IC-MSQUARE 2012: International Conference on Mathematical Modelling in Physical Sciences

Journal of Physics: Conference Series Volume 410

Budapest, Hungary
3-7 September 2012

ISBN: 978-1-62276-988-9
ISSN: 1742-6588

Printed from e-media with permission by:

Curran Associates, Inc.
57 Morehouse Lane
Red Hook, NY 12571

Some format issues inherent in the e-media version may also appear in this print version.

Printed by Curran Associates, Inc. (2013)

For permission requests, please contact the Institute of Physics
at the address below.

Institute of Physics
Dirac House, Temple Back
Bristol BS1 6BE UK

Phone: 44 1 17 929 7481
Fax: 44 1 17 920 0979

techtracking@iop.org

Additional copies of this publication are available from:

Curran Associates, Inc.
57 Morehouse Lane
Red Hook, NY 12571 USA
Phone: 845-758-0400
Fax: 845-758-2634
Email: curran@proceedings.com
Web: www.proceedings.com

TABLE OF CONTENTS

PAPERS

Journal of Physics: Conference Series **410** (2013) 011001 doi:10.1088/1742-6596/410/1/011001

Preface

The first International Conference on Mathematical Modelling in Physical Sciences (IC-MSQUARE) took place in Budapest, Hungary, from Monday 3 to Friday 7 September 2012. The conference was attended by more than 130 participants, and hosted about 290 oral, poster and virtual papers by more than 460 pre-registered authors. The first IC-MSQUARE consisted of different and diverging workshops and thus covered various research fields in which mathematical modelling is used, such as theoretical/mathematical physics, neutrino physics, non-integrable systems, dynamical systems, computational nanoscience, biological physics, computational biomechanics, complex networks, stochastic modelling, fractional statistics, DNA dynamics, and macroeconomics. The scientific program was rather heavy since after the Keynote and Invited Talks in the morning, two parallel sessions ran every day. However, according to all attendees, the program was excellent with a high level of talks and the scientific environment was fruitful; thus all attendees had a creative time.

The mounting question is whether this occurred accidentally, or whether IC-MSQUARE is a necessity in the field of physical and mathematical modelling. For all of us working in the field, the existing and established conferences in this particular field suffer from two distinguished and recognized drawbacks: the first is the increasing orientation, while the second refers to the extreme specialization of the meetings. Therefore, a conference which aims to promote the knowledge and development of high-quality research in mathematical fields concerned with applications of other scientific fields as well as modern technological trends in physics, chemistry, biology, medicine, economics, sociology, environmental sciences etc., appears to be a necessity. This is the key role that IC-MSQUARE will play.

We would like to thank the Keynote Speaker and the Invited Speakers for their significant contributions to IC-MSQUARE. We would also like to thank the members of the International Scientific Committee and the members of the Organizing Committee.

Conference Chairmen

Theocharis Kosmas
Department of Physics, University of Ioannina

Elias Vagenas
RCAAM, Academy of Athens

Dimitrios Vlachos
Department of Computer Science and Technology, University of Peloponnese

Journal of Physics: Conference Series **410** (2013) 011001 doi:10.1088/1742-6596/410/1/011001

International Scientific Committee

Bari Ferenc University of Szeged, Hungary
Medical Informatics

Mark Cross Swansea University, UK
Computational modelling of multi-physics processes on high performance systems

Morten Hjorth-Jensen University of Oslo, Norway
Computational quantum mechanics

Nathan A Baker Pacific Northwest National Laboratory, Richland, WA, USA
Computational biology, biophysics, applied mathematics, signature science, nanotechnology and informatics

Nikolas Markatos National Technical University of Athens, Greece
Computational chemistry

Shiyi Chen Peking University, People's Republic of China
Computational physics

Kostas Kokkotas Eberhard Karls University of Tübingen, Germany
Gravitational waves

Sergey Nazarenko University of Warwick, UK
Non-equilibrium statistical mechanics

Rho Shin Myong Gyeongsang National University, Republic of Korea
Physical modelling

Michel A Van Hove City University of Hong Kong
Computational physics

Jacob Barhen Oak Ridge National Laboratory, Oak Ridge, USA
Global optimization, neural networks

George Karniadakis Brown University, USA
Computational science

J Hernández-Andrés University of Granada, Spain
Optics

M Martcheva University of Florida, USA
Epidemic models

Emilio Elizalde CSIC, Barcelona, Spain
Cosmology, special functions

Angelo Tartaglia Politecnico di Torino, Italy
Gravitation theory

Yoshiharu Yamamoto University of Tokyo, Japan
Physiology

A Mouchet Université de Tours, France
Mathematics and theoretical physics

Mairi Sakellariadou King's College of London, UK
Theoretical physics and cosmology

Kostas Skenderis University of Amsterdam, The Netherlands
String theory

Orfeu Bertolami University of Porto, Portugal
Classical and quantum gravity, noncommutativity, astrophysics, cosmology

Saurya Das University of Lethbridge, Canada
Black hole physics, quantum gravity phenomenology

Theodosios Christodoulakis University of Athens, Greece
Quantum gravity and quantum cosmology

Damianos Sakas University of Peloponnese, Greece
Mathematical modelling in social sciences

Keynote Speaker

Professor Constantino Tsallis
Department of Theoretical Physics, Centro Brasileiro de Pesquisas Fisicas, Brazil

Invited Speakers

Professor Orfeu Bertolami
Departmento de Física e Astronomia, University of Porto, Portugal

Dr Lazaros Gallos
Levich Institute of the City College of New York, USA

Peer review statement

All papers published in this volume of *Journal of Physics: Conference Series* have been peer reviewed through processes administered by the proceedings Editors. Reviews were conducted by expert referees to the professional and scientific standards expected of a proceedings journal published by IOP Publishing.

IC-MSQUARE 2012: International Conference on Mathematical Modelling in Physical Sciences IOP Publishing
Journal of Physics: Conference Series **410** (2013) 012001 doi:10.1088/1742-6596/410/1/012001

Local Model of Arteriovenous Malformation of the Human Brain

Ms Nadezhda Telegina, Mr Aleksandr Chupakhin and Mr Aleksandr Cherevko

Lavrentyev Institute of Hydrodynamics Siberian Branch of Russian Academy of Science, Novosibirsk, Russia

e-mail: nadezhda.telegina@gmail.com, chupakhin@hydro.nsc.ru, cherevko1@ngs.ru

Abstract. Vascular diseases of the human brain are one of the reasons of deaths and people's incapacitation not only in Russia, but also in the world. The danger of an arteriovenous malformation (AVM) is in premature rupture of pathological vessels of an AVM which may cause haemorrhage. Long-term prognosis without surgical treatment is unfavorable. The reduced impact method of AVM treatment is embolization of a malformation which often results in complete obliteration of an AVM. Pre-surgical mathematical modeling of an arteriovenous malformation can help surgeons with an optimal sequence of the operation. During investigations, the simple mathematical model of arteriovenous malformation is developed and calculated, and stationary and non-stationary processes of its embolization are considered. Various sequences of embolization of a malformation are also considered. Calculations were done with approximate steady flow on the basis of balanced equations derived from conservation laws. Depending on pressure difference, a fistula-type AVM should be embolized at first, and then small racemose AVMs are embolized. Obtained results are in good correspondence with neurosurgical AVM practice.

1. Introduction

The human brain is the unique organ which has an expanded complex system of blood supply. Its hemodynamics is of great importance for medicine [1]. Human brain hemodynamics under normal and pathological conditions is difficult to investigate. It requires description of general physiological features of blood circulation for the human brain, special experimental data and further mathematical modeling using these data.

Outstanding complexity of the cardio-vascular system of the human brain, difficult mechanisms of brain circulation regulation and the lack of full experimental data on blood flow explain the absence of the uniform model which could describe blood flowing in brain vessels. Different geometrical and mechanical properties of vessels (i.e. arteries and veins), taking account of the influence of blood corpuscles, and the necessity of local and global blood flow modeling are main difficulties at the development of adequate hemodynamic models [2 – 6].

In recent decades there appeared new high-technology medicine equipment (e.g. various tomography equipment, intravascular surgery equipment) that leads neurosurgery to higher level of development. This equipment allows obtaining and using information on blood flow in real time during surgical operations.

2. Abnormalities of the Vascular System of the Human Brain. Arteriovenous malformations

Vascular diseases of the human brain are one of the reasons of deaths and people's incapacitation not only in Russia, but also in the world. An arteriovenous malformation (AVM) is a congenital pathology which occurs with the interconnection between arterial and venous bed and with the absence of the capillary area in the circulatory system. An AVM is a 'bundle' of small twisted vessels (up to 6 cm in diameter). These bundles do not have a capillary network, so blood is directly shunted from arterial network to the system of superficial and deep veins. Therefore, some parts of the human brain are not fed with oxygen and nutrients which are distributed with blood. The most common malformation type is a racemose AVM which is shown in Figure 1.

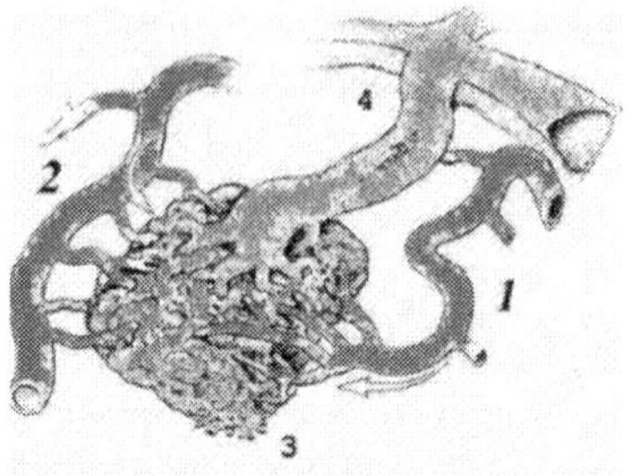

Figure 1. Arteriovenous malformation
1, 2 – supplying arteries; 3 – 'bundle' of modified vessels ('core' of an AVM); 4 – draining vein

The presence of abnormalities in the vascular system causes the change of both hemodynamic (i.e. flow velocity and flow pressure) and strength properties of this vascular system. The danger of an AVM is in premature rupture of pathological vessels of an AVM which causes haemorrhage. Long-term prognosis without surgical treatment is unfavorable.

At present, the process of AVM treatment is a complex staged procedure (e.g. open operation, embolization, and radiosurgery), an AVM should be completely removed without any negative neurological consequences for a patient. The reduced impact method of AVM treatment is embolization of a malformation which often results in complete obliteration of an AVM. Embolization is also used before an open operation for minimizing an existing malformation. Nowadays, information on the procedure of embolization is based on practical experience of surgeons (in other words, professional skills of a surgeon).

Pre-surgical mathematical modeling of an AVM can help surgeons with an optimal sequence of the operation (i.e. optimal sequence of the procedure of embolization), evaluation of the degree of possible risks and possible aftereffects for a patient (e.g. possible repeated haemorrhages).

3. Local Model of Arteriovenous Malformation

The local model of malformation (see Figure 2) is developed as follows: input artery (1) is divided into two vessels (2, 3) with a racemose AVM which is modeled as a 'bundle' of many pipes (4, 5). Then vessels gather into common vein (6). There is also a fistula-type AVM between arterial and venous areas. A fistula-type AVM is a pathologically modified great vessel shunting the most part of flowing blood. Vessels are modeled as pipes with rigid clean walls where water with density of 998 kg/m^3 and viscosity of 4.2 cP flows. Pressure at the input of the AVM is set to 100 mmHg while at the output of the AVM pressure is slightly overrated to 19 mmHg which corresponds to real practice. Flow rate through the AVM is slightly overrated and supposed to be constant. It is set to 747 cm^3/min. Calculations are done with the approximation of one-dimensional steady-state flow on the basis of balance equations derived from conservation laws.

At the first stage embolization of one or two racemose AVMs is performed. As a result, decrease of blood flow rate in afferent vessels (see 'operation 1+2' in Figure 3) is observed.

Using embolization combination for a fistula-type AVM and one smaller racemose AVM, venous blood pressure is normalized to the value of 7 mmHg (see 'operation 1+f' in Figure 3).

Embolization of a fistula-type AVM with one bigger racemose AVM leads to the decrease of flow rate and final normalization of venous blood pressure (see 'operation 2+f', 'operation 1+2+f' in Figure 3).

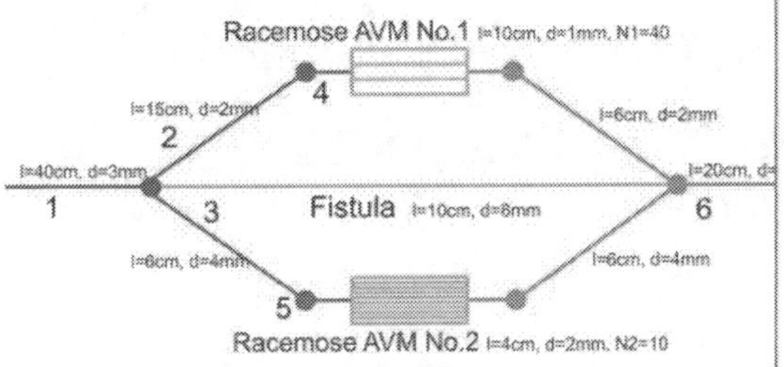

Figure 2. Local model of malformation

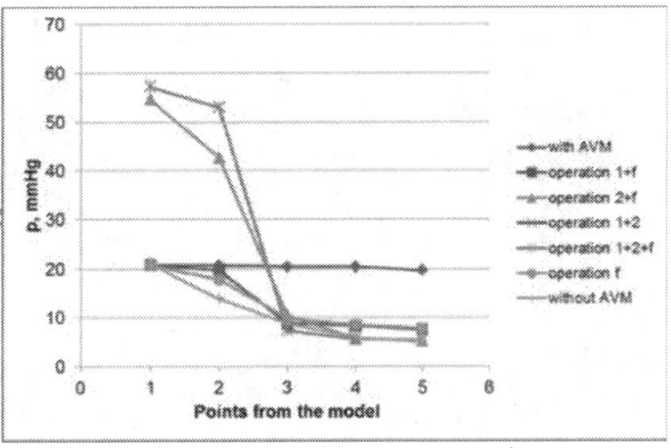

Figure 3. Pressure change curve along an AVM

Figure 3 demonstrates a pressure change curve along a racemose AVM No.1 with various methods of embolization. Depending on pressure difference, a fistula-type AVM should be embolized at first, and then small racemose AVMs are embolized. Obtained results are in good correspondence with neurosurgical AVM practice.

4. Modeling for Patients with AVM

The paper presents the results of modeling of vessel blood circulation for two patients with AVM (the first patient is with AVM located in end lobe, the second patient is with AVM in temporal lobe).

For calculations, data obtained by neurosurgeons of Meshalkin State Research Institute of Circulation Pathology in Novosibirsk during surgical operations is used. Special device provides a surgeon with additional information at any moment of operation. It measures pressure and velocity inside a vessel simultaneously with endovascular operation. This device is equipped with a sensor of 0.36mm diameter which is inserted into a vessel through a catheter, while on the screen display surgeons can observe graphs of velocity and pressure changing in real time. Further, this data can be digitally processed.

Measurements are conducted in various points of a vessel located with a different distance from abnormality. The most important measurements for surgeons are those which are conducted in the same point before and after an AVM embolization.

Output data for pressure and velocity is digitally processed and used for mathematical modeling. In each case with a help of patient angiograms a graph of significant vessels is plotted. Malformation modeling is performed according to methods described in Section 3. Required values for pressure and velocity are achieved by preliminary choice of malformation parameters (i.e. parameters of 'pipes' in an AVM and their quantity).

Figures 4a and 4b show local graphs for significant vessels near AVM for patient 1 (end lobe AVM) and patient 2 (temporal lobe AVM). Sensor positions before embolization are marked with green circles, after embolization – red circles.

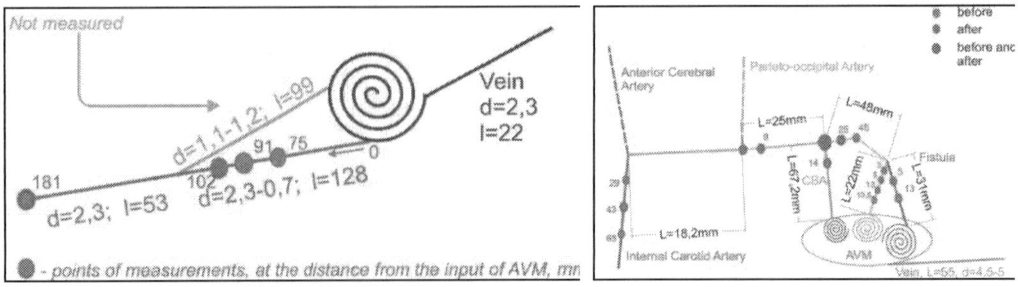

(a) (b)

Figure 4. Local graph for patient 1 (end lobe AVM, figure a) and patient 2 (temporal lobe AVM, figure b)

Calculations confirm experimental data for pressure and velocity before and after an embolization. Comparison of calculated and experimentally obtained parameters is shown in Table 1.

Table 1. Calculated and experimentally obtained parameters for AVM embolization

Diameter of vessel	p, mmHg *experimentally obtained*	p, mmHg *calculated*	v, m/s *experimentally obtained*	v, m/s *calculated*
patient 1: before embolization, sensor – in a vessel with the following diameter, mm:				
d=2,3mm	71	70.91	0.27	0.2631
d=2,5mm	74	71.56	0.24	0.2227
d=3,6mm	73	72.91	0.17	0.1512
patient 1: after embolization, sensor – in a vessel with the following diameter, mm:				
d=3,6mm	79	79.95	0.11	0.1112
patient 2: before embolization, sensor – in a vessel with the following diameter, mm:				
d=3,3mm	50	50.33	0.61	0.5952
d=2,2mm	–	51.99	0.45	0.4716
patient 2: after embolization, sensor – in a vessel with the following diameter, mm:				
d=3,3mm	50	48.72	0.45	0.4531
d=2,2mm	–	48.58	0.57	0.5588
d=3,3mm	48	51.16	0.72	0.7015

5. Conclusions

1. AVM modeling is performed, different types of embolization are studied.
2. Results of AVM modeling for real patients are presented.
3. Obtained results are in good correspondence with neurosurgical AVM practice. This means that even such simple models help us in correct evaluation of risks for surgical intervention. For detailed prognosis, more complex models are required; spatial relationship between vessels, structure of vessel walls and blood rheology should be also taken into account.

Acknowledgments
The work was supported partially by Russian Foundation for Basic Research (RFBR) (grants 11-01-00026, 12-01-09343-mob_z), Program OEMMPU No.2.13.4, Integration project SB RAS No.44, SC-6706.2012.1.

References

[1] "Human brain: fundamental and applied issues", Proceedings of science session of Russian Academy of Science, Vestnik of RAS, No. 5-6, pp. 397-520, 2010.
[2] Pedley T., "The fluid mechanics of large blood vessels", Moscow: Mir, 1983.
[3] Belotserkovsky O.M., Kholodov A.S., "Medicine with relation to information technology", Moscow: Nauka, 2008.
[4] Parashin V.B., Itkin G.P., "Biomechanics of blood circulation", Moscow: Moscow State Technical University, 2005.
[5] Quarteroni A., Tuveri M., Venezani A., "Computational vascular fluid dynamics: problems, models and methods", Comput. Visual. Sci, No. 2, pp. 163-197, 2000.
[6] Ashmetov I.V., Bunicheva A.Ya., Mukhin S.I., Sokolova T.V., Sosnin N.V., Favorsky A.P., "Mathematical modeling of hemodynamics in the human brain and in the systemic circulation" // Computer and brain. New technologies, Moscow: Nauka, p. 231, 2005.

Journal of Physics: Conference Series **410** (2013) 012002 doi:10.1088/1742-6596/410/1/012002

Control of vehicle active suspensions by using PD+PI type fuzzy logic with sliding surface

Yuksel Hacioglu, Nurkan Yagiz

Department of Mechanical Engineering, Faculty of Engineering, Istanbul University, 34320 Avcilar, Istanbul, Turkey

E-mail: yukselh@istanbul.edu.tr

Abstract. A PD+PI type fuzzy logic controller with sliding surface is presented in this study. This controller consists of two parts which are PD type and PI type fuzzy logic units. Inputs to those fuzzy logic units are the sliding surface functions and their derivatives. The integrated controller is applied to two degrees of freedom vehicle active suspension model. Both time and frequency domain analysis are evaluated. Numerical results demonstrate that the proposed controller improves the vibration isolation of the vehicle body, without causing a suspension degeneration problem and without degrading road holding very much.

1. Introduction

Semi-active suspensions in which electrorheological (ER) and magnetorheological (MR) fluid dampers are frequently preferred provide considerable improvements in ride comfort of passengers especially with very small power requirements [1]. On the other hand active suspensions offer significant improvements in ride comfort if compared with passive and semi-active suspensions. Therefore, extensive studies concerning active suspensions have been carried out during last decades. Hrovat [2] applied the optimal control laws on quarter car, half car and full car models and compared their performances with their passive counterparts. D'Amato and Viassolo [3] proposed a fuzzy logic controller (FLC) for a quarter car active suspension system.

By using fuzzy logic (FL) the knowledge coming from experts can be expressed by fuzzy rules. FLC is applicable to systems with uncertain mathematical model and there are two types of FLCs. The first one is the PD type FLC in which the error and its derivative are used as inputs and the control signal is the output. The second one is the PI type FLC in which error and its derivative are used as inputs and the incremental change in control signal is the output. Transient response of the PD type FLC is better than the PI type FLC, but in some cases steady state error can not be removed for PD type FL controlled system [4].

2. Design of the controller

In this section, a PD+PI type FLC with sliding surface is presented (Figure 1). The sliding function σ_N and its derivative $\dot{\sigma}_N$ are used as inputs. Here the sliding function (surface) is defined as $\sigma = \alpha e + \dot{e}$ where e is the error, $\dot{e}$ is derivative of error and α is the negative value of the slope of the sliding surface. In sliding mode control (SMC), by changing the control input according to certain predefined rules, system states are driven to the sliding surface and then forced to remain there. Also it is known that conventional FLCs operate like SMCs with a boundary layer [5]. This is why sliding function and its derivative were chosen as inputs in this study. The outputs of the PD+PI type FLC are control signal u_N and incremental change in control signal Δu_N.

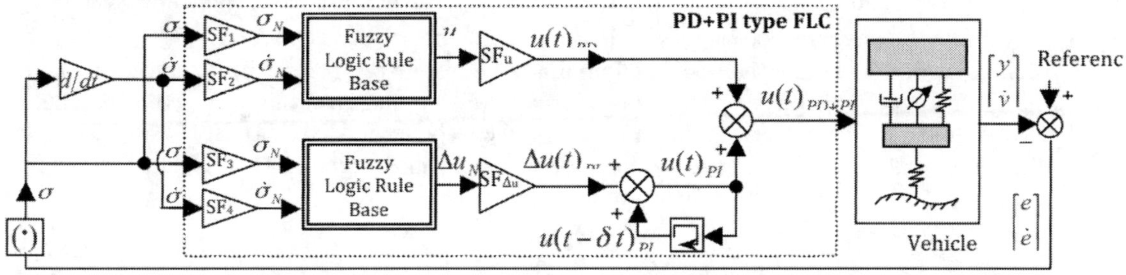

Figure 1 Block diagram of the PD+PI type FLC with sliding surface

Triangular membership functions for the input and output variables are presented in Figure 2. Here *NB*, *NM*, *NS*, *Z*, *PS*, *PM* and *PB* denote negative big, negative medium, negative small, zero, positive small, positive medium and positive big, respectively. SF_i (i=1,2,3,4) are the input scaling factors and SF_u and $SF_{\Delta u}$ are the output scaling factors of the PD+PI type FLC.

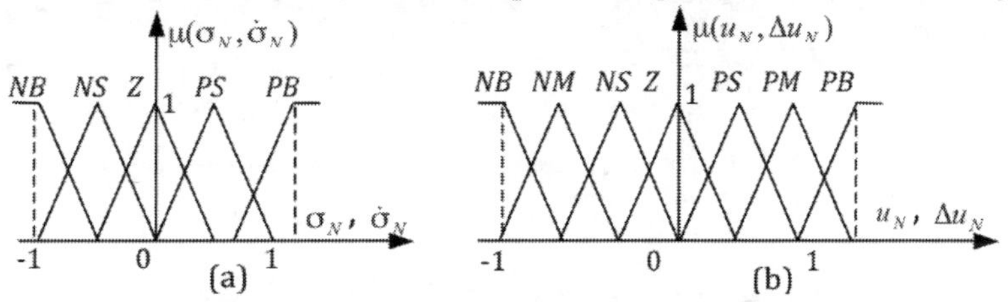

Figure 2 Membership functions for the a) inputs variables b) output variables

Table 1 Rule table for computing u_N and Δu_N

σ_N \ $\dot{\sigma}_N$	*NB*	*NS*	*Z*	*PS*	*PB*
NB	*NB*	*NB*	*NM*	*NS*	*Z*
NS	*NB*	*NM*	*NS*	*Z*	*PS*
Z	*NM*	*NS*	*Z*	*PS*	*PM*
PS	*NS*	*Z*	*PS*	*PM*	*PB*
PB	*Z*	*PS*	*PM*	*PB*	*PB*

For the PD part output of the FLC it is thought that if the states are far from the sliding surface then control input should be big and vice versa in order to bring the states of he system to the that surface. Similarly, for the PI part output of the FLC it is thought that if the states are far from the sliding surface then incremental change in control input should be increased and vice versa. The rule table for the PD+PI type FLC is constructed by using this manner and it is given in Table 1.

3. Vehicle model

Quarter car model, given in Figure 3, has two degrees of freedom which are body bounce y_2 and displacement of the wheel y_1 that are both in vertical directions. In order to have a realistic model, the spring and damper elements used in the suspension and the tire spring have nonlinear characteristics as seen in Figure 3 and the corresponding spring and damper forces are defined as [6]:

$$F_t(\xi_t) = k_t\xi_t + k_{tn}\xi_t^{\ 3};\ F_s(\xi_s) = k_s\xi_s + k_{sn}\xi_s^{\ 3};\quad F_d(\dot{\xi}_s) = b\dot{\xi}_s + b_n\dot{\xi}_s^{\ 2}\,\mathrm{sgn}(\dot{\xi}_s) \tag{1}$$

Here F_t is the tire force, F_s and F_d are the spring and damper forces of the suspension, $\xi_t = y_1$-y_0, $\xi_s = y_2$-y_1 and $\dot{\xi}_s = \dot{y}_2 - \dot{y}_1$ are the tire deflection, suspension deflection and suspension deflection velocity, respectively. In the spring and damper force equations k_t, k_s, b and k_{tn}, k_{sn}, b_n are the coefficients for the linear and nonlinear terms, respectively. The actuator, producing the control input u, is placed between the sprung mass m_2 and unsprung mass m_1 as parallel to the suspension elements. V is the velocity of the vehicle and y_0 is the road surface input. Numerical values of the vehicle parameters are given in Appendix. Equations of motion of the nonlinear quarter vehicle suspension model are as

$$m_1\ddot{y}_1 - b(\dot{y}_2 - \dot{y}_1) - b_n(\dot{y}_2 - \dot{y}_1)^2\,\mathrm{sgn}(\dot{y}_2 - \dot{y}_1) + k_1(y_1 - y_0) + k_{1n}(y_1 - y_0)^3 - k_2(y_2 - y_1) - k_{2n}(y_2 - y_1)^3 = -u \tag{2}$$

$$m_2\ddot{y}_2 + b(\dot{y}_2 - \dot{y}_1) + b_n(\dot{y}_2 - \dot{y}_1)^2\,\mathrm{sgn}(\dot{y}_2 - \dot{y}_1) + k_2(y_2 - y_1) + k_{2n}(y_2 - y_1)^3 = u$$
(3)

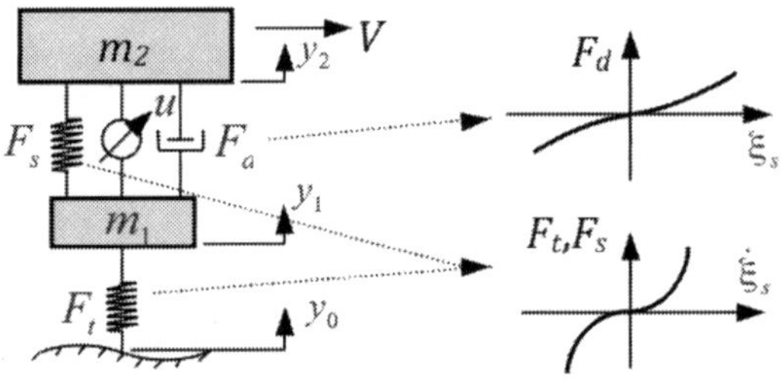

Figure 3 Vehicle suspension model

4. Numerical results

In order to demonstrate the efficiency of the proposed controller, numerical results both in time and frequency domains are presented in this section. For comparison purposes, the numerical results of the vehicle model with PID controller are also presented. Numerical parameters for the designed controllers are given in Appendix.

Time responces for the step input with limited ramp are presented in Figure 4. It is seen in Figure 4.a that passive, PID and fuzzy controlled vehicle body displacements settles on the road surface input height value where the fuzzy controlled one has no overshoot which indicates the efficiency of the proposed controller. The decrease in body acceleration magnitudes assures improved ride comfort as presented in Figure 4.b. If the suspension deflection is checked in Figure 4.c, it is observed that there is not any permanent offset which indicates that the proposed control method is a suitable choice for implementation. The control forces for the PID and fuzzy controllers are also given in Figure 4.d.

Since frequency responses are valid only for linear systems, the nonlinear terms in spring and damper equations are neglected. The frequency plots for the vertical body displacement and acceleration are given in Figure 5.a and 5.b, respectively. It is seen that by using the proposed fuzzy controller the magnitudes for both vertical displacement and acceleration of the vehicle body are decreased in a wide frequency range of interest while PID controller increased the magnitudes of both displacement and acceleration of the vehicle body between 2-4 Hz. These results indicate that the ride comfort of the passengers is improved greatly by using the proposed FL controller. The frequency response for the dynamic tire load is given in Figure 5.c. It seen that especially by using the proposed FLC, the road holding is also

Journal of Physics: Conference Series **410** (2013) 012002 doi:10.1088/1742-6596/410/1/012002

improved for the low frequency ranges and there is not any significant deterioration for the high frequencies. On the other hand, it is seen that with the PID controller the magnitude of the dynamic tire load increased for the frequency range of 2-4 Hz.

5. Conclusion

For the vibration suppression of a vehicle active suspension a PD+PI type fuzzy logic controller was presented in this study. Performance of the controller was demonstrated using both time and frequency responses. Results have shown that proposed fuzzy controller was successful in vibration isolation of the vehicle body without degrading road holding and while preserving suspension working space.

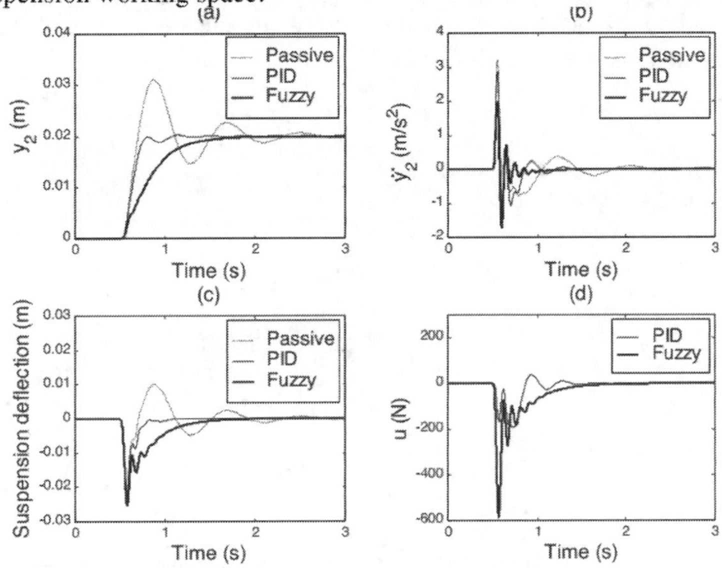

Figure 4 Time responses for step road input a) Bounce motion of the vehicle body b) Vertical acceleration of the vehicle body c) Suspension deflection d) Control force

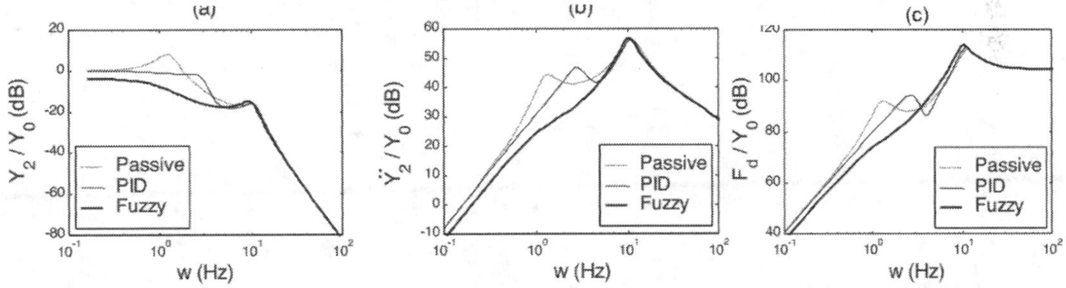

Figure 5 Frequency responses a) Body motion b) Vertical body acceleration c) Dynamic tire force

Appendix

Vehicle parameters		Controller parameters: Fuzzy		PID
$m_1 = 36$ [kg]	$k_t = 160000$ [Nm^{-1}]	$SF_1 = 4$	$SF_u = 700$	$K_p = 83400$ (proportional gain)
$m_2 = 240$ [kg]	$k_{tn} = 16000$ [Nm^{-3}]	$SF_2 = 0.125$	$SF_{\Delta u} = 20$	$T_i = 0.13$ (integral time)
$b = 980$ [Nsm^{-1}]	$k_s = 16000$ [Nm^{-1}]	$SF_3 = 2$	$\alpha = 5$	$T_d = 0.0325$ (derivative time)
$b_n = 402$ [Ns^{2}m^{-2}]	$k_{sn} = 1600$ [Nm^{-3}]	$SF_4 = 0.25$	$\tau = 0.12$	$\tau = 0.12$
$V = 72$ [km/h]				

References

[1] Hrovat D 1997 Survey of advanced suspension developments and related optimal control applications *Automatica* **33(10)** 1781-1817.

[2] Hrovat D 1993 Applications of optimal control to advanced automotive suspension design, *Journal of Dynamic Systems, Measurement, and Control* **115** 328-342.

[3] D'Amato FJ, and Viassolo DE 2000 Fuzzy control for active suspensions *Mechatronics* **10** 897-920.

[4] Lee J 1993 On methods for improving performance of PI-type fuzzy logic controllers, *IEEE Transactions on Fuzzy Systems* **1(4)** 298-301.

[5] Palm R 1992 Sliding mode fuzzy control, *Proceedings of the IEEE International Conference on Fuzzy Systems*, San Diego, CA, 519-526.

[6] Ando Y and Suzuki M 1996 Control of active suspension systems using the singular perturbation method *Control Engineering Practice* **4(3)** 287-293.

Infinite dimensional integrals beyond Monte Carlo methods: yet another approach to normalized infinite dimensional integrals

Jean-Pierre Magnot
Lycée Blaise Pascal, Avenue Carnot, F-63000 Clermont-Ferrand
E-mail: jean-pierr.magnot@ac-clermont.fr

Abstract. An approach to (normalized) infinite dimensional integrals, including normalized oscillatory integrals, through a sequence of evaluations in the spirit of the Monte Carlo method for probability measures is proposed. in this approach the normalization through the partition function is included in the definition. For suitable sequences of evaluations, the ("classical") expectation values of cylinder functions are recovered

Keywords: infinite dimensional integration, means, expectation value, Dirac measures

1. Introduction

Let us first recall two well-developped frameworks.

• Let (X, μ) be a measured space. Following [9], [11], [12], let us fix a vector subspace $\mathcal{F} \subset L^\infty(X,\mu)$ such that $1_X \in \mathcal{F}$. A **mean** on $\mathcal{F}$ is a linear map $\phi : \mathcal{F} \to \mathbb{C}$ such that $\phi(1_X) = 1$. Alternately, if (X, d) is a metric space, given $\mathcal{F} \subset C_b^0(X)$ (space of bounded maps), a **mean** on $\mathcal{F}$ is a linear map $\phi : \mathcal{F} \to \mathbb{C}$ such that $\phi(1_X) = 1$. These two terminologies come from the basic example where μ is a Borel probability measure on a metric space (X, d), for which the mean of a continuous integrable map f is its expectation value and can be approximated by sequences of barycenters of Dirac measures via Monte Carlo method, namely, via some sequences $(x_n)_{n\in\mathbb{N}} \in X^{\mathbb{N}}$ such that

$$\forall f \in L^1(X,\mu), \int_X f d\mu = \lim_{n\to+\infty} \frac{1}{n+1} \sum_{k=0}^{n} f(x_k).$$

The technical condition for such a sequence is the following: for each μ-measurable set A, $\mu(A) = \lim_{n\to+\infty} \frac{1}{n+1} \sum_{k=0}^{n} 1_A(x_k)$.

• the Feynmann-Kac's formula: It is heuristic in the original Feynman's work, and a very difficult question is to give it a mathematical (rigorous) sense. Many approaches have been developedin which the following heuristic integral is central:

$$\frac{1}{\int e^{-iS} d\lambda} \int f e^{-iS} d\lambda$$

where S is the action functional of the physical theory; f is $\mathbb{C}$−valued prescribed map defined on an infinite dimensional vector space of configurations; λ is a heuristic infinite dimensional Lebesgue measure, that is a translation invariant measure on the space of configurations; $\int e^{-iS} d\lambda$ is a so-called "normalization constant", called "partition function" and sometimes noted as $Z(S)$, which can be understood as the total volume of the heuristic measure $e^{-iS}\lambda$, of "density" e^{-iS} with respect to λ; and the whole formula stands as a mean value of f, called "expectation value" because taken with respect to the (heuristic) probability measure $\frac{1}{\int e^{-iS} d\lambda} e^{-iS}\lambda$.

This too short exposition on Feynman-Kac formula is not, of course completely satisfying, compared to the huge litterature on it and the wide variety of tentatives of rigorous definitions. We focuse here on the theory of Fresnel integrals, which are rigorously defined. The reader can refer to [1],[2], [3], [4], [5], [7], [8], [13].

The technical features of the finite measure setting are almost the same as the ones of compact spaces. This suggests that, getting into the setting of infinite dimensional integrals, such as heuristic integrals of the Feynman-Kac formula, where straight way computations of the normalization constant (the partition function integral) lead to divergent approximations that need to be renormalized (see e.g. [2]), one can expect the same kind of problems in generalizing e.g. a Daniell integral to more complex theories as the problems that occur while passing from compact operators to bounded (or even unbounded) operators acting on a Hilbert space. With [10], we began a research program where the normalized integrals with respect to infinite volume measures are seen as particular means (which are means spanned by finite measures), and are not measures in the strictly speaking sense. This approach is coherent with the standard definition of Fresnel integrals (see e.g. [2]) which are defined through sequences of complex measures and normalizing weights. The goal of this communication is to define a similar appraoach with complex measures spanned by Dirac measures, and to compare with the classical approach of integration via cylinder functions.

2. Sequences on a separable space and Dirac means

Let X be a separable topological separable space. Given $x \in X$, the evaluation map $\delta_x : f \in \mathbb{C}^X \mapsto f(x)$ is viewed as a Dirac measure in a probabilistic way. These measures are the extremals of the convex set of Borel probability measures $P(X)$. Let us define the affine space spanned by these elements.

Definition 2.1 *A complex (resp. real)* ***Dirac mean*** *is a linear map $\tau : \mathcal{D}_\tau \subset C^0_b(X) \to C$ which is defined as the limit of barycenters with complex (resp. real) weights of a sequence of Dirac measures on X, i.e. for $\mathbb{K} = \mathbb{R}$ or $\mathbb{C}$,*

$$\exists (x_n, \alpha_n)_{n\in\mathbb{N}} \in (X \times \mathbb{K})^{\mathbb{N}}, \forall m \in \mathbb{N}^*,$$

$$\left\{\sum_{n=0}^{m} \alpha_n \neq 0\right\} \quad \wedge \quad \left\{\forall f \in C^0_b(X), \tau(f) = \lim_{m\to+\infty} \frac{1}{\sum_{n=0}^{m} \alpha_n}\left(\sum_{n=0}^{m} \alpha_n \delta_{x_n}(f)\right)\right\}.$$

We note by $\widetilde{\mathcal{DM}}_{\mathbb{K}}(X)$ the space of $\mathbb{K}-$Dirac means, by $\mathcal{DM}_{\mathbb{K}}(X)$ the set of Dirac means τ such that $\mathcal{D}_\tau = C^0_b(X)$, by $\widetilde{\mathcal{DM}}^+_{\mathbb{R}}(X)$ the means τ obtained by a sequence $(\alpha_n)_{n\in\mathbb{N}} \in \mathbb{R}^*_+$ and by $\mathcal{DM}^+_{\mathbb{R}}(X)$ the space $\widetilde{\mathcal{DM}}_{\mathbb{R}}(X) \cap \widetilde{\mathcal{DM}}^+_{\mathbb{R}}(X)$.

Examples.

- *Linear extensions of the limit :* Let $(x_n)_{n\in\mathbb{N}} \in X^{\mathbb{N}}$ such that $\lim_{n\to+\infty} x_n = x$. Let $(\alpha_n)_{n\in\mathbb{N}} \in (\mathbb{R}^*_+)^{\mathbb{N}}$ such that $\liminf_{n\to+\infty} \alpha_n > 0$. Then, applying the results of [10], we get that the corresponding Dirac mean τ equals to the limit at x. By its definition, τ is defined on a wider class of functions, and then defines a linear extension of the limit.
- *Monte Carlo method extended :* Let $\mu \in P(X)$ and let $(x_n)_{n\in\mathbb{N}}$ such that $\forall f \in L^1(X,\mu)$, $\int_X f d\mu = \lim_{n\to+\infty} \frac{1}{n+1}\sum_{k=0}^{n} f(x_k)$. Then for any finite measure $\mu' = \varphi.\mu$, where $\varphi \in L^\infty$ is the μ-density of μ',

we have

$$\frac{1}{\mu'(X)}\int_X f d\mu' = \lim_{n\to+\infty}\frac{1}{\sum_{k=0}^{n}\varphi(x_k)}\sum_{k=0}^{n}\varphi(x_k)f(x_k).$$

If μ' is only σ-finite, the density φ is no longer bounded but the same formula defines $\tau \in \mathcal{DM}_{\mathbb{R}}(X)$ by

$$\tau(f) = \lim_{n\to+\infty}\frac{1}{\sum_{k=0}^{n}\varphi(x_k)}\sum_{k=0}^{n}\varphi(x_k)f(x_k).$$

For example, if $X = [0;1]$ equipped with the Lebesgue measure λ and $\Psi : X \to \mathbb{R}$ be a smooth diffeomorphismsuch that $\Psi' > 1$. Let $\mu' = \Psi^*(\lambda)$. Then, the density φ of μ' with respect to λ on $[0;1]$ is bounded below by 1, and hence the corresponding mean τ is a linear extension of the limit at infinity [10].

- *Means on a Hilbert space and on its unit sphere:* A Hilbert space has dense sequences $(x_n)_{n\in\mathbb{N}}$ which are mimicking, by their topological properties, the density property of the sequences used in the classical Monte-Carlo method. Then with the same formula as before, one can take:

$$\tau(f) = \lim_{m\to+\infty}\frac{1}{\sum_{n=0}^{m}\alpha_n}\left(\sum_{n=0}^{m}\alpha_n\delta_{x_n}(f)\right)$$

for an adequate function f. We shall discuss a desired class of sequences $(\alpha_n)_{n\in\mathbb{N}}$ next section. But what we have to remark is the following: the unit sphere S of the Hilbert space has the same properies, and then carries also such means. One can even wonder whether radial projections are possible, and emphasis some invariance under the action of an orthogonal group. We also remark that many efforts have been made in the context of metric geometry (see e.g. [9], [12]) to study means on the sphere. We leave open the question of such a spherical integration, and its significance for integration on a ball or on the Hilbert space through a spherical-like procedure.

3. Normalized infinite dimensional integrals as mappings on sequences

Here we come to the heuristic integral $\frac{1}{\int e^{-iS}d\lambda}\int fe^{-iS}d\lambda$ on a Hilbert space H. Here, since the part e^{-iS} stands as a density, we can make the following definition:

Definition 3.1 *We define a function*

$$\begin{array}{rccl} I: & H^{\mathbb{N}} \to & \mathcal{DM}_{\mathbb{C}}(H) & \\ & (x_n)_{n\in\mathbb{N}} \mapsto & \tau = \lim_{m\to+\infty}\frac{1}{\sum_{n=0}^{m}e^{-iS(x_n)}}\left(\sum_{n=0}^{m}e^{-iS(x_n)}\delta_{x_n}\right) & \end{array}$$

Since we have complex coefficients $e^{-iS(x_n)}$, we have to take care about the normalization constant which can be zero for " bad" sequences. So that, the domain of I cannot be $H^{\mathbb{N}}$. Then comes the analysis, beyond the convergence of the Monte Carlo method for classical propability measures. Let us quote some open questions:

- Which kind of sequence lead to invariance by a group acting on H?
- In this general setting, what will be the status of 1-Lipschitz functions that play a crucial role in the approach by metric geometry?
- Which sequences lead to (true) measures on H?
- For which perturbations of the action functionnal do we have some adequate Taylor expansions?

The list of open questions can be very long, but we wish to finish by an answer and an open perspective.

Journal of Physics: Conference Series **410** (2013) 012003 doi:10.1088/1742-6596/410/1/012003

4. Link with the "classical" Monte Carlo method and perspectives

Let us begin with integration of cylinder functions on the infinite cube $[O; 1]^{\mathbb{N}}$ (Daniell integral). (heuristic) Wick rotation in order to get a positive measure. Let us consider now a cylinder function f. Let P be a finite dimensional projection such that $f = f \circ P$ and let $S_P = S \circ P$. Then, adequate sequences for the Monte Carlo method are those whose push-forward on $[0; 1]^{dimImP}$ are also adequate for this method. Taking now a creasing sequence of orthogonal projectors P_k converging (weakly) to identity, the condition on the sequence $(x_n)_{n\in\mathbb{N}}$ is that for each $k \in \mathbb{N}$, the push-forwards of the sequences $(P_k(x_n))_{n\in\mathbb{N}}$ on $[0; 1]^{dimImP_k}$ fit with the desired conditions. A sequence $(x_n)_{n\in\mathbb{N}}$ for such a method exists, through e.g. the powers of π. Taking an action functionnal S and condidering a finite product measure e^{-S} (after Wick rotation), the classical approach of the (classical) Monte-Carlo method is to pull-back the sequences on $[0; 1]^{\mathbb{N}}$ to $\mathbb{R}^{\mathbb{N}}$ through the coordinatewise pull-back of a product measure. This approach carries no additive weight $(\alpha_n)_{n\in\mathbb{N}}$. For Fresnel integrals, the density e^{-iS} is approximated by a density ξe^{-iS} where the unction ξ is chosen to get two convergent integrals $\int \xi e^{-iS} f d\lambda$ and $\int \xi e^{-iS} d\lambda$. Assuming ξ integrable by itself and defined as a product function, there is also (after normalizeation) a possible pull-back of a sequence $(x_n)_{n\in\mathbb{N}}$ adapted for the Monte Carlo method, and the weight we get is only $\alpha_n = e^{-iS(x_n)}$. We recover here an old problem, already quoted in [10]: the "Lebesgue" measures on $\mathbb{R}^{\mathbb{N}}$ have been extensively studied in the 40's but contain very few finite measure subsets (see e.g. [6] for an up-to-date exposition). As for the example described in [10], the theory of means is an attractive candidate to complete the theory of probabilities in such infinite dimensional problems. Unfortunately for applications, analysis on such objects has to be developped, and the topolgies of the space $\mathcal{DM}$ have to be studied.

[1] Albeverio S and Brzeźniak Z (1994) *Acta Appl. Math.* **35** 5-27

[2] Albeverio S, Hoegh-Krohn R and Mazzuchi S (2005) *Mathematical theory of Feynman Path Integrals; an introduction* 2nd edition; Lect. Notes in Math. **523** (Springer)

[3] Albeverio S and Mazzucchi S (2005) *Bull. Sci. Math.* **129** no1 1-23

[4] Albeverio S and Mazzucchi S (2004) *C.R. Acad. Sci. Paris sr. A 338* no3 255-9

[5] Albeverio S and Mazzucchi S (2005) *J. Funct. Anal.* **221** no1 83-121

[6] Baker R (1991) *Proc. AMS* **113** no4 1023-9

[7] Duistermaat J (1974) *Comm. Pure Appl. Math.* **27** 207-281

[8] Elworthy D and Truman A (1984) *Ann. Inst. H. Poincaré Phys. Theo.* **41** (2) 115-142

[9] Gromov M (1999) *Metric structures for Riemannian and Non Riemannian Spaces* Progr. Math. **152** (Birkauser)

[10] Magnot, J-P.; The mean value for infinite volume measures, infinite products and heuristic infinite dimensional Lebesgue measures; Preprint arXiv:1012.2452v3

[11] Paterson A (1988) *Amenability* Math. surveys and Monographs **29** (Amer. Math. Soc., Providence, R.I.)

[12] Pestov V (2006) *Dynamics of Infinite-Dimensional Groups : the Ramsey-Dvoretzky-Milman phenomenon* University Lecture Series **40** (Amer. Math. Soc., Providence, R.I.)

[13] Stein E (1993) *Harmonic analysis: real-variable methods, orthogonality and oscillatory integrals.* Princeton Math. series **43** Monographs in Harmonic analysis III. (Princeton University Pres, Princeton, NJ)

Nuclear Potential Clustering As a New Tool to Detect Patterns in High Dimensional Datasets

V Tonkova[1], D Paulus[2], H Neeb[1]
[1]RheinAhrCampus Remagen, University of Applied Sciences Koblenz, Südallee 2, 53424 Remagen, Germany
[2]University of Koblenz-Landau, Faculty of Computer Science, Universitätsstraße 1, 56070 Koblenz

E-mail: tonkova@rheinahrcampus.de

Abstract. We present a new approach for the clustering of high dimensional data without prior assumptions about the structure of the underlying distribution. The proposed algorithm is based on a concept adapted from nuclear physics. To partition the data, we model the dynamic behaviour of nucleons interacting in an N-dimensional space. An adaptive nuclear potential, comprised of a short-range attractive (strong interaction) and a long-range repulsive term (Coulomb force) is assigned to each data point. By modelling the dynamics, nucleons that are densely distributed in space fuse to build nuclei (clusters) whereas single point clusters repel each other. The formation of clusters is completed when the system reaches the state of minimal potential energy. The data are then grouped according to the particles' final effective potential energy level. The performance of the algorithm is tested with several synthetic datasets showing that the proposed method can robustly identify clusters even when complex configurations are present. Furthermore, quantitative MRI data from 43 multiple sclerosis patients were analyzed, showing a reasonable splitting into subgroups according to the individual patients' disease grade. The good performance of the algorithm on such highly correlated non-spherical datasets, which are typical for MRI derived image features, shows that Nuclear Potential Clustering is a valuable tool for automated data analysis, not only in the MRI domain.

1. Introduction

Cluster analysis is one of the standard techniques to analyze and blindly group data into different categories. However, exploring natural phenomena has often to do with high dimensional correlated non-spherical distributed features, which confine the application of standard clustering methods [1]. A current problem in the neuroscience, for example, is the extraction of image-based biomarkers for the diagnosis of Multiple Sclerosis (MS), a neurodegenerative disease leading to a progressive lifelong handicap. Here, a mismatch between conventional medical imaging and clinical manifestation of the disease is observed [2, 3]. Even though new quantitative magnetic resonance imaging techniques that offer the possibility of measuring multiple parameters simultaneously have emerged [2, 6], no sensitive biomarkers have yet been extracted, since the values measured in a multimodal MRI sequence are biophysically correlated.

In order to detect patterns in high dimensional datasets, we developed a new clustering algorithm, which does not rely on *a priori* assumptions about the distribution of the data, i.e., correlated data do not pose severe problems. The basic idea is to treat the points to be classified as nucleons in an N-dimensional space and model their dynamic behaviour. A nuclear potential is associated to every 'particle', so that, as times goes on, points that are densely distributed in space fuse to form 'nuclei', i.e. clusters.

To validate the algorithm, correlated test datasets of different dimensionality were simulated and analyzed. The method was then applied on a dataset, consisting of image based features derived from spatially normalized T_1-, T_2^*-, and absolute water content maps of 43 MS patients. The results showed a very good performance of the Nuclear Potential Clustering on high dimensional correlated datasets.

2. Materials and Methods

2.1. Nuclear Potential Clustering

The Nuclear Potential Clustering method is based on the modelling of a physical system in which each data point of an N-dimensional dataset represents a single nucleon. As such it possesses: (1) a short range potential well, defined here by a Woods-Saxon approach

$$\varphi^S = \frac{-V_0}{1+e^{\frac{r-R}{a}}}, \qquad (1)$$

that traps nearby nucleons (strong interaction) and (2) at separations larger than the width of the potential well ($r > R$) a positive Coulomb potential $\varphi^C = r^{-1}$ (electromagnetic interaction, where the electric charge is normalized to $4\pi\varepsilon_0$) [5]. Here, r defines the Euclidean distance between two points in feature space, R is the radius at which the strong interaction is effective and a sets the slope of the potential at the border between the two terms [7]. The interaction strength is given by the $|V_0|$. Prior to clustering, features are normalized to have zero mean and unit standard deviation in order to provide an equal weight to each dimension when calculating distances. The basic idea underlying our clustering approach is now the following: while the force between nearby particles is dominated by the strong interaction, resulting in a mutual attraction, particles at larger distances experience only the repelling electromagnetic force. Thus, points that are close enough fuse to larger nuclei (clusters), while single nucleons (outliers) are repelled when dynamically modelling the system. By testing the performance of the algorithm on artificial data sets with different number, shape, density and dimensionality of the clusters, the values characterizing the potential function were empirically set to: $V_0 = 3500$ [a.u.], $R = 10\%$ quantile of the initial distance distribution between all points, and $a = R/50$. The range of the strong interaction is thus adapted to the data set to be clustered.

Starting from the initial spatial distribution of the nucleons, we explore the time evolution of the system. The effective force acting on a single particle is given by the sum of two terms: (1) the negative gradient of the total potential function $\varphi(\vec{x})$ resulting from the interaction with all other points $\vec{F_n} = -\vec{\nabla}\varphi(\vec{x})$, and (2) a friction term proportional to the particles velocity, $\vec{F_f} = -\eta \cdot \vec{v}$. Setting the damping constant to $\eta > 0$ suppresses the scattering of nearby particles if their kinetic energy is large enough to leave the strong potential well. Furthermore, it leads to a quicker transition to equilibrium state, making the algorithm significantly faster. η is set to 200 [a.u.], whereby the algorithm is largely immune against variations of this parameter. Given the total force, the motion of a point $\vec{x}(t)$ is obtained by solving the Newtonian equation of motion, using a fourth-order Runge-Kutta method:

$$m \cdot \vec{a}(t) = \vec{F_n}(t) - \eta \cdot \vec{v}(t), \text{ with } \vec{v}(t) = \dot{\vec{r}}(t) \text{ and } \vec{a}(t) = \dot{\vec{v}}(t). \qquad (2)$$

Here m represents the mass of a particle which we set to 1. After every Runge-Kutta step, the total potential and kinetic energy of every particle are determined.

We have observed that points do not directly form stable nuclei but rather oscillate towards an equilibrium state. In order to choose an objective stopping condition for our algorithm, we determine the system temperature as a function of time, $T(t)$ (an example for a test data set is shown in Figure 1c)). The temperature T is given by the average kinetic energy of all particles. The algorithm is stopped as T reaches a value less than 1% of the absolute maximum reached. Furthermore, the repelling Coulomb force results in a splitting of larger clusters into sub-clusters at locations where the particle density decreases. In order to prevent from such an artificial splitting, we increase the value of R for times later than the time point, where the first temperature maximum is reached:

$$R(t_i) = R(t_0) \cdot (1 + a_i^{-1}), a_i = 4, 5, 6 \dots . \quad (3)$$

The algorithm is repeated five times with different initial particle velocities $\vec{v}(t_o)$ in order to reduce the incidental assignment of noise points to one of the clusters formed. During the first run, particles start at rest. For the next repetitions $\vec{v}(t_o)$ are chosen to have random direction and normal distribution with zero mean and standard deviation given by the average velocity calculated from the previous step.

Finally, the number of clusters is defined as the number of maxima in the potential energy distribution, averaged over all five repetitions. Points with a positive potential energy are classified as outliers. The rest is split into clusters according to the particles' potential energy level.

2.2. Data

A series of test datasets with numbers of clusters and cluster members, as well as shape and dimensionality were generated. Here, three toy problems are presented: set 1 consisting of a 'horseshoe' and two spheroids overlaid by random noise in 3-dimensional space (461, 36, 79 and 57 points, respectively, Figure 1(a)); set 2 consisting of two dense regions positioned around two skew lines overlaid by random noise in 3-dimensional space (200, 320 and 100 points, respectively; Figure 1(b)); and set 3 consisting of two dense regions positioned around two skew lines overlaid by random noise in 100-dimensional space (70, 90 and 35 points, no graphic depiction possible).

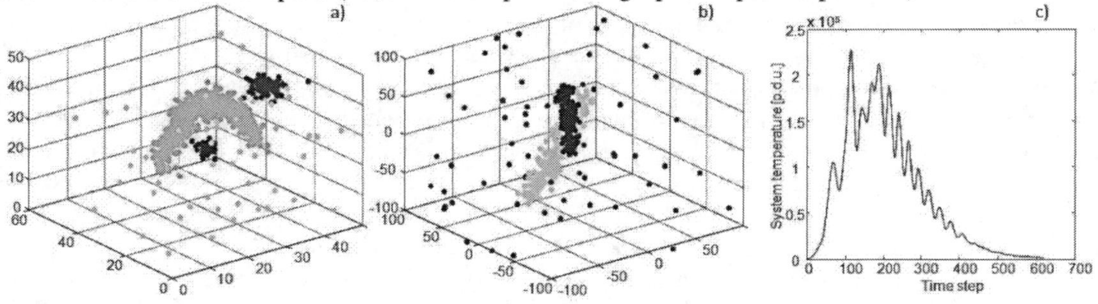

Figure 1. (a) Test data set 1; (b) test data set 2; (c) temperature evolution of test data set 1

In addition, data from 43 MS patients (17 male and 26 female subjects with a mean age of 41±11 years), scanned on a 3T TRIO Siemens scanner, were analyzed. The expanded disability status scale (EDSS) score [4] of the patients at the time of the examination ranged between 1 and 6.5. Quantitative T_1-, T_2^*-, and absolute water content maps were constructed for each subject as described in [6]. The individual maps were spatially normalized to the MNI ICBM452 brain template, so that same anatomical structures appear at the same coordinates in the maps. From the reconstructed maps, a 494-dimensional feature vectors was calculated for each subject. The features include the average and standard deviation of all quantitative MR parameters for the whole brain as well as for each individual slice. Furthermore, the parameter histograms were included as additional features in order to implicitly include higher moments of the respective distributions.

3. Results

Applying the Nuclear Potential Clustering on the test datasets leads to a correct clustering of 96% of the points of set 1, 99% of the points of set 2 and 98% of the points of set 3. Figure 2 demonstrates the final data distribution for the 'horseshoe' test set. As it can be seen in Figure 2(a), nearby nucleons fuse to almost point-like nuclei. Figure 2(b) depicts the corresponding final energy distribution. The different horizontal lines represent the different clusters, as all nucleons in a point-like spherical nucleus have, to a first approximation, the same potential energy. This distribution is used to define the number of groups formed and to assign the points to the appropriate clusters. The corresponding result is shown in Figure 2(c). Similar figures for the second and the third test data sets as well as videos demonstrating their spatial (for the 3D data) and energy evolution can be found at: http://www.rheinahrcampus.de/Nuclear-Potential-Clustering.5142.0.html.

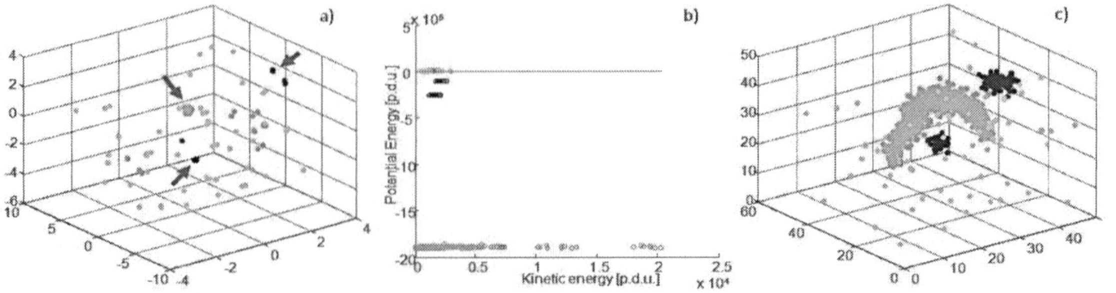

Figure 2. Results for test data set 1: a) final spatial constellation; b) final energy distribution; c) clustered data

Applying the algorithm on the MS data resulted in a splitting of all points into two main categories. The first group consist of 27 subjects with an average EDSS of 0.67±1.58 and an average age of 37 ±10 years. The second group consists of 16 points classified as noise. It contains primarily older patients with a mean age of 46±11 years and a higher EDSS score of 2.50±2.29.

4. Discussion

Nuclear Potential Clustering showed a good performance on all artificial data sets examined. Taking a closer look at our results showed that the falsely classified data are mostly particles on the rim of a cluster where the nucleon density decreases, or noise that is accidentally positioned nearby or inside of a denser group. It is yet important to note that the algorithm is sensitive to variations of R. Omitting the modulation of the short-range strong force leads to building of sub-formations depending on the intra-cluster density variation. However this could even be advantageous, if one is interested in exploring the inner structure of a larger group. Our goal here was to detect major clusters, which was successfully achieved by applying the parameters described in section 2.1.

Analyzing the MS data showed that patients with low grade disease (mean EDSS=0.67) form a homogeneous group, meaning that their image derived features are comparable. On the other hand, patients with a higher EDSS grade (mean EDSS=2.7) are identified as noise. This appears reasonable as higher disease grades are expected to be much more heterogeneous. Indeed our data reveal that this group does not show a consistent pattern, at least with respect to features derived from quantitative T_1-, T_2^*-, and absolute water content maps.

In summary, we have developed a new method for grouping high dimensional data based on a simplified model of nuclear fusion. The new approach allows for the clustering of non-spherical, correlated datasets, which are typically observed in MR derived image features. It therefore presents an ideal tool for the automated analysis of medical imaging data, not only in the MRI domain.

References

[1] Duda RO, Hart PE, Stork DG 2001, *Pattern Classification* (New York: Wiley-Interscience)

[2] Neema M, Stankiewicz J, Arora A, Guss ZD, Bakshi R. MRI in MultipleSclerosis: What'sInside the Toolbox?, Neurotherapeutics 2007;4: 602-617.

[3] Bakshi R, Thompson AJ, Rocca MA, Pelletier D, Dousset V, Barkhof F, Inglese M, Guttmann CRG, Horsfield MA, Filippi M. MRI in multiple sclerosis: current status and future prospects. Lancet Neurol 2008; 7:615-625.

[4] Kurtzke JF. Rating neurologic impairment in multiple sclerosis: an expanded disability status scale (EDSS). Neurology 1983; 33 (11): 1444–52.

[5] Krane KS 1988, *Introductory Nuclear Physics* (New York: Wiley & Sons)

[6] Neeb H Ermer V, Stoecker T, Zilles K, Shah NJ. Rapid high-resolution mapping of cerebral water content with full brain coverage. Neuroimage 2008; 42(3): 1094-1109.

[7] Woods RD, Saxon DS. Diffuse surface optical model for nucleon-nuclei scattering, Phys. Rev. 1954; 95: 577–578

Journal of Physics: Conference Series **410** (2013) 012005 doi:10.1088/1742-6596/410/1/012005

A Computer Library for Ray Tracing in Analytical Media

Eduardo Miqueles*, J.J.S de Figueiredo†, Tiago A. Coimbra‡

$\star$, $\ddagger$ Department of Applied Mathematics, University of Campinas, São Paulo, Brazil
$\dagger$ Geophysics Departament, Federal University of Pará, Pará, Brazil
Grants: $\star$ Supported by FAPESP No. 09/15844-9, $\dagger,\ddagger$ Supported by CnPq, CAPES
E-mail: `miqueles@ime.unicamp.br`, `jadsomjose@gmail.com`, `tgo.coimbra@gmail.com`

Abstract. Ray tracing technique is an important tool not only for forward but also for inverse problems in Geophysics, which most of the seismic processing steps depends on. However, implementing ray tracing codes can be very time consuming. This article presents a computer library to trace rays in 2.5D media composed by stack of layers. The velocity profile inside each layer is such that the eikonal equation can be analitically solved. Therefore, the ray tracing within such profile is made fast and accurately. The great advantage of an analytical ray tracing library is the numerical precision of the quantities computed and the fast execution of the implemented codes. Although ray tracing programs already exist for a long time, for example the `seis` package by Červený, with a numerical approach to compute the ray. Regardless of the fact that numerical methods can solve more general problems, the analytical ones could be part of a more sofisticated simulation process, where the ray tracing time is completely relevant. We demonstrate the feasibility of our codes using numerical examples.

1. The Computer Library `art`

We have developed a free C-library called `art`, standing for *analytical ray tracing*. The 2.5D synthetic medium for ray tracing considered by the library is presented in Figure 1. The main goal of the library is the determination of a ray within the profile π of Figure 1, starting a source point (usually at the surface, identified by the symbol $\star$) and arriving at a receiver or at an interface (identified by ∇). Since the medium is composed by a stack of layers, where each layer has one of the analytical velocities presented in the previous section, the first step in tracing a ray through a profile, is the computation of incidence points along a sequence of layers. Computationally, the union of incidence points determines completely the raypath, because we know exactly (or analytically) the behavior of the curve between such extreme points. The sequence of layers is given a priori, and is known as the *ray code*. In a synthetic profile like the described above, the library allows the following two basic operations: *one-way ray tracing*, given a point source and an initial slowness (inverse of velocity) direction, find the ray through the profile, following a ray code sequence and *two-point ray tracing*, find the ray through the profile that starts at a given point source and finish at a given receiver, obeying a ray code sequence. Others operations are also allowed in the library, but are nothing more than applications of this two basic procedures.

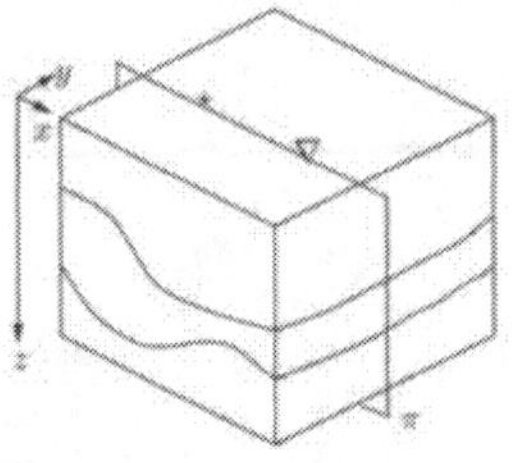

Figure 1. Seismic exploration through plane π.

Published under licence by IOP Publishing Ltd

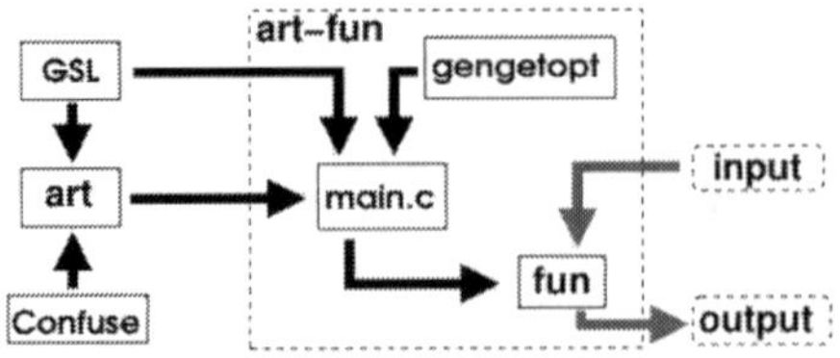

Figure 2. Fluxogram for art.

As other free libraries, like `gsl` [1], our library provide many 'ray tracing functions', so the user can be free to implement his/her own codes; which is basically the idea of `art`. On the other hand, we also provide programs that already contain some standard routines in seismic ray tracing, like two-point ray tracing, one-way ray tracing, common-shot experiments, among others. These functions are contained within `art-fun`. Also, the library depends on other free softwares, the `libConfuse`[2] and the `gengetopt`[3], as shown in the fluxogram (at the left). The library is in continuous expansion and in a near future we will present an expanded article with the whole library, examples, and more sophisticated applications. This paper has the purpose to present the library, comparing with the well-known software `seis`.

2. Seismic Ray Tracing

The seismic wave propagating through an elastic medium obeys the elastodynamic wave equation [1], $\rho\frac{\partial^2 \mathbf{u}}{\partial t^2} = \mathcal{E}(\mathbf{x}, \mathbf{u}, \nabla \cdot \mathbf{u}, \nabla \times \mathbf{u})$, where $\mathcal{E}$ is an operator over the displacement vector $\mathbf{u}$ and ρ is the density of the isotropic medium. In acoustic media, it is easy to show that elastic waves are a superposition of two other waves, the compressional scalar wave $\theta = \nabla \cdot \mathbf{u}$ and the shear vectorial wave $\Omega = \nabla \times \mathbf{u}$, also called P and S waves respectively. For a non-acoustic media, the splitting is no longer possible, although in a high-frequency sense, this is approximately true. Hence, the ray-theory, assume that the zeroth-order approximation $\mathbf{u}(\mathbf{x}, t) = \mathbf{U}(\mathbf{x})F(t - T(\mathbf{x}))$ is a high-frequency solution of the above PDE, where $\mathbf{U}$ is a complex-valued vectorial *amplitude* function, F is a high-frequency signal and T is the real-valued *traveltime* function. So, to estimate the zeroth-order solution, we need to compute the traveltime and amplitude functions. After several operations with the *ansatz* $\mathbf{u}$ and the elastodynamic equation, it can be shown that the traveltime obeys the so-called *eikonal equation* $\mathbf{p}^T\mathbf{p} = \|\mathbf{p}\|^2 = \frac{1}{v(\mathbf{x})^2}$, where $\mathbf{p} = \nabla T$ is the *slowness vector* and v is the wave velocity.

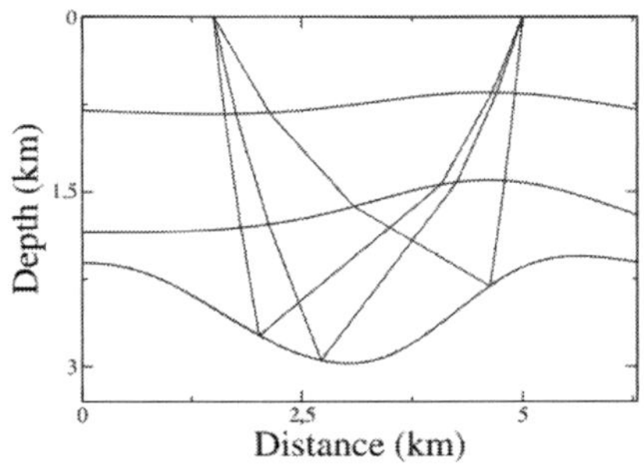

Figure 3. Two-point ray tracing using `art`, between source point $x = 1.5$ and receiver $x = 5$. This corresponds to slice of model depicted in Figure 1.

Usually, we had to solve this equation for P and S waves, so we denote $v = \alpha$ for the first case and $v = \beta$ for the second. Functions α, β are related to the parameters of the medium. For pressure waves the displacement vector is perpendicular to the wavefront and thus, we can write $\mathbf{U}(\mathbf{x}) = \alpha(\mathbf{x})A(\mathbf{x})\mathbf{p}$. Since shear waves propagate in the parallel wavefront direction, we write $\mathbf{U}(\mathbf{x}) = B(\mathbf{x})\mathbf{e}_1(\mathbf{x}) + C(\mathbf{x})\mathbf{e}_2(\mathbf{x})$. In both cases, we need to compute the scalar amplitude functions A, B and C. In the case of P waves, the function A obeys the *transport equation* $2\mathbf{p}^T\nabla\left(\sqrt{\rho\alpha^2}A\right) + \sqrt{\rho\alpha^2}A\nabla^2 T = 0$. Also, the same transport equation applies for S waves, with $v = \beta$.

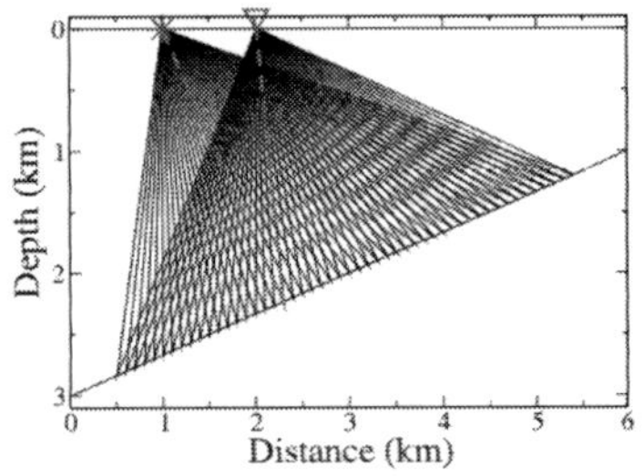

Figure 4. Ray set using `art`, for the Kirchhoff integral using a dipping reflector.

The eikonal equation is associated to a Hamiltonian operator,

[1] `http://www.gnu.org/software/gsl/`

[2] `http://www.nongnu.org/confuse/`

[3] `http://www.gnu.org/software/gengetopt/`

which is related to an index $n \in \mathbb{N}$ - which we describe as the *hamiltonian index*. After using the method of characteristics, which is fully described in [4] and [3], we arrive at a system of ordinary equations. In our case, these equations are given by

$$\left\{ \begin{array}{l} d_u\mathbf{x} = v(\mathbf{x})^{2-n}\mathbf{p}, \quad d_u\mathbf{p} = \frac{1}{n}\nabla\left(\frac{1}{v(\mathbf{x})^n}\right) \\ d_uT = \frac{1}{v(\mathbf{x})^n} \end{array} \right\} \quad \left\{ \begin{array}{l} d_u\mathbf{x} = v(\mathbf{x})^2\mathbf{p}, \quad d_u\mathbf{p} = -\nabla\left(\ln v(\mathbf{x})\right), \\ d_uT = 1 \end{array} \right\} \tag{1}$$

for $n \neq 0$ and $n = 0$ respectively, and with $d_u(\cdot) = d(\cdot)/du$. These are the so-called *ray equations*. The curve $\mathbf{x}\colon \mathbb{R} \to \mathbb{R}^3$ has the slowness vector $\mathbf{p}$ as a tangent vector and the integration parameter u varying over the domain $u \geq u_0$. Usually, we will adopt the convention $u_0 = 0$. For the case $n = 0$, the integration parameter is the traveltime itself. The kinematic part of a ray is mathematically represented by the characteristic curve $(\mathbf{x}, \mathbf{p}, T)$ and the raypath is determinaned by $\mathbf{x}$. Along this ray, the amplitude function A can be computed by $A(u) = \sqrt{\frac{\rho(u_0)v(u_0)J(u_0)}{\rho(u)v(u)J(u)}}$, where $v(u_0) = v(\mathbf{x}(u_0)))$ and J is the ray Jacobian related to a usual orthonomic ray system parameterization, see [4]. The *dynamical ray tracing* provides a theory mainly concerned with the computation of J through a new system of ordinary differential equation that can be decoupled into the so called *in-plane* and *out-of-plane* systems.

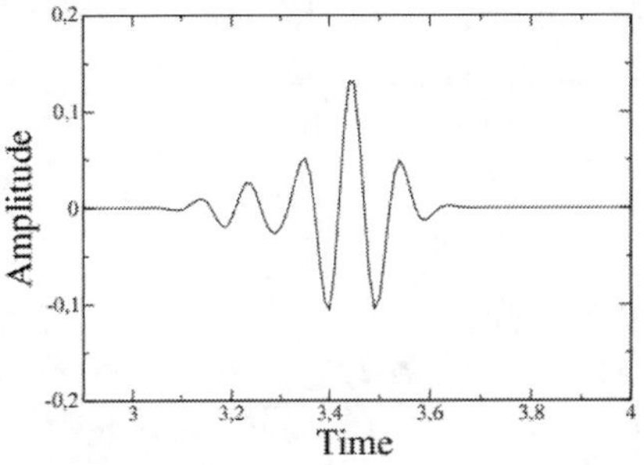

Figure 5. Seismic trace, depending on the amplitude function A, computed along the ray.

3. Analytical Medium

Since a ray is determined by the solution of (1), either for $n \neq 0$ or $n = 0$, we are inspired to choose velocity fields that enable analytical solutions. In this sense, we select particular values for the index n in order to make easier the ray equations. In our case, these values are $n \in \{-1, 0, 1, 2\}$. In a general sense, we consider that velocities can be written in the form $\Phi(v(\mathbf{x})) = \mathbf{x}^T\mathbf{A}\mathbf{x} + \mathbf{a}^T\mathbf{x} + c$ where Φ assume the value of one of the functions $\{v, v^{-2}, \ln v\}$, as ilustrated by the following Table. By a suitable choice of the parameters $\mathbf{A}$, $\mathbf{a}$ and c we can always ensure that $\Phi(v(\mathbf{x})) > 0$.

Id.	(n, u)	$\Phi(v)$
ART_VCONST	$(1, s)$	$v(\mathbf{x}) = c$
ART_VAFF	$(-1, \xi)$	$v(\mathbf{x}) = \mathbf{a}^T\mathbf{x} + c$
ART_VASS, ART_VQSS	$(2, \sigma)$	$\frac{1}{v(\mathbf{x})^2} = \mathbf{x}^T\mathbf{A}\mathbf{x} + \mathbf{a}^T\mathbf{x} + c$
ART_VCGL	$(0, T)$	$\ln v(\mathbf{x}) = \mathbf{a}^T\mathbf{x} + c$

Figure 6. Analytical velocity fields.

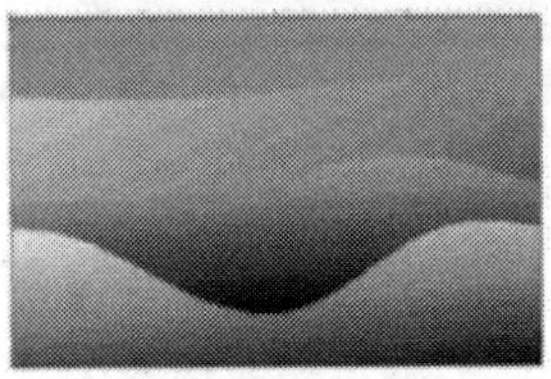

Figure 7. Affine velocity.

To distinguish the velocities, we established the following identification: `vass` standing for affine square of slowness, `vqss` for quadratic square of slowness, `vcgl` for constant gradient of logarithmic velocity, `vaff` for affine velocity and `vconst` for constant velocity. Since we are mainly interested in rays that are confined in a vertical plane along the seismic profile, we can assume that parameters in equation $\Phi(v(\mathbf{x})$ are two-dimensional, i.e, $\mathbf{A} \in \mathbb{R}^{2\times 2}$ and $\mathbf{a} \in \mathbb{R}^2$ where the position vector $\mathbf{x}$ has first component related to distance and second component related to depth. Figure 7 shows a profile with nonconstant velocity. We note that the solution of the kinematic ray tracing system (1) can always be written by $\mathbf{x}(u) = \mathbf{x}_0 + \mathbf{d}_n(u)$ for the position vector, $T(u) = T_0 + \mathbf{t}_n(u)$ for the traveltime and $\mathbf{p}(u) = \mathbf{p}_0 + \mathbf{s}_n(u)$ for the slowness vector; where $\{\mathbf{x}_0, T_0, \mathbf{p}_0\}$ are initial conditions for the ray. The analytical solution varies according to n for each velocity model. Functions $\{\mathbf{t}_n, \mathbf{s}_n, \mathbf{p}_n\}$ can be found explicitly in [1] and [5], also in [3]

4. Numerical Examples

The ray tracing with `art` is presented in Figures 9.a to 9.e. Each figure has depth (in Km) as the vertical axis and distance (in Km) as the horizontal axis. In figure (a), we present a typical common-shot experiment (CS) with source placed at $x = 3$ and receivers equally distributed at the surface; in this case, each reflector branch has dipping angle $\theta = 11,30°$. The seismogram obtained for this example is presented in Figure 8, using seismic traces as the one depicted in Figure 5. Figure (b) and (c) presents the behaviour of a ray family (including Fermat rays), usually needed for the computation of the Kirchhoff integral [3], the first one being at a synthetic anticlinal reflector and the second at a synclinal reflector; both using affine velocity ($n = -1$, see Figure 6) within the first layer (also, source at $x = 1$ and receiver at $x = 3.5$). Figure (d) presents a ray-diagram for a common-shot experiment at a profile with affine velocity ($n = -1$) within second, third and fourth layers, as shown in Figure 7. The associated Kirchhoff seismic section to this velocity model is presented in Figure 10. It is shown in Figures (e) and (f) a comparison between the rays obtained at a common-shot example, with package `art` and `seis`, respectively.

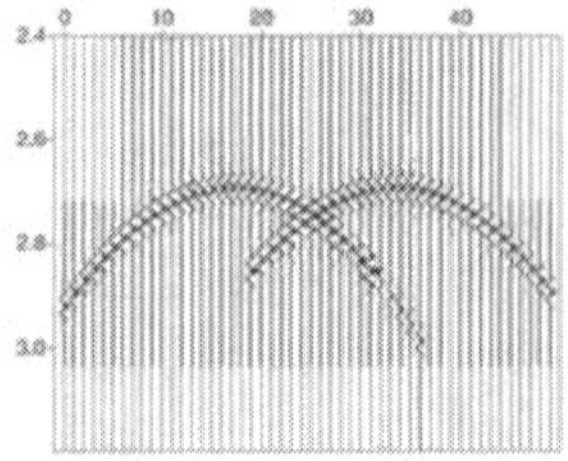

Figure 8. Seismogram obtained with `art`, from rays at Figure 9.a.

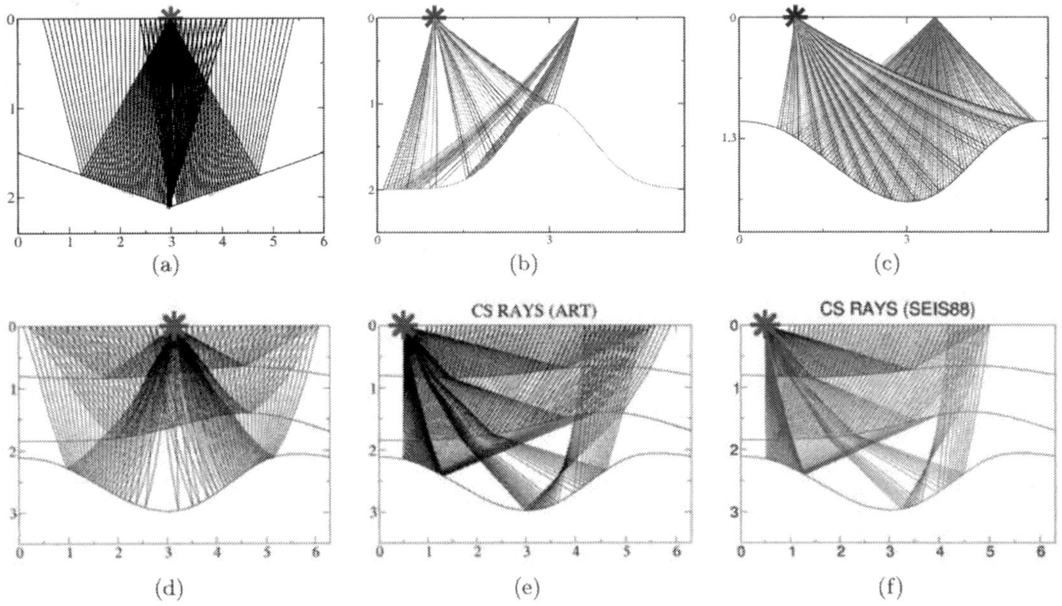

Figure 9. Ray tracing using `art`. See text for more details.

References

[1] V. Červený, *Seismic Ray Theory*, Cambridge Univ. Press, 2001.
[2] P. Hubral, T. Krey, *Interval velocities from Seismic Reflection Time Measurements*, SEG Monograph Series no. 3, 1980.
[3] N. Bleistein, *Mathematics of Modelling, Migration and Inversion with Gaussian Beams* Department of Geophysics, Colorado School of Mines, USA, 2008
[4] N. Bleistein, *Mathematical Methods for wave phenomena*, Academic Press, 1984.
[5] E. Miqueles, *Seismic Modelling in Analytical Media*, Master Thesis (*in portuguese*), Federal University of Paraná, Brazil, 2006

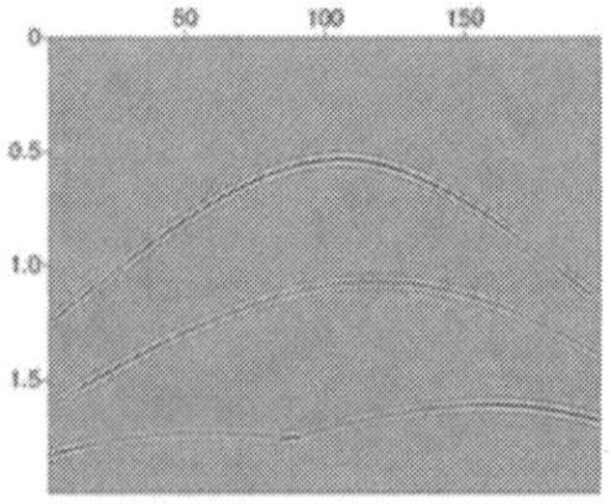

Figure 10. Seismic section for rays from Figure 9.d and velocity model from Figure 7, using `art`.

Journal of Physics: Conference Series **410** (2013) 012006 doi:10.1088/1742-6596/410/1/012006

Fuzzy logic control for active bus suspension system

Mujde Turkkan and Nurkan Yagiz
Department of Mechanical Engineering, Faculty of Engineering, Istanbul University, 34320 Avcilar, Istanbul, Turkey

E-mail: mujdeg@istanbul.edu.tr, nurkany@istanbul.edu.tr

Abstract. In this study an active controller is presented for vibration suppression of a full-bus suspension model that use air spring. Since the air spring on the full-bus model may face different working conditions, auxiliary chambers have been designed. The vibrations, caused by the irregularities of the road surfaces, are tried to be suppressed via a multi input-single output fuzzy logic controller. The effect of changes in the number of auxiliary chambers on the vehicle vibrations is also investigated. The numerical results demonstrate that the presented fuzzy logic controller improves both ride comfort and road holding.

1. Introduction

Vibrations, caused by road roughness affect the ride comfort and drive safety negatively. These vibrations can be minimized by using semi-active [1] and active suspension systems [2].
In this study, an active control strategy was applied to the full-bus suspension system. Fuzzy logic controllers were used on this system with a fixed number of auxiliary chambers. Fuzzy logic control was preferred since it has a model-free design methodology.

2. The bus model with seven degrees of freedom

The physical model of the full-bus is presented in Figure 1.

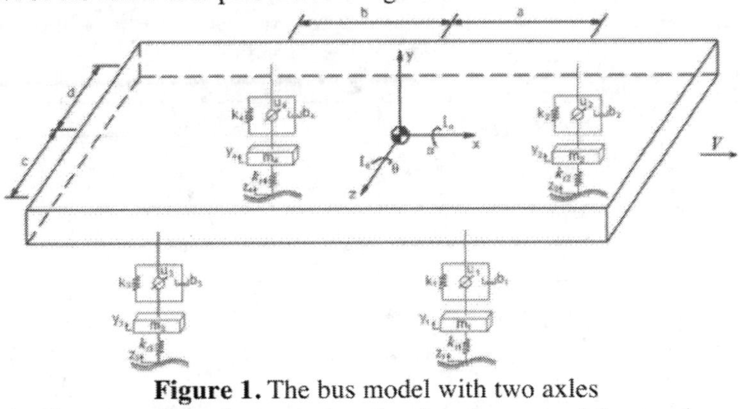

Figure 1. The bus model with two axles

a, b, c and d are the distances from the body inertia of each rear and front axles; z_1, z_2, z_3, z_4 are the front and rear wheel road surface inputs; k_{t1}, k_{t2}, k_{t3}, k_{t4} are the stiffness of front and rear wheels; m_1, m_2, m_3, m_4 are front and rear unsprung masses; y_1, y_2, y_3, y_4 are front and rear wheel hops; k_1, k_2, k_3, k_4 are front and rear suspension spring constants; b_1, b_2, b_3, b_4 are front and rear suspension damping

coefficients; u_1, u_2, u_3, u_4 are control inputs to the front and rear of the vehicle; y is body bounce; θ and α are body pitch and roll angles; I_θ and I_α are body inertias of pitch and roll, respectively.
The air spring model proposed by Bal [3] is used in this study. It can be designed to have different spring constants while carrying the same load. This is achieved by using auxiliary chambers which can be added to or separated from the air spring volume. The air spring model with auxiliary chamber is shown in Figure 2.

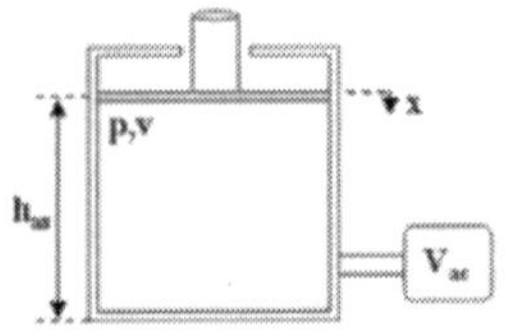

Figure 2. Air spring model with auxiliary chamber

Here p is the pressure and v is the volume of air spring at time t; V_{ac} is volume of auxiliary chamber; has is height of air spring; x is displacement of the air spring. Accordingly, the spring constant of the air spring is

$$k = nMgA_{as}(h_{as}.A_{as} + n_{ac}.V_{ac})^n[(h_{as}.A_{as} + n_{ac}.V_{ac}) \quad A_{as}x]^{-(n+1)} \tag{1}$$

n : Polytropic exponent
M : Body mass
g : Acceleration of gravity
A_{as} : Area of auxiliary chamber
n_{ac} : Number of auxiliary chamber

3. The fuzzy logic controller

In this section the effects of the fuzzy logic controller on ride comfort and drive safety are investigated. The algorithm of the multi-input single-output fuzzy logic controller for the vehicle suspension system uses the errors of the suspension end velocities, their accelerations and suspension gap velocities as the input variables while the control forces are u_1, u_2, u_3 and u_4. The range of membership functions has been chosen as [-1, 1] and scaling factors are used in order to map the crisp input output values to their corresponding fuzzy values. A model of the rule base developed for each suspension is given in Table 1. In this Fuzzy Associative Memory (FAM) table, the rules are given for front right suspension. x_1 denotes the displacement of the front right suspension vehicle connection. The column of with each (P or N) row stand for two rules using the "or" connector. Therefore, there are 75 rules in this table. Identical FAM tables are valid for the other suspensions as well. P, N, ZE, VB, B, M and S represent Positive, Negative, Zero, Very Big, Big, Medium and Small, respectively. The rule base is generated, based on the study of Yagiz and Sakman [2].

Table 1. Fuzzy Associative Memory (FAM) table

$e(\dot{x}_1 - \dot{z}_1)$	$e(\dot{x}_1)$	$e(\ddot{x}_1)$	u_1	$e(\dot{x}_1 - \dot{z}_1)$	$e(\dot{x}_1)$	$e(\ddot{x}_1)$	u_1
PM	PM	ZE	ZE	PM	PM	P or N	PS
PS	PM	ZE	PS	PS	PM	P or N	PM
ZE	PM	ZE	PM	ZE	PM	P or N	PB
NS	PM	ZE	PM	NS	PM	P or N	PB
NM	PM	ZE	PB	NM	PM	P or N	PVB
PM	PS	ZE	ZE	PM	PS	P or N	PS
PS	PS	ZE	PS	PS	PS	P or N	PM
ZE	PS	ZE	PS	ZE	PS	P or N	PM
NS	PS	ZE	PM	NS	PS	P or N	PB
NM	PS	ZE	PM	NM	PS	P or N	PB
PM	ZE	ZE	NS	PM	ZE	P or N	NM
PS	ZE	ZE	ZE	PS	ZE	P or N	NS
ZE	ZE	ZE	ZE	ZE	ZE	P or N	ZE
NS	ZE	ZE	ZE	NS	ZE	P or N	PS
NM	ZE	ZE	PS	NM	ZE	P or N	PM
PM	NS	ZE	NM	PM	NS	P or N	NB
PS	NS	ZE	NM	PS	NS	P or N	NB
ZE	NS	ZE	NS	ZE	NS	P or N	NM

NS	NS	ZE	NS	NS	NS	P or N	NM
NM	NS	ZE	ZE	NM	NS	P or N	NS
PM	NM	ZE	NB	PM	NM	P or N	NVB
PS	NM	ZE	NM	PS	NM	P or N	NB
ZE	NM	ZE	NM	ZE	NM	P or N	NB
NS	NM	ZE	NS	NS	NM	P or N	NM
NM	NM	ZE	ZE	NM	NM	P or N	NS

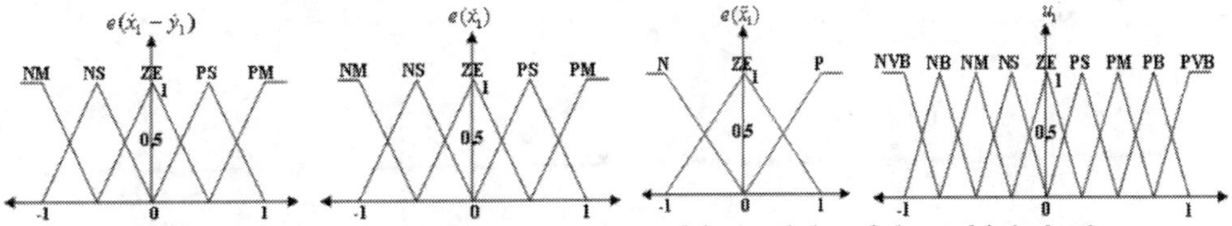

Figure 3. Membership function plots of front right of the vehicle body

4. Numerical results

The bus model is assumed to travel over the bump shown in Figure 4. The road input to the rear wheel is the same but delayed by Δt

$$\Delta t = (a + b)/V \tag{2}$$

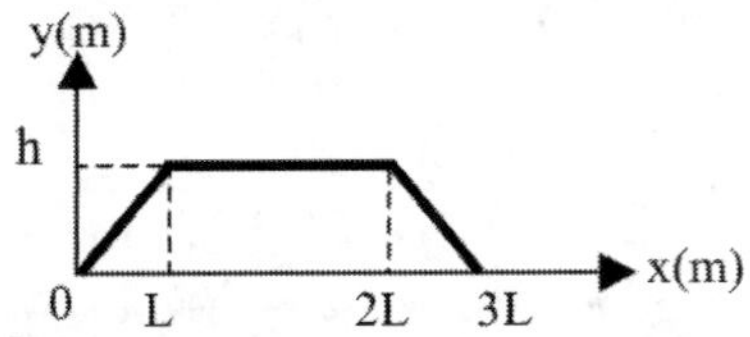

Figure 4. Road disturbance input

The vehicle bounce and pitch motion and their accelerations are shown in Figure 5 for controlled, uncontrolled-without auxiliary chambers and uncontrolled-with 4 auxiliary chambers cases. It is seen that when auxiliary chambers are in effect the ride comfort improves since magnitudes of the displacements are reduced.

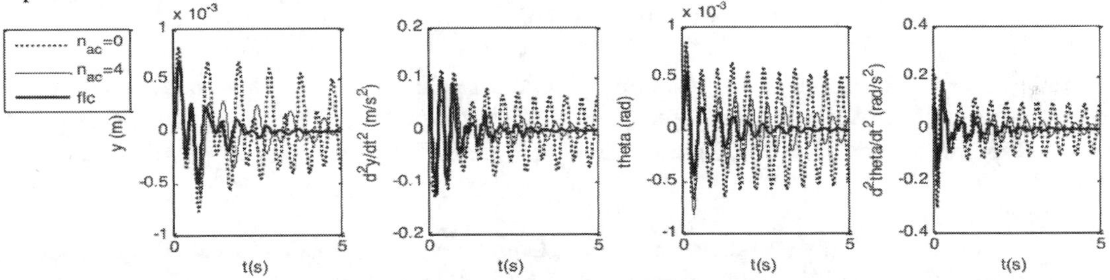

Figure 5. Time responses of the full-bus model

Suspension deflections for the uncontrolled and fuzzy controlled cases are shown in Fig. 6. It is clear from this figure that the maximum suspension deflections for the fuzzy controlled case do not exceed the un-controlled ones with four auxiliary chambers and that there is no permanent deflection, which indicates that the ride comfort is improved without degenerating the suspension working limits.

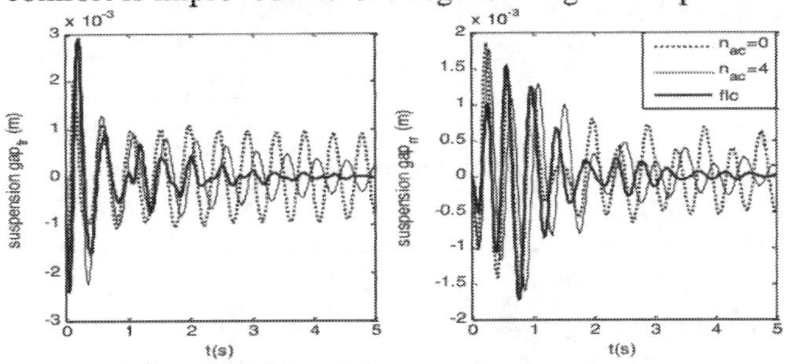

Figure 6. Time history of suspension gap

Frequency response is another way to evaluate the behaviour of a dynamic system. The frequency response of main body displacement and its acceleration are given in Figure 7. The resonances of the body motions belonging to body bounce and its acceleration disappear and the amplitude of the motion over the most of the frequency range becomes smaller effectively. The amplitude of the dynamic wheel load is reduced by fuzzy logic controller.

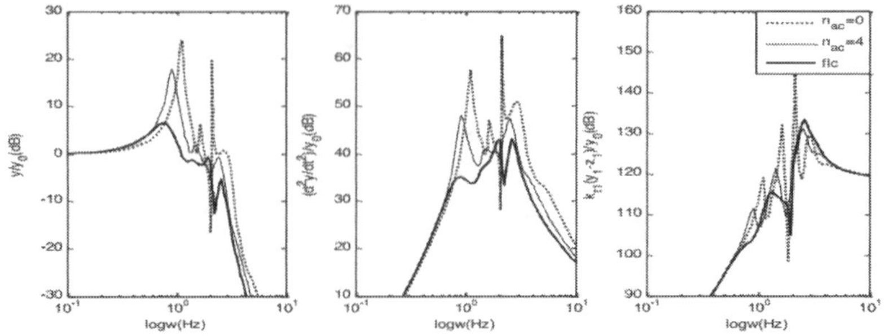

Figure 7. Frequency response of main body, its acceleration and dynamic wheel load

Figure 8 depicts the time history of the control force. The maximum values of the front and rear control forces are about 1200N and 400N, respectively.

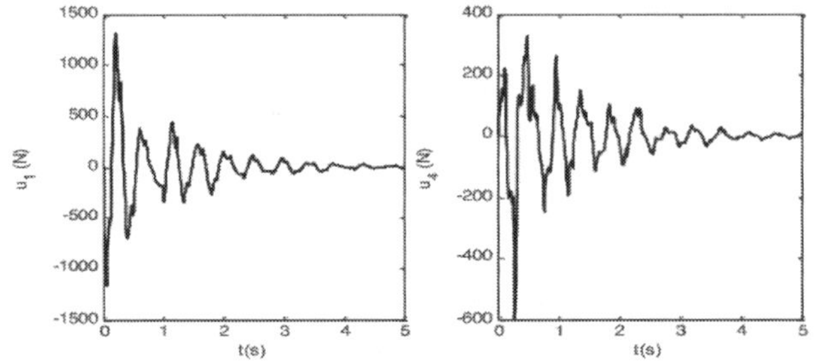

Figure 8. Controller forces of change over time graphs

5. Conclusions

In this study, a controller is designed for comfortable and reliable ride for a bus model using an air spring with auxiliary chambers. A fuzzy logic controller was presented in this. The controller is able to protect the suspension working limits. On the other hand ride comfort can be improved while dynamic wheel load improve.

6. References

[1] N. Al-Holou, A. Bajwa and D. S. Joo, “Computer Controlled Individual Semi-Active Suspension System”, Circuits and Systems, Vol.1, 1993, pp 208 – 211.

[2] N. Yagiz, Y. Hacioglu, Y. Taskin, “Fuzzy Sliding-Mode Control of Active Suspensions”, Industrial Electronics, Vol. 55, 2008, pp 3883 – 3890.

[3] H. Bal, Theoretical Investigation of the Performances of The Semi-active Suspension Systems with Air Springs and Comparison with Mechanical Suspensions, M. Sc. Thesis, Gazi University Institute of Science and Technology, 2006.

[4] Z. Sihong, W. Jiasheng, Z. Ying, “Research on Theoretical Calculation Model for Dynamic Stiffness of Air Spring with Auxiliary Chamber”, Vehicle Power and Propulsion Conference, 2008, pp. 1 – 6.

[5] E. Guglielmino, T. Sireteanu, C. W. Stammers, G. Ghita, M. Giuclea, Semi-active Suspension Control, London: Springer, 2008.

[6] G. Quaglia, M. Sorli, “Air Suspension Dimensionless Analysis and Design Procedure”, Vehicle System Dynamics, Vol. 35, No. 6, 2001, pp 443-475.

[7] K. Toyofuku, C. Yamada, T. Kagawa, T. Fujita, “Study on Dynamic Characteristic Analysis of Air Spring with Auxiliary Chamber”, Journal of Sound and Vibration, Vol. 20, 1999, pp 349–355.

IC-MSQUARE 2012: International Conference on Mathematical Modelling in Physical Sciences IOP Publishing
Journal of Physics: Conference Series **410** (2013) 012007 doi:10.1088/1742-6596/410/1/012007

Efficient gradient and Hessian calculations for numerical optimization algorithms applied to atomistic molecular simulations

Marco Hülsmann[1], Sonja Kopp[1], Markus Huber[1], Dirk Reith[1]

[1]Fraunhofer-Institute for Algorithms and Scientific Computing (SCAI), Schloß Birlinghoven, 53754 Sankt Augustin, Germany

E-mail: dirk.reith@scai.fraunhofer.de

Abstract. Computer simulations of chemical systems can be used to reliably predict physical properties. Accurate molecular models, which are indispensable, are mathematically described by force fields, which have to be parameterized. Recently, an automated gradient-based parametrization procedure was published by the authors based on the minimization of a loss function between simulated and experimental physical properties. The applicability of the utilized algorithms is not trivial at all because of two reasons: First, simulation data is affected by statistical noise and second, the molecular simulations required for the loss function evaluations (involving finite differences approximations of gradients and Hessians) are extremely time-consuming. In this work, a more efficient approach to compute gradients and Hessians is presented. The method developed here is based on directional instead of partial derivatives. It is shown that up to 75% of the simulations can be avoided using this method.

1. Introduction: Why is efficient force field optimization crucial?

A common goal within computational chemistry, be it Newtonian or quantum based methods, is the accurate modeling of physical forces and energetics through mathematics and algorithm design. Through reliable modeling of the underlying forces, molecular simulations can then provide atomistic insights into macroscopic experimental observations. Here, we focus on simulations of systems that consist of several hundreds up to roughly a million atoms, usually representing condensed phase substances. Due to their size, they can only be treated by methods based on Newtons equations of motion, e.g. molecular dynamics or Monte Carlo simulations [1]. For such systems the task of creating a meaningful model, often termed as force field, is not straightforward. That is because the terms involved are then based on classical potentials and, hence, the model parameters are semi-empirical and cannot be derived strictly from first principles any more. Consequently, the only option in the past to reach an acceptable model for a given substance was to manually adjust the involved parameters. This process was, at best, extremely time consuming. Recently, we finished to create two seperate tools in order to accomplish an automated, quick and standardized parameter optimization, both for the intramolecular and intermolecular degrees of freedom [2, 3]. Even though well-defined workflows have been established, the optimization of the intermolecular parameters with the GRadient-based Optimization Workflow (GROW) is still a very demanding task in terms of computational resources. That is because they can only be verified via comparison with experimentally determined observables, e.g. density, thermal and electrical conductivity or diffusion. The mathematical formulation of such an optimization task reads as follows:

$$F(x) = \sum_{i=1}^{n} \sum_{T \in \mathcal{T}} w_{i,T} \left(\frac{f_{i,T}^{\text{exp}} - f_{i,T}^{\text{sim}}(x)}{f_{i,T}^{\text{exp}}} \right)^2 , \tag{1}$$

The iterative calculation of the resulting gradients (see Section 2 for details) represents the resource-intense step: each entry in the derivation of the gradient corresponds to the analysis of a full-blown molecular simulation. On a current single-node multi-core CPU, this can involve real time calculations of several hours up to days. Hence, it is a primary goal to save as many executions of molecular simulations during the optimization process as possible. One way to achieve this goal is to choose the best (combination of) gradient-based algorithm(s) for our specific problem which we already pursued and successfully implemented [4]. Here, we want to investigate another possibility to reach this goal: Within the methodological framework of gradient-based optimization algorithms, we want to re-utilize data points that were already calculated in previous steps of the iterative process.

2. Theory: Gradient and Hessian calculations via directional derivatives

Consider a function $F\colon \mathbb{R}^N \to \mathbb{R}, (x_1,\ldots,x_N)^T \mapsto F(x_1,\ldots,x_N)$, and iterations x^k of a numerical optimization algorithm:

$$x^k = x^{k-1} + t_k d^k, \; k \in \mathbb{N}^{\geq 1}. \tag{2}$$

Thereby, d^k is the current descent direction and t_k the step length. The descent direction d^k is determined using the gradient of F, e.g. $d^k = -\nabla F(x^{k-1})$ in the case of the Steepest Descent method (for more details cf. [3]). The gradient is given by the partial derivatives of F, described by the euclidian scalar product of the gradient and the unit vectors $e_i \in \mathbb{R}^N$, $i = 1,\ldots,N$. Hence, they are considered as the directional derivatives in the direction of e_i, $i = 1,\ldots,N$. The partial derivatives can be numerically approximated using finite differences. This way of computing the gradient is defined as *classical gradient calculation*. Thereby, the gradient is approximated via the determination of $N+1$ function evaluations, i.e. $F(x^k), F(x^k + he_1),\ldots,F(x^k + he_N)$. In this work, the goal is to compute a gradient while avoiding function evaluations for $x^k + he_i$, $i = 1,\ldots,N$. To achieve this, a special base transformation is needed: A set of base vectors $\{v_i^k\}_{i=1}^N$ is considered so that the linear combinations $x^k + v_i^k$, $i = 1,\ldots,N$, lead to vectors for which F has already been evaluated during the iterative process (2). For example, they can be chosen as $v_i^k := x^k - x^{k-1} + he_i$, whenever the vectors $x^{k-1} + he_i$, $i = 1,\ldots,N$, have already been evaluated. It holds

$$F(x^k + v_i^k) = F(x^{k-1} + he_i), \; i = 1,\ldots,N. \tag{3}$$

The $\{v_i^k\}_{i=1}^N$ form a set of new base vectors, and instead of calculating the partial derivatives, the directional derivatives in the direction of v_i^k, $i = 1,\ldots,N$, are calculated. Using a backward base transformation from $\{v_1^k,\ldots,v_N^k\}$ to $\{e_1,\ldots e_N\}$, the gradient of $\nabla F(x^k)$ can be calculated by solving a linear equation system. This sort of gradient approximation is only applicable, if the following two conditions are fulfilled:

- In order to guarantee that the accuracy of the gradient remains in $\mathcal{O}(h)$, there has to exist a constant $C > 0$ so that and the following has to be met: $\|v_i^k\| \leq Ch$, $i = 1,\ldots,N$. Then, $\|v_i^k\| = \mathcal{O}(h)$ for all $i = 1,\ldots,N$.
- It has to be ensured that $\{v_1^k,\ldots,v_N^k\}$ is a base. Therefore, the condition number of A, $\kappa(A)$, must not exceed a certain threshold $\epsilon_\kappa > 0$: $\kappa(A) < \epsilon_\kappa$. If $\kappa(A) \leq 1$, the matrix is well-conditioned.

Table 1. Results of the efficient gradient calculations compared to the classical calculations for the Steepest Descent (SD) and the Conjugate Gradient methods (PR and FR). The numbers shown are the mean values of ten statistically independent random replicates. In the columns, the algorithm (Alg.), the number of successful replicates (# Repl.), the number of iterations (# It.), the number of function evaluations (# Eval.), the number of function evaluations per iteration (# Eval./# It.), and the final parameter vectors fulfilling the stopping criterion (x^{opt}) are indicated. The average number of function evaluations was reduced by a factor of 2–4. In each optimization process, the gradient was calculated no more than twice in the classical way. Hence, in most cases, it was calculated efficiently.

	Classical Gradient Calculation					Efficient Gradient Calculation				
Alg.	# Repl.	# It.	# Eval.	# Eval./# It.	x^{opt}	# Repl.	# It.	# Eval.	# Eval./# It.	x^{opt}
SD	9	6	46–47	7–8	0.3056 0.3285 0.0176 0.1120	9	5	20–21	4–5	0.3054 0.3286 0.0172 0.1110
FR	10	6–7	82	12–13	0.3055 0.3285 0.0174 0.1122	10	6–7	27–28	4–5	0.3051 0.3288 0.0170 0.1114
PR	10	5–6	55–66	10–11	0.3056 0.3284 0.0176 0.1120	9	5–6	18	3–4	0.3054 0.3283 0.0171 0.1109

If one of the above conditions is not fulfilled, the gradient has to be calculated in the classical way. The efficient gradient calculation can directly be transfered to the Hessian, which is needed for example for the Trust Region method, cf. [4, 5]. The efficient gradient calculation can simply be applied to $\nabla F(x^k + he_i)$, $i = 1, \dots, N$, but in this case, function evaluations for $x^k + he_i$, $i = 1, ..., N$, are required and the gradient has to be computed classically. However, in the case of the Hessian, the first condition has to be strengthened: Due to the division by h^2 using finite differences, small errors in the enumerator have an enourmous effect on the accuracy of the Hessian. Hence, $C = 1$ is the only admissible choice guaranteeing that the accuracy of the Hessian remains in $\mathcal{O}(h)$.

3. Results: Applicability of the efficient gradient and Hessian calculations

As molecular simulations require a high amount of computational time, an analytical model replacing them was used for a first assessment of the methodology. In [4], a similar assessment has already been performed for the classical gradient and Hessian calculations. In such a way, physical properties like the saturated liquid density ρ_l and the vapor pressure p_σ are calculated directly by fit functions with minimal computational efforts. The fit functions were determined by [6] by nonlinear regression using a high amount of simulation data for two-centered Lennard–Jones (LJ) fluids [1] with a quadrupolar moment. Here, nitrogen was considered as an example for this kind of fluids. The model parameters to be opimized were the elongation L of the two LJ sites, the quadrupolar moment Q^2, and the two LJ parameters σ and ε. In order to mimic molecular simulations, uniformly distributed artificial noise (0.5% for the density and 3% for the pressure) was added for the optimization. The two properties were optimized simultaneously at six different temperatures. As the simulation data were noisy, ten statistically independent random replicates were performed for each optmization run, whose results were averaged. A replicate was considered as *successful*, if its respective stopping criterion $F(x) \leq \tau$, $\tau > 0$, could be reached. The results of the efficient gradient calculations compared to the classical calculations are shown in Table 1: As can be seen, the number of function evaluations could be decreased considerably, in most cases by a factor of 2 or 4. At the same time the speed of convergency remained almost constant and the improvements that were made with the efficiently calculated gradients were in the same order as for the classical calculations. These results indicate

Table 2. Results of the efficient Hessian calculations compared to the classical calculations for the Trust Region method (TR). The numbers shown are the mean values of ten statistically independent replicates. In the columns, the algorithm (Alg.), the number of successful replicates (# Repl.), the number of iterations (# It.), the number of function evaluations (# Eval.), the number of function evaluations per iteration (# Eval./# It.), and the final parameter vectors fulfillling the stopping criterion (x^{opt}) are indicated. The number of function evaluations was reduced, whereas the number of iterations and successful replicates remained almost constant compared to the classical calculation.

	Classical Hessian Calculation				Efficient Hessian Calculation			
Alg.	# Repl.	# It.	# Eval.	# Eval./# It.	# Repl.	# It.	# Eval.	# Eval./# It.
TR	10	7–8	115–116	15	10	7	65–66	9

that the methodology of the efficient calculation of the gradient can generally be applied with success. If it is calculated efficiently several times successively, the base vectors become more and more linearly dependent because the parameter vectors which belong to last classical gradient calculation are situated very close to each other, while the subsequent iterations move away. At some point, the norm condition and/or the condition number criterion will be violated and the gradient will be calculated in the classical way. Thus, it cannot be expected that the gradient is calculated efficiently at each iteration. At the beginning, the constants C and ϵ_κ were signficantly greater than near the minimum. In a small neighborhood of the minimum, $C \leq 1.0$ and $\epsilon_\kappa \leq 4$ were the only possible choices in order to get a reliable accuracy for the gradient. Because of the strict choice of the conditions, the applicability of the methodology is very limited near the minimum. It is most suitable at the beginning of the optimization process.

Table 2 shows the results of the efficient calculation of the Hessian: As can be seen, the number of function evaluations was reduced as well by a factor of about 2, while the number of successful replicates and iterations remained constant. Thus, also the efficient calculation of the Hessian could be applied successfully. The resulting Hessian was only a rough approximation but suitable for the Trust Region method, where a highly accurate Hessian is not absolutely necessary [4]. The number of simulations could not be decreased as significantly as for gradients because of the stricter norm criterion with $C = 1$.

4. Conclusions

Optimization methods based on gradients and Hessians are fast and robust instruments for the parameterization of force fields using molecular simulations. In this paper, a new methodology to compute gradients and Hessians more efficiently was presented. It increases the applicability of gradient-based numerical optimization methods to force field parameterization problems, as it is capable to decrease significantly the number of molecular simulations required for the loss function evaluations. The method is based on directional instead of classical partial derivatives. It was shown that the number of function evaluations could be reduced by a factor of 2–4, if two main assumptions were fulfilled. However, they were not fulfilled in a close neighborhood of the minimum because they had to be strengthened. Hence, the methodology is not applicable in a small neighborhood of the minimum. However, as their application was shown to be very successful at the beginning of the optimization process, it is highly recommended to use the efficient gradient and Hessian calculations instead of the classical ones.

References

[1] Jensen F 2006 (2nd edition) *Introduction to Computational Chemistry* (John Wiley & Sons)
[2] Reith D and Kirschner K N 2011 *Computer Physics Communications* **182** 2184–2191
[3] Hülsmann M, Köddermann T, Vrabec J and Reith D 2010 *Comput. Phys. Commun.* **181** 499
[4] Hülsmann M, Vrabec J, Köddermann T, Maaß A and Reith D 2010 *Comput. Phys. Commun.* **181** 887
[5] Nocedal J and Wright S J 1999 *Numerical Optimization* (Springer-Verlag)
[6] Stoll J, Vrabec J, Hasse H and Fischer J 2001 *Fluid Phase Eq.* **179** 339

Numerical solution of the time-dependent Ginzburg-Landau equation for a superconducting mesoscopic disk: Link variable method

J. Barba-Ortega[1], J. D. Gonzalez[2], Miryam R.Joya[3]

[1] Departamento de Física, Universidad Nacional de Colombia, Bogotá, Colombia
[2] Departamento de Física, Universidad del Magdalena, Santa Marta, Colombia
[3] Departamento de Física, Universidad Nacional de Colombia, Bogotá, Colombia

E-mail: jjbarbao@unal.edu.co

Abstract. The time dependent GinzburgLandau equations (TDGLE) are used to study the superconducting properties of a disk by taking into account the influence of internal defects. The Link variable algorithm is applied to a circular geometry surrounded by an insulator and immersed in an external magnetic field applied perpendicular to its plane. The TDGLE are used upon taking the magnetic field and the order parameter invariant along z-direction. We show that the magnetic response is substantially modified by the competition between the confinement geometry and the geometric position of the defects leads to vortex configurations which are not compatible with the symmetry of the sample geometry.

1. Introduction

Vortex state in conventional mesoscopically tailored superconducting samples has generate much activity in the wide scientific community. Many topologies have been studied experimentally and theoretically, such as simple loops and circular geometry [1, 2, 3, 4], surface roughness and surface defects [5, 6]. All these superconducting structures have attracted attention as potential new components for low-temperature electronics. By solving the non-linear time dependent Ginzburg-Landau (TDGL) equations, using the Link variable technique, we calculate the magnetization and the Coopers pair density for a disk in presence of an external applied magnetic field.

2. Theoretical Formalism

Superconducting state properties are described in the Ginzburg-Landay Theory by the order parameter ψ, and the potential vector $\mathbf{A}$. The TDGL equations in absence of external currents are given by:

$$\begin{aligned} \frac{\partial\psi}{\partial t} &= -(i\nabla + \mathbf{A})^2\psi + (1-T)\psi(1-|\psi|^2)\,, \\ \frac{\partial\mathbf{A}}{\partial t} &= (1-T)\mathrm{Re}\left[\bar{\psi}(-i\nabla - \mathbf{A})\psi\right] - \kappa^2\nabla\times\mathbf{h}\,, \end{aligned} \tag{1}$$

In the present scenario, T is temperature in critical temperature units T_C, order parameter in ψ_∞ units, length in units of coherence length at zero temperature $\xi(0)$, and fields in $H_{c2}(0)$

units. $\kappa = \lambda(0)/\xi(0)$ is the material-dependent Ginzburg-Landau parameter (for more details, see Ref. [5, 7, 8, 9]). The z-dependence of the order parameter has been ignored. It is convenient to introduce auxiliary vector field $\mathcal{U} = (\mathcal{U}_\rho, \mathcal{U}_\theta)$.

$$\mathcal{U}_\rho(\rho,\theta) = \exp\left(-i\int_{\rho_0}^{\rho} A_\rho(\theta,\xi)\,d\xi\right), \mathcal{U}_\theta(\rho,\theta) = \exp\left(-i\int_{\theta_0}^{\theta} A_\theta(\xi,\rho)\rho\,d\xi\right), \tag{2}$$

where (ρ_0, θ_0) is an arbitrary reference point. We use $\mathbf{D} = -i\nabla - \mathbf{A}$ to write:

$$D_\rho\psi = -i\bar{\mathcal{U}}_\rho \frac{\partial(\mathcal{U}_\rho\psi)}{\partial\rho}, \quad D_\theta\psi = -i\frac{\bar{\mathcal{U}}_\theta}{\rho}\frac{\partial(\mathcal{U}_\theta\psi)}{\partial\theta}. \tag{3}$$

Upon using these two last equations recursively, we obtain

$$D_\rho^2\psi = -\bar{\mathcal{U}}_\rho \frac{\partial^2(\mathcal{U}_\rho\psi)}{\partial\rho^2}, \quad D_\theta^2\psi = -\frac{\bar{\mathcal{U}}_\theta}{\rho^2}\frac{\partial^2(\mathcal{U}_\theta\psi)}{\partial\theta^2}. \tag{4}$$

As a consequence, we obtain for the kinetic term in the first TDGL equation

$$\mathbf{D}\cdot\mathbf{D}\psi = D_\rho^2\psi + D_\theta^2\psi - \frac{i}{\rho}D_\rho\psi = -\frac{\bar{\mathcal{U}}_\rho}{\rho}\frac{\partial}{\partial\rho}\left[\rho\frac{\partial(\mathcal{U}_\rho\psi)}{\partial\rho}\right] - \frac{\bar{\mathcal{U}}_\theta}{\rho^2}\frac{\partial^2(\mathcal{U}_\theta\psi)}{\partial\theta^2}. \tag{5}$$

From equations (3), it can also be easily proved that

$$\mathrm{Re}[\bar{\psi}D_\rho\psi] = \mathrm{Im}\left[\bar{\mathcal{U}}_\rho\bar{\psi}\frac{\partial(\mathcal{U}_\rho\psi)}{\partial\rho}\right], \mathrm{Re}[\bar{\psi}D_\theta\psi] = \mathrm{Im}\left[\frac{\bar{\mathcal{U}}_\theta\bar{\psi}}{\rho}\frac{\partial(\mathcal{U}_\theta\psi)}{\partial\theta}\right], \tag{6}$$

where Im indicates the imaginary part of a complex variable. Finally, using equations (5) and (6), the TDGL equations of (1) can be rewritten as

$$\begin{aligned}
\frac{\partial\psi}{\partial t} &= \frac{\bar{\mathcal{U}}_\rho}{\rho}\frac{\partial}{\partial\rho}\left[\rho\frac{\partial(\mathcal{U}_\rho\psi)}{\partial\rho}\right] + \frac{\bar{\mathcal{U}}_\theta}{\rho^2}\frac{\partial^2(\mathcal{U}_\theta\psi)}{\partial\theta^2} + (1-T)\psi(1-|\psi|^2),\\
\beta\frac{\partial A_\rho}{\partial t} &= (1-T)\mathrm{Im}\left[\bar{\mathcal{U}}_\rho\bar{\psi}\frac{\partial(\mathcal{U}_\rho\psi)}{\partial\rho}\right] - \kappa^2\frac{1}{\rho}\frac{\partial h_z}{\partial\theta},\\
\beta\frac{\partial A_\theta}{\partial t} &= (1-T)\mathrm{Im}\left[\frac{\bar{\mathcal{U}}_\theta\bar{\psi}}{\rho}\frac{\partial(\mathcal{U}_\theta\psi)}{\partial\theta}\right] - \kappa^2\frac{\partial h_z}{\partial\rho}.
\end{aligned} \tag{7}$$

3. Numerical Method

We will discretize the TDGL equations of (7) on a circular sector (for more details see Ref. [1]). Also, Let a superconducting disk domain be given by $\{(x, y, z) \in R^3, |z| < dg(x, y)\}$, for all (x, y). Here, $g(x, y)$ is some function which describes the topology of the top surface of the disk and d is the thickness of the disk. We used the popular link variable formalism to avoid not only the divergence of the derivative of the order parameter at the interface, but also to obtain a discretization of the Ginzburg-Landau equations which preserves the gauge invariance of these equations (see Ref. [7] and [8] for more details). Let us define the following discrete variables.

$$\begin{aligned}
U_{\rho,i,j} &= \bar{\mathcal{U}}_\rho(\rho_i,\theta_j)\mathcal{U}_\rho(\rho_{i+1},\theta_j) = \exp(-ia_\rho A_{\rho,i,j}),\\
U_{\theta,i,j} &= \bar{\mathcal{U}}_\theta(\rho_i,\theta_j)\mathcal{U}_\theta(\rho_i,\theta_{j+1}) = \exp(-i\rho_i a_\theta A_{\theta,i,j}),
\end{aligned} \tag{8}$$

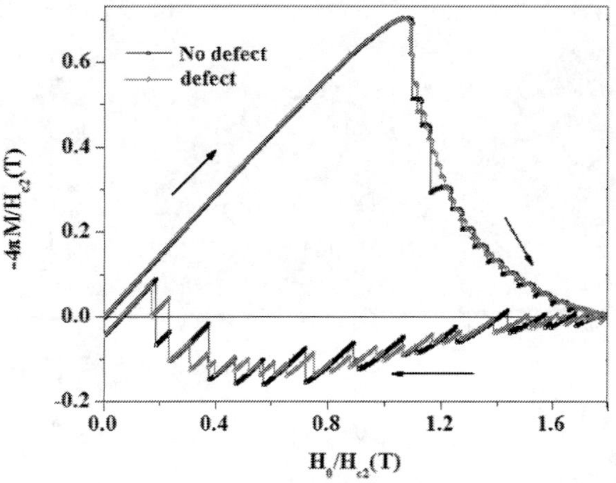

Figure 1. Magnetization curves as a function of external magnetic field for a disk with (and without) a linear defects on its surface.

for all $\{1 \le i \le N_\rho\ , 1 \le j \le N_\theta + 1\}$, and $\{1 \le i \le N_\rho + 1\ , 1 \le j \le N_\theta\}$, respectively. In what follows, it will be important to define the following discrete variable

$$L_{i,j} = \exp\left(-i \oint_{\partial\mathscr{D}} \mathbf{A}\cdot d\mathbf{r}\right) = \exp\left(-i \int_{\mathscr{D}} h_z\, \rho\, d\rho d\theta\right) = \exp\left(-i a_\rho \rho_{i+1/2} a_\theta h_{z,i,j}\right)\ , \qquad (9)$$

for all $\{1 \le i \le N_\rho\ , 1 \le j \le N_\theta\}$, where $\mathscr{D}$ is the domain of a unit cell limited by a closed path $\partial\mathscr{D}$. Stoke's theorem and the midpoint rule for numerical integration have been used. A simple inspection of equation (9) leads to $L_{i,j} = U_{\rho,i,j} U_{\theta,i+1,j} \bar{U}_{\rho,i,j+1} \bar{U}_{\theta,i,j}$. The boundary conditions were as follows. Let $\mathbf{n}$ be a unit vector normal to the interface. It will be assumed that normal current density not vanishes at the interface (i.e. $(-i\nabla - \mathbf{A})\psi \cdot \mathbf{n} = 0$).

4. Results and Discussion

The parameters used in our numerical simulations were: $\kappa = 2.17$, which is a value for to a thin film of Nb with thickness d (assuming $\xi(0) = 380$ nm, $T_c = 3.7K$, $d \approx 60$ nm, $T = 0$, the internal and external radius of the disk were $r_i = 0.3\xi(0)$ and $r_e = 6.5\xi(0)$. Largest unit cell length was taken as being no larger than 0.2×0.2, the applied magnetic field was ramped in steps of $\Delta H = 10^{-4}$. $g(x, y) = 0.8$ for $r_i < r < R_e$ and $\theta = 0$. Figures 1 show the magnetization as a function of increasing and decreasing applied magnetic field for a superconducting disk with (and without) a lineal defect. From these figures, we can see that the both the surface barrier field and the third critical thermodynamic fields are independent of the nature of defects. In the Figure 2 we shown the contour plot of the order parameter for several vorticities. Following the panels from the left to the right, in this order, we can see initially the sample with the defect at zero magnetic field, later three vortices nucleate asymmetrically, one of them at the defect position. Increasing the magnetic field (Figure 3 (up)) four vortices are trying to nucleated, they will form an unusual vortex shape for a circle geometry, it is, the geometry of the defects is imposing the vortex configuration. However, for higher magnetic fields (Figure 3 (down)), we observe a chain of twelve vortices with equal distance from the center but with a space in the equator and poles of the disk due to the defect.

References

[1] E. Sardella and E. H. Brandt, Supercond. Sci. Technol. **23**, 025015 (2010).

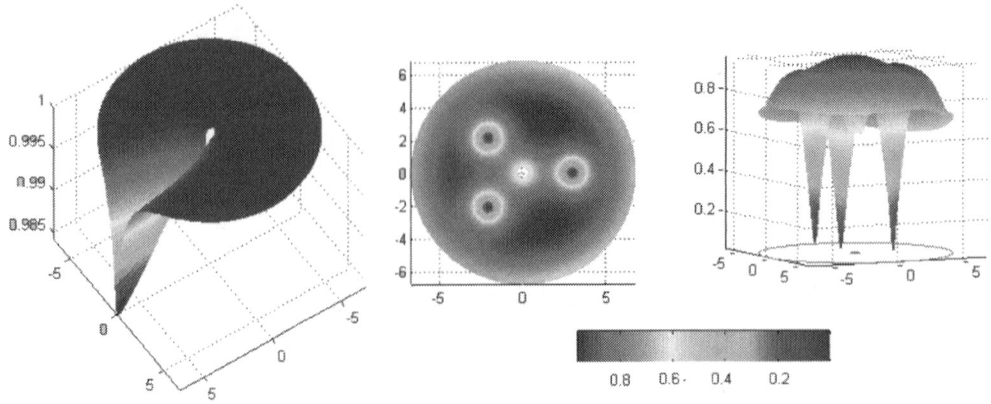

Figure 2. (Color online) Contour plot of the order parameter for a disk with radius $R = 6.5\xi(T)$ and one defect in $r_i < r < r_e$ and $\theta = 0$. Dark and bright regions represent values of the modulus of the order parameter from 0 to 1

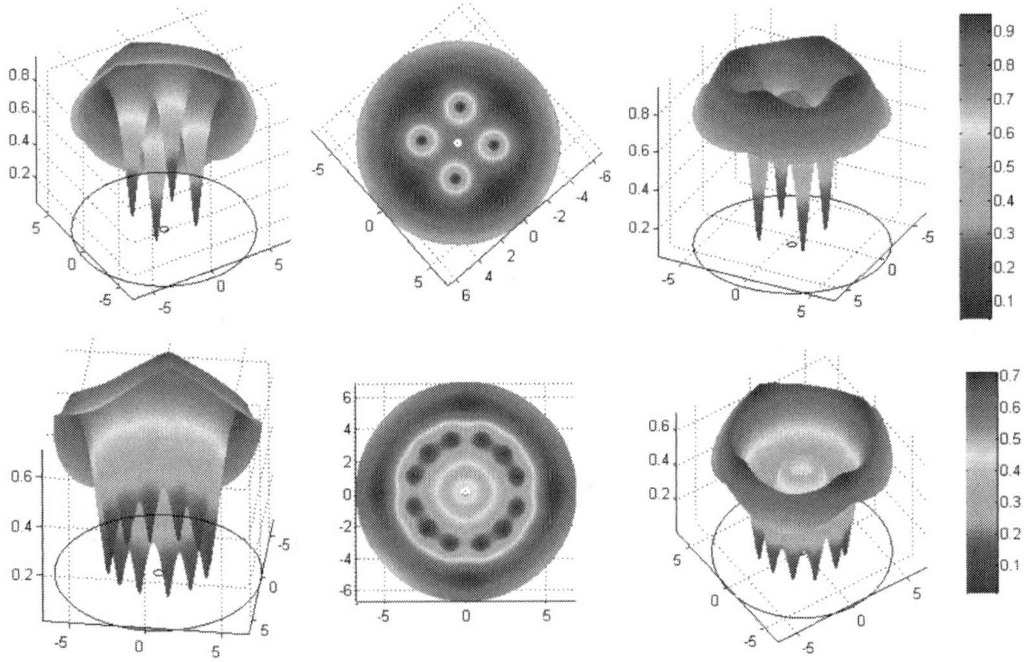

Figure 3. (Color online) Contour plot of the order parameter for a disk with radius $R = 6.5\xi(T)$ and one defect in $r_i < r < r_e$ and $\theta = 0$. Dark and bright regions represent values of the modulus of the order parameter from 0 to 1

[2] J. Barba-Ortega, Miryam R. Joya, J. Albino Aguiar, J. Sup. Nov. Magn, In press (2012).
[3] J. Barba-Ortega, Edson sardella, J. Albino Aguiar, Supercond. Sci. Technol.24, 015001 (2011).
[4] F. M Peeters, V.A Schweigert, B.J Baelus, P.S Deo, Physica C **332**, 255 (2000).
[5] J.R. Clem, in: K.D. Timmerhaus, W.J. O Sullivan, E.F. Hammel (Eds.), Low Temperature Physics, vol. 3, Plenum, New York, 1974, p. 102.

[6] J. Barba-Ortega, Edson sardella, J. Albino Aguiar, E. H. Brandt, Physica C, **479**, 49 (2012).
[7] W. D. Gropp, H. G. Kaper, G. K. Leaf, D. M. Levine, M. Palumbo, and V. M. Vinokur, J. Comput. Phys. **123**, 254 (1996).
[8] J. Berger and J. Rubinstein, Phys. Rev. B. **59**, 8896 (1999).
[9] G. C. Buscaglia, C. Bolech and A. López,*Connectivity and Superconductivity*, J. Berger, and J. Rubinstein (Eds.), Springer,(2000).
[10] V. V. Moshchalkov *et al.* in *Connectivity and Superconductivity*, edit by J. Berger and J. Rubinstein (Springer, Heidelberg, 2000).
[11] V. A. Schweigert and F. M. Peeters, Phys. Rev. B **60**, 3084 (1999).
[12] A. K. Geim, S. V. Dubonos, I. V. Grigorieva, K. S. Novoselov, F. M. Peeters, and V. A. Schweigert, Nature **407**, 55 (2000).

Journal of Physics: Conference Series **410** (2013) 012009 doi:10.1088/1742-6596/410/1/012009

Computational simulation of vortex matter in Type-II mesoscopics superconductors

J. Barba-Ortega[1], J. D. Gonzalez[2], Miryam R.Joya[3]

[1] Departamento de Física, Universidad Nacional de Colombia, Bogotá, Colombia
[2] Departamento de Física, Universidad del Magdalena, Santa Marta, Colombia
[3] Departamento de Física, Universidad Nacional de Colombia, Bogotá, Colombia

E-mail: `jjbarbao@unal.edu.co`

Abstract. In recent years, algorithms and computer simulation methodologies have developed to the stage where simulation can now play a complementary role with experiment, aiding in interpretation of experimental data. Several different approaches to the study of superconductor samples have been developed and their use demonstrated. In this work, we solve numerically the nonlinear time dependent Ginzburg Landau equations to study the vortex dynamics in a mesoscopic type II cylinder superconductor containing one rectangular hole in the presence of an external field applied perpendicular to the surfaces. We calculate the spatial distribution of the superconducting electron density and the phase of the superconducting order parameter using the numerical method based on the technique of gauge invariant variables. It is assumed that the inner hole edge is in contact with a metallic material while the outer edge is in contact with vacuum. The vortex dynamics and the magnetization curves are studied as a function of the external magnetic field.

1. Introduction

In a type-II superconductor; below some critical value H_{c1}, the magnetic field is expelled. Above this value a superconductor forms a lattice of vortices which carry magnetic flux through the system. Only at a higher second critical value, H_{c2} superconductivity is destroyed. These responses are viewed as consequences of the vortex interaction in these systems, the energy cost of a boundary between superconducting and normal states and the thermodynamic stability of vortex excitations. In a type-II superconductor the energy cost of a boundary between the normal and the superconducting state is negative, while the interaction between vortices is repulsive. This leads to a formation of stable vortex lattices in conventional mesoscopics superconducting samples has generate much activity in the wide scientific community. Many topologies have been studied experimentally and theoretically, such as circular and square geometry [1, 2, 3, 4], surface roughness and surface defects [5, 6]. All these superconducting structures have attracted attention as potential new components for low-temperature electronics. By solving the non-linear time dependent Ginzburg-Landau (TDGL) equations, using the Link variable technique, we calculate the magnetization and the superconducting electron density for the samples in presence of an external applied magnetic field.

Published under licence by IOP Publishing Ltd

2. Theoretical Formalism

Superconducting state properties are described in the Ginzburg-Landau Theory by the order parameter ψ, and the potential vector $\mathbf{A}$ (see Ref. [5, 7, 8, 9]). The TDGL equations in absence of external currents are given by:

$$\frac{\partial \psi}{\partial t} = -\frac{1}{\eta}\left[(i\nabla + \mathbf{A})^2 \psi + (1-T)\left(|\psi|^2 - 1\right)\psi\right] \tag{1}$$

$$\frac{\partial \mathbf{A}}{\partial t} = -(1-T)\,Re\left[\bar{\psi}(i\nabla + \mathbf{A})\psi\right] - \kappa^2 \nabla \times \nabla \times \mathbf{A} \tag{2}$$

Eqs. (1) and (2) were rescaled as follows: ψ in units of $\psi_\infty(T)$, lengths in units of $\xi(0)$, time in units of t_0, H_e in units of $H_{c2}(0)$, A in units of $H_{c2}(0)\xi(0)$, temperatures in units of T_c, we use $\eta = 1$. The dynamical equations are complemented with the appropriate boundary conditions for the order parameter $\hat{n}.\left(-i\hbar\nabla - \frac{e^*}{c}\mathbf{A}\right)\psi = \frac{i\hbar}{b}\psi$ where $\hat{n}$ is the unity vector perpendicular to the surface of the superconductor. b is de Gennes parameter, (see Ref. [3]).

3. Numerical Method

The full discretization of the TDGL equations can be found in more detail in Ref. [9]. We used $U\psi$ method [7] for solve the TDGL equations in a discrete grid. Complex link variables $\mathcal{U}^x$ and $\mathcal{U}^y$ are introduced to preserve the gauge-invariant properties of the discretized equations. $\mathcal{U}^x$ and $\mathcal{U}^y$ are related to A by:.

$$\mathcal{U}^x(x,y,t) = \exp\left(-i\int_{x_0}^{x} A_x(\xi,y,t)\,d\xi\right) \Leftrightarrow \mathcal{U}^y(x,y,t) = \exp\left(-i\int_{y_0}^{y} A_y(x,\eta,t)\,d\eta\right) \tag{3}$$

The link variable method is used since a better numerical convergence is obtain in high magnetic fields [10]. The TDGL equations 1 and 2 can be written in the following form:

$$\frac{\partial \psi}{\partial t} = \bar{\mathcal{U}}_x \frac{\partial^2(\mathcal{U}_x\psi)}{\partial x^2} + \bar{\mathcal{U}}_y \frac{\partial^2(\mathcal{U}_y\psi)}{\partial y^2} + (1-T)\psi\left(1-|\psi|^2\right) \Leftrightarrow \mathbf{J}_{\mathbf{s}\alpha} = (1-T)\,Im\left[\bar{\mathcal{U}}_\alpha \bar{\psi}\frac{\partial(\mathcal{U}_\alpha\psi)}{\partial\alpha}\right] \tag{4}$$

where $\alpha = (x,y)$, and Im indicates the imaginary part. We used this method to obtain our results. The outline of this simulation procedure is as follows: the sample is divided in a rectangular mesh consisting of $N_x \times N_y = 72$ cells, with mesh spacing $a_x \times a_y = 0.5$ and a square defect of dimensions $L_x \times L_y = 12$. To derive the discrete equations let us define by $x_i = (i-1)a_x$, $y_i = (i-1)a_y$, an arbitrary vertex point in the mesh and:

$$U^x_{i,j} = \bar{\mathcal{U}}^x_{i,j}\mathcal{U}^x_{i+1,j} = \exp\left(-i\int_{x_i}^{x_i+1} A_x(\xi,y_i)\,d\xi\right) \Leftrightarrow U^y_{i,j} = \exp\left(-i\int_{y_j}^{y_j+1} A_y(x_i,\eta)\,d\eta\right)$$

$$L_{i,j} = U^x_{i,j}U^y_{i+1,j}\bar{U}^x_{i,j+1}\bar{U}^y_{i,j} = \exp\left(-ia_x a_y H_e\right)$$

Then the discretized version of the TDGL equations maintaining second order accuracy in space are gyven by:

$$\frac{\partial \psi_{i,j}}{\partial t} = \frac{U^x_{i,j}\psi_{i,j} - 2\psi_{i,j} + \bar{U}^x_{i-1,j}\psi_{i-1,j}}{\eta a_x^2} + \frac{U^y_{i,j}\psi_{i,j+1} - 2\psi_{i,j} + \bar{U}^y_{i,j-1}\psi_{i,j-1}}{\eta_y^2} - \frac{1-T}{\eta}(\bar{\psi}_{i,j}\psi_{i,j}-1)\psi_{i,j}$$

$$\frac{\partial U^x_{i,j}}{\partial t} = -i(1-T)U^x_{i,j}Im\left(\bar{\psi}_{i,j}U^x_{i,j}\psi_{i+1,j}\right) - \frac{\kappa^2}{a_y^2}U^x_{i,j}\left(\bar{L}_{i,j-1}L_{i,j} - 1\right)$$

$$\frac{\partial U^y_{i,j}}{\partial t} = -i(1-T)U^y_{i,j}Im\left(\bar{\psi}_{i,j}U^x_{i,j}\psi_{i,j+1}\right) - \frac{\kappa^2}{a_x^2}U^y_{i,j}\left(\bar{L}_{i,j}L_{i-1,j} - 1\right)$$

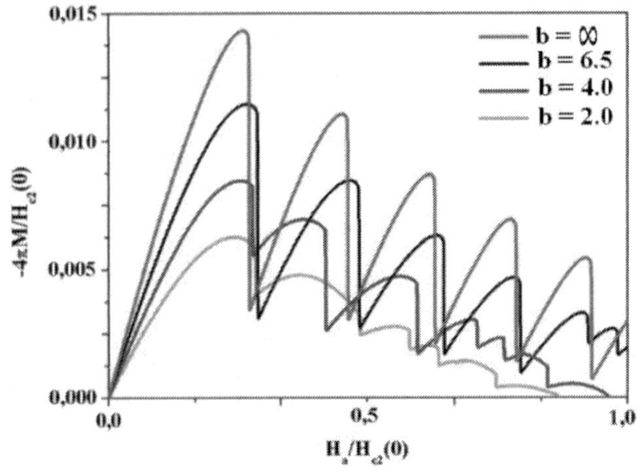

Figure 1. Magnetization curves as a function of external magnetic field for a square with a defects for $b \to \infty, 6.5, 4.5, 2.0$.

4. Results and Discussion

The parameters used in our numerical simulations were: $\kappa = 5.0$, $T = 0.53$,$b = 2.0$ simulating a metallic defect, the applied magnetic field was ramped in steps of $\Delta H = 10^{-4}$. Figures 1 show the magnetization as a function of increasing applied magnetic field for a superconducting square with a centered defect. From these figures, we can see that the first surface barrier field is independent of the metallic quality of the defects, but third critical thermodynamic fields are dependent. The growth of the magnetization with b means that the less the metallic boundary is the more diamagnetic the material is. This effect can be even more enhanced if the interface is a superconductor of higher T_c, $(b < 0)$, see (Ref ??). In the Figures 2 and 3 we shown the contour plot of the order parameter and its phase Θ and the magentic induction for several vorticities. Following the panels from the left to the right, in this order, we can see initially the sample with eight defect,increasing the magnetic field sixteen vortices sit in the sample, eight in the hole and the other eight in the superconductor region. Although they are not visible in the contour plot of the magnitude of order parameter, there is a change in the phase around the hole equal to $\Delta\Theta = 16\pi$. Values of the phase close to zero are given by dark gray regions and close to 2π by light gray regions. The phase allows to determine the number of vortices in a given region, by counting the phase variation in a closed path around this region. If the vorticity in this region is L, then the phase changes by $2\pi L$. At a higher magnetic field forty vortices nucleate the sample, twelve at the defect position and twenty eight in the superconducting sample. In conclusion we investigated theoretically the spatial distribution of the vortices in a square mesoscopic superconductor with one metallic square hole. The presence of this hole affects the vortex entry. If the hole is filled with any metallic material the number of vortices into the hole increases, the first critical field H_{c1} remains constant while the third thermodynamic field decrease with the decrease of the value of b.

References

[1] E. Sardella and E. H. Brandt, Supercond. Sci. Technol. **23**, 025015 (2010).
[2] J. Barba-Ortega, J. Albino Aguiar, Physica C **470**, 225 (2010).
[3] J. Barba-Ortega, Edson sardella, J. Albino Aguiar, Supercond. Sci. Technol.24, 015001 (2011).
[4] F. M Peeters, V.A Schweigert, B.J Baelus, P.S Deo, Physica C **332**, 255 (2000).

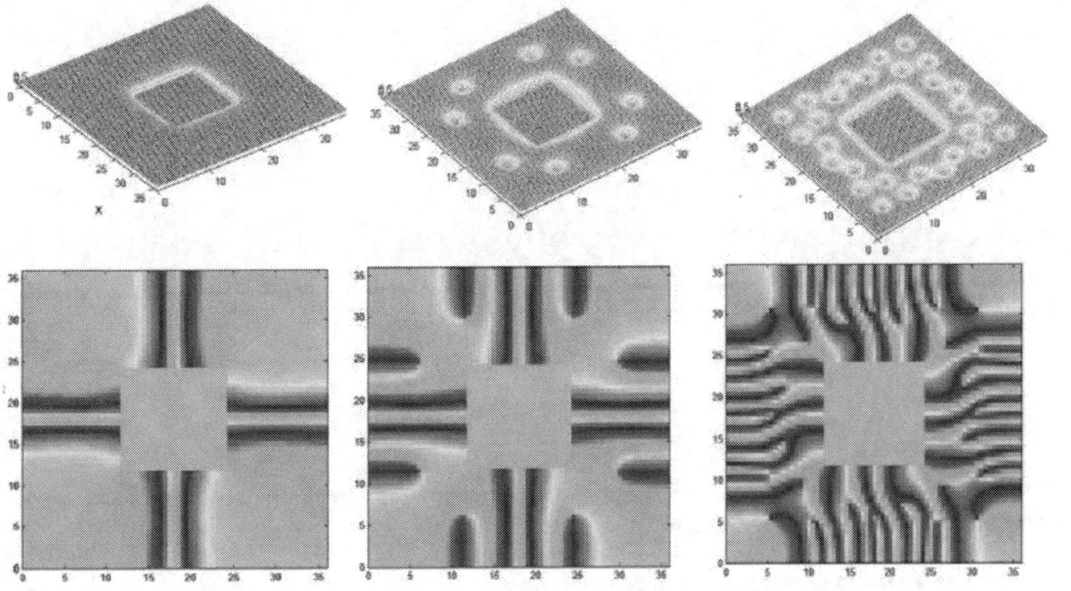

Figure 2. (Color online) Contour plot of the order parameter for a square with dimension $36\xi(T) \times 36\xi(T)$ and one centered metallic defect with dimension $10\xi(T) \times 10\xi(T)$, simulated by $b = 2.0\xi(T)$. Dark and bright regions represent values of the modulus of the order parameter from 0 to 1

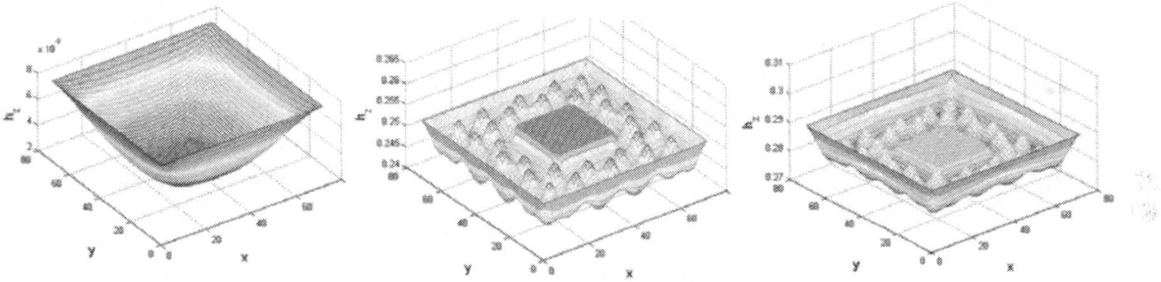

Figure 3. (Color online) Contour plot of the magnetic induction for a square with dimension $36\xi(T) \times 36\xi(T)$ and one centered metallic defect with dimension $10\xi(T) \times 10\xi(T)$, simulated by $b = 2.0\xi(T)$. Dark and bright regions represent values of the modulus of the order parameter ((as well its phase as $\Delta\Theta/2\pi$, from 0 to 1).from 0 to 1)

[5] J.R. Clem, in: K.D. Timmerhaus, W.J. O Sullivan, E.F. Hammel (Eds.), Low Temperature Physics, vol. 3, Plenum, New York, 1974, p. 102.
[6] J. Barba-Ortega, Edson sardella, J. Albino Aguiar, E. H. Brandt, Physica C, **479**, 49 (2012).
[7] W. D. Gropp, H. G. Kaper, G. K. Leaf, D. M. Levine, M. Palumbo, and V. M. Vinokur, J. Comput. Phys. **123**, 254 (1996).
[8] J. Berger and J. Rubinstein, Phys. Rev. B. **59**, 8896 (1999).
[9] G. C. Buscaglia, C. Bolech and A. López,*Connectivity and Superconductivity*, J. Berger, and J. Rubinstein (Eds.), Springer,(2000).
[10] R. Kato, Y. Enomoto, S. Maekawa, Phys. Rev. **B** 47, 8016 (1993).

Self-avoiding walk representation for the Green's function and localization properties

Fumika Suzuki

Department of Physics, The University of British Columbia, Vancouver, BC V6T 1Z1, Canada

E-mail: fumika@physics.ubc.ca

Abstract. The Green's function contains much information about physical systems. Mathematically, the fractional moment method (FMM) developed by Aizenman and Molchanov connects the Green's function and the transport of electrons in the Anderson model. Recently, it has been discovered that the Green's function on a graph can be represented using self-avoiding walks on a graph, which allows us to connect localization properties in the system and graph properties. We discuss FMM in terms of the self-avoiding walks on a general graph with a small number of assumptions.

1. Introduction

We introduce the Anderson model which describes the motion of an electron in a disordered system. There exist many good reviews of the Anderson model such as [3,5] and we follow their notations in this article. In physics, it is common to model the system by a lattice $\mathbb{Z}^n$ or a graph $\mathbb{G}$. In this article, we deal with the Anderson model on a graph $\mathbb{G}$ with the following assumptions.

Let $\mathbb{G} = (V, E)$ be a graph with vertices V and edges E. We assume $\mathbb{G}$ is a connected graph where the number of edges between any pair of vertices is either one or zero. We write $v \sim w$ if an edge connects the vertices v and w. Let $N(v)$ be the degree of a vertex $v \in V$. i.e., $N(v) := \#\{w \in V : w \sim v\}$. We assume that the degree of a vertex is bounded above by some constant N, $N(v) \leq N < \infty$ for all $v \in V$. $d(v, w)$ is the graph distance from v to w, which is the minimum number of edges from v to w on $\mathbb{G}$.

$\mathcal{W}(v, w)$ is the set of self-avoiding walks (sequences of vertices) $[v, v_1, \ldots, v_{d(v,w)}]$ with $d(v, w)$ steps starting at $v_0 = v$. The walks need not end at w.

Furthermore, $\mathcal{W}'(v, w)$ is the set of self-avoiding walks $[v, v_1, \ldots, v_{d(v,w)}]$ with $d(v, w)$ steps starting at $v_0 = v$ and with $v_{d(v,w)}$ connected to w in the graph obtained by deleting all edges attached to $[v, v_1, \ldots, v_{d(v,w)-1}]$.

We define a function $\mathscr{W}$ which measures the maximum number of self-avoiding walks $\mathcal{W}'(v, w)$ with d steps that can happen in $\mathbb{G}$.

$\mathscr{W}(d) = \max\{|\mathcal{W}'(v, w)| : d(v, w) = d\}$, thus $|\mathcal{W}'(v, w)| \leq \mathscr{W}(d(v, w))$.

We write the set of vertices on a sphere (or shell) with radius d from some origin v as $S(v, d)$. We write the set of vertices in a ball with radius d from some origin v as $B(v, d)$. $|S(v, d)|$ and $|B(v, d)|$ are the number of vertices in $S(v, d)$ and $B(v, d)$ respectively. Then, we define $\mathscr{S}(d)$

to be the largest possible value of $|S(v,d)|$ as v ranges over the graph, $\mathscr{S}(d) = \max_v |S(v,d)|$. $\mathscr{S}(d)$ is bounded by the biggest possible value $N(N-1)^d$.

Disordered matter can be described by a *random Schrödinger operator* acting on the Hilbert space $l^2(V)$:

$$H = -\sum_{\{v,w\}:v\sim w} (|v\rangle\langle w| + |w\rangle\langle v|) + \lambda \sum_{v\in V} \omega_v |v\rangle\langle v|$$

where the first term called next neighbour hopping operator describes the kinetic energy, the random potential $\lambda\omega_v$ is a multiplication operator on $l^2(V)$ with a coupling constant $\lambda > 0$. We assume the simplest case where $(\lambda\omega_v)_{v\in V}$ is a set of independent, identically distributed (i.i.d.) real-valued random variables with single-site distribution ρ. We assume ρ is bounded and has compact support.

Large coupling constant $\lambda >> 1$ indicates large disorder (randomness) and small coupling constant $\lambda << 1$ indicates small disorder. As λ increases, the distribution is spread out over larger supports and the random potential can take a wider range of possible random values.

Now we introduce *dynamical localization*, which implies the expectation of the amplitudes that an initial state of the system δ_x evolves into state δ_y after time t is exponentially localized with the distance d for all times.

Definition 1.1. (Dynamical localization): *H exhibits dynamical localization in I if there exist constants $C < \infty$ and $\mu > 0$ such that*

$$\sum_{y\in S(x,d)} \mathbb{E}\left(\sup_{t\in\mathbb{R}} |\langle \delta_y, e^{-itH}\chi_I(H)\delta_x\rangle|\right) \le Ce^{-\mu d} \tag{1}$$

for all $x \in V$.

$\mathbb{E}$ is the expectation with respect to the probability measure for random variables $\lambda\omega_v$. χ_I is the characteristic function of I and so $\chi_I(H)$ is the orthogonal projection onto the spectral subspace of H corresponding to energies in I. i.e., we only deal with the initial states with energy in I. μ is a constant which expresses disorder and often can be written in terms of λ. Larger λ implies larger μ.

2. Dynamical localization and graph properties

The Green's function can be represented using self-avoiding walks on a graph. There are several studies discussing it in different ways [3,6]. In this article, we introduce the self-avoding walk (SAW) representation for the Green's function in the way which gives us physical intuition.

We define a depleted random Schrödinger operator which can be made from H by self-avoiding walk process as follows (Figure 1).

$$H_{[v_0]} = H + \sum_{v_1\sim[v_0]} (|v_0\rangle\langle v_1| + |v_1\rangle\langle v_0|) \text{ (i.e., } H_{[v_0]} \text{ is } H \text{ without edges connected to } v_0.)$$

$$H_{[v_0,\dots v_i]} = H_{[v_0,\dots v_{i-1}]} + \sum_{v_i\sim[v_0,\dots v_{i-1}]} (|v_{i-1}\rangle\langle v_i| + |v_i\rangle\langle v_{i-1}|).$$

where $\sum_{v_i\sim[v_0,\dots v_{i-1}]}$ is summing over every possible vertex v_i which can be reached by taking the

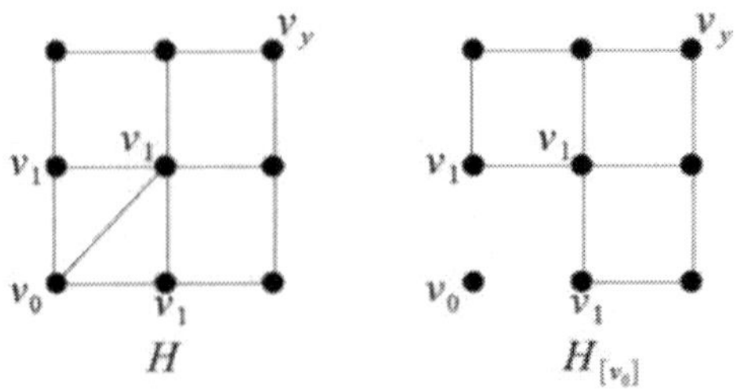

Figure 1. H and $H_{[v_0]}$

next step after the self-avoiding walk $[v_0, \ldots, v_{i-1}]$. Then we obtain the following proposition.[1]

Proposition 2.1. (SAW representation) *Let $x = v_0, y \in V$ and $z \in \mathbb{C} \setminus \mathbb{R}$. Then the Green's function can be written as*

$$G(x,y;z) = \langle x|(H-z)^{-1}|y\rangle = \sum_{[v_0,\ldots,v_{d(x,y)}]} \prod_{i=0}^{d(x,y)-1} \langle v_i|(H_{[v_0,\ldots,v_{i-1}]}-z)^{-1}|v_i\rangle\langle v_{d(x,y)}|(H_{[v_0,\ldots,v_{d(x,y)-1}]}-z)^{-1}|y\rangle \tag{2}$$

where $\sum_{[v_0,\ldots,v_{d(x,y)}]}$ *is summing over all self-avoiding walks starting at v_0 with length $d(x,y)$. When $i=0$, $H_{[v_0,v_{-1}]} = H$. When $i=1$, $H_{[v_0,v_0]} = H_{[v_0]}$.*

Here, self-avoiding walks are sequences of vertices $[v_0, \ldots, v_{d(x,y)}]$. When moving from a vertex to one that is joined by multiple edges, we must delete all the edges joining those vertices (Figure 1). Note that if a walker can not find any edge to walk from v_i since he deleted all edges connected to v_i, then the contribution of that walk to the Green's function is 0 since $\langle v_i|(H_{[v_0,\ldots,v_{i-1}]} - z)^{-1}|y\rangle = 0$.

Now, we discuss the fractional moment bounds for the Green's function represented by self-avoiding walks above.

Firstly, let $\mathcal{W}(x,y)$ be the set of self-avoiding walks $[x, v_1, \ldots, v_{d(x,y)}]$ with $d(x,y)$ steps starting at $v_0 = x$. Then, we can divide $\mathcal{W}(x,y)$ into three subsets:

$\mathcal{Y}(x,y)$: Self-avoiding walks in $\mathcal{W}(x,y)$ with $v_{d(x,y)} = y$.

$\mathcal{X}(x,y)$: Self-avoiding walks in $\mathcal{W}(x,y)$ with $v_{d(x,y)} \neq y$ where $v_{d(x,y)}$ is connected to y in the graph obtained by deleting all edges attached to $[x, v_1, \ldots, v_{d(x,y)-1}]$.

$\overline{\mathcal{X}}(x,y)$: Self-avoiding walks in $\mathcal{W}(x,y)$ with $v_{d(x,y)} \neq y$ where $v_{d(x,y)}$ is not connected to y in the graph obtained by deleting all edges attached to $[x, v_1, \ldots, v_{d(x,y)-1}]$.

Only $\mathcal{Y}(x,y)$ and $\mathcal{X}(x,y)$ contribute to the Green's function since

$$\langle v_{d(x,y)}|(H_{[v_0,\ldots,v_{d(x,y)-1}]} - z)^{-1}|y\rangle = 0 \quad \text{if} \quad [x, v_1, \ldots, v_{d(x,y)}] \in \overline{\mathcal{X}}.$$

Now we can define $\mathcal{W}'(x,y) = \mathcal{Y}(x,y) \cup \mathcal{X}(x,y)$ be the set of self-avoiding walks $[x, v_1, \ldots, v_{d(x,y)}]$ with $d(x,y)$ steps starting at $v_0 = x$ and $v_{d(x,y)}$ is connected to y in the graph obtained by deleting all edges attached to $[x, v_1, \ldots, v_{d(x,y)-1}]$. y is conncetd to itself in the case $v_{d(x,y)} = y$. Then, we obtain the following two theorems using the fractional moment

[1] Proofs of propositions and theorems in this article are given in [5].

method.

Theorem 2.2. **(Fractional moment bounds)** *Let us write the number of* $\mathcal{W}'(x,y)$ *as* $|\mathcal{W}'(x,y)|$. *Let* $0 < s < 1$. *Then, the fractional moment bounds for the Green's function can be written as follows.*

$$\mathbb{E}(|G(x,y;z)|^s) \leq |\mathcal{W}'(x,y)| \left(\frac{C}{\lambda^s}\right)^{d(x,y)+1}$$

Theorem 2.3. *Let* $I \subset \mathbb{R}$, $0 < s < 1$ *and* $\epsilon \in (0, 1/2)$. *Then dynamical localization (1) holds for disorder* λ *which satisfies the following condition.*

$$s\epsilon \ln \lambda > \epsilon \ln C + \sup_{x,d} \left(\frac{\ln |S(x,d)|}{d} \right) \quad \text{and} \quad \sum_{d'=0}^{\infty} \mathscr{S}(d')\mathscr{W}(d') \left(\frac{C}{\lambda^s} \right)^{(1-2\epsilon)d'} < \infty$$

where $C = \|\rho\|_\infty^s 2^{s+1} \frac{2^s s^{-s}}{1-s}$

3. Conclusions and discussion

One of the most important open problems in random operator theory is to understand the transition between the localized regime and the extended states regime. There is an attempt to understand the transition using the *level statistics conjecture* and *random matrix theory* (RMT). This method allows physicists to distinguish the two regimes numerically.

We use the statistical distribution of the eigenvalues of finite volume restrictions of the Anderson model to distinguish two regimes. It is expected that the localized regime and the extended states regime are corresponding to Poisson statistics and Gaussian orthogonal ensemble (GOE) statistics of the eigenvalues respectively.

Some studies in physics communities make use of such method to test the transition. Giraud et al. [2] studied the model of a circular graph with on-site disorder where each vertex is linked with its two nearest-neighbours and also they added shortcut edges between random pairs of vertices. Therefore, this is the one-dimensional Anderson model with extra off-diagonal random elements.

They studied level spacing statistics for Hamiltonian and obtained GOE distribution for small on-site disorder λ and Poisson distribution as they made on-site disorder λ larger.

It might be possible to make a relation between this transition and our theorem 2.3. In our work, the connection between distant regions is reflected in the size of $\mathscr{W}(d)$. When $\mathscr{W}(d)$ is large, the system needs larger disorder to obtain dynamical localization.

Acknowledgments

This article is based on part of the author's master thesis. I am deeply grateful to my advisors Richard Froese and P. C. E. Stamp for their guidance and support. It is a pleasure to thank Joel Feldman for giving me detailed comments and suggestions.

References

[1] Combes G., Germinet F. & Klein A. 2009 *Poisson Statistics for Eigenvalues of Continuum Random Schrödinger Operators* [arXiv:0807.0455]

[2] Giraud O., Georgeot B. & Shepelyansky D.L 2005 *Quantum computing of delocalization in small-world networks Phys. Rev. E.* **72** 036203.

[3] Hundertmark D 2008 *A short introduction to Anderson localization* In Analysis and Stochastics of Growth Processes and Interface Models, Oxford Scholarship Online Monographs, 194-219.

[4] Stolz G 2010 *An introduction to the mathematics of Anderson localization* Lecture notes of the Arizona School of Analysis with Applications

[5] Suzuki F 2012 *Anderson Localization with Self-Avoiding Walk Representation* MSc thesis, UBC.

[6] Tautenhahn M 2011 *Localization criteria for Anderson models on locally finite graphs Journal of Statistical Physics* **144** 60-75.

Exact results for non-integrable systems

Zsolt Gulácsi

Uneversity of Debrecen, Department of Theoretical Physics, H-4010 Debrecen, Egyetem tér 1, POB 5, Hungary

E-mail: Zsolt.Gulacsi@Phys.Unideb.Hu

Abstract. Quantum mechanical many-body systems described by an arbitrary Hamiltonian ($\hat{H}$) are analyzed. It is shown how positive semidefinite operator ($\hat{P}$) properties are able to lead in this case to exact results related to the ground state and the low lying part of the excitation spectrum. This is done independent on dimensionality and integrability. The technique first casts the Hamiltonian in a positive semidefinite form in exact terms ($\hat{H} = \hat{P} + C$, where C is a scalar). Second, the ground state is deduced by constructing the most general Hilbert space vector, on which applying $\hat{P}$, one obtains zero as a result. It is underlined, that the procedure, if applied for variable total number of particles N, allows to obtain information also related to the low lying part of the excitation spectrum. The uniqueness of the ground states can be demonstrated via the study of the kernel of $\hat{H}' = \hat{H} - C$. The physical properties of the obtained phases are deduced based on ground state expectation values calculated in terms of the constructed ground states. Since for a fixed structure of $\hat{P}$, usually the transformation $\hat{H} = \hat{P} + C$ is exact only in a restricted parameter space domain ($\mathcal{D}$), the deduced ground states are present only in $\mathcal{D}$. A global view on the phase diagram is obtained by different transformations of $\hat{H}$ in positive semidefinite form.

1. Introduction

A positive semidefinite operator $\hat{P}$ is defined as the operator which has only non-negative expectation values with all components $|\chi\rangle \in \mathcal{H}$ of the Hilbert space $\mathcal{H}$, i.e. $\langle\chi|\hat{P}|\chi\rangle \geq 0$. From this relation automatically results that all eigenvalues p of $\hat{P}$, $\hat{P}|\psi\rangle = p|\psi\rangle$, are non-negative scalars, that is, $p \geq 0$. This information seems trivial, but can be directly connected to the system Hamiltonian notion as follows: The spectrum of an arbitrary Hamiltonian ($\hat{H}$) describing a physical system is always bounded below, the lower bound being the ground state energy E_g. Consequently $\hat{H} - E_g = \hat{H}'$ is a positive semidefinite operator. This means that always, independent on integrability or dimensionality, an arbitrary $\hat{H}$ can be written as a positive semidefinite operator $\hat{P}$ plus a scalar C, i.e.

$$\hat{H} = \hat{P} + C. \tag{1}$$

Based on (1), the ground state can be deduced by constructing the most general wave vector $|\Psi_g\rangle$ holding the property

$$\hat{P}|\Psi_g\rangle = 0. \tag{2}$$

If $|\Psi_g\rangle$ as a solution of (2) exists, it will be the ground state, while the corresponding ground state energy becomes $E_g = C$. The procedure merits attention since several techniques have been worked out [1-13] for deducing (1) and solving (2).

The basic ideas serving as the background of the method are those presented above. Starting from these, Sect. 2 presents in details the steps of the procedure. Finally, Sect. 3 containing a short summary closes the presentation. The exemplifications are given in the fermionic case, but similar strategies are applicable also for bosonic systems as well.

2. Basic steps of the method

2.1. Transformation of $\hat{H}$ in a positive semidefinite form

The first step of the procedure represents an exact transcription of the Hamiltonian in a positive semidefinite form. For this reason, for example block operators can be used which are a linear combination of creation operators acting on the sites of a given block. In order to emphasize the aspects of this transformation, I present below an example in the case of the simplest two dimensional Hubbard model taken with periodic boundary conditions on a square lattice with Bravais vectors $\mathbf{x}, \mathbf{y}$. The Hamiltonian becomes of the form

$$\hat{H} = \sum_{\mathbf{i},\sigma} (t_x \hat{c}^\dagger_{\mathbf{i}+\mathbf{x},\sigma} \hat{c}_{\mathbf{i},\sigma} + t_y \hat{c}^\dagger_{\mathbf{i}+\mathbf{y},\sigma} \hat{c}_{\mathbf{i},\sigma} + H.c) + \hat{H}_U, \quad \hat{H}_U = U \sum_{\mathbf{i}} \hat{n}_{\mathbf{i},\sigma} \hat{n}_{\mathbf{i},-\sigma}, \tag{3}$$

where $\hat{c}^\dagger_{\mathbf{j},\sigma}$ creates an electron at the site $\mathbf{j}$ with spin projection σ, $\hat{n}_{\mathbf{j},\sigma} = \hat{c}^\dagger_{\mathbf{j},\sigma} \hat{c}_{\mathbf{j},\sigma}$ represents the particle number operator, t_x and t_y are the hopping matrix elements connecting nearest neighbors in x and y directions, while $U > 0$ represents the strength of the on-site Coulomb repulsion. Now for each cell defined at the arbitrary site $\mathbf{i}$ one introduces two block operators

$$\hat{A}_{\mathbf{i},\sigma} = a_2 \hat{c}_{\mathbf{i}+\mathbf{x},\sigma} + a_3 \hat{c}_{\mathbf{i}+\mathbf{x}+\mathbf{y},\sigma} + a_4 \hat{c}_{\mathbf{i}+\mathbf{y},\sigma}, \quad \hat{B}_{\mathbf{i},\sigma} = b_1 \hat{c}_{\mathbf{i},\sigma} + b_2 \hat{c}_{\mathbf{i}+\mathbf{x},\sigma} + b_4 \hat{c}_{\mathbf{i}+\mathbf{y},\sigma}, \tag{4}$$

where a_2, a_3, a_4 and b_1, b_2, b_4 are at the moment unknown block operator parameters, and the in-cell notation of sites by $n = 1, 2, 3, 4$ follows in order the sites $\mathbf{i}, \mathbf{i}+\mathbf{x}, \mathbf{i}+\mathbf{x}+\mathbf{y}, \mathbf{i}+\mathbf{y}$. Considering the total number of electrons N fixed, from (3,4), on the line of (1), one obtains

$$\hat{H}_{AB} = \sum_{\mathbf{i},\sigma} (\hat{A}^\dagger_{\mathbf{i},\sigma} \hat{A}_{\mathbf{i},\sigma} + \hat{B}^\dagger_{\mathbf{i},\sigma} \hat{B}_{\mathbf{i},\sigma}) = \hat{H} - \hat{H}_U + q\hat{N}, \quad \hat{P} = \hat{H}_{AB} + \hat{H}_U, \quad C = -qN, \tag{5}$$

where both $\hat{H}_{AB}$ and $\hat{H}_U$ are positive semidefinite operators, and the unknown block operator parameters are related to the known Hamilton operator parameters by

$$\begin{aligned} &t_x = b_2^* b_1 + a_3^* a_4, \;\; t_y = b_4^* b_1 + a_3^* a_2, \;\; t_{y+x} = t_{y-x} = b_4^* b_2 + a_4^* a_2 = 0, \\ &q = |b_1|^2 + |b_2|^2 + |b_4|^2 + |a_2|^2 + |a_3|^2 + |a_4|^2. \end{aligned} \tag{6}$$

The system of equations (6) represents the *matching conditions*. This system is obtained by effectuating the calculations present in $\hat{H}_{AB}$, and equating term by term the obtained result with the similar components present in $\hat{H}$. For example, in $\hat{H}$ one has $t_x \hat{c}^\dagger_{\mathbf{i}+\mathbf{x},\sigma} \hat{c}_{\mathbf{i},\sigma}$, while from $\hat{H}_{AB}$ one obtains $(b_2^* b_1 + a_3^* a_4) \hat{c}^\dagger_{\mathbf{i}+\mathbf{x},\sigma} \hat{c}_{\mathbf{i},\sigma}$, consequently $t_x = b_2^* b_1 + a_3^* a_4$ holds, etc.

Three observations have to be underlined at this stage: i) The matching conditions are representing usually coupled, complex algebraic non-linear equations. The solutions of this system of equations are present usually in a restricted $\hat{H}$ parameter space region $\mathcal{D}$ (in the present case, for example $t_{y+x} = t_{y-x} = 0$). Consequently, the deduced results will be valid only in a restricted region $\mathcal{D}$. ii) As can be seen from (5,6), the constant C in (1) is not a given number, but usually a scalar with a complicated and implicit structure, depending on $\hat{H}$ parameters. Consequently, the explicit expression of C becomes known only at the end of the calculation. iii) The number of unknown parameters and the number of equations in matching conditions [here Eqs.(6)] are usually different. Depending on the studied case, this difference leads to: a) interdependences between $\hat{H}$ parameters which determine $\mathcal{D}$, b) arbitrary parameters in the solutions of (6) which can be tuned during the process of solving (2) or deduction of E_g, and c) a combination of both a) and b).

Several other decompositions in positive semidefinite operators of the same $\hat{H}$ can be effectuated, each of these placing the results in different regions $\mathcal{D}$. For example (5) usually works in the below system half filling region. Above half-filling, expressions as $\sum_{\mathbf{i},\sigma}(\hat{A}_{\mathbf{i},\sigma}\hat{A}^{\dagger}_{\mathbf{i},\sigma}+\hat{B}_{\mathbf{i},\sigma}\hat{B}^{\dagger}_{\mathbf{i},\sigma})$ must be used (the structure of the corresponding ground states demands that). Furthermore, $\hat{P}$ can mix kinetic and interaction terms from $\hat{H}$. For example, this can be achieved by using instead of $\hat{H}_U$ in the transformed $\hat{H}$, the positive semidefinite operator $\hat{P}_{\mathbf{i}} = \hat{n}_{\mathbf{i},\sigma}\hat{n}_{\mathbf{i},-\sigma} - (\hat{n}_{\mathbf{i},\sigma}+\hat{n}_{\mathbf{i},\sigma})+1$. This last requires for its minimum eigenvalue zero at least one electron on the site $\mathbf{i}$. Similar mixing can be obtained by introducing in the block operators non-linear contributions, e.g. of the form $a_{\mathbf{i}}\hat{c}_{\mathbf{i},\sigma} + b_{\mathbf{i}}\hat{c}_{\mathbf{i},\sigma}\hat{n}_{\mathbf{i},-\sigma}$. I also note that the block operator coefficients can be considered cell variables (see [6,8]), which is useful in the study and description of random systems [6], or textures [8]. For multi-band systems, also the mixing of spin indices inside the same block operator can occur (see [7]). I further note that the number of block operators used in the positive semidefinite transcription of the Hamiltonian does not depend on dimensionality, but rather it is connected to the number and variety of terms present in $\hat{H}$.

2.2. Deduction of the ground states

Now one considers (1) given, and concentrates on solving (2) for $|\Psi_g\rangle$. The solution depends on the structure of $\hat{P}$, hence one analyzes below two of most often emerging cases, considering the block operators (as $\hat{A}_{\mathbf{i},\sigma}$ and $\hat{B}_{\mathbf{i},\sigma}$) linear combinations of fermionic operators.

2.2.1. The $\hat{P}$ operator contains terms of the form $\sum_{\mathbf{i},\sigma}\hat{A}^{\dagger}_{\mathbf{i},\sigma}\hat{A}_{\mathbf{i},\sigma}$, $\sum_{\mathbf{i},\sigma}\hat{B}^{\dagger}_{\mathbf{i},\sigma}\hat{B}_{\mathbf{i},\sigma}$. In this case the ground state is constructed by looking for a block operator $\hat{C}^{\dagger}_{\mathbf{i},\sigma}$ which satisfies the anti-commutation relation

$$\{\hat{A}_{\mathbf{i},\sigma}, \hat{C}^{\dagger}_{\mathbf{j},\sigma'}\} = \{\hat{B}_{\mathbf{i},\sigma}, \hat{C}^{\dagger}_{\mathbf{j},\sigma'}\} = 0, \tag{7}$$

for all possible values of all indices. The reason for this strategy is that defining $|\chi\rangle = \prod_{\mathbf{i},\sigma}\hat{C}^{\dagger}_{\mathbf{i},\sigma}|0\rangle$ where $|0\rangle$ is the bare vacuum, starting from (7), one finds $\hat{H}_{AB}|\chi\rangle = 0$. The reason for this last equality is that based on (7), all $\hat{A}_{\mathbf{i},\sigma}, \hat{B}_{\mathbf{i},\sigma}$ operators present in $\hat{H}_{AB}$ can be anti-commuted with all $\hat{C}^{\dagger}_{\mathbf{j},\sigma'}$ present in $|\chi\rangle$ and pushed in the right side of the expression $\hat{H}_{AB}|\chi\rangle$ just in front of the vacuum state. Then, since $\hat{A}_{\mathbf{i},\sigma}, \hat{B}_{\mathbf{i},\sigma}$ contain only annihilation operators, $\hat{A}_{\mathbf{i},\sigma}|0\rangle = \hat{B}_{\mathbf{i},\sigma}|0\rangle = 0$ holds, consequently $\hat{H}_{AB}|\chi\rangle = 0$ is indeed satisfied. After this stage, usually a restriction $(\mathbf{i},\sigma) \in \mathcal{M}$ introduces $|\chi\rangle$ also in the kernel [14] of the other positive semidefinite operators present in $\hat{P}$, hence the ground state becomes $|\Psi_g\rangle = \prod_{(\mathbf{i},\sigma)\in\mathcal{M}}\hat{C}^{\dagger}_{\mathbf{i},\sigma}|0\rangle$. I note that often, the $\hat{C}^{\dagger}_{\mathbf{i},\sigma}$ operators extend along the whole system (see [7]).

2.2.2. The $\hat{P}$ operator contains terms of the form $\sum_{\mathbf{i},\sigma}\hat{A}_{\mathbf{i},\sigma}\hat{A}^{\dagger}_{\mathbf{i},\sigma}$, $\sum_{\mathbf{i},\sigma}\hat{B}_{\mathbf{i},\sigma}\hat{B}^{\dagger}_{\mathbf{i},\sigma}$. In this case one uses the property valid for all linear combination of fermionic operators

$$\hat{A}^{\dagger}_{\mathbf{i},\sigma}\hat{A}^{\dagger}_{\mathbf{i},\sigma} = 0, \quad \hat{B}^{\dagger}_{\mathbf{i},\sigma}\hat{B}^{\dagger}_{\mathbf{i},\sigma} = 0, \tag{8}$$

and starts the construction of the ground state by $|\chi'\rangle = \prod_{\mathbf{i},\sigma}\hat{A}^{\dagger}_{\mathbf{i},\sigma}\hat{B}^{\dagger}_{\mathbf{i},\sigma}|0\rangle$. Indeed, using (8), one obtains $\hat{H}_{AB}|\chi'\rangle = 0$. Now in most cases (see [9,10]), the introduction of a supplementary $\hat{F}^{\dagger}$ operator in the constructed wave vector pushes $\hat{F}^{\dagger}|\chi'\rangle$ in the kernel of the other operators present in $\hat{P}$, consequently the ground state becomes $|\Psi'_g\rangle = \prod_{\mathbf{i},\sigma} \hat{A}^\dagger_{\mathbf{i},\sigma} \hat{B}^\dagger_{\mathbf{i},\sigma} \hat{F}^\dagger |0\rangle$. I further note that the number of particles N present in the system is given by the number of multiplied creation operators present in the expression of the ground state. Since this is higher in the case of $|\Psi'_g\rangle$, $\sum_{\mathbf{i},\sigma} \hat{A}_{\mathbf{i},\sigma} \hat{A}^\dagger_{\mathbf{i},\sigma}$ and $\sum_{\mathbf{i},\sigma} \hat{B}_{\mathbf{i},\sigma} \hat{B}^\dagger_{\mathbf{i},\sigma}$ type of positive semidefinite operators are used in the above system half filling concentration region. For the case of non-linear Fermi operator combinations in block operators, similar techniques apply [4].

2.2.3. Information regarding the low lying part of the excitation spectrum. If $E_g(N)$ is deduced, the chemical potential $\mu(N+1) = E_g(N+1) - E_g(N)$, $\mu(N) = E_g(N) - E_g(N-1)$, and $\delta\mu = \mu(N+1) - \mu(N)$ can be obtained. Now $\delta\mu \neq 0$, ($\delta\mu = 0$), signals the presence, (absence) of the charge gap, consequently insulating, (conducting) behavior can be tested.

2.3. Demonstration of the uniqueness of the ground states

In demonstrating the uniqueness of the deduced ground states one concentrates on the kernel $Ker(\hat{P})$ [14] of the operator $\hat{H} - C = \hat{P}$. For the most general case, when the deduced ground state $|\Psi_g(m)\rangle$ is M fold degenerate (i.e. $m = 1, 2, ..., M$), the proof of the uniqueness must be done in two steps: i) First one shows that for all possible m values $|\Psi_g(m)\rangle \in Ker(\hat{P})$ holds. ii) Second, one demonstrates that an arbitrary wave vector $|\nu\rangle \in Ker(\hat{P})$ can be expressed as a linear combination of the $|\Psi_g(m)\rangle$ terms (see for exemplification [9,13]). In the non-degenerate case the demonstration is the same, but only the $m = 1$ ground state component exists. As seen, at this step, the uniqueness and non-degenerate notions should be taken distinct: uniqueness means that non-deduced (i.e. unknown) vectors $|v\rangle$, satisfying $\hat{P}|v\rangle = 0$ are not present.

2.4. Study of the physical properties of the deduced ground states

The ground state being deduced, its physical properties are not visible at first view. In order to deduce them, different ground state expectation values must be calculated.

3. Summary and conclusions

It was shown how positive semidefinite operator properties allow the deduction of exact ground states for many-body quantum systems independent on integrability or dimensionality.

Acknowledgments

I kindly acknowledge financial support provided by contracts OTKA-K-100288, TAMOP 4.2.1/B-09/1/KONV-2010-0007, and Alexander von Humboldt Foundation.

References

[1] Gulácsi Z and Orlik I 2001 *Jour. Phys. A.* **34** L359
[2] Gurin P and Gulácsi Z 2001 *Phys. Rev. B.* **64** 045118
[3] Gulácsi Z 2002 *Eur. Phys. Jour. B.* **30** 295
[4] Chalupa I and Gulácsi Z 2007 *Jour. of Phys: Cond. Matter.* **19** 386209
[5] Gulácsi Z 2002 *Phys. Rev. B.* **66** 165109
[6] Gulácsi Z 2004 *Phys. Rev. B.* **69** 054204
[7] Gulácsi Z 2008 *Phys. Rev. B.* **77** 245113
[8] Gulácsi Z and Gulácsi M 2006 *Phys. Rev. B.* **73** 014524
[9] Gulácsi Z and Vollhardt D 2005 *Phys. Rev. B.* **72** 075130
[10] Gulácsi Z and Vollhardt D 2003 *Phys. Rev. Lett.* **91** 186401
[11] Gulácsi Z, Kampf A and Vollhardt D 2007 *Phys. Rev. Lett.* **99** 026404
[12] Gulácsi Z, Kampf A and Vollhardt D 2010 *Phys. Rev. Lett.* **105** 266403
[13] Gulácsi Z, Kampf A and Vollhardt D 2008 *Prog. Theor. Phys. Suppl.* **105** 1
[14] The kernel $Ker(\hat{O})$, of an arbitrary operator $\hat{O}$, is a Hilbert subspace containing all wave vectors $|\phi\rangle$ with the property $\hat{O}|\phi\rangle = 0$.

IC-MSQUARE 2012: International Conference on Mathematical Modelling in Physical Sciences IOP Publishing
Journal of Physics: Conference Series **410** (2013) 012012 doi:10.1088/1742-6596/410/1/012012

Sharpening minimum-phase filters

G Jovanovic Dolecek[1] and A Fernandez-Vazquez[2]

[1] Department of Electronics, INAOE, E. Erro 1, 72740 Puebla , Mexico
[2] School of Computer Engineering, IPN, 07738 Mexico City, Mexico

E-mail: gordana@ieee.org

The minimum–phase requirement restricts that filter has all its zeros on or inside the unit circle. As a result the filter does not have a linear phase. It is well known that the sharpening technique can be used to simultaneous improvements of both the pass-band and stop-band of a linear-phase FIR filters and cannot be used for other types of filters. In this paper we demonstrate that the sharpening technique can also be applied to minimum-phase filters, after small modification. The method is illustrated with one practical examples of design.

1. Introduction

There are some applications where high delay introduced by linear-phase finite impulse response (FIR) filters is not permitted like in communication data systems. In order to obtain much lower delay, the solution is a minimum-phase filter (MP) design which preserves desired magnitude response. There have been proposed different methods to obtain minimum-phase filters.

The existing methods can be classified into two principal groups of methods: methods based on linear-phase (LP) filter design, and methods based on complex cepstrum [1-5]. We consider here the FIR MP design method based on the corresponding LP filter. As a difference to LP filter which generally have zeros in quadruplets [6], a MP filter has all its zeros inside or on the unit circle.

The simplest method to design a MP filter for a given specification consists of two following steps:
1. Design the LP filter which satisfies the specification.
2. Reflect all zeros of LP filter which are outside the unit circle to their reciprocal positions inside the unit circle.

The obtained MP filter has the same magnitude response as the original LP filter [6]. However, the MP filter does not have the linear phase and, as a result, its unit sample impulse response does not have the symmetry.

It is well known that the sharpening technique can be used to simultaneous improvements of both the pass-band and stop-band of a linear-phase FIR filter and is not used for other types of filters including MP filters.

The main objective of this work is to propose a novel procedure which make possible the sharpening of minimum phase filters.

2. Sharpening technique

The sharpening technique is introduced in [7] for simultaneous improvements of both the pass-band and stop-band of a linear-phase FIR filter. The sharpening technique uses the amplitude change function (ACF) which is a polynomial relationship of the form $H_{sh} = f(H)$ between the amplitudes of the sharpened and the prototype filters, H_{sh} and H, respectively. The improvement in the gain response near the passband edge H=1, or near the stopband edge H=0, depends on the order of tangencies m and n of the ACF at H=1, or at H=0, respectively.

We consider here the most important polynomials:

$$\mathbf{P_1}: H_{sh}=2H-H^2, (m=1, n=0);\ \mathbf{P_2}: H_{sh}=3H^2-2H^3, (m=1, n=1);\ \mathbf{P_3}: H_{sh}=4H^3-3H^4, (m=1, n=2). \quad (1)$$

The polynomial $\mathbf{P_1}$ provides the best passband improvement, while $\mathbf{P_3}$ results in best stopband improvement. However, the sharpening technique cannot be applied to minimum phase filters. Next section describes the procedure to apply sharpening also to MP filters.

3. Description of the procedure

The system function of sharpening polynomials (1) can be rewritten in the form:

$$H_{sh}(z) = H^k(z)H_s(z), \quad (2)$$

where $H(z)$ is the original LP filter, k is an integer, z is complex value, and $H_s(z)$ is a LP filter in the form:

$$H_s(z) = K_1 z^{-N/2} - K_2 H(z), \quad (3)$$

where K_1 and K_2 are constants. Note that the delay $z^{-N/2}$ must be introduced in order that the filter $H_s(z)$ impulse response has a symmetry, i.e. the filter $H_s(z)$ is a LP filter. This implies that the order of the filter must be even i.e. the filter must be the Type 1 filter [8]. The values k, K_1 and K_2 for the polynomials (2) are given in Table 1.

From (2) we write the corresponding sharpening polynomials for the corresponding minimum phase filter (MP filter that has the same magnitude response):

$$H_{shm}(z) = H_m^{\ k}(z)H_{sm}(z), \quad (4)$$

where $H_m(z)$ and $H_{sm}(z)$ are obtained from the corresponding LP filters $H(z)$ and $H_s(z)$ by reflecting the zeros from the outside of unit circle to the reciprocal positions inside the unit circle. As a result, the LP filters $H(z)$ and $H_s(z)$ have equal magnitude responses as MP filters $H_m(z)$ and $H_{sm}(z)$, respectively.

Table 1. Parameters of sharpening polynomials.

Polynomial	k	K_1	K_2
P1	1	2	1
P2	2	3	2
P3	3	4	3

The procedure is given in the following steps:

1. Design the LP filter for a given specification.
2. Transform LP filter into MP filter.
3. From (3) find $H_s(z)$ and transform it into MP filter $H_{sm}(z)$.
4. The sharpened MP filter is given in (4).

Using Rouche's theorem [9] we can find that LP filter $H_s(z)$ has $N/2$ zeros inside the unit circle. Being a LP filter, it has other $N/2$ zeros outside the unit circle. As a consequence, the zeros of the MP filter $H_{sm}(z)$ can be easily find by doubling the zeros of $H_s(z)$ which are inside the unit circle, and eliminate that which are outside the unit circle.

The procedure is illustrated in the following example.

Example 1:

Using polynomials (1) sharpen the MP filter satisfying the following specification:

$$\omega_p = 0.1; \omega_a = 0.3; \delta_p = 0.1dB; A_s = 50dB, \tag{5}$$

where ω_p and ω_s are normalized passband and stopband frequencies, δ_p is a maximal passband ripple and A_s is a minimum stopband attenuation.

Step 1: We design a LP filter $H(z)$ using the minmax equiripple method. The order of the filter N=24.

Step 2: We transform the filter $H(z)$ into a MP filter $H_m(z)$ by reflecting zeros of $H(z)$ from their outside position inside the unit circle and converting the roots to polynomial. Figure 1 shows the positions of zeros and impulse responses.

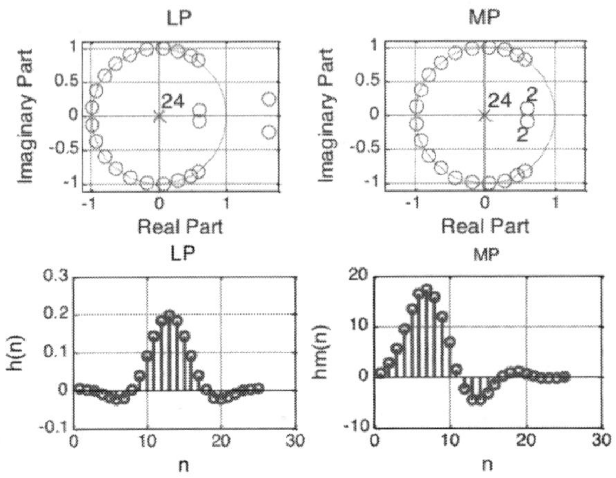

Figure 1. Positions of zeros and impulse responses.

Step 3: From (3) we have:

$$H_{s_i}(z) = K_1 z^{-13} - K_2 H(z) \;.\quad i=1, 2, 3. \tag{6}$$

Denoting the impulse responses of $H(z)$ and $H_s(z)$, as $h(n)$ and $h_s(n)$, respectively, we have:

$$h_{si}(n) = K_1 h(12) - K_2 h(n), n=0,\dots,24 \;, i=1, 2, 3. \tag{7}$$

Taking the values from Table 1 we get:

$$\textbf{P1: } h_{s1}(n) = 2h(13) - h(n); \textbf{ P2: } h_{s2}(n) = 3h(13) - 2h(n); \textbf{ P3: } h_{s3}(n) = 4h(13) - 3h(n) \;. \tag{8}$$

The filter $H_{si}(z)$ is transformed into a minimum system $H_{sim}(z)$ by doubling the zeros which are inside the unit circle, and deleting all others.

Step 4: The corresponding MP sharpened polynomials are:

$$H_{shm1}(z) = H_m(z)H_{sm1}(z); \; H_{shm2}(z) = H_m^{\,2}(z)H_{sm2}(z); \; H_{shm3}(z) = H_m^{\,3}(z)H_{sm3}(z). \tag{9}$$

Figure 2 shows the corresponding magnitude responses of original MP and the sharpened MP filters The passband zooms are also shown.

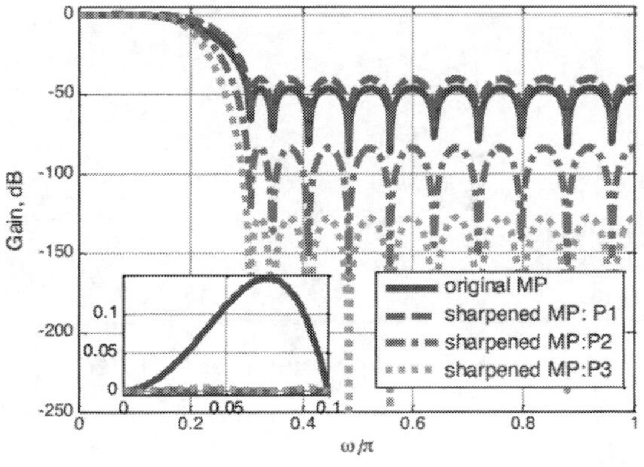

Figure 2. Magnitude responses of MP filters.

4. Conclusion

This paper presents the procedure for the sharpening MP filters. In that way the useful technique, used for FIR filters, can also be used for MP filters. We considered three most important sharpening polynomials. However, the same procedure can also be applied for any sharpening polynomial.

Acknowledgement
This work is supported by Conacyt Mexico under grant No.179585.

References

[1] Kale I, Cauin G.D and Morling R C S 1995 *Electronic Letters* Minimum-phase filter design from linear-phase start point via balanced model truncation **31** 1728-1729.

[2] Stathaki T and Fotinopoulos I 2001 *IEEE Trans. on Circuits and Systems-I: Analog and Digital Signal Processin* Equiripple minimum-phase FIR filter design from linear phase systems using root moments **6** 580-587.

[3] Huang Xiang-dong J Wang Zhao-hua, 2006 *Journal of Tianjin University* A Novel Technique to Design Filter with Discontinuous Frequency Character **39** 614-620.

[4] Okuda M Khan I R Ikehara M and Takahashi S 2004 *IEE Proc. Vis. Image Signal Processing* Quasi-equiripple approximation of minimum phase FIR filters by updating desired response **151**.

[5] Okuda M Ikehara M and Takahashi S 2006 *IEICE Transactions on Fundamentals of Electronics* Design of Equiripple Minimum Phase FIR Filters with Ripple Ratio Control **E89-A**.

[6] Oppenheim A V Schafer R W 1989 *Discrete-Time Signal Processing* (N J: Prentice-Hall International).

[7] Kaiser J F Hamming R W 1977 *IEEE Trans. on Acoustics, Speech, and Signal Processing,* Sharpening the response of a symmetric non-recursive filter **ASSP-25** 415-422.

[8] Mitra S K 2005 *Digital Signal Processing* (N Y: McGraw-Hill).

[9] Churchill R V and Brown J W 1976 *Complex variables and applications* (N Y: Mc Graw Hil).

Cinema Fire Modelling by FDS

J Glasa[1,2], L Valasek[1], P Weisenpacher[1] and L Halada[1]

[1] Slovak Academy of Sciences/Institute of Informatics, Dubravska cesta 9, 845 07 Bratislava, Slovakia

E-mail: jan.glasa@savba.sk

Abstract. Recent advances in computer fluid dynamics (CFD) and rapid increase of computational power of current computers have led to the development of CFD models capable to describe fire in complex geometries incorporating a wide variety of physical phenomena related to fire. In this paper, we demonstrate the use of Fire Dynamics Simulator (FDS) for cinema fire modelling. FDS is an advanced CFD system intended for simulation of the fire and smoke spread and prediction of thermal flows, toxic substances concentrations and other relevant parameters of fire. The course of fire in a cinema hall is described focusing on related safety risks. Fire properties of flammable materials used in the simulation were determined by laboratory measurements and validated by fire tests and computer simulations.

1. Introduction

Advances in computer fluid dynamics (CFD) and significant growth of computational performance of current computers have led to the development of fire models which are capable to describe fire in complex geometries and incorporate a wide variety of physical phenomena related to fire [3, 4]. In zone (multi-zone) models appeared in the seventies, the computational space was divided into two separated, relatively homogeneous parts, significantly more or less affected by fire. The theoretical background of zone models were conservation laws of mass and energy supplemented by models of physical processes describing the plumes spread, gas flow through vents, heat transfer and pyrolysis of solid substances. Their relative physical and computational simplicity has led to their widespread use especially for compartment fires analysis. However, in certain fire scenarios, more detailed spatial distributions of physical properties are required. Therefore, CFD models were introduced in the nineties. Such systems allow to describe fire in spaces with complex geometry and incorporate a broad scale of physical and chemical phenomena [4]. They represent a time-averaged approximation of governing equations of fluid dynamics describing the transfer of mass, momentum and energy by gas flows induced by fire. They are based on RANS (Reynolds averaged Navier-Stokes equations) which model low-speed gas flows driven by chemical heat release and buoyancy forces and are referred to in the combustion community as low Mach number combustion equations [3, 11]. Numerical solving of these equations requires to divide the physical space into 3D rectangular meshes, the density of which markedly affects the computational time and simulation accuracy.

Several fire simulators based on CFD models have been developed, such as for instance the CFX, SMARTFIRE and FDS systems. In this paper, we describe the use of Fire Dynamics Simulator (FDS) [8, 9] for potential cinema fire modelling.

---

[2] To whom any correspondence should be addressed.

2. Fire Dynamics Simulator

FDS is an advanced CFD simulator intended for fire simulation in variable complex environments. The first FDS version was released in 2000 by National Institute of Standards and Technology, USA; at present, the FDS 5.5.3 version is available. FDS numerically solves a form of Navier-Stokes equations for low-speed fire induced flows with emphasis on transport of smoke and heat from fire. The heart of algorithm is an explicit predictor-corrector scheme of the second degree accuracy according to space and time. Turbulence is modelled by LES (Large Eddy Simulation) which is suitable for simulations in larger spaces. Combustion is modelled using mass fractions (ratios of combustible gasses originated in given place corresponding to all main reactants and products which can be derived from mass fractions of mixtures by laboratory analysis and measurements). Radiation heat transport is modelled by FVM (Finite Volumes Method). For solid surfaces and domain boundaries, thermal boundary conditions and reliable information about fire properties of materials must be entered. Fire suppression and activation of sprinklers and detectors of heat and smoke are also included. The methods, models and their implementation in FDS are completely described in [8]. All simulation inputs (the definition of input geometry, computational meshes, material properties, thermal and boundary conditions for solid surfaces and domain boundaries, initiation fire source, fire duration and the required outputs form) must be concentrated in a single text file. In the case of structures with a complex geometry, the input geometry creation can be labourious and time consuming because all objects in the computational space must conform with computational meshes. FDS has confirmed its usefulness for simulation of fire in structures with higher concentration of people such as theatre [13, 14], supermarket [6], room [7] and office [1].

3. Simulation inputs and fire scenario

Let us consider a 3D model of a cinema (see Fig. 1) consisting of the entrance hall, projection room and cinema hall [12]. The cinema hall has a curved ceiling, podium with 2 small stairways and seating space for spectators constructed as a stairway consisting of nine 20 cm high stairs with chairs. 108 chairs are ordered into 9 rows of chairs (3 rows of 10 chairs, 3 rows of 12 chairs and 3 rows of 14 chairs; the rows and chairs are numbered uplink and from the left, respectively). The cinema input geometry shown in Fig. 1 was created by the graphical user interface PyroSim which enables interactive creating of complex structures geometry using ground-plans and efficient drawing walls, repetitious objects, stairways, curved walls and other complex elements of buildings [12, 2].

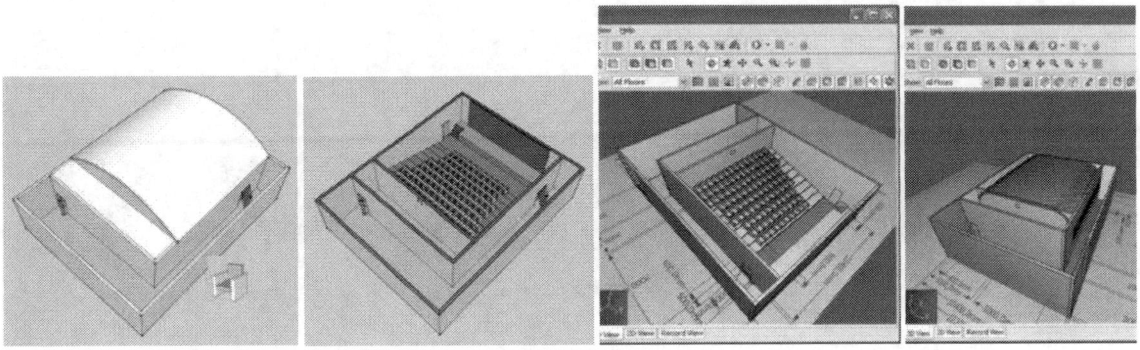

Fig. 1 3D cinema model (scheme of the building and its interior) and cinema input geometry

A simple fire scenario, in which a fire started under the left-most chair in the 5th chair row during the performance, was assumed. The initial fire source, represented by a hot 0,2 x 0,2 m surface (800 kW/m^2 HRRPUA, 12 s duration), was placed on the floor under the chair. All walls and chairs were composed of concrete and upholstery material, respectively. Other surfaces in the cinema (the floor, podium, doors, etc.) had specified material properties of inert material. The upholstery material parameters used in the simulation have been determined by laboratory measurements and validated by fire tests and FDS simulations [5, 10]. For the 12,6 x 16 x 4,8 m cinema, a single computational mesh of 10 cm density was defined consisting of 967 680 cube cells (126, 160 and 48 cells in the direction of x-, y- and z axis, respectively). Simulation would confirm that spectators in the cinema hall would be endangered even during the 1st minute of the fire.

4. Simulation results

Simulation of the 1-minute cinema fire scenario described above was realized on a PC with Intel Core i7 990-X 3.46GHz, 24 GB RAM. The total computational time of was about 2 h 56 min 30 s.

The first plumes appeared already at the first second of fire in front of the 1st chair of the 5th chair row since the floor, chair seat, both side hand rests and the stair, on which the higher chair row was placed, formed a semi-closed space opened forwards. Then, a thin thread of smoke moving quickly up to the ceiling appeared. It reached the ceiling at the 4th s. After the smoke hit the ceiling, it spread under the ceiling and hit the nearer and farther vertical side wall almost immediately and at the 16th s, respectively. At the 7th s, the smoke reached the vertical back wall. The hit of smoke into the vertical and curved vertical obstacles caused its spread downwards, turbulent mixing of quick hot gases with cold fresh air and forming toxic gas clouds at first in the left back corner and the left side part of the cinema hall (see Fig. 2). Later, the toxic gases layer under the ceiling thickened and the toxic turbulent clouds were formed at the back part and under the left curved part of the cinema hall. At the 60th s of fire, spectators sitting in 5 highest chair rows would be endangered by the fire. The simulation captured several specific phenomena appeared during the fire and smoke spread in the cinema hall. It confirmed that the fire of upholstered cinema chairs belongs to very dangerous fires because of its fast development and toxic gases released by the fire.

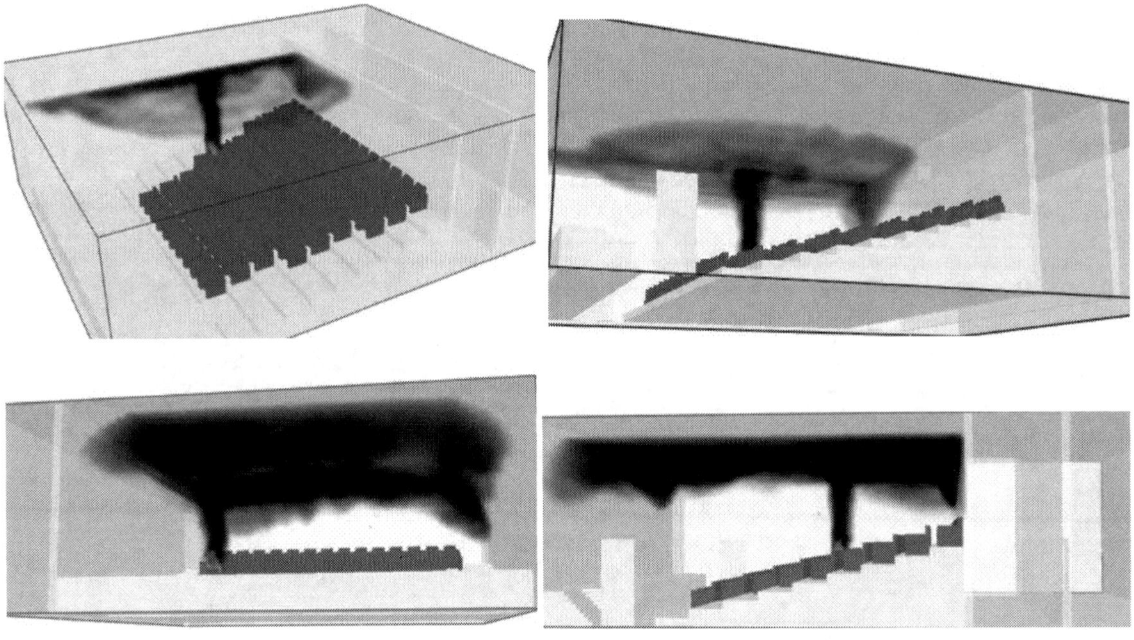

Fig. 2 Fire and smoke spread under the ceiling and formation of toxic gas clouds: the 11th (top) and 22nd second of (bottom) of fire.

5. Conclusions

In this paper, the use of Fire Dynamics Simulator (FDS) for modelling potential cinema fires is illustrated. A great potential of CFD models implemented in FDS for modelling of fire and smoke and safety risks evaluation was confirmed. The simulation results indicate that for the given fire scenario, the highest danger in the cinema hall is in its back and left side parts even during the first minute of the fire and that the curved ceiling increases the danger for spectators. A significant advantage of computer simulation for fire simulation and safety risks evaluation is its ability to vary parameters of tested fire scenarios according to user requirements.

6. Acknowledgment

This work was partially supported by VEGA 2/0216/10 and ASFEU ITMS 26240220060 projects.

7. References

[1] O. Dvorak, J. Angelis, T. Kundrata, H. Matheislova, P. Bursikova, and M. Jahoda, "Computer Simulation of a Fire Test in Mokrsko" (in Czech), *Transactions of the VSB-TU Ostrava*, vol. 5, iss. 2, pp. 45-52, 2010.

[2] J. Glasa, L. Valasak, P Weisenpacher, and L Halada, "Use of PyroSim for simulation of cinema fire", *Int. J. on Recent Trends in Engineering and Technology*, vol. 7, no. 2, 2012.

[3] N. Gourdain, L. Gicquel, G. Staffelbach, O. Vermorel, F. Duchaine, J.-F. Boussuge, and T. Poinsot, "High performance parallel computing of flows in complex geometries: II. Applications", *Computational Science & Discovery*, vol. 2, 28 pp, 2009, doi:10.1088/1749- 4699/2/1/015004.

[4] L Halada, P Weisenpacher, and J Glasa, "Computer modelling of automobile fires" (Chapter XX), IN Fluidodynamics (Ch. Liu, ed.), INTECH Publisher, Rijeka, in press.

[5] J. Hietaniemi, S. Hostikka, and J. Vaari, FDS Simulation of Fire Spread-Comparison of Model Results with Experimental Data, VTT Working Papers 1459-7683, VTT Building and Transport, Finland, 2004.

[6] D. Ling, and K. Kan, "Numerical Simulations on Fire and Analysis of the Spread Characteristics of Smoke in Supermarket", *Communications in Computer and Information Science*, vol. 176, pp. 7-13, DOI: 10.1007/978-3-642-21802-6_2, 2011.

[7] H. Matheislova, M. Jahoda, T. Kundrata, and O. Dvorak, "*CFD simulations of compartment fires*", *Chemical Engineering Trans.*, vol. 21, pp. 1117-1122, DOI: 10.3303/CET1021187, 2010.

[8] K. McGrattan, H. Baum, R. Rehm, W. Mell, R. McDermott, S. Hostikka, and J. Floyd, Fire Dynamics Simulator (Version 5), Technical Reference Guide, NIST Special Publication 1018-5, NIST, Gaithersburg, Maryland, USA, 2010.

[9] K. McGrattan, B. Klein, S. Hostikka, and J. Floyd, Fire Dynamics Simulator (Version 5): User's Guide, NIST Special Publication 1019-5, NIST, Washington, 2009.

[10] K. Prasad, R. Kramer, N. Marsh, M. Nyden, T. Ohlemiller, and M. Zammarano, Numerical Simulation of Fire Spread on Polyurethane Foam Slabs, Fire Research Division NIST, Gaithersburg, MD, USA, 2009.

[11] R. G. Rehm, and H. R. Baum, "The Equations of Motion for Thermally Driven, Buoyant Flows", *Journal of Research of the NBS*, vol. 83, pp. 297–308, 1978.

[12] L. Valasek, The use of Graphical User Interface for Simulation of Fire in Building (in Slovak), thesis, Faculty of Civil Engineering, Slovak Technical University, Bratislava, 2012.

[13] M. Y. Wang, X. Han, G. H. Wu, and Q. Q. Liu, "Simulation analysis of temperature characteristics for a theater fire", *Proc. of the Int. Symp. on Innovations and Sustainability of Structures in Civil Engineering*, vol. 1 and 2, Shanghai, People R. of China, pp. 1145-1152, 2008.

[14] G. H. Wu, X. Han, M. Y. Wang, and Q. Liu, "Simulation analysis of smoke distribution features for a theater fire", *Proc. of the Int. Symp. on Innovations and Sustainability of Structures in Civil Engineering*, vol. 1 and 2, Shanghai, People R. of China, pp. 1153-1159, 2008.

Analysis and Design of Offset QPSK Using Redundant Filter Banks

Alfonso Fernandez-Vazquez[1] and Gordana Jovanovic-Dolecek[2]

[1]School of Computer Engineering, National Polytechnic Institute, ESCOM-IPN, Mexico City, 07738, Mexico.

[2]Electronics Department, National Institute of Astrophysics, Optics and Electronics, INAOE, Tonantzintla, Puebla, 72840, Mexico

E-mail: afernan@ieee.org

Abstract. This paper considers the analysis and design of OQPSK digital modulation. We first establish the discrete time formulation, which allows us to find the equivalent redundant filter banks. It is well known that redundant filter banks are related with redundant transformation of the Frame theory. According to the Frame theory, the redundant transformations and corresponding representations are not unique. In this way, we show that the solution to the pulse shaping problem is not unique. Then we use this property to minimize the effect of the channel noise in the reconstructed symbol stream. We evaluate the performance of the digital communication using numerical examples.

1. Introduction

Digital communication has been one of the successful technologies during the last decades. The main goal of digital communication is to send binary information from one point to another point. Digital modulation plays a key role in digital communication. Different modulations techniques have been proposed in the literature. In particular, offset quadriphase shift keying (OQPSK) provides better performance that QPSK in presence of jitter at the receiver [1].

Multirate signal processing plays an important role in the analysis and design of digital communications. Particularly, filter banks have been used in both Single Input Single Output (SISO) and Multiple Input Multiple Output (MIMO) channels [2, 3]. As an example, Discrete Multitone Modulation (DMT) used in high speed data communication over the twisted pair telephone line is analyzed by using a filter bank. Generally, filter bank transceivers are important tools in channel equalization applications [2, 3].

The analysis and design of Orthogonal Frequency Division Multiplexing with Offset Quadrature Amplitude Modulation (OFDM/OQAM) systems based on filter banks is proposed in [4]. Thus, the authors proposed a filter bank OFDM/OQAM with an even number of subbands. The conditions of discrete orthogonality are also established. OFDM/OAQM is reduced to OQPSK when a single frequency is used and the number of symbols in OQAM is four with equal energy. Unfortunately, this case is not considered in [4]. Therefore, in this paper, we propose the use of filter banks in the analysis and design of OQPSK. We show that the discrete time formulation of OQPSK involves the use of redundant filter banks, which are particular cases of redundant transformation in the frame theory. The main advantage of

Published under licence by IOP Publishing Ltd

using redundant transformation is that its representation is not unique. We use this interesting property to minimize the effect of the channel noise in the reconstructed symbol stream.

This paper is organized as follows. Section 2 gives the analysis of offset QPSK modulation using multirate systems. Section 3 deals with the design of the pulse shaping filters for OQPSK systems.

2. Analysis of Offset QPSK using Redundant Filter Banks

This section introduces the discrete time equivalent of OQPSK signals. We assume that the transmitter OQPSK signal can be written as [1]

$$y(t) = \Re\left\{ \sum_{k=-\infty}^{\infty} \Big(a(k)p(t-kT) + jb(k)p(t-T/2-kT)\Big) e^{j2\pi f_0 t} \right\}, \tag{1}$$

where $a(n)$ and $b(n)$ are the real and imaginary parts of QPSK code words, respectively, f_0 is the carrier frequency, T is the symbol duration, and $\Re\{\cdot\}$ stands for the real part of $\{\cdot\}$. The basedband equivalent signal of $y(t)$ is given by

$$s(t) = \sum_{k=-\infty}^{\infty} \Big(a(k)p(t-kT) + jb(k)p(t-T/2-kT)\Big). \tag{2}$$

In order to obtain a maximum spectral efficiency, we critically sample $s(t)$, i.e., $T_s = T$, where T_s is the sample period. Consequently, we have

$$s(n) = \sum_{k=-\infty}^{\infty} c(k)p_d(2n-kT), \qquad c(n) = \begin{cases} a(n/2), & n \text{ even}, \\ jb((n-1)/2), & n \text{ odd}. \end{cases} \tag{3}$$

Similarly, $p_d(n)$ denotes the oversampled version of the the pulse $p(t)$, that is, $p_d(n) = p(nT/2)$. Using multirate signal processing, we observe that equation (3) represents a decimation filter operating on the input $c(n)$ [5]. The decimation factor in this case is 2.

We now turn our attention to recover the symbol $c(n)$ from $s(n)$. To do that, we recall that the right inverse of a decimation filter is an interpolation filter [6]. Therefore, the symbol stream $s(n)$ can be recovered using the following equation:

$$c(n) = \sum_{k=-\infty}^{\infty} s(k)g(n-2k), \tag{4}$$

where $g(n)$ depends on $p_d(n)$.

Therefore, the problem to recover $c(n)$ from $s(n)$ involves the design of $g(n)$. As we shall see, $g(n)$ is not unique. To do this, consider the Type 1 polyphase components of the z-transform of $p_d(n)$, i.e., $P_d(z) = E_0(z^2) + z^{-1}E_1(z^2)$, where $E_0(z)$ and $E_1(z)$ are the Type I polyphase components of $P_d(z)$. Additionally, using the Type 2 polyphase components $R_0(z)$ and $R_1(z)$ of $G(z)$, we have $G(z) = R_0(z^2) + zR_1(z^2)$, where $G(z)$ is the z-transform of $g(n)$.

Cascading the decimation and interpolation filters, and using the polyphase components of $P_d(z)$ and $G(z)$, we arrive at the structure shown in Figure 1(a), which can be redrawn as shown in Figure 1(b). In order to perfectly recover $c(n)$, the following equation should be held (see Figure 1(b)): $R_0(z)E_0(z) + R_1(z)E_1(z) = z^{-\ell}$, where ℓ is the system delay.

The structure shown in Figure 1(b) is a redundant filter banks. As it is well known the design of redundant filter banks is not unique [7]. From the Frame theory, the polyphase components $R_k(z)$, $k = 0, 1$, can be written as [7, 8, 9];

$$\begin{bmatrix} R_0(z) & R_1(z) \end{bmatrix} = \begin{bmatrix} \widehat{R}_0(z) & \widehat{R}_1(z) \end{bmatrix} + A(z)\begin{bmatrix} E_1(z) & -E_0(z) \end{bmatrix}, \tag{5}$$

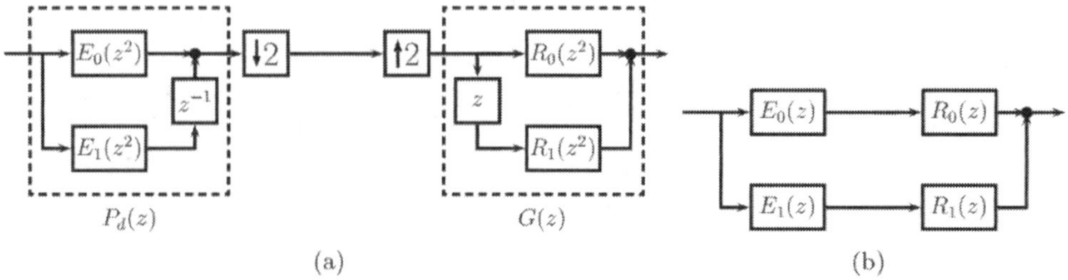

Figure 1. a) cascade of decimation and interpolation b) polyphase form of a), and c) Simplification of part a).

where $A(z)$ is any system function, and $\widehat{R}_0(z)$ and $\widehat{R}_1(z)$ are a particular solution of $\widehat{R}_0(z)E_0(z)+\widehat{R}_1(z)E_1(z) = z^{-\ell}$. Next section illustrates the design of the filters $p_d(n)$ and $g(n)$.

3. Design of pulse shaping filters

Figure 2(a) shows the communication model used to find $A(z)$. The signal $w(n)$ stands for the noise at the communication channel. Using multirate operation, the structure in Figure 2(a) is reduced to the structure shown in Figure 2(b).

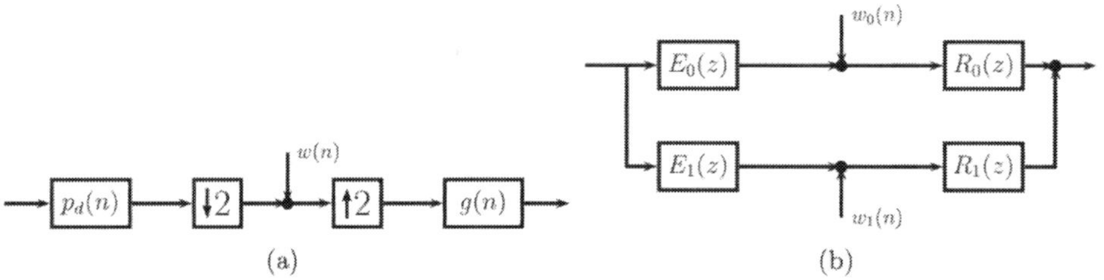

Figure 2. Communication model.

We select the filter $A(z)$, such that it minimizes the noise component at the output of the communication system. Consequently, we find that the optimal $A(z)$ is the Wiener filter given by $A(z) = -S_{uv}(z)/S_{vv}(z)$, where $S_{uv}(z)$ and $S_{vv}(z)$ are respectively the z-transforms of the correlations $R_{uv}(m) = \mathcal{E}(u(n)v^*(n-m))$ and $R_{vv}(m) = \mathcal{E}(v(n)v^*(n-m))$, where $\mathcal{E}(\cdot)$ is the expected value operator. Finally, assuming uncorrelated white noises $w_0(n)$ and $w_1(n)$, we have

$$A(z) = \frac{R_1(z)\widetilde{E}_1(z) - R_0(z)\widetilde{E}_0(z)}{E_0(z)\widetilde{E}_0(z) + E_1(z)\widetilde{E}_1(z)}. \tag{6}$$

We illustrate the design using the following examples.

Example 1. We assume that $p_d(n)$ is the wavelet filter db4, i.e., the number of filter tabs is 6. Since $p_d(n)$ is orthogonal, we select $g(n) = p_d(6-n)$ [10]. In this way, we evaluate the performance of the proposed communication system by using the bit error rate (BER), which have a closed form equation given by $BER = \frac{1}{2}\text{erfc}\,(\sqrt{E_b/N_0})$, where E_b/N_0 is the signal to noise ratio (SNR). Figure 3(a) illustrates the comparison between the simulation results and theoretical value.

Example 2. Now we consider that $p_d(n)$ comes from the root Nyquist filter. Additionally, we applied the Euclidean algorithm to find the corresponding polyphase components $\widehat{R}_0(z)$ and $\widehat{R}_1(z)$ [2]. Figure 3(b) shows the performance of the communication systems. Observe that the

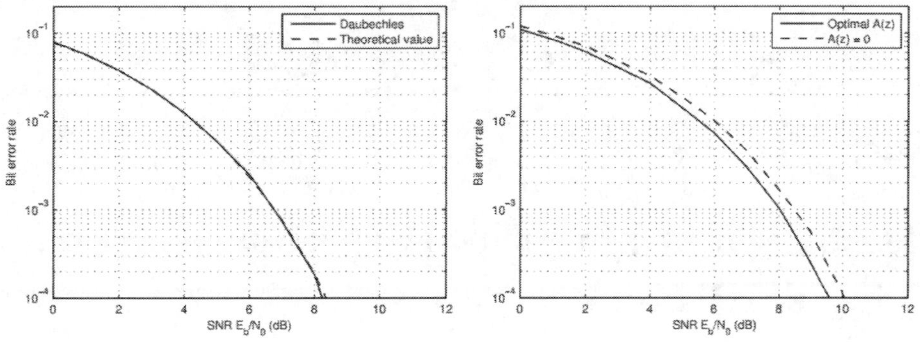

(a) (b)

Figure 3. BER performances of OQPSK transmission a) using Daubechies wavelet filter db4 in Example 1 and b) using multirate filter banks in Example 2.

BER is better using an optimal $A(z)$ than the case where $A(z) = 0$, i.e., $R_0(z) = \widehat{R}_0(z)$ and $R_1(z) = \widehat{R}_1(z)$.

4. Conclusions

This paper addresses the analysis and design of offset QPSK modulation systems using multirate systems. The discrete analysis of offset QPSK modulation involves redundant filter banks, which are special cases of redundant transformation from Frame theory. The main advantage of redundant transformation is that its representation is not unique. Using this interesting property and Wiener theory, we proposed the design of pulse shaping filter of offset QPSK, which minimize the noise component at the output of the communication system. As a future work, we will extend our result to OFDM/OQAM.

Acknowledgments

This work was supported by CONACyT under project 157858 and SIP-IPN project 20121069.

References

[1] Proakis J G 2001 *Digital Communications* 4th ed (Mc Graw Hill)
[2] Vaidyanathan P P, Phoong S M and Lin Y P 2010 *Signal Processing and Optimization for Transceiver Systems* (New York, NY, USA: Cambridge University Press)
[3] Lin Y P, Phoong S M and Vaidyanathan P P 2011 *Filter Bank Transceivers for OFDM and DMT Systems* (New York, NY, USA: Cambridge University Press)
[4] Siohan P, Siclet C and Lacaille N 2002 *IEEE Trans. Signal Process.* **50** 1170–1183
[5] Vaidyanathan P P 1993 *Multirate Systems and Filter Banks* (Englewood Cliffs, NJ: Prentice Hall)
[6] Vaidyanathan P P and Vrcelj B 2001 *IEEE Trans. Signal Process.* **49** 1013–1027
[7] Bolckei H, Hlawatsch F and Feichtinger H G 1998 *IEEE Trans. Signal Process.* **46** 3256–3268
[8] Christensen O 2006 *Modern Mathematical Models, Methods And Algorithms for Real World Systems* ed Siddiqui A H, Duff I S and Christensen O (Anamaya Publishers, New Delhi, India) chap 9, pp 185–221
[9] Li S 1995 *Numer. Funct. Anal. and Optimiz* **16** 1181–1191
[10] Daubechies I 1992 *Ten Lectures on Wavelets* (SIAM, Philadelphia)

Reconstruction of potential part of 3D vector field by using singular value decomposition

Anna Polyakova
Graduate student, Sobolev Institute of Mathematics SB RAS, Novosibirsk, Russia
E-mail: anna.polyakova@ngs.ru

Abstract. In this paper we suggest the method of 3D vector tomography problem solving. The problem consists in determination of potential part of 3D vector field by its known the normal Radon transform. The singular value decomposition of the normal Radon transform operator is obtained. Based on obtained decomposition inversion formula is derived. The decomposition can be the basis for numerical solution of given problem.

1. Introduction

The development both of mathematical approaches and of systems of data measurements and processing induced new mathematical models in tomography such as thermotomography, diffusion tomography, vector and tensor tomography. The models appear due to the necessity of the reconstruction of properties of media with different degree of complication. Thus the tomography of vector fields arises for the description of vector characteristics of currents of fluids, vectors of electromagnetic fields inside the conductor in inhomogeneous media, and many others.[1]

We consider here a method of solving the 3D vector tomography problem in the case of parallel scheme of observation. As the problem of scalar tomography consists in the inversion of the Radon transform for a function, the vector tomography problem is the problem of inversion of normal Radon transform operator applying to potential part of 3D vector fields. In other words, one has to solve operator equations $Af = g$ of the first kind. Here A is a linear, bounded operator. In the operator equation g is a known right hand-side (data of tomographic measurements), and f is an unknown vector field to be determined.

The method of singular value decomposition (SVD) is well known and often used for inversion of compact linear operators. The idea of the approach consists in representing the operator in a form of series of singular numbers and basic elements in the image space. Then the inverse operator is a similar series with the reciprocal of the singular numbers and pre-images of the bases elements. The solution of 3D vector tomography problem, which is proposed in the paper, is based on the possibility of 3D vector field potential part representation by using of potentials, which are constructed as a product of harmonic functions and classical orthogonal polynomials.

2. Definitions and statement of the problem

In this paper we solve the problem of recovery a potential part of vector field, which is given in unit ball $B = \{(x, y, z) \in R^3 \mid x^2 + y^2 + z^2 < 1\}$ with boundary $\partial B = \{(x, y, z) \in R^3 \mid x^2 + y^2 + z^2 = 1\}$. A notification $Z = \{(\xi, s) \mid |\xi| = 1, s \in R\}$ is used for cylinder.

Published under licence by IOP Publishing Ltd

Journal of Physics: Conference Series **410** (2013) 012015 doi:10.1088/1742-6596/410/1/012015

We use the following differential operators:
1) gradient operator $\mathrm{d} : H^k(B) \to H^{k-1}(S^1(B))$, which acts on the potential ψ by formula:

$$\mathrm{d}\psi = \left(\frac{\partial \psi}{\partial x}, \frac{\partial \psi}{\partial y}, \frac{\partial \psi}{\partial z} \right);$$

2) rotor operator $\mathrm{rot} : H^k(S^1(B)) \to H^{k-1}(S^1(B))$, which acts on a vector field $\mathbf{w}$ by next way:

$$\mathrm{rot}\mathbf{w} = \left(\frac{\partial w_3}{\partial y} - \frac{\partial w_2}{\partial z}, \frac{\partial w_1}{\partial z} - \frac{\partial w_3}{\partial x}, \frac{\partial w_2}{\partial x} - \frac{\partial w_1}{\partial y} \right);$$

3) divergence operator $\delta : H^k(S^1(B)) \to H^{k-1}(B)$, which acts on a vector field $\mathbf{w}$ by rule:

$$\delta\mathbf{w} = \frac{\partial w_1}{\partial x} + \frac{\partial w_2}{\partial y} + \frac{\partial w_3}{\partial z}.$$

A vector field $\mathbf{u} \in H^k(S^1(B))$ is a potential vector field, if there is $\phi \in H^{k+1}(B)$ (potential), such as $\mathbf{u} = \mathrm{d}\phi$. A vector field $\mathbf{v} \in H^k(S^1(B))$ is solenoidal, if $\delta\mathbf{v} \in H^{k-1}(B) = 0$. It is obvious that field $\mathbf{u} = \mathrm{rot}\mathbf{v}$ is solenoidal.

It is well known [2] that every vector field $\mathbf{w} \in L_2(S^1(B))$ in R^3 can be decomposed uniquely in a sum of potential and solenoidal parts

$$\mathbf{w} = \mathrm{d}\,\phi + \mathrm{rot}\mathbf{v}.$$

where $\phi \in H^1_0(B)$ and $\mathbf{v} \in H^1(S^1(B))$.

2.1. The Radon transform

The plane $P_{\xi,s}$ in R^3 is given by equation $\langle \xi, \mathbf{x} \rangle - s = 0$ for $\mathbf{x} = (x, y, z)$, $\xi = (\xi^1, \xi^2, \xi^3)$, $|\xi| = 1$. There $|s|$ is distance between the plane and the origin of coordinates, ξ is a normal vector of the plane.

The Radon transform $\mathcal{R}f : L_2(R^3) \to L_2(Z, \rho)$ of function $f(\mathbf{x})$ is given by formula

$$\mathcal{R}f(s, \xi) = \int\limits_{P_{\xi,s}} f(u\mathbf{e_1} + v\mathbf{e_2} + s\xi)\, du\, dv.$$

Integral in the right-hand side does not depend on the choice of the basis $\mathbf{e_1}$, $\mathbf{e_2}$ on the plane of integration.

2.2. The ray transform of a vector field

Let $T = \{(u, v, \xi) \big| u \in [-\sqrt{1 - v^2}, \sqrt{1 - v^2}], v \in [-1, 1], |\xi| = 1\}$.
The ray transform $\mathcal{P} : L_2(S^1(B)) \to L_2(T)$ of a vector field $\mathbf{w}$ is given by formula

$$(\mathcal{P}\mathbf{w})\,(u, v, \xi) = \int\limits_{-\infty}^{\infty} \langle \mathbf{w}, \xi \rangle \;\; dt.$$

Easy to show that the kernel of the operator consist of potential vector fields $\mathrm{d}\phi \in L_2(S^1(B))$ with potential $\phi \in H^1_0(B)$. That is if we know the ray transform of a vector field we can reconstruct only its solenoidal part.

2.3. The normal Radon transform

The normal Radon transform $\mathcal{R}^{\perp} : L_2(S^1(B)) \to L_2(Z,(1-s^2)^{-1})$ of a vector field $\mathbf{w} = \mathbf{w}(x,y,z) = (w_1, w_2, w_3)$ is given by formula

$$\mathcal{R}^{\perp}\mathbf{w} = \iint\limits_{P_{\xi,s}} \left(w_1\xi^1 + w_2\xi^2 + w_3\xi^3\right) du\, dv.$$

To obtain decomposition we need to use following connection between the normal Radon transform of a vector field and the Radon transform of a potential $f \in H_0^1(B)$:

$$\left(\mathcal{R}^{\perp}(\mathrm{d}f)\right)(s,\xi) = \frac{\partial}{\partial s}\left((\mathcal{R}f)(s,\xi)\right). \tag{1}$$

Lemma 1. The kernel of the normal Radon transform consists of solenoidal vector fields, that is, the following equation holds $\mathcal{R}^{\perp}(\mathrm{rot}\mathbf{w}) = 0$ with $\mathbf{w}\big|_{\partial B} = 0$.

In other words if we know the normal Radon transform of a vector field, we can reconstruct only its potential part.

2.4. Statement of the problem

Let we have some potential vector field $\mathrm{d}\phi \in L_2(S^1(B))$, $\phi \in H_0^1(B)$, which is given in a unit ball B. One has to recover this field by its known the normal Radon transform.

3. A singular value decomposition of the normal Radon transform operator

We choose the following system of polynomials as the potentials

$$\Phi_{k,n}(x,y,z) = (1-x^2-y^2-z^2)\, H_k(x,y,z)\, P_n^{(k+2,5,k+1,5)}(x^2+y^2+z^2),\ \ k,n = 0,1,2,\ldots,$$

where $H_k(\mathbf{x})$ are harmonic functions and $P_n^{(p,q)}(\mathbf{x})$ are Yakobi polynomials for the interval $[0,1]$, $k,n = 0,1,\ldots$. In spherical coordinates (r,φ) the functions have the form

$$\Phi_{k,n}(r,\theta,\varphi) = (1-r^2)\, r^k\, P_n^{(k+2,5,k+1,5)}(r^2)\, Y_k(\omega).$$

An application of the operator d leads to a set of potential vector fields

$$\mathbf{T}_{k,n}(x,y,z) \stackrel{def}{=} \mathrm{d}\Phi_{k,n}(x,y,z).$$

Theorem 1. System of potentials (in spherical coordinates)

$$F_{k,n}(r,\theta,\phi) = \frac{\Gamma(n+k+1.5)}{(n+1)!\Gamma(k+1.5)\|Y_k(\omega)\|}\sqrt{\frac{2n+k+2.5}{2}}(1-r^2)\, r^k\, P_n^{(k+2,5,k+1,5)}(r^2)\, Y_k(\omega)$$

forms a system of potential vector fields

$$(\mathbf{F}_{k,n})(x,y,z) = \mathrm{d}F_{k,n}(x,y,z),$$

which is orthonormal in space $L_2(S^1(B))$.

The following proposition is very important to get the images of the potential basis vector fields[3].

Proposition 1. Let $\nu > 0.5$, $k,\ n \geq 0$,

$$\Psi(\omega,\ s) = (1-s^2)^{\nu-0.5}\, C_{2n+k}^{(\nu)}(s)\, Y_k(\omega),$$

with $C^{(\nu)}_{2n+k}(s)$ – Gegenbauer polynomials. Then

$$\Phi = \mathcal{R}^{-1}\Psi = c(n,k,\nu)\left(1-r^2\right)^{\nu-1.5}\; r^k\; P_n^{(k+\nu,k+1.5)}(r^2)\; Y_k(\omega),$$

with $c(n,k,\nu) = \dfrac{(-1)^n\, 2^{1-2\nu}\Gamma(2n+k+2\nu)\Gamma(k+n+1.5)}{\sqrt{\pi}\Gamma(\nu)\,\Gamma(n+\nu-0.5)\Gamma(k+1.5)}$ and $P_n^{(p,q)}(r^2)$— Yakobi polynomials of degree n with indices p, q.

With usage of the Proposition 1 and the formula (1), we obtain

Theorem 2 A system of function

$$G_{k,n} = \frac{(-1)^{n-1}\sqrt{2n+k+2.5}}{\sqrt{(2n+k+3)(2n+k+2)}\|Y_k(\omega)\|}(1-s^2)C^{(1.5)}_{2n+k+1}(s)Y_k(\omega)$$

forms an orthonormal system in space $L_2(Z,\,(1-s^2)^{-1})$.
In other words we have the following relation:

$$(\mathcal{R}^{\perp}\mathbf{F}_{k,n})(s,\,\theta,\,\phi) = \sigma_{k,n}\,G_{k,n}(s,\,\theta,\,\phi),\quad k,n = 0,1,2\ldots,$$

with $\sigma_{k,n} = \dfrac{2\sqrt{2}}{\sqrt{(2n+k+2)(2n+k+3)}}$ – singular values of the operator $\mathcal{R}^{\perp}$. Therefore, a singular value decomposition of the normal Radon transform operator $\mathcal{R}^{\perp}$ has the form

$$\mathcal{R}^{\perp}v = \sum_{k,n=0}^{\infty} \sigma_{k,n}\,(v,\mathbf{F}_{k,n})_{L_2(S^1(B))}\,G_{k,n},$$

but the inverse operator can be calculated by the formula

$$v = \left(\mathcal{R}^{\perp}\right)^{-1} g = \sum_{k,n=0}^{\infty} \sigma^{-1}_{k,n}\,(g,G_{k,n})_{L_2(Z,(1-s^2)^{-1})}\,\mathbf{F}_{k,n}.$$

4. Conclusion

The algorithm can be used for solving of 3D vector tomography problem, namely for reconstruction of a potential part of a 3D vector field if its the normal Radon transforms are known. For this it needs to use truncated singular value decomposition, which is regularization of the problem.

5. Acknowledgments

The work was supported partially by RFBR (grants 11-07-00447-a, 12-01-09336-mob z), SB RAS (grant 2012-32), DMS RAS (grant 2012-1.3.1).

References

[1] Derevtsov E. Yu., Efimov A. V., Louis A. K., Schuster T. Singular value decomposition and its application to numerical inversion for ray transforms in 2D vector tomography. *J. Inverse Ill-Posed Problems*, 2011, Vol. 19, No. 4-5, pp. 689–715.

[2] Weyl H., The method of orthogonal projection in potential theory. *Duke Math. J.*, 1940, No. 7, pp. 411–444.

[3] Louis A.K. Orthogonal function series expansions and the null space of the Radon transform. *Society for industrial and applied mathematics.* Vol. 15, 3 (1984), p.621-633.

Two Spin-1/2 Particles Interacting with a Common Environment

Hyuk-jae Lee, Seok Lee, Deok Ha Woo, Taik Jin Lee, Chul Ki Kim and Jae Hun Kim

Korea Institute of Science and Technology, Seoul, 130-650, Korea

E-mail: jaekim@kist.re.kr

Abstract. We know that the quantum system evolves to different physical phenomena according to initial states. However, the situation is very different in the open system. We consider the system and the environment governed by the Markovian process. The density matrix (the reduced density matrix) of the system satisfies the Kossakowski-Lindblad equation. We calculate the expectation value of the z-direction-magnetization in the system for various types of 1) initial states and 2) interactions between the system and the environment. In all the cases, the values of the magnetization approached asymptotically to the same point even though the different initial states and different interaction types between the system and the environment are applied. The paths to asymptotically approach the final values are different according to the initial states and the interactions. These facts show that the physical results of expectation value are independent of the initial states.

1. Introduction

Every quantum system is unavoidable to the interaction with the environments, and its quantum dynamics has to be considered of the affection of the environment. We know that dissipation and destruction of quantum correlation in the open quantum system result from the environment. The dynamics of the open quantum system is described by the time evolution of a reduced density operator. In particular, the open quantum system, in which a Markovian process proceeds under the influence of the environment, satisfies the Kossakowski-Lindblad equation [1, 2].

Many studies have contributed to drive the Kossakowski-Lindblad equation in mathematical and physical methods. Solving the Kossakowski-Lindblad equation is very hard work in a usual open system and uses numerical calculation or indirect methods to understand the physical phenomena of the open system. Then it is usually hard to know the meaning of the Markovian process and the role of the initial state because we cannot solve the exact solution of the Kossakowski-Lindblad equation. Here we consider the quantum system stated in the thermal environment and calculate the average of the z-component of the total spin. Then, we will check the role of the initial state of the open quantum system and interaction between the open system and the environment. Of course, we will treat the particular open quantum system, which can be solved with the Kossakowski-Lindblad equation.

This paper is organized as follows: Firstly, we drive the Kossakowski-Lindblad equation under the environment, which consists of harmonic modes, for the sake of simplicity, in the zero temperature. Two spin-1/2 particles, in which each particle interacts with the common environment, are considered. Secondly, we solve the time evolution equation of a reduced density operator and calculate the magnetization of the system. Finally, we discuss and summarize our results.

2. Magnetization of two particle system in the Markov Process

The two spin-1/2 particles interacting with the common environment (bath) are described by the Hamiltonian,

$$\mathcal{H} = \mathcal{H}_{S_1} + \mathcal{H}_{S_2} + \mathcal{H}_B + \mathcal{H}_{S_1B} + \mathcal{H}_{S_2B}, \tag{1}$$

where the subscripts S_i and B denote the two subsystems and the bath, respectively. $\mathcal{H}_{S_1B}$ and $\mathcal{H}_{S_2B}$ are the interaction Hamiltonians of each subsystem with bath are given by

$$\mathcal{H}_{S_1B} = \sum_i \sigma_i \otimes I \otimes \Gamma^i, \qquad \mathcal{H}_{S_2B} = \sum_i I \otimes \sigma_i \otimes \Gamma^i, \tag{2}$$

where σ_i is the Pauli's matrix, and Γ^i is the bath operator. Here we assume that there are no direct interactions between two spins. The dynamics of the total system including both the system and the bath can be described by the Liouville equation as following:

$$\frac{\partial \rho_T}{\partial t} = -i[\mathcal{H}, \varrho_T] \tag{3}$$

where ρ_T denotes the density operator of the total system. The density operator of the total system is reduced to an operator acting only in the Hilbert space of the system as

$$\varrho(t) = tr_B \varrho_T(t), \tag{4}$$

where tr_B means the partial trace over the bath space. The operator $\rho(t)$ is called reduced density operator of the system.

We consider the weak-coupling limit that the change rate of the reduced density operator is quite slow compared to that of the bath operators. In this case the Markov approximation is reasonable. We will assume that no correlation between system and bath initially exists. The initial state of the system and bath is factorized as $\rho_T(0) = \rho(0) \otimes \rho_B$, where we assume that the bath is in the equilibrium state of the bath. If the above assumptions are considered, the reduced density operator satisfies the Kossakowski-Lindblad equation [3] described as

$$\frac{d\varrho}{dt} = -\sum_{i,j=1}^{3}[A_{ij}\left(\frac{1}{2}\{(\sigma_j\sigma_i \otimes I), \rho\} - (\sigma_i \otimes I)\rho(\sigma_j \otimes I)\right) + C_{ij}\left(\frac{1}{2}\{(I \otimes \sigma_j\sigma_i), \rho\} - (I \otimes \sigma_i)\rho(I \otimes \sigma_j)\right) + B_{ij}\left(\frac{1}{2}\{(\sigma_i \otimes \sigma_j), \rho\} - (\sigma_i \otimes I)\rho(I \otimes \sigma_j)\right) + B^*{}_{ij}\left(\frac{1}{2}\{(\sigma_i \otimes \sigma_j), \rho\} - (I \otimes \sigma_j)\rho(\sigma_i \otimes I)\right), \tag{5}$$

where the coefficient matrices are defined by $A_{ij} = \frac{1}{\hbar^2}\int_0^\infty d\tau\,(V_{ji} + V_{ij}^*)$, $B_{ij} = \frac{1}{\hbar^2}\int_0^\infty d\tau\,(W_{ji} + W_{ij}^*)$, and $C_{ij} = \frac{1}{\hbar^2}\int_0^\infty d\tau\,(Y_{ji} + Y_{ij}^*)$. V, W, X and Y denote the correlation functions of bath operators Γ^i.

The magnetization of the system at time t is defined by the reduced operator as

$$M_z(t) = \frac{1}{2}\, tr[\rho(t)(\sigma_z \otimes I + I \otimes \sigma_z)] \tag{6}$$

The reduced density operator at time t is obtained by solving eq. (5). The effects of the interaction between the system and bath are included in the coefficient matrices A, B, and C in eq. (5).

We first consider the case that the matrices are given by

$$\mathrm{A} = \mathrm{B} = \mathrm{C} = \frac{1}{2}\begin{pmatrix} \gamma & -i\gamma & 0 \\ i\gamma & \gamma & 0 \\ 0 & 0 & 0 \end{pmatrix}. \tag{7}$$

This case describes the spontaneous transition of the two spin-1/2 particles separated by far distance. Eq. (5) can be analytically solved. First, by using the Laplace transformation, eq. (5) evolves to the algebraic equation, which can be solvable. Then, the inverse Laplace transformation gives the reduced density matrix [4]. For the calculation of the magnetization, we insert the reduced density matrix into eq. (6). Here we can calculate the magnetization of the system for the four cases that has different initial states of the system as

$$M_z(t) = e^{-2\gamma t}(2 - e^{2\gamma t} + 2\gamma t) \quad \textit{for the initial state } |\uparrow\uparrow\rangle,$$

$$M_z(t) = e^{-2\gamma t}(1 - e^{2\gamma t} + \gamma t) \quad \textit{for the initial state } \frac{1}{\sqrt{2}}(|\uparrow\uparrow\rangle + |\downarrow\downarrow\rangle),$$

$$M_z(t) = e^{-2\gamma t}(1 - e^{2\gamma t}) \quad \textit{for the initial state } \frac{1}{\sqrt{2}}(|\uparrow\downarrow\rangle + |\downarrow\uparrow\rangle),$$

$$M_z(t) = -1 \qquad \textit{for the initial state } |\downarrow\downarrow\rangle. \tag{8}$$

We draw the graph of the magnetization versus time at Fig. 1. The magnetization approaches to the same value of -1, independent of each initial state after certain duration in t.

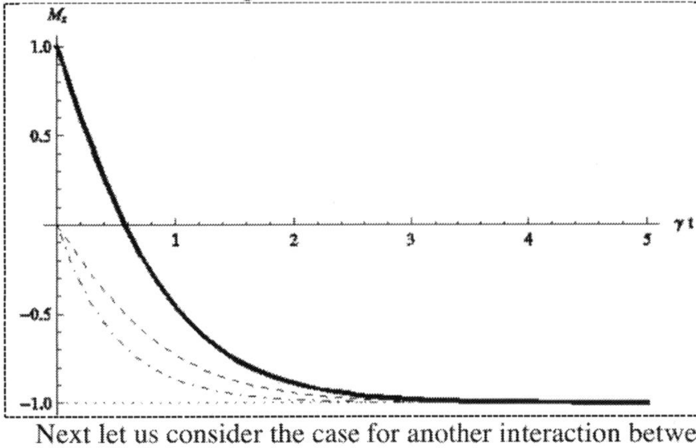

Figure 1. The time dependence of the magnetization for the initial states of two spin-1/2 particles system of eq. (7). The solid line, the dotted line, the dashed line and the dotted-dashed line correspond to the magnetization for $|\uparrow\uparrow\rangle$, $|\downarrow\downarrow\rangle$, $\frac{1}{\sqrt{2}}(|\uparrow\uparrow\rangle + |\downarrow\downarrow\rangle)$ and $\frac{1}{\sqrt{2}}(|\uparrow\downarrow\rangle + |\downarrow\uparrow\rangle)$, repectively.

Next let us consider the case for another interaction between the system and bath as

$$A = C = \begin{pmatrix} 1 & \frac{-i}{\sqrt{2}} & 0 \\ \frac{i}{\sqrt{2}} & 1 & 0 \\ 0 & 0 & 0 \end{pmatrix}, \quad B = \begin{pmatrix} \frac{1}{\sqrt{2}} & 0 & 0 \\ 0 & \frac{-1}{\sqrt{2}} & 0 \\ 0 & 0 & 0 \end{pmatrix} \quad . \tag{9}$$

There is no particular physical meaning for this interaction. However, we think as the counter-example to understand the meaning of the Markov process in the open quantum system. If we give special meaning, this interaction is used to explain the entanglement generation in the system with dissipation [3, 5]. This open quantum system has an exact analytic solution for the reduced density operator. Then, we can obtain the magnetization as the similar way. For each initial state, the final forms of the magnetization are given as follows:

$$M_z(t) = \frac{1}{\sqrt{2}} + \left(1 - \frac{1}{\sqrt{2}}\right)e^{-4t} \qquad \textit{for the initial state } |\uparrow\uparrow\rangle,$$
$$M_z(t) = \frac{1}{\sqrt{2}}e^{-4t}(-1 + e^{4t}) \qquad \textit{for the initial state } \frac{1}{\sqrt{2}}(|\uparrow\uparrow\rangle + |\downarrow\downarrow\rangle),$$
$$M_z = \frac{1}{\sqrt{2}}e^{-4t}(-1 + e^{4t}) \qquad \textit{for the initial state } \frac{1}{\sqrt{2}}(|\uparrow\downarrow\rangle + |\downarrow\uparrow\rangle),$$
$$M_z(t) = \frac{1}{\sqrt{2}} - (1 + \frac{1}{\sqrt{2}})e^{-4t} \qquad \textit{for the initial state } |\downarrow\downarrow\rangle. \tag{10}$$

We calculate eq. (5) for the same initial states as eq. (8) to compare the results. The magnetization converses to the same value of $1/\sqrt{2}$, independent of different initial states as the previous system, eq. (7).

As the third example, we consider the open system with the different magnitude coefficients with following equations, which are the similar interaction between the system and bath as eq. (9):

$$A = C = \begin{pmatrix} 1 & \frac{-i}{4} & 0 \\ \frac{i}{4} & 1 & 0 \\ 0 & 0 & 0 \end{pmatrix}, \quad B = \begin{pmatrix} \frac{\sqrt{15}}{4} & 0 & 0 \\ 0 & \frac{-\sqrt{15}}{4} & 0 \\ 0 & 0 & 0 \end{pmatrix} \quad . \tag{11}$$

By using the same procedure, we can get the magnetization as the function of the time as follows:

$$M_z(t) = \frac{1}{4}(1 + 3e^{-4t}) \qquad \textit{for the initial state } |\uparrow\uparrow\rangle,$$
$$M_z(t) = \frac{1}{4}(1 - e^{-4t}) \qquad \textit{for the initial state } \frac{1}{\sqrt{2}}(|\uparrow\uparrow\rangle + |\downarrow\downarrow\rangle),$$
$$M_z = \frac{1}{4}(1 - e^{-4t}) \qquad \textit{for the initial state } \frac{1}{\sqrt{2}}(|\uparrow\downarrow\rangle + |\downarrow\uparrow\rangle),$$
$$M_z(t) = \frac{1}{4}(1 - 5e^{-4t}) \qquad \textit{for the initial state } |\downarrow\downarrow\rangle. \tag{12}$$

In this case, the magnetization approaches to the same value of 1/4, but obviously different value from the previous case as shown in Fig. 3.

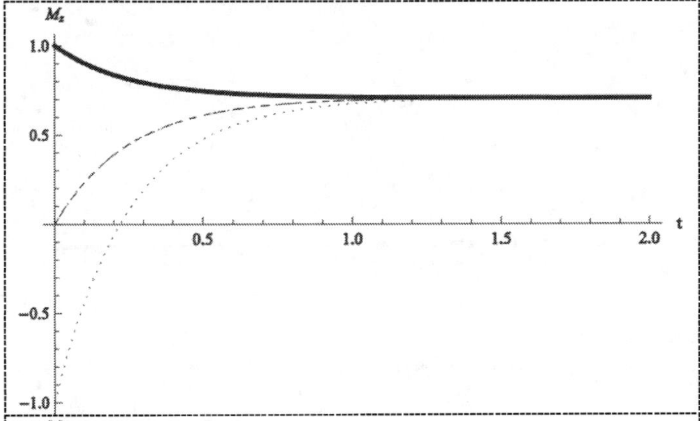

Figure 2. The time dependence of the magnetization for the initial states of two spin-1/2 particles system of eq. (9). The solid line and the dotted line correspond to the magnetization for $|\uparrow\uparrow\rangle$ and $|\downarrow\downarrow\rangle$, respectively. The dashed line and the dotted-dashed line for $\frac{1}{\sqrt{2}}(|\uparrow\uparrow\rangle + |\downarrow\downarrow\rangle)$ and $\frac{1}{\sqrt{2}}(|\uparrow\downarrow\rangle + |\downarrow\uparrow\rangle)$ has the same trajectory.

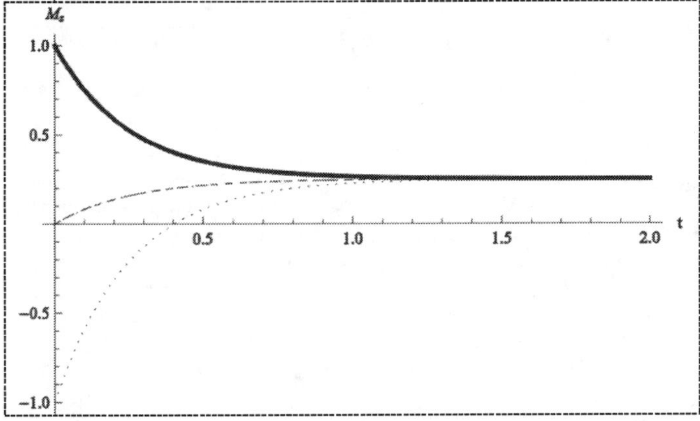

Figure 3. The time dependence of the magnetization for the initial states of two spin-1/2 particles system of eq. (11). The solid line and the dotted line correspond to the magnetization for $|\uparrow\uparrow\rangle$ and $|\downarrow\downarrow\rangle$, respectively. The dashed line and the dotted-dashed line for $\frac{1}{\sqrt{2}}(|\uparrow\uparrow\rangle + |\downarrow\downarrow\rangle$ and $\frac{1}{\sqrt{2}}(|\uparrow\downarrow\rangle + |\downarrow\uparrow\rangle)$ has the same trajectory.

3. Discussion

We can usually consider that the quantum system with different initial states evolves to different final states causing different physical phenomena. We obtained the magnetization of the open systems with the various situations. We calculated the expectation values of the z-direction magnetization of the two spin-1/2 system in several different cases. Even under the different interaction types and values between the system and bath, the magnetization with the different initial states approached asymptotically to the same final value. The paths to approach to the final values were different from the initial states and the interactions.

From our results, we assume that the open quantum system results in dissipation and forgets the old history because of the Markovian process with an environment. According to calculation results of the three different cases in the previous section, we suggest the fact that the magnetization evolves to the same value independent of the initial states.

References

[1] V. Gorini, A. Kossakowski, and E.□C.□G. Sudarshan, J. Math. Phys. (N.Y.) **17** (1976) 821
[2] G. Lindblad, Commun. Math. Phys. 48 (1976) 119.
[3] F. Benatti, R. Floreanini, and M. Piani, Phys. Rev. Lett. 91 (2003) 070402.
[4] L. Jak\'{o}bczyk, J. Phys. A 35 (2002) 6383.
[5] T. Choi and H.-j. Lee, Phys. Rev. A 76 (2007) 012308.

Modeling of the magnetic properties of nanomaterials with different crystalline structure

Yury Kirienko and Leonid Afremov

Department of Theoretical and Experimental Physics, School of Natural Sciences, Far-Eastern Federal University, 8, Sukhanova str., Vladivostok, Russia

E-mail: yury.kirienko@gmail.com

Abstract. We propose a method for modeling the magnetic properties of nanomaterials with different structures. The method is based on the Ising model and the approximation of the random field interaction. It is shown that in this approximation, the magnetization of the nanocrystal depends only on the number of nearest neighbors of the lattice atoms and the values of exchange integrals between them. This gives a good algorithmic problem of calculating the magnetization of any nano-object, whether it is ultrathin film or nanoparticle of any shape and structure, managing only a rule of selection of nearest neighbors. By setting different values of exchange integrals, it is easy to describe ferromagnets, antiferromagnets, and ferrimagnets in a unified formalism. Having obtained the magnetization curve of the sample it is possible to find the Curie temperature as a function of, for example, the thickness of ultrathin film. Afterwards one can obtain the numerical values for critical exponents of the phase transition "ferromagnet – paramagnet". Good agreement between the results of calculations and the experimental data proves the correctness of the method.

1. Introduction

Creation of magnetic materials with predetermined properties – the task of unquestionable importance. While the manipulation of individual atoms becomes a routine, nanotechnologies need appropriate calculation methods, which would allow prototyping magnetic properties of materials without heavy calculations on supercomputers. The method proposed below satisfies such requirements. Moreover, it allows calculus of the properties of magnets of any type (ferromagnets, antiferromagnets, ferrimagnets) to be provided using a uniform formalism.

Another feature of nano-objects is the existence of size effects at their scale. Obviously, such effects are different in two-dimensional films and three-dimensional particles. Furthermore, it is common to use different methods of modeling for nanoparticles of different shape. But if to consider the theory with short-range interaction, the global geometry of the material should not play any role. In this case, there must be a simulation method that is insensitive to the shape of the simulated sample as a whole. We offer in this paper such a method.

2. General model of magnetic material

We assume that magnetic atoms are distributed over N sites of the sample with probability p. According to [1], the distribution function for random interaction fields H on a particle located at the origin can be defined as:

$$W(H) = \int \delta \left(H - \sum_k h_k(\mathbf{r}_k, \mathbf{m}_k) \prod_k F(\mathbf{m}_k)\, \delta(\mathbf{r}_k - \mathbf{r}_{k0}) \right) d\mathbf{r}_k d\mathbf{m}_k, \tag{1}$$

where $\delta(x - x_0)$ – Dirac delta function, $h_k = h_k(\mathbf{m}_k, \mathbf{r}_k)$ – field created by atoms with magnetic moments $\mathbf{m}_k$ located at coordinates $\mathbf{r}_k$, $\mathbf{r}_{k0}$ – the coordinates of lattice sites, $F(\mathbf{m}_k)$ – the distribution function for the magnetic moments, which in the approximation of Ising model for a ferromagnets can be represented as follows:

$$F(\mathbf{m}_k) = (\alpha_k \delta(\theta_k) + \beta_k \delta(\theta_k - \pi)) \left((1-p)\, \delta(m_k) + p\delta(m_k - m_0) \right). \tag{2}$$

Here θ_k – the angle between $\mathbf{m}_k$ and OZ-axis, α_k and β_k – relative probabilities of the spin orientation along and against OZ-axis ($\theta_k = 0$ and $\theta_k = \pi$, respectively); m_0 – magnitude of magnetic moment of a magnetic atom. Probabilities α_k and β_k hold normalization condition $\alpha_k + \beta_k = 1$. In the approximation of nearest neighbors and the direct exchange interaction between magnetic atoms, the equation (1) can be represented as:

$$W_j(H) = \sum_{n=0}^{z} p^{z-n}(1-p)^n \sum_{\nu}^{C_z^{z-n}(k_j)} \sum_{l_\nu \in L(C_z^n(k_j))}^{2^n} \omega_{l_\nu} \delta(H - M_{l_\nu} J_j), \tag{3}$$

where k_j is a set of nearest neighbors of the magnetic atom numbered as j, $z = \dim k_j$ – its coordination number; $C_z^n(k_j)$ – subset of n atoms of the total number of z nearest neighbors of jth atom; $L(\Omega)$ is a binomial set of permutations of an arbitrary set Ω with the amount of elements equal to $2^{\dim \Omega}$. Introducing symmetric notation $\alpha_{-n} \equiv 1 - \alpha_n$ we have got $\omega_{l_j} = \prod_{(\nu=1; l_\nu \in k_j)}^{\nu=z} \alpha_{\pm l_\nu}$ and $M_{l_j} = \sum_{n\in k_j} \pm m_n = m_0 \sum_{n \in k_j} \pm|2\alpha_n - 1|$. Finally, J_j is the constant of exchange interaction (exchange integral).

Using the expression for the distribution function of interaction fields (3), one can obtain equations that determine average relative magnetic moments at each site of lattice:

$$\mu_j = \int \tanh\left(\frac{m_j H}{k_B T}\right) W_j(H)\, dH = \sum_{n=0}^{z} p^{z-n}(1-p)^n \sum_{\nu}^{C_z^{z-n}(k_j)} \sum_{l_\nu \in L(C_z^n(k_j))}^{2^n} \omega_{l_\nu} \tanh\left(\frac{M_{l_\nu} J_j}{k_B T}\right). \tag{4}$$

Expression (4) allows to investigate the dependence of total magnetic moment of the sample $M = \sum_{j=1}^{N} \mu_j$ at the temperature T and concentration p, as well as to determine the dependence on the number N of atoms of the temperature of phase transition and the percolation threshold. System of N equations with N unknowns (4) can be solved numerically using Newton's method[1].

3. Modeling of magnetic materials with different crystalline structure

Equation (4), despite the complicated form, has written in the algorithmically convenient form. This form allows to simulate the magnetic properties of materials, based only on the knowledge of the crystalline structure of the sample and the numerical values of the exchange integrals.

Crucial part of this method is the rule of selection of nearest neighbors for each atom (the way of constructing of the set k_j). By changing this rule of selection we can easily adjust our model to different physical systems. We also can simulate antiferromagnetic and ferrimagnetic materials by reversing signs and values of exchange integrals of individual atoms.

Consider two examples of applying the method described above.

[1] In our research we use `Wolfram Mathematica` for fast lattice prototyping and `C++` and `Fortran` non-linear solvers with `python` wrapper for precise calculus.

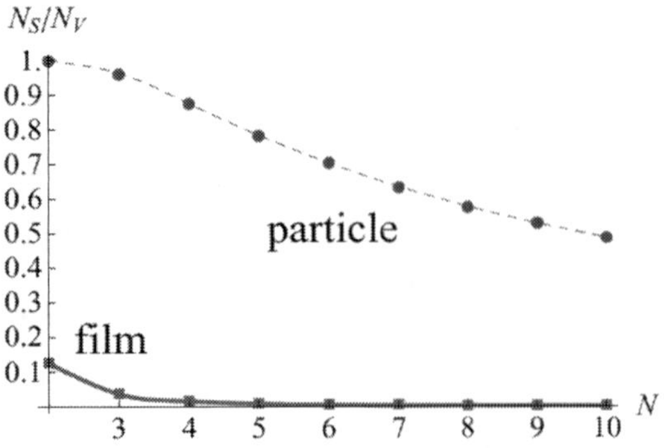

Figure 1. The ratio of the number of atoms on the surface of the sample N_S to the number of atoms in the volume N_V as a function of size (for cubic particles N is the number of atoms on the edge, for films – its thickness).

3.1. Ultrafine particles

Consider a nanoparticle of N atoms. Then (4) is a system of N independent equations with N unknowns. In some cases, when $p = 1$, the symmetry of the particle reduces the number of unknowns, but in general all variables are different.

For example, for cubic-shaped nanoprticle with n atoms on the edge and simple cubic lattice there is a system of n^3 non-algebraic equations with n^3 unknowns. Even relatively small cubic particle with 10 atoms on the edge contains 10^3 atoms, which makes calculation non-trivial and significantly reduces its accuracy.

3.2. Thin films

Now consider ultrathin film that is composed of N infinite monolayers. In a simple case when all atoms from the same layer are equal, (4) turns into a system of N equations with N unknowns. Study of size effects in films is much easier than in the particles, because it is sufficient to simulate a film of 10-15 layers to achieve the bulk properties. (See fig. 1.)

General equation that determines the average relative magnetic moment μ_n in n-th monolayer is given by (4). Replacing in expressions for ω_ℓ and M_ℓ all $\alpha_{\pm n}$ on their average values $\langle\alpha_{\pm n}\rangle = (1 \pm \mu_n)/2$ and substituting (3) into (4), one can obtain the equations that determine μ_n in each monolayer:

$$
\left\{
\begin{aligned}
\mu_1 &= \sum_{l=0}^{z_{1,1}} \binom{l}{z_{1,1}} \langle\alpha_1\rangle^l \langle\beta_1\rangle^{z_{1,1}-l} \sum_{k=0}^{z_{1,2}} \binom{k}{z_{1,2}} \langle\alpha_2\rangle^k \langle\beta_2\rangle^{z_{1,2}-k} \tanh\left(\frac{(2l-z_{1,1})+(2k-z_{1,2})i_{1,2}}{t}\right), \\
\mu_n &= \sum_{l=0}^{z_{n,n}} \binom{l}{z_{n,n}} \langle\alpha_n\rangle^l \langle\beta_n\rangle^{z_{n,n}-l} \sum_{k=0}^{z_{n-1,n}} \binom{k}{z_{n-1,n}} \langle\alpha_{n-1}\rangle^k \langle\beta_{n-1}\rangle^{z_{n-1,n}-k} \times \\
&\qquad \sum_{r=0}^{z_{n,n+1}} \binom{r}{z_{n,n+1}} \langle\alpha_{n+1}\rangle^r \langle\beta_{n+1}\rangle^{z_{n,n+1}-r} \times \\
&\qquad\qquad \tanh\left(\frac{(2k-z_{n-1,n})i_{n-1,n}+(2l-z_{n,n})i_{n,n}+(2r-z_{n,n+1})i_{n,n+1}}{t}\right) \\
\mu_N &= \sum_{l=0}^{z_{N,N}} \binom{l}{z_{N,N}} \langle\alpha_N\rangle^l \langle\beta_N\rangle^{z_{N,N}-l} \sum_{k=0}^{z_{N-1,N}} \binom{k}{z_{N-1,N}} \langle\alpha_{N-1}\rangle^k \langle\beta_{N-1}\rangle^{z_{N-1,N}-k} \times \\
&\qquad \tanh\left(\frac{(2l-z_{N-1,N})i_{N-1,N}+(2k-z_{N,N})i_{N,N}}{t}\right),
\end{aligned}
\right.
\tag{5}
$$

where $z_{n,n}$ is the number of nearest neighbors in the n-th layer, $z_{n-1,n}$ is the number of nearest neighbors of the atom in $(n-1)$-th layer, located in the n-th layer; $i_{nn} = J_{nn}m_n/J_{11}m_1$, $i_{n-1,n} = J_{n-1,n}m_{n-1}/J_{11}m_1$, $i_{n,n+1} = J_{n,n+1}m_{n+1}/J_{11}m_1$, $t = k\,T/J_{11}m_1$. Using (5), one can study the dependence of the average magnetic moment of the film on its temperature and thickness [2], and, as a consequence, the dependence of Curie temperature on the thickness (see fig. 2).

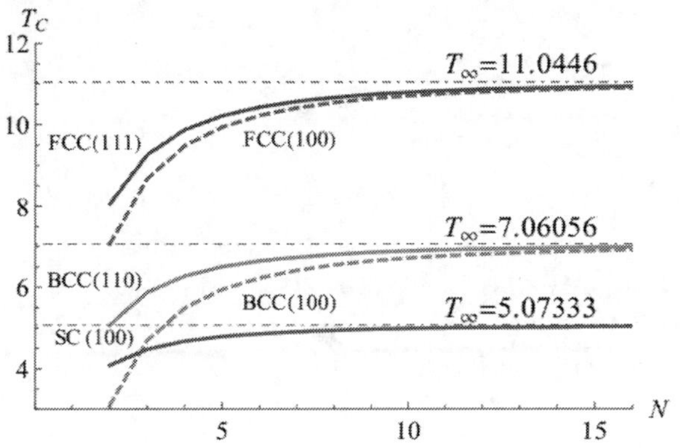

Figure 2. The dependence of reduced Curie temperature T_C on thickness of ultrathin film (N is the number of monolayers) for different crystal lattices (FCC, BCC and SC) and different crystallographic orientations of the surface.

4. Comparison with experimental data
Comparison of the results of modeling with experimental data was carried out directly and indirectly.

4.1. Direct comparison: relative Curie temperature change
In [3], the dependence of Curie temperature $T_C(N)$ on the size N of cubic-shaped particle was calculated. And then the magnitude of the relative change of the Curie temperature $\varepsilon(N) = T_C(N)/T_C(N \to \infty)$ was obtained. Comparison with experimental data [4] revealed the validity of the modeling method.

4.2. Indirect comparison: critical exponents
It is possible to obtain the value of critical exponent ν of spin-spin correlation for phase transition "ferromagnet – paramagnet" by appropriate approximation of $\varepsilon(N)$ defined above. Critical exponents for ultrathin films of different crystalline structures were obtained in [2]. It was shown that the value of ν for three-dimensional Ising model is independent from the type of the lattice. The numerical value of ν is close to the value obtained from RG-calculations [5].

5. Conclusion
Offered method of simulating allows to substitute the real magnetic material with the relatively simple model, where the key role is played by the coordination number and the exchange integrals. All the magnetic geometry lies in these two integral characteristics. By setting only the rules of choice of nearest neighbors and values of corresponding exchange integrals, we make the modeling procedure very effective.

Acknowledgments
The work was supported by grant of Scientific Fund of Far Eastern Federal University (FEFU) № 12-07-13000-FEFU_a.

References
[1] Belokon V and Nefedev K 2001 *Journal of Experimental and Theoretical Physics* **93**(1) 136–142 ISSN 1063-7761 10.1134/1.1391530 URL `http://dx.doi.org/10.1134/1.1391530`
[2] Afremov L and Kirienko Y 2012 *Advanced Materials Research* **378–379** 589–592 (*Preprint* `http://arxiv.org/abs/1108.0745`)
[3] Kirienko Y and Afremov L 2012 *Advanced Materials Research* **472–473** 1827–1830 (*Preprint* `http://arxiv.org/abs/1201.1562`)
[4] Sadeh B, Doi M, Shimizu T and Matsui M 2000 *Journal of the Magnetics Society of Japan* **24** 511–514
[5] Le Guillou J and Zinn-Justin J 1977 *Phys. Rev. Lett.* **39** 95–98

Journal of Physics: Conference Series **410** (2013) 012018 doi:10.1088/1742-6596/410/1/012018

Multi-frequency imaging of perfectly conducting cracks via boundary measurements

Won-Kwang Park[1] **and Dominique Lesselier**[2]

[1]Department of Mathematics, Kookmin University, Seoul, 136-702, Korea

[2]Département de Recherche en Électromagnétisme - Laboratoire des Signaux et Systèmes, UMR8506 (CNRS-Supélec-Université Paris-Sud 11) 91192 Gif-sur-Yvette cedex, France

E-mail: `parkwk@kookmin.ac.kr` and `dominique.lesselier@lss.supelec.fr`

Abstract. Imaging of perfectly conducting crack(s) in a 2-D homogeneous medium using boundary data is studied. Based on the singular structure of the Multi-Static Response (MSR) matrix whose elements are normalized by an adequate test function at several frequencies, an imaging functional is introduced and analyzed. A non-iterative imaging procedure is proposed. Numerical experiments from noisy synthetic data show that acceptable images of single and multiple cracks are obtained.

1. Introduction

One considers the inverse scattering problem of the retrieval of the shape of perfectly conducting cracks buried within a homogeneous medium. Data are scattered field measured at the search domain boundary.

To our knowledge, inverse scattering with a single, perfectly conducting crack in 2-D satisfying a Dirichlet boundary condition has been studied in [4]. In this reference, the retrieval algorithm was based on a Newton-type iterative scheme so that a good initial guess close to the unknown crack was needed to ensure convergence. But an iterative scheme generally requires a large amount of computational time, and it is hard to extend to the reconstruction of multiple cracks unless strong hypotheses are made beforehand.

So, non-iterative reconstruction algorithms have been developed, e.g., MUltiple SIgnal Classification (MUSIC)-type ones [2, 9, 10], topological-derivative-based ones [5, 8]. Recently, a multi-frequency algorithm has been proposed; related works are in [1, 6, 7, 11] and references. But in many cases, the analysis is not fully reliable, in particular since phenomena such as the unexpected appearance of ghost replicas cannot be explained.

Here, one is proposing an improved imaging algorithm for perfectly conducting cracks from scattered field data measured at the search domain boundary, and one analyzes its structure by designing a multi-frequency imaging functional weighted by each frequency. This involves that elements of the Multi-Static Response (MSR) matrix normalized by a test function can be represented in a suitable asymptotic form. Via careful analysis, one can infer a relationship between the proposed imaging functional and Bessel functions of integer order and explain why this imaging functional yields location and shape of the cracks.

The discussion goes as follows. In section 2, the direct scattering problem is summarized and the asymptotic expansion formula introduced. In section 3 the multi-frequency-based imaging functional weighted by each frequency is put forth and its properties discussed. Numerical experiments are in section 4 for single and multiple cracks. Conclusion is in section 5.

2. Direct scattering problem and asymptotic expansion formula

In this section, one considers 2-D electromagnetic scattering by a perfectly conducting crack, Γ, fully buried within a homogeneous domain $\Omega \subset \mathbb{R}^2$. One assumes that Γ is a narrow, linear crack represented as follows

$$\Gamma = \{\vec{z} : \vec{z} = (x, y), -\rho < x < \rho\}, \tag{1}$$

and that it does not touch the boundary $\partial\Omega$, i.e., $\Gamma \cap \partial\Omega = \emptyset$.

Throughout this contribution, we remain with the Transverse Magnetic (TE) polarization case: at a given frequency $\omega_k \gg \rho$, let $u^{(n)}(\vec{x};\omega_k)$ be the time-harmonic total field that satisfies

$$\begin{cases} \Delta u^{(n)}(\vec{x};\omega_k) + \omega_k^2 u^{(n)}(\vec{x};\omega_k) = 0 & \text{for} \quad \vec{x} \in \Omega \backslash \overline{\Gamma}, \\ u^{(n)}(\vec{x};\omega_k) = 0 & \text{for} \quad \vec{x} \in \Gamma, \\ \dfrac{\partial u^{(n)}(\vec{x};\omega_k)}{\partial \vec{\nu}(\vec{x})} = \dfrac{\partial e^{i\omega_k \vec{\theta}_n \cdot \vec{x}}}{\partial \vec{\nu}(\vec{x})} & \text{for} \quad \vec{x} \in \partial\Omega. \end{cases} \tag{2}$$

Here, $\vec{\nu}$ is the outward unit normal vector to $\partial\Omega$ and $\{\vec{\theta}_n : n = 1, 2, \cdots, N\}$ are the N vectors that describe the directions of incidence, assumed equi-distributed on the unit circle $\mathbb{S}^1$. In the same manner, $u_0^{(n)}(\vec{x};\omega_k)$ satisfies (2) without Γ. Then, the following asymptotic relationship of $u^{(n)}(\vec{x};\omega_k)$ and $u_0^{(n)}(\vec{x};\omega_k)$ holds uniformly on $\vec{x} \in \partial\Omega$ (see [2] for instance):

$$u^{(n)}(\vec{x};\omega_k) - u_0^{(n)}(\vec{x};\omega_k) = \frac{2\pi}{\ln(\rho/2)} u_0^{(n)}(\vec{z};\omega_k) \mathcal{N}(\vec{x}, \vec{z};\omega_k) + O\left(\frac{1}{|\ln \rho|^2}\right), \tag{3}$$

where $0 < \rho < 2$ and $\mathcal{N}(\vec{x}, \vec{z};\omega_k)$ is the Neumann function for Ω,

$$\begin{cases} \Delta \mathcal{N}(\vec{x}, \vec{z};\omega_k) + \omega_k^2 \mathcal{N}(\vec{x}, \vec{z};\omega_k) = -\delta(\vec{x}, \vec{z}) & \text{for} \quad \vec{x} \in \Omega, \\ \dfrac{\partial \mathcal{N}(\vec{x}, \vec{z};\omega_k)}{\partial \vec{\nu}(\vec{x})} = 0 & \text{for} \quad \vec{x} \in \partial\Omega. \end{cases} \tag{4}$$

3. Non-iterative multi-frequency imaging algorithm from MSR matrix

One assumes that there exist L different small cracks Γ_l with same length ρ centered at $\vec{z}_l$, $l = 1, 2, \cdots, L$, in Ω. We apply the asymptotic formula (3) to establish an imaging functional of the crack. For this, one uses the eigenvalue structure of the MSR matrix $\mathbb{M}(\omega_k) = \left(M_{nm}^{(k)}\right)_{n,m=1}^{N}$, where $M_{nm}^{(k)}$ denotes the normalized boundary measurement data:

$$M_{mn}^{(k)} := -\frac{1}{N} \int_{\partial\Omega} \left(u^{(n)}(\vec{x};\omega_k) - u_0^{(n)}(\vec{x};\omega_k)\right) \frac{\partial v^{(m)}(\vec{x};\omega_k)}{\partial \vec{\nu}(\vec{x})} dS(\vec{x}) \approx -\frac{2\pi}{N \ln(\rho/2)} \sum_{l=1}^{L} e^{i\omega_k (\vec{\theta}_n - \vec{\theta}_m) \cdot \vec{z}_l}. \tag{5}$$

Here, one eliminates the residue term of (3) and chooses a test function $v^{(m)}(\vec{x};\omega_k) = e^{-i\omega_k \vec{\theta}_m \cdot \vec{x}}$.

Now, one designs an imaging functional. First, one performs the Singular Value Decomposition (SVD) of $\mathbb{M}(\omega_k)$, so that

$$\mathbb{M}(\omega_k) \approx \sum_{l=1}^{L} \vec{U}_l(\omega_k) \sigma_l(\omega_k) \overline{\vec{V}}_l^T(\omega_k),$$

Journal of Physics: Conference Series **410** (2013) 012018 doi:10.1088/1742-6596/410/1/012018

where $\sigma_l(\omega_k)$ are (non-zero) singular values and where $\vec{U}_l(\omega_k)$ and $\vec{V}_l(\omega_k)$ are left and right singular vectors. Then, one introduces a weighted imaging functional at several discrete frequencies $\{\omega_k : k = 1, 2, \cdots, K\}$ as

$$\mathbb{E}(\vec{x}; K) := \sum_{k=1}^{K}\sum_{l=1}^{L} \omega_k \left(\overline{\vec{W}}(\vec{x}; \omega_k) \cdot \vec{U}_l(\omega_k)\right)\left(\overline{\vec{W}}(\vec{x}; \omega_k) \cdot \overline{\vec{V}}_l(\omega_k)\right), \tag{6}$$

where

$$\vec{W}(\vec{x}; \omega_k) = \frac{\vec{D}(\vec{x}; \omega_k)}{|\vec{D}(\vec{x}; \omega_k)|} \quad \text{with} \quad \vec{D}(\vec{x}; \omega_k) = \left(e^{i\omega_k \vec{\theta}_1 \cdot \vec{x}}, e^{i\omega_k \vec{\theta}_2 \cdot \vec{x}}, \cdots, e^{i\omega_k \vec{\theta}_N \cdot \vec{x}}\right)^T.$$

It should be noted that $\vec{W}(\vec{x}; \omega_k)$ satisfies the following properties (see [1])

$$\vec{U}_l(\omega_k) \sim \vec{W}(\vec{z}_l; \omega_k) \quad \text{and} \quad \overline{\vec{V}}_l(\omega_k) \sim \vec{W}(\vec{z}_l; \omega_k). \tag{7}$$

If one assumes that numbers N and K are large enough, applying (7) to (6) yields

$$\begin{aligned}
\mathbb{E}(\vec{x}; K) &\sim \sum_{l=1}^{L}\sum_{k=1}^{K} \omega_k \left(\sum_{s=1}^{N} e^{i\omega_k \vec{\theta}_s \cdot (\vec{z}_l - \vec{x})}\right)\left(\sum_{t=1}^{N} e^{i\omega_k \vec{\theta}_t \cdot (\vec{z}_l - \vec{x})}\right) \\
&\approx \sum_{l=1}^{L} \int_{\omega_1}^{\omega_K} \omega \left(\int_{\mathbb{S}^1} e^{i\omega \vec{\theta} \cdot (\vec{z}_l - \vec{x})} d\vec{\theta}\right)^2 d\omega = \frac{1}{4\pi^2} \sum_{l=1}^{L} \int_{\omega_1}^{\omega_K} \omega J_0(\omega \vec{r}_l)^2 d\omega, \quad \vec{r}_l := |\vec{z}_l - \vec{x}| \\
&= \frac{1}{4\pi^2} \sum_{l=1}^{L} \left[\frac{\omega_K^2}{2}\left(J_0(\omega_K \vec{r}_l)^2 + J_1(\omega_K \vec{r}_l)^2\right) - \frac{\omega_1^2}{2}\left(J_0(\omega_1 \vec{r}_l)^2 + J_1(\omega_1 \vec{r}_l)^2\right)\right],
\end{aligned}$$

where $J_p(\cdot)$ denotes the Bessel function of order p and of the first kind. A full derivation is to appear in an extended version of this contribution. Some properties of $\mathbb{E}(\vec{x}; K)$ are as follows:

- $J_0(\omega|\vec{z}_l - \vec{x}|)^2 + J_1(\omega|\vec{z}_l - \vec{x}|)^2$ reaches its maximum value at $\omega|\vec{z}_l - x| = 0$. Hence, $|\mathbb{E}(\vec{x}; K)|$ plots are of large and small magnitudes at $\vec{x} = \vec{z}_l \in \Gamma_l$ and at $\vec{x} \neq \vec{z}_l \in \Gamma_l$, respectively.
- When K is small, $|\mathbb{E}(\vec{x}; K)|$ plots exhibit replicas at local maxima and minima of J_0.
- N must be large enough. It means that if one applies this algorithm with a small N, poor results follow. Moreover, by looking at (5), this analysis cannot be applied to the limited-view inverse scattering problem.
- Using frequencies high enough guarantees successful imaging performance.
- Based on the Statistical Hypothesis Testing [1], the proposed algorithm is robust vs. random noise when K is large enough.

4. Numerical experiments

In this section, one introduces numerical examples for imaging small and long arc-like cracks. The operation frequency is taken of the form $\omega_k = 2\pi/\lambda_k$; here λ_k, $k = 1, 2, \cdots, K(= 10)$, is the given wavelength. The ω_k are equi-distributed in the interval $[\omega_1, \omega_K]$ with $\lambda_1 = 0.6$ and $\lambda_K = 0.4$. A white Gaussian noise with 20dB signal-to-noise ratio (SNR) is added to the unperturbed data.

Figure 1 displays maps of $|\mathbb{E}(\vec{x}; 10)|$ for three small cracks (*left*), a long single crack (*center*), and two long cracks close to one another (*right*). Notice that the analysis above is restricted to the small crack case, but it can be applied to extended (long) ones. Although some replicas appear since the chosen value of K is rather small ($= 10$), $|\mathbb{E}(\vec{x}; K)|$ plots exhibit maximum amplitudes at the location of the cracks so that one can recognize their shape. Taking them as initial guess could yield more accurate shapes via a Newton-type algorithm [4] or a level-set method [3].

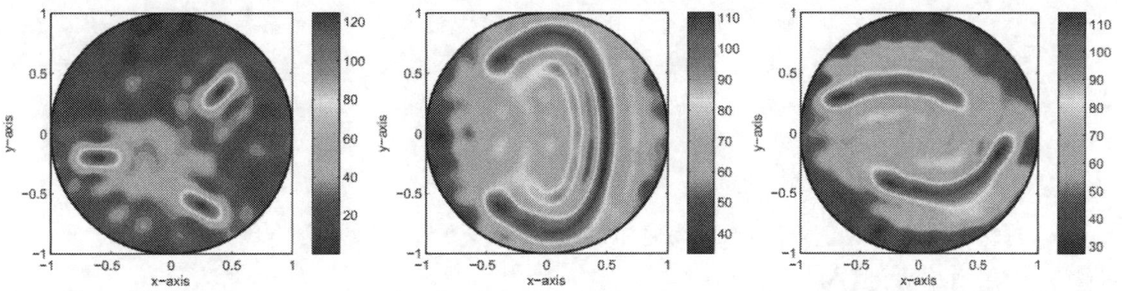

(a) $|\mathbb{E}(\vec{x};10)|$ with $N=12$ (b) $|\mathbb{E}(\vec{x};10)|$ with $N=32$ (c) $|\mathbb{E}(\vec{x};10)|$ with $N=36$

Figure 1. Map of $|\mathbb{E}(\vec{x};K)|$ with $K=10$ for various cracks.

5. Conclusion

Based on the structure of the singular vectors of the MSR matrix at several discrete frequencies of operation, an improved image function motivated from Kirchhoff migration has been proposed to image perfectly conducting cracks. The proposed imaging functional is related to Bessel functions of integer order. Numerical experiments with noisy data are discussed and they show the effectiveness of the corresponding imaging algorithm. The current contribution deals with a full-view inverse scattering problem. Application to a limited-view inverse problem is forthcoming work. Along the same line of thought, extension to cracks with Neumann boundary condition should be an interesting subject.

Acknowledgments

This work was supported by Basic Science Research Program through the National Research Foundation of Korea (NRF) funded by the Ministry of Education, Science and Technology (No. 2012-0003207) and research program of Kookmin University in Korea.

References

[1] Ammari H, Garnier G, Kang H, Park W-K, and Sølna K 2011 Imaging schemes for perfectly conducting cracks *SIAM J. Appl. Math.* **71** 68–91.
[2] Ammari H, Kang H, Lee H, and Park W-K 2010 Asymptotic imaging of perfectly conducting cracks *SIAM J. Sci. Comput.* **32** 894–922.
[3] Dorn O and Lesselier D 2006 Level set methods for inverse scattering, *Inverse Problems* **22** R67–R131.
[4] Kress R 2003 Inverse scattering from an open arc *Math. Methods Appl. Sci.* **18** 267–293.
[5] Ma Y-K, Kim P-S, and Park W-K 2012 Analysis of topological derivative function for a fast electromagnetic imaging of perfectly conducing cracks *Prog. Electromagn. Res.* **122** 311–325.
[6] Park W-K 2010 Non-iterative imaging of thin electromagnetic inclusions from multi-frequency response matrix *Prog. Electromagn. Res.* **110** 237–252.
[7] Park W-K 2010 On the imaging of thin dielectric inclusions buried within a half-space *Inverse Problems* **26** 074008.
[8] Park W-K 2012 Topological derivative strategy for one-step iteration imaging of arbitrary shaped thin, curve-like electromagnetic inclusions *J. Comput. Phys.* **231** 1426-1439.
[9] Park W-K and Lesselier D 2009 Electromagnetic MUSIC-Type imaging of perfectly conducting, arc-like cracks at single frequency *J. Comput. Phys.* **228** 8093–8111.
[10] Park W-K and Lesselier D 2009 MUSIC-type imaging of a thin penetrable inclusion from its far-field multi-static response matrix *Inverse Problems* **25** 075002.
[11] Park W-K and Lesselier D 2012 Fast electromagnetic imaging of thin inclusions in half-space affected by random scatterers *Waves Random Complex Media* **22** 3–23.

Journal of Physics: Conference Series **410** (2013) 012019 doi:10.1088/1742-6596/410/1/012019

New approach to modeling of a local structure of silicate glasses and melts

A A Osipov, L M Osipova

Institute of Mineralogy UB RAS, Miass, Chelyabinsk region, 456317, Russian Federation

E-mail: armik@mineralogy.ru

Abstract. Structural model of silicate melts and glasses obtained by rapid cooling of the melt is far from complete at present. Modern models suggest the existence of the five different types of silicon-oxygen tetrahedra which are distinguish by the ratio of bridging and terminal oxygen atoms per silicon atom (the so-called basic structural blocks or units) in the structure of silicate glasses and melts and also assume that the several types of such units can exist in the glass and melt structures simultaneously. The knowledge of the concentrations of these units depending on the composition and temperature (the so-called Q^n distribution) is a necessary element for the development of an adequate model of silicate systems and to explaining their physical and chemical properties. Three-parameter statistical method to modeling of Q^n distribution in silicate melts and glasses is present in this paper. This method allows calculating the concentration of structural units in a wide composition range at different temperatures. The method is based on a destruction of a preassigned network as a model of depolymerization of the silicon-oxygen framework in the process its interaction with the modifier oxides. Q^n distribution, distribution of silicon-oxygen tetrahedra according to their next-nearest neighbours, and distribution of bridging bonds in alkali-silicate glasses and melts were obtained in frame of the developed approach.

1. Introduction

The silicate glasses are the typical representative of the covalently bonded glasses and their structure is well described by the continuous random network model (CRN) [1]. According to this model, the structure of vitreous SiO_2 is a continuous network of SiO_4 tetrahedra randomly interconnected via the oxygen apexes. In other words, each silicon atom is surrounded by four oxygen atoms and each oxygen atom is shared between two silicon atoms, forming so-called bridging oxygen (BO) bond in structure of glassy SiO_2. Addition of alkali or alkaline-earth oxides to silica leads to break-up of the bridging bonds and formation of the terminal oxygen atoms each of which is connected only to one silicon atom. In case of alkali silicate glasses, such structural reorganization can be schematically described by $Si\text{-}O\text{-}Si + R_2O \Rightarrow 2[Si\text{-}O^-\text{-}R^+]$ reaction, where R^+ is the alkali ion and O^- is the terminal or non-bridging oxygen atom (NBO). At such interaction, four oxygen atoms remain in nearest surrounding of silicon atom but new silicon-oxygen tetrahedra with differ BO/NBO ratio are formed in glass structure. Five types of silicon-oxygen tetrahedra are accepted to distinguish according to BO number per silicon-oxygen tetrahedron (so-called Q^n units, where $n = 0...4$ is the number of BO). The existence of Q^n units in silicate glasses and melts is universally recognized and no any significant deviations from their crystalline counterparts were found.

At present, Q^n species are widely used in interpretation of experimental data such as Raman and ^{29}Si MAS NMR spectroscopy [2-6], and can be obtained by method of thermodynamic simulation [7, 8] and molecular dynamic (MD) method [9]. Experimental determination of dependence Q^n concentrations (Q^n distribution) on composition and temperature are is laborious and the high-temperatures experiments require the special equipment to detect the weak signal against the background of the strong thermal radiation from sample. Thermodynamic modeling allows performing calculation in a wide range of compositions and temperatures but knowledge of all crystalline phases and standard Gibbs free energies of their formation are required as input data [10]. The whole set of such data is not always available for researchers and it limits the possibility of using of this method. In addition, more and more attention is paid to not only Q^n distribution but also connectivity of Q^n species, i.e., so-called $Q^{n,a1,..an}$ distribution ($a1,...,an$ indices denote the type of the structural units which are connected to the given Q^n unit) , lately [11,12]. Such problem does not considered in thermodynamic simulation. The MD allows us to study both Q^n and $Q^{n,a1,..an}$ distributions. To date, such investigations, however, were performed only in a limited compositional range and do not show the complete picture of distribution of large $Q^{n,a1,..an}$ groups as a function of glass composition and temperature. From this point of view, the elaboration of other methods in order to study the above mentioned problems is certain interest. An alternative method of study of the local structure of alkali silicate glasses and melts is reported in this paper. It was tested on Li_2O-SiO_2, Na_2O-SiO_2 and K_2O-SiO_2 glass forming systems.

2. Modeling details

As it is known, the Zachriasen's rules [1] of formation of continuous random glass network are focused only on the local configuration of cation-oxygen polyhedra and their connectivity to each other (via corners, not edges or faces). According to this, it is possible to assume that the most important point in modeling of the above mentioned distributions is a coordination number of glass forming cation rather than the network dimensionality and topology as a whole. Any regular or random network where each knot is linked to four nearest knots and each link is shared only two knots completely reproduces the Q^n, $Q^{n,a1,..an}$ and Q^i-Q^j (various types of Si-O-Si bonds, i,j = 1,..4) distributions in glassy SiO_2 if the knots and links between them are interpreted as silicon atoms and Si-O-Si bridging bonds, respectively. In this case, a break of the link between two knots simulates the depolymerization process of SiO_2 network that can be schematically represented by the follow reaction:

$$Q^i + Q^j + R_2O \rightarrow Q^{i-1} + Q^{j-1}, \quad i,j = 1, \ldots, 4 \qquad (1)$$

Thus, in order to calculate all above distributions it is necessary to break the preassigned number of bonds (determined by the glass composition) and then derived configuration must be analyzed.

It is known, that the Q^n distribution strongly depends on alkali cation type. Therefore, three parameters, w_1, w_2 and w_3, were introduced into the model in order to achieve the best agreement between experimental and calculating data. Figure 1 shows an importance of these parameters in the simulations algorithm. As it is follows from the equation (1), ten different bonds may exist in the system when $R_2O \neq 0$. The choice of the bond that should be broken is absolutely random and does not depend on type of structural units which share this bond when $w_1 = w_2 = w_3 = 1$. The break of the Q^i-Q^j bonds with the highest values of i and j will preferentially occur when $w_k > 1$.

Three different initial networks (2-D regular square network, 3-D quartz-like regular network and network in form of table of the random bonds (TRB)) were used in calculation in order to test the assumption that the Q^n, $Q^{n,a1,..an}$ and Q^i-Q^j distributions are independent from the network dimensionality and topology. The TRB was generated in such a manner that each bond belongs to two knots and each knot was joined to four nearest neighbours. Thus, the condition of four-fold coordination of glass forming cation remained but such network has no any real geometry. Results of all experiments were identical when we used the same w_k parameters. It can be explained by the fact that both 2-D flat square and 3-D quartz-like regular networks are special cases of the TRB. Therefore, the results only for TRB are reported in this paper.

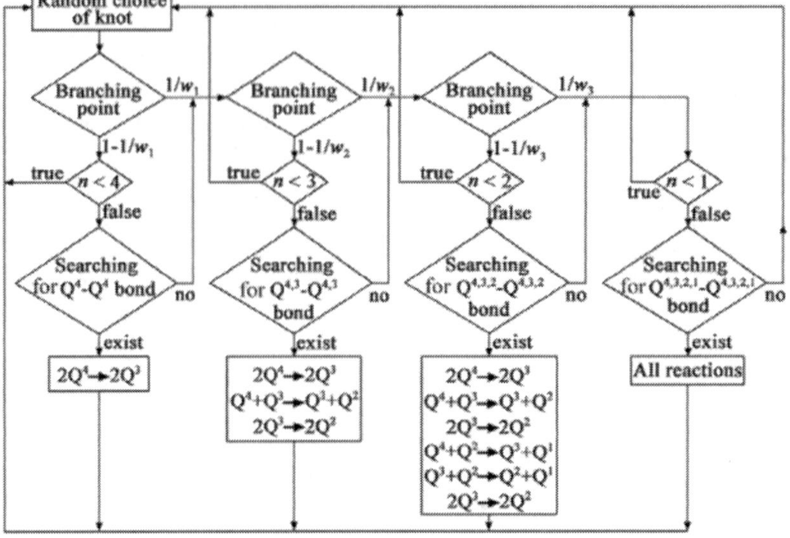

Figure 1. Block-diagram of simulation algorithm.

3. Results

The Q^i-Q^j distributions, several examples of $Q^{n,a1...,an}$ distributions ($Q^{n,a1...,an}$ distributions were calculated as ratio of $Q^{n,a1...,an}$ groups relative the total amount of Q species) and Q^n distributions are shown in figures 2, 3 and

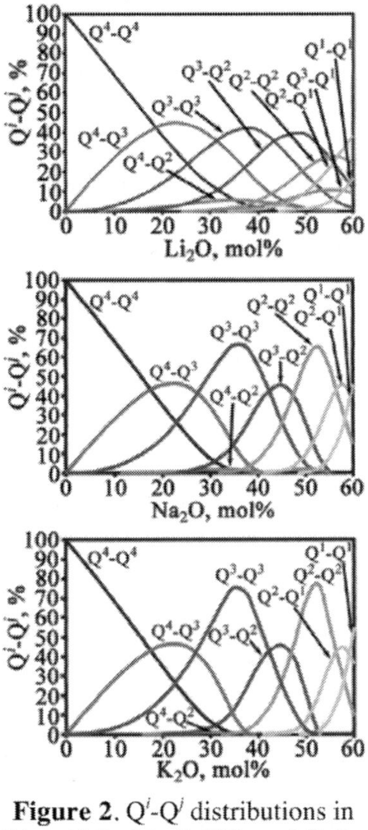

Figure 2. Q^i-Q^j distributions in Li_2O-SiO_2, Na_2O-SiO_2 and K_2O-SiO_2 glasses.

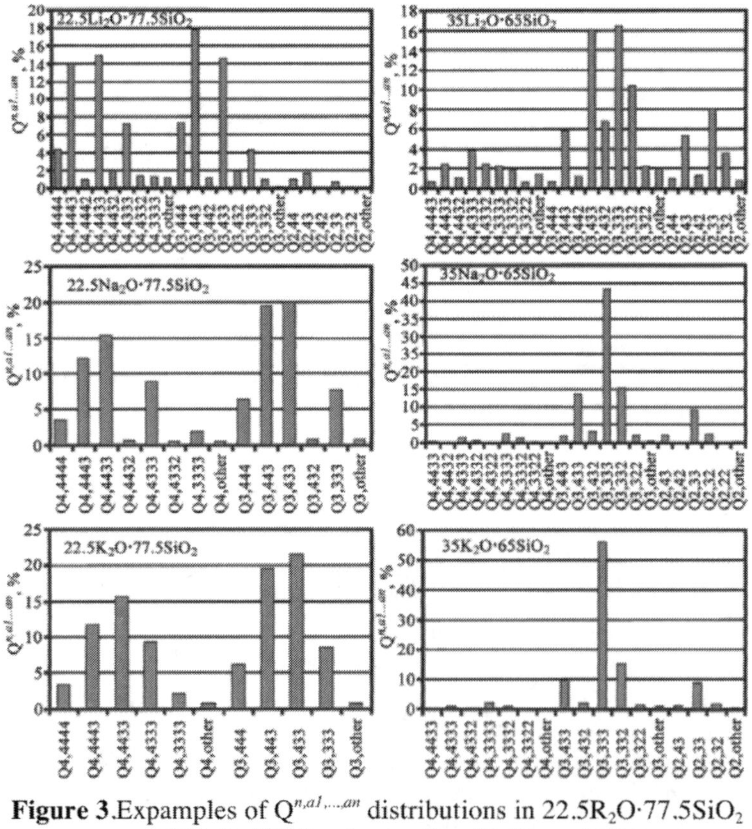

Figure 3.Expamples of $Q^{n,a1,...,an}$ distributions in $22.5R_2O{\cdot}77.5SiO_2$ and $35R_2O{\cdot}65SiO_2$ glasses (R = Li, Na and K).

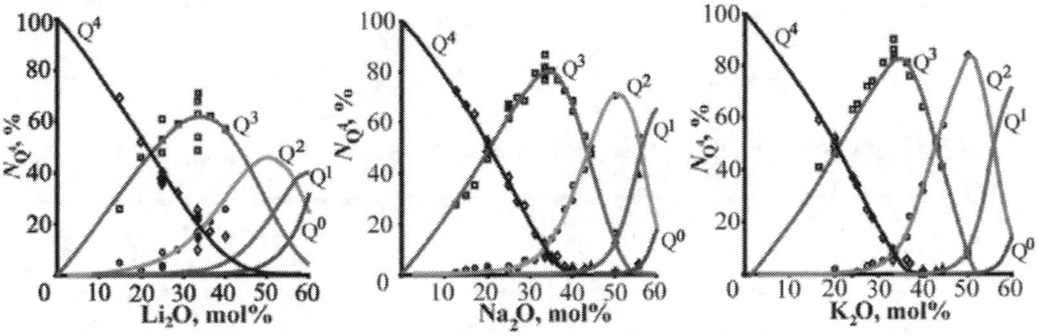

Figure 4. Q^n distributions in glasses of R_2O-SiO_2 systems (R = Li, Na, K). Solid lines are simulation results, symbols are experimental data [2,4-6,14-19].

4, respectively. It is obvious, that dispersion of all distributions increases with decrease in radius of alkali cation. It is known, that dynamic equilibrium between Q^n species in alkali silicate melts can be described by $2Q^n \leftrightarrow Q^{n-1} + Q^{n+1}$ (n = 1, 2,.3) equilibrium reaction. The Q^n distribution is practically independent from the temperature in the Li_2O-SiO_2 system while it strongly depends on the temperature in the Na_2O- and K_2O-SiO_2 systems (above mentioned equilibrium is shifted to the right with increase in temperature). Thus, it is possible to assume that the Q^n, Q^i-Q^j and $Q^{n,a1,...an}$ distributions in the Li_2O-SiO_2 glasses may be considered as analogous of corresponding distributions in sodium and potassium silicate melts.

Acknowledgments

Acknowledgment is made to the Special Federal Program No 753 "Scientific-Pedagogical Staff of Innovative Russia".

4. References

[1] Zachariassen W N 1932 *Amer. Chem. Soc.* **54** 3841
[2] Mysen B O and Frantz J D 1992 *Chem. Geol.* **96** 321
[3] Bykov V N, Osipov A A and Anfilogov V N 2000 *Phys. Chem. Glasses* **41** 10
[4] Malfait W J, Zakaznova-Herzog V P and Halter W E 2008 *Amer. Mineral.* **93** 1505
[5] Maekawa H, Maekawa T, Kawamura K and Yokokawa T 1991 *J. Non-Cryst. Solids* **127** 53
[6] Sen S, Youngman R E 2003 *J. Non-Cryst. Solids* **331** 100
[7] Vedishcheva N M, Shakhmatkin B A and Wright A C 1998 *Glass Phys. And Chem.* **24** 308
[8] Vedishcheva N M, Shakhmatkin B A, Shultz M M, Vessal B, Wright A C, Bachra B, Clare A G, Hannon A C and Sinclair R N 1995 *J. Non-Cryst. Solids* **193** 292
[9] Macháček J and Gedeon O 2003 *Ceramics-Silikáty* **47** 45
[10] Shakhmatkin B A, Vedishcheva N M 1998 *Glass Phys. And Chem.* **24** 229
[11] Macháček J and Gedeon O 2008 *J. Non-Cryst. Solids* **354** 138
[12] Gedeon O, Liška M and Macháček *J. Non-Cryst. Solids* 2008 **354** 1133
[13] Oliver L, Yuan X, Cormack A N, Jäger C 2001 *J. Non-Cryst. Solids* **293-295** 53
[14] Maechara T, Yano T and Shibata S 2005 *J. Non-Cryst. Solids* **351** 3685
[15] Schneider J, Mastelaro V R, Zanotto E D, Shakhmatkin B A, Vedishcheva N M, Wright A C and Panepucci H 2003 *J. Non-Cryst. Solids* **325** 164
[16] Bykov V N, Bobylev I B and Anfilogov V N 1987 *Fiz. Khim. Stekla* **13** 854
[17] Bykov V N, Koroleva O N and Osipov A A 2008 *Rasplavy* **3** 50
[18] Bykov V N, Osipov A A and Anfilogov V N 1998 *Rasplavy* **6** 86
[19] Bykov V N, Koroleva O N and Osipov A A 2009 *Geochemistry International* **47** 1067

IC-MSQUARE 2012: International Conference on Mathematical Modelling in Physical Sciences IOP Publishing
Journal of Physics: Conference Series **410** (2013) 012020 doi:10.1088/1742-6596/410/1/012020

Calculation of a double reactive azeotrope using stochastic optimization approaches

Gustavo Mendes Platt, Roberto Pinheiro Domingos, Matheus Oliveira de Andrade

Instituto Politécnico, Universidade do Estado do Rio de Janeiro, Nova Friburgo, R. Bonfim 25, Brazil

E-mail: gmplatt@iprj.uerj.br, roberto.pinheiro@iprj.uerj.br, oamatheus@gmail.com

Abstract. An homogeneous reactive azeotrope is a thermodynamic coexistence condition of two phases under chemical and phase equilibrium, where compositions of both phases (in the Ung-Doherty sense) are equal. This kind of nonlinear phenomenon arises from real world situations and has applications in chemical and petrochemical industries. The modeling of reactive azeotrope calculation is represented by a nonlinear algebraic system with phase equilibrium, chemical equilibrium and azeotropy equations. This nonlinear system can exhibit more than one solution, corresponding to a double reactive azeotrope. The robust calculation of reactive azeotropes can be conducted by several approaches, such as interval-Newton/generalized bisection algorithms and hybrid stochastic-deterministic frameworks. In this paper, we investigate the numerical aspects of the calculation of reactive azeotropes using two metaheuristics: the Luus-Jaakola adaptive random search and the Firefly algorithm. Moreover, we present results for a system (with industrial interest) with more than one azeotrope, the system isobutene/methanol/methyl-tert-butyl-ether (MTBE). We present convergence patterns for both algorithms, illustrating - in a bidimensional subdomain - the identification of reactive azeotropes. A strategy for calculation of multiple roots in nonlinear systems is also applied. The results indicate that both algorithms are suitable and robust when applied to reactive azeotrope calculations for this "challenging" nonlinear system.

1. Introduction

An azeotrope (in its classical statement) is a thermodynamic phenomenon where a boiling mixture produces a vapor with identical composition (the relative quantities of the mixture components, expressed in molar scales) when compared to the liquid phase. Azeotropic states can introduce confinement regions for separations that explore differences between boiling points of the mixture components. Ordinarily, the azeotrope condition is represented by the equalities of compositions for both phases in equilibrium, i.e.: $x_i = y_i, i = 1, 2, \ldots, c$, where x_i and y_i represent liquid and vapor molar fractions, and c is the number of components. For reactive systems, it is possible to obtain a similar expression, but using a nonlinear transformation on the vector of molar fractions [1], producing a new set of compositional coordinates for liquid and vapor phases, X_i and Y_i, respectively, which are equal in the reactive azeotropic condition:

$$X_i = \left(\frac{x_i - \nu_i^t N^{-1} x_{ref}}{1 - \nu_{tot}^t N^{-1} x_{ref}}\right); Y_i = \left(\frac{y_i - \nu_i^t N^{-1} y_{ref}}{1 - \nu_{tot}^t N^{-1} y_{ref}}\right) \quad (1)$$

Journal of Physics: Conference Series **410** (2013) 012020 doi:10.1088/1742-6596/410/1/012020

In this expression, ν_i refers to the stoichiometric coefficient of component i in the chemical reaction, ν_{tot}^t is a vector with sums of stoichiometric coefficients for chemical reactions, N is a matrix of stoichiometric coefficients for reference components (the number of reference components is equal to the number of equilibrium chemical reactions that are chosen among the set of reacting species), x_{ref} (resp., y_{ref}) is the vector of molar fractions in liquid (resp., vapor) phase for the reference components. Reactive azeotropes can be observed even in ideal mixtures (vapor and liquid ideal phases), whereas ordinary homogeneous azeotropes are consequences of the nonidealities of liquid phase. Moreover, several systems show reactive azeotropes, and, in some cases, more than one azeotrope [2]. The reactive azeotrope calculation is characterized by a nonlinear set of algebraic equations, representing phase coexistence conditions, chemical equilibria and the azeotropy condition. Thus, this nonlinear system can be converted in an optimization problem and solved by classical (deterministic) nonlinear optimization methods or by using metaheuristics. Platt [7] presents an analysis of Differential Evolution algorithm and Luus-Jaakola method in a calculation of a single reactive azeotrope in a quaternary ideal system.

In this work we addressed the calculation of reactive azeotropes in a ternary non-ideal system (with two azeotropes) using two metaheuristics: the Luus-Jaakola (LJ) algorithm (an adaptive random search method) [3] and the Firefly algorithm (FF) [4]. We also studied some strategies to identify further solutions, using a modification of the objective function [5] or through a repulsion technique [6] , in order to avoid solutions previously obtained. The main objective of this work is not to compare efficiency or robustness of LJ and FF algorithms, but to provide an useful framework for "hard"nonlinear systems with multiple roots.

2. Problem Formulation and Numerical Methods

The reactive homogeneous azeotrope is a phase coexistence condition where the compositional coordinates for liquid and vapor phases, in Ung-Doherty sense, are equal. Thus, we can represent an homogeneous reactive azeotrope in a mixture with r equilibrium reactions as [1]: $X_i = Y_i \quad i = 1, 2, \ldots, c-r-1$. Simultaneous chemical and phase equilibrium problem is expressed by the following equations:

$$\begin{cases} Py_i - x_i\gamma_i P_i^{sat} = 0 & i = 1, 2, \ldots, c \\ \sum_{i=1}^{c} x_i - 1 = 0 & \\ \sum_{i=1}^{c} y_i - 1 = 0 & \\ \ln(K) - \nu^t \ln(a) = 0 & i = 1, 2, \ldots, r \end{cases} \quad (2)$$

In Eqs. (2), P is the system pressure, γ_i is the activity coefficient for component i (calculated via Wilson model) and P_i^{sat} is the saturation pressure for specie i (evaluated by Antoine equation). Also in Eqs. (2), K is the vector of chemical equilibrium constants, ν is a matrix of stoichiometric coefficients (dimension $c \times r$). In monoreactive problems, ν is a vector with the stoichiometric coefficients of each component participating in the equilibrium chemical reaction and, obviously, K is a scalar quantity. The vector a is the vector of activities; the activity of component i is defined as $a_i = \gamma_i x_i$. Physical parameters are found in [2].

An analysis of degrees of freedom for this nonlinear system indicates $c + 2 + r$ equations. The thermodynamic coordinates are: x, y, T and P, performing $2c + 2$ variables. Thus, the number of degrees of freedom for the phase coexistence problem is $(2c+2) - (c+r+2) = c-r$. If one consider pressure (or temperature) previously specified, we obtain $c - r - 1$. This quantity is satisfied by the reactive azeotrope conditions. The vector of unknowns will be referred as: $\theta^t = [x_1 \ x_2 \ldots x_c \ y_1 \ y_2 \ldots y_c \ T]$. The nonlinear system appears as: $F_l(\theta) = 0, l = 1, 2, \ldots, 2c+1$, where F_l refers to each nonlinear equation formed by Eqs.(2) and the reactive azeotrope condition. Thus, a related optimization problem can be constructed, by the scalar function $f(\theta) = \sum_{l=1}^{2c+1} (F_l(\theta))^2$. A minimizer θ of the function $f(\theta)$ that produces a null minimum is also a solution of the nonlinear system represented by Eqs. (2) plus reactive azeotrope equations.

Journal of Physics: Conference Series **410** (2013) 012020 doi:10.1088/1742-6596/410/1/012020

Table 1. Reactive azeotropes – isobutene + methanol + MTBE – $K = 49.0$.

Component	x_i^{az1}	y_i^{az1}	T^{az1}	x_i^{az2}	y_i^{az2}	T^{az2}
isobutene (1)	0.0446038	0.1727712	392.2444	0.0138208	0.0748231	391.17559
methanol (2)	0.1197971	0.2378773	-	0.4037263	0.4406101	-
MTBE (3)	0.8355991	0.5893516	-	0.5824529	0.4845668	-

As pointed previously, this nonlinear optimization problem was solved by two metaheuristics: the LJ method [3] and the FF algorithm [4]. The LJ method is an adaptive random search algorithm, with a double-loop structure and a contraction strategy for the search region. Here, we used $n_{int} = 1000$ (internal iterations), $n_{out} = 1000$ (number of contractions of search region), $\epsilon = 0.997$ (contraction factor). The FF algorithm mimetizes the behaviour of fireflies in nature. FF algorithm parameters used are: $\gamma = 2$ (light absorption coefficient), $\beta_0 = 0.5$ (attraction of two collapsed fireflies), $\alpha = 0.05$ (except for temperature, where $\alpha = 5$) (random movement parameter), number of fireflies = 30, generations = 500. More details on LJ and FF algorithms can be found in [3, 4]. In order to calculate multiple roots (minima), some additional steps are necessary, usually a modification in the objective-function. In this work, we evaluate two possible modifications in the objective-function (fitness): the proposal of Hirsch et al. [5] and the repulsion technique employed by Parsopoulos and Vrahatis [6].

3. Results and Discussions

We present results for the reactive system isobutene (1) + methanol (2) $\rightleftharpoons$ MTBE (3), at $P = 8$ atm, considering $K = 49$ [2]. In this situation, this reactive system shows two azeotropes, as presented in Table 1. Here, we consider MTBE (component 3) as the reference component, accordingly to Eqs. (1). Figure 1 shows a typical execution of Firefly algorithm in this problem, in a two-dimensional projection (in fact, the problem is in $\Re^5$). The two azeotropes correspond to the two stationary points in the diagram (a maximum and a minimum). Our computational experiments indicate that the Azeotrope 1 is found more easier than Azeotrope 2. In this scenario, we consider that, in a typical execution, Azeotrope 1 is found.

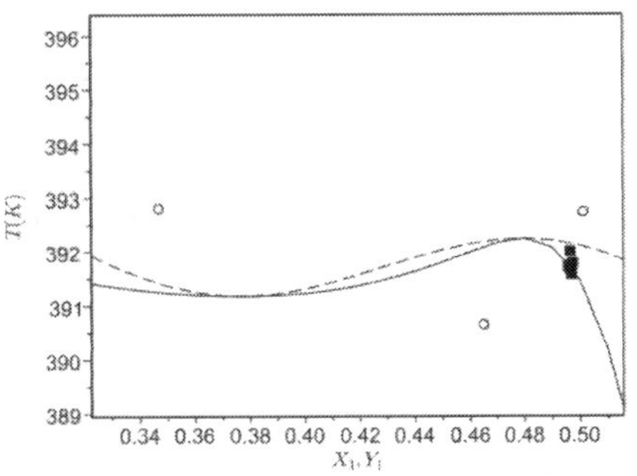

Figure 1. Typical FF execution ($X_1, Y_1 \times T$ projection). ○: initial fireflies; ■: final fireflies.

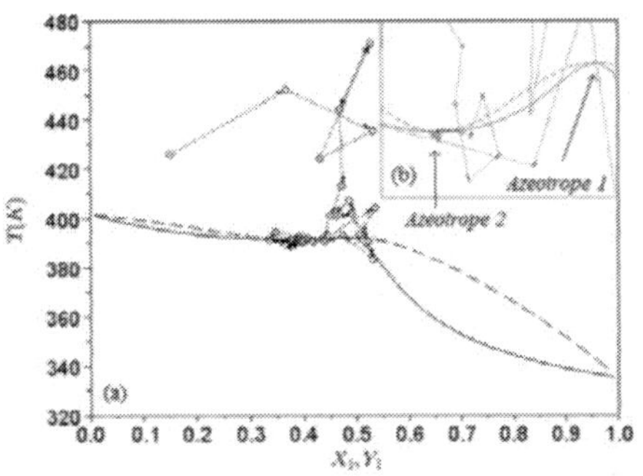

Figure 2. (a) Typical LJ execution ($X_1, Y_1 \times T$ projection). (b) Amplification of the neighbourhood of azeotropes.

Then, some strategy to calculate further solutions is necessary. Here, we tested the modification of the objective-function proposed by [5]. After the localization of k roots, $\{\theta^1, \theta^2, ..., \theta^k\}$, the new objective-function is defined as: $\mathcal{F}(\theta) = \sum (F_i(\theta))^2 + \beta \sum_{j=1}^{k} \exp(-\|\theta - \theta^j\|)\chi_\rho(\|\theta - \theta^j\|)$ where $\chi_\rho(\delta) = 1$, if $\delta \leq \rho$, or $\chi_\rho(\delta) = 0$, otherwise. Qualitatively, in the last expression, β should be a large value and ρ a small one. In these numerical experiments, we used $\beta = 1 \times 10^5$ and $\rho = 0.1$. First, the LJ/FF algorithm is applied with a random estimate. After the localization of the first azeotrope, the objective-function is modified and the algorithm is executed again. The procedure stops when the value of the objective-function in the optimum is larger than a tolerance (in fact, we are searching for null minima). The use of the technique proposed by Hirsch et al. [5] is useful, but, in some instances, the algorithms converged to local minima. Indeed, this kind of approach produces several local minima, in a "mexican hat" form, as pointed by Parsopoulos and Vrahatis [6]. Considering this scenario, the repulsion procedure proposed by these authors [6] is based on the calculation of an Euclidean norm $d_{ij} = \|\theta_i - \theta_j^\star\|$, where $\theta_j^\star$ is a minimizer previously found (in our numerical experiments, we used a Manhattan norm, or ℓ^1 norm). If this norm is less than a specified control parameter r_{ij} (0.1 in our case), the current point is repelled from the center of the repulsion area, using an expression like as $\theta_i \leftarrow \theta_i + p_{ij}(\theta_i - \theta_j^\star)/d_{ij}$, where p_{ij} is a scale factor (equal to one in our computational experiments). In order to avoid the "mexican hat effect", we used the repulsion approach [6] to obtain a second azeotrope. A typical execution of LJ algorithm with this modification, and considering that the first azeotrope identified was Azeotrope 1, is presented in Figure 2. Clearly, the algorithm promotes a convergence to Azeotrope 2. We execute both algorithms for 20 instances using the repulsion technique. For LJ algorithm we observed a mean objective-function equal to 2.03×10^{-3} (best value: 6.26×10^{-5}), and for FF algorithm we obtain a mean value of 1.32×10^{-3} (best value: 9.37×10^{-4}). Obviously, a simple evaluation of mean values does not represent a perfomance or statistical comparison between algorihtms, but it is useful to demonstrate that both methods produce final results in the neighbourhood of Azeotrope 2. Indeed, with a deterministic method applied after the stochastic algorithm (Newton or Simplex methods, for instance) one can easliy obtain an accurate value for Azeotrope 2. For azeotropic temperature, for example, FF founds a mean value of 391.42 K (with standard deviation equal to 0.25) and LJ obtains 391.27 K (and 0.27 for standard deviation). Clearly, we observed a convergence pattern to Azeotrope 2.

4. Conclusions

In this work, we addressed the problem of calculation of reactive azeotropes in a ternary nonideal system, using two metaheuristics – the LJ algorithm [3] and the FF algorithm [4]. In this case, there are two reactive azeotropes, in opposite sides of compositional domain. We applied some repulsion strategies in order to produce a robust framework capable to obtain both azeotropes.

Acknowledgments

Gustavo M. Platt is grateful to Fundação Carlos Chagas Filho de Apoio à Pesquisa do Estado do Rio de Janeiro (FAPERJ), by the financial support for this research. Matheus O. Andrade is also grateful to FAPERJ, by the undergraduate research grant.

References

[1] Ung S and Doherty M F 1995 *AIChE Journal* **41** 2383.
[2] Okasinski M J and Doherty M F 1997 *AIChE Journal* **43** 2227.
[3] Luus R and Jaakola T H I 1973 *AIChE Journal* **19** 760.
[4] Yang X S 2008 *Nature-Inspired Metaheuristic Algorithms* (Cambridge: Luniver).
[5] Hirsch M J , Pardalos P M and Resende M G C. 2009 *Nonlinear Analysis RWA* **10** 2000.
[6] Parsopoulos K E and Vrahatis M N 2004 *IEEE Transactions on Evolutionary Computation* **8(3)** 211.
[7] Platt G M 2012 *Proc. Nat. Congress of Applied and Computational Mathematics (Brazil)* (to appear).

Journal of Physics: Conference Series **410** (2013) 012021 doi:10.1088/1742-6596/410/1/012021

Hyper-heuristic applied to nuclear reactor core design

R.P.Domingos, G.M.Platt

Instituto Politécnico - Universidade do Estado do Rio de Janeiro

E-mail: `roberto.pinheiro@iprj.uerj.br`

Abstract. The design of nuclear reactors gives rises to a series of optimization problems because of the need for high efficiency, availability and maintenance of security levels. Gradient-based techniques and linear programming have been applied, as well as genetic algorithms and particle swarm optimization. The nonlinearity, multimodality and lack of knowledge about the problem domain makes de choice of suitable meta-heuristic models particularly challenging. In this work we solve the optimization problem of a nuclear reactor core design through the application of an optimal sequence of meta-heuritics created automatically. This combinatorial optimization model is known as hyper-heuristic.

1. Introduction

Hyperheuristic is a new methodology that aims to automate the process of selecting or combining simpler heuristics to solve difficult computational problems.The choice of a suitable metaheuristic algorithm is not the only challenge one must tackle, there is also the choice of the proper set of parameters for the specific search space. In fact the task of to adjust the chosen algorithm demands considerable time and effort. In an attempt of reduce this dependency, alternative schemes has been tried. From hybrid metaheuristics aiming to aggregate the strength of distincts algorithms to iteration dependent parameters aiming to adjust automaticaly their values as the execution evolves. In these work we present a model called hyperheuristic which generalizes even more, automaticaly chosing each metaheuristcs to be applied, from a set of available metaheuristics. This model also has the potential of reduce considerably the effect of bad parameterization through the exchange of algorithm on the fly during the optimization process. We applied this model to a known, non-linear and difficult optimization problem of nuclear engineering, the design of nuclear core. Many distincts metaheuristics has been applied to this problem, our results has shown that hyperheuristic reaches comparable solutions for the problem with considerable less work in terms of parameterization, and algorithm choice.

2. Hiperheuristic

Due the complexity of the search space of many problems, one can be interested in find not the absolute optimal solution but a good enough solution with a reasonable computational effort. Metaheuristics, understood here as a type of search algorithm that can be specialized to solve different optimization problems, are normally applied in these cases. It is known [1] that all search process have the same average performance when averaged on all problems defined on a finite search space, but there is also the practical fact that each metaheuristic

Published under licence by IOP Publishing Ltd

has specific limitations and virtues. The idea behind hyperheuristic method is to try to combine distinct metaheuristics in specific phases of the solution process with the hope of compensate the weakness of a metaheuristic A with the strength of metaheuristic B. However the task of finding a combination of metaheuristics able to solve an optimization problem can be understood as a search process itself, one that aims to find the best sequence from a set available metaheuristics. We call this high level problem independent heuristic as hyperheuristic (HH). Moreover metaheuristics must be adjusted to a given problem, and to find the proper parameterization set (population size, mutation rates, cognitive factors) demands considerable effort and skill. On the contrary, hyperheuristics can be directly applied to the problem. In this work a tabu-list HH based on a related work [2] was implemented. It chooses which metaheuristic to used based on how good it has worked before, in order to accomplish this HH builds two lists: a tabu and rank list. Every time a metaheuristic is successful in improve a solution its rank is increased by one. Otherwise it is decreased and it is put in a tabu list, which means it will not be chosen again until another one runs and improve the solution. The next metaheuristic will be chosen between those with higher rank, figure 3. The meta-heuristc set available for the

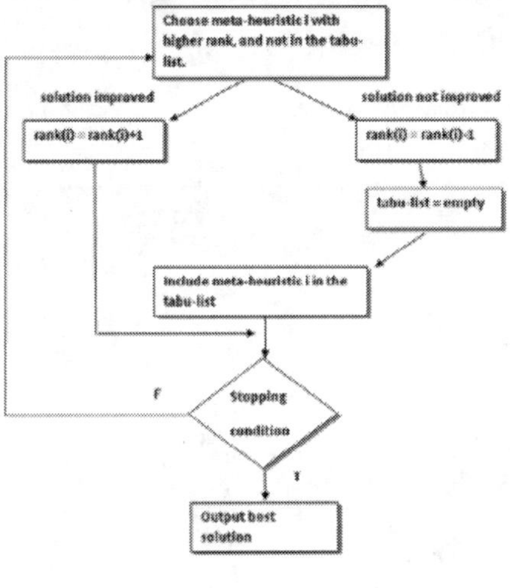

Figure 1. Tabu-list hyperheuristic framework.

Table 1. metaheuristics parameters.

Type	Pop.Size	Iterations	Cognt-Social	Mutation	Int.l/External loop/Constriction	obs
PSO	50	50	2.85/1.26	x	x	Gbest
DE	50	50	x	F*	x	rand/1/bin
DE2	50	50	x	F*	x	best/1/bin
HC	50	100	x	x	x	x
LJ	x	50	x	x	100/20/0.1	x
SA	x	100	x	x	x	Temp. init:10

HH consisted of Particle Swarm Optimization (PSO), Differential Evolution (DE/rand/1/bin), Simulated Annealing (SA), Hill Climbing (HC), Luus Jaakola (LJ) and a variation of DE (DE/best/1/bin) algorithm. The set of parameters for these meta-heuristcs are shown in the table 4, in this table the F* value for the mutation factor parameter of both DE algorithms represents a uniform random value in the interval [-1,-0.4] U[0.4,1].

3. Nuclear reactor core design

We consider a simplified cylindrical three-enrichment-zone pressurized water reactor with a typical cell composed of moderator (light water), cladding, and fuel, figure 2. The design parameters, which may be adjusted in the optimization process, as well as their variation ranges are shown in Table 2. The objective of the optimization problem is to minimize the average power peak factor fp of the proposed reactor for a given average thermal flux f_0, considering as constraints the criticality (keff = 1.0 +/- 0.01) and sub-moderation. So, the optimization problem can be written as a minimization of $f_p(R_f, \Delta_c, R_e, E_1, E_2, E_3, M_f, M_c)$ subject to :

$$\Phi(R_f, \Delta_c, R_e, E_1, E_2, E_3, M_f, M_c) = \Phi_0 \tag{1}$$

$$0.99 \leq k_{eff}(R_f, \Delta_c, R_e, E_1, E_2, E_3, M_f, M_c) \leq 1.01 \tag{2}$$

$$\frac{dk_{eff}}{dV_m} > 0 \tag{3}$$

$$R_{fmin} \leq R_f \leq R_{fmax} \tag{4}$$

$$\Delta_{cmin} \leq \Delta_c \leq \Delta_{cmax} \tag{5}$$

$$R_{emin} \leq R_e \leq R_{emax} \tag{6}$$

$$E_{1min} \leq E_1 \leq E_{1max} \tag{7}$$

$$E_{2min} \leq E_2 \leq E_{2max} \tag{8}$$

$$E_{3min} \leq E_3 \leq E_{3max} \tag{9}$$

$$M_f = UO_2 \quad Umetal \tag{10}$$

$$M_c = (Zircaloy - 2 \quad aluminum \quad Type304stainlesssteel) \tag{11}$$

where V_m is the moderator volume and the min and max subscripts refer to the lower and upper limits of the parameter.The HAMMER system was used to solve cell and diffusion equations. It performs a multigroup calculation of the thermal and epithermal flux distribution from the integral transport theory in a unit cell of the lattice.

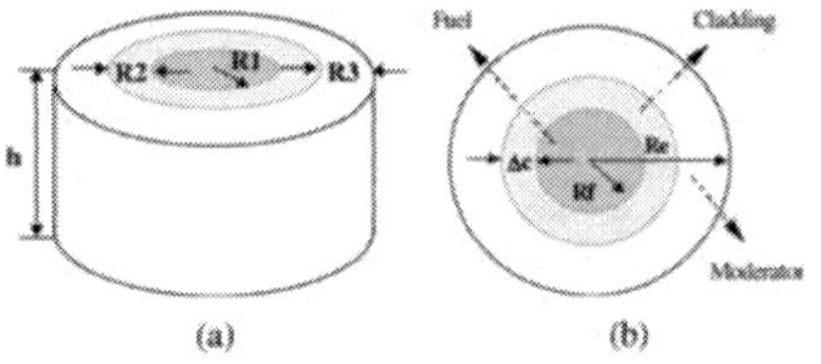

Figure 2. The reactor(a) and its typical cell (b)

4. Results and Conclusion

We performed 10 independent executions of the hyperheuristic model, each one running on Intel I7 under Windows 7 operating system. Table 3 shows the best reactor core parameters obtained along with best configurations from previous works. The results show that hyperheuristic model is able to find a fitness comparable with the best value found by different metaheuristic for this particular problem. However it should be noted that the effort of adjust and fine tunning all these metaheuristics is completely absent in our case. This task normally demands a considerable time when applying a metaheuristic to a given problem together with some experience with the metaheuristic itself. In figure 4 it is shown the sequence of metaheuristics applied when obtaining the best fitness value, it is clear from the figure that PSO is good choice for the initial stage of

Journal of Physics: Conference Series **410** (2013) 012021 doi:10.1088/1742-6596/410/1/012021

Table 2. Comparison with published best configurations.

Parameters	GA^3	PSO^3	DE^3	Hyper
R_f	0.5621	0.7459	0.7281	0.7014
Δ_c	0.1770	0.1647	0.1932	0.0948
Δ_m	0.6581	0.7620	0.7610	0.5666
E_1	2.7560	2.5364	2.7318	2.1961
E_2	4.032	2.6608	2.8657	2.3154
E_3	4.4570	4.6067	4.9686	4.2986
M_f	U-Metal	U-Metal	U-Metal	U-Metal
M_c	SS-304	SS-304	SS-304	SS-304
k_{eff}	1.000	0.990	1.000	1.008
Avg.Φ	$8.02e^{-5}$	$8.07e^{-5}$	$8.08e^{-5}$	$8.02e^{-5}$
$fitness$	1.3100	1.2767	1.2763	1.2762

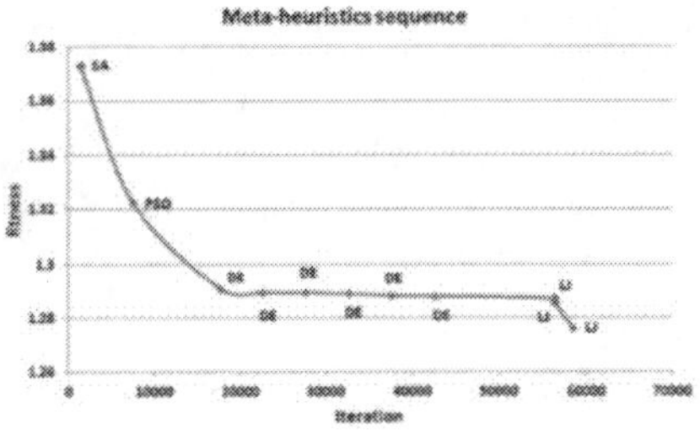

Figure 3. Best metaheuristic sequence

execution and Differential Evolution is very effective in dealing with this problem when in final (difficult) execution stages,being able to conduct to a good solution when followed by Luus Jaakola algorithm. It is also interesting to note that this sequence has started with Simulated annealing, a non-populational algorithm, that has not been the first choice in recent published works for this optimization problem. However immediatly after Simulated annealing three populational metaheuristics were chosen what conducted to an extensive exploratory phase,also each metaheuristic belongs to the set of metaheuristics applied in recent published works to tackle this problem. As an extension to this work we plan to implement a paralel hyperheuristic model and analyse the potential of other frameworks besides tabu-list hyperheuristc.

5. References

[1] Wolpert D., Macready W. 1997 *IEEE Transaction on Evolutionary Computation.* **1** 67
[2] Burke K., Kendall G., Soubeiga A. 2003 *Journal of Heuristics.* **9(6)** 451
[3] Sacco W. et al 2009 *Annals of Nuclear Energy.* **36** 1093

Journal of Physics: Conference Series **410** (2013) 012022 doi:10.1088/1742-6596/410/1/012022

Multiscale Fractal Descriptors Applied to Texture Classification

João Batista Florindo, Odemir Martinez Bruno

Instituto de Física de São Carlos (IFSC), Universidade de São Paulo (USP), Avenida do Trabalhador São-carlense, 400, 13560-970 São Carlos SP Brazil

E-mail: `jbflorindo@ursa.ifsc.usp.br`

Abstract. This work proposes the combination of multiscale transform with fractal descriptors employed in the classification of gray-level texture images. We apply the space-scale transform (derivative + Gaussian filter) over the Bouligand-Minkowski fractal descriptors, followed by a threshold over the filter response, aiming at attenuating noise effects caused by the final part of this response. The method is tested in the classification of a well-known data set (Brodatz) and compared with other classical texture descriptor techniques. The results demonstrate the advantage of the proposed approach, achieving a higher success rate with a reduced amount of descriptors.

1. Introduction

Fractal geometry shows to be an efficient tool to characterize and discriminate complex objects usually found in texture images [1, 2, 3]. This precision of fractal representation is explained by the flexibility of a fractal object, given its nature instrinsically complex. Moreover, fractals are strongly identified by their self-similarity behavior, that is, each part constitutes a copy only linearly transformed of the whole structure. Such characteristic is also widely observed in many objects from the real world, in a greater or smaller degree of intensity.

Among the most known fractal-based techniques, we can cite the fractal descriptors, which demonstrate high efficiency in several works studying texture analysis [6, 7]. Despite its efficiency, fractal descriptors still have problems with the analysis of images presenting a significant amount of patterns and details, which may be observed only at some specific scales. This is explained by the fact that, although fractal descriptors analyzes the image under different scales, the descriptors themselves are represented globally and are not able to highlight any kind of individual scale.

With the aim of overcoming such drawback, this work develops and studies a novel approach where a multiscale transform [8] is applied to the original descriptors. Particularly, we employed a space-scale transform (derivative + Gaussian filter) and applyed a threshold over the filter response in order to discard some irrelevant information. The application of a multiscale transform enabled the use of fractal descriptors as not only a global curve which describes the image, but a set of informations about details, patterns and irregularities observed at a wide range of scales.

2. Fractal Geometry

Fractal geometry is the area which studies fractal objects. These are complex geometrical structures characterized by an infinite degree of self-similarity, that is, each part of the object corresponds to a copy of the whole, only affected by linear geometrical transforms (affine transforms) [9, 10]. The fractal dimension is the most used measure to describe a fractal object.

2.1. Fractal Dimension

Essentially, this metric is obtained by measuring the extension of the object through a rule with a variable side-length. The dimension of an object S is given by the following expression:

$$D = \lim_{\epsilon \to 0} \frac{\log(N_\epsilon(S))}{\log(\epsilon)}, \tag{1}$$

where N is the number of units with length ϵ necessary to cover the whole object.

The above equation is only usable when we know the exact mathematical rule which gave rise to the fractal object. In practical situations like in this work, when we are dealing with objects represented in a digital image, we must turn to estimation methods which provide an approximate fractal dimension value. Most of such methods are based on a generalization of the Equation 1, where, instead of the number of rules N, we use any complexity metric whether in spatial or frequency space.

Among such estimation methods, we opted to use the Bouligand-Minkowski (BM) technique. This choice was based on its good performance in texture analysis problems similar to those showed in this work [11, 7, 6, 12, 13]. In this approach, the gray-level texture imagem is mapped onto a surface, which is dilated through a variable radius r. The dimension is estimated from the power-law relation between the volume $V(r)$ of the dilated surface and the radius.

3. Fractal Descriptors

Despite its importance as a characterizer of objects, fractal dimension is still limitted, mainly when we need to discriminate among a large number of objects with a high degree of similarity. Such difficulty can be justified by two main aspects. The first is that fractal dimension is only a unique value to describe the whole complexity of an object. The second is that we observe a lot of objects with a completely distinct appearance whilst they have the same estimated fractal dimension [12]. In order to overcome such drawback, Bruno et al. [6] developed a novel methodology called fractal descriptors.

This technique consists in composing a set of features (descriptors) from the power-law fractality curve of the analyzed object. In Bouligand-Minkowski method used here, the descriptors are obtained from the power-law relation u between $V(r)$ and r, that is:

$$u : \log(r) \to \log(V(r)). \tag{2}$$

In this way, $V(r)$ acts as a fractality measure while r corresponds to the representation scale.

The function u can be used directly, as in [11] or after some kind of transform. In this work, we propose a multiscale transform to enhance the results obtained with fractal descriptors.

4. Proposed Method

In image analysis, and particularly texture analysis, we are often faced with situations where the behavior of some analysis method varies severely according to the scale at which the image is analyzed. Multiscale transform was developed with the aim of attenuating such difficulty, providing robust information about features of the object which are observed only at specific scales.

The multiscale transform, applied over a curve $u(t)$ like that of fractal descriptors, consists in a mapping function of $u(t)$ onto $U(b, a)$, where b is directly related to t and a is the scale parameter.

Basically, the literature shows three categories of multiscale transform, to know, space-scale, time-frequency and time-scale [14]. Here, we adopted the space-scale approach, given its large use in image analysis and particularly interesting results in works related to the identification of patterns in natural objects, like those studied in this work [7, 6].

In this approach, the function $U(b, a)$ is obtained by derivating the function $u(t)$ and applying a Gaussian smoothing filter g_a, where a is the smoothing parameter. More formally, we have:

$$U(b,a) = U^1(t) * g_a(t) = U^1(t,a), \tag{3}$$

where $U^1(t,a)$ is the convolution of the original signal $u(t)$ with the first derivative of the gaussian g_a^1, that is:

$$U^1(t,a) = u(t) * g_a^1(t). \tag{4}$$

Inasmuch as multiscale is a two-dimensional transform applied over a one-dimensional signal, it adds a subtantial amount of redundance information. To solve this drawback, we employed a technique based on the adjustment of visual system to select objects in a scene [14]. Such solution, named fine-tuning selection consists in selecting a specific value of parameter a and projecting the transform over that point. Here, we select the value of a empirically.

Finally, we applied a threshold over the obtained signal in order to eliminate noises added by the application of the multiscale over the final part of Bouligand-Minkowski curve, where the data becomes more sensible due to the high level of wavefront interference. The threshold point is also determined empirically.

5. Experiments

The efficiency of the proposed method is verified by a comparison with other classical and state-of-the-art methods in the classification of a texture data set.

Here, we used the well-known Brodatz [15] texture data. This corresponds to 1110 images, each one divided into 10 windows. Each image is a class and each window is a sample.

The classification itself is carried out by the Linear Discriminant Analysis (LDA) classifier [16] using a Hold-Out scheme, with one half of the data to train and the other half to validate.

We compare the success rate to the use of Bouligand-Minkowski descriptors (without multiscale) [11], Gabor wavelets descriptors [17], Co-occurrence matrix [18] and a multiscale approach described in [19].

6. Results

The Table 1 shows the success rate of the compared descriptors in the classification task of Brodatz data. We also show the number of descriptors used to provide the best result and the κ index [16], a measure of the reliability of the result. The value of $1 - \kappa$ may also be used as an error measure of the experiment. For the propoed method, we employed a parameter $a = 0.7$ and a threshold after the 51th element.

In summary, we observe that the combination of fractal descriptors and space-scale transform constituted a powerful and robust method to provide texture descriptors. Although fractal descriptors provide a good result, as attested by the success rate of Minkowski method, it is still limitted to an analysis more global of the image, which limits its efficiency in more complex situations as, for instance, when we have a high degree of similarity from different classes and/or a high dissimilarity intra-class. The global fractal descriptors approach also has some dificulties in dealing with noised images. The multiscale approach allows the highlighting of scale-dependent nuances, characterizing the image in a richer way, unveiling important structures which only

Method	Correctness Rate (%)	κ	Number of descriptors
Co-occurrence	92.0721	0.9201	84
Gabor	90.0901	0.9001	20
Multifractal	37.4775	0.3732	101
Minkowski	98.9189	0.9891	85
Proposed	99.4595	0.9945	51

Table 1. Correctness rate for the Brodatz dataset.

appear at some specific scale as well as isolating artifacts and noises inherent to the original image.

7. Conclusion

We compared the proposed method to other well-known texture descriptors and the results showed that the multiscale combination porvided the best success rate even with a reduced number of descriptors. Such performance confirmed the efficiency of multiscale transform associated to fractal geometry, once it allows the study of details which only can be detected at specific scales, turning, in this way, the analysis more complete and robust.

The results also suggest the application of the proposed approach to other problems from the real world, involving the discrimination of objects in complex scenarios.

Acknowledgments

Joao B. Florindo gratefully acknowledges the financial support of CNPq (National Council for Scientific and Technological Development, Brazil) (Grant #140624/2009-0). Odemir M. Bruno gratefully acknowledges the financial support of CNPq (National Council for Scientific and Technological Development, Brazil) (Grant #308449/2010-0 and #473893/2010-0) and FAPESP (The State of São Paulo Research Foundation) (Grant # 2011/01523-1).

[1] Quevedo R, Mendoza F, Aguilera J M, Chanona J and Gutierrez-Lopez G 2008 *Journal of Food Engineering* **84** 509–515
[2] Tian-Gang L, Wang S and Zhao N 2007 *Computerized Medical Imaging and Graphics* **31** 665–671
[3] Millan H and Gonzalez-Posada M 2005 *GEODERMA* **125** 25–38
[4] Bruno O M, de Oliveira Plotze R, Falvo M and de Castro M 2008 *Information Sciences* **178** 2722–2733
[5] Plotze R O, Padua J G, Falvo M, Vieira M L C, Oliveira G C X and Bruno O M 2005 *Canadian Journal of Botany-Revue Canadienne de Botanique* **83** 287–301
[6] Witkin A P 2003 *Proceedings...* ICASSP - IEEE International Conference on Acoustics, Speech, and Signal Processing (Saint Martin d'Hres, France: GRETSI) pp 79–95
[7] Mandelbrot B B 1968 *The Fractal Geometry of Nature* (NY, USA: Freeman)
[8] Peitgen H O, Barnsley M F and Saupe D 1988 *The Science of fractal images / Heinz-Otto Peitgen, Dietmar Saupe, editors ; Michael F. Barnsley ... [et al.] ; with contributions by Yuval Fisher, Michael McGuire* (Berlin : Springer-Verlag, New York :)
[9] Backes A R, Casanova D and Bruno O M 2009 *International Journal of Pattern Recognition and Artificial Intelligence (IJPRAI)* **23** 1145–1160
[10] Florindo J B and Martinez Bruno O 2011 *Chaos* **21** 043112(1–12)
[11] Florindo J B, Backes A R, de Castro M and Bruno O M 2012 *Pattern Recognition Letters* **33** 798–806
[12] da F Costa L and Cesar, Jr R M 2000 *Shape Analysis and Classification: Theory and Practice* (CRC Press)
[13] Brodatz P 1966 *Textures: A photographic album for artists and designers* (New York: Dover Publications)
[14] Duda R O and Hart P E 1973 *Pattern Classification and Scene Analysis* (Wiley)
[15] Manjunath B and Ma W 1996 *IEEE Transactions on Pattern Analysis and Machine Intelligence* **18** 837–842 ISSN 0162-8828
[16] Haralick R M 1979 *Proceedings of the IEEE* **67** 786–804
[17] Parrinello T and Vaughan R A 2002 *International Journal of Remote Sensing* **23** 1799–1825

Journal of Physics: Conference Series **410** (2013) 012023 doi:10.1088/1742-6596/410/1/012023

Efficient numerical solvers for the nonlinear beam and wave equations

M. J. Moreta[1] **and B. Cano**[2]

[1]Departamento de Fundamentos del Análisis Económico I. Facultad de Ciencias Económicas y Empresariales. Universidad Complutense de Madrid. Campus de Somosaguas, Pozuelo de Alarcón, CP 28223, Madrid, Spain.

[2]Departamento de Matemática Aplicada, IMUVA. Facultad de Ciencias. Paseo Belén 7, CP 47011, Valladolid, Spain.

E-mail: [1]`mjesusmoreta@ccee.ucm.es` and [2]`bego@mac.uva.es`

Abstract.

In this paper we give the main ideas about the numerical temporal resolution of second order in time partial differential equations with high order symmetric multistep cosine methods. These methods are efficient numerical solvers when integrating second-order in time partial differential problems subject to periodic boundary conditions, like the nonlinear wave or the beam equations.

1. Introduction

High order symmetric multistep cosine methods [2] are known to be efficient numerical solvers when integrating second-order in time partial differential problems subject to periodic boundary conditions, like

$$\begin{aligned} &u_{tt}(t,x) = Au(t,x) + g(t,u(t,x)), \quad x \in [a,b], \quad t_0 < t < T, \\ &u(t_0,x) = u_0(x), \quad u_t(t_0,x) = v_0(x), \\ &u(t,a) = u(t,b), \quad u_x(t,a) = u_x(t,b), \end{aligned} \tag{1}$$

where g, u_0 and v_0 are smooth enough functions and A is a differential linear operator with negative eigenvalues. For the nonlinear wave equation we have $Au = u_{xx}$ and in the case of the nonlinear beam (or Euler-Bernoulli) equation we have $Au = -u_{xxxx}$. Here, we are assuming hypotheses of enough regularity of the solution, which is interesting in practice, for example, when the solutions are solitons or balanced geophysical flows [5].

Multistep cosine methods can be considered as a subtype of 'exponential integrators' [6, 7, 8, 9, 10], which integrate the linear and stiff part of a problem in an exact way. Because of this, it is possible to obtain methods which are explicit and stable at the same time for linearly stiff problems. In order to construct these methods, we have taken profit of the existence in the literature [11] of standard explicit symmetric multistep methods of high order for second-order systems, which have been optimized with respect to their interval of stability. These methods are used when integrating planetary orbits. In [4] it is shown the detailed construction of methods of orders 8 (SMC8) and 10 (SMC10), which are the ones we are using here in the numerical experiments to numerically solve the beam equation. These methods present a good long-term behaviour when integrating Hamiltonian systems because of its symmetry (see [1] for the study about the conservation of invariants when using these methods to integrate Hamiltonian systems).

The structure of the paper is as follows. In Section 2 we give the main ideas about the spatial discretization we are using. In Section 3 we introduce multistep cosine methods SMC8 and SMC10 and finally, we give some numerical results in Section 4.

2. Problem to solve

In order to discretize in space problem (1), we take, for a natural number M, as an approximation of the solution, $u_M(t,x) = \sum_{j=-M}^{M-1} a_j(t) e^{2\pi i j \frac{x-a}{b-a}}$. Then, we obtain

$$\begin{aligned} \ddot{Y}_M(t) &= -\Omega_M^2 Y_M(t) + G_M(t, Y_M(t)), \\ Y_M(t_0) &= Y_{M,0}, \quad \dot{Y}_M(t_0) = V_{M,0}, \end{aligned}$$

where $Y_M(t) = [a_{-M}(t), \ldots, a_{M-1}(t)]^T$, $\Omega_M = \mathrm{diag}(\lambda_j)_{j=-M}^{M-1}$, with $\lambda_j = \frac{2\pi j}{b-a}$ for the nonlinear wave equation and $\lambda_j = (\frac{2\pi j}{b-a})^2$ for the beam equation. Furthermore,

$$G_M(t, Y_M(t)) = \frac{1}{2M} \mathrm{shift}\Big(F_{2M} g(t, 2M F_{2M}^{-1} \mathrm{shift}(Y_M(t))) \Big),$$

where F_{2M} is the $2M$-dimensional discrete Fourier transform, with shift an application such that for every vector $\vec{v} = [v_0, \ldots, v_{2M-1}]$, $\mathrm{shift}(\vec{v}) = [v_M, \ldots, v_{2M-1}, v_0, \ldots, v_{M-1}]$. Finally, $g(t, \vec{v})$ denotes the vector whose components are $(g(t, v_j))_{j=0}^{2M-1}$ and $Y_{M,0}$ and $V_{M,0}$ contain the initial conditions. This spatial discretization is very accurate, so the values $Y_M(t)$ and $G_M(t, Y_M(t))$ approximate the Fourier coefficients of the regular functions $u(t,x)$ and $g(t, u(t,x))$, respectively.

In what follows, in order to simplify the notation, we will omite the subscript M.

3. Multistep cosine methods

Multistep cosine methods are of the form

$$\begin{aligned} & Y_{n+2k} + \delta_{k-1}(h\Omega)(Y_{n+2k-1} + Y_{n+1}) + \ldots + \delta_1(h\Omega)(Y_{n+k+1} + Y_{n+k-1}) + Y_n \\ & -2\left[\cos(kh\Omega) + \delta_{k-1}(h\Omega)\cos((k-1)h\Omega) + \ldots + \delta_2(h\Omega)\cos(2h\Omega) + \delta_1(h\Omega)\cos(h\Omega)\right] Y_{n+k} \\ = \; & h^2[\gamma_1(h\Omega)(G_{n+2k-1} + G_{n+1}) + \ldots + \gamma_{k-1}(h\Omega)(G_{n+k+1} + G_{n+k-1}) + \gamma_k(h\Omega) G_{n+k}], \end{aligned} \tag{2}$$

where Y_n is the numerical approximation to $Y(t_n)$ and G_{n+j} means $G(t_{n+j}, Y_{n+j})$, $j = 1, \ldots, 2k-1$. Besides, $\delta_j(h\Omega)$, for $j = 1, \ldots, k-1$ and $\gamma_i(h\Omega)$, for $i = 1, \ldots, k$ are real functions. The method is uniquely determined by functions $\delta_j(h\Omega)$, for $j = 1, \ldots, k-1$ and have the desired order of consistency with the only condition of their being bounded on the real axis. In [4] we suggest how to choose them in order to assure stability and we give a detailed and very technical study for general multistep cosine methods. Once they have been calculated, functions $\gamma_i(\epsilon)$, $i = 1, \ldots, k$ are given by

$$\gamma_l(\epsilon) = \int_0^k \epsilon^{-1} \sin((k-u)\epsilon) \hat{I}_{k-l}(u) du + \sum_{j=1}^{k-1} \delta_j(\epsilon) \int_0^j \epsilon^{-1} \sin((j-u)\epsilon) \hat{I}_{k-l}(u) du, \tag{3}$$

with $\hat{I}_l(u)$ the polynomial functions of degree $2k-2$ in u, $l = 0, \ldots, k-1$, such that $\hat{I}_l(-l) = \hat{I}_l(l) = 1$ and $\hat{I}_l(j) = 0$, for $j = -(k-1), \ldots, k-1$, with $j \neq l$.

3.1. Resonances

When we are integrating a problem like (2), sometimes we can observe that the behaviour of the numerical solution is not the expected one and that the error is much more greater than the desirable one. These bad results are caused because there are some particular values of the time step size for which the error is greater than for other values which are well apart from them. This is called as resonances and it is related with the double roots of the first characteristic polynomial.

In [3, 4] resonances are deeply studied. There, it is suggested to considered a filter function

$$\phi(\epsilon) = \begin{cases} 1 & \text{if} \quad |\epsilon| < \delta, \\ 0 & \text{if} \quad |\epsilon| \geq \delta. \end{cases} \tag{4}$$

This value is related with the first value of ϵ for which double roots of the first characteristic polynomial are met. The filter method then reads as (2) but by substituting $G(t_{n+j}, Y_{n+j})$ by $G(t_{n+j}, \phi(h\Omega)Y_{n+j})$, $j = 1, \ldots, 2k-1$.

3.2. Optimal suggested method of eighth order

In this case, we take $k = 4$. Therefore, the method is uniquely determined when functions $\delta_l(h\Omega)$, $l = 1, 2, 3$ are given. For method SMC8 we have the following ones [4]:

$$\delta_1(\epsilon) = 1 - 2\cos(\epsilon), \qquad \delta_2(\epsilon) = 2, \qquad \delta_3(\epsilon) = -2\cos(\epsilon).$$

Functions $\gamma_j(\epsilon)$, $j = 1, \ldots, 4$ have been calculated with Mathematica® by using formulae (3). These functions are given by

$$\begin{aligned}
\gamma_1(\epsilon) &= -\frac{360 - 840\epsilon^2 + 721\epsilon^4 - 288\epsilon^6 + \cos(\epsilon)(-360 + 660\epsilon^2 - 376\epsilon^4 + 72\epsilon^6)}{36\epsilon^8}, \\
\gamma_2(\epsilon) &= -\frac{-120 + 260\epsilon^2 - 192\epsilon^4 + 52\epsilon^6 + \cos(\epsilon)(120 - 200\epsilon^2 + 87\epsilon^4)}{2\epsilon^8}, \\
\gamma_3(\epsilon) &= \frac{-600 + 1240\epsilon^2 - 855\epsilon^4 + 228\epsilon^6 + \cos(\epsilon)(600 - 940\epsilon^2 + 360\epsilon^4 - 8\epsilon^6)}{4\epsilon^8}, \\
\gamma_4(\epsilon) &= \frac{1800 - 3660\epsilon^2 + 2480\epsilon^4 - 612\epsilon^6 + \cos(\epsilon)(-1800 + 2760\epsilon^2 - 1025\epsilon^4 - 18\epsilon^6)}{9\epsilon^8}.
\end{aligned}$$

For this method we have taken $\delta = 1.2$ in (4).

3.3. Optimal suggested method of tenth order

In order to construct the method of tenth order, we have taken $k = 5$. For this method, we have obtained

$$\delta_1(\epsilon) = 5 - 4\cos(\epsilon), \quad \delta_2(\epsilon) = 3 - 4\cos(\epsilon), \quad \delta_3(\epsilon) = 3 - 2\cos(\epsilon), \quad \delta_4(\epsilon) = 1 - 2\cos(\epsilon).$$

For this method, by using again (3) and Mathematica® we have

$$\begin{aligned}
\gamma_1(\epsilon) &= \frac{15120 - 39060\epsilon^2 + 39165\epsilon^4 - 19811\epsilon^6 + 5544\epsilon^8 + \cos(\epsilon)(-15120 + 31500\epsilon^2 - 22785\epsilon^4 + 7085\epsilon^6 - 1008\epsilon^8)}{504\epsilon^{10}} \\
\gamma_2(\epsilon) &= -\frac{60480 - 148680\epsilon^2 + 136500\epsilon^4 - 58013\epsilon^6 + 10584\epsilon^8 + \cos(\epsilon)(-60480 + 118440\epsilon^2 - 74760\epsilon^4 + 15614\epsilon^6 + 504\epsilon^8)}{252\epsilon^{10}} \\
\gamma_3(\epsilon) &= \frac{15120 - 35820\epsilon^2 + 31065\epsilon^4 - 12197\epsilon^6 + 2214\epsilon^8 + \cos(\epsilon)(-15120 + 28260\epsilon^2 - 16305\epsilon^4 + 2846\epsilon^6 - 72\epsilon^8)}{18\epsilon^{10}} \\
\gamma_4(\epsilon) &= -\frac{60480 - 140040\epsilon^2 + 117780\epsilon^4 - 44549\epsilon^6 + 7380\epsilon^8 + \cos(\epsilon)(-60480 + 109800\epsilon^2 - 60360\epsilon^4 + 9710\epsilon^6 + 144\epsilon^8)}{36\epsilon^{10}} \\
\gamma_5(\epsilon) &= \frac{75600 - 173700\epsilon^2 + 144705\epsilon^4 - 54055\epsilon^6 + 9216\epsilon^8 + \cos(\epsilon)(-75600 + 135900\epsilon^2 - 73605\epsilon^4 + 11485\epsilon^6 - 216\epsilon^8)}{36\epsilon^{10}}
\end{aligned}$$

In this case, we have $\delta = 1$ in (4).

4. Numerical experiments

Here, we integrate the following equation, which has periodic boundary conditions

$$\begin{aligned} u_{tt}(t,x) &= -u_{xxxx}(t,x) - 0.4u^3(t,x), \quad x \in [0,2\pi], \\ u(0,x) &= \frac{5}{2}e^{\frac{\cos(x)}{5}} - \frac{5}{2}, \; u_t(0,x) = e^{\frac{\sin(x)}{5}} - \frac{5}{2}, \quad x \in [0,2\pi], \\ u(t,0) &= u(t,2\pi), \; u_x(t,0) = u_x(t,2\pi), \; u_{xx}(t,0) = u_{xx}(t,2\pi), \; u_{xxx}(t,0) = u_{xxx}(t,2\pi). \end{aligned} \tag{5}$$

We have integrated this problem with the spatial discretization described in Section 2, with 32 nodes in space. After this, we have integrated in time, by considering both methods with the filter function (4). In Table (1) we can observe the discrete global L^2-error until time $T = 1$ committed for different values of h and the order achieved by SMC8 method in the left part. In the right part of this table, we have the results corresponding to SMC10 method.

Table 1. Errors and order for problem (5) and SMC8 and SMC10 methods.

h	L^2-error	Order	h	L^2-error	Order
$\frac{1}{40}$	6.61590E-12		$\frac{1}{40}$	9.05268E-13	
$\frac{1}{80}$	2.01536E-14	8.35876	$\frac{1}{80}$	3.48342E-16	11.3436
$\frac{1}{160}$	7.49850E-17	8.07022	$\frac{1}{160}$	2.36999E-19	10.5214
$\frac{1}{320}$	2.89651E-19	8.01614	$\frac{1}{320}$	2.22237E-22	10.0586
$\frac{1}{640}$	1.12577E-21	8.00726	$\frac{1}{640}$	2.14850E-25	10.0146
$\frac{1}{1280}$	4.38926E-24	8.00272	$\frac{1}{1280}$	2.11144E-28	9.99089
	SMC8 method			SMC10 method	

Acknowledgments

This research was supported by MTM 2011-23417

References

[1] Cano B 2012 Conservation of invariants by symmetric multistep cosine methods, published on line in *BIT Numer. Math.* DOI: 10.1007/s10543-012-0393-1.
[2] Cano B and Moreta M J 2010 Multistep cosine methods for second-order partial differential equations, *IMA J. Numer. Anal.* **30** pp 431-61.
[3] Cano B and Moreta M J, Stability and resonances of multistep cosine methods, accepted for publication in *J. Comput. Math.*
[4] Cano B and Moreta M J, High order symmetric multistep cosine methods, submitted for publication.
[5] Leimkuhler B and Reich S 2005 *Simulating Hamiltonian Dynamics* (Cambridge: Cambridge University Press)
[6] Cox S M and Matthews P C 2002 Exponential Time Differencing for Stiff Systems, *Journal of Computational Physics* **176** pp. 430-55.
[7] Calvo M P and Palencia C 2006 A class of explicit multistep exponential integrators for semilinear problems, *Numer. Math.* **102** pp. 367-81.
[8] Hochbruch M and Lubich C 1999 A Gautschi-type method for oscillatory second-order differential equations, *Numer. Math.* **83** pp. 403-26.
[9] Hochbruck M and Ostermann A 2005 Exponential Runge-Kutta methods for parabolic problems, *Applied Numerical Math.* **53** pp. 323-39.
[10] Ostermann A, Thalhammer M and Wright W 2006 A class of explicit exponential general linear methods, *BIT Numer. Math.* **46** pp. 409-31.
[11] Quinlan G D and Tremaine S 1990 Symmetric multistep methods for the numerical integration of planetary orbits, *The Astronomical Journal*, **100** pp. 1694-1700.

Prediction of stability constants of coordination compounds from their molecular graphs

Nenad Raos, Ante Miličević

Institute for Medical Research and Occupational Health, Ksaverska c. 2, P.O.B. 291, HR-1001, Zagreb, Croatia

raos@imi.hr

Abstract. Every molecular structure can be regarded as a graph, and each graph can be transformed into characteristic scalar – topological index. There are many, over 1000 topological indices used by chemists, but they are not equally successful in solving specific problems. For the prediction of stability of coordination compounds valence molecular connectivity index of the 3rd order ($^3\chi^v$) proved best. Stability of these compounds was expressed as log K (K stays for stability constant), which is proportional to the Gibbs free energy of reaction between metal (or metal complex) and ligand. Various molecular graphs were checked, from a simple graph of ligand, to the graphs of various metal-ligand complexes (aqua complex, complex with additional bonds etc.) leading to conclusion that chemically sounder graphs generally yield better models. The best results were obtained for the systems of one metal, mesured at the same experimental conditions. However, our graph-theoretical approach enables development of models for the simulateneous prediction of stability constants of few metals.

1. Topological indices in chemistry

Application of topological indices in chemistry commenced in 1947 when H. Wiener introduced path number W, later known as Wiener number or Wiener index [1,2]. The path number was defined as the number of bonds between all pairs of atom in an acyclic molecule, i.e.:

$$W(G) = \frac{1}{2} \sum D_{ij}(G) \quad (1)$$

where $D_{ij}(G)$ are off-diagonal elements of (topological) distance matrix $\mathbf{D}(G)$. Wiener correlated his *path number* with physical parameters of acyclic hydrocarbons, and this problem occupy the chemists and mathematicians till our days; i.e. most of the papers in *mathematical chemistry* deal with modelling boiling points of hydrocarbons by using a variety of topological indices [3,4]. Later, topological indices (now, there are more than thousand of them) were used for prediction of many physico-chemical properties (QSPR, quantitative structure-properties relationships) [5-7] and biological activity (QSAR, quantitative structure-activity relationship) [8] of many kinds of molecules. Among the hundreds of topological indices [9] very few, however, found its application in chemistry. In our first paper in application of these indices for the estimation of stability of coordination compounds [10], we found the best fit with the Wiener index, W, and three valence connectivity indices, ${}^1\chi^v$, ${}^2\chi^v$, and ${}^3\chi^v$. The last one, the valence molecular connectivity index of the 3rd order, later proved to be the best of all indices, is defined by relation [11,12]:

$$ {}^3\chi^v\,(\mathrm{G}) = \Sigma\,[\delta(i)\delta(j)\delta(k)\delta(l)]^{-0.5} \qquad (2)$$

where $\delta(i)$, $\delta(j)$, $\delta(k)$, and $\delta(l)$ are weights (valence values) of vertices (atoms) i, j, k, and l making up the path of length 3 (three consecutive chemical bonds) in vertex-weighted molecular graph, G.

2. The stability of coordination compounds

The stability of coordination compounds is usually measured by log K values, where K is equilibrium constant of reaction between metal (M) and ligand (L):

$$\mathrm{M} + \mathrm{L} \xrightleftharpoons{K_1} \mathrm{ML}$$

$$\mathrm{ML} + \mathrm{L} \xrightleftharpoons{K_2} \mathrm{ML_2}$$

which are related to the Gibbs energy of reaction, i.e. stability of the complex ($\Delta_r G = -\,RT\ln K$). There are many models aimed at prediction of log K_1, log K_2 and log β_2 (= log K_1 + log K_2) [13], but there is yet no general model for this purpose. Problem is that stability of coordination compounds is determined by many diverse factors which are not easy to calculate, and, as an additional obstacle, stability constants of coordination compounds were measured at different conditions (temperature, ionic strength, solvent composition, etc.). Our approach is, however, based on quantitative comparison of structural (constitutional) similarity of molecules of coordination compounds, via their topological indices, assuming that more related molecules should have closer values of stability constants [14].

3. Influence of molecular graph

There are many possibilities to correlate log K with topological index ${}^3\chi^v$(G), because there are many possibilities to write a graph, G, representing a molecule. G may represent ligand (L), and many forms of complexes ML and ML_2. Exact structure of the complex in solution is not known; it could be generally denoted as $(H_2O)_nML$ or $(H_2O)_nML_2$, where n may be anything from 0 to maximal coordination number (Figure 1). However, more chemically sound structures (graphs) gave better correlations: L < ML < $ML^{aq} \approx ML^{cor}$ [15].

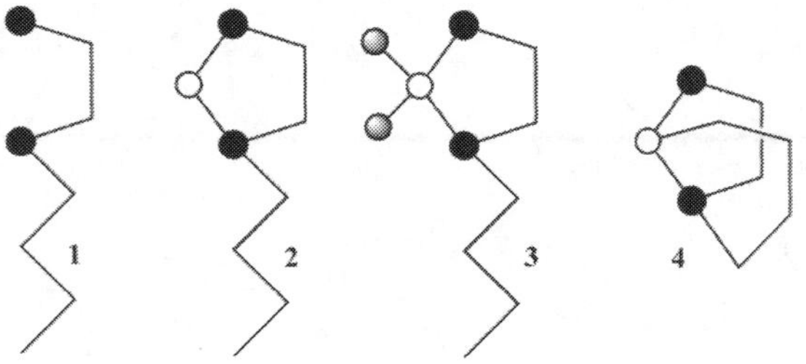

Figure 1. Four kinds of graphs used to estimate stability constants of coordination compounds (M = metal, L = butyldiaminoethane): L (**1**), ML (**2**), ML^{aq} (**3**), and ML^{cor} (**4**).

An another approach to proper graph of coordination compounds is based on the segmentation of molecule (graph) to its fragments (subgraphs). This approach was applied to copper(II) complexes of dipeptides, tripeptides and higher peptides [16,17]. Both approaches are original in graph-theoretical analysis, because researchers usually try many topological indices of the same, well defined graph. We, however, have tried the same topological index, ${}^3\chi^v$, but of the different graphs.

4. Reliability of method

Stability constants predicted from regression of log *K* on ${}^3\chi^v$ usually gave *rms* (log *K*) < 0.3, and even 0.05 [18,19]. In judging these numbers, it has to bear in mind that according to IUPAC recommendation constants determined with SD(log *K*) ≤ 0.05 were regarded as recommended, and the ones determined with 0.05 ≤ SD(log *K*) ≤ 0.2 were denoted as "tentative" [20]. However, it is not uncommon that the constants for the same system, measured at the same experimental conditions, differ more than 0.5 log units.

To check the validity of our method we applied our models on the constants K_1 and β_2 of copper(II) complexes with glycine and alanine, and their *N*-alkylated derivatives [21]. We compared log K_1 and log β_2 values measured by two experimental methods (glass electrode potentiometry, GEP, and square wave voltammetry, SWV) and various linear models based on different sets of stability constants and different molecular graphs (ML and ML^{aq}). It was found that the theoretical method reproduces experimental constants in the error range commensurable to the differences between GEP and SWV values. Moreover, it was found that theoretical values for log K_1 fit better GEP constants, and the opposite is true for SWV log β_2 values. This is in accordance with general statement that K_1 constants are better determined with GEP, and β_2 constants with SWV. That means that our theoretical method should be usable for checking experimental data.

5. Modeling the complex systems

Most of our models deals with one metal, usually copper(II) or nickel(II), and one class of ligands, usually amino acids, their *N*-alkylated derivatives, and diamines. However, in our recent paper [18] we developed models for *mono*- and *bis*-complexes of glycine, alanine, valine, and leucine with bivalent transition metals (Ni, Co, Fe, and Mn) by using the equations (Figure 2):

$$\log K_1 = a_1 [{}^3\chi^v (\mathrm{ML}) - {}^3\chi^v (\mathrm{NiL})]^2 + a_2 [{}^3\chi^v (\mathrm{ML}) - {}^3\chi^v (\mathrm{NiL})] + a_1 {}^3\chi^v (\mathrm{NiL}) + b \quad (3)$$

$$\log \beta_2 = a_1 [{}^3\chi^v (\mathrm{ML_2}) - {}^3\chi^v (\mathrm{NiL_2})]^2 + a_2 [{}^3\chi^v (\mathrm{ML_2}) - {}^3\chi^v (\mathrm{NiL_2})] + a_1 {}^3\chi^v (\mathrm{NiL_2}) + b \quad (4)$$

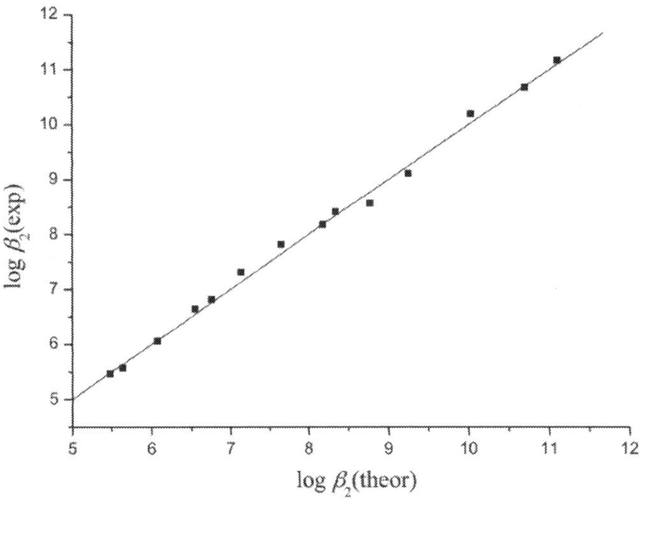

Figure 2. Experimental vs. theoretical (fit) log β_2 for Mn^{2+}, Fe^{2+}, Co^{2+}, and Ni^{2+} complexes with glycine, alanine, valine, and leucine, $r = 0.998$, $rms = 0.11$.

6. Conclusion

Application of topological indices for the prediction of values of stability constants of coordination compounds provides a simple and efficient tool for evaluating experimental data and planning future experiments. We hope that this short introduction will move the chemists, engaged in determination of stability constants, to use our method.

References

[1] H. Wiener, J. Amer. Chem. Soc. 69 (1947) 17.
[2] H. Wiener, J. Amer. Chem. Soc. 69 (1947) 2636.
[3] M. Randić, J. Amer. Chem. Soc. 97 (1975) 6609.
[4] L. H. Hall, L. B. Kier, W. J. Murray, J. Pharm. Sci. 64 (1975) 1974.
[5] E. V. Konstantinova, H. M. Diudea, Croat. Chem. Acta 73 (2000) 383.
[6] B. Lučić, S. Nikolić, N. Trinajstić, Croat. Chem. Acta 68 (1995) 417.
[7] A. T. Balaban, D. Mills, O. Ivancius, S. C. Basek, Croat. Chem. Acta 73 (2000) 923.
[8] L. B. Kier, L. H. Hall, Molecular Connectivity in Chemistry and Drug Research, Academic Press, New York, 1976.
[9] Z. Mihalić, N. Trinajstić, J. Chem. Educ. 69 (1992) 701.
[10] B. Grgas, S. Nikolić, N. Paulić, N. Raos, Croat. Chem. Acta 72 (1999) 885.
[11] M. Randić, J. Mol. Graph. Modell. 20 (2001) 19.
[12] X. Li, I. Gutman, Mathematical Aspects of Randić-type Molecular Structure Descriptors, Mathematical Chemistry Series, Kragujevac, 2006.
[13] N. Raos, A. Miličević, Prediction of stability constants, in: Handbook of Inorganic Chemistry Research (D. A. Morrison, Ed.), Nova Science Publishers, New York, 2010, pp. 267-294.
[14] N. Raos, A. Miličević, Arh. Hig. Rada Toksikol. 60 (2009) 123.
[15] A. Miličević, N. Raos, Polyhedron 25 (2006) 2800.
[16] A. Miličević, N. Raos, Polyhedron 27 (2008) 887.
[17] A. Miličević, N. Raos, Croat. Chem. Acta 82 (2009) 633.
[18] A. Miličević, G. Branica, N. Raos, Molecules 16 (2011) 1103.
[19] A. Miličević, N. Raos, Chin. J. Chem. 29 (2011) 2421.
[20] I. Sóvágó, T. Kiss, A. Gergely, Pure Appl. Chem. 65 (1993) 1029.
[21] N. Raos, G. Branica, A. Miličević, Croat. Chem. Acta 81 (2008) 511.

How is an optimized path of classical mechanics affected by random noise?

Tomoi Koide

Instituto de Física, Universidade Federal do Rio de Janeiro, C.P. 68528, 21941-972, Rio de Janeiro, Brazil

E-mail: tomoikoide@gmail.com

Abstract. The variational principle is one of important guiding principles in physics. Classical equations of motion of particle can be formulated so as to give the optimized path of an action. However, when there exist uncontrollable degrees of freedom such as noise, the optimized path is affected and the original classical equations of motion may not correspond to the optimized path. The stochastic variational method (SVM) is a framework to calculate the modified optimized path by the effect of noise. This method has been developed to show that the Schrödinger equation can be derived from the classical action which leads to Newton's equation of motion by taking into account the modification of the optimized path due to noise. In this work, we will extend this idea to the case of the continuum media and show that the Euler equation of the ideal fluid is converted to the Navier-Stokes equation or the Gross-Pitaevskii equation in SVM.

1. Introduction

As is well-known, the classical dynamics is described by Newton's equation of motion, and the path described by this equation corresponds to the optimized path of an action. However, it is also possible to consider the situation where this optimized path is not necessarily realized.

For example, let us consider to walk on a deck board on a luxury liner. When the ship lays at anchor, it is easy to go wherever we want to. However, when it is on the ocean and the sea runs high, it is difficult to reach there by the most direct way.

Similarly, if there exist uncontrollable degrees of freedom such as random noise, the optimized dynamics should be affected. Then is it possible to estimate the effect? The stochastic variational method (SVM) is a promising approach to discuss optimization under the Gaussian white noise[1].

So far, SVM has been exclusively applied to particle systems and used to discuss the relation between classical dynamics and quantum mechanics. In this work, we briefly summarize the application of SVM to the system of particle and report the extension of the method to the case of continuum media [2].

2. Stochastic Variational Method

In this section, we briefly summarize SVM. See also Ref. [2] and references therein for more details.

2.1. Definition of velocity

In general, there will exist various possibilities for the introduction of the effect of noise in the variational method. In SVM, this effect is taken into account through the definition of velocity as is the case of the Brownian motion.

First, we introduce the forward stochastic differential equation (SDE),

$$d\mathbf{r}(t) = \mathbf{u}(\mathbf{r}(t), t)dt + \sqrt{2\nu}d\mathbf{W}(t), \quad (dt > 0), \tag{1}$$

where ν is the strength of the noise and $\mathbf{W}(t)$ is the Wiener process,

$$E\left[W_j(s) - W_j(u)\right] = 0, \tag{2}$$

$$E\left[(W_j(s) - W_j(u))(W_k(s) - W_k(u))\right] = \delta^{jk}|s-u|, \tag{3}$$

$$E\left[(W_j(s) - W_j(u))W_k(t)\right] = 0. \tag{4}$$

The trajectory $\mathbf{r}(t)$ is not smooth and hence we cannot define the time derivative at t uniquely, as is shown in Fig. 1. Consequently, we further introduce backward SDE which describes the time reversed process of Eq. (1),

$$d\mathbf{r}(t) = \tilde{\mathbf{u}}(\mathbf{r}(t), t)dt + \sqrt{2\nu}d\tilde{\mathbf{W}}(t), \quad (dt < 0) \tag{5}$$

The noise $d\tilde{\mathbf{W}}$ satisfies the same correlation properties as $d\mathbf{W}$ and there is no correlation between $d\mathbf{W}$ and $d\tilde{\mathbf{W}}$.

The backward SDE must describe the time reversed process of Eq. (1). To satisfy this condition, the two Fokker-Planck equations which are derived from Eqs. (1) and (5) should be equivalent. From this condition, $\tilde{\mathbf{u}}$ is related to $\mathbf{u}$, as

$$\mathbf{u} = \tilde{\mathbf{u}} + 2\nu\nabla \ln \rho, \tag{6}$$

where ρ is the particle density which is given by the solution of the Fokker-Planck equation,

$$\partial_t \rho = \nabla(-\mathbf{u} + \nu\nabla)\rho = \nabla(-\tilde{\mathbf{u}} - \nu\nabla)\rho. \tag{7}$$

Correspondingly to the two SDEs, we introduce the two different definitions of the time derivative: one is the mean forward derivative,

$$D\mathbf{r} = \mathbf{u}, \tag{8}$$

and the other the mean backward derivative,

$$\tilde{D}\mathbf{r} = \tilde{\mathbf{u}}. \tag{9}$$

More precise definitions of these two derivatives, see Ref. [3]. The time derivative terms which appear in the kinetic term of the Lagrangian are replaced by using these $D\mathbf{r}$ and $\tilde{D}\mathbf{r}$.

2.2. Various formulas

The most part of the variational procedures in SVM is same as those in the classical variational method. However we cannot use the Taylor expansion and the partial integration formula. These are modified in SVM as follows.

(i) Ito formula

For an arbitrary differentiable function F which is a function of the stochastic variable $\mathbf{r}$ satisfying Eq. (1), there is the following expansion,

$$dF(\mathbf{r}, t) = (\partial_t F + \mathbf{u} \cdot \nabla F + \nu\nabla^2 F)dt + O(dt^{3/2}). \tag{10}$$

(ii) stochastic partial integration formula [4]

For the stochastic variables $\mathbf{X}$ and $\mathbf{Y}$, the partial integration formula is extended as follows,

$$\int_{t_a}^{t_b} dt E[D\mathbf{X}(t) \cdot \mathbf{Y}(t) + \mathbf{X}(t) \cdot \tilde{D}\mathbf{Y}(t)] = E[\mathbf{X}(t_b)\mathbf{Y}(t_b) - \mathbf{X}(t_a)\mathbf{Y}(t_a)]. \tag{11}$$

Journal of Physics: Conference Series **410** (2013) 012025 doi:10.1088/1742-6596/410/1/012025

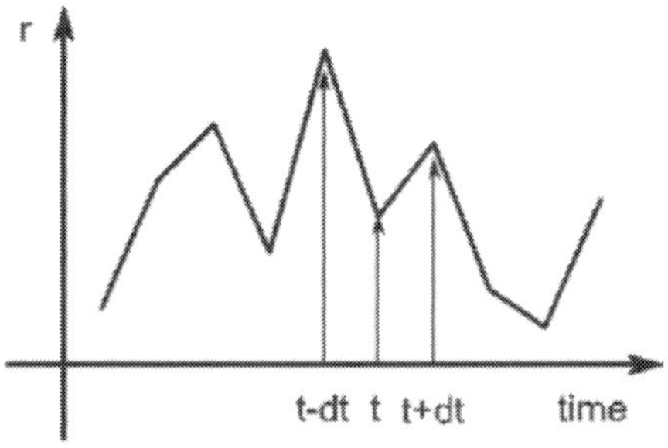

Figure 1. The ambiguity for the definition of the time derivative of $\mathbf{r}$ at t.

3. Application to Systems of Particle and Continuum Media

Let us apply SVM to an action which leads to Newton's equation of motion,

$$I(\mathbf{r}) = \int_{t_a}^{t_b} dt \left[\frac{m}{2} \left(\frac{d\mathbf{r}}{dt} \right)^2 - V(\mathbf{r}) \right], \tag{12}$$

where t_a and t_b are the initial time and the final time of the evolution of dynamics, respectively. As was discussed, there are two time derivatives in SVM and it is not unique to express the kinetic term in terms of the stochastic variable. In this section, we consider the following replacement,

$$I(\mathbf{r}) = \int_{t_a}^{t_b} dt E \left[\frac{m}{2} \frac{(D\mathbf{r})^2 + \left(\tilde{D}\mathbf{r}\right)^2}{2} - V(\mathbf{r}) \right], \tag{13}$$

By considering the variation of the stochastic trajectory $\mathbf{r} \longrightarrow \mathbf{r} + \delta\mathbf{r}$, we obtain

$$(\partial_t + \mathbf{u}_m \cdot \nabla) u_m^i - 2\nu^2 \nabla^i (\rho^{-1/2} \nabla^2 \sqrt{\rho}) = -\frac{1}{m} \nabla^i V, \tag{14}$$

where $\mathbf{u}_m = (\mathbf{u} + \tilde{\mathbf{u}})/2$. As was mentioned, ρ is the solution of the Fokker-Planck equation.

This result can be cast into an another form. The phase variable θ can be introduced by $\nabla\theta = \mathbf{u}_m/(2\nu)$. Then the new variable $\phi = \sqrt{\rho} e^{i\theta}$ satisfies the following equation,

$$i\partial_t \phi = \left[-\nu \nabla^2 + \frac{1}{2\nu m} V \right] \phi. \tag{15}$$

This equation is reduced to the Schrödinger equation by choosing $\nu = \hbar/(2m)$. That is, the Schrödinger equation can be derived form Newton's equation of motion by considering the fluctuation by random noise. It is also easy to see that the particule density is given by $\rho = |\phi|^2$ without introducing the interpretation of the wave function ϕ.

For the case of continuum media, we consider the classical action given by the following equation,

$$I(\mathbf{r}) = \int_{t_a}^{t_b} dt \int d^3\mathbf{R} \left[\frac{\rho_0^m(\mathbf{R})}{2} \left(\frac{d\mathbf{r}(\mathbf{R},t)}{dt} \right)^2 - J\varepsilon \right], \tag{16}$$

where ε is an internal energy density, $\mathbf{J}$ is the Jacobian between $\mathbf{r}$ and $\mathbf{R}$, and the mass density is defined by $\rho_m = m\rho$. Because this action is expressed in the Lagrangian coordinate, the spatial integral is done for the initial distribution of the mass density ρ_m^0, which is a function of $\mathbf{R}$. When we apply the classical variational method to this action, the Euler equation is obtained.

The corresponding stochastic Lagrangian density is given by

$$\begin{aligned} \mathcal{L} &= \frac{\rho_0^m}{2}\left[\left(\frac{1}{2}+\alpha_2\right)\left\{\left(\frac{1}{2}+\alpha_1\right)(D\mathbf{r}(t))^2 + \left(\frac{1}{2}-\alpha_1\right)(\tilde{D}\mathbf{r}(t))^2\right\} + \left(\frac{1}{2}-\alpha_2\right)(\tilde{D}\mathbf{r}(t))(D\mathbf{r}(t))\right] \\ &\quad - J\varepsilon. \end{aligned} \tag{17}$$

Here α_1 and α_2 is arbitrary constants, come from the ambiguity for the stochastic representation of the kinetic term. One can easily check that, in the vanishing noise limit, $\nu \to 0$, the difference between $D\mathbf{r}$ and $\tilde{D}\mathbf{r}$ disappears, and hence the above Lagrangian density always reproduces the classical one, independently of α_1 and α_2.

After applying SVM, we obtain

$$\rho^m \left(\partial_t + \mathbf{u}_m \cdot \nabla\right) u_m^i - \sum_j \partial_j(\mu\rho_m e_{ij}^m) - 2\kappa\rho^m \sum_j \partial_i(\sqrt{\rho^m}^{-1}\partial_j^2\sqrt{\rho^m}) = -\nabla P, \tag{18}$$

where $\eta = \alpha_1(1+2\alpha_2)\nu\rho^m$, $\kappa = 2\alpha_2\nu^2$ and $e_{ij}^m = \partial_j u_m^i + \partial_i u_m^j$. The pressure P is defined by $(\rho_m)^2 d(\varepsilon/\rho_m)/d\rho_m$.

One can see that the second term on the left hand side corresponds to the viscosity and this equation is reduced to the Navier-Stokes (NS) equation when we set $\alpha_2 = 0$ [1]. When $\alpha_2 \neq 0$, the third term is finite and corresponds to the correction term to the NS equation. It is worth mentioning that the same correction term was, later, re-derived from a different phenomenological argument [5].

The above NS equation is reduced to the Gross-Pitaevskii (GP) equation in a special limit. Let us set $(\alpha_1, \alpha_2) = (0, 1/2)$ and $\nu = \hbar/(2m)$. As for the internal energy density, we choose $\varepsilon = V(\mathbf{x})\rho_m/m + U_0/2(\rho_m/m)^2$ where V is a trapped potential and U_0 is the interaction strength. Then the equation for the wave function which is defined as is the case with the previous section is given by the GP equation.

4. Concluding remarks

We discussed the optimization under random noise in the framework of SVM. We showed the application to the systems of particle and continuum media, and derived the Schrödinger, NS and GP equations. SVM further predicted that the NS equation can be modified.

This method has not yet been established. For example, the applicability to field theory and relativistic systems is still open question. The derivation of the uncertainty relation in SVM is now under investigation.

[1] Exactly speaking, the second coefficient of viscosity vanishes in this equation. To obtain a finite value, we have to consider the variation of entropy. See Ref. [2]

References

[1] K. Yasue 1981 J. Funct. Anal. **41** 327; E. Nelson 1985 Quantum Fluctuations (Princeton, NJ: Princeton University Press); Guerra F and Morato L M 1983 Phys. Rev. **D27** 1774; Pavon M 1995 J. Math. Phys. **36** 6774; Nagasawa M 2000 Stochastic Process in Quantum Physics (Bassel:Birkhäuser); Cresson J and Darses S 2007 J. Math. Phys. **48** 072703.
[2] Koide T and Kodama T 2012 J. Phys. **A45** 255204.
[3] Nelson E 1966 Phys. Rev. **150** 1079.
[4] Zheng W A and Meyer P -A 1984 *Sèminarie de Probabilitès XVIII, Lecture Notes in Mathematics Vol 1059, P223*, (Berlin:Springer-Verlag).
[5] Dadzie S K and Reese J M 2012 Phys. Rev. **E85** 041202.

New model describing the dynamical behaviour of penetration rates

Tohru Tashiro, Hiroe Minagawa and Michiko Chiba

E-mail: tashiro@cosmos.phys.ocha.ac.jp
Department of Physics, Ochanomizu University, 2-1-1 Ohtsuka, Bunkyo, Tokyo 112-8610, Japan

Abstract. We propose a hierarchical logistic equation as a model to describe the dynamical behaviour of a penetration rate of a prevalent stuff. In this model, a memory, how many people who already possess it a person who does not process it yet met, is considered, which does not exist in the logistic model. As an application, we apply this model to iPod sales data, and find that this model can approximate the data much better than the logistic equation.

1. Introduction

How does fashion diffuse in our society? Fashion spreads out, although we, members of society, do not aim to do so: Most of us are not out to spread it, while some may have that aim. New stuffs that are somewhat of a curiosity at the beginning will become commonplace before we notice. Some of them may disappear from our life style. This phenomenon is essentially similar to various changes of phase in matter which we cannot imagine from an interaction between atoms or molecules. That is to say, the human being is "the social atom"[1].

We want to clarify the mechanism producing the occurrence of fashion. In order to deal with this as a scientific problem, quantitative data indicating the extent of diffusing are necessary. Here, we shall employ a *penetration rate*. This has the following universality: Generally, penetration rates increase slowly at the beginning and, then, the growth reaches its maximum. Finally, the rates become saturated. This change in time is called the S-shaped curve. The logistic function, solution of the logistic equation, has often been used to analyze the rates.

As is well known, the logistic equation was proposed as an equation describing a population growth with an upper limit by Verhulst [2]. Griliches made the first adoption of the logistic equation for the dynamical behaviour of innovation diffusion [3]. After that, the logistic function has been utilized for dynamical behaviour of penetration rates for various stuffs.

However, what kind of human communication results in such a dynamical behavior? It is not a self-evident question. In this work, we shall unveil it first of all. According to our study, it is clear that the logistic equation applying to penetration rates supposes the following human communication: those who do not have a prevalent stuff start to possess it shortly after they meet people already possessing it.

Hence, a new question arises: Are we influenced by others so easily?, which is the real start-line of this paper. Therefore, we constructed a brand new model supposing more natural human communication, which implies that we extend the logistic equation. Moreover, we adopt the total number of iPod sales as the real penetration data and, then, clarify that our model can describe the behaviour much better than using the logistic equation.

2. Penetration rate in an imitating group

In this section, we unveil the human communication yielding the penetration rate which can be described by the logistic function.

Let us consider a group composed of N people. For this group, we shall apply the following rules: i) At the beginning, some people have a stuff which will diffuse in this society. ii) If those who do not possess the stuff yet (*non-adopters*) meet people who already possessing it (*adopters*) , they start to adopt it at once. iii) A non-adopter is not influenced by more than one adopter and an adopter can not influence more than one non-adopter at the same time. iv) Adopters do not part with it. Such a group can be realized by considering people existing on lattices whose number is $n \times n$ ($< N$). We suppose that he/she moves to one of the next lattices with the same probability at each step whose time interval is Δt. The third rule means that only one adopter and one non-adopter can share a lattice.

Here, we respectively set the number of adopters and non-adopters at ith step as P_i and Q_i. Therefore, it is natural to consider that a probability to meet adopters or non-adopters is proportional to each number of them: The probability at ith step to meet adopters and non-adopters is P_i/n^2 are Q_i/n^2, respectively. Indeed, a probability to meet nobody at each step is $(n^2 - P_i - Q_i)/n^2 = 1 - N/n^2$.

Therefore, $(P_i/n^2) \times Q_i$ people of non-adopters become adopters at the next step, so that we can obtain the following recursion formulae:

$$P_{i+1} = P_i + \frac{P_i}{n^2}Q_i \ , \ Q_{i+1} = Q_i - \frac{P_i}{n^2}Q_i \ . \tag{1}$$

We shall define the number of them at t as $P(t) = P(i \cdot \Delta t) \equiv P_i$ and $Q(t) = Q(i \cdot \Delta t) \equiv Q_i$. Here, we take the limits as $\Delta t \to 0$ and $n \to \infty$ with $n^2 \Delta t$ fixed. By setting the fixed value as a/N and using $P(t) + Q(t) = N$, the following differential equation is derived:

$$\frac{\mathrm{d}P(t)}{\mathrm{d}t} = \frac{a}{N}\left\{N - P(t)\right\}P(t), \tag{2}$$

that is the exact logistic equation. The penetration rate $p(t) \equiv P(t)/N$ satisfies the following logistic equation

$$\frac{\mathrm{d}p(t)}{\mathrm{d}t} = a\left\{1 - p(t)\right\}p(t) \ . \tag{3}$$

3. Hierarchical logistic equation

It is clarified in the previous section that the dynamical behavior of the penetration rate depicted by the logistic function is based on *imitation*: People not possessing a prevalent stuff yet go buying it shortly after they meet people already possessing it. Here, most of us reach the same question: But are we really like that? In that rule, the definitive human psychology, when we imitate, is forgotten. That is, the memory, how many adopters we have met, is essential for us to start to possess the stuff

In order to integrate this feature, hereby, we shall extend the rules of the group of random walker by the following way: We set the number of people starting to process the stuff after they meet μ adopters at i step as Q_i^{μ}, in which we call μ as *remaining adopters number* (RAN). Indeed, if a non-adopter, whose RAN is μ, meet one of adopters, his/her RAN becomes $\mu - 1$ at the next step. We do not alter other rules. Namely, we do not consider interactions between non-adopters despite the fact that the non-adopter gets more varied.

If the maximum of RAN is m, the recursion formulae exchange into

$$P_{i+1} = P_i + \frac{P_i}{n^2} Q_i^1 \ , \ Q_{i+1}^1 = Q_i^1 - \frac{P_i}{n^2} Q_i^1 + \frac{P_i}{n^2} Q_i^2 \ , \ \cdots \ , \ Q_{i+1}^m = Q_i^m - \frac{P_i}{n^2} Q_i^m \ . \tag{4}$$

By the previous continuation of time and space, we can obtain the following differential equations:

$$\frac{\mathrm{d}P(t)}{\mathrm{d}t} = \frac{a}{N} Q^1(t)P(t) \ , \ \frac{\mathrm{d}Q^1(t)}{\mathrm{d}t} = -\frac{a}{N} P(t)Q^1(t) + \frac{a}{N} P(t)Q^2(t) \ , \ \cdots \ , \ \frac{\mathrm{d}Q^m(t)}{\mathrm{d}t} = -\frac{a}{N} P(t)Q^m(t) \tag{5}$$

where $Q^\mu(t) = Q^\mu(i \cdot \Delta t) \equiv Q_i^\mu$.

We shall call this the *hierarchical logistic equation*. Indeed, $P(t) + Q^1(t) + \cdots + Q^m(t) = N$ is always conserved.

We can solve Eq. (5) easily: The solution is $Q^m(t) = Q^m(0) \exp\left[-a \int_0^t \mathrm{d}t' P(t')\right]$. If $Q^m(0) = 0$, $Q^m(t)$ is always 0. Then, the contribution of Q^m into the differential equation of Q^{m-1} disappears, and so Q^{m-1} can be calculated similarly. Therefore, if $Q^m(0) = Q^{m-1}(0) = \cdots = Q^2(0) = 0$, $Q^m(t) = Q^{m-1}(t) = \cdots = Q^2(t) = 0$, which means the *normal* logistic equation is recovered. Namely, the hierarchical logistic equation includes the normal one.

If we use ratios of adopters and non-adopters to the total number N, the differential equations become

$$\frac{\mathrm{d}p(t)}{\mathrm{d}t} = aq^1(t)p(t) \ , \ \frac{\mathrm{d}q^1(t)}{\mathrm{d}t} = -ap(t)q^1(t) + ap(t)q^2(t) \ , \ \cdots \ , \ \frac{\mathrm{d}q^m(t)}{\mathrm{d}t} = -ap(t)q^m(t) \ , \tag{6}$$

where $q^\mu(t) \equiv Q^\mu(t)/N$.

4. Fitting the iPod sales data

Now, let us apply the hierarchical logistic equation to fitting a real data. As this data, we shall employ the iPod, created and marketed by Apple Inc., sales which can be obtained from the official Website, http://www.apple.com/. We shall use the data from November on 2001 to May on 2006, which includes only one peak. We treat N as a fitting parameter, because the number of sales is not saturated and so we cannot obtain N from the data.

Setting November on 2001 as the origin of time, we construct the cumulative sales and, then, we fit the data with the hierarchical logistic equation. If we minimize the residual sum of squares or the sum of the absolute value of error (SAE) when fitting this data, the solution of the hierarchical logistic equation does not match the data with small values.

Therefore, we shall minimize the product of SAE and the sum of the absolute value of relative error (SARE). The results are show in Tab. 1. The parameters minimize the product. SARE diminishes with increase of m: SARE with $m = 4$ reduces to nearly half that with $m = 1$. In other words, the average relative error for the hierarchical logistic model with $m = 4$, 5.3%, is about as half as that for the logistic model, 9.7%.

From the parameters with $m = 4$, we can find the following facts: the market size producing the first peak is about 66 million people; the ratio of the trend-conscious people, $q^1(0)$, is about 33%; the ratio of the cautious people, $q^2(0)$, about 66% and the ratio of the more cautious people, $q^3(0) + q^4(0)$, is about 1%.

Comparing the logistic and the hierarchical logistic model, we show the result with $m = 1$ and $m = 4$ in Fig. 1. The hierarchical logistic function with $m = 4$ matches the data which the logistic model cannot approximate.

Table 1. Parameters minimizing the product of SARE and SAE and SAREs by them. All values are rounded to a three-digit number, and so the sum of $p(0)$ and $p^{\mu}(0)$ is not equal to one.

	$m = 1$ (logistic)	$m = 2$	$m = 3$	$m = 4$
a	1.43	4.42	4.25	4.17
N [$\times 10^6$]	1.28×10^5	64.6	65.4	66.2
$p(0)$	9.26×10^{-7}	0.00207	0.00200	0.00189
$q^1(0)$	1.00	0.288	0.311	0.325
$q^2(0)$	–	0.710	0.675	0.657
$q^3(0)$	–	–	0.0125	0.00619
$q^4(0)$	–	–	–	0.00760
SARE	1.84	1.12	1.02	1.01

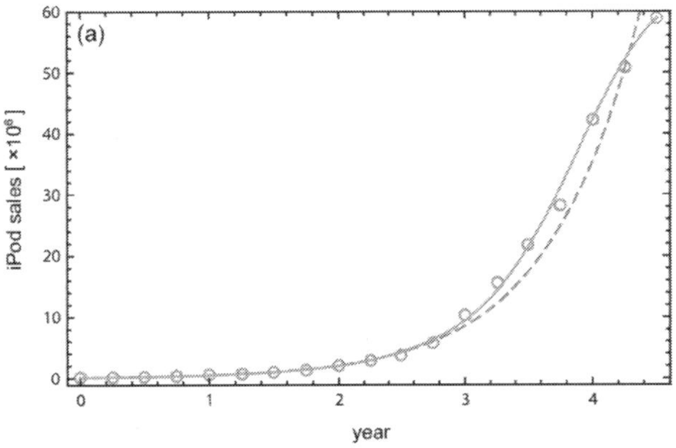

Figure 1. (color online) Cumulative iPod sales expresses by circles and fitting curves. The (purple) dashed and (light blue) full curves are solutions of the logistic and the hierarchical logistic equation ($m = 4$) with parameters minimizing the product of SARE and SAE, respectively.

5. Concluding remarks

In this work, we have unveiled the following fact that the essential human communication within a group, where the dynamical behaviour of the penetration rate can be approximated by the logistic function, is imitation; non-adopters start to process a prevalent stuff shortly after meeting adopters. Indeed, this is not natural. Thereby, we have proposed the extended logistic equation, the *hierarchical logistic equation*, considering the memory of the number of adopters they met. In addition, we have applied this model to the change of iPod sales in time, and so the model has approximated the data much better than the logistic equation.

6. Acknowledgments

We would like to thank Tomo Tanaka and members of astrophysics laboratory at Ochanomizu University for extensive discussions.

References

[1] Buchanan M 2007 *The Social Atom* (New York: Bloomsbury Press)
[2] Verhulst P P 1838 *Corr. Math. Phys* **10** 113
[3] Griliches Z 1957 *Econometrica* **25** 501

Numerical simulation of 2D and 3D compressible flows

Jaroslav Huml[1], Karel Kozel[1], Jaromír Příhoda[2]

[1] Czech Technical University in Prague, Faculty of Mechanical Engineering, Department of Technical Mathematics, Karlovo náměstí 13, 121 35 Prague, CZ
[2] Institute of Thermomechanics, Academy of Sciences of the Czech Republic, Dolejškova 1402/5, 182 00 Prague, CZ

E-mail: jaroslav.huml@fs.cvut.cz[3]

Abstract. The work deals with numerical solutions of 2D inviscid and laminar compressible flows in the GAMM channel and DCA 8% cascade, and of 3D inviscid compressible flows in a 3D modification of the GAMM channel (Swept Wing). The FVM multistage Runge-Kutta method and the Lax-Wendroff scheme (Richtmyer's form) with Jameson's artificial dissipation were applied to obtain the numerical solutions. The results are discussed and compared to other similar results and experiments.

1. Introduction

The goal of the work was to achieve experience and knowledge in the field of numerical simulations of inviscid and viscous (laminar) compressible flows and their future applications in the field of turbulence modelling, etc.

2. Mathematical models

2.1. Navier-Stokes equations

A general two-dimensional flow of a compressible laminar fluid[4] is described by the system of Navier-Stokes equations that was applied in the dimensionless form[5].

$$W_t + F_x + G_y = \frac{Ma_\infty}{Re_\infty}(R_x + S_y) \tag{1}$$

where

$$W = (\rho, \rho u, \rho v, e)^T, F = \left(\rho u, \rho u^2 + p, \rho uv, (e+p)u\right)^T, G = \left(\rho v, \rho uv, \rho v^2 + p, (e+p)v\right)^T \tag{2}$$
$$R = (0, \tau_{xx}, \tau_{xy}, u\tau_{xx} + v\tau_{xy} - q_x)^T, S = (0, \tau_{xy}, \tau_{yy}, u\tau_{xy} + v\tau_{yy} - q_y)^T$$

and

$$p = (\gamma - 1)\left[e - \frac{1}{2}\left(u^2 + v^2\right)\right], \; \tau_{ij} = \mu\left(\frac{\partial u_i}{\partial x_j} + \frac{\partial u_j}{\partial x_i}\right) - \frac{1}{3}\delta_{ij}\frac{\partial u_k}{\partial x_k}, \; q_j = -\frac{\gamma}{\gamma - 1}\frac{\mu}{Pr}\frac{\partial}{\partial x_j}\left(\frac{p}{\rho}\right) \tag{3}$$

[3] If readers have questions or comments, they can write to jar.huml@gmail.com as well.
[4] The authors assumed that the flowing medium was a calorically perfect gas – the Newtonian fluid.
[5] The inflow variables $(.)_\infty$ were used as the reference variable for the transformation of the equations to the dimensionless form (e.g. $\bar{\rho} = \rho/\rho_\infty$).

Journal of Physics: Conference Series **410** (2013) 012027 doi:10.1088/1742-6596/410/1/012027

2.2. Euler equations

Assuming an inviscid fluid ($\mu = 0$ in viscous fluxes R, S in the 2D case), the system of Euler equations is used instead of the system of Navier-Stokes equations.

$$W_t + F_x + G_y + H_z = 0 \tag{4}$$

where a vector of conservative variables W and vectors of inviscid fluxes F, G are extended by terms $\rho w, \rho uw$ and ρvw respectively. The third vector of the inviscid flux in the direction of z-axis is

$$H = \left(\rho w, \rho uw, \rho vw, \rho w^2 + p, (e + p)w\right)^T. \tag{5}$$

3. Numerical methods

For simulation of the flows mentioned previously, two finite volume method numerical schemes[6] were applied to the cell centered form on non-orthogonal structured grids of quadrilateral or hexahedral cells.

- Lax-Wendroff scheme in Richtmyer's predictor-corrector form (LW)
- multistage Runge-Kutta method (RK)

Both schemes were used for modelling the two-dimensional flow in the GAMM channel. For other mentioned simulations only the multistage Runge-Kutta method was used.

Each scheme was extended by including Jameson's artificial dissipation[6] to improve the stability of the method.

4. Formulation of the problems

The authors took into account the numerical simulations of 2D and 3D inviscid compressible flows in the GAMM channel and its 3D modification (denoted as Swept Wing) and 2D laminar compressible flows in a DCA 8% cascade. The outlines of all computational domains are shown in Figures 1 and 2.

Figure 1. GAMM channel and its three-dimensional modification Swept Wing

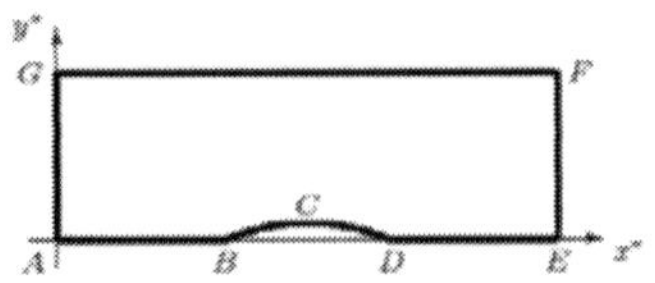

$A = (0,0)$
$B = (1,0)$
$C = (1.5,0.1)$
$D = (2,0)$
$E = (3,0)$
$F = (3,1)$
$G = (0,1)$

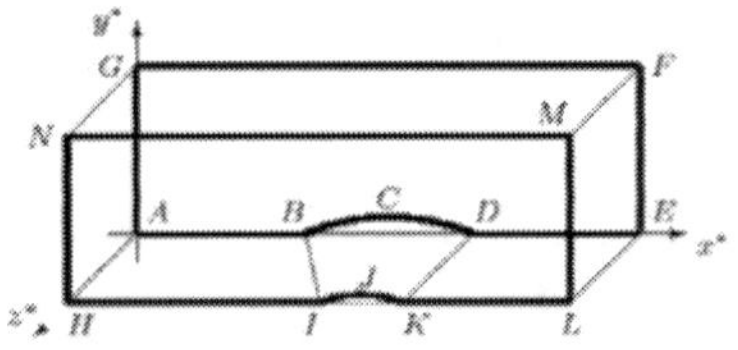

$A = (0,0,0)$ $H = (0,0,1)$
$B = (1,0,0)$ $I = (1.5,0,1)$
$C = (1.5,0.1,0)$ $J = (1.75,0.04,1)$
$D = (2,0,0)$ $K = (2,0,1)$
$E = (3,0,0)$ $L = (3,0,1)$
$F = (3,1,0)$ $M = (3,1,1)$
$G = (0,1,0)$ $N = (0,1,1)$

Figure 2. DCA 8% cascade

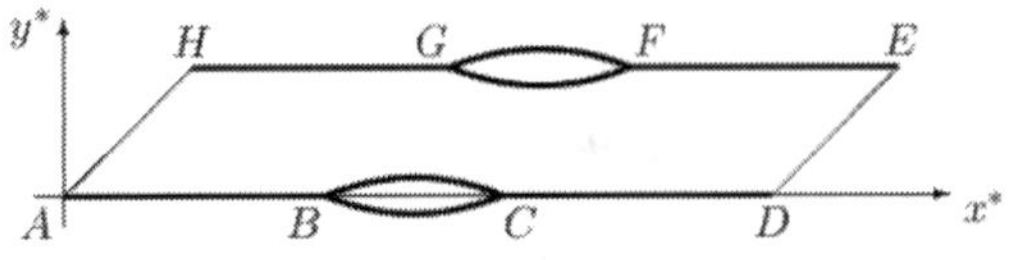

$A = (0,0)$ $E = (4 + \frac{\sqrt{2}}{2}, \frac{\sqrt{2}}{2})$
$B = (1.5,0)$ $F = (2.5 + \frac{\sqrt{2}}{2}, \frac{\sqrt{2}}{2})$
$C = (2.5,0)$ $G = (1.5 + \frac{\sqrt{2}}{2}, \frac{\sqrt{2}}{2})$
$D = (4,0)$ $H = (\frac{\sqrt{2}}{2}, \frac{\sqrt{2}}{2})$

[6] The numerical schemes and artificial dissipation used are presented in detail, e.g. in [4] or in [6].

4.1. Boundary conditions

- **inlet** : $\rho_1 = 1$, $u_1 = Ma_1 \cos\alpha$, $v_1 = Ma_1 \sin\alpha$, $(w_1 = 0,)$ p_1 was extrapolated from the flow field and e_1 was calculated using the equation of state. The angle of attack α was non-zero only for the flows in the DCA 8% cascade.
- **outlet** : p_2 was prescribed and the other components were extrapolated from the flow field or calculated.
- **solid wall**: velocity components were prescribed so that the sum of velocity vectors equals zero (viscous flows – $u = v = 0$) or equals zero in their tangential component (inviscid flows – $(u, v, w)\vec{n} = 0$).
- **periodicity**: the value of the variable in the cell at the bottom/top part of the boundary corresponds to the value in the cell from the flow field near the top/bottom part of the boundary.

5. Numerical results and discussion

For numerical simulation of 2D inviscid compressible flows in the GAMM channel, the authors applied both numerical schemes (*LW* and *RK*) on a structured non-orthogonal mesh with 240×50 cells and compared with J. Fürst's results (e.g. [5]) obtained by *WLSQR* scheme on two different computational grids with 150×50 and 300×100 cells – see Figure 3. If we compared the maximums of Mach number on the lower wall, we would found out that the values $Ma_{max} = 1.42$ (*LW*) and 1.39 (*RK*) are larger than the results by J. Fürst ($Ma_{max} = 1.29$ or 1.32 – the fine mesh) and those presented in literature (1.37).

Figure 3. Inviscid compressible flow in the GAMM channel at $Ma_\infty = 0.675$: Mach number isolines – *RK* scheme, mesh with 240×50 cells (left) and *WLSQR* scheme [5], mesh with 300×100 cells (right).

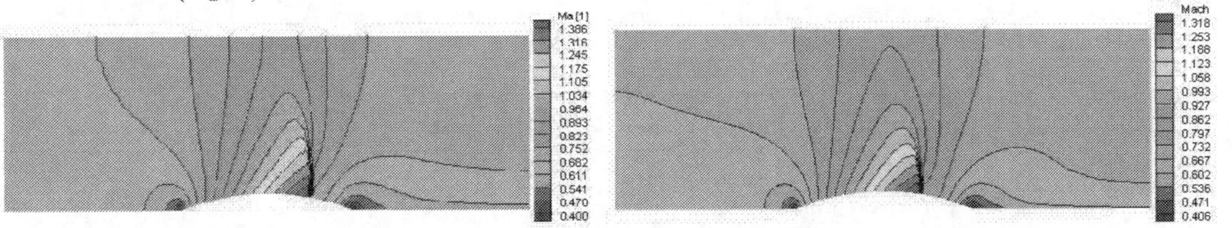

In the case of 3D inviscid compressible flows around the Swept Wing, *RK* scheme and a structured non-orthogonal grid with $110 \times 30 \times 10$ cells were used only and compared with J. Holman's results ([3]) gained by *WLSQR* scheme on a mesh with $180 \times 35 \times 35$ cells (see Figure 5). The comparison shows a finer mesh should have been used for better detection of the Zierep singularity.

The authors applied *RK* scheme and a non-orthogonal structured grids with 170×120 cells for modelling of 2D inviscid and laminar transonic flows in the DCA 8% cascade at different values of inlet Mach Ma_∞, Reynolds numbers Re_∞ and angles of attack α. The results were compared with P. Pořízková results ([7]) and the experiment carried out by R. Dvořák ([1]) at the Institute of Thermodynamics of Academy of Sciences of the Czech Republic.

Based on the comparisons with other similar numerical results or experiments, the software developed by the authors has proved itself (the good mutual agreement in all the simulated cases was achieved) and it can be used in the future.

6. Conclusions

This article presents a few results achieved by using the software with the implemented FVM Lax-Wendroff scheme and multistage Runge-Kutta method with Jameson's artificial dissipation

Figure 4. Inviscid compressible flow around the Swept Wing at $Ma_\infty = 0.675$: Mach number isosurface – RK scheme, mesh with $110 \times 30 \times 10$ cells (left) and $WLSQR$ scheme [3], mesh with $180 \times 35 \times 35$ cells (right).

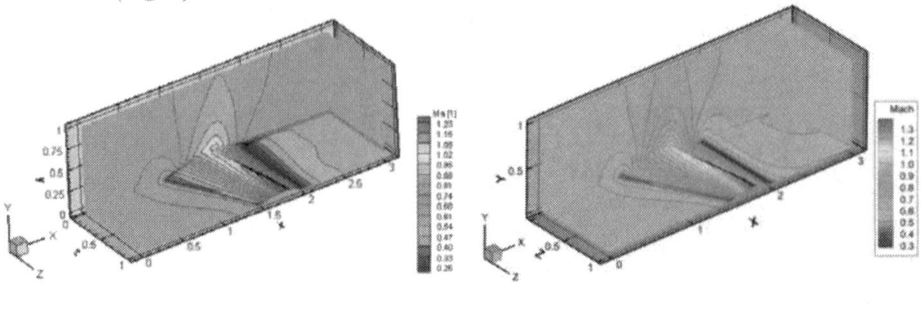

Figure 5. Laminar compressible flow in the DCA 8% cascade at $Ma_\infty = 1.1$, $Re_\infty = 2.1{\cdot}10^6$ and $\alpha = 0°$: Mach number isolines – RK scheme, mesh with 170×120 cells (left) and MacCormack scheme [7], mesh with 150×30 cells (right).

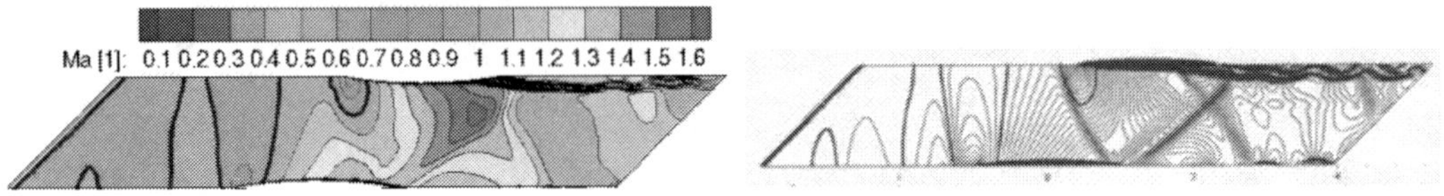

for the simulation of a 2D and 3D transonic flow of the inviscid and laminar compressible fluid in the GAMM channel and its 3D modification, and in the DCA 8% cascade. Numerical results show the good agreement with other numerical results (e.g. J. Holman [3], P. Pořízková [7]) and experimental results carried out at the Institute of Thermodynamics AS CZ.

Acknowledgments
This work was partly supported by grants No. P101/10/1329 and P101/12/1271 of the Grant Agency of the Czech Republic, grant No. IAA200760801 of the Grant Agency of the Academy of Science of the Czech Republic and grant No. SGS 10/243/OHK2/3T/12 of the Student Grant Competition of Czech Technical University in Prague.

References
[1] Dvořák R 1986 *Transonic Flows* (Prague: Academia) [in Czech]
[2] Dvořák R and Kozel K 1996 *Mathematical modelling in aerodynamics* (Prague: CTU in Prague) [in Czech]
[3] Holman J and Fürst J 2008 Comparision of High Order Methods for Transonic Inviscid Flows *Proc. Colloquium Fluid Dynamics 2008* (Prague: Institute of Thermomechanics, AS CR, v.v.i.) pp. 11-12
[4] Huml J Fürst J Kozel K and Příhoda J 2009 Numerical Solution of Subsonic and Transonic Flows in a Channel *Proc. Topical Problems of Fluid Dynamics 2009* (Prague: Institute of Thermodynamics, AS CR, v.v.i.) pp. 45-48 [in Czech]
[5] Huml J Holman J Fürst J and Kozel K 2010 Numerical Solution of 3D Inviscid Flows in a Channel *Proc. Topical Problems of Fluid Dynamics 2010* (Prague: Institute of Thermomechanics, AS CR, v.v.i.) pp. 73-76.
[6] Huml J Kozel K and Příhoda J 2012 Numerical Solution of Inviscid and Viscous Flows in Internal Aerodynamics *Proc. 11th conference on Power System Engineering, Thermodynamics & Fluid Flow 2012* (Pilsen: University of West Bohemia) pp. 73-76.
[7] Pořízková P 2009 *Numerical Solution of Compressible Flows Using Finite Volume Method* (Prague: CTU in Prague) [PhD Dissertation, in Czech]

A sampling algorithm to estimate the effect of fluctuations in particle physics data

Federico Colecchia
Brunel University, Kingston Lane, Uxbridge, Middlesex UB8 3PH, UNITED KINGDOM
E-mail: federico.colecchia@brunel.ac.uk

Abstract. Background properties in experimental particle physics are typically estimated using large data sets. However, different events can exhibit different features both because of the quantum mechanical nature of the underlying physics and due to statistical fluctuations. While signal and background fractions in a given data set can be evaluated using a maximum likelihood estimator, the shapes of the corresponding distributions are traditionally obtained using high-statistics control samples, which normally neglects the effect of fluctuations. On the other hand, if it was possible to subtract background using templates that take fluctuations into account, this would be expected to improve the resolution of observables of interest, and to reduce systematics depending on the analysis. This study is an initial step in this direction. We propose a novel algorithm inspired by the Gibbs sampler that estimates the shapes of signal and background probability density functions from a given collection of particles, using control sample templates as initial conditions and refining them to include the effect of fluctuations. Results on Monte Carlo data are presented, and the prospects for future development are discussed.

1. Introduction

While signal and background fractions in a given data set can be obtained using a maximum likelihood estimator, a traditional approach when analyzing data in particle physics consists in estimating the shapes of signal and background probability density functions (PDFs) from high-statistics control samples. Generally speaking, such templates can be used to subtract background from a set of candidate events containing signatures of a process of interest.

However, even if the physics processes can be exactly the same, both the quantum nature of the underlying physics and statistical fluctuations can produce different features in different events. For this reason, the shapes of signal and background PDFs in a collection of events of interest can deviate from the templates obtained using high-statistics control samples.

When physics analysis leads to the identification of a large enough number of candidate events, such discrepancies are often of no practical relevance. However, in some cases, most notably with reference to searches for new physics, analysis can result in only a few interesting events, and fluctuations can then affect the shapes of signal and background PDFs in ways that cannot be described using high-statistics control samples.

Two major aspects underline the innovative nature of this work. First of all, this article describes the first attempt at estimating the effect of fluctuations on the shapes of probability distributions in particle physics data by using a novel algorithm inspired by the Gibbs sampler [1]. This builds on the results of previous studies presented in [2]. The algorithm starts from control sample templates, and uses the stationary distribution of an appropriately-defined Markov Chain to estimate the effect of fluctuations on the shapes of signal and background PDFs.

Secondly, this paper proposes a new approach to background discrimination in the field. Instead of concentrating on the likelihood for entire events to contain a process of interest, individual particles inside events are assigned a probability for them to originate from signal as opposed to background. This reformulation of background discrimination in terms of a classification problem at the particle level is used to illustrate how the algorithm estimates the effect of fluctuations on the shapes of particle-level PDFs in the data.

However, the two aspects of (i) estimating the effect of fluctuations on PDF shapes and (ii) concentrating on individual particle properties to discriminate signal from background are in principle independent, and can therefore lead to distinct lines of development in the future.

For this reason, this research has two main prospective goals. First of all, this work is an initial step toward the development of techniques to study the effect of fluctuations on particle-level probability distributions inside individual events at the Large Hadron Collider (LHC). The emphasis will be on using event-specific templates to associate individual particles to signal or background on a probabilistic basis, considering the effect of fluctuations on the shapes of particle-level PDFs event by event. We argue that using different templates for different events to subtract background may improve the resolution of observables of interest and may contribute to reduce systematic uncertainties, depending on the analysis.

Secondly, from a broader perspective, this research aims to guide the future development of new background subtraction techniques based on PDFs that take into account the effect of fluctuations in a collection of events of interest. This aspect is more general and not necessarily restricted to particle-level distributions. Both lines of development refer to the LHC at this point, but they are anticipated to be of broader relevance in the field.

Finally, it should be noted that it is not the purpose of this article to present the algorithm for use in a proper physics analysis framework at the LHC. On the other hand, this study is a proof of concept of a novel approach to data analysis, illustrating the use of a data-driven technique that estimates the effect of fluctuations on the shapes of probability distributions, with an emphasis on individual particle kinematics.

2. The algorithm

One of the distinctive features of the proposed approach is a population-based view of high-energy collision events at the particle level. This aspect focuses on decomposing a collection of particles into subpopulations that are associated with different processes and that are described in terms of different PDFs.

The input data set consists of a mixture of particles, some of which originated from a hard scattering of interest, others from background. The algorithm uses properties of individual particles to decompose the mixture, iteratively sampling from posterior distributions that encode information as to which particles are more likely to originate from either process.

As compared to classical mixture models, which typically enforce given subpopulation PDF shapes, our statistical model is based on a nonparametric mixture of the form $F = \sum_{j=1}^{K} \alpha_j f_j(x)$, where the subpopulation fractions α_j ("mixture weights") satisfy $\sum_{j=1}^{K} \alpha_j = 1$. In the present application, x corresponds to particle pseudorapidity η, a kinematic variable related to particle polar angle θ in the laboratory frame in terms of $\eta = -\ln(\tan\theta/2)$, or to p_T i.e. particle transverse momentum with respect to the beam direction.

The PDFs f_j are here defined in terms of regularized histograms of x, namely based on spline interpolation of histogram bin contents[1]. This aspect is further discussed in [4] with reference to the existence of a stationary distribution for the associated Markov Chain. The symbol φ_j

[1] This implementation makes use of the ALGLIB software library [3].

denotes an estimate of PDF f_j corresponding to a splined histogram of x at a given iteration of the sampler. Estimates of f_j obtained from control samples are used to associate individual particles with signal or background on a probabilistic basis at each iteration of the algorithm ("mapping"). The distribution of x in the data set analyzed is ultimately estimated as a splined histogram averaged over iterations.

In general, given the above mixture F of probability distributions and a set of N observations $\{x_i\}_{i=1,\dots,N}$, Markov Chain Monte Carlo (MCMC) techniques can be used to cluster the latter into K groups by associating each observation with a distribution of origin. The Gibbs sampler, which as anticipated directly inspired this work, belongs to this family of numerical methods.

The pseudocode of the proposed algorithm is shown below, subscripts sig and bkg relating to signal and background, respectively. The value of quantity v at iteration t is denoted by $v^{(t)}$ throughout.

(i) **Initialization:** Set $\alpha_{bkg} = \alpha_{bkg}^{(0)} = \alpha_{sig} = \alpha_{sig}^{(0)} = 0.5$, where $\alpha_{bkg} = \alpha_0$ and $\alpha_{sig} = \alpha_1 = 1-\alpha_{bkg}$. Initial estimates $\varphi_j^{(0)}$ of the subpopulation PDFs f_j, $j \in \{1,2\}$, are given by splined one-dimensional histograms of η and p_T obtained from a high-statistics control sample.

(ii) **Iteration** t**:**

(a) Generate $z_{ij}^{(t)}$ ("allocation variables") for all particles i and distributions j according to $P(z_{ij}^{(t)} = 1|\alpha_j^{(t-1)}, \varphi_j^{(0)}, x_i) = \alpha_j^{(t-1)}\varphi_j^{(0)}(x_i)/\sum_{l=0}^{K-1} \alpha_l^{(t-1)}\varphi_l^{(0)}(x_i)$, $K = 2$. Both the nonparametric treatment of the PDFs and the use of $\varphi_j^{(0)}$ instead of $\varphi_j^{(t-1)}$ for mapping distinguish this implementation from the classical Gibbs sampler for mixture models.

(b) Set $\alpha_j^{(t)} = \sum_i z_{ij}^{(t-1)}/N$, $j \in \{1,2\}$.

The sampler estimates the shapes of signal and background subpopulations in an input collection of particles, based on the data as well as on initial conditions on the subpopulation PDFs provided by control samples. As opposed to the traditional Gibbs sampler for mixture models, where the stationary distribution of the corresponding Markov Chain relates to the joint posterior of the allocation variables and of the subpopulation PDF parameters, this algorithm samples from the posterior probability of z_{ij}, and PDFs are kept fixed at control sample estimates.

This article reports results obtained on a Monte Carlo data set generated using Pythia 8.140 [5] [6]. The data set consists of charged particles with 2 GeV/c $< p_T <$ 5 GeV/c from 50 signal $gg \to t\bar{t}$ events from pp interactions at $\sqrt{s} = 14$ TeV, superimposed on 350 soft QCD interactions, so called Minimum Bias events, to simulate background. The data set comprises 1,635 particles, out of which 1,269 originate from signal and 366 from background, corresponding to a fraction of background particles of ~ 0.22. In addition to reducing the correlation between particle η and p_T to a negligible level, the restriction to the kinematic window 2 GeV/c $< p_T <$ 5 GeV/c is necessary for p_T to contribute to discrimination between signal and background. If all p_T values were allowed, the difference between signal and background p_T distributions would in fact be too small for p_T to have any discriminating power.

The higher-statistics Monte Carlo data set ("control sample") that was used to obtain initial estimates $\varphi_j^{(0)}$ of the subpopulation PDFs f_j consists of a total of $\sim 48,000$ particles. It was obtained by generating 1,000 $gg \to t\bar{t}$ events and superimposing 7 Minimum Bias events per signal interaction. Charged particles were considered in the kinematic range 2 GeV/c $< p_T <$ 5 GeV/c. Both samples are exclusively used for illustrative purposes in this article.

The algorithm correctly estimated the fraction of background particles contained in the input data set with a shorter burn-in phase than typically reported in the MCMC literature. In fact, since the sampler essentially refines control sample PDF templates, it already starts from reasonable knowledge of the target PDFs. For this reason, the equilibrium distribution of the

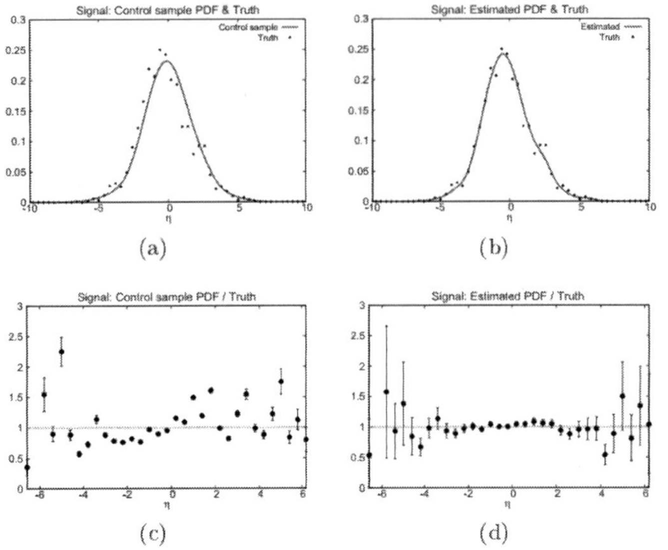

Figure 1. (a) True signal η distribution normalized to unit area (points) compared to the control sample template (curve). (b) The same true distribution (points) compared to the PDF estimated by the algorithm (curve). (c) Ratio between control sample and true distribution. (d) Ratio between estimated PDF and true distribution.

Markov Chain, although corresponding to an improved description of the PDF shapes that takes fluctuations into account, is normally close to the initial conditions.

Estimated mixture weights can be biased because control sample PDFs are used for mapping instead of the unknown true distributions. However, as previously noticed, obtaining the signal and background fractions is not the purpose of this algorithm, and estimated PDFs were observed to be remarkably unaffected by a possible bias on the mixture weights.

The algorithm was run for 1,000 iterations and probabilities were averaged over the last 100. No significant gain was found in this study in choosing a higher number of iterations.

The statistical model is well defined, i.e. the stationary distribution of the Markov Chain exists and is unique, which is a required condition for the estimated signal and background PDF shapes to reflect meaningful properties of the data set analyzed. This was verified explicitly in two ways. First of all, different initial conditions were used for the mixture weights in the range $[0.1, 0.9]$, and it was confirmed that the estimated signal and background PDFs did not vary appreciably, as expected. Secondly, the algorithm was run repeatedly using different pseudorandom seeds, and the estimated PDFs were again observed to be essentially unaffected.

Figure 1 illustrates the effect of fluctuations by comparing the shape of the true signal η distribution to the corresponding control sample template (a) and to the PDF estimated by the algorithm (b). Points correspond to the true distribution normalized to unit area, while curves result from spline interpolation of control sample template (a) and estimated PDF (b).

The actual distribution in the data differs notably from the high-statistics control sample template. Not only is its peak shifted toward lower values, but its shape is not symmetric, as shown by the structure at $\eta \simeq 2$ in figure 1 (b). As anticipated, the algorithm can estimate these features thanks to the use of a nonparametric model. This underlines the importance of abandoning classical mixture models in favor of a more flexible formulation that can in principle accommodate any deviation in the PDF shapes due to fluctuations in the data. Figures 1 (c) and (d) display the corresponding ratios of control sample to true PDF, and of estimated to true PDF, respectively. As it can be noticed, the PDF obtained using the algorithm provides a much more faithful description of the true distribution than the control sample template. A detailed autocorrelation analysis of the Markov Chain is outside the scope of this work and is left for future studies. Results were also cross-checked on a second Monte Carlo data set corresponding to a different fraction of background particles, as well as on toy Monte Carlo data [4].

3. Conclusions and outlook

We have described a novel sampling algorithm that uses a Markov Chain to estimate the shapes of signal and background PDFs from an input data set taking fluctuations into account, starting from initial conditions provided by high-statistics control samples. This article concentrates on the use of discriminating variables associated with particle-level kinematics.

It is worth noticing that the algorithm is presented here neither as a classifier nor as a tool for physics analysis at the LHC. This contribution is a proof of concept of a new approach to data analysis in particle physics that aims to estimate the effect of fluctuations on the shapes of probability distributions, while contextually proposing a reformulation of background discrimination as a classification problem at the particle level.

In particular, this study is a first step toward the development of future mixture decomposition techniques to quantify the effect of fluctuations on the shapes of particle-level probability distributions inside individual events at the LHC. We argue that using event-by-event templates to subtract background may improve the resolution of observables of interest, may increase sensitivity in searches for new physics, and may reduce systematic uncertainties depending on the analysis. Since the number of particles in this study is in line with typical LHC charged particle multiplicities, these results are an encouraging starting point. It should also be noted that, although the present version of the algorithm concentrates on mixtures of one signal and one background subpopulation, this method can in principle be extended to more complex scenarios.

It is worth noticing that efforts to eliminate noise in event-by-event analysis of multiparticle production documented in the heavy ion literature [7] are fundamentally different from this approach. As opposed to traditional techniques whereby functional forms are enforced on the PDFs and parameters are subsequently adjusted within the templates, this approach is based on a nonparametric statistical model that can in principle describe any deviation from the shapes of control sample templates due to fluctuations. Furthermore, this method builds on a novel population-based view of particle physics events, and reformulates background discrimination as a classification problem at the level of individual particles.

For the sake of completeness, it should be mentioned that a conceptual issue can in principle arise when color connection is involved [4], but it is our opinion that this technique can anyway cover many situations of practical relevance. It should also be noted that, due to its iterative nature, the algorithm can be outperformed by established methods in terms of execution time. However, the running time in this study was $\sim$ 20 s on a 2 GHz Intel Processor with 1 GB RAM, and so still reasonable for offline use. Improvements may be possible in this respect [4].

4. Acknowledgments

The author wishes to thank the High Energy Physics Group at Brunel University for a stimulating environment and for many fruitful discussions. Particular gratitude also goes to the High Energy Physics Group at University College London for their hospitality at an earlier stage, and to Prof. Jonathan M. Butterworth for his precious comments. The author also wishes to thank Prof. Trevor Sweeting and Dr. Alexandros Beskos at the UCL Department of Statistical Science for fruitful discussions. Particular gratitude also goes to Prof. Carsten Peterson and to Prof. Leif Lönnblad at the Department of Astronomy and Theoretical Physics at Lund University.

References

[1] Geman S and Geman D 1984 *IEEE T. Pattern Anal.* **6** 721
[2] Colecchia F 2012 *J. Phys.: Conf. Ser.* **368** 012031
[3] ALGLIB (www.alglib.net), Sergey Bochkanov and Vladimir Bystritsky
[4] Colecchia F 2012 (Preprint arXiv:1205.5886v1 [physics.data-an])
[5] Sjöstrand T, Mrenna S and Skands P 2006 *J. High Energy Phys.* JHEP05(2006)026
[6] Sjöstrand T, Mrenna S and Skands P 2008 *Comput. Phys. Comm.* **178**
[7] Voloshin S A, Koch V and Ritter H G 1999 *Phys Rev* C **60** 024901

FDTD simulation of millimeter-wave corrugated waveguides with cylindrical symmetry model

Y. Tamura[1], H. Nakamura[2], T. Okamura[3], N. Kashima[3], S. Fujiwara[4] and S. Kubo[2]

[1] Department of Intelligence and Informatics, Konan University, Okamoto, Higashinada, Kobe, 658-8501, Japan

[2] Department of Helical Plasma Research, National Institute for Fusion Science, Oroshi, Toki, 509-5292, Japan

[3] Department of Energy Engineering and Science, Nagoya University, Frocho, Chikusa, Nagoya, 464-8603, Japan

[4] Department of Macromolecular Science and Engineering, Kyoto Institute of Technology, Matsugasaki, Sakyo, Kyoto, 606-8585, Japan

E-mail: `tamura@konan-u.ac.jp`

Abstract. The purpose of this study is to calculate the relationship between the circular shape of a corrugated waveguide and its electrical transfer characteristics by the finite-difference time-domain (FDTD) method. To improve the waveguide to transport with a lower loss of energy than the previous design, we optimize the waveguide structure by the time evolution of the electromagnetic fields in the waveguide. However, the calculation in a 3D model is difficult because of the large data size and the long calculation run time. Therefore, we used a cylindrical symmetry model. As a result of this simulation, a correlation of the electromagnetic field in the input source position and the output position has been calculated.

1. Introduction

In nuclear fusion, a high-density- and high-temperature environment is required. Therefore, it is necessary to provide the low-loss transmission of electromagnetic waves for the electron cyclotron heating (ECH) system to realize nuclear fusion. As one of the brilliant successes, it was found that corrugating the walls reduces the transmission loss further not only in straight runs but also in bends and that the corrugation makes the waveguide robust against small deformations [1]. However, the distance from the energy source to nuclear fusion reactor is generally very long because a nuclear fusion reactor and peripheral devices are so large that it is difficult to set the energy source, that is, the gyrotron, near the reactor. Therefore, small deformations in the waveguide can have a deleterious effect on energy transmission. In this study, we estimate the influences of structural errors on the corrugated circular waveguide by numerical simulation using the FDTD method [2].

2. Simulation Model

In the FDTD method, the 3D Maxwell equation is solved numerically, in the perfect matched layer (PML) [3] boundary condition, for the same situation as the experimental configuration. However, this equation is difficult to solve directly in the 3D environment because the data size becomes big and the simulation time is quite long. To reduce these costs, we use a 2D model. Equations (1) - (6) show the Maxwell equations in the cylindrical symmetry model.

$$\frac{1}{r}\frac{\partial H_z}{\partial \theta} - \frac{\partial H_\theta}{\partial z} = \epsilon_0 \frac{\partial E_r}{\partial t}, \tag{1}$$

$$\frac{\partial H_r}{\partial z} - \frac{\partial H_z}{\partial r} = \epsilon_0 \frac{\partial E_\theta}{\partial t}, \tag{2}$$

$$\frac{1}{r}\frac{\partial (rH_\theta)}{\partial r} - \frac{1}{r}\frac{\partial H_r}{\partial \theta} = \epsilon_0 \frac{\partial E_z}{\partial t}, \tag{3}$$

$$\frac{1}{r}\frac{\partial E_z}{\partial \theta} - \frac{\partial E_\theta}{\partial z} = -\mu_0 \frac{\partial H_r}{\partial t}, \tag{4}$$

$$\frac{\partial E_r}{\partial z} - \frac{\partial E_z}{\partial r} = -\mu_0 \frac{\partial H_\theta}{\partial t}, \tag{5}$$

$$\frac{1}{r}\frac{\partial (rE_\theta)}{\partial r} - \frac{1}{r}\frac{\partial E_r}{\partial \theta} = -\mu_0 \frac{\partial H_z}{\partial t}, \tag{6}$$

where E is the electric field and H is the magnetic field; the suffixes r, z and θ denote each component in each direction; and ϵ_0 and μ_0 are permittivity and permeability under vacuum, respectively.

Figure 1 shows the corrugated waveguide geometry in this simulation. The walls of the waveguide are perfect conductors and the boundary at both sides of waveguide is the PML boundary. The actual material is aluminum; however, calculating with an aluminum wall is difficult because thin skin depth of the aluminum makes lattice size very small. The inside region of waveguide is under vacuum.

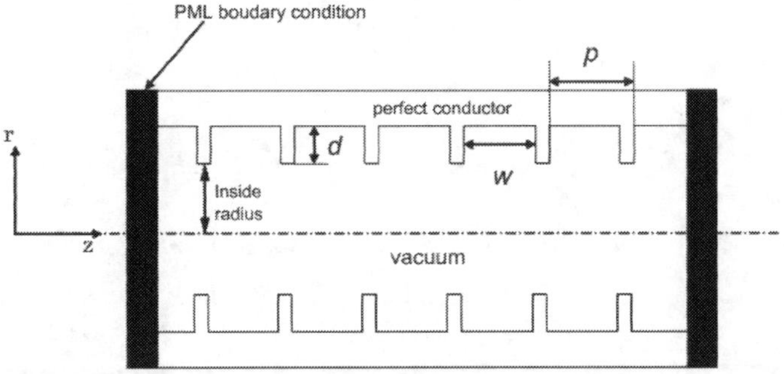

Figure 1. Simulation model of the corrugated waveguide. w is width of a groove, d is depth of a groove, and p is pitch.

3. Results

Table 1 shows the parameters used in this simulation. For estimating the influence of structural error, the shape error described in eq. (7) is appended.

$$d^{error} = d + a\sin(\frac{2\pi}{p^{error}}z), \tag{7}$$

where a is the amplitude of the error and p^{error} is the period of the error. In the simulation, we used several values of a, p^{error}, and d, and inputted the transverse electric mode wave (TM01 mode, electromagnetic field is constant in the θ-direction) at position E^{in} as shown in fig. 2.

Table 1. Parameters in FDTD simulation of the corrugated waveguide

Element	Parameter
Input source frequency	84 GHz
Wave length	3.57 mm
Inside radius	15.9 mm
Depth of groove	0.5 mm, 0.75 mm, 1.0 mm
Width of groove	1.0 mm
Amplitude of error	0.1 mm, 0.3 mm, 0.5 mm, no error
Pitch of error	10 mm, 50 mm, 100 mm
Time step	8.3e-14 sec
Total simulation time	16.7 nsec
Lattice size (r-direction and z-direction)	0.05 mm
PML thickness	1.0 mm

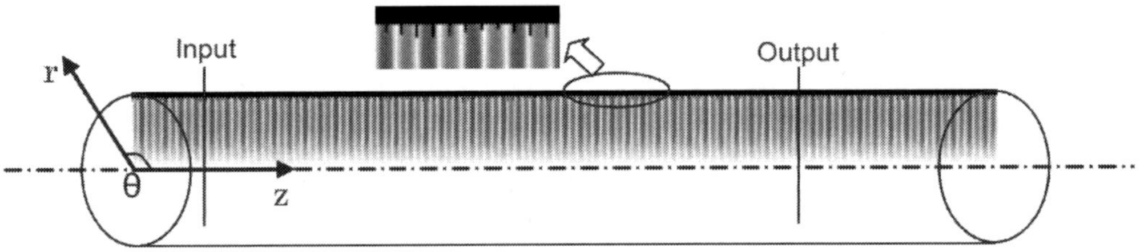

Figure 2. Simulation results in which d is 0.75 mm, a is 0.3 mm, and p^{error} is 10 mm. The length of the z-direction in the simulation model is 200 mm. The "input" and "output" show the lines for calculating the correlation factor.

Figure 2 shows one example of simulation results. Plane waves are transmitted; however small turbulence can occur in local areas.

For estimating simulation results and turbulence, we introduce the correlation factor eq. (8).

$$C = \frac{\sum_{i=1}^{n}(E_i^{in} - \bar{E^{in}})(E^{out} - \bar{E_i^{out}})}{\sqrt{\sum_{i=1}^{n}(E_i^{in} - \bar{E^{out}})^2}\sqrt{\sum_{i=1}^{n}(E_i^{in} - \bar{E^{out}})^2}}, \tag{8}$$

where C is the correlation factor, n is the number of data points (inside radius/lattice size) in the r-direction, and E^{in} and E^{out} are the values of the electric field in the input and output positions, respectively. The output position is apart from the edge of the simulation box to remove the influence of PML. The distance between input and output position is 125 mm, and 35 waves are included in this area. Therefore, the phase is the same for both the input and output positions, and therefore, eq. (8) indicates to calculates the auto correlation factor.

Figure 3 shows the values obtained for the dependency of depth without error, and indicates that $d = 0.75$ mm is the best parameter. This depth parameter is used in later estimations. The results obtained for structural error (Figures 4 and 5) show that the worse the error is, the smaller is the correlation factor. These results can be understood intuitively. However, this estimation is just compared with the error and shape change of the electric field, and therefore, a comparison with experimental results is necessary.

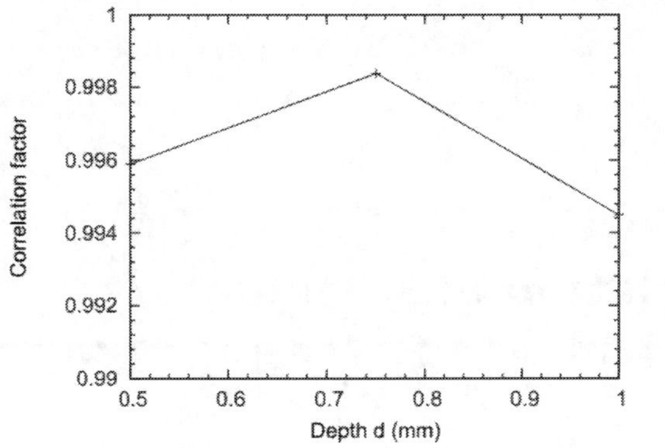

Figure 3. Dependency of depth d without error.

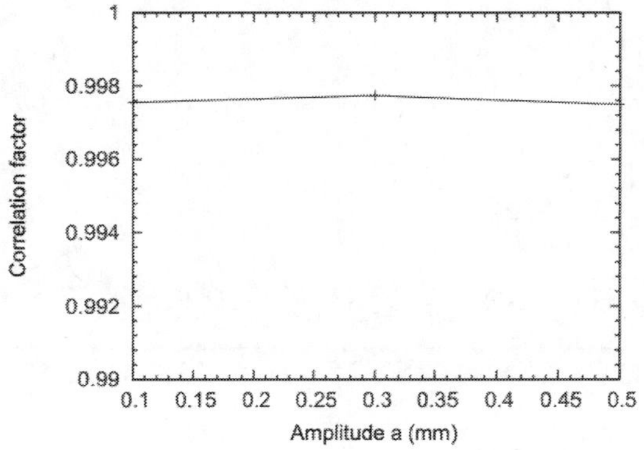

Figure 4. Dependency of error amplitude a. $p = 50$ mm and $d = 0.75$ mm.

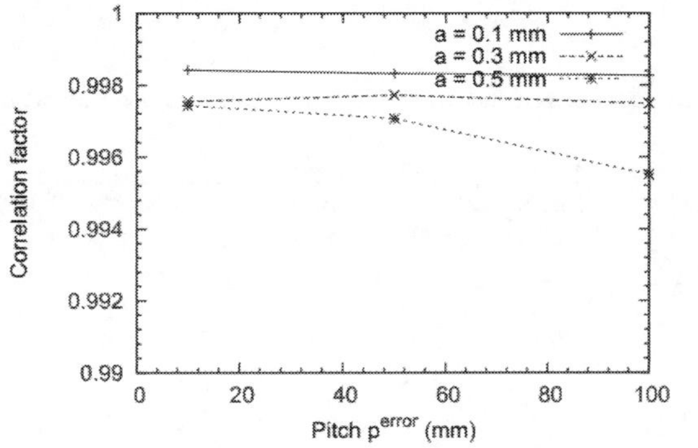

Figure 5. Dependency of pitch p. $d = 0.75$ mm.

4. Conclusion

We applied cylindrical FDTD simulation to a corrugated waveguide using structural error parameters. As the results of this simulation, we established the relationship between structural variance and electromagnetic waveform shape as a correlation factor. In this simulation, we do not show transmission loss directly. In an actual nuclear fusion device, electromagnetic waves are emitted to the nuclear fusion reactor. For an estimate of the validity of this simulation, it is necessary to compare the simulated results with experimental results from the actual system. Moreover, the input source is a TM01 mode wave; however, the actual input source is the HE11 mode. It is difficult to apply this mode wave in a cylindrical symmetry model. Therefore, it will be necessary to simulate it in a 3D model and compare it with a 2D model.

Acknowledgement

This work is supported by KAKENHI (22500114).

References

[1] J. L. Doane 2008 *Fusion Science Technology* **53** 159
[2] K. Kunz and R. Luebbers 1993 The Finite Difference Time Domain Method for Electromagnetics *Florida: CRC Press*
[3] J. P. Berenger 1994 *Journal of Computational Physics* **114** 2

Self-adaptive differential evolution algorithm incorporating local search for protein-ligand docking

Hwan Won Chung[1], Seung Joo Cho[2], Kwang-Ryeol Lee[1] and Kyu-Hwan Lee[1]

[1] Center for Computational Science, Korea Institute of Science and Technology, South Korea
[2] Department of Cellular and Molecular Medicine, University of Chosun, South Korea

E-mail: hwchung@kist.re.kr

Abstract. Differential Evolution (DE) algorithm is powerful in optimization problems over several real parameters. DE depends on strategies to generate new trial solutions and the associated parameter values for searching performance. In self-adaptive DE, the automatic learning about previous evolution was used to determine the best mutation strategy and its parameter settings. By combining the self-adaptive DE and Hooke Jeeves local search, we developed a new docking method named SADock (**S**trategy **A**daptation **Dock**) with the help of AutoDock4 scoring function. As the accuracy and performance of SADock was evaluated in self-docking using the Astex diverse set, the introduced SADock showed better success ratio (89%) than the success ratio (60%) of the Lamarckian genetic algorithm (LGA) of AutoDock4. The self-adapting scheme enabled our new docking method to converge fast and to be robust through the various docking problems.

1. Introduction

During the process to discover a new drug, the molecular docking is used frequently to find new promising hit molecules and also to optimize the binding affinities of lead compounds. The molecular docking consists of two components [1]. The first is a scoring function to define and measure the interactions which a ligand experiences when it approaches to a receptor. The second is a searching method to find the numerous poses that the ligand will take to fit into the active site of the receptor. Many searching methods are based on stochastic optimizations such as genetic algorithm, particle swarm optimization, and DE, etc[1, 2].

DE is an evolutionary strategy that develops a population of real solution vectors using operators such as mutation, crossover, and selection[3]. The performance of DE is significantly impacted by three major parameters: the number of solution vectors in the population (NP), mutation scaling (F), and crossover rate (CR)[3].

The present paper is focused on applying the self-adapting differential evolution [4] to the molecular docking problem using AutoDock4 scoring [1, 5] and estimating new searching method. Subsequently, we will describe the basic DE algorithm, its enhancement by self-adapting parameters, and explanation about the molecular docking problem. After introducing docking experiment, benchmark docking result and its analysis will be explained. At last, some concluding remarks will be shown.

Journal of Physics: Conference Series **410** (2013) 012030 doi:10.1088/1742-6596/410/1/012030

2. Methods

2.1. Basic Differential Evolution

DE evolves a population using evolutionary operators and a scoring function to find optimal solutions for the given problem. The population consists of NP target vectors $x_{i,G}, i = 1, \ldots, NP$ with G generation number and NP population size.

After initialization of the population, a target vector is mutated to a mutation vector $u_{i,G}$ with the difference vector of two target vectors as $u_{i,G} = x_{r1,G} + F(x_{r2,G} - x_{r3,G})$, $r1 \neq r2 \neq r3$ where F is a mutation scaling factor and $r1, r2, r3$ are random indexes. Next, the mutant vector $u_{i,G}$ will undergo a recombination with the current target vector $x_{i,G}$ to generate a trial vector $v_{i,G}$ in the crossover operator as follows: if $rand < CR$ or $i = rnd_j$ then $v_{i,G} = u_{i,G}$ else $v_{i,G} = x_{i,G}$. Here,the CR is a crossover rate, the $rand$ is a random number in $0, 1$, and the rnd_j is one index of the random target vector. The selection operator of DE is greedy; DE will choose the trial vector only if the fitness value of the trial vector is better than that of the current target vector.

2.2. Self-adaptive Differential Evolution

According to A.K. Qin[4], DE can increase performance using the adaptive change of its mutation strategies and the parameter values. SADock was implemented to include such a mutation strategy pool consisting of four mutation methods with diverse behaviors so that each strategy could show its strength at the different stage of the evolution. The probability to choose each strategy was updated from the previous learning of successful trial solutions.

In SADock, the scaling parameter F was generated from a normal distribution $N(0.5, 0.3)$ and the crossover parameter CR was also generated from another normal distribution $N(CR_{m_k}, 0.1)$. With the first normal distribution, random F values were sampled and assigned to the target vectors in the current population. Then, $N(CR_{m_k}, 0.1)$ was used to create random CR values and these values were assigned to the target vectors according to their strategy types. The initial value of the CR_{m_k} for the kth strategy is assigned to 0.5. Qin also holds that CR value is critical for optimal performance and different strategies may need different CR values[4]. Therefore, the CR_m values were also adapted by the previous learning experiences.

The NP was a user-defined parameter because it depends on the difficulty of a given problem. For the more detailed explanation about the self-adaptive DE, see [4].

2.3. Representation and Scoring Function

The molecular docking searches for optimal binding poses with the lowest energy of a ligand in the problem space defined by three movements: translation, orientation, and conformation[1]. At first, the ligand will define its position by three coordinates of t_x, t_y,t_z. Next, the ligand can change its orientation using a rotation axis and a rotation angle. The rotation axis is unit vector of n_x, n_y, $n_z \in \{0, 1\}$ and the ligand will rotate as much as the rotation angle $\alpha \in \{-\pi, \pi\}$ around the rotation axis. Moreover, the ligand can take conformations corresponding to the combination of all these torsion angles (T) with the angle $\in \{-\pi, \pi\}$. So, one pose will have $7 + T$ number of total parameters to be optimized.

Autodock4 scoring function was used to calculate empirical energy to estimate a docked conformation[1]. This scoring function evaluates the binding energy between the ligand and the receptor using the intermolecular energy, the internal energy of the ligand, and the desolvation energy: $E_{total}(X) = E_{inter} + E_{internal} + E_{desolvation}$ where E_{inter} consists of van der Waals force, hydrogen bonding, and electrostatic potential and $E_{internal}$ also comprises these three forces. The last term models a charge-based desolvation and hydrophobic effect[6, 5].

Table 1. Molecular docking result of four algorithms in terms of success number and ratio.

# torsional range[a]	# complex	LGA	SADock(m1)	SADock(m4)	SADock(m4,LS)
0 ~ 5 (A)	39	32 (0.82[b])	**35 (0.89)**[c]	**35 (0.89)**	**35 (0.89)**
6 ~ 10 (B)	40	27 (0.67)	32 (0.80)	**36 (0.90)**	35 (0.87)
11 ~ 15 (C)	6	1 (0.16)	4 (0.66)	**5 (0.83)**	**5 (0.83)**
total	85	60 (0.70)	71 (0.83)	**76 (0.89)**	75 (0.88)

[a]All the complexes are classified into three groups (A,B, and C) according to the number of torsional angles. [b]After docking success number, success ratio are surrounded by parentheses. [c]The cells with the bold type have the best success ratio.

3. Docking Experiment

We used Astex diverse set[7] with 85 protein-ligand complexes in the docking experiment. Gasteiger partial charge was generated for every ligand and protein with AutodockTools. The grid box of each receptor was calculated for its corresponding ligand to be moved freely in its extended state. As a success criterion of the docking, the root mean squared deviation ($RMSD$) was calculated between the reference structure and predicted conformation. When the $RMSD$ was less than 2.0, the associated docking run was considered as a success.

We implemented three different algorithms for comparison: SADock(m1), SADock(m4), SADock(m4,LS). Four mutation strategies used in our strategy adaptation were *DE/rand/1*, *DE/current to best/2*, *DE/best/1*, and *DE/rand/2*. Since the *DE/current to best/2* was the best strategy in a preliminary experiment, SADock(m1) used only this mutation. SADock(m4) had four mutation strategies. In SADock(m4,LS), Hooke Jeeves local search[8] was embedded to refine the current global best. This refinement was applied only when the new global best was found and a random number was less than 0.01.

Each docking run was iterated for 10 times and one conformation with the best (lowest) energy was chosen for comparison. LGA was assigned the population size of 150 and the evaluation number of 1,000,000 and another parameters were chosen with the default values. In the all SADocks, the learning period were 50 and the CR_m learning period was 20. However, the evaluation number for all SADock algorithms was assigned to be one fifth (200,000) of that of LGA.

4. Results and Discussion

Table 1 shows the molecular docking result of four algorithms and the docking result is summarized by three groups according to the number of rotatable bonds. In group A, all SADock algorithms produced the same best success rate (89%) whereas LGA produced the lower success rate (82%) than the others. In group B, SADock(m4) yielded the best success rate of 90%. LGA and SADock(m1) showed the distinct weak performance in this group. In group C, SADock(m4) and SADock(m4,LS) brought about the best result (83%). In a whole, SADock(m4) algorithm produced the best success ratio (89%). SADock(m4,LS) did not show significant improvement than SADock(m4) and it is likely that SADock(m4,LS) will need additional docking experiments using different refinement ratios.

Figure 1 shows that four mutation strategies and associated parameters are adapting to the evolution stage. The probability to select each mutation strategy is drawn in the panel (a) of Figure 1. In the early stage of the evolution, *DE/current to best/2* and *DE/rand/2* showed high selection probability whereas *DE/rand/1* and *DE/best/1* showed low probability. As the evolution proceeds, the selection probability of *DE/current to best/2* increased continuously; On the contrary, the other probabilities decreased continuously. The CR_m value of each strategy

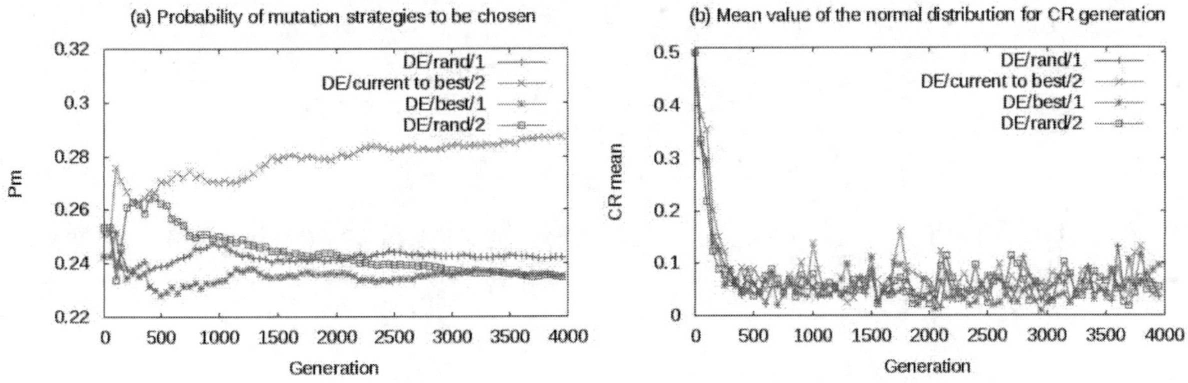

Figure 1. Parameter adaptation during the evolution of SADock(m4) with 1r1h (pdb id)

is drifting about the range between 0.01 and 0.2 in the panel (b) of Figure 1. In most of our docking experiments, the CR_m value was observed in the range between 0.01 and 0.25. It is likely that the molecular docking problem is generally considered to be multi-modal.

The primary idea about SADock was to treat docking problems without adjusting the major parameters manually. Without such manual tuning, our SADock showed better docking success ratio than LGA when only one fifth of the total evaluations of LGA was used. Moreover, SADock(m4) showed better success ratio (89%) than SADock(m1) (83%) without the strategy adaptation. This explains that the strategy adaptation is essential for the efficiency and robustness of SADock.

5. Conclusion

The result of this study indicated that the SADock outperformed AutoDock LGA in the self-docking of the Astex diverse set due to its adaptation of the mutation strategies and parameters. Our SADock was both fast and robust in reproducing the bounded conformation of ligands. These findings means that our SADock can be applied to the virtual screening applications for the drug discovery. In the future, we need to develop more mutation strategies and to test more diverse conditions when embedding the local search.

Acknowledgments

We would like acknowledge an institutional research program of Korean Institute of Science and Technology (KIST) (Grant No. 2E22790).

References

[1] Morris G M, Goodsell D S, Halliday R S, Huey R, Hart W E, Belew R K and Olson A J 1998 *Journal of Computational Chemistry* **19** 1639–1662 ISSN 1096-987X
[2] Thomsen R 2003 *The 2003 Congress on Evolutionary Computation, 2003. CEC '03* vol 4 pp 2354 – 2361 Vol.4
[3] Storn R and Price K 1997 *J. of Global Optimization* **11** 341359 ISSN 0925-5001
[4] Qin A, Huang V and Suganthan P 2009 *Evolutionary Computation, IEEE Transactions on* **13** 398 –417 ISSN 1089-778X
[5] Morris G, Huey R, Lindstrom W, Sanner M, Belew R, Goodsell D and Olson A 2009 *Journal of computational chemistry* **30** 27852791
[6] Huey R, Morris G M, Olson A J and Goodsell D S 2007 *Journal of Computational Chemistry* **28** 11451152 ISSN 1096-987X
[7] Hartshorn M J, Verdonk M L, Chessari G, Brewerton S C, Mooij W T M, Mortenson P N and Murray C W 2007 *J. Med. Chem.* **50** 726–741 ISSN 0022-2623
[8] Hooke R and Jeeves T A 1961 *J. ACM* **8** 212229 ISSN 0004-5411

Closed trajectories of the conformal arclength functional

Emilio Musso

Dipartimento di Scienze Matematiche, Politecnico di Torino, Corso Duca degli Abruzzi 24, I-10129 Torino, Italy

E-mail: `emilio.musso@polito.it`

Abstract. The purpose of this report is to give a brief overview of some unpublished results about the geometry of closed critical curves of a conformally invariant functional for space curves.

1. Introduction

Conformal geometry of curves is a well studied subject in classical differential geometry [4, 15, 16, 17]. In more recent times, the topic has been considered in connection with the regularization of the Kepler problem [6, 9, 14], within the theory of integrable systems and in the topology of knots [3, 1, 2, 5, 8, 10, 11]. In a previous paper, published several years ago [12], I studied the variational problem defined by the *conformal density* of a space curve, improperly called *conformal arc-element.* In that paper I computed the critical curves in terms of elliptic integrals. This was not much surprising because the trajectories of the variational problem can be obtained as projections of the integral curves of a collective completely integrable contact Hamiltonian system, defined on a 13-dimensional momentum space. Recently, my work was taken up and its results have been generalized to curves immersed in Euclidean spaces of arbitrary dimension [13]. The problem that I will consider in this report is the existence and the explicit determination of the *closed trajectories* of the variational problem. A more complete and exhaustive analysis will be the subject of a forthcoming paper.

2. Conformal geometry of space curves and the period map

Consider an oriented, complete bi-regular curve $\gamma : \mathbb{R} \to \mathbb{R}^3$, parameterized by the Euclidean arc-length. A vertex is a point $\gamma(s)$ such that $\dot{\kappa}(s)^2 + \kappa^2(s)\tau(s)^2 = 0$, where κ and τ are the curvature and the torsion respectively. This notion is invariant by conformal transformations of $\mathbb{R}^3$. The *conformal arc element* (or more appropriately the *conformal density*) of γ is the 1-form $\zeta = \sqrt[4]{\dot{\kappa}^2 + \kappa^2\tau^2}ds$. Assuming that the curve is generic (i.e., bi-regular and without vertices), then the conformal density is nowhere vanishing and we can define the two conformal curvatures

$$\begin{aligned} k_1 &= r^5\left(\kappa^2\tau^3 + \kappa\dot{\kappa}\dot{\tau} + \tau(2\dot{\kappa}^2 - \kappa\ddot{\kappa})\right), \\ k_2 &= \frac{1}{2}\left(\dot{r}^2 - 2r\ddot{r} - r^2\kappa^2\right). \end{aligned} \tag{1}$$

Published under licence by IOP Publishing Ltd

where $r = (\dot{\kappa}^2 + \kappa^2\tau^2)^{-1/4}$. The conformal density and the two curvatures determine the shape of the curve, up to conformal transformations. This means that if two curves γ and $\widetilde{\gamma}$ have the same invariants, then there is a conformal transformation of $\mathbb{R}^3$ that sends one trajectory onto the other. The conformal symmetry group of a curve γ is the group G_γ made up of all conformal transformations which leave unchanged its trajectory. From now on we will assume that the curves are bi-regular and without vertices.

Definiton. The critical curves of the functional defined by the integral of the conformal density are called *conformal geodesics*. For simplicity, a closed conformal geodesic with non-constant conformal curvatures is referred to as a *conformal string*.

The critical curves with periodic curvature functions are characterized by the Euler-Lagrange equations [12]

$$dk_1 + (k_1^2 - a)(k_1^2 - b)\zeta = 0, \quad k_2 = -\frac{3}{2}k_1 + \frac{a+b}{a}, \tag{2}$$

where a and b are two real constants, the *natural parameters*, such that $a > 0$, $b \neq 0$ and $a > b$. An analysis of the momentum map shows that the natural parameters of a string belong to the *admissible region*

$$\Sigma = \{(a, b) : a > 1, a^{-1} < b < a\}.$$

Using the conservation laws it can be checked (see [12]) that, up to conformal transformations, the parametrization of a critical curve with natural parameters $(a, b) \in \Sigma$ is given by

$$\begin{aligned} x(t) &= \frac{\sqrt{2}}{r(t)}\mu\sqrt{k(t)^2 - \upsilon^2}\cos(\Theta_2(t)), \\ y(t) &= \frac{\sqrt{2}}{r(t)}\mu\sqrt{k(t)^2 - \upsilon^2}\sin(\Theta_2(t)), \\ z(t) &= \frac{\sqrt{2}}{r(t)}\upsilon\sqrt{\mu^2 - k(t)^2}\sin(\Theta_1(t)), \end{aligned} \tag{3}$$

where $k(t)$ is the Jacobi elliptic function $\sqrt{a} \cdot dn(\sqrt{a}t, a^{-1}(a-b))$, μ and υ are the real constants

$$\mu = \sqrt{\frac{a+b+\sqrt{4+(a-b)^2}}{2}}, \quad \upsilon = \sqrt{\frac{a+b-\sqrt{4+(a-b)^2}}{2}}, \tag{4}$$

$\Theta_1(t)$ and $\Theta_2(t)$ are the incomplete elliptic integrals of the third kind

$$\Theta_1(t) = \int_0^t \frac{\mu}{\mu^2 - k(u)^2}du, \quad \Theta_2(t) = \int_0^t \frac{\upsilon}{\upsilon^2 - k(u)^2}du.$$

and $r(t)$ is the radial function

$$r(t) = \sqrt{\mu^2 - \upsilon^2}k(t) + \upsilon\sqrt{\mu^2 - k(t)^2}\cos(\Theta_1(t)).$$

The curve defined by the formula (3) is said to be the *symmetrical configuration* of the critical curve with natural parameters (a, b). We denote by $\Pi(m, n)$, $-1 < n, m < 1$, the complete elliptic integral of the third kind

$$\Pi(m, n) = \int_0^{\pi/2} \frac{dt}{(1 - n\sin^2(t))\sqrt{1 - m\sin^2(t)}},$$

and, for every $(a, b) \in \Sigma$ we set

$$\begin{aligned}\Phi_1(a,b) &= -\frac{\mu(a,b)}{\pi\sqrt{a}(a-\mu(a,b)^2)}\Pi\left(\frac{a-b}{a-\mu(a,b)^2}, \frac{a-b}{a}\right),\\ \Phi_2(a,b) &= \frac{\upsilon(a,b)}{\pi\sqrt{a}(a-\upsilon(a,b)^2)}\Pi\left(\frac{a-b}{a-\upsilon(a,b)^2}, \frac{a-b}{a}\right).\end{aligned} \tag{5}$$

where $\mu(a, b)$ and $\upsilon(a, b)$ are defined as in (4). The real-analytic function

$$\Phi = (a, b) \in \Sigma \to (\Phi_1(a, b), \Phi_2(a, b)) \in \mathbb{R}^2$$

is called the *period map* of the conformal arc-length functional. With a thorough analysis of the behavior of the period map it can be proved that Φ is a real-analytic diffeomorphism of Σ onto the circular domain

$$\Omega = \{(x, y) \in \mathbb{R}^2 : 1/2 < x < 1/\sqrt{2}, y > 0, x^2 + y^2 < 1/2\}.$$

3. Closed trajectories and their phenomenological invariants

From the previous discussion it follows that a critical curve with natural parameters $(a, b) \in \Sigma$ is closed if and only if $\Phi(a, b)$ is a rational point of the domain Ω. This shows that, up to conformal transformations, the conformal strings are in one to one correspondence with the pairs (q_1, q_2) of rational numbers belonging to Ω. More precisely, for every rational point q of the domain Ω there is a unique conformal string in its symmetrical configuration, with natural parameters $(a, b) \in \Sigma$ such that $q = \Phi(a, b)$. The rational points of Ω are called the *modules of the conformal strings*. To clarify the geometric meaning of the modules we must say a few words about the maximal tori of the conformal group. The conformal group of the Euclidean space is isomorphic to the pseudo-orthogonal group $SO(4, 1)$. Its maximal tori are conjugates to the subgroup $SO(2) \times SO(2) \subset SO(4, 1)$. The infinitesimal generators for the action of $SO(2) \times SO(2)$ on $\mathbb{R}^3$ are

$$\xi_1 = -y\partial_x + x\partial_y, \quad \xi_2 = -\frac{xz}{\sqrt{2}}\partial_x - \frac{yz}{\sqrt{2}}\partial_y + \frac{x^2 + y^2 - z^2 - 2}{2\sqrt{2}}\partial_z.$$

The first vector field generates the rotations around the Oz-axis, while the second one generates the toroidal rotations around the *Clifford circle*, that is the circle of the Oxy-plane, centered at the origin, with radius $\sqrt{2}$. The geometric meaning of the modules is clarified by the following Theorem.

Theorem. *Let $\gamma : \mathbb{R} \to \mathbb{R}^3$ be a conformal string in its symmetrical configuration. Denote by $q_1 = m_1/n_1$ and $q_2 = m_2/n_2$ the modules of γ and assume that $gcd(m_1, n_1) = \gcd(m_2, n_2) = 1$. Let n be the least common multiple of n_1 and n_2. Then, the conformal symmetry group of γ is the cyclic subgroup $G_\gamma \subset SO(2) \times SO(2)$ of order n generated by the composition of a rotation $R_1(2\pi q_1)$ of an angle $2\pi q_1$ around the Oz-axis and a toroidal rotation $R_2(2\pi q_2)$ of an angle $2\pi q_2$ around the Clifford circle. Moreover, if we denote by h_1 and h_2 the coprime integers n/n_1 and n/n_2, then m_1h_1 is the linking number of γ with the Clifford circle and m_2h_2 is the linking number of γ with the Oz-axis.*

This result implies that, up to the action of the conformal group, the conformal strings are are uniquely determined by three numerical invariants: the order of the conformal symmetry group and the linking numbers with the Oz-axis and the Clifford circle.

The inverse of the period map can be calculated by numerical methods. Therefore, once that we know the three phenomenological invariants (order of symmetry and linking numbers), the symmetrical configuration of the string can be determined explicitly. The numerical experiments suggest that the strings are simple curves. However, we do not have a mathematically rigorous proof of this fact. Another interesting problem is to find an asymptotic estimate of the cardinality $\rho(n)$ of the set of the equivalence classes of conformal strings with order of symmetry n. Our expectation is that the growth of $\rho(n)$ is quadratic.

4. Prospectives

It is likely that the techniques and the results obtained so far can be extended to time-like curves in the conformal completion of the Minkowski space (and more generally, in any conformally flat Lorenzian 4-manifold). In this framework, it is important to analyze the physical meaning of the scattering data of the critical curves with non-periodic conformal curvatures.

References

[1] S. Bryson, M. H. Freedman, Z-X. He, Z. Wang, Mœbius invariance of knot energy *Bull. of Am.Math.Soc.*, **28** (1993), pp 99–103.
[2] K. S. Chou, C. Qu, Integrable equations arising from motions of plane curves. II, *J. Nonlinear Sci.* **13** (2003), 487–517.
[3] M. Eastwood, G. Mari-Beffa, Geometric Poisson brackets on Grassmannians and Conformal Spheres, to appear in *Proceedings Royal Society of Edimburg*, (2012), arXiv:1006.5753, 30 June 2010.
[4] A. Fialkow, The conformal theory of curves, *Trans. Amer.Math.Soc.*, **51** (1942), 435–501.
[5] M. H. Freedman, Z-X. He, Z. Wang, Mœbius energy of knots and unknots *Annals of Mathematics*, **139** (1994), 1–50.
[6] V. Guillemin, S. Sternberg, *Variations on a Theme by Kepler*, AMS Colloquium Publications **285**, Providence R. I. (1990).
[7] H. Liebmann, Beiträge zur inversionsgeometrie der Kurven, Mœnchener Berichte (1923).
[8] R. Langevin, J. O'Hara, Conformally invariant energies of knots, *J. Inst. Math. Jussieu*, **4** (2005), 219-280.
[9] C-M. Marle, A property of conformally invariant Hamiltonian vector fields; application to the Kepler problem, *Journal of Geometric Mechanics*, **4** (2012), 181–206.
[10] G. Mari-Beffa, Relative and Absolute Differential Invariants for Conformal Curves, *Journal of Lie Theory*, **12** (2003), 213–245.
[11] G. Marih-Beffa, Poisson brackets associated to the Conformal geometry of curves, *Transaction of the AMS*, **357** (2005), 2799–2827.
[12] E. Musso, The conformal arclength functional, *Math. Nachr.*, **165** (1994), 107–131.
[13] M. Magliaro, L. Mari, M. Rigoli, On the geometry of curves and conformal geodesics in the Moebius space, *Ann. Global Anal. Geom.*, **40** (2011), 133–165.
[14] J.-M. Soriau, Sur la varieté de Kepler, *Sympos. Math.* **14** (1974), 343–360.
[15] R. Sulanke, Submanifolds of the Mœbius space II, Frenet Formulas and Curves of Constant Curvatures, *Math. Nachr.*, **100** (1981), 235–257.
[16] T. Takasu, *Differentialgeometrie in den Kugelräume, Vol. 1, Konforme Differentialgeometrie von Liouville and Mœbius*, Tokyo (1938).
[17] E. Vessiot, Enveloppes de sphéres et courbes gauches, *J, Ecole Polytechnique*, **25** (1925), 43–91.

Time parallelization of advanced operation scenario simulations of ITER plasma

D. Samaddar[1], T.A. Casper[1], S.H. Kim[1], L.A. Berry[2], W.R. Elwasif[2], D. Batchelor[2], W.A. Houlberg[1]

1. ITER Organization, Route de Vinon sur Verdon, 13115 Saint Paul Lez Durance, France
2. Oak Ridge National Laboratory, Oak Ridge, USA

Abstract. This work demonstrates that simulations of advanced burning plasma operation scenarios can be successfully parallelized in time using the parareal algorithm. CORSICA - an advanced operation scenario code for tokamak plasmas is used as a test case. This is a unique application since the parareal algorithm has so far been applied to relatively much simpler systems except for the case of turbulence. In the present application, a computational gain of an order of magnitude has been achieved which is extremely promising. A successful implementation of the Parareal algorithm to codes like CORSICA ushers in the possibility of time efficient simulations of ITER plasmas.

1. Introduction

With the construction of the ITER project well under way, understanding and predicting burning plasma scenarios have gained tremendous importance. These studies are used for designing of components for the world's largest tokamak and contribute towards its successful operation. Plasma scenario studies are carried out by codes such as CORSICA [1, 2]. CORSICA employs a free or fixed boundary 2D equilibrium package that is coupled with various transport and source models. The advanced capabilities of these simulations allow studies with realistic physics and engineering constraints. CORSICA has also been used to predict and analyse plasma operations in the existing tokamak DIII-D[3] which served as a means of validation and verification.

It must be borne in mind that simulations of plasma operation scenarios require extremely long wallclock time. Space parallelization is a common approach to achieve computational speed-up although this technique experiences saturation while reduction in computation time is still desired. The option of time parallelization introduces a new opportunity, although this approach has not been explored much in the past for these kinds of simulations. The parareal algorithm [4] lays out a scheme for the application of temporal parallelization to simulations. The algorithm has attracted significant attention in recent times and has undergone modifications to maximize computational speed-up and efficiency [5, 6, 7, 8].

Except for turbulence [5, 9] which is a complex system with high dimensional chaos, the parareal algorithm has so far been applied to relatively simple problems [10, 11, 12, 13, 14]. This work demonstrates that the parareal algorithm may be successfully applied to CORSICA to attain computational speed-up for plasma scenario simulations. The coupling of the equilibrium solver with the transport and source modules makes this a unique application compared to previous attempts. These simulations are characterized by quasi steady states interrupted by events which may be results of MHD activity, pellet injection or an introduction of ELMs. The success of temporal parallelization in such simulations ushers the possibility of time efficient simulations of ITER relevant plasmas.

The parareal algorithm has been applied using the Parareal framework developed at ORNL as part of the SWIM-IPS project [6].

2. The parareal algorithm

The parareal algorithm allows the splitting of a time series into multiple time slices that are solved in parallel. This unique technique is described in more detail in [5, 4]. Two solvers - one coarse (G) and one fine (F) - are required for this purpose. G is used to evolve the time series to obtain an approximate estimate of the fine solution. G is characterized by high computational speed and is always applied as a serial process. F is always applied in parallel, is computationally slow and yields a solution whose accuracy is higher than that obtained by the coarse solver G.

At the first parareal iteration, G is used to generate an estimate for an initial value for each time slice. F is then applied in parallel to evolve that state. G and F then alternate across successive iterations until parareal convergence is obtained. At each parareal iteration, k, for the (i +1)th time slice, the initial state λ_{i+1}^k is computed using the parareal correction (eq.1) after the state λ_i^k has been evolved using F(λ_i^k) and G(λ_i^k). The parareal correction is defined as

$$\lambda_{i+1}^k = F(\lambda_i^{k-1}) + G(\lambda_i^k) - G(\lambda_i^{k-1}) \tag{1}$$

The Parareal framework is used to apply the event driven parareal technique, which has been described in [6]. Using the framework allows better computational gain and efficiency. Moreover, since the choice of the coarse solver for a case like CORSICA is non trivial, experimenting with different options becomes relatively simpler with the framework.

2.1. Convergence

A parareal convergence is said to occur when the relative error between two successive fine solutions (k and k −1) on a given time slice (i) corresponding to a given processor reduces below a prescribed tolerance. In case of CORSICA, two parameters β_p and li were used to define the convergence metric. β_p is the normalized stored energy of the system and li is the internal inductance whose variation is a measure of the evolution of the plasma current density profile. If $\beta_{p(i)}^{(k)}$ and $li_{(i)}^{(k)}$ represent the variables for the ith time slice at the kth parareal iteration, the relative errors are given by $Err_{\beta(i)}^{(k)} = \frac{\beta_i^k - \beta_i^{k-1}}{\beta_i^{k-1}}$ & $Err_{li(i)}^{(k)} = \frac{li_i^k - li_i^{k-1}}{li_i^{k-1}}$.

The simulation across the processor solving the ith time slice is said to have converged if it satisfies the convergence criterion given by $\xi_i^k \leq tolerance$ where $\xi_i^k = Err_{\beta(i)}^{(k)} + Err_{li(i)}^{(k)}$.

3. Results

The parareal algorithm has been applied to a prescribed boundary simulation of CORSICA with neutral beam injection as a source term. The introduction of the NBI using the Nfreya package makes the simulation realistic in terms of ITER operational scenarios. NBI is the injection of high energy neutral particle beams across the magnetic field to heat the plasma to the temperature required for fusion reaction. The Monte-Carlo method is used to statistically simulate the localized ionisations of the beam particles along its trajectory. The parareal algorithm has been applied to CORSICA simulations without realistic source modules in [15], but this application with the NBI significantly enhances the complexity.

The choice of the coarse solver is always important for the success of the parareal algorithm. In case of CORSICA, selecting the optimum coarse solver becomes critical. Since the ITER NBI system injects 33MW of power, a fast analytic source term was introduced in the coarse solver to simulate power of the same order. This point is illustrated in Fig.(1). Care was also taken to see that the total radial plasma current had a similar profile as is shown in Fig.(2). The use of an analytic source term simplifies the calculation thus requiring lesser computational time. However, a similarity in the current profiles and the total power between the coarse and the fine solvers allow parareal convergence in lesser number of iterations. In addition to using an analytic source model, the size of the timesteps were increased in the coarse solver. The CORSICA solver uses a variable timestep and the size varied between 0.001 and 0.01 seconds for the fine solution. The coarse solver on the other hand allowed a variation of 0.004 to 0.6 seconds.

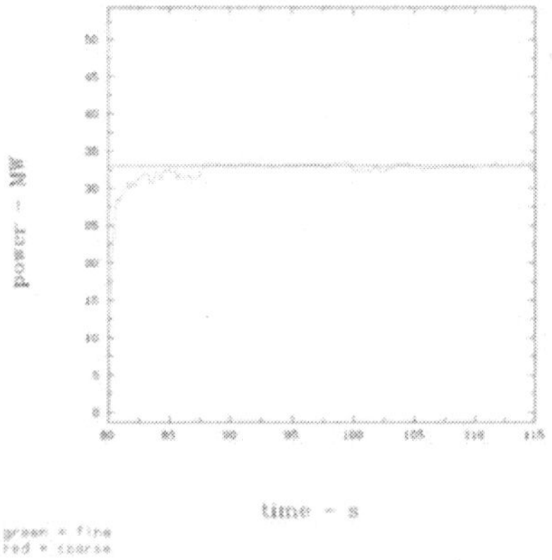

Figure 1. The powers generated by the analytic source term (red) in the coarse solver and the realistic source module (green) in the fine solver are chosen such that their values are similar.

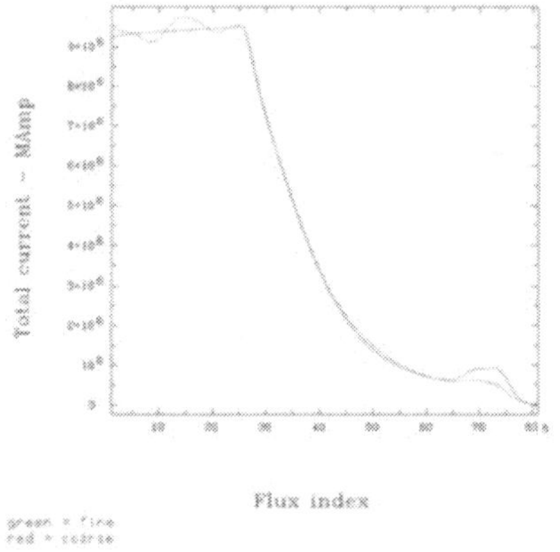

Figure 2. The radial profiles of the total plasma current for the coarse (red) and fine (green) solvers are similar, so convergence is achieved in the least number of parareal iterations.

A typical serial run using the fine solver took 10.5 hours wallclock time to complete a 35 second simulation. A computational gain of 10.13 was achieved with 32 processors on application of the parareal algorithm. The plots in Fig.(3) represent solutions obtained using a single processor with the fine solver as well as that obtained using the parareal scheme. For the parareal solution in Fig.(3), a time slice of size 5 seconds was solved on each processor across a total of 16 processors. The different colors represent solutions across multiple processors. For this case, convergence was attained in 2 parareal iterations with an error tolerance of $5e - 3$.

4. Summary and future work

The parareal algorithm has been successfully applied to simulations of plasma scenarios which are highly complex systems with coupled non linearities. The approach of time parallelization has been applied to this type of simulations for the first time and the results are very promising. The reduced wall time makes studies of burning plasma scenarios at ITER more tractable.

Future studies may include more complex cases of plasma scenarios. The inclusion of free-boundary equilibrium control, ray tracing electron-cyclotron heating and full-wave ion-cyclotron heating serve as examples. Since temporal parallelization adds an extra domain of parallelization, it may be combined with spatial parallelization to attain maximum computational speed-up. The

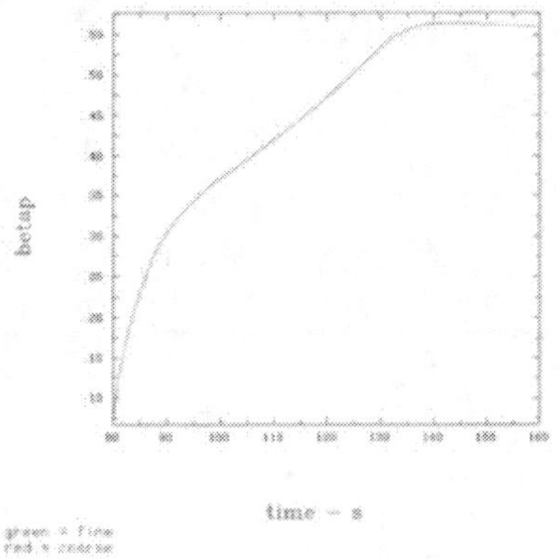

(a) Single processor using the fine solver.

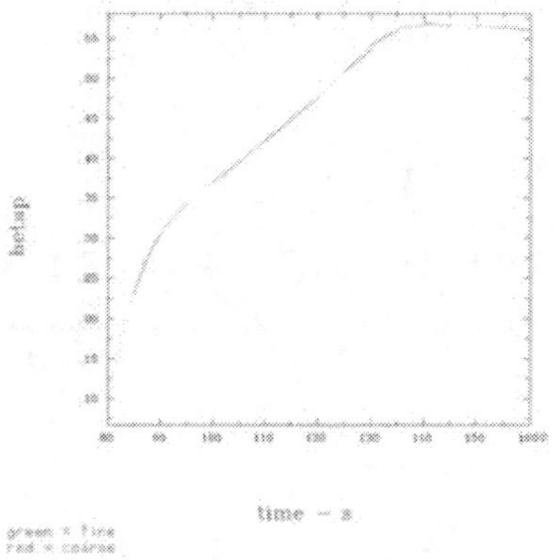

(b) Parareal solution using 16 processors.

Figure 3. Variation of the poloidal beta β_p with time. The different colors on the parareal solution represent solutions obtained across individual processors.

reduced simulation times will allow for further systematic parameter variation studies needed for scenario evolution and discharge development and validation.

5. Acknowledgements

The simulations were performed at the NERSC supercomputing facility Carver. The authors are thankful for the computer time allocated.

6. Disclaimer

The views and opinions expressed here do not necessarily represent those of the ITER Organization.

References

[1] Crotinger J A et al, LLNL Report UCRL-ID-126284; NTIS PB2005-102154. (1997).
[2] Casper T A et al, 23rd Int. Conf. on Fusion Energy (Daejeon, Korea) ITR/P1-19. (2010).
[3] Casper T A et al, Nucl. Fusion 47 (2007) 825-832
[4] Lions J., Maday Y. and Turinici G., CR Acad. Sci. I - Math. 332 (7), (2001) pp. 661-668.
[5] Samaddar D., Newman D. E. and Sanchez R., J. Comp. Phys 229 (18), (2010) pp. 6558-6573.
[6] Berry L. A., W. Elwasif, J.M Reynolds-Barredo, D. Samaddar, R. Sanchez and D. E. Newman, J. Comp. Phys 2012 (In Press).
[7] Elwasif W.R et al, 4th IEEE Workshop on Many-Task Computing on Grids and Supercomputers, MTAGS 2011, (2011).
[8] Aubanel E. et al, Scheduling of tasks in the parareal algorithm, Parallel Computing 37 (3), (2011).
[9] Reynolds-Barredo J.M et al, J. Comp. Phys 2012 (In Press).
[10] Baffico L., Bernard S., et. al, Phys. Rev. E 66 (5), 057706 (2002).
[11] Bal G., Maday Y., Lecture Notes in Computer Science and Engineering, vol. 23, Springer, Berlin, 2002.
[12] Garrido I., Espedal M.S., Fladmark G.E., Proceedings of 15th International Conf. on Dom. Decom. Methods, vol. 40, Springer, Berlin, pp. 469-476.
[13] Fischer P.F., Hecht F., Maday Y., Proceedings of 15th International Conf. on Dom. Decom. Methods, vol. 40, Springer Verlag, 2004, (2004), pp. 433-440.
[14] Gander M.J., Hairer E., Dom. Decom. Methods in Sc. and Eng. XVII,vol. 60, Springer, (2008) pp. 45-56.
[15] Samaddar D. et al, 39th EPS Conf. & 16th Int. Cong. on Plasma Phys., Stockholm, Sweden (2012) P2.140

Heterogeneous patterns enhancing static and dynamic texture classification

Núbia Rosa da Silva[1,2] and Odemir Martinez Bruno[1,3]

[1] Universidade de São Paulo, Av. do Trabalhador Sãocarlense, 400 13560-970 São Carlos, São Paulo, Brasil

[2] Instituto de Ciências Matemáticas e de Computação

[3] Instituto de Física de São Carlos

E-mail: nubiars@icmc.usp.br, bruno@ifsc.usp.br

Abstract. Some mixtures, such as colloids like milk, blood, and gelatin, have homogeneous appearance when viewed with the naked eye, however, to observe them at the nanoscale is possible to understand the heterogeneity of its components. The same phenomenon can occur in pattern recognition in which it is possible to see heterogeneous patterns in texture images. However, current methods of texture analysis can not adequately describe such heterogeneous patterns. Common methods used by researchers analyse the image information in a global way, taking all its features in an integrated manner. Furthermore, multi-scale analysis verifies the patterns at different scales, but still preserving the homogeneous analysis. On the other hand various methods use textons to represent the texture, breaking texture down into its smallest unit. To tackle this problem, we propose a method to identify texture patterns not small as textons at distinct scales enhancing the separability among different types of texture. We find sub patterns of texture according to the scale and then group similar patterns for a more refined analysis. Tests were performed in four static texture databases and one dynamical one. Results show that our method provide better classification rate compared with conventional approaches both in static and in dynamic texture.

1. Introduction

Pattern recognition is the identification and interpretation of patterns in images, in order to extract relevant information on the image to identify and classify your content. Classification of patterns can be used in a variety of applications in different fields such as nanotechnology [1, 2], biology [3, 4], medicine [5] and computer science [6, 7]. Different approaches are developed according to the application, however, most of them analyze the information in a global way, using all the features in an integrated manner. One side in the pattern classification approach uses multi-scale analysis of patterns, that is, different scales of observation are used to perform the analysis and find similar patterns, because important structures in an image usually occur at different spatial scales [8] [9]. Methods based on textons [10, 11] represent each pixel of a texture as the convolution of a multi-scale and multi-orientation filter bank producing a texton vocabulary. Thus, these methods of texture analysis characterize the image homogeneously on the scale to be analyzed. Both the overall analysis, such as multi-scale approach is appropriate for the vast majority of problems in image analysis and pattern recognition. However, in some problems, due to the heterogeneous nature of the composition of the objects under consideration, is needed one more step in the process of pattern recognition, analysis of heterogeneous patterns.

The aim of this paper is to demonstrate how can apply heterogeneous analysis to improve results regarding the homogeneous analysis. To validate our proposal experiments were performed on four static e one dynamic texture databases using the same texture descriptor but with different approaches. In all tests, heterogeneous analysis proved to be better than the homogeneous analysis enhancing the rate classification.

This paper is organized as follows: Section 2 describes the heterogeneous pattern analysis. Section 3 shows the results and discussions and in Section 4 the conclusions.

2. Method

Heterogeneous pattern arises when the object under analysis presents combined patterns in its composition. Similar to the classic definition of heterogeneous compositions in chemical compounds [12], which is the characteristic of presenting a different appearance or composition when analyzed in parts. The same analogy can be applied in recognizing patterns in images. In each image can be found heterogeneous patterns, i. e., it is possible to distinguish the different patterns in each of the texture images. Figure 1(a) shows an image from Brodatz, which analyzed by conventional methods, would be defined only a pattern for this image. Using the approach of heterogeneous patterns, two different types of texture patterns are identified in its formation.

This new view to analyze the various patterns requires a new approach for analysis of similarity between images. The first step is to segment the texture by defining regions where the image belong to a given pattern while defining the patterns in the image. Figures 1(b) and 1(c) illustrate the segmentation of the image of Figure 1(a) in two of patterns.

(a) (b) (c)

Figure 1. Segmentation according to the heterogeneous patterns. (a) Original image. (b) and (c) Two patterns obtained from (a).

To obtain this result the image was divided into smaller sized windows of 8×8 pixels, for windows with texture sufficiently homogeneous, that is, where only one pattern is found. Furthermore, there was obtained a characteristic to represent each window. Windows with similar characteristics remain the same pattern, and those with distinct characteristics were separated into different patterns. Figures 1(b) and 1(c) show the windows stayed grouped using *k-means* algorithm according to their pattern with the remaining windows in each pattern. Textural features from Haralick [13] descriptors were used to characterize each window. Contrast, Correlation, Energy and Homogeneity were computed from resulting co-occurrence matrices to obtain a set of 32 descriptors for each window. Distances of 1 and 2 pixels with angles of $0°$, $45°$, $90°$ and $135°$ were used.

After the windows were divided into patterns, we performed a survival analysis of windows, 25% of the windows with features farthest from the group average were discarded making the representation of pattern more consistent. Each pattern was characterized by a feature vector average that is the average of the characteristics of all windows belonging to the pattern. The next step is to find the best matching among the patterns to define the image similarity as Figures 2(a) and 2(b) exemplify. They show two examples where it has an optimal matching among the image patterns of the same class. When you find the best fit among all the patterns in each image, we have the degree of similarity, defining whether the images belong to the same class. Figure 2(c) shows an example where it is not possible to perform the fitting patterns and all possibilities of fitting are tested. That's why the images belong to different classes.

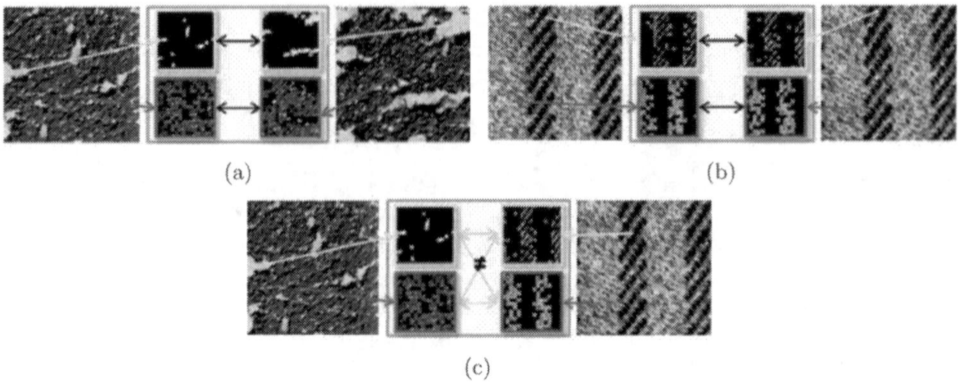

Figure 2. Matching patterns. (a) and (b) matching of two images from the same class. (c) Matching of patterns can not be made because the images belong to different classes.

3. Results and Discussions

To validate the method of analysis for heterogeneous patterns, this technique was applied in four different texture databases, each with its peculiarities: Brodatz, USPTex, Vistex and Outex and one dynamic texture database: Dyntex. Figure 3 shows some samples of these databases. The classification method k-Nearest Neighbor ($k - NN$) with 10-fold cross-validation scheme was used in all experiments.

Brodatz [14] contains 1110 natural textures of 200 × 200 pixels divided into 111 classes.

Vistex [15] contains 864 images of 128 × 128 pixels size with 54 texture classes.

USPTex has 3984 natural texture images of 128 × 128 pixels size divided into 332 classes.

Outex [16] has 1360 images of 128 × 128 pixels size divided into 68 classes.

Dyntex [17] consists of 1230 videos with 250 frames with 400 × 300 pixels size divided into 123 classes of dynamic texture.

Figure 3. (a) Brodatz. (b) Vistex. (c) Usptex. (d) Outex.

The method used as classical approach was Haralick descriptors with the same configurations of characterization of windows. This way we can compare the same descriptor but with different

Table 1. Accuracy rate by comparing the traditional method of analysis with the heterogeneous method.

Database	Brodatz	Vistex	Usptex	Outex	Dyntex
Classical approach	88.92	89.47	69.58	76.62	76.89
Heterogeneous pattern	93.06	93.63	78.49	76.99	98.29

perspectives. One of the most important parameters is the window size that directly interferes in the results. Large windows remain with the general representation of the image, preserving the homogeneous analysis of patterns. However, becomes more homogeneous window, or small windows, the classification rate increases, as the representative pattern of the window increases. For this experiment two patterns were analyzed in each image. Table 1 shows the results obtained when using the conventional method in comparison with heterogeneous patterns, in all cases heterogeneous pattern method improved the classification rate proving to be better than the standard approach in which only one pattern is established for the entire image.

4. Conclusions

Usually, analysis in texture pattern recognition using the homogeneous approach taking all the image information in a global way, that is, the entire image is defined as one pattern. However, it is possible to find more than one texture pattern on the image that we called here, heterogeneous pattern. It is necessary to map the patterns in each image and then check the similarity of patterns among them. This method improved the accuracy rate of classification using the same method to characterize the image showing robustness to assess patterns heterogeneous image.

References

[1] Annampedu V and Wagh M D 2007 *Microelectronics Journal* **38** 430 – 438 ISSN 0026-2692
[2] Oh E H, Song H S and Park T H 2011 *Enzyme and Microbial Technology* **48** 427 – 437 ISSN 0141-0229
[3] Galas D J, Eggert M and Waterman M S 1985 *Journal of Molecular Biology* **186** 117 – 128 ISSN 0022-2836
[4] Isasi A G, Zapirain B G and Zorrilla A M 2011 *Computers in Biology and Medicine* **41** 742 – 755 ISSN 0010-4825
[5] Landeweerd G, Gelsema E, Bins M and Halie M 1981 *Pattern Recognition* **14** 239 – 244 ISSN 0031-3203
[6] Este A, Gringoli F and Salgarelli L 2009 *Comput. Netw.* **53**(14) 2476–2490 ISSN 1389-1286
[7] Canals V, Morro A and Rossell J L 2010 *Pattern Recognition Letters* **31** 2353 – 2356 ISSN 0167-8655
[8] Koenderink J 1984 *Biological Cybernetics* **50**(5) 363–370 ISSN 0340-1200 10.1007/BF00336961
[9] Witkin A P 1983 Scale-space filtering *IJCAI* pp 1019–1022
[10] Julesz B 1981 *Nature* **290** 91–97
[11] Leung T and Malik J 1999 Recognizing surfaces using three-dimensional textons *Computer Vision, 1999. The Proceedings of the Seventh IEEE International Conference on* vol 2 pp 1010–1017
[12] Staley D D, Matta M S and Waterman E L 2004 *Chemistry* (Pearson Prentice Hall)
[13] Haralick R M, Shanmugam K and Dinstein I 1973 *Systems, Man and Cybernetics, IEEE Transactions on* **3** 610 –621 ISSN 0018-9472
[14] Brodatz P 1966 *Textures; a photographic album for artists and designers* (Dover Publications New York,) ISBN 0486216691
[15] 2009 Vision texture database URL `vismod.media.mit.edu/vismod/imagery/VisionTexture/vistex.html`
[16] Ojala T, Mäenpää T, Pietikäinen M, Viertola J, Kyllönen J and Huovinen S 2002 Outex - new framework for empirical evaluation of texture analysis algorithms *Proceedings of the 16 th International Conference on Pattern Recognition (ICPR'02) Volume 1 - Volume 1* ICPR '02 (Washington, DC, USA: IEEE Computer Society) p 10701 ISBN 0-7695-1695-X
[17] Péteri R, Fazekas S and Huiskes M J 2010 *Pattern Recognition Letters* **31** 1627–1632 http://projects.cwi.nl/dyntex/

Journal of Physics: Conference Series **410** (2013) 012034 doi:10.1088/1742-6596/410/1/012034

Phase Diagram of the Blume-Emery-Griffiths-Vannimenus Model on the Cayley Tree

Nasir Ganikhodjaev

Department of Computational and Theoretical Sciences, Faculty of Science, IIUM, 25200 Kuantan, Malaysia

E-mail: nasirgani@hotmail.com

Abstract. We study the phase diagram of the Blume-Emery-Griffiths-Vannimenus model on a Cayley tree with competing nearest-neighbour couplings and next-nearest-neighbour couplings and show that a detailed study of its properties can be carried out with essentially exact results, using rather simple numerical methods. In addition to the expected paramagnetic and ferromagnetic phases, we find modulated phase for Blume-Emery-Griffiths model and hence for Blume-Emery-Griffiths-Vannimenus model also.

1. Introduction

The existence of competing interactions lies at the heart of a variety of original phenomena in magnetic systems, ranging from the spin-glass transitions found in many disordered materials to the modulated phases with an infinite number of commensurate regions, that are observed in certain models with periodic interactions [1]. Vannimenus [1] and Mariz et al [2] examine the local properties of the Ising model on a Cayley tree with nearest-neighbour interactions and competing next-nearest-neighbour interactions. By a process of iteration, Vannimenus found a new modulated phase, in addition to the expected paramagnetic, ferromagnetic and antiferromagnetic phases, which consisted of commensurate (periodic) and incommensurate (aperiodic) regions corresponding to the so-called "devil's staircase", found also in the ANNNI [3] and other competing interaction models [4]. As expanded Ising model, the Blume-Emery-Griffiths (BEG) model, which is characterized by bilinear and biquadratic exchange interactions and crystalfield interaction, has played important role in the development of the theory of tricritical phenomena. The Hamiltonian of the BEG model on the Cayley tree is given by

$$H(\sigma) = -J \sum_{<x,y>} \sigma(x)\sigma(y) - K \sum_{<x,y>} \sigma^2(x)\sigma^2(y) + D \sum_{x} \sigma^2(x) \quad (1)$$

where $\sigma(x) \in \{-1, 0, 1\}$ is the spin at site x, first and second summation run over all the nearest-neighbour pairs, and last summation runs over all the sites. Here J, K, D describe the bilinear exchange, biquadratic interactions, and crystal-field interaction. This Hamiltonian was originally proposed to explain the phase separation and superfluidity in $^3He -^4He$ mixtures [5]. The present paper is devoted to the study of a Blume-Emery-Griffiths-Vannimenus model on a Cayley tree with competing nearest-neighbour couplings and next-nearest-neighbour couplings and to show that a detailed study of its properties can be carried out with essentially exact results, using rather simple numerical methods.

2. The Blume-Emery-Griffiths-Vannimenus model

The model considered consists of spins ($\sigma(x) \in \{-1, 0, 1\}$) on a a semi-infinite Cayley tree $\Gamma_+^2 = (V, L)$ of second order, i.e., an infinite graph without cycles with 3 edges issuing from each vertex except for x^0, so-called a root of the tree, which has only 2 edges, where V is the set of vertices and L is the set of edges. Two kinds of bonds are present: nearest-neigbours interactions and prolonged next-nearest-neighbour interactions, these being restricted to spin belonging to the same branch of the tree.

The Model: The Hamiltonian of the Blume-Emery-Griffiths-Vannimenus (BEGV) model on the Cayley tree is defined by

$$H(\sigma) = -J_p \sum_{\widetilde{<x,y>}} \sigma(x)\sigma(y) - J_1 \sum_{<x,y>} \sigma(x)\sigma(y) - K \sum_{<x,y>} \sigma^2(x)\sigma^2(y) + D \sum_x \sigma^2(x) \quad (2)$$

where $\sigma(x) \in \{-1, 0, 1\}$ is the spin at site $x \in V$, the first summations runs over all prolonged next-nearest-neighbours, the second and third summation runs over all nearest-neighbours pairs, and last summation runs over all the sites. Here J_p, J_1, K, D describe the bilinear exchange, biquadratic interactions and crystal-field interaction respectively. The Hamiltonian (2) with $J_p = 0$ defines the BEG model [5] and with $K = D = 0$ defines Vannimenus model [1] .

3. Recurrence equations

In order to produce the recurrence equations, we consider the relation of the partition function of an $n-$ generation tree V_n to the partition function of its subsystems containing $(n-1)$ generations V_{n-1}.

Let $< x^0, x >= l \in L$ be an edge of semi-infinite Cayley tree Γ_+^2 of second order. The infinite subtree $\Gamma_+^2(l) = (V^l, L^l)$ is called a single-trunk Cayley tree, if from vertex x^0 a single edge l emanates and from any other vertex $x \in V^l, x \neq x^0$ exactly 3 edges emanate. Let $W_1 = \{x_1, x_2\}$ and $< x^0, x^1 >= l_1, < x^0, x^2 >= l_2$ be two edges emanating from x^0. It is evident that semi-infinite Cayley tree Γ_+^2 splits into two components - two single-trunk Cayley trees $\Gamma_+^2(l_i), i = 1, 2$.

Assume $Z^{(n)}(i, j)$ be a partition function on V_n^l with the configuration (i, j) on an edge $l =< x^0, x >$, where $i, j = -1, 0, 1$; and $Z^{(n)}(i_1, i_0, i_2)$ be a partition function on V_n where the spin in the root x^0 is i_0 and the two spins in the proceeding vertices x^1, x^2 are i_1 and i_2, respectively.

There are 27 different partition functions $Z^{(n)}(i_1, i_0, i_2)$ and the partition function $Z^{(n)}$ in volume V_n can the be written as follows

$$Z^{(n)} = \sum_{i_1, i_0, i_2 = -1}^{1} Z^{(n)}(i_1, i_0, i_2). \quad (3)$$

and

$$Z^{(n)}(\sigma(x^1), \sigma(x^0), \sigma(x^2)) = Z^{(n)}(\sigma(x^0), \sigma(x^1)) \cdot Z^{(n)}(\sigma(x^0), \sigma(x^2)). \quad (4)$$

Let

$$a = \exp\left(\frac{J_p}{k_B T}\right), b = \exp\left(\frac{J_1}{k_B T}\right), c = \exp\left(\frac{K}{k_B T}\right), d = \exp\left(-\frac{D}{2k_B T}\right).$$

From (4) one can select only nine new independent variables

$$u_1^{(n)} = \sqrt{Z^{(n)}(-1,-1,-1)}, \quad u_2^{(n)} = \sqrt{Z^{(n)}(-1,0,-1)}, \quad u_3^{(n)} = \sqrt{Z^{(n)}(-1,1,-1)},$$
$$u_4^{(n)} = \sqrt{Z^{(n)}(0,-1,0)}, \quad u_5^{(n)} = \sqrt{Z^{(n)}(0,0,0)}, \quad u_6^{(n)} = \sqrt{Z^{(n)}(0,1,0)},$$
$$u_7^{(n)} = \sqrt{Z^{(n)}(1,-1,1)}, \quad u_8^{(n)} = \sqrt{Z^{(n)}(1,0,1)}, \quad u_9^{(n)} = \sqrt{Z^{(n)}(1,1,1)}.$$

We note that, in the paramagnetic phase (high symmetry phase), $u_1 = u_9$, $u_2 = u_8, u_3 = u_7$ and $u_4 = u_6$. For discussing the phase diagram, the following choice of reduced variables is convenient:

$$x_1 = \frac{u_2+u_8}{u_1+u_9},\quad x_2 = \frac{u_3+u_7}{u_1+u_9},\quad x_3 = \frac{u_4+u_6}{u_1+u_9},\quad x_4 = \frac{2u_5}{u_1+u_9},$$
$$y_1 = \frac{u_1-u_9}{u_1+u_9},\quad y_2 = \frac{u_2-u_8}{u_1+u_9},\quad y_3 = \frac{u_3-u_7}{u_1+u_9},\quad y_4 = \frac{u_4-u_6}{u_1+u_9}.$$

It is straightforward to establish the following recursive relations:

$$\begin{aligned}
x_1' &= \frac{[1+x_2+x_3]^2+[y_1-y_3+y_4]^2}{bcdD}, & y_1' &= \frac{2(a+a^{-1}x_2+x_3)(ay_1-a^{-1}y_3+y_4)}{D} \\
x_2' &= \frac{[a^{-1}+ax_2+x_3]^2+[a^{-1}y_1-ay_3+y_4]^2}{b^2D}, & y_2' &= \frac{2(1+x_2+x_3)(y_1-y_3+y_4)}{bcdD} \\
x_3' &= \frac{[(a+a^{-1})x_1+x_4]^2+[(a-a^{-1})y_2]^2}{bcD}, & y_3' &= \frac{2(a^{-1}+ax_2+x_3)(a^{-1}y_1-ay_3+y_4)}{b^2D} \\
x_4' &= \frac{[2x_1+x_4]^2}{bcdD}, & y_4' &= \frac{2(a-a^{-1})[(a+a^{-1})x_1+x_4]y_2}{bcD},
\end{aligned} \tag{5}$$

where

$$D = (a + a^{-1}x_2 + x_3)^2 + (ay_1 - a^{-1}y_3 + y_4)^2,$$

and primed variables correspond to the $Z^{(n+1)}(i_1, i_0, i_2)$. The system of eight equations (5) are essentially complicated than the similar basic equations of the Ising model [1], [2].
The total partition function is given in terms of $(u_i^{(n)})$ by

$$Z^{(n)} = (u_1^{(n)} + u_4^{(n)} + u_7^{(n)})^2 + (u_2^{(n)} + u_5^{(n)} + u_8^{(n)})^2 + (u_3^{(n)} + u_6^{(n)} + u_9^{(n)})^2. \tag{6}$$

The average magnetization $m^{(n)}$ for the nth generation is given by

$$m^{(n)} = -\frac{4(1+x_2+x_3)(y_1-y_3+y_4)}{(2x_1+x_4)^2+2(1+x_2+x_3)^2+(y_1-y_3+y_4)^2}. \tag{7}$$

4. Phase Diagram

In this chapter we consider the broad features of the phase diagram. This can be achieved numerically in a straightforward fashion. Starting from random initial conditions (with $y_1, y_2, y_3, y_4 \neq 0$), one iterates the recurrence relations (5) and observes their behaviour after a large number of iterations. In the simplest situation a fixed point $(x_1^*, x_2^*, x_3^*, x_4^*, y_1^*, y_2^*, y_3^*, y_4^*)$ is reached. It corresponds to a paramagnetic phase (P) if $y_1^* = y_2^* = y_3^* = y_4^* = 0$, or to a ferromagnetic phase (F) if $y_1^*, y_2^*, y_3^*, y_4^* \neq 0$. Secondary, the system may be periodic with period p, where case $p = 2$ corresponds to antiferromagnetic phase and case $p = 4$ corresponds to so-called antiphase, that denoted $< 2 >$ for compactness. Finally, the system may remain aperiodic. The distinction between a truly aperiodic case and one with a very long period is difficult to make numerically. Below we consider periodic phases with period p where $p \leq 12$. All periodic phases with period $p > 12$ and aperiodic phase we will consider as modulated phase (M). We plot phase diagrams in $(k_BT/J_1, D/J_1)$ plane for some fixed values of $p = -J_p/J_1$ and $k = K/J_1$ with $J_1 > 0$ and $D > 0$.
The phase diagrams of BEG model, i.e., $J_p = 0$, are shown in Fig.1. Here $p = 0$ and $k = \pm 0.5$. One can see that the phase diagram contains modulated phase, in addition to the expected paramagnetic and ferromagnetic ones. As shown by Vannimenus [1] for the Ising model one can reach modulated phase only in presence of competing next-nearest-neighbours interactions, meanwhile for the BEG model with nearest-neighbour interactions we have modulated phase. If $K > 0$ then the phase diagram contains paramagnetic, ferromagnetic and modulated phases and all three phases meet at the multicritical point $(T/J_1 = 1.6, D/J_1 = 1)$, and if $K < 0$ then the phase diagram contains ferromagnetic and modulated phases only.

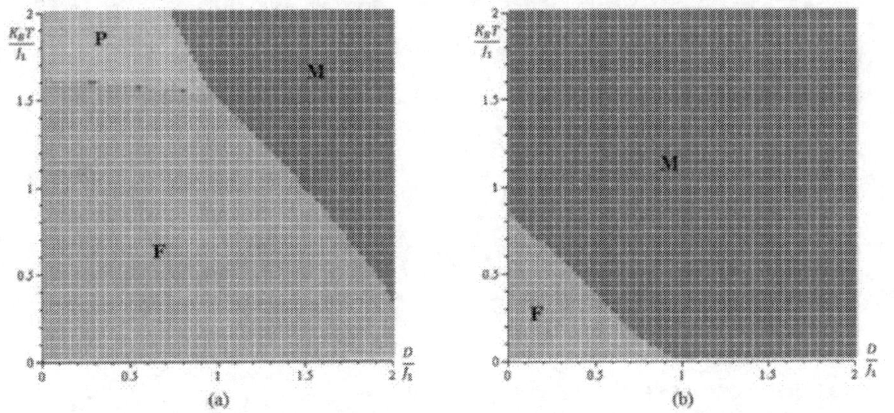

Fig. 1. Phase diagram of BEG model for (a) $K/J_1 = 0.5$ and (b) $K/J_1 = -0.5$

The phase diagrams of BEGV model with $p = 0.6$ and $k = \pm 0.5$ are shown in Fig.2
If $K > 0$ then the phase diagram contains antiphase and modulated phases only, and if $K < 0$ the phase diagram contains antiphase, modulated phase and regions with period 5, 6, and 9.

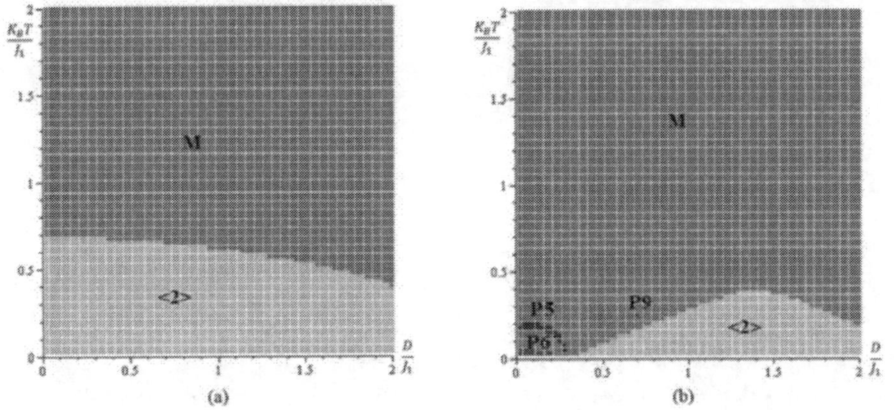

Fig. 2. Phase diagram of BEGV model for (a) $K/J_1 = 0.5$, $-J_p/J_1 = 0.6$ and (b) $K/J_1 = -0.5$, $-J_p/J_1 = 0.6$

5. Conclusion
We show that the local magnetization for BEG model with nearest-neighbour interactions on a Cayley tree is mainly chaotic with oscillatory glasslike behavior (see [1], [3] , [4]) and for BEGV model the local magnetization is fully chaotic. For BEG model on periodic lattice one can expect similar result. The structure of modulated phase and stability of the phases will be considered in another paper.

5.1. Acknowledgments
This work is supported by the FRGS Grant Phase 2/2010, FRGS 10-022-0141.

6. References
[1] Vannimenus J 1981 *Z.Phys.* B **43**, 141
[2] Mariz A M, Tsallis C. and Albuquerque E L 1985 *J.Stat.Phys.* **40**, 577.
[3] Fisher M E , *Physica* A **106** 28
[4] Bak P and von Boehm J 1980, *Phys.Rev.* B **21** 5297
[5] Blume M, Emery V J and R.B. Griffiths R B 1971, *Phys.Rev.* A **4** 1071

Scattering problems for the one-dimensional nonlinear Dirac equation with power nonlinearity

Hironobu Sasaki

Department of Mathematics and Informatics, Chiba University, 263–8522, Japan

E-mail: sasaki@math.s.chiba-u.ac.jp

Abstract. We study scattering problems for the one-dimensional nonlinear Dirac equation $(\partial_t + \alpha\partial_x + i\beta)\Phi = \lambda|\Phi|^{p-1}\Phi$. We prove that if $p > 3$ (resp. $p > 3 + 1/6$), then the wave operator (resp. the scattering operator) is well-defined on some 0-neighborhood of a weighted Sobolev space.

1. Introduction

In this paper, we introduce main results of [7]. We study scattering problems for nonlinear Dirac equations in one spatial dimension

$$(\partial_t + \alpha\partial_x + i\beta)\,\Phi(t,x) = F(\Phi(t,x)), \quad (t,x) \in \mathbb{R}\times\mathbb{R}. \tag{NLD}$$

Here, $\Phi = \Phi(t,x) = {}^t(\phi_1(t,x), \phi_2(t,x))$ is a $\mathbb{C}^2$-valued unknown function defined on $\mathbb{R}\times\mathbb{R}$, $\partial_t = (\partial/\partial t)I$, I is the 2×2 identity matrix, $\partial_x = (\partial/\partial x)I$, $i = \sqrt{-1}$, α and β are 2×2 Hermitian matrices satisfying that $\alpha^2 = \beta^2 = I$ and $\alpha\beta + \beta\alpha = 0$, F fulfills the following condition for some $p > 1$:

(N) $F = F(z_1, z_2)$ is a function from $\mathbb{C}^2$ into itself. Under the identification $\mathbb{C}^2 \ni (z_1, z_2) \simeq (x_1, \cdots, x_4) =: \mathbf{x} \in \mathbb{R}^4$, F is of class C^2 and the following inequalities hold for any multi-index $\gamma = (\gamma_1, \cdots, \gamma_4)$ with $\gamma_1 + \cdots + \gamma_4 \le 2$:

$$|\partial_{\mathbf{x}}^{\gamma} F(\mathbf{x})| \le C|\mathbf{x}|^{p-|\gamma|},$$
$$|\partial_{\mathbf{x}}^{\gamma} F(\mathbf{x}) - \partial_{\mathbf{x}}^{\gamma} F(\mathbf{y})| \le C|\mathbf{x}-\mathbf{y}|(|\mathbf{x}| + |\mathbf{y}|)^{p-|\gamma|-1}.$$

Since the term $F(z) = \lambda|z|^{p-1}z$ ($|z| := |z_1| + |z_2|$, $z = (z_1, z_2) \in \mathbb{C}^2$) satisfies (N), the nonlinear Dirac equation

$$(\partial_t + \alpha\partial_x + i\beta)\Phi(t,x) = \lambda|\Phi(t,x)|^{p-1}\Phi(t,x), \quad (t,x) \in \mathbb{R}\times\mathbb{R}$$

is an example of (NLD).

Nonlinear Dirac equations (NLD) have been used to model many physical systems (see, e.g., Pelinovsky [6] and references therein). For instance, the massive Thirring model is expressed by (NLD) with

$$F(\Phi) = \lambda\begin{pmatrix} |\phi_2|^2\phi_1 \\ |\phi_1|^2\phi_2 \end{pmatrix}, \quad \alpha = \begin{pmatrix} 1 & 0 \\ 0 & -1 \end{pmatrix}, \quad \beta = \begin{pmatrix} 0 & -1 \\ -1 & 0 \end{pmatrix},$$

and (NLD) with

$$F(\Phi) = \lambda|\phi_1|^2|\phi_2|^2\Phi + \lambda|\Phi|^2 \begin{pmatrix} |\phi_2|^2\phi_1 \\ |\phi_1|^2\phi_2 \end{pmatrix}, \quad \alpha = \begin{pmatrix} 1 & 0 \\ 0 & -1 \end{pmatrix}, \beta = \begin{pmatrix} 0 & -1 \\ -1 & 0 \end{pmatrix}$$

occurs in the context of the Feshbach resonance for Bose–Einstein condensates.

By using decay estimates for solutions to the free Dirac equation, we can expect that $p = 3$ is the critical exponent for the long time behavior of solutions to (NLD). In particular, if $p > 3$, then we can also expect that the wave operator W_- for (NLD) is well-defined on some 0-neighborhood of a Hilbert space. Our aim of this paper is to prove the existence of W_- for (NLD) with $p > 3$.

As for the long-time behavior of solutions to the one-dimensional nonlinear Schrödinger equation

$$iu_t + u_{xx} = \lambda|u|^{p-1}u, \tag{1.1}$$

the critical exponent is $p = 3$ (see, e.g., [1, 3]). The existence of W_- for (1.1) with $p > 3$ can be shown by using inequalities associated with $J_S := U_S(t)xU_S(-t)$. Here, $\{U_S(t)\}_{t\in\mathbb{R}}$ is the free Schrödinger evolution group. Since the identity

$$J_S = \exp\left(\frac{ix^2}{4t}\right)\left(2it\frac{\partial}{\partial x}\right)\exp\left(-\frac{ix^2}{4t}\right)$$

holds, the operator J_S is applicable to estimates of the nonlinearity. In the case of the one-dimensional nonlinear Klein-Gordon equation

$$u_{tt} - u_{xx} + u = \lambda|u|^{p-1}u, \tag{1.2}$$

the critical exponent is also $p = 3$ (see, e.g., [4, 5]). Hayashi–Naumkin [5] proved the existence of W_- for (1.2) with $p > 3$ by using inequalities associated with $J_{KG}^{\pm} := U_{KG}^{\pm}(t)xU_{KG}^{\pm}(-t)$, where $U_{KG}^{\pm}(t) = \exp(\mp it\omega)$ and $\omega = \sqrt{1 - (\partial^2/\partial x^2)}$. Since the free Klein-Gordon equation can be expressed by $((\partial/\partial t) - i\omega)((\partial/\partial t) + i\omega)u = 0$, we can call $\{U_{KG}^{+}(t) \oplus U_{KG}^{-}(t)\}_{t\in\mathbb{R}}$ the free Klein-Gorodon evolution group on the Hilbert space $H^s(\mathbb{R}^n) \oplus H^s(\mathbb{R}^n)$ $(s \in \mathbb{R})$. In order to estimate the nonlinearity, the following identity was used:

$$J_{KG}^{\pm} = \pm i\omega^{-1}\left(P_{KG} - \left(\frac{\partial}{\partial t} \pm i\omega\right)x\right),$$

where we define $P_{KG} = t(\partial/\partial x) + x(\partial/\partial t)$, which is directly applicable to estimates of the nonlinearity.

In this paper, we modify $J_{KG}^{\pm}$ and P_{KG} to study Dirac equations and we show that W_- for (NLD) is well-defined on some 0-neighborhood of a Hilbert space if $p > 3$. Furthermore, we prove the existence of the scattering operator S defined on some 0-neighborhood of a Hilbert space if $p > 3 + 1/6$.

In order to state our main results, we list some notation. For $s, w \geq 0$, we define the weighted Sobolev space $\mathbf{H}^{s,w}$ by

$$\mathbf{H}^{s,w} = \left\{\Psi \in L^2(\mathbb{R}, \mathbb{C}^2);\ x^m\partial_x^k\Psi \in L^2(\mathbb{R}, \mathbb{C}^2),\ m = 0, 1, \cdots, w,\ k = 0, 1, \cdots, s\right\}.$$

The $\mathbf{H}^{s,w}$-norm is given by

$$\|\Psi; \mathbf{H}^{s,w}\| = \sum_{m=0}^{w}\sum_{k=0}^{s}\left\|x^m\partial_x^k\Psi; L^2(\mathbb{R}, \mathbb{C}^2)\right\|.$$

Moreover, we denote $\mathbf{H}^{s,0}$ by $\mathbf{H}^s$, which is equal to the usual Sobolev space $H^s(\mathbb{R},\mathbb{C}^2)$. For a positive number δ and for a Hilbert space $(X,\|\cdot\|)$, we denote the closed ball of radius δ centered at the origin in X by $B(\delta;X)$. That is, $B(\delta;X)=\{x\in X;\ \|x\|\leq\delta\}$. We define the linear operator $D(t)$ for $t\in\mathbb{R}$ by

$$D(t)=\cos(t\omega)I-\omega^{-1}\sin(t\omega)(\alpha\partial_x+i\beta).$$

For any $s\geq 0$, we find that $D(t)\Phi_0$ ($\Phi_0\in\mathbf{H}^s$) is a time-global solution to the free Dirac equation and that $\{D(t)\}_{t\in\mathbb{R}}$ is a strongly continuous one parameter unitary group on the Hilbert space $\mathbf{H}^s$. Therefore we call $\{D(t)\}_{t\in\mathbb{R}}$ the free Dirac evolution group. For $s,w\geq 0$ and for $w_1\in[0,w]$, we define the wave operator $W_-:B(\delta;\mathbf{H}^{s,w})\ni\Phi_-\mapsto\Phi(0)\in\mathbf{H}^{s,w_1}$ if the following property holds for some $\delta>0$ and for some Banach space $Y\subset C(\mathbb{R};\mathbf{H}^s)$:

For any $\Phi_-\in B(\delta;\mathbf{H}^{s,w})$, there uniquely exists a time-global solution $\Phi\in Y$ to (NLD) such that $D(-t)\Phi(t)\in C(\mathbb{R};\mathbf{H}^{s,w_1})$ and

$$\lim_{t\to-\infty}\|D(-t)\Phi(t)-\Phi_-;\mathbf{H}^{s,w_1}\|=0.$$

Moreover, we define the inverse wave operator $V_+:B(\delta;\mathbf{H}^{s,w})\ni\Phi_0\mapsto\Phi_+\in\mathbf{H}^{s,w_1}$ if the following property holds:

For any $\Phi_0\in B(\delta;\mathbf{H}^{s,w})$, there uniquely exist a time-global solution $\Phi\in Y$ to (NLD) and a data $\Phi_+\in\mathbf{H}^{s,w_1}$ such that $\Phi(0)=\Phi_0$, $D(-t)\Phi(t)\in C(\mathbb{R};\mathbf{H}^{s,w_1})$ and

$$\lim_{t\to\infty}\|D(-t)\Phi(t)-\Phi_+;\mathbf{H}^{s,w_1}\|=0.$$

If we see that $W_-:B(\delta;\mathbf{H}^{s,w})\to\mathbf{H}^{s,w}$ and $V_+:B(C_0\delta;\mathbf{H}^{s,w})\to\mathbf{H}^{s,w}$ are well-defined and that $W_-(B(\delta;\mathbf{H}^{s,w}))\subset B(C_0\delta;\mathbf{H}^{s,w})$, then we define *the scattering operator* $S:B(\delta;\mathbf{H}^{s,w})\to\mathbf{H}^{s,w}$ by $S=V_+\circ W_-$.

We are ready to state our main results:

Theorem 1.1 *Let F satisfy (N) with $p\geq 5$. Then there exists some positive number δ_0 such that the scattering operator $S:B(\delta_0;\mathbf{H}^1)\to\mathbf{H}^1$ is well-defined.*

Theorem 1.2 *Let F satisfy (N) with $3+1/6<p<5$. Then there exists some positive number δ_0 such that the scattering operator $S:B(\delta_0;\mathbf{H}^{2,1})\to\mathbf{H}^{2,1}$ is well-defined.*

Theorem 1.3 *Let F satisfy (N) with $3<p\leq 3+1/6$. Then there exist some positive numbers δ_0 and C_0 satisfying the following properties:*

(1) The wave operator $W_-:B(\delta_0;\mathbf{H}^{2,2})\to\mathbf{H}^{2,1}$ is well-defined. Moreover, the inverse wave operator $V_+:B(C_0\delta_0;\mathbf{H}^{2,2})\to\mathbf{H}^{2,1}$ is well-defined.

(2) We have $W_-(B(\delta_0;\mathbf{H}^{2,2}))\subset B(C_0\delta_0;\mathbf{H}^{2,2})$ and $V_+(B(C_0\delta_0;\mathbf{H}^{2,2}))\subset B(C_0^2\delta_0;\mathbf{H}^{2,2})$, and hence the composite mapping $V_+\circ W_-:B(\delta_0;\mathbf{H}^{2,2})\to B(C_0^2\delta_0;\mathbf{H}^{2,2})$ is well-defined.

We give some remarks on our main results.

Remark 1.4 *Fix $p>3$. Let Φ_- (resp. Φ_0) belong to the domain of W_- (resp. V_+). We see from the proof of Theorems 1.1–1.3 that $W_-(\Phi_-)$ and $V_+(\Phi_0)$ are expressed by*

$$W_-(\Phi_-)=\Phi_-+\int_{-\infty}^0 D(-t)F(\Phi(t))dt$$

and

$$V_+(\Phi_0)=\Phi_0+\int_0^\infty D(-t)F(\Phi(t))dt.$$

Here, $\Phi(t)$ is a time-global solution to (NLD). Moreover, if $p > 3 + 1/6$, then we obtain

$$S(\Phi_-) = \Phi_- + \int_{\mathbb{R}} D(-t)F(\Phi(t))dt.$$

Remark 1.5 *Let $3 < p \leq 3+1/6$. By S', we denote the composite mapping $V_+ \circ W_-$ appearing in Theorem 1.3. Unfortunately, we do not see whether S' becomes the scattering operator because we do not prove the existence of a time-global solution $\Phi(t)$ satisfying $\|D(-t)\Phi(t) - \Phi_\pm; \mathbf{H}^{2,2}\| \to 0$ as $t \to \pm\infty$. However, we see that S' has the same expression as the scattering operator. That is, we have*

$$S'(\Phi_-) = \Phi_- + \int_{\mathbb{R}} D(-t)F(\Phi(t))dt.$$

Remark 1.6 *Let $p > 3$. Using the above expression of $V_+ \circ W_-$, we can obtain some results of inverse scattering problems for (NLD). For instance, for any $\Phi_- \in \mathcal{S}(\mathbb{R}) \setminus \{0\}$, we have*

$$p = \lim_{\varepsilon \to 0} \frac{\log \|(V_+ \circ W_-)(\varepsilon\Phi_-) - \varepsilon\Phi_-; L^2(\mathbb{R}; \mathbb{C}^2)\|}{\log \varepsilon}$$

and

$$\lambda = \frac{\lim_{\varepsilon \to 0} \varepsilon^{-p} \langle (V_+ \circ W_-)(\varepsilon\Phi_-) - \varepsilon\Phi_-, \Phi_- \rangle_{L^2(\mathbb{R};\mathbb{C}^2)}}{\int_{\mathbb{R}^{1+1}} |D(t)\Phi_-(x)|^{p+1}\, d(t,x)}.$$

For the case of nonlinear Schrödinger equations, see, e.g., [2] and [8].

[1] J.E. Barab, Nonexistence of asymptotically free solutions for a nonlinear Schrödinger equation, J. Math. Phys. 25 (1984) 3270–3273.
[2] R. Carles and I. Gallagher, Analyticity of the scattering operator for semilinear dispersive equations, Comm. Math. Phys. 286 (2009) 1181–1209.
[3] T. Cazenave and F.B. Weissler, Rapidly decaying solutions of the nonlinear Schrodinger equation, Comm. Math. Phys. 147 (1992) 75–100.
[4] N.Hayashi and P.I.Naumkin, The initial value problem for the cubic nonlinear Klein-Gordon equation, Z. Angew. Math. Phys. 59 (2008) 1002–1028.
[5] N.Hayashi and P.I.Naumkin Scattering operator for nonlinear Klein-Gordon equations, Communications in Contemporary Mathematics 11 (2009) 771–781.
[6] D. Pelinovsky, Survey on global existence in the nonlinear Dirac equations in one spatial dimension, RIMS Kokyuroku Bessatsu B26 (2011) 37-50.
[7] H. Sasaki, Small data scattering for the one-dimensional nonlinear Dirac equation with power nonlinearity, preprint.
[8] W.A. Strauss, Nonlinear scattering theory, in: Scattering Theory in Math. Physics, Reidel, Dordrecht, 1974, pp. 53–78.

Wave optics and image formation in gravitational lensing

Yasusada Nambu

Department of Physics, Graduate School of Science, Nagoya University, Chikusa, Nagoya 464-8602, Japan

E-mail: nambu@gravity.phys.nagoya-u.ac.jp

Abstract. We discuss image formation in gravitational lensing system using wave optics. Applying the Fresnel-Kirchhoff diffraction formula to waves scattered by a gravitational potential of a lens object, we demonstrate how images of source objects are obtained directly from waves without using a lens equation for gravitational lensing.

wave optics; image formation; gravitational lens

1. Introduction

Gravitational lensing is one of the prediction of Einstein's general theory of relativity and many samples of images caused by gravitational lensing have been obtained observationally [1]. Light rays obey null geodesics in curved spacetime and they are deflected by the gravitational potential of lens objects. In weak gravitational field with thin lens approximation, a path of a light ray obeys the so called lens equation for gravitational lensing and many analysis concerning the gravitational lensing effect have been carried out based on this equation. Especially, we can obtain images of source objects by solving the lens equation using a ray tracing method. Since the path of light ray is derived as the high frequency limit of electromagnetic wave, wave effects of gravitational lensing become significant when the wavelength is not so much smaller than the size of lens objects and in such a situation, we must take into account of wave effects. For example, when we consider a gravitational wave which is scattered by gravitational lens objects, the wave effect gives significant impact on the amplification factor of intensity for waves [2, 3, 4]. Another example is the direct detection of black holes via their shadows [5, 6]. The apparent angular size of black hole shadows are so small that their detectability depends on the angular resolution of telescopes which is related to the diffraction limit of the image formation system.

Although interference and diffraction of waves by gravitational lensing have been discussed in connection with amplification of waves, little has been discussed about how images by gravitational lensing are obtained based on wave optics. For an electromagnetic wave, E. Herlt and H. Stephani [7] discussed the position of images by a spherical gravitational lens evaluating the Poynting flux of scattered wave at an observer. They claimed that there is a disagreement between wave optics and geometrical optics concerning the position of double images of a point source. But they have not presented a complete understanding of image formation. In this paper, we consider the image formation in gravitational lensing using wave optics and aim to understand how images by gravitational lensing are obtained in terms of waves. For this purpose, we adopt the diffraction theory of image formation in wave optics [8], which explains image formation in optical systems in terms of diffraction of waves.

2. Image formation based on wave optics

2.1. Image formation by a convex lens

To establish relation between the interference pattern of the wave and the images of the source in the gravitational lensing system, we first consider an image formation system composed of a single convex lens (Figure 1). We assume the distribution of the source field with the angular frequency ω on the object plane $z = -a$ as $\Phi_0(\boldsymbol{x}_0)$. Using the Fresnel-Kirchhoff diffraction formula, the amplitude of the wave at $\boldsymbol{x}$ on the $z = b$ plane behind the lens is

$$\Phi_2(\boldsymbol{x}) \propto \int d^2x_0\, d^2x_1\, \Phi_0(\boldsymbol{x}_0)t_L(\boldsymbol{x}_1)\exp\left[i\omega\left\{\frac{1}{2}\left(\frac{1}{a}+\frac{1}{b}-\frac{1}{f}\right)|\boldsymbol{x}_1|^2 - \left(\frac{\boldsymbol{x}_0}{a}+\frac{\boldsymbol{x}}{b}\right)\cdot\boldsymbol{x}_1\right\}\right] \times\exp\left[i\frac{\omega}{2}\left(\frac{|\boldsymbol{x}_0|^2}{a}+\frac{|\boldsymbol{x}|^2}{b}\right)\right] \quad (1)$$

where $\boldsymbol{x}_1$ is the location of the light path on the lens plane, t_L is the aperture function and f represents the focal length of the lens. For a value of b satisfying the lens equation for a convex thin lens, *i.e.* $1/a + 1/b = 1/f$, the wave amplitude becomes

$$\Phi_2(\boldsymbol{x}) \propto \int d^2x_0\Phi_0(\boldsymbol{x}_0)\left(\frac{2J_1(\omega D|\boldsymbol{x}/b+\boldsymbol{x}_0/a|)}{\omega D|\boldsymbol{x}/b+\boldsymbol{x}_0/a|}\right)\exp\left[i\omega\frac{|\boldsymbol{x}_0|^2}{2a}\right]. \quad (2)$$

For $\omega D \to \infty$ limit, the Bessel function, *i.e.* J_1, in (2) becomes the delta function and we obtain the following wave amplitude on $z = b$:

$$\Phi_2(\boldsymbol{x}) \propto \int d^2x_0\, \Phi_0(\boldsymbol{x}_0) \times \delta^2\left[\frac{\boldsymbol{x}_0}{a}+\frac{\boldsymbol{x}}{b}\right] = \Phi_0\left(-\frac{a}{b}\boldsymbol{x}\right). \quad (3)$$

Thus, a magnified image of the source field appears on the $z = b$ plane. This reproduces the result of image formation in geometric optics.

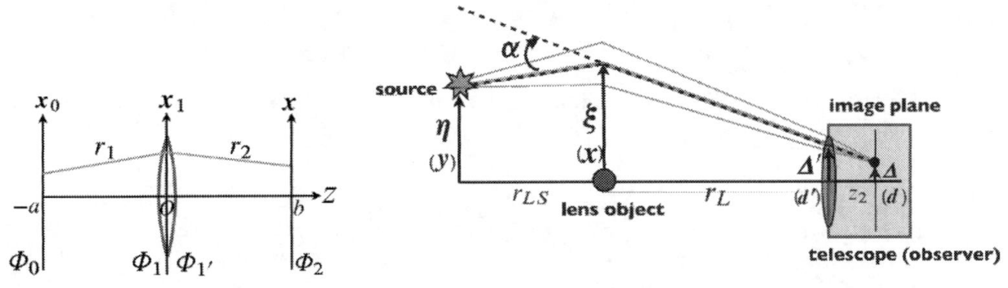

Figure 1. The left panel: one lens image formation system. The right panel: configuration of a gravitational lens with a convex lens system. r_L is the distance from the observer to the lens object and r_S is the distance from the observer to the source. $r_{LS} = r_S - r_L$.

2.2. Image formation in gravitational lens system

As we have observed that a convex lens can be a device for image formation in wave optics, we combine it with a gravitational lensing system and obtain images by gravitational lensing. We consider a configuration of the gravitational lens system shown in the right panel in Figure 1 and examine how the images of the source object appear using wave optics. As source object, we assume a point source of wave with intensity a_0. For a monochormatic wave with angular frequency ω, the wave function obeys

$$(\nabla^2 + \omega^2)\Phi = 4\omega^2 U(\boldsymbol{r})\Phi \quad (4)$$

where U is the gravitational potential of the lens object. Introducing the two dimensional gravitational potential $\hat{\psi}(\boldsymbol{\xi}) = 2\int_{-\infty}^{\infty} dz U(\boldsymbol{\xi}, z)$, the wave amplitude on the image plane becomes

$$\Phi_I(\boldsymbol{y}, \boldsymbol{d}) = \frac{a_0}{r_S}\int_{|\boldsymbol{d}'|\le d_0} d^2d' F(\boldsymbol{y}+\boldsymbol{d}') \quad (5)$$
$$\times\exp\left[i\omega\left\{\frac{r_Lr_S}{2r_{LS}}\left(\frac{1}{r_S}+\frac{1}{z_2}-\frac{1}{f}\right)\boldsymbol{d}'^2 - \left(\boldsymbol{y}+\frac{r_Lr_S}{r_{LS}z_2}\boldsymbol{d}\right)\cdot\boldsymbol{d}' + \frac{1}{2}\left(\frac{r_{LS}}{r_L}\boldsymbol{y}^2 + \frac{r_Lr_S}{r_{LS}z_2}\boldsymbol{d}^2\right)\right\}\right]$$

where F is the amplification factor by a gravitational lensing given by

$$F(\boldsymbol{y},\boldsymbol{d}) = \frac{w}{2\pi i}\int d^2x \exp\left[iw\left(\frac{1}{2}(\boldsymbol{x}-\boldsymbol{y}-\boldsymbol{d})^2 - \psi(\boldsymbol{x})\right)\right] \tag{6}$$

and the following dimensionless variables are introduced

$$\boldsymbol{x} = \frac{\boldsymbol{\xi}}{\xi_0}, \quad \boldsymbol{y} = \frac{r_L}{r_S}\frac{\boldsymbol{\eta}}{\xi_0}, \quad \boldsymbol{d} = \left(1-\frac{r_L}{r_S}\right)\frac{\boldsymbol{\Delta}}{\xi_0}, \quad w = \frac{r_S\xi_0^2}{r_{LS}r_L}\omega, \quad \psi = \frac{r_L r_{LS}}{r_S\xi_0^2}\hat{\psi}$$

and we choose ξ_0 as $\xi_0 = r_L\theta_E, \theta_E = \sqrt{4Mr_{LS}/(r_L r_S)}$. M is the mass of the gravitational source, ξ_0 and θ_E represent the Einstein radius and the Einstein angle, respectively.

If we choose the location of the image plane z_2 to satisfy the following "lens equation" for a convex lens $1/r_S + 1/z_2 = 1/f$, then the wave amplitude on the image plane becomes

$$\Phi_I(\boldsymbol{y},\boldsymbol{d}) = \frac{a_0}{r_S}\int_{|\boldsymbol{d}'|\le d_0} d^2d' F(\boldsymbol{y}+\boldsymbol{d}')\exp\left[-iw\left(\boldsymbol{y}+\frac{r_L r_S}{r_{LS}f}\boldsymbol{d}\right)\cdot\boldsymbol{d}'\right]\times e^{i(w/2)g(\boldsymbol{y},\boldsymbol{d})} \tag{7}$$

where $g = (r_{LS}/r_L)\boldsymbol{y}^2 + (r_L r_S/r_{LS}z_2)\boldsymbol{d}^2$. Thus the wave amplitude on the image plane is the Fourier transform of the amplification factor F which gives the interference fringe pattern. Under the geometrical optics limit $w \gg 1$, $\boldsymbol{x}$ integral in the amplification factor (6) can be approximated by the WKB form $F(\boldsymbol{y}+\boldsymbol{d}') \approx A\, e^{iw\left[\frac{1}{2}(\boldsymbol{x}_*-\boldsymbol{y}-\boldsymbol{d}')^2-\psi(\boldsymbol{x}_*)\right]}$, where $\boldsymbol{x}_*(\boldsymbol{y},\boldsymbol{d}')$ is the solution of the lens equation

$$0 = \boldsymbol{x}-\boldsymbol{y}-\boldsymbol{d}'-\boldsymbol{\nabla}_{\boldsymbol{x}}\psi(\boldsymbol{x}) \approx \boldsymbol{x}-\boldsymbol{y}-\boldsymbol{\nabla}_{\boldsymbol{x}}\psi(\boldsymbol{x}). \tag{8}$$

We have assumed that the aperture of the convex lens is sufficiently smaller than the size of the gravitational lensing system and $|\boldsymbol{d}'| \le d_0 \ll 1$ holds. Then, the wave amplitude on the image plane is

$$\Phi_I(\boldsymbol{y},\boldsymbol{d}) \propto A\, e^{iw\left[\frac{1}{2}(\boldsymbol{x}_*-\boldsymbol{y})^2-\psi(\boldsymbol{x}_*)\right]} e^{iwg(\boldsymbol{y},\boldsymbol{d})/2} \times 2\pi d_0^2\,\frac{J_1(w|\boldsymbol{x}_*+\beta\boldsymbol{d}|d_0)}{w|\boldsymbol{x}_*+\beta\boldsymbol{d}|d_0}, \quad \beta \equiv \frac{r_L r_S}{r_{LS}f}. \tag{9}$$

For $wd_0 \to \infty$ limit (large lens aperture limit or high frequency limit), we obtain $\Phi_I(\boldsymbol{y},\boldsymbol{d}) \propto \delta^2\left[\boldsymbol{x}_*(\boldsymbol{y}) + \frac{r_L r_S}{r_{LS}f}\boldsymbol{d}\right]$ and the image of the point source appears at the location on the image plane determined by the lens equation (8). This reproduces the same result of image formation in the geometrical optics (ray tracing) in terms of the wave optics.

As an example of image formation in a gravitational lensing system using wave optics, we present the wave optical images of a point source by the gravitational lensing of a point mass (Figure 2). These images are obtained by Fourier transformation of the amplification factor F (equation (7)). This procedure corresponds to image formation by a convex lens. We can observe the wave effect in these images.

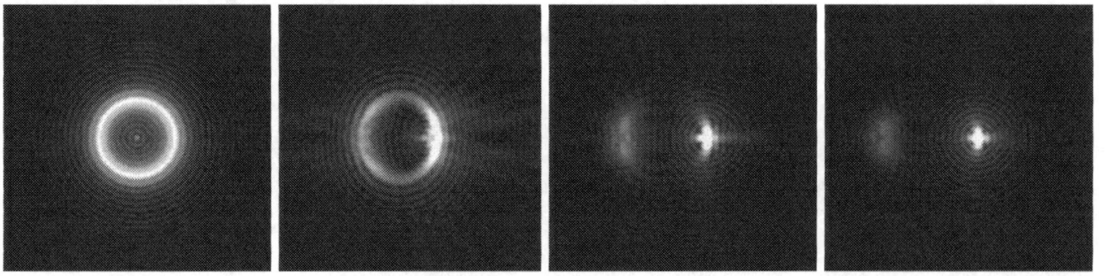

Figure 2. Wave optical images of a point source by the gravitational lensing of a point mass. Parameters are $w = 40, d_0 = 0.5$(aperture of a convex lens), $y = 0, 0.5, 1, 1.5$.

In each one of these images, we observe a concentric interference pattern which is caused by the finite size of the lens aperture and this is not an intrinsic feature of the gravitational lensing system. We can also observe radial non-concentric patterns. These are caused by interference between double images and represent the intrinsic feature of the gravitational lensing system. For $y = 0$ case which corresponds to the Einstein ring in the geometrical optics limit, we can observe a bright spot at the center of the ring, which is the result of constructive interference and does not appear in geometric optics. For sufficiently large values of wd_0, the wave amplitude at the observer coincides with the result obtained by geometric optics.

3. Summary

We have investigated image formation in gravitational lensing system based on wave optics. Instead of using a ray tracing method, we obtained images directly from wave functions at the observer. For this purpose, we introduced a "telescope" with a single convex thin lens, which acts as a Fourier transformer for the interference pattern formed at an observer. The analysis in this article relates the wave amplitude and images by the gravitational lensing directly. In the geometric optics limit of waves, images by lens systems are obtained by a lens equation which determines paths of each light rays. As light rays are trajectories of massless test particles (photons), expressing images in terms of waves is like expressing particles in terms of waves.

As an application and extension of analysis presented in this paper, we plan to investigate gravitational lensing by a black hole and obtain wave optical images of black holes. This is related to the topic of observation of black hole shadows [5, 6]. As the apparent angular sizes of black hole shadows are so small, the diffraction effect on images are crucial to resolve black hole shadows in observation using radio interferometer. For SgrA*, which is the black hole candidate at Galactic center, the apparent angular size of its shadow is estimated to be $\sim 30\mu$ arc seconds and this value is the largest among black hole candidates. For a sub-mm VLBI with a baseline length D, using the resolving power of images $\theta_0 = \lambda/D$ detemined by the diffraction limit, the condition to resolve the shadow requires $D > 1000$ km, and this shows the possibility to detect the black hole shadow of SgrA* using the present day technology of VLBI telescope. Thus, analysis of black hole shadows based on wave optics is important to evaluate detectability of shadows and determination of black hole parameters via imaging of black holes.

References

[1] Schneider P, Ehlers J and Falco E E *Gravitational Lenses* (Springer-Verlag, 1992)
[2] Nakamura T T and Deguchi S 1999 Wave Optics in Gravitational Lensing *Prog. Theor. Phys. Suppl.* **133** 137–153
[3] Baraldo C and Hosoya A 1999 Gravitationally induced interference of gravitational waves by a rotating massive object *Phys. Rev. D* **59** 083001
[4] Matsunaga N and Yamamoto K 2006 The finite source size effect and wave optics in gravitational lensing *JCAP* **01** 023
[5] Falcke H, Melia F and Agol E 2000 Viewing the shadow of the black hole at the galactic center *Astrophys. J.* **528** L13–L16
[6] Miyoshi M, Ishituska K, Kameno S and Shen Z Q 2004 Direct imaging of the black hole, SgrA* *Prog. Theor. Phys. Suppl.* **155** 186–189
[7] Herlt E and Stephani H 1976 Wave Optics of the Spherical Gravitational Lens Part I: Diffraction of a Plane Electromagnetic Wave by a Large Star *Int. J. of Theor. Phys.* **15** 45–65
[8] Sharma K K *Optics: principles and applications* (Academic Press, 2006)

Translating the Concept of Five-Dimensional Tangent Vectors in Space-Time to Internal Gauge Symmetries in Elementary Particle Physics

Alexander B. Krasulin
Institute for Nuclear Research of the Russian Academy of Sciences (former affiliation)
E-mail: krasulin@post.com

Abstract. This paper discusses the analogs of space-time five-dimensional tangent vectors for the case of internal symmetry vector spaces associated with gauge groups in elementary particle physics and their possible application in the Standard Model of quarks and leptons.

1. Introduction

Vector-like objects not directly associated with the space-time manifold are widely used in today's theory of elementary particle. In their properties, such objects resemble ordinary tangent vectors associated with some differentiable, usually complex, manifold endowed with appropriate metric. Since, historically, such nonspacetime vectors have been first introduced to reflect certain symmetries—exact or approximate—observed in the world of elementary particles, physicists usually speak of representations with respect to a symmetry group rather than of nonspacetime vectors of a certain kind; of conjugation and contraction rather than of an antilinear/linear inner (scalar) product; of "gauging" the corresponding symmetry rather than of introducing a new kind of geometry represented by nontrivial rules of parallel transport for the considered type of nonspacetime vectors; and of the corresponding gauge fields rather than of connection coefficients.

Mathematically, both approaches are equivalent. However, when regarded as paradigms within which one searches for new physics beyond the Standard Model, they may offer one different possibilities. One particular advantage of the "geometric" approach—the one where one talks about nonspacetime vectors and their parallel transport—is that within it one may pose the following question: If the nonspacetime analogs of ordinary tangent vectors find such useful application in the theory of elementary particle, could it be that the analogs of five-dimensional tangent vectors in space-time can also be of use?

Let me recall that the so-called five-dimensional tangent vectors (see ref.[1] or ref.[2] for a simpler presentation) are a special kind of geometric objects that can be defined in space-time. Like ordinary tangent vectors, these five-dimensional vectors and the tensors constructed out of them can be used for describing certain local quantities and in this capacity find direct application in physics. For example, such familiar physical quantities as the stress-energy and angular momentum tensors prove to be parts of a single five-tensor. Similar objects can be defined in any other differentiable manifold. In the latter case they will be called $(n+1)$-vectors, where n is the dimension of the manifold. At each point of the latter, such vectors

Published under licence by IOP Publishing Ltd

form a vector space whose dimension is one unit greater than that of the manifold and which is isomorphic to the direct product of the vector space of ordinary tangent vectors and of an abstract one-dimensional vector space. In other words, algebraically, each $(n+1)$-vector is equivalent to a pair consisting of an ordinary tangent vector and a scalar. What makes the latter pair an $(n+1)$-vector is the special way in which its components changes under parallel transport. In the particular case of the space-time manifold, the rules of parallel transport for five-vectors are such that the four-vector in the pair representing the five-vector is transported according to ordinary rules of parallel transport for four-vectors, while the scalar in the pair acquires an additive which at each infinitesimal step of the transport is proportional to the scalar product of the four-vector in the pair with the infinitesimal four-vector representing this step. In the case of $(n+1)$-vectors of any other type the situation is similar: The rules of parallel transport for $(n+1)$-vectors are such that the ordinary (n-dimensional) tangent vector in the pair representing it is transported according to the rules of parallel transport fixed by the corresponding ordinary gauge fields, while the scalar in the pair acquires an additive which at each infinitesimal step of the transport is a linear function of the vector components. In more detail, $(n+1)$-vectors are examined in the last section of ref.[3].

2. The necessary technicalities

It is easier to understand the situation if one writes out the formulae that give the components of the covariant derivative of an $(n+1)$-vector. Let us introduce the following notations. The components of ordinary n-dimensional nonspacetime vectors in some appropriate basis will be denoted as U^i, V^i, etc., where the index i, as all lower-case Roman indices, is assumed to run 1 through n. The components of the covariant derivative of the vector field U^i are given by the usual formula:

$$U^i{}_{;\mu} = \partial_\mu U^i + C^i{}_{j\mu} U^j,$$

where $C^i{}_{j\mu}$ are ordinary gauge fields. The components of $(n+1)$-vectors in a corresponding basis will be denoted as u^Θ, v^Θ , etc., where the index Θ, as all upper-case Greek indices, is assumed to run 1 through $(n+1)$. The components of the covariant derivative of the vector field u^Θ are then given by the following formulae:

$$u^i{}_{;\mu} = \partial_\mu u^i + C^i{}_{j\mu} u^j \tag{1}$$

and

$$u^{(n+1)}{}_{;\mu} = \partial_\mu u^{(n+1)} + C^{(n+1)}{}_{j\mu} u^j + C^{(n+1)}{}_{(n+1)\,\mu} u^{(n+1)}, \tag{2}$$

where i, $j = 1$, ..., n. As one can see, the expression for the first n components of the covariant derivative of an $(n+1)$-vector field is exactly the same as it is for the field whose values are ordinary (n-dimensional) vectors. However, the $(n+1)$-st component of such a covariant derivative has two kinds of terms involving connection coefficients: (i) a term proportional to $u^{(n+1)}$, which reflects the change in the absolute value and/or phase of the $(n+1)$-st component of the field not related to its first n components; and (ii) terms proportional to the components u^i, which reflect the "admixing" of these latter components to $u^{(n+1)}$.

In the particular case of complex vectors for which the inner product is Hermitian and is positively definite, formula (1) can be presented in the following more familiar form:

$$u^i{}_{;\mu} = \partial_\mu u^i + (i/2)\, g\, (t_a)^i{}_j\, C^a_\mu\, u^j + \; ig\, [2n(n+1)]^{-1/2}\, C^0_\mu\, u^i, \tag{3}$$

where the index a runs 1 through n^2-1; the matrices $(t_a)^i{}_j$ are the usual (Hermitian) generators for the fundamental representation of SU(n), normalized by the condition $\mathrm{Tr}(t_a t_b) = 2\delta_{ab}$; the fields C^a_μ and C^0_μ are real; and g is a dimensionless constant, which together with the factors

$1/2$ and $[2n(n+1)]^{-1/2}$ is introduced for convenience. In a similar manner, formula (2) can be cast into the following form:

$$u^{(n+1)}{}_{;\,\mu} = \partial_\mu u^{(n+1)} - ig\,[n/2(n+1)]^{1/2}\,C^0{}_\mu\,u^{(n+1)} + gX_{j\mu}\,u^j, \tag{4}$$

where $X_{j\mu} \equiv g^{-1}C^{(n+1)}{}_{j\mu}$. The expression for the components of the covariant derivative of a field v_i whose values are linear forms (*co*variant vectors) associated with the considered type of $(n+1)$-vectors, can be written down as follows:

$$v_{i\,;\,\mu} = \partial_\mu v_i - (i/2)\,g\,v_j\,(t_a)^j_i\,C^a_\mu - ig\,[2n(n+1)]^{-1/2}\,v_i\,C^0_\mu - g\,v_{(n+1)}X_{i\mu} \tag{5}$$

and

$$v_{(n+1)\,;\,\mu} = \partial_\mu v_{(n+1)} + ig\,[n/2(n+1)]^{1/2}\,v_{(n+1)}C^0{}_\mu. \tag{6}$$

One should note that the interaction with the fields $X_{i\mu}$ is not C-invariant, which means that in this case the charge asymmetry is implemented directly in the nonspacetime degrees of freedom of the fields. A more detailed discussion of this issue can be found in ref.[3].

3. A simplified model

Let me now show how the concept of $(n+1)$-vectors can be applied in the framework of the Standard Model to one generation of leptons. Let us take the first generation, where one has a doublet of left-handed particles (e_L, ν_L), where e is the electron and ν is the electronic neutrino. Following our recipe for constructing $(n+1)$-vectors, one should suppose that these fields are the first two components of a $(2+1)$-vector or of a two-plus-one-vector 1-form. Furthermore, one should suppose that there exists a third left-handed field, say, χ_L, that makes up the third component of this $(2+1)$-vector or this two-plus-one-vector 1-form.

One now has to decide whether this triplet is a $(2+1)$-vector or a two-plus-one-vector 1-form. In the case of ordinary nonspacetime vectors, a similar question was rather trivial, for one could always choose the generators of the gauge group in such a way that the considered n-plet would transform according to the fundamental representation. If need be, one could always redefine the symmetry group so that this n-plet would transform according to its anti-fundamental representation. In the case of $(n+1)$-vectors the choice between a fundamental and an anti-fundamental representation—alias, between a vector and a 1-form—should be made based on the kind of interaction one wishes to have between the fields involved. Indeed, if the triplet (e_L, ν_L, χ_L) were components of a $(2+1)$-vector, then according to equation (4), there should exist an interaction, mediated by the gauge fields $X_{j\mu}$, by means of which the doublet (e_L, ν_L) would decay into χ_L. At the same time, since the triplet of anti-particles $(\overline{e_L}, \overline{\nu_L}, \overline{\chi_L})$ in this case form a two-plus-one-vector 1-form, by virtue of equation (5), the same interaction should result in that the SU(2) singlet $\overline{\chi_L}$ would decay into the doublet $(\overline{e_L}, \overline{\nu_L})$. As a result, one would be left with the anti-particles $\overline{e_L}$ and $\overline{\nu_L}$ and particle χ_L, which is *not* what one observes in reality.

The alternative is that the triplet (e_L, ν_L, χ_L) be a two-plus-one-vector 1-form. In this case, in accordance with equation (5), the particle χ_L would decay into the doublet (e_L, ν_L), while the pair of anti-particles $(\overline{e_L}, \overline{\nu_L})$ would decay into the SU(2) singlet $\overline{\chi_L}$. In the end, one will be left with the particles e_L and ν_L and anti-particle $\overline{\chi_L}$, the latter being a right-handed particle. Considering what one observes in reality, within this simplified model, it is natural to suppose that $\overline{\chi_L}$ is actually the *right-handed electron*, e_R. One therefore comes to the conclusion that the SU(2) doublet (e_L, ν_L) and the anti-particle $\overline{e_R}$, which is an SU(2) singlet, make up a two-plus-one-vector 1-form.

We thus see that the additional interaction that should exist if e_L, ν_L, and $\overline{e_R}$ indeed make up a two-plus-one-vector 1-form, may have very well resulted in the observed particle-antiparticle asymmetry in the Universe. Of course, to demonstrate that this is ineed so, one should construct a more complicated model that would include quarks and other generations.

4. Conclusion

The purpose of this short paper was to demonstrate how the concept of $(n+1)$-vectors can be applied in the framework of the Standard Model to one generation of leptons, and to describe one interesting phenomenon that should occur if the assumptions made are true. The considered case should be regarded only as a toy model that just illustrates the idea, but cannot pretend to be a real model. The inclusion of quarks and of other generations into the scheme will be discussed in a subsequent paper.

References

[1] A.B. Krasulin, Five-dimensional tangent vectors in space-time. I. Introduction and formal theory, Moscow, see e.g. math-ph/9805004 at arxiv.org.
[2] A. Krasulin, Bivector gauge fields from classical electrodynamics, In: Dvoeglazov, V.V. (ed.): Photon and Poincare group, 1999, p.218-230.
[3] A.B. Krasulin, Five-dimensional tangent vectors in space-time. V. Generalization of covariant derivative, Moscow, see e.g. math-ph/9808014 at arxiv.org.

Modeling DNA Dynamics by Path Integrals

Marco Zoli
School of Science and Technology - CNISM
University of Camerino, I-62032 Camerino, Italy

E-mail: marco.zoli@unicam.it

Abstract. Complementary strands in DNA double helix show temporary fluctuational openings which are essential to biological functions such as transcription and replication of the genetic information. Such large amplitude fluctuations, known as the breathing of DNA, are generally localized and, microscopically, are due to the breaking of the hydrogen bonds linking the base pairs (*bps*). I apply imaginary time path integral techniques to a mesoscopic Hamiltonian which accounts for the helicoidal geometry of a short circular DNA molecule. The *bps* displacements with respect to the ground state are interpreted as time dependent paths whose amplitudes are consistent with the model potential for the hydrogen bonds. The portion of the paths configuration space contributing to the partition function is determined by selecting the ensemble of paths which fulfill the second law of thermodynamics. Computations of the thermodynamics in the denaturation range show the energetic advantage for the equilibrium helicoidal geometry peculiar of B-DNA. I discuss the interplay between twisting of the double helix and anharmonic stacking along the molecule backbone suggesting an interesting relation between intrinsic nonlinear character of the microscopic interactions and molecular topology.

1. Hamiltonian Model for DNA

While it has been long recognized that the form of the DNA molecule is key to understand its biological function [1], a considerable amount of work has been recently devoted to explain the DNA dynamics in terms of the microscopic interactions at play in the double helix. A fully atomistic description of DNA, even of a short fragment, would represent a formidable computational task due to the huge number of degrees of freedom. Moreover, the specificity of DNA lies in its capability to store the genetic information and allow a reading of the latter through large amplitude motions which temporarily bring apart portions of the complementary strands. To describe these properties we need mesoscopic models at the scale of the base pair, the fundamental entity in the nucleotide which encodes the information. The one-dimensional Dauxois-Peyrard-Bishop (DPB) model [2] has provided a fundamental tool for the biophysicists working in the field.

The DPB Hamiltonian for a system of N *bps*, with reduced mass μ, assumes the pair mates separation y_n (for the *n-th* base pair) with respect to the ground state position as the relevant degree of freedom. The inter-base pair interactions are modeled by a Morse potential $V_M(y_n)$ whereas the intra-base pair stacking along the molecule backbone is described by an anharmonic potential $V_S(y_n, y_{n-1})$.

Recently, I have proposed to apply the path integral method [3] to a modified DPB Hamiltonian which includes a twist angle θ between adjacent bases, n and $n-1$, along the DNA backbone [4] as shown in Fig. 1. Twisting is described by the angle that the *bps* rotate

Journal of Physics: Conference Series **410** (2013) 012038 doi:10.1088/1742-6596/410/1/012038

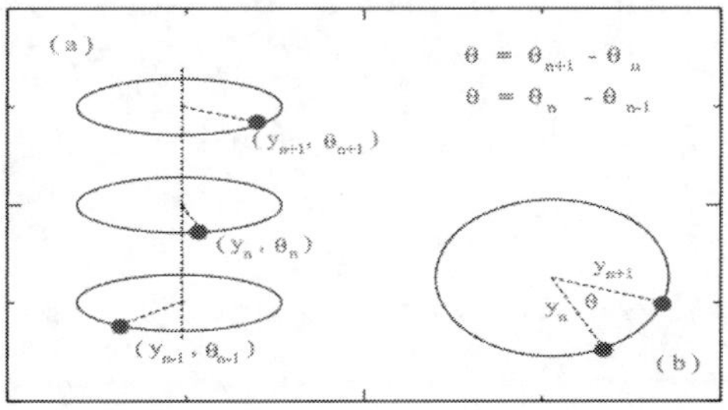

Figure 1. (Color online) (a) Fixed planes picture for the right-handed helicoidal model. The blue filled circles denote the pointlike base pairs stacked along the molecule axis with twist θ. The radial coordinate y_n describes the $n-th$ base pair displacement from the ground state. The dashed vertical axis corresponds to the $y_n \equiv 0$ configuration, the minimum for the one-coordinate potential $V_M(y_n) + V_{sol}(y_n)$ in Eq. (1). (b) The helix plane seen from above.

around the molecule axis. B-DNA at room temperature has a helix repeat of $\sim 35\,\AA$ hosting $h \sim 10\ bps$, hence the equilibrium twist angle is $\theta_{eq} = 2\pi/h \sim 0.6\,rad$. Taking a short fragment, $N = 100$, the (integer) equilibrium twist is $(Tw)_{eq} = N/h = 10$ [1]. The Hamiltonian reads:

$$
\begin{aligned}
H &= \sum_{n=1}^{N} \left[\frac{\mu \dot{y}_n^2}{2} + V_S(y_n, y_{n-1}) + V_M(y_n) + V_{sol}(y_n) \right] \\
V_S(y_n, y_{n-1}) &= \frac{K}{2} \Big[1 + \rho \exp[-\alpha(y_n + y_{n-1})] \Big] (y_n^2 - 2 y_n y_{n-1} \cos\theta + y_{n-1}^2) \\
V_M(y_n) &= D_n (\exp(-a_n y_n) - 1)^2 \\
V_{sol}(y_n) &= -D_n f_s (\tanh(y_n/l_s) - 1)\,.
\end{aligned} \tag{1}
$$

D_n and a_n are the pair dissociation energy and the inverse length setting the hydrogen bond potential range for the n-th base pair. K is the harmonic stacking whereas ρ and α are the anharmonic stacking parameters which are taken independent of the type of base at the n and $n-1$ sites. The homogeneity assumption for the stacking relies on the observation that both types of *bps* contain a purine plus a pyrimidine, the former being larger and heavier. Thus the AT- and GC- *bps* are comparable in size and weight. For the choice of the model potential parameters see Refs.[4]. Eq. (1) also introduces the solvent potential V_{sol} which adds to V_M thus enhancing by $f_s D_n$ the height of the energy barrier above which the base pair dissociates. The factor $f_s = 0.3$ mimics the effect of a high salt concentration which screens the negatively charged phosphate groups. The length $l_s = 3\AA$ defines the range beyond which the Morse plateau is recovered and D_n returns to be the fundamental energy scale. For $y_n > l_s$, the two strands are apart from each other and the hydrogen bond with the solvent is established.

2. Path Integral Method

The imaginary time path integral method [5] is applied to Eq. (1) by introducing the idea that the y_n can be described by paths $x(\tau_i)$, the latter being periodic functions of the imaginary time τ_i, $x(\tau_i) = x(\tau_i + \beta)$ with β being the inverse temperature,. The index i numbers the *bps* along the τ-axis. In fact, there are $N + 1$ base pairs in Eq. (1) but the presence of an extra base pair y_0 is remedied by taking periodic boundary conditions, $y_0 = y_N$, which close the finite chain into a loop. This condition is incorporated in the path integral description as the path is a closed trajectory, $x(0) = x(\beta)$. Hence a molecule configuration is given by N paths and, in the discrete time lattice, the separation between nearest neighbors *bps* is $\Delta\tau = \beta/N$. Then, Eq. (1) transforms onto the time axis by mapping: $y_n \to x(\tau_i)$ and $y_{n-1} \to x(\tau_i - \Delta\tau)$. The paths can be expanded in Fourier series with cutoff M_F

$$x(\tau_i) = x_0 + \sum_{m=1}^{M_F} \Big[a_m \cos(2m\pi\tau_i/\beta) + b_m \sin(2m\pi\tau_i/\beta)\Big] , \tag{2}$$

and this introduces the following physical picture:

a) given a set of coefficients $\{x_0, a_m, b_m\}$, the N *bps* are represented by the configuration $\{x(\tau_i), i = 1, .., N\}$.

b) A set of coefficients corresponds to a point in the path configuration space thus, sampling the latter amounts to build an ensemble of distinct configurations for the system. As this is done for any temperature, we have a tool to describe the base pair thermal fluctuations around the equilibrium ($x(\tau_i) \sim 0$).

c) In principle the configurations ensemble for the DNA fragment is infinite as it may include any possible combination of Fourier coefficients. For practical purposes some physical criteria intervene to select computationally the path coefficients defining a molecule configuration and contributing to the partition function. This poses a restriction on the ensemble size.

Such criteria are of two types: *First*, V_M excludes too negative base pair stretchings due to the hard core which mimics the repulsion between negatively charged sugar-phosphate groups. *Second*, the ensemble of paths has to be consistent with the thermodynamics laws. This means that the numerical code selects, at any temperature, a path ensemble and evaluates the entropy of the DNA fragment. If the entropy is growing versus T, the code proceeds to the next temperature step otherwise a new partition is performed in the Fourier coefficients integration, a new path ensemble is selected and the entropy is recalculated. This is done at any T until the macroscopic constraint of the second law of thermodynamics is fulfilled throughout the whole investigated temperature range. I emphasize that the method does not put any constraint on the shape of the *entropy versus* T- plot aside from the requirement that the entropy derivative has to be positive. It follows that the path ensemble is a dynamical object accounting for the manyfold of molecule configurations which enter the thermodynamical calculation. The size of the ensemble is a measure of the cooperativity degree of the system. By increasing T, some *bps* may open and cooperatively lead to bubble formation along segments of the double helix. Accordingly the ensemble size is expected to grow versus T.

Applying the mapping technique to Eq. (1), the classical partition function for the DNA molecule in the solvent is

$$Z_C = \oint Dx \exp\Bigg[-\beta \sum_{i=1}^{N} \Big[\frac{\mu}{2}\dot{x}(\tau_i)^2 + V_S(x(\tau_i), x(\tau_i - \Delta\tau)) + V_M(x(\tau_i)) + V_{sol}(x(\tau_i))\Big]\Bigg]$$

$$\oint Dx \equiv \frac{1}{\sqrt{2}\lambda_\mu} \int dx_0 \prod_{m=1}^{M_F} \Big(\frac{m\pi}{\lambda_\mu}\Big)^2 \int da_m \int db_m , \tag{3}$$

where $\lambda_\mu = \sqrt{\pi/\beta K}$ is the thermal wavelength. From Eq. (3), I compute the ensemble average for the displacement of the $i-th$ base pair, $< x(\tau_i) >$, which permits to get the fraction f of open *bps*. As the UV signal changes quite abruptly when the *bps* dissociate, f is defined in terms of the Heaviside function $\vartheta(\bullet)$ as: $f = N^{-1} \sum_{i=1}^{N} \vartheta(< x(\tau_i) > -\zeta)$.

The *threshold* ζ yields a criterion to establish whether an average base pair displacement is open, $< x(\tau_i) > \geq \zeta$, or not.

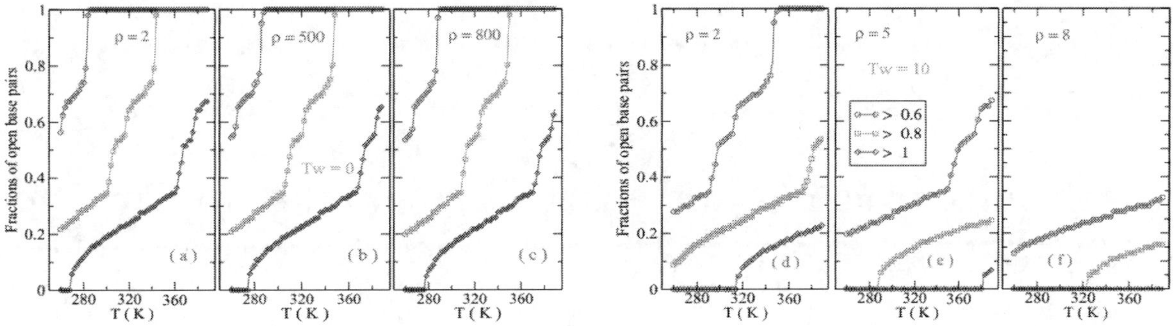

Figure 2. (Color online) Fractions of average displacements larger than $\zeta = 0.6$ (circles), 0.8 (squares), 1 $\AA$ (diamonds) versus temperature in the DPB ladder model ((a) - (c)) and in the equilibrium twist conformation ((d) - (f)). Three anharmonic stacking ρ are assumed both in the left and in the right panel. In (e) and (f), ρ is smaller by a factor 100 than in (b) and (c) respectively.

3. Anharmonic Stacking versus Twist

The computational method is here applied to investigate the interplay between twisting and stacking anharmonicity also in view of the special role given to the latter in the DPB model where a finite ρ induces a sharp denaturation transition driven by sizeable melting entropy [2]. In Fig. 2 the melting profiles are plotted *both* for the DPB model with zero twist (left panel) and *for* the equilibrium twist conformation (right panel). Several ρ values are assumed in both cases. In the $Tw = 0$ system, even very large ρ produce scant variations in the denaturation patterns pointing to a substantial irrelevance of the ρ driven anharmonicity for the DPB ladder model. Quite different is the physical picture emerging from the right panel regarding the $(Tw)_{eq}$ conformation: even slight enhancements over the $\rho = 2$ value *i)* shift upwards along T the opening of the average base pair displacements and *ii)* flatten the melting profiles suggesting that the denaturation becomes more gradual. Note that ρ values in Figs. 2(e),(f) are two orders of magnitude smaller than in Figs. 2(b),(c) respectively. The $\rho = 8$ plot says that, even at $T \sim 400K$, there are no average displacements larger than $1\AA$ while only 35% are larger than $0.6\AA$. Thus it is the twisting that switches on the ρ effect. The latter induces those cooperative interactions along the molecule backbone which are peculiar of the fluctuational openings. Accordingly, the anharmonic stacking renders the double helix flexible hence it does increase the molecule resilience against the whole thermal disruption of the hydrogen bonds. In this sense, anharmonicity is a stabilizing factor for the double helix. Modeling DNA by path integrals offers the advantage to include in the computation a great number of molecule configurations meanwhile accounting for those fluctuational effects which are key to the molecule dynamics, mainly in short fragments.

References

[1] Bates A D, Maxwell A (2009) *DNA Topology* (Oxford University Press, Oxford, UK)
[2] Dauxois T, Peyrard M and Bishop A R (1993) *Phys. Rev. E* **47** R44-R47
[3] Zoli M (2009) *Phys. Rev. E* **79** 041927; Zoli M (2010) *Phys. Rev. E* **81** 051910
[4] Zoli M (2011) *J. Chem. Phys.* **135** 115101; Zoli M (2011) *Eur. Phys. J. E* **34** 68
[5] Feynman R P and Hibbs A R (1965) *Quantum Mechanics and Path Integrals*, (Mc Graw-Hill, New York)

Journal of Physics: Conference Series **410** (2013) 012039 doi:10.1088/1742-6596/410/1/012039

EM field, tracking and scattering simulations for the KATRIN experiment

Ferenc Glück for the KATRIN Collaboration
Karlsruhe Institute of Technology, IKP, Karlsruhe, Germany
E-mail: ferenc.glueck@kit.edu

Abstract. The aim of the KATRIN experiment is to determine the absolute neutrino mass scale in a model independent way, by measuring the electron energy spectrum shape near the endpoint of tritium beta decay. For this purpose, the KATRIN experiment uses two spectrometers with high voltage, and many superconducting and air coils, to create the necessary electric and magnetic fields. In order to design the spectrometer electrodes and the coils, and to investigate and understand the various background processes and systematic effects, one needs various accurate electric and magnetic field, charged particle tracking and scattering computations, for which we use mainly our self-written C and C++ codes.

1. The KATRIN experiment

Neutrino oscillation studies during the past 15 years have proved that the neutrinos have non-zero masses. However, they yield no information on the absolute neutrino mass scale. Neutrinoless double beta decay experiments and cosmological observations are sensitive to the absolute neutrino mass scale, but they are rather model dependent. The KATRIN (Karlsruhe Tritium Neutrino) experiment [1] plans to determine the absolute neutrino mass scale in a direct and model independent way with a sensitivity of 200 meV, by measuring the integral electron energy spectrum shape close to the tritium beta decay endpoint (18.6 keV).

Fig. 1 shows the experimental setup of the KATRIN experiment. The electrons are created in the windowless gaseous tritium source by beta decay of T_2 molecules and are guided by high magnetic field (a few T) of many superconducting coils in the transport system towards the spectrometers. The pumping section reduces the tritium flow towards the spectrometers by many orders of magnitude. The main spectrometer (length: 23 m, diameter: 10 m, magnetic field at center: 0.3 mT) collimates the electrons by the inverse magnetic mirror effect and analyzes their energy by its high negative potential (around -18.6 kV). Inside the main spectrometer a complicated wire electrode system has been installed; this consists of 0.2 and 0.3 mm diameter wires with 40 km total length, and its main task is to reduce the background.

The KATRIN experiment requires various detailed electromagnetic field, charged particle tracking and scattering computations. For this purpose, we use mainly our self-written C and C++ codes developped in the past 10 years. The C++ package KASSIOPEIA [2] is now the standard simulation framework of the KATRIN experiment.

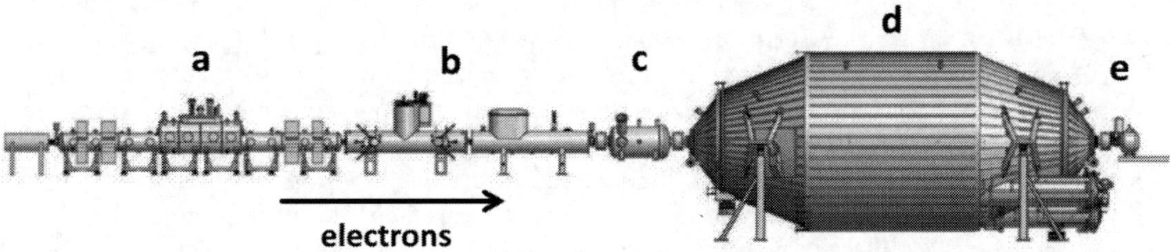

Figure 1. The KATRIN experimental setup: a: tritium source; b: transport and pumping section; c: pre-spectrometer; d: main spectrometer; e: detector.

2. Electromagnetic field computations

2.1. Magnetic field

The KATRIN system contains about 70 superconducting and air coils. In addition, these magnets are surrounded by a lot of magnetic materials. The precise and reliable computation of the magnetic field from these sources is an important task of the KATRIN experiment.

For the calculation of the axisymmetric coils we use the zonal harmonic expansion. This is a special version of the spherical harmonic method, applied for axisymmetric systems. It can be 100–1000 times faster than the more widely known elliptic integral method, and it is more general than the similar radial series expansion. It has not only high speed but also high accuracy, which makes the method especially appropriate for trajectory calculations of charged particles. Due to these properties, no interpolation is necessary when the magnetic field during particle trajectories is computed with the aid of the zonal harmonic method.

The zonal harmonic field series formulas are convergent at field points within the central and remote regions, which have spherical boundaries, and their center, the source point, can be arbitrarily chosen on the symmetry axis. The rate of convergence of the field series depends on the distance of the field and the source point; smaller distance for central field points and larger distance for remote field points correspond to higher convergence rate. The zonal harmonic method is applicable also for special three-dimensional magnetic systems, whose components (coils or magnetic materials) are axially symmetric in their own local coordinate systems. See Ref. [3] for more details.

Some of the KATRIN coils (earth field compensation coils, dipole coils) are not axisymmetric; they are computed by integrated Biot-Savart formulae. In order to accelerate the speed of these calculations we use three-dimensional Hermite interpolation. The field of the magnetic materials in the KATRIN buildings is computed by equivalent (fictitious) magnetic charges.

2.2. Electric field

Some parts of the KATRIN electrode system are approximately axisymmetric. In order to compute the electric field of axisymmetric electrodes, one can use again the zonal harmonic expansion method, like in the case of magnetic field [4]. The sources of the electric field — the charges on the electrode surfaces — are usually not known in the beginning of the field calculation. The most natural way to compute the charge densities is by the boundary element method (BEM). With BEM, one has to discretize only the two-dimensional surface of the electrodes, not the whole three-dimensional space of the electrode system, as it is the case using the finite difference and finite element methods. The electrode system is then approximated by N subelements (e.g. axisymmetric conical elements); in our method, the charge density of each element is assumed to be constant over the element surface. The Coulomb matrix elements C_{ij}, defined as the potential at the centre of element i due to the element j with unit charge density, can be computed by some routines in our codes. The potential U_i of the element i is connected with the charge densities σ_j by the linear equation system: $U_i = \sum_{j=1}^{N} C_{ij}\sigma_j$. The N charge densities can be computed by some direct method (like Gauss-Jordan elimination). The source constants that are needed for the zonal harmonic expansions can be calculated using these charge density values.

The BEM is also useful for general 3-dimensional electric field computations. It is especially advantageous for electrodes having small scale structures within large volumes, like the wire electrode system of KATRIN. In order to get a sufficient discretization of the KATRIN wire electrodes, more than 1 million subelements are necessary. On the other hand, using the above scheme, the number of different charge densities N cannot be more than several 10000, otherwise the Coulomb matrix does not fit into the computer main memory (and the charge density computation time becomes also extremely large). One method to solve this problem is by using the approximate rotational symmetry of the wire electrodes. For example, in the cylindrical part of the KATRIN main spectrometer there are 1200 wires in the azimuthal direction, and all of them have approximately the same charge density. Using this symmetry, the number of geometrical subelements can be much larger than the number of different charge densities.

Another possibility to overcome the charge density number limit is by using an iterative method to solve the above linear algebraic equation system. The Robin Hood method [5], which is similar to the well-known Gauss-Seidel iterative algorithm, has several advantages relative to the direct method: i, the memory requirement grows only linearly with N, instead of quadratically; ii, it can be easily parallelized; iii, the computation time grows quadratically with N, instead of the third power.

3. Relativistic charged particle tracking with explicit Runge-Kutta method

The first-order differential equation system $\dot{\mathbf{y}}(t) = \mathbf{f}(\mathbf{y}, t)$, $\mathbf{y}(t_0) = \mathbf{y}_0$ (the $\mathbf{y}$, $\mathbf{y}_0$ and $\mathbf{f}$ vectors having generally n components) can be solved by the explicit s-stage Runge-Kutta method defined by

$$\mathbf{y}(t_0 + h) = \mathbf{y}_0 + h \sum_{j=1}^{s} b_j \mathbf{k}_j, \tag{1}$$

$$\mathbf{k}_j = \mathbf{f}(\mathbf{y}_0 + h \sum_{l=1}^{j-1} a_{jl}\mathbf{k}_l, t_0 + hc_j) \quad (j > 1), \quad \mathbf{k}_1 = \mathbf{f}(\mathbf{y}_0, t_0). \tag{2}$$

The coefficients b_j, c_j, a_{jl} ($j = 1, \ldots, s$; $l = 1, \ldots, j-1$) can be found in various Runge-Kutta tables (see e.g.: [6, 7, 8, 9]). The number of function evaluations during a Runge-Kutta step is proportional to the stage s. The local and the global errors are proportional to h^{p+1} and h^p, respectively, where p is the order of the Runga-Kutta method. The classical Runge-Kutta method has $p = 4$ and $s = 4$. The Runga-Kutta methods of [6, 7, 8] have $p = 8$ (eighth-order) and require $s = 13$ stages. According to our experience, these eighth-order Runga-Kutta methods are much more efficient (have much better precision for a given computation time) than the fourth-order classical Runge-Kutta method.

The relativistic charged particle motion can be expressed by the first order differential equations $\dot{\mathbf{x}} = \mathbf{v}$, $\dot{\mathbf{p}} = \mathbf{F}_L$, where $\mathbf{x}$, $\mathbf{v}$ and $\mathbf{p}$ are the space coordinate, velocity and relativistic momentum 3-vectors, and $\mathbf{F}_L = Q\ (\mathbf{E} + \mathbf{v} \times \mathbf{B})$ denotes the Lorentz force (Q is the charge of the particle, $\mathbf{E}$ and $\mathbf{B}$ are the electric and magnetic field vectors). In this case the vectors $\mathbf{y}$ and $\mathbf{f}$ both have 6 components: the first 3 components are defined by $y_i = x_i$ and $f_i = v_i$ ($i = 1, \ldots, 3$); the second 3 components are then $y_i = p_{i-3}$ and $f_i = \mathbf{F}_{L,i-3}$ ($i = 4, \ldots, 6$).

The time step h should be automatically controlled by the computer code. For this control we use several different methods. For example, $h = P_c/m$, where P_c denotes the cyclotron motion period, and the m integer is around 20 to 50 (higher m value corresponds to higher accuracy level, but more computation time). Another simple possibility is $h = s/v$, where s is some user-defined distance and v the particle velocity. The widely used time step control is by local error estimation and embedded Runge-Kutta formulae [10].

In the case of time-independent fields, the accuracy of our trackings can be tested by energy conservation. Due to the high accuracy of our fields and to the high order of our Runge-Kutta method, the relative change of the energy of our tracked charged particles was in many cases order of 10^{-10} or smaller, even for very large tracking time values. In addition, in order to estimate the global error of a charged particle trajectory, one can perform the tracking with two different time step series (e.g. the second tracking can have twice smaller steps than the first one), and then one compares the coordinate and velocity values of the two tracks corresponding to the same tracking time. Often, these coordinate and velocity values agree quite well, even for large tracking time, certifying again the high accuracy of our field and tracking computations. On the other hand, in the case of higher energy stored electrons in the KATRIN main spectrometer, it occurs often that after some time value the two trajectories start to deviate (diverge) strongly from each other. It turned out that these divergences of the trajectories with different time steps are due to chaotic motion of non-adiabatic electrons; in this case, it is not possible to improve significantly the long-time accuracy of the numerical tracking method.

In addition to exact tracking described above, we use also adiabatic tracking. In this case we compute first the motion of the guiding center in the adiabatic approximation, assuming that for the motion along the magnetic field line the first adiabatic invariant is constant [11], and using also the magnetron drift perpendicularly to the magnetic field line. The gyration of the charged particle along the guiding centre is also taken into account. The adiabatic tracking can be used only if the charged particle motion is approximately adiabatic. But in this case the adiabatic tracking can be 100 times faster than the exact tracking.

4. Electron–hydrogen molecule scattering

The KATRIN source contains molecular tritium with a few times 10^{-3} mbar pressure, and the beta decay electrons can have collisions with T_2 molecules. In addition, H_2 molecules are expected to provide the main part of the residual gas in the KATRIN vacuum system, and stored electrons can collide with these H_2 molecules even if the vacuum pressure is as low as 10^{-11} mbar. Therefore, to investigate and understand various background processes and systematic effects, one has to perform electron-$H_2(T_2)$ scattering simulations. In our codes, we use total and angular differential cross sections for elastic, electronic excitation, ionization, vibrational and rotational excitation scatterings [12, 13, 14], for electron kinetic energy range from a few meV up to a few times 100 keV.

References

[1] J. Angrik et al. (KATRIN Collaboration), "KATRIN Design Report 2004", Wissenschaftliche Berichte FZKA 7090. `http://bibliothek.fzk.de/zb/berichte/FZKA7090.pdf`.
[2] M. Babutzka et al., "The Comprehensive Guide to KASSIOPEIA", version 1.00.00, KATRIN internal report.
[3] F. Glück, Progress In Electromagnetics Research B 32 (2011) 351.
[4] F. Glück, Progress In Electromagnetics Research B 32 (2011) 319.
[5] J. Formaggio et al., Progress In Electromagnetics Research B 39 (2012) 1.
[6] J. H. Verner, SIAM Journal on Numerical Analysis 15 (1978) 772.
[7] P. Prince, J. Dormand, Journal of Computational and Applied Mathematics 7 (1981) 67.
[8] C. Tsitouras, S. N. Papakostas, SIAM Journal on Scientific Computing 20 (1999) 2067.
[9] G. Engeln-Müllges, F. Uhlig: Numerical algorithms with C (Springer, Berlin, 1996).
[10] J. R. Dormand: Numerical methods for differential equations (CRC-Press,1996).
[11] J. D. Jackson, Classical Electrodynamics (John Wiley & Sons, New York, 1999).
[12] W. Hwang, Y.-K. Kim, M. E. Rudd, The Journal of Chemical Physics 104 (1996) 2956.
[13] S. Trajmar, D. Register, A. Chutjian, Physics Reports 97 (1983) 219.
[14] H. Tawara et al., Journal of Physical and Chemical Reference Data 19 (1990) 617.

Journal of Physics: Conference Series **410** (2013) 012040 doi:10.1088/1742-6596/410/1/012040

Bayesian optimal experimental design for the Shock-tube experiment

G Terejanu, C M Bryant and K Miki

Institute for Computational Engineering and Sciences, The University of Texas at Austin, 1 University Station C0200, Austin, TX, 78712, USA

E-mail: cbryant@ices.utexas.edu

Abstract. The sequential optimal experimental design formulated as an information-theoretic sensitivity analysis is applied to the ignition delay problem using real experimental. The optimal design is obtained by maximizing the statistical dependence between the model parameters and observables, which is quantified in this study using mutual information. This is naturally posed in the Bayesian framework. The study shows that by monitoring the information gain after each measurement update, one can design a stopping criteria for the experimental process which gives a minimal set of experiments to efficiently learn the Arrhenius parameters.

1. Introduction

A mathematical description of the gas-phase combustion kinetic problem is very challenging in context of Bayesian optimal experimental design. The system is composed of very stiff nonlinear ordinary differential equations (ODEs). Relevant studies on Bayesian optimal experimental design applied to such a problem are still sparse in literature, see Ref's [1, 14, 7], and often test their models using "manufactured" data, which is generated using the model with artificial noise added.

Terejanu et al.[4] have shown that Bayesian experimental design can be formulated as an information-theoretic sensitivity analysis, where the optimal design maximizes the mutual information between the parameters and the future observations.

In this study, we present the application of the sequential BOED as presented in [4] to the ignition delay problem in the hydrogen-oxygen combustion. We explore the behavior of BOED when used to calibrate the model in the presence of model inadequacy and real data. It is shown that by closely monitoring the sequential experimental process using information theoretic measures such as information gain, one can perform a more efficient calibration using fewer experiments.

2. The Proposed Experimental Design Methodology

Given a set of observations $D_{n-1} = \{\tilde{\mathbf{d}}_1, \tilde{\mathbf{d}}_2, \ldots, \tilde{\mathbf{d}}_{n-1}\}$ we are concerned with finding the next experimental design $\boldsymbol{\xi}_n \in \Xi$ such that the model parametric uncertainty is reduced after the experiment is performed and the associated measurement data $\tilde{\mathbf{d}}_n$ is collected. The mathematical models used in this paper are generally represented by the following abstract model:

$$\hat{\mathbf{r}}(\mathbf{u}, \boldsymbol{\theta}, \boldsymbol{\xi}_n) = \mathbf{0} \tag{1}$$

$$\mathbf{d}_n = \mathbf{d}(\mathbf{u}, \boldsymbol{\theta}, \boldsymbol{\xi}_n) + \boldsymbol{\epsilon}_{model}(\boldsymbol{\xi}_n) + \boldsymbol{\epsilon}_{meas} \tag{2}$$

Here $\mathbf{u}$ is the state of the system which obeys the governing equations defined by $\mathbf{r}(\cdot)$, $\boldsymbol{\xi}_n \in \boldsymbol{\Xi}$ is the control variables associated with the experimental scenario, $\boldsymbol{\Xi}$ is the design space and $\boldsymbol{\theta} \in \Theta$ are the model parameters, where Θ is the parameter space of the model. Here, the model prediction $\mathbf{d}_n$ calculated using the measurement model $\mathbf{d}(\cdot)$ is comparable with the experimental data $\tilde{\mathbf{d}}_n$ for a particular scenario input $\boldsymbol{\xi}_n$. The random vector or random field $\boldsymbol{\epsilon}_{meas}$ is the observation noise due to sensor imprecision or the discretization of computer simulations. The random vector or random field $\boldsymbol{\epsilon}_{model}(\boldsymbol{\xi}_n)$ captures discrepancy between model predictions and experimental data, and is due to missing physics in the governing equations.

When the objective of the experiment is to infer the parameters $\boldsymbol{\theta}$, then the optimal experiment is given by maximizing the mutual information between the parameters and the model predictions, $\tilde{\mathbf{d}}_n$, see Ref. [4]. In other words we would like to sample where the expected observations have the highest impact on model parameters.

$$\boldsymbol{\xi}_n^* = \arg\max_{\boldsymbol{\xi}_n \in \boldsymbol{\Xi}} \underbrace{\int_D \int_\Theta p(\boldsymbol{\theta}, \mathbf{d}_n | D_{n-1}, \boldsymbol{\xi}_n) \log \frac{p(\boldsymbol{\theta}, \mathbf{d}_n | D_{n-1}, \boldsymbol{\xi}_n)}{p(\boldsymbol{\theta}|D_{n-1}) p(\mathbf{d}_n | D_{n-1}, \boldsymbol{\xi}_n)} \mathrm{d}\boldsymbol{\theta} \mathrm{d}\mathbf{d}_n}_{\mathrm{I}(\boldsymbol{\theta}; \mathbf{d}_n | D_{n-1}, \boldsymbol{\xi}_n)}$$

Compared with maximum entropy sampling for parameter estimation proposed in Ref.[16], the sampling based on maximizing the mutual information is more general as no assumption is made about the independence between model discrepancy and design variables. For this study, the estimation of the mutual information is done using the kNN (Nearest Neighbor) method proposed in Ref. [9].

3. Physical Model and Experimental data

Assuming that the temperature and pressure behind the reflected shock waves are constant during the reaction, the chemical reactions can be expressed in ordinary differential equations for each species (i.e. a well-stirred reactor model), see Ref.[6]. The forward rate coefficient of the r-th reaction, $k_{f,r}$, is calculated using the modified Arrhenius equation: $k_{f,r} = A_r\, T^{m_r} \exp(-\frac{\Theta_r}{T})$, $r = 1, \ldots, N_r = 35$, where A_r and m_r are the pre-exponential Arrhenius parameters and Θ_r is the characteristic temperature of the activation energy. We use the detailed chemical kinetic mechanism previously verified in [6] to model the ignition delay of $O_2/H_2/Ar$ mixture behind the reflected shock wave. This mechanism considers 35 elementary reactions and 11 species.

Experimental data were recently collected at the High Temperature Gasdynamics Laboratory at Stanford University [6]. The data is composed of six profiles of the H_2O concentration history behind reflected shock waves over the temperature range 1100 - 1472 K. All six experimental scnearios have an initial O_2 concentration of 0.001%. Three have an initial H_2 concentration of 0.029% with $T_5 = \{1100, 1197, 1256\}$K and corresponding $P_5 = \{1.95, 1.84, 2.01\}$ atm. The other three have initial H_2 concentration of 0.009% and $T_5 = \{1317, 1448, 1472\}$K and $P_5 = \{1.91, 1.85, 1.83\}$ atm. Given our assumption that the model discrepancy is not time dependent we choose five data points where the level of the error is assumed to be constant. The detailed description of the shock tube experiment is provided in [5].

4. Description of Stochastic Models

In the current study we treat the pre-exponential Arrhenius parameters, A_r and m_r, and the characteristic temperature, Θ_r, as uncertain and thus opened for calibration. The prior distribution for these parameters is given by $A_r \sim \mathcal{U}(.64289, 1.21429)$, $m_r \sim \mathcal{U}(0,1)$, $\Theta_r \sim \mathcal{U}(0,2)$. The measurement model for the reaction is assumed to have the following probabilistic formulation,

$$\tilde{d}(t) = d(A_r, m_r, \Theta_r, t) + \epsilon_{model} \tag{3}$$

where $d(A_r, m_r, \Theta_r, t)$ is the concentration of H_2O given by the kinetic mechanism and $\tilde{d}(t)$ is the experimental observation given in Ref.[6]. The discrepancy between model predictions and observations, ϵ_{model}, accounts for the small experimental uncertainty and any missing physics in the governing equation. Kennedy & O'Hagan [8] have emphasized the importance of model discrepancy, ϵ_{model}, in Bayesian calibration when dealing with approximate models of complex mathematical models. In this work, we model this model discrepancy using an additive Gaussian model, $\epsilon_{model} \sim \mathcal{N}(0, \sigma^2)$. The hyper-parameter σ of the Gaussian model will also be inferred from the data with prior distribution for σ given by $\sigma \sim \mathcal{U}(0,1)$.

5. Results and Discussions

The estimated mutual information between model parameters and the model predictions are shown in Table 1. At some stages, stage 1 in particular, the mutual information is nearly identical for multiple experiments. This may indicate that any one of these experiments could be added to inform the parameters.

Table 1. Mutual information at each design stage and the optimal sequence of designs. In bold are the maximums found at each stage of the design process.

	$\mathrm{I}(\boldsymbol{\theta}; \mathbf{d}_n \mid D_{n-1}, \boldsymbol{\xi}_n)$					
Index	stage 1	stage 2	stage 3	stage 4	stage 5	stage 6
1	**0.454**					
2	0.452	0.429	1.500	1.375	**1.212**	
3	0.441	0.306	**1.517**			
4	0.434	**0.466**				
5	0.422	0.411	1.060	1.071	0.897	**0.882**
6	0.451	0.458	1.074	**1.399**		
Info.Gain	4.9	7.7	9.4	9.6	9.6	9.1

Additionally Table 1 also lists the information gain after each stage of the process. A large number signifies that the posterior has dramatically changed with respect to the prior given the experimental data. We can see that there is a convergence in the information gain which is reached after the fourth stage. This convergence signals that the inference process is slowing down and that additional data provides less information compared with the first data points.

Posterior means and marginal 95% confidence intervals for the four parameters are shown in Figure 1. Note that for Θ_r and σ the mean value remains relatively unchanged after stage four, see Figs 1(a),1(b). In addition the confidence intervals for both parameters narrow initially and then remain relatively stable as additional experiments are added. This indicates a convergence of the distributions to a stable representation of the uncertainty for these parameters.

On the other hand, the results for A_r and m_r are not as stable, see Figs1(c),1(d). While the mean values seem to fluctuate only slightly after two stages, the confidence intervals only narrow significantly after the sixth stage of the design process. Even though in the 5th stage of optimal experimental deign, there is an increase in the marginal uncertainty for A_r and m_r, see Figs.1(c),1(d), there is also an increase in the correlation between these two parameters, see

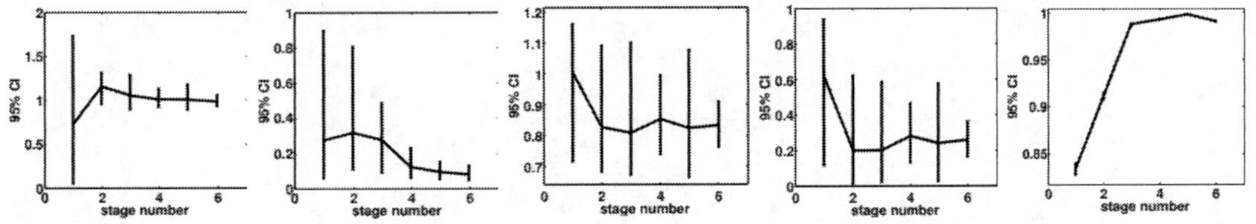

(a) Θ_r (b) σ (c) A_r (d) m_r (e) $Corr(A_r, m_r)$

Figure 1. Posterior uncertainties after each experimental design stage

Fig.1(e), which results in an overall decrease in the uncertainty given by the joint distribution of A_r and m_r.

6. Conclusions

In this paper the sequential Bayesian optimal experimental design as presented by Terejanu et al. 2012 is applied to an ignition delay problem in hydrogen-oxygen combustion. Monitoring the sequential experimental process using information theoretic measures we are able to determine an efficient calibration using fewer experiments. Model inadequacy is also taken into account in the methodology. More work needs to be done to propagate this uncertainty to the reaction rate of interest.

Acknowledgments

Acknowledgment This work has been supported by the Department of Energy [National Nuclear Security Administration] under Award Number [DE-FC52-08NA28615].

References

[1] Blau G, Lasinski M, Orcun S, Hsu S H, Caruthers J, Delgass N and Venkatasubramanian V. *Comput. Chem. Eng.* 32 (2008) 971–989.
[2] Cheung S H and Beck J L. *Computer-Aided Civil and Infrastructure Engineering* 25 (2010) 304–321.
[3] Franceschini G and Macchietto S. *Chem. Eng. Sci.* 63 (2008) 4846–4872.
[4] Terejanu G, Upadhyay R R and Miki K. *Exp. Therm. Fluid Sci* 36 (2012) 178–193.
[5] Hong Z, Barbour E A, Davidson A and Hanson R K. *J. Chem. Phys.* 113 (2009) 12919–12925.
[6] Hong Z, Davidson D E, Barbour E A and Hanson R K. *Proc. Combust. Inst.* 33 (2011) 309–316.
[7] Huan X and Marzouk Y M. *J. Comput. Phys.* (2011) under review.
[8] Kennedy M C and O'Hagan A. *J. Royal Statisti. Soc.: Series B (Statistical Methodology)* 63(3) (2001) 1467-9868.
[9] Kraskov A, Stogbauer H and Grassberger P. *Phys. Rev. E* 69(6) (2004) 066138.
[10] Lindley D V. *Ann. Math. Statist.* 27(4) (1956) 986-1005.
[11] Manion J A, et al. NIST chemical kinetics database, NIST standard reference database 17, version 7.0 (web version), release 1.4.3, data version 2008.12, National Institute of Standards and Technology, Gaithersburg, Maryland, 20899-8320. Web address: http://kinetics.nist.gov/.
[12] Masten D A, Hanson R K, Bowman C T. *The Journal of Physical Chemistry* 91 (1990) 7119–7128.
[13] Miki K, Cheung S H, Prudencio E E and Varghese P L. *International Journal of Chemical Kinetics* (2012), to appear.
[14] Mosbach S, Braumann A, Man P L W, Kastner C A, Brownbridge G P E and Kraft M. *Combust. Flame* doi:10.1016/j.combustflame.2011.10.019 (2011).
[15] Pirraglia A N, Michael J V, Sutherland R B and Klemm J. *The Journal of Physical Chemistry* 93 (1989) 282–291.
[16] Sebastini P and Wynn H P. *J. R. Statist. Soc. B* 62(1) (2000) 145-157.
[17] Sheen D and Wang H. *Combust. Flame* 158 (2011) 2358–2374.
[18] Smith G P, et al. GRI-Mech 3.0. available at http://www.me.berkeley.edu/grimech.

Journal of Physics: Conference Series **410** (2013) 012041 doi:10.1088/1742-6596/410/1/012041

LS-SVM based solution for delay differential equations

Siamak Mehrkanoon and Johan A. K. Suykens

KU Leuven, ESAT-SCD, Kasteelpark Arenberg 10, B-3001 Leuven (Heverlee), Belgium

E-mail: {Siamak.Mehrkanoon,Johan.Suykens}@esat.kuleuven.be

Abstract. In this paper a new approach based on Least Squares Support Vector Machines (LS-SVMs) is proposed for solving delay differential equations (DDEs) with single-delay. The proposed method provides a closed form approximate model based solution without using any interpolation techniques. The result of this paper can be extended for DDE with multi-lags. The results of some numerical experiments are presented and compared with analytic solutions to confirm the validity and applicability of the proposed method.

1. Introduction

Delay differential equations (DDEs) can be found in the mathematical formulation of real life phenomena in a wide variety of applications especially in science and engineering such as population dynamics, infectious disease and control problems [1]. In contrast with ordinary differential equations (ODEs) where the unknown function and its derivatives are evaluated at the same time instant, in a DDE the evolution of the system at a certain time, depends on the state of the system at an earlier time. A general form of first order multi-delay DDE is given by:

$$\begin{aligned} \dot{x}(t) &= f(t, x(t), x(t-\tau_1), \ldots, x(t-\tau_n)), \quad t \geq t_{in}, \\ x(t) &= g(t), \quad t \leq t_{in} \end{aligned} \tag{1}$$

where $g(t)$ is the initial function and $\{\tau_i\}_{i=1}^{n}$ are the delays or lags which are non-negative and can be constant, time dependent or state dependent. The term $x(t-\tau_i)$ is called the delay term. Due to the presence of the delay term, numerical methods that provide only discrete solutions at the grid points are not suitable for DDE and one needs to apply interpolation techniques, for approximating the delay term, in order to advance the solution of the given DDE.

Least Squares Support Vector Machines (LS-SVMs) is a methodology which has been applied in a wide variety of fields such as regression and classification [3]. Like in support vector machines [6], in this method one maps the data into a high dimensional feature space using a feature map. Solving ordinary differential equations (ODEs) and differential algebraic equations (DAEs) using LS-SVMs have been discussed in [4, 5]. In this paper the method developed in [4] is extended to approximate the solution of a scalar DDE. The method provides a closed form approximate model based solution without using any interpolation techniques.

Making use of Mercer's Theorem [6], derivatives of the feature map can be written in terms of derivatives of the kernel function.

Let us define the following differential operator which will be used in the subsequent section

$$\nabla_n^m \equiv \frac{\partial^{n+m}}{\partial p^n \partial q^m}. \tag{2}$$

When $\boldsymbol{\varphi}(p)^T\boldsymbol{\varphi}(q) = K(p,q)$, one can show then that

$$[\boldsymbol{\varphi}^{(n)}(p)]^T\boldsymbol{\varphi}^{(m)}(q) = \nabla_n^m[\boldsymbol{\varphi}(p)^T\boldsymbol{\varphi}(q)] = \nabla_n^m[K(p,q)] = \frac{\partial^{n+m}K(p,q)}{\partial p^n \partial q^m}. \tag{3}$$

Using formula (3), it is possible to express all derivatives of the feature map in terms of the kernel function itself (provided that the kernel function is sufficiently differentiable).

Suppose that $t_i \in \varrho$ and $v_j \in \tau$ are two arbitrary points and ϱ and $\tau \subseteq \mathbb{R}$. Now for notational convenience let us list the following notations which are used in the following sections.

$$\Big[\nabla_n^m K\Big](t,y) = \Big[\nabla_n^m K(p,q)\Big]\Big|_{p=t,q=y}, \quad \Big[{}^{\varrho}\Omega^{\tau}\Big]_{i,j} = \nabla_0^0[K(p,q)]\Big|_{p=t_i,q=v_j} = K(t_i,v_j),$$

$$\Big[{}^{\varrho}_{n}\Omega^{\tau}_{m}\Big]_{i,j} = \nabla_n^m\Big[K(p,q)\Big]\Big|_{p=t_i,q=v_j} = \frac{\partial^{n+m}K(p,q)}{\partial p^n \partial q^m}\Big|_{p=t_i,q=v_j},$$

where $\Big[{}^{\varrho}_{n}\Omega^{\tau}_{m}\Big]_{i,j}$ denotes the (i,j)-th entry of matrix ${}^{\varrho}_{n}\Omega^{\tau}_{m}$. In the case that $\varrho = \tau$, we denote the matrix by ${}^{\varrho}_{n}\Omega_{m}$.

2. Formulation of the method for DDE

Consider a scalar linear DDE with single-delay of the form

$$\begin{aligned} &\dot{x}(t) = a(t) + b(t)x(t) + c(t)x(t-\tau(t)), \quad t_{in} \le t \le t_f, \\ &x(t) = g(t), \quad t \le t_{in}, \end{aligned} \tag{4}$$

where $\dot{g}(t_{in}^-) = \dot{x}(t_{in}^+)$. Assume that a general approximate solution to equation (4) is of the form of $\hat{x}(t) = \boldsymbol{w}^T\boldsymbol{\varphi}(t) + d$, where $\boldsymbol{\varphi}(\cdot) : \mathbb{R} \to \mathbb{R}^h$ is the feature map and h is the dimension of the feature space. To obtain the optimal value of $\boldsymbol{w}$ and d, a collocation method is used [2] with a discretization of the interval $[t_{in}, t_f]$ into a set of points $\{t_i\}_{i=1}^N$. The estimated solution follows from

$$\begin{aligned} \min_{w,d,e_i} \quad & \frac{1}{2}\boldsymbol{w}^T\boldsymbol{w} + \frac{\gamma}{2}\sum_{i=1}^m e_i^2 \\ \text{s.t.} \quad & \boldsymbol{w}^T\left[\boldsymbol{\varphi}'(t_i) - b_i\boldsymbol{\varphi}(t_i) - c_i\boldsymbol{\varphi}(s_i)\right] - (b_i + c_i)d = \\ & a_i + e_i, \; i = 2, \ldots, N \\ & \boldsymbol{w}^T\boldsymbol{\varphi}(v_k) + d = g(v_k), \; k = 1, \ldots, M \end{aligned} \tag{5}$$

where $s_i = t_i - \tau(t_i)$ and $v_k = \{s_k \,|\, s_k \le t_{in}\}$. N is the number of collocation points (which is equal to the number of training points). Furthermore $a_i = a(t_i), b_i = b(t_i)$ and $c_i = c(t_i)$. Problem 5 is obtained by combining the LS-SVM cost function with constraints constructed by imposing the approximate solution $\hat{x}(t) = \boldsymbol{w}^T\boldsymbol{\varphi}(t) + d$, given by the LS-SVM model, to satisfy the given differential equation with corresponding initial function at collocation points $\{t_i\}_{i=1}^N$. Problem (5) is a quadratic minimization under linear equality constraints, which enables a straightforward solution.

Lemma 2.1. *Given a positive definite kernel function $K : \mathbb{R} \times \mathbb{R} \to \mathbb{R}$ with $K(t,s) = \varphi(t)^T\varphi(s)$ and a regularization constant $\gamma \in \mathbb{R}^+$, the solution to (5) is obtained by solving the following dual problem:*

$$\left[\begin{array}{c|c|c} \mathcal{K} + I_{N-1}/\gamma & \mathcal{U} & -\mathcal{H} \\ \hline \mathcal{U}^T & {}^{\mathcal{V}}_{0}\Omega_0 & 1_M \\ \hline -\mathcal{H}^T & 1_M & 0 \end{array}\right] \left[\begin{array}{c} \boldsymbol{\alpha} \\ \hline \boldsymbol{\beta} \\ \hline d \end{array}\right] = \left[\begin{array}{c} \boldsymbol{F_a} \\ \hline \boldsymbol{G} \\ \hline 0 \end{array}\right] \tag{6}$$

with

$\boldsymbol{\alpha} = [\alpha_2, \ldots, \alpha_N]^T, \boldsymbol{\beta} = [\beta_2, \ldots, \beta_N]^T, \boldsymbol{F_a} = [a(t_2), \ldots, a(t_N)]^T \in \mathbb{R}^{N-1}$,
$\boldsymbol{F_b} = [b(t_2), \ldots, b(t_N)]^T \in \mathbb{R}^{N-1}, \boldsymbol{F_c} = [c(t_2), \ldots, c(t_N)]^T \in \mathbb{R}^{N-1}, \boldsymbol{G} = [g(t_2), \ldots, g(t_N)]^T \in \mathbb{R}^{N-1}$,
$\mathcal{K} = \bar{D}\,\bar{\Omega}\,\bar{D}^T$, $\mathcal{U} = \bar{D}\,\Delta^T$, $\mathcal{H} = \boldsymbol{F_b} + \boldsymbol{F_c}$

$$\bar{\Omega} = \left[\begin{array}{c|c|c} {}^{\mathcal{T}}_{1}\Omega_1 & {}^{\mathcal{T}}_{0}\Omega_1 & {}^{\mathcal{S}}_{0}\Omega_1^{\mathcal{T}} \\ \hline {}^{\mathcal{T}}_{1}\Omega_0 & {}^{\mathcal{T}}_{0}\Omega_0 & {}^{\mathcal{S}}_{0}\Omega_0^{\mathcal{T}} \\ \hline {}^{\mathcal{T}}_{1}\Omega_0^{\mathcal{S}} & {}^{\mathcal{T}}_{0}\Omega_0^{\mathcal{S}} & {}^{\mathcal{S}}_{0}\Omega_0 \end{array}\right] \in \mathbb{R}^{3(N-1)\times 3(N-1)},\ \bar{D} = [\mathbb{1}, -D_b, -D_c],\ \Delta = [{}^{\mathcal{V}}_{0}\Omega_1^{\mathcal{T}}, {}^{\mathcal{V}}_{0}\Omega_0^{\mathcal{T}}, {}^{\mathcal{V}}_{0}\Omega_0^{\mathcal{S}}]$$

D_b and D_c are diagonal matrices with the elements of $\boldsymbol{F_b}$ and $\boldsymbol{F_c}$ on the main diagonal respectively. $\mathbb{1}$ is the identity matrix. In addition $\mathcal{S} = \{s_i\}_{i=2}^N$, $\mathcal{V} = \{v_i\}_{i=1}^M$ and $\mathcal{T} = \{t_i\}_{i=2}^N$.

The solution in the dual form becomes:

$$\hat{x}(t) = \sum_{i=2}^{N} \alpha_i \Big([\nabla_1^0 K](t_i, t) - b(t_i)[\nabla_0^0 K](t_i, t) - c(t_i)[\nabla_0^0 K](s_i, t) \Big) + \sum_{k=1}^{M} \beta_k \, [\nabla_0^0 K](v_k, t) + d,$$

where $[\nabla_0^0 K](t,s) = K(t,s) = \varphi(t)^T\varphi(s)$ is the kernel function and $[\nabla_1^0 K](t,s) = \varphi'(t)^T\varphi(s)$ is its derivative. Here the Gaussian RBF kernel is used. α_i and β_k are Lagrange multipliers associated with (3). The result of this paper can be extended for linear DDE with multi-lags.

3. Numerical Result

Problem 3.1: Consider the following scalar DDE,

$$\dot{x}(t) = -x(t - 1 + \exp(-t)) + \sin(t - 1 + \exp(-t)) + \cos(t),\ t \geq 0,$$
$$x(t) = \sin(t),\ t \leq 0$$

The problem is solved on domain $t \in [0, 40]$ for $N = 200$. Figure 1 shows the residuals $e(t) = x(t) - \widehat{x}(t)$.

Problem 3.2: Consider the delay equation with asymptotically vanishing-lag:

$$\dot{x}(t) = \frac{t^4 - 3}{(t^5 + t)\ln(t - t^{-3} + [t - t^{-3}]^{-3})} x(t - t^{-3}), \quad t \in [2, 30],$$
$$x(t) = \ln(t + t^{-3}), \quad t \in [1.5, 2]$$

whose exact solution is $x(t) = \ln(t + t^{-3})$. The result obtained by the proposed method method, with $N = 400$, is depicted in Figure 2.

4. Conclusion

In this paper, we have presented a method based on LS-SVMs for the numerical solution of delay differential equations. The proposed method provides a close form approximate solution for the problem without using any interpolation techniques.

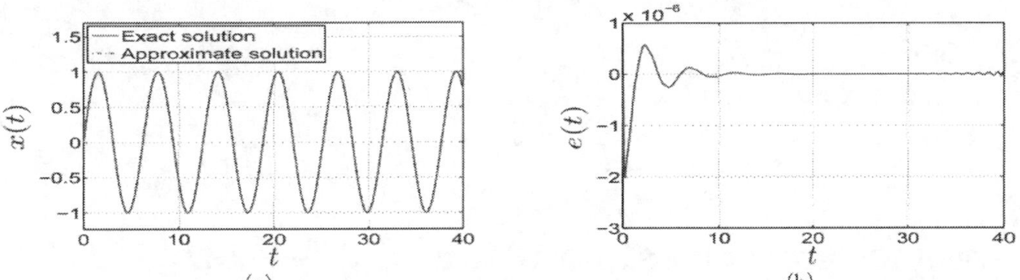

(a) (b)

Figure 1. (a) Numerical results for Problem 3.1, when 200 equidistant points in [0,40] are used for training. (b) Obtained model errors for problem 3.1.

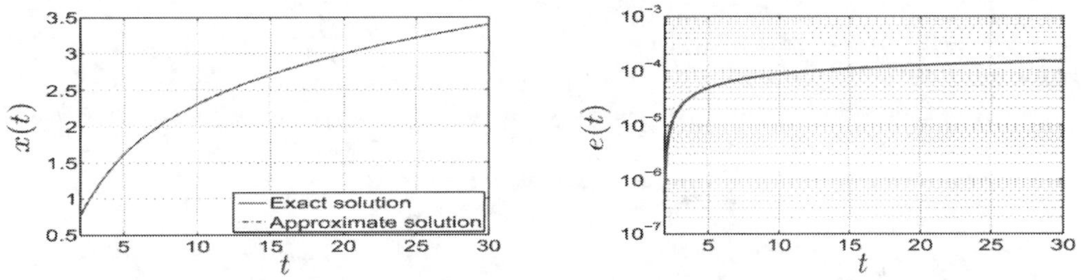

(a) (b)

Figure 2. (a) Numerical results for Problem 3.2, when 400 equidistant points in [2,30] are used for training. (b) Obtained model errors for problem 3.2.

Acknowledgments
This work was supported by: • Research Council KUL: GOA/10/09 MaNet, PFV/10/002 (OPTEC), several PhD/postdoc & fellow grants • Flemish Government: ∘ IOF: IOF/KP/SCORES4CHEM; ∘ FWO: PhD/postdoc grants, projects: G.0320.08 (convex MPC), G.0558.08 (Robust MHE), G.0557.08 (Glycemia2), G.0588.09 (Brain-machine), G.0377.09 (Mechatronics MPC); G.0377.12 (Structured systems) research community (WOG: MLDM); ∘ IWT: PhD Grants, projects: Eureka-Flite+, SBO LeCoPro, SBO Climaqs, SBO POM, O&O-Dsquare • Belgian Federal Science Policy Office: IUAP P7/ (DYSCO, Dynamical systems, control and optimization, 2012-2017) • IBBT • EU: ERNSI, FP7-EMBOCON (ICT-248940), FP7-SADCO (MC ITN-264735), ERC ST HIGHWIND (259 166), ERC AdG A-DATADRIVE-B • COST: Action ICO806: IntelliCIS • Contract Research: AMINAL • Other: ACCM. Johan Suykens is a professor at the KU Leuven, Belgium.

References

[1] Driver R D 1977 *Ordinary and Delay Differential Equations* (New York: Springer-Verlag New York Inc).
[2] Kincaid D R and Cheney E W 2002 *Numerical Analysis*, (Mathematics of Scientific Computing, third ed., Brooks/Cole, Pacific Grove, CA).
[3] Suykens J A K, Van Gestel T, De Brabanter J, De Moor B and Vandewalle J 2002 *Least Squares Support Vector Machines* (World Scientific, Singapore.)
[4] Mehrkanoon S, Falck T and Suykens J A K 2012 Approximate Solutions to Ordinary Differential Equations Using Least Squares Support Vector Machines *IEEE Transactions on Neural Networks and Learning Systems*, **23** 1356-67.
[5] Mehrkanoon S and Suykens J A K 2012 LS-SVM approximate solution to linear time varying descriptor systems *Automatica*, **48**, 2502-2511.
[6] Vapnik V 1998 *Statistical learning theory* (New York: Wiley)

Slice-by-slice numerical solution of $3D$-vector tomography problem

Dr. Ivan Svetov

Researcher, Sobolev Institute of Mathematics SB RAS, Novosibirsk, Russia

E-mail: svetovie@math.nsc.ru

Abstract. We suggest a numerical solution of the problem by reconstruction of solenoidal part of the vector field defined in the unit ball. That is, approximation of solenoidal part of vector field is built by known ray transform, calculated along the lines parallel to one of the coordinate planes. Test calculations have shown good results of reconstruction solenoidal vector fields by using the proposed method.

1. Introduction

In this paper the $3D$-vector tomography problem is considered in the following formulation. Let a bounded domain in R^3 be filled by a medium without refraction (probing rays permeate along straight lines). In the domain some vector field v is given. It is require to find this field by its known ray transform.

We consider R^3 as the Euclidean vector space with the standard scalar product $\langle\cdot,\cdot\rangle$ and norm $|\cdot|$. The ray transform $\mathcal{I}$ on R^3 maps a vector field $v = (v_j)$ to the function $[\mathcal{I}v]$ on the manifold of oriented lines $l = \{x + t\xi \mid x, \xi \in R^3,\ |\xi| = 1,\ \langle x, \xi\rangle = 0,\ t \in R\}$ by the formula

$$[\mathcal{I}v](x,\xi) = \sum_{j=1}^{3} \int_{-\infty}^{\infty} \xi_j v_j(x + t\xi)\, dt.$$

In calculating the ray transform along the ray only projection of the desired field on the ray makes contribution to the result. Therefore problem of reconstruction of vector fields v by its ray transform $[\mathcal{I}v]$ has no unique solution. Namely, the operator $\mathcal{I}$ possesses not trivial kernel consisting of potential vector fields with vanishing potentials on the boundary [1]. Therefore only the solenoidal part ${}^s v$ of the field v can be recovered from $[\mathcal{I}v]$.

The problem of recovering vector field ${}^s v$ from $[\mathcal{I}v]$ is underdetermined in the number of unknown functions. Namely, three unknown functions ${}^s v_j$ are connected by one differential equation $\delta\, {}^s v = \sum\limits_{j=1}^{3} \dfrac{\partial({}^s v_j)}{\partial x_j} = 0$. This means that ${}^s v$ depends on two functions. We need to recover these functions by one function $[\mathcal{I}v]$. This underdetermined does not complicate the vector field recover problem as compared with scalar field recover problem. On the other hand, the problem is overdetermined in dimension, because we try to recover functions ${}^s v_j(x)$, where $x \in R^3$, from function $[\mathcal{I}v]$ on the four-dimensional manifold of oriented lines (variables x, ξ have the dimension 6 in total, but there are two conditions: $|\xi| = 1$, $\langle x, \xi\rangle = 0$). Therefore it is

Published under licence by IOP Publishing Ltd

natural to pose the problem of recovering ${}^s v$ from incomplete data $[\mathcal{I}v]|_{M^3}$, where M^3 is some three-dimensional submanifold of manifold of oriented lines.

Investigation of scalar incomplete data problems is a classical subject of mathematical tomography. As far as studying incomplete data problems of vector tomography is concerned, this direction is in its very beginning. To the author's knowledge, there are very few mathematical papers in the direction. Denisjuk in his article [2] proved the possibility of recovery of a solenoidal part of m-tensor field, which is given in R^3, from incomplete data with usage of cone-beam scheme of observations. Namely, it is possible to recover a solenoidal part of m-tensor field by its known ray transforms computed along oriented lines from three-dimension manifold. Sharafutdinov in his article [3] proposes formulas for recovery of a solenoidal part of vector field by three-dimension data too, but with usage of parallel scheme of observations.

2. Definitions and statement of problem

In this paper we numerically solve the problem of recovery of a solenoidal part of vector field, which is given in unit ball $B = \{x \mid x_1^2 + x_2^2 + x_3^2 < 1\} \subset R^3$, from incomplete data. Namely, data have three dimensions. The algorithm of solving is based on the inversion formulas from paper [3], in which the problem of recovering a solenoidal part of a vector field on R^3 from ray transforms known over all lines parallel to one of the coordinate planes is considered.

We used coordinate planes $\pi_j = \{x_j = 0\} \subset R^3$ $j = 1, 2, 3$ only. We introduce coordinates (θ, s, z) on

$$M^3(\pi_j) = \{(\xi, x) \in R^3 \times R^3 \mid |\xi| = 1,\ \xi_j = 0,\ \langle \xi, x\rangle = 0\}$$

such that

$$\begin{aligned} \xi &= \cos\theta\, e_{j+1} + \sin\theta\, e_{j+2}, \\ x &= s(-\sin\theta\, e_{j+1} + \cos\theta\, e_{j+2}) + z e_j, \end{aligned}$$

where (e_1, e_2, e_3) is the standard basis.

For vector field v a functions $[\mathcal{I}_{(j)}v] = [\mathcal{I}v]|_{M^3(\pi_j)}$ are defined by formula

$$[\mathcal{I}_{(j)}v](\theta, s, z) = \int_{-\infty}^{\infty} \big(\cos\theta\, v_{j+1}(x + t\xi) + \sin\theta\, v_{j+2}(x + t\xi)\big)\, dt.$$

We consider simultaneously two problems that are referred to as the $2P$-problem and $3P$-problem respectively.

$2P$-problem: One has to recover the solenoidal part of a vector field v given two functions $[\mathcal{I}_{(1)}v]$ and $[\mathcal{I}_{(2)}v]$.

$3P$-problem: One has to recover the solenoidal part of a vector field v given three functions $[\mathcal{I}_{(j)}v]$, $j = 1, 2, 3$.

Thus, data space have dimension 3 for both formulations of the vector tomography problem. For $2P$-problem $M^3 = \bigcup_{j=1}^{2} M^3(\pi_j)$, but for $3P$-problem $M^3 = \bigcup_{j=1}^{3} M^3(\pi_j)$.

Two different coordinate planes are sufficient for the uniqueness ($2P$-problem), but three coordinate planes are needed for the stable reconstruction of a solenoidal part of vector field ($3P$-problem) (see, [3]).

The main ingredients of the reconstruction algorithms are the $3D$ operators of the direct and inverse Fourier transform and the $2D$ back-projection operator.

3. Schemes of algorithms of solution $2P$- and $3P$-problems

$3P$-problem. Let for vector field v three functions $[\mathcal{I}_{(j)}v]$, $j = 1, 2, 3$ are given. The following steps should take to restore the solenoidal part of ${}^s v$ field v on these data.

Step 1. Calculate a value $\left(B[\mathcal{I}_{(j)}v]'_s\right)(x)$ of the $2D$ back-projection operator, $j = 1, 2, 3$ by formula

$$\left(B[\mathcal{I}_{(j)}v]'_s\right)(x) = \frac{1}{2\pi}\int_0^{2\pi} [\mathcal{I}_{(j)}v]'_s(\theta, -x_{j+1}\sin\theta + x_{j+2}\cos\theta, x_j)\, d\theta.$$

Step 2. Find a functions $\lambda_{(j)}(y)$, $j = 1, 2, 3$ by formula

$$\lambda_{(j)}(y) = \frac{i}{2}(y_{j+1}^2 + y_{j+2}^2)^{1/2} F_3[B[\mathcal{I}_{(j)}v]'_s](y),$$

where $F_3[\cdot]$ is operator of the direct $3D$-Fourier transform.

Step 3. Calculate $u(y) = F_3[{}^s v](y)$ by formula

$$u_i(y) = |y|^{-2}(\lambda_{(i+1)}(y)y_{i+2} - \lambda_{(i+2)}(y)y_{i+1}).$$

Step 4. Find ${}^s v(x)$ by applying the inverse $3D$-Fourier transform.

$2P$-problem. Let for vector field v two functions $[\mathcal{I}_{(j)}v]$, $j = 1, 2$ are given. The following steps should take to restore the solenoidal part of ${}^s v$ field v on these data.

Step 1. Calculate $\left(B[\mathcal{I}_{(j)}v]'_s\right)(x)$, $j = 1, 2$ by formula

$$\left(B[\mathcal{I}_{(j)}v]'_s\right)(x) = \frac{1}{2\pi}\int_0^{2\pi} [\mathcal{I}_{(j)}v]'_s(\theta, -x_{j+1}\sin\theta + x_{j+2}\cos\theta, x_j)\, d\theta.$$

Step 2. Find a functions $\lambda_{(j)}(y)$, $j = 1, 2$ by formula

$$\lambda_{(j)}(y) = \frac{i}{2}(y_{j+1}^2 + y_{j+2}^2)^{1/2} F_3[B[\mathcal{I}_{(j)}v]'_s](y).$$

A function $\lambda_{(3)}$ are found by formula

$$\lambda_{(3)} = -\frac{1}{y_3}(\lambda_{(1)}y_1 + \lambda_{(2)}y_2).$$

Step 3. Calculate $u(y) = F_3[{}^s v](y)$ by formula

$$u_i(y) = |y|^{-2}(\lambda_{(i+1)}(y)y_{i+2} - \lambda_{(i+2)}(y)y_{i+1}).$$

Step 4. Find ${}^s v(x)$ by applying the inverse $3D$-Fourier transform.

4. Numerical implementation of algorithms

Input date. We use discretization 100 for $z \in [-1, 1]$ and 100, 200, 300, 400, 500 for $s \in [-1, 1]$, and $\theta \in [0, 2\pi]$.

Regularization for solution of $2P$-problem. If we have $\{y_3 = 0\}$ ($|(y_3)_q|$ is very small), then $\lambda_{(3)}((y_1)_m, (y_2)_k, (y_3)_q)$ are calculated with usage regularization.

1) Cut: $((y_3)_q)^{-1}$ are replaced $((y_3)_{q-1})^{-1}$ in formula for finding $\lambda_{(3)}(y)$.

2) Linear interpolation:

$$\lambda_{(3)}((y_1)_m, (y_2)_k, (y_3)_q) = \frac{\lambda_{(3)}((y_1)_m, (y_2)_k, (y_3)_{q-1}) + \lambda_{(3)}((y_1)_m, (y_2)_k, (y_3)_{q+1})}{2}.$$

Without regularization for small $|(y_3)_q|$ components ${}^s v_1(y)$ and ${}^s v_2(y)$ can not restore.

Numerical simulation. We reconstruction continuous solenoidal vector field with components

$$v_i(x) = \begin{cases} 48e^{-24|x|^2}(x_{i+2} - x_{i+1}), & \text{for } |x| < 1, \\ 0, & \text{else.} \end{cases}$$

Dependence relative error of reconstruction ($\%, L_2$-norm) on discretization of input data for s, θ is given in the Table 1.

Table 1. Relative error of reconstruction ($\%, L_2$-norm).

problem, regularization	100	200	300	400	500
$2P$-problem, regularization 1	36.2	34.7	33.1	32.5	31.9
$2P$-problem, regularization 2	21.9	8.4	4.7	3.2	2.5
$3P$-problem	4.3	2.4	1.3	0.9	0.7

5. Conclusion

Numerical simulations have shown good accuracy of the approximation by using the proposed algorithms for reconstruction of a solenoidal part of vector field, which is given in the unit ball, with incomplete data. It is sufficient to solve the $2P$-problem for the uniqueness of a solenoidal part of the vector field reconstruction. On the other hand, the solution of $2P$-problem is unstable. However, good choice of regularization allowed to recover a solenoidal part of the vector field with accuracy close to the accuracy in $3P$-problem solving algorithm realization.

Acknowledgments

The work was supported partially by RFBR (grants 11-07-00447-a, 12-01-09335-mob_z), SB RAS (grant 2012-32), DMS RAS (grant 2012-1.3.1)

References

[1] V. A. Sharafutdinov, *Integral Geometry of Tensor Fields*, Utrecht: VSP, 1994.

[2] A. Denisjuk, Inversion of the X-ray transform for 3D symmetric tensor fields with sources on a curve. *Inverse Problems*. No.**22** (2006), pp. 399–411.

[3] V. A. Sharafutdinov, Slice-by-slice reconstruction algorithm for vector tomography with incomplete data. *Inverse Problems*. No.**23** (2007), pp. 2603–2627.

IC-MSQUARE 2012: International Conference on Mathematical Modelling in Physical Sciences IOP Publishing
Journal of Physics: Conference Series **410** (2013) 012043 doi:10.1088/1742-6596/410/1/012043

Models for biaxial nematic liquid crystals

Lorenzo Nicolodi

Dipartimento di Matematica e Informatica, Università degli Studi di Parma, Parco Area delle Scienze 53/A, I-43124 Parma, Italy

E-mail: `lorenzo.nicolodi@unipr.it`

Abstract. In the Landau-de Gennes theory, the order parameter describing a biaxial nematic liquid crystal is, at each point of the region occupied by the system, a symmetric, traceless 3×3 matrix with three distinct eigenvalues. In the constrained case of matrices with constant eigenvalues, the order parameter space identifies with an eightfold quotient of the 3-sphere, and a configuration of a biaxial nematic liquid crystal is described by a map into such a quotient. We express the (simplest form of the) Landau-de Gennes elastic free-energy density of biaxial nematics as a density on maps into the 3-sphere, whose functional dependence is restricted by the requirements that it is well-defined on the class of configuration maps (residual symmetry) and is independent of arbitrary superposed rigid rotations (frame indifference). This is a report on joint work with D. Mucci [18], to which we refer for more details and applications.

1. The Landau–de Gennes Q-tensor model for nematic liquid crystals

A liquid crystal is a state of matter, called mesomorphic, intermediate between the crystal state and the liquid state, in which the molecules retain preferential orientations relative to one another over large distances. There are many different types of liquid crystals, the main classes being nematics, cholesterics and smectics. Nematic liquid crystals consist of rod-like molecules which have a locally preferred direction of orientation.

In the Landau–de Gennes theory [7, 16], the state of alignment of a nematic liquid crystal occupying a region Ω is described, at each point of Ω, by an order parameter, $\mathbf{Q}$, the so-called $\mathbf{Q}$-*tensor*, which is a symmetric, traceless 3×3 matrix. By definition, $\mathbf{Q}$ vanishes in the isotropic phase and thus measures the extent to which the system is ordered in the region Ω. In a general nematic phase, $\mathbf{Q}$ has five degrees of freedom, two of them specify the degree of order, while the remaining three are the angles needed to specify the principal directions. A nematic phase is said *biaxial* when $\mathbf{Q}$ has three distinct eigenvalues, *uniaxial* when $\mathbf{Q}$ has two non-zero equal eigenvalues. While uniaxial nematics has been known to exist for more than a century, the experimental evidence of biaxial nematics is more recent [13]. A general $\mathbf{Q}$-tensor can be written as

$$\mathbf{Q} = S_1\Big(\mathbf{n} \otimes \mathbf{n} - \frac{1}{3}\mathbf{I}\Big) + S_2\Big(\mathbf{m} \otimes \mathbf{m} - \frac{1}{3}\mathbf{I}\Big), \quad S_1, S_2 \in \mathbb{R}, \ \mathbf{n}, \mathbf{m} \in \mathbb{S}^2, \tag{1}$$

where $\mathbf{n}$, $\mathbf{m}$ are orthonormal eigenvectors of $\mathbf{Q}$ and S_1, S_2 are scalar order parameters given by $S_1 = \lambda_1 - \lambda_3 = 2\lambda_1 + \lambda_2$, $S_2 = \lambda_2 - \lambda_3 = \lambda_1 + 2\lambda_2$ in terms of the eigenvalues $\lambda_1, \lambda_2, \lambda_3$ of $\mathbf{Q}$. We may assume $\lambda_1 \leq \lambda_2 \leq \lambda_3$ and $\lambda_1, \lambda_2, \lambda_3 \in (-\frac{1}{3}, \frac{2}{3})$ (see [2]). In the uniaxial case, $\mathbf{Q}$ takes the form

$$\mathbf{Q} = s\Big(\mathbf{r} \otimes \mathbf{r} - \frac{1}{3}\mathbf{I}\Big), \quad s \in \mathbb{R}, \ \mathbf{r} \in \mathbb{S}^2, \tag{2}$$

where s is the only scalar order parameter. The Landau–de Gennes free-energy functionals are nonlinear integral functionals of $\mathbf{Q}$ and its spatial derivatives. In general, it is required that any energy density $\Psi = \Psi(\mathbf{Q}, \nabla\mathbf{Q})$ satisfy the condition of *frame indifference* which amounts to

$$\Psi(\mathbf{Q}, \nabla\mathbf{Q}) = \Psi\left(M\mathbf{Q}M^T, \mathbf{D}^*\right) \qquad \forall M = (M_j^i) \in SO(3)\,, \tag{3}$$

where $\mathbf{D}^*$ denotes a third order tensor such that $\mathbf{D}^*_{ijk} = M_l^i M_m^j M_p^k \mathbf{Q}_{lm,p}$, and $\mathbf{Q}_{ij,k} = \frac{\partial}{\partial x_k}\mathbf{Q}_{ij}$ denote the first order partial derivatives of $\mathbf{Q}$ (compare, for example, [1]).

A commonly used expression for the free energy of a nematic liquid crystal is [7, 6, 16, 20]

$$\mathcal{F}(\mathbf{Q}, \Omega) := \int_\Omega \left[\psi(\mathbf{Q}, \nabla\mathbf{Q}) + f_B(\mathbf{Q})\right]\, dx\,, \tag{4}$$

where $f_B(\mathbf{Q}) = f_B(\mathrm{tr}(\mathbf{Q}^2), \mathrm{tr}(\mathbf{Q}^3))$ is the bulk part (a function of the principal invariants of $\mathbf{Q}$) and

$$\psi(\mathbf{Q}, \nabla\mathbf{Q}) = L_1 I_1 + L_2 I_2 + L_3 I_3 + L_4 I_4 \tag{5}$$

is the elastic free-energy density; the L_i are material constants and the I_i are elastic invariants,

$$I_1 = \mathbf{Q}_{ij,j}\mathbf{Q}_{ik,k}\,, \quad I_2 = \mathbf{Q}_{ik,j}\mathbf{Q}_{ij,k}\,, \quad I_3 = \mathbf{Q}_{ij,k}\mathbf{Q}_{ij,k}\,, \quad I_4 = \mathbf{Q}_{lk}\mathbf{Q}_{ij,l}\mathbf{Q}_{ij,k} \tag{6}$$

(summation over repeated indices is assumed). The bulk energy $f_B(\mathbf{Q})$ is invariant under the $SO(3)$-action by conjugation on the five-dimensional space of $\mathbf{Q}$-tensors, so that the critical points of the bulk energy form an orbit of solutions in the space of $\mathbf{Q}$-tensors. Clearly, a $\mathbf{Q}$-tensor taking values in a group orbit has constant scalar order parameters S_1, S_2. In many applications, it suffices to work in the *constrained Landau-de Gennes theory*, in which $\mathbf{Q}$ has constant S_1, S_2, independent of $x \in \Omega$, and hence constant eigenvalues [3, 14]. In the constrained theory, the bulk energy is constant and one only has to consider the elastic free energy.

2. The Oseen–Frank vector model for uniaxial nematics

In the *Oseen–Frank theory* [10, 19], a configuration of a liquid crystal is described by a unit vector field $\mathbf{r} : \Omega \to \mathbb{S}^2$, which represents the direction of preferred molecular alignment. This theory has received much analytical and computational attention, but is not sufficiently general to represent biaxial states. In this model, the elastic energy density associated to $\mathbf{r}$ is given by

$$w(\mathbf{r}, \nabla\mathbf{r}) = K_1(\mathrm{div}\,\mathbf{r})^2 + K_2(\mathbf{r}\cdot\mathrm{curl}\,\mathbf{r})^2 + K_3|\mathbf{r}\times\mathrm{curl}\,\mathbf{r}|^2 + (K_2+K_4)\left[\mathrm{tr}[(\nabla\mathbf{r})^2] - (\mathrm{div}\,\mathbf{r})^2\right],$$

where the K_i are elastic constants. This energy density satisfies the *invariance properties*

$$w(\mathbf{r}, \nabla\mathbf{r}) = w(-\mathbf{r}, -\nabla\mathbf{r}); \quad w(\mathbf{r}, \nabla\mathbf{r}) = w(H\mathbf{r}, H\,\nabla\mathbf{r}\,H^T), \quad \forall\, H \in O(3), \tag{7}$$

so that the corresponding functional is well defined on vector fields in Ω, regardless of the orientation. The first equation in (7) accounts for the lack of polarity of nematics, while the second expresses the frame indifference of the energy density and the material symmetry condition corresponding to the lack of chirality of nematics. Requiring that the second condition in (7) hold for $SO(3)$ only, is equivalent to the frame indifference condition (3) for constrained uniaxial $\mathbf{Q}$-tensors. The vector approach has been further developed in [8, 9, 12]. The Oseen-Frank model does not respect the "head-to-tail" symmetry, in which $\mathbf{r}$ and $-\mathbf{r}$ represent the same state; instead, $\mathbf{r}$ should take values in $\mathbb{S}^2$ with antipodal points identified, i.e., in the projective plane $\mathbb{RP}^2$. The constrained uniaxial $\mathbf{Q}$-tensor representation (2) clearly respects this symmetry.

In the constrained uniaxial case, formally calculating the energy density (5) in terms of $\mathbf{r}$ and $\nabla\mathbf{r}$, it can be checked (see [3, 16, 18]) that *the Landau-de Gennes elastic free-energy density*

$\psi(\mathbf{Q}, \nabla\mathbf{Q})$ *reduces to the Oseen-Frank energy density* $w(\mathbf{r}, \nabla\mathbf{r})$, that is, it is possible to choose the L_i and the K_i, $i = 1, 2, 3, 4$, so that

$$\psi(\mathbf{Q}, \nabla\mathbf{Q}) = w(\mathbf{r}, \nabla\mathbf{r}). \tag{8}$$

However, according to [3], although the elastic energies can be taken to be the same in the two theories, the result of the energy minimization might be different. For the related problems of line field orientability and map lifting in the Sobolev setting, we refer the reader to [3, 17].

3. Constrained theory of biaxial nematic liquid crystals

Let $\mathcal{Q}(\lambda_1, \lambda_2, \lambda_3)$ be the set of all constrained biaxial $\mathbf{Q}$-tensors of the form (1) with distinct constant eigenvalues $\lambda_1, \lambda_2, \lambda_3$. Any element $\mathbf{Q} \in \mathcal{Q}(\lambda_1, \lambda_2, \lambda_3)$ can be written in the form $\mathbf{Q} = \mathbf{GAG}^T$, for some $\mathbf{G} \in SO(3)$, where $\mathbf{A} = \mathrm{diag}(\lambda_1, \lambda_2, \lambda_3)$ is the diagonal matrix of the eigenvalues. Thus, $\mathcal{Q}(\lambda_1, \lambda_2, \lambda_3)$ coincides with the orbit of $\mathbf{A}$ with respect to the $SO(3)$-action by conjugation on the space of $\mathbf{Q}$-tensors, and can be identified with the homogeneous space $SO(3)/D_2$, where D_2 is the abelian four-element dihedral group ([15]). Using the identification of $\mathbb{S}^3$ with the Lie group of unit quaternions, $Sp(1)$, and the 2:1 covering map Φ: $\mathbb{S}^3 \to SO(3)$, the order parameter space of constrained biaxial nematics is then diffeomorphic to the homogeneous manifold $\mathbb{S}^3/\mathcal{H}$, where $\mathcal{H} = \{\pm 1, \pm \mathsf{i}, \pm \mathsf{j}, \pm \mathsf{k}\}$ is the non-abelian eight-element quaternion group. In this model, a configuration of a biaxial nematic liquid crystal is described by a map fromΩ to $\mathbb{S}^3/\mathcal{H}$, as opposed to the constrained uniaxial case where the order parameter space is $\mathbb{RP}^2$.

4. A vector model for the elastic energy of constrained biaxial nematics

Motivated by (8) in the constrained uniaxial case, our purpose is to express, in the constrained biaxial case, the Landau–de Gennes elastic free-energy density $I_3(\mathbf{Q}, \nabla\mathbf{Q})$ as a density on maps $q : \Omega \to \mathbb{S}^3$. For, we need first to identify the conditions on a generic energy density $f : \mathbb{S}^3 \times \mathbb{M}_{4\times 3} \to [0, +\infty)$, depending on $q : \Omega \to \mathbb{S}^3$ and its first derivatives, in order that: (a) f is independent of arbitrary superposed rigid rotations (frame indifference condition); (b) f is well defined on the class of configuration maps$\Omega \to \mathbb{S}^3/\mathcal{H}$ (residual symmetry condition), that corresponds to the "head-to-tail" symmetry in the uniaxial case. As for condition (a), we say that $f : \mathbb{S}^3 \times \mathbb{M}_{4\times 3} \to [0, +\infty)$ satisfies the *frame invariance condition* if, for any $q \in \mathbb{S}^3 \cong Sp(1)$,

$$f(w, H) = f(qw, L(q)H\Phi(q)^T) \quad \forall (w, H) \in \mathbb{S}^3 \times \mathbb{M}_{4\times 3},$$

where $L(q)$ is the matrix of the $\mathbb{R}$-linear map $w \mapsto qw$ on the algebra of quaternions $\mathbb{H}$, relative to $\{1, \mathsf{i}, \mathsf{j}, \mathsf{k}\}$, and $\Phi(q)$ is the rotation induced by q via the 2:1 covering map Φ: $\mathbb{S}^3 \to SO(3)$. We prove that this invariance condition is indeed equivalent to the frame indifference condition in the sense of $\mathbf{Q}$-tensors (see [18]). The proof uses the identification $\mathbb{S}^3 \cong Sp(1)$ and the relations between quaternions and rotations in $\mathbb{R}^3$ and $\mathbb{R}^4$. As for condition (b), we say that $f : \mathbb{S}^3 \times \mathbb{M}_{4\times 3} \to [0, +\infty)$ satisfies the *residual symmetry property* if, for any $q \in \mathcal{H}$, one has

$$f(w, H) = f(qw, L(q)H) \quad \forall (w, H) \in \mathbb{S}^3 \times \mathbb{M}_{4\times 3}.$$

Given that, we express the Landau-de Gennes energy density I_3 in terms of maps $(u, \mathbf{v}) : \Omega \to \mathbb{S}^3$, namely, we explicitly compute $f_3 : \mathbb{S}^3 \times \mathbb{M}_{4\times 3} \to [0, +\infty)$, so that

$$I_3(\mathbf{Q}, \nabla\mathbf{Q}) = f_3((u, \mathbf{v}), \nabla(u, \mathbf{v})),$$

provided that $\mathbf{Q}$ corresponds to $(u, \mathbf{v})$, that is, $\mathbf{Q} = \mathbf{GAG}^T$ with $\mathbf{G} = \Phi((u, \mathbf{v}))$. For $(u, \mathbf{v}) \in \mathbb{S}^3$ and $H \in \mathbb{M}_{4\times 3}$, the function $f_3 : \mathbb{S}^3 \times \mathbb{M}_{4\times 3} \to [0, +\infty)$ is given by

$$f_3((u, \mathbf{v}), H) := (k_1 - k_2 + k_3)|H|^2 - k_1\, A((u, \mathbf{v}), H)^2 + k_2 B((u, \mathbf{v}), H)^2 - k_3 C((u, \mathbf{v}), H)^2,$$

where $H = (H_i^j)$, $j = 0,1,2,3$; $i = 1,2,3$, $|H|^2 := \sum_{i=1}^3 \sum_{j=0}^3 (H_i^j)^2$, $k_1 := 8S_1(S_1 - S_2)$, $k_2 := 8S_2(S_1 - S_2)$, $k_3 := 8S_1S_2$ ($k_1, k_2, k_3 > 0$ and $(k_1 - k_2 + k_3) = 8(S_1^2 + S_2^2 - S_1S_2) > 0$), and

$$\begin{aligned} A((u,\mathbf{v}),H)^2 &:= \sum_{i=1}^{3} \left(-v_1H_i^0 + uH_i^1 + v_3H_i^2 - v_2H_i^3\right)^2, \\ B((u,\mathbf{v}),H)^2 &:= \sum_{i=1}^{3} \left(-v_2H_i^0 - v_3H_i^1 + uH_i^2 + v_1H_i^3\right)^2, \\ C((u,\mathbf{v}),H)^2 &:= \sum_{i=1}^{3} \left(-v_3H_i^0 + v_2H_1^1 - v_1H_i^2 + uH_i^3\right)^2. \end{aligned}$$

In particular, the energy density model $f_3((u,\mathbf{v}),\nabla(u,\mathbf{v}))$ is proved to satisfy the above mentioned frame indifference and residual symmetry conditions. Therefore, f_3 may be interpreted as the elastic energy density model for the configuration maps $(u,\mathbf{v}) : \Omega \to \mathbb{S}^3/\mathcal{H}$ of a constrained biaxial nematic liquid crystal. (In principle, similar calculations can be performed also for the elastic invariants I_1, I_2 and I_4, see [6, 18, 21].) The corresponding elastic energy functional

$$\mathcal{F}_3((u,\mathbf{v}),\Omega) := \int_\Omega f_3((u,\mathbf{v})(x),\nabla(u,\mathbf{v})(x))\,dx$$

is well defined, for instance, on Sobolev maps $(u,\mathbf{v}) : \Omega \to \mathbb{S}^3/\mathcal{H}$, where Ω is a bounded domain of $\mathbb{R}^3$. This representation is also used to prove the coercivity of the energy density. This yields, for instance, that the class of measurable and a.e. weakly differentiable functions from Ω to $\mathbb{S}^3/\mathcal{H}$ with finite $\mathcal{F}_3$-energy agrees with the Sobolev class $W^{1,2}(\Omega,\mathbb{S}^3/\mathcal{H})$. A general lifting technique described by Bethuel and Chiron in [5], combined with a strong density result of Bethuel [4], are then used to prove results on lower semicontinuity and strong density of smooth maps (see [18] for more details).

References

[1] Ball J M 2010 Lecture slides, Benin Summer School, 28 June–2 July
[2] Ball J M and Majumdar A 2010 *Mol. Cryst. Liq. Cryst.* **525**, 1–11
[3] Ball J M and Zarnescu A 2011 *Arch. Ration. Mech. Anal.* **202**, 493–535
[4] Bethuel F 1991 *Acta Math.* **167**, 153–206
[5] Bethuel F and Chiron D 2007 *Contemp. Math.* **446**, 125-152
[6] Davis T A and Gartland E C Jr 1998 *SIAM J. Numer. Anal.* **35**, 336-362
[7] de Gennes P G and Prost J 1995 *The Physics of Liquid Crystals*, 2nd edn. (Oxford: Oxford University Press)
[8] Ericksen J L: Equilibrium theory of liquid crystals. In: *Advances in liquid crystals* **2**, G.H. Brown (ed.), Academic Press, New York, 233–299 (1976)
[9] Ericksen J L 1990 *Arch. Ration. Mech. Anal.* **113**, 97-120
[10] Frank F C 1958 *Discuss. Faraday Soc.* **25**, 19–28
[11] Kinderlehrer D: Recent developments in liquid crystals theory. In: *Frontiers in Pure and Applied Mathemathics*, North-Holland, Amsterdam, 151–178 (1991)
[12] Leslie F C: Theory of flow phenomena in liquid crystals. In: *Advances in liquid crystals* **2**, G.H. Brown (ed.), Academic Press, New York, 1–81 (1976)
[13] Luckhurst G R 2004 *Nature* **430**, 413-414 (2004)
[14] Majumdar A and Zarnescu A 2010 *Arch. Ration. Mech. Anal.* **196**, 227-280
[15] Mermin N D 1979 *Rev. Modern Phys.* **51**, 591-648
[16] Mottram N J and Newton C 2004 *Introduction to Q-tensor theory*, Technical report, University of Strathclyde, Department of Mathematics
[17] Mucci D 2012 *Discrete Contin. Dyn. Syst. Ser. B* **17**, 597-635
[18] Mucci D and Nicolodi L 2012 *Arch. Ration. Mech. Anal.* **206**, 853–884
[19] Oseen C W 1933 *Trans. Faraday Soc.* **29**, 883-899
[20] Schopohl N and Sluckin T J 1987 *Phys. Rev. Lett.* **59**, 2582–2584
[21] Vissenberg N, Stallinga S and Vertogen G 1997 *Phys. Rev. E* **55**, 4367–4377

Journal of Physics: Conference Series **410** (2013) 012044 doi:10.1088/1742-6596/410/1/012044

How optimal synchronization of oscillators depends on the network structure and the individual dynamical properties of the oscillators

R Markovič[1], M Gosak[1,2,3] and M Marhl[1,2]

1 Faculty of Natural Sciences and Mathematics, University of Maribor, Koroška cesta 160, SI-2000 Maribor, Slovenia

2 Faculty of Education, University of Maribor, Koroška cesta 160, SI-2000 Maribor, Slovenia

3 Faculty of Civil Engineering, University of Maribor, Smetanova ulica 17, SI-2000 Maribor, Slovenia

E-mail: rene.markovic@uni-mb.si

Abstract. The problem of making a network of dynamical systems synchronize onto a common evolution is the subject of much ongoing research in several scientific disciplines. It is nowadays a well-known fact that the synchronization processes are gradually influenced by the interaction topology between the dynamically interacting units. A complex coupling configuration can significantly affect the synchronization abilities of a networked system. However, the question arises what is the optimal network topology that provides enhancement of the synchronization features under given circumstances. In order to address this issue we make use of a network model in which we can smoothly tune the topology from a highly heterogeneous and efficient scale-free network to a homogeneous and less efficient network. The network is then populated with Poincaré oscillators, a paradigmatic model for limit-cycle oscillations. This oscillator model exhibits a parameter that enables changes of the limit cycle attraction and is thus immediately related to flexibility/rigidity properties of the oscillator. Our results reveal that for weak attractions of the limit cycle, intermediate homogeneous topology ensures maximal synchronization, whereas highly heterogeneous scale-free topology ensures maximal synchronization for strong attractions of the limit cycle. We argue that the flexibility/rigidity of individual nodes of the networks defines the topology, where maximal global coherence is achieved.

Complex networks are nowadays used for the description of several natural and artificial systems. Since topological features of interactions between individual units characterize the global properties of a given system, the research of structural properties is increasingly gaining on attention. To qualitatively analyze the local and global structural properties of a network, numerous techniques have been developed, which have been utilized in various disciplines and diverse circumstances [1]. One of the basic measures that describes the properties of a network is its degree distribution. Scale-free networks, for instance, are known to have a highly heterogeneous power-law degree distribution [1], which can foster the synchronization abilities of a network of coupled oscillators [2], especially due to their hight global efficiency. Furthermore, it has been shown that a high level of heterogeneity can suppress synchronization, even thought it also reduces the average distance between the nodes. Another importing network measure is the average clustering coefficient C characterizing the cliquishness of a network. Remarkably, high clustering coefficients were detected in several real-life networks [1]. On the other hand, high clustering implies the existence of many transitive connections and can thus hinder global synchronization [3]. There are several other topological factors, such as for example the existence of a community structure [4], which also have an impact on synchronization processes. Nevertheless, even though the dynamical behavior of networks is inevitable and non-trivially connected with structural properties of the underlying networks, there are also several other important factors having an impact. With this in mind, in the present study we focus on the analysis of the role of dynamical properties of individual oscillators. In a previous theoretical study on cellular oscillators it has been argued that the local dissipation rate crucially determines the coupling ability of cellular oscillators [5]. Notably, the dissipation rate is directly correlated with flexibility/rigidity of an oscillator. A low dissipation rate stands for rigid oscillators whereas on the other hand, oscillators with a near zero dissipation rate have attractors that are very susceptible to external perturbations and are thus flexible. Therefore, the main goal of the present work is to explore how the relation between flexibility/rigidity properties of individual oscillators on one hand and the structure of the network topology that characterizes the connectivity patterns between them on the other hand impacts the synchronization of oscillators.

By applying the algorithm described previously [6, 7] we can generate networks whose topologies can smoothly be altered between a highly heterogeneous scale-free network and a more homogeneous network. First, N nodes are randomly distributed in a unit square and to each node a fitness value f_i is prescribed. Two nodes are connected if:

$$\theta < \frac{f_i f_j}{I_{ij}^{\delta}}, \tag{1}$$

where f_i and f_j are fitness values of the i-th and j-th node, I_{ij} is the Euclidean distance between them and the δ is used to alter the topology of the network. The parameter θ is used as a threshold to control the average node degree of the network $\langle k \rangle$. In order to quantify topological features of the network and its degree of heterogeneity, the standard deviation of the node degrees STD(k) (Fig. 1a), the standard deviation of the clustering coefficients STD(C) (Fig. 1b) and the average clustering coefficient C (Fig. 1c) are plotted as a function of δ. Both, STD (k) and STD(C) are decreasing monotonically with increasing values of the topology parameter δ. Higher values of STD(k) and STD(C) are indicators of diverse local topological properties indicating a high degree of network heterogeneity.

Characteristic examples of generated networks are shown in Fig. 1d-f. It can be observed that for low values of δ indeed very heterogeneous networks are generated, in which mostly long-range connections exist. On the other hand, if $\delta >> 1$ mostly nearby nodes are connected and, in addition, there are no expressive differences in individual node degrees. Finally, intermediate values of δ, i.e. $\delta = 1.1$, result in an intermediate heterogeneous network with both long- and short-range connections. In other words, Fig. 1d-f provide a visual assessment that the network heterogeneity indeed decreases with increasing values of δ.

The dynamics of individual nodes is driven by the paradigmatic Poincaré oscillator:

$$\dot{x}_i = -\gamma\,(r_i - A)\,x_i - \omega_i y_i - \epsilon \sum_{i=1}^{N} d_{ij}\,(x_i - x_j)\,, \tag{2}$$

$$\dot{y}_i = -\gamma\,(r_i - A)\,y_i + \omega_i x_i - \epsilon \sum_{i=1}^{N} d_{ij}\,(y_i - y_j)\,, \tag{3}$$

where x_i and y_i are the phase space coordinates of the i-th oscillator, A is the limit cycle radius, $r_i = \sqrt{x_i^2 + y_i^2}$ is the distance between the origin of the phase space and the position of the i-th

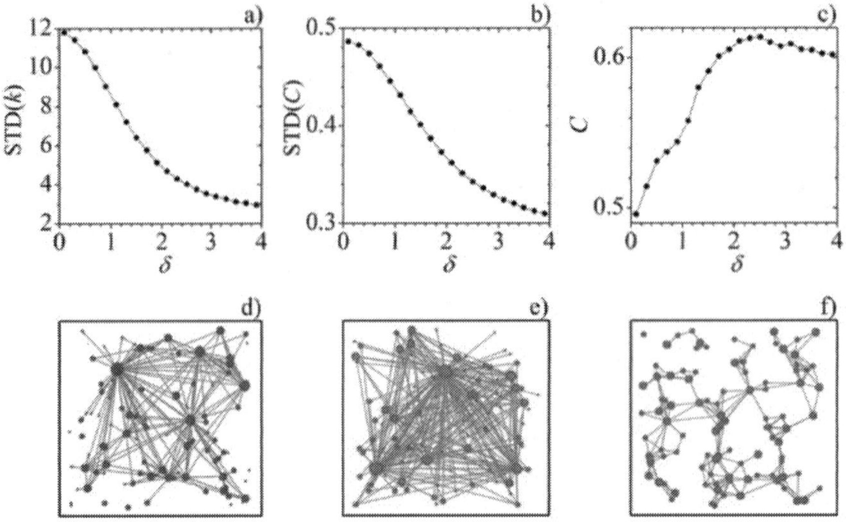

Figure 1. Standard deviation of node degrees (a), standard deviation of the local clustering coefficient (b), the average clustering coefficient (c) and characteristic network structures obtained for $\delta = 0.1$ (d), $\delta = 1.1$ (e) and $\delta = 4.0$ (f). Note that in lower panels the size of the nodes is proportional to their degrees. The number of nodes is $N = 100$ and the average node degree is $\langle k \rangle = 5.0$

node in the phase space, ω_i is the angular velocity of the i-th oscillator and ϵ is the coupling strength and d_{ij} is the connectivity matrix with values of 1 if the i-th and j-th node are connected and 0 otherwise. Angular velocities ω_i were distributed according to a normal distribution with mean angular velocity $\overline{\omega} = 1.0$ with standard deviation of 0.2. The limit cycle radius was set to $A = 1.0$ and the coupling strength to $\epsilon = 0.25$.

The parameter γ determines the relaxation towards the limit cycle and is immediately related with the dissipation rate of the oscillators [5, 8]. In order to investigate the interplay between structural and dynamical features of the network, we simulated how the overall synchronization in the systems changes according to the initial rigidity of the oscillators and structural properties of the network. For that purpose we analyze the synchronization behavior as a function of the relaxation rate γ and network topology parameter δ. For the quantification of the global network synchronization we calculate the average correlation coefficient $\overline{R}$. Accordingly, we have to construct the $N \times N$ correlation matrix, whose ij-th element is defined as:

$$R_{ij} = \frac{1}{M} \sum_{t=1}^{M} \frac{[x_i(t) - \overline{x_i}][x_j(t) - \overline{x_j}]}{S_i S_j}, \qquad (4)$$

where M is the number of integration steps, $x_i(t)$ and $x_j(t)$ are the time series of the i-th and j-th oscillator, $\overline{x_i}$ and $\overline{x_j}$ are the average values of the time series and S_i and S_j the corresponding standard deviations. The average correlation coefficient $\overline{R}$ is then obtained as the average over all non-diagonal elements of the matrix. Its values range between 0 (uncorrelated dynamics) and 1 (complete synchronized motion).

Numerical simulations were carried out on a network of $N = 100$ Poincaré oscillators, with an average node degree $\langle k \rangle = 5.0$. Fig. 2a features the results showing the average correlation as a function of the dissipation rate γ and the network structure δ. It can be observed that both parameters mutually affect the synchronization behavior. For rigid oscillations where $\gamma > 100$, maximal synchronization is achieved in the highly heterogeneous scale-free network.

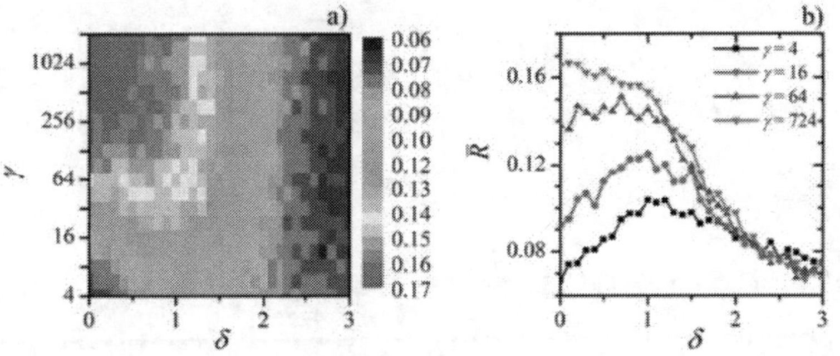

Figure 2. a) The average correlation $\overline{R}$ as a function of the topology parameter δ and flexibility parameter γ. b) Average correlation values for four different values of γ as a function of δ.

On the other hand, in case the network is populated with flexible oscillators ($\gamma < 100$), maximal synchronization is achieved for intermediate values of the topology parameter δ, where the network is less heterogeneous as the scale-free network and is constituted by long- as well as short-range links (see Fig. 1). Notably, for this intermediate values of δ the network exhibits a higher average clustering coefficient. Furthermore, in the homogeneous network ($\delta > 2$) with mostly short-range interactions, the degree of synchronization does not change by changing γ, thus indicating that in this case the synchronization does not depend on flexibility/rigidity properties of oscillators. In order to provide a better inside into the reported phenomena, we additionally plotted characteristic cross-sections of the color-contour plots for different values of γ. The results presented in Fig. 2b additionally confirm the existence of a resonant response due to changes in network topology for flexible oscillators, whereas for rigid oscillators the level of synchronization decreases monotonically with increasing δ.

In sum, we have shown that the network topology ensuring most synchronized response in an ensemble of coupled oscillators depends on individual oscillator properties. While rigid oscillators synchronize best in heterogeneous scale-free networks, flexible oscillator exhibit the most coherent collective response when they are connected in a less heterogeneous network. Our findings provide novel insights into synchronization behavior of coupled oscillators, which may be of importance especially from biological point of view. Signal transduction systems have to respond sensitively to weak external stimuli, thus indicating that cellular oscillators in general should behave like flexible oscillators. On the other hand it is known that different biological oscillators differ in their rigidity and, moreover, that the coupling influences the flexibility/rigidity properties of an oscillator [8]. Apparently, the optimal structural organization of the intercellular communication networks is determined also by characteristics of individual cells. Interestingly, the intermediate network structure that ensures best synchronizability of flexible oscillators is very economic, since it represents a good compromise between efficiency, wiring economy and robustness - a desirable attribute of several real-life systems.

References

[1] Albert R and Barabási A L 2002 *Rev. Mod. Phys.* **74**(1) 47–97
[2] Hong H, Kim B J, Choi M Y and Park H 2004 *Phys. Rev. E* **69**(6) 067105
[3] McGraw P N and Menzinger M 2005 *Phys. Rev. E* **72**(1) 015101
[4] McGraw P N and Menzinger M 2008 *Phys. Rev. E* **77**(3) 031102
[5] Perc M and Marhl M 2004 *Bioelectrochemistry* **62** 1–10
[6] Gosak M, Korošak D and Marhl M 2011 *New Journal of Physics* **13** 013012
[7] Gosak M, Markovič R and Marhl M 2012 *Physica A: Statistical Mechanics and its Applications* **391** 2764 – 2770 ISSN 0378-4371
[8] Abraham U, Granada A E, Westermark P O, Heine M, Kramer A and Herzel H 2010 *Molecular System Biology* **6** 13

Journal of Physics: Conference Series **410** (2013) 012045 doi:10.1088/1742-6596/410/1/012045

Navigating catastrophes: Local but not global optimisation allows for macro-economic navigation of crises

Michael S. Harré

School of Engineering and IT, The University of Sydney, Sydney, Australia

E-mail: michael.harre@sydney.edu.au

Abstract. Two aspects of modern economic theory have dominated the recent discussion on the state of the global economy: Crashes in financial markets and whether or not traditional notions of economic equilibrium have any validity. We have all seen the consequences of market crashes: plummeting share prices, businesses collapsing and considerable uncertainty throughout the global economy. This seems contrary to what might be expected of a system in equilibrium where growth dominates the relatively minor fluctuations in prices. Recent work from within economics as well as by physicists, psychologists and computational scientists has significantly improved our understanding of the more complex aspects of these systems. With this interdisciplinary approach in mind, a behavioural economics model of local optimisation is introduced and three general properties are proven. The first is that under very specific conditions local optimisation leads to a conventional macro-economic notion of a global equilibrium. The second is that if both global optimisation and economic growth are required then under very mild assumptions market catastrophes are an unavoidable consequence. Third, if only local optimisation and economic growth are required then there is sufficient parametric freedom for macro-economic policy makers to steer an economy around catastrophes without overtly disrupting local optimisation.

1. Optimisation in Localised and Global Economies

In this study the Maximum Entropy (MaxEnt) technique is applied to a simple economy in which production output is set through a competitive financial market mechanism. A justification for the use of MaxEnt in economics is given in [1] but an alternative is introduced here. MaxEnt provides a way of finding the *least structured distribution over output choices* for a given set of economic constraints. As entropy over production choices decreases, this lowers the level of output for products with a lower profit and the minimum entropy is achieved when only the most profitable products are produced. Across an entire economy this may not be desirable as variety in economic output provides a level of economic security. MaxEnt provides a mechanism to achieve this: Given a minimum Gross Domestic Product (GDP) target to reach (which may increase over time), produce the most diverse range and largest quantity of different products across the economy. If economic targets allow for it, producers should tend to produce anything and everything in equal proportion. However, in practice economic constraints require choosing to manufacture some products and not others, and shifting economic targets and competition may mean some products are not economically viable. This study considers the most basic elements of a model that instantiates these ideas.

Journal of Physics: Conference Series **410** (2013) 012045 doi:10.1088/1742-6596/410/1/012045

1.1. Global Optimisation of a Simple Economy

Following [1] there are two groups of economic agent, i and $-i$, in a joint strategy economy $x_j, y_k,\ j,k \in \{1,2\}$. Each group contributes to economic output (GDP) based on their choices:

$$u^i_{x,y} = \begin{bmatrix} u^i_{x_1,y_1} & u^i_{x_1,y_2} \\ u^i_{x_2,y_1} & u^i_{x_2,y_2} \end{bmatrix}, \quad u^{-i}_{x,y} = \begin{bmatrix} u^{-i}_{x_1,y_1} & u^{-i}_{x_1,y_2} \\ u^{-i}_{x_2,y_1} & u^{-i}_{x_2,y_2} \end{bmatrix}$$

The expected GDP is found via the probability distributions q_i and q_{-i} over choices:

$$E(u^i) = GDP_i = \sum_{x,y} q_i(x) q_{-i}(y) u^i_{x,y}\,, \quad E(u^{-i}) = GDP_{-i} = \sum_{x,y} q_i(x) q_{-i}(y) u^{-i}_{x,y}$$

The constrained optimisation is to maximise the entropy subject to a minimum GDP:

$$\max_{f(q_i,q_{-i})} S(f(q_i,q_{-i})) = \max_{f(q_i,q_{-i})} \Big(-\sum_{x,y} f(q_i,q_{-i}) \log(f(q_i,q_{-i})) \Big) \tag{1}$$

subject to the constraints:

$$f(q_i,q_{-i}) \geq 0 \;\forall\, x,y \tag{2}$$

$$\sum_{x,y} f(q_i,q_{-i}) = 1 \tag{3}$$

$$\sum_{x,y} f(q_i,q_{-i}) u^i_{x,y} + \sum_{x,y} f(q_i,q_{-i}) u^{-i}_{x,y} = \sum_{x,y} f(q_i,q_{-i})(u^i_{x,y} + u^{-i}_{x,y}) \geq \text{GDP} \tag{4}$$

where total $GDP = GDP_i + GDP_{-i}$. The solution to this optimisation is to form the Lagrangian $\mathcal{L}(f(q_i,q_{-i}))$, find the stationary point and solve for $f(q_i,q_{-i})$:

$$\mathcal{L}(f(q_i,q_{-i})) = S(f(q_i,q_{-i})) + \beta\Big(\sum_{x,y} f(q_i,q_{-i})(u^i_{x,y} + u^{-i}_{x,y})\Big) + \beta_0 \sum_{x,y} f(q_i,q_{-i}) \tag{5}$$

$$f(q_i,q_{-i}) = \mathcal{Z}^{-1} \exp\Big(\beta \sum_{x,y} (u^i_{x,y} + u^{-i}_{x,y})\Big) \tag{6}$$

where $\mathcal{Z}^{-1}$ is the normalisation factor and β enforces the minimum GDP constraint. In a free market with no collusion between participants q_i and q_{-i} are independent: $f(q_i,q_{-i}) = q_i q_{-i}$ and the distributions are separable: $q_i q_{-i} \propto \exp\left(\beta \sum_{x,y} u^i_{x,y}\right) \exp\left(\beta \sum_{x,y} u^{-i}_{x,y}\right)$. To find the joint probability when it is in a global equilibrium state simply substitute in all the values for $u^i_{x,y}$ and $u^{-i}_{x,y}$, which are known from the problem definition. However, to then ask "What is the probability that q_i chooses $x = x_j$" is an ill-posed question, setting $x = x_j$ and using $q_i(x = x_j) \propto \exp(\beta \sum_y u^i_{x_j,y})$ has no unique solution as y is still undetermined, i.e. global optimisation has insufficient information for local decisions. This is fixed by assuming that i is able to estimate q_{-i}, learned through previous interactions or some other means. The problem of finding a clearly defined local distribution is now well-posed: $q_i(x = x_j | q_{-i}) \propto \exp(\beta \sum_y q_{-i} u^i_{x_j,y})$.

1.2. Local Optimisation Within a Simple Economy

Now we start from the other extreme, given a local optimisation problem what is the form of q_i and q_{-i}? Each set of economic agents has to maximise an entropy function subject to the constraint that their contribution to the $GDP = GDP_i + GDP_{-i}$ reaches a minimum level:

$$\max_{q_i} S(q_i) = \max_{q_i} \Big(-\sum_x q_i \log(q_i) \Big) \tag{7}$$

subject to the constraints:

$$q_i \geq 0 \,\forall\, x \tag{8}$$

$$\sum_x q_i = 1 \tag{9}$$

$$\sum_{x,y} q_i q_{-i} u^i_{x,y} \geq \mathrm{GDP}_i \tag{10}$$

Following an analogous process to the global case we have:

$$\mathcal{L}(q_i) = S(q_i) + \beta_i\Big(\sum_{x,y} q_i, q_{-i}u^i_{x,y}\Big) + \beta_0 \sum_x q_i \tag{11}$$

$$q_i = \mathcal{Z}_i^{-1} \exp\Big(\beta_i \sum_{x,y} q_{-i}u^i_{x,y}\Big), \quad q_{-i} = \mathcal{Z}_{-i}^{-1} \exp\Big(\beta_{-i} \sum_{x,y} q_i u^{-i}_{x,y}\Big) \tag{12}$$

Note that the need to post-hoc stipulate an estimate for q_{-i} (when we optimise for i) disappears as q_{-i} does not vanish when we differentiate with respect to q_i as it does when we differentiated with respect to $f(q_i, q_{-i})$. We also note that Equation 12 is identical with the Quantal Response Equilibrium (QRE) of MacKelvey and Palfrey [2] although the derivation used here was only introduced recently [1] and that Equation 6 is a macroscopic (MaxEnt) and previously unreported equivalent of the derivation of the QRE.

In the next section a *self-consistent* economic solution is one whereby both locally optimising economic agents as well as globally optimising economic agents can find numerical values for their probability distributions. A *parametric path* through an optimised economy is a slow variation in β for Equation 6 or a joint variation in β_i and β_{-i} in Equation 12 and its q_{-i} counterpart. The QRE (Equation 12) contains the Nash Equilibria (NE) as a special case where both $\lim_{\beta_i \to \infty} q_i$ and $\lim_{\beta_{-i} \to \infty} q_{-i}$ [2]. A *disordered economic state* is when $\beta = 0$ (global case) or $\beta_i = 0$ and $\beta_{-i} = 0$ (local case), at which point there is only ever one equilibrium point [2] and it is a uniform distribution over choices. An *Economic Catastrophe* [3] is where a disproportionate shift in the optimised solution occurs for a relatively small change in a macroscopic economic parameter [1] such as β, β_i or β_{-i}, illustrations of this are shown below.

2. Main Results

Theorem 1. *Local Maximum Entropy optimisation contains self-consistent global Maximum Entropy optimisation as a special case.*

Proof: From the global optimisation case, Equation 6, we have:

$$\begin{aligned} f(q_i, q_{-i}) &= q_i q_{-i} = \mathcal{Z}^{-1} \exp\Big(\beta \sum_{x,y} (u^i_{x,y} + u^{-i}_{x,y})\Big) \\ &= \mathcal{Z}^{-1} \exp\Big(\beta \sum_{x,y} u^i_{x,y}\Big) \exp\Big(\beta \sum_{x,y} u^{-i}_{x,y}\Big) \quad \text{not self-consistent} \\ \to q_i q_{-i} &= \mathcal{Z}^{-1} \exp\Big(\beta \sum_{x,y} q_{-i}u^i_{x,y}\Big) \exp\Big(\beta \sum_{x,y} q_i u^{-i}_{x,y}\Big) \quad \text{self-consistent} \end{aligned} \tag{13}$$

From the local optimisation case, Equation 12, we have:

$$q_i q_{-i} = \mathcal{Z}_i^{-1} \mathcal{Z}_{-i}^{-1} \exp\Big(\beta_i \sum_{x,y} q_{-i}u^i_{x,y}\Big) \exp\Big(\beta_{-i} \sum_{x,y} q_i u^{-i}_{x,y}\Big) \tag{14}$$

For the special case where $\beta_i = \beta_{-i}$ the two exponentials in Equation 14 are equivalent to those in Equation 13. As both of these representations of $q_i q_{-i}$ are normalised then $Z = Z_i Z_{-i}$ and therefore self-consistent global optimisation is a special case of local optimisation.

Theorem 2. *If an equilibrium point is not a Pareto Efficient equilibrium point then there is no parametric path in global optimisation that can reach a Pareto Efficient point without going through an Economic Catastrophe.*

Proof: If there are 3 unique NE then there is always a bifurcation point on the 45° path $\beta_i = \beta_{-i} = \beta$ at which multiple equilibria appear. At least one of these points is a Pareto Efficient point and at least one other is not, if the economy is not at the Pareto Efficient point for a fixed β, no variation in β can arrive at a Pareto Efficient point without a discontinuous step change in Q_i, see Figure 1.

Theorem 3. Theorem 2 *is not always true for local optimisation.*

Proof: By inspection the left plot of Figure 1 shows a path is possible from the lower surface to the upper surface that does not require a discontinuous step in Q_i.

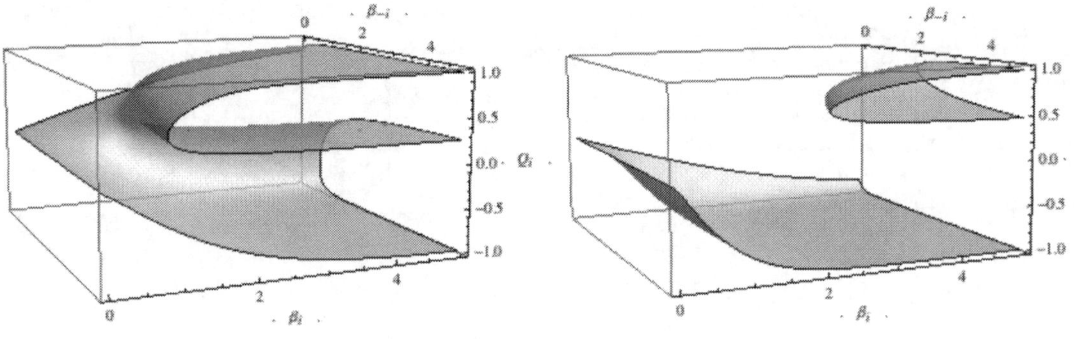

Figure 1. Two Equilibrium surfaces for two different economies each with 3 NE. The vertical axis is a rescaled probability: $Q_i = 1 - 2q_i$. The 45° path always passes through a bifurcation if there are 3 NE. The left figure has a path from the top surface to the bottom surface for suitable variation β_i and β_{-i}, this is not possible for the right figure.

3. Discussion

Bifurcations occur in financial markets [4] and macro-economics [3]. Recent evidence from econophysics has shown signatures of phase-transitions in both the macroscopic [5] and microscopic [6] interactions in financial markets. This work shows that the way in which such systems are optimised has significant impact on the flexibility and manageability of an economy. Importantly, early warning signatures are detectable before a catastrophic crash [7] and so the possibility of 'steering' an economy around such phase transitions as described in this work is possible.

[1] D. Wolpert, M. Harré, E. Olbrich, N. Bertschinger, and J. Jost. Hysteresis effects of changing the parameters of noncooperative games. *Physical Review E*, 85(3), 2012.

[2] R.D. McKelvey and T.R. Palfrey. Quantal response equilibria for normal form games. *Games and Economic Behavior*, 10(1):6–38, 1995.

[3] T. Poston and I. Stewart. *Catastrophe theory and its applications*, volume 2. Dover Pubns, 1996.

[4] V. Plerou, P. Gopikrishnan, and H.E. Stanley. Two-phase behaviour of financial markets. *Nature*, 421(6919), 2003.

[5] D. Sornette and A. Johansen. Large financial crashes. *Physica A: Statistical Mechanics and its Applications*, 245(3):411–422, 1997.

[6] M. Harré and T. Bossomaier. Phase-transition–like behaviour of information measures in financial markets. *EPL (Europhysics Letters)*, 87:18009, 2009.

[7] M. Scheffer, J. Bascompte, W.A. Brock, V. Brovkin, S.R. Carpenter, V. Dakos, H. Held, E.H. Van Nes, M. Rietkerk, and G. Sugihara. Early-warning signals for critical transitions. *Nature*, 461(7260):53–59, 2009.

Optimization of a corrugated millimeter-wave waveguide and a miter bend by FDTD simulation

H. Nakamura[1,2], N. Kashima[2] A. Takayama[1], K. Sawada[3], Y. Tamura[4], S. Fujiwara[5], and S. Kubo[1,2]

[1] Department of Helical Plasma Research, National Institute for Fusion Science, Oroshi-cho 322-6, Toki-city, GIFU, Japan
[2] Department of Energy Engineering and Science, Graduated School of Engineering. Nagoya University, Oroshi-cho 322-6, Toki-city, GIFU, Japan
[3] Faculty of Engineering, Shinshu University, Wakasato 4-17-1, Nagano-city, NAGANO, 380-8553, Japan
[4] Department of Intelligence and Informatics, Konan University, Okamoto 8-9-1, Higashinada-ku, Kobe-city, HYOGO, Japan
[5] Department of Helical Plasma Research, National Institute for Fusion Science, Matsugasaki, Sakyo-ku, Kyoto-city 606-8585, KYOTO, Japan

E-mail: hnakamura@nifs.ac.jp

Abstract. We estimate the transmission efficiency of the electromagnetic wave through the system composed of waveguide and miter bend by Finite-Difference Time-Domain (FDTD) simulation. As the fisrt approach of this estimation, we choose the case that the input wave is TE_{11} mode. In this case, the efficiency is estimated as 99.65 % for the system without grooves and 76.48 % for the system with grooves (which is called as "corrugate"). Comparing the distributions of the input electric field with that of the output electric field, the effect of the grooves is found as follows: Because the TE_{11} mode has an anisotropy, its shape is changed by the miter bend. This property appears in the corrugated system more strongly than the non-corrugated system.

1. Introduction

Electromagnetic waves to heat plasma in the nuclear fusion devices have been studied both in experiment and in theory. These studies are mainly classified into three fields, that is (i) the interaction between plasma and electromagnetic wave, (ii) the transmission of electromagnetic wave in waveguide, and (iii) the generation of electromagnetic wave. In this paper, we analyze the electromagnetic field in a miter bend (Fig.1) by Finite-Difference Time-Domain (FDTD) simulation to improve the transmission efficiency of electromagnetic wave.

The FDTD method was introduced by Yee in 1966 [1] to solve numerically Maxwell's equation, that is $\partial \boldsymbol{E}/\partial t = -\boldsymbol{j}/\epsilon + \boldsymbol{\nabla} \times \boldsymbol{H}/\epsilon$, $\partial \boldsymbol{H}/\partial t = -\boldsymbol{\nabla} \times \boldsymbol{E}/\mu$, where $\boldsymbol{E} = (E_x, E_y, E_z)$, $\boldsymbol{H} = (H_x, H_y, H_z)$ and $\boldsymbol{j} = (j_x, j_y, j_z)$ denotes an electric field, a magnetic field and an electric current respectively. The parameters ϵ and μ are electric permittivity and magnetic permeability. In our previous works[2, 3, 4], we have developed the FDTD simulation with the Drude-Lorentz model to investigate optical phenomena related to the electromagnetic response of metals using near-field scanning optical microscope (NSOM) [5]. We adopted our FDTD-simulation method, which we have developed in the research of NSOM, into the research of the transmission through the waveguide for the nuclear fusion.

The transmission system of the electromagnetic wave is composed mainly of two components, that is cylindrical corrugated-waveguide and miter bend. Grooves are engaged in both of them, which prevent eddy currents from being generated on the surface of them. The depth, the pitch, and the width of the grooves are designed by the phenomenological theory [6] (see Table 1). Our aim to use the FDTD simulation is to improve the propagation loss of the components.

In the present paper, we pick up the miter bend as the target of analyzation. (As for the corrugated waveguide, we report in another paper[7].) The miter-bend bends the direction of the electromagnetic wave as Fig.2. We estimate the propagation loss of the millimeter-wave through the miter bend.

2. Simulation Model

Figure 2 illustrates the cross-sectional view in the xz plane at $y = 0$ of the three-dimensional FDTD geometry of the miter bend. The cross-sectional view in the xy plane of the cylindrical waveguide is Fig.3.

The input electromagnetic wave with TE_{11} mode enters from the top-left size of the waveguide and it is bended by the miter bend. After all, the electromagnetic wave goes out from the lower-right side (Fig.2). The waveguide and the miter bend (which are made of aluminum) are approximated as the perfect electric conductor (PEC). The whole system is surrounded by the simulation box (67mm×34mm×67mm) with Mur's first-order absorbing boundary condition[8] as Fig.2. The parameters for the FDTD simulation are summarized in Table 1.

Figure 1. Photshop of miter bend. Grooves are engraved in the inside of miterbend. The width, the depth and the pitch of grooves are 1.0 mm, 0.76 mm, and 1.3 mm, respectively.

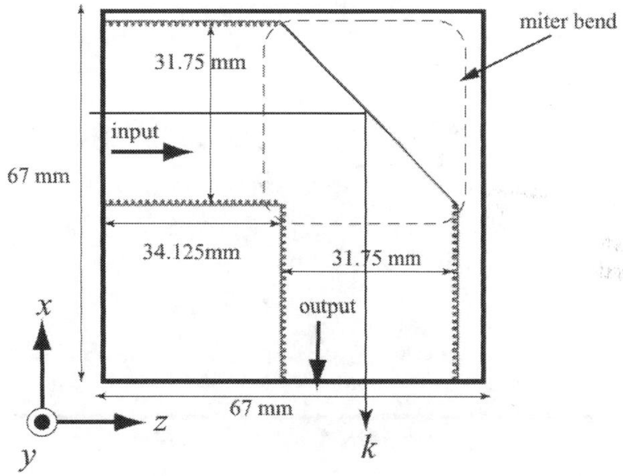

Figure 2. Schematic picture of wave guide system. The direction of electromagnetic traveling-wave is bended by miter bend. The guiding axis k is defined in the center of the wave guide along the traveling-wave. The input wave enters at $k = 0$ and go out of the wave guide at $k = 100$mm.

3. Results

To consider the effect of the groove engraved on the surface of the miter bend, we simulated two types of the miter bends as shown in Fig.4. In Fig.4, distributions in the xz plane at

Table 1. Parameters in FDTD simulation of the corrugated waveguide and the miter bend

Input source frequency	84 GHz	Wave length	3.57 mm
Inside radius	15.9 mm	Depth of groove	0.76 mm
Pitch error	1.3 mm	Width of groove	1.0 mm
Grid size	$\Delta x = \Delta y = \Delta z = 0.357$ mm	Number of meshes	$188 \times 96 \times 188$

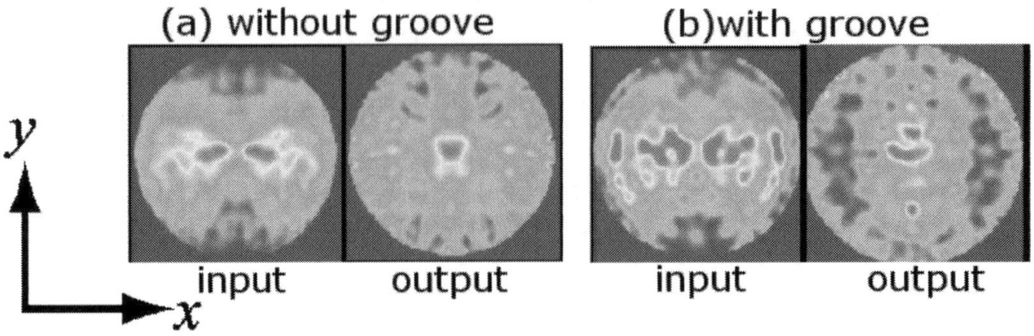

Figure 3. Simulated distributions in the xy plane of the averaged electric-field intensity $|\bar{E}(x,y)|_{k=k_0}$ for two types of miter bends. We plotted the distributions $|\bar{E}(x,y)|_{k=k_0}$ at the input position ($k_0/\Delta z = 14$) and the output position ($k_0/\Delta z = 258$) for each miter bend without grooves (a) and with grooves (b), respectively.

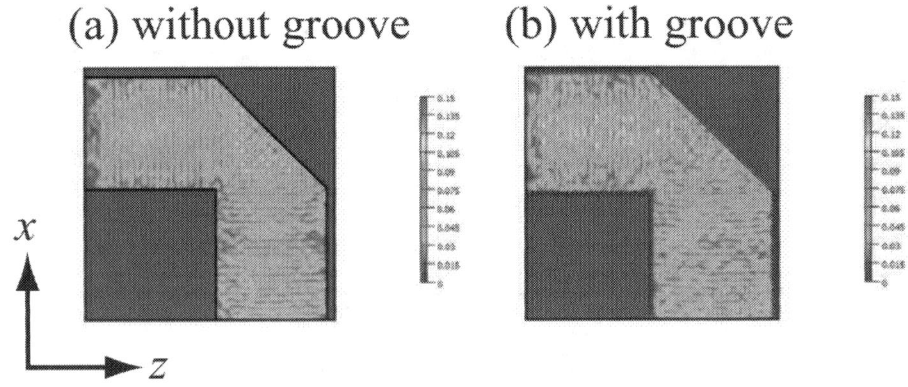

Figure 4. Simulated distribution in the xz plane at $y = 0$ of the averaged electric-field intensity $|\bar{E}(x,z)|_{y=0}$ for two types of miter bends. (a) miter bend without grooves, (b) miter bends with grooves.

$y = 0$ of the averaged electric-field intensity $|\bar{E}(x,z)|_{y=0}$ in the steady state are plotted. Here $|\bar{E}(x,z)|_{y=0} := (1/T_0)\int_t^{t+T_0} |\boldsymbol{E}(x, y=0, z; t_1)|dt_1$, where $T_0 = 5 \times 1/(88\text{GHz}) \sim 5.68 \times 10^{-11}$sec.

Moreover, we plotted the $|\bar{E}(x,y)|_{k=k_0}$ distributions for the steady state in Fig. 3, where $|\bar{E}(x,y)|_{k=k_0} := (1/T_0)\int_t^{t+T_0} |\boldsymbol{E}(x,y,z;t_1)|dt_1$ for each position (x,y) in the $k = k_0$ plane. Here k axis is defined in Fig.2.

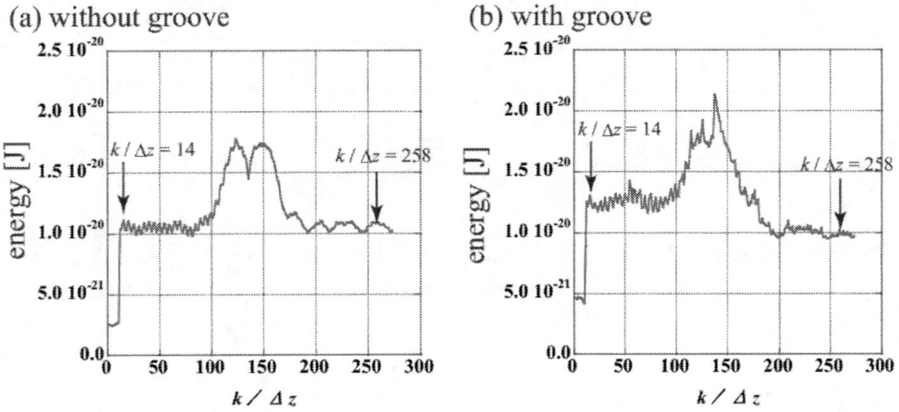

Figure 5. The k dependence of the averaged electromagnetic energy $\bar{\varepsilon}(k)$ for the waveguides without grooves (a) and with grooves (b).

In Fig.5, the k dependence of the averaged electromagnetic energy $\bar{\varepsilon}(k)$ is plotted to estimate the transmission efficiency. $\bar{\varepsilon}(k)$ is given by the time average for 5 periods of $\varepsilon(k) = \sum_{(x,y,z)\in S(k)}(\epsilon|\boldsymbol{E}(x,y,z)|^2 + \mu|\boldsymbol{H}(x,y,z)|^2)/2$, where $S(k)$ is the cross-section area of the waveguide at k. We define the transmission efficiency as the ratio $\bar{\varepsilon}(258\Delta z)/\bar{\varepsilon}(14\Delta z)$, which becomes 99.65 % for the waveguide without grooves and 76.48 % for the waveguide with grooves.

4. Discussion and Conclusion

The transmission efficiency for the system composed of waveguide and miter bend is obtained by the FDTD simulation. Comparing the two cases (without grooves and with grooves), it is found that the grooves does not increase the efficiency for the input wave with TE_{11} mode. The reason of this fact is as follows: Because the TE_{11} mode is not the eigenmode of the corrugated waveguide and miter bend (*i.e.*, with grooves), the input wave changes its shape through the system with grooves more strongly than without grooves. In the present simulation, we chose the TE_{11} mode as the first step of the analysis with the miter bend. To improve the estimation of the efficiency, it is necessary to choose the eigenmode of the corrugated waveguide and miter bend as the input wave (e.g. HE_{11} mode).

Acknowledgment

This work is supported by KAKENHI (24656560).

References

[1] Yee K S 1966 *IEEE Trans. Antennas Propagat.* **14** 302
[2] Nakamura H, Sawada K, Kambe H, Saiki T and Sato T 2000 *Prog. Theor. Phys. Suppl.* **138** 173
[3] Nakamura H, Saiki T, Kambe H, Sawada 2001 *Comp. Phys. Com.* **142** 464
[4] Sawada K, Nakamura H, Maruoka T, Tamura Y, Imura K, Saiki T, Okamoto H 2010 *Plasma and Fusion Research* **5** S2110
[5] Furukawa H and Kawata S 1996 *Opt. Commun.* **132** 170
[6] Doane J L 1985 *Infrared and Millimeter Waves* **13** 123
[7] Tamura Y, Nakamura H, Okamura T, Kashima N, Fujiwara S and Kubo S 2012 *International Conference on Mathematical Modeling in Physical Sciences* (IC-MSQUARE 2012) September 3-7, 2012 Budapest, Hungary
[8] Mur G 1981 *IEEE Trans. Electromagn. Compat.* **23** 377

Adaptive control in multi-threaded iterated integration

Elise de Doncker[1] **and Fukuko Yuasa**[2]

[1]Department of Computer Science, Western Michigan University, Kalamazoo, MI 49008, U.S.A.

[2]High Energy Accelerator Research Organization (KEK), 1-1 Oho, Tsukuba, Ibaraki 305-0801, Japan

E-mail: `elise.dedoncker@wmich.edu`, `fukuko.yuasa@kek.jp`

Abstract. In recent years we have developed a technique for the direct computation of Feynman loop-integrals, which are notorious for the occurrence of integrand singularities. Especially for handling singularities in the interior of the domain, we approximate the iterated integral using an adaptive algorithm in the coordinate directions. We present a novel multi-core parallelization scheme for adaptive multivariate integration, by assigning threads to the rule evaluations in the outer dimensions of the iterated integral. The method ensures a large parallel granularity as each function evaluation by itself comprises an integral over the lower dimensions, while the application of the threads is governed by the adaptive control in the outer level. We give computational results for a test set of 3- to 6-dimensional integrals, where several problems exhibit a loop integral behavior.

1. Introduction and background

Automatic integration can be viewed as a black-box procedure for the computation of an integral

$$\mathcal{I}f = \int_{\mathcal{D}} f(\vec{x})\, d\vec{x},$$

which takes a user-specification of the problem as input, including the integrand function $f(\vec{x})$, the domain $\mathcal{D}$, the dimension N and a requested accuracy. The latter may be determined by a relative and/or absolute error tolerance, t_r and t_a, respectively, in order to impose a criterion of the form $|\,Qf - \mathcal{I}f\,| \;\leq\; Ef \;\leq\; \max\{\,t_a\,,\, t_r\,|\,\mathcal{I}f\,|\,\}$, where Qf represents the obtained integral approximation and Ef is an estimate of the actual absolute error $|Qf - \mathcal{I}f|$.

Let us consider an *adaptive* integration procedure that adheres to the meta-algorithm of Figure 1. By domain partitioning it is intended to sample the integrand intensively in the vicinity of problematic function behavior. This type of algorithm maintains a priority queue data structure (such as a heap or a linked list) on the set of subregions, which is generally keyed with the estimated error over the subregions. At each step, the highest priority region is selected, deleted from the priority queue and subdivided. The local integration is performed over the new subregions, which in turn are added to the priority queue.

The 1D general adaptive QUADPACK algorithms [1] use Kronrod rules for the local integration, and generate local error estimates based on the Kronrod approximation and its corresponding

```
Initialize results and priority queue
while (evaluation limit not reached and estimated error too large)
    Retrieve region from priority queue
    Split region
    Evaluate new subregions and update results
    Insert new subregions into priority queue
```

Figure 1. Adaptive integration meta-algorithm

Guass rule sum. As another example, the multivariate adaptive routine DCUHRE [2] calculates a nested family of rules for the integration and error estimation on each subregion.

We calculate an iterated integral, as formulated in Section 2, where the numerical integration on each level of the iterated integral is performed adaptively and according to the algorithm description of Figure 1. A parallel algorithm is obtained by evaluating the integration rule points in parallel in the outer level. In Section 3 we give results and discuss the parallel performance of the method for a set of sample problems, including the "DICE" integrals which characterize the singular behavior of some classes of Feynman loop integrals. The latter occur in perturbation calculations for the cross section of particle interactions in high-energy physics. This paper presents an extension of the work described in [3] (which was restricted to three dimensions).

2. Parallel iterated integration

Consider an *iterated (repeated)* integral

$$\mathcal{I}f = \int_{\mathcal{D}_1} d\vec{x}^{(1)} \ldots \int_{\mathcal{D}_\ell} d\vec{x}^{(\ell)}\; f(\vec{x}^{(1)}, \ldots, \vec{x}^{(\ell)})$$

over the N-dimensional product region $\mathcal{D} = \mathcal{D}_1 \times \ldots \times \mathcal{D}_\ell$, where the $\mathcal{D}_j,\ j = 1, \ldots, \ell$ may be 1D or multidimensional regions with associated (for example, adaptive) integration methods.

For an iterated integration over $\mathcal{D} = \mathcal{D}_1 \times \mathcal{D}_2$, the integral over a selected subregion $S \subseteq \mathcal{D}_1$ is approximated numerically by a (scaled) integration rule with points $\vec{x}_k$ and weights w_k, $k = 1, \ldots, K$, of the form

$$\int_S d\vec{x}^{(1)}\; F(\vec{x}^{(1)}) \;\approx\; \sum_{k=1}^{K} w_k\, F(\vec{x}_k^{\,(1)}), \quad \text{where} \quad F(\vec{x}_k^{\,(1)}) \;=\; \int_{\mathcal{D}_2} d\vec{x}^{(2)}\; f(\vec{x}_k^{\,(1)}, \vec{x}^{(2)}).$$

The values of F are themselves integrals and they are independent from each other; hence they can be evaluated in parallel. The weighted sum referred to as the (outer) integration rule sum is thus calculated by parallel threads, over each subregion $S \subseteq \mathcal{D}_1$ generated by the adaptive partitioning strategy in the outer level of the iterated integral.

Our current parallel implementation uses successive one-dimensional quadratures, over the given regions (intervals) $\mathcal{D}_j$, for $j = 1, \ldots, \ell = N$. For the integration in each coordinate direction we use a general adaptive program (DQAGE) from the QUADPACK package [1]. The interval in each coordinate direction is partitioned adaptively according to the algorithm of Figure 1. For the local integral and error estimates over the subintervals, a pair of quadrature formulas is applied consisting of a Gauss rule and its corresponding Kronrod formula.

The test results in the next section are obtained by evaluating the outer integral in parallel using the 21-point Gauss-Kronrod pair over the subintervals generated by the adaptive strategy. For the inner integrations, the 21-point or the 15-point pair can be selected by the user.

3. Parallel integration results

A set of simple test integrals of dimensions 1 through 6 is given in formulas (1)-(10) below. Integrals (1)-(3) are 3-dimensional, (4)-(6) are 4-dimensional, (7) and (8) are 5-dimensional, (9) and (10) are 6-dimensional sample problems. Various problem characteristics are represented, including (very) high-degree polynomials in (4) and (10), and an oscillating function in (8). The integrands of (3) and (6) have a corner singularity which requires intensive local partitioning around a single point of the domain.

$$\mathcal{I}f_1 = \int_{-1}^{1} dx_1 \int_{-1}^{1} dx_2 \ldots \int_{-1}^{1} dx_N \, \frac{\delta\, r\, \theta(1-r^2)}{(r^2-\alpha^2)^2+\delta^2}, \tag{1}$$

$$\text{where } r = \Big(\sum_{j=1}^{N} x_j^2\Big)^{1/2}, \; N=3, \; \delta = 10^{-6}, \; \alpha = 0.8$$

$$\text{and } \theta(\chi) = 1 \; (\text{if } \chi > 0); \; 1/2 \; (\text{if } \chi = 0); \; 0 \; (\text{if } \chi < 0)$$

$$\mathcal{I}f_2 = \int_{-1}^{1} dx_1 \int_{-1}^{1} dx_2 \ldots \int_{-1}^{1} dx_N \, \frac{2\,\delta\, x_2}{(\sum_{j=1}^{N} x_j - 1)^2 + \delta^2}, \quad \text{with } N=3, \; \delta = 10^{-6} \tag{2}$$

$$\mathcal{I}f_3 = \int_0^1 dx \int_0^1 dy \int_0^1 dz \, \frac{1}{(x+y+z)^2} \tag{3}$$

$$\mathcal{I}f_4 = \int_0^1 dx \int_0^1 dy \int_0^1 dz \int_0^1 du \; (0.66 \times 10^7)\, x^{54}\, y^{39}\, z^{59}\, u^{49} \tag{4}$$

$$\mathcal{I}f_5 : \; \text{as } \mathcal{I}f_2 \text{ for } N=4, \; \text{with } \delta = 10^{-6} \tag{5}$$

$$\mathcal{I}f_6 = \int_0^1 dx \int_0^1 dy \int_0^1 dz \int_0^1 du \, \frac{1}{(x+y+z+u)^2} \tag{6}$$

$$\mathcal{I}f_7 : \; \text{as } \mathcal{I}f_1 \text{ for } N=5, \; \text{with } \delta = 10^{-1} \tag{7}$$

$$\mathcal{I}f_8 = \int_0^{\pi} dx \int_0^{\pi} dy \int_0^{\pi} dz \int_0^{\pi} du \int_0^{\pi/2} dv \; \cos(6(x+y+z+u+v)) \tag{8}$$

$$\mathcal{I}f_9 : \; \text{as } \mathcal{I}f_1 \text{ for } N=6, \; \text{with } \delta = 10^{-1} \tag{9}$$

$$\mathcal{I}f_{10} = \int_0^1 dx \int_0^1 dy \int_0^1 dz \int_0^1 du \int_0^1 dv \int_0^1 dw \; (3.78 \times 10^9)\, x^{49}\, y^{39}\, z^{44}\, u^{29}\, v^{39}\, w^{34} \tag{10}$$

The integral (1) was used in [4] with $N = 3$, as a test problem (named "DICE3") for the Monte Carlo integration program DICE, and (2) is a 3D version of the problem "DICE1", treated for $N = 2$ in [4]. A two-dimensional version of the integrand f_1 (as a function of x_1 and x_2 over $[-1,1]^2$) is depicted in the plot of Figure 2 for $\delta = 0.1$ and $\alpha = 0.8$ [5]. The function has a discontinuity on the unit circle (outside of which it falls to zero), and a ridge at $r^2 = \alpha^2 = 0.64$, with a peak height of α/δ. Thus the ridge inside the circle is steep and narrow when delta is small (e.g., $\delta = 10^{-6}$ in (1)). Both the discontinuity and the ridge present numerical integration difficulties, occurring in the interior of the domain. The integrand of problem (2) has a peaked behavior at $\sum_{j=1}^{N} x_j = 1$, which is more pronounced for smaller δ (note $\delta = 10^{-6}$ in (2)). The peaked behavior of the integrand functions in (1) and (2) mimics the irregular behavior of the integrands of some classes of Feynman loop integrals in high energy physics.

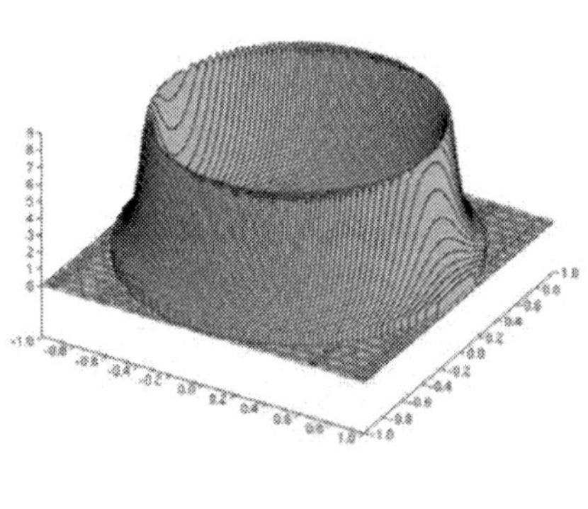

Figure 2. Plot of integrand $f_1(\vec{x})$ of (1) for $N = 2, \; \delta = 0.1, \; \alpha = 0.8$

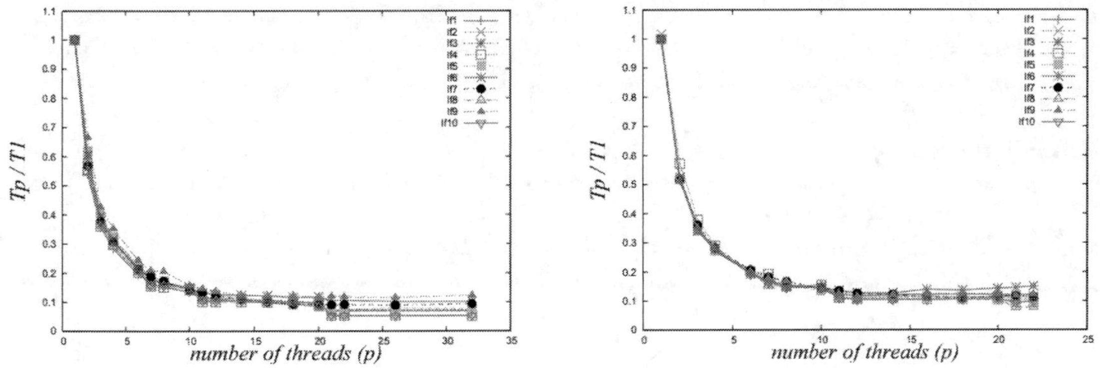

(a) (b)

Figure 3. ρ_p vs. number of threads p : (a) on SR16000; (b) on Intel Xeon workstation

Particularly in relatively low dimensions ($N = 2, 3, 4$, say), the ability of the iterated adaptive integration method to achieve high accuracy for very small δ is often remarkable, compared to, e.g., slowly converging Monte-Carlo (MC) methods and standard multivariate adaptive subdivision strategies. In [5], results were given for DICE3 ($N = 3$, $\delta = 10^{-m}$, $1 \le m \le 6$). For example, a simple MC method with antithetic variates (RCRUDE [6]), given a maximum of $100\,M = 10^8$ points, yields results to relative accuracies of 0.013 and 0.017 for $\delta = 10^{-5}$ and $\delta = 10^{-6}$, respectively. For the same δ values, the standard global adaptive (GA) algorithm of DCUHRE [2] using $250\,M$ points has no correct digits in the results, with relative errors of 0.65 and 0.95. However, the iterated integration method can achieve high requested accuracies (e.g., 10 correct digits) for these parameter values, at the expense of function evaluations but with only a fraction of the memory usage of standard GA.

Parallel performance results of our multi-threaded (OpenMP) implementation of the code for the test problems (1)-(10) are shown in Figures 3(a) and 3(b). Each curve represents the ratio $\rho_p = T_p/T_1$ as a function of the number of threads p. Here T_1 is the sequential time and T_p is the execution time when using p threads; thus ρ_p is the parallel time scaled by T_1.

Figure 3(a) is obtained from timing runs on a node of the Hitachi SR16000 system at KEK (with a Power7@3.83GHz CPU), and using the f90 compiler. Correspondingly, Figure 3(b) shows results obtained on a 12-core workstation (with dual 6-core Intel Xeon E5645 processors), and using gfortran for the compilation. Generally, good speedups are observed for our functions on both systems. The difficulty of the corner point singularity is apparent in the timing results for (4D) problem (6) on the 12-core Intel machine. Timings on this system show the 12-core limit; however, speedup is achieved for some functions to higher numbers of threads through hyper-threading. On the Hitachi SR16000 node, the current version of our code shows the effect of the parallelization based on the sample 21-point rule in the outer layer of the iterated integral, even though some of the times still decrease slightly for more threads.

The parallelization can easily be extended to the simultaneous integration over the two (or more) subregions obtained in the subdivision (*Split*) step of the meta-algorithm in Figure 1. We will further investigate an extension of the scheme allowing for massive parallelism by implementing parallelization in more than one, e.g., the outer two levels of the iterated integral.

4. Conclusions

We presented a novel parallel algorithm for iterated adaptive multivariate integration. Iterated multivariate integration in relatively low dimensions is superior to MC and GA with respect to accuracy, for problems exhibiting the behavior of some classes of Feynman loop integrals. Good speedups of the parallel scheme are achieved on the SR16000 Hitachi system at KEK, Japan, and on a 12-core Intel (Xeon processor) based workstation. In future work we plan to target the parallelization to systems with many cores.

Acknowledgments

We acknowledge the support from the National Science Foundation under MRI Award Number 1126438, and from Western Michigan University and the College of Engineering and Applied Sciences (CEAS). In Japan, this work is supported by Grant-in-Aid for Scientific Research (24540292) of JSPS and the Large Scale Simulation Program No.12-07 of KEK.

References

[1] Piessens R, de Doncker E, Überhuber C W and Kahaner D K 1983 *QUADPACK, A Subroutine Package for Automatic Integration* Springer Series in Computational Mathematics (Springer-Verlag)
[2] Berntsen J, Espelid T O and Genz A 1991 *ACM Transactions on Mathematical Software* 17 452–456 URL `http://www.sci.wsu.edu/math/faculty/genz/homepage`
[3] de Doncker E and Yuasa F 2012 *Proc. of the Conference on Computational Physics, Oak Ridge* (The Journal of Physics: Conference Series) to appear
[4] Tobimatsu K and Kawabata S 1998 Multi-dimensional integration routine DICE Tech. Rep. 85 Kogakuin University
[5] de Doncker E, Kaugars K, Cucos L and Zanny R 2001 *Proc. of Computational Particle Physics Symposium (CPP 2001)* pp 110–119
[6] Genz A 1998 MVNDST URL `http://www.sci.wsu.edu/math/faculty/genz/homepage`

Pattern recognition tool based on complex network-based approach

Dalcimar Casanova[1], André Ricardo Backes[2] and Odemir Martinez Bruno[1]

[1] Instituto de Física de São Carlos (IFSC), Universidade de São Paulo, Av. Trabalhador São Carlense, 400, 13560-970 - São Carlos, SP, Brasil
[2] Faculdade de Computação, Universidade Federal de Uberlândia, Av. João Naves de Ávila, 2121, 38408-100, Uberlândia, MG, Brasil

E-mail: `arbackes@yahoo.com.br, dalcimar@gmail.com` and `bruno@ifsc.usp.br`

Abstract. This work proposed a generalization of the method proposed by the authors: 'A complex network-based approach for boundary shape analysis'. Instead of modelling a contour into a graph and use complex networks rules to characterize it, here, we generalize the technique. This way, the work proposes a mathematical tool for characterization signals, curves and set of points. To evaluate the pattern description power of the proposal, an experiment of plat identification based on leaf veins image are conducted. Leaf vein is a taxon characteristic used to plant identification proposes, and one of its characteristics is that these structures are complex, and difficult to be represented as a signal or curves and this way to be analyzed in a classical pattern recognition approach. Here, we model the veins as a set of points and model as graphs. As features, we use the degree and joint degree measurements in a dynamic evolution. The results demonstrates that the technique has a good power of discrimination and can be used for plant identification, as well as other complex pattern recognition tasks

1. Introduction

In recent years, many real-world problems and phenomenas are being studied with the use of complex network theory [6]. Between these problems the computer vision is one field that has benefited of these new developments. In this way, the work of [1] uses a complex network to measure features of generic shapes and use it in a classification problem.

In fact, this method have several interesting properties as (1) rotation and scale invariance: shapes with different sizes or positions have the same properties; (2) noise tolerance: the network model allows to deal with different levels of noise; (3) robustness: the network model do not need space or sequential information about the shape points. This properties allows the method to be extended for a widely applications. As example [14] use the same idea to measure the similarity of graphs on a problem of 3D objects classification.

In same way, here we generalize the technique to apply the method in a plant identification problem based on venation of their leaves. We explore the fact of the method does not require sequential informations, to model the leaf veins as a graph and measure the degree features. These features will be used in classical pattern recognition problem.

On the other hand, we propose a method to solve the problem of automated plant taxonomy [4], more specifically identify plants by analysis of their leaves.

After their shape, the next most studied aspect of leaves is the vein structure, also referred to as the venation [10]. Leaf venation corresponds to the blood vessel of organisms and provide the transport mechanism for water, minerals, sugars and other substances. The venation structure can be used to identify a plant. Although the fine detail may vary, the overall pattern of veins is conserved within many species [4].

So, in this work we use the complex network method to characterize the leaf venation. Few attempts are made to analyze or compare the venation in literature [11, 13, 12], the majority of studies is devoted to the process of extracting venation [4].

2. Complex network model and measurements of the vein leaf network

We can describe the research in complex networks as the intersection between graph theory and statistical mechanics. It integrates computer sciences, mathematics and physics [5], so that, this area presents a truly multidisciplinary nature.

In the last few years, many researches have focused on applying the developed concepts of the complex networks to many real data and situations. This is motivated because of the great flexibility and generality of the complex networks to model virtually any problem.

Here we propose to apply the complex network theory in a classification problem. Let P a set of N points in the R^n space. Each point is a typical vector in the form of $p_i = [x_i, \ldots, x_n]$. To apply the Complex Networks Theory we must build a representation of this set P as a network $G(V, E)$. To accomplish this, we consider each point in the set P as a vertice in the network (i.e., $P = V$). Then, we add an edge binding each pair of vertices, $e_{i,j} \in E : V \times V$, where its weight $w(e_{i,j})$ is the Euclidean distance between them:

$$w(e_{i,j}) = |p_i - p_j| \tag{1}$$

To ensure scale invariance of the network, we normalize the weight into the interval [0, 1],

$$w(e_{i,j}) = \frac{w(e_{i,j})}{\max_{w(e_{i,j}) \in E}} \tag{2}$$

Initially, all vertices in the network are connected to each other, so that, all vertices present the same number of connections. This is a regular network and it does not present properties relevant for classification. Thus, in order to extract relevant properties of this network, we propose to apply a dynamic evolution process to this network [5]. To achieve this, we apply a threshold t over the edges E to achieve a new set of edges E^*, where

$$E^* = \delta_t(E) = \{e_{i,j} \in E | w(e_{i,j}) \leq t\}. \tag{3}$$

From this new set of edges E^*, $E^* \subseteq E$, arises a new network $G^* = (V, E^*)$ which can be interpreted as a intermediate step in the evolution of the network G and it owns relevant properties of the network topology. In our approach, we characterize each network G^* by using two degree measurements: *maximum degree*

$$k_\kappa = \max_i k_i \tag{4}$$

and *average degree*

$$k_\mu = \frac{1}{N} \sum_{i=1}^{N} k_i \tag{5}$$

where k_i is the degree of the vertice v_i in the network G^*. By considering different threshold values t, it is possible to build a feature vector that characterizes the evolution of the network G and, consequentely, the original set of points P:

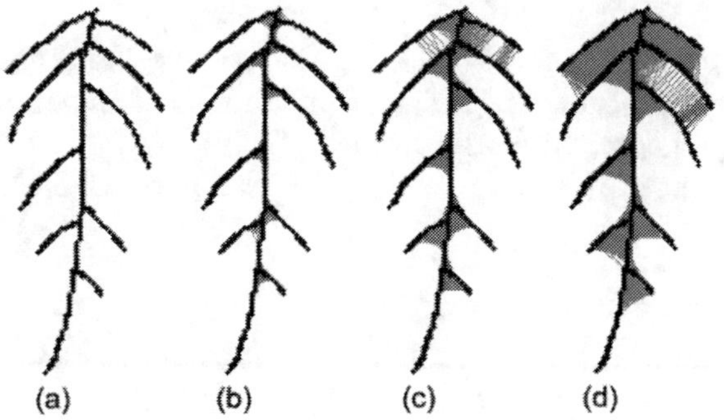

Figure 1. (a) leaf nerv; (b)-(d) complex network computed for different t values, $t = 0.050, 0.075, 0.100$.

$$[k_\mu(t_1), k_\kappa(t_1), k_\mu(t_2), k_\kappa(t_2), \ldots, k_\mu(t_m), k_\kappa(t_m)]\,. \tag{6}$$

The complete procedure is showed in Figure 1.

3. Experiments and Results

Since [10], in his pioneer work on classification of Dicotyledonous leaves, the venation structure is used to help the plant identification and taxonomy. In fact, after their shape, the next most studied aspect of leaves is the vein structure.

Although the fine detail may vary, the overall pattern of veins is conserved within many species. Several studies also show properties superintendent this taxonomic character. According [2], for example, the angles between the veins are well defined and are directly related to the segment size. These properties are then used to build a mathematical model based on fractals to model the leaf venation.

The big challenge with automated venation analysis, however, is the extraction of the vein networks (segmentation), where we have several works dedicated to solve this problem [13, 8, 9, 3, 12].

Here this complex task is carried out in five steps, as in [13]: (i) high pass filtering; (ii) segmentation; (iii) morphological filters; (iv) thinning; (v) and outline subtraction.

In the first stage, high frequency signals were intensified through a high pass filtering of the Fourier transform. The second stage segmentation was done using the morphological Laplacean method. An morphological filters applied, removing the isolated pixel groups and leaving only the veins and the leaf outlines. A thinning algorithm was adopted for reducing the vein map to 1 pixel width and finally the last step is the subtraction of the leaf outline. This process results in a database with 44 leaf veins grouped into 11 classes.

In the experiments, we computed an initial network G for each leaf vein. Then, we generate different manifestations of G, G^*, by applying the δ transformation for $t = 0.050, 0.175, 0.300, 0.425, 0.550$. In the sequence, we computed the maximum and average degree from each network G^*. This results in a feature vector containing 10 features for each sample. We used these feature vectors to describe each leaf vein.

We evaluated the method by applying the Linear Discriminant Analysis (LDA), a supervised statistical classification method, over the feature vectors [7]. We also used the leave-one-out cross-validation scheme to evaluate the samples. The method was able to correctly classify 41 of 44 samples in their respective classes, which gives us a success rate of 93.18%.

4. Conclusions

The results show that is possible generalize the complex network method in order to characterize the leaf venation. Due the properties of rotation and scale invariance, noise tolerance and robustness is possible build a graph model from the leaf venation and evaluate their similarity using measurements of their graph model. We believe that the same approach can be used on others similar problems with good results too.

References

[1] A. R. Backes, D. Casanova, and O. M. Bruno. A complex network-based approach for boundary shape analysis. *Pattern Recogn.*, 42(1):54–67, 2009.

[2] S. Bohn, B. Andreotti, S. Douady, J. Munzinger, and Y. Couder. Constitutive property of the local organization of leaf venation networks. *Phys. Rev. E.*, 65(6):061914, 2002.

[3] O. M. Bruno, R. de Oliveira Plotze, M. Falvo, and M. de Castro. Fractal dimension applied to plant identification. *Inform. Sci.*, 178(12):2722–2733, 2008.

[4] J. S. Cope, D. Corney, J. Y. Clark, P. Remagnino, and P. Wilkin. Plant species identification using digital morphometrics: A review. *Expert Syst. Appl.*, 39(8):7562–7573, 2012.

[5] L. F. Costa, F. A. Rodrigues, G. Travieso, and P. R. Villas Boas. Characterization of complex networks: A survey of measurements. *Adv. Phys.*, 56(1):167–242, 2007.

[6] L. F. Costa, O. N. Oliveira Jr., G. Travieso, F. A. Rodrigues, P. R. Villas Boas, L. Antiqueira, M. P. Viana, and L. E. C. Rocha. Analyzing and modeling real-world phenomena with complex networks: A survey of applications. *Adv. Phys.*, 60(3):329–412, 2008.

[7] B. S. Everitt and G. Dunn. *Applied Multivariate Analysis.* Arnold, 2nd edition edition, 2001.

[8] H. Fu and Z. Chi. Combined thresholding and neural network approach for vein pattern extraction from leaf images. *IEE Proc.-Vis. Image Signal Process.*, 153(6):881–892, 2006.

[9] F. Gouveia, V. Filipe, M. Reis, C. Couto, and J. Bulas-Cruz. Biometry: the characterization of chestnut-tree leaves using computer vision. In *ISIE*, 1997.

[10] L. J. Hickey. Classification of the architecture of dicotyledons leaves. *Amer. J. Bot.*, 60: 17–33, 1973.

[11] Y. Nam, E. Hwang, and D. Kim. Clover: A mobile content-based leaf image retrieval system. In *ICADL*, pages 139–148, 2005.

[12] R. O. Plotze and O. M. Bruno. Automatic leaf structure biometry: Computer vision techniques and their applications in plant taxonomy. *Int. J. Pattern Recogn.*, 23(2):247–262, 2009.

[13] R. O. Plotze, M. Falvo, J. G. Pádua, L. C. Bernacci, M. L. C. Vieira, G. C. X. Oliveira, and O. M. Bruno. Leaf shape analysis using the multiscale minkowski fractal dimension, a new morphometric method: a study with passiflora (passifloraceae). *Can. J. Bot.*, 83(3): 287–301, 2005.

[14] J. Tang, B. Jianga, C.-C. Chang, and B. Luo. Graph structure analysis based on complex network. *Digit. Signal Process.*, 22(5):713–725, 2012.

Ab-initio study of the structural and electronic properties of osmium under high pressure

A. Rubio-Ponce[1], D. Olguín[2] and R. de Coss[3]

[1]Departamento de Ciencias Básicas, UAM-Azcapotzalco, Av. San Pablo 180, 02200 México, D. F. MEXICO

[2]Departamento de Física, Centro de Investigación y de Estudios Avanzados, A.P. 14740, C.P. 07300 México, D. F. MEXICO

[3]Departamento de Física Aplicada, Centro de Investigación y de Estudios Avanzados - Mérida, A.P. 73 Cordemex 97310, Mérida, Yucatán. MEXICO

E-mail: arp@correo.azc.uam.mx

Abstract. The structural and electronic properties of osmium (Os) have studied using the full potential linearized augmented plane wave method and the generalized gradient approximation for the exchange-correlation energy. The calculations were done incluiding the spin-orbit (SO) coupling and for hydrostatic pressures up to 400 GPa. The total-energy as a function of the cell volume was computed assuming four different crystal structures, namely *hcp*, *fcc*, $hcp - \omega$ and *bcc*. Contrary to previous non-relativistic LDA calculations our study shows that the equilibrium phase of Os correspond to the *hcp* structure and that remain stable in the studied range of pressures and no structural transition to the *fcc*, $hcp - \omega$ or *bcc* phases are obtained.

Keywords: Density functional theory, electronic band structure, hydrostatic pressure.

1. Introduction

The transition metal osmium (Os) has an hexagonal close packed (*hcp*) structure and the highest density among the elements under ambient conditions. Recently, there has been a lot of interest in the high pressure properties of Os after that the first measurement of the bulk modulus (B_0) and first derivative (B_0') were made few years ago by Cynn *et al.*, [1]. These authors reported diamond anvil cell compression studies up to 60 GPa for Os. They found that Os is less compressible than diamond ($B_0 = 446(1)$ GPa and $B_0' = 3.0(1)$)[2], and has a B_0 and B_0' equal to 462(12) GPa and 2.40(5), respectively.

In the same direction, first principles calculations for Os under pressure, using the pseudopotential plane wave method and the local density approximation (LDA) for the exchange and correlation energy density functionals done by Hebbache and Zemzemi [3], predicted that in the limit of very low temperatures the face centered cubic (*fcc*) structure is more stable than the *hcp* structure. Also, they claimed that, for small compression pressures, i.e., $p < 115$ GPa, Os is harder than diamond, their calculated value for B_0 was larger than the corresponding value measured for diamond (see Table II and Fig. 3 on Ref.[3]).

Recent high–pressure and low temperature measurements have shown that Os is more compressible than diamond. Values reported by Takemura ($B_0 = 395(15)$GPa and $B_0' = 4.5(5)$)[4] and Occelli *et al.*, ($B_0 = 411(6)$GPa and $B_0' = 4.0(2)$)[5], show these facts. On the other hand, using pulse–echo ultrasound and resonance ultrasound spectroscopy, Pantea, *et al.*, [6] have measured the low temperature B_0. They found that at any temperature B_0 does not exceeds the diamond value, the reported values are 410 GPa and 405 GPa for zero–tempertature and 300 K, respectively.

Sahu and Kleinman [7] using the fully relativistic linear augmented plane–wave method calculated the B_0 and its B_0' for Os. In their calculations they used the LDA ($B_0 = 453.59$GPa and $B_0' = 5.26$), as well as the generalized gradient approximation (GGA) ($B_0 = 401.30$GPa and $B_0' = 5.47$) to compute the exchange and correlation energy density functional. From their results concluded that Os is softer than diamond.

2. Computational strategy

The structural and electronic properties of Os were obtained by using the full potential linearized augmented plane wave (FLAPW) method as it is implemented in the WIEN2k code [8], the exchange and correlation energy was evaluated within the GGA[9], and we have included the SO coupling as well. In the calculations, a step analysis was carefully performed to ensure the convergence of the total energy in terms of the variational parameters, here we chose the next values: muffin-tin radii (RMT) of 2.1 a.u., and the plane wave expansion parameter $R_{mt}K_{\max} = 9.0$ (where R_{MT} is the MT radius and $K_{\max}$ is the plane wave cutoff), while the energy cutoff parameter was $G_{\max} = 16$. At the same time an apropiate set of $k-$points in the irreducible part of the Brillouin zone was used. Convergence was assumed when the energy difference was less than 10^{-5} Ry in all cases.

In our study, we have considered the *hcp* ($P6_3/mmc = D_{6h}^4$, space group 194) as the stable crystal struture of Os. Then, three other hypotetical crystal structures were also considered in our calculations, namely: the $hcp - \omega$ (p6/mmm=D_{6h}^1, space group 191), the fcc (Fm3m = O_h^5, space group 225) and the *bcc* (Im3m = O_h^9, space group 229). The number of k–points used in the full irreducible part of the Brillouin zone (FBZ) was 240 for the *hcp* and $hcp - \omega$ phase, and 220 for the fcc and *bcc* structures. The Os atomic electronic configuration used in the calculations was: [Xe] $4f^{14}$, $5d^6$, $6s^2$, where the $5p$ orbitals were included by using the local orbital extension of the FLAPW method [8].

The structural parameters are obtained by fitting the calculated total energy of each primitive cell to the third–order Birch–Murnaghan equation of state (EOS)[10]

3. Results

Our calculated lattice parameters for Os–*hcp*, considering SO coupling, are $a = 2.7693$Å and $c = 4.3676$Å, where the c/a ratio was optimized for each unit cell volume ($c/a = 1.5771$). This value is in excellent agreement with the experimental data (namely, $c/a = 1.5794$ [4]), which means that there is not change in the total energy.

Then, our calculated value for B_0 and B_0' was 390.74 GPa and 4.57, respectively, in agreement with previous experimental reports [4, 5, 6, 11] and other calculations [7, 12, 13, 14, 15, 16, 17].

For the $hcp-\omega$ structure we optimized the lattice parameters a and c, as was done in the *hcp* case, the found values were $a = 4.3420$Å, $c = 2.7281$Å, and $c/a = 0.6283$, while $B_0 = 370.97$GPa and $B_0' = 4.55$. For the other hypothetical two cases, the found structural parameters were $a = 3.8716$Å, $B_0 = 385.64$GPa, and $B_0' = 4.72$ for fcc; and $a = 3.1139$Å, $B_0 = 316.09$GPa, and $B_0' = 4.80$ for *bcc*. In our calculations, all the found values for B_0 are lower than the corresponding value for diamond, in agreement with previous theoretical results [15, 16, 17, 18].

For the different studied crystal structures, the *hcp* phase has the lowest total energy (see the inset in Fig. 1). With respect to the *hcp* curve, the minimum of the energy for the hypothetical crystal structures fcc, $hcp - \omega$, and *bcc* is located at 137, 338, and 778 meV, respectively. Our calculations confirm the interpretation that the *hcp* is the most stable crystal structure for Os, in disagreement with the pseudopotential calculations carried out by Hebbache and Zemzemi [3].

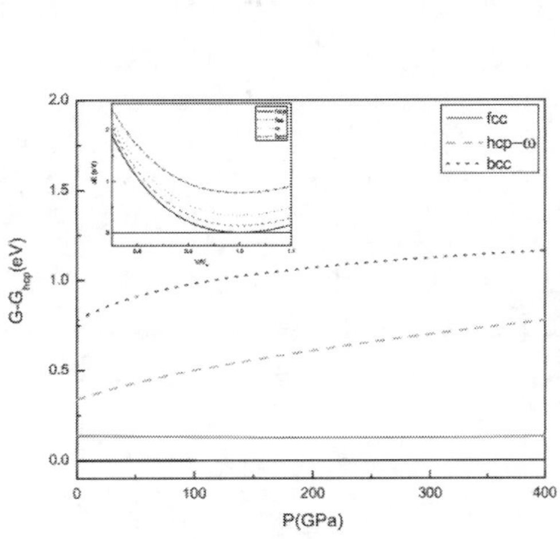

Figure 1. (Color online) Gibbs free energy difference at 0K *vs.* pressure for the diffrent Os structures. Note that there is not predicted phase transition for the calculated hydrostatic pressures up to 400 GPa.

Loking for a possible phase transition between the different crystal structures considered in this work, we have calculated the Gibbs free energy $G = E + PV - TS$ per atom at 0K, where G is equivalent to the enthalpy. The intersecting point of the different functions $G_{hcp}(P)$, $G_{fcc}(P)$, $G_{hcp-\omega}(P)$ or $G_{bcc}(P)$ defines the pressure of the rigid lattice transition between the different structures. According to our calculations, no phase transition is obtained for the *hcp*, fcc, $hcp - \omega$ or *bcc* crystal structures for these pressures, as it is clearly shown in Fig. 1, where no crossing for the different curves is obtained.

As it is well known, to describe properly the electronic properties of the heavy metals it is very important to take into account, in the calculations, the SO coupling effects [19, 20]. Thus, we compare in Fig. 2 the calculated density of states (DOS), solid lines show the calculation including SO coupling, while dashed show the calculation without SO. As we have found, energies just below E_F, the $6s$ (Fig. 2a) electrons contribute in an important way to the electronic properties of Os. Fig. 2b depicts the contribution of the $5p$ orbitals, the states are located around −45eV. With the SO interaction these states split in two new bands with a gap of 10.7eV, and are the more affected states. Fig. 2c shows the $5d$ states, as in the case of the $6s$ orbitals, the introduction of SO coupling produces not changes on these orbitals. Then, the states around −47eV are derived mainly from the $4f$ orbitals, as it is shown in Fig. 2d. The SO coupling splits these states in two new bands with a gap of 2.8 eV. Our results show that the electronic properties of Os at E_F are mainly governed by the $5d$ electrons, while the contribution of the $5p$ and $4f$ electrons is almost two orders or magnitude lower. These facts mean that at E_F the electronic hybridization is mainly obtained from the $6s$ and $5d$ states, while hybridization of the $5p$ and $4f$ states is almost negligible.

4. Conclusions

Our calculated P-V equation of state for Os is in good agreement with recent experimental data. We found that Os has a B_0 and B'_0 equal to 390.74 GPa and 4.57 respectively and is more compressible than diamond. In contradiction with the previous non-relativistic LDA-calculation[3], we found that Os in the *hcp* phase has lower energy than the fcc, $hcp - \omega$ and *bcc* structures. The *hcp* structure remains stable for pressures up to 400 GPa and structural transition to the fcc, $hcp - \omega$ or *bcc* was not found. On the other hand, the SO coupling should

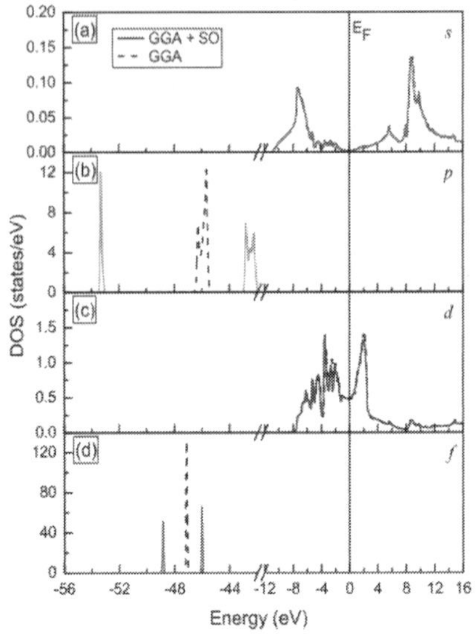

Figure 2. (Color online) Partial contributions to the calculated DOS. The splitting of the lower bands occurs only for the p and f states.

be included to properly describe the Os electronic band structure. At this moment, diamond is the hardest known material and also has the highest known bulk modulus $B_0 = 443$ GPa.

References

[1] H. Cynn, J.E. Klepeis, C.-S. Yoo, and D.A. Young. Phys. Rev. Letters **88**, 135701 (2002).
[2] F. Occeli, P. Loubeyre, and R. Le Toullec. Nature Materials **2**, 151 (2003).
[3] M. Hebbache and M. Zemzemi. Phys. Rev. B **70**, 224107 (2004).
[4] K. Takemura. Phys. Rev. B, **70**, 012101 (2004).
[5] F. Occeli, *et al.* Phys. Rev. Letters **93**, 095502 (2004).
[6] C. Pantea, *et al.* Phys. Rev. B **80**, 024112 (2009).
[7] B. R. Sahu and L. Kleinman. Phys. Rev. B **72**, 113106 (2005).
[8] P. Blaha, K. Schwarz, G. K.H.Madsen, D. Kvasnicka and J. Luitz,WIEN2k, An Augmented Plane Wave + Local Orbitals Program for Calculating Crystal Properties (Karlheinz Schwarz, Techn. Universität Wien, Austria), 2001. ISBN 3-9501031-1-2.
[9] J. P. Perdew, K. Bourke, and M. Ernzerhof. Phys. Rev. Lett. **77**, 3865 (1996).
[10] F. Birch. Phys. Rev. **71**, 809 (1947).
[11] C. Kittel, *Introduction to Solid State Physics* (John Wiley & Sons. Inc, 2005).
[12] Jin-Cheng Zheng. Phys. Rev. B **72**, 052105 (2005).
[13] Yongcheng Liang and Zhong Fang. J. Phys.: Condens. Matter **18**, 8749 (2006).
[14] Y. Ma, T. Cui, L. Zhang, Yu Xie, G. Zou, J. S. Tse, X. Gao and D. D. Klug. Phys. Rev. B **72**, 174103 (2005).
[15] Michael J. Mehl and Dimitrios A. Papaconstantopoulos. Phys. Rev B **54**, 4519 (1996).
[16] Liu Bo, *et al.* Phys. B **19**, 026301 (2010).
[17] Lars Fast, J. M. Wills, Börje Johansson and O. Ericksson. Phys. Rev. B **51**, 17341 (1995).
[18] M. Sigalas, D. A. Papaconstantopoulos and N. C. Bacalis. Phys. Rev B **45**, 5777 (1992).
[19] D. Legut, M. Friák, and M. Šob. Phys. Rev. Lett. **99**, 016402 (2007); D. Legut, M. Friák, and M. Šob. Phys. Rev. B **81**, 214118 (2010) and references inthere.
[20] A. Rubio–Ponce, J. Morales, and D. Olguín. Advances in the Theory of Quantum Systems in Chemistry and Physics: Progress in Theoretical Chemistry and Physics, 2012, Vol. 22, Part 3, p. 119-127. Springer Netherlands, ISBN: 978-94-007-2076-3.

Ising model on $L \times L$ square lattice with free boundary conditions up to $L = 19$

Seung-Yeon Kim

School of Liberal Arts and Sciences, Korea National University of Transportation, Chungju 380-702, Korea

E-mail: `sykimm@ut.ac.kr`

Abstract. The exact partition functions of the Ising model on $L \times L$ square lattices with free boundary conditions are evaluated up to $L = 19$ using microcanonical transfer matrix. The critical behavior of the Ising antiferromagnet is analyzed using the partition function zeros for free boundary conditions and compared with the results for periodic boundary conditions.

1. Introduction

The square-lattice Ising model has played a central role in our understanding of phase transitions and critical phenomena [1]. The exact solution of the square-lattice Ising model with periodic boundary conditions is well known both in the thermodynamic limit [2] and in finite-size systems [3]. However, its exact solution with free boundary conditions is not known for systems of arbitrary size. To overcome this, Ferdinand and Fisher [4] constructed the exact partition functions of the Ising model on $L \times L$ square lattices with free boundary conditions up to $L = 4$. But their lattice sizes were too small and their results were not useful. Bhanot [5] computed the exact partition functions up to $L = 10$ and began obtaining some useful results. Stosic *et al.* [6] calculated the exact partition functions up to $L = 12$ and tested finite-size scaling theory. Stodolsky and Wosiek [7] obtained the exact partition function for $L = 13$, and studied phase transition based on microcanonical entropy. Recently, Kim obtained the exact partition functions of the Ising model for $L = 14$ [8] and $L = 15$ [9], and investigated its properties. In this work, for the first time, we evaluate the exact partition functions of the Ising model on $L \times L$ square lattices with free boundary conditions for $L = 16 \sim 19$ using microcanonical transfer matrix [5, 6, 7, 10]. Currently, we are evaluating its exact partition functions up to $L = 25$.

2. Density of states

The Ising model on a lattice with N_s sites and N_b bonds is defined by the Hamiltonian

$$\mathcal{H} = -J \sum_{\langle i,j \rangle} \sigma_i \sigma_j, \tag{1}$$

where J is the coupling constant, $\langle i, j \rangle$ indicates a sum over all nearest-neighbor pairs of lattice sites, and $\sigma_i = \pm 1$. We define the density of states, $\Omega(E)$, with a given energy

$$E = -\sum_{\langle i,j \rangle} \sigma_i \sigma_j, \tag{2}$$

Table 1. The density of states $\Omega(E)$ for the Ising model on the 4×4 square lattice with free boundary conditions. Here, E denotes the given energy values.

E	$\Omega(E)$	E	$\Omega(E)$	E	$\Omega(E)$	E	$\Omega(E)$
−24	2	−10	1216	2	9984	14	224
−20	8	−8	2638	4	7344	16	72
−18	32	−6	4928	6	4928	18	32
−16	72	−4	7344	8	2638	20	8
−14	224	−2	9984	10	1216	24	2
−12	584	0	11472	12	584		

where E is integers $(-N_b \leq E \leq N_b)$. Then the partition function of the Ising model (a sum over 2^{N_s} possible spin configurations)

$$Z = \sum_{\{\sigma_n\}} e^{-\beta \mathcal{H}}, \tag{3}$$

where $\beta = 1/k_B T$ (k_B is the Boltzmann constant and T is temperature), can be written as

$$Z(T) = \sum_{E=-N_b}^{N_b} \Omega(E) e^{-\beta J E}. \tag{4}$$

Microcanonical transfer matrix [5, 6, 7, 10] is used to evaluate the *exact* integer values for the density of states $\Omega(E)$ of the Ising model on $L \times L$ (up to $L = 19$) square lattices with free boundary conditions ($N_s = L^2$ and $N_b = 2L^2 - 2L$). Here, microcanonical transfer matrix is briefly described for the Ising model on the 4×4 square lattice. First, an array $\omega^{(1)}$, which is indexed by energy E and spin variables $\sigma_i^{(1)}$ $(1 \leq i \leq 4)$ for the first row, is initialized as

$$\omega^{(1)}(E; \sigma_1^{(1)}, \sigma_2^{(1)}, \sigma_3^{(1)}, \sigma_4^{(1)}) = \delta(E + \sigma_1^{(1)}\sigma_2^{(1)} + \sigma_2^{(1)}\sigma_3^{(1)} + \sigma_3^{(1)}\sigma_4^{(1)}), \tag{5}$$

where δ is the Kronecker delta. Next, by introducing a new spin variable $(\sigma_1^{(2)})$ from the second row, each spin in the first row is traced over in turn

$$\tilde{\omega}^{(2)}(E; \sigma_1^{(2)}, \sigma_2^{(1)}, \sigma_3^{(1)}, \sigma_4^{(1)}) = \sum_{\sigma_1^{(1)}} \omega^{(1)}(E + \sigma_1^{(1)}\sigma_1^{(2)}; \sigma_1^{(1)}, \sigma_2^{(1)}, \sigma_3^{(1)}, \sigma_4^{(1)}). \tag{6}$$

This procedure is repeated until all the spins in the first row have been traced over, leaving a new function of the four spins $(\sigma_1^{(2)}, \sigma_2^{(2)}, \sigma_3^{(2)}, \sigma_4^{(2)})$ in the second row. Now, the horizontal bonds connecting the spins in the second row are then taken into account by shifting the energy,

$$\omega^{(2)}(E; \sigma_1^{(2)}, \sigma_2^{(2)}, \sigma_3^{(2)}, \sigma_4^{(2)}) = \tilde{\omega}^{(2)}(E + \sigma_1^{(2)}\sigma_2^{(2)} + \sigma_2^{(2)}\sigma_3^{(2)} + \sigma_3^{(2)}\sigma_4^{(2)}; \sigma_1^{(2)}, \sigma_2^{(2)}, \sigma_3^{(2)}, \sigma_4^{(2)}). \tag{7}$$

These two procedures, Eqs. (6) and (7), are then applied to each row in turn until the final (4th) row is reached. Finally, the density of states is given by

$$\Omega(E) = \sum_{\sigma_1^{(4)}} \sum_{\sigma_2^{(4)}} \sum_{\sigma_3^{(4)}} \sum_{\sigma_4^{(4)}} \omega^{(4)}(E; \sigma_1^{(4)}, \sigma_2^{(4)}, \sigma_3^{(4)}, \sigma_4^{(4)}), \tag{8}$$

as shown in Table 1. The sum over all densities of states is equal to 2^{N_s} (all possible spin configurations), for example, $2^{16} = 65536$ for $L = 4$.

We evaluate the exact integer values for the density of states $\Omega(E)$ of the Ising model on $L \times L$ square lattices with free boundary conditions for $L = 16 \sim 19$ using microcanonical transfer matrix. The largest densities of states are as follows:

$$\begin{aligned}\Omega_{\max}(L=16) \;=\;& 422717712806305626915190216446266421777790439457026923297936 \\ & 5969506075458048\end{aligned} \tag{9}$$

(approximately 4.227×10^{75}),

$$\begin{aligned}\Omega_{\max}(L=17) \;=\;& 340989849864650344940632877971629373806476542617613543966009 \\ & 923245953630254995011346966\end{aligned} \tag{10}$$

(approximately 3.410×10^{85}),

$$\begin{aligned}\Omega_{\max}(L=18) \;=\;& 110436968889888645233777742673235427292106386519331404321433 \\ & 7764044519276950524948647188833880536\end{aligned} \tag{11}$$

(approximately 1.104×10^{96}), and

$$\begin{aligned}\Omega_{\max}(L=19) \;=\;& 143544606619440356957791141868826926778768964447532779063754 \\ & 4543890911238153755873632608552272496208460396 24\end{aligned} \tag{12}$$

(approximately 1.435×10^{107}).

3. Partition function zeros

Given the density of states $\Omega(E)$, the partition function of the Ising model is a polynomial in $a = e^{2\beta J}$. For antiferromagnetic interaction ($J < 0$), the physical interval is $0 \leq a \leq 1$ ($0 \leq T \leq \infty$), while for ferromagnetic interaction ($J > 0$), it is $1 \leq a \leq \infty$ ($\infty \geq T \geq 0$). Its specific heat (per volume) is given by

$$C(a) = (N_s k_B T^2)^{-1} \frac{\partial^2}{\partial \beta^2} \ln Z(a) = \frac{k_B}{N_s} (\ln a)^2 (\langle E^2 \rangle - \langle E \rangle^2). \tag{13}$$

In the thermodynamic limit, the specific heat of the square-lattice Ising model diverges at the antiferromagnetic critical point $a_c^{\text{af}} = \sqrt{2} - 1$ and at the ferromagnetic critical point $a_c^{\text{fm}} = \sqrt{2} + 1$.

The partition function can also be expressed as its zeros $\{a_i\}$ in the complex temperature (a) plane:

$$Z(a) = A \prod_{i=1}^{N_b} (a - a_i), \tag{14}$$

where A is constant [5, 8, 10]. In terms of the partition function zeros, the specific heat is given by

$$C(a) = \frac{k_B}{N_s} (\ln a)^2 \sum_{i=1}^{N_b} \left[\frac{a}{a - a_i} - \left(\frac{a}{a - a_i} \right)^2 \right]. \tag{15}$$

In finite sizes, no zero cut the positive real axis in the complex temperature plane, but some zeros (for a system showing a phase transition) approach the positive real axis as the system size increases, determining the critical point and exponents and resulting in the singularity of the specific heat $C(a)$ in the thermodynamic limit. It is clear from $C(a)$ that the leading behavior of such a singularity is due to the pair of partition function zeros closest to the positive real axis,

Table 2. The first zeros (a_1) of the Ising antiferromagnet on $L \times L$ square lattices for free boundary conditions and periodic boundary conditions.

L	a_1 (free)	a_1 (periodic)
4	$0.26982689 + 0.23832440i$	$0.44443953 + 0.18729421i$
5	$0.30440253 + 0.19570711i$	$0.44508165 + 0.33653188i$
6	$0.32534009 + 0.16622629i$	$0.43618425 + 0.12068694i$
7	$0.33943334 + 0.14458498i$	$0.43921166 + 0.22325545i$
8	$0.34959422 + 0.12799881i$	$0.43135614 + 0.08937469i$
9	$0.35728151 + 0.11486764i$	$0.43516252 + 0.16855597i$
10	$0.36330846 + 0.10420557i$	$0.42825730 + 0.07104314i$

called the first zeros (a_1 and its complex conjugate). Therefore, by calculating the partition function zeros and examining the behavior of the first zero in the thermodynamic limit, the critical behavior can be much more accurately analyzed than examining the behavior of the specific heat, which is plagued by the noise due to the subleading terms containing zeros other than the first ones.

Table 2 shows the first zeros of the square-lattice Ising antiferromagnet for free and periodic boundary conditions. For periodic boundary conditions, even and odd sizes show different finite-size behaviors. Therefore, we have to choose even sizes in the extrapolation to the infinite size. On the other hand, there is no different behavior between even and odd sizes for free boundary conditions, enabling us to use the whole data. Using the data in the table for free boundary conditions and Bulirsch-Stoer extrapolation method [11], we have obtained the critical point $a_c = 0.414213(16)$ and the thermal scaling exponent $y_t = 1.0000(81)$ in the infinite size, in excellent agreement with the exact values $a_c = \sqrt{2} - 1$ and $y_t = 1$. The error estimates are twice the difference between the $(n-1,1)$ and $(n-1,2)$ approximants. Similarly, from the even-size data in the table for periodic boundary conditions, the extrapolated values are $a_c = 0.414538(2255)$ and $y_t = 0.9940(252)$. Our results show clearly that free boundary conditions are better than periodic boundary conditions in the study of antiferromagnet.

Acknowledgments
This research was supported by Basic Science Research Program through the National Research Foundation of Korea (NRF) funded by the Ministry of Education, Science and Technology (grant number 2012-0004527).

References

[1] Domb C 1996 *The Critical Point* (London: Taylor and Francis)
[2] Onsager L 1944 *Phys. Rev.* **65** 117
[3] Kaufman B 1949 *Phys. Rev.* **76** 1232
[4] Ferdinand A E and Fisher M E 1969 *Phys. Rev.* **185** 832
[5] Bhanot G 1990 *J. Stat. Phys.* **60** 55
[6] Stosic B, Milosevic S and Stanley M E 1990 *Phys. Rev.* B **41** 11466
[7] Stodolsky L and Wosiek J 1994 *Nucl. Phys.* B **413** 813
[8] Kim S-Y 2004 *Phys. Rev. Lett.* **93** 130604
[9] Kim S-Y 2011 *World Academy of Science, Engineering and Technology* **59** 701
[10] Creswick R J 1995 *Phys. Rev.* E **52** R5735
[11] Press W H, Teukolsky S A, Vetterling W T and Flannery B P 1992 *Numerical Recipes in Fortran 77*, 2nd edition (New York: Cambridge University Press) p 104

Computing sparse derivatives and consecutive zeros problem

B V Ravi Chandra[1] **and Shahadat Hossain**[2]

[1] Birla Institute of Technology and Sciences (BITS), Pilani, India
[2] Department of Mathematics and Computer Science, University of Lethbridge, Alberta T1K 3M4, Canada

E-mail: [1] `ravichandra363@gmail.com`, [2] `shahadat.hossain@uleth.ca`

Abstract. We describe a substitution based sparse Jacobian matrix determination method using algorithmic differentiation. Utilizing the a priori known sparsity pattern, a compression scheme is determined using graph coloring. The "compressed pattern" of the Jacobian matrix is then reordered into a form suitable for computation by substitution. We show that the column reordering of the compressed pattern matrix (so as to align the zero entries into consecutive locations in each row) can be viewed as a variant of traveling salesman problem. Preliminary computational results show that on the test problems the performance of nearest-neighbor type heuristic algorithms is highly encouraging.

1. Introduction

We consider the problem of computation or estimation of the sparse Jacobian matrix $J \equiv \left\{\frac{\partial f_i}{\partial x_j}\right\}, 1 \leq i \leq m, 1 \leq j \leq n$ of a mapping $f : \mathbb{R}^n \mapsto \mathbb{R}^m$. Mathematical derivatives constitute an important ingredient in important algorithms e.g., Newton's method in nonlinear optimization. In this paper we assume that the matrix has a priori known fixed sparsity pattern in the domain of interest. Furthermore, for computational efficiency (there are many common and computationally expensive subexpressions in the vector function which should be evaluated only once) and/or because the function is only available as a subroutine call, we compute the whole column of J or J^T, rather than computing each nonzero unknown separately.

A group of columns in which no two columns have nonzero elements in the same row position is known as *structurally orthogonal* [3]. If columns j and l are structurally orthogonal, then for each row index i, at most one of $J(i,j)$ and $J(i,l)$ can be nonzero. In general, $\sum_j J(i,j) = J(i,l)$ for some l (l will depend on i) where the sum is taken over a group of structurally orthogonal columns. An estimate of the nonzero elements in the group can be obtained in

$$\left.\frac{\partial f(x+ts)}{\partial t}\right|_{t=0} = f'(x)s = As \approx \frac{1}{\epsilon}[f(x+\epsilon s) - f(x)] \equiv b, \tag{1}$$

with a forward difference (one extra function evaluation) [3] where b is the finite difference approximation and $s = \sum_j e_j$ with e_j denoting the jth coordinate vector. Using the forward automatic differentiation [5] the unknown elements in the group are obtained as the product $b = f'(x)s$ accurate within working accuracy. Unfortunately, finding a structurally orthogonal column partition with minimum number of column groups and equivalently coloring the vertices of the column intersection graph is NP-hard [2, 8]. Therefore, many computationally effective algorithms employ heuristic techniques to find a "good" partition. Vector b in (1) holds the nonzero unknowns in a orthogonal group of columns. If the columns of A are partitioned into p orthogonal groups, (1) takes the form,

$$b_j \equiv As_j,\ j = 1, \cdots, p \text{ or } B \equiv AS,$$

where the jth column of S, s_j, is defined by the column indices in jth orthogonal group, $j = 1, 2, \ldots, p$. It is clear that since the location of nonzero unknowns of A are assumed to be known, these values can simply be read from the compressed matrix B. Consider the matrix

$$A = \begin{pmatrix} 0 & a_{12} & 0 & a_{14} & a_{15} & a_{16} \\ a_{21} & a_{22} & a_{23} & 0 & 0 & a_{26} \\ a_{31} & 0 & a_{33} & a_{34} & a_{35} & 0 \end{pmatrix},$$

where no two columns are structurally orthogonal. Therefore, the seed matrix S is simply the 6×6 identity matrix; the compressed matrix B will have 6 columns, and hence no savings in AD or FD computations can be realized. Now, with the seed matrix:

$$S = \begin{pmatrix} 1 & 0 & 0 & 0 & 0 \\ 1 & 1 & 0 & 0 & 0 \\ 0 & 1 & 1 & 0 & 0 \\ 0 & 0 & 1 & 1 & 0 \\ 0 & 0 & 0 & 1 & 1 \\ 0 & 0 & 0 & 0 & 1 \end{pmatrix}, \text{ we have, } B = AS = \begin{pmatrix} a_{12} & a_{12} & a_{14} & a_{14}+a_{15} & a_{15}+a_{16} \\ a_{21}+a_{22} & a_{22}+a_{23} & a_{23} & 0 & a_{26} \\ a_{31} & a_{33} & a_{33}+a_{34} & a_{34}+a_{35} & a_{35} \end{pmatrix}.$$

We can recover the nonzero unknowns by solving a linear system of equations associated with each row. For example, for row 2, we obtain (after removing zero column from the matrix at the left and zero entry from vector b)

$$x^T \begin{pmatrix} 1 & 0 & 0 & 0 \\ 1 & 1 & 0 & 0 \\ 0 & 1 & 1 & 0 \\ 0 & 0 & 0 & 1 \end{pmatrix} = \begin{pmatrix} a_{21}+a_{22} & a_{22}+a_{23} & a_{23} & a_{26} \end{pmatrix}, \tag{2}$$

which is a triangular linear system and can be solved cheaply (i.e., without multiplication or division) for the nonzero unknowns. Hossain and Steihaug [6] show that if each row of matrix A contains at least d consecutive zero entries, then a seed matrix S that results triangular linear systems for each row of A as discussed above, can be constructed with only $n - d$ columns. However, reordering the columns so as to maximize d is NP-hard [6]. In the next section we present a column reordering heuristic and computational test results.

2. Column reordering

The example matrix A in the preceding section contains two zero entries in each row, and except for row 2, the zeros are not consecutive. However, applying the permutation $\Pi = (1, 3, 5, 4, 6, 2)$ to the columns, the zero entries in each row of the resulting matrix $A_\Pi = AP$ where the jth column of P is the $\Pi(j)$th coordinate vector, become consecutive.

> **Consecutive Zeros Problem (CZP).** Given the sparsity pattern of a matrix $A \in \mathbb{R}^{m\times n}$ find a permutation Π of the columns of A that maximizes $d = \min_i\{\rho_i | \rho_i$ is the maximal sequence of consecutive zero entries in row i of $A_\Pi\}, i = 1, 2, \ldots, n$.

The central idea of our heuristic method is to represent the matrix ordering problem as a variant of traveling salesman problem (TSP). For every pair of columns j, l of matrix A, we define

$$c_{jl} = \sum_{i=1}^{n} A(i,j) \odot A(i,l),$$

where,

$$A(i,j) \odot A(i,l) = \begin{cases} 0, & \text{if } (A(i,j) = A(i,l) = 0) \text{ or } (A(i,j) \neq 0 \text{ and } A(i,l) \neq 0) \\ 1, & \text{otherwise} \end{cases}.$$

Our objective is to find a permutation Π of the columns that minimizes the cost $C(A_\Pi) = \sum_{j=1}^{n-1} c_{\Pi(j)\Pi(j+1)}$. For the example matrix from the preceding section, we have $C(A) = 8$ and $C(A_\Pi) = 4$. The best approximation guarantee of 1.5 for a polynomial time heuristic for TSP is due to Christofides [1] and can be implemented in $O(n^3)$. For the CZP we have implemented a greedy heuristic "nearest-neighbor (NN)" and an improvement heuristic" 2-OPT". In NN we start at an arbitrary column as the first column in the permutation ordering i.e., $\Pi(1)$ is the column selected. At each iterative step $j > 1$, we choose column l such that $c_{\Pi(j)l}$ is minimum and set $\Pi(j+1) = l$.This process is continued until all the columns are ordered. In our implementation of the 2-OPT improvement heuristic the input is the permuted column indices obtained from the NN heuristic. At each iterative step, a pair of column indices j, l are chosen arbitrarily. A new permutation is created by exchanging their position in the ordering and its cost is computed. If the new cost is an improvement over the current best, the current best is updated. Otherwise, no exchange is made. This step is repeated for a user-defined maximum number of iterations.

Table 1. Nearest-neighbor heuristic test results

Name	m	n	$z_{\min}$	d_1	d_2
abb313.mtx	313	10	4	2	2
ash219.mtx	219	4	2	1	1
ash331.mtx	331	6	4	2	2
ash608.mtx	608	6	4	2	2
ash958.mtx	958	6	4	2	2
will199.mtx	199	7	1	1	1
af23560.mtx	23560	41	20	3	3
cage11.mtx	39082	62	31	4	4
cage12.mtx	130228	68	35	4	4
e40r0100.mtx	17281	69	7	1	1
fs_541_1.mtx	541	13	2	1	1
fs_541_2.mtx	541	13	2	1	1
fs_541_3.mtx	541	13	2	1	1
impcol_b.mtx	59	11	4	2	2
impcol_d.mtx	425	11	1	1	1
impcol_e.mtx	225	21	9	4	4
lp_ken_11.mtx	14694	125	3	2	2
lp_ken_13.mtx	28632	171	1	1	1
lp_maros_r7.mtx	3136	83	35	3	3
lund_a.mtx	147	13	1	1	1
lund_b.mtx	147	13	1	1	1

3. Computational experiments

Table 1 displays test results for NN and 2-OPT algorithms on matrix instances drawn from University of Florida Test Collection [4]; $m, n, z_{\min}, d_1, d_2$ denote, respectively, number of rows, number of columns, minimum number of zeros in any row, value of d (as in CZP) with NN, value of d (as in CZP) with 2-OPT. The pattern matrices in the test suite have been pre-compressed by using structurally orthogonal partitions (or vertex coloring); for 2-OPT algorithm we perform 200 exchange iterations. For the studied test instances, 2-OPT does not yield any improvement in d. In fact, using an exhaustive search (by testing all permutations for smaller problems indicated in **boldface**) we have verified that the NN heuristic yields optimum d. For larger problems, we set a time limit of one hour and exhaustive search provided no improvement over NN heuristic. We observe that for a few of the test instances $z_{\min}$ is significantly larger than the value of d computed by the NN heuristic; for test instances with $z_{\min} > 2$, except for problem `lp_ken_11.mtx`, the savings (with respect to extra function evaluation in FD or AD forward evaluation) can be as high as 50%. The test instances for which the optimality of d can be verified, we get about 19% savings, on average, in computational cost.

4. Concluding remarks

In this paper we have proposed a formulation for the consecutive zeros problem arising in the computation of sparse derivative matrices. The test results indicate savings in computational cost when the Jacobian matrix to be determined is preprocessed with structurally orthogonal partitioning of columns. In iterative schemes e.g., Newton's method for solving nonlinear system of equations, the most expensive computation at each iteration is the evaluation of the Jacobian matrix. For large-scale problems the overall computational time is greatly reduced by savings in Jacobian evaluation. Our test results indicate that the savings can be as large as 67% over structurally orthogonal column partitioning alone. However, for certain test instances, elimination methods as outlined in [7, 9] can be employed to fully exploit the available sparsity in the matrix.

Acknowledgments

B V Ravi Chandra was supported by a MITACS Globalink Summer Fellowship and Shahadat Hossain's work was partially supported by a NSERC Discovery Grant (individual).

References

[1] Christofides N 1976 Worst-case analysis of a new heuristic for the travelling salesman problem *Management Sciences Research Report* **388**, CMU

[2] Coleman T. F and J. J. Moré J. J 1983 Estimation of sparse Jacobian matrices and graph coloring problems *SIAM J. Numer. Anal.*, **20(1)**:187–209

[3] Curtis A. R, Powell M. J. D, and Reid J. K 1974 On the estimation of sparse Jacobian matrices *J. Inst. Math. Appl.*, **13**:117–119

[4] Davis T. A and Hu Y The University of Florida sparse matrix collection *ACM Trans. Math. Softw.* (to appear)

[5] Griewank, A and Walther, A 2008 *Evaluating Derivatives: Principles and Techniques of Algorithmic Differentiation*, 2nd edn. No. **105** in Other Titles in Applied Mathematics. SIAM, Philadelphia, PA

[6] Hossain S and Steihaug T 2002 Sparsity issues in the computation of jacobian matrices In Mora T, editor, *Proceedings of the International Symposium on Symbolic and Algebraic Computing (ISSAC)*, pages 123–130, New York. ACM

[7] Hossain S and Steihaug T 2006 Computing sparse Jacobian matrices optimally In Bcker M, Corliss G, Naumann U, Hovland P, Norris B, Barth T J, Griebel M, Keyes D E, Nieminen R M, Roose D, and Schlick T, editors, *Automatic Differentiation: Applications, Theory, and Implementations*, volume **50** of *Lecture Notes in Computational Science and Engineering*, pages 77–87. Springer Berlin Heidelberg

[8] Hossain S and Steihaug T 2012 Optimal direct determination of sparse Jacobian matrices *Opt. Meth. and Soft.* doi: 10.1080/10556788.2012.693927

[9] Newsam G N and Ramsdell J D. 1983 Estimation of sparse Jacobian matrices. *SIAM J. Alg. Disc. Meth.*, **4(3)**:404–417

A CLASS OF DEDUCTIVE THEORIES THAT CANNOT BE DETERMINISTIC

Iegor Reznikoff
Professor Emeritus, Département de Philosophie,
Université de Paris-Ouest, 92001 Nanterre, France

E-mail: dominiqueleconte@yahoo.fr

Abstract. Although the concept of an observing device with memory is very simple and definable in terms of algorithmic machine, it is not compatible with a predictive theory involving this concept. This is sufficient to show that Physics as a whole cannot be deterministic; moreover, using the Conway-Kochen Free Will Theorem [1] and the simple logical proof of it we gave in [2], we show, without any need of free will or observer's freedom, that neither Quantum Mechanics nor any extension of this theory can be deterministic.

I - Definitions

1). By *deductive* theory, we mean a theory that is based on – not necessarily completely formal – logical reasoning, as e.g. mathematics, logic itself of course, or more casually physical science, theoretical or applied. If F is the theory and $\vdash$ the symbol of deduction, F $\vdash$ A means that A is deducible from (or can be proved in) the theory F.

2). By *deterministic* theory, we mean that in the deductive theory a time variable t is defined, t belonging as usual to an interval of real numbers (however an ordered set is sufficient here), and that any true event E(t) expressible in the theory and occurring at time t can be deduced 'in advance' in the theory i.e. there exists $t' < t$ such that

(a) $F(t') \vdash E(t)$

where in F appear only events that occurred at time $\leq t'$ (this can be formulated precisely in logical terms, but is not necessary here). In such case the event E(t) is said to be predicted or (pre)determined at time t' by theory F.

3). By an *observing* machine M with *memory* we mean a machine e.g. a Turing machine, a computer or a human brain, that can *observe* (or read) any finitely coded sentence or formula expressible in the theory, that appears on a tape or in the coded observational field of the machine; moreover, the machine stores this sentence or a part of it in its *memory*, i.e., the finite set of already stored sentences since a given time (the machine, of course, can observe only a finite number of times). In particular, the machine M can observe the second member of formula (a) above relative to the event E(t) as soon as the formula has appeared among (or in the list of) deducible formulas in F i.e. at time $t' < t$, and stores the predicted event E(t) in its memory.

4). Let us write M(t, x(t), m(t)) for "at time t M observes x(t) and stores it in its memory m(t) ". By definition above, memory has obviously the following property:

(b) (b) if $t' < t$ then m(t') Ì m(t).

This corresponds to the very meaning of the notion of memory.

II. Main Results

Theorem *The existence of an observing machine with memory based on a deductive theory* F *is inconsistent with the theory* F *being deterministic.*

Proof. Suppose the contrary. Since the theory F is deterministic, in particular in what concerns M, at some time $t' < t$ the behaviour of the machine M at time t is predetermined in the theory, i.e., F(t') $\vdash$ M(t, x(t), m(t)). Then already at time t', M observes the predicted event and stores it in its memory m(t'), thus

(c) M(t, x(t), m(t)) Î m(t').

Now, when later on, time t indeed occurs, if the deterministic prediction is correct, the expected event M(t, x(t), m(t)) should occur as well, but since $t' < t$, because of (b) we have m(t') Ì m(t), and therefore, M(t, x(t), m(t)) Î m(t), which is impossible since m(t) is finitely coded.

Commentary. What we need is that the interval of time between t' and t, be large enough for M to have observed and stored (possibly at time t'' with $t' < t'' \leq t$) the predicted sentence M(t, x(t), m(t)) before (or at the latest at) time t. Technically for such simple a task, a machine nowadays needs a very small fraction of time, while a deterministic theory, if all needed information is available, should calculate and predict an event, in principle, 'sufficiently' in advance. Otherwise, if there is no time to know a predicted result before it happens, there is of course no prediction and the determinism would have no more value than the religious belief that 'everything is already written' in some Book for ever unknown. In any case, if we accept determinism in the ordinary meaning of the word, so that the predicted event can be recorded as such, the theorem shows that it is not compatible with observation and memory.

Corollary 1 *In the deterministic part of classical physics, an observing machine with memory cannot be defined (and therefore constructed).*

Now, since an observing machine with memory certainly exists in our physical world e.g. as the brain of human being or, more elementary, as a modern computer, we have

Corollary 2 *Actual Physics cannot be completely deterministic.*

Corollary 3 *Quantum Mechanics, if consistent, are not deterministic.*

Proof. In [2], we have given a formal proof (improving the Conway – Kochen *Free Will Theorem* [1]) that if an observer of spin components is *free* in this observation, then Quantum Mechanics are not deterministic; but our main theorem here proves that the behaviour of an observer or machine with memory cannot be predetermined, i.e., is necessarily free.

The proof shows that no extension, e.g. by *hidden variables*, of Q.M. can be a deterministic theory.

If observed, Nature cannot be deterministic; actually, when observed, Nature is observing itself and clearly, since it includes the observing device, it has memory; but this, as we have shown, yields non-determinism. Extending Einstein's famous saying, we should say: *God doesn't play dice, but lets the dice play.*

References

[1] J.Conway and S.Kochen, The Strong Free Will Theorem, *AMS*, vol 56/2, p.226-232, Providence, February 2009.

[2] I.Reznikoff, A Logical Proof of the Free Will Theorem, arXiv: 1008.3661v1 [quant-ph] 21 Aug 2010 (http://arxiv.org/abs/1008.3661).

[3] I.Reznikoff, The Consistency of Quantum Mechanics Implies Its Non-Determinism, arXiv: 1010.4020v1 [quant-ph] 19 Oct 2010 (http://arxiv.org/abs/1010.4020). (In this paper some proofs are not clearly explained and a logical assumption is actually unsatisfactory; the paper given here proves, in a much more powerful way, the same main conclusion).

Nitriding modelling in nanoscale

R Politano

IPEN/CNEN-SP - Nuclear and Energy Research Institute/National Nuclear Energy Commission - Av. Prof. Lineu Prestes, 2242, Cidade Universitária, São Paulo, SP, Brazil.
CCTM - Materials Science and Technology Center

politano@ipen.br

Abstract. Modelling on Plasma and gas nitriding of austenitic and low alloy steels have strong influence of interfacial - nanoescale phenomena. Gamma prime and epsilon nitrogen-iron phases evolution during the nitriding process wasn't found explicitly simulated on literature. On present work we simulated nitrind process on low alloy steels - with their precipitation phenomena and nitrides moving interfaces - using diffusional models plus cellular automata. Surface effects and heterogeneities and local transition phenomena considered on our models show that some experimental results considered "errors" on literature are predicted by our simulations. In present work parameters like diffusion coefficient and surface conditions and gradient of nitrogen concentrations between different phases was measured.

1. Introduction

Nitriding process is a thermochemical method for increasing wear and corrosion resistance, and surface hardness of metals. Gas and plasma nitriding are commercial process mostly used for low alloy and austenitic steels. Efforts in modeling have either been solving the diffusion equation analytically or by numerical methods in one dimension [1]. But it`s not so often have found simulations in two or three dimensions [2]. Because of this, some important effects are missed, mainly surface imperfections effects [18] which have an important effect on global diffusion. Surface effects and nitride layers have a thickness of about a few microns or smaller. So multidimensional simulations on this scale have been done numerically and have necessary grids with nanometer increments. Results of simulations are very close to the experimental results found in most experimental works, but because of the nanometer scale, some unusual numerical effects became important – mainly due to the diffusion process.

Some old concepts are revisited, like "diffusion distance", stability of numerical methods of partial differential equations solutions, and local equilibrium. So despite this work being a solution of an old technological problem with old mathematical tools, looking at that in a nanometer scale caused some useful conclusions to be made for nanomaterial`s and material`s diffusional process in this scale, and how it can have an important influence on bigger scales (micrometer or millimeter lenghs).

The simulations of present work were made on a computer program, developed by the author in Object Pascal language in a open source compiler (Lazarus), with a visual and user-friendly interface. Results can be saved as raw data and as a visual representation in colored 2D images.

2. Theory

2.1.Surface phenomena and diffusion:

In general, the process of nitriding is inserted in the reactions between gases and metals [3-5]. This type of reaction has been attracting interest for a long time due to their applications not only to surface treatments, but also in studies related to oxidation and corrosion of metals. Under this generic

approach, is the work of Fromm and Horz [3], which divides this kind of reaction in the following steps:
I) Transport of the gas molecules to the metal surface.
II) dissociation of the molecules and chemisorption thereof.
III) Transfer of gas atoms or ions through the surface.
IV) Diffusion of atoms, or ions, the lattice of the metal.

Stages I and II (transport and chemisorption of gas molecules) is strongly influenced by the type of process employed. The gas nitriding, which in most cases uses mixtures-of N2, H2 and NH3, where partial pressure will be an important variable, which influences the formation (or not) of the nitrided layer. This is due to the mechanism of interaction between the gaseous atmosphere and the surface, and the number of collisions between molecules of gases. The reaction of nitrogen absorption can be written as:

$$NH_3 = N_{(diss)} + \frac{3}{2}H_2 \tag{2.1}$$

The potential of nitrogen in gas nitriding is given by:

$$K_N = \frac{p_{NH_3}}{p_{H_2}^{3/2}} \tag{2.2}$$

The formation of the nitride layer depends on the nitrogen potential in the surface (Eq. 2.2) , it`s condition, and the presence of stable non-metallic compounds like oxides. The nitrides of Fe-N system are the phases γ '(Fe4N) and ε (Fe2N1-z). Finally, the nitrogen atom diffuses into the material - Step IV.

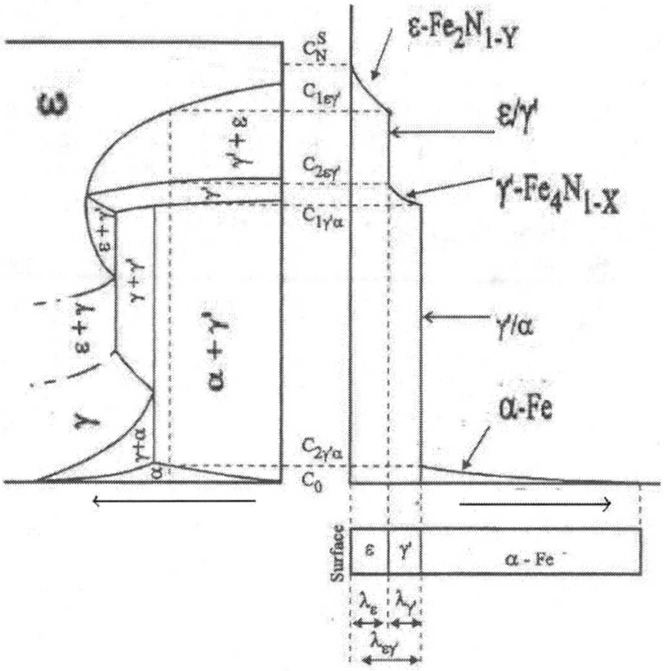

Figure 1: Relationship between Fe-N phase diagram and diffusion profile.[6]

The relationship between the phase diagram, diffusion, formation of nitrides, and their respective growth is shown in Figure 1[6]. In this figure the limits for concentrations of each phase and their respective profiles are schematized. The advancement of each layer depends on the diffusion process between them. It's possible to determine the speed in which each of these layers will advance. Using the symbols of Figure 1 we have that the advancement of the layers and their interfaces are determined by the total flow of nitrogen through the interfaces and the mass balance[6]:

Interface ε / γ ':

$$(C_{1\varepsilon\gamma} - C_{2\varepsilon\gamma})\frac{\partial\lambda_\varepsilon}{\partial t} = \left(-D^\varepsilon \nabla_{x=\lambda_\varepsilon} C_\varepsilon\right) - \left(-D^{\gamma'}\nabla_{x=\lambda_\varepsilon} C_{\gamma'}\right) \tag{2.3}$$

Interface γ '/ α:

$$(C_{1\gamma'\alpha} - C_{2\gamma'\alpha})\frac{\partial\lambda_{\varepsilon\gamma'}}{\partial t} = \left(-D^{\gamma'}\nabla_{x=\lambda\varepsilon\gamma'} C_{\gamma'}\right) - \left(-D^\alpha \nabla_{x=\lambda_{\varepsilon\gamma'}} C_\alpha\right) \tag{2.4}$$

By means of such equations it is possible to determine the speed of advancement of the layers and their respective positions by numerical procedures which are discussed below. The diffusion on each layer obey the Fick's 2 Law

$$\frac{\partial C}{\partial t} = D\nabla^2 C \tag{2.5}$$

2.2. Precipitation of alloy elements nitrides:

In the case of steels containing alloying elements, they can be influenced during the process in three ways: by changing (in some cases dramatically) the boundaries of the phase diagrams, by changing the diffusion coefficient of nitrogen present in phases - mainly in the ferrite, or by reacting with nitrogen to form their nitrides. The last point refers to the presence or absence of the precipitation of nitrides of alloying elements (forming the "strong" are nitrides of Al, Ti, V, Cr, Mo, Nb, W and Mn [1]. The formation of these precipitates influences in the diffusion of nitrogen along the matrix phase, and it`s behavior is studied extensively by some authors [7-10], the resulting microstructures were analyzed [11], showing the formation of precipitated coherent or semi-coherent finely distributed through the matrix. It is precisely these precipitates which confer the positive effects of increased strength resulting from treatment. This precipitation occurs already in the initial stage. Even when it is expected a formation of the nitride layer, the formation of nitrides occurs, delaying thus the kinetics of the process. The formation of nitrides obeys the following reaction:

$$\boldsymbol{y}N(x,t) + \boldsymbol{x}M(x,t) \leftrightarrow M_xN_y \tag{2.6}$$

whose equilibrium constant is given by

$$K_e = \frac{1}{[a_N]^y[a_M]^x} \tag{2.7}$$

Assuming that Henry's Law is obeyed, and the solubility product using NMR, it follows that:

$$K_{M_xN_y} = {}^1/_{K_e} = [N(x,t)]^y[M(x,t)]^x \tag{2.8}$$

where M and N concentrations of alloying and nitrogen in solution, respectively. Occurs when the precipitation in a given region and at any given time:

$$[N(x,t)]^y[M(x,t)]^x > K_{MN} \tag{2.9}$$

All these assumptions are valid if one takes into account the local equilibrium in each time interval analyzed, what will be conceptually and mathematically proved later on.

3. Numerical schemas and simulations.

3.1. *Numerical procedures*:

Discrete mathematics and the finite difference method for the manipulation of variables and solving differential equation, which are present in the theoretical model, were used. As the mathematical approach will be done in a space of three dimensions - x, y and t, using the notation: x = i Δx ; y = j Δy and t = k Δt

So is the space (x, y) for discrete mesh M x N, where M and N are the total numbers of points in x and y, respectively. Concentrations are represented by:

CFN (i, j, k): concentration of nitrogen in f phase

CM (i, j, k): concentration of alloying element M

If (i, j, k) position of the interface nitride f

Df: diffusion coefficient of nitrogen in phase f

The usual operators for the finite difference method are defined from the differential diffusion process used in [12-17] have applied to the solution of equation 2.5 for the two-dimensional case:

$$\frac{c_{i,j,k+1} - c_{i,j,k}}{\Delta t} = \frac{D}{(\Delta x)^2}\left(c_{i-1,j,k} - 2c_{i,j,k} + c_{i+1,j,k}\right) + \frac{D}{(\Delta y)^2}\left(c_{i,j-1,k} - 2c_{i,j,k} + c_{i,j+1,k}\right) \tag{3.1}$$

Where the stability condition for the solution is:

$$D\left(\frac{1}{\Delta x^2} + \frac{1}{\Delta y^2}\right)\Delta t \leq \frac{1}{2} \tag{3.2}$$

The boundary conditions can be expressed in the same way:

For the plasma nitriding to α CN (0, j, k) <γ C2 'α:

(CN α represented by c)

$$c_{0,j,k+1} = \frac{2\Delta t}{\Delta x^2}\left[\frac{A\Delta x}{\rho} + D\,c_{1,j,k}\right] + \left[1 - \frac{D\Delta t}{\Delta x^2}\right]c_{0,j,k} \tag{3.3}$$

where A and ϱ are process parameters of plasma treatment (total surface area and density of current, respectively). For the gas nitriding:

$$c_{0,j,k+1} = c_{0,j,k} + \frac{2\Delta t}{(\Delta x)^2} D\left(c_{1,j,k} - c_{0,j} - k\Delta x\left(c_{0,j,k} - c_{eq}\right)\right) \tag{3.4}$$

where Ceq is indicated by Lehrer diagram for the respective nitriding potential, and k is given by:

k = 9x10-4 pH2 exp (-64220/RT) (3.5)

where R is gas constant and T is the temperature in K). At the opposite end surface, the concentration gradient will be zero. Hence the boundary condition for j = N:

$$c_{i,N,k+1} = c_{i,N,k} + \frac{2D\Delta t}{\Delta x^2}\left(c_{i,N-1,k} - c_{i,N,k}\right) \tag{3.6}$$

Implicit Cranck-Nickolson method or alternating direction implicit method can also be used for bi-dimensional schema solution where the constant of tridiagonal matrix of the linear equations set is:

$$\alpha = \frac{\Delta t D}{2\Delta x^2} \qquad \beta = \frac{\Delta t D}{2\Delta y^2} \tag{3.7}$$

Theoretically these methods are stable in most cases.

Finally, the diffusivity D of interstitial nitrogen in α-iron (diffusion layer) is given by[6]:

$$D_N^{\alpha} = 6.6\times10^{-7} exp\left(\frac{-77900}{RT}\right) m^2 s^{-1} \tag{3.8}$$

3.2 Simulations:

Initial conditions are set, such as:

- surface conditions (proportion of surface barriers and their fragmentation/relative size);

- alloying composition;
- atmosphere conditions of treatment (partial pressures of molecular gases)
- time and temperature of treatment.
- size of region to be simulated, number of points in the mesh (M and N) and Δx. The time increment Δt is calculated from stability condition form equation 3.2.

The steps of calculations can be summarized as:

1) surface calculations: length of nitride layers (eqs 2.3 and 2.4). Concentration on diffusion layer boundaries (eqs 3.3 or 3.4 and 3.5)
2) diffusion calculations (eq. 3.1)
3) precipitation calculation from 2.9 condition. Decrement on nitrogen and alloy elements solution concentrations CFN(i,j,k) and CM(i,jk).

4.Results and discussions

4.1 Choosing parameters for nanometer scale simulations.

A useful concept for diffusive phenomena is "diffusion distance" or "diffusion lenght":

$$< x >= (Dt)^{1/2} \tag{4.1}$$

For a temperature of 823K, D=7,6x10^{-12}m^{2}s^{-1}. In one hour (3600 s) <x> would be 1,65x10-4 m, or 165μm - three orders of magnitude greater than a nanometer scale. So, is necessary determine how many time is involved in one nanometer distance of diffusion:

$$t = \frac{\Delta x^2}{D} \tag{4.2}$$

for Δx=1nm , t=1,3x10-7 s, 10nm would be t=1,3x10-5s! So, times to be choose must be very short to not cover transient phenomena on this scales. Indeed, the mathematical condition for stability of explicit numerical solutions (eq. 3.6), when rearranged for $\Delta x=\Delta y$ allows :

$\Delta t \leq \frac{\Delta x^2}{4D}$ what means that with this time interval transients are not mathematically covered, but local equilibrium is reached in 4 steps (eq 4.2). These results, which are almost obvious, are important for the assumptions used on the modelling as a whole, mainly about the precipitation of nitrides and reactions on the surface, when considerate on nanoscale. A general remark is that when talking about nanomaterials, there is a high surface and interface proportioned material but also a material with very high diffusional processes.

Even though literature says that alternating direction implicit methods in 2 dimensions are always stable, instable behaviour of solutions with this method was found for α and β of 3.11 greater than 0.5. Therefore, explicit method - with clear stability criteria - was chosen.

4.2. Simulations results:

Figures 2, 3 and 4 show a quantitative and semi quantitative simulations for a 0.1%V,0.1%Ti,0.1%Al Fe alloy , nitrated in a 0.1atmN_2 and 0.1atmH_2 atmosphere, for 200s at 823K, with 20% of the surface with some barrier with a size of 700nm and $\Delta x=\Delta y$=10nm. These bidimentional figures show a typical microstructure with a wavy frontiers of nitrided regions, which traditional simulations don't show, as a result of the presence of proportionally tiny barriers on the surface of the experimental samples. Usually the non-diffusional profile show in Figure 2 found on most of other author`s experimental data is considered as an experimental error of the determination on either the local nitrogen concentrations or microhardness measurements. It's clear that it is an effect of the sample surface and their composition [18].

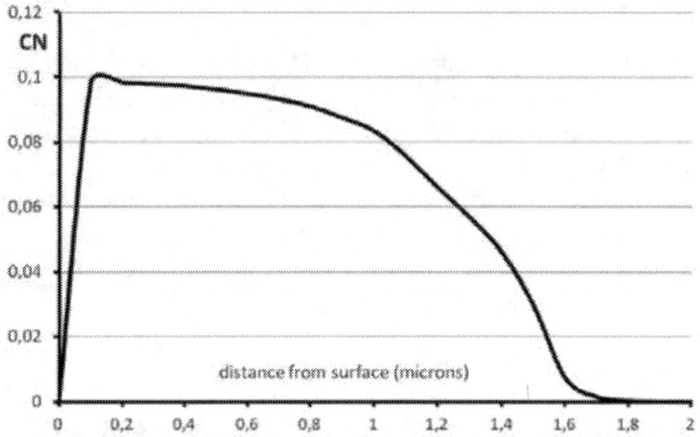

Figure 2: result of simulation - average total concentration of nitrogen (CN) in %mass with distance from surface in micrometers and respective.

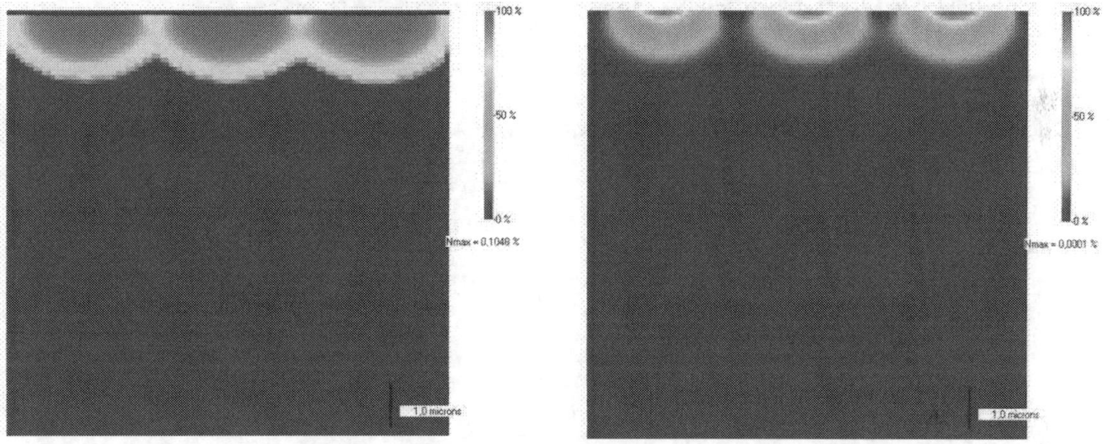

Figure 3: bidimentional image of simulation show in Figure 2, with normalized scale of colors. In the left side, total nitrogen (precipited and in solid solution), in right side, nitrogen in solid solution. Each side of squares have 10 micrometers.

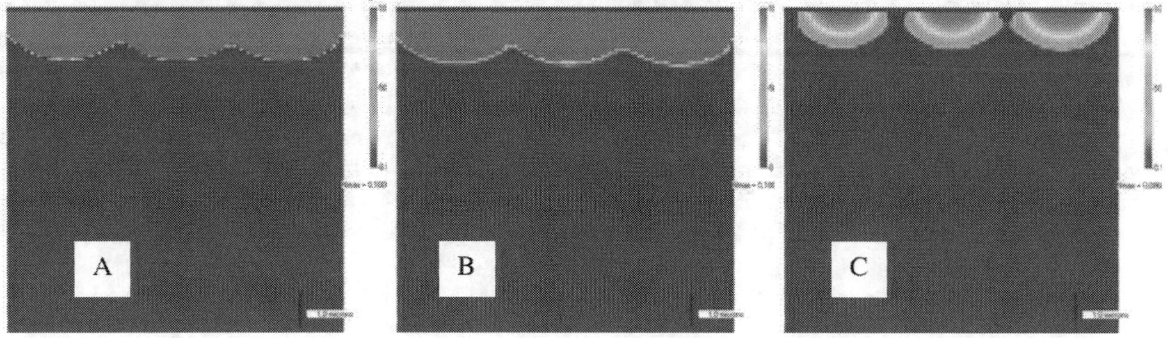

Figure 4: Result of simulation where a) is amount of nitrogen precipitated in AlN, b) TiN and c)VN.

5.Conclusions

The present work makes the relationship between nanometer scale and velocity of transformations, especially diffusion controlled process, clear. Some attractive properties of nanomaterials have come from this. But fast reactions in metal surfaces in a nanometer scale make clear the importance to pay attention to that in their planning. It's clear for 2.3 and 2.4 equations, where time scales for reach the local equilibrium in hundred nanometers makes the usual theory approach for gas-solid –specifically nitrogen – iron (steel) reactions inappropriate. Another point is the importance to take account of nanoscale objects in simulations of micro and milimeter scale process. The results of present work showed the importance of computer simulations, a mathematical and phenomenological approach for understanding technological problems.

6.References

[1] SUN, Y., BELL, T. A numerical model of plasma nitriding of low alloy steels. **Mater. Aci. Eng**, A A224 ,1997, p. 33-47.
[2] POLITANO,R, ROSSI, J, The nitring of ferrous powder alloys - mathematical modelling and simulation. **Mat. Science Forum** , v. 498/499,2004, p. 285-291
[3] FROMM, E., HÖRZ, G. Hydrogen, nitrogen, oxygen and carbon in metals. **Int. Met. Rev**..5/6 ,1980, p. 269-311.
[4] JACK, K.H., Nitriding, **Heat Treat**. ,1973, p. 39-50
[5] GRABKE, H. J. Adsorption, segregation and reactions of non-metal atoms on iron surfaces. **Mat. Sc. Eng**.,1980, v. 42 p. 91-99.
[6] TORCHANE, L., BILGER, P., DULCY, J., GANTOIS, M. Control of Iron nitride layers grow kinetics in the binary Fe-N system**. Metall. Mater. Trans. A** V. 27A.July,1996, p. 1823-1835
[7] BIGLARI, M.H., BRAKMAN, C.M., MITTEMEIJER, E.J., VAN DER ZWAAG, S., The Kinetics of internal nitriding of Fe-2 At. Pct Al alloy. ***Metall. Mater. Trans. A*** V. 26A.April,1995, p. 765-776.
[8] BORGES, C.F.M., HENNECKE, S., PFENDER, E. Decreasing chromium precipitation in AISI 304 stainless steel during the plasma-nitriding process. ***Surf. Coat. Technol***. V. 123 ,2000, p. 112-121.
[9] STRAVER, W.T.M., ROZENDAAL, H.C. F, MITTEMEIJER, E.J. Consequences of the heterogeneous nitriding of alfa-Iron: dislocation production and oriented precipitation. ***Metall. Trans.A*** V.15.April ,1984, p. 627-637.
[10] DAHMEN, U., FERGUSON, P., WESTMACOTT, K. H. A Tem study of Fe-16N2 and Fe4N precipitation in iron-nitrogen. **Acta Metall**. V. 35.5 ,1987, p. 1037-1046.
[11] USTINOVSHIKOV,Y., RUTS, A., BANNYYKH, O., BLIKOV, V. The microstructure of Fe-18%Cr alloys with high N contents. ***Acta Mater***. V. 44.3 1996, p. 1119-1125.
[12] K. KACZMARSKI, ANTOS D. Fast Finite Difference Method for Solving Multicomponent Adsorption-Chromatography models. **Computers chem. Eng**. V. 20, 1996, p.1271-1276.
[13] BONGARTZ, K, QUADAKKERS, W.J, SCHULTEN, R, NICKEL H. A Mathematical Model Describing Carburization in Multielement Alloy Systems. **Met. Trans. A**,1989, V. 20A. p. 1021-1028.
[14] BONGARTZ, K., LUPTON, D. F., SCHUSTER, H. A Model to predict carburization profiles in high temperature alloys. **Met. Trans. A**, November 1980, V.11A. p. 1883-1893.
[15] BELMONTE, T., GOUNÉ, M., MICHEL H,. Numerical modeling of interstitial diffusion in binary systems. Aplication to iron nitriding. **Materials Science & Engeenering A 302**, 2001, p. 246-257.
[16] AGREN, J., DU HONG. Theoretical treatment of nitriding and nitrocarburizing of Iron. **Metallurgical and Materials Transactions A.** April 1996, V. 27A. p. 1073-1080.
[17] DEHGHAN MEHDI. Fully implicit finite differences methods for two-dimensional diffusion

with a non-local boundary condition. **Journal of Computational and Applied mathematics 106**.,1999, P.255-269.

[18] HAASE, B.; DONG, J.; IRRETIER, O.; BAUCKHAGE, K. Influence of steel surface composition on gas nitriding mechanism **Surface Engineering**, Vol. 13, 1997 , pp. 251-256(6)

Numerical Simulations of Photon Echoes for both Space and Time using Density Matrix and Iterative Approach

Salman Abdullah and Byoung S. Ham*

Centre for Photon Information Processing and School of Electrical Engineering, Inha University, 253 Younghyun-Dong, Nam-gu, Incheon 402-751, South Korea

E-mail: *bham@inha.ac.kr

Abstract. We present a three-dimensional computational tool for atom-field interactions such as optically locked photon echoes in a three level system for both space and time axes, where time-dependent density matrix calculations are combined with iterative methods. We provide an exact numerical simulation tool without any approximations. To the best of our knowledge, this is the first computational tool for short-pulse light interactions with an optical medium, whose space dimension along the pulse propagation direction is simultaneously cared for density matrix calculations.

1. Introduction

Quantum Optical data storage using a reversible inhomogeneous broadening has been a main stream recently [1-3]. For a two level atomic system, following the photon echo generation, the echo intensity in a forward propagation scheme must suffer from reabsorption because of the remaining non interacted atom distribution along the propagation direction. This situation leads to degradation of the echo efficiency, where normally echo efficiency in rare-earth doped solids drops down to less than 1% [4]. According to Beer`s law, the optical depth dependent data pulse absorption [5] along the propagation direction is governed by an exponential decay of the coherence. For a data pulse whose initial intensity is I (0), the length dependent transmitted intensity is

$$I(x) = I(0)e^{-\alpha x} \tag{1}$$

Here $\propto$ is absorption coefficient. The exponential decay behavior governed by this relation will be used later on to combine the output from the density matrix with the iterative approach to transfer into space time domain. The optical coherence excited by the data pulse is obtained from the density matrix operator ρ obtained from the Schrodinger equation,

$$\frac{d\rho}{dt} = -\frac{i}{\hbar}[H,\rho] - \frac{1}{2}\{\Gamma,\rho\} \tag{2}$$

Where $\{\Gamma,\rho\} = \Gamma\rho + \rho\Gamma$ and $[H,\rho] = H\rho - \rho H$. H is Hamiltonian and ρ_{ij} is density matrix element. The initial condition for Eq. (2) is ρ_{11}=1, ρ_{22}=0 and ρ_{33}=0.

Using equation (2), we derive nine time dependent density matrix equations, where the following equations serve as the main coupled equations for time dependent numerical solutions and are solved using an explicit Runge-Kutta (4,5) solver (Ode45) in MATLAB [6].

$$\frac{d\rho_{13}}{dt} = i\Omega_{13}(\rho_{33} - \rho_{11}) - i\Omega_{23}\rho_{12} - i(\delta_1 + \gamma_{13})\,\rho_{13} \tag{3.a}$$

$$\frac{d\rho_{23}}{dt} = i\Omega_{23}(\rho_{33} - \rho_{22}) - i\Omega_{13}\rho_{21} - i(\gamma_{23} + \delta_1)\rho_{23} \tag{3.b}$$

$$\frac{d\rho_{12}}{dt} = i\Omega_{13}\rho_{32} - i\Omega_{23}\rho_{23} - i\Omega_{13}\rho_{21} - i(\delta_1 - \delta_2)\rho_{12} - \gamma_{12}\rho_{12} \tag{3.c}$$

Here, Ω_{13} is the Rabi frequency of the Data and the rephasing pulse, Ω_{23} is the Rabi frequency of the locking (unlocking) pulse B1 (B2) as shown in the energy level diagram in figure.1. δ_1 is the detuning of the atoms from the line center and γ_{ij} is the phase coherence decay rate for the corresponding transition. Equations (1) and (2) are solved together where the value I(0) is derived from the $\rho_{ij.\ (i=1,\ j=3)}$ depending on the population distribution, the Echo intensity, I(x) either increases or decays down. In order to get the corresponding density matrix entries, ρ_{ij}, in space and time we use the following model [4, 5]:

$$\rho_{ij}(x,t) = \rho_{ij}(0,t)\,e^{(\rho_{22}(x,t) - \rho_{11}(x,t)(L-x))} \tag{4}$$

According to equation (4), the Echo efficiency inside the medium depends on the population distribution. Following an echo generation, the non-interacted population distribution along a medium of length L, for a forward propagating echo scheme determines the resultant echo intensity. The initial parameter for the echo emission is the density matrix element of ρ_{13}. We iteratively solve each entry of the density matrix for all values of points in the space domain using equation (4).

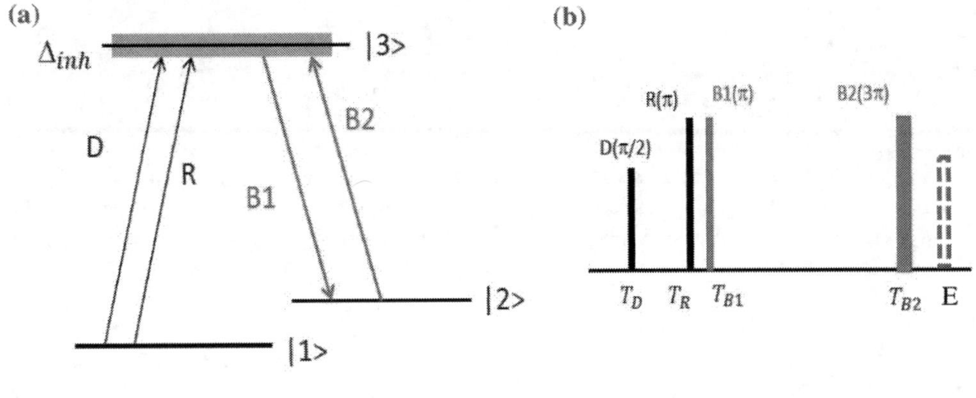

Figure 1 (a) Energy level diagram for the three level lambda system. The grey line on the excited state |3> shows Inhomogeneous broadening. The FWHM of the Gaussian distributed atomic ensemble is 680 kHz. (b) Programmed pulse sequence: T_D=4 μs. T_R=50 μs. T_{B1}=50.1 μs. T_{B2}=180 μs. Ω_D=392 kHz, Ω_R= 785 kHz. $\Omega_{B1} = \Omega_{B2} = \Omega_D \cdot \gamma_{13} = \gamma_{23} = \gamma_{12} = 0$. The pulse area is defined by $\int dt$. Duration of the Data pulse is 2μs. Duration of R, B1 and B2 is 200 ns each.

2. Results and Discussions

In this section we show numerical calculations for equations (3) according to the programmed pulse sequence applied to a three level lambda system as shown in figure 1. For better visualization, we use a maximum coherence excitation by the data pulse, i.e. $\pi/2$. The coherence excited by the Data pulse quickly decays at a speed determined by the width of inhomogeneous broadening. For our numerical simulations we divide 680 kHz inhomogeneously broadened atomic ensemble into 161 groups. The phase reversal as a result of the π rephasing pulse is shown in figure 2 (a), where there is an abrupt colour change. In figure 2(a), the coherence ($Im\rho_{13}$) for all groups of atoms is shown as a function of time and detuning. The white dashed line shows the time instant of photon echo generation.

Figure 2 (b) shows the coherence evolution as a function of space and time for the 170 kHz detuned group of atoms represented by the black dashed line in figure 2 (a). The speed of the coherence evolution for a particular group of atoms depends on the detuning from the line center. The coherence change for every group of atoms along the medium is governed by Eq. (4) where the remnant population distribution for the forward propagating photon echo determines intensity of the generated echo signal. The behavior of each group of atoms in space-time domain is shown in figure. 2 (b). The blue color shows maximum excitation by the data pulse. After the rephasing process is intiated by the rephasing pulse, the coherence for the particular group of atoms evloves along the time axis, and correpondingly changes along the space domain.

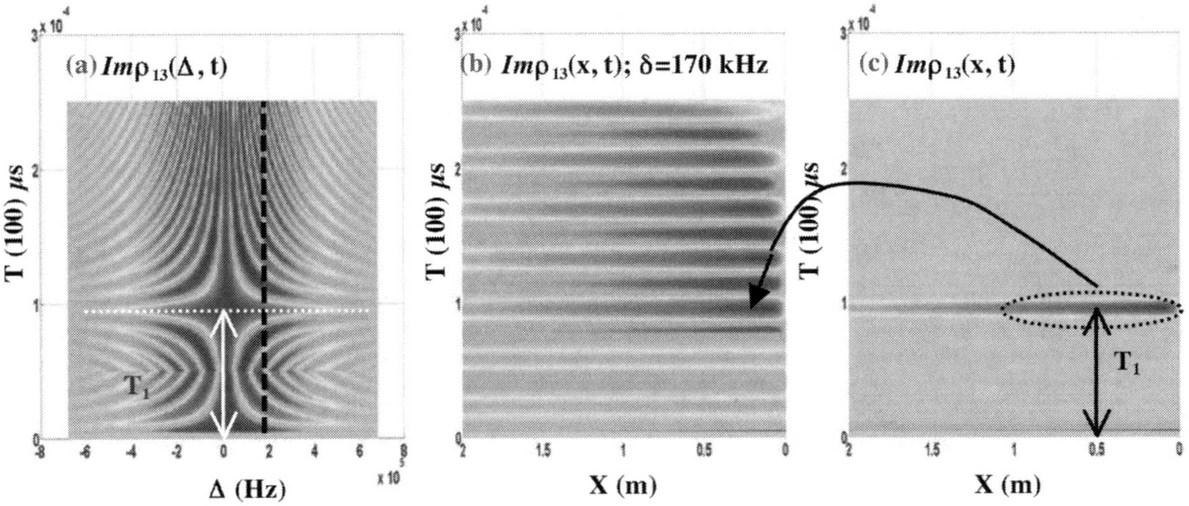

Figure 2 (a) Coherence evolution as a function of time and detuning δ. Rephasing pulse induced phase change is clear from the change of color for all groups of atoms. (b) Coherence evolution for a δ=170 kHz detuned atoms group as a function of space and time. The rephasing initiated phase reversal applies for all groups of atoms. (c) Coherence for all groups of atoms in space time domain. All parameters same as in figure 1.

To represent the entire Gaussian distributed atomic ensemble in space-time domain, we sum up all the groups of atoms along the detuning domain. The sum of all such groups is shown in figure 2 (c). the contribution from all other groups sum up to zero leaving out the echo signal as pointed by the dashed circle showing the contributing atoms correspondingly in figure 2 (b). The time between the two strips in figure 2 (c) marks the storage time, which, later on in the case of an optically locked echo will be extended. In figure 2 (c) we note an interesting phenomena, even we set the decay rates to zero (ideal efficiency) along the time axis, there is a decay along the x axis, which points to the fact that the remnant population distribution along the length of the medium changes, which leads to the degradation of the echo efficiency in the forward direction.

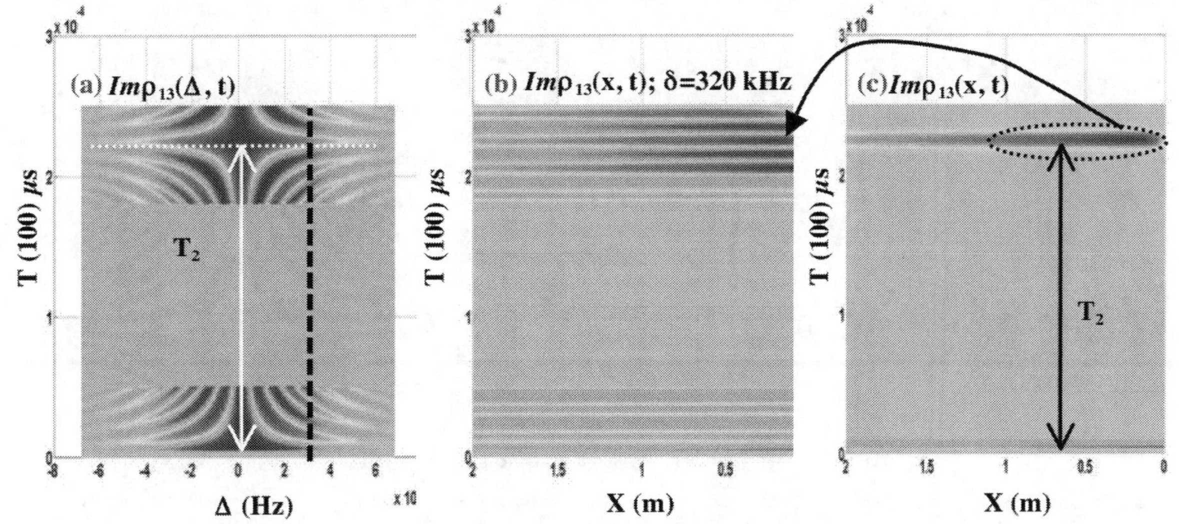

Figure 3 (a) Coherence evolution as a function of time and detuning for an optically locked echo. (b) Coherence evolution for a 320 kHz detuned atom group as a function of space and time. (c) Coherence for all groups of atoms for optically locked echo as a function of space and time.

Figure 3 show numerical calculations for optically locked photon echo [4]. Optical locking pulse pair follows the rephasing pulse in figure 3(a), where storage time extension is expected. The coherence evolution for a particular group of atoms in space and time also depends on the detuning from the line center. In figure 3 (a) the black dashed line indicates a particular group of atoms, detuned by 320 kHz, presented in space and time domain in figure 3 (b). We deliberately select a highly detuned atoms group in order to show the coherence evolution more clearly for space and time axes in figure 3 (b). The farther the atoms from the line center, faster is the speed of their phase evolution and more pronounced are the color changes in figure 3 (b). The sum of all such groups of atoms is shown in figure 3 (c), with the summing up of all groups of atoms explained in figure 2 (c).

The storage time extension as shown by T_2-T_1, is shown in figure 3(c). Because of the optical locking pulse pair, the coherence is locked to zero both in space and time domains. This means that during the locking interval there is no change in the population distribution as would otherwise be experienced by the generated echo signal in a non-optically locked echo case.

Acknowledgement
This work was supported by Creative Research Initive Program by the Korean government (MEST) and National Research Foundation. The authors gratefuly acknowledge fruitful discussions with Torsten Hennig from Fraunhofer UMSICHT, Germany.

References

[1] C. Clausen et al., "Quantum storage heralded polarizatioin qubits in birefringent and anisotropically absorbing material," Phys. Rev. Lett. **108**, 190503 (2012)
[2] M. P. Hedges, J. J. Longdell, and M. J. Sellars, "Efficient quantum memory for light," Nature **465**, 1052 (2010).
[3] B. S. Ham, "Ultralong quantum optical storage using reversible inhomogeneous spin ensembles", Nature Photonics **3**,518 (2009).
[4] J. Hahn and B. S. Ham, "Rephasing halted photon echoes using controlled optical deshelving," New J. Phys. **13**, 193011 (2011).
[5] B. S. Ham, "A contradictory phenomena of deshelving pulses in a dilute medium used for lengthening photon storage time", Optics Express, **18**, 17749 (2010).
[6] P. Bogacki, L.F. Shampine, "An efficient Runge-Kutta (4,5) pair", Computers Math Applic. **32**,15 (1996).

A gauged baby Skyrme model and a novel BPS bound

C Adam[1], C Naya[1], J Sanchez-Guillen[1] and A Wereszczynski[2]

[1] Departamento de Física de Partículas, Universidad de Santiago, and Instituto Galego de Física de Altas Enerxías (IGFAE) E-15782 Santiago de Compostela, Spain
[2]Institute of Physics, Jagiellonian University, Reymonta 4, Kraków, Poland

E-mail: adam@fpaxp1.usc.es, carlos.naya87@gmail.com, joaquin@fpaxp1.usc.es, wereszczynski@th.if.uj.edu.pl

Abstract. The baby Skyrme model is a well-known nonlinear field theory supporting topological solitons in two space dimensions. Its action functional consist of a potential term, a kinetic term quadratic in derivatives (the "nonlinear sigma model term") and the Skyrme term quartic in first derivatives. The limiting case of vanishing sigma model term (the so-called BPS baby Skyrme model) is known to support exact soliton solutions saturating a BPS bound which exists for this model. Further, the BPS model has infinitely many symmetries and conservation laws. Recently it was found that the gauged version of the BPS baby Skyrme model with gauge group U(1) and the usual Maxwell term, too, has a BPS bound and BPS solutions saturating this bound. This BPS bound is determined by a superpotential which has to obey a superpotential equation, in close analogy to the situation in supergravity. Further, the BPS bound and the corresponding BPS solitons only may exist for potentials such that the superpotential equation has a global solution. We also briefly describe some properties of soliton solutions.

1. Introduction

The Skyrme model was originally introduced by Skyrme as a novel way to describe nucleons and nuclei [1], and has ever since served both as a paradigmatic nonlinear field theory supporting topological solitons [2] and as a successful complementary approach to strong interaction physics [3]. Concretely, the primary fields in the Skyrme model are the Goldstone bosons of quantum chromodynamics (pions) taking values in the group manifold SU(2), whereas the nucleons and nuclei emerge as collective excitations of the former, i.e., topological solitons. Due to the complexity of the resulting nonlinear field theory, lower-dimensional theories containing similar topological solitons have been studied, too (e.g., the so-called baby Skyrme model in 2+1 dimensions [4]), both as toy models for the 3+1 dimensional theory and as independent objects of study with their proper applications, e.g., in condensed matter physics and in brane cosmology.

If the action is required to be Poincare invariant and to allow for a proper Hamiltonian formulation (no higher than second powers in time derivatives) then the possible Lagrangians for the Skyrme and baby Skyrme models are, in fact, quite restricted. For the Skyrme model with target space SU(2) it consists (in ascending order of powers of first derivatives) of a potential term, the "non-linear sigma model" term (quadratic), the Skyrme term (quartic in first derivatives) and a "topological" term (the square of the topological or baryon current, sextic in first derivatives). For the baby Skyrme model with target space S^2 (two-sphere) we have a potential term, the quadratic non-linear sigma term, and a quartic term which is, at the same time, topological (the square of the topological current). Topological soliton solutions have been found numerically both for the full models and for various restrictions obtained by suppressing some of the terms described above.

It has been found recently that the specific restrictions to the potential and topological terms only (the so-called BPS Skyrme [5] and BPS baby Skyrme models [6]) possess both a BPS bound and exact soliton solutions saturating this bound (the full Skyrme models have BPS bounds, too, but solitons in general do not saturate these bounds). In addition, these BPS submodels exhibit the property of generalized integrability and have, therefore, infinitely many symmetries and conservation laws.

A last relevant piece of information consists in the fact that both the Skyrme [7] and baby Skyrme models [8] may be coupled to a U(1) gauge field (i.e., to electromagnetism) in a natural way, and that soliton solutions of these gauged Skyrme [9] and baby Skyrme models [10] have been found numerically. This leads to the natural question whether the electromagnetically gauged versions of the BPS submodels share some of the nontrivial properties (saturation of BPS bounds, infinitely many symmetries and conservation laws) with the ungauged theories. It is the purpose of these notes to briefly demonstrate that for the BPS baby Skyrme model this is indeed the case. Specifically, we want to describe a rather novel, nontrivial BPS bound (for more details and a more exhaustive list of references we refer to [11]).

2. The gauged BPS baby Skyrme model

2.1. Baby Skyrme models

The fields of the baby Skyrme model form a three-component vector $\vec{\phi} = (\phi_1, \phi_2, \phi_3)$ taking values in the unit two-sphere S^2, $\vec{\phi}^2 = 1$. The corresponding baby Skyrme Lagrangian is

$$L = E_0 \int d^2x \left(-\mu^2 V(\vec{\phi}) + \frac{\nu^2}{2}(\partial_\alpha \vec{\phi})^2 - \frac{\lambda^2}{8} K_\alpha^2 \right) \tag{1}$$

where E_0 is a constant with the dimension of energy, and (μ, ν, λ) are constants with the dimensions of (l^{-1}, l^0, l^1), respectively (l stands for "length"). Here, the first term on the r.h.s. is the potential term, the second one the sigma model term, and the third term is the square of the topological current $K^\alpha = \epsilon^{\alpha\beta\gamma}\vec{\phi}\cdot(\partial_\beta \phi \times \partial_\gamma \vec{\phi})$. Finite energy field configurations are required to approach a unique value $\vec{\phi}_0$ at spatial infinity (where necessarily the potential must obey $V(\vec{\phi}_0) = 0$), therefore the base space $\mathbb{R}^2$ is effectively compactified to a S^2, and finite energy field configurations may be classified by an integer winding number $k = (1/8\pi)\int d^2x K^0$. For the potential we now assume that it only depends on ϕ_3 and that its unique vacuum is at $\vec{\phi}_0 = (0,0,1)$, i.e., $V = V(\phi_3)$, $V(\phi_3 = 1) = 0$. This implies that the potential and, therefore, the full Lagrangian, is still invariant under rotations of the field vector $\vec{\phi}$ about the three axis (i.e. rotations in the ϕ_1-ϕ_2 plane). A natural coupling to electromagnetism consists in gauging this residual U(1) symmetry. To achieve this, partial derivatives have to be replaced by covariant derivatives $D_\alpha\vec{\phi} \equiv \partial_\alpha\vec{\phi} + A_\alpha\vec{\phi}_0 \times \vec{\phi}$ (where A_α is the gauge potential, and $F_{\alpha\beta} \equiv \partial_\alpha A_\beta - \partial_\beta A_\alpha$), and the standard Maxwell term has to be added. The resulting Lagrangian is

$$L = E_0 \int d^2x \left(-\mu^2 V(\phi_3) + \frac{\nu^2}{2}(D_\alpha \vec{\phi})^2 - \frac{\lambda^2}{8} \tilde{K}_\alpha^2 - \frac{1}{4g^2} F_{\alpha\beta}^2 \right) \tag{2}$$

where $\tilde{K}^\alpha = \epsilon^{\alpha\beta\gamma}\vec{\phi}\cdot(D_\beta\phi \times D_\gamma\vec{\phi}) = K^\alpha + 2\epsilon^{\alpha\beta\gamma}A_\beta\partial_\gamma(\vec{\phi}_0\cdot\vec{\phi})$.

Journal of Physics: Conference Series **410** (2013) 012055 doi:10.1088/1742-6596/410/1/012055

2.2. BPS bounds

From now on we restrict to the BPS submodels, i.e., we set $\nu = 0$ in the above Lagrangians. The energy functional of the ungauged BPS submodel is $(q \equiv (1/2)K_0)$
$E = \frac{1}{2}E_0 \int d^2x \left[\lambda^2 q^2 + 2\mu^2 V(\phi_3)\right]$, and obeys the BPS bound

$$E = \frac{1}{2}E_0 \int d^2x \left(\lambda q \pm \mu\sqrt{2V}\right)^2 \mp E_0\lambda\mu \int d^2xq\sqrt{2V} \geq \mp E_0\lambda\mu \int d^2xq\sqrt{2V} \tag{3}$$

with equality if the BPS equation $\lambda q \pm \mu\sqrt{2V} = 0$ is satisfied. The bound is genuinely topological (i.e., for a given potential only depends on the winding number k), because the two-form d^2xq is the pullback (under $\vec{\phi}$) of the target space area two-form $d\Omega$, and, therefore, $d^2xq\sqrt{2V(\phi_3)}$, too, is the pullback of a two-form on target space [12], whose integral results in

$$\int d^2xq\sqrt{2V(\phi_3)} = 4\pi k \langle\sqrt{2V}\rangle_{S^2}\ , \quad \langle\sqrt{2V}\rangle_{S^2} \equiv \frac{\int d\Omega\sqrt{2V}}{\int d\Omega}. \tag{4}$$

The static energy of the gauged BPS submodel is $(Q \equiv (1/2)\tilde{K}_0 = q + \epsilon_{ij}A_i\partial_j\phi_3)$

$$E = \frac{1}{2}E_0 \int d^2x \left[\lambda^2 Q^2 + 2\mu^2 V(\phi_3) + \frac{1}{g^2}B^2\right] \tag{5}$$

where B is the magnetic field, and we used the known fact [10] that for static finite energy solutions of the gauged baby Skyrme model the electric field must be zero. To find the BPS bound, we now start with a non-negative expression and try to re-express it as the energy minus a topological term, such that the topological term provides the bound. Concretely, we start from

$$0 \leq \frac{1}{2}E_0 \int d^2x \left[\lambda^2(Q - W'(\phi_3))^2 + \frac{1}{g^2}(B + g^2\lambda^2 W(\phi_3))^2\right]. \tag{6}$$

Expanding the squares, we get for the mixed terms $-2\lambda^2(q + \epsilon_{ij}A_i\partial_j\phi_3)W' + 2\lambda^2 W\epsilon_{ij}\partial_i A_j = -2\lambda^2 qW' + 2\lambda^2\epsilon_{ij}\partial_j(A_iW)$ and, therefore, a total derivative (which integrates to zero) plus a topological term. The remaining contributions from the squares produce the energy provided that the function $W(\phi_3)$ (which has not yet be determined) obeys the first order ODE

$$\lambda^2 W'^2 + g^2\lambda^4 W^2 = 2\mu^2 V. \tag{7}$$

Assuming this we find the BPS bound $E \geq E_0\lambda^2 \int d^2xqW' = 4\pi E_0\lambda^2 k\langle W'\rangle_{S^2}$, with equality if the two (first order) BPS equations $Q - W' = 0$ and $B + g^2\lambda^2 W = 0$ hold.

2.3. Remarks

- Equation (7) is well-known in supergravity where it provides the relation between potential and superpotential, so we call W the "superpotential" and eq. (7) the "superpotential equation".
- Eq. (7) is of first order and seems to have a one-parameter family of solutions. This is, however, not the case. At the vacuum value $\vec{\phi} = \vec{\phi}_0$, i.e., $\phi_3 = 1$ (where $V(1) = 0$), $W(\phi_3)$ must take the unique value $W(1) = 0$. The solution is, therefore, unique near the vacuum. The global existence of this solution in the whole interval $1 \geq \phi_3 \geq -1$ is a different and rather nontrivial problem.
- It can be proved that the BPS equations imply the static second order field equations. The existence of nontrivial BPS solitons solutions can therefore be expected.

- The global existence of the superpotential is a necessary condition for the existence of BPS soliton solutions. For some potentials like, e.g., $V = 1 - \phi_3$ or $V = (1 - \phi_3)^2$ (and probably for the whole family $V = (1 - \phi_3)^a$, $a \geq 1$) the superpotential exists globally, whereas for others, like the two-vacua potential $V = 1 - \phi_3^2$, it doesn't.
- In all cases where soliton solutions have been calculated numerically, their energies saturate the BPS bound, and these solitons are, therefore, BPS solutions. For potentials without a globally existing superpotential, numerical soliton solutions have not been found. This motivates the conjecture that all solitons are, in fact, BPS solitons.
- For some potentials, exact superpotentials and exact BPS solitons can be calculated.
- The gauged BPS baby Skyrme model shares all the infinitely many symmetries and conservation laws with its ungauged version.
- For more details, see [11].

2.4. Open problems and further research

- For wich potentials the superpotential exists globally? Which further conditions the potential must satisfy such that BPS solitons exist?
- Can the present construction be generalized to the gauged BPS Skyrme model in 3+1 dimensions?
- Can the gauged (BPS) baby Skyrme model be supersymmetrized and can the BPS bound we found be related to supersymmetry (we remark that the baby Skyrme model can be supersymmetrized [13])?
- Can the novel BPS bound discussed in these notes be used for other field theories? Etc.

Acknowledgments

The authors acknowledge financial support from the Ministry of Education, Culture and Sports, Spain (grant FPA2008-01177), the Xunta de Galicia (grant INCITE09.296.035PR and Conselleria de Educacion), the Spanish Consolider-Ingenio 2010 Programme CPAN (CSD2007-00042), and FEDER. CN thanks the Spanish Ministery of Education, Culture and Sports for financial support (grant FPU AP2010-5772). Further, AW was supported by polish NCN grant 2011/01/B/ST2/00464.

References

[1] T.H.R. Skyrme, Proc. Roy. Soc. Lon. **260**, 127 (1961); Nucl. Phys. **31**, 556 (1962).
[2] R.A. Battye, P.M. Sutcliffe, Phys. Rev. Lett. **79**, 363 (1997); Phys. Rev. Lett. **86**, 3989 (2001).
[3] G. Adkins, C. Nappi, E. Witten, Nucl. Phys. B**228**, 552 (1983); G. Adkins, C. Nappi, Nucl. Phys. B**233**, 109 (1984).
[4] B.M.A.G. Piette, B.J. Schoers and W.J. Zakrzewski, Z. Phys. C **65** (1995) 165; Nucl. Phys. B **439** (1995) 205.
[5] C. Adam, J. Sanchez-Guillen, A. Wereszczynski, Phys. Lett. B**691**, 105 (2010); Phys. Rev. D**82**, 085015 (2010).
[6] C. Adam, T. Romanczukiewicz, J. Sanchez-Guillen, A. Wereszczynski, Phys. Rev. D**81**, 085007 (2010).
[7] E. Witten, Nucl. Phys. B**223**, 422 (1983); Nucl. Phys. B**233**, 433 (1983); C.G. Callan, E. Witten, Nucl. Phys. B**239**, 161 (1984).
[8] B.J. Schroers, Phys. Lett. B**356** 291, 1995.
[9] B.M.A.G. Piette, D.H. Tchrakian, Phys. Rev. D**62**, 025020 (2000).
[10] J. Gladikowski, B.M.A.G. Piette, B.J. Schroers, Phys. Rev. D**53** 844, 1996.
[11] C. Adam, C. Naya, J. Sanchez-Guillen, A. Wereszczynski, arXiv:1205.1532; to be published in PRD.
[12] J.M. Speight, J. Phys. A**43**, 405201 (2010).
[13] C. Adam, M. Queiruga, J. Sanchez Guillen, A. Wereszczynski, Phys. Rev. D**84**, 025008 (2011).

Journal of Physics: Conference Series **410** (2013) 012056 doi:10.1088/1742-6596/410/1/012056

A Finsler geometric model for membranes on triangulated surfaces

Hiroshi Koibuchi

Ibaraki National College of Technology, Nakane 866, Hitachinaka, Ibaraki 312-8508, Japan

E-mail: koibuchi@mech.ibaraki-ct.ac.jp

Abstract. In this paper, we study a new surface model for anisotropic membranes. The model is characterized by an interaction between the surface and a three-dimensional tilt order, of which projected components on the surface are recognized as a vector filed. This vector field defines a Finsler metric on the surface, and consequently the length unit on the surface becomes dependent on the position and direction. The induced interaction dynamically alters the surface tension and the bending rigidity. We numerically find that the tilts form the Kosterlitz-Thouless and low temperature configurations, which correspond to two different anisotropic phases such as disk and tubular on both of the connection-fixed and dynamically triangulated surfaces. The internal tilt order plays an important role for those anisotropic configurations.

1. Introduction

To construct a surface model, we usually assume Helfrich-Polyakov (HP) Hamiltonian, which is composed of the bond potential S_1 and a curvature energy S_2 [1, 2]. Since the HP Hamiltonian is symmetric under three-dimensional rotations, the symmetric shape transformations including the crumpling transition have long been studied in the framework of HP theory [3, 4]. To the contrary, there is a lot of rotationally non-symmetric phases in membranes, such as tubular and disk phases. The prolate-oblate transition is a shape transformation between two-different non-symmetric phases [6]. Those non-symmetric states have been studied with a model called the area-difference bilayer one [7]. An external flow field can also be an origin of non-symmetric surface states [8]. Hexatic membrane model is also for an interaction between the surface and internal tilts order [9]. However, the geometric origin of the non-symmetric states still remains to be studied. In this paper, we show that a Finsler geometric (FG) surface model has three different phases including the disk and tubular phases at high bending region. The FG model is simply obtained by replacing the Riemannian metric with a Finsler metric in the HP model [10].

2. Model

The model is defined by a mapping $X : M \to \mathbf{R}^3$ from a two-dimensional surface M to $\mathbf{R}^3$ (Fig. 1(a)). A smooth triangulation of M corresponds to a linear triangulation in $\mathbf{R}^3$. The triangulated surface in $\mathbf{R}^3$ is obtained by splitting the faces of icosahedron, and it is characterized by the three numbers (N, N_B, N_T), which are the total number of vertices, the total number of bonds, and the total number of triangles.

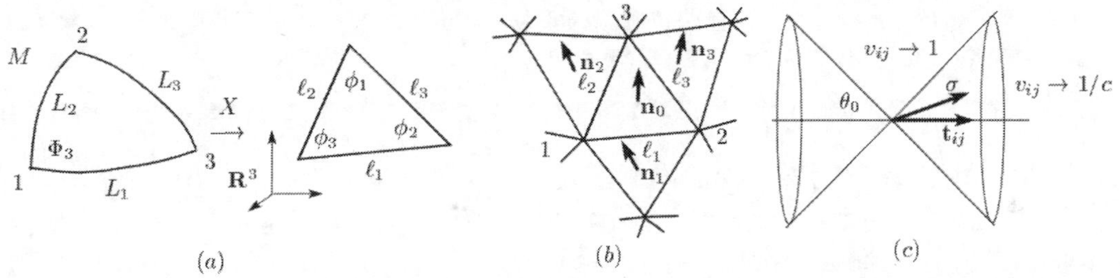

Figure 1. (a) A mapping $X : M \to \mathbf{R}^3$, (b) a triangle 123 and the three neighboring triangles in $\mathbf{R}^3$, and (c) a vector v_{ij} along $\mathbf{t}_{ij}$ is defined by the tilt σ_i such that $v_{ij} = 1/c$ (or $v_{ij} = 1$) if $|\sigma_i \cdot \mathbf{t}_{ij}| \geq \cos\theta_0$ (or $|\sigma_i \cdot \mathbf{t}_{ij}| < \cos\theta_0$).

The Hamiltonian is given by

$$S(X, \sigma) = \lambda S_0 + S_1 + bS_2, \tag{1}$$

where S_0 is an energy for the directors (or tilts) $\sigma_i(\in \mathbf{S}^2)$. The tilt is defined at the vertex i and represents some external degrees of freedom.

S_1 the Gaussian bond potential, and S_2 the bending energy. These energies are defined as follows:

$$\begin{aligned} &S_0 = \sum_{ij} (1 - \sigma_i \cdot \sigma_j), \quad S_1 = \sum_{\Delta} S_{1\Delta}, \quad S_2 = \sum_{\Delta} S_{2\Delta}, \\ &S_{1\Delta} = \frac{1}{6}\left(a_1\ell_1^2 + a_2\ell_2^2 + a_3\ell_3^2\right), \\ &S_{2\Delta} = \frac{1}{6}\left[b_1(1 - \mathbf{n}_0 \cdot \mathbf{n}_1) + b_2(1 - \mathbf{n}_0 \cdot \mathbf{n}_2) + b_3(1 - \mathbf{n}_0 \cdot \mathbf{n}_3)\right], \end{aligned} \tag{2}$$

where $\ell_i(i = 1, 2, 3)$ is the bond length of the triangles Δ, and $\mathbf{n}_i(i = 0, 1, 2, 3)$ denote the unit normal vectors of triangles (Fig. 1(b)). $\sum_{\Delta}$ denotes the sum over all Δ.

The coefficients a_i and b_i in Eq. (2) for model 2 are defined by

$$\begin{aligned} &a_1 = v_{13}^{-1} + v_{23}^{-1}, \quad a_2 = v_{12}^{-1} + v_{32}^{-1}, \quad a_3 = v_{21}^{-1} + v_{31}^{-1}, \\ &b_1 = v_{12}^{-1} + v_{21}^{-1}, \quad b_2 = v_{13}^{-1} + v_{31}^{-1}, \quad b_3 = v_{23}^{-1} + v_{32}^{-1}, \\ &v_{ij}^{-1} = \begin{cases} 1 & (|\sigma_i \cdot \mathbf{t}_{ij}| < \cos\theta_0), \\ c\ (< 1) & (|\sigma_i \cdot \mathbf{t}_{ij}| \geq \cos\theta_0). \end{cases} \end{aligned} \tag{3}$$

This definition implies that a_i and b_i have values in $\{2, 2c, 1{+}c\}$. If all σ_i satisfy $|\sigma_i \cdot \mathbf{t}_{ij}| < \cos\theta_0$, then we have $a_i = b_i = 2$ and therefore S_1 and S_2 reduces to the ordinary ones such as $S_1 = \sum_{ij} \ell_{ij}^2$, $S_2 = \sum_{ij} (1 - \mathbf{n}_i \cdot \mathbf{n}_j)$ up to a multiplicative factor 2/3 in both models. We should note that a_i becomes an effective surface tension while bb_i becomes an effective bending rigidity at the bond i. Thus, the surface tension and bending rigidity dynamically changes depending on the position and the direction on the surface.

The partition function $Z_{\rm fix(flu)}$ for the connection-fixed (or fluid) surface model is given by

$$Z_{\rm fix(flu)}(\lambda, b) = \left(\sum_{\mathcal{T}}\right) \sum_{\sigma} \int' \prod_{i=1}^{N} dX_i \exp\left[-S(X, \sigma)\right], \tag{4}$$

where $\int' \prod_{i=1}^{N} dX_i$ is the multiple three-dimensional integrations, which are performed by fixing the center of mass of the surface to the origin of $\mathbf{R}^3$. The $\sum_{\sigma}$ denotes the sum over tilt variables, and $\sum_{\mathcal{T}}$ denotes the sum over all possible triangulation $\mathcal{T}$. By eliminating $\sum_{\mathcal{T}}$ from $Z_{\rm flu}$, we have $Z_{\rm fix}(\lambda, b)$ for the connection-fixed surface model.

The assumed Finsler metric g_{ab} is given by $g_{ab} = \begin{pmatrix} v_{12}^{-2} & 0 \\ 0 & v_{13}^{-2} \end{pmatrix}$. This metric comes from the Finsler function $L(x(t), y(t)) = \sqrt{\sum_i y_i^2}/|\mathbf{v}|$, where $x=(x_1, x_2)$ and $y=(y_1, y_2)=(dx_1/dt, dx_2/dt)$ are a point and a tangential vector at x on an edge of triangle in M, and $\mathbf{v}$ is a vector along the edge [10]. This $\sqrt{\sum_i y_i^2}/|\mathbf{v}|$ corresponds to v_{12}^{-1} or v_{13}^{-1}.

Both of the discrete Hamiltonians S_1 and S_2 in Eq. (2) with the coefficients in Eq. (3) are obtained from the continuous Hamiltonians

$$S_1 = \int \sqrt{g} d^2 x g^{ab} \partial_a X^\mu \partial_b X^\mu, \quad S_2 = \frac{1}{2} \int \sqrt{g} d^2 x g^{ab} \partial_a n^\mu \partial_b n^\mu. \tag{5}$$

Indeed, the partial derivatives in S_1 are replaced as $\partial_1 X^\mu \to X_2^\mu - X_1^\mu$, $\partial_2 X^\mu \to X_3^\mu - X_1^\mu$, and those in S_2 are $\partial_1 n^\mu \to \mathbf{n}_0 - \mathbf{n}_2$, $\partial_2 n^\mu \to \mathbf{n}_0 - \mathbf{n}_1$, where $\mathbf{n}_i (i=0,1,2,3)$ are shown in Fig. 1(a). The integration symbol $\int \sqrt{g} d^2 x$ is simply replaced by $\sum_\Delta 1$, then we have a_i and b_i in Eq. (3). It is also possible to use $\sum_\Delta v_{12}^{-1} v_{13}^{-1}$ for a discretization of $\int \sqrt{g} d^2 x$. Including two additional terms, which are cyclic under permutations of vertex positions $1 \to 2, 2 \to 3, 3 \to 1$, and multiplying a factor 1/3 to them, we have the discretization of a_i and b_i in Eq. (3).

3 Monte Carlo results

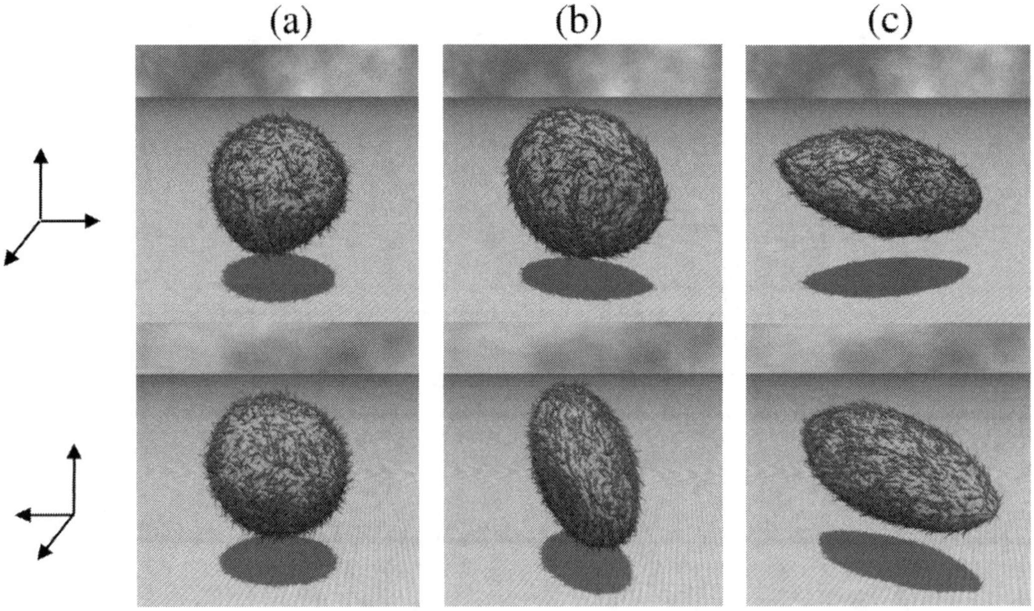

Figure 2. Snapshots obtained at (a) $\lambda=0.56$ (spherical), (b) $\lambda=0.68$ (disk), (c) $\lambda=1$ (tubular). $N=2562$, $b=320$, and $(c, \theta_0)=(0.2, \pi/4)$ on connection-fixed surfaces. The view point of the lower snapshot is rotated $\pi/2$ around the vertical axis of the upper one in (a)–(c).

Snapshots of connection-fixed surfaces of size $N=2562$ are shown in Figs. 1(a)–(c). We see that there are three different phases; spherical phase, disk phase and tubular phase. Brushes on the surface denote the tilts. The directions of tilts are at random in the spherical phase (Fig. 1(a)), and aligned in the tubular phase (Fig. 1(c)). In the disk phase in Fig. 1(b), the tilts form a vortex-like configuration just like the Kosterlitz-Thouless phase in the two-dimensional Heisenberg spin model.

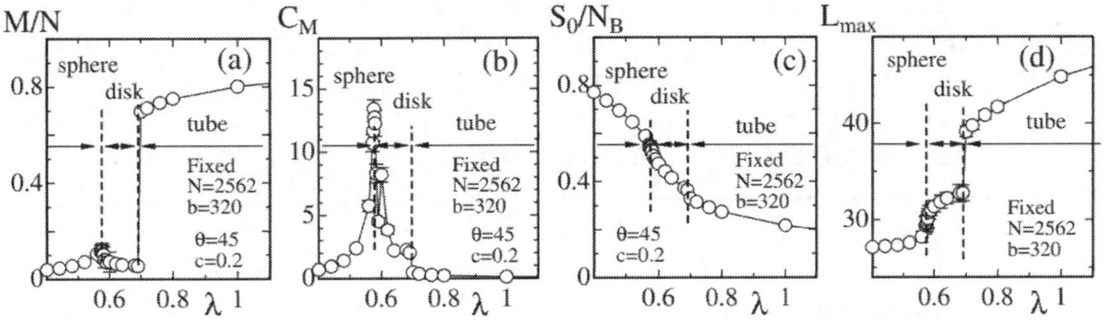

Figure 3. (a) M/N vs. λ, (b) C_M vs. λ, (c) S_0/N_B vs. λ, and (d) the maximal linear extension $L_{\rm max}$ vs. λ. $N\!=\!2562$ of connection-fixed surface, $b\!=\!320$, and $(c,\theta_0)\!=\!(0.2, 45^\circ)$.

The magnetization M defined by $M = |\sum_i \sigma_i| = |(\sum_i \sigma_i^x, \sum_i \sigma_i^y, \sum_i \sigma_i^z)|$ discontinuously changes against λ at the phase boundary between the disk and tubular phases as shown in Fig.3(a). This indicates that the tilt configuration is reflected in the surface shape. The variance $C_M = (1/N)\langle (M - \langle M \rangle)^2 \rangle$ has a peak at the boundary between the disk and spherical phases (Fig. 3(b)). The correlation energy S_0/N_B rapidly changes at the same boundary, and this implies a phase transition (Fig. 3(c)). Indeed, the specific heat $C_{S_0} = (\lambda^2/N)\langle (S_0 - \langle S_0 \rangle)^2 \rangle$ has a peak at the transition. The maximal linear extension $L_{\rm max}$ of the surface changes discontinuously (almost discontinuously) at the boundary between the disk and tubular (spherical) phases (Fig. 3(d)). Thus, we confirm that an internal in-plane or external tilt order is a possible origin of anisotropy. We should note that the fluid surface model on dynamically triangulated surfaces has almost the same phase structure as the connection-fixed model.

4. Summary and conclusion

By performing Monte Carlo simulations, we have demonstrated that a Finsler geometric model has a variety of phases including anisotropic ones such as disk and tubular. The anisotropic surfaces emerge depending on the configurations of internal tilt order, which makes the surface tension and bending rigidity anisotropic. This allows us to understand that the internal tilt order is an origin of surface anisotropy. The interaction between the tilt order and surface should be studied more extensively. The author acknowledges Hideo Sekino for discussions, and he also acknowledges Hiroki Mizuno for computer analyses.

References

[1] W. Helfrich, Z. Naturforsch **28c** (1973) 693.
[2] A.M. Polyakov, Nucl. Phys. B **268** (1986) 406.
[3] K.J. Wiese, *Phase Transitions and Critical Phenomena 19*, C. Domb, and J.L. Lebowitz Eds. (Academic Press, 2000) p.253.
[4] M. Bowick and A. Travesset, Phys. Rep. **344** (2001) 255.
[5] L. Radzihovsky, in Statistical Mechanics of Membranes and Surfaces, Second Edition, edited by D. Nelson, T. Piran, and S. Weinberg, (World Scientific, 2004) p.275.
[6] H.-G. Döbereiner and U. Seifert, Europhys. Lett. **36**, 325 (1996).
[7] Ling Miao, Udo Seifert, Michael Wortis, and Hans-Günther Döbereiner, Phys. Rev. E **49** (1998) 5389.
[8] Martin Kraus, Wolfgang Wintz, Udo Seifert, and Reinhard Lipowsky, Phys. Rev. Lett. **77** (1994) 3685.
[9] D.R. Nelson and L. Peliti, J. Physique **48** (1987) 1085.
[10] H. Koibuchi and H. Sekino, "Monte Carlo studies of a Finsler geometric surface model", http://arxiv.org/abs/1208.1806.

On a family of linear recurrences

C M Wilmott

Faculty of Informatics, Masaryk University, Šumavská 416/15, 602 00 Brno, Czech Republic

E-mail: wilmott@fi.muni.cz

Abstract. We concern ourselves with the family of linear recurrence relations $a_j = a_{j-1}+a_{j-d}$ with the initial conditions $a_0 = \ldots = a_{d-1} = 1$. We discuss the periodicity evaluation of such recurrences for prime powers d, and demonstrate that a key feature of our evaluation method relates to an instance of Shor's algorithm for factoring. As an application, we discuss how efficient quantum circuit designs may be completely recast as a problem relating to linear recurrence relations.

1. Introduction

Recurrence relations possess a rich history with applications extending from biology to economics. Indeed, many algorithms have time complexities that can be modelled by recurrence relations. In this paper, we will discuss a family of dth-order linear recurrence relations before revealing a novel application to quantum circuit designs. Our main result focuses on the task of evaluating the periodicity of a family of recurrence relations.

Let d be a positive integer and consider the recurrence relation

$$a_j = a_{j-1} + a_{j-d} + [j = 0] \pmod d \tag{1}$$

where $[j = 0]$ adds 1 when $j = 0$. The solution to this recurrence relation is given by the binomial summation

$$a_j = \sum_{i=0}^{j/d} \binom{j-(d-1)i}{i}, \tag{2}$$

and, furthermore, its closed-form is given by

$$a_j \quad = \quad \sum_{l=1}^{d} \beta_l \alpha_l^j, \tag{3}$$

where α_l denote the reciprocals of the roots of $B(z) = 1 - z - z^d$ and $\beta_l = -\alpha_l/B'(1/\alpha_l)$ [1]. For d prime, the periodicity of the recurrence relation is $d^2 - 1$. For the prime power case $d = p^m$ with p prime, the periodicity of this recurrence relation remains an open problem, however, it is conjectured to be $p^{m-1}(p^{2m} - 1)$ [1].

Interestingly, Wilmott [1] established that the periodic property of recurrence relations can be used to study the computational problem associated with the construction of certain quantum circuits possessing an exact number of constituent gates; in particular, quantum SWAP gates with a minimal number of CNOT gates. Crucially, however, the answer to this question directly depends on knowing the periodicity for the family of recurrence relations presented in Eq. (1). We now present some preliminary material before progressing to our main result.

2. Recurrence relations and Shor's Algorithm

Consider a d-dimensional complex Hilbert space $\mathbb{C}^d$ and fix each orthonormal basis state of the d-dimensional space to correspond to an element of ring $\mathbb{Z}_d$ of integers modulo d. The basis elements $\{|0\rangle, |1\rangle, \ldots, |d-1\rangle\} \subset \mathbb{C}^d$ correspond to the column vectors of the identity matrix $\mathbb{I}_d$ and is called the computational basis. A qudit is then a d-dimensional quantum state $|\psi\rangle \in \mathbb{C}^d$ written $|\psi\rangle = \sum_{i=0}^{d-1} \alpha_i |i\rangle$ where $\alpha_i \in \mathbb{C}$ and $\sum_{i=0}^{d-1} |\alpha_i|^2 = 1$. An n-qudit state is the tensor product of the basis states of the single system $\mathbb{C}^d$; $(\mathbb{C}^d)^{\otimes n}$, and its basis states are $|i_1\rangle \otimes |i_2\rangle \otimes \ldots \otimes |i_n\rangle = |i_1 i_2 \ldots i_n\rangle$ where $i_j \in \mathbb{Z}_d$. Further, the CNOT gate is a two-qudit quantum gate whose action on the basis states $|i\rangle \otimes |j\rangle \in \mathbb{C}^d \otimes \mathbb{C}^d$ is given by

$$\text{CNOT}\,|i\rangle \otimes |j\rangle = |i\rangle \otimes |j \oplus i\rangle, \qquad i, j \in \mathbb{Z}_d, \tag{4}$$

with $\oplus$ denoting addition modulo d. We will now outline how Shor's algorithm [2] can be used to confirm the conjecture relating the period of the recurrence relation presented in Eq. (1).

Consider an initial state written as a composition of two registers;

$$|\Psi_0\rangle = |0\rangle\,|1, 1, \ldots, 1\rangle. \tag{5}$$

The first register is prepared in the zero state while the second register is prepared as a collection of d one states. For $d = p^m$, let $r = p^{m-1}(p^{2m} - 1)$ be the period of the recurrence relation in Eq. (1), and choose an $n \in \mathbb{N}$ such that r divides d^n. Placing the first register in a uniform superposition of states representing integers a mod d^n, we have

$$|\Psi_1\rangle = \frac{1}{\sqrt{d^n}} \sum_{a=0}^{d^n-1} |a\rangle\,|1, 1, \ldots, 1\rangle. \tag{6}$$

Focusing on the second register of Eq. (6), let us define a set of states $|\mathrm{u}_s\rangle$ by

$$|\mathrm{u}_s\rangle := \frac{1}{\sqrt{r}} \sum_{k=0}^{r-1} e^{-2\pi\imath ks/r} \left| \sum_{l=1}^{d} \beta_l \alpha_l^k, \sum_{l=1}^{d} \beta_l \alpha_l^{k+1}, \ldots, \sum_{l=1}^{d} \beta_l \alpha_l^{k+d-1} \pmod d \right\rangle \tag{7}$$

with $0 \le s < r$. Following the style style presented in Nielsen & Chuang [3], we exploit the fact that a suitable combination of such states $|\mathrm{u}_s\rangle$ yields the state of the second register. In particular,

$$\begin{aligned} \frac{1}{\sqrt{r}} \sum_{s=0}^{r-1} |\mathrm{u}_s\rangle &= \frac{1}{r} \sum_{s=0}^{r-1} \sum_{k=0}^{r-1} e^{-2\pi\imath ks/r} \left| \sum_{l=1}^{d} \beta_l \alpha_l^k, \sum_{l=1}^{d} \beta_l \alpha_l^{k+1}, \ldots, \sum_{l=1}^{d} \beta_l \alpha_l^{k+d-1} \pmod d \right\rangle \\ &= \sum_{k=0}^{r-1} \delta_{k,0} \left| \sum_{l=1}^{d} \beta_l \alpha_l^k, \sum_{l=1}^{d} \beta_l \alpha_l^{k+1}, \ldots, \sum_{l=1}^{d} \beta_l \alpha_l^{k+d-1} \pmod d \right\rangle \\ &= |1, 1, \ldots, 1\rangle. \end{aligned} \tag{8}$$

Next, we define a unitary transformation $\Lambda(U) := I \otimes U^a$ which acts according to the circuit of Fig. 1. The unitary transformation $\Lambda(U)$ is a controlled-U^a operation and acts on the state of

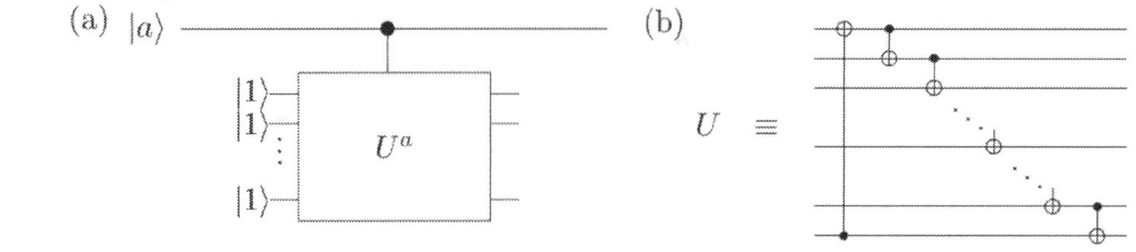

Figure 1. (a) The unitary transformation $\Lambda(U)$ is represented as a controlled-U^a operation. When the state of the first register is $|a\rangle$, an a-fold product of U is applied to the second register. (b) The transformation U consist of d CNOT gates which produce clock cycles of length d of the recurrence relation given in Eq. (1).

the second register. The effect of $\Lambda(U)$ is to produce a set of disjoint cycles of coefficients of length d related to the recurrence relation of Eq. (1). For example, when the state of the first register is one, $\Lambda(U)$ will act on the state $|1\rangle\,|\mathrm{u}_s\rangle$ by applying a set of d CNOT gates conditioned on successive states of the target register. The effect of these gates mimics the action of the recurrence relation transforming the second register from $|a_0, a_1 \ldots, a_{d-1}\rangle$ to $|a_d, a_{d+1}, \ldots, a_{2d-1}\rangle$;

$$\begin{aligned}
\Lambda(U)\,|1\rangle\,|\mathrm{u}_s\rangle &= |1\rangle\, U^1\,|\mathrm{u}_s\rangle \\
&= \frac{1}{\sqrt{r}}\sum_{k=0}^{r-1} e^{-2\pi\iota ks/r}\,|1\rangle\, U \left|\sum_{l=1}^{d}\beta_l\alpha_l^k, \sum_{l=1}^{d}\beta_l\alpha_l^{k+1}, \ldots, \sum_{l=1}^{d}\beta_l\alpha_l^{k+d-1} \pmod d\right\rangle \\
&= \frac{1}{\sqrt{r}}\sum_{k=0}^{r-1} e^{-2\pi\iota ks/r}\,|1\rangle \left|\sum_{l=1}^{d}\beta_l\alpha_l^{k+d}, \sum_{l=1}^{d}\beta_l\alpha_l^{k+d+1}, \ldots, \sum_{l=1}^{d}\beta_l\alpha_l^{k+2d-1} \pmod d\right\rangle. \qquad (9)
\end{aligned}$$

When the shift invariance property of the Fourier transform is applied to Eq. (9), it is readily established that input state $|1\rangle\,|\mathrm{u}_s\rangle$ is an eigenstate of $\Lambda(U)$;

$$\begin{aligned}
&\frac{1}{\sqrt{r}}\sum_{k=0}^{r-1} e^{-2\pi\iota(k-d)s/r}\,|1\rangle \left|\sum_{l=1}^{d}\beta_l\alpha_l^k, \sum_{l=1}^{d}\beta_l\alpha_l^{k+1}, \ldots, \sum_{l=1}^{d}\beta_l\alpha_l^{k+d-1} \pmod d\right\rangle \\
&= e^{2\pi\iota ds/r}\,|1\rangle\,|\mathrm{u}_s\rangle. \qquad (10)
\end{aligned}$$

More generally, it can be shown that $|a\rangle\,|\mathrm{u}_s\rangle$ is also an eigenstate of $\Lambda(U)$ and, as such, we have it that $\Lambda(U)\,|a\rangle\,|\mathrm{u}_s\rangle = |a\rangle\, U^a\,|\mathrm{u}_s\rangle = e^{2\pi\iota ads/r}\,|a\rangle\,|\mathrm{u}_s\rangle$. Noting this, we apply $\Lambda(U)$ to state $|\Psi_1\rangle$;

$$\begin{aligned}
\Lambda(U)\,|\Psi_1\rangle &= \Lambda(U)\left(\frac{1}{\sqrt{d^n}}\sum_{a=0}^{d^n-1}|a\rangle\,|11\ldots1\rangle\right) \\
&= \Lambda(U)\left(\frac{1}{\sqrt{d^n r}}\sum_{a=0}^{d^n-1}\sum_{s=0}^{r-1}\sum_{k=0}^{r-1} e^{-2\pi\iota ks/r}\,|a\rangle \left|\sum_{l=1}^{d}\beta_l\alpha_l^k, \sum_{l=1}^{d}\beta_l\alpha_l^{k+1}, \ldots, \sum_{l=1}^{d}\beta_l\alpha_l^{k+d-1} \pmod d\right\rangle\right) \\
&= \frac{1}{\sqrt{d^n r}}\sum_{a=0}^{d^n-1}\sum_{s=0}^{r-1}\sum_{k=0}^{r-1} e^{-2\pi\iota ks/r}\,|a\rangle\, U^a \left|\sum_{l=1}^{d}\beta_l\alpha_l^k, \sum_{l=1}^{d}\beta_l\alpha_l^{k+1}, \ldots, \sum_{l=1}^{d}\beta_l\alpha_l^{k+d-1} \pmod d\right\rangle \\
&= \frac{1}{\sqrt{d^n r}}\sum_{a=0}^{d^n-1}\sum_{s=0}^{r-1}\sum_{k=0}^{r-1} e^{-2\pi\iota ks/r}\,|a\rangle \left|\sum_{l=1}^{d}\beta_l\alpha_l^{k+ad}, \sum_{l=1}^{d}\beta_l\alpha_l^{k+ad+1}, \ldots, \sum_{l=1}^{d}\beta_l\alpha_l^{k+(a+1)d-1} \pmod d\right\rangle. \\
&\qquad (11)
\end{aligned}$$

The sequence values recorded in the second register is periodic as the recurrence relation is reversible and must repeat as soon as d consecutive terms, of which there are only finitely many possibilities, are repeated. Therefore, there exists an $a = jr + l$ for some $1 \leq l < r$ and $0 \leq j < d^n/r$ such that Eq. (11) can be written as

$$\frac{1}{\sqrt{d^n r}} \sum_{j=0}^{d^n/r-1} \sum_{l=0}^{r-1} \sum_{s=0}^{r-1} \sum_{k=0}^{r-1} e^{-2\pi\iota ks/r} \left|jr+l\right\rangle \left| \sum_{l=1}^{d} \beta_l \alpha_l^{k+(jr+l)d}, \ldots, \sum_{l=1}^{d} \beta_l \alpha_l^{k+(jr+l+1)d-1} \pmod d \right\rangle . \quad (12)$$

Applying the shift invariance property of the Fourier transform yields

$$\frac{1}{\sqrt{d^n r}} \sum_{j=0}^{d^n/r-1} \sum_{l=0}^{r-1} \sum_{s=0}^{r-1} \sum_{k=0}^{r-1} e^{-2\pi\iota(k-(jr+l)d)s/r} \left|jr+l\right\rangle \left| \sum_{l=1}^{d} \beta_l \alpha_l^{k}, \ldots, \sum_{l=1}^{d} \beta_l \alpha_l^{k+d-1} \pmod d \right\rangle \quad (13)$$

and, by definition of $|\mathrm{u}_s\rangle$, we can simplify Eq. (13) to

$$\frac{1}{\sqrt{d^n r}} \sum_{j=0}^{d^n/r-1} \sum_{l=0}^{r-1} \sum_{s=0}^{r-1} e^{2\pi\iota(jr+l)ds/r} \left|jr+l\right\rangle |\mathrm{u}_s\rangle. \quad (14)$$

Implementing the inverse Fourier transform on the first register transforms Eq. (14) to

$$\frac{1}{d^n\sqrt{r}} \sum_{m=0}^{d^n-1} \sum_{j=0}^{d^n/r-1} \sum_{l=0}^{r-1} \sum_{s=0}^{r-1} e^{2\pi\iota j(ds-mr/d^n)} e^{2\pi\iota l(ds/r-m/d^n)} |m\rangle |\mathrm{u}_s\rangle. \quad (15)$$

Now, noting that

$$\sum_{j=0}^{d^n/r-1} \left(e^{2\pi\iota(ds-mr/d^n)}\right)^j = \begin{cases} d^n/r & \text{if } e^{2\pi\iota(ds-mr/d^n)} = 1 \\ 0 & \text{otherwise} \end{cases} . \quad (16)$$

However, $e^{2\pi\iota(ds-mr/d^n)} = 1$ if and only if $ds - mr/d^n$ is an integer. Therefore, m must be a multiple of d^n/r. Thus, the state of the system can be given as

$$\frac{1}{r\sqrt{r}} \sum_{c=0}^{r-1} \sum_{l=0}^{r-1} \sum_{s=0}^{r-1} e^{2\pi\iota l(ds-c)/r} |cd^n/r\rangle |\mathrm{u}_s\rangle. \quad (17)$$

Again, since $\sum_{l=0}^{r-1} e^{2\pi\iota l(ds-c)/r} = r\delta_{ds-c,0}$, it follows that Eq. (17) can be written as

$$\frac{1}{\sqrt{r}} \sum_{s=0}^{r-1} \left|sd^{n+1}/r\right\rangle |\mathrm{u}_s\rangle . \quad (18)$$

Finally measuring the first register, we find an m equal to sd^{n+1}/r for some integer s with $0 \leq s < r$. Since both m and d^{n+1} are known, the value r can be confirmed by writing m/d^{n+1} in its lowest form.

3. Conclusion

We discussed the problem of determining the period of a family of linear recurrences and related this problem to a particular instance of Shor's algorithm for factoring.

References

[1] Wilmott C M 2008 *On quantum codes and networks* Ph.D. Thesis Royal Holloway University of London
[2] Shor P W 1997 *Polynomial-Time Algorithms for Prime Factorization and Discrete Logarithms on a Quantum Computer* SIAM Journal on Computing **26** 5 1484-1509
[3] Nielsen M A and Chuang I L 2000 *Quantum Computation and Quantum Information* Cambridge University Press

Exact solution of Random Graph fragmentation and physical, chemical and biological applications

P. Désesquelles, D.-T. Nga, A. Cholet

Univ. Paris-Sud and CSNSM CNRS/IN2P3, 15 rue G. Clémenceau, 91405 Orsay, France

E-mail: Pierre.Desesquelles@in2p3.fr

Abstract. Random graphs (RG), also called mean field percolation in the frame of mesoscopic physics, are a basic model composed solely of connected entities called nodes. The connections, called bonds, can be active of broken. When the number of broken bonds is too large, the system of entities separates into a set of fragments called a partition. The exact solution for the micro-canonical partition probabilities of finite size systems was yet unresolved, thus a series of fundamental questions about the model could not be answered. We have established the exact equations of RG partition probabilities as a function of the number of nodes and of the number of broken bonds. From these probabilities, it is also possible to deduce intrinsic properties of RG. Many actual networks, while composed of complex interactions, behave like RG. We show examples where information was deduced, using RG, from systems consisting of sets of nucleons, atoms or termite nest chambers.

keywords: Random Graphs, partitions.

1. Introduction

Random Graphs (RG) [1] are the most general and the most studied model of sets of connected entities. Its importance is due to its simplicity, to the richness of its properties and to the fact that it has shown its usefulness to describe and analyze real-world systems. When the size of the system tends to infinity, it shows a phase transition and its evolution near the critical point has been described precisely via a set of universal parameters [2, 3]. New interest has been triggered by the study of mesoscopic objects, such as nano-objects or atomic clusters, that cannot be considered as infinite and that are not necessarily close the critical points. Hence, we have calculated the exact equations of microcanonical RG for finite systems [4], using the tools of statistical physics. Indeed, the nodes of RG can be seen as particles or molecules and the bonds as binding energies. The resulting equations give the probabilities of fragment partitions.

Many real-life networks, while composed of complex interactions, show a behavior close RG This model can therefore be used as a tool to evidence properties or correlations in physical systems submitted to fragmentation or to nucleation. Different examples are shown where physical information was deduced, using RG, from systems consisting of sets of nucleons [5] (nuclear spinodal decomposition [6, 7]) of atoms (cluster fragmentation [8, 9]), of molecules [10] and from the topology of insect nests [11].

2. Implementation of the model equation

In the following the number of nodes will be noted S (size of the system) and the number of broken bonds E^* as we consider that the breaking of a bond requires a unit excitation energy.

Published under licence by IOP Publishing Ltd

A fragment is a set of connected nodes such that it is possible to go from any node to any other node, directly or indirectly, following active bonds. A fragment partition [12] will be noted as a vector $\mathbf{n}$, whose component n_s indicates the number of fragments with size s. The maximum number of bonds is $S\,(S-1)/2$. When enough bonds are broken, the system can separate into fragments.

The microcanonical (obtained for a given E^*) weight of a fragmentation partition depends both on the number of ways to arrange the nodes among the fragments and on the number of ways to arrange the broken bonds, i.e. the excitation energy, among the fragments.

The first combinatory factor accounts for the number of ways to allocate S nodes to the fragments. There are $S!$ ways to arrange the nodes. However, the permutation of nodes inside a fragment does not change the partition nor does the permutation of equal size fragments. Thus the combinatory factor reads:

$$\omega_{\mathrm{comb}}(\mathbf{n}) = \frac{S!}{\prod_s s!^{n_s}\; n_s!}\,. \tag{1}$$

A part of the energy is used to separate the fragments, the remaining energy is distributed among the fragments. For a given fragment i, the density of states $\rho(s_i, E_i^*)$ is equal to the number of ways to choose E_i^* broken bonds among $s_i\,(s_i-1)/2$ bonds without breaking the fragment. Hence, the product of the fragments level densities has to be taken into account in the evaluation of the partition weight. Moreover, there are different ways to sort the excitation energy among the fragments. The energy sorting is accounted for by a vector $\mathbf{E}$ which components are the E_i^* and $\sum_i E_i^*$ is the remaining energy. Thus, the partition weights read:

$$\omega(\mathbf{n}, E^*) = \omega_{\mathrm{comb}}(\mathbf{n}) \sum_{\mathbf{E}} \prod_i \rho(s_i, E_i^*)\,. \tag{2}$$

All the observables and the thermodynamic features of the model can be deduced from this equation.

The density of states $\rho(S, E^*) = \omega((0, \ldots, 0, 1), E^*)$ can easily be computed noticing that the total number of ways to sort E^* broken bonds in the system is equal to the sum of all the partition weights. Hence we have:

$$\rho(S, E) = \binom{S(S-1)/2}{E^*} - \sum_{\mathbf{n}\neq(0,\ldots,0,1)} \omega_{\mathrm{comb}}(\mathbf{n}) \sum_{\mathbf{E}} \prod_{i=1}^{M} \rho(s_i, E_i^*)\,. \tag{3}$$

Here s_i and E_i^* are smaller than S and E thus the equation can be used to calculate recursively the densities of states. The results are shown in figure 1. As can be seen, the density of states reaches very high values even for relatively small systems. For fifty nodes, the maximum is of the order of 10^{367}.

Equation (2), gives the microcanonical probabilities. The canonical RG model corresponds to the breaking of bonds according to a breaking probability p playing the role of a temperature. The distribution of the number of broken bonds for a given p is binomial, thus:

$$\omega_{\mathrm{cano}}(\mathbf{n}, E^*) = \omega_{\mathrm{comb}}(\mathbf{n}) \sum_{E^*=0}^{S\,(S-1)/2} \binom{S(S-1)/2}{E^*} p^{E^*} (1-p)^{S\,(S-1)/2-E^*} \sum_{\mathbf{E}} \prod_i \rho(s_i, E_i^*)\,. \tag{4}$$

Another way to calculate this quantity was also proposed recently in [13].

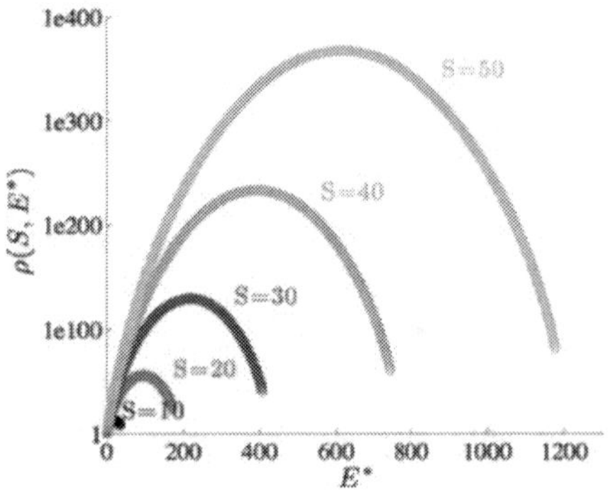

Figure 1. Level density (number of stable configurations) as a function of the excitation energy for different system sizes.

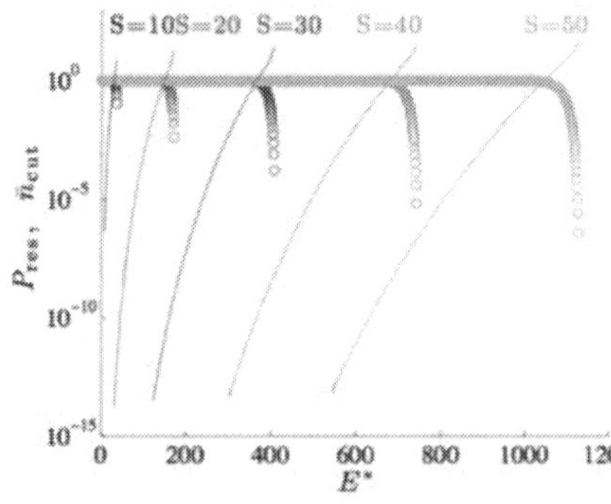

Figure 2. Open dots: resilience probability, lines: mean number of cutting bonds as a function of the excitation energy for different system sizes.

Many real life systems are composed of different types of "nodes". For T types, the combinatory factor is given by [4]:

$$\omega_{\rm comb}(\mathbf{n}) = \frac{\prod_{t=1}^{T} S_t!}{\left[\prod_{s=1}^{\max(S_1,\ldots,S_T)} (s!)^{\sum_t n_t(s)}\right] \left[\prod_{s_1=0}^{S_1} \cdots \prod_{s_T=1}^{S_T} n_{s_1 \ldots s_T}!\right]} . \quad (5)$$

where S_t the number of type t nodes, $n_t(s)$ the number of fragments containing s type t nodes and $n_{s_1 \ldots s_T}$ the number of fragments with s_1 type 1 nodes, …, s_T type T nodes.

3. Applications

In [4] we have shown how the preceding equations could be used to evidence an over counting of the fragmentation of the carbon nucleus into three alpha particles, which could sign the Hoyle state [14].

3.1. Networks

From equation (2), different features of the model can be deduced. For example, the resilience probability (probability that the system remains connected if one bond is broken) and the mean number of cutting bonds (bonds such that if they are broken, the system splits into two fragments), which play important roles in the frame of network stability studies, are shown in figure 2. The resilience probability $P_{\rm r}$ is the ratio of the density of states to the sum of all partition weights and the mean number of cutting bonds is given by the following equation [4]:

$$\bar{n}_{\rm cut}(S, E^*) = \left(\frac{S\,(S-1)}{2} - E^*\right)\left(1 - \frac{P_{\rm r}(S, E^*+1)}{P_{\rm r}(S, E^*)}\right) . \quad (6)$$

3.2. Distillation process

The simplest two-colour graph is composed of, say, red and blue nodes and of bonds that connect nodes with different colours only. When bonds are broken, two "phases" appear. The "gas" gathers the free (unbounded) nodes and the "liquid" is composed of the connected nodes. The results (figure 3) illustrate the distillation property that appears when a majority of entities belong to a given species: the relative proportion of the different species in the disordered phase (gas) is systematically more asymmetric than in the ordered phase (liquid). A similar result has been evidenced in atomic nuclei decay [15].

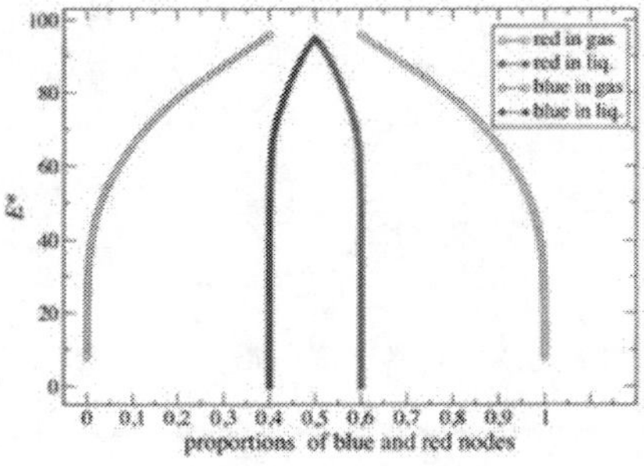

Figure 3. Distillation process in a 2 node-type RG model. The system is composed of 8 blue and 12 red nodes. The curves give the mean proportion of nodes in each phase at a given excitation energy. At low energy, all the nodes are in the "liquid". At the evaporation threshold, $E^* = 8$, the "gas" is composed of red nodes only. Over the threshold, the "gas" is always more asymmetric than the "liquid".

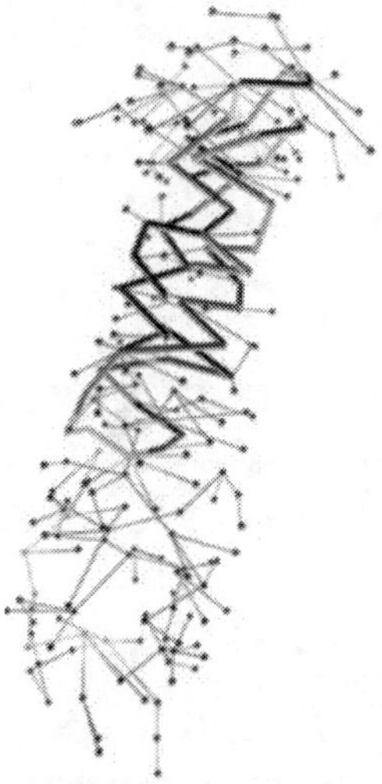

Figure 4. Upper part of a cubitermes nest. The center of the chambers are indicated by dots, cutting corridors by thin lines and non-cutting corridors by bold lines.

3.3. Cubitermes nests
Cubitermes, like ants and termites, build complex nests composed of corridors and chambers (figure 4) [11]. One way to evidence their topological properties is to compare them to RG. As an example, the mean number of cutting corridors is much larger in the nests than in RG which is interpreted in terms of mobility management and of protection against intruders.

4. Conclusions
We have established the exact expressions of the RG partition probabilities and shown that a fruitful way to use the result is to apply RG as a null hypothesis to evidence properties of real-life systems.

References
[1] Erdös P, Rényi A 1959 *Publicationes Mathematicae* **6** pp 290-297.
[2] Albert R, Barabási A-L 2002 *Rev. Mod. Phys.* **74** pp 47-97.
[3] Bollobàs B, Janson S, Riordan O 2007 *Random Struct. Alg.* **31** pp 3-122.
[4] Désesquelles P 2011 *Physics Letters* **B 698** pp 284-287.
[5] Lleres A *et al.* 1994 *Phys. Rev.* **C 50** pp 1973-1981.
[6] Désesquelles P 2002 *Phys. Rev.* **C 65** p 034603.
[7] Borderie B *et al.* 2004 *Nucl. Phys.* **A 734** pp 495-503.
[8] Diaz-Tendero S 2006 *International Journal of Mass Spectrometry* **252** pp 126-132.
[9] Beroff K 2009 *Nucl. Instr. and Meth.* **A 267** p 866.
[10] Van-Oanh N-T *et al.* 2006 *The Journal of Physical Chemistry* **A 110** pp 5592-5598.
[11] Perna A *et al.* 2008 *Naturwissenschaften*, **95** pp 877-884.
[12] Désesquelles P 2002 *Phys. Rev.* **C 65** p 034604.
[13] Allard A, Hébert-Dufresne L, Noël V, Marceau P-A and Dubé LJ 2012 *Europhysics Letters* **98** p 16001.
[14] Schuck P *et al.* 2007 *Progress in Particle and Nuclear Physics* **59** pp 285-304.
[15] Ducoin C, Chomaz P, Gulminelli F 2006 *Nuclear Physics* **A 771** pp 68-92.

Leak Air in a Double-Wall Chimney System

Klaus Lichtenegger, Babette Hebenstreit, Christian Pointner
Bioenergy 2020+, Gewerbepark Haag 3, A 3250 Wieselburg-Land
E-mail: klaus.lichtenegger@bioenergy2020.eu

Abstract. Operating biomass stoves in modern buildings with tight shells often requires a room-independent air supply. One possibility to arrange this supply is to use a double-wall chimney with fresh air entering through the annular gap. For this setup, a mathematical model has been developed and checked with experimental data. It turned out that for commercially available chimneys, leakage is not negligible and inclusion of leak air in the calculation is crucial for reproduction of the experimental data. Even with inclusion of this effect, discrepancies remain which call for further investigations and a refinement of the model.

1. Introduction

Biomass is an important source of renewable energy, in particular used for heating purposes on small and large scale. Biomasse stoves have to be equipped with a chimney, through which the flue gases are led to the environment, as well as a combustion air supply. Traditionally the combustion air was directly taken from the room where the heating appliance was located.

During the last decades, building shells got much tighter, advancing in the direction of passive houses. Therefore building shells of modern houses are often too tight to enable enough air supply through leakage. Consequently, modern stoves are constructed room sealed with separate air supply in order to enable the operation of stoves in almost air-tight buildings.

One possibility to arrange air supply is by using a double wall chimney: The flue gas exits through the inner pipe, while fresh air enters through the annular gap. Due to preheating, the buoyant force of the air reduces the natural draft of the chimney; an effect which can be significantly reduced by proper insulation. This setup has been investigated in a research project conducted in cooperation with an industrial partner [1]. The project consisted both of measurements performed on different experimental setups and the creation of a mathematical model for the chimney which is outlined in sec. 2.

2. Outline of the Model

The model is quasi-static and effectively one-dimensional, making use of the wealth of heuristic relationships for heat and mass transfer established in engineering [2]. The cylindric chimney is discretized by introducing $N_{\rm cyl}$ slices, for which mass and energy balances are employed to calculate pressure and temperature. Some relevant physical quantities in a single cylinder slice are illustrated in fig. 1, where also the notation is fixed.

Convective heat transfer coefficients are determined from the Nußelt numbers, which depend on the properties of the gas flow, characterized by Reynolds and Prandtl number. Pressure in pipe and annular gap is determined to a large extent by buoyancy. In addition, dynamic pressure and friction have some influence, all taken into account in Bernoulli's equation.

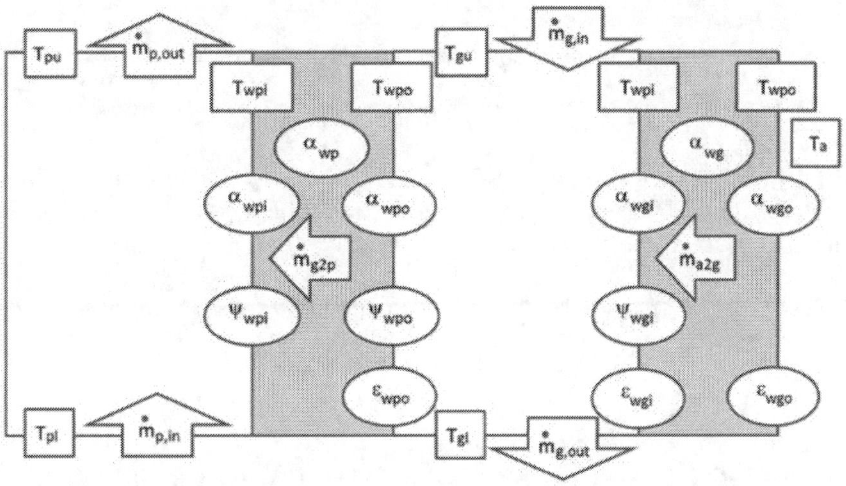

Figure 1. Some quantities used to describe a cylinder slice in a double-wall chimney: T denotes temperature, $\dot{m}$ the mass flow, α the coefficent of convective or conductive heat transfer, ε the coefficient of radiative emission and ψ the dimensionless number characterizing wall friction. Subscripts p denote the pipe section, g the gap section, a the ambient. Quantities defined on the lower/upper boundary of the slice carry an additional subscript l/u. The subscripts wp, wpi and wpo denote the pipe wall, its inner and its outer surface, wg, wgi and wgo indicate the gap wall in the same way. Mass flows entering/leaving the slice have a subscript in/out, mass flows between regions within the cyclinder slice are written as $\dot{m}_{\rm a2g}$ and $\dot{m}_{\rm g2p}$.
Other important physical and geometrical quantities not explicitly shown in this figure are the area A of the wall barrel, volume flows $\dot{V}$, gas density ρ, gas velocity v and gas pressure p, characterized by the difference Δp to the ambient pressure $p_{\rm a}$.

2.1. Energy balance for heat transfer at the outside of the pipe wall

As an example, we set up the balance equation for the heat transfer for the outside of the pipe wall: The heat conducted through the pipe wall is partially transfered to the gas in the gap und partially converted to radiation which is emitted to the inner gap wall,

$$(\alpha A)_{\rm wp}\,(T_{\rm wpi}-T_{\rm wpo}) = (\alpha A)_{\rm wpo}\left(T_{\rm wpo}-\frac{T_{\rm gu}+T_{\rm gl}}{2}\right)+\Sigma_{\rm p2g}\,(T_{\rm wpo}-T_{\rm wgi}), \tag{1}$$

where we have employed a condensed notation for subscripts of products, $(\alpha A)_{\rm wpo}=\alpha_{\rm wpo}A_{\rm wpo}$. In addition we have performed a factorization of the radiation term and defined

$$\Sigma_{\rm p2g} := \frac{\sigma_{\rm SB}A_{\rm wpo}}{\frac{1}{\varepsilon_{\rm wpo}}+\frac{A_{\rm wpo}}{A_{\rm wgi}}\left(\frac{1}{\varepsilon_{\rm wgi}}-1\right)}\left(T_{\rm wpo}^3+T_{\rm wpo}^2T_{\rm wgi}+T_{\rm wpo}T_{\rm wgi}^2+T_{\rm wgi}^3\right), \tag{2}$$

with $\sigma_{\rm SB}$ denoting the Stefan-Boltzmann constant. The prefactor is chosen to account also for reflection and re-absorption of radiation within the annular gap.

Here, as in the whole model, arithmetic means are used instead of more accurate expressions based on logarithmic temperature differences. This introduces a controllable systematic error which is small compared to other uncertainties. The benefit is that the equation is formally linear with respect to the unknown temperatures.

2.2. Inclusion of Leak Air

The walls are modelled as homogenously porose; for the volume flows of leak air we set

$$\dot{m}_{\mathrm{a2g}} = \sigma_{\mathrm{g}} \rho_{\mathrm{g}} A_{\mathrm{wg}} \Delta p_{\mathrm{g}}, \qquad \dot{m}_{\mathrm{g2p}} = \sigma_{\mathrm{p}} \rho_{\mathrm{p}} A_{\mathrm{wp}} (\Delta p_{\mathrm{p}} - \Delta p_{\mathrm{g}}) \tag{3}$$

with the (initially unknown) porosity coefficients σ_{p} and σ_{g}. These mass flows change the gas velocities, determined from the continuity equation which, for the (quasi-)stationary case, simplifies to $\operatorname{div} \boldsymbol{j} = 0$. Including leak air, one arrives at

$$v_{\mathrm{p,out}} \rho_{\mathrm{p,out}} = v_{\mathrm{p,in}} \rho_{\mathrm{p,in}} + \dot{m}_{\mathrm{g2p}}, \qquad v_{\mathrm{g,out}} \rho_{\mathrm{g,out}} = v_{\mathrm{g,in}} \rho_{\mathrm{g,in}} - \dot{m}_{\mathrm{g2p}} + \dot{m}_{\mathrm{a2g}} \tag{4}$$

with gas densities calculated from the ideal gas equation.

2.3. Structure of the Complete Model

The resulting equation for the unknown temperatures (T_{pu}, T_{wpi}, T_{wpo}, T_{gl}, T_{wgi}, T_{wgo}) is formally linear and can be solved by simple matrix inversion. However, some entries of the coefficient matrix and of the inhomogenity depend directly or indirectly on the unknown temperatures. In addition, the equations for the cylinder slices are coupled, since for example $T_{\mathrm{pu}}^{(k)} = T_{\mathrm{pl}}^{(k+1)}$ and $v_{\mathrm{p,out}}^{(k)} = v_{\mathrm{p,in}}^{(k+1)}$ with k denoting the index of the current cylinder slice.

Obtainig the final solution requires an iterative approach, during which also velocities, pressure values, leak air mass flow and heat transfer coefficients are re-calculated several times. Boundary values like T_{a}, p_{a}, $T_{\mathrm{pl}}^{(1)}$ and $v_{\mathrm{p,in}}^{(1)}$ are obtained from experiment.

3. Comparison to Experimental Data and Outlook

The model works reasonably well for some setups, if a sufficient amount of leak air is permitted, as shown in figs. 2 and 3. For $\sigma_{\mathrm{p}} = \sigma_{\mathrm{g}} = 0.0 \, \frac{\mathrm{m}^3/\mathrm{s}}{\mathrm{m}^2\,\mathrm{Pa}}$ the simulation yields flue gas temperatures which are up to 100°C too high. Changing this to $\sigma_{\mathrm{p}} = 0.00015 \, \frac{\mathrm{m}^3/\mathrm{s}}{\mathrm{m}^2\,\mathrm{Pa}}$, which results in a volume flow of $\dot{V}_{\mathrm{g2p}} \approx 5 \, \frac{\mathrm{m}^3}{\mathrm{h}}$, significantly improves the agreement between simulation and experiment. (Further parameter-tuning could be done to obtain an even better agreement.)

The remainig deviations can be mainly attributed to the fact that the quasi-stationary simulation responds faster to changes than the real system, which possesses some thermal inertia.

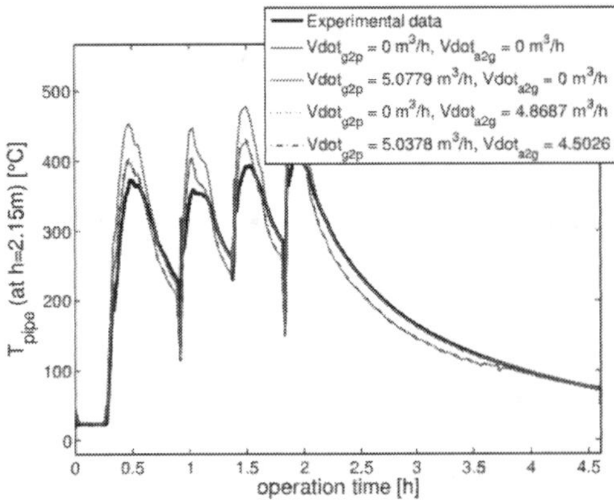

Figure 2. Flue gas temperatur $T_{\mathrm{p}}(2.15\,\mathrm{m})$ for double-wall chimney with insulated liner

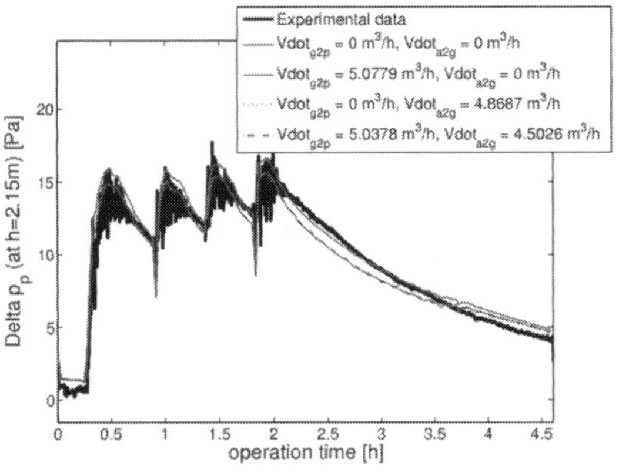

Figure 3. Pipe pressure difference $\Delta p_{\mathrm{p}}(2.15\,\mathrm{m})$ for double-wall chimney with insulated liner

In other setups, the discrepancy between simulation results and experimental data is unacceptably large. This happens in particular for setups with insulated case, as shown in figs. 4 and 5. For this setup also numerical problems (in particular instabilities) have been repeatedly encountered.

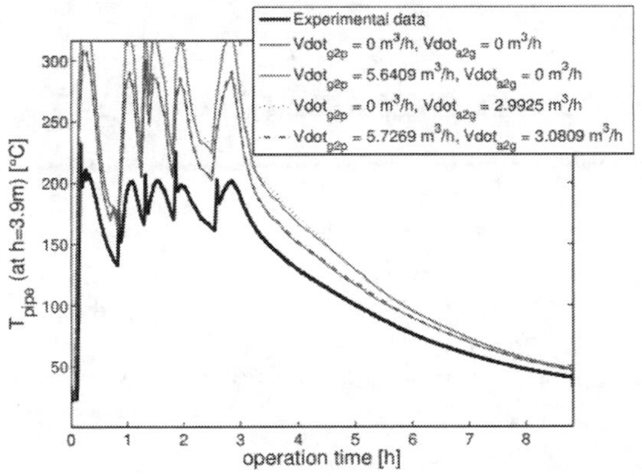

Figure 4. Flue gas temperatur $T_\mathrm{p}(3.9\,\mathrm{m})$ for double-wall chimney with insulated case

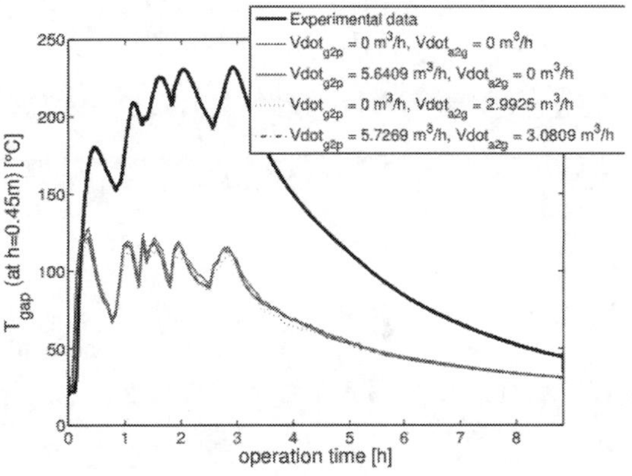

Figure 5. Air temperatur $T_\mathrm{g}(0.45\,\mathrm{m})$ for double-wall chimney with insulated case

Also here, some deviations can be attributed to the quasi-stationary modelling and could be removed by setting up a fully dynamical simulation. A more serious issue is the yet unsolved problem of heat transfer coefficients for the asymetrically heated annular gap [3]. For the present studies, the expressions for an annular gap with thermal insulation have been used as an approximation, but this introduces an uncontrollable systematic error, which is a serious source of uncertainty for the results of the model. Here, some progress on theoretical grounds is required to improve the model.

Other improvements may be necessary as well in order to obtain a model with predictive power, but already at the present stage it is clear that for commercially available chimneys leakage is not negligible and inclusion of leak air in the corresponding simulation is a crucial point.

Acknowledgements

The authors are grateful to Schiedel AG, in particular to Gerald Steinecker, for the excellent cooperation during the research project. They are grateful to Erich Essletzbichler for valuable aid during the measurements and to Christoph Schmidl, Ernst Höftberger, Walter Haslinger and Manfred Wörgetter for support and useful advice concerning model building, data analysis and publication of results.

Development and dissemination of the model have been supported by the Klima- und Energiefonds and the Österreichische Forschungsförderungsgesellschaft FFG.

References

[1] Pointner C, Lichtenegger K, Hebenstreit B 2012 *Flows – Future Log Wood Stove: Untersuchung eines doppelwandigen Edelstahl-Schornsteinsystems für den raumluftunabhängigen Ofenbetrieb* final report for project C-I-1-5-3, (Graz Wieselburg: Bioenergy 2020+)

[2] Verein Deutscher Ingenieure VDI-Gesellschaft Verfahrenstechnik und Chemieingenieurwesen (ed) 2002 *VDI-Wärmeatlas*, 9th edition (Berlin Heidelberg: Springer)

[3] Mitrovic J 2005 *Wärmeübergang im konzentrischen Ringspalt bei unterschiedlichen Wandtemperaturen Ch. Ing. Tech.* **77** 1618-20

Application of quadratures and operator splitting to partial differential equations

Francisco J. Sánchez-Bernabe [1] **and J. Salcedo** [2]

[1] Department of Mathematics, Universidad Autónoma Metropolitana-Iztapalapa, San Rafael Atlixco 186, Colonia Vicentina, México, D.F., 09340, Mexico
[2] E. S. I. M. E., Culhuacán, I.P.N., Av. Santa Ana 1000, México D. F., 04430, Mexico

E-mail: `fjsb@xanum.uam.mx`

Abstract. In this work we study the stability of a method for the numerical solution of initial value problems, that combines finite differences with Simpson's rule. This method is applied to a one spatial dimension, convection-dominated transport problem. To solve the same problem in two spatial dimensions, the proposed method was used in combination with Strang's operator decomposition method.

1. Introduction

In this paper we study a simple method to find numerical approximations to the solutions of initial value problems. This method transforms such problems into an equivalent problem of solving systems of integral equations, and then approximates the involved time-dependent definite integrals by trapezoidal rule and the spatial definite integrals by Simpson's rule.

To prove the effectiveness of the proposed scheme, we first applied it to solve a convection-dominated transport problem in one spatial dimension [1]. To solve the same problem in two spatial dimensions, the proposed method was used successfully in combination with Strang's operator decomposition method [2].

2. Application to a one-dimensional convection diffusion problem

Given the following initial value problem:

$$\frac{\partial \phi}{\partial t} + a\,\frac{\partial \phi}{\partial x} = \nu\,\frac{\partial^2 \phi}{\partial x^2}, \qquad x \in (0, L), \qquad t \in (0, T], \tag{1}$$

where $L, T > 0$, with the initial condition

$$\phi(x, 0) = g(x) = \frac{2.5}{\sigma}\,exp\left[-\frac{(x - x_0)^2}{2\sigma^2}\right], \tag{2}$$

and the boundary conditions:

$$\phi(0, t) = \phi_1(t) = 0$$

$$\phi(L,t) = \phi_2(t) = \frac{2.5}{\sqrt{\sigma^2 + 2\nu t}}\, exp\left[-\frac{(150 - x_0 - t)^2}{2(\sigma^2 + 2\nu t)}\right], \tag{3}$$

where: $L = 150$, $\Delta x = 0.1$, $\Delta t = 0.4$, $a = 1.0$, $\nu = 0.01$, $\sigma = 3.5$, $x_0 = 20$, $Pe = Peclet\ number = 5.0$.

To solve (1)-(3) is necessary, for each $n = 0, 1, 2, \ldots, q - 1$, to solve the following system of equations:

$$\int_{x_i}^{x_{i+2}} \int_{t^n}^{t^{n+1}} \frac{\partial \phi}{\partial t}\, dt\, dx + a \int_{t^n}^{t^{n+1}} \int_{x_i}^{x_{i+2}} \frac{\partial \phi}{\partial x}\, dx\, dt$$

$$= \nu \int_{t^n}^{t^{n+1}} \int_{x_i}^{x_{i+2}} \frac{\partial^2 \phi}{\partial x^2}\, dx\, dt\,, \tag{4}$$

where $i = 0, 1, 2, \ldots, m - 2$. Next, calculating the second iterated integrals for each side of the equations (4), it follows:

$$\int_{x_i}^{x_{i+2}} [\phi(x, t^{n+1}) - \phi(x, t^n)]\, dx + a \int_{t^n}^{t^{n+1}} [\phi(x_{i+2}, t) - [\phi(x_i, t)]\, dt$$

$$= \nu \int_{t^n}^{t^{n+1}} \left[\frac{\partial \phi}{\partial x}(x_{i+2}, t) - \frac{\partial \phi}{\partial x}(x_i, t)\right] dt\,, \tag{5}$$

$$i = 0, 1, 2, \ldots, m - 2\,.$$

3. Stability of proposed method

We take $a = 0$, and approximating the left hand side integrals by using Simpson's rule, and by substituting the partial derivatives of right hand side by centered differences, we obtain

$$\int_{x_i}^{x_{i+2}} [\phi(x, t^{n+1}) - \phi(x, t^n)]\, dx$$

$$= \nu \int_{t^n}^{t^{n+1}} \left[\frac{\partial \phi}{\partial x}(x_{i+2}, t) - \frac{\partial \phi}{\partial x}(x_i, t)\right] dt\,, \tag{6}$$

$$\frac{x_{i+2} - x_i}{6}[\phi(x_{i+2}, t^{n+1}) + 4\phi(x_{i+1}, t^{n+1}) + \phi(x_i, t^{n+1})] -$$

$$\frac{x_{i+2} - x_i}{6}[\phi(x_{i+2}, t^n) + 4\phi(x_{i+1}, t^n) + \phi(x_i, t^n)]$$

$$= \nu \int_{t^n}^{t^{n+1}} \left[\frac{\phi(x_{i+3}, t) - \phi(x_{i+1}, t)}{2\Delta x} - \frac{\phi(x_{i+1}, t) - \phi(x_{i-1}, t)}{2\Delta x}\right] dt\,, \tag{7}$$

by defining now

$$\lambda = \frac{3\nu\Delta t}{4(\Delta x)^2}$$

we obtain

$$\phi(x_{i+2}, t^{n+1}) + 4\phi(x_{i+1}, t^{n+1}) + \phi(x_i, t^{n+1}) - \phi(x_{i+2}, t^n)$$

$$-4\phi(x_{i+1}, t^n) - \phi(x_i, t^n)$$

$$= \lambda \left[\phi(x_{i+3}, t^n) + \phi(x_{i+3}, t^{n+1}) - 2\phi(x_{i+1}, t^n) - 2\phi(x_{i+1}, t^{n+1})\right] +$$

$$\lambda \left[\phi(x_{i-1}, t^n) + \phi(x_{i-1}, t^{n+1})\right], \qquad (8)$$

shifting the index i one unit to the left, and by considering the approximate solution ϕ_i^n,

$$-\lambda\phi_{i+2}^{n+1} + \phi_{i+1}^{n+1} + (4+2\lambda)\phi_i^{n+1} + \phi_{i-1}^{n+1} - \lambda\phi_{i-2}^{n+1}$$

$$= \lambda\phi_{i+2}^{n} + \phi_{i+1}^{n} + (4-2\lambda)\phi_i^{n} + \phi_{i-1}^{n} + \lambda\phi_{i-2}^{n}, \qquad (9)$$

next, we substitute

$$\phi_i^n = w_n e^{rx_i\sqrt{-1}},$$

to obtain

$$-\lambda w_{n+1} e^{r(x_i+2h)\sqrt{-1}} + w_{n+1} e^{r(x_i+h)\sqrt{-1}} + (4+2\lambda) w_{n+1} e^{rx_i\sqrt{-1}}$$

$$+w_{n+1} e^{r(x_i-h)\sqrt{-1}} - \lambda w_{n+1} e^{r(x_i-2h)\sqrt{-1}} = \lambda w_n e^{r(x_i+2h)\sqrt{-1}} + w_n e^{r(x_i+h)\sqrt{-1}}$$

$$+(4-2\lambda) w_n e^{rx_i\sqrt{-1}} + w_n e^{r(x_i-h)\sqrt{-1}} + \lambda w_n e^{r(x_i-2h)\sqrt{-1}}, \qquad (10)$$

where we have used the notation $h = \Delta x$. Next, by the exponential function properties, and by the definition of cosine function,

$$\left[4\lambda \sin^2(\frac{1}{2}2rh) + 4\cos^2(\frac{1}{2}rh) + 2 \right] w_{n+1}$$

$$= \left[-4\lambda \sin^2(rh) + 4\cos^2(\frac{1}{2}rh) + 2 \right] w_n, \qquad (11)$$

if we define

$$\kappa = \frac{1 + 2\cos^2(\frac{1}{2}rh) - 2\lambda\sin^2(rh)}{1 + 2\cos^2(\frac{1}{2}rh) + 2\lambda\sin^2(rh)} \qquad (12)$$

Thus,

$$w_{n+1} = \kappa\, w_n \tag{13}$$

Therefore, the method is stable because the absolute value of κ is less than 1, [3].

4. Applications to a two-dimensional convection-diffusion problem

Let $\phi = \phi(x, y, t)$ be a function of two spatial variables, and A is a spatial differential operator that can be decomposed in the form:

$$A = A_1 + A_2\,, \tag{14}$$

where A_1 includes only *x-derivatives* and A_2 includes only *y-derivatives*. More specifically,

$$A_1\phi = -\,a\,\frac{\partial\phi}{\partial x} + \nu\,\frac{\partial^2\phi}{\partial x^2}\,, \qquad A_2\phi = -\,a\,\frac{\partial\phi}{\partial y} + \nu\,\frac{\partial^2\phi}{\partial y^2}\,. \tag{15}$$

Thus, our problem [1], [4] can be written in the following way:

$$\frac{\partial\phi}{\partial t} = -\,a\,\frac{\partial\phi}{\partial x} + \nu\,\frac{\partial^2\phi}{\partial x^2} - a\,\frac{\partial\phi}{\partial y} + \nu\,\frac{\partial^2\phi}{\partial y^2},$$

$$x \in (0, L_x), \qquad y \in (0, L_y), \qquad t \in (0, T], \tag{16}$$

where $L_x, L_y, T > 0$. The initial condition is given by

$$\phi(x, y, 0) = g(x, y) = \frac{2.5}{\sigma}\, exp\left[-\frac{(x-x_0)^2 + (y-y_0)^2}{2\sigma^2}\right], \tag{17}$$

and the boundary conditions can be found in [5].

5. Conclusions

The one-dimensional convection diffusion problem is solved directly, while the two-dimensional convection diffusion problem is solved by applying the Strang's operator decomposition method. More details can be found in [5]. Actually, the combination of the proposed method with Strang's operator decomposition method provides great advantages when it is used to solve three spatial dimension problems.

References

[1] Donea J, Roig B and Huerta A 2000 High-order accurate time-stepping schemes for convection-diffusion problems *Comput. Meth. Appl. Mech. Engrg.* **182** 249-75

[2] Strang G 1968 On the construction and comparison of differences schemes *SIAM J. Num. Analysis* **5** 506-17.

[3] Holmes M H 2007 *Introduction to numerical methods in differential equations* (New York: Springer)

[4] Glowinski R 2003 *Numerical methods for fluids, Part 3* (Handbook of numerical analysis vol IX) ed P G Ciarlet and J L Lions (Amsterdam: Elsevier Science)

[5] Salcedo J, Salcedo B and Sánchez-Bernabe F J 2011 The use of quadratures for solving convective and highly stiff transport problems *Computers and Math. with Applic.* **61** 586-91.

Statistical Distribution of a Completely Open System Based on Incomplete Shannon Entropy

Heling Li, Bin Yang, Tao Yang, Ying Xiong

School of Physics Electronics Information Engineering,
Ningxia University, Yinchuan 750021, China.

E-mail: nigxiayclhl@163.com

Abstract. Micro canonical, canonical and grand canonical systems are special cases of completely open systems. Within the framework of non-extensive and incomplete statistics, we derive the statistical distribution for a completely open system on the basis of incomplete Shannon entropy, using the maximum entropy principle. We calculate the physical properties of a linear filament using this distribution. The results are the same as those calculated using the incomplete *E-V* distribution, except that average values of the thermodynamic variables replace the corresponding constant values. However, the relative fluctuations in thermodynamic variables are completely different. In the canonical and grand canonical distributions, relative fluctuations are proportional to $1/\sqrt{N}$. In the incomplete statistical distribution for a completely open system, the relative fluctuations are proportional to 1. This result can explain some phenomenon that traditional and current statistical mechanics cannot explain, such as the large fluctuations of critical, super-cooled and overheated states.

1. Introduction

We deduced the statistical distribution of a completely open system and a few thermodynamic formulas by applying Tsallis entropy [1] and the maximum entropy principle in our previous paper [2]. However, even with these tools it is still complicated to study the thermodynamic properties of a completely open system with the Tsallis statistical distribution. In this paper, we will use another approach. In 2001 French physicist Q.A. Wang proposed a modification to Tsallis statistics called incomplete statistics [3]. Wang also discussed some physical systems on the basis of incomplete entropy [3-5] and incomplete Shannon entropy [6,7]. Others made a contribution to the field [8-10].

It is well known that multi-particle systems with complex interactions can have fractal or chaotic phase spaces. The macroscopic behavior of such systems might be influenced by the edge effect and other factors related to the structure of the phase space. In such cases, we cannot accurately calculate the Hamilton canonical equation (for classical systems) or the Schrödinger equation (for quantum systems).Therefore, we cannot characterize all the microstates of the system and their corresponding probabilities. This problem is referred to information imperfection.

We denote the total number of possible microstates for a system as v. Owing to information imperfection, the number of microstates that we can record is w (w can be greater or less than v). Let $\{p_i\}$ denote the probability distribution for microstates of the system (p_i is the probability of microstate i, i=1,2,3,...,w). If $w=v$, the distribution is complete; if $w\neq v$, the distribution is incomplete. Since the normalizing condition $\sum_i^v p_i = 1$ holds for a complete distribution, the statistical average of a physical quantity O is simply $\overline{O} = \sum_i^v O_i p_i$, where O_i is the value of O in state i.

For an incomplete distribution, $\sum_i^w p_i = Q \neq 1$. To normalize the incomplete distribution, we assume that $\sum_i^w \frac{p_i}{Q} = \sum_i^w p_i^q$, and that $\sum_i^w p_i^q = 1$. Since each $p_i \in [0,1]$, we have $q \in [0,+\infty)$. The statistical average of a physical quantity O is given by $\overline{O} = \sum_i^w O_i p_i^q$. (1)

The statistical physics based on Eq. (1) is called incomplete statistics.

2. Theory

2.1. The probability distribution and the partition function

The Shannon entropy is defined as the statistical average of $-k \ln p_i$ (where k is the Boltzmann constant). Within the framework of complete statistical physics, its formula is $S = -k\sum_{i=1}^{v} p_i \ln p_i \quad S_0$. We usually express the Shannon entropy as an absolute entropy with constant S_0=0. The arbitrary entropy constant can be nonzero, and derives from the Clausius entropy. We should emphasize that selecting a non-zero entropy constant does not violate the first, second, or third laws of thermodynamics. BG statistical mechanics assumes that entropy is extensive, so in that framework S_0 should also be extensive.

In non-extensive systems with long-range interactions, we take S_0 as a non-extensive, nonzero constant. Within the framework of incomplete and non-extensive statistical physics, the statistical average is based on Eq. (1), so the Shannon entropy can be written in the following form [6-7]:

$$S_q = -k\sum_{i=1}^{w} p_i^q \ln p_i \quad S_0 \qquad (2)$$

We call this quantity as *incomplete Shannon entropy*.

Here we consider a single-component system which only does expansion work (systems that do work in other forms can be analyzed analogously) and is surrounded by a heat-particle source. The system can do work by changing volume, and it can exchange both heat and particles with the source. This system is completely open. The source is so enormous that these interactions do not influence its macroscopic states, and each microscopic state of the system can have a different particle number N_i, mechanical energy E_i and volume V_i, in which the volume V_i is a microscopic quantity, and represents a new concept, and it is different from purely mechanical quantities such as coordinates and momentum, while having some common ground with the mechanical energy particle number, For example, a determined value of the volume V_i may correspond more microscopic states such as the mechanical energy E_i and particle number V_i. The volume is alterable in a completely open system, when the volume change the coordinates of particles can also change (i.e., the microscopic states change). So, different microscopic states may correspond with different volume, the energy and particle number.

After the system and the source reach equilibrium, the mean values of N, E=U and V will be definite. They are given by the formulas $\sum_{i=1}^{w} p_i^q E_i = U$, $\sum_{i=1}^{w} p_i^q V_i = V$, $\sum_{i=1}^{w} p_i^q = 1$ (3)

We can introduce a Lagrangian function

$$F = \sum_{i=1}^{w} p_i^q \ln p_i + \alpha \sum_{i=1}^{w} p_i^q N_i + \beta \sum_{i=1}^{w} p_i^q E_i + \kappa \sum_{i=1}^{w} p_i^q V_i + \gamma \sum_{i=1}^{w} p_i^q \quad S_0/k$$

. Where α, β, ϰ and γ are Lagrangian multipliers. From this we can calculate the conditional extrema, the probability distribution of microstates pi (called the N-E-V distribution, because the particle number N, energy E and volume V are all alterable), and the partition function Zq:

$$p_i = \exp(-\alpha N_i - \beta E_i - \kappa V_i - \gamma - 1/q) = \exp(-\alpha N_i - \beta E_i - \kappa V_i)/Z_q(\alpha,\beta,\kappa) \tag{4}$$

$$Z_q(\alpha,\beta,\kappa) = \exp(\gamma + 1/q) = \left[\sum_{i=1}^{w}\exp(-\alpha q N_i - \beta q E_i - \kappa q V_i)\right]^{1/q} \tag{5}$$

Inserting Eq. (4) into the incomplete Shannon entropy (2) with $S_0=0$, we obtain $S_q = k(\gamma + 1/q + \alpha N + \beta U + \kappa V)$. The constraint that E=0, V=0 and Sq=0 should hold when N=0 leads to the rule $\gamma+1/q=0$, so $Z(\alpha, \beta, \varkappa)=1$ always. In other words, the partition function is normalizing. We cannot calculate any thermodynamic quantities on this basis. However, if we select an arbitrary constant $S0\neq0$ (non-extensive entropy), then for N=0 we have $\gamma+1/q\neq0$. The function $Z(\alpha, \beta, \varkappa)$ in Eq. (5) is not always equal to 1, and becomes a practical partition function.

2.2. *The thermodynamic formulas*

Substituting (4) and (5) into (3), we obtain

$$N = -\frac{\partial \ln Z_q(\alpha,\beta,\kappa)}{\partial\alpha},\quad U = -\frac{\partial \ln Z_q(\alpha,\beta,\kappa)}{\partial\beta},\quad V = -\frac{\partial \ln Z_q(\alpha,\beta,\kappa)}{\partial\kappa}. \tag{6}$$

And substituting (4) and (5) into (2), we have

$$S_q = -k\left(\alpha\frac{\partial \ln Z_q}{\partial\alpha} + \beta\frac{\partial \ln Z_q}{\partial\beta} + \kappa\frac{\partial \ln Z_q}{\partial\kappa}\right) \tag{7}$$

Notice (6), we have now $S_q = k(\alpha N + \beta U + \kappa V)$ (8)

Obviously there are

$$k\alpha = \left(\frac{\partial S_q}{\partial N}\right)_{U,V},\quad k\beta = \left(\frac{\partial S_q}{\partial U}\right)_{N,V},\quad k\kappa = \left(\frac{\partial S_q}{\partial V}\right)_{N,U} \tag{9}$$

Contrast Eq. (8) with the fundamental thermodynamic equation for open systems, TdS=dU+PdV−μdN. By analogy, we can select the constants

$$\alpha = -\mu/kT,\quad \beta = 1/kT,\quad \kappa = P/kT. \tag{10}$$

Moreover, the fluctuations in the thermodynamic variables are

$$\frac{\overline{N_i^2}-N^2}{N^2} = \frac{1}{qZ_q^2N^2}\left[Z_q\frac{\partial^2 Z_q}{\partial\alpha^2} - \left(\frac{\partial Z_q}{\partial\alpha}\right)^2\right],\quad \frac{\overline{E_i^2}-U^2}{U^2} = \frac{1}{qZ_q^2U^2}\left[Z_q\frac{\partial^2 Z_q}{\partial\beta^2} - \left(\frac{\partial Z_q}{\partial\beta}\right)^2\right],\quad \frac{\overline{V_i^2}-V^2}{V^2} = \frac{1}{qZ_q^2V^2}\left[Z_q\frac{\partial^2 Z_q}{\partial\kappa^2} - \left(\frac{\partial Z_q}{\partial\kappa}\right)^2\right] \tag{11}$$

3. Applications

Consider a linear filament composed of independent units), each of which has two (or more, for example DNA) possible states: short (length a and energy εa) or long (length b and energy εb). T is the temperature of the system, and J is the tension in the filament. The partition function for the N-E-V distribution for this system is

$$Z_q(\alpha,\beta,\kappa) = \left(\sum_i e^{-\alpha q N_i - \beta q E_i - \kappa q L_i}\right)^{1/q} = \left(\sum_{N=0}^{\infty} e^{-\alpha q N}\sum_{E_i,L}\Omega e^{-\beta q E_i(L) - \kappa q L}\right)^{1/q}, \tag{12}$$

Where $\alpha = -\mu/kT$, $\beta = 1/kT$, $\varkappa = -J/kT$, and Ω is the degeneracy of the system with energy level Ei and length L. If there are Na units in state a and Nb units in state b, then the energy and length of the system are

$$E_i(L) = N_a\varepsilon_a + N_b\varepsilon_b,\quad L = N_a a + N_b b,\quad (N = N_a + N_b) \tag{13}$$

The degeneracy Ω=N!/（Na! Nb! ） is the combinatorial number for selecting Na from N units.

Inserting Eqs. (13) into Eq. (12), we have:

$$Z_q(\alpha,\beta,\kappa) = \left\{\sum_{N=0}^{\infty}\left[e^{-\alpha q}\left(e^{-\kappa q a - \beta q \varepsilon_a} + e^{-\kappa q b - \beta q \varepsilon_b}\right)\right]^N\right\}^{1/q}$$

We define $y \equiv e^{-\alpha q}\left(e^{-\kappa q a - \beta q \varepsilon_a} + e^{-\kappa q b - \beta q \varepsilon_b}\right)$. When y<1, the series becomes $Z_q(\alpha,\beta,\kappa) = (1-y)^{-1/q}$.

The average unit number is

$$\overline{N} = -\frac{\partial \ln Z(\alpha,\beta,\kappa)}{\partial\alpha}\ \frac{y}{1-y}. \tag{14}$$

The average length is

$$\overline{L} = -\frac{\partial \ln Z(\alpha,\beta,\kappa)}{\partial\kappa} = \overline{N}\frac{ae^{(Ja-\varepsilon_a)q/kT} + be^{(Jb-\varepsilon_b)q/kT}}{e^{(Ja-\varepsilon_a)q/kT} + e^{(Jb-\varepsilon_b)q/kT}} = \overline{Nl}. \tag{15}$$

Where $\bar{l} \equiv \dfrac{a e^{(Ja-\varepsilon_a)q/kT} + b e^{(Jb-\varepsilon_b)q/kT}}{e^{(Ja-\varepsilon_a)q/kT} + e^{(Jb-\varepsilon_b)q/kT}}$ is the average length of a single unit in the filament.

The internal energy is

$$U = -\frac{\partial \ln Z(\alpha,\beta,\kappa)}{\partial \beta} = \overline{N}\frac{\varepsilon_a e^{(Ja-\varepsilon_a)q/kT} + \varepsilon_b e^{(Jb-\varepsilon_b)q/kT}}{e^{(Ja-\varepsilon_a)q/kT} + e^{(Jb-\varepsilon_b)q/kT}} = \overline{N}\overline{\varepsilon}\ . \qquad (16)$$

Where $\overline{\varepsilon} \equiv (\varepsilon_a e^{(Ja-\varepsilon_a)q/kT} + \varepsilon_b e^{(Jb-\varepsilon_b)q/kT})/(e^{(Ja-\varepsilon_a)q/kT} + e^{(Jb-\varepsilon_b)q/kT})$ is the average energy of a unit.

The results have an extra parameter q compared to the BGS theory, since k/q replaces the Boltzmann constant k. The average unit number also replaces the constant unit number. The solution in BGS theory (i.e., q=1) is compared to two solutions for values of the deformed parameter q>1 and q<1 in Figures 1. The entropy is

$$S_q = \overline{N}\left(\overline{\varepsilon} - \mu - J\right)/T\ . \qquad (17)$$

The effect of the incompleteness can be estimated using the energy difference:

$$\Delta U = U - U_0 = \overline{N}\frac{(\varepsilon_a - \varepsilon_b)e^{(Ja-\varepsilon_a)q/kT+(Jb-\varepsilon_b)/kT} + (\varepsilon_b - \varepsilon_a)e^{(Jb-\varepsilon_b)q/kT+(Ja-\varepsilon_a)/kT}}{\left(e^{(Ja-\varepsilon_a)q/kT} + e^{(Jb-\varepsilon_b)q/kT}\right)\left(e^{(Ja-\varepsilon_a)/kT} + e^{(Jb-\varepsilon_b)/kT}\right)}, \qquad (18)$$

where U_0 is the internal energy of the linear filament when q=1.
The squared relative fluctuations are

$$\frac{\overline{E_i^2} - U^2}{U^2} = 1 + \frac{\overline{\varepsilon^2}}{\overline{\varepsilon}^2\overline{N}} \propto 1, \quad \frac{\overline{L^2} - \overline{L}^2}{\overline{L}^2} = 1 + \frac{\overline{l^2}}{\overline{l}^2\overline{N}} \propto 1, \quad \frac{\overline{N^2} - \overline{N}^2}{\overline{N}^2} = 1 + 1/\overline{N} \propto 1 \qquad (19)$$

where $\overline{\varepsilon^2} \equiv \dfrac{\varepsilon_a^2 e^{(Ja-\varepsilon_a)q/kT} + \varepsilon_b^2 e^{(Jb-\varepsilon_b)q/kT}}{e^{(Ja-\varepsilon_a)q/kT} + e^{(Jb-\varepsilon_b)q/kT}}$ is the mean squared energy of an individual unit, $\overline{l^2} \equiv \dfrac{a^2 e^{(Ja-\varepsilon_a)/kT} + b^2 e^{(Jb-\varepsilon_b)/kT}}{e^{(Ja-\varepsilon_a)/kT} + e^{(Jb-\varepsilon_b)/kT}}$ is the mean squared length of an individual unit.

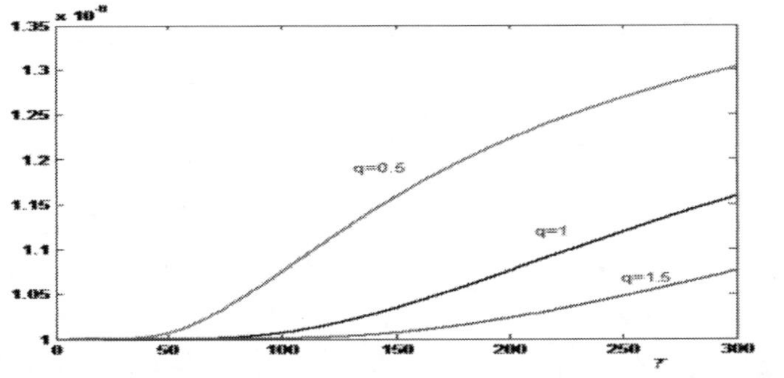

Figure1. The deviation (*U* or *L*) from *q*=1 for values of the deformed parameter *q*>1 and *q*<1..

4. Conclusions

By introducing the concepts of non-extensive entropy constant and microscopic volume in incomplete statistics, we deduce the statistical distribution of a completely open system using the maximum entropy principle in Eqs.(4-5) and the corresponding thermodynamic formulas in Eqs.(6-11).When the deformation parameter q□1,both results reduce to the statistical distribution deduced by T. L. Hill by the method of ensemble transformation in small systems [11]. However, our approach belongs to the family of non-extensive statistical mechanics, and is suitable for systems with any size.

The case of a linear filament system (comparable to biomacromolecules such as DNA) has never before been discussed in incomplete statistics. In this paper, we derive analytical results for the internal energy, average length, and entropy of the linear filament in Eqs. (15-17).

The order of magnitude of the relative fluctuations in thermodynamic quantities is 1. T. L. Hill found the same result for classical systems with very small numbers of particles (10^2-10^3), but our results can apply to the systems of any size. This result can account for the large fluctuations and instability observed in critical phenomena, supercooled states, and overheated states that are not explained by traditional or current statistical mechanics.

Acknowledgment
The work is supported by the National Natural Science Foundation of China (No.11047024), the Key Project of Chinese Ministry of Education (No.211194), the Natural Science Foundation of Ningxia (No.NZ1023) and the Science Research Project for Colleges and Universities in Ningxia (2011).

References:

[1].C. Tsallis, *J. Stat. Phys.* **52**, 479(1988).
[2].Li Heling, XiongYing, Li Yaya, *Physica A* **390**, 2769 (2011). doi:10.1016/j.physa.2011.03.023 .
[3]. Q. A. Wang, *Chaos,Solitons and Fractals* **12**, 1431 (2001).
[4]. Q. A. Wang, *Euro. Phys, J. B* **26**, 357(2002).
[5]. Q. A. Wang, *Chaos,Solitons and Fractals* **19**, 639(2004).
[6]. Q. A. Wang, *Euro. Phys. J. B* **31**, 75(2003).
[7].Q. A. Wang, *Entropy* **5**, 220(2003).
[8].A.S. Parvan, T.S. Biro, *Phys. Lett. A***375**, 372 (2011).
[9]. L. Nivanen, M. Pezeril, Q.A. Wang, A. Le Mehaute, *Chaos Solitons Fractals* **24**, 1337(2005).
[10]. C. Ou, J. Chen, Q.A. Wang, *Chaos Solitons Fractals* **28**, 518(2006).
[11]. T. L. Hill, R. V. Chamberlin, *Nano Lett.***2**, 609（2002）

Bosonization and Vertex Operator of Supersymmetry $U_q(\widehat{sl}(N|1))$ for Level k

Takeo Kojima
Faculty of Engineering, Yamagata University, Jonan 4-3-16, Yonezawa 992-8510, Japan
E-mail: kojima@yz.yamagata-u.ac.jp

Abstract. We construct a bosonization of the quantum superalgebra $U_q(\widehat{sl}(N|1))$ for an arbitrary level k. We construct the screening that commutes with the quantum superalgebra for an arbitrary level $k \neq -N+1$. We propose a bosonization of the vertex operator that gives the intertwiner among the Wakimoto realization and the typical representation.

1. Introduction

There have been many developments in exactly solvable models. Various methods were invented to solve models. The bosonization provides a powerful method to study exactly solvable models. We review recent developments in bosonizations of the $U_q(\widehat{sl}(N|1))$ [7, 8, 10]. The trace of our bosonizations of the vertex operators gives the correlation function of the higher level(spin) k and rank N generalization of the supersymmetric t-J model ($U_q(\widehat{sl}(2|1))$ for level $k=1$) [1].

2. Bosonization of $U_q(\widehat{sl}(N|1))$

We recall Drinfeld generators of the superalgebra $U_q(\widehat{sl}(N|1))$ [3]. We fix a complex number $q \neq 0, |q| < 1$. We use the notations $[x,y] = xy - yx$, $\{x,y\} = xy + yx$, $[a]_q = \frac{q^a - q^{-a}}{q - q^{-1}}$. The Cartan matrix $(A_{i,j})_{0 \leq i,j \leq N}$ of the affine Lie algebra $\widehat{sl}(N|1)$ is given by

$$A_{i,j} = (\nu_i + \nu_{i+1})\delta_{i,j} - \nu_i \delta_{i,j+1} - \nu_{i+1}\delta_{i+1,j}.$$

Here we set $\nu_1 = \cdots = \nu_N = +, \nu_{N+1} = \nu_0 = -$. The Drinfeld generators of the quantum superalgebra $U_q(\widehat{sl}(N|1))$ are $x^{\pm}_{i,m}$, $h_{i,m}$, c, $(1 \leq i \leq N, m \in \mathbf{Z})$. Defining relations are

$$c : \text{central}, \ [h_i, h_{j,m}] = 0,$$

$$[a_{i,m}, h_{j,n}] = \frac{[A_{i,j}m]_q [cm]_q}{m} q^{-c|m|} \delta_{m+n,0} \quad (m,n \neq 0), \ [h_i, x^{\pm}_j(z)] = \pm A_{i,j} x^{\pm}_j(z),$$

$$[h_{i,m}, x^{+}_j(z)] = \frac{[A_{i,j}m]_q}{m} q^{-c|m|} z^m x^{+}_j(z), \quad [h_{i,m}, x^{-}_j(z)] = -\frac{[A_{i,j}m]_q}{m} z^m x^{-}_j(z) \quad (m \neq 0),$$

$$(z_1 - q^{\pm A_{i,j}} z_2) x^{\pm}_i(z_1) x^{\pm}_j(z_2) = (q^{\pm A_{j,i}} z_1 - z_2) x^{\pm}_j(z_2) x^{\pm}_i(z_1) \quad \text{for } |A_{i,j}| \neq 0,$$

$$x^{\pm}_i(z_1) x^{\pm}_j(z_2) = x^{\pm}_j(z_2) x^{\pm}_i(z_1) \quad \text{for } |A_{i,j}| = 0, (i,j) \neq (N,N), \ \{x^{\pm}_N(z_1), x^{\pm}_N(z_2)\} = 0,$$

$$[x^{+}_i(z_1), x^{-}_j(z_2)] = \frac{\delta_{i,j}}{(q - q^{-1}) z_1 z_2} \left(\delta(q^{-c} z_1/z_2) \psi^{+}_i(q^{\frac{c}{2}} z_2) - \delta(q^{c} z_1/z_2) \psi^{-}_i(q^{-\frac{c}{2}} z_2) \right),$$

Published under licence by IOP Publishing Ltd

Journal of Physics: Conference Series **410** (2013) 012062 doi:10.1088/1742-6596/410/1/012062

$$
\begin{aligned}
&\text{for} \quad (i,j) \neq (N,N),\\
&\{x_N^+(z_1), x_N^-(z_2)\} = \frac{1}{(q-q^{-1})z_1z_2}\left(\delta(q^{-c}z_1/z_2)\psi_N^+(q^{\frac{c}{2}}z_2) - \delta(q^c z_1/z_2)\psi_N^-(q^{-\frac{c}{2}}z_2)\right),\\
&\left(x_i^\pm(z_1)x_i^\pm(z_2)x_j^\pm(z) - (q+q^{-1})x_i^\pm(z_1)x_j^\pm(z)x_i^\pm(z_2) + x_j^\pm(z)x_i^\pm(z_1)x_i^\pm(z_2)\right)\\
&+(z_1 \leftrightarrow z_2) = 0 \quad \text{for } |A_{i,j}| = 1,\ i \neq N,
\end{aligned}
$$

where we have used $\delta(z) = \sum_{m\in\mathbf{Z}} z^m$. Here we have used the generating function $x_j^\pm(z) = \sum_{m\in\mathbf{Z}} x_{j,m}^\pm z^{-m-1}$, $\psi_i^\pm(q^{\pm\frac{c}{2}}z) = q^{\pm h_i} e^{\pm(q-q^{-1})\sum_{m>0} h_{i,\pm m} z^{\mp m}}$ and the abbreviation $h_i = h_{i,0}$.

We construct bosonizations of superalgebra $U_q(\hat{sl}(N|1))$ for an arbitrary level [7]. We fix the level $c = k \in \mathbf{C}$. We introduce the bosons and the zero-mode operators a_m^j, Q_a^j $(m \in \mathbf{Z}, 1 \leq j \leq N)$, $b_m^{i,j}, Q_b^{i,j}$ $(m \in \mathbf{Z}, 1 \leq i < j \leq N+1)$, $c_m^{i,j}, Q_c^{i,j}$ $(m \in \mathbf{Z}, 1 \leq i < j \leq N)$ which satisfy

$$
\begin{aligned}
&[a_m^i, a_n^j] = \frac{[(k+N-1)m]_q [A_{i,j}m]_q}{m}\delta_{m+n,0} \quad (m,n \neq 0), \quad [a_0^i, Q_a^j] = (k+N-1)A_{i,j},\\
&[b_m^{i,j}, b_n^{i',j'}] = -\nu_i\nu_j \frac{[m]_q^2}{m}\delta_{i,i'}\delta_{j,j'}\delta_{m+n,0} \quad (m,n \neq 0), \quad [b_0^{i,j}, Q_b^{i',j'}] = -\nu_i\nu_j\delta_{i,i'}\delta_{j,j'},\\
&[c_m^{i,j}, c_n^{i',j'}] = \frac{[m]_q^2}{m}\delta_{i,i'}\delta_{j,j'}\delta_{m+n,0} \quad (m,n \neq 0), \quad [c_0^{i,j}, Q_c^{i',j'}] = \delta_{i,i'}\delta_{j,j'},\\
&[Q_b^{i,j}, Q_b^{i',j'}] = \delta_{j,N+1}\delta_{j',N+1}\pi\sqrt{-1} \qquad ((i,j) \neq (i',j')).
\end{aligned}
$$

Other commutation relations are zero. In what follows we use the standard symbol of the normal orderings :: and use the following abbreviations $b^{i,j}(z), c^{i,j}(z), b_\pm^{i,j}(z), a_\pm^j(z)$ and $\left(\frac{\gamma_1}{\beta_1}\frac{\gamma_2}{\beta_2}\cdots\frac{\gamma_r}{\beta_r}\, a^i\right)(z|\alpha)$ given by

$$
\begin{aligned}
b^{i,j}(z) &= -\sum_{m\neq 0}\frac{b_m^{i,j}}{[m]_q}z^{-m} + Q_b^{i,j} + b_0^{i,j}\log z, \quad c^{i,j}(z) = -\sum_{m\neq 0}\frac{c_m^{i,j}}{[m]_q}z^{-m} + Q_c^{i,j} + c_0^{i,j}\log z,\\
b_\pm^{i,j}(z) &= \pm(q-q^{-1})\sum_{\pm m>0} b_m^{i,j}z^{-m} \pm b_0^{i,j}\log q, \quad a_\pm^j(z) = \pm(q-q^{-1})\sum_{\pm m>0} a_m^j z^{-m} \pm a_0^j \log q,\\
\left(\frac{\gamma_1}{\beta_1}\frac{\gamma_2}{\beta_2}\cdots\frac{\gamma_r}{\beta_r}\, a^i\right)(z|\alpha) &= -\sum_{m\neq 0}\frac{[\gamma_1 m]_q\cdots[\gamma_r m]_q}{[\beta_1 m]_q\cdots[\beta_r m]_q}\frac{a_m^i}{[m]_q}q^{-\alpha|m|}z^{-m} + \frac{\gamma_1\cdots\gamma_r}{\beta_1\cdots\beta_r}(Q_a^i + a_0^i\log z).
\end{aligned}
$$

The generating functions $x_i^\pm(z)$, $\psi_i^\pm(z)$, $(1 \leq i \leq N)$ of $U_q(\widehat{sl}(N|1))$ for an arbitrary level k are realized by the bosonic operators as follows. This is main result of [7].

$$
\begin{aligned}
x_i^+(z) &= \frac{1}{(q-q^{-1})z} : \sum_{j=1}^{i} e^{(b+c)^{j,i}(q^{j-1}z) + \sum_{l=1}^{j-1}(b_+^{l,i+1}(q^{l-1}z) - b_+^{l,i}(q^l z))} \times\\
&\times \left\{ e^{b_+^{j,i+1}(q^{j-1}z) - (b+c)^{j,i+1}(q^j z)} - e^{b_-^{j,i+1}(q^{j-1}z) - (b+c)^{j,i+1}(q^{j-2}z)} \right\} :,\\
x_N^+(z) &= : \sum_{j=1}^{N} e^{(b+c)^{j,N}(q^{j-1}z) + b^{j,N+1}(q^{j-1}z) - \sum_{l=1}^{j-1}(b_+^{l,N+1}(q^l z) + b_+^{l,N}(q^l z))} :,\\
x_i^-(z) &= q^{k+N-1} : e^{a_+^i(q^{\frac{k+N-1}{2}}z) - b^{i,N+1}(q^{k+N-1}z) - b_+^{i+1,N+1}(q^{k+N-1}z) + b^{i+1,N+1}(q^{k+N}z)} :\\
&+ \frac{1}{(q-q^{-1})z} : \left\{ \sum_{j=1}^{i-1} e^{a_-^i(q^{-\frac{k+N-1}{2}}z) + (b+c)^{j,i+1}(q^{-k-j}z) + b_-^{i,n+1}(q^{-k-n}z) - b_-^{i+1,n+1}(q^{-k-n+1}z)} \right.
\end{aligned}
$$

$$
\begin{aligned}
\times\ & e^{\sum_{l=j+1}^{i}(b_-^{l,i+1}(q^{-k-l+1}z)-b_-^{l,i}(q^{-k-l}z))+\sum_{l=i+1}^{N}(b_-^{i,l}(q^{-k-l}z)-b_-^{i+1,l}(q^{-k-l+1}z))}\\
\times\ & \left(e^{-b_-^{j,i}(q^{-k-j}z)-(b+c)^{j,i}(q^{-k-j+1}z)}-e^{-b_+^{j,i}(q^{-k-j}z)-(b+c)^{j,i}(q^{-k-j-1}z)}\right)\\
+\ & e^{a_-^{i}(q^{-\frac{k+N-1}{2}}z)+(b+c)^{i,i+1}(q^{-k-i}z)+\sum_{l=i+1}^{N}(b_-^{i,l}(q^{-k-l}z)-b_-^{i+1,l}(q^{-k-l+1}z))+b_-^{i,N+1}(q^{-k-N}z)-b_-^{i+1,N+1}(q^{-k-N+1}z)}\\
-\ & e^{a_+^{i}(q^{\frac{k+N-1}{2}}z)+(b+c)^{i,i+1}(q^{k+i}z)+\sum_{l=i+1}^{N}(b_+^{i,l}(q^{k+l}z)-b_+^{i+1,l}(q^{k+l-1}z))+b_+^{i,N+1}(q^{k+N}z)-b_+^{i+1,N+1}(q^{k+N-1}z)}\\
-\ & \sum_{j=i+1}^{N-1} e^{a_+^{i}(q^{\frac{k+N-1}{2}}z)+(b+c)^{i,j+1}(q^{k+j}z)+b_+^{i,N+1}(q^{k+N}z)-b_+^{i+1,N+1}(q^{k+N-1}z)+\sum_{l=j+1}^{N}(b_+^{i,l}(q^{k+l}z)-b_+^{i+1,l}(q^{k+l-1}z))}\\
\times\ & \left(e^{b_+^{i+1,j+1}(q^{k+j}z)-(b+c)^{i+1,j+1}(q^{k+j+1}z)}-e^{b_-^{i+1,j+1}(q^{k+j}z)-(b+c)^{i+1,j+1}(q^{k+j-1}z)}\right)\Big\}:.\\
x_N^-(z) =\ & \frac{1}{(q-q^{-1})z}:\left\{\sum_{j=1}^{N-1} e^{a_-^{N}(q^{-\frac{k+N-1}{2}}z)-b_+^{j,N+1}(q^{-k-j}z)-b^{j,N+1}(q^{-k-j-1}z)-\sum_{l=j+1}^{N-1}(b_-^{l,N}(q^{-k-l}z)+b_-^{l,N+1}(q^{-k-l}z))}\right.\\
\times\ & q^{j-1}\left(e^{-b_+^{j,N}(q^{-k-j}z)-(b+c)^{j,N}(q^{-k-j-1}z)}-e^{-b_-^{j,N}(q^{-k-j}z)-(b+c)^{j,N}(q^{-k-j+1}z)}\right)\Big\}:\\
+\ & q^{N-1}:\left(e^{a_+^{N}(q^{\frac{k+N-1}{2}}z)-b^{N,N+1}(q^{k+N-1}z)}-e^{a_-^{N}(q^{-\frac{k+N-1}{2}}z)-b^{N,N+1}(q^{-k-N+1}z)}\right):.
\end{aligned}
$$

$$
\begin{aligned}
\psi_i^{\pm}(q^{\pm\frac{k}{2}}z) &= e^{a_\pm^{i}(q^{\pm\frac{k+N-1}{2}}z)+\sum_{l=1}^{i}(b_\pm^{l,i+1}(q^{\pm(l+k-1)}z)-b_\pm^{l,i}(q^{\pm(l+k)}z)}\\
&\quad\times e^{\sum_{l=i+1}^{N}(b_\pm^{i,l}(q^{\pm(k+l)}z)-b_\pm^{i-1,l}(q^{\pm(k+l-1)}z)+b_\pm^{i,N+1}(q^{\pm(k+N)}z)-b_\pm^{i+1,N+1}(q^{\pm(k+N-1)}z)},\\
\psi_N^{\pm}(q^{\pm\frac{k}{2}}z) &= e^{a_\pm^{N}(q^{\pm\frac{k+N-1}{2}}z)-\sum_{l=1}^{N-1}(b_\pm^{l,N}(q^{\pm(k+l)}z)+b_\pm^{l,N+1}(q^{\pm(k+l)}z))}.
\end{aligned}
$$

3. Vertex Operator

We construct the screening $\mathcal{Q}_i$ that commutes with the quantum superalgebra $U_q(\widehat{sl}(N|1))$ for an arbitrary level $k \neq -N+1$ [10]. The Jackson integral with parameter $p \in \mathbf{C}$ ($|p| < 1$) and $s \in \mathbf{C}^*$ is defined by $\int_0^{s\infty} f(z)d_pz = s(1-p)\sum_{m\in\mathbf{Z}} f(sp^m)p^m$. We introduce the screening operators $\mathcal{Q}_i$ ($1 \le i \le N$) by using the Jackson integral. This is one of main result of [10].

$$\mathcal{Q}_i = \int_0^{s\infty} : e^{-\left(\frac{1}{k+N-1}a^i\right)\left(z\middle|\frac{k+N-1}{2}\right)}\widetilde{S}_i(z) : d_pz, \quad (p = q^{2(k+N-1)}).$$

Here we have set the bosonic operators $\widetilde{S}_i(z)$ $(1 \le i \le N)$ by

$$
\begin{aligned}
\widetilde{S}_i(z) &= \frac{1}{(q-q^{-1})z}\sum_{j=i+1}^{N} :\left(e^{-b_-^{i,j}(q^{N-1-j}z)-(b+c)^{i,j}(q^{N-j}z)}-e^{-b_+^{i,j}(q^{N-1-j}z)-(b+c)^{i,j}(q^{N-j-2}z)}\right)\\
&\quad\times e^{(b+c)^{i+1,j}(q^{N-1-j}z)+\sum_{l=j+1}^{N}(b_-^{i+1,l}(q^{N-l}z)-b_-^{i,l}(q^{N-l-1}z))+b_-^{i+1,N+1}(z)-b_-^{i,N+1}(q^{-1}z)}:\\
&\quad+ q : e^{b^{i,N+1}(z)+b_+^{i+1,N+1}(z)-b^{i+1,N+1}(qz)} : \qquad (1 \le i \le N-1),\\
\widetilde{S}_N(z) &= -q^{-1} : e^{b^{N,N+1}(z)} :.
\end{aligned}
$$

The screening $\mathcal{Q}_i$ commutes with the quantum superalgebra.

$$[\mathcal{Q}_i, U_q(\widehat{sl}(N|1))] = 0 \quad (1 \le i \le N).$$

For $p_a^i \in \mathbf{C}$ $(1 \le i \le N)$, $p_b^{i,j} \in \mathbf{C}$ $(1 \le i < j \le N+1)$, $p_c^{i,j} \in \mathbf{C}$ $(1 \le i < j \le N)$, we set the vector $|p_a, p_b, p_c\rangle$ which satisfies the following conditions.

$$
\begin{aligned}
& a_m^i|p_a,p_b,p_c\rangle = b_m^{i,j}|p_a,p_b,p_c\rangle = c_m^{i,j}|p_a,p_b,p_c\rangle = 0 \quad (m>0),\\
& a_0^i|p_a,p_b,p_c\rangle = p_a^i|p_a,p_b,p_c\rangle,\ b_0^{i,j}|p_a,p_b,p_c\rangle = p_b^{i,j}|p_a,p_b,p_c\rangle,\ c_0^{i,j}|p_a,p_b,p_c\rangle = p_c^{i,j}|p_a,p_b,p_c\rangle.
\end{aligned}
$$

The boson Fock space $F(p_a, p_b, p_c)$ is generated by the bosons $a_m^i, b_m^{i,j}, c_m^{i,j}$ on the vector $|p_a, p_b, p_c\rangle$. We set the space $F(p_a)$ by

$$F(p_a) = \bigoplus_{\substack{p_b^{i,j}=-p_c^{i,j}\in\mathbf{Z}\ (1\leq i<j\leq N)\\ p_b^{i,N+1}\in\mathbf{Z}\ (1\leq i\leq N)}} F(p_a, p_b, p_c).$$

We would like to construct the vertex operator $\Phi^*(z)$ [10] which gives the intertwiner among $F(p_a)$ and the typical representation [13].

$$\Phi^*(z) : F(p_a) \longrightarrow F(p_a + l_a + x_a) \otimes V_z^{*S}, \quad \Phi^*(z) = \sum_j \Phi_j^*(z) \otimes v_j^*.$$

Here V_z^{*S} is the dual evaluation representation of the typical representation with the weight $l_a = (l_a^1, l_a^2, \cdots, l_a^N)$ [13]. The basis of V^{*S} is given by $\{v_j^*\}$. The coefficients are linear maps : $\Phi_j^*(z) : F(p_a + l_a + x_a) \longrightarrow F(p_a)$. For $l_a = (l_a^1, l_a^2, \cdots, l_a^N)$ and $x_a = (x_a^1, x_a^2, \cdots, x_a^N)$, we set the highest element $\Phi_1^*(z)$ of the vertex operator by

$$\Phi_1^*(z) =: \prod_{j=1}^N \mathcal{Q}_j^{x_j} \quad e^{-\sum_{i,j=1}^N \left(\frac{l_a^i}{k+N-1} \frac{\mathrm{Min}(i,j)}{N-1} \frac{N-1-\mathrm{Max}(i,j)}{1} a^j\right)(q^k z|-\frac{k-N+1}{2})} :,$$

where $(x_1, x_2, \cdots, x_N)$ is related to x_a^i by $x_a^i = \sum_{j=1}^N A_{i,j} x_j$. Other elements $\Phi_j^*(z)$ $(j \neq 1)$ of the vertex operators are determined by the intertwining property. We conjecture that this bosonization of the vertex operator $\Phi^*(z)$ gives the intertwiner among our bosonization and the typical representation. Using the Gelfand-Zetlin basis [13], we have checked this conjecture in some cases for $N = 2, 3, 4$ [10]. We balance the "background charge" of the vertex operators by using the screening $\mathcal{Q}_i$. Sometimes we have to multiply nontrivial product of the screenings $\mathcal{Q}_i$ inside the vertex operator in order to have non-zero correlation functions (trace of vertex operators). At the end of this paper, we would like to give some comments on relating works [8, 12]. In [8] we study how to get the Wakimoto realization from our bosonization by ξ-η system. In [12] we study how to construct the elliptic deformed algebra $U_{q,p}(\widehat{sl}(M|N))$ from the quantum group $U_q(\widehat{sl}(M|N))$. Using deformation method developed in [12] we obtain a bosonization of the elliptic deformed algebra $U_{q,p}(\widehat{sl}(N|1))$ for an arbitrary level k.

Acknowledgement
This work is supported by the Grant-in-Aid for Scientific Research **C** (21540228) from Japan Society for Promotion of Science.

References

[1] Zhang F and Rice T 1988 *Phys.Rev.* **B37** 3759
[2] Kac V 1977 *Adv.Math.* **26**(1) 8
[3] Yamane H 1999 *Publ.Res.Inst.Math.Sci.* **35** 321
[4] Wakimoto M 1986 *Commun.Math.Phys.* **104** 605
[5] Awata H, Odake S and Shiraishi J 1994 *Commun.Math.Phys.* **162** 61
[6] Awata H, Odake S and Shiraishi J 1997 *Lett. Math. Phys.* **42** 271
[7] Kojima T 2011 *J.Math.Phys.* **53** 013515.
[8] Kojima T 2011 The q-Wakimoto realization of the superalgebras $U_{q,p}(\hat{sl}(N|1))$ and $U_{q,p}(\widehat{sl}(N|1))$ to appear in *Springer Proceedings : Statistics and Mathematics* (*Preprint* arXiv.1201.4222)
[9] Zhang Y and Gould M 2000 *J.Math.Phys.* **41** 5577
[10] Kojima T 2012 *J.Math.Phys.* **53** 083503
[11] Jimbo M, Konno H, Odake S and Shiraishi J 1999 *Commun.Math.Phys.* **199** 605
[12] Kojima T 2011 *J.Phys.* **A44** 485205
[13] Palev T and Tolstoy V 1991 *Commun.Math.Phys.* **141** 549

Applications of Complex Networks on Analysis of World Trade Network

Jae Woo Lee[1], Seong Eun Maeng, Gyeong-Gyun Ha, Moon Hyeok Lee, Eun Seong Cho

[1] Department of Physics, Inha University, Incheon 402-751 Korea
E-mail: jaewlee@inha.ac.kr

Abstract. We consider the wealth and the money flow of the world trade data. We analyze the world trade data from year 1948 to 2000 which include the total amounts of the import and export for every country per year. We apply the analyzing methods of the complex networks to the world trade network. We define the wealth as the gross domestic products (GDP) of each country. We defined the backbone network of the world trade network. We generate the backbone network keeping the link with the largest wealth flowing out each country by the import and deleting all remaining links. We observed that the wealth was transferred from the poorer countries to the wealthier countries. We found the asymmetry of the world trade flow by the disparity of the networks. From the backbone network of the world trade we can identify the regional economic connections and wealth flow among the countries.

1. Introduction

The economic and financial systems are complex systems which show many emerging properties. Many stylized facts reported on the time series of the stock market and foreign exchange [1,2]. The complex networks were generated by the cross-correlations between the return time series of the stock indices and the foreign exchange rates [1,2]. The minimal spanning trees of the correlation networks showed the scale-free behaviours. The world trade networks (WTN) are extracted from the world trade relationship among the countries [3,4]. In WTN the nodes of the network corresponds to the countries and the size of the node is equal to the relative magnitude of the each country's GDP. We connect the two countries if there are trading relationships such as the import or export. The width of the links corresponds to the relative size of the trading magnitude.

The world trade networks are almost fully connected networks because the country trade with other countries over the world. We consider two types of the network: the backbone networks and the minimal spanning trees of the world trade network. These sub-networks extracted from the original world trade networks. We can understand the main stream of the world trade from these networks. In the backbone networks of WTN shows the flowing dynamics of the money and inter-dependence among the countries by the world trade. We indicate the change of the trading structures when we plot the backbone networks of WTN each year. We also observe the economic zone built by the main stream of the world trade.

We consider the wealth asymmetry of the world trade in the section 2. We extract the backbone network of the WTN in section 3. We generate the minimal spanning trees of the WTN in the section 4. We gave the conclusion remarks in the section 5.

2. Wealth Asymmetry of World Trade Network

[1] To whom any correspondence should be addressed.

We consider the global domestic products (GDP) and data of the world trade from year 1950 to year 2000[5]. We define the wealth of a country as the GDP of the country. In the world trade network the node is the country and the size of the node is equal to the relative magnitude of the wealth. The wealth of a country, i, is defined by the its GDP, $G_i(t)$, where t is a year. The money flows from country j to country i by the export of the country i. We denote that the amount of the money, $f_{ij}(t)$, corresponds to the money flowing into the country i from country j by the export of the country i to the country j at the year t. We observed the asymmetric wealth transfers measured by the wealth fractions belonging to the top 20% country and the bottom 20% country. Figure 1 shows the wealth fraction for the rich country and poor country over the year. In the observed period the wealth of the top 20% rich country increase monotonically, but the wealth of the bottom 20% poor country decrease. Although the total wealth of the world are increasing monotonically over the year, the portions of the wealth owned by the rich country increase more quickly.

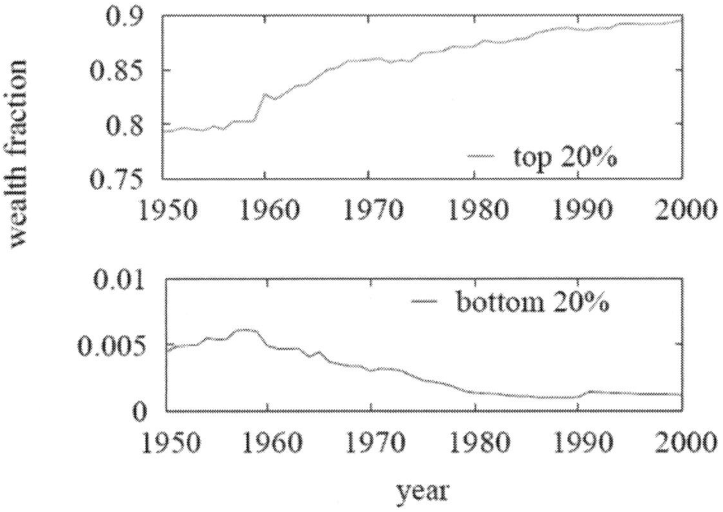

Figure 1. Wealth fractions over the year belong to the top 20% rich country and bottom 20% poor country. The wealth is transferring from the poorer country to the richer country in the observed period.

3. Backbone Network

We generate the backbone networks of the world trade network. We keep the link of the largest money transfer of each country by the import and delete all other connections. In Figure 2 we present a backbone network of the WTN in year 2000. In the network the size of the node corresponds to the relative magnitude of the GDPs and the width corresponds to the relative magnitude of the money flowing out from the country. From the backbone network we can identify the main stream of the money flow by the import. We can also observe the economic zone by the money flows. These economic zones are very similar to the regional economic zone such as EU economic zone, USA economic zone, Asia economic zone etc.

In the backbone network of year 2000 we observe that USA, Japan (JPN) and German Federal Republic (GFR) are the sink country of the wealth. These countries absorb the wealth by the world trade. An interesting pattern exists in east-Asian economic zone. The money circulation occurs at the countries: Japan-United Arab Emirate-China-Korea-Japan cycle. Japan is one of the largest money absorbing country all over the world.

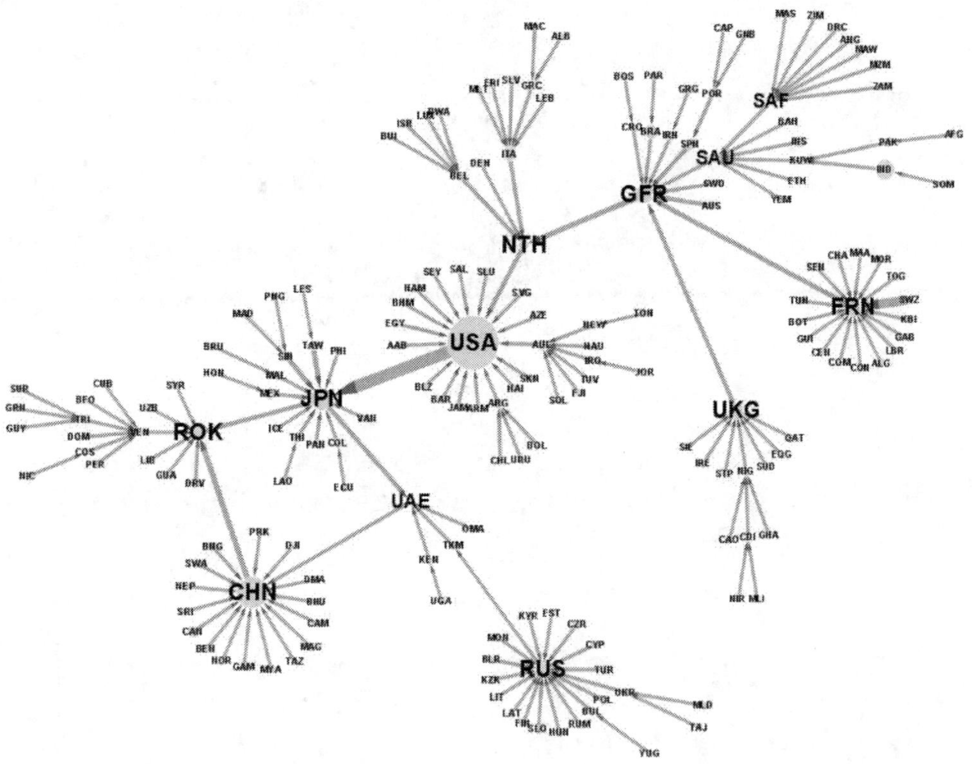

Figure 2. Backbone network of the world trade network in year 2000. We keep the link of the largest import of a country and delete all other connections. The money flows out from the country to the target country because the country transfers the money to buy the products. The size of node corresponds to the relative magnitude of the GDPs. The width of the link corresponds to the relative magnitude of the money flowing out. The name of the node is the country's name, for example, USA (the united state of America), JPN (Japan), GFR (German Federal Republic), CHN (China), etc.

4. Minimal Spanning Tree

Minimal Spanning Tree (MST) generated from the world trade network. We use the Kruskal algorithm to extract the MST in the WTN. In the MST we keep the largest trading links up to (N-1) links. Figure 3 presents the MST in year 2000. This MST is similar to the backbone network of Figure 2. However, there is big difference between two networks. In the backbone network the circular loop can exist, but in the MST the network is just a tree and there is no loop structure. In the MST we also observe the economic zone by the world trade. In the MST we can identify the most influential countries: USA, China, Japan, German Federal Republic, etc. German Federal Republic is the most important trading country in the European countries.

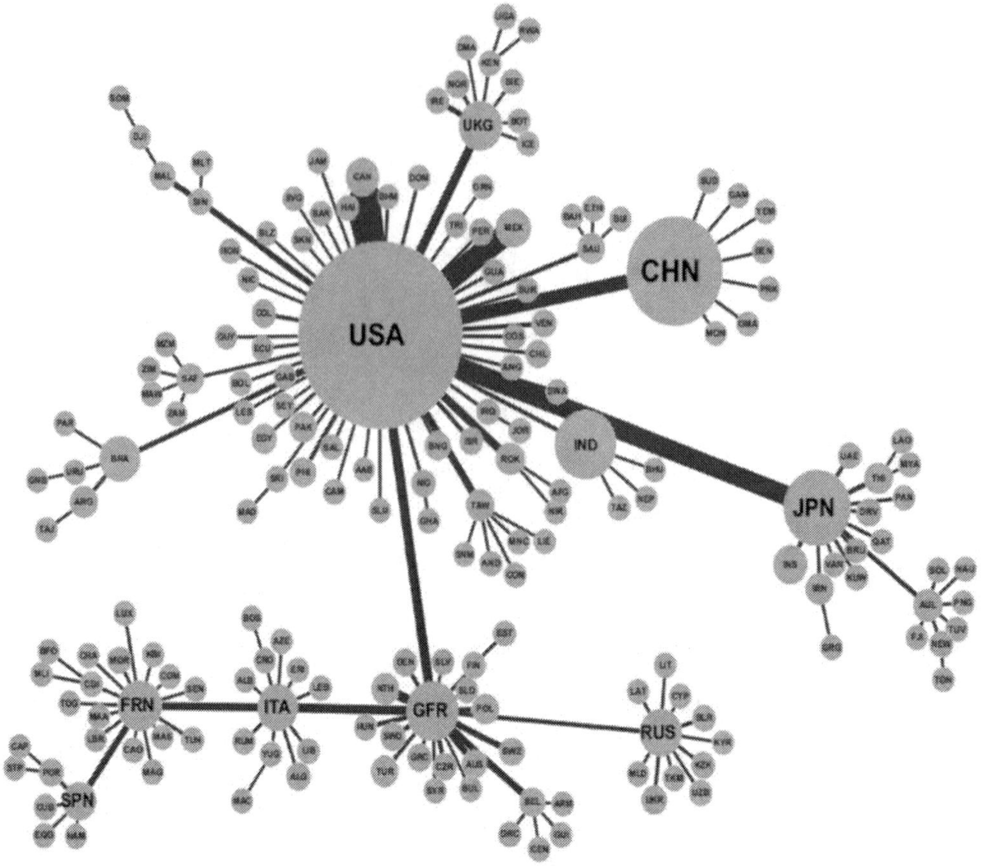

Figure 3. Minimal spanning tree (MST) generated by the import relationship in year 2000. The size of node corresponds to the relative magnitude of the GDPs. The width of the link corresponds to the relative magnitude of the money flowing out.

5. Conclusions

We observed the asymmetry of the world trade network. We generated the backbone network and the minimal spanning trees of the world trade network. We can identify the wealth flow and the formation of sub-economic zone in these networks.

Acknowledgments

This work was supported by the Korea Foundation for the Advancement of Science & Creativity(KOFAC) grant funded by the Korean Government(MEST).

References

[1] Mantegna R N and Stanley H E *An Introduction to Econophysics: Correlations and Complexity in Finance* (Cambridge University Press, New York, 2000) p 20
[2] Voit J *The Statistical Mechanics of Financial Markets* (Springer, Berlin, 2001) P 59.
[3] Fagiolo G, Reyes J and Schiavo S 2009 *Phys. Rev. E* **79** 036115
[4] Garlaschelli D, Di Matteo T, Aste T, Caldarelli G and Loffredo M. I 2007 *Eur. Phys. J. B* **57** 159
[5] Gleditsch K S 2002 *J. Conflict Resolut.* **46** 712

Damage growth in fibre bundle models with localized load sharing and environmentally-assisted ageing

S Lennartz-Sassinek[1], Z Danku[2], F Kun[2], I G Main[1] and M Zaiser[1,3]

[1] School of Engineering and School of Geosciences, University of Edinburgh, Edinburgh, UK
[2] Department of Theoretical Physics, University of Debrecen, Debrecen, Hungary
[3] Department of Materials Science, University of Erlangen-Nürnberg, Fürth, Germany.

E-mail: `sabine.lennartz@ed.ac.uk`

Abstract. One of the most fundamental models for the complex behaviour of damage accumulation in earth materials is the fibre bundle model. One distinguishes between models with localized load sharing (LLS) and models with equal load sharing (ELS). While ELS models can be treated by mean field theory, the behaviour of LLS models is usually more complicated. Here, we consider a fibre bundle model with LLS where, in addition, we introduce a time scale by incorporating time dependent ageing of the fibres due to the accumulation of damage driven by the locally acting stress in a chemically active environment. If the accumulated damage exceeds a random threshold, the fibres fail. The non-trivial time dependence of the cumulative damage in the system can be attributed to different mechanisms that dominate at different time scales. We include this information into an analytical description of the damage accumulation process and show that the analytical description is in agreement with numerical results.

1. Introduction

The study of subcritical fracture processes is an important issue in physics, engineering and geosciences. One of the most fundamental models for the failure of heterogeneous materials is the fibre bundle model (FBM) [1, 2] which envisages a material as a set of parallel fibres with identical elastic properties but with stochastic failure thresholds [1–5]. If a fibre breaks as the local stress exceeds its failure threshold σ_{th}, the stress carried by this fibre is redistributed. In equal load sharing (ELS) models, stress is equally redistributed over all remaining intact fibres, which allows to treat these models by mean field theory. In localized load sharing (LLS) models, on the other hand, redistribution is locally to the close neighborhood of the broken fiber and the behavior is more complicated to analyze. In physical terms, LLS fiber bundle models describe materials failure in situations where the range of internal stresses is limited, e.g. by the vicinity of a free surface [6,7]. To model subcritical failure, the FBM has been extended by incorporating an additional ageing of the fibres [8–12] which accounts for chemical weakening effects associated with the environment. To this end one introduces a damage variable c, and fibres now break either when their stress reaches the threshold σ_{th} or when their damage reaches a threshold c_{th}.

In the present study we consider a FBM on a square lattice with LLS, redistributing the load equally over all intact nearest neighbors of a broken fiber. Regarding the stress-driven ageing process, we assume for the accumulated damage in each fibre a time-dependence of the form

$$c = a \int_0^t \sigma(t')dt' \qquad (1)$$

where $\sigma(t')$ is the time dependent local stress. For σ_{th} we assume a uniform distribution between 0 and 1, and for c_{th} a uniform distribution between $1 - W$ and $1 + W$. By varying W in the range $0 \leq W \leq 1$ we control the amount of damage disorder.

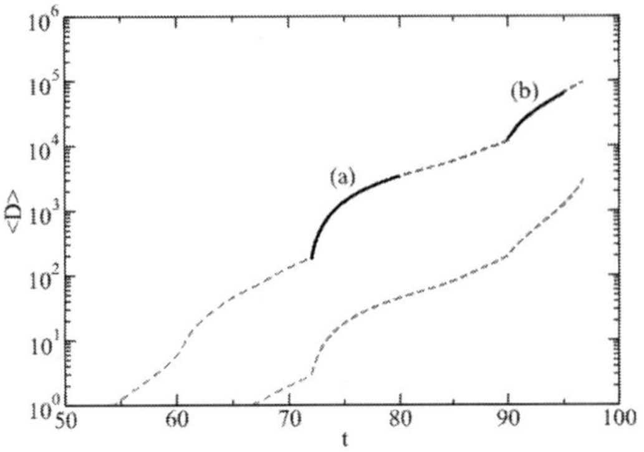

Figure 1. Average damage $\langle D \rangle$ initiated by ageing (dashed blue line) or by local stress redistribution (dashed red line) for a 401×401 sized lattice with $a = 1$, $W = 0.1$ and $\sigma = 0.01$ as a function of time t. The full black curves show the corresponding analytical results from (a) Eq. (6) and (b) Eq. (11) with $d = 146$ and $d' = 2150$ for the respective time periods.

Figure 1 shows the time dependent average damage on a $N = 401 \times 401$ lattice with applied constant stress $\sigma = 0.01$ and constants $a = 1$ and $W = 0.1$. Failures triggered by ageing (dashed blue line) and by local stress redistribution (dashed red line) are plotted separately. One clearly sees that the damage makes significant transient jumps and the functional form of the damage curve seems to be nontrivial. One also sees that in the considered time window failures triggered by local stress redistribution only play a minor role in comparison to those triggered by the ageing process. In the following we focus on this dominant part of the damage and derive analytical equations describing the onset and the functional form of two of the observed jumps. These jumps originate from a sequence of different mechanims that dominate the damage process: In the very beginning of a simulation, a certain fraction of fibres break as we apply the constant stress. Their neighbours thus experience higher local stresses and fail first. Later, also fibres with intact neighbourhood start to fail due to ageing and this process then dominates the system.

2. Breaking of fibres with one broken neighbour

If the number of fibres is large, the smallest threshold for failure by aging has the approximate value $c_{th} \approx 1 - W$. The local stress acting on a fibre with one broken neighbour is $5\sigma/4$. Since the local stress will not increase until other fibres break, the onset time t_1 for failure of fibres with one broken neighbour is given by $1 - W = 5a\sigma t_1/4$, which is equivalent to

$$t_1 = 4(1 - W)/(5a\sigma). \tag{2}$$

Since the critical threshold values for the local stresses are equally distributed between 0 and 1, the number of initially broken fibres can be approximated by σN. Each of these broken fibres has 4 neigbours to which the local stress is redistributed. Thus the total number of fibres with one broken neighbor can initially be approximated by $4N\sigma$. On average all these fibres will have equal distributed thresholds between $1 - W$ and $1 + W$. Thus the average distance between two consecutive thresholds is $W/(2N\sigma)$. After the breaking of i fibres, on average each remaining fibre with one broken neighbour is then carrying the stress $\sigma(i) = \sigma(N/(N - i) + 1/4)$. Between two consecutive breakings, the local stresses do not change. This yields for the time $\tau(i)$ between events i and $(i + 1)$: $W/(2N\sigma) = a\sigma(i)\tau(i)$, which is equivalent to

$$\tau(i) = \frac{W}{a2N\sigma^2(1/4 + N/(N - i))} \approx \frac{2W(N - i)}{5aN^2\sigma^2} \tag{3}$$

where we used that $5N \gg i$. The time of the i-th event is then given by:

$$t_i = t_1 + \sum_{j=1}^{i-1} \tau(j) \approx t_1 + \frac{2W}{5N^2 a\sigma^2}(N(i-1) - (i-1)^2/2) \tag{4}$$

where in the last step $N \gg 1$ has been used. It follows for the cumulative damage D which is equal to the event number i:

$$\begin{aligned} D &= i = 1 + N\left(1 - \sqrt{1 - 5a\sigma^2(t_i - t_1)/W}\right) \\ &= 1 + N\left(1 - \sqrt{1 + 4\sigma(1/W - 1)}\sqrt{1 - t_i\frac{5a\sigma^2}{W + 4\sigma(1-W))}}\right) \end{aligned} \tag{5}$$

where we have used $t_1 = 4(1-W)/(5a\sigma)$ (Eq. (2)). If the system aready contained an initial number d of failed fibers, this has to be added to the damage caused by the newly breaking fibres. Thus, the number of available fibres N has to be replaced by $N - d$, and the stress σ has to be replaced by the effective stress $\sigma N/(N-d)$ leading to

$$\begin{aligned} D &= d + i \\ &= d + 1 + (N-d)\left(1 - \sqrt{1 + \frac{4\sigma(1/W-1)}{(1-d/N)}}\sqrt{1 - t_i\frac{5a\sigma^2}{(1-d/N)^2(W + \frac{4\sigma(1-W)}{(1-d/N)})}}\right) \\ &\approx N - (N-d)\sqrt{1 + \frac{4\sigma(1/W-1)}{(1-d/N)}}\sqrt{1 - t_i\frac{5a\sigma^2}{(1-d/N)^2(W + \frac{4\sigma(1-W)}{(1-d/N)})}} \end{aligned} \tag{6}$$

This equation is plotted for a $N = 401 \times 401$ lattice with $a = 1$, $W = 0.1$, $\sigma = 0.01$ and $d = 146$ together with the numerical results in Fig. 1 and labelled (a) on the diagram. The analytical and numerical results are in very good agreement.

3. Breaking of fibres with intact neigbourhood

In this case the smallest critical threshold for the local damage has again the approximate value $c_{th} \approx 1 - W$. However, the local stress is now given by σ and the onset time t'_1 is given by $1 - W = a\sigma t'_1$ which is equivalent to

$$t'_1 = (1-W)/(a\sigma). \tag{7}$$

On average all N fibres have equal distributed thresholds between $1 - W$ and $1 + W$. Thus the average distance between two consecutive thresholds is $2W/N$. After the breaking of i fibres, on average each fibre carries a stress $\sigma'(i) = \sigma N/(N-i)$, i.e. the stress which was carried by N fibres at the very beginning, is now carried by $(N-i)$ fibres. Between two consecutive breakings, the local stresses do not change. This yields for the time $\tau'(i)$ between events i and $(i+1)$: $2W/N = a\sigma'(i)\tau'(i)$, which is equivalent to

$$\tau'(i) = 2W(N-i)/(N^2 a\sigma). \tag{8}$$

The time of the i-th event is given by:

$$\begin{aligned} t_i &= t'_1 + \sum_{j=1}^{i-1} \tau(j) = t'_1 + \frac{2W}{N^2 a\sigma}(N(i-1) - (i-1)i/2) \\ &\approx t'_1 + \frac{2W}{N^2 a\sigma}(N(i-1) - (i-1)^2/2) \end{aligned} \tag{9}$$

where in the last line $N \gg 1$ has been used. It follows for the cumulative damage D' which is equal to the event number i:

$$D' = i = 1 + N\left(1 - \sqrt{1 - a\sigma(t_i - t_1')/W}\right) = 1 + N\left(1 - \sqrt{1/W}\sqrt{1 - a\sigma t_i}\right) \tag{10}$$

where we have used $t_1' = (1 - W)/(a\sigma)$ (Eq. (7)). If the system aready contained an initial number d' of failed fibers, this has again to be added to the damage caused by the newly breaking fibres. Thus, N has to be replaced by $N - d'$, and the stress σ has to be substituted by the effective stress $\sigma N/(N - d')$ leading to

$$\begin{aligned} D' &= d' + i = d' + 1 + (N - d')\left(1 - \sqrt{\frac{1}{W}}\sqrt{1 - \frac{t_i a\sigma}{(1 - d'/N)}}\right) \\ &\approx N - (N - d')\sqrt{\frac{1}{W}}\sqrt{1 - \frac{t_i a\sigma}{(1 - d'/N)}}. \end{aligned} \tag{11}$$

This equation is also plotted in Fig. 1 (labelled (b)) with $d' = 2150$ and N, a, W and σ as above. Again, the analytical and numerical results are in very good agreement.

4. Discussion

At early times the nontrivial behaviour of the time dependent cumulative damage in a FBM including time dependent ageing can be described analytically by considering a sequence of different dominating mechanisms acting in the system. Explicitly, we have derived analytical equations for the two cases of (i) failure of fibres with one damaged neighbour and (ii) failure of fibres with intact neighbourhood. In a similar manner, earlier jumps in the observed damage can be attributed to the case of two damaged neighbours, etc. We note that the above calculations and approximations are only valid for large systems and small applied stresses when the initial damage is limited and damage accumulation due to cascades or other correlated failure triggered by local stress redistribution can be neglected. In these situations our analytical equations are in very good agreement with the numerical simulations. We have not yet determined the times where correlated damage becomes dominant.

Acknowledgments

This work was supported by the ERA Complexity-NET pilot project LOCAT.

References

[1] Daniels H E 1945 *Proc. R. Soc. London, Ser.* A **183** 405
[2] Coleman B D 1956 *J. Appl. Phys.* **27** 862
[3] Kloster M, Hansen A and Hemmer P C 1997 *Phys. Rev.* E **56** 2615
[4] Kun F, Raischel F, Hidalgo R C and Herrmann H J 2006, in *Modelling Critical and Catastrophic Phenomena in Geoscience: A Statistical Physics Approach*, edited by Bhattacharyya P and Chakrabarti B K, Lecture Notes in Physics (Springer-Verlag, Berlin), pp. 57-92
[5] Pradhan S, Hansen A and Chakrabarti B K 2010 *Rev. Mod. Phys.* **82** 499
[6] Fyffe B and Zaiser M 2004 *Cold Reg. Sci. Tech.* **40** 229
[7] Zaiser M, Moretti P, Konstantinidis A and Aifantis E C 2009 *J. Stat. Mech.* P02047
[8] Kun F, Costa M H, Filho R N C, Andrade J S, Soares J B and Herrmann H J 2007 *J. Stat. Mech.* P02003
[9] Carmona H A, Kun F, Andrade J S and Herrmann H J 2007 *Phys. Rev.* E **75** 046115
[10] Kun F, Carmona H A, Andrade J S and Herrmann H J 2008 *Phys. Rev. Lett.* **100** 094301
[11] Kun F, Halász Z, Andrade J S and Herrmann H J 2009 *J. Stat. Mech.* P01021
[12] Halász Z, Danku Z and Kun F 2012 *Phys. Rev.* E **85** 016116

A model-based approach to monitor complex road-vehicle interactions through first principles

T Chakravarty[1], K Srinivasarengan [2], S Roy[3], S Bilal[1], Balamuralidhar P[2]

[1] TCS Innovation Labs, BIPL, Salt Lake City Sector V, Kolkata, India
[2] TCS Innovation Labs, Plot No 96, EP-IP, Whitefield Road, Bangalore, India
[3] (Purdue University), Apt7, 645 Waldron St., West Lafayette, IN 47906

E-mail: tapas.chakravarty@tcs.com

Abstract. The increasing availability of portable computing devices and their interaction with physical systems ask for designing compact models and simulations to understand and characterize such interactions. For instance, monitoring a road's grade using accelerometer stationed inside a moving ground vehicle is an emerging trend in city administration. Typically the focus has largely been to develop algorithms to articulate meaning from that. But, the experimentation cannot provide with an exhaustive analysis of all scenarios and the characteristics of them. We propose an approach of modeling these interactions of physical systems with gadgets through first principles, in a compact manner to focus on limited number of interactions. We derive an approach to model the vehicle interaction with a pothole on a road, a specific case, but allowing for selectable car parameters like natural damped frequency, tire size etc, thus generalizing it. Different road profiles are also created to represent rough road with sharp irregularities. These act as excitation to the moving vehicle and the interaction is computed to determine the vertical/ lateral vibration of the system i.e vehicle with sensors using joint time-frequency signal analysis methods. The simulation is compared with experimental data for validation. We show some directions as to how simulation of such models can reveal different characteristics of the interaction through analysis of their frequency spectrum. It is envisioned that the proposed models will get enriched further as and when large data set of real life data is captured and appropriate sensitivity analysis is done.

1. Introduction

Portable computing devices are becoming ubiquitous and their interactions with the physical systems are increasing. These interactions can help analyze the characteristics of the interactions and their effects on the interacting systems. This would need modeling of specific interactions, but at the same time generalizing to allow for different types of interacting systems. To illustrate this modeling approach, we model the interaction between a road vehicle and a pothole on the road. Models are obtained from first principles due to the need for capturing of physical characteristics. We allow for variations in the parameters of a car like tire size, natural frequency, etc. We use this data to analyse the interaction's frequency components.

Interaction analysis between a road vehicle and infrastructures have been analysed before, but in a different context. For instance, in [1], [2] the authors present a method to assess dynamic forces that occur after a vehicle passes a road surface irregularity and obtain the contact forces and its effect on bridge vibration. The usage of smart phone accelerometer for pothole identification and road grade analysis have been widely experimented. See for instance, [3], [4]. These works were motivation to look at this specific interaction.

2. Modeling and Simulation of Vehicle road Interaction

In this section, we briefly describe the modeling approach for the pothole vehicle interaction. We consider that the pothole affects only one wheel and hence there is a vehicle tilt. The potholes modeled as a step deformation and the personal computing device accelerometer location is considered to be aligned to the roll center of the front axle. The computation of the acceleration has three phases, (i) the descent into the pothole (ii) the bounce inside the pothole (iii) the ascent from the pothole to the road.

The tire motion is modeled as a projectile. The initial descent is modeled as a horizontally launched projectile followed by a bounce inside the pothole and then by a projectile motion with an ascend and descend. This projectile based modeling is extensible to cover different step pothole scenarios with different length and internal gradient. In this paper we derive the idea of using a single bounce case.

2.1. Mathematical Model

For the equation of motion of horizontally launched projectile, we apply modification to account for the tilt between the two front wheels due to the descent. Apart from this, we also modify the 'g' value appropriately to take care of the characteristics of vehicle body and wheel. This results in vertical and lateral displacements (refer Figures 1 and 2).

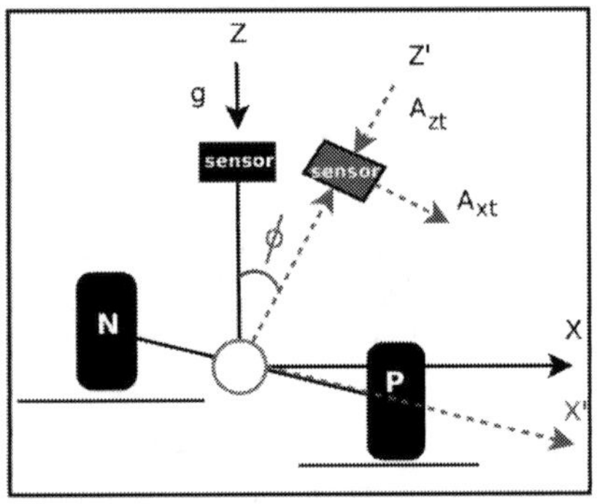

Figure 1. Smartphone Location during pothole traversal (front/rear view)

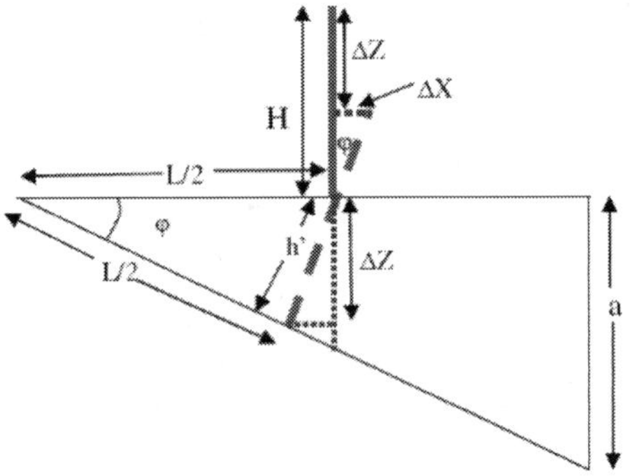

Figure 2. The geometrical parameters of tilt during pothole traversal

Descent The position of the center of the tire, the instantaneous tilt of the wheel and the vertical and lateral displacements respectively, are given by,

$$S_t = \frac{g_{db}}{2}(t-t_i)^2, \quad \phi = tan^{-1}(\frac{S_t}{L}), \quad \Delta Z = \frac{L}{4}sin(2\phi), \quad \Delta X = (H - \frac{L}{2sin\phi})sin\phi \tag{1}$$

where, we g_{db} is the g value corrected for the body natural frequency as, $g_{db} = g - 2\pi f_n M_R$, where M_R is the Motion ratio factor of the vehicle body and f_n is its natural frequency. Here t_i is the initial time. Here L is the axle length and H is the position of the accelerometer above roll axis. The accelerations in the two directions are given by,

$$A_z = \frac{2\Delta Z cos\phi}{(t-t_i)^2} \qquad A_x = \frac{2\Delta X sin\phi}{(t-t_i)^2} \tag{2}$$

Ascent Similar to the descent case, we have the ascending of the tire expressed as,

$$S_t = Depth - (V_{zfb}(t - t_i)) - \frac{g_{db}}{2}(t - t_i)^2 \quad (3)$$

with the rest of the equations as given by (1). Here $V_{zfb} = R_{vel}V_{zf}$, is corrected final velocity after the tire hits the pothole, where R_{vel} is the translation factor between velocity final velocity on the tire to that as experienced at the pothole (after accounting for tilt and other factors).

Bounce and Transmissibility The transmissibility factor is computed with a sinsusoidal road profile ($p = Psin(\frac{2\pi}{L})z)$ to quarter car model as discussed in [5]. We use the transmissibility factor given by [5], and obtain the A_z measured at the phone as,

$$T_R = \frac{\sqrt{1 + (2\zeta\beta)^2}}{\sqrt{(1 - \beta^2)^2 + (2\zeta\beta)^2}} \qquad A_{z-bounce} = \frac{2T_R V_{zf} cos\phi}{\delta t} \quad (4)$$

where, $\beta = \omega/\omega_n$, with ω is the frequency of the road profile is considered and ω_n is the natural frequency of the vehicle. ζ is the body damping factor of the vehicle. We consider the radius of tire (R_{tire}) and the velocity of the vehicle (U_x) to compute the impact interval ($\delta t = K\pi R_{tire}/U_x$), where K is an empirical value changing from very small for smooth and hard surface to close to 1 muddy surface.

Roadgrade modeling and Simulation To represent a road with irregularities, the modeled pothole should be augmented with the profile of a road. The ISO 8608:1995 classifies road profiles based on their degree of roughness given by the Power Spectral Density of the surface roughness. We use the sinusoidal approximation as described in [6], [7] and the generated road surface is input to a quarter car model to obtain the acceleration as discussed in [8].

2.2. Results
In this section, we briefly discuss the results obtained from simulations and their comparisons with experimental results (performed using a TATA Nano car). A comparison of the Z-direction acceleration of the measured and the generated pattern are shows in Figure 3. To illustrate the different modes of this approach, the Figure 4 we illustrate how different cars going on a 50m Type C road with a randomly placed pothole, will give out signatures in the Z-axis acceleration, with a sampling frequency of 20 Hz.

3. Analysis of Modeled Interaction
In this section we briefly discuss how the simulated data can be put to use. In Figure 5, we compare the PSD of experimental and simulated data of the jerk (m/s^3). The amplitude difference and absence of second peak is due to the modeling limitations, which can be overcome using more accurate modeling. In Figure 6, we look at the frequency response impact for an increasing pothole size for a medium saloon car (sampled at 50 Hz), which shows a unique response signature for different potholes in the form of number of peaks. Such analysis can give a deeper understanding of the interaction without performing elaborate experiments.

4. Concluding Remarks
In this paper, we discussed the preliminary results of a modeling approach to suit the increasing interaction of mobile computing device with various physical and engineering systems. We have illustrated this by modeling the interaction between a vehicle and a pothole on a road and analysing the characteristics of this interaction. Indirectly, the data generated through this

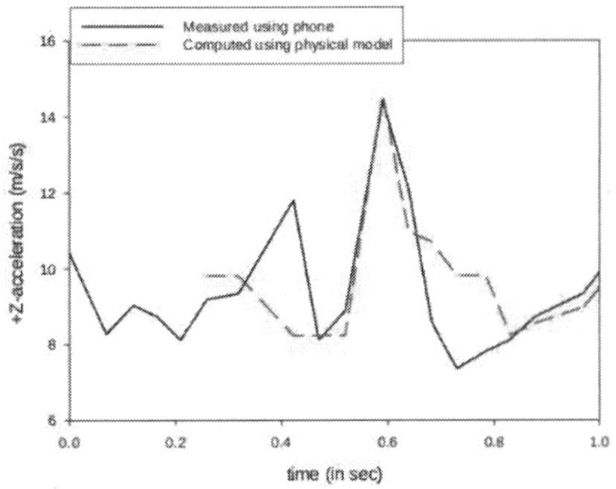

Figure 3. Z-direction acceleration at 30 kmph over a 10cm deep pothole (18 Hz)

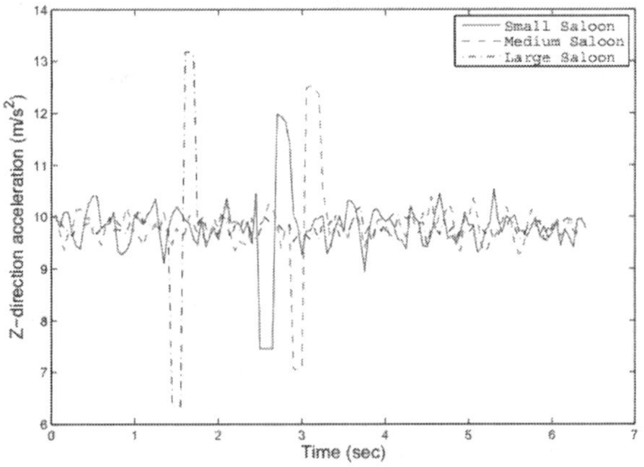

Figure 4. Simulation of a vehicle on road with different type of cars

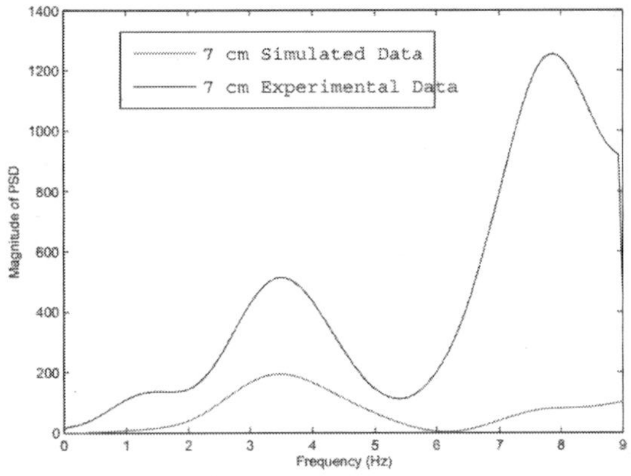

Figure 5. Comparison of PSDs of Simulated and Experimental Data for a 7cm pothole

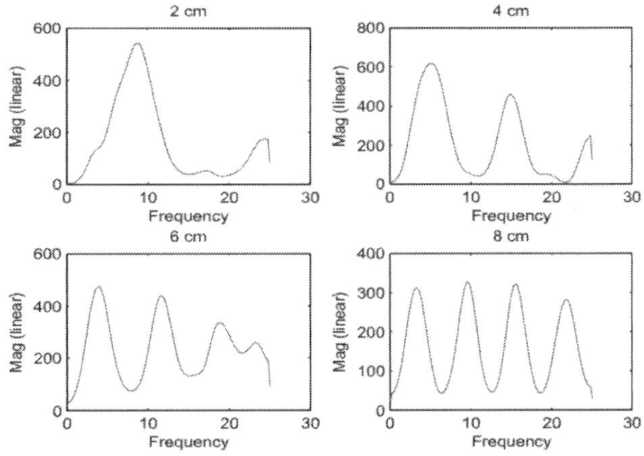

Figure 6. PSD of Jerk (simulated at 50Hz) for different pothole depths

simulation can also aid in evaluating algorithms used to identify potholes. This work can be enhanced by modeling the interaction with further details and analysing their implications. This approach can be used in different applications where a mobile computing device are put to use.

References

[1] Pesterev A, Bergman L, Tan C and Yang B 2005 *Journal of sound and vibration* **279** 817–841
[2] Pesterev A, Bergman L and Tan C 2004 *Journal of sound and vibration* **275** 127–149
[3] Mednis A, Strazdins G, Zviedris R, Kanonirs G and Selavo L 2011 *Distributed Computing in Sensor Systems and Workshops (DCOSS), 2011 International Conference on* (IEEE) pp 1–6
[4] Mohan P, Padmanabhan V and Ramjee R 2008 *Proceedings of the 6th ACM conference on Embedded network sensor systems* (ACM) pp 357–358
[5] Nader S Frequency response principle lecture Notes, http://www.me.gatech.edu/nader.sadegh/ME3015/frqresp.pdf
[6] González A, O'brien E, Li Y and Cashell K 2008 *Vehicle System Dynamics* **46** 483–499
[7] Tyan F, Hong Y, Tu S and Jent W 2009 *Journal of Advanced Engineering* 1373–1378
[8] Wong J 2001 *Theory of ground vehicles* (Wiley-Interscience)

Journal of Physics: Conference Series **410** (2013) 012066 doi:10.1088/1742-6596/410/1/012066

Partial differential equations and fractal analysis to plant leaf identification

Bruno Brandoli Machado, Dalcimar Casanova, Wesley Nunes Gonçalves, Odemir Martinez Bruno

Physics Institute of São Carlos, University of São Paulo, Av. Trabalhador São Carlense, 400 - 13560-970 - São Carlos, SP, Brasil

E-mail:
brandoli@icmc.usp.br,dalcimar@gmail.com,wnunes@ursa.ifsc.usp.br,bruno@ifsc.usp.br

Abstract. Texture is an important visual attribute used to plant leaf identification. Although there are many methods of texture analysis, some of them specifically for interpreting leaf images is still a challenging task because of the huge pattern variation found in nature. In this paper, we investigate the leaf texture modeling based on the partial differential equations and fractal dimension theory. Here, we are first interested in decomposing the original texture image into two components $f = u + v$, such that u represents a cartoon component, while v represents the oscillatory component. We demonstrate how this procedure enhance the texture component on images. Our modeling uses the non-linear partial differential equation (PDE) of Perona-Malik. Based on the enhanced texture component, we estimated the fractal dimension by the Bouligand-Minkowski method due to its precision in quantifying structural properties of images. The feature vectors are then used as inputs to our classification system, based on linear discriminant analysis. We validate our approach on a benchmark with 8000 leaf samples. Experimental results indicate that the proposed approach improves average classification rates in comparison with traditional methods. The results suggest that the proposed approach can be a feasible step for plant leaf identification, as well as different real-world applications.

1. Introduction

The plant cataloging activity is essential to taxonomical studies, especially in Ecology and Botany. Traditionally, it is first necessary: i) to collect and dry some branch of the plant, ii) to process the voucher, iii) to compare it with other vouchers in the herbaria using diagnostic keys, and then iv) to send the material to specialists for confirmation [1]. The main advantage of using leaves in such activity is related to three main facts: (1) in contrast to flowers and fruits, they are found during all seasons, (2) the morphological analysis is considered easy-handling and can provide important features for taxonomy purposes, (3) they are posed to present greater differentiation scope by exploring geometric aspects.

In this paper, we investigate the use of partial differential equations and fractal dimension theory to describe the texture present in leaves. As a first step, the leaf image f is decomposed using the non-linear partial differential equation of Perona-Malik [2] into two components $f = u+v$, where u, v represent the cartoon component and the oscillatory or textural component, respectively. This decomposition enhances the texture component and successfully filter noise while preserving relevant edge details in the image. Then, we estimated the fractal dimension of the enhanced texture component by using the Bouligand-Minkowski method [3, 4]. We have chosen this fractal method rather than other ones due to its precision in quantifying structural properties of images. Finally, the feature vector, which describe the texture image, is used as input to our classification system.

In order to validate the proposed approach, we have used a database containing 20 leaf samples of 40 species of plants. According to the experimental results, the proposed approach improves the average classification rate compared to traditional texture methods. Thus, we conclude that our approach can be used as a feasible step for plant leaf identification, as well as different real-world applications.

2. Partial Differential Equations to Image Texture Enhancement

The two-dimensional Perona and Malik equation is the most conventional PDE in the literature [2]. The discrete formulation can be implemented by:

$$I_{t+1}(x,y) = I_t(x,y) + \frac{1}{4}\sum_{i=1}^{4}\left[g(\nabla I_t^i(x,y)).\nabla I_t(x,y)\right] \tag{1}$$

where (x, y) denotes the spatial position of each pixel, and $\nabla I_t(x,y), i = 1, 2, 3$ and 4 correspond to the brightness gradients of neighbors north, south, east and west. A possible coefficient function, updated at each iteration by the gradient function, is given by the inverse quadratic:

$$g(\nabla I) = \frac{1}{1 + \left(\frac{\|\nabla I\|}{K}\right)^2} \tag{2}$$

where the parameter K is a constant that acts as an edge strength threshold. The computational scheme described in this section generates a scale-space representation, that is, it creates from the input image a family of derived images. After creating the multiscale representation, we obtain two components of every new image: cartoon and texture component. The texture component is determined by the subtraction of the original image and cartoon approximation.

Next section, the fractal method of Bouligand-Minkowski is briefly reviewed. Then, we present our strategy and we point what is different to the conventional approach.

3. Bouligand-Minkowski Fractal on Image Analysis

Let us consider a gray level image $f(x, y)$ represented as a surface $S(x, y, z)$, where (x, y) are the spatial position of the image and z corresponds to the intensity of the pixel [3]. Given the surface S, we can measure the influence of volume $V(r)$ by the dilatation of each point $p \in S$ by using a sphere of radius r. We show an example of this process in Figure 1. The surface of the image is created (Figure 1 (a)), and then it is dilated by sphere of different radius r, as can be seen in Figure 1 (b) and (c).

Here, the arrangement of points of the surface changes the process of expansion. While the value of r grows, the spheres produced by different points on the surface begin to interact up. It causes effects on the final amount of $V(r)$. Thus, for each r value we calculate the volume V to compose a feature vector [5].

4. A Hybrid Approach for Texture Analysis

A traditional strategy is to compute feature vectors by means of a fractal analysis method. However, it does not explore the richness of the feature description task. Instead of obtaining a single value captured from the image, we create a multiscale representation by using the partial differential equations. Such strategy enable us to enhance structures perceived in certain levels of observation. The diagram of Figure 2 summarizes the approach proposed here.

(a) Surface (b) $r = 5$ (c) $r = 7$

Figure 1. Example of the dilation process for two diferent radiuses.

To clarify, we divide our approach in three main steps: (1) color samples from a plant leaf species are obtained and converted to gray-level, (2) we represent an input image by using Perona and Malik equation before straight adopt fractal analysis. After decomposing an image, we obtain two components: cartoon (u) and texture (v) component. The enhanced texture component is determined by the subtraction of the original image and cartoon approximation, and finally in (3), we employ the fractal analysis of Bouligand-Minkowski.

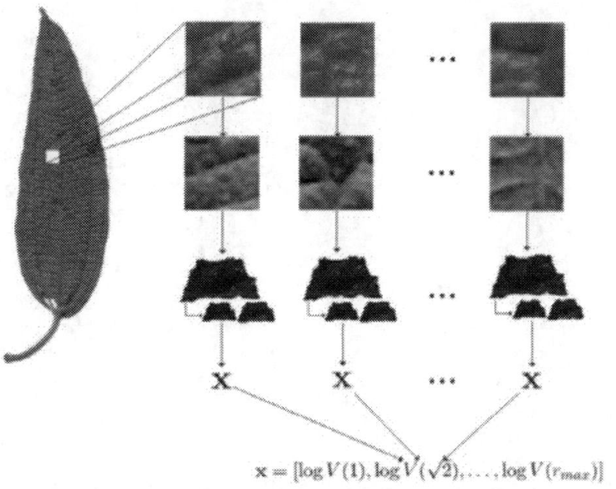

Figure 2. The process used to classify a leaf.

5. Experimental results

In this section, we describe the leaf database used in the experiments and the results of the proposed approach. Given the leaf image, it is possible to identify different plant species using the extracted features of the image. However, the direct classification was not used in this work. First, we randomly extract 10 texture samples from each leaf and, therefore, 10 feature vectors are available for each leaf. To minimize problems with noise and texture variation we have calculated the mean of these 10 feature vectors. With a single feature vector for each leaf, we have chosen to use the Naive Bayes classifier in the 10-fold cross validation.

5.1. Leaf database

Each leaf was digitalized with a traditional scanner (HP-Scanjet 3800) with resolution of 1200dpi (dots per inch). The high resolution was used to access the details in image texture. To avoid rotation problems, all leaves were oriented in the same direction while assembled on the scanner. Both surfaces of the leaf epidermis were digitalized (abaxial and adaxial ones). The adaxial surface of some species can be seen in Figure 3. In the final database, we have used 40 species of plants with 20 leaves for each specie.

Figure 3. Two species of leaf textures used in the experiments.

5.2. Results

The experimental results are shown in Table 1. The Boulingand-Minkowski fractal dimension (FD), which extract features from the original texture image, achieved a correct classification rate of 85.20. Using the texture component, the correct classification rate improves from 85.20 to 87.00, which demonstrates the importance of the PDE on the process. As expected, the cartoon component did not provided satisfactory results since homogeneous regions does not present useful information of textures.

Table 1. Classification results for the leaf database.

Method	N. of Features	Correct Classification Rate
Boulingand-Minkowski FD	85	85.20 (±1.05)
Cartoon Component + FD	85	74.37 (±1.82)
Our approach (Texture Component + FD)	85	**87.00** (±1.11)

6. Conclusion

The results suggest that the computer-aided plant classification is an area that can provide new useful tools for experimental taxonomists and plant morphologists, improving their work, bringing also new forms to asses informative characters that can be useful in systematic and phylogenetic studies. The method proposed here may represent a new, important tool for non-taxonomic botanists or ecologists, working with plant species with very similar morphology, since it requires easily available equipment, such as a regular computer and an optical scanner.

Acknowledgements

This work was supported by FAPESP grants 2011/02918-0, 2008/57313-2 and 2010/08614-0 (to B.B.M., D.C. and W.N.G.), and grants from CNPQ to O.M.B.

References

[1] Rossatto D, Casanova D, Kolb R and Bruno O M 2011 *Plant Systematics and Evolution* **291**(1) 103–116
[2] Perona P and Malik J 1990 *IEEE Transactions on Pattern Analysis and Machine Intelligence* **12**(7) 629–639
[3] Mandelbrot B B 1983 *The Fractal Geometry of Nature* (New York: W. H. Freeman and Company)
[4] Bruno O M, de Oliveira Plotze R, Falvo M and de Castro M 2008 *Information Sciences* **178**(12) 2722–2733
[5] Backes A R, Casanova D and Bruno O M 2009 *International Journal of Pattern Recognition and Artificial Inteligence* **23**(6) 1145–1160

Journal of Physics: Conference Series **410** (2013) 012067 doi:10.1088/1742-6596/410/1/012067

Weighted Scale-Free Network Properties of Ecological Network

Jae Woo Lee[1], **Seong Eun Maeng**
[1] Department of Physics, Inha University, Incheon 402-751 Korea
E-mail: jaewlee@inha.ac.kr

Abstract. We investigate the scale-free network properties of the bipartite ecological network, in particular, the plant-pollinator network. In plant-pollinator network, the pollinators visit the plant to get the nectars. In contrast to the other complex network, the plant-pollinator network has not only the trophic relationships among the interacting partners but also the complexities of the coevolutionary effects. The interactions between the plant and pollinators are beneficial relations. The plant-pollinator network is a bipartite and weighted network. The networks have two types of the nodes: plant and pollinator. We consider the visiting frequency of a pollinator to a plant as the weighting value of the link. We defined the strength of a node as the sum of the weighting value of the links. We reported the cumulative distribution function (CDF) of the degree and the strength of the plant-pollinator network. The CDF of the plants followed stretched exponential functions for both degree and strength, but the CDF of the pollinators showed the power law for both degree and strength. The average strength of the links showed the nonlinear dependence on the degree of the networks.

1. Introduction

The ecological networks are good examples of the complex networks. The scaling invariance is observed in the ecological networks. There are many types of ecological networks such prey-predator networks, plant-pollinator networks, host-parasite networks, plant-seed disperser networks etc. In the food web the degree distribution function of the network followed some functional forms like the power law, stretched exponential function, exponential function, uniform function etc.[1,2]. Many studied are oriented to the network properties of the binary ecological networks[1,2]. However, some ecological networks are bipartite networks with two types of the components such as the mutualistic network and host-parasite networks [3-11].

In this work we consider the mutualistic networks, in particular, the plant-pollinator networks. In these networks, the pollinators obtain their diet from the plants and help pollinate the plants. Such connections give the two species beneficial mutualistic relations. In these networks, the degree and strength distribution function for the pollinators followed the different functional dependence in comparison to those for the plants. In the plant-pollinator network, a pollinator visits frequently some plants, it visits rarely other plants. The ecologists had recorded the visiting frequencies of a pollinator to a plant at a region in a fixed season by the field experiments [3]. We consider the visiting frequencies as the interacting strength between two species. Therefore, the plant-pollinator networks act like the weighted complex networks.

In this work we analyze the seven plant-pollinator networks. It includes boreal forest, alpine forest, montane forest, beach forest, meadow, and upland grassland food web. We reported the network properties in section 3. We consider the nonlinear dependence of the plat-pollinator networks in section 4. We give the conclusions in section 4.

[1] To whom any correspondence should be addressed.

Table 1. Quantitative properties of the seven mutualistic plant-pollinator networks.

Food Web	Type	N_{pl}	N_{po}	C	<k>	<s>
Bar1987	Boreal forest	12	102	0.136	27.83	91.66
Elb1999	Alpine forest	23	118	0.104	20.7	33.3
Ino1988	Montane forest	42	91	0.073	13.38	69.48
Kat1999	Beach forest	93	679	0.019	25.94	51.44
Mem1999	Meadow	25	79	0.151	23.92	174.64
Oll2003	Upland grassland	9	56	0.204	22.89	132
Sma1976	Montane forest	13	34	0.319	21.69	152.62

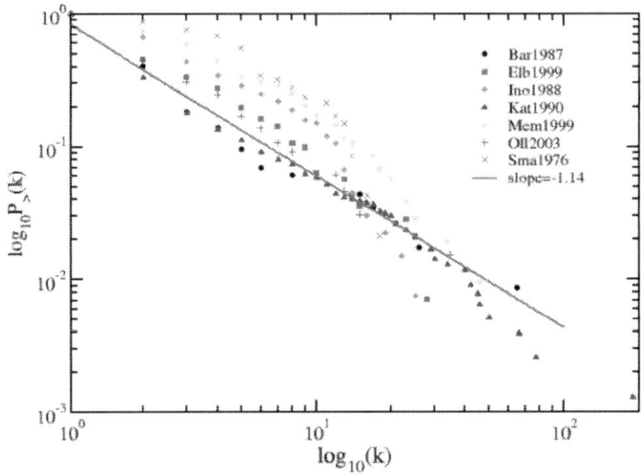

Figure 1. The cumulative probability distribution functions of the degree for the seven mutualistic ecological networks. The CDFs of the plant-pollinator networks follow the power law, $P_{>}(k) \approx k^{-\gamma_d}$, with the average degree exponent, $\gamma_d = 1.14(3)$. The solid line indicates the least square fits.

2. Network Properties of Plant-Pollinator Networks

We consider seven plant-pollinator mutualistic ecological networks as shown in table 1. The plant-pollinator networks are small number of nodes because the food web is limited in an area of the observing field. The seven networks include many types of ecological food web such as boreal forest, alpine forest, montane forest, beach forest, meadow, and upland grassland. The number of pollinators is bigger than the number of the plant in all seven ecological networks. The total number of species, $N = N_{pl} + N_{po}$, lines in the range from N=45 to N=772 in the seven plant-pollinator networks where N_{pl} is the number of the plants and N_{po} is the number of the pollinators. The pollinators visit the plants to get their diets. We define the visiting frequencies, w_{ij}, of a pollinator j, to a plant i. This visiting frequencies act as the weights in the network. We summarize the network properties of the seven networks in table 1 [11-17]. The connectance of the network is defined by $C = \sum_i k_i / N_{pl} N_{po}$, where k_i is the total number of connecting links of the node i. The average degrees of each network are large and the connectances are large compared to the other networks [1,2]. The food web, Ino1988, has the small average degree <k>=13.38 among the networks. The strength of the node i in the network is defined as $s_i = \sum_{i=1}^{N_i} w_{ij}$, where Ni is the number of nearest neighbours of the node i. The average strength of the networks is also very large. However, the average strength lies in the range between 1.609 for food web, Elb1999 and 7.301 for food web, Mem1999. The average shortest path length between two nodes is around 3.0 in the networks. Figure 1 shows the cumulative probability density function (CDF) of the node degree for seven mutualistic ecological networks. The CDF follows a power law, $P_> \sim k^{-\gamma_d}$, up to the cutoff degree $k < k_c$ with $\gamma_d = 1.1(1)$ for Bar1987, $\gamma_d = 1.33(7)$ for Elb1999, $\gamma_d = 0.9(1)$ for Ino1988, $\gamma_d = 1.25(8)$ for Kat1990, $\gamma_d = 1.04(5)$ for Mem1999, and $\gamma_d = 1.22(7)$ for Oll2003, $\gamma_d = 1.15(8)$ for Sma2002. We obtained the average exponent of the degree distribution as $\overline{\gamma_d} = 1.14(3)$. When the total numbers of the nodes are increase, the range of the power law extends to the large value of the degree, i.e. the cutoff k_c increases. The plant-pollinator networks show the scale-free network. The complex networks of the ecological bipartite systems give the beneficial relationships among the species and guarantee the stability and the robustness of the ecological networks.

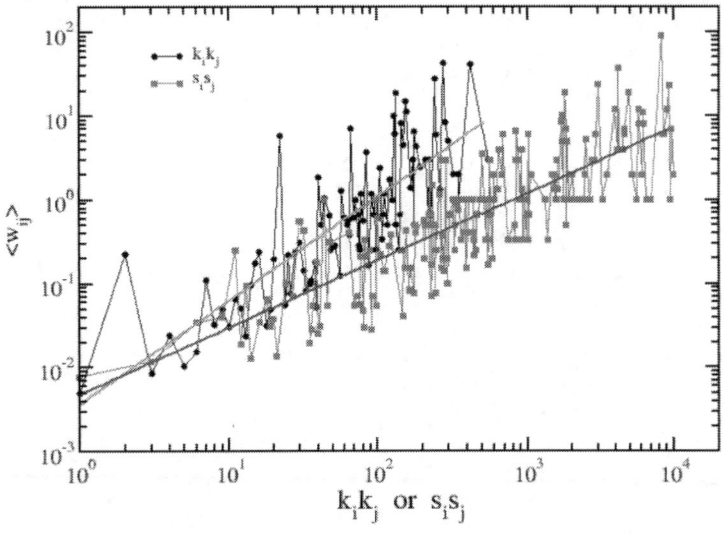

Figure 2. The average weights, $< w_{ij} >$, depend on the degree or strength of the end-point node. The average weights followed the power law, $< w_{ij} > \sim (k_i k_j)^{\theta}$ for the degree (circle) and $< w_{ij} > \sim (s_i s_j)^{\delta}$ for the strength (square). We obtained the exponents as $\theta = 1.22(9)$ and $\delta = 0.77(8)$.

3. Nonlinearity of Network

Figure 2 represents the dependence of the average weight, $< w_{ij} >$ on the degree of the end-point node k_i and k_j, and the strength of the end-point node s_i and s_j. We observed the power law dependence of the average weights, $< w_{ij} > \sim (k_i k_j)^{\theta}$ for the degree and $< w_{ij} > \sim (s_i s_j)^{\delta}$ for the strength. The power law of the average weights indicates that the degree and strength of the end-point node are strongly correlated each other. We observed the average exponents of the power law as $\theta = 1.22(9)$ for the degree and $\delta = 0.77(8)$ for the strength.

4. Conclusions

The network properties of seven plant-pollinator ecological networks were studied. The CDFs followed power law for both degree and strength. Such different distributions are a unique property of ecological networks. The plants adopt a different strategy compared with the pollinators in the interactions between them. The average strengths of the nodes showed nonlinear dependence on the degree or the strength of the mutualistic network.

Acknowledgments

This research was supported by Basic Science Research Program through the National Research Foundation of Korea(NRF) funded by the Ministry of Education, Science and Technology(Grant number: 2012R1A1B3001183).

References

[1] Boccaletti S, Latora V, Moreno Y, Chavez M and Hwang D U 2006 *Phys. Rep.* **424**, 175
[2] Pimm S L, Lawton J H and Cohen J E 1991 *Nature* **350**, 669
[3] Dunne J A, Williams R J and Martinez N D 2002 *Ecol. Lett.* **5**, 558
[4] Onnela J-P, Chakraborti A, Kaski K, Kertesz J and Kanto A, 2003 *Phys. Rev. E* **68**, 656110
[5] Bascompte J, Jordano P, Melian C J and Olesen J M 2003 *PNAS* **100**, 9383
[6] Jordano P, Bascompte J and Olesen J M, 2003 *Ecol. Lett.* **6**, 69
[7] Montoya J M, Pimm S L, and Sole R V 2006 *Nature* **442**, 259
[8] Vazquez D P, Melian C J, Williams N M, Bluthen N, Kransnov B R and Poulin R 2007 *Oikos* **116**, 1120
[9] Maeng S E and Lee J W 2011 *J. Korean Phys. Soc.* **58**, 851
[10] Barrat A, Barthelemy M, Pastor-Satorras R and Vespignani A 2004 *PNAS* **101**, 3747
[11] Barrett S C H and Helenurm K 1987 *Canadian J. Bot.* **65**, 2036
[12] Elberling H and Olesen J M 1999 *Ecography* **22**, 314
[13] Kato M, Makutani T, Inoue T and Itino T 1990 *Contr. Biol. Lab. Kyoto Univ.* **27**, 309
[14] Inouye D W and G H 1988 *Australian J. Ecol.* **13**, 191
[15] Memmott J 1999 *Ecol. Lett.* **2**, 276
[16] Ollerton J, Johnson S P, Cranmer L, and Kellie S 2003 *Ann. Bot.* **92**, 807
[17] Small E 1976 *Canadian Field Naturalist*, **90**, 22

Numerical modeling of the plasma response to local cooling

Mikhail Z. Tokar and Mikhail Koltunov

Institute for Energy and Climate Research - Plasma Physics (IEK-4), Research Center Jülich, 52428, Jülich, Germany

E-mail: m.tokar@fz-juelich.de

Abstract. Thermal instabilities can lead to a sudden drop of local plasma temperature by orders of magnitude and numerical modeling of the global plasma response to such phenomena is performed. An approach to operate with the plasma heat conduction κ, being a non-linear function both of the plasma temperature and of its gradient, if the heat flux limit is taken into account, is developed. It is demonstrated that the deviation of κ from Spitzer-Härm approximation normally assumed is essential to explain experimental findings by massive gas injection experiments carried out to mitigate plasma disruptions in tokamaks.

1. Introduction

There are diverse causes for abrupt cooling of small localized regions in fusion and astrophysical plasmas. This can happen, e.g., by the penetration of impurity particles from intensive sources when electrons and ions in the background hydrogen plasma lose their thermal energy in elastic coulomb collisions and in inelastic excitation processes with impurities. Owing to thermal instabilities the local plasma temperature can drop by orders of magnitude during a very short time in a microsecond range [1]. Modeling of the plasma response to a localized cooling is essential for interpreting and predicting global consequences of such events. It is, however, a challenging numerical task due to some reasons. First, the problem is non-stationary and multi-dimensional. Second, the heat conduction κ is a strongly non-linear function of the plasma temperature T. In addition, because of the very sharp temperature gradient in the vicinity of the plasma area initially cooled down, the local Spitzer-Härm approximation for κ [2] fails and κ becomes depending on the temperature gradient due to heat flux limit phenomenon [3].

In this paper a numerical approach allowing coping with the problems outlined is presented. As the first step in the original heat transport equation we discretize the partial derivatives with respect to time and spatial variables except the one indicated as x henceforth. As a result one obtains coupled ordinary differential equations (ODE) of the second order for $T_j(x)$ with j counting the sets of discretized variables. These equations are solved in an iterative procedure by the method elaborated in Ref.[4] where in the vicinity of grid knots the ODE obtained are approximated by ODE with constant coefficients and exact analytical solutions of these equations are conjugated by requiring the continuity of the solution and its first x-derivative.

Computations are done for the conditions of massive gas injection into the tokamak JET performed to mitigate plasma current disruptions [5]. The radial transport across magnetic surfaces is approximated by a source term proportional to the deviation of the temperature from that before the injection and non-stationary two-dimensional heat transfer on the surface is modeled. It is demonstrated that only if the deviation from the local Spitzer-Härm approximation to the heat conduction due to the heat flux limit is taken into account, calculations give a decay time of the temperature far from the injection position in an agreement with the experimentally measured level, in sub-millisecond range.

2. Basic equations and results of calculations

We proceed from a non-stationary heat balance equation in the coordinate system on the magnetic surface with the coordinates l and s aligned along and across the magnetic field, correspondingly:

$$3\partial_t (nT) + \partial_l q = \nu n (T_0 - T) \tag{1}$$

where n and T are the plasma density and temperature, respectively; $q = -\kappa_{||}\partial_l T$ is the parallel heat flux density with $\kappa_{||}$ being the plasma heat conduction along the magnetic field, and the source term in the right hand side characterizes the heat transport in the radial direction across the magnetic surfaces induced by the deviation from the temperature before the local cooling; the frequency ν is introduced in this study as a free parameter and the sensitivity of the results to it will be investigated. In spite of the fact that only one spatial coordinate is involved into equation (1), this advantage is fictitious since no boundary conditions can be posed to equation (1) in the toroidal geometry. Therefore one has to transfer to the toroidal and poloidal coordinates x and y with respect to which all physical quantities are periodic with periods $2L$ and 2Δ, respectively. The transformation between two coordinate system is straightforward, $l = x\cos\psi + y\sin\psi$, $s = -x\sin\psi + y\cos\psi$, where ψ is the pitch-angle of the magnetic field with respect to the toroidal direction. By using formulas for the differentiation of functions of many variables one gets instead of Eq.(1) the following one:

$$\begin{aligned} & 3\partial_t (nT) - \cos^2\psi\partial_x \left(\kappa_{||}\partial_x T\right) - \sin\psi\cos\psi\partial_y \left(\kappa_{||}\partial_x T\right) - \\ & - \sin\psi\cos\psi\partial_x \left(\kappa_{||}\partial_y T\right) - \sin^2\psi\partial_y \left(\kappa_{||}\partial_y T\right) \\ = & \; \nu n (T_0 - T) \end{aligned} \tag{2}$$

The time and y derivatives are discretized with finite increments τ and h_y, respectively, by taking into account that h_y can vary with y_j counted with the subscript $j_{\min} \le j \le j_{\max}$. In this study we assume the density n as constant in space and time. As a result an ordinary differential equation of the second order is obtained which governs the x dependence of $T_j \equiv T(t, x, y_j)$:

$$d_x^2 T_j + a \cdot d_x T_j = bT_j - f \tag{3}$$

with

$$a = d_x \ln\chi_{||,j},\; b = \frac{3/\tau + \nu}{\chi_{||,j}\cos^2\psi} + \left(1 + \frac{\chi_{||,j+1}/\chi_{||,j}}{1 + h_{y,j}/h_{y,j-1}} + \frac{\chi_{||,j-1}/\chi_{||,j}}{1 + h_{y,j-1}/h_{y,j}}\right)\frac{\tan^2\psi}{h_{y,j}h_{y,j-1}}$$

$$f = \frac{3T_{-,j}/\tau + \nu T_0}{\chi_{||,j}\cos^2\psi} + \frac{\tan\psi}{h_{y,j-1} + h_{y,j}}\left\{\begin{array}{c} \left(1 + \frac{\chi_{||,j+1}}{\chi_{||,j}}\right)\left(\frac{\tan\psi}{h_{y,j}}T_{j+1} + d_x T_{j+1}\right) + \\ + \left(1 + \frac{\chi_{||,j-1}}{\chi_{||,j}}\right)\left(\frac{\tan\psi}{h_{y,j-1}}T_{j-1} - d_x T_{j-1}\right) + \\ + a\left(T_{j+1} - T_{j-1}\right) \end{array}\right\}$$

where $\chi_{||} = \kappa_{||}/n$ and $T_{-,j} = T(t - \tau, x, y_j)$.

In the region $|x - L| \leq \lambda$ and $|y - \Delta| \leq \delta$ the plasma temperature is maintained at a very low level $T_c \ll T_0$ for any time $t \geq 0$; out of this area $T = T_0$ at $t = 0$. Due to the symmetry of the problem the computational domain can be reduced to $0 \leq x \leq L$ and $0 \leq y \leq \Delta$ where the boundary conditions for equations (3) are $d_x T_j\left(x = 0, L\right) = 0$ for $y_j < \Delta - \delta$, and $d_x T_j\left(x = 0\right) = 0, T_j\left(x = L\right) = T_c$ for $\Delta - \delta \leq y_j \leq \Delta$; $T_{j+1}\left(x\right) = T_{j-1}\left(x\right)$ and $d_x T_{j+1}\left(x\right) = d_x T_{j-1}\left(x\right)$ for $j - 1 = j_{\min}$ and $j + 1 = j_{\max}$ corresponding to $y_{j_{\min}} = 0$ and $y_{j_{\max}} = \Delta$, respectively. As the first approximation to $T_j\left(x\right)$ we assume $T_{-,\,j}\left(x\right)$ for all $j_{\min} \leq j \leq j_{\max}$. With $T_{-,\,j}\left(x\right)$ used for $T_{j+1}\left(x\right)$, $T_{j-1}\left(x\right)$ and by calculating $\chi_{||}$, since this is a function of the temperature, equations (3) are solved numerically by the method outlined in Ref.[4]. The found solutions are mixed with the previous approximation to T to get the next one. This procedure continues till the error of calculations becomes much smaller than the iteration mixing coefficient.

Figure 1 demonstrates the results of numerical calculations with the Spitzer-Härm approximation to $\chi_{||}$ [2], $\chi_{||}^{SH}\left[m^{-1}s^{-1}\right] \approx 10^{18} T^{2.5}\left[eV\right]/n\left[m^{-3}\right]$, where the parameter units are in square brackets. Computations have been done for the conditions of the massive gas injection to mitigate plasma disruption in the tokamak JET [5] with $L \approx 9.42m$, $\Delta \approx 3.14m$, $\psi = 0.1$, $\lambda/L = \delta/\Delta = 0.1L$, the initial plasma temperature at the radial position in question $T_0 = 250eV$, $T_c = 1.5eV$ and $n = 5 \cdot 10^{19} m^{-3}$. Different curves in figure 1a, demonstrating the time evolution of the temperature at $x = y = 0$, T_s, correspond to different magnitudes of ν; figure 1b shows the final stationary temperature profile for $\nu = 10^3 s^{-1}$. These calculations were performed with the time step $\tau = 1\mu s$, spatial grid increments decreasing in geometrical progression by approaching to the cooled plasma region, $|L - x| \leq \lambda$ and $|\Delta - y| \leq \delta$, the iteration mixing coefficient equal 1 and the accuracy 10^{-10} at each time step. The correctness of the solutions has been proven by comparing the results obtained with different τ. In figure 1 one can see that in the final stationary state the temperature far away from the cooled area depends on the magnitude of the radial transport characteristic ν, varying less than by factor 2 when ν changes by an order of magnitude. On the contrary, the characteristic rate of the temperature decay is practically independent of ν. However, the characteristic transient time of $100\mu s$ is much smaller than that of $0.5ms$ found in experiments.

A possible reason for discrepancy above, between the calculated and experimental decay rates of the temperature on a magnetic surface locally cooled down by impurity injection, may be a violation of the Spitzer-Härm approximation to $\kappa_{||}$ in the region of extremely sharp temperature gradient. If the characteristic scale for the temperature change along field lines, $L_T = 1/|\partial_l \ln T|$, becomes much less than the mean free path length of electrons between coulomb collisions, $\lambda_c\left[m\right] \approx 10^{16} T^2\left[eV\right]/n\left[m^{-3}\right]$, a significantly reduced free-streaming heat flux, $q = \xi n T^{3/2}/\sqrt{m}$, may be expected. Interpretation of laser fusion experiments [3], experiments with heating of magnetic islands in tokamaks [6], Fokker-Planck simulations of laser produced plasmas [7], modeling of tokamak scrape-off layer with strong recycling [8], has provided $0.002 \leq \xi \leq 0.1$. The fact that $\xi \ll 1$ is explained by non-local effects in collisionless plasma, reducing the perturbation in the distribution function caused by the temperature gradient, and braking of light electrons with the ambipolar electric field. A smooth transition between collisional and collisionless limits can be described by the formula [3]:

$$\chi_{||} \approx \chi_{||}^{SH} / \left(1 + \frac{3\lambda_c}{\xi L_T}\right) \tag{4}$$

One can see that this formula results in an even stronger reduction of the heat conduction near the cooled region as one expects from the temperature dependence of $\chi_{||}^{SH}$. This should result in an increase of the decay time and bring it closer to the experimental observations. However, the dependence of $\chi_{||}$ on the temperature gradient, prescribed by equation (4), may be a source of numerical troubles and a significant reduction of the iteration mixing coefficient, down to

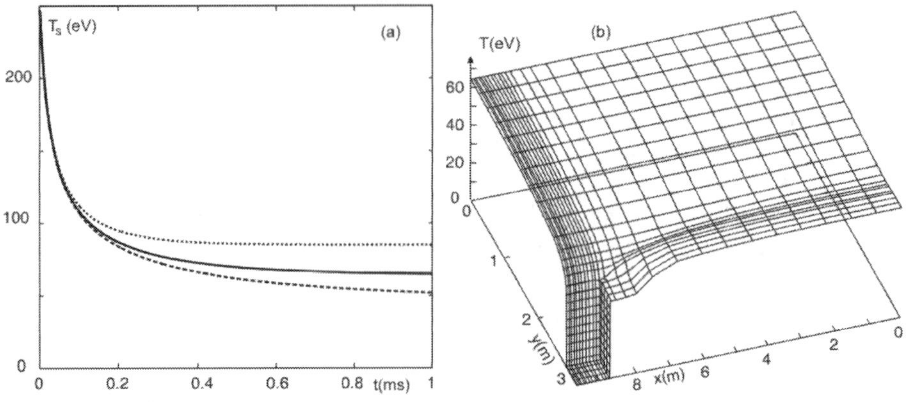

Figure 1. Time evolution of the plasma temperature far from the cooled area for $\nu = 1000$ (solid line), 300 (dashed line) and $3000s^{-1}$ (dotted line) (a) and final stationary temperature profile (b) computed with Spitzer-Härm heat conduction.

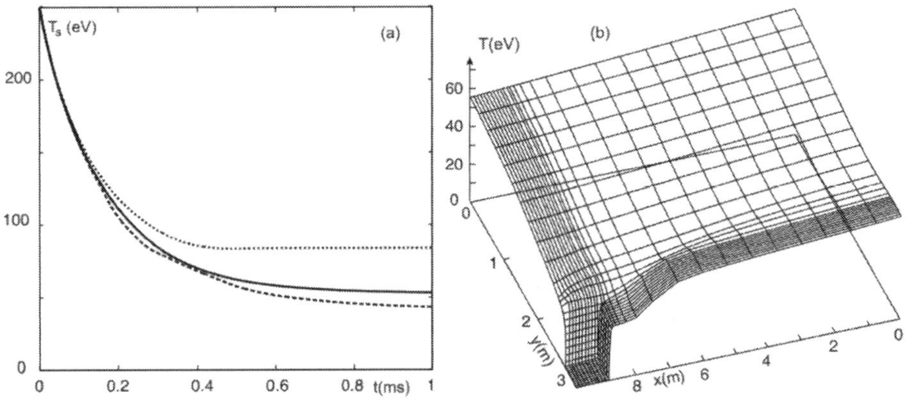

Figure 2. The same parameters as in figure 1 computed with the heat flux limit taken into account.

10^{-3}, is necessary to get converged solutions. This increases the computation time immensely. Figure 2 shows the time variation of T_s found with $\xi = 0.03$ and different magnitudes of ν and the temperature profile in the final stationary state for $\nu = 10^3 s^{-1}$. One can see that with the deviations from the Spitzer-Härm approximation to the heat conduction, see equation (4), the decay time of the temperature far from the injection position agrees significantly better with the experimentally measured level [5] in the sub-millisecond range.

References

[1] Tokar M Z and Koltunov M 2012 *Phys. Plasmas* **19** 042502
[2] Spitzer L and Härm R 1953 *Phys. Rev.* **89** 977
[3] Malone R C, McCrory R L and Morse R L 1975 *Phys. Rev. Lett.* **34** 721
[4] Tokar M Z 2010 *J. Comp. Phys.* **229** 2625
[5] Lehnen M, Alonso A, Arnoux G, et al. 2011 *Nucl. Fusion* **51** 123010
[6] Tokar M Z and Gupta A 2007 *Phys. Rev. Lett.* **99** 225001
[7] Luciani J F, Mora P and Virmont J 1983 *Phys. Rev. Lett.* **51** 1664
[8] Chodura R 1992 *Contrib. Plasma Phys.* **320** 219

Shear Propagation in Sand System

Sasuke Miyazima, Yuko Yamada and Keizo Yamamoto*

Faculty of Engineering, Chubu University, Kasugai, Aichi 487-8501, Japan
*Faculty of Engineering, Setsunan University, Neyagawa, Osaka 572-8508 , Japan

miyazima@isc.chubu.ac.jp

Abstract. Shear propagation and formation of sheared region in a sand system are investigated by a simulation. Our experimental apparatus for sand system is one of the simplest sets in order to obtain shear pattern, called as linear split-bottom cell. In this paper we give an analysis based on simulation for the above-mentioned system. Although a sand particle has a variety of forms from sphere to needle in general, but here we assume an ensemble of the simplest shape such as cubes in a rectangular parallelpipe split into two halves. One of the halves is moved along the split quasi-statically. We obtain a pattern of shear region and various quantities obtained from our simulation are compared with experiments.

1. Introduction

Properties in sand system has recently been studied from basic processes[1]-[3] and also actual applications to technological fields[4]. As is well-known, sand systems consist of sand particles. Generally, forms of sand particles diverge from sphere to needle. Ensemble of such particles forms an example of typical random system. When stress is given at a site, along a line, along a surface and/or to bulk body, propagation of stress and strain becomes very complicated. It is very important to investigate the quasi-static propagation of stress and strain in the sand system.

Therefore we built up an experimental apparatus to study how the shear propagates and is formed as simple as possible which is shown in Fig.1. And our obtained results of sheared width which depends on height from the bottom is shown in Fig.2.

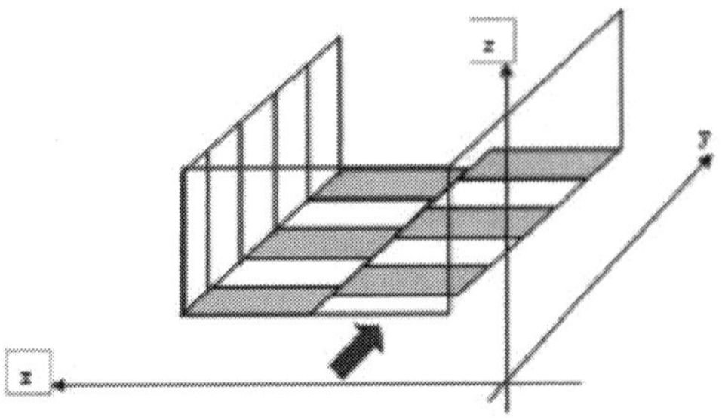

Fig.1 Experiment aparatus

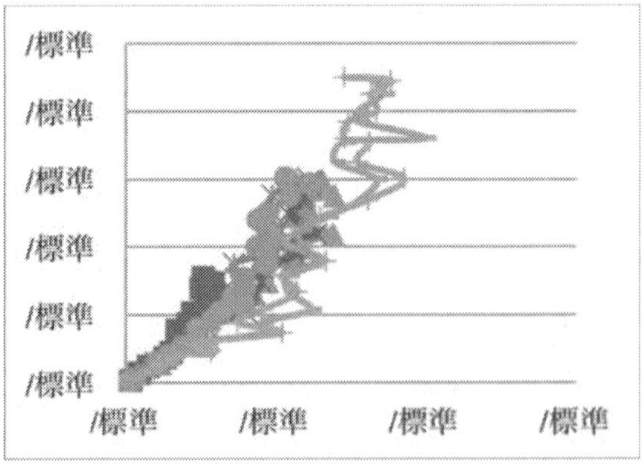

Fig.2. Height-width diagram

In Fig.1, The right hand side of outer box can be moved, where stress propagate from bottom to top. Figure 2 shows that the sheared width which depends on the height of sheared region in a sand system are investigated by a simulation. Our experimental apparatus for sand system is one of the simplest sets in order to obtain shear pattern, called as linear split-bottom cell. In this paper we give an analysis based on simulation for the above-mentioned system. Although a sand particle has a variety of forms from sphere to needle in general, but here we assume an ensemble of the simplest shape such as cubes in a rectangular parallel-pipe split into two halves. One of the halves is moved along the split quasi-statically shown in Fig.1. We obtain a pattern shown in Fig.2 and various quantities obtained from our simulation are compared with experiments.

2. A simulation corresponding the present experiment

We assume that sand particles are approximated by small cubes and our sand ensemble is regularly piled 100×86 cubes which approximates a random piling. When one side of the box is under the force which is shown an arrow in Fig.1, the sand particles at the bottom move together with the bottom plate and many particles at the top share the slip between the right and left half boxes. By assuming the frition force between two particles we have obtained the following equation.

$$\begin{gathered}\left[\mu_x\big(f(x,z)-f(x-1,z)\big)-d_{shr}\right]+\left[\mu_x\big(f(x,z)-f(x+1,z)\big)-d_{shr}\right]\\ +\left[\mu_z\big(f(x,z)-f(x,z+1)\big)(z_{top}-z)-d_{shr}\right]\\ +\left[\mu_x\,(f(x,z)-f(x,z-1))(z_{top}-z+1)-d_{shr}\right]=0\end{gathered}$$

Here, $f(x,z)$ is displacement of a particle at x and z and $\boldsymbol{\mu_x(f(x,z)-f(x-1,z))}$ is friction between the particles at x, z and x-1, z, where $\boldsymbol{\mu_x}$ and $\boldsymbol{\mu_z}$ are the friction coefficient, as shown in Fig.3. $[G-d_{shr}]$ means $G-d_{shr}$ if $G-d_{shr}>0$, and 0 if $G-d_{shr}<0$. Once we know the numeric values of four neighbour sites of $f(x, z)$, we can obtain the value of $f(x, z)$ by using this equation.

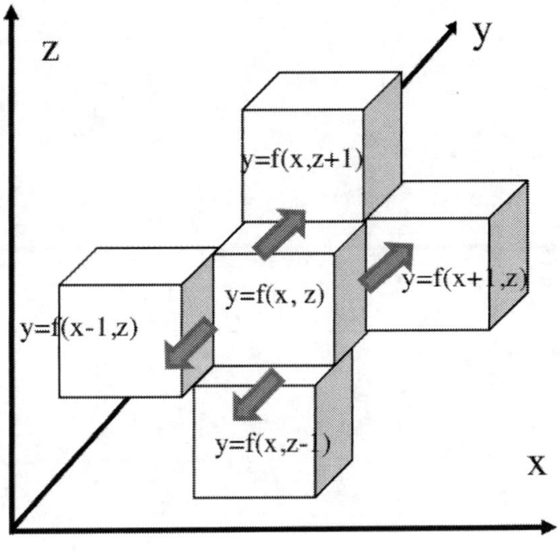

Fig.3. A sand particle is under the force from Four nearest neighbour cubes.

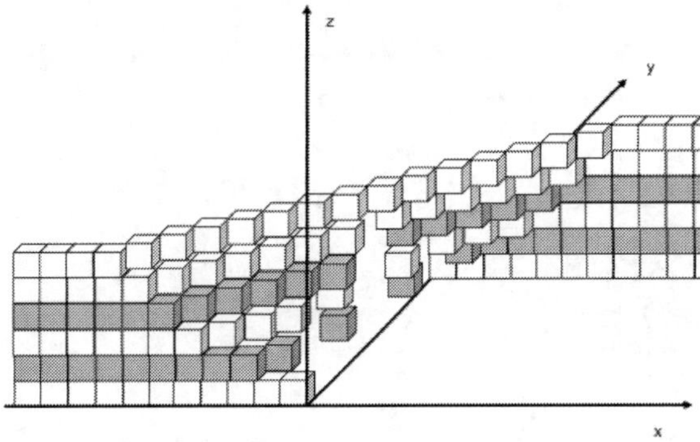

Fig. 4. A sketch of sheared portion.

3. Some results of calculation

Figure 4 shows a movement of particles near the sliding straight plate at the center. Here we adopt that a sliding distance is 7mm as experiment where a side box is moved in a quasi-static manner. The particles at the bottom move together with the plate, but next upper particles near the sliding plate move a little less distance. A graphic is shown in Fig.5, looked from the top b by adopting $\mu_{x'} = 0.2$, $\mu_{z'} = 0.1$ and $d_{shr'} = 0$.

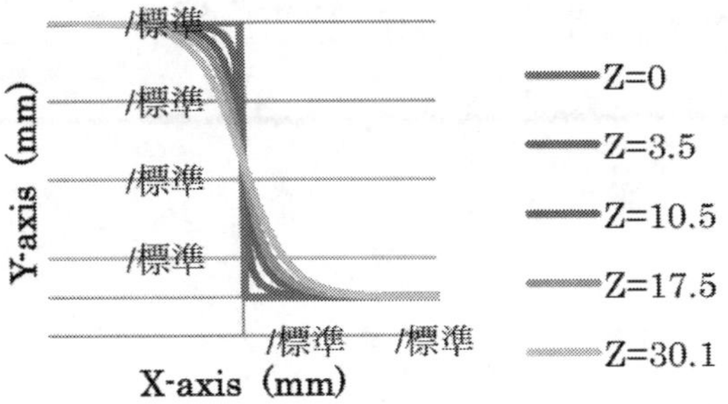

Fig.5. Shape of sheared region.

The width of sheared region is depicted against height in Fig.6.

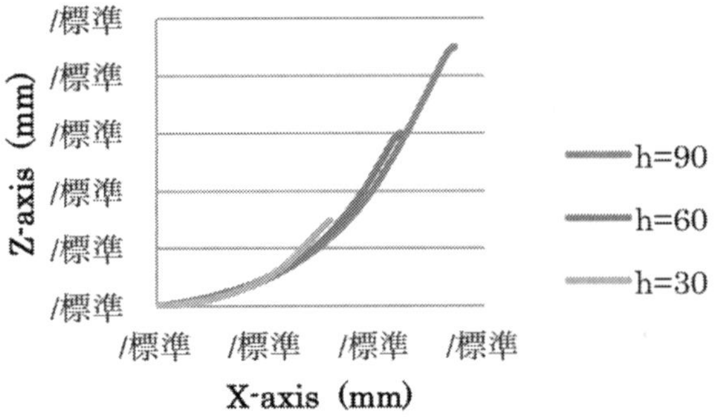

Fig.6. Sheared width for 3 different height of sand pile.

4. Concluding remarks

Shear pattern in the sand system is examined by experiment by many scientists and here is a trial for simulating the experiment. Some results of experiment are realized by the present simulation. The sheared region becomes wider with the height. The present simulation is the first trial and some modification is possible without difficulty. For instance, random packing of sand particles can be introduced. It is also possible to adopt a sphere for a particle. When we assume random packing and/or a spherical form for a sand particle, it is equivalent to introduce spaces for a particle to move with ease. Furthermore when we introduce irregular shape for sand particle, it may connect with avalanche of sliding or moving. Finally these simulation may be useful for investigation of dislocation in the earth

References

[1] Ries A Wolf D E and Unger T 2007 Physical Review E **76** 051301

[2] Fenistein D, van de Meent J W, and van Hecke M Physical Review Letters **92** Number 9

[3] Torok J, Unger T, Kertesz J and Wolf D E 2007 Physical Review E **75**, 011305

[4] Advanced Powder Technology, Elsevier

Journal of Physics: Conference Series **410** (2013) 012070 doi:10.1088/1742-6596/410/1/012070

Periodic Strategies and Rationalizability in Perfect Information 2-Player Strategic Form Games

V. K. Oikonomou[1] * and J. Jost[1,2] †

1 Max Planck Institute for Mathematics in the Sciences
Inselstrasse 22, 04103 Leipzig, Germany
2 Santa Fe Institute, New Mexico, USA

E-mail: *Vasilis.Oikonomou@mis.mpg.de, †jost@mis.mpg.de

Abstract. We define and study periodic strategies in 2-player finite strategic form games. and exploit their connection to non-Nash rationalizable strategies. The result of our findings is that, non-Nash rationalizable strategies are always periodic, but periodic strategies are not necessarily rationalizable.

Non-cooperative game theory [1, 2] has developed to be the most useful tool in strategic decision making and social sciences the last decades, especially when the outcomes of the players are strongly interdependent. This aforementioned attribute renders non-cooperative game theory different from problems of single agent decision theory. The concepts of rationality of the players and also the common belief in rationality of the players that participate in the game are foundational principles of invaluable importance. In the literature rationality is somehow a controversial feature therefore in this paper we assume that rationality means, that each player forms beliefs about his own and his opponents possible strategies. The beliefs, are materialized in the game quantitatively in terms of subjective probabilities and acts according to those probabilities. The word act embodies the fact that a player chooses an optimal strategy based on the beliefs of the player for his opponents strategies. We shall assume that each player acts rationally and additionally that he believes in his opponents rationality. Additionally we shall work within the context of common knowledge [3], which means that either the game and rationality of the players are assumed to be common knowledge. We also confine ourselves to outcomes that are measured by von Neumann-Morgenstern utilities functions [4], so optimality means utility maximization. In this article we shall consider finite strategic form games with complete information that are played simultaneously and only for one time. Strategic form games are very important in game theory since even extensive games can be reduced in strategic form games. A strategic game consists of a finite set of players, a finite set of actions (often so-called pure-strategies), and a utility function for each player. One of the most fundamental concepts in non-cooperative game theory is the Nash equilibrium. A pure-strategy Nash equilibrium is an action profile for which no single player can obtain a higher payoff by deviating unilaterally from this strategy profile. However, as Bernheim notes in his paper [5], Nash equilibrium is neither a necessary consequence of rationality nor a reasonable empirical proposition. A very important refinement of the Nash equilibrium is the rationalizability solution concept and rationalizable strategies [5, 6]. The rationalizability concept appeared independently in Bernheim's [5] and Pearce's work [6]. In this article we shall study periodic strategies in two player strategic form games. With the term periodic is meant that there exist maps between the two players strategy spaces, such that they constitute an automorphism on each strategy space d, which has the property $d^n = 1$ for some $n \in N$. The periodicity concept appeared in Bernheim's paper [5] in a different context and not under that name. We shall prove that periodic strategies exist in two player strategic games and some of these indeed are rationalizable too. In particular our findings prove that the set of periodic strategies include all the rationalizable strategies that are not Nash equilibria. In addition there are Nash equilibria that can be periodic and also there are periodic strategies that are neither Nash strategies nor rationalizable. Being a superset of non-Nash rationalizable strategies, the periodic strategies could serve as a more general concept than rationalizability. This paper is based on the working paper [7], and due to the constraints of space, we omit the proofs and also the study of mixed strategies. These can be found in [7].

We consider a simultaneous move game with two players A and B, with perfect information and we assume that the game is played only once. The actions are considered to be finite in number, as a first approach. Each player can have a finite number of actions but the actions of the two players can be different in number. Denote with $M(A)$ the strategy space of all A's actions and with $N(B)$ the strategy space of all B's actions. The strategic form game is then defined as:

- The set of players: $I = 1, 2$
- The strategy spaces of players A, B, namely $M(A)$ and $N(B)$, and the total strategy space $\bar{G} = M(A) \times N(B)$
- The payoff functions $u_i(\bar{G}) : \bar{G} \to \Re^2$, $i = 1, 2$

We define two continuous maps, ϕ_1 and ϕ_2, between the strategy spaces $M(A)$ and $N(B)$, to act as follows:

$$\phi_1 : M(A) \to N(B) \quad \phi_2 : N(B) \to M(A) \tag{1}$$

These two maps are considered to be injective (monomorphisms). Lets define the action of these maps to each strategy space. With the payoff being defined as:

$$u_i : M(A) \times N(B) \to \Re^2 \tag{2}$$

the actions of the aforementioned maps $\phi_{1,2}$ are defined in such a way that the following hold at each level:

$$\begin{aligned} & u_1(x, \phi_1(x)) \geq u_1(x, y_1) \quad \forall y_1 \in N(B) \\ & u_2(\phi_2 \circ \phi_1(x), \phi_1(x)) \geq u_2(x_1, \phi_1(x)) \, \forall x_1 \in M(A) \\ & u_1(\phi_2 \circ \phi_1(x), \phi_1 \circ \phi_2 \circ \phi_1(x)) \geq u_1(\phi_2 \circ \phi_1(x), y_1) \forall \, y_1 \in N(B) \\ & \vdots \end{aligned} \tag{3}$$

Let us clarify the meaning of the above inequalities. We start with an action $x \in M(A)$, upon which we act with the map ϕ_1. The map is acting in such a way that the inequality $u_1(x, \phi_1(x)) \geq u_1(x, y_1) \quad \forall y_1 \in N$ holds true. What this means is that ϕ_1 maps x to B's strategy space such that this action of B's yields the best payoff for player A playing the action x. So we could say that the map yields the actions of player B to which the action x of A is a best response. At the next step, the map ϕ_2 acts on the strategy space of player B and yields an action $\phi_2 \circ \phi_1(x) \in M(A)$ such that $u_2(\phi_2 \circ \phi_1(x), \phi_1(x)) \geq u_2(x_1, \phi_1(x)) \forall x_1 \in M(A)$. So we could say that $\phi_2 \circ \phi_1(x)$ is an action of player A to which $\phi_1(x)$ is a best response for player B. If we proceed this way, it is possible to end up to the initial x action of player A. Let us depict this procedure with a one dimensional chain of actions, as follows:

$$x \underset{\rightarrow}{P} \phi_1(x) \underset{\rightarrow}{P} \phi_2 \circ \phi_1(x) \underset{\rightarrow}{P} \cdots \underset{\rightarrow}{P} x \tag{4}$$

where with the letter P we denote the procedure described in relation (3) above. Therefore it is possible to construct a chain of continuous action of the maps $\phi_{1,2}$ so that the final action is identical to the action at the beginning of the game. We shall call that action periodic. Notice that for periodic actions, the operator $d = \phi_2 \circ \phi_1$, has the property that $d^n = \mathrm{id}_M$ for some number $n \in N$. This is a very important property of the space of actions of player A. It means that we can find an operator that acts as an automorphism on a subset of $M(A)$ and leaves this subset invariant under this map. It is exactly this subset of the total strategy space of player A that constitutes the set of periodic strategies of players A. So we can formally define the periodic strategies and actions as follows.

DEFINITION 1: In a 2-player simultaneous move strategic form game with finite actions, we define periodic strategies for a player A to be a subset of his available strategies $M(A)$ such that there exists an operator d: $M(A) \rightarrow M(A)$, such that $\exists\ n \in N$ for which $d^n = \mathrm{id}_M$. It is presupposed that the map consists of maps that act in such a way that the inequalities of relation (3) are fulfilled.

There is a clear conceptual distinction between periodic strategies and rationalizable strategies. We will make this clear with some characteristic examples that we analyze in detail. As a comment we have to say that to our knowledge the number n defined previously, depends on the payoff details of the game. So there is no direct connection of this number , to the number of actions and the number of players, at least for finite 2-player strategic form games. There is also some sort of connection between the total number of Nash equilibria and the number n, a relation which at the moment we fail to understand from what arguments it originates. We hope to address these problems along with the case of continuous games in the future. According to our findings, all non-Nash rationalizable actions are periodic. Moreover there are Nash actions that are periodic and some Nash strategies that are not periodic. The most important outcome of our work is materialized in the following theorem:

THEOREM In a 2-player finite action strategic form game, the rationalizable actions that do not belong to a Nash equilibrium, are periodic. The converse is not true.

The proof can be found in [7]. Denote with $P(A)$ the set of all the periodic actions that are contained in player's A strategy space, $N(A)$ the set of player A actions that belong to some Nash equilibrium, $R(A)$ be the set of all rationalizable actions and $R'(A)$ the set of non-Nash rationalizable actions. So $R'(A) = R(A)\backslash N(A)$. The theorem states that all actions that belong to $R'(A)$ also belong to $P(A)$. We could say that the following hold true:

$$R'(A) \subseteq P(A) \tag{5}$$

It is obvious that the set of periodic actions contains the set of non-Nash rationalizable actions. Therefore finding the periodic actions in a finite game but with many actions, is a first step towards finding the set of all rationalizable strategies. Before closing we present an example that will enlighten the concepts that we presented in this section and show us that the theorem finds application. Consider game 1. According to the inequalities of relation (3), we can construct the following periodicity cycles, namely:

$$\begin{aligned} &a_1 \overset{P}{\rightarrow} b_3 \overset{P}{\rightarrow} a_3 \overset{P}{\rightarrow} b_1 \overset{P}{\rightarrow} a_1 \\ &a_3 \overset{P}{\rightarrow} b_1 \overset{P}{\rightarrow} a_1 \overset{P}{\rightarrow} b_3 \overset{P}{\rightarrow} a_2 \end{aligned} \tag{6}$$

It is obvious in the above two examples that $n = 2$ for both the actions, a_1 and a_3. Moreover for the actions that constitute a Nash equilibrium it is not possible to construct such a cycle. Now let us consider the rationalizability cycles. The actions a_1 and a_3 are both rationalizable and neither of them is a Nash strategy. Therefore the theorem we presented previously applies to them. So these two actions are both rationalizable and periodic, that is there exist a rationalizability cycle for these two. By rationalizability cycle is meant a cycle based on rationality and by rationality

	b_1	b_2	b_3	b_4
a_1	0,7	2,5	7,0	0,1
a_2	5,2	3,3	5,2	0,1
a_3	7,0	2,5	0,7	0,1
a_4	0,0	0,-2	0,0	10,-1

Table 1. Game 1

is meant acting optimally under some beliefs about the opponents actions. Indeed such a cycle exists and it looks like:

$$\begin{aligned} &a_1 \underset{\rightarrow}{R} b_3 \underset{\rightarrow}{R} a_3 \underset{\rightarrow}{R} b_1 \underset{\rightarrow}{R} a_1 \\ &a_3 \underset{\rightarrow}{P} b_1 \underset{\rightarrow}{P} a_1 \underset{\rightarrow}{P} b_3 \underset{\rightarrow}{P} a_2 \end{aligned} \tag{7}$$

The reasoning behind this cycle is based on this system of beliefs: Player A considers action a_1 rational if he believes that player B will play b_3, which is rational for player B if he believes that player A will play a_3. Accordingly, A will consider playing a_3 rational if he believes that player B will play b_1, which would be rational for player B if he beliefs that player A will play a_1. So we obtain a cycle of rationalizability based on pure utility maximization rationality. For the Nash action a_2 it is not easy to construct such a cycle because A will be forced to play a_2 since B would never play b_1 or b_3 as a best response to a_2. So the Nash strategy is forced to be rationalizable. In this game the non-Nash rationalizable actions are periodic actions and actually are the only periodic strategies. So the theorem finds direct application in this case.

We have exploited an intrinsic property of 2-player finite simultaneous strategic form games, which we called periodicity of strategies. Additionally we provided a theorem which states that every action that is rationalizable but does not belong to a Nash equilibrium is periodic. Additionally we found that the set of periodic actions is a larger set than these of non Nash rationalizable actions. Clearly there could be some reasoning behind this that resides in some fixed point argument of the appropriate operator d^n, for some $n \in N$, but why this operation excludes some of the Nash equilibria is still a mystery for us. Probably there exists an underlying pattern strongly related to the specific payoff of the games or the combination of actions in a specific way, that we still fail to understand. The periodicity feature for finitely many actions of strategic form games can be very useful. Indeed, all the periodic actions can be found using some simple program code, a fact that is clearly a good step in finding all the rationalizable actions that are non- Nash equilibria. This can be very useful for games that have, as we mentioned, finitely many actions. In [7] we have included into the study the mixed strategies for 2×2 games and also commented about continuous 2-player games. In addition the reference list is more complete in the working paper [7]

References

[1] Fudenberg, Drew; Tirole, Jean (1991), *Game theory*, MIT Press
[2] Osborne, Martin J (2004),*An introduction to game theory*, Oxford University Press
[3] Geanakoplos J, "Common Knowledge", (1992), *Journal of Economic Perspectives*, **6** pp 53-82
[4] *Theory of Games and Economic Behavior*, Princeton University Press, John von Neumann, Oskar Morgenstern (2004)
[5] Bernheim, B D (1984), Rationalizable strategic behavior, *Econometrica* **52**, 1007
[6] Pearce, D (1984), Rationalizable strategic behavior and the problem of perfection, *Econometrica* **52**, 1029-1050
[7] Oikonomou V K, Jost J *working paper*

Exact Enumeration of Protein Conformations from Fragment Assembly

Julian Lee

Department of Bioinformatics and Life Science, Soongsil University, Seoul, Korea

E-mail: jul@ssu.ac.kr

Abstract. Efficient sampling of protein conformations is very important in computational studies of protein folding. The search space of the protein conformations can be reduced by utilizing bioinformatics. The frament assembly is one such method, where the local structures are selected from the fragment sets prepared using bioinformatics tool. The conformational space of proteins is reduced to a finite discrete set by using fragment assembly. The search space can be further reduced by introducing a consistency condition at the junction of the fragments. An algorithm for exact enuemration of such conformations are introduced, which can be used for generating cadidates of the native structure for a relatively short protein.

1. Introduction

Prediction of the native structure of a protein is an important and challenging problem in computational biophysics. According to Anfinsen's thermodynamic hypothesis[1], the native structure is the conformation minimizing the free energy under physiological condition. However, computation of protein native structure by purely physical means has not been so successful, not only because it is difficult to construct a free energy function with reasonable accuracy, but also because the conformational space that has to be sampled is huge. The fragment assembly method, which has exhibited excellent performance in protein structure prediction for sequences with little overal sequence similarity with those in the sutructural database, is a method that uses bioinformatics method to complement the physico-chemical calculations[2, 3, 4, 5, 6, 7, 8, 9, 10, 11, 12, 13]. In a fragment assembly method of protein structure prediction, candidates of possible local structure are constructed for each part of the protein, by collecting fragments from structural database using similarity of sequence features. The conformations are generated by assembling these fragments and only the global tertiary packing of fragments is determined by energy optimization, and hence the name fragment assembly method. By using fragments, not only the burden of modeling interactions between neighboring residues is eliminated, but also the conformational space is reduced to a finite set of discrete conformations, making the conformational search much more tractable.

Despite remarkable successes of fragment assembly methods[2, 3, 4, 5, 6, 7, 8, 9, 10, 11, 12, 13], the conformational space of a protein has been mostly sampled using stochastic methods coupled with appropriate score function to select the model structure. In this work, we develop an algorithm for exact enumeration of all possible conformations generated by fragment assembly, after achieving a further reduction of the conformational space by implementing addtional constraints for fragment assembly. The exact enumeration algorithm is then tested for a small protein 1m2zB .

The exact enumeration of protein conformations, even for short proteins, is useful in several aspects. Since all the conformations are examined for a given protein, the exact global minimum of a free energy function can be found. For a training set consisting of proteins of known structures, this information can be used to develop an accurate free energy function so that a native-like conformation is produced as the global minimum. Also, method for collecting candidate fragments from structural database may be developed, so that there are many native-like conformations in the search space. Lastly, benchmark tests on various stochastic sampling methods can be performed.

2. Methods

2.1. Structural Database

A list of chains with mutual sequence identities less than 25 %, X-ray resolution lower than 3 Å and R factor lower than 1.0 was taken from the protein sequence culling server PISCES [14]. The chains with unknown structure regions were removed to obtain the final database consisting of 1,927 protein chains with a total of 378,149 residues. The fragments are prepared from this database.

2.2. Fragment Set

For each residue of a protein sequence, we collect a set of k fragments from the structural database, corresponding to the most probable conformations of the local neighborhood. By tuning the parameter k, one can control the size of the conformational space. The criterion for the fragment selection is the similarity of sequence feature. However, instead of directly comparing the raw sequences, the profile obtained from PSI-BLAST[15] search is compared.

When PSI-BLAST[15] search is performed against a protein *sequence* database, the sequences similar to the query is retrieved and aligned, and the rate of substitution of each residue of the query protein to other amino acids is then calculated, which is called the sequence profile. The PSI-BLAST sequence profile is a matrix of size (protein length)$\times$20, which contains evolutionary information that cannot be obtained from the raw sequence only.

For each residue of a query sequence, a window of size 15 centered on this residue is constructed. The part of the sequence profile lying within this window is considered as the sequence feature vector. Thus, the feature vector is the matrix of size 15$\times$20, except for windows centered on terminal residues where the windows have to be truncated[16, 17, 18, 19, 20].

Similar construction is performed for each window of each protein sequence in the reference dataset. Then k such windows whose feature vectors are closest to the query window is collected. Since k windows are collected from the reference dataset with known structures, the corresponding fragments of size 15 form the fragment set for the query window. The distance measure used for the sequence feature comparison is

$$d_{ij} = \sum_{i=1}^{15}\sum_{j=1}^{20} |P_{ij} - P'_{ij}| \tag{1}$$

where P_{ij} and P'_{ij} are the elements of the feature vectors[16, 17, 18, 19, 20], except for small details such as fragment sizes or similarity measures used for the sequence feature comparison.

2.3. Fragment Assembly

We assemble fragments to construct a protein conformation. Since a protein chain of length N is assmbled from N fragments, and there are k cadidate fragments centered on each residue, the total number of possible conformations is k^N if the fragments are assembled without any restriction. In the current work, the fragments are joined only when they overlap and share at least one residue with similar values of dihedral angles. Two sets of dihedral angles (ϕ_1, ψ_1) and (ϕ_2, ψ_2) are considered to be similar to each other if either

$$|\phi_1 - \phi_2| + |\psi_1 - \psi_2| \leq 2°. \tag{2}$$

If we find such a residue, then the second fragment is joined smoothly to the first one starting from this residue. This additional constraints prevent appearance of any local structure around fragment junction that is not present in the structural database, and reduce the conformational space substantially, while still allowing for generating native-like structures. The fragments that can be joined are precomputed, and then all possible conformations are enumerated.

3. Results

To test the exact enumeration algorithm, we performed exact enumeration of the conformations of 1m2z chain B, with sequence length 21 consisting of mostly of α helices, a protein included in the structural database. For fairness of the test, it is removed from the structural database before the fragment selection, so that the remaining homology with the structural database is sequence identity less than 25 % by construction.

The total number of conformations are given in table I for various values of k. It is a non-decreasing function of k, as it should be, since the conformational space for a given value of k always includes the space for smaller value of k as a subset.

Table 1. The total number of fragment-assembled protein conformations, the number of conformations with RMSD < 4.0 Å, their ratio, the smallest value of RMSD found among the conformations, and CPU time, as functions of number of candidate fragments k, for the protein 1m2z chain B.

k	total number	< 4.0	fraction	RMSD(smallest)	CPU time (sec)
1	0	0	–	–	0.06
2	0	0	–	–	0.06
3	381	41	0.11	3.89	0.11
4	1360	66	0.049	3.83	0.24
5	3232	196	0.061	3.70	0.46
6	50414	4083	0.081	3.57	6.34
7	164232	13950	0.085	3.57	20
8	418175	36225	0.086	2.72	52
9	668856	55811	0.086	2.72	84
10	1082077	94664	0.086	2.72	136

As the number of conformations increases, the number of native-like conformations will also increase. It is crucial for the success of a fragment-based protein structure prediction method, that a sufficient number of native-like conformations are generated by fragment assembly. The native-like property of a protein conformation is usually measured by root-mean-square-deviation(RMSD) of its backbone atom coordinates from those of the native structure. The smallest RMSD value found among all the conformations, and the number of conformations with RMSD < 4.0 Å, are displayed in table I. Note that for $k = 10$, the total number of conformation would be 10^{21} if the fragments were assembled without imposing any constraint. In the current algorithm, the number of conformations is reduced to about 10^6, achieving a

10^{15} fold reduction, but conformations with RMSD as low as 2.72 Å are generated. The CPU time spent for computation in Intel i7 64 bit Hexacore was 136 seconds for generating these conformations.

4. Discussions

In this work, we developed an algorithm for generating all possible conformations assembled from structural fragments. The crucial elements are the additional reduction of the conformational space by imposing constraints at the fragment junctions, and the precomputation of fragment that can be joined together. The algorithm was tested for a small protein 1m2zB. For $k = 10$ the number of conformations are reduced by 10^{15} fold, but conformations with RMSD as small as 2.72 Å were generated. By combining the current algorithm with accurate score function to select the model structure, a novel protein structure predicton algorithm may be developed.

4.1. Acknowledgments

The author thanks Seung-Yeon Kim for useful discussions. This work was supported by the Soongsil University Research Fund.

References

[1] Anfinsen C B 1973 *Science* **181** 223
[2] Jones D T 2005 *Proteins* **61 (S7)** 143
[3] Rohl C, Strauss C, Misura K and Baker D 2004 *Methods Enzymol.* **383** 66
[4] Chikenji G, Fujitsuka Y and Takada S 2006 *J. Proc. Acad. Natl. Sci. U.S.A.* **103** 3141
[5] Lee J, Kim S-Y, Joo K., Kim I and Lee J 2004 *Proteins* **56** 704
[6] Lee J, Kim S-Y and Lee J 2005 *Biophys. Chem.* **115** 209
[7] Lee J, Kim S-Y and Lee J 2005 *J. Korean Phys. Soc.* **46** 707
[8] Kim S-Y, Lee W and Lee J 2006 *J. of Chem. Phys.* **125** 194908
[9] Kim T K and Lee J 2008 *J. Korean Phys. Soc.* **52** 137
[10] Lee D-S, Seok C and Lee J 2008 *J. Korean Phys. Soc.* **52** 1137
[11] Lee J, Lee D-S, Park H, Coutsias E A and Seok C 2010 *Proteins* **78** 3428
[12] Ko J, Lee D-S, Park H, Coutsias E A, Lee J and Seok C 2011 *Nucleic Acid Res* **39** W210
[13] Park H, Ko J, Joo K, Lee J, Seok C and Lee J 2011 *Proteins* **79** 2725
[14] Wang G and Dunbrack R L 2003 *Bioinformatics* **19** 1589
[15] Altschul S F, Madden T L, Schaffer A A, Zhang J , Zhang Z, Miller W and Lipman D J 1997 *Nucleic Acid Res* **25** 3389
[16] Kim S-Y, Sim J and Lee J 2006 *Lecture Notes in Bioinformatics* **4115** 562
[17] Sim J, Kim S-Y and Lee J 2005 *Bioinformatics* **21** 2844
[18] Joo K, Lee J, Kim S-Y , Kim I, Lee S J and Lee J 2004 *J. Korean Phys. Soc.* **44** 599
[19] Joo K, Kim I, Lee J, Kim S-Y, Lee S J, and Lee J 2004 *J. Korean Phys. Soc.* **45** 1441
[20] Lee J 2009 *J. Korean Phys. Soc.* **54** 1

IC-MSQUARE 2012: International Conference on Mathematical Modelling in Physical Sciences IOP Publishing
Journal of Physics: Conference Series **410** (2013) 012072 doi:10.1088/1742-6596/410/1/012072

Remarks on point interactions in quantum mechanics

José T Lunardi[a], Luiz A Manzoni[b] and Wagner Monteiro[a]

[a] Departamento de Matemática e Estatística, Universidade Estadual de Ponta Grossa
Avenida General Carlos Cavalcanti, 4748. 84030-000, Ponta Grossa, PR, Brazil
[b] Department of Physics, Concordia College, 901 8th St. S., Moorhead, MN 56562, USA

E-mail: jttlunardi@uepg.br, manzoni@cord.edu, wmonteiro137@gmail.com

Abstract. When dealing with singular potentials, such as point interactions, the term $V(x)\psi(x)$ in Schrödinger's equation is, in general, mathematically ill-defined. The traditional way of dealing with this difficulty, by employing regularization techniques, has led to some ambiguous results for the $\delta'(x)$ potential in the literature. Here we propose a mathematically consistent approach to deal with the one-dimensional version of this problem by considering from the beginning the distributional nature of the whole interaction term. We show that the interaction is univocally determined from two simple mathematical requirements on the interaction distribution, together with the physical requirement of probability flux conservation and considerations of symmetry.

The solution of Schrödinger's equation with point (or zero-range) interactions has been widely considered in literature due to both its usefulness in modeling interactions in atomic, nuclear and solid state physics (see, e.g., [1] and references therein) and its value as a "theoretical laboratory" to investigate regularization and renormalization in simple scenarios [2, 3]. However, it is often the case that these point interactions lead to regularization dependent and contradictory results (see, for example, [4, 5] and references therein). The origin of such difficulties lies in the fact that point interactions are represented by singular potentials which are not ordinary functions but distributions (generalized functions). Therefore, the interaction term in Schrödinger's equation $[V(x)\psi(x)]$ is generally mathematically ill-defined, since the product of singular distributions is in general not defined.

In this paper we report on an approach developed by taking into account the distributional nature of the interaction term in a consistent way and using probability flux conservation and symmetry as guides to univocally define the interaction term. Before we proceed in presenting such an approach below we briefly review some basic concepts of distribution theory which will be directly related to this work. For a systematic exposition of the theory we refer to [6].

A distribution d is a linear and continuous functional on a suitable space of infinitely differentiable test functions $\phi : \mathbb{R} \to \mathbb{C}$. Usually the space of test functions is C_0^∞, the space of infinitely smooth functions which vanish outside some finite interval, or the Schwartz's space $\mathcal{S}$ of infinitely smooth functions of rapid descent. Distributions can be infinitely differentiable and integrable, they can be added among themselves, or multiplied by well-behaved functions, but in general distributions cannot be multiplied by themselves. The *support* of a distribution d, denoted by supp d, is the smallest closed set of the real axis outside of which d equals zero. Every locally integrable function f corresponds to a *regular* distribution. Distributions which are not regular are called *singular*. It is said that a distribution has *order* $r = -2$ if it corresponds to a continuous but not differentiable function. Under distributional differentiation or integration the order r changes by $+1$ or -1, respectively. For instance, the Dirac delta $\delta(x)$ is a distribution whose support is concentrated at the origin and it is *singular* with order $r = 0$.

The Dirac's delta potential. Let us start by considering the distributional solution of the one-dimensional Schrödinger's equation with a Dirac's delta potential, $V(x) = \gamma\delta(x)$ (γ gives the strength of the interaction). Although this case does not present any of the ambiguities widespread in more general point interactions and its solution is well-known, it is suitable for the introduction of the basic ideas of our distributional approach and to lay the foundations of the method. Using units such that $\frac{\hbar}{2m} = 1$, the time independent Schrödinger equation is written as $-\psi''(x) + \gamma\delta(x)\psi(x) = E\psi(x)$ [7]. In this case, by working in the subspace of distributions defined on a space of test functions which are continuous but not differentiable at $x = 0$, the distributional product $\delta(x)\psi(x)$ can be identified as being $\psi(0)\delta(x)$. In alternative, our method starts by rewriting the Schrödinger equation in the form

$$\psi''(x) + E\psi(x) = s(x)\,, \tag{1}$$

where the interaction term was substituted by a distribution $s(x)$, yet to be determined. In order to obtain the specific form of the distribution $s(x)$ corresponding to the potential $V(x)$ one will require that (i) supp $s \subseteq$ supp V and that (ii) the distributions $s(x)$ and $V(x)$ have the same order. Both of these requirements are naturally satisfied when $V(x)$ corresponds to an ordinary function – here we are just extending them to the case of potentials described by distributions.

For the present case, $V(x) \propto \delta(x)$, and requirements (i) and (ii) imply that $s(x)$ must be a zero order distribution concentrated at $x = 0$. However, by a general theorem from distribution theory we have that any distribution concentrated at a single point is necessarily a linear combination of delta distributions (concentrated at the same point) and its derivatives up to the order of the distribution (see [6], page 96 ff.). In our case this theorem implies that

$$s(x) = \sum_{m=0}^{r_s} \alpha_m\, \delta^{(m)}(x) = \alpha\, \delta(x), \tag{2}$$

where α is a complex coefficient whose specification completely determines the interaction term.

With the interaction distribution $s(x)$ given by (2), the differential equation (1) can be solved by standard methods of distribution theory [6] and its general solution can be written as

$$\psi(x) = \theta(-x)\left[Ae^{ikx} + \left(B + \frac{\alpha}{2ik}\right)e^{-ikx}\right] + \theta(x)\left[\left(A + \frac{\alpha}{2ik}\right)e^{ikx} + Be^{-ikx}\right], \tag{3}$$

where A and B are arbitrary constants. By taking the limits of the above solution from the right and from the left of the origin, we obtain that the distribution ψ satisfies the following boundary conditions around $x = 0$:

$$\psi\left(0^-\right) = \psi\left(0^+\right) = \psi(0) \qquad ; \qquad \psi'\left(0^+\right) = \psi'\left(0^-\right) + \alpha, \tag{4}$$

where in the first equation we have used the fact that the distribution ψ is continuous at origin [it has $r = -2$, according to Eq. (1)]. It follows that the boundary conditions are completely specified once the coefficient α is determined.

As it is clear by now, the mathematical requirements (i) and (ii) are not enough to unambiguously specify $s(x)$, therefore it is necessary to impose some general physical constraints on the solution of (1). Since we are considering a stationary problem, it is natural to require that ψ be such that the *probability flux is conserved.*

The distribution (3) corresponds to an ordinary continuous and differentiable function on any interval not containing the origin. Thus, the products of ψ and its first derivative occurring in the expression for the probability flux are well defined for $x \in (-\infty, 0) \cup (0, \infty)$, and the requirement of the equality of the flux in both sides of the origin is equivalent to the requirement that $\alpha\psi^*(0)$ be real. If $\psi(0) \neq 0$ this condition can be written as

$$\alpha = \gamma\psi(0), \qquad \gamma \in \mathbf{R}, \tag{5}$$

where the remaining real constant γ is fixed by specifying the strength of the interaction. The case $\psi(0) = 0$ implies that the flux is conserved (and it is actually zero) regardless of the value of α, which becomes arbitrary. This implies that the solutions in the two sides of the origin are disconnected [because $\alpha = \psi'(0^+) - \psi'(0^-)$ from Eq. (4)]. This last case corresponds to a totally reflective barrier concentrated at the origin and it could be also obtained, as usual, by taking the limit $\gamma \to \infty$ in (5).

The above result, as expected, recovers the well-known prescription for the interaction with a Dirac's delta potential, usually obtained through the direct use of the identity $\gamma\psi(x)\delta(x) \equiv \gamma\psi(0)\delta(x)$, valid for continuous functions $\psi(x)$. More importantly, it illustrates the fact that our approach, which is based on requirements (i) and (ii) together with simple physical constraints, allowed us to specify univocally, and in a mathematically consistent manner, the singular point interaction. This approach will be explored below in the investigation of the $\delta'(x)$ interaction.

The first derivative of Dirac's delta potential. Consider the one-dimensional Schrödinger's equation in the presence of a potential $V(x) = \lambda\delta'(x)$, $(\lambda \in \mathbf{R})$. In this case, an attempt to define the interaction term as the product $V(x)\psi(x)$ soon runs into difficulties as can be seen in the following reasoning: if $\psi(x)$ and its derivative $\psi'(x)$ were continuous at the origin, the product $V(x)\psi(x)$ would be well defined in the theory of distributions and it would be given by $s(x) = \lambda\psi(x)\delta'(x) = \lambda\left[\psi(0)\delta'(x) - \psi'(0)\delta(x)\right]$. However, if we substitute this expression into the Schrödinger equation, a quick analysis of the order of the distributions in (1) shows that ψ must be of order $r = -1$, and therefore ψ cannot be continuous at the origin – a contradiction with the initial assumptions.

Let us now address this problem according to the ideas developed in the treatment of the Dirac's delta interaction. Since the potential is now of order $r = 1$ and concentrated at the origin, the most general form of the distribution $s(x)$ is

$$s(x) = \alpha\delta(x) + \beta\delta'(x), \qquad \text{with} \quad \beta \neq 0, \tag{6}$$

and specifying the coefficients α and β completely determines the interaction. The solution of eq. (1) with the above interaction term has the general form

$$\psi(x) = \theta(-x)\left[Ae^{ikx} + \left(B - i\frac{\alpha}{2k} - \frac{\beta}{2}\right)e^{-ikx}\right] + \theta(x)\left[\left(A - i\frac{\alpha}{2k} + \frac{\beta}{2}\right)e^{ikx} + Be^{-ikx}\right], \tag{7}$$

with A and B arbitrary constants. The boundary conditions at the origin are given, in terms of the still undetermined coefficients α and β, by

$$\psi\left(0^+\right) = \psi\left(0^-\right) + \beta \qquad ; \qquad \psi'\left(0^+\right) = \psi'\left(0^-\right) + \alpha. \tag{8}$$

Again, conditions (i) and (ii) are not enough to specify α and β and we need to impose additional physical conditions. Once more, it is natural to require probability flux conservation across the origin, which is equivelent to the requirement that the quantity $\alpha\psi^*(0^-) - \beta\psi'^*(0^-) + \alpha\beta^*$ be real.

For a moment we consider scattering solution with incidence from the left. In this case we can write $\alpha = \alpha_1\psi(0^-) + \alpha_2\psi'(0^-)$ and $\beta = \beta_1\psi(0^-) + \beta_2\psi'(0^-)$, with $\alpha_1, \alpha_2, \beta_1, \beta_2 \in \mathbf{R}$. Upon substitution of these expressions into (8) we obtain

$$\begin{bmatrix} \psi(0^+) \\ \psi'(0^+) \end{bmatrix} = \begin{bmatrix} 1+\beta_1 & \beta_2 \\ \alpha_1 & 1+\alpha_2 \end{bmatrix} \begin{bmatrix} \psi(0^-) \\ \psi'(0^-) \end{bmatrix} = \Lambda \begin{bmatrix} \psi(0^-) \\ \psi'(0^-) \end{bmatrix}, \tag{9}$$

and the condition of flux conservation now reads $(\det\Lambda - 1)\, j(0^-) = 0$. For $j(0^-) \neq 0$ this condition implies $\det\Lambda = 1$. On the other hand, if $j(0^-) = 0$ any value for $\det\Lambda$ is possible, but it can be shown that only $\det\Lambda = 0$ allows for nontrivial solutions [for scattering from the right, Eq. (9) also holds if $\det\Lambda = 1$, whereas a similar equation holds if $\det\Lambda = 0$]. However, conditions (i) and (ii), together with the requirement of flux conservation across the singularity, are not enough to completely determine all the coefficients α_j and β_j, although they reduce the number of free parameters to three. Additional physical requirements, such as considerations of symmetry, are necessary to completely specify the interaction. In this case, we can consider the behavior of the interaction term under parity transformations.

Given that $\delta'(x)$ is an odd distribution, we require that under parity transformations its associated interaction term $s(x)$ behave in the same way as in the case in which the potential corresponds to an ordinary odd function. From this condition we obtain $\alpha_1 = \alpha_2 = 0$. Combined with the condition $\det\Lambda = 1$ this gives $\beta_1 = 0$, while β_2 remains a free parameter of the model. In this case we have $s(x) = \beta_2 \psi'(0^-)\, \delta'(x)$, where $\psi'(0^+) = \psi'(0^-)$, with β_2 being the interaction strength. On the other hand, when $\det\Lambda = 0$ one obtains $\beta_1 = -1$ and β_2 becomes arbitrary. It can be shown that irrespective of the side of incidence (left or right, or any superposition of scattering solutions) this last case implies that $s(x) = [\psi(0^+) - \psi(0^-)]\, \delta'(x)$, with $\psi'(0^\pm) = 0$ and arbitrary $\beta = [\psi(0^+) - \psi(0^-)]$, which can be promptly identified with a completely reflective barrier, with disconnected solutions in the two sides of the barrier.

The results obtained so far are in agreement with those obtained by the (also mathematically consistent) method of self-adjoint extensions [8] for singular operators. However, while in the latter the interaction is determined *from* the specification of the boundary conditions, with no particular form specified for the interaction term, in our approach such a term is univocally determined from basic mathematical and sound physical requirements, such as probability flux conservation and behavior under symmetry transformations, with the boundary conditions being a consequence of the specific form of interaction. In this approach it becomes clear that the boundary conditions usually associated with the delta prime potential emerge as a consequence of the odd parity of this potential. A systematic presentation of this approach, including its relativistic generalization and the investigations of the possibilities of bound states, will appear elsewhere.

Acknowledgments

JTL and WM thank Fundação Araucária for financial support. LAM thanks the Office of Undergraduate Research, Concordia College, for partial financial support.

References

[1] Schmidt A G M, Cheng B K, and da Luz M G E 2002 *Phys. Rev. A* **66** 062712
[2] Manuel C and Tarrach R 1994 *Phys. Lett. A* **328** 113
[3] Nyeo S-L 2000 *Am. J. Phys.* **68** 571
[4] Zolotaryuk A V 2010 *Phys. Lett. A* **374** 1636
[5] Arnbak H, Christiansen P L and Gaididei Yu B 2011 *Phil. Trans. R. Soc. A* **369** 1228
[6] Zemanian A H 1987 *Distribution Theory and Transform Analysis: An Introduction to Generalized Functions, with Applications* (New York: Dover)
[7] Merzbacher E 1970 *Quantum Mechanics, 2nd ed.* (New York: Wiley)
[8] Albeverio S, Gesztesy F, Høegh-Krohn R and Holden H 2004 *Solvable Models in Quantum Mechanics 2nd ed* (AMS Chelsea Publishing)

Supersymmetry breaking, conserved charges and stability in $N = 1$ Super KdV

Adrián Sotomayor[1], **Alvaro Restuccia**[2]

[1] Departamento de Matemáticas, Universidad de Antofagasta, Antofagasta, Chile

[2] Departamento de Física, Universidad de Antofagasta, Antofagasta, Chile

E-mail: asotomayor@uantof.cl, arestu@usb.ve

Abstract. We analyse the non-abelian algebra and the supersymmetric cohomology associated to the local and non-local conserved charges of N=1 SKdV under Poisson brackets. We then consider the breaking of the supersymmetry and obtain an integrable model in terms of Clifford algebra valued fields. We discuss the remaining conserved charges of the new system and the stability of the solitonic solutions.

1. Introduction

Local and non-local conserved charges are relevant in order to analyze perturbative and non-perturbative aspects of a field theory. It is known that the local conserved charges of $N = 1$ SKdV are related to the superconformal algebra. It is then interesting to look for an extension of the abelian algebra, spannned by the local conserved charges of $N = 1$ SKdV, in terms of the non-local conserved charges. We obtain such extension as a non-abelian algebra realized in terms of Poisson brackets and discuss its relation to the supersymmetric structure of $N = 1$ SKdV. We then consider the susy breaking on the symmetries associated to the higher dimensional conserved charges. The resulting system, although has less symmetries than the original SKdV one, has solitonic solutions with nice stability properties.

2. Non-abelian algebra of conserved charges and supersymmetric cohomology for $N = 1$ SKdV

The susy $N = 1$ Korteweg-de Vries equation in superfield form is given by

$$\Phi_t = D^6\Phi + 3D^2(\Phi D\Phi), \tag{1}$$

where $\Phi = \xi + \theta u$ and ξ, u are fields with values on the odd and even parts respectively of a given Grassmann algebra Λ with n generators and a distinguished one θ, and $D = \partial_\theta + \theta\partial_x$ is the covariant derivative with the property $D^2 = \partial_x$ acting on superfields.

In components (1) is equivalent to

$$\begin{aligned} u_t &= u''' + 6uu' - 3\xi\xi'' \\ \xi_t &= \xi''' + 3(\xi u)'. \end{aligned} \tag{2}$$

The system (2) is integrable in the sense that it admits an infinite number of local conserved charges. It is invariant under the supersymmetric transformation defined by $\delta\Phi = \eta Q\Phi$, where

η is an odd parameter and $Q = -\partial_\theta + \theta\partial_x$. It is also known to have an infinite sequence of odd and even non-local conserved charges. They can be obtained using simultaneously an auxiliary equation, the so called super-Gardner equation

$$\chi_t = D^6\chi + 3D^2(\chi D\chi) - \varepsilon^2 3(D\chi)D^2(\chi D\chi) \tag{3}$$

and the super-Gardner transformation $\Phi = \chi + \varepsilon D^2\chi - \varepsilon^2\chi D\chi$, relating (1) with (3) and specifically solutions between them. The infinite sequence of local conserved charges of the $N = 1$ Super KdV equation were obtained by Mathieu [1] from the one conserved charge of the super-Gardner equation

$$G_1 = \int dx d\theta\chi = \sum_{n=0} \epsilon^{2n} H_{2n+1}, \tag{4}$$

using the inverse Gardner transformation. The infinite sequence of fermionic non-local conserved charges were obtained by Dargis and Mathieu [2] and Kersten [3]. This infinite sequence can be obtained also from one non-local conserved charge of the super-Gardner equation [4]

$$G^{NL}_{\frac{1}{2}} = \int dx d\theta \left(\frac{\exp(\epsilon D^{-1}\chi) - 1}{\epsilon}\right) = \sum_{n=0} \epsilon^n H^{NL}_{n+\frac{1}{2}}. \tag{5}$$

And finally, the infinite sequence of bosonic non-local conserved charges was derived [5] from the non-local conserved charge for super-Gardner given by

$$H_G = \frac{1}{2}\int dx d\theta D^{-1}\left\{D\left[\frac{\exp(\epsilon D^{-1}\chi) - 1}{\epsilon}\right]D^{-1}\left[\frac{\exp(-\epsilon D^{-1}\chi) - 1}{\epsilon}\right]\right\} = \sum_{n=0} \epsilon^n H^{NL}_{n+1}. \tag{6}$$

In (4),(5) and (6) the subscripts of the right members denote the dimension of the respective conserved charges.

Given a conserved charge for equation (1) of the form $H = \int dx d\theta h, h \in C^\infty_I$, where C^∞_I is the ring of integrable superfields, we define

$$\delta_Q H := \int dx d\theta Q h, \tag{7}$$

where δ_Q satisfies $\delta_Q\delta_Q = 0$. (7) defines a cohomology on the conserved charges of super-KdV [5]. In fact, for example we have

$$\delta_Q H_{2n+1} = 0, n = 0, 1, \ldots, \tag{8}$$

hence H_{2n+1} must be δ_Q of some other conserved charge, in particular for $n = 1$ we have

$$\delta_Q H^{NL}_{\frac{1}{2}} = H_1. \tag{9}$$

In the same way we obtain a relation between the non-local odd and even conserved charges, for example

$$\delta_Q H^{NL}_1 = H^{NL}_{\frac{3}{2}} - \frac{1}{2}H_1 H^{NL}_{\frac{1}{2}}. \tag{10}$$

The general formula is

$$\sum_{n=0} \epsilon^n \delta_Q H^{NL}_{n+1} = \sum_{n=0} \epsilon^{2n} H^{NL}_{2n+\frac{3}{2}} - \frac{1}{2\epsilon}\left[\exp\left(\sum_{n=0} \epsilon^{2n+1} H_{2n+1}\right) - 1\right]\left[\sum_{n=0} (-\epsilon)^n H^{NL}_{n+\frac{1}{2}}\right]. \tag{11}$$

We then consider the Poisson bracket of superfields Φ at two different points in superspace as given in [1]:

$$\{\Phi(x_1,\theta_1),\Phi(x_2,\theta_2)\} = P_1\Delta, \tag{12}$$

where $P_1 = D_1^5 + 3\Phi_1 D_1^2 + (D_1\Phi_1)D_1 + 2(D_1^2\Phi_1)$ and $\Delta = \delta(x_1-x_2)(\theta_1-\theta_2)$. With this bracket, using the supersymmetric cohomology and after some calculations, it is possible to derive the Poisson algebra of local and non-local conserved charges, in particular we obtain the following relations:

$$\left\{H^{NL}_{\frac{1}{2}}, H^{NL}_{\frac{1}{2}}\right\} = H_1, \tag{13}$$

$$\left\{H^{NL}_{\frac{1}{2}}, H^{NL}_{\frac{3}{2}}\right\} = \frac{1}{2}H_1^2, \tag{14}$$

$$\left\{H^{NL}_{\frac{1}{2}}, H^{NL}_{1}\right\} = -\frac{1}{2}H_1 H^{NL}_{\frac{1}{2}} + H^{NL}_{\frac{3}{2}}, \tag{15}$$

and

$$\left\{H^{NL}_{\frac{3}{2}}, H^{NL}_{\frac{3}{2}}\right\} = -H_3 + \frac{1}{3}(H_1)^3. \tag{16}$$

This gives the first steps in order to construct the complete non-abelian algebra of conserved charges for $N=1$ super-KdV equation. This problem is under development.

The local and non-local conserved charges are related to symmetries of the $N=1$ SKdV equation. We are now interested in considering the breaking of supersymmetry which will induce a breaking of the symmetries associated with the conserved charges.

3. Supersymmetry breaking of SKdV and stability of the ground state and the one-soliton solutions

In order to break the supersymmetry we consider the system [6]

$$\begin{aligned} u_t &= -u''' - uu' - \tfrac{1}{4}(\mathcal{P}(\xi\bar{\xi}))' \\ \xi_t &= -\xi''' - \tfrac{1}{2}(\xi u)', \end{aligned} \tag{17}$$

where the fields u and ξ takes values on a Clifford algebra instead of being Grassmann algebra valued. In order to compare with the stability analysis of KdV in Benjamin [7] we have redefined the u and ξ fields in (2) with convenient factors. We thus take u to be a real valued field while ξ to be an expansion in terms of the generators $e_i, i=1,\ldots$ of the Clifford algebra:

$$\xi = \sum_i \varphi_i e_i + \sum_{ij} \varphi_{ij} e_i e_j + \sum_{ijk} \varphi_{ijk} e_i e_j e_k + \cdots \tag{18}$$

where

$$e_i e_j + e_j e_i = -2\delta_{ij} \tag{19}$$

and $\varphi_i, \varphi_{ij}, \varphi_{ijk}, \ldots$ are real valued functions. We define $\bar{\xi} = \sum_{i=1}^{\infty} \varphi_i \bar{e}_i + \sum_{ij} \varphi_{ij} \bar{e}_j \bar{e}_i + \sum_{ijk} \varphi_{ijk} \bar{e}_k \bar{e}_j \bar{e}_i + \cdots$ where $\bar{e}_i = -e_i$. We denote as in superfield notation the body of the expansion those terms associated with the identity generator and the soul the remaining ones. Consequently the body of $\xi\bar{\xi}$, denoted by $\mathcal{P}(\xi\bar{\xi})$, is equal to $\Sigma_i \varphi_i^2 + \Sigma_{ij}\varphi_{ij}^2 + \Sigma_{ijk}\varphi_{ijk}^2 + \cdots$ In what follows, without loss of generality, we rewrite $\mathcal{P}(\xi\bar{\xi}) = \Sigma_i \varphi_i^2 + \Sigma_{ij}\varphi_{ij}^2 + \Sigma_{ijk}\varphi_{ijk}^2 + \cdots$ simply as $\mathcal{P}(\xi\bar{\xi}) = \Sigma_i \varphi_i^2$. The system (17) is not invariant under supersymmetric transformations and has not a conserved charge of dimension 7, that is there is no analogue of H_7 in SKdV or KdV systems but has the following conserved charges:

$$\begin{aligned} &\hat{H}_{\frac{1}{2}} = \textstyle\int_{-\infty}^{+\infty} \xi dx, \quad \hat{H}_1 = \int_{-\infty}^{+\infty} u dx \\ &V \equiv \hat{H}_3 = \tfrac{1}{2}\textstyle\int_{-\infty}^{+\infty} \left(u^2 + \mathcal{P}\left(\xi\bar{\xi}\right)\right) dx, \\ &M \equiv \hat{H}_5 = \tfrac{1}{2}\textstyle\int_{-\infty}^{+\infty} \left(-\tfrac{1}{3}u^3 - \tfrac{1}{2}u\mathcal{P}\left(\xi\bar{\xi}\right) + (u')^2 + \mathcal{P}\left(\xi'\bar{\xi}'\right)\right) dx. \end{aligned} \tag{20}$$

It is interesting to remark that the following non-local conserved charge of Super KdV [5] is also a non-local conserved charge for the system (17), in terms of the Clifford algebra valued field ξ,

$$\int_{-\infty}^{\infty} \xi(x) \int_{-\infty}^{x} \xi(s) ds dx.$$

However the non-local conserved charges of Super KdV in [2] are not conserved by the system (17). For example,

$$\int_{-\infty}^{\infty} u(x) \int_{-\infty}^{x} \xi(s) ds dx$$

is not conserved by (17).

We now consider the stability of solutions for the system (17) in the sense of Liapunov [6]. In particular we take the same definition as in [7]: $(\hat{u}, \hat{\xi})$, a solution of (17), is stable if given ϵ there exists δ such that for any solution (u, ξ) of (17), satisfying at $t = 0$

$$d_I \left[(u, \xi,), \left(\hat{u}, \hat{\xi}, \right) \right] < \delta \tag{21}$$

then

$$d_{II} \left[(u, \xi,), \left(\hat{u}, \hat{\xi}, \right) \right] < \epsilon \tag{22}$$

for all $t \geq 0$.

d_I and d_{II} denote two distances to be defined.

We denote by $\| \ \|_{H_1}$ the Sobolev norm

$$\|(u, \xi)\|_{H_1}{}^2 = \int_{-\infty}^{+\infty} \left[\left(u^2 + \Sigma_{i=0}^{\infty} \varphi_i^2 \right) + \left(u'^2 + \Sigma_{i=0}^{\infty} \varphi_i'^2 \right) \right] dx. \tag{23}$$

It can be proved that

$$\|(u, \xi)\|_{H_1}{}^2 \leq V + M + \frac{1}{\sqrt{2}} V \|(u, \xi)\|_{H_1}, \tag{24}$$

and this gives a priori bound for solutions to the system (17). The stability of the ground state solution $\hat{u} = 0, \hat{\xi} = 0$ is a direct consequence of (24) taking d_I and d_{II} to be the Sobolev norm $\|(u - \hat{u}, \xi - \hat{\xi})\|_{H_1}$.

The system (17) has also solitonic solutions, for example: $u(x,t) \equiv \phi(x,t) = 3\mathcal{C} \frac{1}{\cosh^2(z)}, z \equiv \frac{1}{2} \mathcal{C}^{\frac{1}{2}} (x - \mathcal{C}t), \xi(x,t) = 0$. $\phi(x,t)$ is the one-soliton solution of KdV equation. We now consider the stability of the one-soliton solution $\hat{u} = \phi, \hat{\xi} = 0$. The proof of stability is based on estimates for the u field which are analogous to the one presented in [7, 8] while a new argument will be given for the ξ field. The distances we will use are

$$d_I \left[(u_1, \xi_1), (u_2, \xi_2) \right] = \| (u_1 - u_2, \xi_1 - \xi_2) \|_{H_1} \tag{25}$$

$$d_{II} \left[(u_1, \xi_1), (u_2, \xi_2) \right] = \inf_{\tau} \| (\tau u_1 - u_2, \xi_1 - \xi_2) \|_{H_1} \tag{26}$$

where τu_1 denotes the group of translations along the x-axis. d_{II} is a distance on a metric space obtained by identifying the translations of each $u \in H_1(\mathbb{R})$. d_{II} is related to a stability in the sense that a solution u remains close to $\hat{u} = \phi$ only in shape but not necessarily in position.

We first assume that

$$V(u, \xi) = V(\hat{u}, \hat{\xi}) = V(\phi, 0) \tag{27}$$

and

$$\int_{-\infty}^{+\infty} \xi dx = \int_{-\infty}^{+\infty} \hat{\xi} dx = 0 \tag{28}$$

and at the end of the argument we relax this conditions to get the most general formulation of the stability problem. It is then possible to prove, after long reasoning that

$$\Delta M \equiv M(u,\xi) - M(\phi,0) \leq \left[\max(1,\mathcal{C}) + \frac{1}{3\sqrt{2}}\delta\right] \parallel (h,\xi) \parallel_{H_1}^2 \tag{29}$$

and that

$$\Delta M \geq \frac{1}{6} l \{d_{II}\left[(u,\xi),(\phi,0)\right]\}^2, \tag{30}$$

and these two inequalities, using that ΔM is a conserved quantity, are sufficient to prove the stability of the one-soliton solution for the system (17). To relax the conditions (27) and (28) we use essentially the triangle inequality.

4. Conclusions

We discussed the algebraic structure of the local and non-local conserved charges of $N = 1$ SKdV. We presented a rigorous construction of the Poisson brackets of these non-local conserved charges, this is in general an open problem in field theory.

The non-abelian algebra has been worked out partially, we hope to provide the complete algebraic structure in a future work. Besides we considered the supersymmetry breaking of $N = 1$ SKdV, this is always an interesting problem to study. The resulting system is integrable, in the sense that it has solitonic solutions, but, of course, it is not supersymmetric. Moreover the infinite sequence of local and non-local conserved charges becomes certainly reduced, possibly only to a finite set. Consequently the associated symmetries have also been broken. In spite of the fact that the new integrable system has less symmetries, its solitonic solutions are stable in the Liapunov sense. The problem of stability of the solutions of the supersymmetric KdV system is also an open problem. Our result is a contribution in this sense.

Acknowledgments

A. R. and A. S. are partially supported by Project Fondecyt 1121103, Chile.

References

[1] P. Mathieu, J. Math. Phys. 29, 2499 (1988).
[2] P. Dargis and P. Mathieu, Phys. Lett. A 176, 67-74 (1993).
[3] P. H. M. Kersten, Phys. Lett. A 134, 25 (1988).
[4] S. Andrea, A. Restuccia and A. Sotomayor, J. Math. Phys. 46, 103517 (2005).
[5] S. Andrea, A. Restuccia and A. Sotomayor, Phys. Lett. A 376, 245-251 (2012).
[6] A. Restuccia and A. Sotomayor, arXiv: 1207.5870v1[math-ph] (2012).
[7] T. B. Benjamin, Proc. R. Soc. Lond. A. 328, 153-183 (1972).
[8] J. Bona, Proc. R. Soc. Lond. A. 344, 363-374 (1975).

Dissociation of multiple hydrogen molecules on the non-planar aluminium cluster: stationary points on the $Al_6 + nH_2$ potential energy surface

Jerzy Moc[1]
Faculty of Chemistry, Wroclaw University, F. Joliot-Curie 14, 50-383 Wroclaw, Poland

E-mail: jmoc@wchuwr.pl

Abstract. Compared to clusters of transition metal atoms, H-H bond activation by main Group metal clusters is much less known. Here, we have examined a potential new way of obtaining a novel alane Al_6H_8 through addition of multiple H_2 molecules to the aluminium hexamer, i.e., $Al_6 + nH_2$ (n=1-4) reactions. To this end, systematic high level quantum chemical modeling calculations using density functional theory (DFT) and coupled-cluster singles-doubles-perturbative triples (CCSD(T)) method in conjunction with the aug-cc-pVTZ basis set were performed to identify the lowest energy barrier paths for the consecutive dissociation of several hydrogen molecules on Al_6, the smallest aluminium cluster with a three-dimensional ground-state structure. These computational results are relevant to the issues of hydrogen storage and novel stable alanes.

1. Introduction

Although studying the processes of adsorption and dissociation of molecular hydrogen on the surface of transition metal clusters has been an active area of research for the last twenty years [1-3], such reactions involving main group metal clusters have been investigated on a much smaller scale. For the light metals like aluminium this type of reactions is of potential technological importance for hydrogen storage. In the early gas phase experimental studies of the reactions of aluminium clusters with molecular hydrogen of Cox et al [4], it was observed that the onset of reactivity of Al_n towards H_2 occured for hexamer Al_6 (n=6). Recently, by using mass spectrometry and photoelectron spectroscopy (PE) techniques, Li et al. [5-6] studied the series of new aluminium hydride anions $Al_nH_{n+2}^-$ ($4 \leq n \leq 8$) and, by analyzing the recorded PE spectra, they drew conclusions on the electronic properties and stability of the corresponding neutral alanes Al_nH_{n+2}. These authors in particular suggested Al_6H_8 species to have an increased stability relative to that of the other Al_6H_m hydrogenated clusters. In other words, Li et al. proposed the former species to be a novel stable "hydrogen rich" alane. Importantly, Al_6 hexamer is the smallest aluminium cluster whose ground state structure is three-dimensional, with the octahedron- and prism-like geometries lying close in energy [7]. If one models multiple H_2 dissociation reactions on the relatively big cluster like Al_6, a systematic exploration of the respective potential energy surfaces (PESs) is crucial. This will ensure that the low-lying transition states (TSs)

[1] To whom any correspondence should be addressed.

for H-H bond dissociation are located. Identifying the "right" H-H dissociation transition states will have implications, e.g., for the kinetic modelling of these processes.

Below, we will briefly describe the results of the high-level quantum chemical calculations of the reaction paths for the consecutive dissociation of multiple H_2 on the aluminium hexamer Al_6. The Al_6H_8 hydrogenated cluster is intended to be the "final" hydrogenation product, consistent with the PE results for the Al_nH_m species with n=6 mentioned above. We note that in the experimental studies of Li et al. [5-6], the anion clusters were made by the reactions of the Al_n^- clusters with H atoms (not H_2 molecules). The present theoretical results are relevant to the issues of hydrogen storage and novel stable alanes.

2. Computational methods

The correlation consistent triple-zeta quality aug-cc-pVTZ basis set [8] was used throughout. All the structures were optimized and verified as minima or transition states on the respective potential energy surfaces using density functional theory (DFT) with the hybrid B3LYP functional [9-10], and the improved energetics were subsequently obtained via the single-point calculations at the B3LYP/aug-cc-pVTZ structures using the coupled-cluster singles, doubles and perturbative triples (CCSD(T)) method [11]. Connectivity of the transition states for H-H bond dissociation with the hydrogenated products was confirmed by determining the intrinsic reaction coordinates [12]. The harmonic vibrational frequencies computed at the B3LYP/aug-cc-pVTZ level were used to estimate the zero-point energy (ZPE) and thermal and entropic corrections. All the computations were performed using the Gaussian 09 program [13].

3. Outline of results

3.1. The scope and labelling of the stationary points

In the subsections below, the low energy barrier paths computed for the hydrogenation reactions $Al_6H_{2n-2} + H_2 \rightarrow Al_6H_{2n}$ with n=1-4 will be summarized. The dissociation of the first H_2 on Al_6 (n=1) was recently discussed in detail [7], thus only a brief account is given below. The bulk of this presentation is devoted to the n=2-4 hydrogenation steps. Transition states involved in the dissociation of the second (S) (n=2), third (T) (n=3) and fourth (F) (n=4) H_2 are denoted as **TSH-HSj**, **TSH-HTj** and **TSH-HFj**, respectively, with **j**=1,2,3,... indicating the TS number for each dissociation type. The respective hydrogenated clusters are termed **Sj**, **Tj** and **Fj**. In order to save space, for each n, only the energetically most favourable structures of the transition states and corresponding hydrogenated clusters are shown, although all the calculated H-H dissociation TSs/hydrogenated clusters are included in the respective enthalpy and free energy profiles.

3.2. n=1

The hydrogenation of the octahedral-based Al_6 to form Al_6H_2(**1**) dihydride (Figure 1) is thermodynamically favourable, both in terms of enthalpy, ΔH(298K)=-25.3 kcal/mol, and Gibbs free energy, ΔG(298K)=-25.3 kcal/mol (Figure 2) [7]. Relative to the ground state reactants $Al_6(D_{3d}, {}^1A_{1g})$ + H_2 [7], no enthalpic barrier $\Delta H^{\ddagger}$(298K) is predicted to be involved in the path that passes through the octahedral-like transition state **TSH-H** (Figure 1), with the corresponding free energy barrier $\Delta G^{\ddagger}$(298K) being 5.3 kcal/mol (Figure 2).

3.3. n=2

Among the three low-lying transition states located on the $Al_6H_2 + H_2$ potential energy surface, the energetically most preferred one is the four-centered **TSH-HS3**, where the H-H bond is broken on the Al-Al edge of the Al_6 prism (Figure 1). Upon dissociating the second H_2 via **TSH-HS3**, the new terminal and new bridging Al-H bonds are formed in the Al_6H_4 **S3** hydrogenated cluster. **TSH-HS3** competes energetically with the octahedral-like **TSH-HS1** transition state, bearing a strong

Journal of Physics: Conference Series **410** (2013) 012074 doi:10.1088/1742-6596/410/1/012074

resemblance (the "active site") to **TSH-H** (n=1). The enthalpic and free energy profiles in Figure 2 show that for n=2 there are indeed two kinetically competitive hydrogenation paths, via **TSH-HS3** and **TSH-HS1** at 8.9 (17.1) and 9.5 (18.0) kcal/mol, in terms of $\Delta H^{\neq}(298K)$ ($\Delta G^{\neq}(298K)$), relative to the Al_6H_2(**1**) + H_2 asymptote. The lower-energy path via **TSH-HS3** affords exothermically the Al_6H_4 cluster **S3,** with the enthalpy (free energy change) of –18.0 (-9.7) kcal/mol (Figure 2).

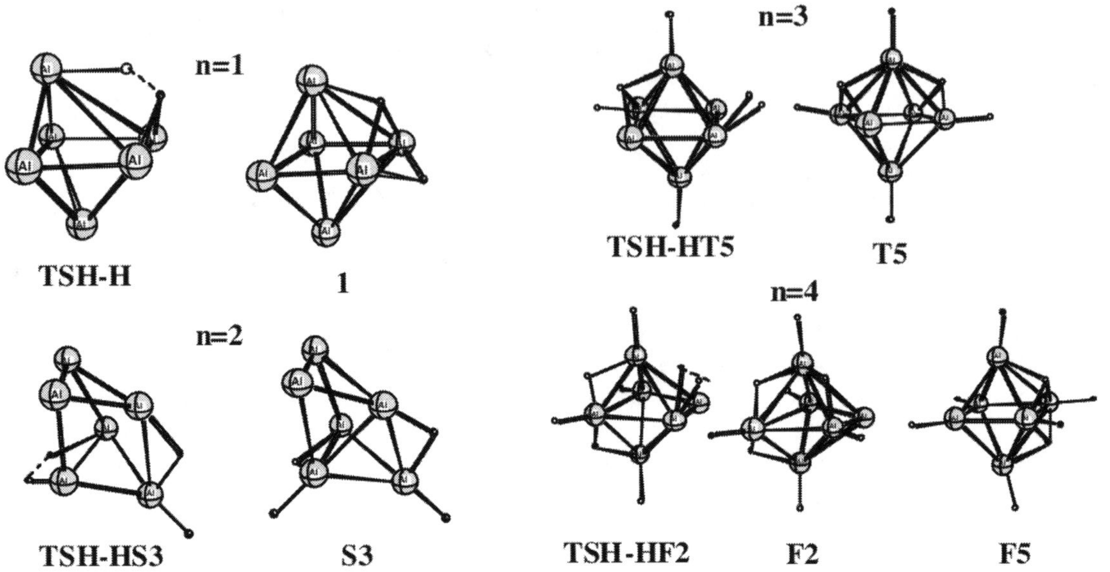

Figure 1. The lowest lying transition states for the hydrogenation reactions $Al_6H_{2n-2} + H_2 \rightarrow Al_6H_{2n}$ with n=1-4 along with the resulting hydrogenated Al_6 clusters.

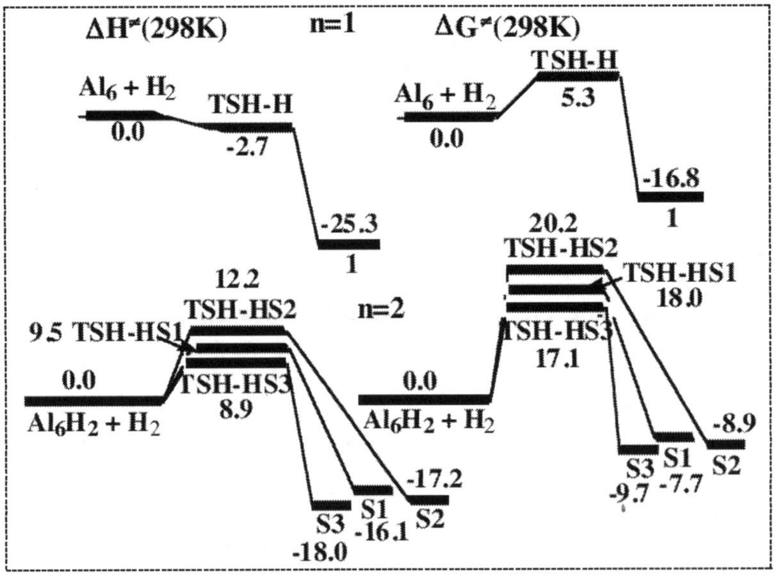

Figure 2. Schematic enthalpic and free energy profiles (kcal/mol) for the hydrogenation reactions $Al_6H_{2n-2} + H_2 \rightarrow Al_6H_{2n}$ with n=1,2 calculated at the CCSD(T)/aug-cc-pVTZ//B3LYP/aug-cc-pVTZ

level.

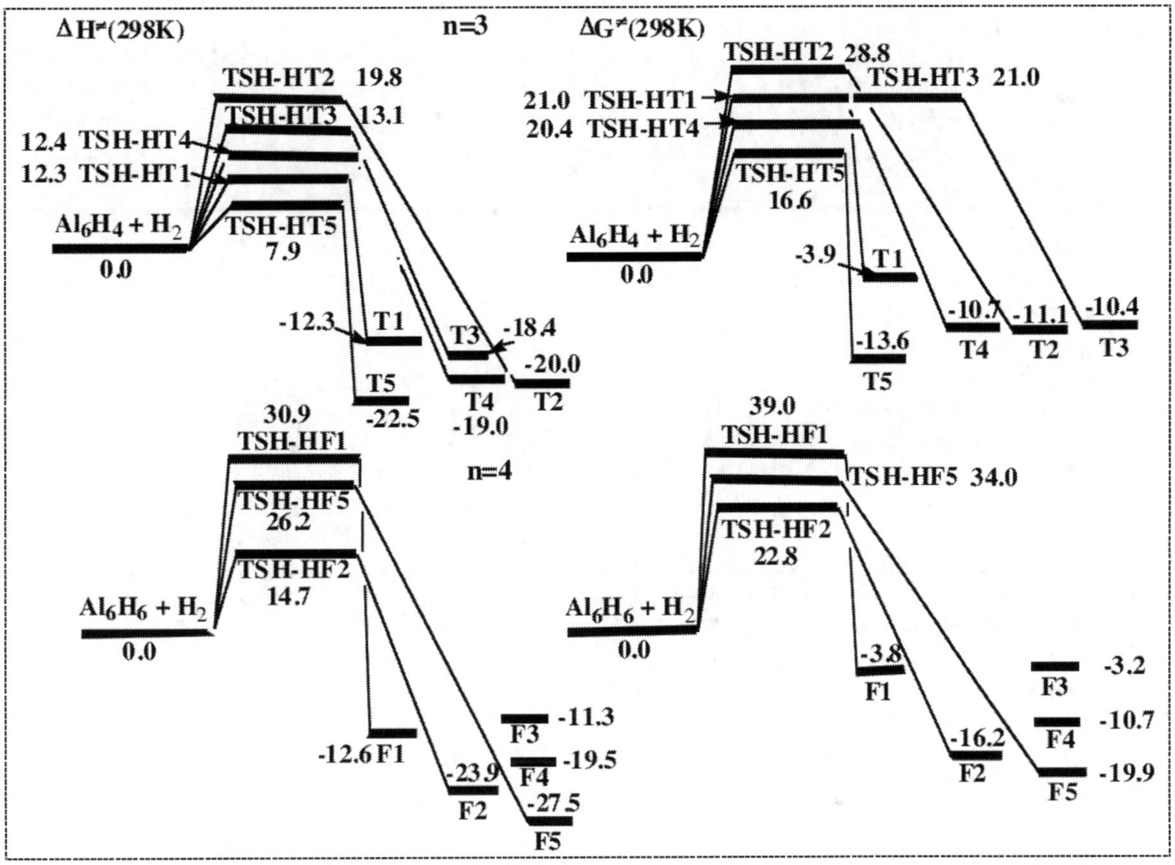

Figure 3. Schematic enthalpic and free energy profiles (kcal/mol) for the hydrogenation reactions $Al_6H_{2n-2} + H_2 \rightarrow Al_6H_{2n}$ with n=3,4 calculated at the CCSD(T)/aug-cc-pVTZ//B3LYP/aug-cc-pVTZ level.

3.4. n=3

For the n=3 hydrogenation step, we have identified five transition states for the H-H bond dissociation, with the octahedral-based **TSH-HT5** (Figure 1) lying significantly lower in energy compared to the other TSs (cf. Figure 3). All he transition states relevant to the third H_2 dissociation are either of the octahedral- or prism-like type. The lowest energy H_2 hydrogenation path, via **TSH-HT5**, leads to the Al_6H_6 **T5** cluster of C_2 symmetry, which features four hydrogens in the terminal sites and two Hs occupying the three-fold bridged positions of the Al_6 octahedron (Figure 1). The current hydrogenation step via **TSH-HT5** requires overcoming the energy barrier comparable to that encountered for the n=2 case, i.e. 7.9 (ΔH·(298K)) kcal/mol and 16.6 kcal/mol (ΔG·(298K)), with respect to the Al_4H_6(**S3**) + H_2 asymptote (Figure 3). Relative to this asymptote, the third H_2 dissociation is exothermic as indicated by the computed enthalpy (free energy change) of –22.5 (-13.6) kcal/mol (Figure 3).

3.4. n=4

Among the three transition states calculated for the fourth H_2 dissociation, the lowest lying is the four-centered **TSH-HF2** one with the H-H bond being broken on the Al-Al edge of the distorted Al_6 octahedron (Figure 1). Upon dissociating the fourth H_2 via **TSH-HF2**, the new terminal and new bridging Al-H bonds are formed in the resulting Al_6H_8 **F2** hydrogenated cluster. The energy barriers corresponding to the fourth H_2 hydrogenation step via **TSH-HF2** are 14.7 and 22.8 kcal/mol in terms of ΔH(298K) and ΔG(298K), respectively, relative to the Al_4H_6(**T5**) + H_2 asymptote (Figure 3). The kinetically most preferred hydrogenation path for n=4 does not lead however to the thermodynamically most favourable Al_6H_8 product. This is because the thermodynamically most favourable Al_6H_8 isomer is **F5** (Figure 1). The latter species of C_{2v} symmetry contains six terminal and two three-fold bridged hydrogens. Relative to the Al_4H_6(**T5**) + H_2 asymptote, the fourth H_2 dissociation step is again exothermic, with the computed enthalpy (free energy change) of –27.5 (-19.9) kcal/mol (Figure 3).

Acknowledgement
The author acknowledges a generous support of computer time at the Wroclaw Center for Networking and Supercomputing, WCSS.

References

[1] Kubas G J 2007 *Chem. Rev.* **107** 4152
[2] Fayet P, Kaldor A and Cox D M 1990 *J. Chem. Phys.* **92** 254
[3] Moc J, Musaev D G and Morokuma K M 2000 *J. Phys. Chem. A* **104** 11606
[4] Cox D M, Trevor D J, Whetten R L and Kaldor A 1988 *J. Phys. Chem. A* **92** 421
[5] Li X, Grubisic A, Stokes S T, Cordes J, Gantefor G F, Bowen K H, Kiran B, Willis M, Jena, P, Burgert R and Schnockel H 2007 *Science* **315** 356
[6] Grubisic A, Li X, Stokes S T, Cordes J, Gantefor G F, Bowen K H, Kiran B, Willis M, Jena, P, Burgert R and Schnockel H 2007 *J. Am. Chem. Soc.* **129** 5969
[7] Moc J 2008 *Chem. Phys. Lett.* **466** 116
[8] Dunning T H 1989 *J. Chem. Phys.* **90** 1007
[9] Becke A D, 1993 *J. Chem. Phys.* **98** 5648
[10] Lee C, Yang W and Parr R G 1988 *Phys. Rev. B* **37** 785
[11] Pople J A, Head-Gordon M and Raghavachari K 1987 *J. Chem. Phys.* **87** 5968
[12] Gonzalez C and Schlegel H B 1989 *J. Chem. Phys.* **90** 2154
[13] Frisch M J *et al* 2009 *Gaussian 09 revision C.01* (Gaussian Inc, Wallingford, CT.)

Journal of Physics: Conference Series **410** (2013) 012075 doi:10.1088/1742-6596/410/1/012075

2D hybrid meshes for direct simulation Monte Carlo solvers

N Sengil[1] and U Sengil[2]

[1]Astronautical Engineering Department, *University of Turkish Aeronautical Association*, Turkkusu Campus 06790 Etimesgut-Ankara-Turkiye

[2]Department of Mathematics, *Koc University*, Rumelifeneri Yolu 34450 Sariyer-Istanbul-Turkiye

E-mail:usengil@ku.edu.tr

Abstract. The efficiency of the direct simulation Monte Carlo (DSMC) method decreases considerably if gas is not rarefied. In order to extend the application range of the DSMC method towards non-rarefied gas regimes, the computational efficiency of the DSMC method should be increased further. One of the most time consuming parts of the DSMC method is to determine which DSMC molecules are in close proximity. If this information is calculated quickly, the efficiency of the DSMC method will be increased. Although some meshless methods are proposed, mostly structured or non-structured meshes are used to obtain this information. The simplest DSMC solvers are limited with the structured meshes. In these types of solvers, molecule indexing according to the positions can be handled very fast using simple arithmetic operations. But structured meshes are geometry dependent. Complicated geometries require the use of unstructured meshes. In this case, DSMC molecules are traced cell-by-cell. Different cell-by-cell tracing techniques exist. But, these techniques require complicated trigonometric operations or search algorithms. Both techniques are computationally expensive. In this study, a hybrid mesh structure is proposed. Hybrid meshes are both less dependent on the geometry like unstructured meshes and computationally efficient like structured meshes.

1. Introduction

Direct simulation Monte Carlo (DSMC) is a molecular based stochastic method to analyze gas flows in the rarefied flow regimes where continuum based solvers are found unsatisfactory. In DSMC method, one particle can represent many physical molecules. Additionally, molecular collisions and movements are decoupled to increase the efficiency in terms of the solution time [1].

Although gridless DSMC methods exist in the literature [2], most of the current DSMC solvers depend on the grids. First DSMC solvers are embedded with the structured grids [3] because of the simplicity of the code maintenance and efficiency. To deal with more complicated geometries, unstructured grids are also implemented in the DSMC method [4]. Currently both structured and unstructured meshes are used in current DSMC solvers.

Firstly, an ideal DSMC grid should be computationally efficient. Secondly, these grids should be fitted to complex geometries. Next, grid sizes should be correlated with the mean-free-path and local gradients. Grid based DSMC solvers display a certain compromise between solution time efficiency and geometry dependence. In this study we propose a hybrid grid system. This new grid system is computationally as efficient as the structured grids. But at the same these grids are adaptable to complex geometries like unstructured grids [1].

2. DSMC 2D grid cell area qualities

The quality of the grid system is defined such that, DSMC cell area values should be as equal as possible and aspect ratios should be close to unity. The number of cells is chosen around 60.

2.1. Structured grids

If grid connections are defined by a general rule, grids are considered as structured [5]. The simplest DSMC solvers are constructed on the structured grids. These kinds of grids are computationally efficient. Additionally, DSMC solvers developed with structured codes are quick to develop and easily maintainable. But these grids are geometry dependent.

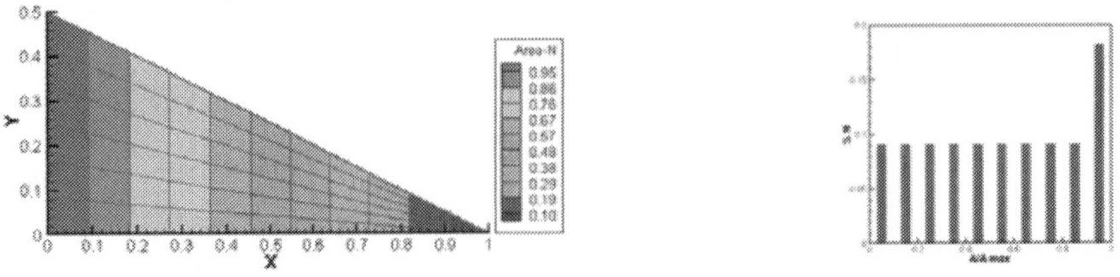

(a) (b)

Figure 1. (a) Normalized cell area values of a triangle with structured grids (b) Normalized cell area distribution of 66 structured cells.

2.2. Unstructured grids

When grid connections different at each grid point, these types of grids are called unstructured grids. Unstructured grids are more capable to mesh complex geometries comparing to the structured grids. The cell area values and their aspect ratios are more uniform compare to the structured grids as shown in figure 2.

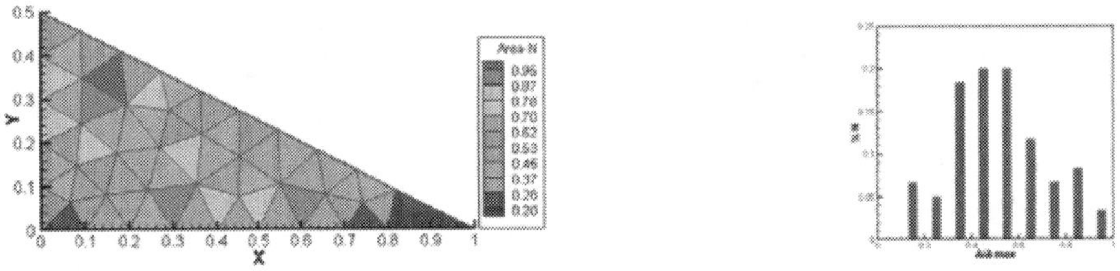

(a) (b)

Figure 2. (a) Normalized cell area values of a triangle with unstructured grids (b) Normalized cell area distribution of 60 unstructured cells.

2.3. Hybrid grids

In this study, a hybrid type grid is proposed to mesh a triangular geometry. After the grid size is determined, the triangle is first sliced along x-direction. The edges of the slices are parallel to y-axis. These slices divided again, along the y- direction according to the predetermined grid size as shown in figure 3. The cell area values and their aspect ratios are more uniform compare to the structured grids.

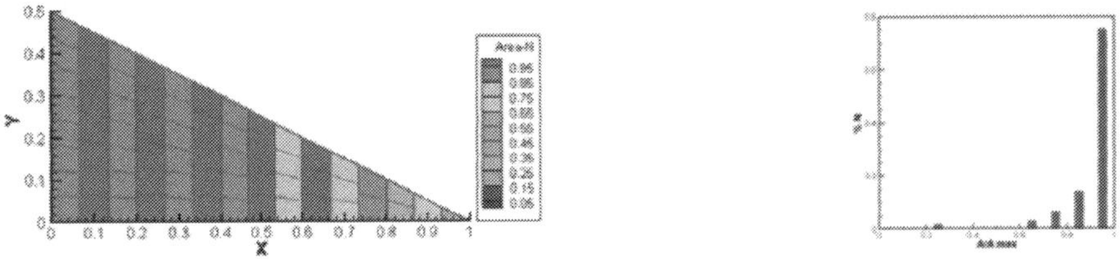

(a) (b)

Figure 3. (a) Normalized cell area values of a triangle with hybrid grids (b) Normalized cell area distribution of 64 hybrid cells.

3. DSMC 2D grid computational efficiencies

In this part of the study we measure the computational efficiency figures of the structured, unstructured and hybrid grids. Cell dimensions are scaled with the local mean-free-path. After these grids are constructed, one DSMC cell is filled with 10.000 DSMC molecules. Each DSMC molecule is positioned randomly in this cell and assigned velocity components sampled from the Maxwellian distribution. The time step is configured such that DSMC molecules travel no more than one cell length which is physically consistent. In "movement" step, these DSMC molecules change their positions. They either position in the original cell or move to the neighbor cells. After completion of the "movement" step, computer clock is resettled. In "indexing" step all the DSMC molecules are assigned to their new cells in which they positioned. When "indexing" step is completed, computer clock is recorded. Comparisons of the time differences during "indexing" step enables us to assess how efficient the grid systems are.

3.1. Structured grids

The simplest DSMC solvers are limited with structured grids and rectangular geometries. In these types of solvers, molecule "indexing" according to the positions can be handled very fast using simple arithmetic operations. When geometry is not a rectangular, non-rectangular physical domain should be mapped to a square computational domain geometry using coordinate transformation techniques [6] as shown in figure 4. If direct spatial relations between two domains are exist, DSMC "movement" step is carried out in the physical domain. Following, "indexing" step is realized in the computational domain. Tailored grids should be generated in the physical domain in order to calculate molecule cell data fast in the square computational domain with structured grids. Cell data calculation of all 10.000 DSMC molecules takes 0.002 Sec.

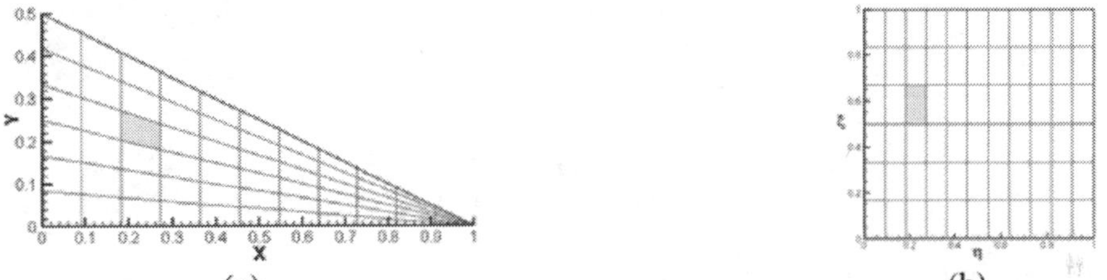

(a) (b)

Figure 4. (a) Physical domain (b) Computational domain.

3.2. Unstructured grids

DSMC molecules should be traced cell-by-cell to calculate the cell data if unstructured grids are employed as shown in figure 5. The cell-by-cell tracing technique is less efficient in comparison to the coordinate transformation technique. To calculate the efficiency of the DSMC method based on unstructured meshes, first 10000 DSMC molecules are randomly placed inside the cell marked with cell number 59. After "movement" step, all the cells in which DSMC molecules positioned are calculated. Although many methods are existed to calculate the cell data, we only use three different techniques. In the first technique, we searched all the cells sequentially to find out in which cells DSMC molecules positioned. The calculation time for this technique is 0.125 Sec. Secondly; we searched the cells starting from the neighbor cells. This technique takes only 0.011 Sec. Our calculations show that %92,1 of the DSMC molecules either positioned in the originated (59) or neighbor cells on the edges (55,57,58). Finally, we trace the DSMC molecules in these cells only. If DSMC molecules are not located these cells, they are searched in other cells. Calculation time of this third technique is 0.006 Sec.

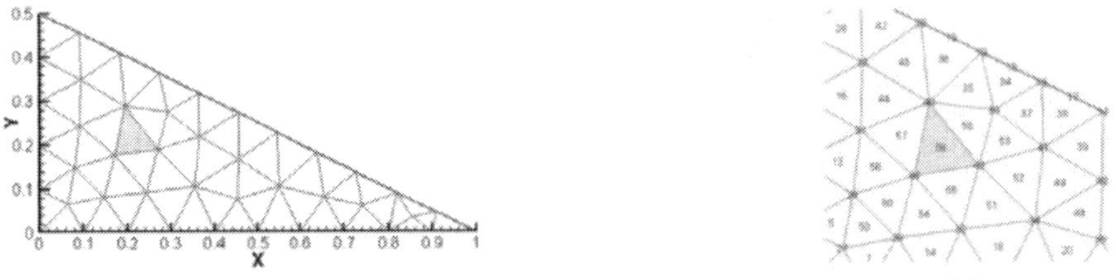

(a) (b)

Figure 5. (a) Physical domain (b) Detailed part.

3.3. Hybrid grids

Hybrid grids can be used with the coordinate transformation technique unlike the unstructured grids in order to increase the efficiency. Primarily, DSMC "movement" step is implemented in the physical domain. Later, molecule positions are transformed to computational domain using analytical relations between two domains. Following, in which slice DSMC molecule positioned is calculated. Knowing the cell numbers in each slice, calculation of the cell data is also simplified. As a result DSMC "indexing" step is carried out in the computational domain. In figure 6, both hybrid grids in the physical domain and the computation domain are displayed. It takes only 0.002 Sec to calculate in which cells DSMC molecules travel.

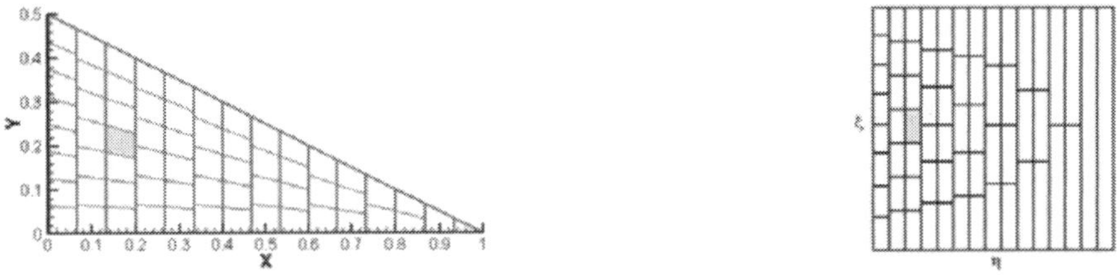

(a) (b)

Figure 6. (a) Physical domain (b) Computational domain.

4. Conclusion

In this study different grid techniques used in 2-dimensional DSMC method are compared in terms of the cell area quality and the computational efficiency. Structured grids are computationally efficient. Moreover they are simple to develop and easily maintainable in the DSMC solvers. But their cell area quality can be significantly degraded when it comes to the complex geometries. Consequently new DSMC solvers embedded with unstructured grids are developed. In these DSMC solvers, molecules are traced cell-by-cell. However this technique is quite inefficient in terms of the computational time. To take advantage of the best parts of these grid systems, we proposed a hybrid type of grid. It is shown that hybrid grids display high cell area qualities like unstructured grids. Besides hybrid grids exhibit efficiency figures as high as the structured grids. In the future, the implementation of the hybrid grids will be further extended to the 3-dimensional DSMC solvers.

References

[1] Bird G A 1994 *Molecular Gas Dynamics and the Direct Simulation of Gas Flows* (Oxford: Clarendon)
[2] Olson S E and Christlieb A J 2008 *J. Comput. Phys.* **227** 8035-64
[3] Bird G A 1976 *Molecular Gas Dynamics* (Oxford: Clarendon)
[4] Piekos E S and Breuer K S 1996 *Trans. ASME J. Fluids Engrs.* **118** 464-9
[5] Liseikin V D 2007 *A Computational Differential Geometry Approach to Grid Generation* (Berlin: Springer)
[6] Abe T 1989 *J. Comput. Phys.* 83 424-32

$\mathcal{PT}$ Symmetric Hamiltonian Model and Exactly Solvable Potentials

Özlem Yeşiltaş
Department of Physics, Faculty of Science, Gazi University, 06500 Ankara, Turkey
E-mail: yesiltas@gazi.edu.tr

Abstract. Searching for non-Hermitian, $\mathcal{PT}$-symmetric Hamiltonians [1] with real spectra has been acquiring much interest for fourteen years. In this article, we have introduced a $\mathcal{PT}$ symmetric non-Hermitian Hamiltonian model which is given as $\hat{\mathcal{H}} = \omega(\hat{b}^\dagger\hat{b} + \frac{1}{2}) + \alpha(\hat{b}^2 - (\hat{b}^\dagger)^2)$ where ω and α are real constants, $\hat{b}$ and $\hat{b}^\dagger$ are first order differential operators. Because the Hamiltonian $\mathcal{H}$ is pseudo-Hermitian, we have obtained the Hermitian equivalent of $\mathcal{H}$ which is in Sturm- Liouville form leads to exactly solvable potential models which are effective screened and hyperbolic Rosen-Morse II potentials. Using convenient sinilarity transformations, we have obtained a physical Hamiltonian h for each case. Then, the Schrödinger equation is solved exactly using Shape Invariance method of Supersymmetric Quantum Mechanics [2].

1. Introduction

There is growing interest in the study of non-Hermitian Hamiltonian operators [1, 3, 4, 5, 6] particularly those which possess $\mathcal{PT}$ symmetry. Ten years ago, pseudo-Hermiticity concept is introduced in a parallel development [3]. A Hermitian operator η and a pseudo-Hermitian Hamiltonian $\mathcal{H}$ satisfy $\mathcal{H}^\dagger = \eta\mathcal{H}\eta^{-1}$. We note that η corresponds to $\mathcal{PC}$ in the case of $\mathcal{PT}$ symmetric Hamiltonians. It is also shown that a Hamiltonian is equivalent to a Hermitian Hamiltonian h under a similarity transformation $\mathcal{H} = \rho^{-1}h\rho$ [3]. In this paper we wish to explore a pseudo-Hermitian Hamiltonian which is introduced by us and which is a special form of the Swanson Hamiltonian in [7] and also in [8, 9], a $\mathcal{PT}$ symmetric non-Hermitian model with two parameters,

$$\hat{\mathcal{H}} = \omega(\hat{b}^\dagger\hat{b} + \frac{1}{2}) + \alpha(\hat{b}^2 - (\hat{b}^\dagger)^2) \tag{1}$$

where $\hat{b}, \hat{b}^\dagger$ are bosonic annihilation and creation operators, ω, α are real constants and $\dagger$ is Hermitian adjoint, $\hat{b}$ is the annihilation operator given in a general form $\hat{b} = A(x)\frac{d}{dx} + B(x)$ and $A(x)$, $B(x)$ are real functions. In terms of differential operators, our Hamiltonian operator is written as

$$\begin{aligned}\hat{\mathcal{H}} &= -\omega A(x)^2\frac{d^2}{dx^2} + (4\alpha A(x)B(x) - 2\omega A(x)A(x)')\frac{d}{dx} \\ &-(\omega - 2\alpha)A(x)B(x)' - (\omega - 2\alpha)A(x)'B(x) + \omega B(x)^2 - \alpha(A(x)A(x)'' + (A(x)')^2) + \frac{\omega}{2}.\end{aligned} \tag{2}$$

We may write the eigenvalue equation as $\hat{\mathcal{H}}\psi = \varepsilon\psi$ and here, the pseudo-Hermitian Hamiltonian $\mathcal{H}$ can be mapped into a Hermitian operator form by using a mapping function ρ where

$\rho(x) = e^{-\frac{2\alpha}{\omega}\int^x dy \frac{B(y)}{A(y)}}$. Here we note that $h\psi = \varepsilon\psi$, $\psi = \rho^{-1}\xi$. So we can introduce operator h which is Hermitian equivalent of $\mathcal{H}$ as

$$h = -\omega\frac{d}{dx}A(x)^2\frac{d}{dx} + U_{eff}(x) \tag{3}$$

$$U_{eff}(x) = \frac{\omega}{2} - \omega(A(x)B(x))' - \alpha\left((A'(x))^2 + A(x)A''(x)\right) + (\omega + \frac{4\alpha^2}{\omega})B^2(x) \tag{4}$$

where the primes denote the derivatives. Then (3) can be mapped into a Schrödinger-like form by using $\xi(x) = \frac{1}{A(x)}\Phi(x)$ which leads to $-\Phi'' + \bar{U}_{eff}(x)\Phi = \frac{\varepsilon}{\omega A^2(x)}\Phi$. Hence, $\bar{U}_{eff}(x)$ becomes

$$\bar{U}_{eff}(x) = \frac{A''(x) + 1/2 - (A(x)B(x))'}{A(x)} - \frac{\alpha}{\omega}\frac{(A'(x))^2 + A''(x)}{A(x)} + \frac{\omega^2 + 4\alpha}{\omega^2}\frac{B^2(x)}{A^2(x)}. \tag{5}$$

Let us take the mapping $\xi(x) = e^{\int dx \frac{4\alpha B(x) - \omega A(x)')}{2\omega A(x)}}\chi(x)$, then eigenvalue equation for (3) becomes

$$\begin{aligned} &-\omega A(x)^2\chi(x)'' - \omega A(x)A(x)'\chi(x)' + [\frac{\omega}{2} - (\omega + 2\alpha)(A(x)B(x))' + (\omega + \frac{8\alpha^2}{\omega})B^2(x) + \\ &\omega/2 + (\omega/4 - \alpha)(A(x)')^2 + (\omega/2 - \alpha)A(x)A''(x)]\chi(x) = \varepsilon\chi(x). \end{aligned} \tag{6}$$

If we use $y = \int^x \frac{dx}{A(x)}$ in (6), we obtain

$$\begin{aligned} -\omega\frac{d^2\chi}{dy^2} + [\frac{\omega}{2} - (\omega + 2\alpha)(A(x)B(x))' + (\omega + \frac{8\alpha^2}{\omega})B^2(x) + (\omega/4 - \alpha)(A(x)')^2 + \\ (\omega/2 - \alpha)A(x)A''(x)]_{x\to y}\chi(y) = \varepsilon\chi(y). \end{aligned} \tag{7}$$

2. Models

Let us choose some special cases for both $A(x)$ and $B(x)$ to show that the real energies can be found for the corresponding systems. If we take $A(x) = \cosh x$ and a constant $B(x) = 1$, then we have

$$\frac{d^2\chi}{dy^2} + \left(\frac{\varepsilon}{\omega} - \frac{7}{4} + \frac{\alpha(1 - 8\alpha)}{\omega} + \frac{\omega + 2\alpha}{\omega}\tanh y - \frac{1}{4}\sec hy\right)\chi(y) = 0 \tag{8}$$

In [2], hyperbolic Rosen Morse potential is given as

$$V(z) = a^2 + \frac{b^2}{a^2} - a(a+1)\sec^2 hx + 2b\tan hx. \tag{9}$$

If we compare our system (8) and (9), it can be seen that

$$b = \frac{\omega + 2\alpha}{\omega}, \qquad a = -1 \pm \frac{\sqrt{5}}{4} \tag{10}$$

where the sign of a can be chosen as the condition on the parameters $b < a^2$ [2], and the corresponding energy eigenvalues are

$$\varepsilon_n = a^2 - (a-n)^2 - \frac{b^2}{(a-n)^2} + \frac{b^2}{a^2}, \qquad b < a^2. \tag{11}$$

Now let us use $A(x) = \frac{e^{-x}}{e^{-x}-q} = B(x)$ in (7), then we have

$$\frac{d^2\chi}{dy^2} + \left(\frac{\varepsilon}{\omega} - \frac{1}{2} + \frac{11(8\alpha^2 + 2q\alpha\omega + q\omega^2)}{4\omega^2}\frac{e^{-2y}}{(e^{-y}-q)^2} - \left(\frac{1}{2} + \frac{3\alpha}{\omega}\right)\frac{e^{-y}}{e^{-y}-q}\right)\chi = 0. \quad (12)$$

In [10], the deformed four-parameter exponential type potential which is

$$V(r) = \frac{V_0 e^{\alpha x}}{C + q' e^{\alpha x}} - \frac{V_1 e^{\alpha x}}{(C + q' e^{\alpha x})^2} \quad (13)$$

where V_0, V_1, C, q' are real parameters. The spectrum of (13) is given in [10] as

$$E_n = -\left(\frac{\frac{V_0}{q'}}{-(2n+1) + \sqrt{1 + \frac{4V_1}{q'C}}} - \frac{1}{4}\left(-(2n+1) + \sqrt{1 + \frac{4V_1}{q'C}}\right)\right)^2. \quad (14)$$

Comparing (15) and (12) leads to the real spectrum of (12),

$$\varepsilon_n = \frac{\omega}{2} - \omega\left(\frac{\frac{1/2+3\alpha/\omega}{q}}{-(2n+1) + \sqrt{1 + \frac{11(8\alpha^2+2q\alpha\omega+q\omega^2)}{q\omega^2}}} - \frac{1}{4}\left(-(2n+1) + \sqrt{1 + \frac{11(8\alpha^2 + 2q\alpha\omega + q\omega^2)}{q\omega^2}}\right)\right)^2, \quad (15)$$

where the parameters used for our system are

$$q = -q', \quad C = 1, \quad \frac{V_0}{q'} = \frac{\frac{1}{2} + \frac{3\alpha}{\omega}}{q}, \quad V_1 = \frac{11(8\alpha^2 + 2q\alpha\omega + q\omega^2)}{4\omega^2}. \quad (16)$$

Hereafter, we will look at the system (5) which may lead to more complicated results. Then we will discuss two potential models as hyperbolic Rosen Morse II and deformed effective screen potential whose pseudo-Hermiticity is discussed and results for the spectrum given above.

3. Rosen-Morse II potential

Let us choose $A(x)$ and $B(x)$ in these forms: $A(x) = a\cos h\mu x$, $B(x) = -1 + \gamma A(x)'$. Here a, μ, γ are real parameters. These choices in (5) lead to

$$\frac{\varepsilon}{\omega A^2(x)} - \bar{U}_{eff}(x) = -\mu\tanh\mu x + \left(2\gamma^2\mu^2 + \frac{\varepsilon - 4\alpha}{\omega a^2} - \frac{2-a}{2a^2}\right)\sec h\mu x - \mu^2\left(1 + \frac{\alpha}{\omega} + 2\gamma^2\right) \quad (17)$$

if and only if the parameters are given as

$$\gamma = \frac{\alpha}{2\omega}, \quad \mu = -\frac{4\alpha}{\omega^2}. \quad (18)$$

Thus, the eigenvalue equation becomes

$$\Phi''(x) + \left(E - \mu\tanh\mu x + \left(2\gamma^2\mu^2 + \frac{\varepsilon - 4\alpha}{\omega a^2} - \frac{2-a}{2a^2}\right)\sec h\mu x\right)\Phi(x) = 0 \quad (19)$$

where E is the constant and given as $E = -\mu^2\left(1 + \frac{\alpha}{\omega} + 2\gamma^2\right)$. If we put the super-potential $W(x)$ in the form of $W(x) = A + B\tanh(\mu x)$, then we have the ground-state wave-function

$\Phi_0(x) = e^{-Ax}(\cosh(\mu x))^{-B/\mu}$ where $B > 0$ and $|A| < B$ because of the boundary conditions $x \to \pm\infty$, $\Phi_0(x) \to 0$. Now we can give the partner potentials $V_-(x)$ and $V_+(x)$ as

$$V_-(x) = A^2 + B^2 + 2AB\tanh(\mu x) - B(B+\mu)\sec h^2(\mu x) = \frac{\varepsilon}{\omega A^2(x)} - \bar{U}_{eff}(x) - E_0 \quad (20)$$

$$V_+(x) = A^2 + B^2 + 2AB\tanh(\mu x) - B(B-\mu)\sec h^2(\mu x) \quad (21)$$

and we match (17) and (20), then we have

$$2AB = \mu \quad (22)$$

$$B = -\frac{\mu}{2} + \frac{1}{2\sqrt{\omega}}\sqrt{4v_1 + \mu^2}, \quad v_1 = 2\gamma^2\mu^2 + \frac{\varepsilon - 4\alpha}{\omega a^2} - \frac{2-a}{2a^2} \quad (23)$$

$$E_0 = -\left(B^2 + \left(\frac{\mu}{2B}\right)^2\right). \quad (24)$$

Shape invariance relationship between the partner potentials is known as $V_+(x, a_0) = V_-(x, a_1) + R(a_1)$ where $R(a_1)$ is the reminder independent of x and $a_0 = B$, $a_1 = 2B - \mu$.Because the energy spectrum is given by $E_0^- = 0$, $E_n^- = \sum_{k=1}^{n} R(a_k)$, we can obtain the spectrum of (19) as

$$E_n^- = -\left(\frac{\mu}{2B - n\mu}\right)^2 - (B - n\mu)^2, \quad n = 0, 1, 2, ... \quad (25)$$

To find corresponding eigenfunctions of the system, we will use E_n^- in eigenvalue equation (19) then we obtain un-normalized Φ_n,

$$\Phi_n(x) = B_n(1 + \tanh(\mu x))^{-r}(1 - \tanh(\mu x))^{-s} P_n^{(-2r,2s)}(-\tanh(\mu x)) \quad (26)$$

where B_n is the normalization constant, $r = \frac{1}{2}(n + \nu - \frac{1}{2\mu}\frac{1}{n+\nu})$, $s = \frac{1}{2}(n + \nu + \frac{1}{2\mu}\frac{1}{n+\nu})$, $\nu = B/\mu$ and $P_n^{(-2r,2s)}(-\tanh(\mu x))$ stands for Jacobi polynomials. For the normalization, one may use the $\rho = e^{-\frac{2\alpha}{\omega}\int^x dz \frac{-1+\gamma A(z)'}{A(z)}} dz$ used in Hermitian inner product $\ll \Phi|\Phi \gg = < \Phi|\rho^2\Phi >$. The boundary conditions require $r > 0, s > 0$.

4. Deformed Effective Screened Potential

Now we can choose $A(x)$ and $B(x)$ in these forms: $A(x) = 1$, $B(x) = \frac{e^{-\delta x+\tau}}{ae^{-\delta x+\tau} - q}$ in (5). These choices lead to

$$\frac{\varepsilon}{\omega A^2(x)} - \bar{U}_{eff}(x) = \frac{\epsilon}{\omega} - \frac{1}{2} + \delta\frac{e^{-\delta x+\tau}}{ae^{-\delta x+\tau} - q} - \left(1 + a\delta + \frac{4\alpha}{\omega^2}\right)\frac{e^{-2\delta x+2\tau}}{(e^{-\delta x+\tau} - q)^2} \quad (27)$$

where δ, a, τ, q are real parameters. If we give the super-potential $W(x) = A + B\frac{e^{-\delta x+\tau}}{e^{-\delta x+\tau}-q}$, then we have the ground-state function in the form of $\Phi_0(x) = e^{-(A+B)x}(e^{\tau} - e^{\delta x}q)^{B/\delta}$. The boundary conditions require $(A + B) > 0$. If we consider the partner potentials $V_{\mp}(x)$ for this case,

$$V_{\mp}(x) = A^2 + (B^2 \mp \delta B)\frac{e^{-2\delta x+2\tau}}{(e^{-\delta x+\tau} - q)^2} + B(2A \pm \delta)\frac{e^{-\delta x+\tau}}{e^{-\delta x+\tau} - q} \quad (28)$$

then we have,

$$B = \frac{\delta}{2} + \frac{\delta}{2}\sqrt{4(1 + a\delta + \frac{4\alpha}{\omega^2}) + \delta^2} \quad (29)$$

$$A = -\frac{\delta}{2} - \frac{\delta}{2B} \quad (30)$$

$$E_0 = -A^2. \quad (31)$$

It is noted that $(a_0, b_0) = (A, B)$ and $(a_1, b_1) = \left(\frac{B(A-\delta)-\frac{\delta^2}{2}}{B+\delta},\ B+\delta\right)$. Then, we have obtained the energy spectrum as

$$\varepsilon_n^- = \frac{1}{2} - \left((B+n\delta) - \frac{B^2+7B\delta+\delta(1+3\delta)}{2(B+n\delta)}\right)^2, \quad n = 0, 1, 2, ... \tag{32}$$

Following the same procedure shown in the last section, the corresponding wave-functions are obtained as

$$\Phi_n(x) = B_n s^{\varepsilon}(q-s)^{\frac{q+\sqrt{q^2-4\gamma}}{2q}} P_n^{(2\varepsilon, \frac{\sqrt{q^2-4\gamma}}{q})}(q-2s) \tag{33}$$

where B_n is the normalization constant and $s = e^{-\delta x+\tau}, \nu = \sqrt{q^2 - 4(1 + a\delta + \frac{4\alpha}{\omega^2})/\delta^2}$. The mapping function $\rho = e^{-\frac{2\alpha}{\omega}\int^x dz \frac{e^{-\delta z+\tau}}{ae^{-\delta z+\tau}-q}}$ can be given for this case for the specific values of $A(x)$ and $B(x)$ given above.

5. Conclusions

We have found the real spectrum of a non-Hermitian, parity-time-symmetric Hamiltonian (1) which is introduced in terms of the differential operators for the hyperbolic Rosen-Morse and the deformed effective screen potentials using convenient wave-function transformations within the point transformations such as $y = \int \frac{dx}{A(x)}$. In sections 3 and 4, we have generated two specific potentials using (5) which is a different effective potential given in section 2, hyperbolic Rosen Morse and deformed effective screened potentials namely, then, using shape invariance method we have seen that real energy is guaranteed if the inside of the square root is positive in (23) that leads to restrictions as $a < 2$, $\mu^2 > \frac{4\omega(2-a)-8(\varepsilon-4\alpha)}{2\omega a^2}$. The real spectrum is also guaranteed for the deformed effective screen potential. We have also obtained the corresponding wave functions. It is also shown that the ϵ which is the eigenvalue of the non-Hermitian Hamiltonian can be obtained using point transformations for the equivalent Hamiltonian h, however it may appear as a potential parameter in (17). In figure 1, for the first potential example, it is shown that when μ gets smaller, singularity appears for the large n quantum numbers.

[1] C. M. Bender and S. Boettcher, Phys. Rev. Lett. 80 5243 (1998); J. Phys. A 31 L273 (1998).
[2] Cooper F, Khare A, Sukhatme U 2001 *Supersymmetry in Quantum Mechanics* (Singapore: World Sci. Pub.).
[3] A. Mostafazadeh, J. Math. Phys. 43 205-214 (2002); J. Math. Phys.43 2814 (2002); J. Math. Phys. 43 3944 (2002).
[4] A. Sinha, J. Phys. A-Math. Theo. 45(18) (2012).
[5] Z. Ahmed, J. Phys. A: Math. Theo. 45(3) (2012).
[6] Bender C M, Brody D C and Jones H F Phys. Rev. Lett. 89 270401 (2002).
[7] M. S. Swanson, J. Math. Phys. 45 585 (2004).
[8] A. Sinha, P. Roy, J. Phys. A: Math. Theo. 41 335306 (2008)
[9] Ö. Yeşiltaş, J. Phys. A: Math. Theo. 44 305305 (2011).
[10] H. Eĝrifes et al, Phys. Lett. A 275 2000 229

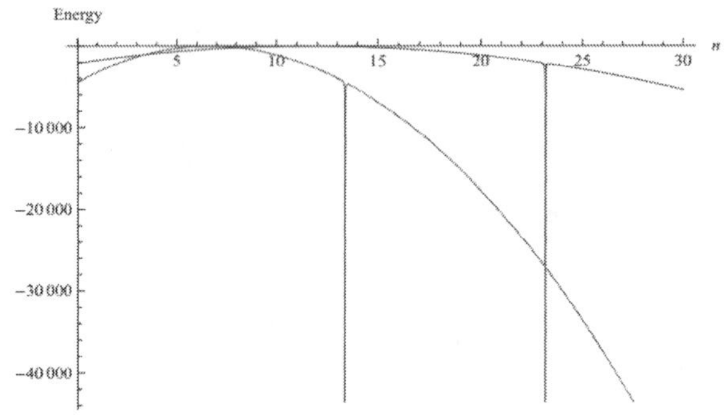

Figure 1. E_n versus n for (25). Parameters for the pink curve: $\mu = 10, \omega = 1/15, \alpha = 1/2; a = 1; \gamma = 1; \epsilon = 10$ and for the blue curve: $\mu = 4, \omega = 1/15, \alpha = 1/2, a = 1, \gamma = 1, \epsilon = 10$.

Principle of completeness: genesis of hyperdynamics within the theory of reactive motion

V Yu Tertychny-Dauri
Saint-Petersburg State University of Information Technologies, Mechanics & Optics
Department of Physics and Engineering
49 Kronverkskiy Prospect, Saint-Petersburg 197101, Russia

E-mail: `tertychny-dauri@mail.ru`

Abstract. The paper deals with a hyperreactive model of motion, which is based on a new approach to the fundamental momentum concept applied to a variable-mass point. In this model, the equations of motion differ from known Meshchersky-Tsiolkovsky's equations. The latter equations cannot be considered as satisfactory ones for analysis of many problems because of their inconsistency with some real processes. The suggested concept of hyperreactive motion allows determining the dependence of parameters of motion upon the time-dependent mass and, in particular, allows finding a solution to a basic problem of the highest absolute velocities that can be reached in space.

1. Introduction

If one assumes that generalized (Lagrange) coordinates and mass $M(t)$ of a point are independent at a given instant $t \in [t_0, t]$, the mass being thus considered as one more Lagrange coordinate [1], a total momentum (or measure of motion) of the point can be written in the form

$$Q(t) = M(t)\frac{dR(t)}{dt} + \frac{dM(t)}{dt}R(t), \tag{1}$$

where $R(t) = r(t) - \rho(t)$ is the reactive vector, $dr(t)/dt = v(t)$ is the absolute velocity of the point, $d\rho(t)/dt = u(t)$ is the absolute velocity of the particle outflow from the radiating center, which is a vector function of time given on $[t_0, t]$. We shall distinguish the quantity $Q(t)$ from the classical momentum $Q_c(t) = M(t)dR(t)/dt$, where the velocity $u(t)$ is, as a rule, assumed to be stationary $(du(t)/dt = 0)$. From relation (1), one obtains a new universal differential law of dynamics (the principle of completeness [2])

$$\frac{dS(t)}{dt} = Q(t),$$

where the vector quantity $S(t) = M(t)R(t)$ will be referred to as the motion composition. The principle of completeness, or the theorem of changing the motion composition, can be treated as a starting fundamental dynamic law which implies all the rest.

Let the law of changing the momentum be applied to $Q(t)$. Then the nonstationary equation of point motion will take the form

$$M\frac{dv}{dt} = F + M\frac{du}{dt} - 2\frac{dM}{dt}\frac{dR}{dt} - \frac{d^2M}{dt^2}R, \tag{2}$$

where F is the applied force, $-(dM/dt)(dR/dt) = (dM/dt)\,V$ is the standard reactive force [3], $V = u - v$ is a relative velocity of particle outflow, $-R(d^2M/dt^2)$ is the hyperreactive force. Equation (2) essentially differs from the known Meshchersky equation, whose derivation is founded on an improper hypothesis of short-range interaction. It is essential that a new model of motion (2) takes into account a hyperreactive d^2M/dt^2-containing summand in addition to the M- and dM/dt-containing terms. Basing on the new concept of variable-mass point motion, the present work investigates a number of specific cases of such motion, reveals its characteristic features and corresponding dynamic laws.

2. The first problem and Tsiolkovsky's hypothesis

Let us consider a variable-mass point that moves straightly in a nonresisting medium with no forces applied, the relative velocity of particle outflow V being constant, collinear to the reactive vector R, and directed opposite to the point motion. The whole set of these conditions forms, as known, Tsiolkovsky's hypothesis. The point movement is to be determined.

Note that the logarithmic relation [4]

$$v = v_0 + V \ln (M_0/M), \qquad v_0 = v(t_0), \qquad M_0 = M(t_0),$$

which results from integrating the Meshchersky equation

$$M\,\frac{dv}{dt} = -V\,\frac{dM}{dt},$$

is not self-consistent, since $V = \text{const}$, $u = \text{const}$, and $v = \text{const}$.

Let us use a scalar form of equation (2). We obtain

$$\frac{dR}{dt} = -\frac{R\,d^2M/dt^2}{2\,dM/dt}. \tag{3}$$

On integrating (3) with respect to R, we have

$$R = R_0 \exp\left\{-\frac{1}{2}\ln\frac{dM/dt}{(dM/dt)_0}\right\},$$

where $R_0 = R(t_0)$, $(dM/dt)_0 = (dM/dt)_{t=t_0}$. So it appears that

$$\left(\frac{R}{R_0}\right)^2 = \frac{(dM/dt)_0}{dM/dt}. \tag{4}$$

Substitute (4) into (3). We obtain

$$\frac{dR}{dt} = -\frac{R_0\,(dM/dt)_0^{1/2}\,(d^2M/dt^2)}{2\,(dM/dt)^{3/2}}. \tag{5}$$

To be definite, the point mass $M(t)$ will be considered a continuous, monotonously descreasing function of time ($dM/dt < 0$, $d^2M/dt^2 > 0$, $[\,(dM/dt)_0/(dM/dt)\,]^{1/2} > 0$). Therefore, the right side of relation (5) is positive, and the velocity of the point, v, increases obeying the relation

$$v = u - \frac{K_0\,d^2M/dt^2}{(dM/dt)^{3/2}}, \tag{6}$$

where $K_0 = R_0\,(dM/dt)_0^{1/2}/2$. The fulfilment of relation (6) is ensured by the mass change following a certain law $M = M(t)$ given beforehand.

3. The second Tsiolkovsky problem

Let us analyze the motion of a variable-mass point in a homogeneous gravity field when the point moves vertically upwards. The problem is to find the law of mass and distance variation as a function of time and to find the maximal altitude reached by the point. The relative velocity V of the emitted particles is constant and directed vertically downwards.

In this case, the equation of motion has the form

$$\frac{dR}{dt} = -\frac{R\,d^2M/dt^2}{2\,dM/dt} - \frac{Mg}{2\,dM/dt}. \tag{7}$$

We integrate (7) and obtain

$$R(t) = \frac{(dM/dt)_0^{1/2}}{(dM/dt)^{1/2}}\left[R_0 - \frac{g}{2\,(dM/dt)_0^{1/2}}\int_{t_0}^{t}\frac{M\,dt}{(dM/dt)^{1/2}}\right]. \tag{8}$$

After differentiating (8), we obtain the expression for $dR(t)/dt$ as a function of $M(t)$, $d\,M(t)/dt$, and $d^2\,M(t)/dt^2$, which may be written in the form

$$v = u - \frac{(dM/dt)_0^{1/2}\,(d^2M/dt^2)}{2\,(dM/dt)^{3/2}}\left[R_0 - \frac{g}{2\,(dM/dt)_0^{1/2}}\int_{t_0}^{t}\frac{M\,dt}{(dM/dt)^{1/2}}\right] - \frac{Mg}{2\,dM/dt},$$

where $u(t)$, $M(t)$ are given functions of time. If the relative velocity V is chosen constant, then equation (7) provides a differential law of mass variation in the form

$$\frac{d^2M}{dt^2} - \frac{2V}{R_0 - V\,(t-t_0)}\,\frac{dM}{dt} - \frac{gM}{R_0 - V\,(t-t_0)} = 0. \tag{9}$$

Equation (9) is not integrable in closed form. It is to be noted that (9) is the equation of relative equilibrium for a variable-mass point in a gravity field provided $V = u = \text{const}$.

4. Optimal conditions of motion in Tsiolkovsky problems

According to the previous theory, the more fuel mass of the object, the greater the velocity it reaches, the time dependence of the quantities $dm(t)/dt$, $d^2m(t)/dt^2$ being not important. This is, of course, wrong: taking into accuont the fuel mass and at the same time ignoring these quantities essentially means taking into account the impact effects only (a spread or distributed impact). But in fact, as shown by the new theory, the small fuel mass properly utilized can race the point to considerable absolute velocities.

Let us once again turn to the first Tsiolkovsky problem. When the fuel mass $m(t)$ becomes zero at $t = t_*$, the length of an active section of the trajectory will be

$$r(t_*) = \rho_0 + \int_{t_0}^{t_*} u(s)\,ds + \frac{R_0\,(dM/dt)_0^{1/2}}{(dm/dt)_*^{1/2}},$$

where t_* is determined from the equation $M(t_*) = N$. If the relative velocity $V = \text{const}$ is given, then t_* and $(dM/dt)_*$ are estimated as

$$t_* = t_0 + \frac{m_0 R_0}{m_0 V - R_0\,(dm/dt)_0}, \qquad \left(\frac{dM}{dt}\right)_* = \left(\frac{dm}{dt}\right)_* = \frac{R_0^2\,(dm/dt)_0}{[\,R_0 - V\,(t_* - t_0)\,]^2}.$$

The case of an instantaneous fuel discharge (a concentrated single impact) corresponds to $t_* = t_0$, i. e. $r(t_*) = r(t_0) = \rho_0 + R_0$,

$$\frac{V}{R} - \frac{(dm/dt)_0}{m_0} \to \infty.$$

But if we choose

$$V \to \frac{R_0\,(dm/dt)_0}{m_0},$$

then $t_* \to \infty$: one observes an infinitesimal fuel-mass variation, and in addition, $r(t_*) \to \infty$.

Let us turn to the second Tsiolkovsky problem. We are to find a maximal altitude of an active section of the trajectory followed by a variable-mass point which moves in the homogeneous gravitational field vertically upwards, the law of changing $M(t)$ being given in advance. We have (for $V = \text{const}$ and not given a priori)

$$H = \rho_0 + \int_{t_0}^{t_*} u(s)\,ds + R(t_*), \qquad M(t_*) = N,$$

$$R(t_*) = \frac{(dm/dt)_0^{1/2}}{(dm/dt)_*^{1/2}} \left[R_0 - \frac{g}{2\,(dm/dt)_0^{1/2}} \int_{t_0}^{t_*} \frac{M\,dt}{(dm/dt)^{1/2}} \right].$$

This problem will have an analytical solution if the $M(t)$ function of time is known. If the value $V = \text{const}$ is given in advance, then equation (16) cannot be integrated in closed form with respect to $M(t)$. However, the time t_* for which $m(t_*) = 0$ can be found approximately by numerical methods, and the maximum altitude of the active section can be approximated as well.

5. Conclusions

Let us formulate the main conclusions which are based on the results achieved by using the hyperreactive theory and presented in the paper.

(1) The conceptual basis of Meshchersky-Tsiolkovsky model is flawed. First, this model is inconsistent at a differential level; second, it does not permit taking into account the terms containing $d^2\,M(t)/dt^2$, and, finally, this model does not comply with the Lagrange scheme describing a dynamical object using generalized independent coordinates.

(2) The hyperreactive model allows gaining necessary absolute velocities mainly by choosing the rate and acceleration of the object mass change.

(3) The previous theory puts special emphasis on setting a relative velocity of particle outflow, V, and the new one gives preference to setting an absolute velocity of particle outflow, u. In the latter case the dynamics of the object is completely defined.

References

[1] Tertychny-Dauri V Yu 2005 Variational dynamic problems with parameters and its adaptive interpretation *Automat. and Telemech.* **9** 114–128

Tertychny-Dauri V Yu 2008 *Galamech* vol 1 *Adaptive Mechanics* (Moscow: Fizmatlit) 544 p

[2] Tertychny-Dauri V Yu 2004 *Hyperreactive Mechanics* (Moscow: Fizmatlit) 560 p

Tertychny-Dauri V Yu 2006 *Varied Mechanics* (Moscow: Factorial Press) 272 p

Tertychny-Dauri V Yu 2011 Hyperreactive model in dynamics of a variable-mass point *Eprint Arxiv* 1109.3597 10 p (http:// arxiv.org/abs/1109.3597)

[3] Meshchersky I V 1897 Dynamics of a Variable-Mass Point (Saint-Petersburg) *Works on Mechanics of Variable-Mass Bodies* 1952 (Moscow: Gostechizdat) 280 p

Levi-Civita T 1928 Sul moto di un corpo di massa variable *Rendiconti della Reale Academiae dei Lincei* 329–333, 621–622

[4] Tsiolkovsky K E 1914 *Research of Space by Jet Engines* (Kaluga) (1967 *Research of Space by Jet Engines* (Moscow: Mashinostroenie) 375 p)

Population Induced Instabilities in Genetic Algorithms for Constrained Optimization

D.S. Vlachos and K.J. Parousis-Orthodoxou

Department of Computer Science and Technology,
Faculty of Sciences and Technology,
University of Peloponnese,
GR-221 00 Tripolis, Greece

E-mail: dvlachos@uop.gr

Abstract. Evolutionary computation techniques, like genetic algorithms, have received a lot of attention as optimization techniques but, although they exhibit a very promising potential in curing the problem, they have not produced a significant breakthrough in the area of systematic treatment of constraints. There are two mainly ways of handling the constraints: the first is to produce an infeasibility measure and add it to the general cost function (the well known penalty methods) and the other is to modify the mutation and crossover operation in a way that they only produce feasible members. Both methods have their drawbacks and are strongly correlated to the problem that they are applied. In this work, we propose a different treatment of the constraints: we induce instabilities in the evolving population, in a way that infeasible solution cannot survive as they are. Preliminary results are presented in a set of well known from the literature constrained optimization problems.

1. Introduction

During the last years, genetic algorithms have received much attention regarding their potential as global optimization techniques. Since one of the most interesting and challenging field in global optimization is the class of constrained problems (both linear and nonlinear), a lot of effort has been recorded from researcher in this direction and several techniques have been developed and embedded in standard genetic algorithms ([1], [2], [3], [4], [5], [6], [7]). A nice short review of proposed methods with selected references can be found in [8].

In general, there are two ways in handling the constraints. The first one is the penalty function methods, in which a penalty term is added to the objective function. This term should be zero in the case of feasible solutions while in the case of constrains violation, the penalty increases. The second method is to take extra actions in order to limit the members of the population in the feasible region of the search space. There are several techniques that have been proposed, from modified mutation and crossover operation with the property that they only produce feasible offsprings from feasible parents to the death penalty method in which a member is destroyed if it violates a certain constrain.

Both methods have their pros and cons. Penalty methods are simple and self-evident. On the other hand, as far as the second approach is concerned, searching in the feasible region of the search space can be a very complicated task. The method is equivalent to transform the constrained problem to a constrain-free one. In any case, the basic requirement is that infeasible

Published under licence by IOP Publishing Ltd

members cannot survive for long. It is important that the method allows for infeasible members in order to preserve the searching capabilities of the genetic algorithm, but sooner or later, these members must be replaced by feasible ones.

this work we develop a new approach to obtain this. Assuming a binary coding of the chromosomes, the representation of the population of the genetic algorithm can be viewed as an array of binary digits that oscillate with time (generations). As the solution is approached, these oscillations are decreased until they are disappear when the genetic algorithm reaches the global minimum. Instead of replacing infeasible solutions with feasible ones (modifying and thus oscillate the binary digits), we could obtain the same result by inducing instabilities (oscillations) to the infeasible members of the populations without modifying the objective function. In the following, we shall investigate this idea by tuning the mutation and crossover probabilities according to the feasibility of a member of the population.

2. Formulation

Consider the general nonlinear programming problem

$$Minimize\ f(x),\ x \in R^n \tag{1}$$

subject to q equality constraints

$$h^i(x) = 0\ ,\ i = 1...q \tag{2}$$

and $m - q$ inequality constraints

$$g^j(x) \geq 0\ ,\ j = q + 1...m \tag{3}$$

We can now define a measure of infeasibility c^i of constrain i by

$$c^i(x) = w^i \cdot \begin{cases} \delta_b(h^i(x)) & ,\quad i = 1...q \\ u_a(g^i(x)) & ,\quad i = q+1...m \end{cases} \tag{4}$$

where w^i is a weight in order to bring the constrains in a comparable between them range of values and a, b are positive constants. The functions u_a and δ_b are given:

$$u_a(x) = \frac{\lambda(ax)}{1 - \lambda(ax)} \tag{5}$$

and

$$\delta_b(x) = \frac{\lambda(1 - b|x|)}{1 - \lambda(1 - b|x|)} \tag{6}$$

where $\lambda(x)$ is a continuous and at least k-times differentiable function (the value of k will be calculated later) with the property that is zero if $x \geq 0$, and for $x < 0$ is increasing and

$$\lim_{x \to -\infty} \lambda(x) = 1 \tag{7}$$

It can be easily proved now that $c^i(x)$ equals to zero for feasible solutions and $c^i(x)$ are everywhere C^k functions if this is true for the functions $h^i(x)$ and $g^j(x)$. Moreover, a very important property of $u_a(x)$ and $\delta_b(x)$ is that their n-th derivative at $x = 0$ is zero for every $n \leq k$. Fig. 1 shows the general shape of the functions $u_a(x)$ and $\delta_b(x)$ for different values of a and b. The values of the functions $c^i(x)$ for a specific member x of the population, measure how

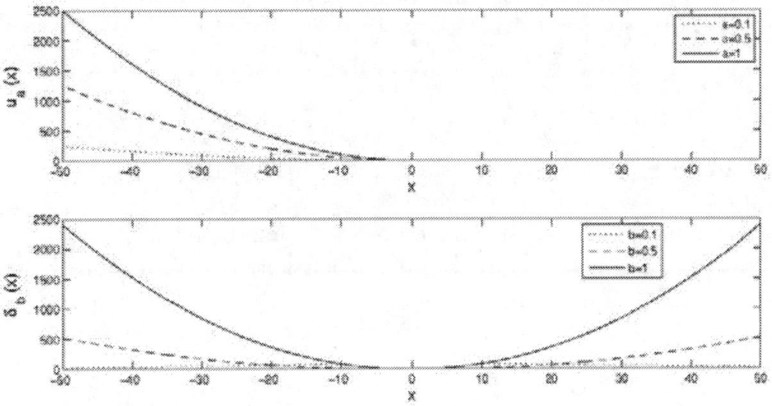

Figure 1. Functions $u_a(x)$ and $\delta_b(x)$ for different values of a and b.

deep in the infeasible region, the member lies. Consider now a feasible member x. If we induce a change Δx in the member, the infeasible measure becomes

$$c^i(x+\Delta x) = \frac{\partial^{k+1} c^i}{\partial x^{j_1} \ldots \partial x^{j_{k+1}}} \Delta x^{j_1} \cdot \ldots \cdot \Delta x^{j_{k+1}} + \mathcal{O}(|\Delta x|^{k+2}) \tag{8}$$

while for an infeasible member, the infeasibility measure becomes

$$c^i(x+\Delta x) = c^i(x) + \frac{\partial c^i}{\partial x^j} \Delta x^j + \mathcal{O}(|\Delta x|^2) \tag{9}$$

The vector $\frac{\partial c^i}{\partial x^j}$ now shows us the direction of the steepest descent (or ascent) of the infeasibility measure c^i. An important conclusion of the above analysis, is that, if we let k to be big enough, then any small change in a feasible member, will produce a feasible member again, since from eq. (8) the infeasibility measure will be negligible. The ideal case is if we can make $k \to \infty$. Fortunately, this can be obtained if we select

$$\lambda(x) = \begin{cases} e^{-\frac{1}{x^2}} & , \quad x < 0 \\ 0 & , \quad x \geq 0 \end{cases} \tag{10}$$

It can be easily proved that $\lambda(x)$ is a $\mathcal{C}^\infty$ function with $\lambda^{(n)}(x) = 0$ for every n. Using now eq. (10) to define the infeasibility measure, we have ensured that (i) every infeasible member will have a measure which increases with increasing distance of the member from the feasible part of the search space, (ii) every feasible member will have a null measure and most important that (iii) a small change of a feasible member will produce again a feasible one. We can now use these properties and define mutation and crossover probabilities in order to induce instabilities in the infeasible members.

2.1. Mutation

For every member x of the population, we construct the vector s_j by

$$s_j = \sum_{i=1}^{m} \frac{\partial c^i}{\partial x^j} \tag{11}$$

If the $|s|$ equals to zero, then we use the standard mutation probability, otherwise we produce a random integer number j between 1 and m with probability

$$p(j) = \frac{s_j}{|s|} \tag{12}$$

and mutate a random bit in the binary encoding of the j-th part of the chromosome representing the member x. Here we assume that the chromosome uses a Gray binary code to encode the n-variables of the problem. Each variable occupies a continuous part of bits in the chromosome.

The result of this operation is that every infeasible member will mutate, making impossible the stabilization of the solution in the infeasible area of the search space. But for a feasible member, this mechanism produce no effect. Mutation will occur as in the standard genetic algorithm in order to span the search space.

2.2. Crossover

Consider now the case where a member x of the population is candidate for crossover. There are two possibilities. The first one is that x is a feasible solution. In this case we should enhance crossover with the hope that offsprings from x will inherit the feasibility. The other case is when x is an infeasible solution. Suppose that the other parent is member y. If y is a feasible solution, then again we use the standard crossover probability. If y is infeasible, then it might violate the same or different constraints as x. If the same constraints are violated from both x and y, then it is likely that offsprings will inherit this property. But if x and y violate different constraints, it is possible that an offspring inherits the good part of the genetic information from each parent. In order to quantify this idea, we modify the standard crossover probability p_c by

$$p'_c = p_c/(1 + s_x \cdot s_y) \tag{13}$$

where s_x and s_y are the vectors defined in eq. (11) for the parents x and y. It is clear that if any of the members is feasible, then the crossover probability is that of the standard genetic algorithm, but if both x and y are infeasible, the crossover probability is decreased with increasing number of common violated constrains.

2.3. Selection

The selection mechanism is important for two reasons: first it enhances the best individuals to crossover and second, it selects between the individuals the best ones to survive. Since the searching mechanism of a genetic algorithm is based on crossover and mutation, the selection mechanism forces the searching to be performed in areas of the search space which are close to the best individuals. By enhancing crossover of the best individuals, the selection mechanism results in searching near these individuals and by allowing them to survive, this searching is continued. Selection can be performed only by defining an ordering relation between the members of the population, and this is achieved by the fitness function and in the case of constrained optimization by both the fitness and penalty functions.

Consider now a different mechanism which leads to the same result as the selection mechanism described before. Every member of the population has its own clock which counts how much time this member has lived. A member dies if the number of ticks of its clock cross a predefined level. The frequency of the clock is proportional to the value of the objective function. Thus, the smaller the value of the objective function for a member, the slower the frequency of the clock of this member and the longer this member will live. Let

$$F = f(x) \; and \; C = \sum_{i=1}^{m} c^i(x) \tag{14}$$

be the value of the objective function and a measure of the infeasibility for a given member x respectively. Now, combine these two measures to create the clock frequency of the member

$$\Omega = F + iC \tag{15}$$

The member x continues to survive as long as

$$\frac{|e^{i\Omega t}|}{winding\ number\ of\ curve\ e^{i\Omega t}} > threshold \tag{16}$$

where the winding number of a closed curve in the plane around a given point is an integer representing the total number of times that curve travels counterclockwise around the point and *threshold* is a predefined value. For a feasible member, the numerator of equation (16) is always 1 and that member dies when

$$\frac{F \cdot t}{2\pi} > \frac{1}{threshold} \tag{17}$$

The condition now that best members live more (and thus crossover and mutate more) is fulfilled. On the other hand, for an infeasible member, the numerator of equation (16) decreases with time and thus this member will die sooner.

2.4. Technical details

It is clear from the analysis of the proposed algorithm that the number of the individuals in the population is variable. In the first few generations, every member of the population survives, so it is a good practice to start with a small size of the population. Since the value of the objective function may change for a specific member (due to mutation), we have to update at every generation and for all members the phase Ωt. Finally, it is a natural selection to transform every calculated quantity in the population according to its average value. Since evolution operators depend on probabilities, only relevant probabilities are of interest.

3. Experimental results

The performance of the proposed constraint handling method has been evaluated using a set of 11 test cases ([9],[10]). These test cases include various forms of objective function (linear, quadratic, cubic, polynomial, nonlinear), and each test case has a different number of variables (n). The test problems also pose a range of constraint types and number of constraints [linear inequalities (LI); nonlinear equalities (NE); and nonlinear inequalities (NI)]. The general form of each test case is given in Table 1, which also indicates the number of constraints active at the optimum solution (a). Note here that every optimization problem is transformed in the form of eq. (1). For example, in $G3$ instead of maximizing $\sqrt{(n)}^n \prod_{i=1}^{n} x_i$, we minimize the function $1 - \sqrt{(n)}^n \prod_{i=1}^{n} x_i$.

Fig. 2 shows the result of the application of the new method in problems $G1$ to $G11$. The error is the distance of the solution calculated from the best known solution. In this error, we have to include the quantization error, since for the needs of this experiment we use an 8-bit encoding of each variable, so for variable i we have an extra error e_i given

$$e_i = \frac{range\ of\ variable_{\ i}}{256} \tag{18}$$

The starting size of the population in all test problems was 50 members and this size changed during the experiment as it was explained before. The probabilities for standard crossover and mutation was 1 and 0.06 respectively. The horizontal axis of Fig. 2 corresponds to the number of generations and of course this is in a one to one correspondence with the computational time

Case	n	Objective	LI	NE	NI	a
G1	13	quadratic	9	0	0	6
G2	k	nonlinear	0	0	6	1
G3	k	polynomial	0	1	0	1
G4	5	quadratic	0	0	6	2
G5	4	cubic	2	3	0	3
G6	2	cubic	0	0	2	2
G7	10	quadratic	3	0	5	6
G8	2	nonlinear	0	0	2	0
G9	7	polynomial	0	0	4	2
G10	8	linear	3	0	3	6
G11	2	quadratic	0	1	0	1

Table 1: Summary of Test Cases

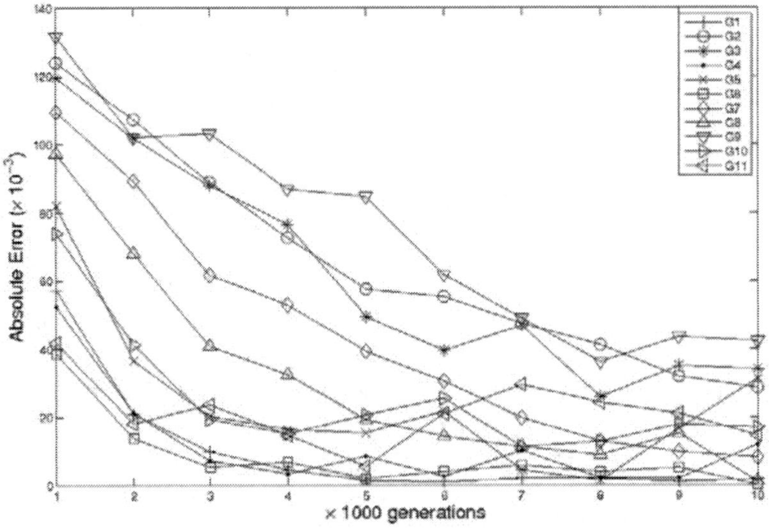

Figure 2. Absolute error vs number of generations for the eleven problems of table 1. In problems $G2$ and $G3$ the number of the unknown variables was 10.

of the method. The method seems to behave well in the case of linear constraints, although it exhibits a slow convergence in the case of non linear ones, especially $G2, G3$ and $G9$. On the other hand, the method treats successfully the case of boundary searching ($G1, G7$ and $G10$ for example).

Finally, it is interesting to see the dynamics of the method. First, the population size increases during the first generations. But later, there is an almost stable population size because the number of new offsprings are equal to the number of members that cannot survive. This is shown if Fig. 3a and for problem $G1$. On the other hand, the frequency of the bit changes in the binary representation of members is shown in Fig. 3b for the feasible and infeasible members and

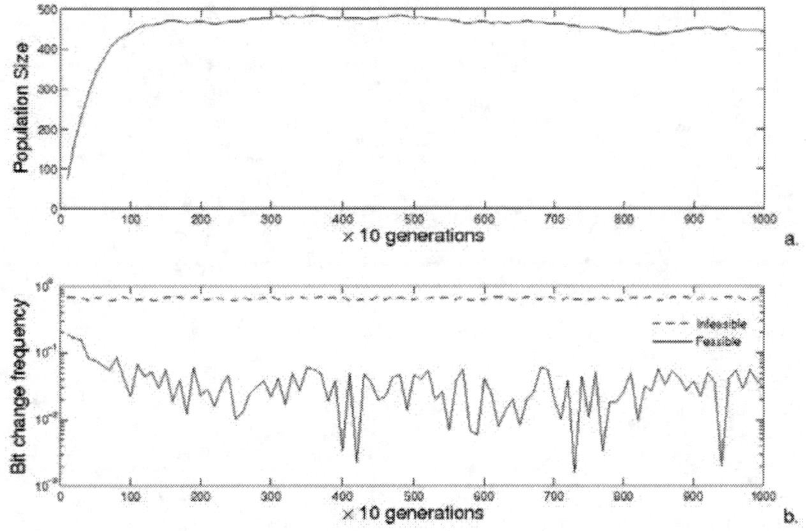

Figure 3. (a) The population size vs the number of generations. (b) The frequency of bit changes in the binary representation of infeasible (dashed line) and feasible (solid line) members vs the number of generations.

for the same problem $G1$. It is clear that bits in infeasible members oscillate with a frequency which is one order of magnitude greater than the bit oscillation caused by standard mutation and crossover.

4. Conclusion

A new approach for handling constrains in optimization using genetic algorithms is presented in this letter. The new method induces instabilities in the infeasible members of the population by mutation and selective crossover, while it introduces a new mechanism for selection by letting every member of the population to survive a number of generations which depends on both the fitness and the feasibility of the member. The preliminary results from the application of the method to a well known academic set of constrained problems is very promising. Since the method is parameter free, it might be used for complex constrained optimization problems.

References

[1] Hamida S B and Schoenauer M 2000 *Proc. Parallel Problem Solving from Nature* **VI** 5296–538
[2] Coit D W, Smith A E and Tate D M 1996 *INFORMS J. Comput.* **8** 1736–182
[3] Deb K 2000 *Computer Methods in Applied Mechanics and Engineering* **186** 3116–338
[4] Hadj-Alouane A B and Bean J C 1997 *Oper. Res.* **45 (1)** 926–101
[5] Powell D and Skolnick M M 1993 *Proceedings of 5th Int. Conf. Genetic Algorithms*
[6] Runarsson T P and Yao X 2000 *IEEE Trans. Evol. Comput.* **4** 2846–294
[7] Wright J A and Farmani R 2001 *Proc. Genetic and Evolutionary Computation Conf.* (San Francisco, CA) pp 7256–732
[8] Farmani R and Wright J A 2003 *IEEE Trans. Evol. Comput.* **7** 4456–455
[9] Michalewicz Z and Fogel D B 2000 *How to Solve It: Modern Heuristics* (Berlin, Germany: Springer-Verlag)
[10] Koziel S and Michalewicz Z 1999 *Evolutionary Computation* **7 (1)** 196–44

Global vaccination strategies in Modular Networks

K.J. Parousis-Orthodoxou, M.M. Stamos and D.S. Vlachos

Department of Computer Science and Technology,
Faculty of Sciences and Technology,
University of Peloponnese,
GR-221 00 Tripolis, Greece

E-mail: dvlachos@uop.gr

Abstract. We study the effect of vaccinating networks with different growing strategies, using various techniques that require the complete knowledge of the network. The goal is to restrain the epidemic before it spreads throughout the network and target the few key nodes that will help contain it. Our target networks are chosen to have relatively large modularity index and various immunization techniques are applied to them.

1. Introduction

The spread of epidemics can be considered to occur on networks that describe contacts between individuals. The process of epidemic spreading throughout large networks is being researched widely the last years. The spreading dynamics are closely related to the structure of networks. For this reason network epidemiology has become one of the most vibrant subdisciplines of complex network studies [1],[2] and [3].

The form of epidemic spreading can also be acknowledged as information flow in a network. If we consider the virus spreading in an epidemic, we can safely say that an information will spread the same way. In this case, we want the opposite results than the virus spreading. The fortunate news are that with successful vaccination techniques, we can control the information flow, by immunizing or not certain nodes that are points of strategic importance in the network's topology.

Disease propagation has also been studied in scale-free networks [4], [5] and the problem of vaccination strategies has been specifically addressed [6]. Many times, it is not possible to vaccinate the whole population, due to high price of the vaccine. Some vaccines have severe side effects and for this reason one may also want to keep the number of vaccinated individuals low.

In this paper, in search of a better vaccination technique, we study two different approaches in selecting the group of nodes to be immunized. These techniques require the full knowledge of the network and their characteristics that define them. The networks that we use for this study, are of medium scale (approximately 2000 nodes) and are created with specific models throughout the bibliography of complex networks.

2. Network Models and Parameters

As indicated earlier in the paper, the main parameter, taken into consideration to select the networks in which we will experiment, is the modularity factor. This factor indicates the degree in which the network can be divided into modules or communities. It is simply the probability

Published under licence by IOP Publishing Ltd

that a link begins at a node i, a_i, multiplied by the fraction of links that end at a node in i, a_i. So the expected number of intra-community links is just $a_i\, a_i$. On the other hand we know that the real fraction of links exclusively within a partition is e_{ii}. So, we can compare the two directly and sum over all the partitions in the graph. Modularity fluctuates between the values of zero and one with one being the completely modular network. The equation for the modularity index is $Q = \sum(e_{ii} - {a_i}^2)$. The range following this pattern, four models of growing networks are chosen, as to satisfy that parameter. These four models are of different parameters, but the common parameter for all of them is the number of nodes they are containing, which is 2000 (medium-scale).

The first model for growing network which examine, introducing as basic content the fact that new nodes are attached to the network through one existing node and then explore the network through the links of the visited nodes [?]. At each time step, a new node is introduced into the preexisting network. This new new node, chooses at random, a node in then network. The new node follows the path of neighbors, starting from the random chosen node. As it passes through the node trail, it decides if it will connect or not. In our network that probability is p=0.2. This network is directed.

The second network based on Volkan Sevin and Per Arne Rikvold model [8]. In this model we use the preferential-attachment scheme which is primarily used in the Barabasi Albert model [9]. The model, has a user-defined number of initial nodes. These nodes have only outgoing links, in a way that they provide incoming links to the nodes that join at each time step. The new node attaches to an existing node with probability $\Pi(k_i) = k_i - 1$ where k_i, is the number of outgoing links at i. This model creates directed networks.

The next network is based on a modified BA model [9] in which all the edges connected to the new nodes are made locally to the creator and its nearest neighbors [10]. This means that, once again, at each time step a new node is introduced in the network. This node chooses among the preexisting nodes of the network, with the logic of preferential-attachment. After it has chosen a node, it uses the same PA scheme to choose the next, but only among the first node's neighbors. This model is used primarily for the generation of networks that simulate various natural food webs in certain ecosystems. In the network used in this paper, the new node makes two new connections with existing nodes. In contrary with the previous networks, this network is undirected, and the epidemic is spread differently.

The final network is based in the creation of triangles [11]. The newly introduced node, chooses at random among the preexisting edges of the network and it then connects to both ends of this edge. The control parameter q decides if the old edge between the two preexisting end-nodes, stays or not. This model operates in such a way as to create network with embedded small-world property. In the network examined in this paper, the control parameter q has the value of 0.5, which means that approximately half of the edges in question, are erased. This network is also undirected.

3. Vaccination Techniques

Describing the vaccination techniques in this section, first we will analyze the way the infection spreads over the networks. The epidemic model we choose for this occasion is the SIR model. The specific model of epidemic spreading operates with basic principle that there are three pools which contain all the nodes of the network. At the beginning of the spreading all the nodes (except from the vaccinated ones) are in the "Susceptible" pool. As the spreading progresses, the nodes are transferred with probability $p_i = 0.5$ to the "Infected" pool. Then with probability $p_c = 0.1$ are transferred to the Recovered pool. The key component of the SIR model is the fact that, as soon as a node has recovered from the infection, it can no longer carry it again. This is why the three states of the model are Susceptible Infected Recovered. The vaccinated nodes are considered to be in the Recovered state as soon as the infection starts to spread.

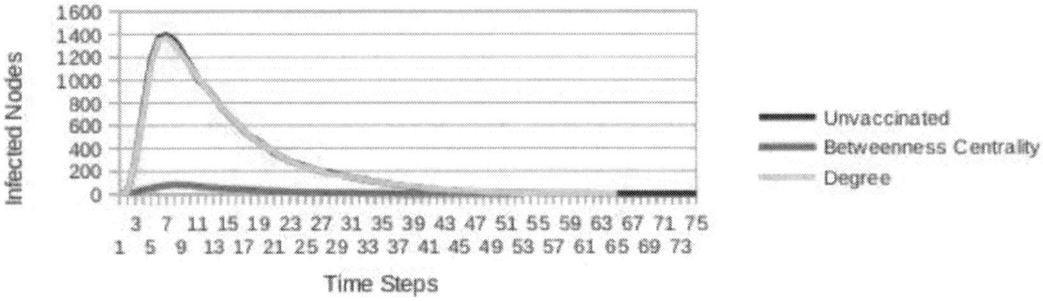

Figure 1. The epidemic spreading of the first network.

The first vaccination technique that we use in this paper, is the most commonly used in the papers that study immunization in complex networks. The nodes that are selected to be immunized before the spread of the epidemic are the ones with the highest degree of the network. For this vaccination, we target the highest 1% of the total population.

The other vaccination technique which yielded the most promising results is the vaccination based on the betweenness centrality factor of the network nodes. This metric indicates the number of shortest paths passing through a node. The value fluctuates between zero and one (0-1) with the values closer to one showing higher number of shortest paths passing through the specific node. The equation of the betweenness centrality factor is

$$C_B(v) = \sum_{s \neq v \neq t \in V} \frac{\sigma_{st}(v)}{\sigma_{st}}$$

For the purpose of this experiment we locate the 1% of the total nodes that have the higher value of betweenness centrality factor.

4. Results and Discussion

The results of the vaccination and epidemic spreading will be presented in the following graphs. These graphs show the number of infected nodes per time step. For reference purposes, we give in each graph, the values of the epidemic spreading without vaccination. As we expect, in all occasions, the number of infected nodes, as well as the speed of the spreading are higher than the vaccinated ones.

The first network that we examine has a modularity factor of 0.753 which is considered to be fairly modular. This network is created from the model of Alexei Vazquez paper [7]. The chart below **Figure 1** shows the progression of the infection in the three occasions.

This network is not created using preferential-attachment and this is the main reason that the basic characteristic of the network is the existence of triangles. This is why the network is very susceptible to the epidemic spreading, due to the great number of alternative routes that exist. As it is shown in **Figure 1**, the vaccination with respect to the betweeness centrality has

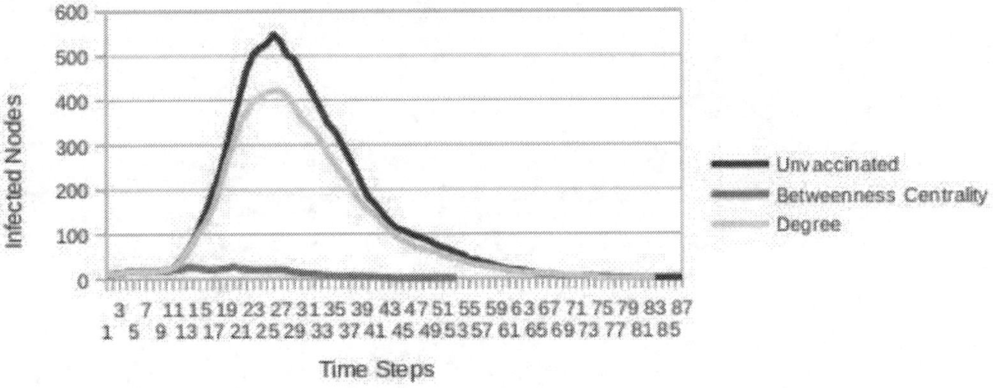

Figure 2. The epidemic spreading of the second network.

the best performance. This result makes sense because the network is fairly modular, but the scale of it is an important outcome.

The second network has a modularity factor of 0.939 for which we can safely say that is very modular. It is generated with the model introduced in the paper of Volkan Sevim and Per Arne Rikvold [8].It is fairly obvious in **Figure 2** that this network shows great resilience against epidemic spreading. The technique in which this network is generated, since it is a food-web based model, has minimal number of connections per node, and the existence of hubs is close to zero. The unvaccinated spreading reaches up to 25% of the population being simultaneously infected (approximately 550 nodes peaking at the time step 26). This proves the resilience of the network, even if there is no immunization. If we proceed to the next line of the chart which shows the vaccination with the degree, we will see even more resilience, but the difference is not very significant, due to the structure of the network as described above. The final line in the chart, shows the epidemic spreading with the vaccination of some nodes using the betweenness centrality value of the nodes. The results here are even more impressive than the previous network, because the modularity factor of the network is even higher than the previous one.

The third network has a modularity factor of 0.93 and it is also considered very modular. It is generated as a variation of the Barabasi Albert model [9], used to give small world networks [10]. This network was firstly introduced in the paper of XU Xin-Ping, LIU Feng and Li Wei [10] and the network we examine here, is created in relation to the model described in that paper. Results as shown in **Figure 3**.Even though this network is also very modular, it allows the spreading to reach higher levels than the previous network. This is because the structure of the network is very different than the previous one but quite similar to the first one. This network has many triangles which allow to the epidemic to spread more easily throughout the network. The fact that the vaccination technique using the degree of the nodes behaves better than the first network, which also has triangular structure, is caused by the use of the preferential-attachment in the model. Again due to high modularity and the network structure the vaccination technique using betweenness centrality value performs significantly better than the other two.

The fourth and final network has a modularity factor of 0.931 which makes this network very modular as well. The model from which this network was created is an attempt to create

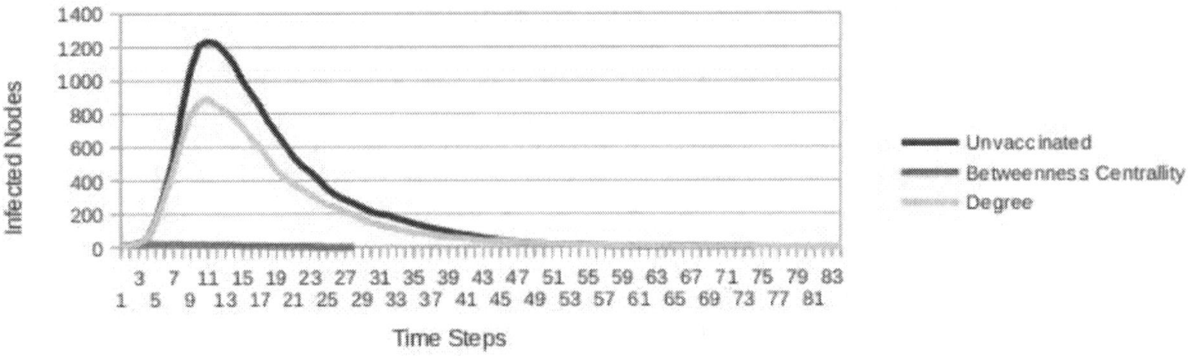

Figure 3. The epidemic spreading of the third network.

small-world networks [11]. The results in **Figure 4** show an interesting outcome.This network's behavior in epidemic spreading is quite similar to the previous networks. Yet we observe an important differentiation. The spreading in the procedure with vaccination based on the degree is significantly delayed than the other two. This means that the spreading has a small fluctuation in the beginning before it reaches its high peak (approximately at the time steps 77) in comparison with the other two techniques which have their peak earlier in the procedure.For the purpose of this experiment we locate the 1% of the total nodes that have the higher value of betweenness centrality factor.

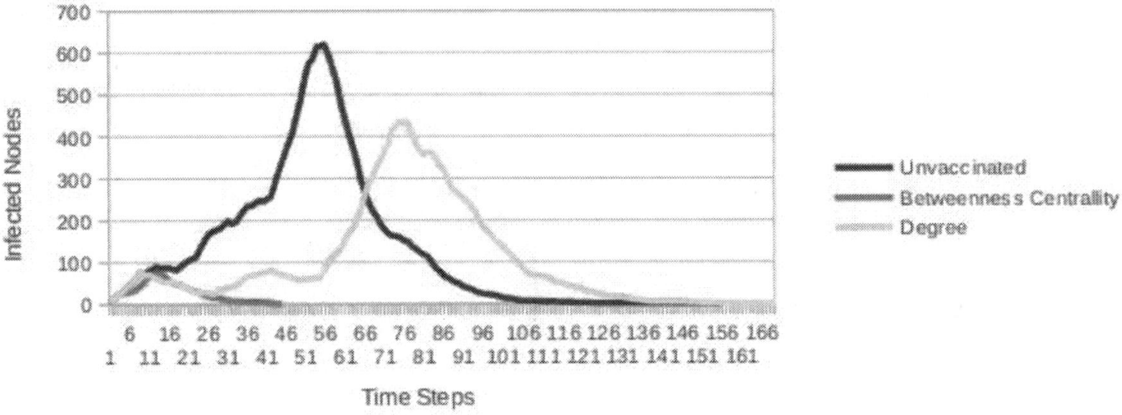

Figure 4. The epidemic spreading of the fourth network.

5. Conclusions

Summarizing this paper's results, it is safe to say that networks with high modularity index are more susceptible to the vaccination technique using the betweenness centrality of the nodes. Yet we observe that the networks that have triangles in their basic structure are more resilient in the epidemic spreading even without vaccination. This is a safe conclusion when we the knowledge of the networks are in global scale.

Acknowledgment

The project is implemented under the Operational Program "Education and Lifelong Learning" and funded by the European Union (European Social Fund) and by Greek national resources.

References

[1] J.Giesecke, Modern infectious disease epidemiology, Arnold , London, 2 ed., 2002.
[2] F. Liljeros, C. R. Edling, and L. A. Nunes Amaral, Sexual networks: Implication for the transmission of sexually transmitted infection, Microbes Infect. 5 (2003), pp. 189-106.
[3] M. E. J. Newman, The structure and function of complex networks, SIAM Rev. 45 (2003), pp. 167-256.
[4] R. Pastor-Satorras and A. Vespignani, Phys. Rev. Lett. 86 (2001) 3200-3203.
[5] R. Pastor-Satorras and A. Vespignani, Phys. Rev. E 63 (2001) 066117.
[6] R. Pastor-Satorras and A. Vespignani, Optimal immunisation of complex networks, cond-mat/0107066.
[7] Alexei Vazquez, Knowing a network by walking on it: emergence of scaling, cond-mat/0006132v4 (2000)
[8] V. Sevim and P. A. Rikvold. Effects of preference for attachment to low-degree nodes on the degree distributions of a growing directed network and a simple food-web model. Phys. Rev. E, 73:056115, 2006.
[9] Barabsi, A. -L. & Albert, R. Emergence of scaling in random networks. Science 286, 509512 (1999).
[10] XU Xin-Ping;LIU Feng;LI Weim, Growing Small-World Networks Based on a Modified BA Model, Chinese Physics Letters 2006 23 (3): 750-753.
[11] Zhongzhi Zhang, Shuigeng Zhou, Zhen Shen, From regular to growing small-world networks, Physica A 385 (2007) 765-772.

Journal of Physics: Conference Series **410** (2013) 012080 doi:10.1088/1742-6596/410/1/012080

The concept of roles in complex networks

K.J. Parousis-Orthodoxou, M. Stamos and D.S. Vlachos

Department of Computer Science and Technology,
Faculty of Sciences and Technology,
University of Peloponnese,
GR-221 00 Tripolis, Greece

E-mail: dvlachos@uop.gr

Abstract. This paper is an analytic process of introducing the concept of roles in a complex network. Each node has a specific connectivity pattern to the rest of the nodes in the network. There are many nodes that have the same number of connections and also the same type. In this case we say that two nodes can have the same role. The work described here, shows the process of identifying the roles of all the nodes in a network and grouping them accordingly.

1. Introduction

Numerous systems in modern day life can be assigned to complex networks and complex network's properties. From the protein network within the human body to the configuration of the Internet or the World Wide Web, all can be treated with respect to their topological configuration of nodes connected to eachother with the use of distinct edges. The variety and diversity of these networks (and many more) are what prompted the scientific community to investigate the mechanisms that explore their topology [1], [2], [3], and [4].

In modern day Computer Science as well as various different fields, the use of complex network modeling and experimentation is widespread. Within the past decade, various efforts have taken place for the modeling and analysis of complex systems. These efforts have been very fruitful, both in manufacturing networks by following certain rules (such as the preferential attachment rule) [5], as well as analyzing real data and mining emerging phenomena such as the small-world property, which implies that the average topological distance between any two nodes within a network, increases very slowly along with the evolution of the network.

By taking a closer look into a complex network, it is noticeable that each node has a specific position in the graph, not necessarily topologically, but with respect to the other nodes of the network. It is also easy to see that there are many nodes that may have the same positioning within the graph. The algorithm presented here will adopt the methodology to define the positioning of each node in the network, newly introducing roles.

2. The use of roles

The concept of *roles* is explained by analyzing the function of a node within the network, in accordance with the other nodes. Every node has a role in the network, whereas many nodes can have the same role, with regard to their own connectivity.

In order to find the role that can be assigned to every node, we must first divide the nodes into groups, and create each *connectivity vector* to the other groups. Each connectivity vector

Journal of Physics: Conference Series **410** (2013) 012080 doi:10.1088/1742-6596/410/1/012080

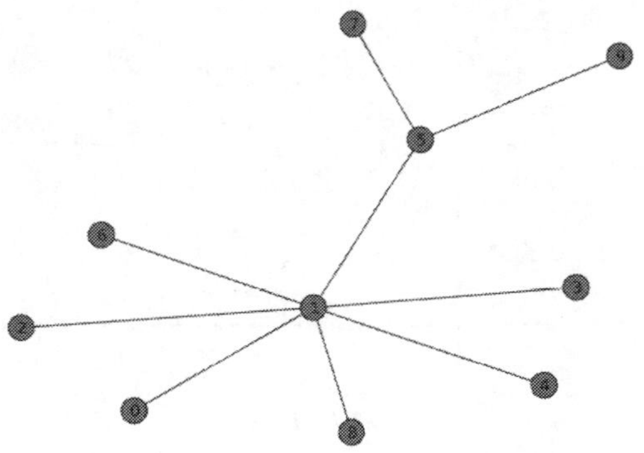

Figure 1. An example network for role identification.

records the number of connections that the group has, to all the rest (including itself) and also the number of nodes that reside in that group (the final metric of the vector).

For better understanding of the entire process, an example network is shown in **Figure 1**. The algorithm tries first to categorize the nodes into groups. The process begins by creating N distinct groups and assigning a single node to each one. Then the connectivity vectors of the network are generated for each group (initially for each node). This instance of the connectivity vectors of the example network is shown in **Table 1**.The second step compares the connectivity vectors of the groups and if it finds that two vectors are the same (apart from the last metric that shows the number of nodes in the group), it merges the two groups into one. After the merging of the two groups takes place, the algorithm recalculates the connectivity vectors and starts searching for the next groups that may share a common connectivity vector. When there can be no more merging among the groups, the algorithm has finished the grouping stage.

CV No	Grp 0	Grp 1	Grp 2	Grp 3	Grp 4	Grp 5	Grp 6	Grp 7	Grp 8	Grp 9	No
0	0	1	0	0	0	0	0	0	0	0	1
1	1	0	1	1	1	1	1	0	1	0	1
2	0	1	0	0	0	0	0	0	0	0	1
3	0	1	0	0	0	0	0	0	0	0	1
4	0	1	0	0	0	0	0	0	0	0	1
5	0	1	0	0	0	0	0	1	0	1	1
6	0	1	0	0	0	0	0	0	0	0	1
7	0	0	0	0	0	1	0	0	0	0	1
8	0	1	0	0	0	0	0	0	0	0	1
9	0	0	0	0	0	1	0	0	0	0	1

Table 1: The connectivity vectors of the example network, before the first iteration.

The example network has 10 nodes and 9 edges. The first step of the process, creates 10 groups and places a single node of the network into them. Then, it checks the connections of each group (node for now) and creates the connectivity vector for each group. The starting vectors are taken unchanged from each line of the adjacency matrix of the network, with the addition of the number of nodes for each group at the end of the vector (*one* at first). After the initial step, the algorithm identifies that the groups with number *0* and *2*, which initially

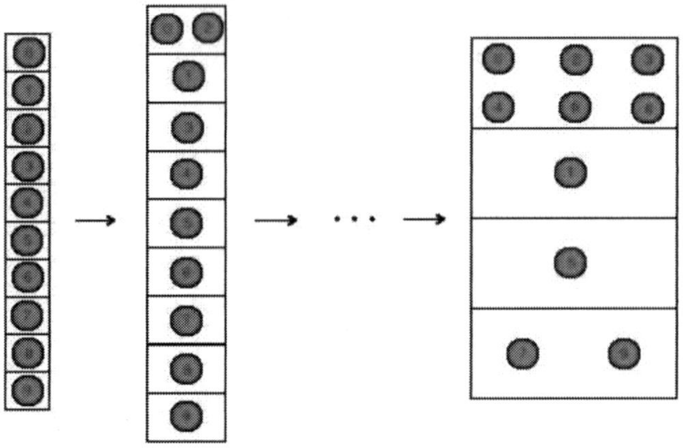

Figure 2. The process of the grouping algorithm.

contain the nodes 0 and 2 respectively, have the same connectivity vector. The two groups have only one connection with the group *1*. Then it increments the final metric of the group *0* and deletes group *2*. After that, it recalculates the connectivity vectors with one group less. On the next iteration, the groups *0* and *3* are found to have the same connectivity vector and so, the group *3* is deleted and the counter of group *0* is incremented by 1. This process is continued on the same pattern until no further grouping can be performed.

So, at the end of the final iteration, the number of groups is reduced to 4. These groups are shown in **Figure 2** and the nodes that each group contains. Note that the numbering of the nodes here is for explanatory purposes and the algorithm only focuses on the number of nodes in each group. The final connectivity vector of the example graph is shown in **Table 2**.

CV No	Grp 0	Grp 1	Grp 2	Grp 3	No
0	0	6	0	0	6
1	6	0	1	0	1
2	0	1	0	2	1
3	0	0	2	0	2

Table 2: The connectivity vectors of the example network at the end of the process.

3. Testing the algorthm

The network that will be used for the testing of the algorithm is undirected and unweighted and it is provided by Duncan Watts and Steven Strogatz [6]. The graph presents information about the topology of the Western States Power Grid of the United States of America. The network has 4941 nodes and 6594 edges. The mean degree of the graph is 2.67 whereas the maximum degree that can be found is 19.

After performing the aforementioned operation to our test network, the result was the creation of table (of similar form to **Table 2**, but significantly larger). This table contains the connectivity vectors of all the groups that can be created that would hold information about each role that might exist in the network. The specific network generated a table with 3954 groups and equivalent roles. The number of nodes that assume these roles are shown in **Figure 3**

It is fairly clear from the illustration of the connectivity vector of the specific network that

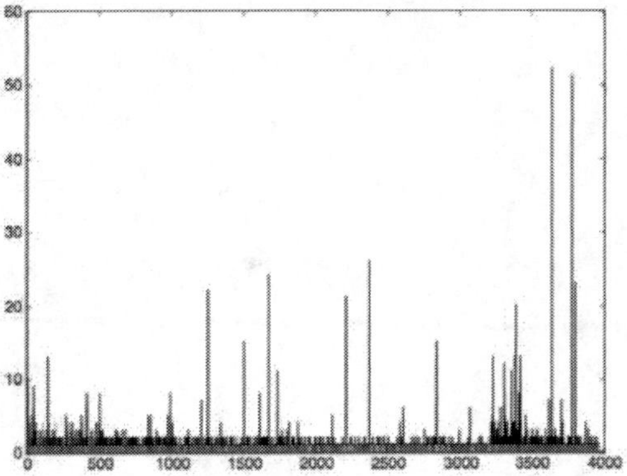

Figure 3. The role distribution of the test network.

there is a significant number of roles that are assumed by a single node (and that is why so many roles can be found in the network) and also that there are certain roles that are assumed by many nodes, illustrated by the spikes that emerge from the figure.

4. Future work

The concept of roles and their distribution over a network can provide useful information regarding the connectivity of the nodes. This algorithm allows the user to manipulate a network further when analyzing its role distribution, in several ways such as replicating or scaling the network. This part can be extensively studied in a different project an may yield very promising results.

Acknowledgment

The project is implemented under the Operational Program "Education and Lifelong Learning" and funded by the European Union (European Social Fund) and by Greek national resources.

References

[1] L.A.N. Amaral, A. Scala, M. Barthlemy, H.E. Stanley, Proc. Natl. Acad. Sci. USA, 97 (2000), p. 11149
[2] S.N. Dorogovtsev, J.F.F. Mendes, **Evolution of Networks: From Biological Nets to the Internet and WWW**, Oxford University Press, Oxford (2003)
[3] R. Pastor-Satorras, A. Vespignani, **Evolution and Structure of the Internet: A Statistical Physics Approach**, Cambridge University Press, Cambridge (2004)
[4] M.E.J. Newman, Phys. Rev. Lett., 89 (2002), p. 208701
[5] R.Albert, A.-L. Barabasi, Rev.Mod.Phys., 74(2002), p.47
[6] D.J. Watts, S.H. Strogatz, Nature, 393 (1998), p. 440

Evolutionary Processes in Complex Networks and Small Worlds

D.S. Vlachos and K.J. Parousis-Orthodoxou

Department of Computer Science and Technology,
Faculty of Sciences and Technology,
University of Peloponnese,
GR-221 00 Tripolis, Greece

E-mail: dvlachos@uop.gr

Abstract.
In recent years, researchers discovered by empirical studies that the majority of networked systems that occurring in nature and society, exhibit an emerging behavior which is caused by the statistical properties of these systems. Two of the most important factors that characterize those complex systems is the degree distribution of the underlying network and the so called *small world* property. In this work, we examine the application of evolutionary processes on these systems in order to derive useful results for human dynamics modeling. Simulation experiments are used to explain universal properties observed recently in the vote distribution of proportional elections. Moreover, it has been founded that the hierarchical or modular nature of complex networks accelerates the convergent of a hybrid genetic algorithm applied in a rather complex search space.

1. Introduction

Complex systems are those groups or organizations which are made up of a large number of non identical elements connected by diverse interactions. Some representative examples of complex systems include the global climate, economies, ant colonies, immune systems, the Web, the Internet (routers and domains), electronic circuits, computer software, movie actors, coauthorship networks, sexual web, citations, phone calls, metabolic, protein interaction, protein domains, brain function web, linguistic networks, international trade, bank system, encryption trust net, energy landscapes, earthquakes, astrophysical network etc. In such systems the individual parts and the interaction between them often lead to large-scale behaviors which are not easily predicted from a knowledge only of the behavior of the individuals. Such collective effects are called *emergent* behaviors [1]. Examples of such emergent behaviors include short and long-term climate changes, price fluctuations in markets, foraging and building by ants, and the ability of immune systems to distinguish 'self' from 'other' and to protect the former and eradicate the latter.

A straightforward representation of complex systems is by using graph theory. The interacting parts of a complex system are represented by vertices, while the interactions between them with edges. Such networks are called *complex networks* and may be directed (citations) or undirected (sexual web), weighted (international trade), multi-graphed (coauthorship or movie actor network) etc. The size and complexity of the complex networks along with the statistical behavior of the system that they represent, brought network research close to statistical physics and this new approach has been driven largely by the availability of computers and communication networks. Traditional tools used in graph theory like the adjacency and the incidence matrix, the Laplacian and the set of eigenvalues [2], have been replaced by the characterization of networks by statistical properties.

Of great importance in the field of complex network and modeling of complex systems is their dynamical behavior. The spreading of a virus in human or computer networks is a common example. The range of possible dynamical processes in complex networks is very wide. Watts in [3] studied the impact of clustering on several processes, including games, cooperation, the Prisoner's Dilemma, cellular automata while Lago-Fernadéz in [4] studied synchronization. Wang and Chen in [5] demonstrated that the inhomogeneous scale-free topology plays an important role in determining synchronization in complex networks. Search and random walks in complex networks is another well studied example. But one of the most important, although unexpected, result presented by Pastor-Satorras and Vespignani in [6] and [7] concerning the effect of network topology in the spreading of a disease. They showed that while for random networks a local infection spreads to the whole network only if the spreading rate is higher than a critical value, for scale-free networks any spreading rate leads to the infection of the whole network. This property motivates the work of this paper as it will be explained in the following.

Dynamical process in general can be modeled by genetic algorithms. Genetic algorithms are one class of agent based modeling techniques that were designed to capture the essence of evolution and adaptation and yet to be simple enough to be mathematically tractable. In genetic algorithm methods one studies the evolution of simple strings of symbols on a computer, or fragments of computer code, rather than attempting to simulate the behavior of real organisms. An early result from research on genetic algorithms was the mathematical characterization of adaptation as a near-optimal trade off between exploitation of traits that have been already been found to be useful and exploration for new useful traits. Genetic algorithm research has also led to mathematical characterization of the roles of mutation, sexual recombination, diploidy, and other genetic processes and characteristics. In addition, genetic algorithms have a practical use as computational search and learning methods inspired by evolution. In general, in a genetic algorithm, a set of individuals-members constitute a population in which genetic operation like mutation and crossover produce offsprings which populates the next generation through a process of selection. There are many flavors and implementations of genetic algorithms which try to optimize either the effect of genetic operators or the coding of information inside each member of the population. On the other hand, a number of techniques have been developed for the implementation of the selection operation, mainly through a fitness function, capable to handle even large number of constraints. A nice short review of proposed methods with selected references can be found in [8].

One of the most important challenges one has to face, when implementing a genetic algorithm, is the so called exploration-exploitation trade-off. Exploration is needed to ensure every part of the search space is searched thoroughly in order to provide a reliable estimate of the global optimum. Exploitation is important since the refinement of the current solution will often produce a better solution. Population-based heuristics (where genetic algorithms [9] and estimation of distribution algorithms [10] are found) are powerful in the exploration of the search space, and weak in the exploitation of the solutions found.

The purpose of this work is two folded; first, we design and implement a special hybrid genetic algorithm, the population members of which 'live' on the vertices of a complex network. Crossover now is restricted only between members that correspond to adjacent vertices of the network. The small world property of the network assures that finally, information from every member will be available to the whole population, while the degree distribution and the transitivity of the network will enhance the exploration capabilities of the algorithm. On the other hand, we present a model for the evolution and opinion formation in a human network using a special designed genetic algorithm. This model will be used to explain the recently observed property of the distribution of votes received by candidates in proportional elections which is the same function in different countries and years, when the number of votes is rescaled according to the strength of the party to which each candidate belongs [11].

2. Hybrid genetic algorithm on complex networks

(i) Assign at each vertex of the network a chromosome. This can be done randomly or we can load to the network a pre calculated population.

(ii) Crossover each vertex with all the adjacent ones. Here we can use the crossover probability used in standard genetic algorithms. In the case that no crossover is selected, the offsprings are copies of the parents.

(iii) Make a copy of each offspring to both parents.

(iv) After finishing with crossover, every vertex of the network has the chromosome from the previous population plus a number of offsprings. Using this set of chromosomes, we estimate the distribution of genes in the chromosomes and finally produce a chromosome according to that distribution. This chromosome is the one who enters the next generation.

Since the chromosome who finally survives to the next generation is produced randomly (following of course a given distribution), there is no need for mutation.

3. Simulation results

Since the aim of this work is to examine the effect of network topology in the exploration-exploitation dynamics of the genetic algorithm, we use a special designed fitness function with several local minima. The used function is taken from [12] and is given by

$$f(x,y) = \sum_{j,k=1}^{5} j \cdot k \cdot cos\left((j+1) \cdot x + j\right) \cdot cos\left((k+1) \cdot y + k\right) \tag{1}$$

4. Experimental results

In figure 1 the results of the dynamics of the genetic algorithm are shown. As a first test, we study the entropy of the population as a function of the evolution time. The networks that have been tested vary in degree distribution (scale free networks with different exponent), modularity, clustering and assortativity.

5. Conclusion

In this work, the dynamics of a hybrid genetic algorithm has been studied for different topological properties of the population. The population has been modeled with a complex network and the entropy has been measured as a function of the evolution time. The results show that the diversity of the population is decreased with decreasing exponent in the degree distribution, with increasing modularity, clustering and assortativity.

References

[1] Mitchell M and Newman M 2002 *Encyclopedia of Evolution, M. Pagel (ed.)* (New York: Oxford University Press)
[2] Biggs N 1993 *Algebraic Graph Theory* (Cambridge: Cambridge University Press)
[3] Watts D 1999 *Small Worlds: The dynamics of networks between order and randomness* (Princeton, NJ: Princeton University)
[4] LF Lago-Fernádez R Huerta F C and Siguenza J 2000 *Phys. Rev. Lett.* **84** 2758–2761
[5] Wang X and Chen G 2001 *preprint cond-mat/0105014* 445–455

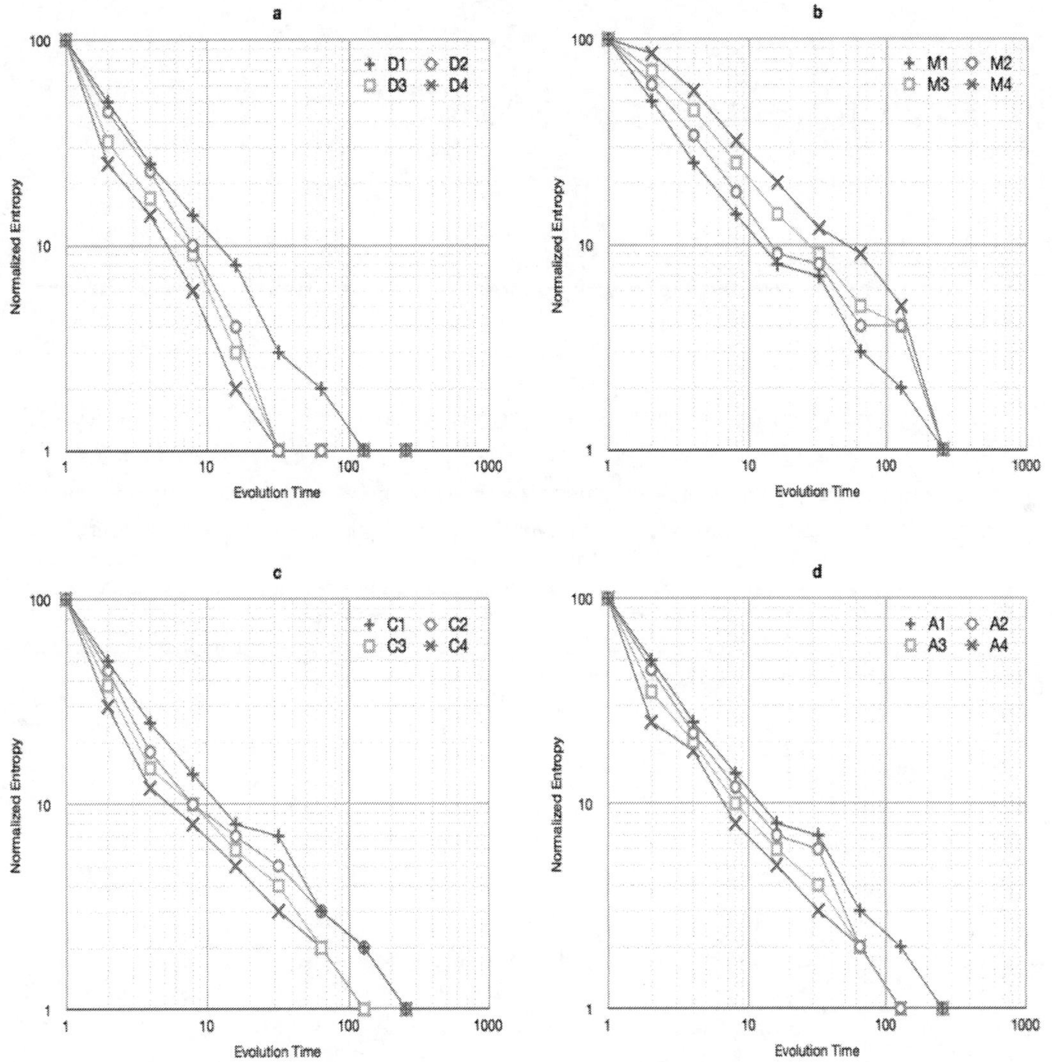

Figure 1. (a). Evolution of entropy for various exponents of degree distribution (decreasing exponent from D1 to D4). (b) The same for various modularities (increasing modularity from M1 to M4). (c) The same for various clustering coefficients (increasing from C1 to C4). (d) The same for various assortativity coefficients (increasing from A1 to A4).

[6] Pastor-Satorras R and Vespignani A 2001 *Phys. Rev. Lett.* **86** 3200–3203
[7] Pastor-Satorras R and Vespignani A 2001 *preprint cond-mat/0102028*
[8] Farmani R and Wright J A 2003 *IEEE Trans. Evol. Comput.* **7** 445–455
[9] Holland J 1975 *Adaption in natural and artificial systems* (Ann Harbor, MI: The University of Michigan Press)
[10] Larranaga P and Lozano J 2001 *Estimation of Distribution Algorithms. A New Tool for Evolutionary Computation* (Norwell, MA: Kluwer Academic Publisher)
[11] Fortunato S and Castellano C 2007 *Phys. Rev. Lett.* **99** 138701
[12] W Huyer A N Global optimization by multilevel coordinate search Technical report Institut fr Mathematik Universitt Wien

IC-MSQUARE 2012: International Conference on Mathematical Modelling in Physical Sciences IOP Publishing
Journal of Physics: Conference Series **410** (2013) 012082 doi:10.1088/1742-6596/410/1/012082

Probabilistic methods for physics

G. Cirier
LSTA. Paris VI University, France

Email: guy.cirier@gmail.com

Abstract. We present an asymptotic method giving a probability of presence of the iterated spots of R^d by a polynomial function f. We use the well-known Perron Frobenius operator (PF) that lets certain sets and measure invariant by f. Probabilistic solutions can exist for the deterministic iteration. If the theoretical result is already known, here we quantify these probabilities. This approach seems interesting to use for computing situations when the deterministic methods don't run. Among the examined applications, are asymptotic solutions of Lorenz, Navier-Stokes or Hamilton's equations. In this approach, linearity induces many difficult problems, all of whom we have not yet resolved.

1. Presentation of the problem

The Perron-Frobenius's equation (PF) is a very difficult functional equation. Ulam and Von Neumann gave the first and only known asymptotic solution for the logistic $f(a) = 4(a - a^2)$ in R.

Here, we study an iteration f, which is a polynomial function of R^d in R^d. The main new result for physics is the random behaviour of solutions of special differential equations.
It is impossible to expose all our results and demonstrations in this 4-pages paper. The detailed references and proofs can be found in the preprint hal-00691097, v3[1] after some little corrections.
Here, we insist only on the leading points to understand the process we used to obtain these results.

1.1 The main hypothesis H0

The polynomial iteration f applies a compact set C *of* R^d *in* C. $f(0) = 0$. *Let* λ *be the real eigenvalues of the linear part of* f *at* 0. *There is no resonance if we have* $1 \neq \lambda^n$ *for all* $n \in N^d$.

We iterate indefinitely this application $f^{(k)} = f \circ f^{(k-1)}$. Then, we obtain infinity of spots in C. When $k \to \infty$, these spots can converge toward some points:

- Fixed points: $f(\alpha) = \alpha$;
- Cycles such as $f^{(p)}(\alpha_p) = \alpha_p$ but $f \circ f^{(p)}(\alpha_p) \neq \alpha_p$;
- Sets B where exists an **invariant probability measure** P called Perron Frobenius (PF) (see[3]):

$$\boxed{P(B) = P \circ f^{-1}(B)}$$

In this paper, we want to find the measure P solution of this functional equation on C.

We note the generic collection of $f^{(p)}$ as $f_\bullet \in C(f)$; $Fix(f_\bullet)$ is the set of the fixed points of $f_\bullet$.

2 - The resolving equation L and the resolving deviation $e^n(y)$

First, we transform all measure P by Stieltjes-Laplace's L in order to capture the fixed point 0:

$$\phi(y) = \mathrm{L}(P) = \lim_{\varepsilon \to 0} \int_{-\varepsilon} e^{yx} dP(x)$$

As in [3], for all f and P, we have: $P_f = P \circ f^{-1}$ and

$$\phi_f(y) = \int e^{yx} dP_f(x) = \int e^{yf(x)} dP(x)$$

Second, we compute $\phi_f(y)$ in 2 steps:

- As C is compact, we can expand in convergent series: $\phi(y) = \Sigma_n b_n y^n$,
- We translate the origin with a: $x = a + u$; then; for all measure P on C, we have:

$$\phi(y) = E(e^{yx}) \text{ becomes } \phi_a(y) = \phi(y)e^{ya} = \Sigma_n b_n \partial^n (e^{ya}) / \partial a^n,$$

$$\phi_f(y) = E(e^{yf(x)}) \text{ becomes } \phi_f(y,a) = \Sigma_n b_n \partial^n e^{yf(a)} / \partial a^n,$$

Thence, if P is the PF measure for f, then $\theta_f(\phi_a) = \phi_f(y,a) - \phi(y,a) = 0$ is the **resolving equation** L. If $a = 0$, we have $\theta_f(\phi) = \Sigma_n b_n (y^n - H_n(y)) = 0$ with $H_n(y) = \partial^n e^{yf(a)} / \partial a^n |_{a=0}$.

2.1. The resolving deviation

We call **resolving deviation** $e^n(y)$, the polynomial defined by: $\boxed{e^n(y) = y^n - H_n(y)}$

Then: $\theta_f(\phi) = \Sigma_n b_n e^n(y)$

2.2. The ideal E_n

If $Z(e^n)$ is the set of zeros of $e^n(y)$, let $Z(E_n) = Z(e^n) \bigcap_{\ell=1}^{\ell=d} Z(e^{n+l_1})$ $\ell = 1,..d$ be the set of these common zeros. E_n is the ideal generated by these $d+1$ polynomials e^n, e^{n+l_1}, $\ell = 1,2,..,d$.

If $\varphi(y)$ is a solution of L and $\phi(y)$ the solution of (PF), then, the relation between $\varphi(y)$ and $\phi(y)$ is: $\phi(y,a) = \partial^2 \varphi(y,a) / \partial y \partial a = \partial(y\varphi(y,a)) / \partial y$ *at* $a = 0$. ($\theta(\phi) = 0$ is an identity for all y and a).

2.3. Theorem 1

When $n \to \infty$ with $1 \neq \lambda^n$ (no resonance):

- **1.** There exists an integer m such as $(\varphi_n(y) - 1)^m \in E_n$. With a probability 1, the distribution defined by $\varphi_n(y)$ is constant on the algebraic affine real manifolds defined by E_n.
- **2.** These results are valid for all $f_\bullet \in C(f)$.

The demonstration needs the following steps: First, in order to vanish $\theta_f(\varphi_n)$ for a solution $\varphi_n(y)$, $e^n(y)$ must be null when $n \to \infty$. Second, if we have $e^n(y) = 0$, by translation of the origin, we must have $e^{n+1_\ell}(y) = 0$ $\ell = 1,..d$; then, immerging the problem in the algebraically closed field of the complex numbers, the Hilbert's theorem of zeros (nullstellensatz) says $(\varphi_n - 1)^m \in E_n$.

3. Approximation of the zeros by the Riemann-Debbie's steepest descent method

The main idea is the following:

• The leading term of $e^n(y)$ is $y^n(1-\lambda^n)$. If $|\lambda^n| > 1$, y^n is negligible with respect to $H_n(y)$. So, the asymptotic distribution of the real zeros of $H_n(y)$ gives us the solution when $y \to \infty$ with n and $y_\ell = n_\ell s_\ell$. But we can search this distribution for all λ^n. As Plancherel and Rotach[4] did for the Hermitian polynomials, we represent $H_n(y)$ in the field of the complexes with the Cauchy's integral :

$$H_{n-1}(y) = \partial^{n-1} e^{yf(a)} / \partial a^{n-1} \big|_{a=0} = c \oint_\Gamma \frac{e^{yf(a)} da}{a^n}$$

We call the Plancherel-Rotach's function PRF: $\gamma(a) = yf(a) - n \log a$. Γ is a closed poly-disk around the fixed point 0 of f, $a \in C^d$. $n-1 = (n_1 - 1, .., n_d - 1)$. c is a finite non-null constant.

• The steepest descent's method says: if a is critical point of $\gamma(a)$, $ce^{\gamma(a)}$ gives an approximation of $H_{n-1}(y)$ when $y \to \infty$. As $yf(a)$ is polynomial in a, we can refer to Delabaere [2] for a formulation of the conditions (general position). The definite negative Hessian of $\gamma(a)$ at a is sufficient condition.

• As only the real zeros of $H_{n-1}(y)$ interest us, we can neglect all factors $\neq 0$ in the representation of $H_{n-1}(y)$. Then, we can write this equivalence $H_{n-1}(y) \underset{0}{\approx} e^{\gamma(a)}$ at a. But, $H_{n-1}(y) = 0$ iff the critical point $a = a(y)$ has complex coordinates. Then, $H_{n-1}(y) \underset{0}{\approx} e^{\mathrm{Re}(\gamma(a))} \sin(\mathrm{Im}(\gamma(a)) \underset{0}{\approx} \sin(\mathrm{Im}(\gamma(a))$ and $\mathrm{Re}(\gamma(a))$ must be maximum.

3.3. Theorem 2

If the critical point $a \in Z(J_n)$ is in general position, then, when $n \to \infty$:

- **1.** $\gamma(a)$ have one common index of derivation n for all coordinates of y and $y = ns$.
- **2.** If a has p complex coordinates, the $d-p$ dimensional manifolds solutions are indexed by a code $k \in [1, n-1]^p$ and verify : $s_\ell \,\mathrm{Im}\, f_\ell(a) - \vartheta_\ell = k_\ell \pi / n$, with the code $k_\ell \in (1, ..., n-1)$. Between two zeros at time $n-1$ there is one zero at time n : then we have a Rolle's foliation.
- **3.** If $\kappa_\ell = \lim_{n\to\infty} k_\ell / n$ for $\ell = 1, .., p$, then κ is asymptotically a random uniform vector on the unit cube $]0,1[^p$. Then, the $d-p$ dimensional manifolds are indexed by the random vector $\kappa \in]0,1[^p$ and verify the random equation $\boxed{\pi\kappa_\ell = s_\ell \,\mathrm{Im}\, f_\ell(a) - \vartheta_\ell}$, $\ell = 1, 2, .., p$.

Demonstrations use the facts that a is critical, and, not only $H_{n-1}(y) = 0$, but also $H_{n-1+1_\ell}(y) = 0$, $\ell = 1, 2, .., d$. Unfortunately, in many problems, the Hessian is not definite (as in the Henon's case[1]).

4. Applications to ODE and PDE

Many physical problems can be written with a differential equation as $\partial a / \partial t = F(a)$, where $F(a)$ is an application of R^d in R^d. In this equation, the unknown variables are a vector a of R^d and the known variables are a vector t of R^{+k}. Here, the number of equations equals the number of unknown variables a. We write the differential equation in terms of iteration:

4.1. A differential iteration is the application $f(a,\delta)$ of R^d in R^d, defined by $f(a,\delta) = a + \delta F(a)$ for all fixed path $\delta_0 > \delta > 0$ of R^{+k}.

In order to use the previous results, we assume the *hypothesis H*:

The polynomial application F iterates a compact set $C \subset R^d$ in C and we can apply the steepest descent's method to $f(a,\delta)$. But here, the steepest descent's method will be managed, especially if the Hessian is degenerate.

4.2. The iteration $f(\boldsymbol{a},\delta)$ is said partly linear in $\boldsymbol{a} = (a,b)$ if $\boldsymbol{F}$ linear in b.

We denote: $\boldsymbol{F(a)} = (A(a)b + B(a), C(a)b + D(a))$ where $\boldsymbol{a} = (a,b)$ with $a \in R^p$ and $b \in R^{d-p}$. For all each zero $\boldsymbol{\alpha} = (\alpha,\beta)$ of $\boldsymbol{F}$ we note $\boldsymbol{a} = \boldsymbol{u} + \boldsymbol{\alpha}$ and $\boldsymbol{u} = (u,v)$ in R^p and $\boldsymbol{y} = (x,y) \in R^d$. For every zero $\boldsymbol{\alpha}$ of $\boldsymbol{F}$, we define a partly linear iteration: $\boldsymbol{G}_\alpha(\boldsymbol{u}) = (u + \tau A(u+\alpha)v, v)$ at $\boldsymbol{\alpha}$.

4.3. Theorem 3

If the differential iteration $f(a,\delta) = a + \delta F(a)$, applying $C \subset R^d$ in C for all $\delta_0 > \delta > 0$ with $\delta = |t|\tau / n \rightarrow 0 \in R^{+k}$, verify the hypotheses H of *(4.1)* for every fixed point $\alpha \in C$:

- **1.** The invariant distribution dominates the fixed point according to $\lambda\tau$ will be positive or negative. In the EDO case, the condition $\Sigma\lambda < 0$ implies generally the convergence toward the fixed point. We have resonance with $\Sigma\lambda = 0$. For the PDE, the planes $\lambda\tau = 0$ get big gaps.
- **2.** If the Hessian of $yF(a)$ is definite, there exists, besides the fixed points of $\boldsymbol{F}$, only cyclical orbits with uniform density and period T such as $\int_0^T F(a(t+t_0)dt = 0$ for all t_0.
- **3.** If $\boldsymbol{F}$ is partly linear, then asymptotically:
  the critical point $\boldsymbol{a} = (a,b)$ de $\gamma_f(\boldsymbol{a})$ can tends to the critical point of either $\gamma_G(\boldsymbol{a})$ at $\boldsymbol{0}$ or γ_{G_α} at $\boldsymbol{\alpha}$. Thence, we find the asymptotic invariant solution among the solutions of each iteration $\boldsymbol{G}_\alpha(\boldsymbol{a})$. As $\boldsymbol{G}_\alpha$ is partly linear, we have families of random manifolds by iteration of the critical point.
- **4.** If we have two fixed points $\boldsymbol{0}$ or $\boldsymbol{\alpha}$, we shall take the distribution at $\boldsymbol{0}$ or at $\boldsymbol{\alpha}$ according to $\mathrm{Re}(x\tau A(a)\beta)$ is positive or negative at the critical point. The dominating distribution must get the greatest contribution of the steepest descent's method. The critical points realising $\mathrm{Re}(x\tau A(a)\beta) = 0$ induce a route of communication between the two sets.

But, owing to the degeneracy of the Hessian, it is clear that the present description doesn't give the complete solution for the points • **3** and • **4**, as for the Lorenz or Navier Stokes attractor where the main solutions are *random ovals* [1]. We don't have studied cycles and their zones of domination. We observe that many terms of $\boldsymbol{F}$ vanish asymptotically.

5.Conclusion

Some asymptotic solutions of a differential equation can be randomized. The probabilistic theory brings information whereas the deterministic approach cannot explain everything.
- Many ideas on chaos must be corrected: the definition of chaos, as a situation between determinism and probability, seems devoid of interest. The multiple errors of approximation making unpredictable the behaviour does not give a good mathematical argument. Notions of sensibility to the initial conditions have to be seen differently. We have only zones of domination or attraction.
- On the other hand, the degeneracy of the Hessian remains a difficult problem.

References

[1] G. Cirier 2012 *Probabilistic and asymptotic methods with the Perron Frobenius's operator*.
Preprint, hal-00691097, v. 2 .

[2] E. Delabaere and C. J. Howls 2002 *Global asymptotic for multiple integrals with boundaries*,

[3] A. Lasota, M.C. Mackey1991 *Chaos, Fractals and Noise*, Sec. Ed., Berlin, Springer.

[4] M. Plancherel et W. Rotach 1929 *Sur les valeurs asymptotiques des polynômes d'Hermite*
$\mathrm{H}_n(x) = (-1)^n e^{x^2/2} d^n e^{-x^2/2} / dx^n$, Com. Math. Helvetici **1**, EMS, p 227 - 254.

Journal of Physics: Conference Series **410** (2013) 012083 doi:10.1088/1742-6596/410/1/012083

Study of two elements constituting turbulence by a multifractal theory for probability density functions — Through the analyses of 4096^3 DNS —

T Arimitsu[1], **N Arimitsu**[2], **K Takechi**[3], **Y Kaneda**[4] **and T Ishihara**[4]

[1] Faculty of Pure and Applied Sciences, University of Tsukuba, Tsukuba 305-8571, Japan
[2] Faculty of Environment and Information Sciences, Yokohama National University, Yokohama 240-8501, Japan
[3] Graduate School of Pure and Applied Sciences, University of Tsukuba, Tsukuba 305-8571, Japan
[4] Graduate School of Engineering, Nagoya University, Nagoya 464-8603, Japan

E-mail: arimitsu.toshi.ft@u.tsukuba.ac.jp

Abstract. The probability density functions (PDFs) of energy dissipation rates created from the whole of the DNS region of the size 4096^3 (named *whl-PDFs*) for various sizes of coarse graining areas are analyzed by multifractal PDF theory (MPDFT). Furthermore, other PDFs named *max-PDF* and *min-PDF* for various sizes of coarse graining areas, created respectively from the partial DNS regions of the size 512^3 with maximum and minimum enstrophy, are also analyzed by making use of the whl-PDF. It is shown that there is a whl-PDF with an appropriate size of coarse graining area, whose slope of the tail part can adjust the slope of the tail-part of a max-PDF. On the other hand, there is a whl-PDF whose center part can adjust quite accurately the center part of a min-PDF. Note that the partial DNS region with maximum (minimum) enstrophy is the region of maximally dense (rare) vortexes, and therefore contributes mainly to the tail (center) part of the whl-PDF. It is found that these adjustments are systematic with respect to the sizes of coarse graining areas.

We analyze in this paper the probability density functions (PDFs) for energy dissipation rates created from the snapshot raw data of 4096^3 DNS for fully developed turbulence [1] by means of multifractal PDF theory (MPDFT) [2, 3, 4, 5, 6]. The DNS is specified by the Taylor micro-scale Reynolds number $R_\lambda = 1132$ and the integral length $L/\eta = 2.130 \times 10^3$ with the Kolmogorov length scale $\eta = 5.12 \times 10^{-4}$. The inertial range is estimated to be in the region $126 \lesssim r/\eta \lesssim 448$ [1].

MPDFT is a statistical mechanical ensemble theory for analyzing intermittent behavior providing fat-tail PDFs [4, 5]. It was constructed following Frisch and Parisi [7] by assuming that, in the limit of high Reynolds number and therefore in the situation in which the contribution of the dissipation term in the Navier-Stokes (N-S) equation can be neglected, the singularities due to the invariance of N-S equation under a scale transformation with an arbitrary real number α distribute themselves in a multifractal way in real physical space. This distribution of singularities is specified by a multifractal spectrum introduced [3] by assuming that the distribution of α is given by Tsallis-type PDF [8]. The multifractal spectrum is a manifestation of self-similar hierarchical structure (a multi-scale Cantor set) producing intermittent fluid motion of turbulence to which we refer to as a *coherent turbulent motion*. The finiteness of Reynolds

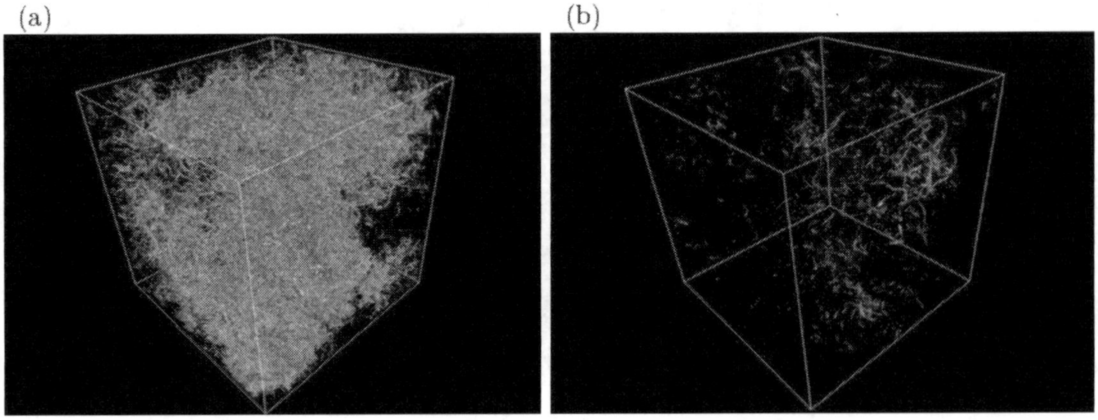

Figure 1. 3D-snapshots of the partial regions with (a) maximum and (b) minimum enstrophy. Presented are the iso-surfaces of the value $\omega_{\rm iso} = \langle|\omega|\rangle + 12\sigma_\omega$ of vorticity with the average $\langle|\omega|\rangle$ of the absolute values of vorticity and their variance $\sigma_\omega^2 = \langle|\omega|^2\rangle_{\rm c}$.

number due to the dissipation term violating the invariance under the scale transformation manifests itself as an *incoherent fluctuating motion* of fluid. Therefore, in the study of fully developed turbulence, we have to look for the characteristics of two elements in turbulent fluid motion, i.e, one is the coherent and the other is the incoherent fluid motion.

Within MPDFT [4, 5, 6], it is assumed that the fat-tail PDF can be divided into two parts, i.e., the *tail part* and the *center part*. The tail part PDF and the center part PDF are connected at a specific point ε_n^* of the energy dissipation rates ε_n under the conditions that both of the PDFs have the common value and the common log-slope there (see [6] for the theoretical formula of the PDF witin MPDFT). As the tail part is responsible for the intermittent character of turbulence, at the tail part the coherent contribution may dominate over the incoherent contribution. On the other hand, the center part consists of both coherent and incoherent contributions. With the above consideration, we neglect within A&A model of MPDFT the incoherent contribution to the tail part in deriving the theoretical PDF formula, and conjecture that the coherent contribution to the tail part can be determined solely by the multifractal spectral for α.

In order to prove the above assignment for the center and the tail parts of PDF, we create PDFs for energy dissipation rates as follows. In the first place, we construct the PDFs from the whole of 4096^3 DNS region, which we refer to as *whl-PDFs*. For the extraction of the intermittent character of turbulence, it is necessary to have information of self-similar hierarchical structure of the system. This is realized by producing a series of PDFs for responsible singular quantities, such as the energy dissipation rates, with different lengths $\ell_n = \ell_0 \delta^{-n}$ $(n = 0, 1, 2, \cdots)$ which are the diameters of the regions where microscopic energy dissipation rates are coarse-grained. The value of the magnification δ (> 1) is chosen freely by observers. In the next place, we create the PDFs from partial regions of the size 512^3 which are obtained by cutting the whole of 4096^3 DNS region into 512 pieces, and analyze each of them with the help of the center part or the rail part of appropriate whl-PDF. We will refer to the PDF created from the partial region as *p-PDF* in the following. Among 512 partial DNS regions, we select in this paper two DNS regions, i.e., one has the maximum enstrophy and the other the minimum enstrophy (see Fig. 1). We call the p-PDF created from the partial DNS region with maximum (minimum) enstrophy *max-PDF* (*min-PDF*).

In [6], we analyzed PDFs for energy dissipation rates for 4096^3 DNS turbulence given in Fig. 11 of [1]. However, the number of data points used in drawing the PDFs of the figure

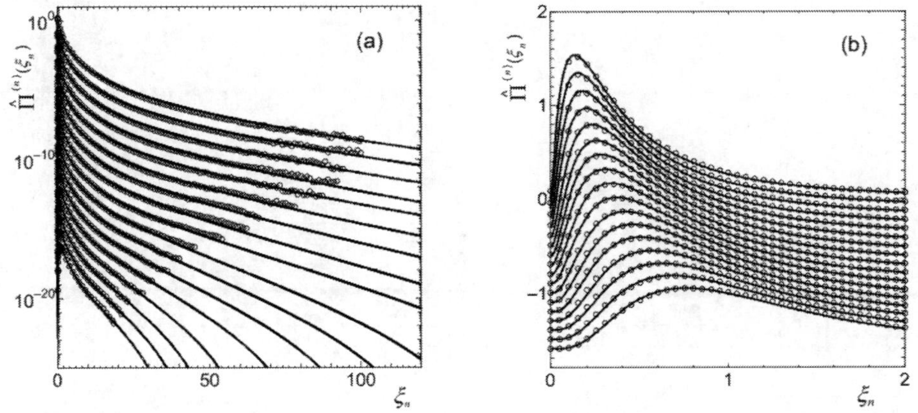

Figure 2. Whl-PDFs for the normalized energy dissipation rates $\xi_n = \varepsilon_n/\langle\langle\varepsilon_n^2\rangle\rangle_c^{1/2}$. The average $\langle\langle\cdots\rangle\rangle$ is taken with $\Pi^{(n)}(\varepsilon_n)$. Open circles represent the whl-PDFs extracted out from the whole of 4096^3 DNS region. For better visibility, the PDFs in each figure are shifted along the vertical axes by the amount (a) -1 in log-scale and (b) 0.1 in linear-scale. They are displayed according to the values of $r/\eta = \ell_n/\eta = 27.5$, 32.8, 38.9, 46.3, 55.1, 65.6, 78.0, 92.7, 110, 131, 156, 186, 221, 264, 314, 374 and 442 successively in the sequence that the PDF with the smallest value of r/η is placed at the top and the one with the largest value at the bottom. Solid lines represent the curves given by theoretical PDFs of MPDFT with the intermittency exponent $\mu = 0.345$.

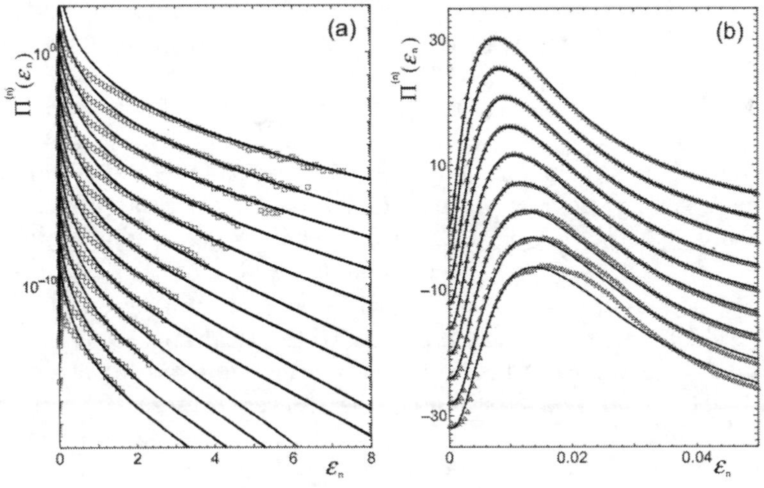

Figure 3. (a) Max-PDFs ($\square$) and (b) min-PDFs ($\triangle$). The PDFs in each figure are shifted along the vertical axes by the amount (a) -1 and (b) -4 according to the values (a) r/η = 27.5, 32.8, 38.9, 46.3, 55.1, 65.6, 78.0, 92.7, 110, 131, 156, 186 and 221, (b) r/η = 27.5, 32.8, 38.9, 46.3, 55.1, 65.6, 78.0, 92.7 and 110, successively, from the top to the bottom. Solid lines are the whl-PDFs cooked someway (see the main text).

turned out to be not enough for the more precise analyses of the central part of the PDFs under the above motivation. Therefore, making use of the the snapshot *raw* data of the 4096^3 DNS [1], we have created PDFs for energy dissipation rates with higher enough resolution even for their central parts. In Fig. 2, thus created whl-PDFs for $\delta = 2^{1/4}$ (open circles) from the whole of 4096^3 DNS region are analyzed in a high precision with the theoretical PDF within MPDFT (solid lines).

In Fig. 3 (a) and (b), the max-PDFs ($\square$) and the min-PDFs ($\triangle$) of energy dissipation rates for $\delta = 2^{1/4}$ are given, respectively. It was revealed from Fig. 3 (a) that the tail-part of max-PDFs can be adjusted with the slope of the tail-part of whl-PDF with a specific value of r/η, magnified vertically by some factor χ, and that the value of $\varepsilon_n^\dagger$ at which the max-PDF ($\square$) and

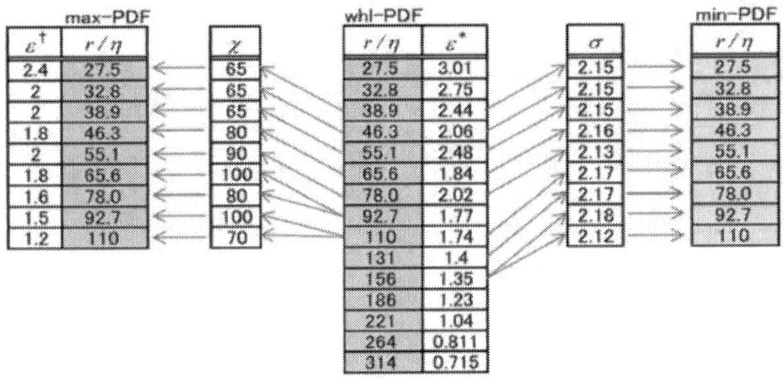

Figure 4. Recipe to analyze the max- and min-PDFs with the whl-PDFs. The whl-PDF with a specific r/η at the origin of an arrow can adjust the max- or min-PDF at the point of the arrow by the magnification factor χ and by the scaling factor σ. $\varepsilon^{\dagger}$ is the point from which a max-PDF and a whl-PDF multiplied by χ start to merge. The values of the connection points ε^{*} for theoretical whl-PDFs are listed.

the corresponding whl-PDF (solid line) start to merge is quite close to the connection point ε_n^{*} associated with the whl-PDF (see also Fig. 4). It was found from Fig. 3 (b) that the center-part of min-PDFs can be adjusted by the scaled PDF, $\breve{\Pi}^{(n)}(\breve{\varepsilon}_n)$, which is introduced through $\breve{\Pi}^{(n)}(\breve{\varepsilon}_n)\, d\breve{\varepsilon}_n = \Pi^{(n)}(\varepsilon_n)\, d\varepsilon_n$, $\breve{\varepsilon}_n = \varepsilon_n/\sigma$ with an appropriate scaling factor σ (see Fig. 4).

In Fig. 4, we listed the values of the magnification χ which is necessary to analyze the max-PDFs in Fig. 3 (a), and of the scaling factor σ which is necessary for the scaled PDF $\breve{\Pi}^{(n)}(\breve{\varepsilon}_n)$ to analyze the min-PDFs in Fig. 3 (b). The arrows indicate that which whl-PDF with a specific value r/η at the origin of an arrow can adjust the max-PDF or the min-PDF at the point of the arrow. In between arrows, there inserted the magnification factor χ for the case of max-PDF or the scale factor σ for min-PDF.

Summarizing the above investigation, we have the following outcomes. 1) We can find a whl-PDF whose tail part can adjust the slope of the tail-part of a max-PDF with appropriate magnification factor χ. 2) The center part of the min-PDFs can be adjusted quite accurately by the whl-PDFs scaled with the factor σ which has almost common value around 2.15. 3) The value of $\varepsilon_n^{\dagger}$ at which the tail part of the whl-PDF multiplied by χ starts to merge the tail part of the max-PDF coincides well with the connection point ε_n^{*} for the theoretical whl-PDF. These findings support the assumption of the existence of the two elements in turbulent flow, i.e., coherent and incoherent fluid motions, and of the division of the PDF within MPDFT into two parts, i.e., the tail part and the center part, is reasonable. We conjecture that the above notion about the characteristics of turbulence may provide us with a strong conceptual tools for the separate visualization of the two elements constituting fully developed turbulence. The study in this direction is now in progress and will be reported elsewhere in the near future.

References

[1] Aoyama T, Ishihara T, Kaneda Y, Yokokawa M, Itakura K, Uno A 2005 *J. Phys. Soc. Jpn.* **74** 3202
[2] Arimitsu T and Arimitsu N 2000 *Phys. Rev.* E **61** 3237
[3] Arimitsu T and Arimitsu N 2000 *J. Phys. A: Math. Gen.* **33** L235 [CORRIGENDUM: 2001 *ibid* **34** 673]
[4] Arimitsu N and Arimitsu T 2002 *J. Korean Phys. Soc.* **40** 1032
[5] Arimitsu T and Arimitsu N 2002 *Physica* A **305** 218
[6] Arimitsu N and Arimitsu T 2011 *Physica* A **390** 161 [CORRIGENDUM: 2012 *ibid* **391** 1908)]
[7] Parisi G and Frisch U 1985 *Turbulence and predictability in geophysical fluid dynamics and climate dynamics* ed Ghil M, Benzi R and Parisi G (New York: North-Holland) p 84
[8] Tsallis C 1988 *J. Stat. Phys.* **52** 479

Prediction of atomic arrangement of Pt-Cu nanoalloy by genetic algorithm

J S Oh[1], H-S Nam[2], J-H Choi[1] and S-C Lee[1]

[1] Future Convergence Technology Research Division, Korea Institute of Science and Technology, Seoul, 136-791, Korea
[2] Center for Materials and Processes of Self-Assembly and School of Advanced Materials Engineering, Kookmin University, Seoul, 136-702, Korea

E-mail: leesc@kist.re.kr

Abstract. The prediction of atomic arrangements for a nanoalloy is of critical importance in developing a novel catalyst for clean energy applications because these arrangements determine the catalytic activities of a nanoalloy. So, the development of a reliable method for predicting atomic arrangements has become increasingly important. In this study, the atomic arrangements of the Pt-Cu in a truncated octahedron (TOh) shaped nanoalloy of various Pt-Cu compositions were predicted. We developed a genetic algorithm (GA) code to predict atomic arrangements, and this code was combined with classical molecular dynamics (MD) simulations to evaluate the optimality of each atomic arrangement. The GA calculation predicted a predominantly multilayered core-shell structure for the Pt-Cu nanoalloy, regardless of the Pt-Cu composition ratio. The cause of a multilayered core-shell structure in the Pt-Cu nanoalloy can be interpreted by the interaction between the intrinsic properties of Pt and Cu, such as their surface energies and miscibility.

1. Introduction

It has been widely accepted that platinum (Pt) is the best catalyst in many practical applications such as fuel cells, lithium-based batteries and etc. However, the limited amount of Pt reserves and the high cost of Pt have demanded the search for other materials having comparable catalytic activity of Pt. Since Pt cannot be completely replaced in current applications, the alloying of Pt with other metals has been widely studied. Reducing the Pt load and at the same time, maximizing the catalytic activity, Pt alloys have been synthesized as nanoparticles with sizes ranging from 3 to 5 nanometers in diameter. From these sizes, high surface-to-volume ratio can be achieved. Considering that the catalytic activity of a nanoparticle depends critically on the atomic configurations of its constituent metals, it is necessary to understand the atomic configurations of nanoparticles. Unfortunately, neither experimental nor single computational method is effective for observing the atomic configurations of nanoparticles with those sizes.

In this paper, we have approached the prediction of atomic arrangement in a fixed shape from the point of an optimization problem based on the genetic algorithm (GA). The atomic arrangement with the lowest total energy is regarded as the optimal arrangement, which is searched by an optimization method. For such a search, GA was adopted for its outstanding searching ability and generality [1]. In order to search for, or predict, the optimal atom arrangement, the 'pyGA' code written in Python language [2] was developed. It was then combined with the molecular dynamics (MD) code XMD [3] and the newly developed embedded atom method (EAM) potential to estimate the total energy of the designated atomic arrangement. With the pyGA, we successfully predicted the atomic arrangements of Pt-Cu nanoalloy in the truncated octahedron (TOh) shape with 1,654 atoms and observed their distinctive structural features in the predicted atomic arrangements regardless of the Pt-Cu composition.

2. Structural Features of Pt-Cu nanoalloy

The TOh has been known to be one of the most stable structures observed for an Face-Centered Cubic (FCC) based nanoalloy [4], and the FCC stacking sequence is preferred for nanoalloys larger than 3 nm in diameter [5,6]. An isotropic TOh nanoalloy with 1,654 atoms was constructed, which is approximately 3.5 nm in diameter [7].

Finding the optimal atomic arrangement of an A-B binary nanoalloy of a given TOh structure corresponds to finding an arrangement having the lowest total energy in the solution space, which contains all feasible atomic arrangements. The solution space size is defined as the number of feasible atomic arrangements of a:b composition for the binary nanoalloy, without consideration of the atomic symmetry in the nanoalloy . It is estimated based on the formula $N!/(A!B!)$ [8], where $A+B=N$, A, B are the numbers of the corresponding atomic species in the compositional fraction a, b and N is the total number of atoms in the nanoalloy. The solution space size of various compositions with 1,654 atoms are tabulated in the Table 1.

Table 1. The solution space size, the number of searching to reach an atomic arrangement having the lowest total energy and the lowest energy itself at each composition. The prediction for the 1:1 composition was obtained from Ref [7].

Pt:Cu	Solution space size	Number of searching	Total energy [eV]
1:3	1.13×10^{402}	1.4×10^{6}	-6571.34
2:3	4.32×10^{482}	1.6×10^{6}	-7156.00
1:1	1.57×10^{497}	2.0×10^{6}	-7502.38
3:2	4.32×10^{482}	1.3×10^{6}	-7835.75
3:1	1.13×10^{402}	1.5×10^{6}	-8323.15

3. Calculation Methods

3.1. Genetic algorithm for the searching method

The GA is a very well known searching and optimization method based on the natural selection and survival of the fittest [9]. It mimics the biological evolution in the natural world. Following the basic concept of GA, pyGA was developed to predict the atomic arrangement in a bimetallic nanoalloy [7]. In pyGA, the total energy represented the optimality of an atomic arrangement in the solution space.

The input parameters of the pyGA code used in this study were as follows. The population size was 100 and the initial population was generated at random. The 'Roulette Wheel' selection operator and Partially Matched Crossover (PMX) [9] crossover operator were chosen. The probability of mutation was 0.0015. In order to pass down the dominant properties to the next generation, elitism was also implemented in pyGA. The stopping criterion was also prescribed like other input parameters. When a value of the lowest total energy had not changed in 1,000 consecutive generations, then pyGA terminated the evolution and regarded the atomic arrangement having the lowest total energy as the optimal one.

3.2. Molecular dynamics for the total energy calculation

For calculating the total energy of an atomic arrangement in the searching procedure, the MD simulation program XMD with the EAM potential was adopted. In order to include the effect of atomic position relaxation to yield a more realistic result, the conjugate gradient method was additionally implemented in the XMD. The analytic formula of EAM was originally proposed by A. Voter [10], but some parameters describing the inter-atomic interactions and thermodynamic properties were adjusted for this prediction [11]. The crucial properties of the nanoalloy, such as the surface energy, cohesive energy, miscibility, which is the heat of mixing, and the melting temperature were well reproduced by using the newly developed EAM potentials.

4. Results and Discussion

The search performances for the prediction at each composition were almost equivalent in spite of the variations in the composition. In the pyGA calculation, 10^6 atomic arrangements were commonly searched to find the lowest energy configuration, which quantitatively equals 1.2 ~ 1.5 % of the solution space. This pyGA prediction required smaller number of searching than the Monte Carlo simulation [11]. Table 1 lists the number of searching to predict an atomic arrangement having the lowest total energy predicted from pyGA. The corresponding atomic arrangements are figured in Fig 1.

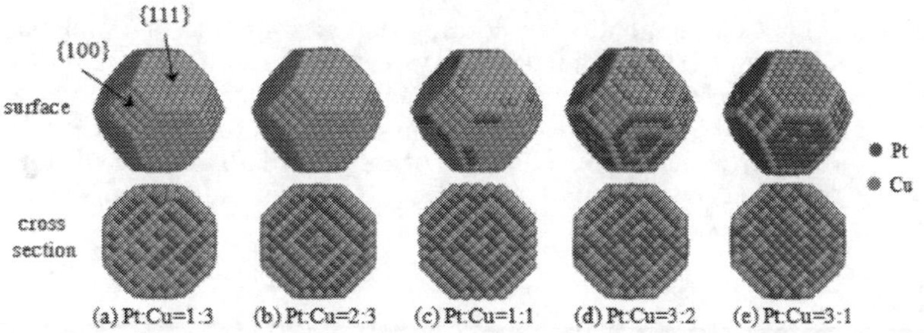

Figure 1. Predicted atomic arrangements in the Pt-Cu nanoalloy. (dark/blue : Pt, light/red : Cu) The first row shows the surface configurations, and the second row shows the cross sectional arrangements along the <010> direction.

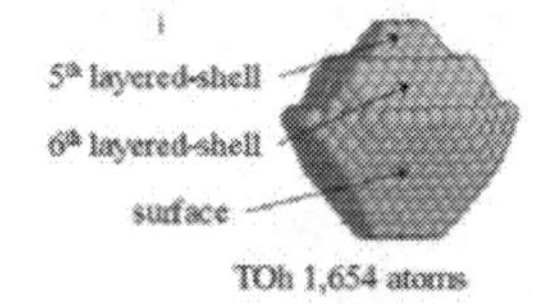

Figure 2. The layered-shell structure in the TOh with 1,654 atoms.

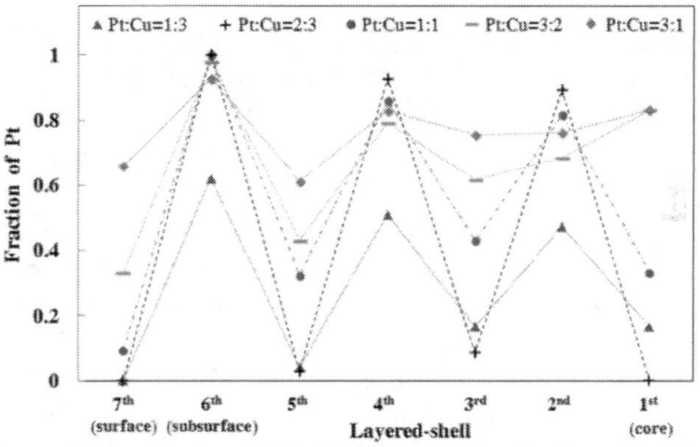

Figure 3. Prediction of Pt fraction in the Pt-Cu nanoalloy according to layered-shell. The fraction of Pt show a zigzag pattern, which starts from the surface to the core.

Fig. 1 shows the predicted ones from the pyGA code according to composition. The upper and lower rows are snapshots of the surface and cross-sectional views, respectively. The proportion of atom species on the surface changed according to composition. At the composition ratio of Pt:Cu = 3:1, only Cu species existed all along the edges of the surface. At the composition ratio of 1:1, entire Cu {100} facets and Cu-rich {111} facets were predicted. However, at the composition ratios of 1:3 and 2:3 = Pt:Cu, no Pt atoms existed on the {100} facets or on the {111} facets. This indicated that on the surface, Cu occupied firstly on the edges, then on the {100} and {111} facets with increasing composition of Cu. This phenomenon can be interpreted by the difference of the surface energy as described in the EAM potential [11]. The relations between the surface energies of Pt and Cu is as follows: Pt{100} > Cu{100}, Cu{100}/Pt{100} = 0.76 and Pt{111} > Cu{111}, Cu{111}/Pt{111} = 0.80. Due to the smallest coordination number on the edges in the FCC shape, the atoms on the edges are greatly influenced by the surface energy. Therefore, preoccupancy of Cu on the edges is explained by the lower surface energies of both Cu {100} and {111} facets. Consequently, the Cu species is regarded as the dominant atom species on the overall surface.

For the further analysis, TOh with 1,654 atoms was divided into 7 layered-shells like an onion, as partially shown in the Fig. 2. The fraction of Pt in each layered-shell was calculated and drawn in a graphical form in Fig. 3. The fraction of Pt drastically oscillated along the x-axis in all composition ratios. Analyzing the oscillation tendency from Fig. 3, it was concluded that the Pt-Cu nanoalloy had a remarkable structural feature that of an onion-like multilayered core-shell with a Cu dominant surface for all composition ratios. But closer to the core, the onion-like multilayered core-shell structure lost its form. Considering the relatively smaller number of atoms on the inner shells than on the surface or subsurface, the structural tendency appeared to be stronger on the outer layered-shells than on the inner shells.

5. Conclusions

Atomic arrangements of Pt-Cu nanoalloy according to various Pt:Cu composition ratios were predicted by using the GA and MD combined calculations. We developed the pyGA code written in Python language. The combined simulation predicted an onion-like multilayered core-shell structure of TOh Pt-Cu nanoalloy with 1,654 atoms, regardless of the composition ratio. This core-shell structure resulted from the surface energies and heat of mixing of Pt and Cu. In further studies, bulk alloy structure will be compared with nanoalloy structure to obtain more details and to understand the relations between size and morphology.

Ackknowledgments

This work was partly supported by Nano R&D program (2012-0006197) through the National Research Foundation (NRF) of Korea funded by the Ministry of Education, Science and Technology (MEST) of Korea, and partly supported by Future Research program (2E22731) funded by Korea Institute of Science and Technology (KIST). H.-S Nam also acknowledges the support of the ERC program of MEST/KOSEF, Grant No. R11-2005-048-00000-0 and the Priority Research Centers Program (2009-0093814) through the NRF of Korea funded by the MEST of Korea.

References

[1] Johnston R L 2003 *Dalton Trans.* **2003** 4193-207
[2] Python Software Foundation, Python Programming Language 1990 http://www.python.org
[3] xmd-Molecular Dynamics for Metals and Ceramics 2004 http://xmd.sourceforge.net
[4] Kuo C-L and Clancy P 2005 *J. Phys. Chem.* B **113** 13743-54
[5] Nakin A V and Khmelevskaya V S 2006 *Crystallogr. Rep.* **51** 322-328
[6] Baletto F, Ferrando R, Fortunelli A, Montalenti F and Mottet C 2002 *J. Chem. Phys.* **116** 3856–62
[7] Oh J S, Nam H-S, Choi J-H and Lee S-C 2012 *Metal. Mater. Int.* in press
[8] Ferrando R, Fortunelli A and Johnston R L 2008 *Phys. Chem. Chem. Phys.* **10** 640-9
[9] Moon B-R 2005 *Genetic algorithm*, 2nd *ed.* (Seoul: Doo Yang Sa) pp 44-50 (in Korean)
[10] Voter A F 1993 Los Alamos National Laboratory Report LA-UR-93-3901
[11] Yun K, Cho Y-H, Cha P-R, Lee J, Nam H-S, Oh J S, Choi J-H and S-C Lee 2012 *Acta* Materialia **60** 4908-16

A numerical procedure to model and monitor CO_2 sequestration in aquifers

Juan E. Santos[1], **Gabriela B. Savioli**[2], **José M. Carcione**[3], **and D. Gei**[3]

[1] CONICET, Inst. del Gas y del Petróleo, Fac. Ingeniería, Universidad de Buenos Aires, Av. Las Heras 2214 Piso 3, Buenos Aires, Argentina (C1127AAR) and Universidad Nacional de La Plata and Department of Mathematics, Purdue University, U.S.A (santos@math.purdue.edu)
[2] Inst. del Gas y del Petróleo, Fac. Ingeniería, Universidad de Buenos Aires, Argentina (gbsavioli@yahoo.com.ar)
[3] Istituto Nazionale di Oceanografia e di Geofisica Sperimentale (OGS), Borgo Grotta Gigante 42c, 34010 Sgonico, Trieste, Italy (jcarcione@inogs.it)

Abstract. Carbon Dioxide (CO_2) sequestration into geologic formations is a means of mitigating greenhouse effect. In this work we present a new numerical simulation technique to model and monitor CO_2 sequestration in aquifers. For that purpose we integrate numerical simulators of CO_2-brine flow and seismic wave propagation (time-lapse seismics). The simultaneous flow of brine and CO_2 is modeled applying the Black-Oil formulation for two phase flow in porous media, which uses the Pressure-Volume-Temperature (PVT) behavior as a simplified thermodynamic model. Seismic wave propagation uses a simulator based on a space-frequency domain formulation of the viscoelastic wave equation. In this formulation, the complex and frequency dependent coefficients represent the attenuation and dispersion effect suffered by seismic waves travelling in fluid-saturated heterogeneous porous formations. The spatial discretization is achieved employing a nonconforming finite element space to represent the displacement vector. Numerical examples of CO_2 injection and time-lapse seismics in the Utsira formation at the Sleipner field are analyzed. The Utsira formation is represented using a new petrophysical model that allows a realistic inclusion of shale seals and fractures. The results of the simulations show the capability of the proposed methodology to monitor the spatial distribution of CO_2 after injection.

1. Introduction

Storage of CO_2 in geological formations is a procedure employed to reduce the amount of greenhouse gases in the atmosphere to slow down global warming [1].

Geologic sequestration involves injecting CO_2 into a target geologic formation at depths typically greater than 1000 m where pressure and temperature are above the critical point for CO_2 (31.6C, 7.38 MPa).

The CO_2 injection operation at the Sleipner gas field in the North Sea is the world first industrial scale CO_2 injection project [1]-[2]. CO_2 separated from natural gas produced at Sleipner is injected into the Utsira Sand, a saline aquifer some 26000 km^2 in area. Injection started in 1996 and is planned to continue for about twenty years, at a rate of about one million tonnes per year.

Time-lapse seismic surveys aim to monitor the migration and dispersal of the CO_2 plume. Recent papers [3]-[4] applied seismic modeling using synthetic generated CO_2 saturation fields. In this work we combine numerical simulations of CO_2 injection and seismic modeling using a viscoelastic model that takes into account attenuation effects due to the presence of heterogeneities in the solid and fluid properties.

2. The Black-Oil formulation of two-phase flow in porous media

The simultaneous flow of brine and CO_2 is described by the well-known Black-Oil formulation applied to two-phase, two component fluid flow [5]. In this model, CO_2 may dissolve in the brine but the brine is not allowed to vaporize into the CO_2 phase. This formulation uses as a simplified thermodynamic model, the PVT data: CO_2 solubility in brine (R_s) and CO_2 and brine formation volume factors (B_{CO2}, B_b). They are determined using the Hassanzadeh's correlations [7]. The nonlinear system of partial differential equation is,

$$\nabla \cdot (\underline{k}(\frac{k_{rCO2}}{B_{CO2}\mu_{CO2}}(\nabla p_{CO2} - \rho_{CO2} g \nabla D) + \frac{R_s k_{rb}}{B_b \mu_b}(\nabla p_b - \rho_b g \nabla D))) + q_{CO2} = \frac{\partial \left[\phi \left(\frac{S_{CO2}}{B_{CO2}} + \frac{R_s S_b}{B_b} \right) \right]}{\partial t}, \tag{1}$$

$$\nabla \cdot (\underline{k}(\frac{k_{rb}}{B_b \mu_b}(\nabla p_b - \rho_b g \nabla D)) + q_b = \frac{\partial \left[\phi (\frac{S_b}{B_b}) \right]}{\partial t}. \tag{2}$$

The unknowns are the fluid pressures p_β and saturations S_β for the β-phases, with $\beta = CO2, b$. The parameters k and ϕ are the absolute permeability and porosity. Also, the funcions $k_{r\beta}$, μ_β and ρ_β are the relative permeability, viscosity, and density of the β-phase, respectively.

Besides, phase saturations add to one and phase pressures are related by the capillary pressure (P_C) relation $p_{CO2} - p_b = P_C(S_b)$.

The solution of the Black-Oil fluid-flow model was obtained employing the BOAST simulator [8], which solves the flow equations with the IMPES finite difference technique [9].

3. A VISCOELASTIC MODEL FOR WAVE PROPAGATION

The propagation of waves is described using a viscoelastic model that takes into account the dispersion and attenuation effects due to the presence of heterogeneities in the fluid and solid phase properties.

The equation of motion in a 2D isotropic viscoelastic domain Ω with boundary $\partial\Omega$ can be stated in the space-frequency domain as

$$-\omega^2 \rho u - \nabla \cdot \sigma(u) = f(x, \omega), \quad \Omega \tag{3}$$

$$-\sigma(u)\nu = i\omega \mathcal{D} u, \quad \partial\Omega, \tag{4}$$

where $u = (u_x, u_y)$ is the displacement vector. Here ρ is the bulk density and (4) is a first-order absorbing boundary condition using the positive definite matrix $\mathcal{D}$.

The stress tensor $\sigma(u)$ is defined in the space-frequency domain by

$$\sigma_{jk}(u) = \lambda_G(\omega) \nabla \cdot u \delta_{jk} + 2\mu_m(\omega)\varepsilon_{jk}(u), \tag{5}$$

where $\sigma_{jk}(u)$ and $\varepsilon_{jk}(u)$ are the stress and strain tensors, λ_G and μ_m are the complex and frequency dependent Lamé parameters. The Lamé coefficients, computed using White's model [6] for patchy saturation, take into account attenuation and dispersion of waves due to the presence of CO_2 after injection.

The solution of (3)-(4) was obtained using an iterative finite element domain decomposition procedure employing a nonconforming finite element space, since it generates less numerical dispersion than the standard bilinear elements [3] .

4. Model of the Utsira Formation and CO_2 Injection.

We consider a 2D model of the Sleipner field constructed using an initial porosity at hydrostatic pressure and the clay content of the formation. The model has 400 m thickness (top at 700 m and bottom 1100 m b.s.l.). Within the formation, there are several mudstone layers which act as barriers to the vertical motion of the CO_2 and can be observed in Figure 1 (left). The viscosity, density and bulk modulus of CO_2 were obtained from the Peng-Robinson equations as a function of temperature and pore pressure [7].

CO_2 is injected at a constant flow rate of one millon tons per year. The injection point is located at the bottom of the Utsira formation: $x = 400$ m, $z = 1060$ m. Figure 1 (right) shows the saturation field after 2 years of CO_2 injection computed using the BOAST flow simulator.

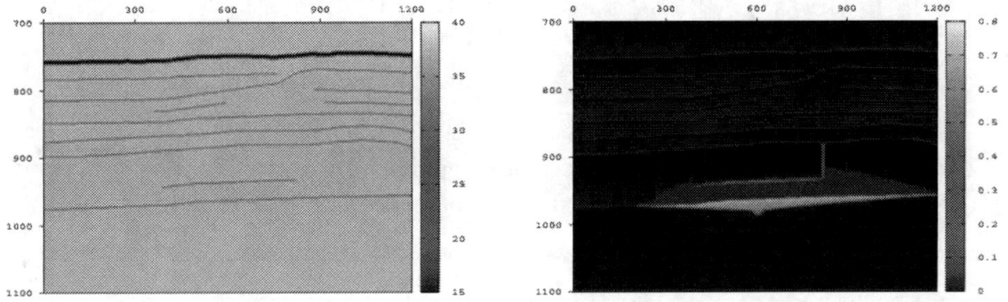

Figure 1. Porosity map (left) and CO_2 saturation distribution (rigth) after 3 years of injection. The injection point is located at x= 600 m, z= 1060 m

5. Time-Lapse Seismic Monitoring

To analyze the capability of seismic monitoring to identify zones of CO_2 accumulation, the media is excited with a compressional point source located at $x = 400$ m, $z = 710$ m before and after 2 years of CO_2 injection. Time histories measured near the surface are shown in Figures 2 before CO_2 injection (left) and after 2 years of CO_2 injection (right). The upper reflection in both figures is due to the direct wave coming from the point source. The other reflection in (b) is due to the CO_2 accumulations below the deepest mudstone layer. Figure 3 displays the traces measured at x= 750 m, z = 710 m shown in Figure 2.

6. Conclusions

In this work we presented a new numerical methodology to model and monitor CO_2 sequestration. For that purpose we integrated numerical simulators of CO_2-brine flow and seismic wave propagation. The numerical experiment shows the capability of seismic monitoring to identify the spatial distribution of CO_2 after injection. Therefore, this approach constitute a valuable tool to analyze storage integrity, provide early warning should any leakage occur, and monitor the migration and dispersal of the CO_2 plume.

7. References

[1] Arts R, Chadwick A, Eiken O, Thibeau S and Nooner S, 2008 Ten years of experience of monitoring CO_2 injection in the Utsira Sand at Sleipner, offshore Norway *First break* **26** pp 65-72

[2] Chadwick A, Arts R and Eiken O 2005 4D seismic quantification of a growing CO_2 plume at Sleipner, North Sea. In: Dore A G and Vincent B (Eds) *Petroleum Geology: North West Europe and Global Perspectives - Proc. 6th Petroleum Geology Conference* (Geological Society, London) pp 1385-99

[3] Santos J, Rubino J and Ravazzoli C 2008 Modeling mesoscopic attenuation in a highly heterogeneous Biot's medium employing an equivalent viscoelastic model *Proc. 78th Annual International Meeting SEG (Las Vegas)* pp 2212-15

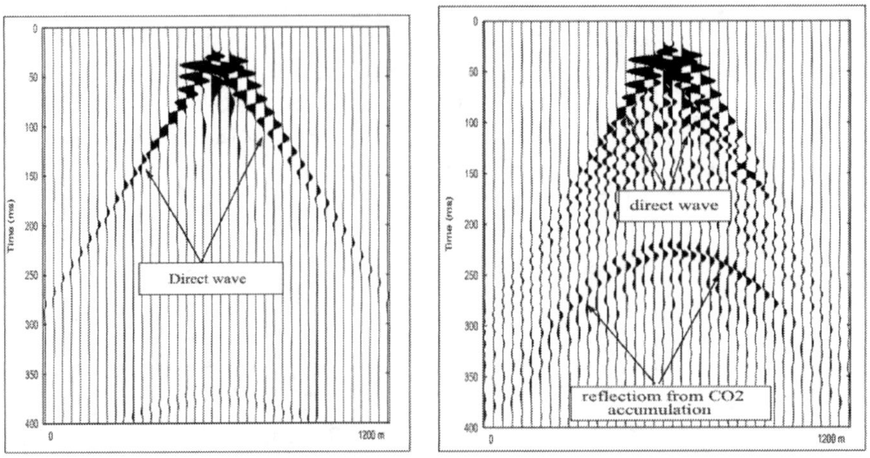

Figure 2. Traces of the z-component of the displacement before (left) and after (right) 2 years of CO_2 injection. The second reflection is due to the deepest CO_2 accumulation .

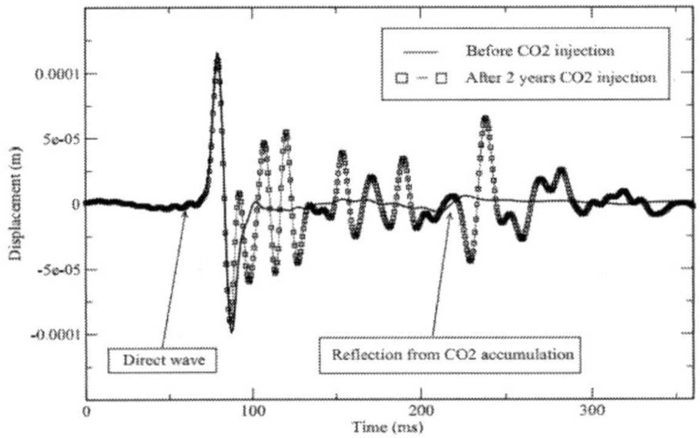

Figure 3. Traces of particle velocity of the solid phase before and after 2 years of CO_2 injection.

[4] Carcione J, Picotti S, Gei D, and Rossi G 2006 Physics and Seismic Modeling for Monitoring CO_2 storage *Pure and Applied Geophysics* **163** pp 175-207
[5] Aziz K and Settari A 1985 *Petroleum Reservoir Simulation* (Great Britain: Elsevier Applied Science Publishers)
[6] White J.E., Mikhaylova N G, Lyakhovitskiy F M 1975 Low-frequency seismic waves in fluid-saturated layered rocks *Izvestija Academy of Siences USSR, Physics of Solid Earth* **10** pp 654-659
[7] Hassanzadeh H, Pooladi-Darvish M, Elsharkawy A, Keith D and Leonenko Y 2008 Predicting PVT data for CO_2-brine mixtures for black-oil simulation of CO_2 geological storage *International Journal of Greenhouse Gas Control* **2** pp 65-77
[8] Fanchi J 1997 *Principles of Applied Reservoir Simulation* (Houston, Texas: Gulf Professional Publishing Company)
[9] Savioli G and Bidner M S 2005 Simulation of the oil and gas flow toward a well - A stability analysis *Journal of Petroleum Science and Engineering* **48** pp 53-69

Journal of Physics: Conference Series **410** (2013) 012086 doi:10.1088/1742-6596/410/1/012086

Modelling of strongly coupled particle growth and aggregation

F Gruy and E Touboul

Ecole Nationale Supérieure des Mines, 158 Cours Fauriel, 42023 Saint-Etienne France

gruy@emse.fr

Abstract. The mathematical modelling of the dynamics of particle suspension is based on the population balance equation (PBE). PBE is an integro-differential equation for the population density that is a function of time t, space coordinates and internal parameters. Usually, the particle is characterized by a unique parameter, e.g. the matter volume v. PBE consists of several terms: for instance, the growth rate and the aggregation rate. So, the growth rate is a function of v and t. In classical modelling, the growth and the aggregation are independently considered, i.e. they are not coupled. However, current applications occur where the growth and the aggregation are coupled, i.e. the change of the particle volume with time is depending on its initial value v_0, that in turn is related to an aggregation event. As a consequence, the dynamics of the suspension does not obey the classical Von Smoluchowski equation. This paper revisits this problem by proposing a new modelling by using a bivariate PBE (with two internal variables: v and v_0) and by solving the PBE by means of a numerical method and Monte Carlo simulations. This is applied to a physicochemical system with a simple growth law and a constant aggregation kernel.

1. Introduction

Particle or crystal synthesis involves several stages: nucleation, growth and aggregation. It is of great importance to model these phenomena in order to manage the dynamics of the particle formation.
At a given time one considers a population of particle. Each particle is characterized by its radius or its matter volume. The population can evolve by growth of particle, i.e. by mass transfer from the surrounding medium to the particle and by aggregation or coalescence, i.e collision of particles possibly followed by merging. Generally it is assumed that the growth law of the particle is a function of the radius or the matter volume:

$$\frac{dv}{dt} = G(v) \tag{1}$$

The coalescence is a particular case of aggregation phenomena. So, a particle with a volume v is coming from the collision between a particle with a volume $\bar{v}$ and another one with a volume $v - \bar{v}$. Let us introduce the population density $n(t, v)$. Its change with time obeys the Von Smoluchowski equation [1]:

$$\frac{\partial n}{\partial t} = \frac{1}{2}\int_0^v K(\bar{v}, v - \bar{v})n(\bar{v})n(v - \bar{v})d\bar{v} - n(v)\int_0^\infty K(v, \bar{v})n(\bar{v})d\bar{v} \tag{2}$$

$K\left(\bar{v}, v-\bar{v}\right)$ is a kinetic constant or kernel that is a function of the volume of the colliding particles. If one takes into account at once the growth and the coalescence, the equation (2) becomes:

$$\frac{\partial n}{\partial t}+\frac{\partial\left(G(v)n\right)}{\partial v}=\frac{1}{2}\int_0^v K\left(\bar{v}, v-\bar{v}\right)n\left(\bar{v}\right)n\left(v-\bar{v}\right)d\bar{v}-n(v)\int_0^\infty K\left(v,\bar{v}\right)n\left(\bar{v}\right)d\bar{v} \qquad (3)$$

It has to be underlined that the growth rate only depends on the particle volume at the given time. There is no memory effect. Equation (3) assumes that the growth and aggregation are independent phenomena. However there are several cases where the growth rate at a given time explicitly depends on events that have previously took place. So, the growth law is allowed to depend on the actual volume v and the initial volume v_0 of the particle.

$$\frac{dv}{dt}=G\left(v,v_0\right) \qquad (4a)$$

If the growth is linked to the coalescence, then the growth rate will change at each coalescence event, i.e. at the birth of a new particle. As a consequence, the dynamics of the particle population does not obey the equation (3). For instance, if the colliding particles are facetted crystals, the resulting crystal has a surface area smaller than the sum of the initial crystals surface areas because some crystalline face is shared as a contact area. In this case, the growth law obeys the expression:

$$\frac{dv}{dt}=B\left(A-\alpha A_0\right)=C\left(v^{2/3}\quad \alpha v_0^{2/3}\right) \qquad (4b)$$

B and C are functions of the physicochemical parameters (solute concentrations, temperature ...) depending on the intrinsic growth mechanism. α is a dimensionless constant within the [0; 1] range. Higher the α-value stronger is the effect of the face joint.
The second section of the paper is dedicated to the modelling of such coupled growth and aggregation. The third section describes a numerical method to solve the corresponding population balance equation. The fourth one presents a comparison of the results coming from this numerical method and Monte Carlo simulations.

2. Modelling

In order to take into account the new form of the equation (4a) one considers two internal variables v and v_0 instead of the unique v. Then one defines the two population densities: $N\left(t,v,v_0\right)$ and $n(t,v)$ with:

$$n(t,v)=\int_0^v N\left(t,v,v_0\right)dv_0 \qquad (5)$$

and

$$N\left(t,v<v_0,v_0\right)=0 \qquad (6)$$

One introduces:

$$J_{v0}=\frac{1}{2}\int_0^{v0} K\left(\bar{v}, v_0-\bar{v}\right)n\left(t,\bar{v}\right)n\left(t,v_0-\bar{v}\right)d\bar{v} \qquad (7)$$

The population balance equation is written as:

$$\frac{\partial N\left(t,v,v_0\right)}{\partial t}+\frac{\partial G\left(v,v_0\right)N}{\partial v}=\delta\left(v-v_0\right)J_{v0}-N\left(t,v,v_0\right)\int_0^\infty K\left(v,\bar{v}\right)n\left(t,\bar{v}\right)d\bar{v} \qquad v\geq v_0 \qquad (8)$$

In order to avoid the use of the Dirac function in the equation (8), this one is transformed into [2]:

$$\frac{\partial N\left(t,v,v_0\right)}{\partial t}+\frac{\partial G\left(v,v_0\right)N}{\partial v}=-N\left(t,v,v_0\right)\int_0^\infty K\left(v,\bar{v}\right)n\left(t,\bar{v}\right)d\bar{v} \qquad v \quad v_0 \qquad (9)$$

With the boundary condition:

$$G(v_0, v_0)N(t, v_0, v_0) = J_{v0} \tag{10}$$

If the growth law G does not depend on v_0, one can check that the integration of equation (8) between v_0 =0 and *v(-)* leads to the equation (3).

3. Numerical method

Among several numerical methods [3] a finite difference method has been selected and fitted. The finite difference scheme is based on upwind derivative for the convection term, and simple rectangle method for the evaluation of the integrals. The hyperbolic integro-differential equation holds on a triangular two dimensional space ($v \geq v_0$). The flux $G(v, v_0)N(v, v_0)$is directed in the v direction. The coupling in the v_0 direction occurs through the source term, and the boundary conditions on the line $v = v_0$. The numerical mesh (figure 1) covers a triangular domain, half of a square: $(v, v_0) \in [0, v_{max}] \times [v_0, v_{max}]$ divided into squares of side length δv, indexed by $(i, j) \in (1...p) \times (1...p)$. $N_{i,j}^k \approx N(t_k, v_i, v_{0j})$ is the discrete population density, and thus $n_i^k = \sum_{j=1}^{p} N_{ij}^k$. $G_{i,j} \approx G(v_i, v_{0j})$is the discrete growth function, $K_{i,l} \approx K(v_i, v_l)$ is the discrete kinetic constant. The time step and volume step are noted δt and $\delta v (= \delta v_0)$.

The numerical scheme based on equation (8) can be written as:

$$N_{i,j}^{k+1} = N_{ij}^k + \frac{\delta t}{\delta v}\left(N_{i-1,j}^k G_{i-1,j}^k - N_{i,j}^k G_{i,j}^k\right) + \delta t (\delta v)^2 \left(-N_{ij}^k \sum_{l=1}^{p} K_{i,l} n_l^k \quad \frac{1}{2}\delta_{ij} \sum_{l=1}^{i-1} K_{l,i-l} n_l^k n_{i-l}^k\right) \tag{11}$$

where δ_{ij} is the Kronecker symbol. The initial condition is set to : $N_{ij}^1 = 0$ except for i=j=1: $N_{11}^1 = N_0^1$. The algorithm is of p^2 complexity since it goes trough the couples of cells for the coalescence process.

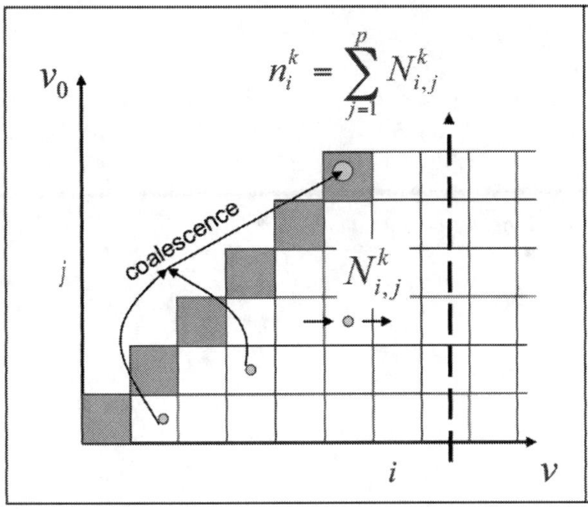

Figure 1: Numerical mesh

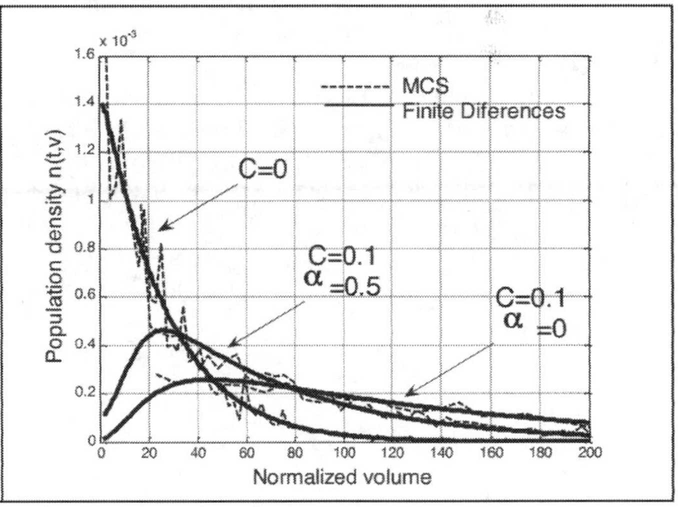

Figure 2: Population density versus volume, Time = 0.1

4. Results

In order to study the consistency of the modelling (equations (8), (9) and (10)) we have compared its numerical solution with Monte Carlo Simulations (MCS). The latter are based on the direct application of the involved elementary processes, i.e. growth and aggregation. Here we use the algorithm suggested by Tandon and Rosner [4] when they studied aggregation coupled with surface area reduction. In our case, the sintering kinetic law was replaced by the matter volume expansion law. The aggregation kernel has been taken as a constant. The choice of the time step in the MCS needs a particular consideration. As a consequence we used a refined method proposed by Smith and Matsoukas [5] in which the number of particles is kept constant. To start the simulations we consider a set of N (N=5000) particles, each of unit volume. During aggregation each aggregate is characterized by v and v_0.

Figure 2 shows the comparison between MCS and finite difference method. The kinetic parameter was constant: K=0.5. The growth law (4b) was applied, with different values for parameters C and α. MCS and finite difference method are in agreement. This shows the consistency of the equations (8), (9) and (10) to represent these strongly coupled growth-aggregation processes. As expected the growth of particles shifts the population density to the large volume values whereas the coupling, as in equation (4b), delays the shift.

5. Conclusion

Strongly coupled growth and aggregation processes have been studied by MCS and by a finite difference method applied to the equations of a dedicated model. Results show the validity of the modelling. Nowadays the authors are investigating various coupled growth-aggregation processes for which the kinetics are available: perikinetic and orthokinetic aggregation, growth in presence of impurity or adsorbed polymer. The next step of this work will be to apply this methodology to other phenomena that strongly depend on the collisions, e.g. restructuring or morphological dynamics of aggregating particles [6].

6. References

[1] Elimelech M, Gregory J, Jia X and Williams R 1995 *Particle Deposition & Aggregation* (Oxford: Butterworth-Heinemann Ltd)

[2] Randolph A.D. and Larson M.A. 1988 *Theory of Particulate Processes* (New York: Acad. Press)

[3] Ramkrishna D 2000 *Population balances: theory and application to particulate systems in Engineering* (San Diego: Acad. Press)

[4] Tandon P and Rosner D.E. 1999 *J. Coll. Interf. Sci.* **213** 273

[5] Smith M and Matsoukas T 1998 *Chem. Eng. Sci.* **53** 1777

[6] Gruy F 2011 *Colloids and Surfaces A: Physicochem. Eng. Aspects* **374** 69

A Simulation Based Study of Maximum Development of Extensive Air Showers in Highest Energies

P. Davoudifar[1], **K. Rowshan Tabari**[1]
Research Institute for Astronomy and Astrophysics of Maragha,
P.O.B 55177-36698, Vali Asr Town, Eastern Azarbayjan, Iran

dfpantea@riaam.ac.ir; p_davoudifar@yahoo.com

Abstract. The maximum development of an Extensive Air Shower (EAS) is an important parameter which is possible to be studied with EAS simulation programs. This parameter is also related to hadronic interaction models in high and low energy ranges. A Detailed study been considered to investigate the effects of zenith angle and interaction models as well as thinning parameters on the maximum development of an EAS and its mean values. Diagrams of Maximum shower development vs. energy being used to recognize the masses of the primary particles initiated those EASs. An empirical formula presented to remove the effects of zenith angle.

1. Introduction

The question of cosmic rays type has been essential to many fields of physics. Knowing the composition of very high energy particles (EeV range and above) been researched alongside the development of cosmic ray large area detectors. For primaries with energies of ~ 10^{15} eV and above, due to their low rate of hitting the upper atmosphere (1 $m^{-2}yr^{-1}$ to 1 $km^{-2}yr^{-1}$ and lower), their direct observation is practically impossible and they can only be investigated through their interactions with the atmosphere, where they produce Extensive Air Showers (EASs).

Till now, many experimental techniques were used to detect the features of these EASs amongst them is to measure the "elongation rate" [1, 2, 3] which is made essentially by detection of Cherenkov light [4]. Hence, the mean values of "maximum longitudinal shower development", $<X_{max}>$, is possible to be estimated using "elongation rates". But the random nature of shower development in the atmosphere, causes uncertainties arise against the nature of those primaries which are declared through experimental techniques.

Development of simulation programs, like as CORSIKA (Cosmic Ray Simulation for Kascade)[5, 6, 7], made it possible to explore the random features of EASs and it appeared that to compare the results of a simulation based study and experimental values, high statistic is essential.

2. CORSIKA simulation

We started to study the mass composition of higher energy particles using CORSIKA 6.98[5], were different primary particles (gamma, proton, CNO group, and Iron) were used. As the maximum

[1] This work has been supported by Research Institute for Astronomy & Astrophysics of Maragha (RIAAM) under research project No. 1/2522.

development of an EAS, X_{max}, is used in mass composition studies we considered a detailed analysis of the zenith angle dependence of X_{max}.

Due to this study using CORSIKA simulation, it was seen that there are considerable fluctuations in $<X_{max}>$ with respect to its mean values when zenith angles were changed up to 60 degrees. The fluctuations seemed to be related in primary type as there were lower for Iron primaries. For all primary types the fluctuations were higher in lower zenith angles.

Then we considered our simulation for QGSJETII [8, 9] as high and FLUKA 2011[10, 11] as low energy interaction models, thinning parameter of 10^{-4}[6, 7], for Yakutsk array (observation level of 105 gr/cm^2, 62° North latitude and 129° East longitude [12]).

To calculate mean values of $<X_{max}>$, the zenith angles were ranged in 30 ranges by separations of 2 degrees (i.e. 0-2, 2-4, 4-6, 6-8… and 58-60), 100 showers in each range and total of 3000 showers for each energy were produced. Energy bins were ranged from EeV to 10^3 EeV (i.e. 1EeV, 2EeV, …, 10 EeV, 20 EeV, …, 100 EeV, 200 EeV, …, and 1000 EeV), total of 28 energy bins and 252000 showers were simulated.

In a given energy, the distribution of the shower maximum development in different zenith angles is shown in figure 1:

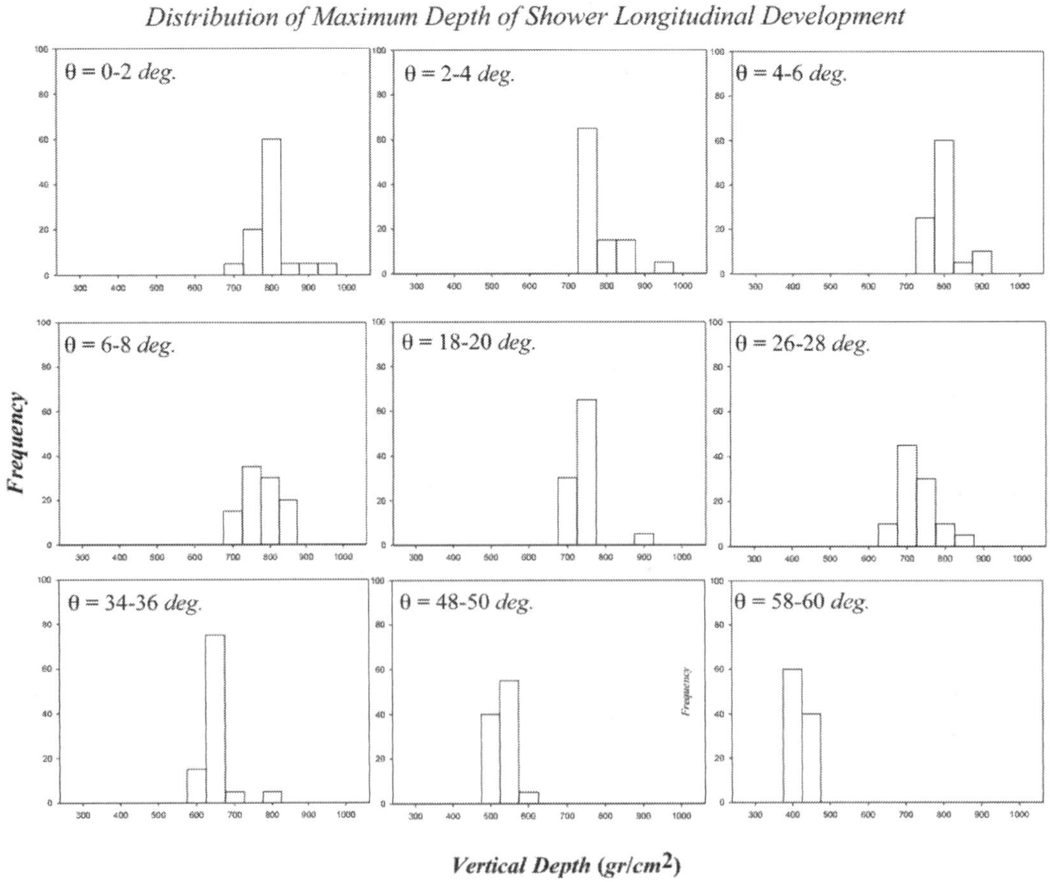

Figure 1. Distribution of shower development maxima in different zenith angles for a 100 EeV proton (this simulation).

in which, it is seen that not only the frequency, but the values of X_{max} differs when the zenith angle changes.

Due to later result, we considered a path length distribution function to estimate the average values in each zenith angle range. The path length distribution function considered to be in the form of:

$$f(x) = polynomial(2) \cdot \begin{cases} e^{(p[0].x+p[1])} & \text{lower limit} < x < x_{max} \\ p[2] & x = x_{max} \\ e^{(p[3].x+p[4])} & x_{max} < x < \text{upper limit} \end{cases} \quad (1)$$

in which lower and upper limits been considered as the lower and upper ranges of distribution of shower development maxima, where no more event were recorded. The common form of a function of this type is shown in figure 2.

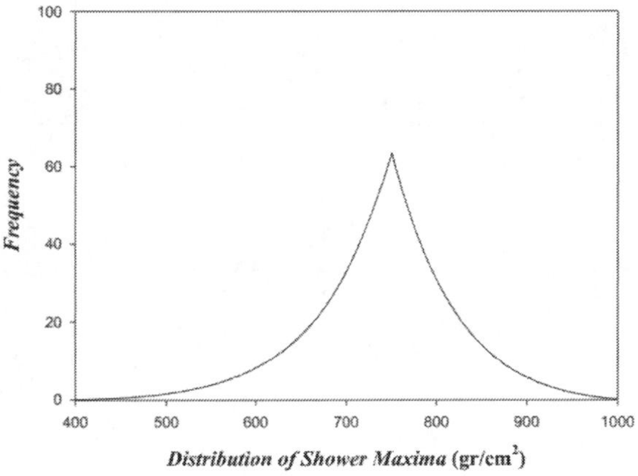

Figure 2. Common form of the function of formula 1.

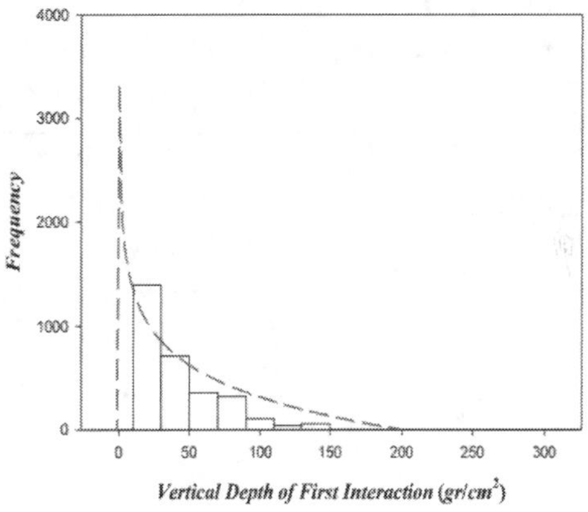

Figure 3. Distribution of Vertical Depth of First Interaction for a 100 EeV Proton.

These functions were fitted to the frequencies and then normalized to get the probability functions:

$$P(x) = \frac{f(x)}{\int_{\text{lower limit}}^{\text{upper limit}} f(x).dx} \quad (2)$$

The comparison of these probabilities shows that in lower zenith angles (i.e. less than 10 deg.), the shape of probability function does not change with zenith angle increase, but this change is obvious in larger zenith angles up to about 30 deg. (i.e. the width of probability function increases) and then the width of probability function slowly decreases without drastic changes in its maxima.

The result for primaries with larger mass numbers is different as for example, for an Iron primary there are fewer changes in all zenith angle ranges and the value of maxima decreases with zenith angle very smoothly. In this stage no more could be down. In each angle range, the mean value of X_{max} is then equal to:

$$\langle X_{max} \rangle = \int_{\text{lower limit}}^{\text{upper limit}} x.P(x).dx \quad (3)$$

As the width of this probability function is more at lower zenith angles, this treatment will affect the calculated $< X_{\max} >$ at lower angles. The results are shown in table 1.

To remove the effect of zenith angle, a function of type:

$$\Theta(\theta) = a.\cos\theta + b \tag{4}$$

been used. Coefficients were estimated to be a=753.1315 and b=11.5003. Eliminating the effect of this function:

Table 1. Result of simulated showers. Distribution of $<X_{max}>$ v.s. zenith angle.

θ (deg)	1	3	5	7	9	11	13	15	17	19
$<X_{max}>$	773	721	789	751	782	776	738	742	711	723
θ(deg)	21	23	25	27	29	31	33	35	37	39
$<X_{max}>$	735	744	700	630	651	630	642	639	622	579
θ(deg)	41	43	45	47	49	51	53	55	57	59
$<X_{max}>$	583	559	543	529	517	481	479	439	426	397

the value of $< X_{\max} >$ for a primary proton of 100 EeV is estimated:

$$\langle X_{\max} \rangle = 764.7 \tag{5}$$

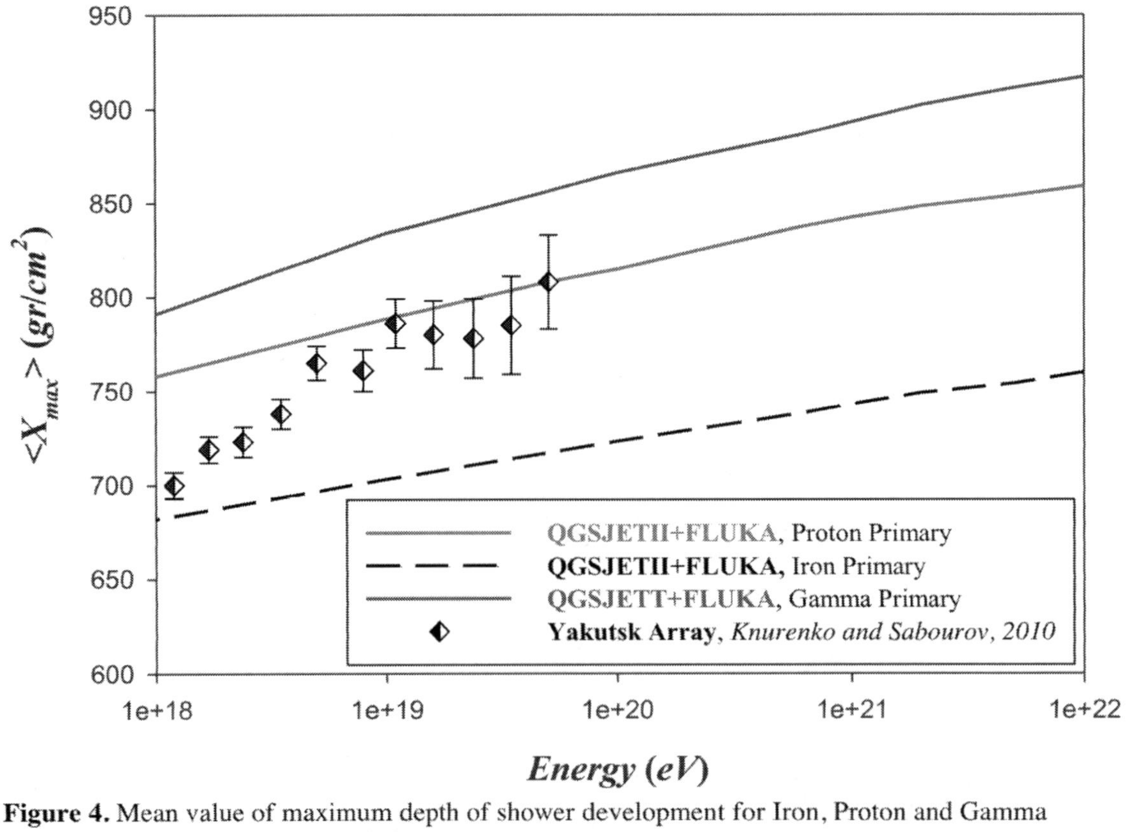

Figure 4. Mean value of maximum depth of shower development for Iron, Proton and Gamma primaries as resulted in this simulation.

To produce a diagram of $<X_{max}>$ v.s. energy (figure 4), this procedure been repeated for Proton, Iron and Gamma primaries in different energies (i.e. 10 EeV up to 1000 EeV).

A function of the logarithmic form (figure 3), were adopted to consider the effect of first interaction height:

$$H(x) = c + d.\ln x \tag{6}$$

The function were fitted to our results, in which x denoted the atmospheric depth in gr/cm^2. For a primary proton of 100 EeV, these parameters were estimated to be c= 2394.2031 and d= -451.8835. From equation 7:

$$\langle H \rangle = \frac{\int_{\text{lower limit}}^{\text{upper limit}} x.(c + d.\ln x).dx}{\int_{\text{lower limit}}^{\text{upper limit}} (c + d.\ln x).dx} \tag{7}$$

the average value of H were obtained ~50. More studies showed that this average value for a proton primary remains unchanged (or have small changes) when the interaction models are not varied.

3. Results

These results are collected in figure 3, in which a set of resent experimental data of Yakutsk array are also shown [4]. Though the most of these primaries at higher energies seems to be protons but the effect of other interaction models should be considered in detail.

The effect of zenith angle on the depth of shower maximum was studied. It is shown that in lower zenith angles, the values of shower maximum development, X_{max} , have drastic changes and a statistical study of $<X_{max}>$ values always should be considered to remove these effects. We introduced a procedure to remove the effect of zenith angle. As a result it shown that in a simulation, for practical means, it is essential to consider this effect when use the data from an arrays with lower resolution. For higher resolution arrays the accuracy of zenith angle estimation is a few degrees and this resolution determines the zenith angles ranges should be considered in the simulation.

4. References

[1] Linsley J 1977 *Proceedings of 15th International Conference on Cosmic Rays* **12** 89-96
[2] Linsley J and Watson A A 1981 *Physical Review Letters* **46** 459-463
[3] Badea A F, et al. 2001 *Astroparticle Physics*, **15** 19-28
[4] Knurenko S P and Sabourov A 2011 *Astrophysics and Space Science Transactions* **7** 251-255
[5] Heck D and Pierog T 2011 *Extensive Air Shower Simulation with CORSIKA: A User's Guide* (Version 6.980)
[6] Heck D and et al. 1998 *FZKA 6019*
[7] Knapp J, Heck D and Schatz G 1996 *FZKA 5828*
[8] Kalmykov N N, Ostapchenko S S, and Pavlov A I 1994 *Bulletin of Russian Academy of Science (Physics)* **58** 1966
[9] Ostapchenko S S 2006 *Nuclear Physics B (Proceedings Supplements)* **151** 143, 147; 2006 *Physical Review D* **74** 014026
[10] Fass`o A, Ferrari A, Ranft J, and Sala P R 2005 *Report CERN-2005-1*
[11] **http://www.fluka.org/**
1988 *"Catalogue of Highest Energy Cosmic Rays, No. 3, Yakutsk"* World Data Center C2 for Cosmic Rays Institute of Physical and Chemical Research

Modeling of a self-healing process in blast furnace slag cement exposed to accelerated carbonation

Serguey V. Zemskov[1], Bilal Ahmad[2], Oguzhan Copuroglu[2], Fred J. Vermolen[1]

[1]Faculty of Electrical Engineering, Mathematics and Computer Science,
Delft University of Technology, Mekelweg 4, 2628 CD Delft, The Netherlands
[2]Faculty Civil Engineering & Geosciences,
Delft University of Technology, Stevinweg 1, 2628 CN Delft, The Netherlands

E-mail: s.zemskov@tudelft.nl

Abstract. In the current research, a mathematical model for the post-damage improvement of the carbonated blast furnace slag cement (BFSC) exposed to accelerated carbonation is constructed. The study is embedded within the framework of investigating the effect of using lightweight expanded clay aggregate, which is incorporated into the impregnation of the sodium mono-fluorophosphate (Na-MFP) solution. The model of the self-healing process is built under the assumption that the position of the carbonation front changes in time where the rate of diffusion of Na-MFP into the carbonated cement matrix and the reaction rates of the free phosphate and fluorophosphate with the components of the cement are comparable to the speed of the carbonation front under accelerated carbonation conditions. The model is based on an initial-boundary value problem for a system of partial differential equations which is solved using a Galerkin finite element method. The results obtained are discussed and generalized to a three-dimensional case.

1. Introduction

Blast furnace slag cement (BFSC) is the most important product of the cement industry in the northern European countries, particularly in the Netherlands, Belgium and Germany. Apart from the low-heat application, the superior durability of BFSC against aggressive environment makes this material a suitable binder for the concrete exposed to chloride, acid and sulfate attack.

In general, the microstructure of Portland-slag cement pastes is not too different from that of ordinary Portland cement (OPC) [1]. However, it has been found that cements containing a high amount of BFS have a poor resistance against carbonation. It is generally known that the content of calcium hydroxide ($Ca(OH)_2$) in concrete plays an important role for carbonation resistance. Calcium carbonate ($CaCO_3$), which is a major product of $Ca(OH)_2$ carbonation, blocks the capillary pore in concrete leading to the reduction of gas diffusivity, thereby reducing the rate of carbonation. However, BFSC concrete tends to suffer much more from carbonation than the OPC mixture does due to its significantly lower $Ca(OH)_2$ content. One way to improve this disadvantage of BFSC concrete is to use sodium-monofluorophosphate (Na-MFP) as an addition to the cement mixture [2].

The solution of Na-MFP is encapsulated by use of cement paste to produce a *self-healing* system in the BFSC. The content of the capsule (Na-MFP) dissolves and diffuses through water which occupies the pores of both uncarbonated and carbonated parts of the cement specimen. The chemical reaction between Na-MPF and certain components of the carbonated cement matrix takes place such that fluoroapatite ($Ca_5(PO_4)_3F$) forms according to the following reaction equation:

$$5Ca(OH)_2 + 3Na_2PO_3F \quad \rightarrow \quad Ca_5(PO_4)_3(F, OH) + 4NaOH + 2NaF + 3H_2O \ .$$

Fluoroapatite is a highly insoluble compound which improves the micro-structure of BFSC and increases its tensile resistance.

In this paper we are going to consider a mathematical model for the post-damage improvement of the carbonated BFSC using expanded clay spherical capsules filled with Na-MFP.

2. The mathematical model

To construct a mathematical model of the healing process, we consider one spherical capsule of radius R within a cement cube with side length $L \in [4R, 10R]$. The following assumptions are used:

- the cement cube is subject to carbonation and the ingress of gaseous carbon dioxide into the cement matrix takes place through the upper edge of the cube;
- the reaction of carbonation takes place in the narrow zone which can be approximated by a surface (a plane) referred to as *the carbonation front*; so, after some definite time, the cube is supposed to be divided by the carbonation front into two rectangular cuboids consisting respectively of carbonated and uncarbonated cement;
- both parts of the cement cube are penetrable for water to a different extent;
- the pores of both parts of the cement cube are filled with water, which facilitates diffusion of Na-MFP from the capsule through pores; we further neglect diffusive transport through solid cement particles;
- the carbonation front position l is assumed to move downwards according to the relation $l(t) = L/2 - \kappa\sqrt{T+t}$ [4], where $l(0)$ is the initial z-coordinate of the horizontal carbonation front and T is the known time of carbonation process which is necessary to reach the position $l(0)$.

The vertical rectangular section of the cement cube passing through the center of the capsule provides us with a representation for the two-dimensional mathematical model (Fig. 1 (*left*)). Due to symmetry reasons, we consider half of this cross-section (Fig. 1 (*right*)).

The domain of computation Ω consists of three parts, Ω_1, Ω_2, and Ω_3 representing, respectively, the carbonated cement matrix, the capsule and the uncarbonated part of the cement matrix. The boundary $\partial\Omega$ of Ω is a rectangle of dimensions $L/2 \times L$.

Since the vast majority of the reacting chemicals in the cement is calcium hydroxide ($Ca(OH)_2$), we neglect the contribution of other cement components.

We denote the density of Na-MFP and $Ca(OH)_2$ by $c_1 = c_1(r, z, t)$ and $c_2 = c_2(r, z, t)$, respectively. The following differential equations hold for c_1 and c_2:

$$\frac{\partial c_1}{\partial t} = \nabla \cdot (D \nabla c_1) - 3\, k\, c_1 \frac{1000\, c_2}{\mathrm{M}_{\mathrm{Ca(OH)_2}}}, \quad (r, z) \in \Omega, \ t > 0 \ , \tag{1}$$

$$\frac{dc_2}{dt} = -5\, k\, c_2 \frac{1000\, c_1}{\mathrm{M}_{\mathrm{Na_2PO_3F}}}, \quad (r, z) \in (\Omega_1 \cup \Omega_3), \ t > 0 \ , \tag{2}$$

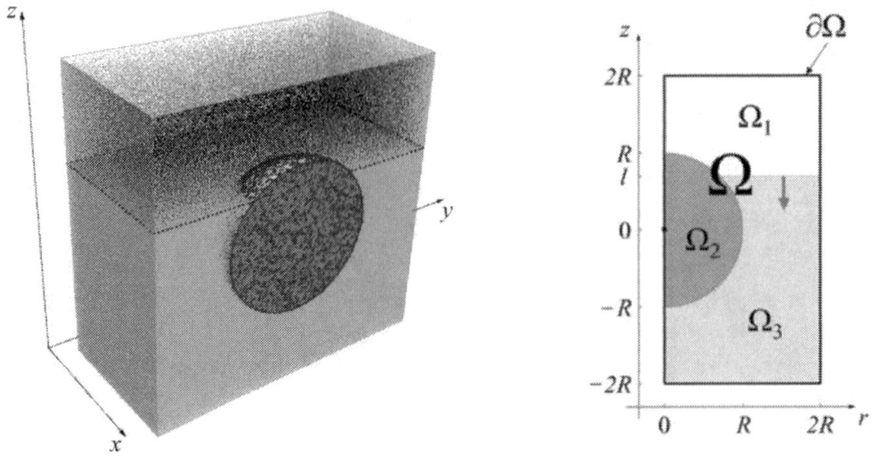

Figure 1. (*left*) Representation of a half of the concrete cube containing the capsule and consisting of uncarbonated (below) and carbonated (above) parts. (*right*) Computational domain for a two-dimensional mathematical model.

where M_X are molar masses of corresponding chemicals, k is a proportionality factor which determines the reaction rate, and the diffusion coefficient D changes over space and time:

$$D = D(r, z, t) = D_0\, \varepsilon_i / \tau_i, \ (r, z) \in \Omega_i, \ t > 0, \ i = 1, 2, 3 \ .$$

Here, D_0 is the actual diffusion coefficient of Na-MFP in water, $\varepsilon_i = \varepsilon_i(r, z, t)$ and $\tau_i = \tau_i(r, z, t)$ are, respectively, the effective porosity and tortuosity of the i-th medium. The capsule and both parts of the cement cube are penetrable for water to a different extent: the carbonated cement possesses a higher porosity which implies, according to the Kozeny-Carman equation [3], a higher permeability, whereas the porosity and, hence, permeability of the uncarbonated cement is much lower. The values of τ_i can be defined empirically or one can invoke the approximation $\tau_i \approx 1 - (\ln \varepsilon_i)/2$ described in [5], which we use in our calculations.

No transport of Na-MFP is assumed through the boundary of Ω, hence we have a zero flux boundary condition.

The initial densities of Na-MFP and $Ca(OH)_2$ are defined as follows:

$$c_1(r, z, 0) = \begin{cases} c_1^{\max}, & (r, z) \in \Omega_2 ; \\ 0, & (r, z) \in \Omega_1 \bigcup \Omega_3 \end{cases} \quad \text{and} \quad c_2(r, z, 0) = \begin{cases} c_2^{\max}, & (r, z) \in \Omega_1 ; \\ 0, & (r, z) \in \Omega_2 ; \\ c_2^{\min}, & (r, z) \in \Omega_3 \ . \end{cases} \tag{3}$$

Equations (1), (2), zero flux boundary condition, and initial conditions (3) form an initial boundary value problem in Ω and for $t > 0$.

After solving the boundary value problem on each time step, the density and the volume of fluoroapatite formed in Ω_1 and Ω_3 as well as values of ε_1 and ε_3 are corrected with regard to amounts of reacted chemicals.

3. Results and discussion

Equations (2) and (1) are solved using Galerkin finite element method over a structured triangle grid. Some results obtained through a computer simulation are presented in figure 2. In this example, we consider a half of the capsule section with the radius of 1 mm.

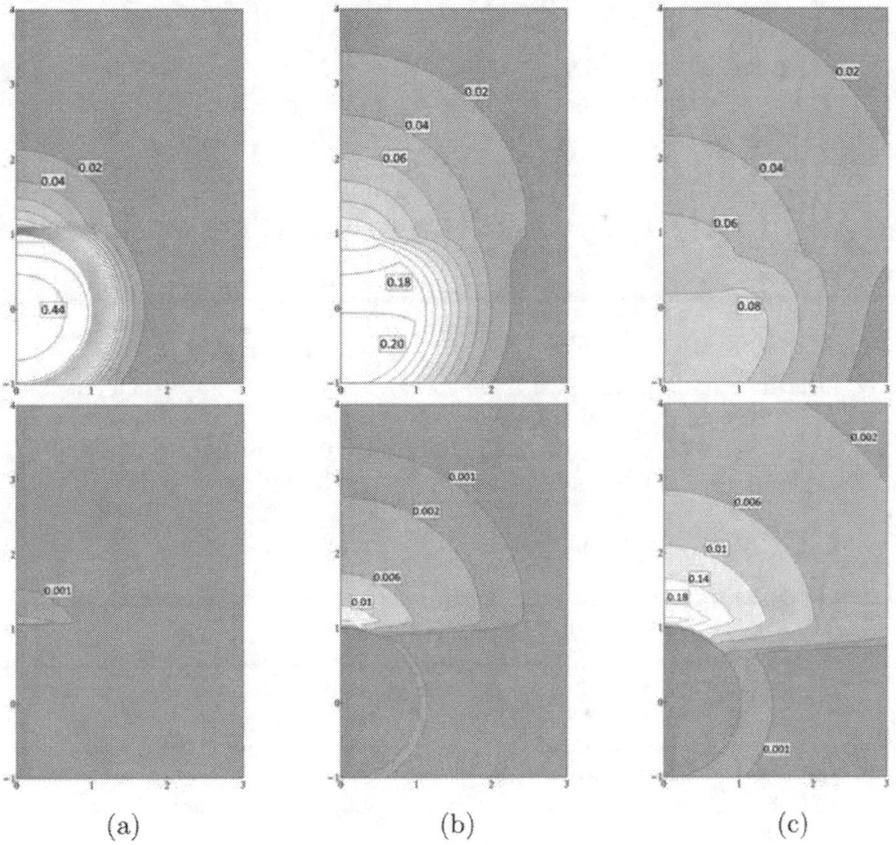

Figure 2. Concentration profile of Na-MFP (above) and density profile of fluoroapatite (below) after (a) 10, (b) 40 and (c) 80 days of healing process; $\varepsilon_1 = 0.15$, $\varepsilon_2 = 0.21$, $\varepsilon_3 = 0.02$.

In figure 2, the first row of illustrations represents the evolution of the Na-MFP concentration profile, whereas the second row shows the corresponding changes of fluoroapatite density around the capsule. The columns (a), (b) and (c) correspond to the state of both profiles at the certain moments of the healing process: after 10, 40 and 80 days, respectively.

It can be seen how the movement of the carbonation front influences the process of diffusion of Na-MFP and, hence, forming of fluoroapatite. It is possible also to retrace the process of growing of the fluoroapatite layer and, in particular, the layer in which the density of the reaction product exceeds 0.01 g/cm^3. This density is considered as sufficient to improve the cement structure. According to the results observed, one can make definite conclusions about necessary amount and concentration of Na-MFP in the capsule and the number of capsules per unit volume of cement.

A three-dimensional representation of this model can be obtained from symmetry considerations in a straight-forward manner.

References

[1] Odler I 2000 *Special inorganic cements* (London, New York: Taylor & Francis) p 416
[2] Copuroglu O, Fraaij A L A and Bijen J M J M 2006 *Cem. Concr. Res.* **36** 1475–82
[3] McCabe W L, Smith J C, Harriott P 2005 *Unit operations of chemical engineering* (New York: McGraw-Hill)
[4] Aiki T, Muntean A 2010 *Comm. Pure Appl. Anal.* **9** 1117–1129
[5] Weissberg H L 1963 *J. Appl. Phys.* **34** 2636-2639

Lumped parameter identification of a quarter car test rig

Y Taskin[1,3], N Yagiz[1] and I Yuksek[2]

[1] Department of Mechanical Engineering, Faculty of Engineering, Istanbul University, 34320 Avcilar, Istanbul, Turkey
[2] Department of Mechatronics Engineering, Faculty of Mechanical Engineering, Yildiz Technical University, Besiktas, Istanbul, Turkey

E-mail: ytaskin@istanbul.edu.tr

Abstract. Quarter car model is a simple and widely used mathematical model to analyze the vibration and control problem of vehicles. In this study, a quarter car test rig is modeled as a lumped parameter system. Model parameters of the system are determined by measurements and experiments. Forced vibration method is used to identify the stiffness and damping parameters of the lumped model. A modal shaker is used to generate the road input in the test rig. The accelerations of the road input, sprung and unsprung masses are measured by piezoelectric accelerometers. The frequency response functions are obtained by using acceleration data. The identified parameters of the test rig are adjusted by comparing the experimental and simulation results.

1. Introduction

Quarter car model represents only the vertical motion of vehicle body and wheel. Thus, vertical dynamics of vehicles can be investigated easily by using this model. A great deal of research has been carried out on improving ride comfort and road holding with quarter car models during last three decades [1-3]. In addition to theoretical and numerical studies, there have been several experimental studies realized during the same period of time [4-9].

In this paper a quarter car test rig which is constructed in laboratory environment, is modeled as a lumped parameter system to determine system parameters. Model parameters of the system are investigated by forced vibration method experiments using acceleration data.

2. Quarter car test rig and lumped parameter model

Quarter car test rig consists of sprung mass, suspension elements, unsprung mass, wheel, excitation plate, support springs, modal shaker and linear sliding bearings that can be seen in figure 1 (a). Sprung mass and unsprung mass are connected together with suspension elements which consist of spring and damper. Wheel is also a part of the unsprung mass above the excitation plate and all parts are supported by springs that are under the excitation plate to avoid damages to modal shaker. Also, all moving plates are guided on columns by linear sliding bearings. Modal shaker is on the ground and connected to excitation plate for generating the sinusoidal road input.

Test rig is modeled as a lumped parameter system shown in figure 1 (b). The model parameters; m_e, m_u and m_s are mass of excitation plate, unsprung and sprung masses, respectively. These

[3] To whom any correspondence should be addressed.

Journal of Physics: Conference Series **410** (2013) 012089 doi:10.1088/1742-6596/410/1/012089

parameters are determined by measurements on a digital scale and are listed in table 1. b_w, b_{ss}, b_u, b_s and b_e are damping coefficients of wheel, suspension system, linear sliding bearings on unsprung mass, sprung mass and excitation plate, respectively. k_s, k_w and k_{ss} are stiffness coefficients of support springs, wheel and suspension system, respectively. Damping and stiffness parameters are determined by forced vibration method which is explained in the following section. $\tilde{P}$ represents the sinusoidal excitation force that can be generated by modal shaker.

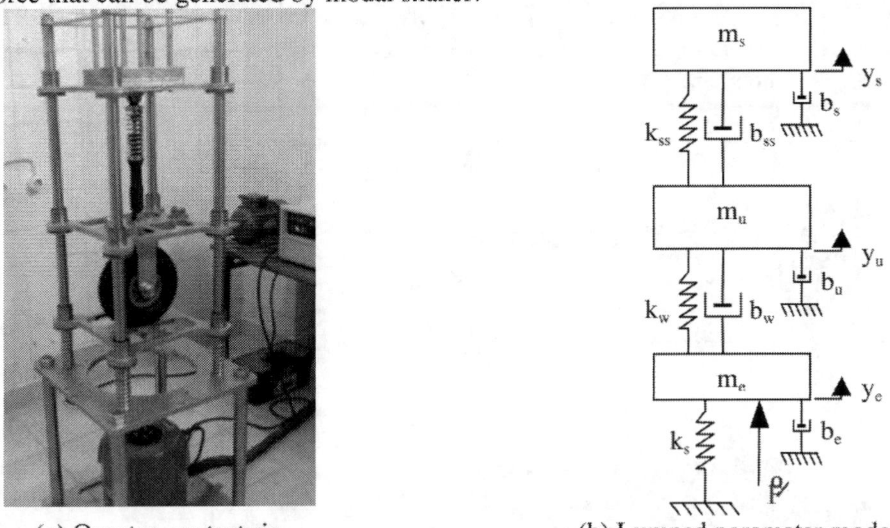

(a) Quarter car test rig (b) Lumped parameter model

Figure 1. Test rig and model

3. Forced vibration method and experimental results

Forced vibration method can be applied easily on a single degree of freedom (sdof) system. The frequency response function (FRF) of the system can be obtained by using two piezoelectric accelerometers to collect input and output vibration data simultaneously during forced vibration motion is generated by a modal shaker. Damping and stiffness parameters of sdof system can be determined from FRF if mass parameter of system is known. Although lumped parameter model of the test rig has three degrees of freedom as seen in figure 1 (b), the test rig can be divided into sdof parts to apply forced vibration method as shown in figure 2.

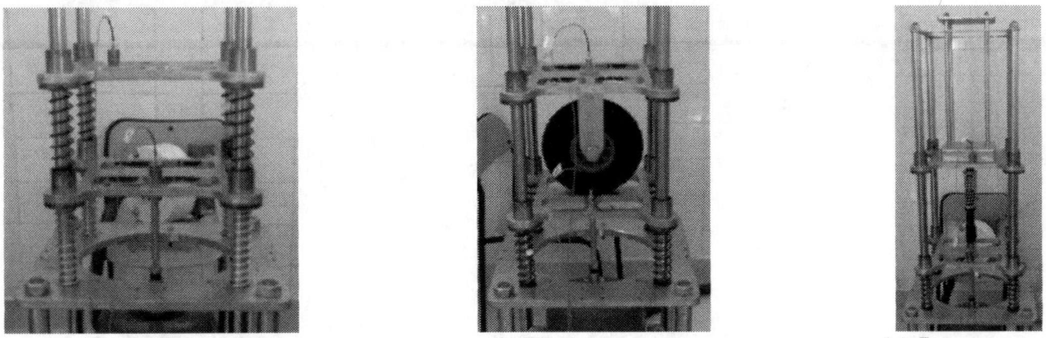

(a) Excitation plate (b) Unsprung mass (c) Sprung mass

Figure 2. Sdof parts of test rig on test setup

A typical forced vibration method application on a sdof excitation plate with springs and linear sliding bearings shown in figure 2 (a). Excitation plate is located over another plate with supporting springs and linear sliding bearings. The combination of components forms a sdof system for the excitation plate and can be modeled as a lumped parameter model seen in figure 3 (b). y_1 and y_2 represent the displacement of input and output points respectively. While modal shaker excites the sdof system, excitation acceleration is measured by a piezoelectric accelerometer as an input data. Simultaneously, acceleration of the sdof system is measured by another piezoelectric accelerometer as an output data. FRF is calculated from acquired input and output vibration data. The input and output vibration data and the FRF that are obtained during the forced vibration method application are shown in figure 3 (a) and (c). Damping and stiffness coefficients are determined from FRF diagram by magnification-factor method [10]. The same procedure is also applied to the other parts and all model parameters are determined and listed in table 1.

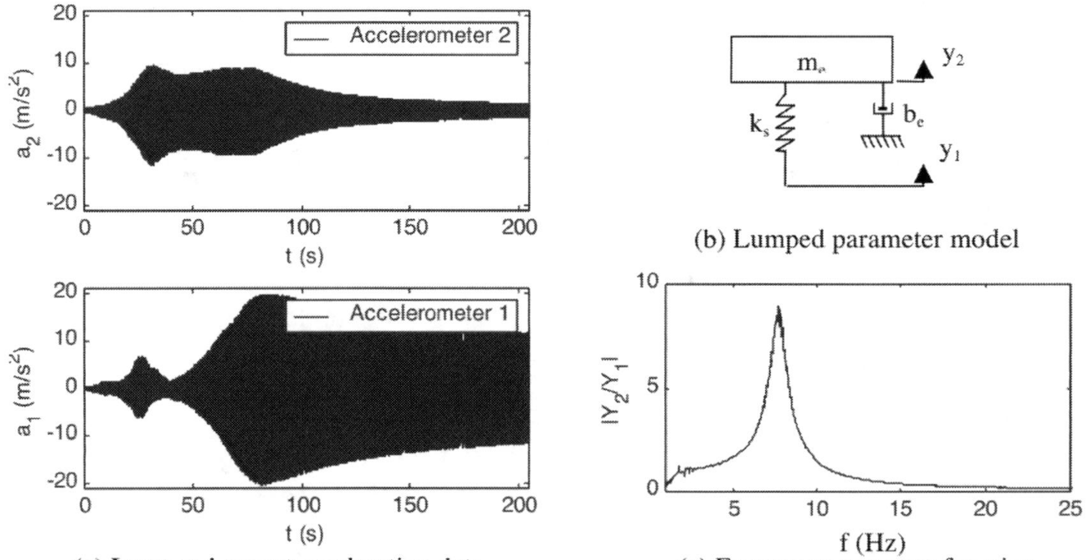

(a) Input and output acceleration data (c) Frequency response function

Figure 3. Typical forced vibration method application for parameter identification of a sdof excitation plate

The equation of motion can be written for the lumped parameter model seen in figure 3 (b) as follows;

$$m_e \ddot{y}_2 + b_e \dot{y}_2 + k_s y_2 = k_s y_1 \tag{1}$$

FRF of the lumped parameter model can be calculated from the equation of motion and can be simulated with the identified parameters.

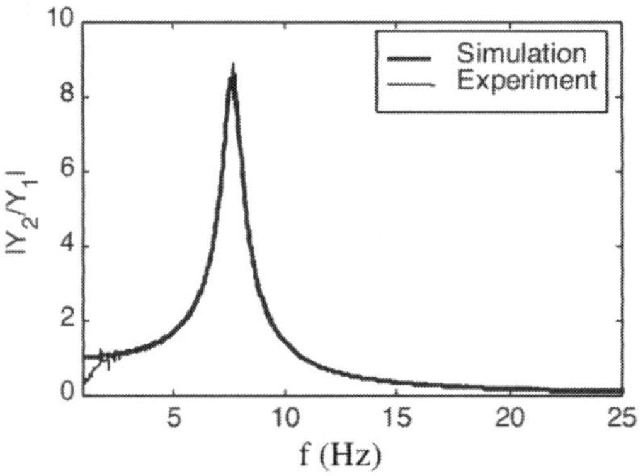

Figure 4. Comparison of experimental and simulation FRF diagrams

In figure 4, the comparison of experimental and simulation FRFs are drawn. The results indicate that the model and the test setup are compatible with the entire frequency band.

Table 1. Identified model parameters

Parameter	Value	Parameter	Value
m_e	9,869 kg	b_e	56,340 N.s/m
m_u	14,355 kg	b_w	80,650 N.s/m
m_s	29,560 kg	b_{ss}	672,250 N.s/m
k_s	23241,550 N/m	b_u	56,340 N.s/m
k_w	75026,420 N/m	b_s	56,340 N.s/m
k_{ss}	30959,910 N/m		

4. Conclusion

In this paper, a quarter car test rig is modeled as a lumped parameter model. Forced vibration method is applied to sdof parts of test rig to determine model parameters. The simulated and experimental FRF of sdof excitation plate is compared for identified parameters. It is concluded that the forced vibration method is practical and preferable method to identify lumped parameters of sdof systems.

References

[1] Margolis D L 1983 A procedure for comparing passive, active, and semi-active approaches to vibration isolation *Journal of the Franklin Institute* **315** 225-38

[2] Huisman R G M, Veldpaus F E, Voets H J M and Kok J J 1993 An optimal continuous time control strategy for active suspensions with preview *Vehicle System Dynamics* **22** 43-55

[3] Taskin Y, Hacioglu Y and Yagiz N 2007 The use of fuzzy-logic control to improve the ride comfort of vehicles *Strojniski vestnik - Journal of Mechanical Engineering* **53(4)** 233-40

[4] Ahmadian M and Pare C A 2000 A quarter-car experimental analysis of alternative semiactive control methods *Journal of Intelligent Material Systems and Structures* **11** 604-12

[5] Huang S J and Chao H C 2000 Fuzzy logic controller for a vehicle active suspension system *Proc Instn Mech Engrs* Part D **214** 1-12

[6] Yoshimura T, Kume A, Kurimoto M and Hino J 2001 Construction of an active suspension system of a quarter car model using the concept of sliding mode control *Journal of Sound and Vibration* **239(2)** 187-99

[7] Anakwa W K N, Thomas D R, Jones S C, Bush J, Green D, Anglin G W, Rio R, Sheng J, Garret S and Chen L 2002 Development and control of a prototype pneumatic active suspension system *IEEE Transactions on Education* **45(1)** 43-9

[8] Lian R J, Lin B F and Sie W T 2005 Self-organizing fuzzy control of active suspension systems *International Journal of Systems Science* **36(3)** 119-35

[9] Sung K G, Han Y M, Cho J W and Choi S B 2008 Vibration control of vehicle er suspension system using fuzzy moving sliding mode controller *Journal of Sound and Vibration* **331** 1004-19

[10] de Silva CW 2000 *VIBRATION Fundamentals and Practice* (London: CRC Press LLC)

Characterization of Rössler and Duffing maps with Rényi entropy and generalized complexity measures

B. Godó and Á. Nagy

Department of Theoretical Physics, University of Debrecen, Debrecen, Hungary

E-mail: godobence1990@gmail.com

Abstract.
Rényi entropy and generalized complexity measures are used to describe the chaotic behaviour of dynamical systems. These measures are found to be sensitive to the fine details of the Rössler and the Duffing maps. They are good descriptors of chaotic behaviour. Periodic windows and the fractal character of the chaotic dynamics are nicely detected.

1. Introduction

There exist several quantities to study the chaotic behaviour of dynamical systems. Complexity measures have proved to be especially efficient in this respect. One of these measures is the LMC (Lopez-Ruiz - Mancini - Calbet) statistical complexity [1]. A couple of years ago, a one- and a two-parameter extension [2] of this measure were put forward. These generalizations are based on the Rényi entropy. First, some simple quantum systems (H-atom, harmonic oscillator and square well) were studied with these measures. Recently, it has been demonstrated [3] that these generalized complexity measures are suitable to describe chaotic behavior. The logistic and Tinkerbell maps were analyzed.

In this work the Rössler and the Duffing maps are studied with the Rényi entropy and the generalized complexity measures.

2. Rényi Entropy and Generalized Statistical Complexity Measures

Consider a set of discrete probabilites p_1, ..., p_N with $\sum_{i=1}^{N} p_i = 1$. The Rényi entropy of order α has the form

$$R^{(\alpha)} = \frac{1}{1-\alpha} \ln \sum p_i^{\alpha}, \quad 0 < \alpha < \infty, \ \alpha \neq 1. \tag{1}$$

The limit $\alpha \to 1$ gives the Shannon entropy:

$$S = -\sum p_i \ln p_i. \tag{2}$$

The LMC complexity was defined as the product of two important information-theoretical quantities: $C = HQ$, where $H = e^S$ is the Shannon entropy power, while $Q = e^{-D} = e^{-R^{(2)}}$ is the logarithm of the Rényi entropy or order 2. The disequilibrium D quantifies the deviation of the probability distribution from uniformity. The Shannon entropy S, on the other hand, is a measure of uncertainty. A one-parameter extension of the generalized statistical measure of complexity [2] is $C^{(\alpha)} = e^{R^{(\alpha)} - R^{(\beta=2)}}$. If $\alpha \to 1$ we obtain the LMC complexity.

Published under licence by IOP Publishing Ltd

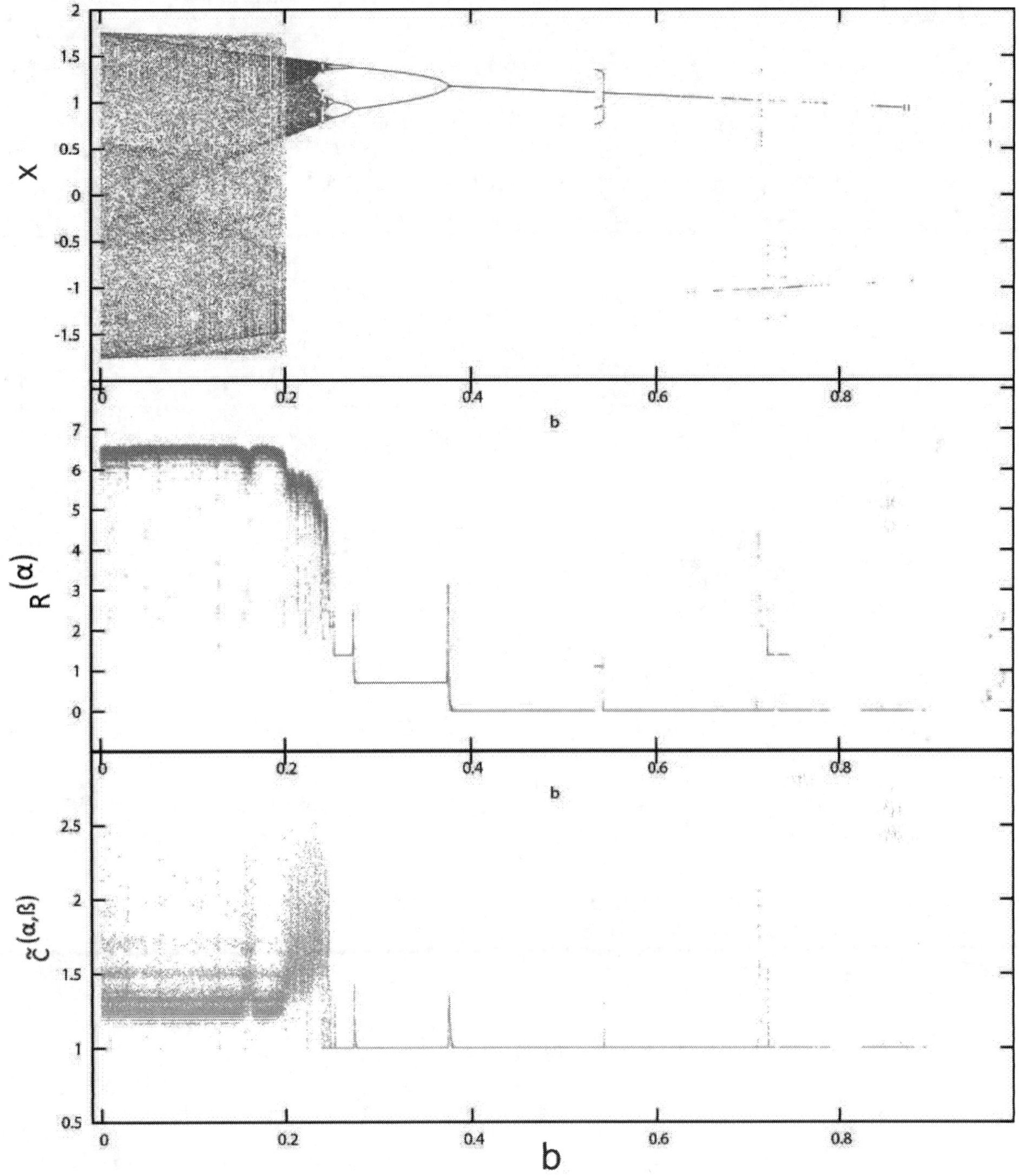

Figure 1. Bifurcation diagram, Rényi entropy($\alpha = 6$) and generalized complexity($\alpha = 3$, $\beta = 6$) for the Duffing map.

In the two-parameter extension, on the other hand, the generalized statistical measure of complexity [2] has the form

$$\tilde{C}^{(\alpha,\beta)} = e^{R^{(\alpha)}-R^{(\beta)}}, \qquad 0 < \alpha, \beta < \infty . \tag{3}$$

Certainly, the special case $\alpha \to 1$ and $\beta = 2$ gives back the LMC complexity. Important properties of the generalized complexity are detailed in [2]. It has been shown that the generalized complexity extends the complexity measure to any kind of well behaved distribution.

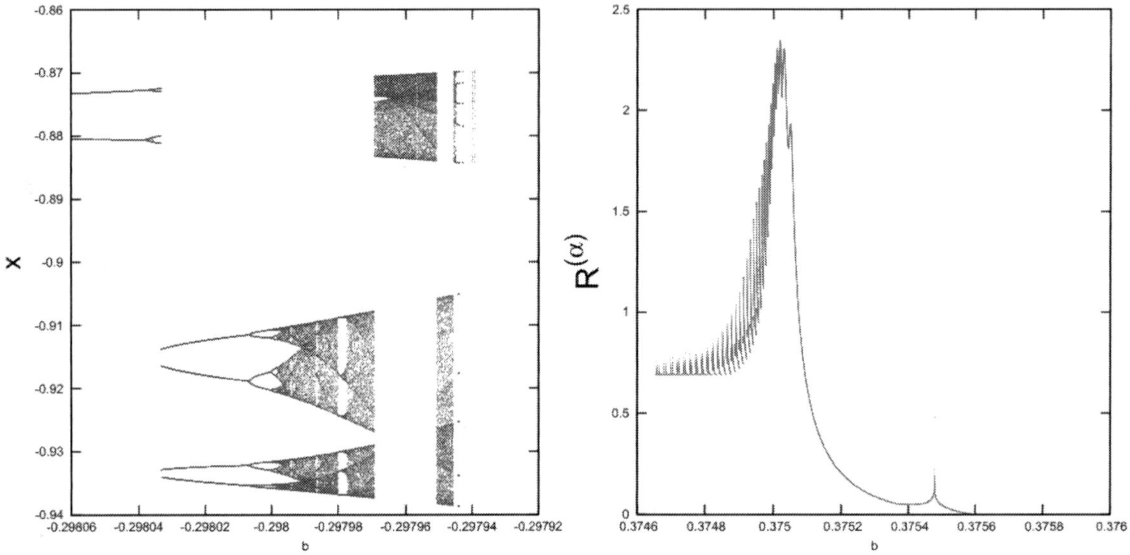

Figure 2. Enlarged bifurcation diagram of the Duffing map for $-0.2981 < b < -0.2979$ and the Rényi entropy in the vicinity of a bifurcation point.

3. Application: Duffing and Rössler maps

Now, we apply the generalized complexity measure to characterize the Duffing and Rössler maps. The Duffing map has the form:

$$x_{n+1} = y_n, \qquad y_{n+1} = -bx_n + ay_n - y_n^3. \tag{4}$$

The parameter a is taken as $a = 2.75$ and the parameter b is selected as a control parameter. The initial coordinates were: $x = 0.1$ and $y = 0.1$. Fig.2 shows the x coordinate. (y behaves similarly.) The probabilities p_i were determined [4] by subdividing the interval $[-2, 2]$ into 10000 equal bins. The number of iterates falling within a bin divided by the total number of iterations (10^4) gives the probability. For an n-periodic dynamics there are only n probabilities that are not zero. As these probabilities are all equal, the Rényi entropy is $\ln n$, independent from the parameter α, therefore the complexity is 1. From the definition (3) follows that $\tilde{C}^{(\alpha,\beta)} \geq 1$ if $\alpha < \beta$ and $\tilde{C}^{(\alpha,\beta)} \leq 1$ if $\alpha > \beta$. As one expects that complexity is larger for a more complex behaviour, the case $\alpha < \beta$ is selected.

The upper panel of Fig. 1 presents the bifurcation diagram. (The values of x are plotted against the parameter b.) Fig. 1 also shows the Rényi entropy for $\alpha = 6$ (middle panel) and the

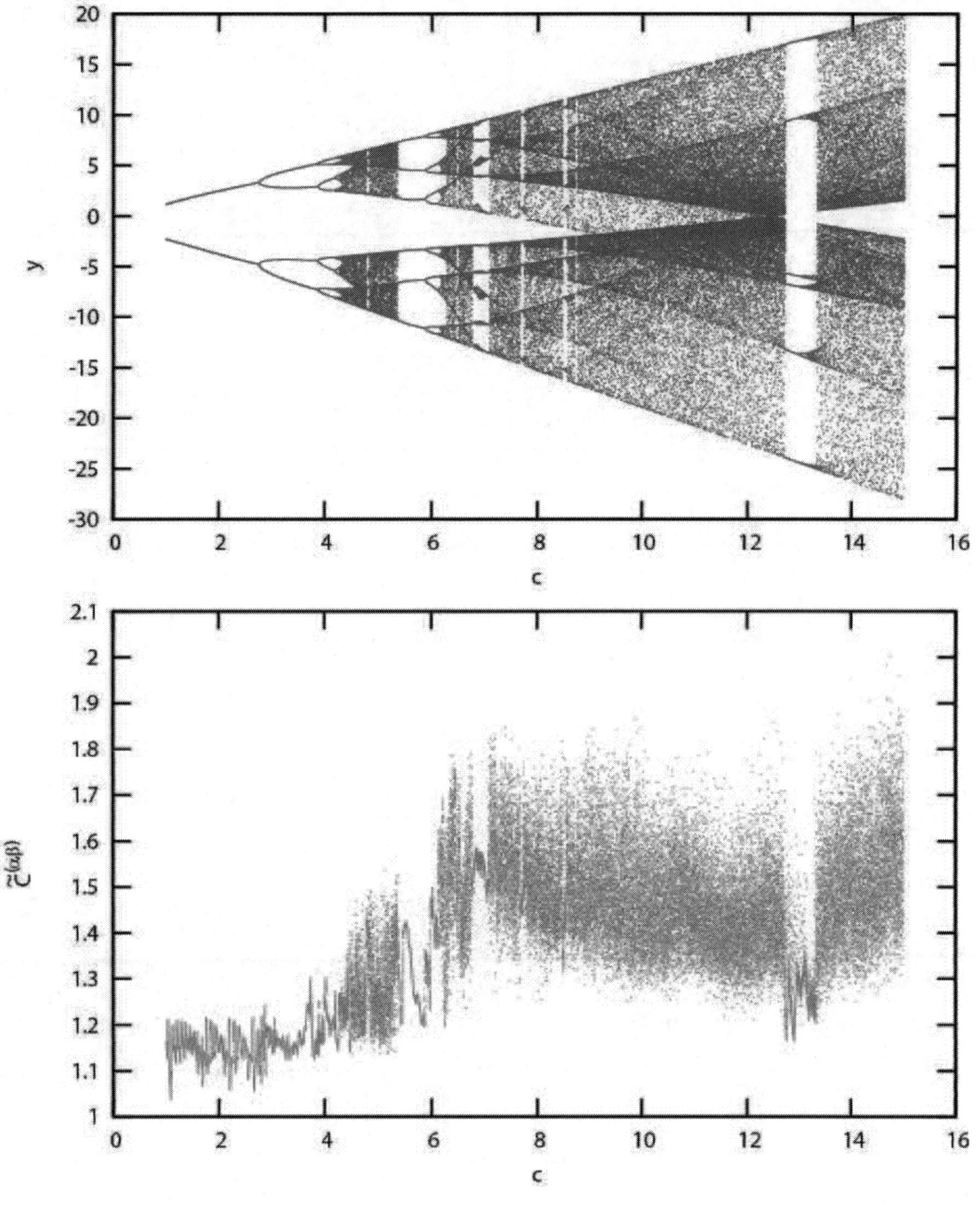

Figure 3. Rössler bifurcation diagram and generalized complexity($\alpha = 3$, $\beta = 6$).

generalized complexity for $\alpha = 3$ and $\beta = 6$ (lower panel) for the interval $0 < b < 1$. Periodic and chaotic behaviour can be seen in the bifurcation diagram, and can also be detected by the

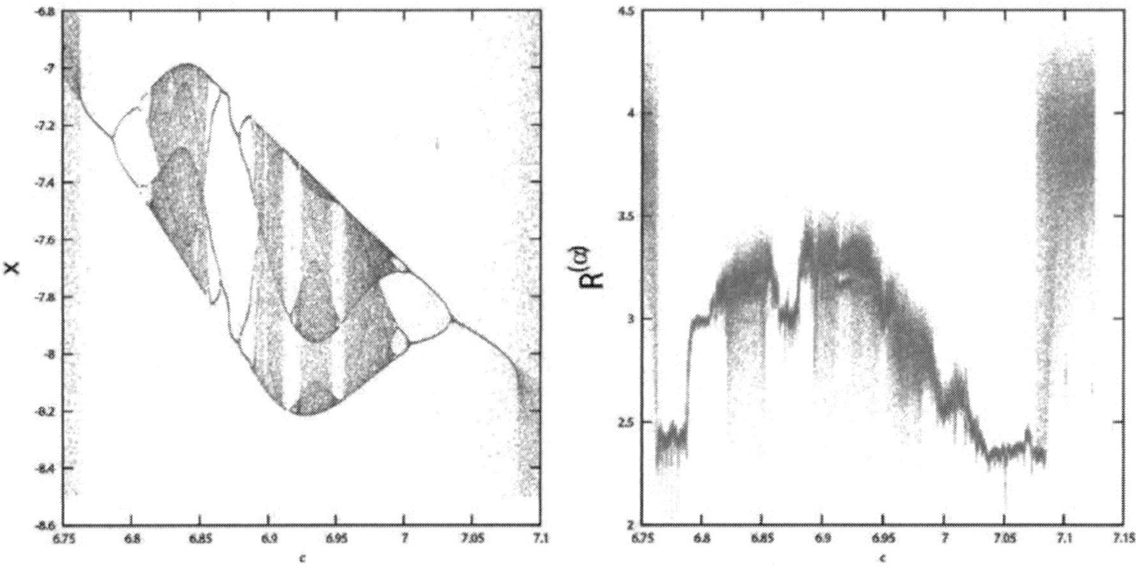

Figure 4. Enlarged Rössler bifurcation diagram and Rényi entropy.

Rényi entropy and the generalized complexity. In the bifurcation points both the Rényi entropy and the generalized complexity icreases abruptly. Fig. 2b enlarges the Rényi entropy in the vicinity of a bifurcation point. Fig. 2a shows an enlargement of the bifurcation diagram: a very interesting behaviour in the intervals $-0.86 < x < -0.94$ and $-0.2981 < b < -0.2979$. At $b = -0.298075$ the diagram is shifted, at $b = -0.29801$ it goes back to the original position. There is another shift in the interval $-0.29799 < b < -0.297985$. A similar behaviour can be observed for other values of x. These shifts can not be detected in the Rényi entropy and the generalized complexity, because the values of the probabilities do not change.

The Rössler model is given by

$$\frac{dx}{dt} = -y - z, \qquad \frac{dy}{dt} = x + ay, \qquad \frac{dz}{dt} = b + z(x - c). \tag{5}$$

The parameters a and b were taken as $a = 0.2$, $b = 0.2$ and c is the control parameter. The initial coordinates were: $x = 0$, $y = -5$ and $z = 0$. The differential equations were solved numerically by the Runge-Kutta (second order). Poincaré sections were taken at $x = 0$ and the figures show the coordinate y. Fig. 3 presents the bifurcation diagram and the generalized complexity($\alpha = 3$, $\beta = 6$) for $1 < c < 15$. Fig. 4 shows the enlarged bifurcation diagram and the Rényi entropy for $6.75 < c < 7.1$. It is a very rich structure, the bifurcation diagram and the Rényi entropy reflects different aspects. The regular and chaotic parts can be clearly distinguished. When periodic windows appear, the Rényi entropy decreases. Further enlargements (not presented here) would reveal additional fine details and the fractal character of the chaotic dynamics.

In summary, we used the Rényi entropy and the generalized complexity measures to describe Rössler and the Duffing maps. These measures nicely show the regular and the chaotic behaviour of dynamical systems. Periodic windows and the fractal character of the chaotic dynamics are clearly detected.

Acknowledgments

The work is also supported by the TAMOP 4.2.1/B-09/1/KONV-2010-0007 and the TAMOP 4.2.2/B-10/1-2010-0024 projects. The project is co-financed by the European Union and the European Social Fund. Grant OTKA No. K 100590 is also gratefully acknowledged.

References

[1] R. Lopez-Ruiz, H. L. Mancini, and X. Calbet, Phys. Lett. A 209, 321 (1995); R.G. Catalan, J. Garay, and R. López-Ruiz, Phys. Rev. E **66**, 011102 (2002).

[2] E. Romera, R. Lopez-Ruiz, J. Sanudo and Á. Nagy, Int. Rev. Phys. 3, 207 (2009); R. Lopez-Ruiz, Á. Nagy, E. Romera and J. Sanudo, J. Math. Phys. 50, 123528 (2009).

[3] B. Godó and Á. Nagy, Chaos 85, 023118 (2012).

[4] G. L. Ferri, I. Pennini and A. Plastino, Phys. Lett. A **373**, 2210 (2009).

Fractal Measures of Complex Networks Applied to Texture Analysis

João Batista Florindo, Dalcimar Casanova, Odemir Martinez Bruno
Instituto de Física de São Carlos (IFSC), Universidade de São Paulo (USP), Avenida do Trabalhador São-carlense, 400, 13560-970, São Carlos, SP, Brazil

E-mail: jbflorindo@ursa.ifsc.usp.br

Abstract. This work proposes a novel technique for the extraction of descriptors of a texture image. The method consists in modeling the image through a complex network, based on the distribution of gray-level intensities. Therefore, we process the adjacency matrix as a geometrical object and compute its fractal dimension. Thus we apply a multiscale transform over the power-law fractal curve, obtaining the descriptors of the image. The method is tested in a classification task with a comparison to other classical texture descriptors.

1. Introduction

We observe that complex network modeling [1] has called a significative attention of computer vision community. Complex networks turn possible the analysis of an object represented in an image as being a system from which we can extract measures capable of characterizing precisely an intrinsic structure in the original image. In this way, this modeling can extract features representative of aspects like spatial and pixel intensity distribution, texture patterns and irregularities, etc. Finally, these features allow an accurate description and discrimination of textures, as demonstrated in works like [2, 3, 4, 5].

An important challenge in complex network modeling is what kind of measure could be extracted from the network. Although works like [4, 5] use simple metrics, it is noticeable that more complicated problems require the employment of more elaborate measures [6]. In this context, we propose the use of fractal dimension as a valuable measure of a complex network. This use is explained by the fact that the fractal dimension estimates the complexity of the network. In the texture modeling, this metric captures the propagation of details under different scales in the network.

The proposed method consists in modeling the texture image through a complex network, representing each pixel intensity value as a vertex and connecting two vertices if there are pixels with the corresponding intensities in an 8-neighborhood. Therefore, we analyze the adjascency metrix as an object of interest in a binary image and estimate the fractal dimension of this object. In the following, we apply a multiscale transform over the dimension obtaining the texture descriptors.

We verified the efficiency of the proposed methodology in the classification of a well-known texture dataset, that is, Brodatz data [7]. The results are compared to other classical texture descriptors [8], to know, Gabor wavelets, Fourier descriptors, Laws and a multifractal approach.

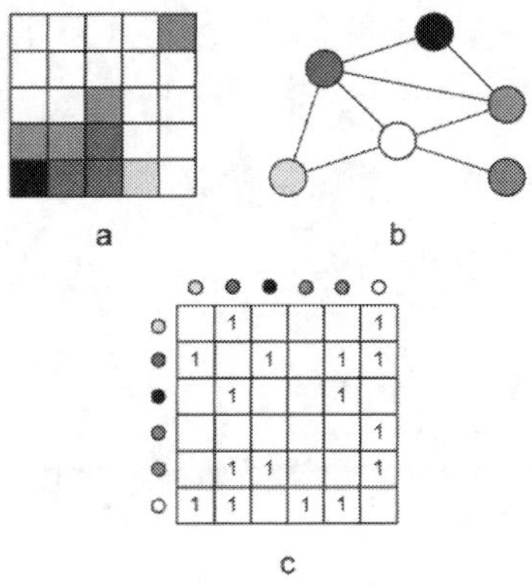

Figure 1. Simple example of color/vertex modeling. (a) 5x5 image; (b) Resultant CAG for 8-neighborhood rule; (c) Adjacency matrix

2. Complex Networks

Given an image I, consisting of a set of pixels $I = \{p_1, p_2, \ldots, p_N\}$, before using measures of networks, it is necessary the definition of a way to turn it into a complex network. This requires the definition of the set of vertices V and edges E and their properties.

A simple way to accomplish this transformation is when each node v is a color component of the image in a given color space. An edge is then added between two vertices if the colors represented by them are adjacent in an 8-neighborhood rule in the image (self connections are ignored). Figure 1 exemplifies this transformation. In the Figure 1a, we suppose that each colored square corresponds to a pixel with a specific RGB color. In 1b, we show the corresponding network and in 1c, the adjascency matrix is depicted, where each value 1 corresponds to a connection.

This is not a novel form of modeling. The work [9] shows the first use of this type of transformation calling it as *Color Adjacency Graph* (CAG). Nevertheless, in that work, the authors use this type of modeling in an application to scene objects locating and recognization, while in our work the use is for texture analysis.

In this representation, the number of nodes is equal to the number of colors, thus we need make an image quantization to reduce the number of nodes obtained and reduce the computational cost. Furthermore, as we are dealing with gray-level images, we consider each color as being a gray-level intensity. In this study, we approximate the image gray-levels by using minimum variance quantization. The quantized image has at most 64 levels.

The resulting network with n vertices can be easily represented by an adjacency matrix A of nxn (Figure 1c):

$$a_{ij} = \begin{cases} 1, & if\{i,j\} \in E, \\ 0, & otherwise, \end{cases} \quad (1)$$

where i and j are two vertices.

3. Fractal Descriptors

Fractal descriptors [10, 11, 12] method is a technique to extract features of an image based on the power-law relation of fractal dimension estimation methods.

Generally, the fractal dimension of an object in a digital image is estimated by the following generic expresion:

$$D = \lim_{\epsilon \to 0} \frac{\log(\mathfrak{M})}{\log(\epsilon)}, \quad (2)$$

being $\mathfrak{M}$ a measure depending on the dimension method and ϵ the scale at which this measure is taken. In practice, this limit is estimated by plotting the values of $\mathfrak{M}$ and ϵ in a log-log scale, as described in the following.

The fractal descriptors are provided from the power-law function u:

$$u : \log(\epsilon) \to \log(\mathfrak{M}). \tag{3}$$

In the proposed work, we extractted the descriptors from the Bouligand-Minkowski dimension method [10]. In this approach, the object of interest in the image is dilated by a variable radius r. For each r, we compute the influence area A, corresponding to the area of the dilated object. Thus the value of $A(r)$ is taken as the measure $\mathfrak{M}$ in the Equation 3 and the scale ϵ is represented by r.

In order to extract the descriptors, the function u can be used directly or summited to some sort of transform to emphasize some aspect relevant to the specific application.

4. Proposed Method

Here, we employed the fractal descriptors technique over the adjascency matrix of the complex network used to model each texture image.

Initially, we model the original gray-level texture image into a complex network using the approach described in the Section 2.

In the following, we extract fractal descriptors $u(t)$ from the modelled network. Here, we used the function u, followed by a multiscale transform. Essentially, the multiscale tranform of a signal $u(t)$ is an operation which maps into a bidimensional function $U(b, a)$, where b is related to t and a corresponds to the scale of analysis. This operation turns possible a local analysis of the fractal power-law, providing a more complete description of the fractal behavior along the scales in the network and, as a consequence, allows a richer modeling of the original texture.

We employed a specific type of multiscale, named time-scale approach. This is based on the wavelets transform. The result of the operation $U(b, a)$ is given through:

$$U(b, a) = \frac{1}{\sqrt(a)} \int_{\Re} \psi^*(\frac{t-b}{a})u(t)dt, \tag{4}$$

where ψ is the wavelet basis function.

Finally, we apply a threshold over the multiscale response, obtaining the final descriptors used in the discrimination of the textures. The threshold point is obtained empirically, from the training data set, as being that which provided the most precise result.

5. Experiments

The proposed technique was tested in the classification of the well-known Brodatz data set [7]. This is composed by 1110 images divided into 111 classes with 10 images in each one. The classification itself was performed by the Linear Discriminant Analysis (LDA) technique and the result was compared to other classical texture descriptors methods [8], that is, Gabor, Fourier, Laws and a multifractal approach.

The classification is carried out in a hold-out scheme, with one half of the data to train and the another one to test. The results are compared in terms of the correctness rate.

6. Results

The Table 1 shows the correctness rates of each technique in a comparison to the proposed method. We see that the novel descriptors have overcome the classical methods in the classification results.

Method	Correctness Rate (%)	Number of descriptors
Gabor	90.0901	20
Fourier	74.5946	100
Laws	87.0270	15
Multifractal	37.4775	101
Proposed method	92.0721	39

Table 1. Correctness rate for Brodatz dataset.

Such good performance was expected, given the fact that we are summing up the richness of two efficient approaches, that is, fractal descriptors and complex networks modelling. The method combines the distribution of pixel intensities along the image with the complexity of the spatial distribution of patterns and irregularities in the texture. These are fundamental features for the description and discrimination of texture images, allowing a precise and robust result for the proposed task.

We also notice that the proposed technique uses a reasonable number of descriptors. Although it provides more values than Gabor method, in this case, the difference is not significant given the size of the dataset. Therefore, the proposed method presents a low computational classification cost and avoids problems like the dimensionality curse or classification bias.

In the Figure 2, we show the respective confusion matrices of each compared method, illustrating the performance per each class in the analyzed data set.

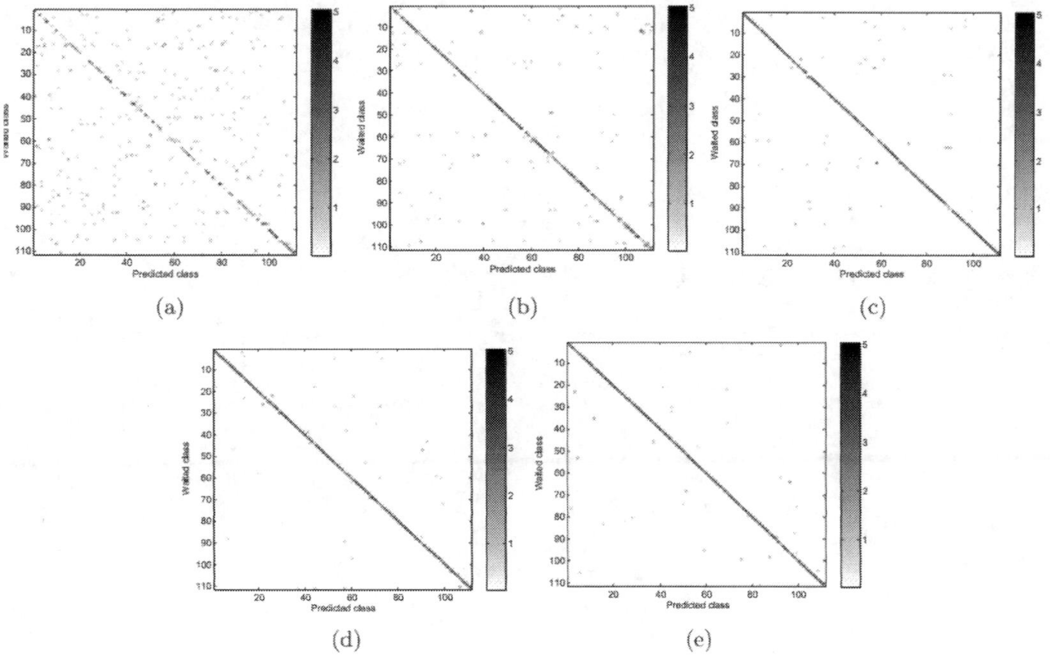

Figure 2. Confusion matrices of the compared methods. a) Multifractal. b) Fourier. c) Laws. d) Gabor. e) Proposed.

The matrices corroborate the efficiency of the proposed descriptors, as this technique showed few gray points outside the diagonal and a more continuous diagonal, showing that the proposed descritors achieved a great performance in all the tested classes.

7. Conclusions

This work proposed a novel methodology for the extraction of descriptors from a texture gray-scale image. The method is based on the modeling of the image by a complex network. The network is composed in a way that each vertex corresponds to a gray-level intensity and the edges are connected if two pixels with the same intensity are 8-connected. In the following, we obtain the descriptors by applying a method which combines the fractal dimension power-law with a multiscale tranform over the adjascency matrix.

The results of a comparison to other classical texture descriptors illustrated the power of the proposed method. The novel technique achieved the best success rate in the classification of a well-known texture data set.

8. Acknowledgements

João B. Florindo gratefully acknowledges the financial support of CNPq (National Council for Scientific and Technological Development, Brazil) (Grant #140624/2009-0). Dalcimar Casanova is grateful to FAPESP (Proc. #08/57313-2) for his doctorate grant. Odemir M. Bruno gratefully acknowledges the financial support from CNPq (National Council for Scientific and Technological Development, Brazil) (Grant #308449/2010-0 and #473893/2010-0) and FAPESP (The State of São Paulo Research Foundation) (Grant # 2011/01523-1).

[1] Costa L F, Oliveira Junior O N, Travieso G, Rodrigues F A, Villas Boas P R, Antiqueira L, Viana M P and Rocha L E C 2011 *Advances in Physics* **60** 329–412
[2] Chalumeau T, Costa L F, Laligant O and Meriaudeau F 2006 *Proceedings of the Second International Conference on Signal-Image Technology and Internet-Based Systems* p 543:550
[3] Costa L F 2004 Complex networks, simple vision URL `http://arxiv.org/abs/cond-mat/0403346`
[4] Chalumeau T, Costa L F, Laligant O, and Meriaudeau F 2008 *Electronic Letters on Computer Vision and Image Analysis* **7** 93–100
[5] Backes A R, Casanova D and Bruno O M 2010 *Progress in Pattern Recognition, Image Analysis, Computer Vision and Applications* (*Lecture Notes in Computer Science* vol 6419) ed Bloch I and Cesar R (Springer Berlin / Heidelberg) pp 354–361
[6] Costa L F, Rodrigues F A, Travieso G and Villas Boas P R 2007 *Advances in Physics* **56** 167–242
[7] Brodatz P 1966 *Textures: A photographic album for artists and designers* (New York: Dover Publications)
[8] Petrou M and García-Sevilla P 2006 *Image processing - dealing with texture* (Wiley)
[9] Matas J, Marik R and Kittler J 1995 *Proceedings of Fifth International Conference on Computer Vision* 726–732
[10] Florindo J B, Backes A R, Castro M and Bruno O M 2012 *Pattern Recognition Letters* **33** 798–806
[11] Florindo J B and Bruno O M 2011 *Chaos* **21** 043112(1–12)
[12] Bruno O M, Plotze R O, Falvo M and de Castro M 2008 *Information Sciences* **178** 2722–2733

Journal of Physics: Conference Series **410** (2013) 012092 doi:10.1088/1742-6596/410/1/012092

Random Walks and Optimal Stopping Strategies as a Model for Evolving Networks

D.S. Vlachos, K.J. Parousis-Orthodoxou and M.M. Stamos

Department of Computer Science and Technology,
Faculty of Sciences and Technology,
University of Peloponnese,
GR-221 00 Tripolis, Greece

E-mail: `dvlachos@uop.gr`

Abstract. In this work, a model for evolving networks is presented based on a Brownian particle. Each time, the Brownian particle enters the network through a randomly selected node. The random walk is terminated after having created m-links with visited nodes. Two strategies have been tested: in the first one, we used a generalized algorithm for the secretary problem in order to maximize the degree of the node with which the new node is connected, while in the second strategy, the Brownian particle creates links with nodes that meets twice. In all cases, scale free, modular, dissasortative networks are created.

1. Introduction

During the last years, there is an increasing interest for the study of complex networks from researchers working in diverse areas of science [1], [2], [3], [4]. Many real world complex systems can be described by networks of nodes or agents connected with edges or interactions. These structures are characterized by both the nature of nodes and the underlying interaction that is represented by the edges. Among the various classes of models used to describe complex networks, *evolving random networks* are by far more realistic. In these models, at each time, a node is added to the network and using a strategy, one or more links are created with other nodes. Let $N(k,t)$ be the number of nodes in the network that have degree less or equal to k at time t. Then, one can easily verify that

$$\frac{\partial N(k,t)}{\partial t} = 1 - q_k^t \left(N(k,t) - N(k-1,t)\right) \tag{1}$$

where q_k^t is the probability of a node with degree k to acquire an edge at time t and it is assumed that $k > m$, where m is the number of edges that each node carries when it is entered the network. The second part of equation (1) says that at each time a new node is entered the network with degree m (thus $N(k,t)$ is increased by 1) but those nodes who have degrre k (they are exactly $N(k,t) - N(k-1,t)$) can acquire an edge and thus the number of nodes with degree $\leq k$ is decreased. Taking the stationary case $t \to \infty$ and noting that $N(k,t) = tP(k)$, where $P(k)$ is the cumulative probability distribution of node degrees, we have

$$\frac{dP(k)}{dk} = \frac{1 - P(k)}{tq_k^t} \tag{2}$$

where he have used the fact that

$$N(k,t) - N(k-1,t) \sim tdP(k)/dk \tag{3}$$

For scale free networks and for k large enough, the cumulative degree distribution must behave as $1 - Ck^{-\gamma+1}$ and replacing in equation (2) we get

$$q_k^t = \frac{k}{(\gamma-1)t} \tag{4}$$

Since $t \sim \sum k$ (because t is equal to the total network nodes and each node carries the same number of edges) we have

$$q_k^t \sim \frac{k}{\sum k} \tag{5}$$

which says that nodes with higher degree are more likely to acquire new edges than the others. This property is known as *preferential attachement* and equation (5) shows its relation to the scale free character of random evolving networks. Since preferential attachement and scale free behavior seem interdependent, how realistic is the fact that when a node is entering the network, knows the degree distribution of all the other nodes? Fortunately, there are models like the *duplication-divergence* one, in which the preferential attachement rule emerges at an effective level from local principles [5]. Following this result, it is of great interest to investigate how local principles and rules based only on limited knowledge of the network topology, can emerge other characteristic properties of real wolrd networks (for example like assortativity or disassortativity, modularity and self-similarity). This question gives the cause for reflection in the current work. The current paper examines the formation of networks under the assumption that random walkers decide to create links with network nodes applying a specific strategy. This strategy can be the optimization of a certain outcome or it can be a simple heuristic rule. The paper is organized as follows: in section 2 general results about random walks and the formulation of optimal stopping strategies are given. In section 3 experimental tests are performed and in section 4 the results from these tests are discussed.

2. Optimal Stoping Strategies in Random Walks

Since many real networks are not static, in order to build a model for them, we adopt the hypothesis that the network grows by addition of new nodes and there is a set of rules govering the way new nodes establish connections with old ones. The proposed model can be summarized as follows: at each time step, a new node is entered in the network and connects with a candidate link to a random node of the network. Then a random walk inside the network is fired, and the final connections are decided during this walk. But before we present the details of the model, we summarize some basic results about the random walks in networks and we give a general formalism for optimal stoping strategies.

2.1. Random Walk in Networks

Consider the case where a new node is entered in a given network and this node has to decide what kind of connections should establish with other nodes. In order to make this decision, the new entered node must have some information about the other nodes, their connectivity patterns etc. But what might be its perspective of the network's topology? An obvious answer is to *walk* the network for some time. For simplicity, we assume that the network is connected, undirected and all edges have the same weight. Let N be the number of nodes before the enrty of the new one and A the adjacency matrix of the network ($A^i_j = 1$ iff there is an edge between nodes i and j, which means that i, j are nearest neighbors). The number of nearest neighbors of node i is its degree k_i. If a t a given time t a Brownian particle is located at node i, then at time $t+1$ it has the same probability to be in any of the nearest neighbors of node i. One can easily verify that after a large amount of movements (compared to the total number of edges), the probability p^i for the Brownian particle to be at node i is given:

$$p^i = \frac{k^i}{k^j \cdot e_j} \tag{6}$$

For the Brownian particle, the observed distance between two nodes (the average number of moves that the particle has to make in order to reach from one node to the other) depends not only on the direct distance (the shortest path) but also on all the possible paths that lead from one node to the other. Let d_j^i be the observed distance between nodes i and j. Then

$$d_j^i = 1 + \sum_{l, l \neq j} P_l^i d_j^l \tag{7}$$

Let now $B(j)_m^l = P_m^l$ if $m \neq j$ and $B(j)_j^l = 0$, then the observed distances are given

$$[I - B(j)] \cdot \begin{pmatrix} d_j^1 \\ d_j^2 \\ \vdots \\ d_j^N \end{pmatrix} = \begin{pmatrix} 1 \\ 1 \\ \vdots \\ 1 \end{pmatrix} \tag{8}$$

Equation 8 can be effectively solved in the case of sparse matrices (see for example [6] and [7]). The matrix d_j^i contsins all the necessary information about how far is node i from node j (assuming Brownian motion). Note here that $d_j^i \neq d_i^j$ (for example if the nodes i, j are connected and the degree of i is $k_i = 1$ while the degree of j is $d_j > 1$ the $d_j^i = 1$ but $d_i^j > 1$). It can be easily seen that nodes with high degree have a pronounced effect on d_j^i and if the clustering coefficient of the network is high enough, a random walk in the network for the majority of the time is located on nodes with high degree.

One important feature of a random walk in a network is the probability that since the random walk is started from an arbitrary node, after m steps, the current node has a specified degree. Thus, we introduce the *modified degree distribution* $P^{(m)}(k)$ which is given:

$$P^{(m)}(k) = \sum_{k_1, k_2, \ldots, k_m} P(k_1) \cdot P(k_2|k_1) \cdot \ldots \cdot P(k|k_m) \tag{9}$$

where

- $P(k)$ is the degree distribution (the probability that a node has degree k).
- $P(k|k')$ is the conditioned degree distribution (the probability that a node with degree k is connected to a node with degree k').

Equation 9 can be rewritten using the joint degree distribution $P(k, k')$ (the probability that two connected nodes have degrees k, k'):

$$P^{(m)}(k) = \sum_{k_1, k_2, \ldots, k_m} P^{(0)}(k_1) \frac{< k >^m}{k_1 \cdot k_2 \cdot \ldots \cdot k_m} \frac{P(k_1, k_2) \cdot P(k_2, k_3) \cdot \ldots \cdot P(k_m, k)}{P(k_1) \cdot P(k_2) \cdot \ldots \cdot P(k_m)} \tag{10}$$

where we have used the identity

$$P(k'|k) = \frac{< k >}{k} \frac{P(k, k')}{P(k)} \tag{11}$$

and $< k >$ is the mean degree. Obviously $P^{(0)}(k) = P(k)$.

2.2. Optimal Stopping Strategies

Let $X_1, X_2, \ldots$ be the outcomes of random identical independent experiments and assume that all of them follow the distribution $F(x)$ ($F(x)$ is the propbability that the outcome $X \leq x$). Suppose that a *Decision Maker* (DM) observes the sequence of up to n outcomes, and at each time has two choices:either to accept or reject the current outcome. Once a decision is made (an outcome is selected), the problem terminates. If reached the nth outcome must be selected, and once rejected, an outcome cannot be recalled. At each time, the DM only observes an indicator of X_t which says whether each observed oucome is the best so far. The DM's payoff for selecting the tth outcome with $X_t = x_t$ is itself x_t, while in the classical secretary problem this payoff is either 1 if the best outcome is selected and 0 otherwise [8], [9].

Let

$$I_t = \begin{cases} 1 & , \quad \text{if} X_t = \max\{X_1, X_2, \ldots, X_{t-1}, X_t\} \\ 0 & , \quad \text{otherwise} \end{cases} \tag{12}$$

The probability that the outcome $X_t \leq X$ given that X_t is the maximum observed one is given:

$$\begin{aligned} P(X_t \leq X | I_t = 1) &= F(X)^{t-1} \cdot F(X) \\ &= F(X)^t \end{aligned} \tag{13}$$

and the density function of the variable X_t given that it is the maximum observed is given:

$$f(X_t | I_t = 1) = t \cdot F(X_t)^{t-1} \cdot f(X_t) \tag{14}$$

where $f(X_t)$ is the unconditional density function of X_t ($f(x) = F'(x)$). Let now E_t be the mean value of X_t given that it is the maximum observed, that is $E_t = E(X_t | I_t = 1)$. Then:

$$\begin{aligned} E_t &= \int_0^1 t \cdot x \cdot F(x)^{t-1} \cdot f(x) \cdot dx \\ &= \int_0^1 x \cdot dF(x)^t \end{aligned} \tag{15}$$

Consider now the case where the DM is called to make multiple selections in order to maximize a given function of them. Let m be the number of selections and $G(X_1, X_2, \ldots, X_m)$ is the function to be maximized. Assume that G ia a non decreasing function in any of its m-arguments. The adopted strategy is again the obvious one: the first $c-1$ outcomes are skiped and then the m maximal outcomes are selected (until the number of tries n is reached). In order to calculate averages of the given function G, one has to calculate first the joint probability density function $f_c^{n,m}(x_{t_1}, x_{t_2}, \ldots, x_{t_m}) = f(x_{t_1}, x_{t_2}, \ldots, x_{t_m} | I_{t_1} = 1, I_{t_2} = 1, \ldots, I_{t_m} = 1, t_1 < t_2 < \ldots < t_m)$ with the parameters c and n for the selected strategy. The time t_1 that the first outcome is selected can be any between c and $n-m$. The time t_2 that the second outcome is selected can be any between $t_1 + 1$ and $n - m + 1$ and finally the time t_m that the last outcome is selected can be any between $t_{m-1} + 1$ and n. Let

$$r_{t_1}^{t_2} = (1 - p_{t_1}) \cdot (1 - p_{t_1+1}) \cdot \ldots \cdot (1 - p_{t_2-1}) \cdot p_{t_2}, \; t_1 \leq t_2 \tag{16}$$

be the probability that after bypassing $t_1 - 1$ outcomes, the t_2 outcome is the first with $I_{t_2} = 1$ (p_t is the probability that $I_t = 1$ as it was defined earlier). We can argue now that:

$$\begin{aligned} f_c^{n,m}(x_{t_1}, x_{t_2}, \ldots, x_{t_m}) &= \sum_{t_1=c}^{n-m} r_c^{t_1} \cdot f(x_{t_1} | I_{t_1} = 1) \cdot f_{t_1+1}^{n,m-1}(x_{t_2}, \ldots, x_{t_m}) \\ &+ \frac{r_c^{n-m}}{p_{n-m}} \prod_{i=1}^{m} f(x_{t_i}) \end{aligned} \tag{17}$$

Similarly,

$$f_{t_1+1}^{n,m-1}(x_{t_2},\ldots,x_{t_m}) = \sum_{t_2=t_1+1}^{n-m+1} r_{t_1+1}^{t_2}\cdot f(x_{t_2}|I_{t_2}=1)\cdot f_{t_2+1}^{n,m-2}(x_{t_3},\ldots,x_{t_m}) + \frac{r_{t_1+1}^{n-m+1}}{p_{n-m+1}}\prod_{i=2}^{m} f(x_{t_i}) \quad (18)$$

and recursively we can calculate $f_c^{n,m}$. Then, the mean value of the function G is given

$$E_G = \int f_c^{n,m}(x_{t_1},x_{t_2},\ldots,x_{t_m})\cdot G(x_{t_1},x_{t_2},\ldots,x_{t_m})\cdot dx_{t_1}\cdot dx_{t_2}\cdot\ldots\cdot dx_{t_m} \quad (19)$$

Finally, setting $\frac{\theta E_G}{\theta c}=0$ we can obtain the optimal strategy (the value of c).

3. Experimental results

3.1. Networks using optimal stopping strategies

Two networks have been constructed using the following strategies: The first network was build using a maximum number of steps equal to 5 and a cutoff parameter equal to 2 (using the results of the previous section). The second network was build using the same strategy, but now the Brownian particle is allowed to travel for more time by using a maximum number of steps equal to 20. In both networks, the Brownian particle stops its motion after making 2 links ($m = 2$). The degree distribution of these two networks is drawn in figure 1. Both networks follow a power law in degree distribution with exponents 2.4 and 2 respectively.

3.2. Networks using the m-Crossing strategy

Two networks have been constructed using the following strategy (m-Crossing strategy): a Brownian particle is entered in the network and makes a link with the first node that it meets twice. The motion can be continued until m-links have been created. The first network was created using the m-Crossing strategy with $m = 2$, while in the second network we used $m = 1$. Both networks follow a power law in degree distribution with exponents 2.4 and 1.9 respectively. The degree distribution of these two networks is drawn in figure 1.

4. Conclusions

We have tested a model based on a Brownian particle for evolving networks. All networks created are scale free which means that there is an emerging preferential attachment rule for the creation of links. This can be easily verified since random walks on randomly selected network edges, visit nodes with higher degree more often. In the case of the optimal stopping strategy, if the number of allowed steps is high enough, the Brownian particle can reach the most connected hubs and thus the exponent of the power law is low. On the other hand, if the number of allowed steps is low, then a percentage of links are created randomly and thus the exponent in the power law is higher [11].

Concerning the degree correlation, these networks are dissasortative. Concerning the modularity of these networks, it must be quite high, since if a Brownian particle enters in the neighborhood of a hub, and having in mind the strong dissasotativity of the network, it is difficult to escape. Moreover, the clustering coefficient of these networks must be quite high if the number of links created by newly entered node is > 1. Again this can be verified by the strong dissasortativity of the network. In the case where only one link is created, the network must have very low clustering coefficient, since the probability for the formation of a triangle is negligible.

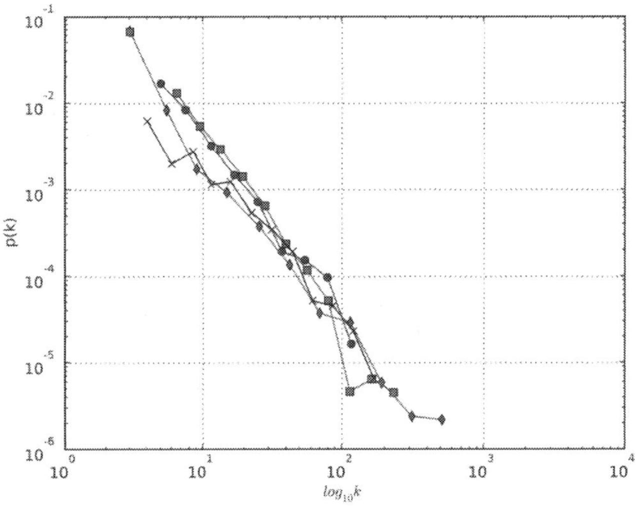

Figure 1. Degree distribution for the following networks: ○-network using optimal stopping strategy with maximum steps 5 and cutoff paramete 2 ($\gamma \sim 2.4$), ◇ -network using optimal stopping strategy with maximum steps 20 and cutoff paramete 2 ($\gamma \sim 2.$), □- network using the m-Crossing strategy with $m = 2$ ($\gamma \sim 2.4$) and ×-network using the m-Crossing strategy with $m = 1$ ($\gamma \sim 1.9$).

Acknowledgement

The project is implemented under the Operational Program "Education and Lifelong Learning" and funded by the European Union (European Social Fund) and by Greek national resources.

[1] Albert, R., Barabasi, A.-L.: Statistical Mechanics of Complex Networks. Rev. Mod. Phys. 74, 47 (2002)
[2] Dorogovtsev, S. N., Mendes, J. F. F..: Evolution of Networks: From Biological Nets to the Internet and WWW. Oxford University Press, Oxford (2003)
[3] Pastor-Satorras, R., Vespignani, A.: Evolution and Structure of the Internet: A Statistical Physics Approach. Cambridge University Press, Cambridge (2004)
[4] Cohen, R., Havlin, S.: Complex Networks: Structure, Stability and Function. Cambridge University Press, Cambridge (2008)
[5] Vázquez, A.: Disordered networks generated by recursive searches, Europhysics Letters 54, 430 (2001)
[6] Tewarson, R.P.: Sparse Matrices. Academic Press, New York (1973)
[7] Davis, T.A., Duff, I.S..: A combined unifrontal/multifrontal method for unsymmetric sparse matrices, ACM Trans. Math. Softw. 25, 1–3 (1999)
[8] Ferguson, T.S..: Who solved the secretary problem?. Statistical Science 4, 282–296 (1989)
[9] Samuels, S.M.: Secretary problems. In: Gosh, B.K., Sen, P.K. (eds.) Handbook of sequential analysis, pp. 381–405. Marcel Dekker, New York (1991)
[10] Bearden, J. Neil.: A new secretary problem with rank-based selection and cardinal payoffs. Journal of Mathematical Psychology 50, 58–59 (2006)
[11] Xenides, D., Vlachos, D.S., Simos, T.E.: Synchronization in complex systems following a decision based queuing process: rhythmic applause as a test case, Journal of Statistical Mechanics: Theory and Experiment P07017 (2008)

Evolution of Correlations in Complex Networks

M.M. Stamos, K.J. Parousis-Orthodoxou, and D.S. Vlachos

Department of Computer Science and Technology,
Faculty of Sciences and Technology,
University of Peloponnese,
GR-221 00 Tripolis, Greece

E-mail: dvlachos@uop.gr

Abstract. We investigate in detail the mechanisms under which degree correlations evolve in complex networks. We consider the case where a vertex is entering the network at each time, carrying a predefined number of edges. We prove in this work, that the same elementary interactions which is responsible for emerging of scaling in complex networks, can give several patterns of degree corellations. As a test case, the effect of preferential attachement rule in degree correlations is studied in detail.

1. Introduction

During the past years there has been a vast progress in the field of complex networks [1], [2]. Various networks in nature, science and technology display, so called, complex phenomena. Networks such as the Internet, at the autonomous level or even the World Wide Web, the protein correlations network or social networks such as the friendship network or the sexual network [3], [4].

Strikingly, many of these networks have complex topological properties and dynamical features that cannot be accounted for by classical graph modelling [5]. For example, small-world properties [6] and scale-free (SF) degree distributions [7] (where the degree of a node is defined as the number of other nodes to which it is attached) seem to influence significantly the topology of real world networks. These global properties show a large connectivity and a short average distance between nodes, which have considerable impact on the behaviour of physical processes taking place on the network.

Interactions in networks may be described in terms of graphs, consisting of vertices and edges, where they represent the connections in the graph (their interactions). This approach was initiated by Erds and Rnyi (ER) [5]. In the ER model, the number of vertices is fixed, while edges connecting one vertex to another occur randomly with certain probability. However, the ER model is too random to describe real complex systems. Earlier studies concentrated on simple random networks, and it was recently discovered that many complex networks are hierarchically organised. Recently, Watts and Strogatz (WS) [6] introduced a small-world network, where a fraction of edges on a regular lattice is rewired with probability p_{WS} to other vertices. More recently, Barabsi and Albert (BA) [7], [9], [10] and [11], introduced a new approach of an evolving network where the number of vertices N increases linearly with time (rather than fixed), and a newly introduced vertex connects to the preexisting network with m edges, following the so-called preferential attachment (PA) rule. The PA rule means that the probability for the new vertex to connect to an already existing vertex is proportional to the degree k of the selected vertex. Then the degree distribution $P_D(k)$ follows a power law $P_D(k) \sim k^{-\gamma}$ with $\gamma = 3$ for the BA model, while for the ER and WS models, it follows a Poisson distribution. In SF networks, one may wonder if the exponent γ is universal in analogy with the theory of critical phenomena. However, the exponent γ turns out to be sensitive to the detail of network structure. Thus, a universal quantity for SF networks is yet to be found.

In this work, we derive a master equation for the degree correlation of an evolving network. The derivation is presented in section 2, while in section 3, simulating results are produced in the case of preferentail attachement rule. Finally, in section 4, conclusions are drawn and open questions for future work are posed.

2. Evolution of Correlations

Suppose that p_k is the degree distribution of our network, i.e., the fraction of vertices in the network with degree k, or equivalently the probability that a vertex chosen uniformly at random from the network will have degree k. The vertex at the end of a randomly chosen edge in the network will have degree distributed in proportion to kp_k , the extra factor of k arising because k counts the available edges of a node with degree k [14], [15] and [16]. In the absence of correlations, the probability that a randomly selected edge has endpoints with degrees (k, k') is proportional to $k \cdot p_k \cdot k' \cdot p_{k'}$. On the other hand, degree correlations dictate that there is a conditional degree distribution $P(k'|k)$ which is the probability that a vertex with degree k is connected with a vertex with degree k'.

Assume that initially there are m_0 vertices with no edges in the network. At each time $t > m_0$, a node is entering the network and makes m connections with other vertices already in the network (suppose that $m \leq m_0$). We define as the k-*island* the set of vertices that have degree k and let $E^t_{k,k'}$ be the number of edges connecting the k and k' islands (that is the number of edges that their endpoints have degrees (k, k')) . Let N^t_k be the number of vertices in the k-island at time t, $P^t(k'|k)$ the conditional degree distribution at time t, $< k >^t$ the mean degree at time t, N^t the number of nodes in the network at time t (which is equal to t), $P^t(k)$ the degree distribution at time t, $P^(k, k')$ the joint degree distribution at time t which is given by

$$P^t(k, k') = \frac{E^t_{k,k'}}{< k >^t N^t} \tag{1}$$

and q^t_k the probability that a vertex with degree k gains an edge at time t. Since at each time m-edges are created

$$\sum_{k=1}^{\infty} q^t_k N_k = m \tag{2}$$

The number of edges connecting the k and k' islands at time $t + 1$ is given:

$$\begin{aligned} E^{t+1}_{k,k'} = E^t_{k,k'} & + q^t_{k-1} N^t_{k-1} P^t(k'|k-1)(k-1) \\ & + q^t_{k'-1} N^t_{k'-1} P^t(k|k'-1)(k'-1) \\ & - q^t_k N^t_k P^t(k'|k) k \\ & - q^t_{k'} N^t_{k'} P^t(k|k') k' \\ & + q^t_{k'-1} N^t_{k'-1} \delta_{k,m} \\ & + q^t_{k-1} N^t_{k-1} \delta_{k',m} \end{aligned} \tag{3}$$

where δ is the Kronecker symbol. The first term in equation 3 says that a new vertex connects to another vertex with degree $k - 1$ with probability $q^t_{k-1} N^t_{k-1}$ since we have N^t_{k-1} vertices with degree $k - 1$. This vertex has $P^t(k'|k - 1)(k - 1)$ links with vertices with degree k. Since the connected vertex with degree $k - 1$ has now degree k, those links must be counted now to $E^{t+1}_{k,k'}$

. The second term in equation 3 has the same meaning (interchange k and k'). The third term says that a new vertex connects to another vertex with degree k with probability $q_k^t N_k^t$. This vertex has $P^t(k'|k)k$ links with vertices with degree k'. Since the connected vertex with degree k will have now degree $k+1$, these links are not counted now to $E_{k,k'}^t$. The fourth term has the same meaning. Finally, the fifth and sixth term counts the new edges that connect the m-island with the other islands (note here that the new vertex enters the m-island). It can be easily verified that

$$\sum_{k,k'>0} E_{k,k'}^{t+1} = \sum_{k,k'>0} E_{k,k'}^t + 2m \tag{4}$$

as expected. For the initial condition at time $t = m_0$ we have:

$$N_k^{m_0} = m_0 \delta_{k,0}, \quad P^{m_0}(k) = \delta_{k,0}, \quad and\ P^{m_0}(k,k') = 0 \tag{5}$$

Equation 3 along with the initial conditions 5 can produce $E_{k,k'}$ at any time given q_k^t. Since

$$P(k,k') = \frac{E_{k,k'}}{<k> N} \tag{6}$$

and

$$kP(k)P(k'|k) =<k> P(k,k') \tag{7}$$

we can divide 3 with $<k>^{t+1} N^{t+1} =<k>^{t+1} (t+1$. The result, after some rearrangements, reads :

$$\begin{aligned} t\left(\frac{<k>^{t+1}}{<k>^t}P^{t+1}(k,k') - P^t(k,k')\right) + \frac{<k>^{t+1}}{<k>^t}P^{t+1}(k,k') + \left(Q^t(k) + Q^t(k')\right)P^t(k,k') = \\ Q^t(k-1)P^t(k-1,k') + Q^t(k'-1)P^t(k,k'-1) + \\ Q^t(k-1)\frac{P^t(k-1)}{<k>^t}\delta_{k',m} + Q^t(k'-1)\frac{P^t(k'-1)}{<k>^t}\delta_{k,m} \end{aligned} \tag{8}$$

where $Q^t(k) = tq_k^t$.

Taking now the limiting case where $t \to \infty$, we have $<k>^{t+1} =<k>^t$ and $P^{t+1}(k,k') = P^t(k,k')$ thus

$$\begin{aligned} P(k,k') = & \frac{Q(k-1)P(k-1,k') + Q(k'-1)P(k,k'-1)}{1+Q(k)+Q(k')} + \\ & \frac{Q(k-1)P(k-1)\delta_{k',m} + Q(k'-1)P(k'-1)\delta_{k,m}}{2m(1+Q(k)+Q(k'))} \end{aligned} \tag{9}$$

where $Q(k) \overset{t\to\infty}{=} Q^t(k)$. Since

$$P(k) = \frac{<k>}{k} \sum_{k'>0} P(k,k') \tag{10}$$

and summing equation 9 for $k' > 0$ we have

$$P(k) = \frac{Q(k-1)P(k-1)}{1+Q(k)}, \ k > m \tag{11}$$

and

$$P(m) = \frac{1}{1+Q(m)} \tag{12}$$

Equation 11 can be rewritten as

$$P(v) + Q(v)P(v) = Q(v-1)P(v-1) \tag{13}$$

Summing for $v = m+1$ until k we have

$$F(k) - 1 + Q(k)P(k) = 0 \tag{14}$$

where $F(k)$ is the cumulative degree distribution and we have used equation 12. In the case of scale free networks, for large k we must have

$$F(k) = 1 - Ck^{1-\gamma} \tag{15}$$

where after substitution in 14 and taking into account that $P(k) = F'(k)$ we have

$$Q(k) \stackrel{k>>1}{=} \frac{k}{\gamma - 1} \tag{16}$$

3. Simulation Results

As an example, we examine the evolution of correlations in the well studied Barabasi-Albert network [7], In this network, the probability of a vertex with degree k to gain an edge at time t is given:

$$q_k^t = m\frac{k}{\sum_{j=1}^{N^t} k_j} \tag{17}$$

where the sum in the denominator is running on all network vertices and thus

$$q_k^t = m\frac{k}{2mt} \tag{18}$$

since the sum of all degrees is twice the number of edges and $N^t = t$. For $Q(k)$ we have

$$Q(k) = \frac{k}{2} \tag{19}$$

and thus the produced network is scale free with exponent equal to 3. Equation 8 now becomes

$$\begin{aligned} t\left(P^{t+1}(k,k') - P^t(k,k')\right) + P^{t+1}(k,k') + \frac{k+k'}{2}P^t(k,k') = \\ \frac{k-1}{2}P^t(k-1,k') + \frac{k'-1}{2})P^t(k,k'-1) + \\ \frac{k-1}{2}\frac{P^t(k-1)}{m}\delta_{k',m} + \frac{k'-1}{2}\frac{P^t(k'-1)}{m}\delta_{k,m} \end{aligned} \tag{20}$$

with initial conditions

$$P^{m_0}(k,k') = 0$$
$$P^{m_0}(k) = \delta_{k,0}$$

Figure 1 shows the calculated joint degree distribution for $t = 10^2,\ 10^3,\ 10^4$ and ∞. The disassortative neture of the Barabasi-Albert network is obvious for large times, while this is not the case when the network is small enough. This happens because initially, the edges are quite uniformly distributed among the vertices, but as the time increases, some vertices collect the majority of new edges (emergence of scaling). These veritces, have small probability that they are connected due to the sparse nature of the network. On the other hand, as time increases, since neighbors of vertices with high degree have small degree, it is expected that the clustering coefficient will decrease with increasing node degree. This is shown in figure 2 where the clustering coefficient is calculated for $t = 10^2,\ 10^3,\ 10^4$.

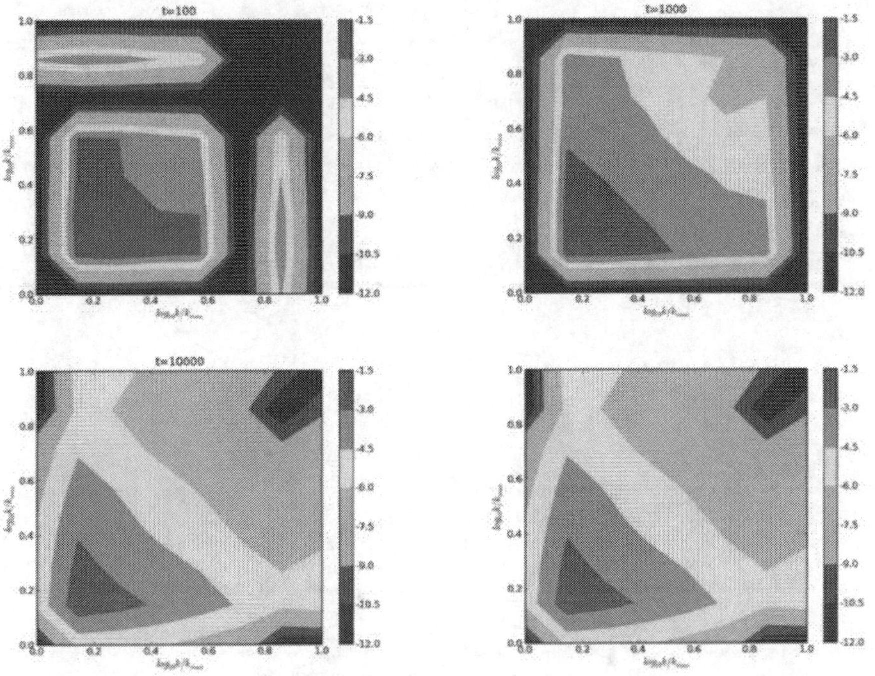

Figure 1. Calculated joint degree distribution for $t = 10^2,\ 10^3,\ 10^4$ and ∞ for an evolving network following the preferential attachement rule.

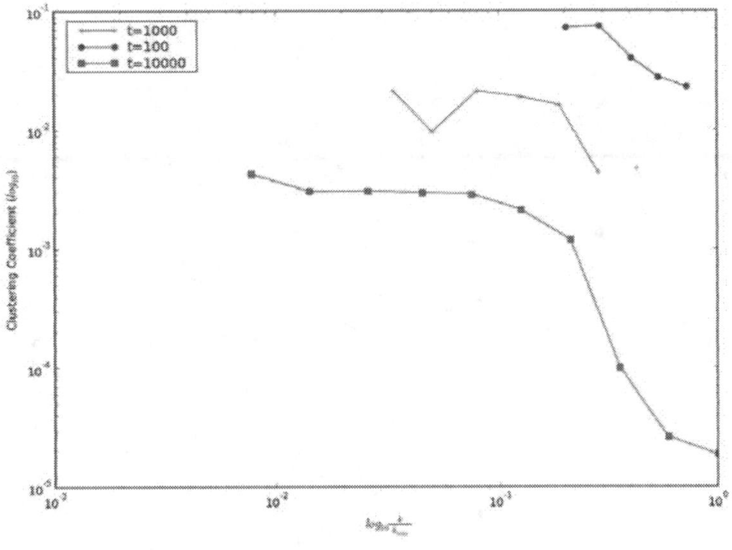

Figure 2. Calculated distribution of clustering coefficient for $t = 10^2,\ 10^3,\ 10^4$ for an evolving network following the preferential attachement rule.

4. Conclusions

It has been shown in this work, that the same elementary processes that are responsible for the scale free structure of real world complex networks, can produce different patterns of degree correlations. This result comes from the construction of a master equation for the evolution of degree correlations in an evolving network in which the only parameter is the probability that a vertex with a predefined degree gains an edge in a specific time ($Q^t(k)$). The master equation was tested in the Barabasi-Albert network where the well studied behaviour of these networks was reproduced. The master equation for the evolution of correlations can be used to model various kikds of networks by simply changing the probability $Q^t(k)$.

Acknowledgement

The project is implemented under the Operational Program "Education and Lifelong Learning" and funded by the European Union (European Social Fund) and by Greek national resources.

[1] R. Albert and A L. Barabasi, *Rev. Mod. Phys.* 74, 47 (2002)
[2] S.N. Dorogovtsev and J.F.F. Mendes, *Adv. Phys.* 51, 1079 (2002)
[3] M.E.J. Newman and Juyong Park, *Physical Review E* 68, 036122 (2003)
[4] Andrea Capocci, Guido Caldarelli, and Paolo De Los Rios, *Physical Review E* 68, 047101 (2003).
[5] P. Erds and P. Rnyi, *Publ. Math. Inst. Hang. Acad. Sci.* 5, 17 (1960).
[6] D.J. Watts and S.H. Strogatz, *Nature* (London) 393, 440 (1998).
[7] A-L. Barabasi and R. Albert, *Science* 286, 509 (1999).
[8] M.E.J. Newman, Stephanie Forest, and Justin Balthorp, *Physical Review E* 66, 035101 (2002).
[9] P.L. Krapivsky, S. Render and F. Leyvraz, *Phys. Rev. Lett.* 85, 4629 (2000).
[10] P.L. Krapivsky, S. Render, *Physical Review E* 63, 066123 (2001).
[11] S.N. Dorogovtsev, J.F.F. Mendes and A.N. Samunkhin, *Phys. Rev. Lett.* 85, 4633 (2000).
[12] L.A. Amaral, A. Scala, N. Barthelemy, and H.E. Stanley, *Proc. Natl. Acad. Sci.* U.S.A. 97, 11 149 (2000).
[13] G. Caldarelli and A. Vespignani (eds), Large Scale Structure and Dynamics of Complex Networks, World Scientific (2007).
[14] S. Feld, *Am. J. Sociol.* 96, 1464 (1991).
[15] M. Molloy and B. Reed, *Random Struct. Algorithms* 6, 161 (1995).
[16] M.E.J. Newman, S.H. Strogatz, and D.J. Watts, *Phys. Rev. E* 64, 026118 (2001).

Emergence of scale-free networks from optimization process

F. L. Forgerini[1,2], **S. N. Dorogovtsev**[1,3] **and J. F. F. Mendes**[1]

[1]Departamento de Física and I3N, Universidade de Aveiro, 3810-193 Aveiro, Portugal
[2]ISB, Universidade Federal do Amazonas, 69460-000 Coari-AM, Brazil
[3]A. F. Ioffe Physico-Technical Institute, 194021 St. Petersburg, Russia

E-mail: `fforgerini@ua.pt`

Abstract. Preferential attachment is a standard mechanism producing power-laws in growing networks. Thanks to its simplicity this mechanism is realized in most of the models of scale-free network, but, unfortunately, it rather mimics scale-free networks and not explains them. Optimization based mechanisms have a much greater potential to explain the evolution of scale-free networks. We consider one of the simplest optimization based models generating power-law growing networks. Our model is defined as follows. At each time step, a new node is created and connected to m previous nodes in the network, which are selected to minimize the product $s^{\alpha}r$, where s is the birth time of the node and r is a random number drawn from some distribution. In the case of complete optimization, the networks generated by this model have a power-law degree distribution with the exponent $\gamma = 1 + 1/\alpha$ for a wide range of the random number distributions. For partial optimization, including a finite fraction of nodes in a network, we observe an exponential degree distribution.

1. Introduction

The joint feature of numerous real-world networks including the Internet, scientific collaborations, WWW, protein and gene interaction networks, etc., is their scale-free organization. In all these examples, the number of nodes of a degree q follows a power-law distribution, $P(q) \sim q^{-\gamma}$ with the exponent γ typically in the range $2 \leq \gamma \leq 3$. One of the most studied mechanisms producing power-laws in networks is the preferential attachment mechanism [1, 2]. In this simple mechanism, more popular nodes with many connections receive new connections preferentially [3, 4]. Preferential attachment can also be considered as a consequence of various processes on networks, e.g., optimization processes aiming at an optimal network design. Moreover, optimization can be treated as an alternative mechanism explaining complex network architectures [5, 6, 7, 8, 9].

Although the idea of preferential attachment is simple and elegant, often the preferential attachment itself cannot be explained. In addition, sometimes it is simply impossible to know about the degrees of all vertices in a network. In this respect, standard preferential attachment models are often not quite realistic. It was shown recently that a refined optimization model, incorporating trade-off between popularity and similarity of nodes, can describe real-world network architectures remarkably well [10]. In the present work we demonstrate that even within a very simple optimization based evolution model one can obtain a scale-free network with exponent γ in a wide range of values. This result is valid only for complete optimization, in which information about all nodes in the network is taken into account. For partial optimization, including a finite fraction of the network at each step, or even a few randomly selected nodes, the degree distribution of a growing network has an exponential form.

2. Optimization based model for growing networks

Our model of a growing network is formulated as follows. At each time step we create a new node t and attach it to m previous nodes in the network. Each of these m nodes is selected from the set of all existing nodes by minimization of the product $s^\alpha r_s$, where the label s is the birth time of a node, $1 \leq s \leq t$, r_s is a random number taken from a given distribution $p(r)$, and $\alpha \geq 0$ is a new exponent. Here the random variable r actually plays the role of multiplicative noise.

Our simulations demonstrate that if the optimization process incorporates all existing nodes at each step (complete optimization), then the growing network is scale-free. In Fig. 1 we show the cumulative degree distribution of a network generated by the complete optimization model, in which α is set to 1 and $m = 1, 2, 5, 10$, after averaging over 100 samples. The random numbers are uniformly distributed and the probability density function is

$$p(r) = \begin{cases} 1 & \text{for } 0 < r < 1 \\ 0 & \text{otherwise.} \end{cases}$$

One can see that the exponent γ of the power-law node degree distribution in our model approaches 2 for any m. Introducing $\alpha < 1$ leads to $\gamma > 2$. For the sake of brevity here we mostly discuss the simplest case of $\alpha = 1$.

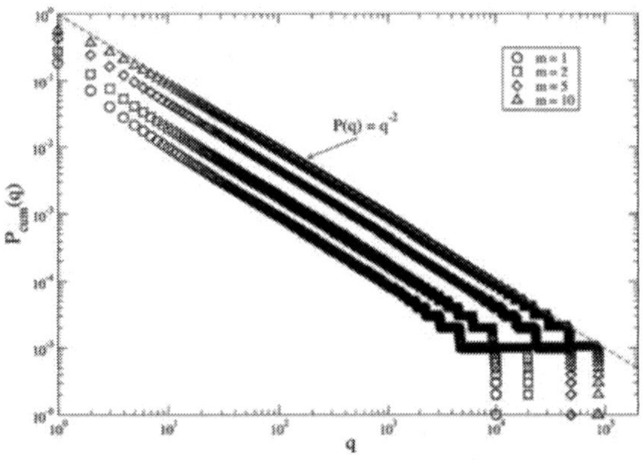

Figure 1. Cumulative degree distribution fora complete optimization process with various values of m, $\alpha = 1$ generating a network of 10^5 nodes averaged over 100 different samples. The distribution $p(r)$ is uniform, $p(0 < r < 1) = 1$ and $p(r > 1) = 0$. The resulting power law has exponent $\gamma = 2$ (dashed line).

3. Generation of power-law degree distributions

Let us derive the degree distribution for our model. Here we only present a simple estimate in the case of $\alpha = 1$, $m = 1$, and for a random number uniformly distributed from 0 to 1. A straightforward derivation for an arbitrary α and other distributions $p(r)$ will be given elsewhere.

In the case of the uniform $p(r)$ defined above, the distribution of the product $\mathcal{P}(sr_s)$ is also uniform, namely $\mathcal{P}(0 < sr_s < s) = 1/s$ and $\mathcal{P}(sr_s > 0) = 0$. Consequently, for small values of sr_s, we have $\mathcal{P}(sr_s) = 1/s$. This means that the probability that node s has the smallest product sr_s is proportional to $1/s$. Therefore, the mean degree $\langle q \rangle(s,t)$ of node s increases according to the following relation:

$$\frac{\partial \langle q \rangle(s,t)}{\partial t} \sim \frac{1}{s \ln t}. \tag{1}$$

So we have $\langle q\rangle(s) \sim 1/s$. In the continuum approximation which is applicable to scale-free networks, this corresponds to the degree distribution

$$P(q) = -\frac{1}{t}\Big(\frac{\partial q(s)}{\partial s}\Big)^{-1}\Big|_{s=s(q)\sim 1/q} \sim \frac{1}{q^2}, \tag{2}$$

where, as is usual in the continuous approximation, we set $\langle q\rangle(s,t) = q(s,t)$. This result agrees with our simulation, Fig. 1. Generally, for $\alpha \leq 1$, assuming $p(r = 0) \neq 0$, one can obtain the relation $\gamma = 1 + 1/\alpha$.

In the work [11], attachment to a node of the maximal degree selected from a random sample of a few nodes was studied. The degree distributions of the resulting networks was found to be rapidly decaying. Inspired by these ideas we modify our model and consider a partial optimization process, in which at each step, the optimal node for attachment is selected from a finite fraction of the existing networks, namely from a uniformly randomly chosen fraction f of all nodes.

In Fig. 2 we show a linear-log plot of the cumulative degree distribution for the result of the partial optimization process, $f = 0.01$. One can see that for various values of m, the degree distribution decays exponentially. In Fig. 3 we show the cumulative degree distribution obtained for various values of f, where $f = 1$ corresponds to complete optimization resulting in the scale-free network having $\gamma = 2$, while $f = 0$ actually corresponds to the standard random recursive tree in which new nodes are attached to uniformly randomly chosen existing nodes. The random recursive graph is well-known to have an exponentially decaying degree distribution. On the other hand, we find that $0 < f < 1$ produces an exponential cutoff of the power-law degree distribution, Fig. 3.

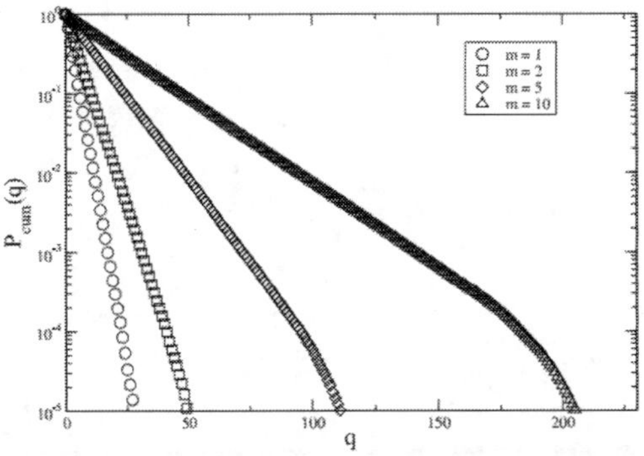

Figure 2. Linear-log plot of the cumulative degree distribution of the network of 10^5 nodes generated by a partial optimization process, $f = 0.01$, $m = 1, 2, 5, 10$, $\alpha = 1$.

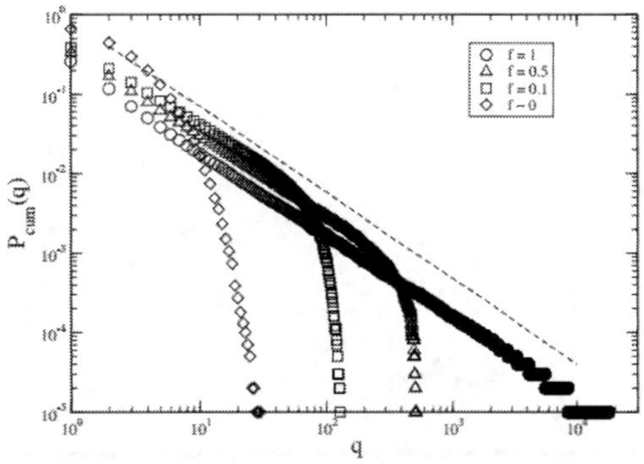

Figure 3. Log-log plot of the cumulative degree distribution of the network of 10^5 nodes generated by a partial optimization process, $f = 0, 0.1, 0.5, 1$, $m = 1$, $\alpha = 1$.

4. Conclusions

In summary, we considered, maybe, the simplest example of the optimization driven evolution of complex networks. The resulting networks are scale-free if at each step, the optimization involves all existing nodes in a network. If the optimization is partial, i.e., it includes only a finite fraction of a network or a few nodes, the result is an exponential cut-off of a power-law degree distribution or even an exponential degree distribution. We suggest that the optimization driven evolution is a widespread mechanism generating complex networks architectures.

Acknowledgments
F. L. Forgerini was supported by FCT project SFRH/BD/68813/2010. This work was partially supported by projects PTDC/FIS/108476/2008, PTDC/SAU-NEU/103904/2008, and PTDC/MAT/114515/2009.

References
[1] Barabási A-L and Albert R 1999 *Science* **286** 509.
[2] Albert R and Barabási A-L 2002 *Rev. Mod. Phys.* **74** 47.
[3] Krapivsky P L, Redner S and Leyvraz F 2000 *Phys. Rev. Lett.* **85** 4629.
[4] Dorogovtsev S N, Mendes J F F and Samukhin A N 2000 *Phys. Rev. Lett.* **85** 4633.
[5] Ferrer i Cancho R and Solé R V 2003 *Statistical Mechanics of Complex Networks* (Optimization in complex networks) eds Pastor-Satorras R *et al* (Berlin: Springer) chapter 7 pp 114-125.
[6] Dorogovtsev S N, Mendes J F F and Samukhin A 2001, *Phys. Rev.* E **63** 062101.
[7] Caldarelli G, Capocci A, Rios P D L and Muñoz M A 2002 *Phys. Rev. Lett.* **89** 258702.
[8] Fabrikant A, Koutsoupias E and Papadimitriou C H 2002 *Lect. Notes Comp. Sci.* **5632** 110.
[9] D'Souza R M, Borgs C, Chayes J T, Berger N and Kleinberg R D 2007, *Proc. Natl. Acad. Sci. USA* **104** 6112.
[10] Papadopoulos F, Kitsak M, Serrano M A, Bogũña M and Krioukov D, arXiv:1106.0286.
[11] D'Souza R M, Krapivsky P L and Moore C 2007 *Eur. Phys. J.* B **59** 535.

Economic consideration of optimal vaccination distribution for epidemic Spreads in complex networks

Bing Wang*[,1], Hideyuki Suzuki[1,2], Kazuyuki Aihara[1]

[1]Institute of Industrial Science, The University of Tokyo, 4-6-1 Komaba, Meguro-ku, Tokyo 153-8505, Japan,

[2]PRESTO, Japan Science and Technology Agency (JST), 4-1-8 Honcho Kawaguchi, Saitama 332-0012, Japan

E-mail: bingbignmath@gmail.com

Abstract. The main concern of epidemiological modeling is to implement an economical vaccine allocation to the population. Here, we investigate the optimal vaccination allocation in complex networks. We find that the optimal vaccine coverage depends not only on the relative cost of treatment to vaccination but also on the vaccine efficacy. Especially with a high cost of treatment, nodes with high degree are prioritized to vaccinate. These results may help us understand factors that may impact the optimal vaccination distribution in the control of epidemic dynamics.

1. Introduction

The concept of basic reproduction number R_0, defined as the average number of secondary cases generated by a 'typical' infectious case, throughout the infectious period, in a totally susceptible population [1], is often used to evaluate the ability of a disease to invade the population. If $R_0 < 1$, the infectious individual is unable to sustain itself in the population; if $R_0 > 1$, there is a positive probability for a large epidemic to occur [2]. In order to efficiently control the outbreak of an infectious disease, a vaccine strategy has to be planned and the efficacy of it has to be evaluated. Designing a vaccine strategy for the control of an outbreak is a key goal for epidemiological modeling. Traditionally, such kind of strategy has been formed to stop an outbreak as quickly as possible regardless of the costs that will be taken, yielding the critical vaccination coverage p_c, i.e., the fraction of population that one needs to vaccinate in order to avoid a major outbreak, $p_c = 1 - \frac{1}{R_0}$. If $p > p_c$, the prevalence will be eradicated. However, from an economic point of view, there is a growing requirement for a successful strategy that minimizes the total cost as lower as possible [3, 4, 5].

Our purpose in this paper is to investigate the optimal vaccination distribution in complex networks with the consideration of cost, which is assumed to be the sum of the cost for treating infected individuals and that for vaccinating susceptible individuals. Since vaccines rarely provide full protection from diseases [6, 7, 8], imperfect vaccines are assumed in the present paper, as observed in the vaccines that are currently being developed against malaria [9, 10]. We find that for homogeneous networks, the minimal cost is relevant to the relative cost of treatment to vaccination. For heterogeneous networks, since it is difficult to determine the fraction of susceptible and infected individuals at equilibrium due to the uncertainty of vaccine allocation and the increased number of variables, optimal vaccination distributions are numerically determined with an optimization routine. In light of our findings, optimal vaccination distributions are found closely relevant to the relative cost. For example, for a high cost of treatment, nodes with high degree have to be prioritized to vaccinate. The results have significant implications for the successs control of infectious diseases that spread via contact and help us understand nodes' roles in the implement of an economical vaccine allocation strategy for public health service.

An SIR model with imperfect vaccines in homogeneous networks

To carry out the epidemiological analysis, a classical SIR model with birth and death processes is used. Assume totally x_0 percent of susceptible individuals can receive vaccines. Unvaccinated susceptible individuals get infected by contacting infected individuals with rate β_I, while vaccinated susceptible individuals get part protection from infection with a decreased infection rate β_v, described by $\beta_v = (1-\alpha)\beta_I$, while infected individuals recover with rate r. Without economic consideration, the critical vaccine coverage, i.e., the fraction of population, necessary to vaccinate in order to avoid a major outbreak is given by $1 = R_0(1-\alpha x_0)$, where R_0 is the basic reproduction number defined as $R_0 = \frac{\beta_I}{\mu+r}$ [1]. Thus, the critical vaccine coverage x_{0c} is given by $x_{0c} = \frac{1}{\alpha}(1-\frac{1}{R_0})$.

Let us consider the cost required for the model. The total cost is assumed to be the sum of the cost for vaccinating susceptible individuals and that for treating infected individuals. For simplicity, we assume that the cost of vaccination is exponentially dependent on the vaccine coverage as has been done in Ref. [4] and the cost of treatment is proportional to the prevalence at equilibrium. Thus, the total cost is defined as

$$f(x_0) = c_v e^{x_0 \bar{S}} + c_I \bar{I}, \tag{1}$$

where $\bar{S}$ and $\bar{I}$ are the stationary fractions of susceptible and infected individuals, c_v and c_I are per-capita costs of vaccination and treatment. Usually we have $c_v < c_I$ described by $c_I = \gamma c_v$ with $\gamma > 1$. The optimal solution of Eq. (1), x_{0c}^* is achieved by checking the first-order condition $\frac{df(x_0)}{dx_0} = 0$ and the second order condition $\frac{df^2(x_0)}{dx_0^2} = 0$.

The total costs as a function of γ and α can are shown in Fig. 1. We compared the total costs for different vaccine efficacy $\alpha = 1.0$ and $\alpha = 0.2$. We see that with a perfect vaccine efficacy $\alpha = 1$, if the cost of treatment γ is not so high, i.e., $\gamma \leq 6$, then it is unnecessary to vaccinate any one at all. By increasing γ, the optimal vaccine coverage x_0 increases up to the value of x_{0c}, during which process, a smaller optimal value x_{0c}^* is observed. For a lower vaccine efficacy $\alpha = 0.2$, vaccine allocation is unnecessary due to its low efficacy. Therefore, the results suggest us that the optimal vaccine coverage is determined by a number of factors such as vaccine efficacy and the cost of treatment, which should be taken into account in the design of vaccine allocation by the public health service .

2. The SIR model with imperfect vaccines in heterogeneous networks

In order to stress the role of heterogeneity on epidemic process with vaccination, we consider the SIR model with imperfect vaccines in heterogeneous networks, which is described by a given degree distribution $p(k) \sim k^{-3}$ with a finite average connectivity $\langle k \rangle = \sum_k kp(k)$. Thus, the problem is equivalent to allocate x_k to nodes with degree k with the minimal total cost constrained by the total amount of vaccine resources $\sum_k p(k)x_k = x_0$.

To find the optimal vaccination distribution, we here directly implement an optimization routine to determine the optimal vaccination distribution x_k in degree k. With the tabu

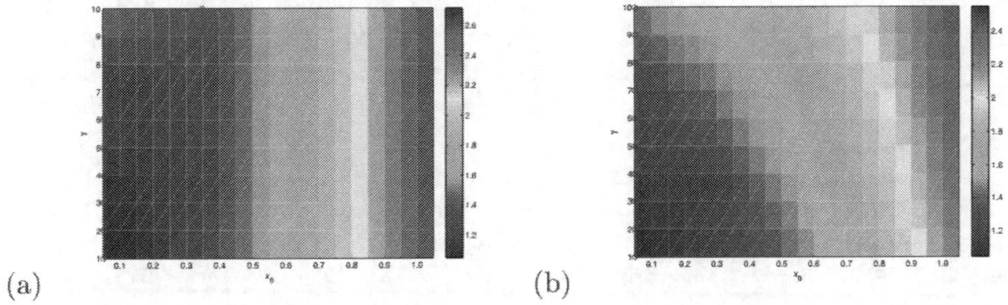

Figure 1. Total costs versus γ and x_0 for different vaccine efficacies α in homogeneous networks. (a)$\alpha = 1.0$; (b) $\alpha = 0.2$. Parameters are set as $R_0 = 1.5$, $\mu = \frac{1}{80}$, and $r = \frac{1}{14}$. The basic per-capita cost is set as $c_v = 1$.

search [12], we determine the optimal vaccination distribution x_k with the minimal cost with combinations of parameters such as the relative cost γ and the vaccine efficacy α. Incorporating these factors into the mode, we implement the tabu search for more than 10 times with different initial solutions, and take the one with the lowest cost as the optimal solution.

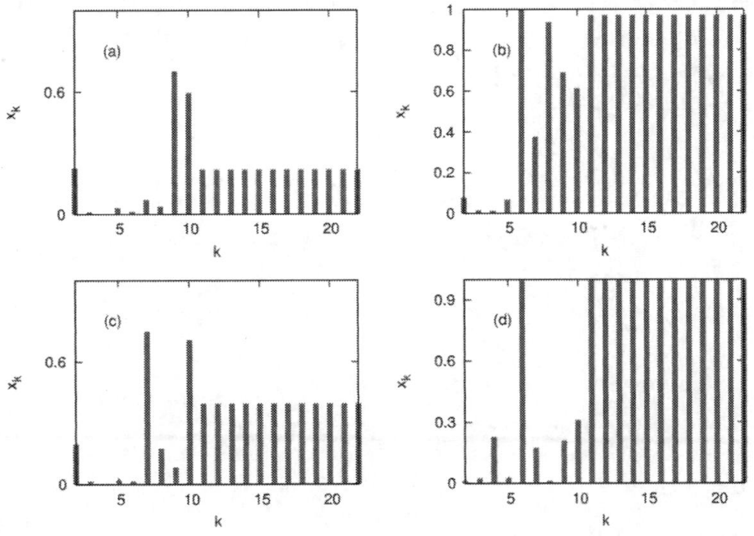

Figure 2. Optimal distributions of x_k versus k for different combinations of γ and α with very limited amount of vaccine resources $x_0 = 10\%$ that is not large enough to eradicate the disease. (a) $\gamma = 1$ and $\alpha = 0.5$; (b) $\gamma = 10$ and $\alpha = 0.5$; (c) $\gamma = 1$ and $\alpha = 1.0$; (d) $\gamma = 10$ and $\alpha = 1.0$. Parameters are set as $c_v = 1$, $R_0 = 1.5$, $\mu = 1/80$, and $r = 1/14$. Nodes with degree $k > k_c = 11$ are taken as one group.

Figure 2 shows the optimal vaccination distribution x_k, as a function of k, at the level of a very limited vaccine resources with $x_0 = 1\%$ and $R_0 = 1.5$, for the comparison between lower and higher relative costs ($\gamma = 1$ in panels (a) and $\gamma = 10$ in panels (b)). It demonstrates that the role of high degree nodes since all nodes with degree larger than $k_c = 11$ are almost prioritized to vaccinate. The role of nodes with medium degrees are found fundamentally important to the decrease of the cost at a high relative cost ($\gamma = 10$), in which the optimal vaccination distribution switches to vaccinate more nodes with medium degrees. For example, degrees less than 5 are not found to vaccinate at all (panels (c) and (d)). At a high relative cost, the optimal vaccination distribution favors the nodes that could efficiently reduce the prevalence at equilibrium, and thereby reducing the cost of treatment.

The results presented above indicate that whether a vaccination strategy is optimal or not depends on some factors such as the available amount of vaccine resources and the cost of treatment as we tested. From a realistic point of view, these results have important implications for the success of economical allocations of vaccines for the diseases that spread via contact, typically characterized by sexually transmitted diseases (STDs), by which people get infected through sexual contact with an infected person. Since the human sexual contact network has been found to follow a power-law distribution [11], according to our results, we guess that if STDs vaccines are very limited, then persons who have broad sexual partners have to be prioritized to vaccinate in order to reduce the perceived cost possibly caused by the potential infection of them.

Conclusions

In this study, we investigate the optimal vaccine allocation to the population with the consideration of economic factors. The study is applicable to diseases that spread through air, such as influenza, and diseases that spread through contact, such as STDs. Our study demonstrates that the design of economical vaccination strategies should be incorporated with a number of factors that involved in the model such as the cost of treatment as we tested and the vaccine efficacy. Achieving more complete control of STDs may be possible if advances in our understanding of nodes' roles in the epidemiology and transmission dynamics can be integrated into new interventional strategies.

Acknowledgments

This research is supported by the "Funding Program for World-Leading Innovative *R&D* on Science and Technology (FIRST Program)" initiated by the Council for Science and Technology Policy (CSTP) (B.W. and K.A.) as well as by Japan Science and Technology Agency (JST) Precursory Research for Embryonic Science and Technology (PRESTO) program (H.S.),

References

[1] Diekmann O. and Heesterbeek J.A.P. and Metz J.A.J. 1990 *J. Math. Biol.* **28** 1365
[2] van den Driessche P. and Watmough J. 2002 *Math. Biosci.* **180** 29
[3] Medlock J. and Galvani A.P. 2009 *Science* **325** 1705
[4] Klepac P. and Laxminarayan R. and Grenfell B.T. 2011 *Proc. Natl. Acad. Sci. USA* **108** 14366
[5] Rowthorn R.E. and Laxminarayan R. and Gilligan C.A. 2009 *J. R. Soc. Interface* **6** 1135
[6] Gandon S and Mackinnon M.J. and Nee S. 2001 *Nature* **414** 751
[7] Gandon S. and Mackinnon M and Nee S. and Read A. 2003 *Proc. R. Soc. B* **270** 1129
[8] Williams P. D. and Day T 2008 *Mol. Ecology* **17** 485
[9] Ballou W.R. and Alo-Herrea M.A. and Carucci D. 2004 *Am. J. Trop. Med. Hyg.* **71** 239
[10] Smith T. and Killeen G.F. and Maire N. 2006 *Am. S. Trop. Med. and Hyg.* **75** 1
[11] Liljeros F. and Edling C.R. and Amaral L. A. and Stanley H. E. and Aberg Y. 2001 *Nature* **411** 907
[12] Glover F 1990 *Interfaces* **20** 74

Journal of Physics: Conference Series **410** (2013) 012096 doi:10.1088/1742-6596/410/1/012096

Entropy growth in emotional online dialogues

J Sienkiewicz[1], M Skowron[2], G Paltoglou[3] and Janusz A. Hołyst[1]

[1] Faculty of Physics, Centre of Excellence for Complex Systems Research, Warsaw University of Technology, Koszykowa 75, 00-662 Warszawa, Poland

[2] Interaction Technologies Group, Austrian Research Institute for Artificial Intelligence, Freyung 6/3/1a, A-1010 Vienna, Austria

[3] School of Technology, University of Wolverhampton, Wulfruna Street, Wolverhampton WV1 1LY, United Kingdom

E-mail: julas@if.pw.edu.pl

Abstract. We analyze emotionally annotated massive data from IRC (Internet Relay Chat) and model the dialogues between its participants by assuming that the driving force for the discussion is the entropy growth of emotional probability distribution.

1. Introduction

Large amounts of data offer new possibilities of examining communication between humans expressed in face-to-face encounters [1], mobile telephone calls [2], surface-mail [3] short messages [4] as well as typical Internet activities such as e-mail correspondence [5], forum postings [6] or web browsing [7]. In the most fundamental part, the communication is based on a *dialogue* - an exchange of information and ideas between two people [8]. As compared to the off-line communication, the exchange of information in the Internet is claimed to be more biased toward the emotional aspect [9]. However, the very Internet that gives the opportunity to acquire massive data, thus making it possible to perform a credible statistical analysis of common habits in communication. As the recent research shows, it is already possible to spot and model certain phenomena of the Internet discussion participants while looking just at the emotional content of their posts [10, 11, 12, 13, 14]. One of them is the collective emotional behavior [10, 13], the other is clear correlation between the length of discussion and its emotional content [10, 11]. Here we present an approach based on the observation of entropy of emotional probability distribution during the conversation that can arguably serve as an indicator of an ending discussion.

2. Data description

As a source of data for analysing online dialogues we chose the Internet Relay Chat (IRC) [15] logs. The analysis is limited only to one of the channels, namely **#ubuntu** [16] in the period 1st January 2007 - 31st December 2009. We focused on dialogues that included only two participants. The final output, after several levels of data processing consists of $N = 93329$ dialogues with the length L between $L_{min} = 11$ and $L_{max} = 339$ each. Each dialogue can be represented as a chain of messages where all odd posts are submitted by one user and all even by another one.

The emotional classifier program that was used to analyze the emotional content of the discussions is based on a machine-learning (ML) approach. The algorithm functions in two

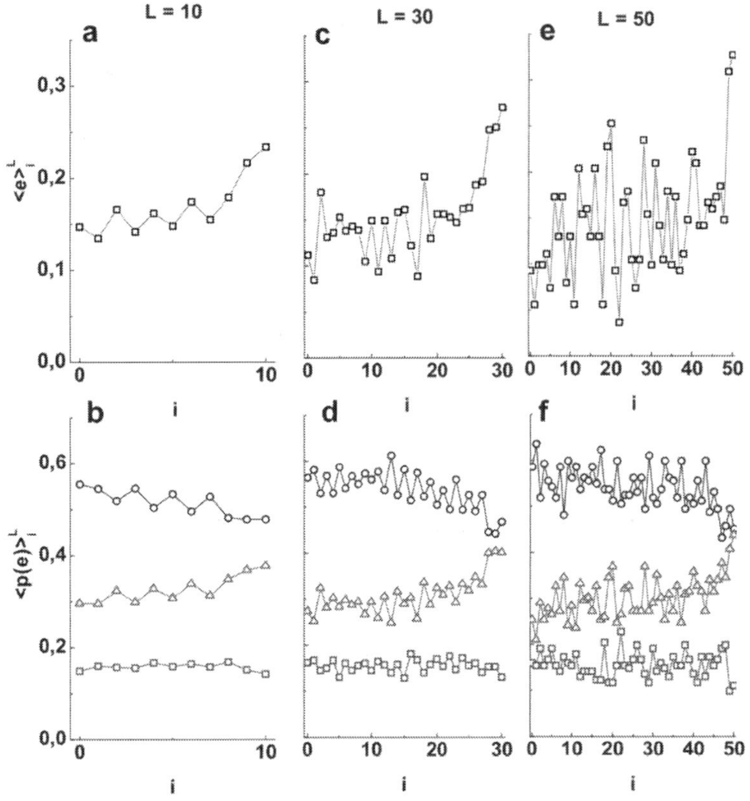

Figure 1. Average emotional value $\langle e\rangle_i^L$ (top row) and average emotional probabilities (bottom row) $\langle p(-)\rangle_i^L$ (squares), $\langle p(0)\rangle_i^L$ (circles), $\langle p(+)\rangle_i^L$ (triangles) in the i-th timestep for dialogues of specific $L = 10$ (first column), $L = 30$ (second column) and $L = 50$ (third column).

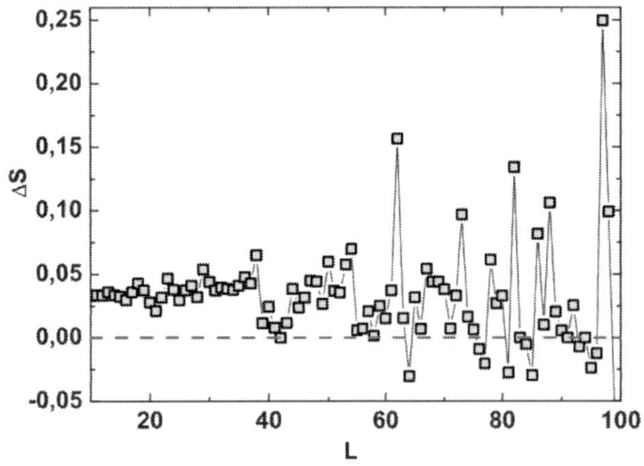

Figure 2. Difference between terminal and initial entropy value ΔS versus the dialogue length L.

phases: during the training phase, it is provided with a set of documents classified by humans for emotional content (positive, negative or objective) from which it learns the characteristics of each category. Then, during the application phase, the algorithm applies the acquired sentiment classification knowledge to new, unseen documents. In our analysis, we trained a hierarchical Language Model [17, 18] on the Blogs06 collection [19] and applied the trained model to the extracted IRC dialogues, during the application phase. The algorithm is based on a two-tier solution, according to which a post is initially classified as objective or subjective and in the latter case, it is further classified in terms of its polarity, i.e., positive or negative. Each level of classification applies a binary Language Model [20, 18]. Posts are therefore annotated with a

Table 1. Properties of the dataset: number of comments C, number of dialogues N, shortest dialogue L_{min}, longest dialogue L_{max}, average valence $\langle e \rangle$, probability of finding negative, neutral or positive emotion (respectively $p(-)$, $p(0)$ and $p(+)$).

C	N	L_{min}	L_{max}	$\langle e \rangle$	$p(-)$	$p(0)$	$p(+)$
1889120	93323	11	339	0.17	0.15	0.53	0.32

single value $e = -1, 0$ or 1 to quantify their emotional content (to be more precise - their valence [21]) as negative, neutral or positive, respectively. The accuracy of the classifier checked for 950 humanly annotated comments in IRC data is 62.49 % for subjectivity detection and 70.25 % for polarity detection.

3. Results

The obtained dialogues have been divided into groups of constant dialogue length L. For such data we follow the evolution of mean emotional value $\langle e \rangle_i^L$ and average emotional probabilities $\langle p(e) \rangle_i^L$ ($\langle e \rangle_i^L$. In both cases the $\langle ... \rangle_i^L$ symbol indicates taking all dialogues with a specific length L and averaging over all comments with number i, thus, for example, $\langle p(-) \rangle_i^L$ is the probability that at the position i in all dialogues of length L there is a negative statement. The characteristic feature observed regardless of the dialogue length is that the $\langle e \rangle_i^L$ at the end of the dialogue is higher than at the beginning (upper row in Fig. 1). In fact, there is especially a rapid growth close the very end of the dialogue, which is probably caused by participants who acknowledge others' support issuing comments like "thank you", "you were most helpful", etc.

The direct reason for such behavior is shown in the bottom row of Fig. 1, which presents the evolution of the average emotional probabilities $\langle p(-) \rangle_i^L$, $\langle p(0) \rangle_i^L$ and $\langle p(+) \rangle_i^L$. The observations can be summarized in the following way:

- the negative emotional probability $\langle p(-) \rangle_i^L$ remains almost constant,
- $\langle p(+) \rangle_i^L$ increases and $\langle p(0) \rangle_i^L$ has an opposite tendency,
- $\langle p(+) \rangle_i^L$ and $\langle p(0) \rangle_i^L$ tend to equalize in the vicinity of dialogue end.

Other manifestation of the system's features can be spotted by examining the level of the entropy S of the emotional probabilities $\langle p(e) \rangle_i^L$. Basing on entropy, it has also been shown how the coherent structures in the e-mail dialogues arise [5] or how to predict conversation patterns in face-to-face meetings [1]. It is also possible to use other information theoretic quantities as Kullback-Leiber or Jensen-Shannon divergence (see e.g., [22]), however due to an established connection to its thermodanymical equivalent as well as its clear role for poiting the direction of physical processes we use entropy (after Shannon's definition [23])

$$S_i^{sh} = - \sum_{e=-1,0,1} \langle p(e) \rangle_i^L \ln \langle p(e) \rangle_i^L. \tag{1}$$

Here, taking into account the fact that $\langle p(-) \rangle_i^L$ is constant in the course of dialogue, we paid attention only to $\langle p(+) \rangle_i^L$ and $\langle p(0) \rangle_i^L$, thus the observed entropy had a form of

$$S_i = - \left[\langle p(0) \rangle_i^L \ln \langle p(0) \rangle_i^L + \langle p(+) \rangle_i^L \ln \langle p(+) \rangle_i^L \right]. \tag{2}$$

Plotting the difference between terminal and initial entropy ΔS versus the length of the dialogue L it is possible to see that for the dialogues up to $L \approx 50$ this difference is always above zero (see Fig. 2). Over that point the system we observe high fluctuations due to small statistics of number of dialogues (dialogues with length $L > 50$ make around 1% of the total number of dialogues N) . It implies a following likely scenario for the dialogue: it evolves in the direction of growing entropy. In the beginning of the dialogue, the probabilities $\langle p(0)\rangle_i^L$ and $\langle p(+)\rangle_i^L$ are separated from each other, contributing to low value of initial entropy S_p. However, then the entropy grows, the probabilities $\langle p(0)\rangle_i^L$ and $\langle p(+)\rangle_i^L$ equalize leading to high value entropy (i.e., higher then the initial one) at the end of the dialogue.

4. Conclusions

Analysis performed on the emotionally annotated dialogues extracted from IRC data demonstrate that following such simple metrics as probability of specific emotion can be useful to predict the future evolution of the discussion. Moreover, all the analyzed dialogues share the same property, i.e., the tendency to evolve in the direction of a growing entropy. Those observations may be of help for designing the next generation of interactive software tools [24]

Acknowledgments

This work was supported by a European Union grant by the 7th Framework Programme, Theme 3: Science of complex systems for socially intelligent ICT. It is part of the CyberEmotions (Collective Emotions in Cyberspace) project (contract 231323). J.S. and J.A.H. also acknowledge support from Polish Ministry of Science Grant 1029/7.PR UE/2009/7. The work was partially supported from polish Ministry of Science and Higher Education (grant no 0490/IP3/2011/71).

References

[1] Takaguchi T, Nakamura M, Sato N, Yano K and N. Masuda 2011 *Phys. Rev.* X **1** 011008.
[2] Onnela J-P, Saramäki J, Hyvönen J, Szabó G, Lazer D, Kaski K, Kertész K and Barabási A-L 2007 *Proc. Net. Acad. Sci. U.S.A.* **104** 7332.
[3] Oliveira J G and Barabási A-L 2005 *Nature (London)* **437** 1251.
[4] Wu Y, Zhou Ch, Xiao J, Kurths J, and Schellnhuber H J 2010 *Proc. Nat. Acad. Sci. U.S.A.* **107** 18803.
[5] Eckmann J-P, Moses E and Sergi D 2004 *Proc. Nat. Acad. Sci. U.S.A.* **101** 14333.
[6] Kujawski B, Hołyst J A and Rodgers G J 2007 *Phys. Rev.* E **76** 036103.
[7] Chmiel A, Kowalska K and Hołyst J A 2009 *Phys. Rev.* E **80** 066122.
[8] Sacks H, Schegloff E A and Jefferson G 1974 *Language* **50** 696.
[9] Sobkowicz P and Sobkowicz A 2010 *Eur. Phys. J.* B **73**, 633.
[10] Chmiel A, Sienkiewicz J, Thelwall M, Paltoglou G, Buckley K, Kappas A and Hołyst J A 2011 *PLoS ONE* **6** e22207.
[11] Chmiel A, Sobkowicz P, Sienkiewicz J, Paltoglou G, Buckley K, Thelwall M and Hołyst J A 2011 *Physica* A **390** 2936.
[12] Weroński P, Sienkiewicz J, Paltoglou G, Buckley K, Thelwall M and Hołyst J A 2012 *Acta Phys. Pol.* A **121** B-128.
[13] Garas A, Garcia D, Skowron M and Schweitzer F 2012 *Sci. Rep.* **2** 402.
[14] Garcia D, Garas A and Schweitzer F 2012 *EPJ Data Science* **1** 3.
[15] http://en.wikipedia.org/wiki/Internet_Relay_Chat.
[16] https://help.ubuntu.com/community/InternetRelayChat.
[17] Sebastiani F 2002 *ACM Computing Surveys* **34**.
[18] Mitrović M, Paltoglou G and Tadić B 2010 *Eur. Phys. J.* B **77**, 597.
[19] Ounis I, Macdonald C, Soboroff I 2008 in *Proceedings of the Second International Conference on Weblogs and Social Media.*
[20] Peng F, Schuurmans D, and Wang S 2003 in *NAACL '03* 110.
[21] Feldman L A 1995 *Journal of Personality and Social Psychology* **69** 153.
[22] Masucci A P, Kalampokis A, Eguíluz V M and Hernández-García E 2011 *Phys. Rev. E* **83**, 026103.
[23] Shannon C E 1948 *Bell Syst. Tech. J.* **27** 379.
[24] Skowron M, Rank S, Theunis M and Sienkiewicz J 2011 *LNCS* **6974** 337.

Journal of Physics: Conference Series **410** (2013) 012097 doi:10.1088/1742-6596/410/1/012097

Tailored Random Graph Ensembles

ES Roberts[1,2,3], **A Annibale** [1,2,3] **and ACC Coolen**[1,2,3,4]

[1] Institute for Mathematical and Molecular Biomedicine, Hodgkin Building, King's College London, London SE1 1UL, United Kingdom

[2] Department of Mathematics, The Strand, King's College London, London WC2R 2LS, United Kingdom

[3] Randall Division of Cell and Molecular Biophysics, New Hunts House, King's College London, London SE1 1UL, United Kingdom

[4] London Institute for Mathematical Sciences, 35a South St, Mayfair, London W1K 2XF, United Kingdom

E-mail: ekaterina.roberts@kcl.ac.uk

Abstract. Tailored graph ensembles are a developing bridge between biological networks and statistical mechanics. The aim is to use this concept to generate a suite of rigorous tools that can be used to quantify and compare the topology of cellular signalling networks, such as protein-protein interaction networks and gene regulation networks. We calculate exact and explicit formulae for the leading orders in the system size of the Shannon entropies of random graph ensembles constrained with degree distribution and degree-degree correlation. We also construct an ergodic detailed balance Markov chain with non-trivial acceptance probabilities which converges to a strictly uniform measure and is based on edge swaps that conserve all degrees. The acceptance probabilities can be generalized to define Markov chains that target any alternative desired measure on the space of directed or undirected graphs, in order to generate graphs with more sophisticated topological features.

1. Introduction

Consider looking at a network not as a set of nodes and edges, but as a collection of topological properties. Any observed set of topological features can be used to define a family of networks. Specifying a sufficient number of topological features will eventually define a network uniquely. If we meaningfully choose our topological features to relate to our understanding of the processes which formed the network, then the associated ensemble is interesting in its own right.

In this paper we will call such a family of networks a 'Tailored Random Graph Ensemble' (TGE). We study several aspects of TGEs. Results are derived analytically, in leading order in N. We look to measure the size of the ensemble via its Shannon Entropy S per bond. This allows us to quantify the topological properties we have imposed from the point of view of how severely they restrict the TGE. We can also compare two TGEs via a Kullback-Leibler distance. Our approach gives a rigorous and quantitative basis to compare two networks based on their topological properties.

Finally, we consider how we might sample without bias from a TGE. To achieve this, we begin by drawing attention to (and providing a correction) to a bias in a common graph randomisation algorithm. We then show that our approach can be generalised to sample networks with desired topological properties.

2. Quantifying the topology of a network via tailored random graph ensembles

[1] and [2] derive closed form solutions for the Shannon entropy per bond

$$S = -\frac{1}{N\langle k\rangle}\sum_{\mathbf{c}} p(\mathbf{c})\log p(\mathbf{c})$$

of the maximum entropy ensemble associated with particular constraints, where k_i is the degree (number of links) of node i, $\langle k\rangle$ is the average number of links per node, $p(\mathbf{c})$ is the degree distribution and N is the number of nodes. We choose to describe degree-degree correlation via the degree statistics of connected nodes

$$W(k,k'|\mathbf{c}) = \frac{1}{N\langle k\rangle}\sum_{ij} c_{ij}\delta_{k,k_i(\mathbf{c})}\delta_{k',k_j(\mathbf{c})}$$

We can define the maximum-entropy ensemble of non-directed graphs constrained by a degree distribution $p(k)$ and a degree-degree correlation $W(k,k')$ by setting the probability of observing a network $\mathbf{c}$ to be

$$p(\mathbf{c}) = \int\prod_i\left[\mathrm{d}k_i\ p(k_i)\right]\frac{\prod_i\delta_{k_i,k_i(\mathbf{c})}}{Z}\prod_{i<j}\left[\frac{\langle k\rangle}{N}\frac{W(k_i,k_j)}{p(k_i)p(k_j)}\delta_{c_{ij},1}+\left(1-\frac{\langle k\rangle}{N}\frac{W(k_i,k_j)}{p(k_i)p(k_j)}\right)\delta_{c_{ij},0}\right] \quad (1)$$

The partition function Z is defined in the usual way. $\pi(k)$ is the Poissonian distribution with the same average degree, and $W(k)$ indicates the marginal. In this case [1] find the Shannon entropy per bond of this ensemble to be

$$S = \frac{1}{2}[1+\log(\frac{N}{\langle k\rangle})] - \underbrace{\frac{1}{\langle k\rangle}\sum_k p(k)\log[\frac{p(k)}{\pi(k)}]}_{\textit{degree complexity}} - \underbrace{\frac{1}{2}\sum_{k,k'}W(k,k')\log\left[\frac{W(k,k')}{W(k)W(k')}\right]}_{\textit{wiring complexity}} + \epsilon_N \quad (2)$$

The ensemble constrained by the degree distribution only can be found by substituting $W(k,k') = p(k)p(k')$ into equation 1. Substituting this into equation 2 can be easily demonstrated to be equivalent to removing the last term. That is, the three terms of equation 2 can be associated with the progressive tightening of the constraint on the ensemble from average degree, then to degree distribution, then additionally to degree-degree correlation.

[2] extended [1] to find a similar expression in the case of directed graphs.

3. Structural dissimilarity of graphs c_A and c_B, based on information-theoretic distance between associated ensembles

[1] uses the same methods as were used to derive equation 2 to determine a Kullback-Leibler distance between ensemble A (defined through p_A and W_A) and ensemble B (defined through p_B and W_B)

$$D_{AB} = \frac{1}{2N}\sum_{\mathbf{c}\in G}\left\{p(\mathbf{c}|p_A,W_A)\log\left[\frac{p(\mathbf{c}|p_A,W_A)}{p(\mathbf{c}|p_B,W_B)}\right] + p(\mathbf{c}|p_B,W_B)\log\left[\frac{p(\mathbf{c}|p_B,W_B)}{p(\mathbf{c}|p_A,W_A)}\right]\right\} \quad (3)$$

This is evaluated to be

$$
\begin{aligned}
D_{AB} &= \frac{1}{2}\sum_k p_A(k)\log\left[\frac{p_A(k)}{p_B(k)}\right]+\frac{1}{2}\sum_k p_B(k)\log\left[\frac{p_B(k)}{p_A(k)}\right] \\
&+\frac{1}{4\langle k\rangle_A}\sum_{kk'} p_A(k)p_A(k')kk'\Pi_A(k,k')\log\left[\frac{\Pi_A(k,k')}{\Pi_B(k,k')}\right] \\
&+\frac{1}{4\langle k\rangle_B}\sum_{kk'} p_B(k)p_B(k')kk'\Pi_B(k,k')\log\left[\frac{\Pi_B(k,k')}{\Pi_A(k,k')}\right] \\
&+\frac{1}{2}\sum_k p_A(k)k\log\rho_{AB}(k)+\frac{1}{2}\sum_k p_B(k)k\log\rho_{BA}(k)
\end{aligned} \tag{4}
$$

with

$$\Pi(k,k') = W(k,k)/W(k)W(k'), \qquad \rho_{AB}(k)=\sum_{k'}\Pi_B(k,k')W_A(k')\rho_{AB}^{-1}(k')$$

The directed case is covered in [2].

4. Constrained Markovian graph dynamics

This section will demonstrate the existence of a bias which applies to the commonly used edge-swapping algorithm for edge-preserving graph randomisation. This bias can be corrected, and a natural generalisation of the correction allows us to sample networks with more finely tailored topological properties.

The aim is to sample networks from a subset $G[\star]\subseteq G$ without bias, where G is the space of all networks and $G[\star]$ is the subset that satisfies some constraint $\star$. To do this, we look to define a Markov Chain process with transition probabilities $W(\mathbf{c}|\mathbf{c}')$ such that $p_{t+1}(\mathbf{c}) = \sum_{\mathbf{c}'\in G[\star]} W(\mathbf{c}|\mathbf{c}')p_t(\mathbf{c}')$ based on some set of auto-invertible allowed moves Φ (excluding identity) such that Φ is ergodic on $G[\star]$ and the process converges to $p_\infty(\mathbf{c}) = Z^{-1}e^{-H(\mathbf{c})}$ on $G[\star]$ for some target $H(\mathbf{c})$. Setting $H(\mathbf{c}) = const$ would sample each member of the set with equal probability. Choosing a different form of $H(\mathbf{c})$ allows us to favour certain networks.

We can express our transition probability as

$$W(\mathbf{c}|\mathbf{c}') = \sum_{F\in\Phi} q(F|\mathbf{c}')\Big[\delta_{\mathbf{c},F\mathbf{c}'}A(F\mathbf{c}'|\mathbf{c}') + \delta_{\mathbf{c},\mathbf{c}'}[1-A(F\mathbf{c}'|\mathbf{c}')]\Big]$$

where $A(\mathbf{c}|\mathbf{c}')$ is the move acceptance probability and $q(F|\mathbf{c})$ is the move proposal probability. If all allowed moves F are equally probable $q(F|\mathbf{c})$ can be written as $q(F|\mathbf{c}) = I_F(\mathbf{c})/n(\mathbf{c})$ where $I_F(\mathbf{c})$ indicates if F is a valid move on $\mathbf{c}$, and $n(\mathbf{c}) = \sum_{F\in\Phi} I_F(\mathbf{c})$ is the number of allowed moves on $\mathbf{c}$.

As a sufficient condition for convergence we impose a detailed balance condition. This is equivalent to

$$(\forall F\in\Phi)(\forall \mathbf{c}\in G[\star]): \qquad q(F|\mathbf{c})A(F\mathbf{c}|\mathbf{c})\mathrm{e}^{-H(\mathbf{c})} = q(F^{-1}|F\mathbf{c})A(\mathbf{c}|F\mathbf{c})\mathrm{e}^{-H(F\mathbf{c})}$$

It then follows that

$$(\forall F\in\Phi)(\forall \mathbf{c}\in G_F[\star]): \qquad \frac{1}{n(\mathbf{c})}A(F\mathbf{c}|\mathbf{c})\mathrm{e}^{-H(\mathbf{c})} = \frac{1}{n(F\mathbf{c})}A(\mathbf{c}|F\mathbf{c})\mathrm{e}^{-H(F\mathbf{c})}$$

which is satisfied by

$$A(\mathbf{c}|\mathbf{c}') = \frac{n(\mathbf{c}')\mathrm{e}^{-\frac{1}{2}[H(\mathbf{c})-H(\mathbf{c}')]}}{n(\mathbf{c}')\mathrm{e}^{-\frac{1}{2}[H(\mathbf{c})-H(\mathbf{c}')]}+n(\mathbf{c})\mathrm{e}^{\frac{1}{2}[H(\mathbf{c})-H(\mathbf{c}')]}}$$

A common example would be to sample from the set of networks with a certain degree sequence by defining a process based on degree-preserving edge swaps [3]. Most common implementations of this scheme accept all proposed moves, i.e. $A(\boldsymbol{c}|\boldsymbol{c}') = const.$ This corresponds to $H(\boldsymbol{c}) = -\log n(\boldsymbol{c})$, so would give $p_\infty(\boldsymbol{c}) = \frac{n(\boldsymbol{c})}{\sum_{\boldsymbol{c}' \in G[\star]} n(\boldsymbol{c}')}$ - a sampling bias.

To sample without bias from the ensemble of all networks which satisfy the constraint $\star$, it is necessary to set $H(\boldsymbol{c}) = const$, which leads to

$$A(\boldsymbol{c}|\boldsymbol{c}') = \frac{n(\boldsymbol{c}')}{n(\boldsymbol{c}') + n(\boldsymbol{c})}$$

Suppose, however, we wanted to study an ensemble which was also tailored to certain soft constraints, such as was seen in equation 1. Using the relation $p_\infty(\boldsymbol{c}) = Z^{-1}e^{-H(\boldsymbol{c})}$, we can do this by setting

$$H(\boldsymbol{c}) = -\sum_{i<j} \log \left[\frac{\langle k \rangle}{N} \frac{W(k_i,k_j)}{p(k_i)p(k_j)} \delta_{c_{ij},1} + \left(1 - \frac{\langle k \rangle}{N} \frac{W(k_i,k_j)}{p(k_i)p(k_j)}\right) \delta_{c_{ij},0} \right]$$

[4] and [5] derive practical expressions for $n(\boldsymbol{c})$ for the most commonly seen degree-preserving process based on swapping pairs of edges.

5. Discussion

In this paper we have summarized some recently developed tools for studying the topology of directed and non-directed binary networks.

Firstly, section 2 provides exact closed form expressions which allow us to measure topological properties with reference to the Shannon entropy of the associated constrained maximum entropy ensembles. In a particular example, calculating a large value indicates that the constraint is not very restrictive, whereas finding a low value indicates that the constraint is strong, and networks with this collection of properties are more rare in the space of all graphs. Section 3 provides expressions that can be used to compare how similar the topology of two networks is based on the Kullback-Leibler distance between the associated ensemble. This provides a far more natural and intuitive basis for comparing networks than the more commonly seen Hamming distance.

Section 4 discusses a complementary approach to studying the topology of networks: how do we generate an example network from our ensemble in an unbiased way? This is an important consideration for numerical work, for example in comparing the frequency of motifs in a real network compared to a suitably randomised test network. We show that one of the most commonly used approaches to randomising networks is biased, and provide a simple correction to eliminate this bias.

6. Acknowledgements

Author ESR thanks BBSRC for financial support.

References

[1] A. Annibale, A. C. C. Coolen, L. P. Fernandes, F. Fraternali, and J. Kleinjung. Tailored graph ensembles as proxies or null models for real networks I: tools for quantifying structure. J. Phys. A, 42(48):485001, 2009.
[2] E S Roberts, T Schlitt, and A C C Coolen. Tailored graph ensembles as proxies or null models for real networks ii: results on directed graphs. J. Phys. A, 44(27):275002, 2011.
[3] R Taylor. Combinatorial Mathematics VIII. Springer, 1981.
[4] A. C. C. Coolen, A. De Martino, and A. Annibale. Constrained Markovian dynamics of random graphs. J. Stat. Phys., 2009.
[5] E. S. Roberts and A. C. C. Coolen. Unbiased degree-preserving randomization of directed binary networks. Physical Review E, 85:046103+, 2012.

A Chaotic Ordered Hierarchies Consistency Analysis Performance Evaluation Model

Wei-Chang Yeh
Integration & Collaboration Laboratory
Advanced Analytics Institute
University of Technology Sydney
PO Box 123 Broadway, NSW 2007, Australia

Email: yeh@ieee.org

Abstract. The Hierarchies Consistency Analysis (HCA) is proposed by Guh in-cooperated along with some case study on a Resort to reinforce the weakness of Analytical Hierarchy Process (AHP). Although the results obtained enabled aid for the Decision Maker to make more reasonable and rational verdicts, the HCA itself is flawed. In this paper, our objective is to indicate the problems of HCA, and then propose a revised method called chaotic ordered HCA (COH in short) which can avoid problems. Since the COH is based upon Guh's method, the Decision Maker establishes decisions in a way similar to that of the original method.

1. Introduction

A fundamental goal of management is to measure and assess the performance of an organization and/or a system. Such evaluation of performance consists of two parts: the choice of the principle and the establishment of evaluation model [1,2]. The question of how to accurately and correctly estimate the performance has always been the most basic topic of discussion in the field of management. Matters of concern such as system stability, the monitoring of the manufacturing process, and quality of service or the performance evaluation itself in fields such as health work, business and so forth all involve performance evaluation.

It is always a major concern for decision-makers (DMs) to establish an effective model for objective assessment in the field of performance evaluation [2]. Many researchers have built performance evaluation models as a hierarchy of goals and means in the shape of a systematic diagram composed of all the criteria elements [3-7]. Hierarchical structure is a good approach for describing a complicated system in which the relationships among all the related criteria are complicated.

However, in most problems, a single framework of an analysis model would provide only one stand point of evaluation [8]. This would have a likely chance in producing an error in evaluation particularly for large-scale complex processes. Hence, a different viewpoint or specific point's emphasis will result in different hierarchy structures in the analysis of large-scale problems [1,9].

Moreover, with the integration and compromise of analysis results, a more accurate and objective result would surface. Thus, Guh [1] proposed a new weighting methodology called Hierarchies Consistency Analysis (HCA) based on the fact that when a DM transfers a multiple criteria problem into a hierarchy structure, several hierarchy structures may actually be produced from emphasizing different viewpoints [1,9]. Guh et al. improved the HCA furthermore and named it AHP-HCA [10]. AHP-HCA uses AHP to acquire the initial weights and then employs HCA to establish a multiple structures model for the AHP.

HCA and AHP-HCA can provide us with a comprehensive framework for solving hierarchy evaluation problems to aid the DM to make decision more reasonable and rational. Generally, the process of determining the evaluation index for a HCA includes the following steps [10]: The lowest level criteria of the problem are defined and collected, next based on some specified management viewpoints these lowest level criteria are further grouped and clustered into multiple level and multiple criteria (hierarchy structure) models, and the DM then assigns directly, or obtains indirectly, the weights for the criteria in all levels of the hierarchy structure. Finally, we calculate the aggregated index by multiplying the weights and the observed values of the criteria for the evaluated unit. If the solutions do not satisfy all DMs, then a new solution is obtained by justifying the weights for the criteria. The above procedure is repeated until all obtained weight are reasonable acceptable, and consistent [1,8-11].

HCA focuses on "reasonable, acceptable, and consistent solution", the way the DM makes his/her final decision with the upmost accuracy possible. What makes this model more reasonable and acceptable than the previous one is the higher level of acceptability it requires. As DMs who follow the old method (e.g., AHP) employ only one set of results, DMs who apply HCA use two sets of results that can be compared with and sifted out to obtain the more correct results. This need

The HCA is a straightforward method and was proven to be very effective to find a reasonable comprised solution especially for complex and large-scale problem performance evaluation without suffering numerical difficulties and reduce the computational burdens (e.g., eigenvectors) [10]. However, the final result derived from HCA may cause inconsistent with our intuitions about the ranking of the weights for the criteria (see Section 3). In the real world, we can illustrate many examples like the above situations again. Thus, there is still a lots room and need to improve HCA (AHP-HCA) to easy help the DM find acceptability and reasonable solution.

Chaos refers to a state lacking order or predictability. It has been studied from many perspectives [12-16]. To the best of the author's knowledge, the concept of chaos has never been investigated in the context of performance evaluation models. Thus, in this research, chaos is incorporated into HCA to construct a new method called the chaotic ordered AHP-HCA (COH), where seek consistency for the weights of the criteria of each hierarchy ability of HCA and chaotic searching behavior are reasonably combined. Example results and comparisons demonstrate the effectiveness and efficiency of the proposed COH.

2. Acronym, notations and assumptions

2.1 Acronym

AHP : Analytical Hierarchy Process
HCA : Hierarchies Consistency Analysis
COH : Chaotic Ordered HCA
CI : Consistency Index
CR : Consistency Ratio

2.2 Notations

Pr(•) : the probability of the occurrence of •.
Max(•) : the maximal number of the set •.
Min(•) : the minimal number of the set •.
$\bullet_1 \rightarrow \bullet_2$: the path from nodes $\bullet_1$ to $\bullet_2$ in the corresponding hierarchy.
|•| : the number of the set •.
n : the number of changes made
i : the ith type (hierarchy), i=1,2,…,s.
t : the tth iteration.
n_i : the total number of criteria in the ith hierarchy.
I : $I=l_1l_2l_3\ldots l_s$ denotes the position in the l_sth sub-criteria of the (l_{s-1})th sub-criteria of the (k_{s-2})th sub-criteria …l_1th sub-criteria.
Ij : $Ij=l_1l_2l_3\ldots l_sj$ if $I=l_1l_2l_3\ldots l_s$.
$P_{i,I}$: the performance index of node I with respective to ith type (hierarchy).
$P_{i,I}^+$: the parent node of $P_{i,I}$ with respective to ith type (hierarchy).
P_{i,I^*} : the bottom node with $P_{i,I}$ is one of its ancestors with respective to ith type (hierarchy).
$w_{i,I}^t$: $w_{i,I}^t=\Pr(P_{i,I+} \rightarrow P_{i,I})$ is the weight corresponding to the sub-criteria in the position I in the tth iteration.
$\overline{w_{i,I}^t}$: the counter-weight corresponding to $w_{i,I}^t$. It is defined as $\overline{w_{i,I}^t}$ =[the average of

$\Pr(P_h \rightarrow P_{h,I})$ in the tth iteration for all $i \neq h$ and I is the bottom node], and $\overline{w^t_{i,Ij}}$ =[the average of $\Pr(P_{i,J} \rightarrow P_{i,J^*} \mid P_{i,J+} \rightarrow P_{i,J^*})$ in the tth iteration for all $i \neq h$].

k_t : The combination parameter in the convex combination

$$w^{t+1}_{i,I} = k_t w^t_{i,I} + (1 - k_t)\overline{w^t_{i,I}} \tag{1}$$

with $0<k<1$. Note that $k_t = k = .5$ in the traditional HCA and AHP-HCA.

CR : $CR = CI/RI$ [3, 4, 5].

$(CI)^t$:

$$(CI)^t = \frac{\sum_i \sum_j |w^t_{i,j} - \overline{w^t_{i,j}}|}{\sum_i n_i}$$

. Note that when $CI=0$, it indicates that the DM has reached consistency of judgment. Higher CI suggests more serious inconsistence. Hence, it was suggested that the error can be accepted with $CI \leq 0.1$ [3, 4, 5].

2.3 Assumption [3]

1. If attribute A is absolutely more important than attribute B and is rated at 9, then B must be absolutely less important than A and is valued at 1/9.
2. The DM is rational, i.e. if A is preferred to B and B is preferred to C, then A is preferred to C.

3. The proposed COH

To enrich the negotiable, reasonable, and practicable behavior of finding comprised weights, a new HCA-based method called the chaotic ordered HCA (COH in short) is proposed. There are three important assumptions in the proposed COH: 1) each hierarchy is always tried to find possible combination for the comprised weights, 2) each hierarchy is always protected its own profit, 3) the DM must have the ability to discriminate which criterion (sub-criterion) is more important than the others under same parent node. The above concerns are reasonable and practical which are three significant factors need to be included in Guh's HCA-based method [1].

The above important assumptions are all based on the psychology of the decision making process [2]. The chaotic random number is employed to k in Eq.(1) achieve the first assumption. A restriction is imposed to k also in Eq.(1) to express that the each hierarchy is always protected its own profit. The more important is a novel method to readjust weights is proposed to keep the order of weights of criteria help the DM. The proposed COH approach is discussed in detail in the rest of this section.

3.1 The Chaotic Random Number

Each hierarchy is always tried to find possible k in Eq.(1) for the best comprised weights to achieve the maximal profit. Theoretically, the optimal k for the best comprised weights can be identified by means of an exhaustive search over the all possible values of k. In practice, however, we are not able to check all possible strategies because of the vast size of weights and k. Therefore, k is set to a fixed number, say k=.5, in all traditional HCA or AHP-HCA [1, 8, 9, 10]. To correct the shortcoming of using fixed k, the chaotic random number is incorporated in the proposed COH to improve the negotiable space of finding comprised weights.

In recently years, growing interests from physics, chemistry, biology and engineering have stimulated the studies of chaos for control, synchronization, and optimization [18, 19]. Due to the easy implementation and special ability to avoid being trapped in local optima, chaos has been a novel optimization technique and chaos-based searching algorithms have aroused intense interests [20]. In general, the chaotic random number is a special ergodicity, pseudo-randomness and irregularity variable [19]. The well-known logistic equation is adapted here to replace k in Eq.(1) such that a slight difference in the initial value of the chaotic random number would result in a significant difference in its long time behavior as follows [18].

$$k_{t+1} = \mathrm{m} k_t (1 - k_t), \tag{2}$$

where m≠4 is the control parameter, $0<k_t<1$ and $k_0 \neq 0.25$, 0.5, and 0.75. The logistic equation has the basic characteristic of chaos, i.e., it reveals the sensitive dependence on initial conditions and includes infinite unstable periodic motions.

3.2 A limitation to k

According to the characteristic of the convex combination shown in Eq.(1), the new generated number $w_{i,I}^{t+1}$ is between $w_{i,I}^{t}$ and $\overline{w_{i,I}^{t}}$. If k>.5, then $w_{i,I}^{t+1}$ is closer to $w_{i,I}^{t}$ than to $\overline{w_{i,I}^{t}}$ and vice versa. Since the priority of each hierarchy is always to protect its own profit in real-life situation. Therefore, to be more reasonable, the k must not less than .5 in this study, i.e., Eq.(1) is revised as:

$$w_{i,I}^{t+1} = \begin{cases} k_t w_{i,I}^{t} + (1-k_t)\overline{w_{i,I}^{t}} & \text{if } k_t \geq .5 \\ (1-k_t) w_{i,I}^{t} + k_t \overline{w_{i,I}^{t}} & \text{otherwise} \end{cases}. \quad (3)$$

3.3 The Novel Method to Readjustment Weights

Experienced DM may not be able to provide a 100% correct value to each weights of criterion. However, they can tell which criterion is more important. That is why Guh's AHP-HCA [1] resulted in failed to apply in real-life. To avoid such flaw of AHP-HCA discussed in Section 3.4, a novel equation is proposed to replace Eqs.(1) and (3) to readjustment each weight by a new convex combination after sorting the weights under same parent in decreasing order as follows:

$$\omega_{i,I_j}^{t+1} = \overline{k_t}\omega_{i,I_j}^{t} + (1-\overline{k_t})\Delta_t. \quad (4)$$

where

$$w_{i,I_j}^{t} > w_{i,I_l}^{t} \text{ if and only if } j<l, \quad (5)$$

$$\Delta_t = \begin{cases} Min\{\overline{w_{i,I_j}^{t}}, \omega_{i,I_{j-1}}^{t+1}\} & \text{if } w_{i,I_i}^{t} < \overline{w_{i,I_i}^{t}} \\ Max\{\overline{w_{i,I_j}^{t}}, w_{i,I_{j+1}}^{t}\} & \text{otherwise} \end{cases}, \quad (6)$$

$$\overline{k_t} = Max\{k_t, (1-k_t)\}. \quad (7)$$

To satisfy the total weight under the same parent is equal to one, a normalization is implemented to Eq.(4):

$$w_{i,I_l}^{t} = \frac{\omega_{i,I_l}^{t}}{\sum_j \omega_{i,I_j}^{t}}. \quad (8)$$

The major concept behind Eqs.(4)-(8) is that the order of the importance of criteria can never be change, i.e. the following important relationship must be satisfied for each new generated w_{i,I_j}^{t+1}:

$$w_{i,Ia}^{t} < w_{i,Ib}^{t+1} < w_{i,Ic}^{t+1}, \quad (9)$$

if

$$w_{i,Ia}^{0} < w_{i,Ib}^{0} < w_{i,Ic}^{0}. \quad (10)$$

3.4 Procedures of the Proposed COH

The procedure of the Proposed COH is presented in detail using the following steps:

Purpose: Find the comprised weights for the DM.

Input: The performance evaluation reference index.

Output: The comprised weights satisfied by the DM.

STEP 0. Design types, qualitative and quantitative criteria of evaluation hierarchy structures for a problem by the experienced DMs.

STEP 1. Implement the AHP to find the prior values of weights to qualitative and quantitative criteria of each type of hierarchy structure constructed in STEP 0 if necessary.

STEP 2. Let t=0.

STEP 3. Compute counter-weight of each criterion based on Eqs.(3) or (4).

STEP 4. Compute $(CI)^t$ of the various indexes.

STEP 5. If $(CI)^t$<0.001, then the performance evaluation model would produce a comprised solution based on the current weight values and halt.

STEP 6. Let t=t+1 and calculate k_t based on Eq.(2).

STEP 7. Adjustment the weight of each criterion based on Eq.(4) , and go to STEP 3.

4. An example

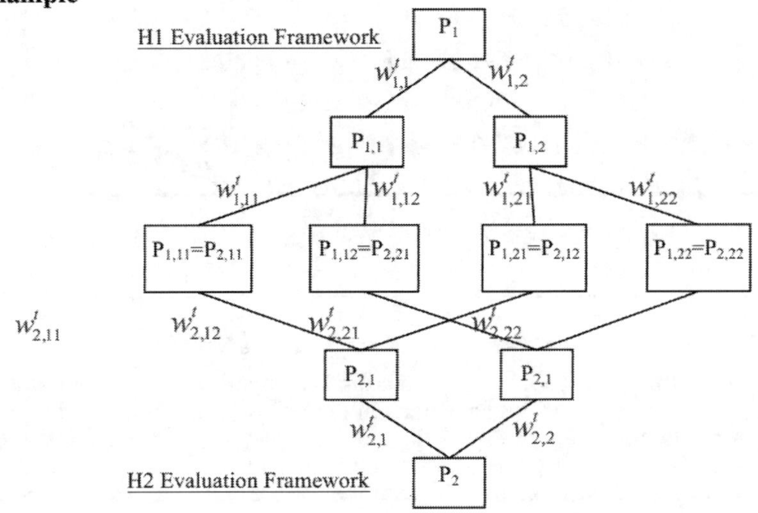

Figure 1. The hierarchy structure schema for the benchmark problem.

The general procedure is best illustrated with an example. Since the total running time and computational difficulty will grow exponentially with the number of hierarchies, levels, criteria, and sub-criteria in the worst case. Owing to this inherent problem, instead of presenting practically large problem, the famous benchmark problem shown in Figure 1 is selected to demonstrate the proposed COH as in all existing known algorithms [1,8-11].

Table 1. The detail of 10 runs of the proposed COH for Figure 1.

	1	2	3	4	5	6	7	8	9	10	Avg	Max	Min	Std
$w^t_{1,1}$	.499	.499	.495	.482	.493	.498	.495	.489	.500	.499	.495	.500	.482	.006
$w^t_{1,2}$	.501	.501	.505	.518	.507	.502	.505	.511	.500	.501	.505	.518	.500	.006
$w^t_{1,11}$	.702	.711	.692	.702	.695	.708	.729	.680	.707	.708	.703	.729	.680	.013
$w^t_{1,12}$	.298	.290	.308	.298	.306	.292	.271	.320	.294	.292	.297	.320	.271	.013
$w^t_{1,21}$	.537	.526	.543	.525	.531	.529	.502	.541	.551	.529	.531	.551	.502	.013
$w^t_{1,22}$	.463	.474	.457	.475	.469	.471	.498	.459	.449	.471	.469	.498	.449	.013
$w^t_{2,1}$	.619	.618	.617	.610	.612	.618	.614	.610	.629	.619	.616	.629	.610	.006
$w^t_{2,2}$	.381	.382	.383	.390	.388	.382	.386	.391	.371	.382	.384	.391	.371	.006
$w^t_{2,11}$	.567	.575	.557	.556	.560	.572	.589	.548	.563	.572	.566	.589	.548	.012
$w^t_{2,21}$	.434	.426	.443	.444	.440	.428	.411	.453	.437	.428	.434	.453	.411	.012
$w^t_{2,12}$	.389	.378	.397	.368	.387	.380	.347	.400	.395	.382	.382	.400	.347	.016
$w^t_{2,22}$	.611	.623	.604	.632	.613	.620	.653	.600	.605	.618	.618	.653	.600	.016
t	11	12	11	12	11	11	13	12	14	12	11.9	14	11	.994
$10^5(CI)^t$	61	79	94	98	63	80	91	70	86	62	78	98	61	14

Table 2. The compromised weights under k=0,.1,...,1.

k	.0	.1	.2	.3	.4	.5	.6	.7	.8	.9	1.	Avg	Max	Min	Std
$w^t_{1,1}$	.60	.583	.565	.545	.525	.505	.484	.462	.441	.420	.400	.503	.600	.400	.067

$w_{1,2}^t$	.40	.417	.435	.455	.475	.495	.517	.538	.559	.580	.600	.497	.600	.400	.067
$w_{1,11}^t$	.80	.785	.769	.751	.731	.710	.688	.666	.643	.621	.600	.706	.800	.600	.068
$w_{1,12}^t$	.20	.215	.231	.250	.269	.290	.312	.334	.357	.379	.400	.294	.400	.200	.068
$w_{1,21}^t$	.30	.353	.400	.442	.482	.518	.554	.589	.624	.661	.700	.511	.700	.300	.129
$w_{1,22}^t$	.70	.647	.600	.558	.519	.482	.446	.411	.376	.339	.300	.489	.700	.300	.129
$w_{2,1}^t$	.60	.605	.608	.610	.613	.615	.619	.624	.633	.644	.660	.621	.660	.600	.018
$w_{2,2}^t$	.40	.395	.392	.390	.388	.385	.381	.376	.367	.356	.340	.379	.400	.340	.018
$w_{2,11}^t$	.80	.757	.714	.671	.627	.583	.538	.493	.448	.405	.364	.582	.800	.364	.146
$w_{2,21}^t$	.20	.243	.286	.329	.373	.418	.462	.507	.552	.595	.636	.418	.636	.200	.146
$w_{2,12}^t$	.30	.317	.333	.349	.365	.380	.396	.412	.429	.448	.471	.382	.471	.300	.055
$w_{2,22}^t$	.70	.683	.667	.651	.635	.620	.605	.589	.571	.552	.529	.618	.700	.529	.055
t	2	3	4	4	4	4	4	4	4	3	2	3.46	4	2	.82

There are 10 runs for the proposed COH to Figure 1, and the result is shown in Table 1. From Table 1, the number of iterations is higher that Guh's HCA (see Table 2), but it can be ignored since the running times all are less than 0.1 second. Moreover, all the order of weights is still held in the proposed COH. According to the results, we find that there is no significant change when we increase the iteration number of COH.

5. Conclusions

Performance evaluation method has been applied extensively in many real-world systems. In general, the structure of evaluation models established for large-scale problem resolution seems to be the multiple-criterion and multiple-level type models that are more practical and simulate real problems in management. Hence, the more knowledge the DM possesses, the better the advantage and success will be.

To the best of the author's knowledge, the concept of chaos has never been investigated in the context of performance evaluation models. In this paper, we improved Guh's HCA-based method and presented a revised method named chaotic ordered HCA (COH in short). The proposed COH has the advantage is that the ranking order of the proposed COH is more consistent with our intuitions than original one. Thus, utilizing the proposed COH is more reasonable than using HCA-based method for ranking criteria, and the DM gains greater understanding of the structure of the problem. Furthermore, the proposed COH not only includes all the situations of AHP and HCA, but also it improves both AHP and HCA as well.

References

[1]. Y. Y. Guh, Determining weight by combining different hierarchy structure-hierarchies consistency analysis, International Journal of Information and Management Science. 7 (1996) 63-80.

[2]. W.-C. Yeh, An interactive augmented max-min model for the multi-objective system reliability problem, International Journal of Systems Science. 38 (2007) 87-99.

[3]. T. L. Saaty, The analytic hierarchy process, McGraw-Hill, New York, 1980.

[4]. T. L. Saaty, Decision making for leaders: The analytic hierarchy process for decisions in a complex world, Lifetime Learning, New York, 1982.

[5]. T. L. Saaty, Axiomatic foundation of the analytical hierarchy process. Management Science. 32 (1986) 841-855.

[6]. E. N. Weiss, V. R. Rao, AHP design issues for large scale systems, Decision Science. 18 (1987) 43-61.

[7]. F.S. Fogliatto, S.L. Albin, A hierarchical method for evaluating products with quantitative and sensory characteristics, IIE Transactions. 33 (2001) 1081-1093.

[8]. Y. Y. Guh, M.S. Yang, R.W. Po, E.S. Lee, Establishing performance evaluation structures by fuzzy relation-based cluster analysis, Computers and Mathematics with Applications. 56 (2008) 572-582.

[9]. Y. Y. Guh, Introduction to a new weighting method - hierarchies consistency

analysis, European Journal of Operational Research. 102 (1997) 215-226.

[10]. Y. Y. Guh, R.W. Po, Establishing a multiple structures analysis model for AHP, International Journal of Information and Management Science. 15 (2004) 35-51.

[11]. Y. Y. Guh, E. S. Lee, K. M. Wang, Hierarchies consistency analysis for non-transitive problems, Computers and Mathematics with Applications. 32 (1996) 67-77.

[12]. S. Gakkhar, R. K. Naji, Order and chaos in predator to prey ratio-dependent food chain, Chaos, Solitons and Fractals. 18 (2003) 229-239.

[13]. Z. Jing, J. Yang, Bifurcation and chaos in discrete-time predator-prey system, Chaos, Solitons and Fractals. 27 (2006) 259-277.

[14]. X. Liu, D. Xiao, Complex dynamic behaviors of a discrete-time predator-prey system, Chaos, Solitons and Fractal. 32 (2007) 80-94.

[15]. R. K. Naji, A.T. Balasim, Dynamical behavior of a three species food chain model with BeddingtonDeAngelis functional response, Chaos, Solitons and Fractals. 32 (2007) 1853-1866.

[16]. H. Cheng, A. Sandu, Efficient Uncertainty Quantification with the Polynomial Chaos Method for Stiff Systems, Computers and Mathematics with Applications. 79 (2009) 3278-3295.

[17]. Z.H. Che, Using fuzzy analytic hierarchy process and particle swarm optimisation for balanced and defective supply chain problems considering WEEE/RoHS directives, International Journal of Production Research. 48 (2010) 3355- 3381.

[18]. R. May, Simple mathematical models with very complicated dynamics, Nature. 261 (1976) 459-467.

[19]. B. Liu, L. Wang, Y.H. Jin, F. Tang, D.X. Huang, Improved particle swarm optimization combined with chaos, Chaos, Solitons and Fractals. 25 (2005) 1261-1271.

[20]. L.Wang, D.Z. Zheng, Q.S. Lin, Survey on chaotic optimization methods, Comput Technol Automat. 20 (2001) 1–5.

The internet differential equation and fractal networks

R G V Baker[2]
[2] BCSS, Faculty of Arts and Sciences, University of New England, Armidale, NSW 2351, Australia

Email: rbaker1@une.edu.au

Abstract. The Internet is an example of a general physical problem dealing with motion near the speed of light relative to different time frames of reference. The second order differential equation (DE) takes the form of 'time diffusion' near the speed of light or alternatively, considered as a complex variable with real time and imaginary longitudinal components. Congestion waves are generated by peak global traffic from different time zones following the Earth's revolution defined by spherical harmonics and a day/night bias. The DE is essentially divided into space and time operators constrained by the speed of light c, band capacity w and a fractal dimension Z (Hausdorff dimension). This paper explores the relationship between the dynamics and the network including the addition of fractal derivatives to the DE for regional networks for $0 < Z < 1$.

1. Introduction

The Internet forms the physical network of connectivity (such as, optical cables and phone wires), where there are nodes or 'routers' that navigate packets of data from one computer to another. The location of these computers around the Earth means that pings of information packets can be transmitted to and from different time zones relative to the arbitrary International Date Line. There will never be instantaneous transfers of packets because of the fundamental limit of the speed of light. The time taken to transmit this data between computers is termed 'latency'. Packet loss is a proxy for peak demand, since in heavy traffic periods, too many packets arrive and the routers then hold them in buffers until the traffic decreases. If these buffers fill during these times of congestion, the routers drop packets and this is measured as 'packet loss'. Sites that are not well-connected in the network, suffer high packet loss and this can be a proxy for congestion. However, with the significant increase in band capacity over the last decade, such congestion has dissipated within the global network. Nevertheless, packet loss still remains a very useful proxy for modelling and such data complied during the Stanford Linear Accelerator (SLAC) internet experiments for 2000 and 2004 [1] are a valuable resource to test physical modelling.

There are a number of approaches employed to study internet traffic, such as, network analysis [2] [3] [4], the fluid dynamics of congestion flows through the system [5] [6], stochastic evolution [7] [8] and the measurement of ping times and geographical distance relative to capacity [9]. An internet differential equation has been developed to describe the dynamics of pings around a global network where information can move either forwards or backwards between relative time zones [7] [10].

Published under licence by IOP Publishing Ltd

This paper summarises and develops the modeling dynamics by looking at two further aspects raised briefly by [10]. Firstly, since we are dealing with relative time, where ping directions can be either forwards or backwards relative to the Earth's rotation, the concept of time for the internet can be framed as a complex variable with real and imaginary components. The meridians of longitude can be visualized as imaginary time lines relative to a real time rotation and can be described as such mathematically. Secondly, the link between the dynamics of global traffic movement and regional network contributions to congestion is critical to any physical description of internet traffic. There is a possibility of a connection between the Gaussian nature of internet traffic movement and the fractal nature of a bound Gaussian network, when the Hausdorff dimension is Z = 2 [10]. This has not proved to be very fruitful because there no easy way to mathematically link congestion patterns and the network fractal. This paper looks briefly at introducing a fractal calculus F^Z derivative in the internet DE for the condition $0 < Z < 1$ where the congestion measurement is proportional to the Hausdorff dimension Z.

2. Background

Random walk modeling has been used to develop a set of internet equations between the Earth's time zones for pings between monitoring machines and remote hosts located within different time zones [7]. Traffic can move forwards and backwards relative to the arbitrary International Date Line with the novelty that one can experience their birthday twice traveling from Sydney to Los Angeles. There is real time movement from the Earth's revolution, but pings traveling near the speed of light can access imaginary time lines elsewhere on the revolving Earth designated by meridians of longitude. This is the difference between information transfer by the internet in the 21st century and telegraphic transfers by cables in the late 19th century, since spatial diffusion then was a major problem for trans-Atlantic cabling.

Internet traffic within and between time zones can be defined by a set of equations for the exchange rates $E(x,t)$, continuity conditions for origin-destination exchanges and the congestion transfers as [7]:

$$E(x,t) = -(w/c^2)(\partial\phi/\partial t + v\phi) \tag{1}$$

$$\partial\phi/\partial x = -\partial E/\partial t \tag{2}$$

$$\partial\phi/\partial x = (w/c^2)\left(\partial^2\phi/\partial t^2 - v\partial\phi/\partial t\right) \tag{3}$$

where $\phi_o\ (x,\ t)$ is the initial site density for a time distance p between pairs assumed to be equal for remote hosts and the ith monitoring site (located at x and having meridians of longitude as time bands between $t_i = ip$); w the transfer band width; c the speed of light and υ as bias in traffic from the Earth's 24-hour revolution. The ping latency Δt and the partitioning of distance Δx are constrained by the inequality [7]:

$$\left(2c^2/w\right)\Delta x \le \Delta t^2 \tag{4}$$

This inequality states that there will never be an instantaneous transfer of information because of the speed of light but improvement in capacity will see pings approach the event horizon (figure 1). For the time wave on a sphere, Eq.3 can be expressed by spherical polar coordinates for the day/night rotation (l =1, m =1). This means that packet loss, as a density proxy at any of the monitoring sites, can be modelled with longitudinal change on the sphere.

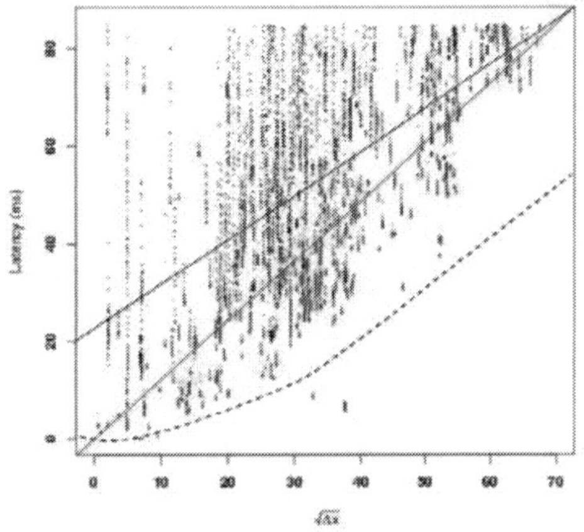

Figure 1. Latency shift towards the limit of transmission for global traffic, SLAC 2000 & 2004

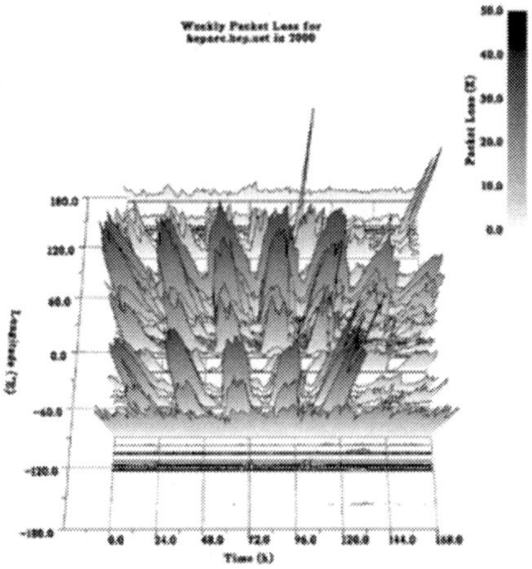

Figure 2. The wave nature of global demand congestion for hepnrc.hep.net, Chicago US

The novelty for the physics is that information can come to a computer located at x from positive and/or negative time directions, because the time zone band Δt is squared, making travel time Gaussian in nature. Also, the return ping to the monitoring site requires a convergence without any fluctuations. This will occur if all the coefficients in the exchange have the same sign and are positive in a Taylor expansion at location x for a modulus λ stopped at the first order Δx. Therefore, for positive coefficients only, the ratio λ of space to time in any network must obey the inequality $0 \leq \lambda \leq \frac{1}{2}$ [7].

Equations (1)-(4) have been applied to the Stanford Linear Accelerator Centre internet experimental data set [1]. The SLAC experiments in 2000 used 27 monitoring sites on five continents and up to 216 remote hosts in a global network. In 2004 there were 25 monitoring sites and up 269 remote hosts. The monitoring sites pinged the network every 30 minutes and SLAC recorded the average ping latency and packet loss between each origin-destination pairs. The raw hourly ping latency data was collated and averaged for each hour of the week. The data begins Monday 00:00 local time of the remote host and is truncated to extract the days (Monday to Sunday 00:00 to 24:00 local time). This is then standardised to Greenwich Mean Time. Using the actual ping time and the distance at which the interaction between the monitoring site and remote host occurs (using latitude and longitude references), plots are generated for all yearly ping time averages.

The key aspects of the differential equation have been compared for each operator subset of the equation to the SLAC global network [10]. The limit of internet information transfer for the whole network ($k\Delta x \leq \Delta t^2$) is shown for 2000 and 2004 (figure 1). The wave metric of congestion ($\partial^2 / \partial t^2$) is demonstrated for the hepnrc.hep.net monitoring machine in Chicago US (figure 2). The distance metric ($\partial / \partial x$) yields a negative exponential distance metric. Regression of the residuals of the phase metric χ of the network gives a proxy for global connectivity of the monitoring site for $0 < \chi \leq 1$. It can be calculated for each monitoring site, where, for $\chi \rightarrow 0$, the site is increasingly well-connected globally. For example, for the well connected hepnrc.hep.net site, $\chi = 0.184$, whereas for the isolated caia.swin.edu.au site at Melbourne Australia, $\chi = 0.764$ (figures 7 and 8). The slope of the regression line is calibrated as unity from traffic in phases with the rotation of the Earth, so there is no significance testing. The day-night divide can also be described through spherical polar coordinates for $l = 1$ and $m = 1$ (figure 4) and this can be used to animate the SLAC data [10].

2. Imaginary Time Zones and Internet Pings

The model in Eq.3 has been applied to the Internet where pings can travel near the speed of light to and from different time zones relative to the Earth's rotation. Whilst the rotation is real, the longitudes are imaginary in the sense that motion can be forwards in time or backwards in time, relative to the arbitrary International Date Line. In recognition of this internal time relativity between longitudes of imaginary time, Eq.1 can be rewritten including the imaginary time lines as:

$$(iD/2\pi)\partial^2\phi/\partial t^2 - v\ \partial\phi/\partial t = \mu\partial\phi/\partial x \tag{5}$$

This can be derived simply by using the complex wave form of $\Phi(t,x) = \exp 2\pi i(kt - \beta x)$, differentiating time twice and space once, recognising that traffic can travel with or against the Earth's rotation. This is the Schrödinger equation in one spatial dimension with the second order operator in time, where the time quantums are just the time zones defined by meridians of longitude. Time is framed as complex with real (physical time) and imaginary components (time relative to the arbitrary International Date Line). As in the case of quantum mechanics, probability densities can be constructed from wave functions between 0 and T of the Earth's 24-hour rotation $\int_0^{24} |\Phi|^2 dt$ and the expectation value of packet loss $\Psi(t)$:

$$\overline{\Psi}(t) = \int_0^{24} \Phi^* \Psi(t) \Phi\ dt \tag{6}$$

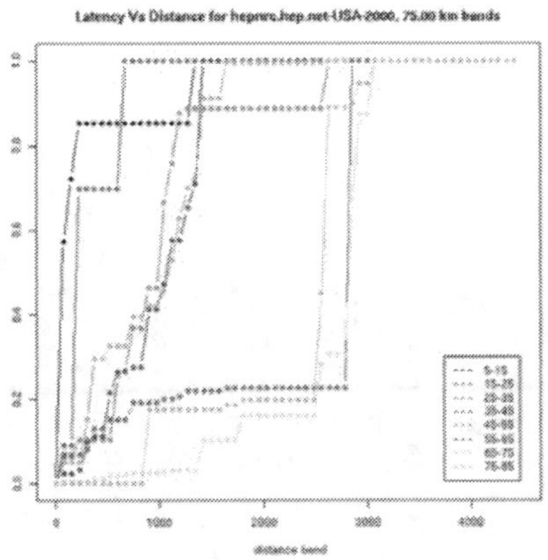

Figure 3. Cumulative plot of demand relative to distance from hepnrc.hep.net Chicago

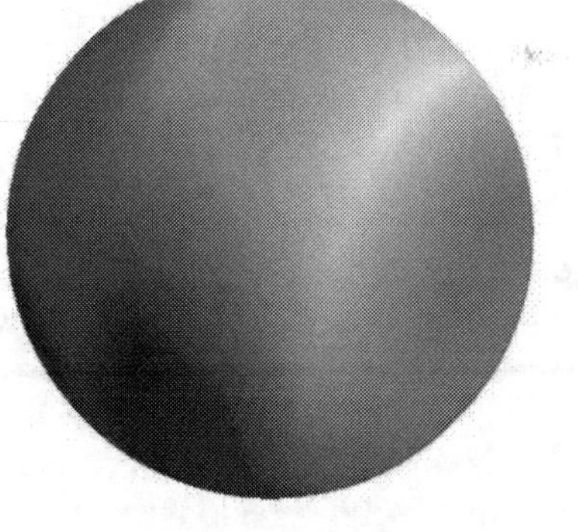

Figure 4. Spherical harmonic solution ($l=1$, $m=1$) for day/night rotation of the Earth

If $\Psi(t)$ varies with location, then there is still an expectation value, since $\overline{\Psi}(t)$ must be evaluated at a particular time as Φ is also a function of location. Therefore, averages in packet loss can be assigned to any monitoring machine measuring internet traffic supporting the data smoothing in averaging the ping data for each hour per week.

The solution of a time Schrödinger equation in spherical co-ordinates is interesting since the separation constant is the rotational period of the Earth. The distribution of global internet traffic suggests (figure 5) three principle axes of demand (Europe, US and Asia) in three time zones. An interesting idea is whether the solutions for time zones according to quantum numbers *l* and *m* in the imaginary time Schrödinger equation are self-similar (figure 6) in that the demand loops for an individual computer in the inner circle have the same fractal dimension of global traffic as the outer circle. If this is the case, then the oscillation with a real rotation between three imaginary time zones would constitute a Julia Set (figure 10).

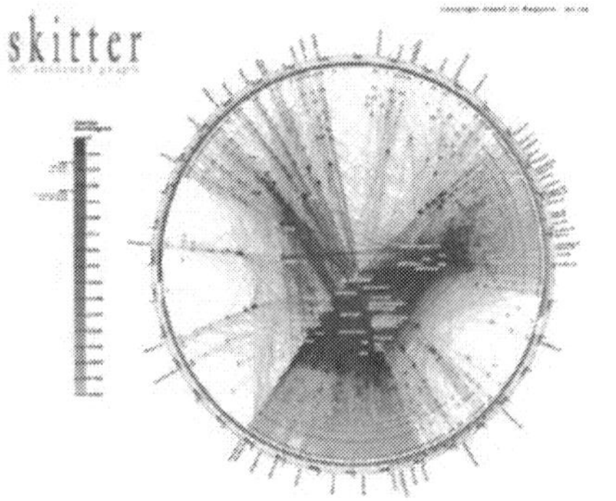

Figure 5. Internet traffic 2000 (source CADA)

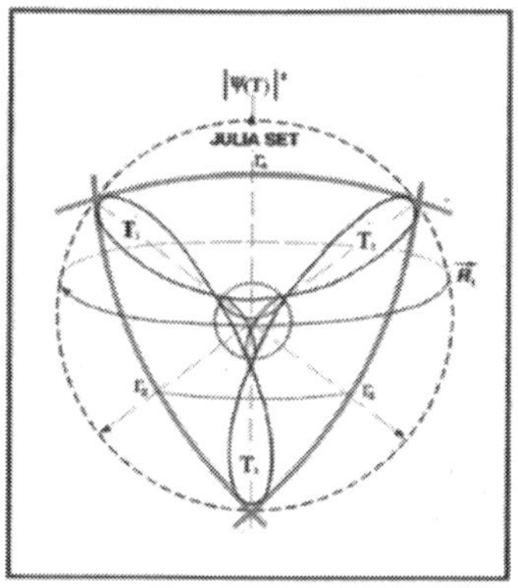

Figure 6. Self-similar oscillation in a Julia Set

4. Fractal Dimensions and Network Analysis

The contribution of the regional network configuration to congestion is an important element of understanding the character of internet traffic. This has been recognized in the origin-destination exchange equation [10] with the addition of the regional network contribution $-Z\Phi$ to congestion resulting in the differential equation, namely:

$$E(x,t) = -(w\chi / c^2)(\partial\phi/\partial t + v\phi - Z\phi) \tag{7}$$

$$\partial\phi/\partial x = (w\chi / c^2)\{\partial^2\phi/\partial t^2 - (1-Z)\partial\phi/\partial t\} \tag{8}$$

The measure of phase connectivity in the network involves two elements, firstly a global phase measurement and secondly a regional network contribution to congestion. The global index χ uses a linear regression plot of the residuals of the wave against the standardised global line for the range (0,1). For well-connected sites, χ approaches 0 and the amplitude of the wave diminishes accordingly (figure 7). This phase line was constructed for points distributed relative to a 360° manifold with a boundary. Conversely, a regional κ-statistic is defined by a slope of zero and is set for residuals of packet loss that are in phase with local congestion. The residuals are distributed along a standardised horizontal line relative to the local time zone of the monitoring machine (figure 8). As κ approaches 0, there is substantial phase congestion from local traffic dropping packets in what could be inferred as poorly connected sites. The relationship between the χ- and κ -statistics are summarised below:

$Z = 0$ $Z = 1$(wave only) $Z = 2$ (bound Gaussian)

← $\chi = 1$ Global Phase 2χ

κ → κ= 1

Regional Phase

If Z is a fractal dimension, for Z = 2, the probability of network jumps for a time-space similarity λ of the ping distribution can be defined through a Lévy distribution as [11]:

$$P(\eta) = 2\, exp\,(-\, b \mid \lambda \mid^{Z}) \qquad 0 < Z \leq 2 \tag{9}$$

When $Z = 2$, the network cluster of computers is a bound Gaussian of Brownian motion from the pings from the monitoring machine. In this case, equation 9 simplifies to equation 3 and so there is an implicit Gaussian network within the time Gaussian. For $Z = 1$, there is zero congestion from the regional networks and no bias from the Earth's rotation. There is only a pure wave of congestion and this is when the global index χ^2 is calculated for the monitoring computer network.

For the global phase regression for a well-connected hepnrc.hep.net monitoring machine in Chicago US (figure 7) in 2000 $\chi = 0.184$ and the regional phase $\kappa = 0.564$ whilst for a less well-connected machine in Melbourne Australia (caia.swin.edu.au) in 2000 $\chi = 0.764$ and $\kappa = 0.115$. For well-connected regional networks for the US in 2000, the mean value for the 14 monitoring sites was 0.488 and for the 11 monitoring machines in 2004 it was 0.485 (figure 9). However, the regional median for US monitoring sites 2004 was 0.395, possibly due to the substantial reduction in periodic congestion measurements used in the regression, due to improvements in capacity. Why would a regional network that is well-connected approach $\kappa \approx 0.5$? It could simply be the relationship between the ratio of the time-space network to global phase is limited to the ratio of space to time in any network mesh which must obey the inequality $0 \leq \lambda \leq ½$; otherwise it fluctuates and is chaotic [7]. The

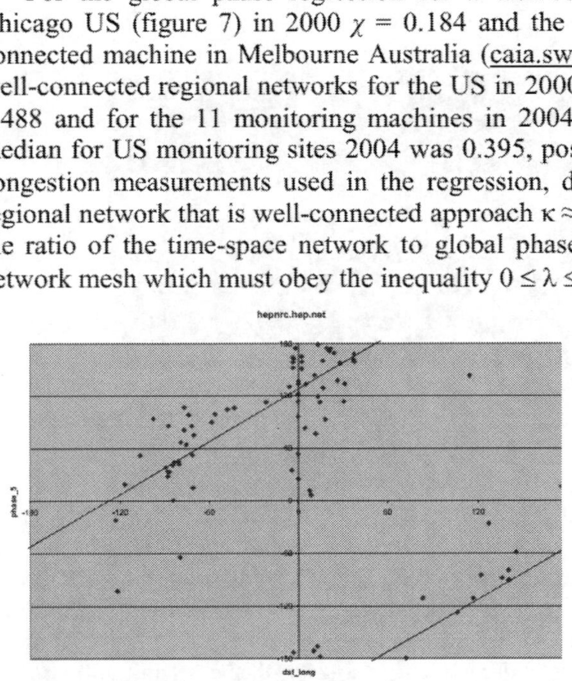

Figure 7. Global phase regression for the well-connected hepnrc.hep.net site Chicago US hepnrc.hep.net site Chicago 2000

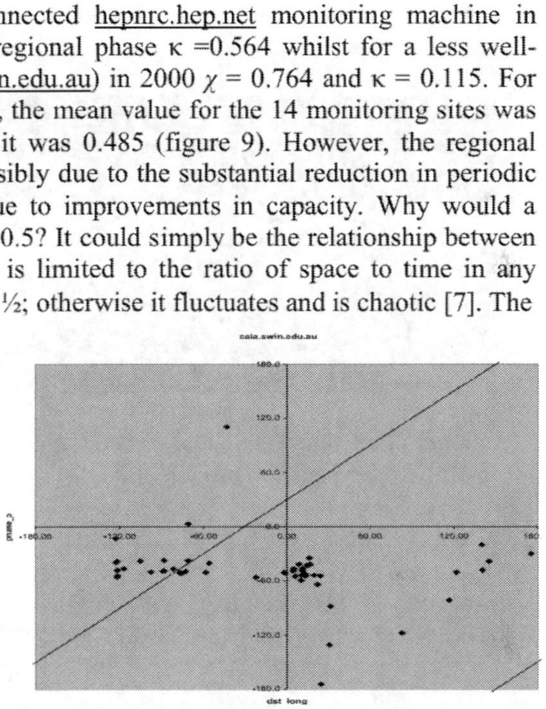

Figure 8 Global phase regression for the poorly connected caia.swin.edu.au site Melbourne Australia 2004

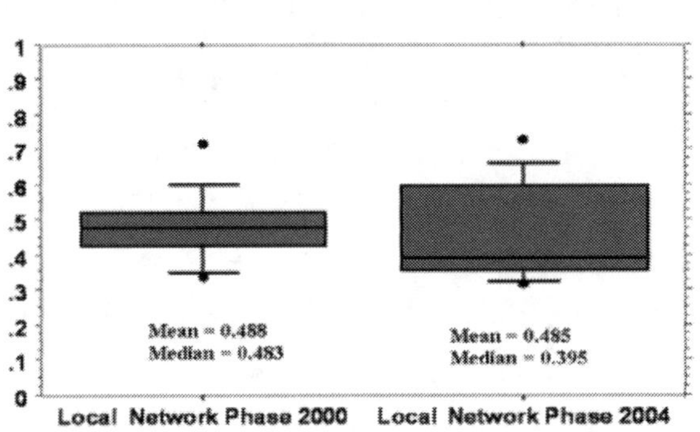

Figure 9. Box and whisker plot of regional phase congestion κ in the US sites

Figure 10. Generalised Cantor set for λ =0.48 (source: Wolfram Demonstrator)

US sites are in phase with global traffic and well-connected regionally, suggesting only positive coefficients in a Taylor expansion around Δx. If this ratio is self-similar for every computer within this regional network (figure 11), then κ may be proportional to a fractal dimension and be non-chaotic for $0 \leq \kappa \leq ½$.

If the well-connected US and European sites do indeed have a fractal Z dimension of ~0.5, then the exchange of traffic between the US and Europe may cascade through a doubling of the time zones as a Feigenbaum attractor where the traffic across the Atlantic is purely the wave (Z =1), whilst the network in the US and Europe may be chaotic for Z > 0.5 (figure 11). The Hausdorff dimension for the Feigenbaum attractor is calculated at 0.538, which is close to the Z boundary for stability. If Z ~ 0.5 for regional phase (and $\chi \rightarrow 0$), equation 9 reduces to the continuity condition (equation 2) for a return ping across the Atlantic.

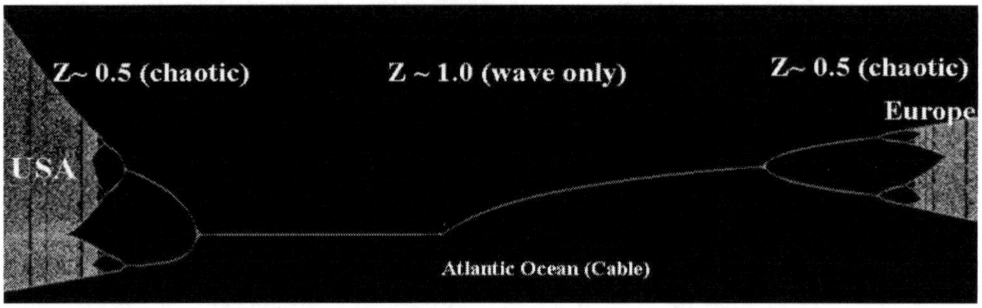

Figure 11. The US and European networks and cable link as a Feigenbaum attractor of a logistic mapping function with a period doubling cascade for deterministic Z =1 and chaotic for Z = 0.538

An alternative idea to model the network contribution to congestion is to introduce a fractal derivative in equation 3. The use of fractional derivatives as a fractal subset of ordinary calculus is an area of growing interest (see [12] for a review). The Φ^z-derivative is best suited for functions like the Cantor staircase which change only on the fractal boundary. The cumulative distance demand graphs for each monitoring computer in the SLAC 2000 network (such as the hepnrc.hep.net monitoring machine in Chicago, figure 3) can be normalised into a Cantor staircase distribution with the Φ^z-integral and Φ^z-derivative characteristics [12]. There is a mass function $E^z(\Phi\ x,t)$ that is proportional to the Hausdorff dimension and because there are finite sub-divisions in regional networks, κ may be analogous to a separation constant for a Φ^z-derivative of congestion from the network. It is fortuitous that the phase of the time derivative from the regression plot of residuals from pings in a regional network is (0,1) and this in the range for the Φ^z-derivative of 0 < Z < 1. When Z = 1 we have to only consider the global wave contribution. Equation 8 can therefore be defined for a constant network congestion threshold Φ_o as:

$$E(x,t) + E^Z(x,t) = -w/c^2 \left(\partial\phi/\partial t + v\phi \ - Z\phi\right) + \Phi_o \tag{10}$$

We assume there is a fractal continuity condition between the fractional derivatives in space and time as:

$$\partial^Z \phi^Z / \partial^Z x^Z = \partial^Z E^Z / \partial^Z t^Z \tag{11}$$

Applying this assumption to equation 11 and noting that the Φ^z-derivative of a constant function is zero, equation 10 simplifies for $\upsilon = 1$ to:

$$\partial\phi_o/\partial x + \partial^z\phi^z/\partial^z x^z = (w/c^2)\{\partial^2\phi/\partial t^2 - (1-Z)\partial\phi/\partial t\} \tag{12}$$

Equation 12 is stating that changes in distance constraint on congestion and the configuration of the spatial network to transmit this congestion is equivalent to the periodic congestion wave of demand with bias relative to whether the local network is in phase with the global network.

We may define the regional network as a generalised Cantor set –log(2)/log(1-k/2) (figure 10). For example, the stability condition $0 \le \kappa \le ½$, the hepnrc.hep.net monitoring site in Chicago with a ratio for network demand of k =0.48 yields a Cantor set in figure 11 for pings coming in and out of the network for a standardised distance.

This paper has summarised the work on dynamic modelling of internet pings as time-based random walks where time lines can be imaginary relative to the International Date Line. There are various ideas presented on how to account the contribution of the network to traffic congestion on Earth. The definition of a regional network as a Cantor set allows for the possibility of introducing Φ^z-derivatives into the dynamics of the internet differential equation and a description of the network from the cumulative patronage as a standardised function of distance.

5. References

[1] Cottrell L, Matthews W and Logg C 2001 http://www-iepm.slac.stanford.edu/pinger/.
[2] Szabo G, Alava M and Kertesz M 2002 *Phys. Rev. E* **66**, 026101/1-026101/8
[3] Barabasi A and Albert R 1999 *Science*, **286**, 509
[4] Cohen R, Erez K, Be-Avraham D and Havlin S 2000 *Phys. Rev. Let.* **85**, 4626
[5] Steger J, Vaderna P, and Vattay G 2006 *Phys. A* **359** 784
[6] Chainais P 2006. *Eur. Phys. J. B* **51** 229
[7] Baker R G V 2005 *J. Geog. Sys.* **7** 361
[8] Eun J J 2007 *Tel. Sys.*, **34**, 3
[9] Ziviani A, Fdida S, De Rezende J F and Duarte O C M B 2005 *Com. Net.* **47,** 503
[10] Baker R G V 2012 *Phys. A* **391** 1133
[11] Montroll E W and West B J 1979 Stochastic Processes eds E Montroll and J L Lebowitz *Fluctuation Phenomena* (North-Holland, Amsterdam) p 92.
[12] Parvate S and Gangal A D 2011 *Fractals* **19** 271

Community detection using Kernel Spectral Clustering with memory

Rocco Langone and Johan A.K. Suykens

Kasteelpark Arenberg 10, 3001, Leuven (Belgium)

E-mail: {rocco.langone,johan.suykens}@esat.kuleuven.be

Abstract. This work is related to the problem of community detection in dynamic scenarios, which for instance arises in the segmentation of moving objects, clustering of telephone traffic data, time-series micro-array data etc. A desirable feature of a clustering model which has to capture the evolution of communities over time is the temporal smoothness between clusters in successive time-steps. In this way the model is able to track the long-term trend and in the same time it smooths out short-term variation due to noise. We use the Kernel Spectral Clustering with Memory effect (MKSC) which allows to predict cluster memberships of new nodes via out-of-sample extension and has a proper model selection scheme. It is based on a constrained optimization formulation typical of Least Squares Support Vector Machines (LS-SVM), where the objective function is designed to explicitly incorporate temporal smoothness as a valid prior knowledge. The latter, in fact, allows the model to cluster the current data well and to be consistent with the recent history. Here we propose a generalization of the MKSC model with an arbitrary memory, not only one time-step in the past. The experiments conducted on toy problems confirm our expectations: the more memory we add to the model, the smoother over time are the clustering results. We also compare with the Evolutionary Spectral Clustering (ESC) algorithm which is a state-of-the art method, and we obtain comparable or better results.

1. Introduction

Community detection of evolving networks recently received much attention in the science community [2, 3, 4, 5, 6]. We aim to cluster networks whose community structure change over time and we wish to obtain a meaningful clustering result over the whole time range taken into account. To this purpose we use the kernel spectral clustering with memory effect (MKSC) model, which is based on a Least Squares Support Vector Machines (LSSVM) [7] formulation. We incorporate the temporal smoothness [8] between consecutive partitioning as a prior knowledge at the primal level, resulting in the dual problem to a set of linear systems. Moreover, the out-of-sample extension is a key feature of our model: we can predict the membership of a new node thanks to the model learned during the training phase. Finally, a systematic model selection scheme is used in order to carefully tune all the necessary parameters.

This paper is organized as follows. In Section 2 we summarize the basic MKSC model (i.e. with one snapshot of memory) and we describe how we can add more memory, which represent the new contribution present in this work. Section 3 presents some simulation results on 4 toy problems. We compare the results obtained by the MKSC model (considering different amounts of memory) with the kernel spectral clustering (KSC, see [9]) applied separately on each snapshot and the evolutionary spectral clustering (ESC) algorithm [10], which is a state-of-the-art method for clustering evolving graphs. Finally in Section 4 we draw some conclusions.

Published under licence by IOP Publishing Ltd

2. MKSC with arbitrary memory

The primal problem of the MKSC model has been stated as follows [11]:

$$\min_{w^{(l)},e^{(l)},b_l} \frac{1}{2}\sum_{l=1}^{k-1} w^{(l)^T} w^{(l)} - \frac{\gamma}{2N}\sum_{l=1}^{k-1} e^{(l)^T} D_{\text{Mem}}^{-1} e^{(l)} - \nu \sum_{l=1}^{k-1} w^{(l)^T} w_{\text{old}}^{(l)}$$
$$\text{such that } e^{(l)} = \Phi w^{(l)} + b_l 1_N.$$

where $e^{(l)} = [e_1^{(l)}, \ldots, e_N^{(l)}]^T$ are the projections, $l = 1, \ldots, k-1$ indicates the score variables needed to encode the k clusters to find, $D_{\text{Mem}}^{-1} \in \mathbb{R}^{N \times N}$ is a weighting matrix, Φ is the $N \times d_h$ feature matrix $\Phi = [\varphi(x_1)^T; \ldots; \varphi(x_N)^T]$, ν and γ are regularization constant. Here $w_{\text{old}}^{(l)} = \sum_{i=1}^{M} w_{\text{prev,i}}^{(l)} = \sum_{i=1}^{M} \Phi_{\text{prev,i}}^{(l)} \alpha_{\text{prev,i}}^{(l)}$ represents the l_{th} binary models developed for the M previous snapshots of the evolving network, not only one snapshot like in the original MKSC model proposed in [11]. The term $w^{(l)^T} w_{\text{old}}^{(l)}$ describes the correlation between the actual and the previous models, which we want to maximize. In this way we are able to introduce temporal smoothness in our formulation, such that the current partition does not deviate too much from the recent history. In particular, the more memory we add (i.e. the more M increases) the smoother will be the clustering results. The clustering model is expressed by:

$$e_i^{(l)} = w^{(l)^T} \varphi(x_i) + b_l, i = 1, \ldots, N.$$

where $\varphi : \mathbb{R}^d \to \mathbb{R}^{d_h}$ is the mapping to a high-dimensional feature space, b_l are bias terms, $l = 1, \ldots, k-1$. The projections $e_i^{(l)}$ represent the latent variables of a set of $k-1$ binary clustering indicators given by $\text{sign}(e_i^{(l)})$. The binary indicators are combined to form a codebook $\mathcal{CB} = \{c_p\}_{p=1}^{k}$, where each codeword is a binary string of length $k-1$ representing a cluster.

The out-of-sample extension to new nodes is done by an Error-Correcting Output Codes (ECOC) decoding procedure.

The dual model is:

$$(D_{\text{Mem}}^{-1} M_{D_{\text{Mem}}} \Omega - \frac{I}{\gamma})\alpha^{(l)} = -\nu D_{\text{Mem}}^{-1} M_{D_{\text{Mem}}} \Omega_{\text{new-old}} \alpha_{\text{old}}^{(l)} = \nu D_{\text{Mem}}^{-1} M_{D_{\text{Mem}}} \sum_{i=1}^{M} \Omega_{\text{new-prev,i}} \alpha_{\text{prev,i}}^{(l)}.$$

where Ω is the kernel matrix with ij-th entry $\Omega_{ij} = K(x_i, x_j) = \varphi(x_i)^T \varphi(x_i)$, $\Omega_{\text{new-old}}$ captures the similarity between the nodes of the current graph and the nodes of each of the previous M snapshots, $\alpha^{(l)}$ are dual variables, $K : \mathbb{R}^d \times \mathbb{R}^d \to \mathbb{R}_+$ is the kernel function, $M_{D_{\text{Mem}}}$ is the centering matrix equal to $M_{D_{\text{Mem}}} = I_N - \frac{1}{1_N^T D_{\text{Mem}}^{-1} 1_N} 1_N 1_N^T D_{\text{Mem}}^{-1}$.

3. Simulation Results

We conducted experiments on the following toy problems, related to both binary and multiway clustering:

- **two moving Gaussians**: from time-steps 1 to 10 two Gaussian clouds move towards each other until they overlap almost completely. This situation can be also described by considering the points of the 2 Gaussians as nodes of a weighted network where the weights of the edges change over time.
- **three moving Gaussians**: three merging Gaussian clouds.
- **switching network**: we build up 9 snapshots of a network of 1000 nodes formed by 2 communities. At each time-step some nodes switch their membership between the two clusters (the software described in [12] is used to generate this benchmark).
- **expanding/contracting network**: a network with 5 communities experiences over-time 24 expansion events and 16 contractions of its communities.

For the two moving Gaussians data-set, in Fig. 1 we show how increasing the memory of the MKSC model can give smoother and better clustering results over time, measured in terms of Normalized Mutual Information (NMI) between consecutive partitioning and the smoothed ARI ([11]). We also compare with the Evolutionary Spectral

Clustering (ESC) algorithm [10] which is a state-of-the-art method for community detection of evolving graphs. We obtain comparable or better performances, depending on the amount of memory added to our model. For what concerns the switching network (see Fig, 2), the NMI plot shows, as we expect, that the more memory we add the more similar are the clustering results over time. Moreover it seems that one snapshot of memory is enough to have good performances ($MKSC_{M1}$ performs better of all the other models). Finally in Table 1 we summarize the results obtained on all the data-sets, showing also the memory requirement and the computation time.

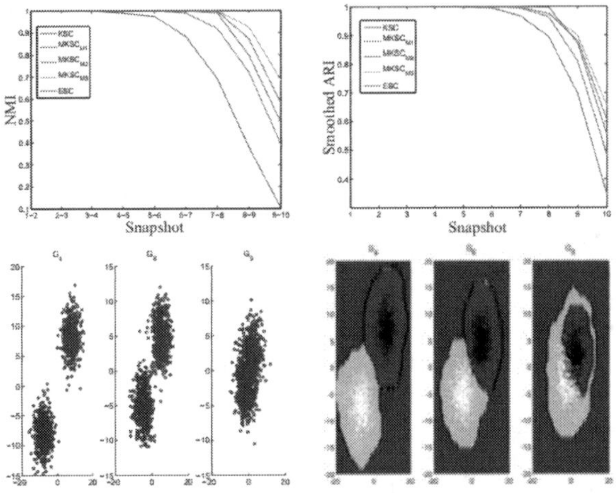

Figure 1. First row: performance of MKSC with 1, 2 and 3 snapshots of memory ($MKSC_{M1}$, $MKSC_{M2}$, $MKSC_{M3}$), KSC and ESC in the two moving Gaussian experiment in terms of the the NMI between consecutive partitions (left) and the smoothed ARI index [11] (right). The smoothed ARI plot and the NMI trend tell us that, as expected, the models with temporal smoothness are more able than KSC to produce clustering results which are more similar to the ground truth and also more consistent and smooth over time. Moreover the more memory we add to the MKSC model, the better the results. In the second row we show the true partitioning (left) and the prediction of $MKSC_{M3}$ (right), where only snapshots 4, 6 and 9 are depicted.

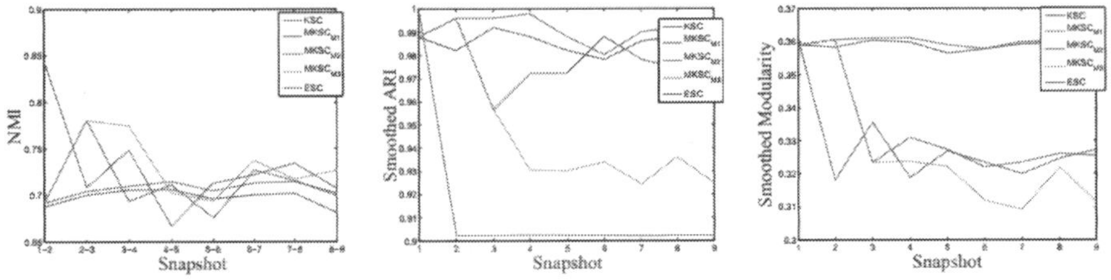

Figure 2. Performance of MKSC with 1, 2 and 3 snapshots of memory ($MKSC_{M1}$, $MKSC_{M2}$, $MKSC_{M3}$), KSC and ESC for the switching network data-set in terms of the the NMI between consecutive partitions, the smoothed ARI index and the smoothed Modularity [11]. The NMI plot shows, as we expect, that the more memory we add the more consistent are the clustering results over time. Moreover $MKSC_{M1}$ performs better of all the other models in terms of smoothed ARI and smoothed Modularity.

4. Conclusions

In this paper we presented an extension of our previous work introduced in [11]. In the latter we proposed the kernel spectral clustering with memory effect (MKSC), a clustering model in the Least Squares Support Vector Machines (LS-SVM) framework with primal-dual formulation and out-of-sample extension. The MKSC model is useful to cluster evolving networks (represented as a sequence of graphs), where a desirable feature of the clustering results over time is their consistency and smooth evolution. This temporal smoothness has been introduced in the primal formulation via a memory term of one snapshot in the past. In this work we explored the consequence of increasing this memory to more than one snapshot. We tested the new model on 4 toy problems. We observed that in some cases we can obtain clusters of better quality and more consistent over time by adding more memory to the basic MKSC model.

EVALUATION MEASURE	MKSC	ESC
Smoothed ARI	**0.88** ($M3$)	0.87
NMI	**0.86** ($M3$)	0.85
CPU time (s)	**0.53** ($M1$)	5.36
Memory requirement (s)	**0.29** ($M1$)	2.92

EVALUATION MEASURE	MKSC	ESC
Smoothed ARI	**0.92** ($M3$)	0.81
NMI	**0.91** ($M3$)	0.75
CPU time (s)	**1.30** ($M1$)	18.95
Memory requirement (s)	**0.73** ($M1$)	10.77

EVALUATION MEASURE	MKSC	ESC
Smoothed ARI	**0.99** ($M1$)	0.91
Smoothed Modularity	**0.36** ($M1$)	0.31
NMI	0.73 ($M3$)	0.73
CPU time (s)	52.23 ($M1$)	**5.56**
Memory requirement (s)	4.31 ($M1$)	**3.06**

EVALUATION MEASURE	MKSC	ESC
Smoothed ARI	0.76 ($M1$)	**0.78**
Smoothed Modularity	**0.51** ($M1$)	0.50
NMI	0.45 ($M1$)	**0.61**
CPU time (s)	51.83 ($M1$)	**5.08**
Memory requirement (s)	**2.57** ($M1$)	2.79

Table 1. Summary of the results for the 4 data-sets (top row refers to the Gaussians and bottom row to the synthetic networks). For each evaluation measure the values represent an average over time (i.e. the mean value per snapshot). Regarding the MKSC model, between parenthesis we indicate with which amount of memory we obtained the best results (1, 2 or 3 snapshots of memory, that is $M1$, $M2$, or $M3$).

Acknowledgements

This work was supported by Research Council KUL: ERC AdG A-DATADRIVE-B, GOA/11/05 Ambiorics, GOA/10/09 MaNet , CoE EF/05/006 Optimization in Engineering(OPTEC), IOF-SCORES4CHEM, several PhD/postdoc & fellow grants;Flemish Government:FWO: PhD/postdoc grants, projects: G0226.06 (cooperative systems and optimization), G0321.06 (Tensors), G.0302.07 (SVM/Kernel), G.0320.08 (convex MPC), G.0558.08 (Robust MHE), G.0557.08 (Glycemia2), G.0588.09 (Brain-machine) G.0377.12 (structured models) research communities (WOG:ICCoS, ANMMM, MLDM); G.0377.09 (Mechatronics MPC) IWT: PhD Grants, Eureka-Flite+, SBO LeCoPro, SBO Climaqs, SBO POM, O&O-Dsquare; Belgian Federal Science Policy Office: IUAP P6/04 (DYSCO, Dynamical systems, control and optimization, 2007-2011); EU: ERNSI; FP7-HD-MPC (INFSO-ICT-223854), COST intelliCIS, FP7-EMBOCON (ICT-248940); Contract Research: AMINAL; Other:Helmholtz: viCERP, ACCM, Bauknecht, Hoerbiger. Johan Suykens is a professor at the KU Leuven, Belgium. The scientific responsibility is assumed by its authors.

References

[1] Fortunato S 2010 *Physics Reports* **486** 75–174

[2] Mucha P J, Richardson T, Macon K, Porter M A and Onnela J P 2010 *Science* **328** 876–878

[3] Lin Y R, Chi Y, Zhu S, Sundaram H and Tseng B L 2009 *ACM Trans. Knowl. Discov. Data* **3**

[4] Bóta A, Krész M and Pluhár A 2011 *Acta Cybernetica* **20** 35–52

[5] Asur S, Parthasarathy S and Ucar D 2009 *ACM Trans. Knowl. Discov. Data* **3** 16:1–16:36

[6] Palla G, lszl Barabsi A, Vicsek T and Hungary B 2007 *Nature* **446** 2007

[7] Suykens J A K, Van Gestel T, De Brabanter J, De Moor B and Vandewalle J 2002 *Least Squares Support Vector Machines* (World Scientific, Singapore)

[8] Chakrabarti D, Kumar R and Tomkins A 2006 *Proceedings of the 12th ACM SIGKDD international conference on Knowledge discovery and data mining* KDD '06 (New York, NY, USA: ACM) pp 554–560 ISBN 1-59593-339-5

[9] Alzate C and Suykens J A K 2010 *IEEE Transactions on Pattern Analysis and Machine Intelligence* **32** 335–347

[10] Chi Y, Song X, Zhou D, Hino K and Tseng B L 2007 *KDD* pp 153–162

[11] Langone R, Alzate C and Suykens J A K 2012 *Submitted*

[12] Greene D, Doyle D and Cunningham P 2010 *Proceedings of the 2010 International Conference on Advances in Social Networks Analysis and Mining* ASONAM '10 (Washington, DC, USA: IEEE Computer Society) pp 176–183

A social network model for the development of a 'Theory of Mind'

Michael S. Harré
School of Engineering and IT, The University of Sydney, Sydney, Australia
E-mail: michael.harre@sydney.edu.au

Abstract. A "Theory of Mind" is one of the most important skills we as humans have developed; It enables us to infer the mental states and intentions of others, build stable networks of relationships and it plays a central role in our psychological make-up and development. Findings published earlier this year have also shown that we as a species as well as each of us individually benefit from the enlargement of the underlying neuro-anatomical regions that support our social networks, mediated by our Theory of Mind that stabilises these networks. On the basis of such progress and that of earlier work, this paper draws together several different strands from psychology, behavioural economics and network theory in order to generate a novel theoretical representation of the development of our social-cognition and how subsequent larger social networks enables much of our cultural development but at the increased risk of mental disorders.

1. Introduction

Imagine if every new-born child had to learn from scratch how to tie a knot, ride a bicycle or even speak the local language without the help of their parents and teachers. What we know, how we learn and how we reason is not just a matter of sitting down and using our brains, however large they may be, to solve a problem, without the supportive network of relationships that guide our learning there are just too many different ways for us to go astray. But what is the relationship between our innate abilities, conferred by our genes, and the development of the social networks in which we live? At the level of the species, Robin Dunbar suggested that the average social group size an individual lived in is related to the size of the neocortex by the following relationship [1]:

$$\log_{10}(N) \quad = \quad 0.093 + 3.89 \log_{10}(C_r) \tag{1}$$

where N is the mean group size across a species and C_r is the ratio of the neocortex size to the rest of the brain across the species. The ratio of neocortex to the rest of the brain was used to allow for those species that need comparatively large brains for the management of their (relatively) large bodies, e.g. whales have large brains in part because they have large bodies that require considerable brain power to co-ordinate. A consequence of this model was the prediction that the average human social group size should be somewhere between 100 and 230 (95% CI), commonly approximated at about 150. At the level of the individual though, rather than at the species level, Dunbar and his colleagues have recently shown that the size of the Orbital Pre-Frontal Cortex (oPFC) region of the human brain linearly correlates with the size of an individual's social network [2].

2. Developing a Theory of Mind

ToM in infants is often measured in several different ways, but the example I will use, called the Sally-Anne test [3], is one of the first developed and is sufficient for the ideas I want to illustrate by example. First used in 1983, the Sally-Anne test uses two dolls that a child, typically between 3 and 9 years old, is introduced to: Sally and Anne. The child witnesses a scene that contains the two dolls, a basket and a box and then watches while Sally places a marble in the basket in clear view of Anne. Sally then goes for a walk (leaves the scene) and Anne removes the marble from the basket and places into the box. When Sally returns the child is then asked: Where will Sally look for the marble? The correct answer, i.e. in the basket and not in the box, requires the child to recognise that Sally has different states of knowledge to that of the child (or that Sally can has inaccurate knowledge of the world). Autistic children often have difficulty passing this test because they are unable to attribute mental states to Sally that are different from their own, this is often interpreted as a failure to properly develop a ToM.

This test requires a child to have some model of the internal workings of another person's mind and that the model needs to accurately reflect that person's state of mind rather than the state of mind of the child. Figure 1 represents four different levels of modelling that a child might have. These are the basic steps by which a child forms a 'link' between themselves, and another through an increasingly better understanding of the other person's point of view.

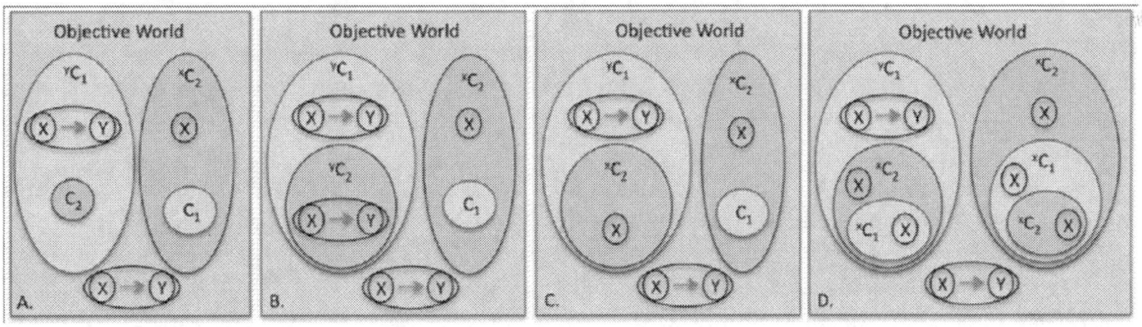

Figure 1. Increasing levels of complexity in the development of a ToM. **A.** C_1 sees the world as simple objects. **B.** C_1 attributes an incorrect state of knowledge to C_2. **C.** C_1 attributes a correct state of knowledge to C_2. **D.** Both C_1 and C_2 attribute states of knowledge to each other, including states of knowledge regarding each other.

3. Network Theory for Small to Medium Societies

In this paper I use the Erdös-Rényi [4] network model where we have a set of N nodes and L links whereby a link forms between two nodes with probability p. The following relationships are well established for this model (ln is the natural log): $\ln(N) = l$ and $N^{-1}\ln(N) = p$ where l is the minimum expected number of links a single node needs to make and p is the minimum probability that a link forms in order for the network to be fully connected[1]

We want to describe the minimum necessary conditions under which a link is formed between two individuals in a small to medium society. Every individual i has a set of m behavioural attributes denoted by a vector $b_i = [x_1, x_2, \ldots, x_m]$, $x_k \in \{0, 1\}$ containing binary behaviours (present-not present). The probability that two people i and j have matching behavioural vectors is a probability $p = q^m$ where q is the probability that a single behavioural attribute is a match. The probability that a person is able to match their b_i to that of l others is $p^l = q^{ml}$.

[1] Throughout we assume the caveat: with high probability

In a network of size N, each person needs to have minimum number of links given by $l = \ln(N)$ in order for a group of individuals to become a fully connected community, so each individual needs to connect with a minimum probability $p = q^{\ln(N)m}$. But this minimum is also known to be given by $p = N^{-1}\ln(N)$, so we have: $N^{-1}\ln(N) = q^{\ln(N)m}$ and solving for m gives:

$$m = \frac{\ln\left(\ln(N)N^{-1}\right)}{\ln(q)\ln(N)} \quad (2)$$

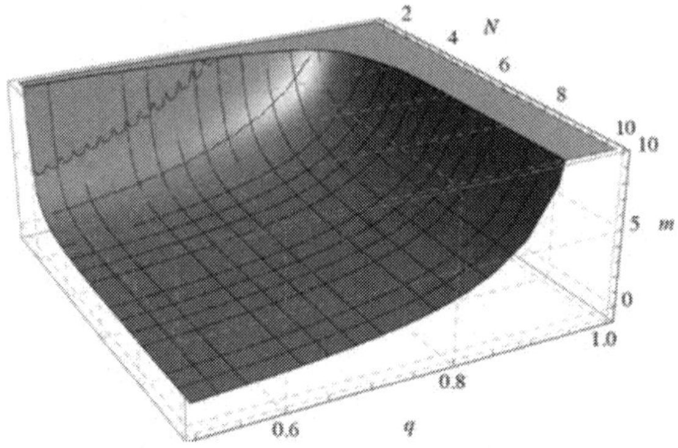

Figure 2. Behavioural complexity m as a function of minimum connected network size N and probability q of accepting any single behavioural attribute.

This equation relates the maximum size of the behavioural repertoire m (the behavioural complexity of each individual) to the group size N and the minimum probability q that a single behavioural trait is 'acceptable' between two members of the social group. If these conditions are met the group becomes a *community* in which all members of the group are connected. A plot of this function is shown in Figure 2. Note that $\frac{-\ln(\ln(N)N^{-1})}{\ln(N)}$ is a smoothly varying function with a limit of 1 as $N \to \infty$. In the discussion that follows this function is essentially a scaling factor over the N of interest with a nearly constant value in the range of $c \in [0.63, 0.76] \simeq 0.7$ for $N \in [4, 3 \times 10^3]$ and so we focus on $m = -c\ln(p)^{-1}$. This is expressed as an equality but it represents a threshold, for m larger than $-c\ln(q)^{-1}$ a connected network is highly unlikely to form, for m smaller than $-c\ln(q)^{-1}$ a connected network is highly likely to form so the region of interest is $m \leq -c\ln(q)^{-1}$. A connected network is a social community of individuals whereby all members have at least one connection with another individual, a community has no loners.

4. Discussion

We appear to have some of the largest and most diverse social networks of any species on earth but we also have some of the most complex behaviour. This work argues for a mentalising ability that mediates the relationship between behavioural complexity and social network size. This mentalising is expressed as the probability that we can 'comprehend' each of the behaviours of another and the size of the social networks that can then form.

This is achieved by considering another person's state of knowledge and constraints and imagining how you might act in the same situation. Such mentalising reduces the probability of a misunderstanding and has been shown to be key in ToM and social network size [2]. The left hand side of Table 1 shows how the number of link formations ($\ln(N)$), which is dependent on an individuals ability to accurately mentalise regarding other people's behaviours, dictates the maximum social group size (N).

N	$\ln(N)$	$\ln(N)N^{-1}$	$[N_1,$	$N_2]$	$[d\ln(N_1),$	$d\ln(N_2)]$
7	1.94	0.278	[3,	5]	[1,	1.46497]
20	3.00	0.150	[9,	15]	[2,	2.46497]
55	4.01	0.073	[27,	45]	[3,	3.46497]
150	5.01	0.033	[81,	135]	[4,	4.46497]
400	5.99	0.015	[243 ,	405]	[5,	5.46497]
1100	7.00	0.006	[729 ,	1215]	[6,	6.46497]
1.30×10^{52}	120	$\simeq 0$	[2187,	3645]	[7,	7.46497]

Table 1. Left: For a fully connected network of size N, the minimum expected links per node $\ln(N)$ and the minimum probability of a link being formed $N^{-1}\ln(N)$ are shown in approximate discrete steps of links. **Right:** A connectivity table scaled by a constant $d = \ln(3)^{-1}$, for maximum group sizes in the range $[N_1, N_2]$ the scaled expected connectivity per person is in the range $[d\ln(N_1), d\ln(N_2)]$

The right hand side of this table shows the scaled connectivity required for networks of maximum sizes between N_1 and N_2. This table reflects the work of Dunbar and colleagues in understanding the multi-layer nature of human social networks [5]. In this work multilayered social networks increase in size with constant scaling factor slightly larger than 3. A range of group sizes they discuss are approximately of size (N): 3-5 (support clique), 9-15 (sympathy group), 30-45 (overnight camps), 150 (clan), 500 (megaband) and 1000-2000 (tribe).

Other numbers were given based on a qualitative analysis of the data: $S_0 = 1$ (the self), $S_1 = 4.6$, $S_2 = 14.3$, $S_3 = 42.6$, $S_4 = 132.5$, $S_5 = 566.6$ and $S_6 = 1728$ with an average scaling between groups of $S_i/S_{i-1} = 3.52$. Note that the first four groupings are a reasonable approximation with N_2 in the table above, offering a potential solution to a question Dunbar et. al. ask at the end of their paper, why discrete scales and a factor of 3? Because individuals that can, on average, be expected to maintain between $d\ln(N)$ and $d\ln(N) + 0.5$ links and as $d\ln(N)$ is in discrete steps it has a maximum fully connected social group size of N that increases stepwise.

Finally, Steven Pinker [6] has recently commented on an on-going debate regarding traditional evolutionary theory where genes act solely at the individual organism level and 'group selection' where genes act at the level of benefiting a group of individuals and not the individual itself. For example group selection argues that self-less altruism can be positively selected for because it benefits the group while disadvantaging the individual organism. The current paper suggests that evolutionary pressure comes to bear on our mentalising abilities, and that the resultant larger groups are probably competitively advantageous. If this means understanding that someone needs help, and that we would expect help were we in the same position, then this 'perspective taking' ability can be selected for at the individual level.

[1] R.I.M. Dunbar. Neocortex size as a constraint on group size in primates. *Journal of Human Evolution*, 22(6):469–493, 1992.

[2] J. Powell, P.A. Lewis, N. Roberts, M. García-Fiñana, and RIM Dunbar. Orbital prefrontal cortex volume predicts social network size: an imaging study of individual differences in humans. *Proceedings of the Royal Society B: Biological Sciences*, 2012.

[3] H. Wimmer and J. Perner. Beliefs about beliefs: Representation and constraining function of wrong beliefs in young children's understanding of deception. *Cognition*, 13(1):103–128, 1983.

[4] P. Erdős and A. Rényi. *On the evolution of random graphs.* Akad. Kiadó, 1960.

[5] W.X. Zhou, D. Sornette, R.A. Hill, and R.I.M. Dunbar. Discrete hierarchical organization of social group sizes. *Proceedings of the Royal Society B: Biological Sciences*, 272(1561):439–444, 2005.

[6] S. Pinker. The false allure of group selection. *Edge*, June 18, 2012.

Principal Component Analysis Based Measure of Structural Holes

Shiguo Deng, Wenqing Zhang, Huijie Yang*
Biz School, University of Shanghai for Science and Technology, Shanghai 200093, China
E-mail: hjyang@ustc.edu.cn

Abstract. Based upon principal component analysis, a new measure called compressibility coefficient is proposed to evaluate structural holes in networks. This measure incorporates a new effect from identical patterns in networks. It is found that compressibility coefficient for Watts-Strogatz small-world networks increases monotonically with the rewiring probability and saturates to that for the corresponding shuffled networks. While compressibility coefficient for extended Barabasi-Albert scale-free networks decreases monotonically with the preferential effect and is significantly large compared with that for corresponding shuffled networks. This measure is helpful in diverse research fields to evaluate global efficiency of networks.

1. Introduction

Structural hole [1] is a novel concept for structural analysis of social networks. Let us consider two nodes or clusters between which there exist not direct linkage or indirect redundancy. This kind of structural pattern will induce resistance to information exchange between the two parts. This hinder effect is called structural hole. As shown in Fig.1, three rectangles and triangle Y connect directly with triangle X, while the three rectangles and Y do not link with each other, namely, X is the single route by which information exchange can occur. X is much more competitive in collecting and controlling information. There exists a typical structural hole. As for the triangle Y, it has an equivalent position with X, but the direct linkages between the three rectangles that are connected with Y introduce redundancy edges besides the routes bridged by Y, which reduces significantly the control ability of Y. Structural hole does not exist here.

Burt [1] introduces the concept of network constraint index to measure quantitatively structural hole. Denoting with $P_{i,j}$ the cost of node i on maintaining its relation with node j, the network constraint index of i reads,

$$H_i = \sum_{j=1}^{N}(W_{i,j}P_{i,j} + \sum_{q=1}^{N} W_{i,q}P_{i,q}P_{q,j}W_{q,j})^2, \tag{1}$$

where W is the adjacency matrix of a social network, namely, $W_{i,j} = W_{j,i} = (0)1$ if node i and node j are (dis-)connected, respectively. A large value of H_i means that the node i is positioned in a dense local pattern, which leads to a low degree of structural hole. But this measure is limited to the contributions from the nearest and the secondary neighbors. Especially, the redundancy from linkages of its neighbors is not included explicitly.

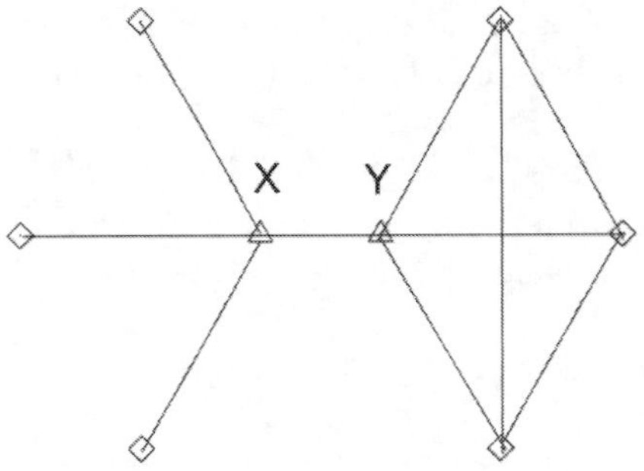

Figure 1. Illustration of structural hole. X is the only way for its neighbor's exchanging information. X is positioned at a typical hole. As for Y, linkages between its neighbors induce redundancy. Y is not positioned at a structural hole.

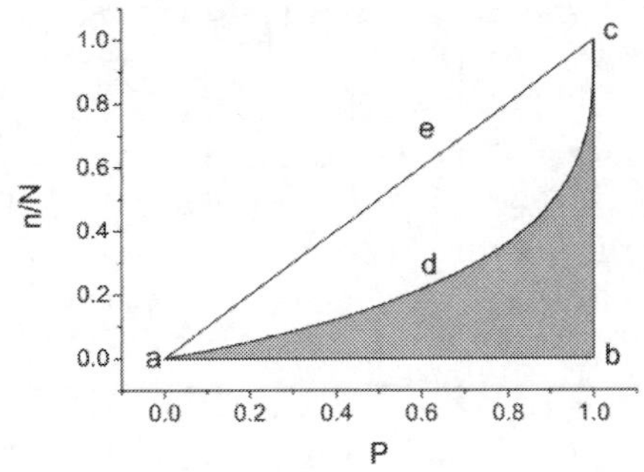

Figure 2. A typical curve for percent explained versus $\frac{n}{N}$. $N = 2000$. Each pair of nodes is linked with probability $\frac{5}{2000}$.

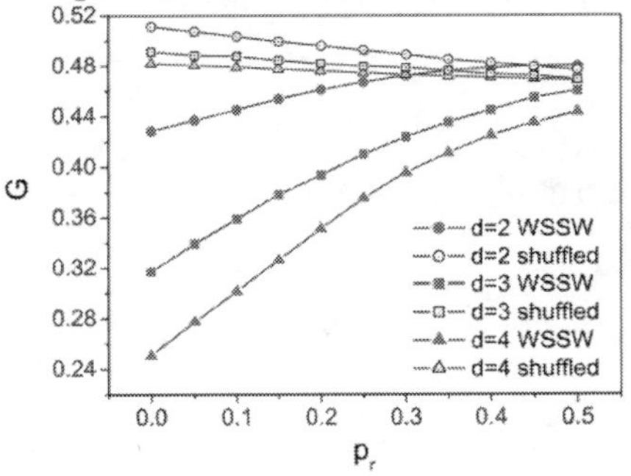

Figure 3. Compressibility coefficients for WSSW networks. $N = 2000$.

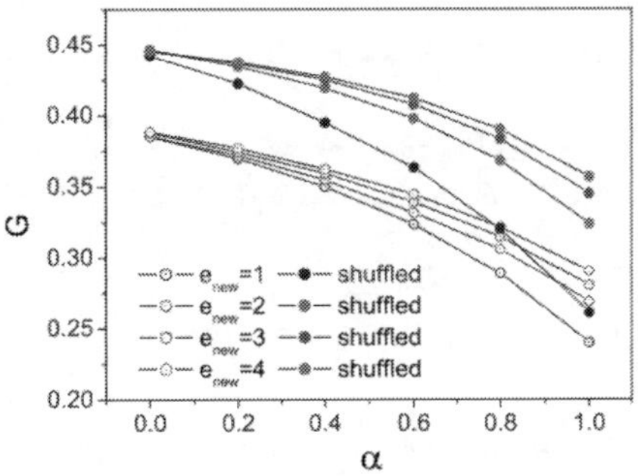

Figure 4. Compressibility coefficients for EBA networks. $N = 2000$.

Besides the contributions of direct linkages, indirect redundant linkages, and linkages between neighbors, there exists actually another important effect from identical patterns in networks. In a social network, optimal strategies in one node is widely duplicated by others. And common environments may force similar nodes to employ the same strategies. Strategies of a node determine its relationships with other nodes. Hence, there may exist some structural patterns occurring with high frequencies. This kind of symmetry may decrease significantly the cost of a node, because the node can use the same strategies in keeping its relationships with structurally identical groups. Once one group obtains information from other nodes, its response action will also be duplicated by the other groups with the same patterns. By this way, information is transferred to a large size in the social network. As a result, degree of structural hole will decrease. This effect is not included in the original definition of structural hole. In the present paper, by means of principal component analysis (PCA) we take into account this effect.

2. Methods and Materials

2.1. Compressibility Coefficients of Networks

Principal component analysis (PCA) [2] is a powerful tool to eliminate redundancy in data, by using of which raw data is compressed. Let us consider a social network described with adjacency matrix A with size N. The element $A_{i,j}$ is 1 and 0 if i and j are connected or disconnected, respectively. Redundant edges between two nodes, i and j means the two nodes share many nodes as their neighbors. At the same time, if the neighbors of node i are densely connected, they will share much more nodes as neighbors, also. These two kinds of redundancy induce high degree of symmetry of the nodes in the social network. The corresponding data can be compressed much more effectively. Hence, compressibility of adjacency matrix can be used to measure structural holes of the social network.

From adjacency matrix A, one can calculate the covariance matrix Cov, whose elements read,

$$Cov(i,j) = \frac{\sum_{k=1}^{N}(A_{k,i} - \frac{\sum_{s=1}^{N} A_{s,i}}{N}) \cdot (A_{k,j} - \frac{\sum_{s=1}^{N} A_{s,j}}{N})}{\sqrt{\sum_{k=1}^{N}(A_{k,i} - \frac{\sum_{s=1}^{N} A_{s,i}}{N})^2 \cdot \sum_{k=1}^{N}(A_{k,j} - \frac{\sum_{s=1}^{N} A_{s,j}}{N})^2}}, \quad (2)$$
$$i,j = 1,2,\cdots,N.$$

Ranking the eigenvalues of matrix Cov in ascending order, $\lambda_1 \geq \lambda_2 \geq, \cdots, \geq \lambda_N$, percent explained by the first n eigenvalues reads,

$$P(n) = \frac{\sum_{s=1}^{n} \lambda_s}{\sum_{s=1}^{N} \lambda_s} \times 100\%. \quad (3)$$

A typical curve for $P(n)$ versus $\frac{n}{N}$ is illustrated in Fig.2. Compressibility coefficient of the social network is defined as,

$$G \equiv \frac{S_{a\to b\to c\to d\to a}}{S_{a\to b\to c\to e\to a}} = 2S_{a\to b\to c\to d\to a}, \quad (4)$$

where $S_{a\to b\to c\to d\to a}$ and $S_{a\to b\to c\to e\to a}$ are the surface areas of the domains surrounded by curves $a \to b \to c \to d \to a$ (filled with grey) and $a \to b \to c \to e \to a$, respectively. A larger value of G tells us that to reach the same percent explained, much more nodes are required to be included. Consequently the corresponding network has low level of redundancy and there exist much more structural holes.

2.2. Data

We consider the Watts-Strogatz small-world (WSSW) model [3]. One can construct initially a regular lattice with periodic boundary condition, in which each node is linked with its right-handed d neighbors. Randomness is then introduced by rewiring with probability p_r destination of each link to another randomly selected node. Double or degeneration of linkages are strictly forbidden.

We consider also an extended version of Barabasi-Albert scale-free (EBA) model [4]. Starting from several connected nodes as seed, at each step a new node is connected to the existing network with e_{new} edges. The probability for an existing node i being linked with the new comer is proportional to k_i^{α}, where k_i is the degree of node i. For the case of $\alpha = 1$, the EBA model degenerates to the standard Barabasi-Albert scale-free model, while for $\alpha = 0$ new comer will be attached randomly to the existing network without any preferential effect.

Detailed calculations show that the results are robustness for changing of size. Results for $N = 2000$ are shown in present paper. The structural properties are obtained by average over 100 realizations. All the networks are shuffled by means of the so-called edge-exchange algorithm [5].

3. Results

Fig.3 shows the compressibility coefficients for WSSW networks with $d = 2, 3$ and 4. One can find that with the increase of rewiring probability p_r, G increases monotonically. When p_r is large enough (e.g. $p_r \approx 0.5$), G saturates to values very close to that for the corresponding shuffled networks. While with the increase of p_r, G for shuffled networks decreases monotonically but slightly. At the same time, increase of d leads to significant decrease of G. The significant low degree of structural hole comes from non-trivial patterns rather than degree distribution. On the initial regular lattice, each node has $2d$ duplications and the two sets of d duplications separated by it form two sub-complete graphs. Tightness and high symmetry of structure induces low degree of structural hole.

Fig.4 presents the compressibility coefficients for EBA scale-free networks. One can find that with the increase of α, i.e., increase of preferential effect, G decreases monotonically. For $e_{new} = 1$, with the increase of α the difference between G values for the initial and the shuffled networks becomes nondistinctive. While for $e_{new} = 2, 3, 4$, the difference keeps significant large with the increase of α. G values for original networks are large compared that for the shuffled ones. Hence, the high degree of structural hole comes from nontrivial patterns in original networks.

4. Conclusions

In the present work we propose the principal component analysis to evaluate degree of structural hole in a global way. For WSSW networks with the increase of rewiring probability degree of structural hole increases monotonically and saturates to a value very close to that for corresponding shuffled networks. While increase of connection density decreases significantly degree of structural hole. For EBA networks, with the increase of α, namely the preferential effect, degree of structural hole decreases monotonically. For $e_{new} = 1$, G tends to a value very close to that for corresponding shuffled networks, while for $e_{new} > 1$, G keeps significant larger compared with that for shuffled networks.

PCA-based measure of structural hole provides a powerful tool to evaluate in a global way social networks. Comparing with the initial definition of structural hole, it takes account of a new effect in generating mechanism of structural hole, namely, identical patterns occurring in networks.

Acknowledgements

The work is supported by the National Science Foundation of China under Grant Nos. 10975099 and 50974052, the Program for Professor of Special Appointment (Eastern Scholar) at Shanghai Institutions of Higher Learning, the Innovation Program of Shanghai Municipal Education Commission under Grant No.13YZ072, and the Shanghai leading discipline project under grant No.S30501. One of the authors (W. Zhang) thank the support of the Innovation Fund Project For Graduate Student of Shanghai under Grant No. JWCXSL1102.

References

[1] R. S. Burt, The social structure of competition (Cambridge: Harvard University Press,1992).
[2] K. Pearson, Philosophical Magazine 2,559(1901).
[3] D.J. Watts, S.H. Strogatz, Nature 393,409(1998).
[4] A.-L. Barabasi, and R. Albert, Science 286(5439),509(1999).
[5] S. Maslov, K. Sneppen, Science 296,910(2002).

A New Method for Detecting Anti-community Structures in Complex Networks

Qiang Yu[1], Ling Chen[12]
1.Institute of Information Science and Technology, Yangzhou University , Yangzhou , China
2.National Key Lab of Novel Software Tech, Nanjing University, Nanjing, China
1.E-mail:77724954@163.com
2.E-mail: yzulchen@gmail.com

Abstract. In this paper, we present a label propagation algorithm named ACD for anti-community detection. Experimental results on some real world networks show that our algorithm can obtain higher quality results than other methods.

1. Introduction

In the past decade, identification of community structure has attracted much attention in various scientific fields. Many methods have been proposed and applied successfully to some specific complex networks. Those community structures studied in the past are also called "homophyily"or "assortative mixing", where vertexes are divided into groups such that the members of each group are densely connected to other members of the same group [1,2]. Recently, another type of community is called "disassortative mixing" or "anti-community" has also been discussed to a less extent [3-5]. In an anti-community, vertexes have most of their connections outside their group and has no or fewer connection with the members within the same group.

Several methods for detecting the anti-community in the complex networks have been proposed recently. Newman et. al. [6] proposed a method based on the modularity matrix to detect both communities and anti-communities. While communities are detected by the eigenvectors corresponding to the largest eigenvalue of the modularity matrix, the anti-communities can be detected by the eigenvectors corresponding to the most negative eigenvalue. Newman et. al. [7] also proposed a method based on the mixed model to detect both communities and anti-communities. The method uses the machinery of probabilistic mixture models and the expectation maximization algorithm. It can detect communities, anti-communities and other types of structure in networks without prior

*Corresponding author

Published under licence by IOP Publishing Ltd

knowledge of what one is looking for. Wang Fan [8] proposed a method detecting the anti-communities using the most negative eigenvalue of the Laplacian matrix of the graph.

In this paper, a label propagation algorithm ACD (Anti-Community Detection) is proposed. We test the algorithm ACD on some real networks and compare the results with other methods [7,8]. Experimental results show that our algorithm can obtain higher quality results than other methods.

2. A label propagation method for detecting the anti communities

Recently, Raghavan et al. [9] introduce a very innovative and promising algorithm called label propagation algorithm (LPA) for community detection. The most striking feature of LPA is that it can find meaningful community structure at very high speed, which is faster than what is possible so far. In this paper, we improve the LPA to make it more suitable for anti-community detection in large scale networks. Note that we define anti-community structure in networks as groups of sparsely connected nodes, in our algorithm the label should be propagated to the non-connected nodes.

In our label propagation algorithm, there is a label for each node in a network and communities are defined as the sets of nodes of the same labels. Let $G=(V,E)$ be a graph with n vertexes. Let $v \in V$ be a vertex in graph G, we define the set $H(v) = \{v' | v' \in V, (v,v') \in E\}$, which is the set of vertexes having connections with v, and the set $F(v) = \{v' | v' \in V, (v,v') \notin E\}$, which is the set of vertexes having no connection with vertex v. The algorithm tries to assign each vertex v a label $L(v)$ which indicates the class v belongs to. Initially each all the vertexes are assigned different labels: $L(v_i)= i$. The algorithm propagates the label $L(v_i)$ of vertex v to the vertexes in $F(v)$. The iterative relabeling process continues until every node has a label. In this fashion, sparsely connected group of nodes can reach a consensus on a unique label and form an anti-community quickly.

If two disconnected vertexes v_1 and v_2 are both connected to a third vertex v, this means they probably can be assigned an identical label. In this case, we assign them $\min\{L(v_1), L(v_2)\}$ as the new label. Since two connected vertexes can not be grouped to the same anti-community, they should not be assigned the same label. Therefore, when a vertex v being assigned a new label L, the algorithm checks the labels of all its neighbor vertexes in $H(v)$ to see if there is any vertex in $H(v)$ which has the same label L. Vertex v can be assigned a new label L, only when there is no vertex in $H(v)$ labeled L. Such label propagation process should continue until no node in the network changes its label.

Since label of each vertex is replaced only by a smaller one, and the label of each vertex is an integer less than n, which is the number of the vertexes in the graph, it takes $O(n)$ time for a label to converge to its minimal value. Therefore, the time complexity of the algorithm is $O(n^2)$. Our ACD algorithm can obtain the anti-communities directly, unlike the bipartition-based method, which requires a recursive procedure of bipartitioning which requires a large amount of computation time. ACD algorithm does not require a predefined number of clusters k. In addition, it can obtain a result where there is no connection between the vertexes within each cluster, and maximize the number of edges between clusters while the number of clusters is minimized.

3. Experimental results

In this section, we empirically demonstrate the effectiveness of our algorithm ACD for detecting the anti-communities in networks on some real world networks. All experiments have been run on Pentium IV, Windows XP, P1.7G, and visualize the results on Matlab 6.0.

We apply our method to the real network shown in Figure 1. The network of the most frequent 50 English adjectives and 50 nouns selected from the novel "*The Old Man and the Sea*" with edges connecting any pair of words that appear adjacent to each other at any point of the text. Because adjectives typically occur next to nouns in English and very few pairs of nouns or adjectives appear in the text, most edges connect an adjective to a noun. The network is thus approximately bipartite or disassortative. In Figure 1 the red nodes represent the adjectives and the blue nodes represent the nouns. In the graph, most edges are observed to run between two nodes of different colors.

We first test on this network using the spectral based algorithm[8], the result is shown in Figure 2. From the figure, we can see that the network has been divided into two anti-communities, of which one consists of most of the red nodes and the other consists of most of the blue ones. But there are 6 blue nodes are assigned into the red anti-community, and 8 red nodes are assigned into blue anti-community. Therefore, the precision of the partitioning is 86%.

We also test our algorithm ACD on this network, the result is shown in Figure 3. From the figure, we can see that the network also has been divided into two anti-communities. There are 3 blue nodes are assigned into the red anti-community, and 5 red nodes are assigned into blue anti-community. Therefore, the precision of the partitioning is 92%. Therefore, ACD algorithm can obtain higher quality results than the spectral based algorithm.

To measure the quality of the anti-community partition, we use the Newman-Girvan modularity for community partition [10]. Since we detect the anti-community, smaller value of such modularity indicates higher quality of the partition result.

We test the on the networks of adjective-noun, Southern women [11], and Scotland Corporate Interlock [12] using the mixture models based algorithm [7], the spectral based algorithm [8], and algorithm ACD. The Southern women data set was collected by Davis et al. [11] in and around Natchez, Mississippi during the 1930s as part of an extensive study of class and race in the Deep South. Scotland Corporate Interlock [12] is a data set on corporate interlocks in Scotland in the early twentieth century. Modularity of the results by the algorithms are as shown in Table 1. From the table, we can see that the result of ACD has lower modularity than other methods. This indicates that algorithm ACD can get better anti-community partition than other methods.

4. Conclusions

A label propagation based algorithm ACD for anti-community detection is proposed. Our ACD algorithm can obtain the anti-communities directly in $O(n^2)$ time. The ACD algorithm does not require a predefined number of clusters k, and can obtain a result where there is no connection between the vertexes within each cluster, and maximize the number of edges between clusters while the number of clusters is minimized. Experimental results show that our algorithm can obtain higher quality results than other methods.

Acknowledgements

This research was supported in part by the Chinese National Natural Science Foundation under grant Nos. 61070047, 61070133 and 61003180. State Key Fundamentals Research (973) Project under contract 2012CB316003, Natural Science Foundation of Jiangsu Province under contracts BK2010318, BK21010134.

References

[1] M. Girvan and M. E. J. Newman, "Community structure in social and biological networks," Proc. Natl. Acad. Sci. U.S.A., vol. 99, no. 12, pp. 7821-7826, 2002.
[2] Danon L, Duch J, Diaz-Guilera A, Arenas A (2005) J Stat Mech, P09008.
[3] M. E. J. Newman, "Finding community structure in networks using the eigenvectors of matrices," Phys. Rev. E, vol. 74, no. 3, pp. 036104, 2006.
[4] P. Holme, F. Liljeros, C. R. Edling, and B. J. Kim, Net-work bipartivity. Phys. Rev. E 68, 056107 (2003).
[5] Estrada E, Rodrı´guez-Vela´zquez JA (2005) Phys Rev E 72:046105.
[6] M.E.J. Newman, Finding community structure in networks using the eigenvectors of matrices, Proc Natl Acad Sci USA, 2007, 104(23) : 9564- 9569
[7] Newman M E, Leicht E A. Mixture Models and Exploratory Analysis in Networks. Proc Nat l Acad Sci USA, 2007, 104(23) : 9564 – 9569
[8] Wang Fan, Detecting "Anti-communities" of Networks Based on Spectral Method, Thesis for the Master Degree, Huazhong University of Science and Technology Wuhan, Hubei 430074, China, June, 2008
[9] U. N. Raghavan, R. Albert, and S. Kumara, "Near linear time algorithm to detect community structures in large-scale networks," Phys. Rev. E, vol. 76, no. 3, 036106, Sep. 2007, doi: 10.1103/PhysRevE.76.036106.
[10] M. E. J. Newman and M. Girvan, Phys. Rev. E 69, art. no. 026113 (2004).
[11] A. Davis, B. B. Gardner, and M. R. Gardner, Deep South, (University of Chicago Press, 1941).
[12] J. Scott and M. Hughes, The anatomy of Scottish capital: Scottish companies and Scottish capital, 1900–1979, (Croom Helm, London, 1980).

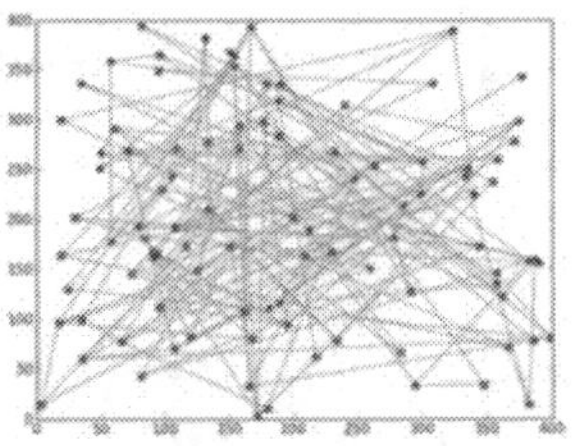

Figure 1. The network of adjectives and nouns

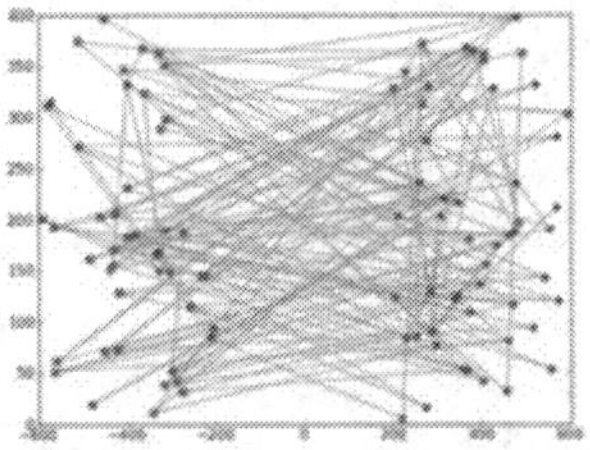

Figure 2. The anti- communities detected by the spectral based algorithm

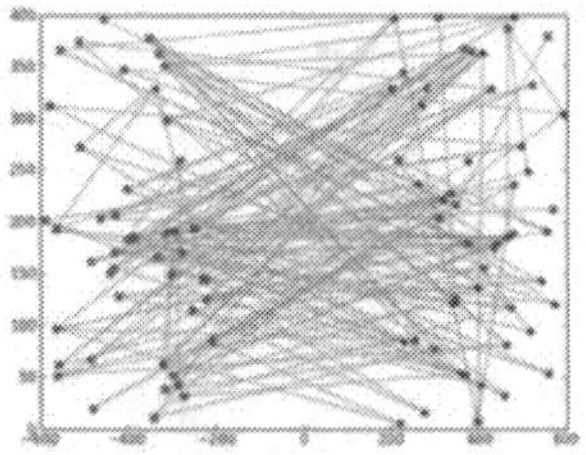

Figure 3. The anti-communities detected by the algorithm ACD

Table 1 Comparison of the modularity of the results by the algorithms

Algorithm / Data set	ACD	spectral based	mixture models based
adjective-noun	-0.3448	-0.2988	-0.3199
Southern women	-0.4778	-0.4386	-0.4561
Scotland Corporate Interlock	-0.4113	-0.3569	-0.3815

IC-MSQUARE 2012: International Conference on Mathematical Modelling in Physical Sciences IOP Publishing
Journal of Physics: Conference Series **410** (2013) 012104 doi:10.1088/1742-6596/410/1/012104

Analysis on controlling complex networks based on dominating sets

Jose C. Nacher‡[1], Tatsuya Akutsu*[1]

‡ Department of Information Science, Faculty of Science, Toho University, Miyama 2-2-1, Funabashi, Chiba, 274-8510, Japan

E-mail: nacher@is.sci.toho-u.ac.jp

* Bioinformatics Center, Institute for Chemical Research, Kyoto University, Uji, 611-0011, Japan

E-mail: takutsu@kuicr.kyoto-u.ac.jp

Abstract. Recent works have emphasized the importance of controlling complex networks and have provided novel analytical approaches based on the maximum matching among others. The analysis leads to the conclusion that the driver nodes tend to avoid the high-degree nodes. Therefore, it implies that apparently inhomogeneous networks are more difficult to control. Here we present an approach to control complex networks from the perspective of the minimum dominating set (MDS). We apply the definition of MDS to complex networks with scale-free structure $P(k) \propto k^{-\gamma}$ and assume that a node controls itself and each of its adjacent nodes through its links. Our theoretical calculations and simulations demonstrate that the more heterogeneous a network's degree distribution is, the smaller the fraction of individuals, devices or molecules required to control the entire system is. Here, we have extended our recent analytical derivations and we now provide a more accurate analysis for the case $\gamma < 2$. Moreover, we have also performed additional computer simulations to study in more detail the dependence of the MDS with the network size and the average degree below the phase transition at $\gamma = 2$.

1. Introduction

The emergence of large complex networks structures in physical, biological and technological systems poses the question as to whether we can effectively control and direct the system's entire dynamics at will. Recent works have investigated the controllability in complex networks [1, 2]. In particular, Liu et al., has used a maximum matching algorithm to investigate network control. As a result, homogeneous (i.e., random networks) can be controlled easily because the number of nodes that need to be controlled externally is small. However, there are several ways to define controllability in complex networks. Nepusz and Vicsek [3, 4] have analyzed this problem from the angle of edge dynamics. In parallel, we have recently introduced the minimum dominating set (MDS) approach to control complex networks, which conceptually has similarities with the controlling link dynamics [3]. First, our theoretical findings suggest that scale-free networks with small scaling exponent values ($\gamma < 2$), where high-degree nodes are present, require relatively few nodes to be controlled [5]. Conversely, networks with large degree exponent values ($\gamma > 2$) or faster exponential decay, where hubs are weakly connected or almost absent, require more nodes to be fully regulated. In this work, we extend our previous analytical derivations and provide a more accurate analysis for the case $\gamma < 2$. Moreover, we also perform

[1] JCN and TA are corresponding authors

additional computer simulations to study in more detail the dependence of the MDS with the network size and the average degree below $\gamma = 2$.

2. Computation of the MDS

A set $S \subseteq V$ of nodes in a graph $G = (V, E)$ is a dominating set if every node $v \in V$ is either an element of S or is adjacent to an element of S [6]. Since the domination problem is a classical NP-complete decision problem in computational complexity theory [6], it is not plausible that there exists a polynomial time algorithm that finds a smallest dominating set for arbitrary graphs. Here, we reduce this problem to the computation of a binary integer programming problem. We then minimize a linear function $f(x)$ subject to linear constraints [5]. Each node i is assigned to a binary integer variable x_i that takes values 1 or 0. A node i that belongs to the dominating set will take the value x_i=1.

3. Theoretical analysis

3.1. Mathematical analysis of MDS size in scale-free networks

We here briefly mention the main results of our previous anaysis and present our latest findings for the case $\gamma < 2$.

Let $G(V, E)$ be an undirected random graph with node set V and edge set E such that the degree distribution follows a power-law $k^{-\gamma}$. We let $n = |V|$. Our results show the following:

- if $\gamma > 2$, the size of the minimum dominating set is $\Theta(n)$,
- if $\gamma < 2$, the expected size of the minimum dominating set is $o(n)$.

This result means that a phase transition occurs at $\gamma = 2$. Here we remind that $f(n)$ is $\Theta(g(n))$ if it is both $O(g(n))$ and $\Omega(g(n))$, that is, $f(n)$ is $\Theta(g(n))$ if $k_1 \cdot g(n) \leq f(n) \leq k_2 \cdot g(n)$ holds for sufficiently large n where k_1 and k_2 are some positive constants. Moreover, $f(n)$ is $o(g(n))$ if for every $\epsilon > 0$, $|f(n)| \leq |g(n)| \cdot \epsilon$ holds for sufficiently large n. Hereafter, we focus on the case $\gamma < 2$, and provide a more accurate analysis.

3.2. On Upper Bound of the Size of the Minimum Dominating Set

Here we give a more accurate estimate of an upper bound of the size of MDS in a scale-free network. We first assume that the degree distribution follows $\alpha k^{-\gamma}$ with cut off at $k = n$, where $1 < \gamma < 2$. From $\alpha n \int_1^n k^{-\gamma} dk = n$, we have $\alpha \approx \gamma - 1$.

We select the nodes with degree from n^β to n as a core part of a dominating set (this core part is denoted by DS), where $\beta < 1$ is a positive constant depending on γ. Then, the number of nodes N_{DS} in DS is given by

$$\begin{aligned} N_{DS} &= \alpha n \int_{n^\beta}^{n} k^{-\gamma} dk \\ &= O(n^{1-\beta(\gamma-1)}). \end{aligned}$$

On the other hand, the total number of edges E_G is given by

$$\begin{aligned} E_G &= \frac{\alpha n}{2} \int_1^n k \cdot k^{-\gamma} dk \\ &= \frac{\gamma - 1}{2(2-\gamma)} \cdot n \cdot (n^{2-\gamma} - 1). \end{aligned}$$

Let E_{DS} be the number of edges covered by DS. A lower bound is estimated by

$$\begin{aligned} E_{DS} &\geq \frac{\alpha n}{2} \int_{n^\beta}^{n} k \cdot k^{-\gamma} dk \\ &= \frac{\gamma - 1}{2(2-\gamma)} \cdot n \cdot (n^{2-\gamma} - n^{\beta(2-\gamma)}). \end{aligned}$$

Therefore, the probability that an arbitrary edge is not covered by DS is

$$\begin{aligned}\frac{E_G - E_{DS}}{E_{DS}} &\leq \frac{n^{\beta(2-\gamma)} - 1}{n^{2-\gamma} - n^{\beta(2-\gamma)}} \\ &\approx \frac{n^{\beta(2-\gamma)}}{n^{2-\gamma}} \\ &= n^{(\beta-1)(2-\gamma)}.\end{aligned}$$

Let $G \ominus DS$ denote the set of nodes in G that are not dominated by DS. Since a node is dominated by DS if at least one edge connecting to the node is covered by DS, the expected number of nodes (denoted by $N_{G\ominus DS}$) of $G \ominus DS$ is bounded as

$$N_{G\ominus DS} \leq O(n \cdot n^{(\beta-1)(2-\gamma)}) = O(n^{1+(\beta-1)(2-\gamma)}).$$

Here, we note that $DS \cup (G \ominus DS)$ becomes a dominating set. Therefore, the size of a minimum dominating set is bounded by

$$|DS \cup (G \ominus DS)| \leq O(n^{1-\beta(\gamma-1)} + n^{1+(\beta-1)(2-\gamma)}).$$

In order to find β minimizing this order (recall that we can choose β depending on γ), we let

$$1 - \beta(\gamma - 1) = 1 + (\beta - 1)(2 - \gamma),$$

which results in $\beta = 2 - \gamma$. Therefore, an upper bound of the size of the dominating set is estimated as

$$O(n^{1+(2-\gamma)(1-\gamma)}).$$

For example, if $\gamma = 1.5$, this number is $O(n^{0.75})$, which means that the size of MDS is $o(n)$. It is interesting to note that this value takes the minimum when $\gamma = 1.5$

4. Computational experiments on MDS size in scale-free networks

To analyze in detail the dependence of the scaling exponent and mean degree on MDS size, we generated a variety of scale-free networks by using an algorithm that generates random simple connected graphs with prescribed degree sequence [7]. Here, we extend our previous results by investigating the behaviour of the MDS size on the network size and the average degree below $\gamma = 2$. Our previous results indicate that the MDS size increases as γ increases. Moreover, the MDS generally decreases when the average degree increases for values of $\gamma > 2$. Conversely, when $\gamma < 2$, the MDS slightly increases or becomes constant as the average degree increases. In particular, it seems to take a minimum value at $\gamma = 1.5$. Here, we study simultaneously the effect of average degree and network size on the MDS size and how it changes for values around $\gamma = 1.5$ and below $\gamma = 2$. Figure 1 shows the dependence of the MDS size on the systems size. The results show that, for $\gamma < 1.5$, denser networks are more difficult to control. This tendency is also observed when the system size grows. However, the networks show a different behaviour for $\gamma > 1.5$. In this case, our results show that the MDS size tend to decreases when average degree increases (data not shown due to limited space).

5. Conclusion

In sharp contrast to a recent work on controllability that uses a maximum matching [1], our analysis demonstrates that the MDS tends to target highly connected nodes, whereas the previous study suggested that driver nodes tend to avoid high-degree nodes. Our approach, however, can be connected to Liu et al,. by assuming that every edge in a network is bi-directional

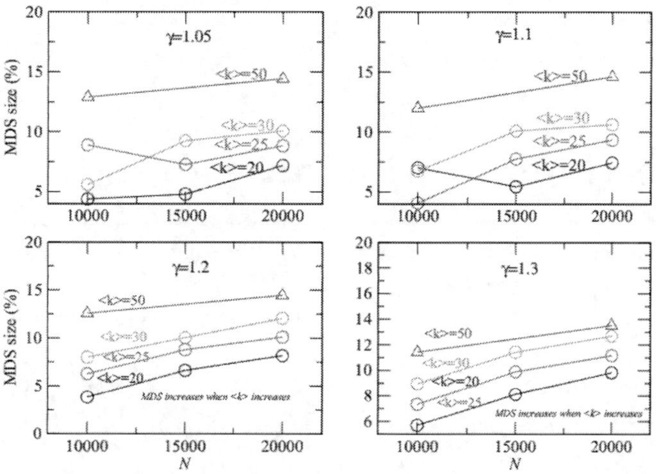

Figure 1. Computer simulations of MDS size as a function of the network size with different average degree $< k >$ in simulated scale-free networks. Scaling exponents are indicated in figure.

and every node in MDS can control all of its outgoing links separately. Then, the network is structurally controllable by selecting the nodes in MDS as the driver nodes [5]. Though Müller and Schuppert [2] recently suggested that iPS cells can be controlled by a few driver nodes, they did not show general results or consider structural properties of networks. Moreover, our approach has some conceptual similarities with controlling edge dynamics [3], although we are tackling the problem from MDS view points rather than using switchboard-dynamics and its mapping to the line graph [8].

Acknowledgments
T.A. was partially supported by MEXT, Japan (Grant-in-Aid 22650045) and J.C.N. was partially supported by MEXT, Japan (Grant-in-Aid 23700352).

References

[1] Liu Y-Y, Slotine J-J and Barabási A-L (2011) Controllability of complex networks. *Nature* **473**, 167-173.
[2] Müller F-J and Schuppert A (2011) Few inputs can reprogram biological networks. *Nature* **478**, E4.
[3] Nepusz T and Vicsek T (2012) Controlling edge dynamics in complex networks *Nature Physics* **8**, 568-573.
[4] Slotine J-J and Liu Y-Y (2012) The missing link *Nature Physics* **8**, 512.
[5] Nacher JC and Akutsu T (2012) Dominating scale-free networks with variable scaling exponent: heterogeneous networks re not difficult to control *New Journal of Physics* **14**, 073003 (24 pages).
[6] Haynes TW, Hedetniemi ST and Slater PJ (1998) *Fundamentals of Domination in graphs* (Chapman and Hall/CRC Pure Applied Mathematics, New York)
[7] Viger F and Latapy M (2005) Efficient and simple generation of random simple connected graphs with prescribed degree sequence, *In Proc. of COCOON'05 (Computing and Combinatorics Conference)*, 1-15.
[8] Nacher JC, Yamada T, Goto S, Kanehisa M and Akutsu T (2005) Two complementary representations of a scale-free network. *Physics A* **349**, 349-363.

Information communication on complex networks

Akito Igarashi, Hiroki Kawamoto, Takahiro Maruyama, Atsushi Morioka and Yuki Naganuma

Department of Applied Mathematics and Physics, Kyoto University, Kyoto 606-8501 JAPAN

Abstract. Since communication networks such as the Internet, which is regarded as a complex network, have recently become a huge scale and a lot of data pass through them, the improvement of packet routing strategies for transport is one of the most significant themes in the study of computer networks. It is especially important to find routing strategies which can bear as many traffic as possible without congestion in complex networks. First, using neural networks, we introduce a strategy for packet routing on complex networks, where path lengths and queue lengths in nodes are taken into account within a framework of statistical physics. Secondly, instead of using shortest paths, we propose efficient paths which avoid hubs, nodes with a great many degrees, on scale-free networks with a weight of each node. We improve the heuristic algorithm proposed by Danila *et. al.* which optimizes step by step routing properties on congestion by using the information of betweenness, the probability of paths passing through a node in all optimal paths which are defined according to a rule, and mitigates the congestion. We confirm the new heuristic algorithm which balances traffic on networks by achieving minimization of the maximum betweenness in much smaller number of iteration steps. Finally, We model virus spreading and data transfer on peer-to-peer (P2P) networks. Using mean-field approximation, we obtain an analytical formulation and emulate virus spreading on the network and compare the results with those of simulation. Moreover, we investigate the mitigation of information traffic congestion in the P2P networks.

1. Introduction

Since the discovery of small world and scale free networks[1,2], where average lengths of shortest paths and degree distribution with a power law, respectively, have characteristic properties, their structures and various dynamics on the networks have been investigated actively in many fields. These networks are called complex networks. Because the Internet has such properties and is regarded as a complex network, information communication on complex networks is one of the most important problems, for examples, packet routing, peer-to-peer (P2P) communication and so on. In routing strategies, hubs, which are nodes with a great many links and characteristic in scale-free networks, should be avoided for mitigation of congestion. We propose an efficient algorithm for avoidance of hubs on communication routes. On the other hand, in P2P networks, there are two notable aspects. One is that a peer searches for a file and downloads it from a "host peer", which has the requested file. The other aspect is that peers leave and join in the network repeatedly. Thus, the topology of the network changes gradually with time and the behavior of the density of infected peers and data transfer become complicated. We appropriately simulate virus spreading and data transfer on the networks under the change of its topology by the effect of leaves and compare the results with mean field approximations. We also consider the mitigation of traffic congestion on the P2P networks.

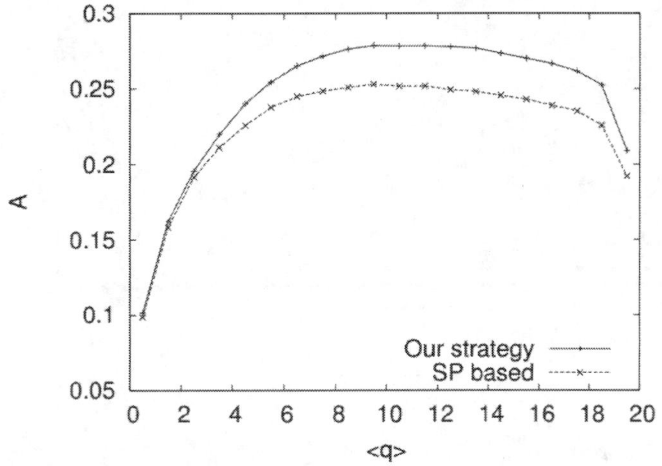

Figure 1. A is depicted as functions of $\langle q \rangle$. Results with our strategy and the strategy based on shortest paths (SP) are compared on the Barabási and Albert (BA) [2] model for total node number 100 and the average degree 10.

2. Packet routing strategy using neural networks on scale-free networks

We investigate routing strategies on complex networks. Using neural networks, we introduce a routing strategy where path lengths and packet-queue lengths are taken into account within a framework of statistical physics. The performance of this strategy becomes more efficient from adjustment of parameters. At the same time, we analyze how the properties of networks influence the performance of this strategy. Moreover, we propose a routing strategy where connection weights in neural networks are adjusted by local information. We also confirm how the distance term and the properties of networks influence the performance of this adjustive strategy.[3]

In Fig.1, the average number of packets arriving at their destinations per transmission time A is depicted as functions of the average packet-queue length in a node $\langle q \rangle$. Since A with our strategy is smaller than that with the strategy based on shortest paths (SP), we confirm that our proposed strategy is more effective than the SP strategy.

3. Efficient packet routing strategy in complex networks

We investigate new packet routing strategies which mitigate traffic congestion on complex networks. Instead of using shortest paths, we propose efficient paths which avoid hubs on scale-free networks with the use of a weight of each node. We improve the heuristic algorithm [6] which improves step by step routing properties on congestion by using the information of the betweenness of each node in every step. We propose a new heuristic algorithm which balances traffic on networks by achieving minimization of the maximum betweenness in the much smaller number of iteration steps [4] than that in [6].

In Fig.2, We compare the average traveling time of a packet $T_{\rm avg}$ as functions of the packet generation rate λ for Barabási and Albert (BA) model [2] with the total node number 500 and the average degree 8 for the shortest path strategy (SPS), the strategy proposed by Yan *et. al.* [5] (DEPS), the strategy proposed by Danila *et. al.* [6] (OPS) and our algorithm (NOPS).

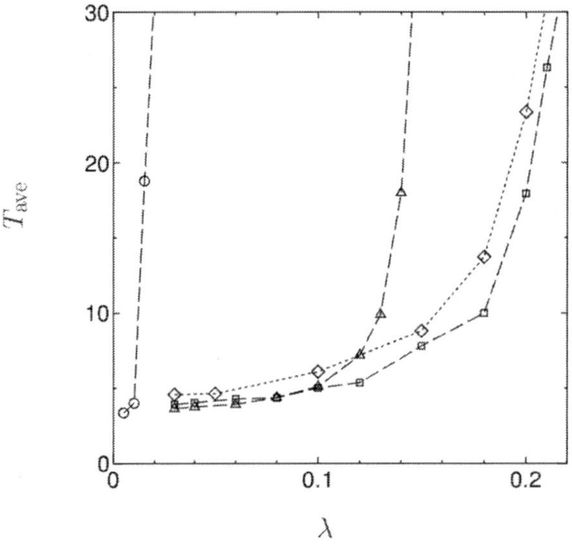

Figure 2. The results with SPS (circles), DEPS (triangles), OPS (diamonds) and NOPS (squares) for T_{avg} are compared as functions of λ for the BA model.

4. Virus spreading and data transfer model in P2P networks

Recently peer-to-peer (P2P) file-sharing systems have been a new communication paradigm. P2P networks have two notable aspects. One is that a peer searches for a file and downloads it from a "host peer", which has the requested file. To determine whether a peer has the requested file or not in modeling of the search and download process, we introduce a parameter P_j, which expresses the normalized amount of files stored in the peer j. It is assumed that the more a peer has files, the more possibly it becomes a host peer, that is, if P_j is large, the peer j has many files and can respond other peers' requests with high probability. The other aspect is that peers leave and join in the network repeatedly. The topology of the network, therefore, changes gradually with time and the behavior of ρ (the fraction of infected peers) becomes complicated. We propose models appropriate to P2P networks and simulate virus spreading on the networks under the change of its topology by the effect of leaves and joins. Since peers empirically seem not to leave the network randomly, the two directions of separation from the networks are examined. One is that peers randomly leave (called "RA-separation") and the other is that the peer j leaves at the rate which is proportional to $1 - P_j$ (called "F2-separation"). We devise F2-separation to take into account the existence of "file suppliers", who hold many files and principally supply files to other peers and rarely leave the network, and "free riders", who possess few files and hardly supply files and quite often leave the network in a moment after joining.

Moreover, using mean-field approximation, we obtain an analytical formulation and emulate virus spreading on the network and compare the results with those of simulation. In practice, an infected computer is recovered when being outside the network. Thus, we introduce an extended SIS model, where infected peers never turn to susceptible in the network, susceptible peers may turn to infected only in the network and infected peers always turn to susceptible while leaving the network. Fig.3 represents the results for two directions of separation. The fraction of infected peers ρ is not zero for F2-separation even when the infection rate μ is too small, whereas for RA-separation, μ_c, the epidemic threshold value of μ, under which ρ is 0, has a positive value. In Fig.3, dashed curves represent simulation results and solid curves represent the results from the theoretical analysis with mean field approximations. Although the simulation results of

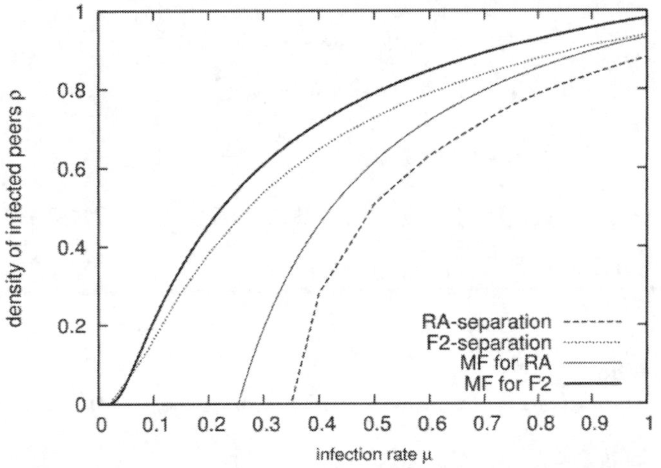

Figure 3. The fraction of infected peers ρ is plotted as a function of μ with extended SIS model for leaving rate, the fraction of leaving node, 0.2. Dashed curves indicate simulation results and solid curves represent the theoretical results with mean-field approximation. For RA-separation, the epidemic threshold μ_c, where $\rho = 0$ for theoretical analysis is smaller than that for simulation results.

the behavior of ρ for F2-separation are in good agreement with the theoretical analysis, the agreement for RA-separation is not so good, especially near the epidemic thresholds μ_c. The reason is that the number of peers in the network is finite in simulation, whereas mean-field approximation theory is effective for infinite systems. The theoretical results for F2-separation show that the epidemic threshold is 0 and ρ is proportional to $-\mu \log \mu$ for quite small μ.

We attain the fact that viruses spread on the network even for quite small infection rate when peers leave in accordance with F2-separation. In other words, the network is more vulnerable to virus in the case of F2-separation than RA-separation. We also investigate traffic congestion for file transport on the P2P network and confirm the effectiveness of the strategy propose in section 3.

5. Conclusion

We devise efficient packet routing strategies on communication networks in order to avoid traffic congestion. With neural networks and the heuristic algorithm the above purpose is achieved. On the other hand, P2P networks are modeled with their notable properties, and investigated virus spreading on them theoretically and with simulation. We also confirm that the similar strategy proposed in section 3 is effective for mitigation of congestion on the P2P network.

Our methods are also applied to treat traffic properties of other communication networks. Various characteristic properties of such networks will be investigate in near future.

[1] Watts D J and Strogatz S H 1998 *Nature* **393** 440
[2] Barabási A L and Albert R 1999 *Science* **286** 509
[3] Naganuma Y and Igarashi A 2010 *Physica* A **389** 623
[4] Kawamoto H and Igarashi A 2012 *Physica* A **391** 895
[5] Yan G, Zhou T, Hu B, Fu Z and Wang B 2006 *Phys. Rev.* E **73** 046108
[6] Danila B, Yu Y, Marsh J A and Beassler K E 2006 *Phys. Rev.* E **74** 046106

Journal of Physics: Conference Series **410** (2013) 012106 doi:10.1088/1742-6596/410/1/012106

A study for system subjected to random shocks with accumulated damage

Jih-An Chen[1]
Kao-Yuan University, Department of Marketing and Distribution Management

E-mail: jachen@cc.kyu.edu.tw

Abstract. System could be continuously operating over an indefinitely long operation cycle, where each operation causes a random amount of damage to the system, and these damages are accumulated. We also propose a maintenance policy to optimize the expected cost rate for an infinite time span. Some useful properties and result discussions are presented, which indicate that the optimal maintenance policy is to perform preventive maintenance only depend on the level of accumulated damage, it is unnecessary to depend on the number of operation. Several special cases of such a maintenance policy are also presented and discussed.

1. Introduction

Maintenance policies for systems that are subject to stochastic failures have been treated extensively in literature. A comprehensive set of models review and discussion on the maintenance policies are included in Nakagawa (2005, 2007). The main aim of optimal maintenance policies is to provide optimum system reliability, availability and safety performance at the lowest possible maintenance cost. Maintenance can also be further divided into several categories according to the degree to which the operating conditions of an item are restored by maintenance.

A perfect maintenance means a maintenance action that restores system operating conditions to as good as new. That is, upon perfect maintenance, a system has the same operating conditions as a new one. A complete overhaul of an engine with a broken connecting rod is an example of perfect maintenance.

Generally, the replacement of a failed system with a new one is a perfect maintenance or perfect repair. A minimal maintenance means that after it is conducted, the system's operating state is as bad as old. Changing a flat tire on a car, or changing a flat or broken fan belt on an engine are examples of minimal maintenance, as the overall state of the operating condition is essentially unchanged.

An imperfect maintenance is an action that does not make a system like new, but less. It is usually assumed that imperfect maintenance restores the system operating state to somewhere between as good as new and as bad as old.

Clearly, imperfect maintenance is a general maintenance that can include two extreme cases, namely, minimal and perfect. An engine tune-up is an example of imperfect maintenance, as a tune-up may not make an engine as good as new, but its performance would be greatly improved. For more details about the classification of maintenance degree, please refer to Pham and Wang (1996). Most optimum maintenance problems are based on a continuous time process where a system is operating.

[1] jachen@cc.kyu.edu.tw

However, in failure studies, the time to item failure is often measured by the number of operational (cycles) to failure, and therefore, operating a system in a discrete time process might be more appropriate. Also, cumulative damage models play an important role in reliability and maintenance theory. These models are considered as a sequence of shocks which occur randomly in time as an event in accordance with a stochastic process and give some amount of damage to a system.

A discrete-time order replacement model was first considered by Kaio & Osaki (1979). Later, they analyzed discrete-time ordering policies in Kaio & Osaki (1981) by taking account of minimal repair. Dohi et al. (2004) treated a generalized model with the similar complex cost structure to Dohi et al. (2004).

The damage suffered for the system is accumulated to the current damage level and weakens the system gradually. The system fails when the total damage exceeds a failure level. A variety of maintenance models subjected to shocks were studied extensively in Wortman et al. (1994), Sheu et al. (1996), Sheu (1998) and Sheu et al. (2002). The reliability properties and optimal maintenance policies for various damage models were also summarized sufficiently in Nakagawa (2007).

Besides, another application of the cumulative process in the maintenance related problems is the cumulative repair-cost limit policy, which concept uses the information of all repair costs to decide whether a system is repaired or replaced. For more details about the cumulative repair-cost limit policy on the maintenance model, please refer to Chien et al. (2009, 2010).

It is importance to avoid failures of a complex system during the actual operation when such an event is costly or dangerous. In this paper, we consider a system continuously operating over an indefinitely long operational cycle. Also, the cumulative damage model is applied. A maintenance policy, where the timing to perform PM is depends on the operational number as well as the accumulated damage level, is determined and investigated. The expected cost rate for an infinite time span is adopted as criterion of optimality.

2. Model Development

Consider a system operating over the time with an indefinitely long operation cycle n ($n = 1, 2, \cdots$) and each operation causes a random amount of damage to the system. These damages are accumulated to the current damage to the system.

2.1. CM and PM

A system fails when the total damage exceeds failure level, then a corrective maintenance (CM) is immediately conducted. Preventive maintenance (PM) action should be performed to prevent such a failure. The maintenance policy that considered and investigated for such a system is that the PM is carried out when the accumulated damage exceeds a pre-specified PM level δ (but less than the failure level), or it is performed at the completion of Nth ($N = 1, 2, \cdots$) operation after the system installation, whichever comes first.

That is, the system undergoes CM when the total damage exceeds a failure level ζ, and undergoes PM at damage δ ($0 \le \delta \le \zeta$) or at operation N ($N = 1, 2, \cdots$), whichever occurs first.

In additional to the CM and PM, regular maintenance (RM) actions also should be performed at every completion of operation in order to maintain the system for the next use.

2.2. Damage Amount

Let a random variable Y_j ($j = 1, 2, \cdots$) be the amount of damage due to jth operation that has an distribution function

$$G(y) \equiv \mathrm{P_r}(Y_j \le y).$$

These damages are accumulated to the current damage level.

Then, the total damage Z_j up to the jth operation has a distribution function

$$G^{(j)}(w) = \Pr(Z_j \le w) = \Pr\left(\sum_{i=1}^{j} Y_i \le w\right).$$

The above distribution function can be represented the j-fold Stieltjes of $G(w)$ with itself, and

$$G^{(0)}(w) \equiv 1 \text{ for } w \ge 0 .$$

Under the maintenance policy that described above, it is obviously that the probability that a PM is performed at the completion of N th operation is $G^{(N)}(\delta)$.
If $N = \infty$, then the probability that a PM is performed at the completion of j th operation is given by

$$\int_0^{\delta} G(\xi - y) dG^{(j-1)}(y) - G^{(j)}(\delta) \tag{1}$$

and the probability that a CM is performed at j th operation is given by

$$G^{(j-1)}(\delta) - \int_0^{\delta} G(\xi - y) dG^{(j-1)}(y). \tag{2}$$

2.3. *Minimal Maintenance*
CM and PM are assumed to be a perfect maintenance, while RM is assumed to be minimal maintenance. That is, a system is as good as new after CM or PM, and after RM the system is as bad as old.
Therefore, the perfect maintenance scheme for such a system is that it is undergoes scheduled PM at operation N th, or unscheduled PM at damage δ, or unscheduled CM at failure, whichever occurs first.
The time between successive perfect maintenance (i.e., CM or PM) can be regarded as a renewal cycle, and the behaviour in each renewal cycle repeats.
From the renewal reward theorem, the expected cost rate for an infinite time span is the expected total cost per renewal cycle, divided by expected renewal cycle length (Ross, 1970).

2.4. *Cost Rate and expected total cost*
To develop the expected cost rate for the maintenance policy in the model framework, the following costs are introduced.
Let c_0 be the fixed operating cost for each operation, c_{rm} be the fixed cost for each RM, C_{PM} be the fixed PM cost, and C_{CM} be the fixed CM cost, where $C_{CM} > C_{PM} > c_{rm}$ is assumed.
For a renewal cycle, we have the expected total cost *ETC*

$$c_0 \cdot \sum_{j=0}^{N-1} G^{(j)}(\delta) + c_{rm} \cdot \sum_{j=1}^{N-1} G^{(j)}(\delta) + C_{PM} \cdot \left\{ \sum_{j=1}^{N} P_j^{(\mathrm{PM})} + G^{(N)}(\delta) \right\} + C_{CM} \cdot \sum_{j=1}^{N} P_j^{(\mathrm{CM})} \tag{3}$$

and the expected renewal cycle length *ECL*

$$\sum_{j=1}^{N} j \cdot \left[G^{(j-1)}(\delta) - G^{(j)}(\delta) \right] + N \cdot G^{(N)}(\delta). \tag{4}$$

The expected cost rate $CR(\delta, N)$ is given by

$$CR(\delta, N) = \frac{ETC}{ECL}. \tag{5}$$

3. Optimization

For further analytical study, we consider that the amount of damage due to each operation has an exponential distribution with a mean of μ.

That is,

$$G(y) = 1 - \exp\left(-\frac{y}{\mu}\right), \tag{6}$$

$$G^{(j)}(y) = \sum_{i=j}^{\infty} \frac{\left(\frac{y}{\mu}\right)^i}{i!} \cdot \exp\left(-\frac{y}{\mu}\right). \tag{7}$$

The expected total cost ETC in (3) can be reduced to

$$(c_0 + c_{rm}) \cdot \sum_{j=0}^{N-1} G^{(j)}(\delta) + (C_{PM} - c_{rm}) + (C_{CM} - C_{PM}) \cdot \overline{G}(\zeta - \delta) \cdot \overline{G}^{(N)}(\delta). \tag{8}$$

The following theorem regarding the properties of the optimal pair (δ^*, N^*) that satisfies

$$\underset{0 \le \delta \le \zeta;\ N=1,2,\cdots}{\text{Min}} CR(\delta, N) = CR(\delta^*, N^*).$$

The above result is presented based on the expected cost rate given in (5).

Theorem 1. For the cost rate $CR(\delta, N)$ given in (5) where $0 \le \delta \le \zeta$ and $N = 1, 2, \cdots$. If

$$\frac{C_{CM} - C_{PM}}{C_{PM} - c_{rm}} \le \frac{\zeta}{\mu},$$

then

$$(\delta^*, N^*) = (\zeta, \infty).$$

Otherwise, $0 < \delta^* < \zeta$ and $N^* = \infty$.

The theorem indicates that the problem of determining the optimal pair (δ^*, N^*) to minimize the expected cost rate $CR(\delta, N)$ is equivalent to only determining the δ^* to minimize $CR(\delta, \infty)$.

4. Conclusion

A preventive maintenance policy for a continuously running system over an indefinitely long operation cycle are proposed in this study,. Each operation causes a random amount of damage to the system, and these damages are accumulated to the current damage to a system, which fails when the total damage exceeds a pre-specified failure level, and then corrective maintenance should be conducted.

To prevent such a failure, a PM action should be carried out at suitable time. In the maintenance policy, the timing to perform PM depends on the operational number as well as the accumulated damage level.

Avoid failures of a operating system during the actual operation when such an event is costly or dangerous is importance. In this study, we have considered a continuously operating system over an indefinitely long operational cycle. The cumulative damage model is also applied.

A maintenance policy, where the timing to perform PM is depends on the operational number as well as the accumulated damage level, is determined and investigated in this research. The expected cost rate for an infinite time span is adopted as criterion of optimality. Some useful properties and result discussions are also presented, which indicate that the optimal maintenance policy is to perform preventive maintenance only depend on the level of accumulated damage. And, it is unnecessary to depend on the number of operation.

5. Acknowledgment

This research was supported by the National Science Council of Taiwan, under Grant No. NSC 101-2221-E-244-014.

6. References

[1] Chien Y.H.; Sheu S.H.; Chang C.C. 2009 *International Journal of Systems Science* **40** 703
[2] Chien Y.H.; Chang C.C.; Sheu S.H. 2010 *Annals of Operations Research* **181** 723
[3] Cox D.R. (1962). Renewal Theory, Metheun Press, London.
[4] Dohi T., K. Iwamoto, N. Kaio and S. Osaki 2004 *IMA Journal of Management Math.* **15(2)** 125
[5] Kaio N. and S. Osaki 1979 *IEEE Transactions on Reliability* **28** 405
[6] Kaio N. and S. Osaki 1980 *R.A.I.R.O. Operational Research* **14** 257
[7] Nakagawa T. 2005 Maintenance Theory of Reliability, Springer Press, London.
[8] Nakagawa T. 2007 Shock and Damage Models in Reliability Theory, Springer Press, London.
[9] Pham P.; Wang H. 1996 *European Journal of Operational Research* **94** 425
[10] Ross S.M. 1970 Applied Probability Models with Optimization Applications, Holden-Day, San Francisco
[11] Sheu S.H.; Griffith W.S. 1996 *Naval Research Logistics* **43** 319
[12] Sheu S.H. 1998 *European Journal of Operational Research* **108** 345
[13] Sheu S.H.; Griffith W.S. 2002 *European Journal of Operational Research* **140** 50
[14] Wortman M.A.; Klutke G.A.; Ayhan H. 1994 *IEEE Transactions on Reliability* **43** 439

A Study on Financial Reporting Standards and Accounting Quality– Evidence from China

Cheng-Hwai Liou
National Taichung University of Science and Technology, Department of Accounting Information, Taichung, Taiwan

jhliou@nutc.edu.tw

Abstract. According to institutional theorists, the forms and business models of corporation are mainly shaped by factors such as politics, regulations, social norms and cultures. This paper examines how the International Financial Reporting Standards (IFRS) and institutional environment influence the accounting quality, in response to the threat of political extraction in China. We took mainland China as an example instead in our study, following the accounting quality definition of Barth et al. [2], we found that the developments of Chinese government performance audit are conspicuously different by region; to reflect such differences, we elaborated our research by dividing mainland China into 31 categories (provinces or cities). We set 2003-2010 as the time horizon for this study. After testing the Regression model, our empirical research achieved two conclusions: 1) IFRS adoption in China should significantly improve the accounting quality, and 2) IFRS and institutional environment should synthetically influence the quality of accounting as well.

1. Introduction

Due to the different legislation system, tax regulation, political and economic environment, the accounting systems are varies across countries. Recently, many important changes in the accounting standards around the world, the most obvious is to adopt international accounting standards (International Financial Reporting Standards, IFRSs), and will be in a world of a single era of financial accounting standards. IFRS[1] have now become the most popular reporting standards languages around the world as numerous listed-companies from many countries have voluntarily adopted the standards over the last decades. International accounting literatures provide evidences that accounting quality has economic consequences, such as reduce the cost of capital and open new opportunities for diversification and improved investment returns [2,7], efficiency of capital allocation, and international capital mobility [4]. On February 15, 2006, the Ministry of Finance of the People's ROC (China), Parliament passed a regulation that requiring all firms listed on stock exchanges of China member mandatory adoption IFRS when preparing their financial statements for fiscal years beginning on or after January 1, 2007.Widespread application of IFRS will result in a significant change in the business environment.

Despite the rapid growth and development of China's economy in the past three decades, legal institutions and economic environment in China are still far from ideal [12]. The weak enforcement of property rights has given chances to rampant rent-seeking activities by government bureaucrats, as highlighted by numerous academic studies in recent years. In particular, firms owned by private entrepreneurs are more likely to be the subject of political extraction than state-owned enterprises. This problem is exacerbated by the variation in legal institutions across different provinces in China. This motivates us to systematically examine the implication of political extraction on firms' corporate policies. The question we address is

[1] .In2001,TheIASB (International Accounting Standards Board)succeeded the IASC(International Accounting Standards Committee) and assumed its standard-setting responsibilities from the IASC to harmonize the financial reporting practices to reduce diversity and facilitate cross-listings and cross-border investment.

whether application of IFRS is associated with higher accounting quality than application of local domestic standards. Furthermore, we also investigate whether institution environment that after apply IFRS reveal more timely loss recognition, and higher value relevance than application of domestic standards.

2. Institutional Background, Literature Review and Hypothesis Development

The Chinese stock market also was organized by the central government as a means to transform its socialist planned economy into a market economy. In order to prevent the mass privatization of state-owned enterprise (SOE) and maintain their dominant role, ownership structure is divided into state-owned shares, institutional shares, employee shares, A shares, B shares, H shares, and other shares. The majority of shares is still controlled by the state and institutional shareholders, and the shares are non-tradable. Such a highly concentrated ownership structure determines the unique characteristics of the agency problem in Chinese corporations; in addition, it has become one of the key corporate governance problems in China [11]. When ownership is concentrated in the hands of a few controlling ownerships, as is the case in most countries around the world, the nature of the agency problem turns into conflicts between the controlling and minority shareholders. The majority shareholders for most listed SOE, are state agencies that lack experience in monitoring and controlling public corporations. Consequently, these controlling shareholders are likely to proscribe the minority shareholders because the undersized shareholdings of outside shareholders provide them with limited power to effectively monitor the controlling shareholders [3, 5-6]. Securities regulations are less rigorous than in many Anglo-American countries. Investors, financial analysts, and government officials are often uncertain about the real meanings and the true enforceability of regulatory codes and decrees. In addition, the audit market in China is not mature. Because of SOEs' lack of demand for international auditors, small local auditors' superior local knowledge, and SOEs' collusion incentives [10].

Accounting is closed linked with the economic and political environments. Existing a variety of parties, according to their economic benefit, are interested in the development of accounting standards. As a matter of facts, institutional theory has come to be regarded as a dominant theoretical perspective in organizational theory research. Accounting in real cannot be isolated from the social processes operating in and around organization. Accounting as a social and institutional practice has come to view the phenomenon of accounting as a symbol of legitimacy. Adopting IFRS appears to reduce information asymmetry between managers and stockholders. Previous literature finds a reduction of information asymmetry as evidenced by lower earnings management, lower costs of capital, and lower forecast errors. We examine each of these economic consequences below. Barth et al. [2] suggest that accounting quality could be improved with elimination of alternative accounting methods that are less reflective of firms' performance and are used by managers to manage earnings. They compare earnings management for firms that voluntarily switch to IFRS with firms that use domestic accounting standards. They find that after IFRS adoption, firms have higher variance of changes in net income, a higher ratio of variance of changes in net income to variance of changes in cash flows, higher correlation between accruals and cash flows, lower frequency of small positive net income, and higher frequency of large losses. Barth et al. [2] also investigate the value relevance of earnings by comparing the R-squared from two regressions: 1) price regressed on book value and earnings; and 2) earnings regressed on positive and negative returns. They find that R-squared increases after IAS adoption, providing evidence of greater value relevance for IFRS earnings.

Based on the above arguments, we propose the following research hypothesis:

H1: There is a positive association among mandatory adoption IFRS, institution environment and the income smoothing.

H2: There is a positive association among mandatory adoption IFRS, institution environment and the value relevance.

3. Source of Data and Variables Description

Our data comprises of provincial-level institutional indices and firm-level corporate governance as well as financial attributes. We collect firm-level financial data and the ultimate controlling shareholder data of state-controlled and entrepreneur-controlled Chinese firms that are listed on the Shanghai or Shenzhen Stock Exchange from the Taiwan Economic

Journal (TEJ) Finance database for the 2005 to 2008 periods are selected for the study. As prior research, accounting quality is proxied by income smoothing and value relevance. Higher accounting quality is expected to reveal less income smoothing and higher value relevance of earnings and equity book value (Barth et al., [2]). To detect a change in accounting quality attributable to changes in accounting standards rather than changes in incentives, we include controls associated with firms' voluntary accounting decisions such as growth, leverage, size, cash flow and so on as Barth et al. [2]. The two research model is as following:

$$Smooth_{it} = \beta_0 + \beta_1 IFRS_{it} + \beta_2 Inst_{it} + \beta_3 IFRS \times Inst_{it} + \beta_4 Eissue_{it} + \beta_5 CF_{it} + \beta_6 Lev_{it}$$

$$+\beta_7 ROA_{it} + \beta_8 Loss_{it} + \beta_9 Second_{it} + \beta_{10} L\arg e_{it} + \beta_{11} Size_{it} + \beta_{12} Big10_{it} + \varepsilon_{it} \qquad (1)$$

$$\Pr ice_{it} = \beta_0 + \beta_1 IFRS_{it} + \beta_2 Inst_{it} + \beta_3 IFRS \times Inst_{it} + \beta_4 Eissue_{it} + \beta_5 CF_{it} + \beta_6 Lev_{it}$$

$$+\beta_7 ROA_{it} + \beta_8 Loss_{it} + \beta_9 Second_{it} + \beta_{10} L\arg e_{it} + \beta_{11} Size_{it} + \beta_{12} Big10_{it} + \varepsilon_{it} \qquad (2)$$

where for firm i in year t, Smooth is net income increment deflate by beginning assets; Price is the year-end market price; Inst is the provinces or cities score where the home office of the listed company registered. Reissue is the incremental amount of the outstanding common stock; CF is the cash flow from operating activities scaled by total assets; Lev is total liabilities divided by shareholders' equity; ROA is the ratio of return on assets; Loss is the prior year's operating, 1 if operating loss, otherwise 0, a dummy variable; Second indicate that if the company haves issues H-share or others DR, also a dummy variable; Large is the largest stockholder, 1 if exist, 0 otherwise; Size is the natural logarithm of total assets; Big10 is a dummy variable taking the value of 1 when the firm's auditor is one of the Big 10 accounting firms, that is, international Big 4 plus the top 6 local audits firms and 0 otherwise.

3. Empirical Results and Discussions

The empirical results show that earnings smoothing is associated with IFRS and the institutional environment, IFRS coefficient of 0.004 and 1% significance level, and expected the same as [], the results that China will adoption the IFRS after the guidelines can improve the volatility of the surplus earnings smoothing phenomenon has improved to support the hypothesis of this study. The institutional environment (INST) coefficient of -0.005, 1% significance level, although contrary to the expected direction of the study, regardless of the institutional environment is good or bad, earnings smoothing is very serious, does not support hypothesis 1. Morck, Yeung and Yu [8] found that emerging capital markets, corporate level information content is very low in the 40 sample countries, the severity of China's capital market ranks second. The inference may be due to local institutional environment in China are not fairly perfect. The variable of IFRS × INST coefficient -0.004, 10% significance level, this article is expected the opposite, but the coefficient is smaller than the INST, on behalf of IFRSs into earnings smoothing has also affected due to new accounting standards adoption, the effect of its influence local institutional environment, so the coefficient is still negative, empirical results support the hypothesis.

In summary, the empirical study reveals significantly higher quality in reported accounting measures after the mandatory adoption of substantially IFRS-convergent accounting standards in China with decreased earnings smoothing and increased value relevance to stock price. The findings also empirically confirm previous researchers' argument that IFRS can be relevant to countries like China where economic and social environments and accounting needs become similar to those of advanced economies [9].

4. Conclusions

In summary, the empirical study reveals significantly higher quality in reported accounting measures after the mandatory adoption of substantially IFRS-convergent accounting standards in China with decreased earnings smoothing and increased value relevance to stock price and return. We found that the developments of Chinese government performance audit are conspicuously different by region; to reflect such differences; we elaborated our research by dividing mainland China into 31 categories (provinces or cities). We set 2003-2010 as the time horizon for this study. After testing the Regression model, our empirical research achieved two conclusions: 1) IFRS adoption in China should significantly improve the accounting quality, and 2) IFRS and institutional environment should synthetically influence the quality of accounting as well.

References

[1]. Ashbaugh, H., and P. Olsson. 2002. An exploratory study of the valuation properties of cross-listed firms' IAS and U.S. GAAP earnings and book values. *The Accounting Review* 77: 107-126.

[2]. Barth, M. E., W. R. Landsman, and M. H. Lang. 2008. International Accounting Standards and Accounting Quality. *Journal of Accounting Research* 46: 467-98.

[3]. Claessens, S., S. Djankov, and L. Lang. 2000. The separation of ownership and control in East Asia corporations. Journal of Financial Economics 58: 81-112.

[4]. Covrig, V., M. DeFond, and M. Hung 2007. Home bias, foreign mutual fund holdings, and voluntary adoption of international accounting standards. *Journal of Accounting Research*, 45, 41-70.

[5]. Faccio, M., Lang, L.H.P., 2002. The ultimate ownership of Western European corporations. Journal of Financial Economics 65, 365-395.

[6]. La Porta, R. Lopez-de-Silanes, F. and Shleifer, A., 1999. Corporate ownership around the world. Journal of Finance, 54, 471-517.

[7]. Leuz, C., and R. E. Verrechia. 2000. The economic consequences of increased disclosure. *Journal of Accounting Research* 38: 91-124.

[8]. Morck, R., B. Yeung, and W Yu, 2000. The information Content of Stock Market: Why Do Emerging Market Have Synchronous Stock Price Movements? Journal of *Financial Economics* 58: 215-260.

[9]. Tyrrall, D., D. Woodward, and A. Rakhimbekova. 2007. The relevance of International Financial Reporting Standards to a developing country: Evidence from Kazakhstan. *The International Journal of Accounting* 42 (1):82-110.

[10]. Wang, Q., T.J. Wong, and L. Xia. 2008. State Ownership, the Institutional Environment, and Auditor Choice: Evidence from China. *Journal of Accounting and Economics* 46: 112-134.

[11]. Xu, X, and Y, Wang, 1999, Ownership structure and corporate and governance in Chinese stock companies, China Economic Review 10, 75-98.

[12]. Yao Y. and L. Yueh, 2009. Law, Finance and Economic Growth in China: An Introduction, World Development 37 (4), 753-762.

On a Stochastic Failure Model under Random Shocks

Ji Hwan Cha[1]
Department of Statistics, Ewha Womans University,
Seoul, 120-750, Rep. of KOREA

E-mail: jhcha@ewha.ac.kr

Abstract. In most conventional settings, the events caused by an external shock are initiated at the moments of its occurrence. In this paper, we study a *new classes of shock model,* where each shock from a nonhomogeneous Poisson processes can trigger a failure of a system not immediately, as in classical extreme shock models, but with delay of some random time. We derive the corresponding survival and failure rate functions. Furthermore, we study the limiting behaviour of the failure rate function where it is applicable.

1. Introduction

Consider an orderly point process (without multiple occurrences) $N(t), t \geq 0$ of some 'initiating' events (IEs) with arrival times $T_1 < T_2 < T_3 < \ldots$. Let each event from this process triggers the 'effective event' (EE), which occurs after a random time (delay) $D_i, i = 1,2,\ldots$, since the occurrence of the corresponding IE at T_i. Obviously, in contrast to the initial ordered sequence $T_1 < T_2 < T_3 < \ldots$, the EEs $\{T_i + D_i\}, i = 1,2,\ldots$ are now not necessarily ordered. This setting can be encountered in many practical situations, when, e.g., initiating events start the process of developing the non-fatal faults in a system and we are interested in the number of these faults in $[0,t)$. Alternatively, effective events can result in fatal, terminating faults (failures) and then we are interested in the survival probability of our system. Therefore, the latter setting means that the *first EE ruins our system.* The corresponding stochastic survival model will be considered in this paper. When there are no delays, each shock (with the specified probability) results in the failure of the survived system and the described model obviously reduces to the classical extreme shock model (Gut and Hüsler, 2005; Finkelstein, 2008; Cha and Finkelstein, 2009).

The IEs can often be interpreted as some external shocks affecting a system, and for convenience, we will often use this wording (interchangeably with the "IE"). We will consider the case of the nonhomogeneous Poisson process (NHPP) of the IEs. The approach can, in principle, be applied to the case of renewal processes, but the corresponding formulas are too cumbersome. However, the obtained results for the NHPP case are in simple, closed forms that allow intuitive interpretations and proper analyses.

[1] To whom any correspondence should be addressed.

The paper is organized as follows. In Section 2, the corresponding lifetime distribution and the failure rate are obtained for the model with delays. In Section 3, the limiting behaviour of the failure rate for $t \to \infty$ is analyzed for different limiting properties of the rate of the NHPP of shocks. Finally, in Section 4, concluding remarks are given.

2. Lifetime Distribution

Consider a system subject to the nonhomogeneous Poisson process of IEs $N(t), t \geq 0$ to be called shocks. Let the rate of this process be $\nu(t)$ and the corresponding arrival times be denoted as $T_1 < T_2 < T_3 ....$ Assume that the ith shock is 'harmless' to the system with probability $q(T_i)$, and with probability $p(T_i)$ it triggers the failure process of the system which results in its failure after a random time $D(T_i)$, $i = 1,2,...$, where $D(t)$ is a non-negative, semi-continuous random variable with the point mass at "0" (at each fixed t). Note that this 'point mass' at 0 opens the possibility of 'immediate failure' of the system at the occurrence of a shock, which is practically very important, and, furthermore the case of the 'full point mass' of $D(t)$ at 0 reduces to the ordinary 'extreme shock model'. Obviously, without point mass at 0, we arrive at an absolutely continuous random variable. The distributions of $D(t)$ having point masses at other values could be considered similarly.

Let $G(t,x) \equiv P(D(t) \leq x)$, $\bar{G}(t,x) \equiv 1 - G(t,x)$, and $g(t,x)$ be the Cdf, the survival function and the pdf for the 'continuous part' of $D(t)$, respectively. Then, in accordance with the terminology in the Introduction, the failure in this case is the *Effective Event (EE)*.

First of all, we are interested in describing the lifetime of our system T_S. The corresponding conditional survival function is given by

$$P(T_S > t \mid N(s), 0 \leq s \leq t; D(T_1), D(T_2), ..., D(T_{N(t)}); J_1, J_2, ..., J_{N(t)})$$
$$= \prod_{i=1}^{N(t)} \left(J_i + (1 - J_i) I(D(T_i) > t - T_i)\right), \tag{1}$$

where the indicators are defined as

$$I(D(T_i) > t - T_i) = \begin{cases} 1, \text{ if } D(T_i) > t - T_i \\ 0, \text{ otherwise} \end{cases},$$

$$J_i = \begin{cases} 1, \text{ if the } i\text{th shock does not trigger the subsequent failure process}, \\ 0, \text{ otherwise.} \end{cases}$$

Assume the following conditions regarding 'conditional independence':

(i) Given the shock process, $D(T_i), i = 1,2,...$, are mutually independent.

(ii) Given the shock process, J_i, $i = 1,2,...$, are mutually independent. (It means that whether each shock triggers the failure process of the system or not is 'independently determined').

(iii) Given the shock process, $\{D(T_i), i = 1,2,...\}$ and $\{J_i, i = 1,2,...\}$, are mutually independent.

Integrating out all conditional random quantities in (1) under the basic assumptions described above results in the following theorem.

Theorem 1. *Suppose that* $\nu(+0) > 0$. *Then*

$$P(T_S \ge t) = \exp\left\{-\int_0^t G(x;t-x)p(x)\nu(x)dx\right\}, t \ge 0,$$

and the failure rate function of the system is

$$\lambda_S(t) = \int_0^t g(x;t-x)p(x)\nu(x)dx + G(t,0)p(t)\nu(t), t \ge 0.$$

Proof.
The proof is omitted due to the page limit.

Remark 1. It has been recognized that many queuing systems are most appropriately described by nonstationary queuing models, in which "both the arrival and service rates" are functions of time. Therefore, there is a growing literature on methods for calculating time-dependent performance measures in these models. It can be seen that in a queuing context, when $G(t,x) = G(x)$, our system's failure time can be interpreted as the first departure time in the $M_t / G / \infty$ system that starts empty (Ross, 1996). The model considered in this section is a time variant of it and, accordingly, the failure time can be interpreted in terms of the first departure time from the corresponding $M_t / G_t / \infty$ system that starts empty.

Remark 2. Speaking formally, the split of effects to effective and ineffective shocks does not add any mathematical complexity due to the NHPP nature of the arrival process. This means that the result would be the same if we had only one type of effects and the NHPP with the rate function $p(t)\nu(t)$. However, from the practical point of view and keeping in mind that we are generalizing here the classical extreme shock model with two types of effects, this splitting seems to be reasonable. Furthermore, we can consider the case of multi-type delayed consequences of shocks ($n > 1$), where the shock that occurs at time t causes the delayed (with distribution $G_i(t,x)$) effect of type i with probability $p_i(t)$, whereas the probability of 'no effect' is $1 - \sum_{i=1}^{n} p_i(t)$. Obviously, this model is the same as the single-type model with $G(t,x) = \sum_{i=1}^{n} p_i^*(t)G_i(t,x)$ and $p(t) = \sum_{i=1}^{n} p_i(t)$, where $p_i^*(t) = p_i(t) / \sum_{i=1}^{n} p_i(t)$. Therefore, similar to Theorem 1,

$$P(T_S \ge t) = \exp\left\{-\int_0^t (\sum_{i=1}^{n} p_i(x)G_i(x,t-x))\nu(x)dx\right\}, t \ge 0$$

and

$$\lambda_S(t) = \int_0^t (\sum_{i=1}^{n} p_i(x)g_i(x,t-x))\nu(x)dx + (\sum_{i=1}^{n} p_i(t)G_i(t,0))\nu(t).$$

3. Limiting behaviour

In this subsection, we study the limiting behavior of the failure rate $\lambda_S(t)$. Without loss of generality, assume that $p(t)$ and $\nu(t)$ are continuous functions with $p(t) > 0$, for all $t \geq 0$. We further assume that

$$D(t) \to D(\infty) \equiv D \text{ in distribution as } t \to \infty,$$

where $D(t)$, $t > 0$, and D are semi-continuous random variables with supports $[0,\infty)$ and point masses at "0", and

$$\frac{g(t,x)}{g(0,x)} \text{ is bounded for all } t,x \geq 0. \tag{2}$$

Obviously, (2) is a rather weak condition. The simplest example of a continuous random variable is

$$g(t,x) = (1/\mu(t))\exp\{-(1/\mu(t))x\},\ x \geq 0, \tag{3}$$

where $\mu(t)$ is continuous and decreasing with

$$\mu(t) \to \mu, \text{ as } t \to \infty.$$

Observe that the distribution which corresponds to the pdf in (3) is an exponential distribution for each fixed t.

Theorem 2. *In addition to the assumptions of Theorem 1, let* $\lim_{t\to\infty} p(t) \equiv p(\infty)$ *and* $\lim_{t\to\infty} \nu(t) \equiv \nu(\infty) \leq \infty$ *exist. Then*

$$\lim_{t\to\infty} \lambda_S(t) \equiv p(\infty)\nu(\infty).$$

Proof.
The proof is omitted due to the page limit.

4. Concluding remarks

One can find a lot of different shock models in the vast literature on this topic (see, e.g., Nakagawa, 2007 and references therein). These models usually deal with systems that are subject to shocks of random magnitudes at random times.

In this paper, we study a new shock model. The model generalizes the extreme shock model to the case when each shock from a nonhomogeneous Poisson processes can be fatal only after some time delay. We derive the corresponding survival probabilities and failure rates in a rather simple, meaningful form that allows probabilistic analysis and prompts for further generalizations.

References

[1] Cha J H and Finkelstein M 2009 On a terminating shock process with independent wear increments *Journal of Applied Probability* **46** 353-362
[2] Gut A and Hüsler J 2005 Realistic variation of shock models *Statistics & Probability Letters* **74** 187-204
[3] Finkelstein M 2008 *Failure Rate Modelling for Reliability and Risk*. Springer, London.
[4] Nakagawa T 2007 *Shock and Damage Models in Reliability Theory*. Springer, London.

Healthcare system simulation using Witness

Masoud Khakdaman, Milad Zeinahvazi, Bahareh Zohoori, Fardokht Nasiri, Kuan Yew Wong

Department of Manufacturing and Industrial Engineering, Universiti Teknologi Malaysia (UTM), Skudai, 81310, Johor, MALAYSIA

masoudkhakdaman@gmail.com

Abstract: Simulation techniques have a proven track record in manufacturing industry as well as other areas such as healthcare system improvement. In this study, simulation model of a health center in Malaysia is developed through the application of WITNESS simulation software which has shown its flexibility and capability in manufacturing industry. Modelling procedure is started through process mapping and data collection and continued with model development, verification, validation and experimentation. At the end, final results and possible future improvements are demonstrated.

1. Introduction

Over time, simulation tools and techniques are applied in problem detection and process improvement of many manufacturing and service systems. Popularity and flexibility of the simulation programs enable organizations to monitor the whole aspect of their systems and plan for business development prior to facing the real behavior of their industry or service [1].

Among various service organizations, healthcare industry has a fundamental and significant role in upgrading lifestyle standards. In this regard, usage of different tools and methods to enhance the capability of healthcare service providers to keep up with the growing changes of their industry is being increased. However, the process of migrating to the new or improved environment or methodology is not easily acceptable by healthcare staff and managers. Simulation allows for various types of assessments and examinations to be done. It does not burden managers with large amounts of money and it reduces any possible risk and danger in the healthcare performance [2]. Thus, this paper aims to assess the benefits and drawbacks of the current healthcare system of a health center in Malaysia and suggest recommendations to enhance the healthcare level in the supposed case study.

In the rest of this paper, a review of the current literature and WITNESS simulation software is presented. Then the case study is introduced and the simulation process is demonstrated. Next, model verification and validation is mentioned. After that, model experimentation and results are depicted as well as final conclusions.

1.1. Computer simulation of healthcare industry

Although simulation modelling has been widely used in manufacturing and logistics during the past four decades, healthcare is an issue that has not benefitted from the advancements and capabilities of simulation modelling [3, 4]. According to [5], application of simulation techniques in healthcare has noticeably increased after 2000 while before that only a handful of studies were conducted in this area. In this regard, application of simulation techniques in healthcare can be categorized into several areas including decision making for medical, administrative and operational issues [6], investigating control programs, monitoring medical progress, forecasting future incidences, designing health care systems [7], improving the factors which affect the patients' flow such as waiting time, stay length, patient throughput and clinical overtime and allocating resources [8]. As an example, [9] applied lean manufacturing concepts through simulation techniques in healthcare and demonstrated its tangible results such as decreasing costs, errors and waiting time of patients. In addition, [10] presented some intangible results, for instance, increasing the customer and employee satisfaction as well as staff motivation. Assessing the impact of several scenarios, [11] reengineered a healthcare system using simulation techniques. [12] and [13] applied mathematical programming for scheduling of doctors of an emergency room and healthcare resource allocation of a health center respectively. [14] developed a simulation approach using multi-objective optimization for treatment of cancer in a health center. By combining design of experiments and simulation, [15] estimated the capacity of an emergency room. [16] merged discrete-event simulation and lean manufacturing concepts to develop a new approach for improving healthcare systems and depicted the impact of their approach through three examples.

1.2. Overview of WITNESS simulation software

In this paper, WITNESS simulation software is applied because of its strengths and flexible features as one of the powerful and popular Visual Interactive Simulation Modeling (VISM) systems [17] for discrete-event simulation. WITNESS is capable of designing complex processes of manufacturing enterprises and service industries through quick and accurate modeling and detailed mathematical support as well as comprehensive assessment and evaluation reports [18].

The modeling procedure in WITNESS is started using different basic entities called elements and after defining appropriate specifications, it is continued with relating the elements through using graphical and coded connectors. In addition, each element can be programmed using built-in codes.

2. Case study: a health center in Malaysia

A health center in the Johor Bahru city of Malaysia is considered. This medical center provides health services including medication, laboratory, dental services etc from 8:00 A.M to 10:00 P.M (regular time) and also during emergency time (from 10:00 P.M to 8:00 A.M). The scope of this case study includes the queuing systems associated with the treatment services (only Medical services) and supportive services (Pharmacy and Laboratory) which are provided by four doctors in regular time.

2.1. Process mapping and data collection

In the first step, the conceptual model of the case study, as one of the initial steps in a simulation process, is used to deeply understand the actual workflow of the system and determine how to design the simulation model (Figure 1). There are three counters in the arrival area of the health center that put the patients into a queue for scanning their barcode number and waiting for visiting doctors. Here, as patients go to the shortest queue behind the three counters, when they arrive in the health center, a decision-making entity has been defined before the counters' queues. In addition, based on the statistical study of patients, almost 13 % of them are advised to go to the laboratory. Thus, another decision-making entity is defined after seeing the doctors in order to separate patients which must go to the laboratory from the rest. Finally, all patients go to the pharmacy to receive their medicines.

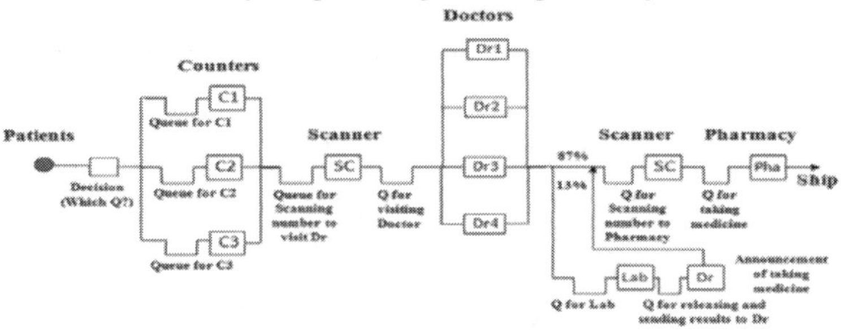

Figure 1. Conceptual model of the health center

Using the stop watch method, required data on cycle times, inter arrival times, setup times, waiting times, break down times, queue capacities and number of servers are collected from around 500 samples (Patients). For analyzing data, Easy Fit software is applied to estimate which distribution is best fitted to the collected data. In this software, the goodness of fit which is ranked by the Anderson Darling test is considered and among the results, the first distribution which exists in Witness software is accounted as the distribution of the data. As an example, the distribution function of the counters' service time is estimated as a Lognormal distribution with $\mu = 0.3$ minute and $\sigma = 0.05$ minute.

2.2. Model development

Based on the conceptual model, collected data, distribution functions of simulation parameters and actual behavior of the system, the simulation model of the health center is built using WITNESS simulation software. Figure 2 shows the simulation model of the health center.

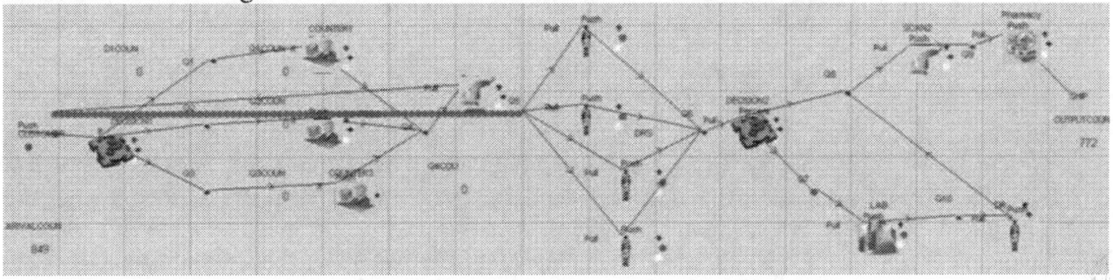

Figure 2. Simulation Model for the Health Center

2.3. Verification and validation of the model

To be confident whether a simulation model and its results are "correct" or not, model verification and validation methods must be applied. Referring to [19], the process of ensuring that the coding and logic of a computerized model and its execution are correct is named model verification; however, model validation is belonged to the concept of ensuring the consistency of the processes and results of the simulation model with the actual behavior of the real system. Using Witness simulation software, the next two parts present model verification and validation of the case study.

2.3.1. *Model verification.* In general, verification must be performed to ensure about the accuracy of the programmed model, its specifications and algorithms, and also its implementation. There are several ways to verify a simulation model such as incremental model building, expert evaluation, internal evaluation, consistency test and so on. In this study incremental model building is applied to verify the simulated model.

In the incremental model building method, the model should be built in small sizes and checked whether it is rightly and logically coded or not. Then it should be expanded to the final model. In this study, four steps are used to demonstrate the process of verification through the incremental model building method. First, the model is built from arrival of patients to taking numbers from counters and then checked whether the number of entered patients (e.g. 262) and output of the three counters (e.g. 262) are equal or not (Figure 3). After verifying the first step, the model is simulated from patients' arrival to visiting doctors (Figure 4). Here, 274 patients have arrived in which 254 of them have been visited by doctors, 4 of them are being visited and 16 of them are waiting for visiting doctors. The third step is building the model up to the stage of taking medicine without considering patients that should go to the laboratory (Figure 5). In the last step, the complete model is built.

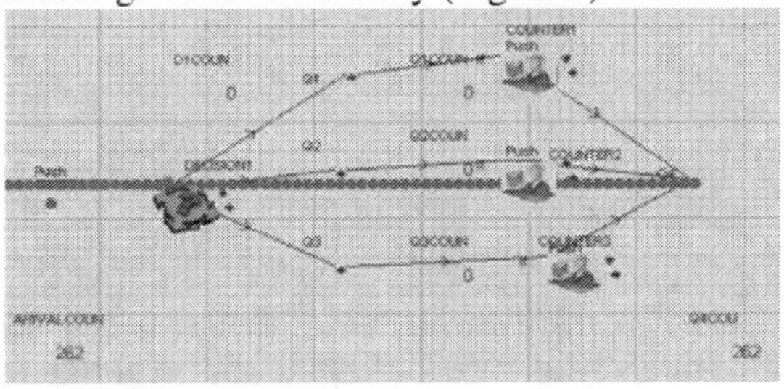

Figure 3. First Step of Verification

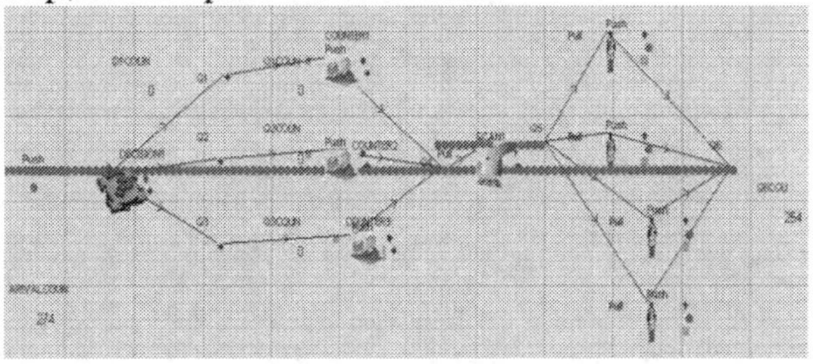

Figure 4. Second Step of Verification

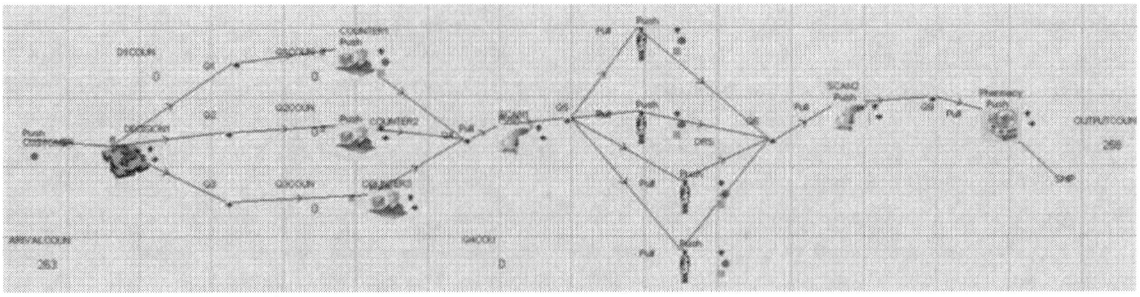

Figure 5. Third Step for Verification of the simulation model

In addition to the incremental model building method, most of the transactions of the simulated model are tested to check whether they are the same with what we have expected or not. For instance, the results of one random run of the model are depicted in Table 1 to demonstrate that the busy time of the pharmacy is 35.11 % when serving 257 patients in 14 hours (840 minutes). On the other hand, the mean value of the distribution function of the pharmacy service time is 1.13 minutes, and the busy time of the pharmacy is 294.92 minutes (840* 0.3511) which brings the expectation of serving almost 261 patients into account (294.92/1.13). Comparison of the results of the simulation model (257) and statistical calculation (261) depicts very little difference (around 1.5 %) which shows the strength of the model in terms of verification.

Table 1. Verification data of the simulated model

Parameter	Pharmacy
Idle Time (%)	64.89%
Busy Time (%)	35.11%
Number of Operations	257

2.3.2. *Model Validation.* The crucial role of model validation can be stated as making the model development team confident about the adoption of the simulated model to the actual situation of the system and its problems in a sensible manner. According to [19] there are four main approaches for model validation. The first approach is making the decision by simulation team members. The second approach is checking the model by different tests and evaluations during the simulation model building process to reach a subjective decision. The third approach which is frequently called "independent verification and validation" (IV&V), validates the model by applying the decision of a third party team. Using a scoring model can be mentioned as the fourth approach.

In this study, the process of checking whether or not the model represents the actual situation of the health center is made by comparing the average number of patients served by the health center during a 14 hours working time and the average number of patients served through the simulation model during a run of 840 minutes. This length of the run is used because the simulation model is a terminating system. In the next step, we should determine the number of replications, which should be large enough to make us confident about the results. Since the distribution of the study's population is not clear, we use the t distribution to determine the number of replications with a 95% confidence level. For estimating number of the replications, the formula presented by [20] , $n_m = \left((s * t_{m-1,\frac{(1-\alpha)}{2}})/(\bar{x} * k)\right)^2$, is applied where n =number of simulation runs to achieve the desired accuracy level, $\bar{x}$ = the mean estimate of an initial m number of runs, s = the standard deviation estimate of m number of runs, α = confidence level, k =permissible percentage of error, and $t_{m-1,\frac{(1-\alpha)}{2}}$= critical value of the two-tailed t-distribution at a level of significance, given m-1 degrees of freedom. Here, the model has been run for six times and the mean and standard deviation are calculated (Table 2). Using $t_{6-1,0.025} = 2.571$ and $k = 0.05$, the number of replications should be 3 or above ($n = 2.26$). The final step is comparing the results of at least 3 replications with the actual average number of patient arrivals which is 281 (Table 3). It is clear that the average number of patient arrivals using the simulation model (276) is very near to the actual one (281).

Table 2. Initial runs for estimating the number of replications

Simulation runs	Patient arrivals
1	273
2	284
3	270
4	263
5	281
6	268
Mean ($\bar{x}$)	273.17
Standard deviation (s)	7.99

Table 3. Validation data of the simulation model

Simulation		Average Actual arrivals
Replications	Patient arrivals	
1	273	281
2	284	
3	270	
Mean ($\bar{x}$)	275.67	
Standard deviation (s)	7.37	

3. Model experimentation and results analysis

After completing the verification and validation of the model, the results of the model experimentation are obtained. In the service sector, one of the most important factors in performance measurement is the average time each customer is in the system. In this study, the average time each patient spends in the health center is around 44 minutes which is a long time. In addition, the average number of patients in the system is around 25 patients which is relatively large (Table 4). However, analysis of data in Table 4 shows that the average utilization of the system is more than 86 percent.

Table 5 depicts the percentage of working and inactiveness of the different entities of the simulation model. As shown, one counter is 100% idle and the other two are idle most of the time. The utilization of doctors in the system is an important factor. As we can see, all of the doctors are almost always busy (more than 98 %). However, Laboratory and Pharmacy are busy for around 65 % and 35% of the working hours.

Table 4. Simulation report of the verified and validated model

Performance factor		Value
Number of entered patients		284
Number of shipped patients		257
Number of patients in the system		27
Average number of patients in the system		25.06
Average time spent by each patient in the system (Minute)		43.21
System Utilization	Mean	86.74 %
	Standard Deviation	13.64 %

Table 5. Report of simulation results for various model entities

Simulation Entity		Idle Time (%)	Busy Time (%)	Operations Number
Counters	1	83.38	16.62	282
	2	99.87	0.13	2
	3	100	0	0
Doctors	1	0.84	99.16	64
	2	1.02	98.98	63
	3	1.85	98.15	65
	4	1.66	98.34	65
Scanners	1	96.74	3.26	284
	2	97.05	2.95	257
Laboratory		35.53	64.47	42
Pharmacy		64.89	35.11	257

4. Conclusions

In this study, the queuing system of a Malaysian health center is studied to determine the strengths and weaknesses of services offered to patients. According to the scope of study, it can be concluded from the simulation model that although the system utilization is more than 86%, the average time each patient spends in the system (around 44 minutes) and the average number of patients in the system (25 patients) are large amounts that reveal the inappropriate service level in the health center. One reason is a long queue of patients waiting to get visited by doctors which often bothers patients. Here, adding one doctor, if possible, can be considered as an improvement to decrease the waiting time of patients in the whole queuing system. Furthermore, there are some amounts of idle time in the counter service which can be reduced by the deduction of one counter.

References

[1] P. L. Market and M. H. Mayer, WITNESS simulation software a flexible suite of simulation tools, Proceedings of the IEEE Simulation Conference, Ohio, USA, 1997, pp.711–717.
[2] C. D. Barnes , J. L. Quiason , C. Benson and D. McGuiness, Success Stories In Simulation In Health Care, Proceedings of the Winter Simulation Conference, San Francisco, USA, 1997, pp.1280-1285.
[3] S. M. Sanchez, D. M. Ferrin, T. Ogazon, J. A. Sepu´lveda and T. J. Ward, Emerging issues in healthcare simulation, Proceedings of the Winter Simulation Conference, New York, USA, 2000, pp.1999-2003.
[4] J. B. Jun, S. H. Jacobson and J. R. Swisher, Application of discrete-event simulation in health care and clinics: a survey, Journal of the Operational Research Society, Vol.50, No.2, 1999, pp.109-123.
[5] MASHnet, Modelling and simulation in healthcare, Available at www.mashnet.org.uk, 2005
[6] T. Eldabi, R. J. Paul and T. Young, Simulation Modelling in Healthcare: Reviewing Legacies and Investigating Futures, *The Journal of the Operational Research Society*, Vol. 58, No. 2, Special Issue: Operational Research in Health, 2007, pp. 262-270
[7] M. Lagergren, What is the role and contribution of models to management and research in the health services? A view from Europe, *European Journal of Operational Research*, Vol. 105, No. 2, 1998, pp. 257-266.
[8] L. P. Baldwin, T. E, Ray J. Paul, Simulation in healthcare management: a soft approach (MAPIU), *Simulation Modelling Practice and Theory,* Vol. 12, No. 7-8, 2007, pp. 541–557
[9] K. Silvester, R. Lendon, H. R. S. Bevan and P. Walley, Reducing waiting times in the NHS: Is lack of capacity the problem?, *Clinician in Management*, Vol. 12, No. 3, 2004, pp. 105–111.
[10] Z. J. Radnor and R. Boaden, Lean in public services: Panacea or paradox?, *Public Money and Management,* Vol. 28, No. 1, 2008, pp. 3-7.
[11] D. Sinreicha and Y. Marmor, Emergency department operations: The basis for developing a simulation tool. *IIE Transactions,* Vol. 37, 2005, pp. 233-245.
[12] H. Beaulieu, J. A. Ferland, B. Gendron, and P. Michelon, A mathematical programming approach for scheduling physicians in the emergency room, *Health Care Management Science*, Vol. 3, No. 3, 2000, pp. 193-200.
[13] S. Flessa, Where efficiency saves lives: A linear programme for the optimal allocation of health care resources in developing countries, *Health Care Management Science,* Vol.3, No. 3, 2000, pp. 249-267.
[14] F. F. Baesler and J. A. Sepúlveda, Multi-objective simulation optimization for a cancer treatment center, *Proceedings of the Winter Simulation Conference*, Arlington, USA, 2001, pp. 1405-1411.
[15] F. F. Baesler, H. E. Jahnsen and M. DaCosta, Emergency departments I: the use of simulation and design of experiments for estimating maximum capacity in an emergency room, *Proceedings of the Winter Simulation Conference,* 2003, New Orleans, USA, pp. 1903-1906.
[16] S. Robinson, Z. J. Radnor, N. Burgess and C. Worthington, SimLean: Utilising simulation in the implementation of lean in healthcare, *European Journal of Operational Research,* Vol. 219, No. 1, 2012, pp. 188-197.
[17] M. Pidd, *Computer simulation in management science (4th ed),* Wiley, 1998
[18] A. P. Waller, *WITNESS for Six Sigma*, Lanner group, 2009
[19] R. G. Sargent, Verification and validation of simulation models, *Proceedings of the IEEE Winter Simulation Conference,* New York, USA, 2010, pp. 166-183.
[20] K. Ahmed, Modeling Drivers' Acceleration and Land Changing Behavior, *PhD thesis*, Massachussettes Institute of Technology, Cambridge, USA, 1999.

Journal of Physics: Conference Series **410** (2013) 012110 doi:10.1088/1742-6596/410/1/012110

Pore structure and effective permeability of metallic filters

Vladimír Hejtmánek[1*], Martin Veselý[2], and Pavel Čapek[2]

[1]Institute of Chemical Process Fundamentals ASCR, v.v.i., Rozvojová 135, 165 02 Prague 6, Czech Republic, hejtmanek@icpf.cas.cz

[2]Institute of Chemical Technology, Prague, Technická 5, 166 28 Prague 6, Czech Republic, pavel.capek@vscht.cz, veselyr@vscht.cz

[*]hejtmanek@icpf.cas.cz

Abstract. The pore structures (microstructures) of two metallic filters were reconstructed using the stochastic reconstruction method based on simulated annealing. The following microstructural descriptors were included in the description of the real microstructures: the two-point probability function, the lineal-path functions for the void or solid phases, i.e. simulated annealing was constrained by all low-order statistical measures that were accessible through the analysis of images of polished sections. An effect of the microstructural descriptors on the course of reconstruction was controlled by modifying two parameters of the reconstruction procedure [1]. Their values resulted from repeated reconstruction of two-dimensional microstructures in such a way that the reference (experimental) and calculated two-point cluster functions deviated negligibly. It was tacitly assumed that the parameters adjusted during two-dimensional reconstruction had the same influence on the formation of the three-dimensional microstructures. Since connectivity of phases is a critical property of the stochastically reconstructed media, clusters of pore and solid voxels were determined using the Hoshen-Kopelman algorithm. It was found that the solid phase formed one large cluster in accordance with the physical feasibility. The void phase created one large cluster and a few small clusters representing the isolated porosity. The percolation properties were further characterised using the local porosity theory [2]. Effective permeability of the replicas was estimated by solving the Stokes equation for creeping flow of an incompressible liquid in pore space. Calculated permeability values matched well their experimental counterparts.

1. Introduction

The replicas like artificial representative 3D models of real microstructure of porous solid material are created by means of several classes of methods. The serial sectioning method [3] or focused ion beam lift-out technique [4], X-ray computed microtomography [5], process-based methods [6,7], and stochastic reconstruction methods can be categorized into basic list. The last considered class belongs to the group of statistical methods and appears as more economical alternative to the other reconstruction methods.

The stochastic reconstruction method based on the simulated annealing (SRSA) [8] generally provides quite realistic information on microstructure of porous materials. Nevertheless, the connectivity of pore space in some generated replicas is not reproduced in long-range scale sense so well; see [9,10]. This is the case of the metallic porous filters where the improved SRSA method was used to enhance long-range pore connectivity of their replicas.

2. Experimental

2.1. Porous metallic filters and preparation of samples

Porous metallic filters from Mott Corporation, USA, in the form of flat disc made from stainless steel were used in two variants differing in their textural properties [11]. The cross-sections in the appropriate various plane orientations were prepared from hardened epoxy resin blocks (vacuum impregnation method) by their cutting, lapping, grounding, and polishing.

2.2. Image acquisition and image processing

The resulting polished cross-sections were monitored in the scanning electron microscope (SEM JSM–5500LV, JEOL, Japan). Thereafter, series of back-scatter electron images from various locations of the considered cross-sections in uniform size and resolution were collected. The grey-scale images were underwent to the nonlinear filtering through the adaptive median and percentile filters. The Otsu's segmentation algorithm of global thresholding was then applied to separate void (pore) and solid phases, respectively. Value of threshold was adjusted on the basis of information obtained from textural analysis (helium pycnometry, mercury porosimetry).

2.3. Microstructure and microstructural descriptors

Binary images of randomly selected parts of porous media represent 2D digitized medium in form of the indicator phase function; the void phase is labeled as 1 and the solid phase as 0. The relation between various shapes of both phases depicted on every 2D plane is captured in the microstructural descriptors which were derived from series of 2D binary images. The two-point probability function for the void phase, $S_2(u)$, the lineal-path function for both phases, $L^p(u)$, $L^m(u)$, respectively, were calculated for mentioned porous metallic filters. The detailed definition of all used microstructural descriptors can be found in [12]. The course of selected microstructural descriptors derived from the images of polished sections is common in 2D and 3D spaces. Therefore, they were used as starting data (reference descriptors having an origin from the real porous material) for the computational reconstruction algorithm.

2.4. Stochastic reconstruction of porous metallic filters

Stochastic reconstruction of porous material by means of simulated annealing method constrained by three microstructural descriptors, $S_2(u)$, $L^p(u)$, $L^m(u)$ was based on minimization of the objective function E, i.e. "energy" of the digitized system

$$E = e_0 + e_1 + e_2 \frac{\eta}{\eta + e_0 + e_1} \quad (1)$$

$$e_s = \sum_{i=1}^{n} \sum_{u=0}^{U_s} \left(M^{\exp}(u) - M_i(u) \right)^2 \; s = 0, 1, 2 \quad (2)$$

where M designs arbitrary microstructural descriptor $S_2(u)$, $L^p(u)$, $L^m(u)$, and superscript exp denotes the referential one. The parameter η was given to alter the weight of lineal-path function for solid phase during time of minimization (1). All used microstructural descriptors were calculated in principal and diagonal directions, under consideration of periodic boundary conditions.

In the principle, the random configuration of solid and void voxels (volume elements) was generated in 3D space on a simple cubic lattice (320×320×320). Consequently, this high-energy uncorrelated microstructure was gradually transformed into a more correlated microstructure with minimal energy content, see figure. The pair of voxels, the first voxel from one and the second from opposite phase, was selected randomly in each iteration step k and their phases were mutually changed. The energy difference ΔE of two states of whole microstructure (before and after this change, respectively) was determined. Thereafter, the more suitable energetic state was accepted with the probability resulting from Metropolis rule [13]. The decision to accept/reject is dependent on ΔE and system "temperature" ϑ_k. The considered temperature was kept constant, when sufficient number of new states had been generated. The temperature was decreased according to the annealing schedule. The two-point cluster function was used indirectly for adjustment of two special parameters η, ζ of our reconstruction procedure. The second parameter ζ controls the voxels selection. Other details on the modified method and on the optimal adjustment of parameters η, ζ are detailed described in [1].

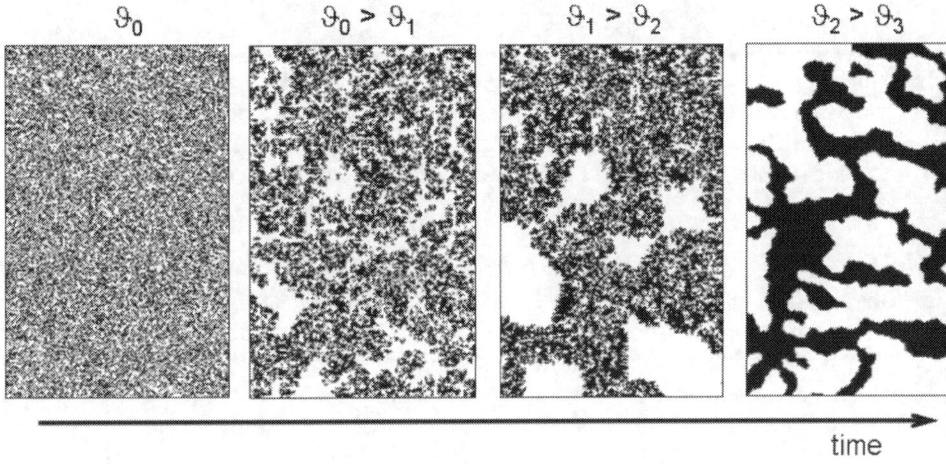

Figure. Relation between the temperature of porous digitized system ϑ_k and elapsed time during stochastic reconstruction on simulated annealing (increased correlation of voxels, cross-sections).

3. Results and discussion

Two real samples of the porous metallic filter (labeled as F1/F2) differed in total porosity (17.8 %, 20.1 %) and pore size distribution (median pore diameter 3.55/2.70 microns) were reconstructed by means of the modified SRSA method. Five replicas were generated for each porous material. Their percolation properties were characterized through the 3D descriptors defined in terms the local porosity theory [2]. Considered descriptors were local porosity distribution, the local percolation probability and the total fraction of percolating cells in the selected directions.

It can be summarize that both kinds of replicas formed large percolating clusters in the void (pore) and solid phases, respectively. The volume fraction of non-percolating void phase was very small (for instance, its value for the sample F1 is 0.0103 versus calculated total porosity 0.1849).

[9] who employed in their study the standard SRSA method, mentioned the existence of non-percolating clusters and poor connectivity of pores in sandstone rock samples as the result of artificial anisotropy in replicas. On the other hand, the adaptation of the objective function as well as the dynamics of the reconstruction of porous material (performance of this algorithm during time of reconstruction) played a significant role in the modified SRSA method. Additive constraint of the lineal-path function for metallic phase $L^m(u)$ into the objective function E (see equation (1)) had a positive impact on the pore connectivity of both metallic replicas. These replicas became more penetrable without small isolated clusters in the metallic phase. The choice of microstructural descriptors calculation in the more orientations (principal and diagonal) supported the formation of metallic microstructures free of the rectangular-like shapes typical for using of descriptors evaluated only in principal directions. The next benefit for the improving of the percolation properties of reconstructed replicas consists in a deceleration of cooling schedule during time of the SRSA, already suggested by [9]. In the case of low-porosity porous material, above described modified SRSA method

led to improving of percolation in all considered principal directions compared with the commonly used SRSA (parameter $\eta = 0$, thus the objective function $E = e_0 + e_1$).

The effective transport properties of the replicas of porous metallic filters were estimated for creeping flow of an incompressible liquid in pore space. Together with the balance (equation of continuity), no-slip boundary condition and considered pressure difference on the opposite walls of replica, the Stokes equation was solved with aid "permsolver", the public domain software [14]. The effective permeability β was derived from the mean survival time calculated as average time of random walker travelling in the pore space of metallic replica [1].

Calculated permeability values matched well their experimental counterparts obtained from permeation measurements with inert gases in Wicke-Kallenbach cell working under quasi-stationary conditions [11]. The correlation between pore connectivity and effective transport properties was clearly demonstrated on the replicas generated by means of the modified SRSA method. The results following from the local pore analysis and the cluster statistics were then confirmed.

4. Conclusion

The reconstruction of real porous metallic filters verified assumption of the pore connectivity improving in low-porosity materials. The both replicas constituted large clusters of the porous and metal phases, whereas non-percolating isolated clusters took only negligible fractions of both phases. Consequently, the modified SRSA method was capable to reproduce good connectivity. The 3D replicas generated from limited statistical information obtained from 2D micrographs can serve for evaluation of various effective properties included elastic modulus, fluid diffusivity, and permeability.

Acknowledgement

The authors are greatly appreciated for financial support from the Czech Science Foundation, Project P204/11/1206. We are grateful to Dr. L. Brabec for microstructure images.

References

[1] Čapek P, Hejtmánek V, Kolafa J, Brabec L 2011 *Transp Porous Media* **88** 87
Čapek P, Hejtmánek V, Brabec L, Zikánová A, Kočiřík M 2009 *Transp Porous Media* **76** 179
[2] Hilfer R 2000 Local Porosity Theory and Stochastic Reconstruction for Porous Media *Statistical Physics and Spatial Statistics, Lecture Notes in Physics* eds K Mecke and D Stoyan vol 254 Springer-Verlag pp 203–241
[3] Lymberopoulos D P and Payatakes A C 1992 *J Colloid Interface Sci* **150**, 61
[4] Shearing P R, Golbert J, Chater R J, Brandon N P 2009 *Chem Eng Sci* **64** 3928
[5] Flannery B P, Deckman H W, Roberge W G, D'Amico K L 1997 *Science* **237** 1439
[6] Øren P E and Bakke S 2002 *Transp Porous Media* **46** 311
[7] Politis M G, Kikkinides E S, Kainourgiakis M E, Stubos A K 2008 *Microporous Mesoporous Mat* **110** 92
[8] Yeong C L Y and Torquato S 1998 *Phys Rev E* **58** 224
Yeong C L Y and Torquato S 1998 *Phys Rev E* **58** 495
[9] Manwart C, Torquato S, Hilfer R 2000 *Phys Rev E* **62** 893
[10] Talukdar M S and Torsaeter O 2002 *J Pet Sci Eng* **33** 265
[11] Hejtmánek V and Čapek P 2011 *Proc 7th Int Conf on Porous Metals and Metal Foams (Busan)*, Korea, September 18–21 p 547
[12] Torquato S 2002 *Random heterogeneous materials: microstructure and macroscopic properties* New York: Springer Verlag
[13] Aarts E and Korst J 1989 *Simulated annealing and Boltzmann machines: A stochastic approach to combinatorial optimization and neural computing* Chichester: John Wiley & Sons
[14] Bentz D P and Martys N S 2007 Tech Rep No 7416 NIST Gaithersburg, USA ftp://ftp.nist.gov./pub/bfrl/bentz/permsolver

IC-MSQUARE 2012: International Conference on Mathematical Modelling in Physical Sciences IOP Publishing
Journal of Physics: Conference Series **410** (2013) 012111 doi:10.1088/1742-6596/410/1/012111

Some control variates for exotic options

J C Ndogmo
School of Mathematics
University of the Witwatersrand
Private Bag 3, Wits 2050
South Africa

E-mail: jean-claude.ndogmo@wits.ac.za

Abstract. There are no known exact formulas for the valuation of a number of exotic options, and this is particularly true for options under discrete monitoring and for American style options. Therefore, one usually recourses to a Monte Carlo Simulation approach, amongst other numerical methods, to estimate the value of these options. The problem which then arises with this method is one of variance reduction. Control variates are often used, and we present some results for the optimization of these control variables, for the valuation of Asian and lookback options. An inequality on functions of correlations useful for comparing estimators in variance reduction procedures is also provided.

1. The problem of exotic options valuations

One important class of exotic options consists of path-dependent (or history-dependent) derivatives. They have a payoff that depends not just on the final price of the underlying asset, but also on the path followed by this price up to maturity. Asian options and lookback options are typical examples of path-dependent options. In some instances, numerical methods are the only means available to value these exotic options, and in addition to Monte Carlo methods, other commonly used numerical methods include the finite difference and the finite element methods, the trinomial tree method, and some times the Markov Chain method (see [1, 2, 3, 4, 5], and the references therein). Until the last decade, Monte Carlo method has been considered as costly and unreliable, but based on innovative techniques which are now a topic of current research, they've yielded more promising results. In this context, Barraquand and Martineau [3], amongst others, obtained some interesting results.

Consider an option on a given security having n days to maturity. Let $S_d(j)$ denote the closing price of the security at the end of day j (for $j = 1, \dots, n$). Assume also that the risk free interest rate is r per year, and denote by N the number of trading days in a year. With these notation, the payoff from an asian option is either given by

$$Y = e^{-rn/N} \left(S_d(n) - \sum_{j=1}^{n} \frac{s_d(j)}{n} \right)^+ , \tag{1.1}$$

where the strike price K is taken as the average of all prices up to the expiry date, or

$$Y = e^{-rn/N}\left(\sum_{j=1}^{n}\frac{S_d(j)}{n} - K\right)^{+}, \tag{1.2}$$

when the underlying's terminal price is taken as the average of all prices up to maturity. The payoff from a lookback option is given by

$$Y = e^{-rn/N}\left(S_d(n) - \operatorname*{Min}_{j=1,\dots,n} S_d(j)\right)^{+}. \tag{1.3}$$

The main problem with using Monte Carlo Simulation to value path-dependent derivatives is that the computation time necessary to achieve a reasonable level of accuracy can be excessively high. Control variate techniques are some of the tools available for implementing the necessary variance reduction procedures, which can lead to dramatic savings in computation time. Our focus in this paper will be on the applications of these techniques to the valuation of Asian and lookback options.

2. Monte carlo simulations and variance reduction procedures

Suppose that the price $S(t)$ of a given security follows a risk neutral geometric Brownian motion with constant parameters μ and σ^2. This means in particular that for all $y \geq 0$, and $t \geq 0$, $\ln(S(t+y)/S(y))$ has a normal distribution with mean μt and standard deviation $\sigma\sqrt{t}$, and that $\mu = r - \sigma^2/2$, where r represents the risk-free interest rate.

We denote as usual by $S(0)$ the initial price of the security, and by $S_d(j)$ the price of the security at the end of day j (for $j = 1, \dots, n$), where n is the number of days to maturity. Set

$$X(i) = \ln\left(\frac{S_d(i)}{S_d(i-1)}\right), \quad (i = 1, \dots, n). \tag{2.1}$$

The random variables $X(1), \dots, X(n)$ are independent and identically distributed with mean μ/N and variance σ^2/N. That is, $X(i) \sim \Phi(\frac{\mu}{N}, \frac{\sigma^2}{N})$. A straightforward calculation using (2.1) shows that

$$S_d(i) = S_d(i-1)e^{X(i)} \tag{2.2}$$

$$= S(0)e^{X(1)+\cdots+X(i)} \tag{2.3}$$

Values of $X(1), \dots, X(n)$ can be generated by a computer and Eq. (2.2)) or (2.3) can be used to sample a random path for the price S of the security. The valuation simulation for an asian option whose payoff Y is as in Eq. (1.2) can thus be implemented by generating for each simulation run j the corresponding payoff $Y_j = e^{-rn/N}\left(\sum \frac{S_d(i)}{n} - K\right)^{+}$ of the derivative. The next step then consists in computing the average mean $\bar{Y} = \frac{1}{R}\sum_1^R Y_j$, where R is the total number of simulation runs. The value of $\bar{Y}$ represents an estimate of the exact cost Y of the derivative. We have

$$\mathrm{E}[\bar{Y}] = \mathrm{E}[Y] \tag{2.4}$$

and

$$\mathrm{Var}(\bar{Y}) = \mathrm{Var}(Y)/R. \tag{2.5}$$

The last equality, Eq. (2.5), shows that the accuracy of $\bar{Y}$ improves with the number of simulations.

The simulation procedure is similar for lookback and for most of the path-dependent derivatives.

The drawbacks with Monte Carlo simulation is that the amount of time necessary to achieve a reasonable accuracy can be unacceptably high. As remedial measures, a number of variance reduction techniques are available, and they can lead to dramatic savings in computation time. These include the antithetic variable technique, the importance sampling, the stratified sampling, and the control variate technique. All these procedures and many others are described by Hull [6], Clewlow and Strickland [7], and Boyle [8].

2.1. Control variates

Suppose that the payoff Y from an option on a given security is of any of the forms given by the Eqs. (1.1)-(1.3). With the notation of the previous section, set

$$V = \sum_{i=1}^{n} \alpha_i X(i) \tag{2.6}$$

where α_i are some weights to be determined for the control variate V, and take

$$W = Y + c(V - \mathrm{E}(V)) \tag{2.7}$$

as the new estimator for Y. We have

$$\mathrm{E}[V] = \sum_{i=1}^{n} \alpha_i \,\mathrm{E}[X(i)] = \sum_{i=1}^{n} \frac{\alpha_i}{N}\left(r - \frac{\sigma^2}{2}\right),$$

so that

$$W = Y + \sum_{i=1}^{n} \beta_i \left[X(i) - (r - \frac{\sigma^2)}{2}/N\right], \qquad \text{where } \beta_i = c\,\alpha_i. \tag{2.8}$$

On the other hand, we have

$$\mathrm{Var}(W) = \mathrm{Var}(Y) + \sum_{i=1}^{n} \beta_i^2 \,\mathrm{Var}(X(i)) + 2\sum_{i=1}^{n} \beta_i \,\mathrm{Cov}(Y, X(i))$$

and it readily follows that the values of β_i that minimize W are given by

$$\beta_i^* = -\frac{\mathrm{Cov}(Y, X(i))}{\mathrm{Var}(X(i))} \tag{2.9}$$

In terms of these optimal values of the β_i, the smallest possible value for $\mathrm{Var}(W)$ is given after simplification by

$$\frac{\mathrm{Var}(W)}{\mathrm{Var}(Y)} = 1 - \sum_{i} \mathrm{Corr}^2(Y, X(i)) \tag{2.10}$$

We have thus proven the following result

Theorem 1. *Let $V = \sum_i \alpha_i X(i)$, for some arbitrary weights α_i and for $i = 1, \ldots, n$. Let $W = Y + c(V - \mathrm{E}[V])$ be an estimator of Y for some constant c.*

(a) *The optimal variance reduction is achieved with this control variable V for the values of α_i and c such that*

$$c\alpha_i = -\frac{\mathrm{Cov}(Y, X(i))}{\mathrm{Var}(X(i))}, \qquad (i = 1, \ldots, n)$$

(b) *The optimal variance reduction thus obtained is given by* (2.10).

Remark .

(i) It follows from part (a) of the theorem that a control variate of the form $V = \sum_i \alpha_i X(i)$ gives rise to an optimal reduction if and only if $\alpha_i = -\lambda \frac{cov(Y, X(i))}{\mathrm{Var}(X(i))}$ for all $i = 1, \ldots, n$ and for a nonzero constant λ.

(ii) This theorem implies that for all $\alpha_1, \ldots, \alpha_n$ and for every constant c

$$\frac{\mathrm{Var}(W)}{\mathrm{Var}(Y)} \geq 1 - \sum \mathrm{Corr}^2(Y, X(i))$$

where

$$W = Y + c\left[\sum_i \alpha_i X(i) - \alpha_i(r - \sigma^2/2)\right]$$

Ross [9] considers the problem of finding the values of α_i (for $i = 1, \ldots, n$) for the best control variate of the form $V = \sum_i \alpha_i X(i)$, and alternatively, the problem of finding the values of $c_1, \ldots, c_n$ for the best estimator of Y of the form $Y + \sum_{i=1}^n c_i \left(X(i) - \frac{r - \sigma^2/2}{N}\right)$. Theorem 1 gives an answer to this question, by determining the values of α_i and c_i and by indicating precisely the optimal variance reduction that can be achieved.

Theorem 2. *Let Y be a random variable and let $X_1, \ldots, X_n$ be n independent random variables. Then for all numbers $\alpha_1, \ldots, \alpha_n$*

$$\mathrm{Corr}^2\left(Y, \sum_i \alpha_i X_i\right) \leq \sum_i \mathrm{Corr}^2(Y, X_i).$$

Proof. Let $V = \sum_i \alpha_i X_i$ and let $W = Y + c(V - \mathrm{E}[V])$ be an estimator of Y for some constant c. For any fixed values of the α_i the optimal variance reduction can be shown similarly to (2.10) to be given by

$$\frac{\mathrm{Var}(W)}{\mathrm{Var}(Y)} = 1 - \mathrm{Corr}^2(Y, V)$$

This is not smaller than the optimal variance reduction obtained for all possible values of the α_i and c given by Eq. (2.10). Consequently, $1 - \sum_i \mathrm{Corr}^2(Y, X_i) \leq 1 - \mathrm{Corr}^2(Y, \sum_i \alpha_i X_i)$; that is, $\mathrm{Corr}^2(Y, \sum_i \alpha_i X_i) \leq \sum_i \mathrm{Corr}^2(Y, X_i)$. □

The result stipulated by this theorem is certainly very important, but it is likely to be unknown. Indeed in statistics books that present the most comprehensive material on the topic of correlation coefficient between random variables, there is rarely any discussion of a relationship between functions of correlations (see [10] and [11]).

2.2. Applications

(i) If $V = \sum_i \alpha_i X_i$ where the α_i are constants independent of X_i, then no estimator of Y of the form $W = Y + \sum_i \alpha_i X_i - \mathrm{E}[V]$ using V as control variable can lead to an optimal variance reduction, since by Theorem 1, the α_i must depend on X_i.

(ii) Let $V = \ln(s_d(n)/S(0))$. By the definition of the $X(i)$, it readily follows that $V = \sum_i X(i)$. Thus by the preceding remark, taking V as control variable cannot give rise to the best estimator.

Journal of Physics: Conference Series **410** (2013) 012111 doi:10.1088/1742-6596/410/1/012111

3. Conclusion

A generalization of this type of inequalities between functions of correlation coefficients is desirable, to compare for instance an estimator obtained with a control variable of the form $V = \sum_i \beta_i S_d(i)$, with that obtained using a control variable of the form $\sum_i \alpha_i X(i)$. In this instance, one would need a general relationship between functions of correlation coefficients of the type $\mathrm{Corr}^2(Y, \sum_i \beta_i S_d(i))$ and $\mathrm{Corr}^2(Y, \sum_i \beta_i \ln(S_d(i)/S_d(i-1)))$. As stated in this paper, no result on such type of relationship seems to be available in the literature. Finally, the results obtained in this paper clearly apply to any random variable for which variance reduction is required, and not only to the specific case of exotic options considered in this paper.

References

[1] Duan J, Dudley E 2003 *J. Derivatives* **10** 9
[2] Broadie M, Glasserman P and Kou S 1997 *Math. Finance* **7** 325
[3] Barraquand J and Martineau D 1995 *J Financ Quant Anal* **30** 383
[4] Ndogmo J C and Ntwiga D B 2011 *Appl. Math. Comput.* **218** 2210
[5] Green R, Fusai G, and Abrahams I D 2010 *Math. Finance* **20** 259
[6] Hull J C 2000 *Options, Futures, & Other derivatives, Fourth edition* (New Jersey: Prentice Hall)
[7] Clewlow L and Strickland C 1998 *Implementing derivatives models* (London: John Wiley and Sons)
[8] Boyle P P 1997 *Journal of financial economics* **4** 323
[9] Ross S M 1999 *An Introduction to Mathematical Finance* (Cambridge: Cambridge University Press)
[10] Ostle B 1972 *Statistics in research: Basic concepts and techniques for research workers, 2nd Edition* (Iowa: Iowa State University Press)
[11] Rees D G 2001 *Essential Statistics, Fouth Edition* (Boca Raton: Chapman & Hall/CRC)

Comparative analysis of the machine repair Problem with imperfect coverage and service pressure condition

Kuo-Hsiung Wang[a,*], Cheng-Dar Liou[b] and Yu-Hsueh Lin[c]

[a]Department of Computer Science and Information Management, Providence University, Taichung 43301, Taiwan
[b]Department of Business Administration, National Formosa University, Huwei, Yunlin, 63201, Taiwan
[c]Department of Applied Mathematics, National Chung-Hsing University, Taichung 402, Taiwan
khwang@pu.edu.tw

Abstract. We analyze the warm-standby M/M/R machine repair problem with multiple imperfect coverage which involving the service pressure condition. When an operating machine (or warm standby) fails, it may be immediately detected, located, and replaced with a coverage probability c by a standby if one is available. A recursive method is used to develop the steady-state analytic solutions. The total expected profit function per unit time is derived to determine the joint optimal values at the maximum profit. We utilize the direct search method to measure the various characteristics of the profit function followed by Quasi-Newton method to search the optimal solutions.

1. Introduction

This paper studies the warm-standby M/M/R machine repair problem (MRP) with multiple imperfect coverage which involving the service pressure condition. The faults that switch a failed machine in a spare machine are called to be not covered. The coverage factor c includes the probabilities of successful detection, location, and recovery (see Trivedi [1]). The service pressure coefficient is a positive constant and indicates degree to which the servers increase the service rate in order to reduce the number of failed machines in the system.

The concept of coverage factor in a repairable system has been introduced by several researchers such as Dugan and Trivedi [2], Trivedi [1], Wang and Chiu [3], etc. Dugan and Trivedi [2] considered the coverage modeling for dependability analysis of fault-tolerant systems. Wang and Chiu [3] considered the cost benefit analysis of availability systems with warm standby units and imperfect coverage for three availability models. Profit and cost analysis in the MRP have been investigated by a variety of authors including Sivazlian and Wang [4], Wang, et al. [5], and so on. The Quasi-Newton method (QNM) has been used by many researchers such as Wang, et al. [5], Wang, et al. [6] and so on. Wang, et al. [7] considered the M/M/1 MRP with working vacation. They constructed a cost model to derive the optimal number of operating machines by the direct search method (DSM) and find the optimal values of two different repair rates by the QNM. Wang, et al. [6] considered the profit analysis of the M/M/R MRP with balking, reneging, and standby switching failures. They used the two methods DSM and steepest descent method to find the maximum profit until the availability, balking, and reneging constraints are satisfied.

The purpose of this paper is two-fold. We first use a recursive method for developing steady-state analytic solutions. Next, we develop a profit model to determine the joint optimal values at the maximum profit. We use the DSM to find the optimal number of servers R and the optimal number of warm standby S, and then apply the QNM to obtain the optimal service rate μ^* after R^* and S^* are fixed.

2. Problem statement

We consider a machine repair model with $N=M+S$ identical machines and R servers in the repair facility. As many as M of these can be operating simultaneously in parallel, the rest of the S machines are warm-standby spares. Each of the operating machines fails independently of the state of the others and has an exponential time-to-failure distribution with parameter λ.Each of the available warm standbys fails independently of the state of all the others and has an exponential time-to-failure distribution with parameter α $(0 < \alpha < \lambda)$.When an operating machine (or warm standby) fails, it may be immediately detected, located, and replaced with a coverage probability c by a standby if one is available. The coverage factor is the same for an operating machine failure as that for a warm standby failure and is denoted by c. We define the unsafe failure state of the system as any one of the breakdowns is not covered. Operating machine failure (or warm standby failure) in the unsafe failure state is cleared by a reboot. Reboot delay occurs at rate β for an operating machine (or warm standby) which is exponentially distributed. Each of the servers has an exponential time-to-repair distribution with parameter μ until the number of failed machines n is greater than or equal to the number of servers R in the system, and the mean service rate μ_n is define as $\mu_n = n\mu$, where $1 \le n \le R-1$. When the number of failed machines n is greater than or equal to the number of servers R, the conditions of service pressure coefficient will be considered, and the mean service rate μ_n is defined as $\mu_n = R\mu\left[n(R+1)/R(n+1)\right]^{\theta}$, where $R \le n \le N$, and θ is the pressure coefficient. When an operating machine (or warm standby) fails, it is immediately sent to a repair facility where it is repaired in the order of their breakdowns; that is, the first-come, first-served discipline. The server can repair only one operating machine (or warm standby) at a time.

3. Steady-state results

We obtain the steady-state solutions given by

$$P_n = \frac{\prod_{j=0}^{n-1}\left[M\theta_\lambda + (S-j)\theta_\alpha\right]}{n!} P_0, \quad 1 \le n \le R-1 \tag{1}$$

$$P_n = \frac{\prod_{k=R}^{n}\left[\frac{k(R+1)}{R(k+1)}\right]^{-\theta} \prod_{j=0}^{n-1}\left[M\theta_\lambda + (S-j)\theta_\alpha\right]}{R!R^{n-R}} P_0, \quad R \le n \le S-1 \tag{2}$$

$$P_n = \frac{M!\theta_\lambda^{n-S} \prod_{k=R}^{n}\left[\frac{k(R+1)}{R(k+1)}\right]^{-\theta} \prod_{j=0}^{S-1}\left[M\theta_\lambda + (S-j)\theta_\alpha\right]}{(N-n)!R!R^{n-R}} P_0, \quad S \le n \le N \tag{3}$$

$$P_{uf_1} = \frac{(M\lambda + S\alpha)(1-c)}{\beta} P_0 \tag{4}$$

$$P_{uf_n} = \frac{\left[M\lambda + (S-(n-1))\alpha\right](1-c)\prod_{j=0}^{n-2}\left[M\theta_\lambda + (S-j)\theta_\alpha\right]}{\beta(n-1)!} P_0, \quad 2 \le n \le R-1 \tag{5}$$

$$P_{uf_n} = \frac{\left[M\lambda + \left(S-(n-1)\right)\alpha\right](1-c)\prod_{k=R}^{n-1}\left[\frac{k(R+1)}{R(k+1)}\right]^{-\theta}\prod_{j=0}^{n-2}\left[M\theta_\lambda + (S-j)\theta_\alpha\right]}{\beta R!R^{(n-1)-R}} P_0,$$

$$R \le n \le S-1 \tag{6}$$

$$P_{uf_n} = \frac{\left[M+S-(n-1)\right]\lambda(1-c)M!\theta_\lambda^{(n-1)-S}\prod_{k=R}^{n-1}\left[\frac{k(R+1)}{R(k+1)}\right]^{-\theta}\prod_{j=0}^{S}\left[M\theta_\lambda + (S-j)\theta_\alpha\right]}{\beta\left(N-(n-1)\right)!R!R^{(n-1)-R}} P_0,$$

$$S \le n \le N-1 \tag{7}$$

where $\theta_\lambda = \frac{\lambda}{\mu}$ and $\theta_\alpha = \frac{\alpha}{\mu}$.

The steady-state solutions P_n and P_{uf_n} always exist because the number of states is finite. For (1)-(7), P_0 can be solved from the following normalizing equation

$$\sum_{n=0}^{N} P_n + \sum_{n=1}^{N-1} P_{uf_n} = 1$$

4. Numerical results

Let $E[O]$ be the expected number of operating machines in the system, $E[S]$ be the expected number of spare machines in the system, $E[I]$ be the expected number of idle servers in the system, $E[B]$ be the expected number of busy servers in the system, $M.A.$ be machine availability, and $O.U.$ be operative utilization fraction. Let p represent revenue per unit time when one machine is in an operating state, C_1 be cost per unit time when one machine is in an operating state, C_2 be cost per unit time when one machine is functioning as a warm standby, C_3 be cost per unit time when one server is busy, C_4 be cost per unit when one server is idle, and C_5 denote cost per unit time of providing the service rate μ. Then the total expected profit function is given by

$$F(R,S,\mu) = (p - C_1)E[O] - C_2E[S] - C_3E[B] - C_4E[I] - C_5\mu. \tag{8}$$

The profit and cost parameters in (8) are assumed to be linear in the expected number of the indicated quantity. We first use DSM to find the optimal value of the number of servers R^* and warm standbys S^*, followed by QNM to search the optimal value of μ^*.

The profit maximization problem can be presented mathematically as

$$Z^* = F(R^*, S^*, \mu^*) = \underset{R,S,\mu}{Maximize}\ F(R,S,\mu).$$

Since R and S are discrete variables, we use direct substitution of successive values of R and S into the profit function until the maximum value of Z^*, say $F(R^*, S^*, \mu^*)$ is obtained. The following numerical results are provided by considering the revenue and cost parameters as follows:

$$p = \$300,\ C_1 = \$175, C_2 = \$30, C_3 = \$100, C_4 = \$60, C_5 = \$50.$$

We fix M=15, vary R from 1 to 10 and vary t S from 1 to 15.

Table 1. System performance measures of the warm-standby M/M/R MRP with multiple imperfect coverage and service pressure coefficient (α =0.05, μ =3.0, c=0.7, β =2.4).

(λ, θ)	(1.0, 0.8)	(2.0, 0.8)	(3.0, 0.8)	(1.0, 0.2)	(1.0, 0.4)	(1.0, 0.6)
(R^*, S^*)	(5, 8)	(9, 15)	(10, 15)	(5, 8)	(5, 8)	(5, 8)
Z^*	1111.12	776.82	436.52	1088.86	1096.92	1104.32

Table 2. QNM and PSO algorithm in searching the global best solution (α = 0.05, β = 2.4)

(λ, θ, c)	QNM					QNM			
	R^*	S^*	μ^*	Z_{QNM}^*		R^*	S^*	μ^*	Z_{QNM}^*
(1.0,0.6,0.7)	5	8	3.4850	1123.53	(2.0,0.8,0.7)	6	12	4.8788	936.85
(1.0,0.6,0.9)	5	9	3.4179	1110.15	(2.0,0.8,0.9)	6	12	4.9821	909.19
(1.0,0.8,0.7)	5	8	3.4140	1126.49	(3.0,0.6,0.7)	8	11	5.8885	782.92
(1.0,0.8,0.9)	5	9	3.3690	1103.31	(3.0,0.6,0.9)	8	11	6.0440	747.18
(2.0,0.6,0.7)	6	12	4.9430	932.08	(3.0,0.8,0.7)	7	15	5.9305	795.07
(2.0,0.6,0.9)	6	12	5.0484	904.54	(3.0,0.8,0.9)	7	15	6.0282	760.44

5. Conclusions

In this paper, we studied the warm-standby M/M/R MRP with multiple imperfect coverage and service pressure coefficient. We first used state-transition-rate diagram to set up the steady-state equations. Next, we used recursive method to develop the steady-state analytical solutions which are used to calculate various system performance measures, such as the expected number of failed machines, the expected number of idle servers, machine availability and operative utilization, etc. Finally, we derived the expected profit function per unit time to determine the optimal number of servers R^*, the optimal number of warm standbys S^*, and the optimal service rate μ^* at the maximum profit by using QNM.

References

[1] K.S. Trivedi, Probability and Statistics with Reliability, Queuing and Computer Science Applications, 2nd ed., John Wiley and Sons, New York, 2002.

[2] J.B. Dugan, K.S. Trivedi, Coverage modeling for dependability analysis of fault-tolerant systems, IEEE Trans. on Comput. 38 (1989) 775-787.

[3] K.H. Wang, L.W. Chiu, Cost benefit analysis of availability systems with warm standby units and imperfect coverage, Appl. Math. and Comput. 172 (2006) 1239-1256.

[4] B.D. Sivazlian, K.-H. Wang, Economic analysis of the M/M/R machine repair problem with warm standbys, Microelectron Reliab. 29 (1989) 25-35.

[5] K.H. Wang, W.L. Chen, D.Y. Yang, Optimal management of the machine repair problem with working vacation: Newton's method, J. Comput. Appl. Math. 233 (2009) 449-458.

[6] K.H. Wang, J.B. Ke, J.C. Ke, Profit analysis of the M/M/R machine repair problem with balking, reneging, and standby switching failures, Comput. Oper. Res. 34 (2007) 835-847.

Statistical techniques for modeling extreme price dynamics in the energy market

L N Mbugua[1] and P N Mwita[1]

[1]Jomo Kenyatta University of Agriculture and Technology, Department of Statistics and Actuarial Sciences, P.O Box 62000-00200, Nairobi, Kenya

Email: mbugualevi@yahoo.co.uk

Key words: Extreme value theory, Peak over threshold, Quantile regression, Value at risk

Abstract. Extreme events have large impact throughout the span of engineering, science and economics. This is because extreme events often lead to failure and losses due to the nature unobservable of extra ordinary occurrences. In this context this paper focuses on appropriate statistical methods relating to a combination of quantile regression approach and extreme value theory to model the excesses. This plays a vital role in risk management. Locally, nonparametric quantile regression is used, a method that is flexible and best suited when one knows little about the functional forms of the object being estimated. The conditions are derived in order to estimate the extreme value distribution function. The threshold model of extreme values is used to circumvent the lack of adequate observation problem at the tail of the distribution function. The application of a selection of these techniques is demonstrated on the volatile fuel market. The results indicate that the method used can extract maximum possible reliable information from the data. The key attraction of this method is that it offers a set of ready made approaches to the most difficult problem of risk modeling

1. Introduction

The extreme rate of change of production factors and their corresponding factor payments expose both the producers and consumers to significant risks, which they can hedge provided that they have good modeling tools of which they can forecast these changes well enough [1]. One major determinant that affects production capacity and the market mechanism structure is the cost of energy [2]. There exists a wide range of mathematical approaches to modeling and forecasting costs as well as price dynamics, but often do perform poorly in case of extreme events [3]. In this paper we focus on regression quantiles to capture the complete underlying structure of a time series data and model the observations that exceed central limit, focusing on the tail of the distribution. We apply Extreme Value Theory (EVT) to capture extremism [4]. Value at risk, the minimal loss which may occur under extra ordinary market circumstances is also discussed.

2. Empirical quantiles

When data has non-constant mean and variance, typically skewed and contains some outliers, it creates difficulties in empirical modelling [5]. More of interest is where the data pattern shows heteroscedasticity and asymmetries. In such data, quantile regression an order statistic is more explicable and robust than the mean regression [6].

2.1 Definition

The qth quantile estimate of the distribution function F that is $\hat{F}_k^{-1}(q)$ is defined as

$$(q) = inf\{x|\hat{F}_k(x) \geq q\},\ 0 < q < 1 \quad (1)$$

where, $\hat{F}_k$ is the empirical distribution function that put mass $1/k$ at each $X_i, 1 \leq i \leq k$ that is

$$\hat{F}_k(x) = \frac{1}{k}\sum_{i=1}^{k} I_{\{X_i \leq x\}},\quad -\infty < x < \infty$$

with I being the indicator function and inf denotes the smallest real number x satisfying $F_k(x) \geq q$. If a random variable $K(k)$ of $F_k(x)$ is continuous, then $q = Pr[K(k) \leq F_k^{-1}]$.

Applying the Rosenblatt-Parzen kernel estimators of the density function f we get

$$\hat{f}(x) = \frac{1}{k}\sum_{i=1}^{k}\frac{1}{h}w\left(\frac{x-X_i}{h}\right),\quad -\infty < x < \infty \quad (2)$$

where w is a kernel which is non negative and that $\int_{-\infty}^{\infty} w(x)dx = 1$. h, is a positive integer which governs the smoothness of the fluctuations. A relatively large value of h gives too much smoothness and a relatively small value of h give big fluctuations. Consequently, with $W(x) = \int_{-\infty}^{x} w(y)dy$ the distribution function corresponding to the density w is given by

$$\hat{F}_w(x) = \frac{1}{k}\sum_{i=1}^{k} W\left(\frac{x-X_i}{h}\right),\quad -\infty < x < \infty \quad (3)$$

which is a kernel estimator of the distribution function F. A smooth alternative to equation (1) may be defined by

$$\hat{F}_W^{-1}(q) = inf\left\{x\left|\frac{1}{k}\sum_{i=1}^{k} W\left(\frac{x-X_i}{h}\right) \geq q\right.\right\},\quad 0 < q < 1. \quad (4)$$

Let the observations Y_t be as a realization of the model;

$$Y_t = \hat{q}_\theta^t + \hat{\epsilon}_\theta^t \quad (5)$$

where $\hat{q}_\theta^t$ is an estimate of the quantile defined in equation (4) and $\hat{\epsilon}_\theta^t = Y_t - \hat{q}_\theta^t$ are the standardized quantile residuals. It turns out that

$$\frac{\hat{\epsilon}_\theta^t}{\hat{q}_\theta^t} = \frac{Y_t - \hat{q}_\theta^t}{\hat{q}_\theta^t} = \frac{Y_t}{\hat{q}_\theta^t} - 1 \quad (6)$$

θ, corresponds to a reasonably low probability for which the quantile can be estimated nonparametrically. Let p denote a very small probability of interest, reformulating the definition of the $p-th$ of the variable rate of change in terms of the $\theta - th$ quantile yields

$$P[Y_t < q_p^t] = P[Y_t < q_\theta^t - q_\theta^t + q_p^t] \quad (7)$$

Assuming that q_p^t is a negative number, we switch the inequality sign to obtain

$$P\left[\frac{Y_t}{q_\theta^t} - 1 > \frac{q_p^t}{q_\theta^t} - 1\right] = \theta \quad (8)$$

Let $\frac{q_p^t}{q_\theta^t} - 1 \equiv z_p$ denote the $p - th$ quantile of the standardized residuals. Then

$$\hat{q}_p^t = (\hat{z}_p + 1)\hat{q}_\theta^t\,. \quad (9)$$

Since data sparseness is more severe in extreme quantiles, we embark upon refining nonparametric quantile regression methods with extreme value theory so as to be proficient to model extreme quantiles accurately. To obtain $\hat{z}_p$ in equation (9) we apply the idea of the standard peaks over threshold (POT) methods [7].

3. Modeling Extreme Changes

The distribution function of the excesses over the threshold u can be defined by

$$F_u(x) = Pr\{X - u \leq x|X > u\} = \frac{F(x+u) - F(u)}{1 - F(u)} \quad (10)$$

Theorem: For a large class of underlying distribution function F, the conditional excess distribution function $F_u(x)$, for u large is well approximated by Pickands and Balkema [8], that is

$$\lim_{u\to x_0} \sup_{0\le x\le x_0-\mu}\left|F_u(x) - H_{\xi,\sigma}(x)\right| = 0 \tag{11}$$

where

$$H_{\xi,\sigma}(x) = \begin{cases} 1 - \left(1+\xi\frac{x}{\sigma}\right)^{-1/\xi} & \xi \neq 0 \\ 1 - e^{(-x/\sigma)} & \xi = 0 \end{cases} \tag{12}$$

ξ is the shape parameter and $\frac{1}{\xi}$ is the tail index. The tail index sets the rate at which the tails of $H_{\xi,\sigma}(x)$ decay away. $\sigma > 0$, is the scale parameter. The tail index as well as the scaling parameter has to be determined by fitting equation (12) to the actual data and estimate the parameters with the maximum likelihood method.

4. Peak over a high threshold (POT)

Denote the random variable of interest by r_t. Let u be a specified high threshold. Suppose that the $i-th$ exceedance at time t_i is $r_{ti} \le u$, we shall focus on the data $(t_i, r_{ti} - u)$ where $r_{ti} - u$, are the excesses over the threshold u. The occurrence times $\{t_i\}$ provides useful information about the intensity of the occurrence of important "rare events". A cluster of t_i indicates a period of large declines and the exceeding amount provides the actual quantity of interest. Different choices of the threshold u leads to different estimates of the tail index $\frac{1}{\xi}$ and is based on risk tolerance. Since different institutions and investors have different risk tolerance, the choice of u is a statistical problem as well as an institutional one. Too low u value and the asymptotic theory breaks down, too high threshold value and one does not have enough data points to estimate the parameters [9]. Rearranging equation (10) and using $F_u(\cdot) \approx H_{\xi,\sigma}(r_t)$, it holds that

$$1 - F(r_t + u) \approx [1 - F(u)][1 - H_{\xi,\sigma}(r_t)]$$

We estimate $1 - F(u) = z(u)$ by use of empirical distribution function and

$$z(u + r_t) = \frac{N_u}{n}\left(1 + \xi\left(\frac{r_t}{\beta}\right)\right)^{-\frac{1}{\xi}} \tag{13}$$

where, N_u denotes the number of excesses over the threshold u. Employing the change in variables $y = u + r_t$ and fixing the distribution value at the probability of interest, $F(y) = \theta$, we estimate the quantile estimator $\hat{q}_\theta$ by inverting equation (13) to obtain

$$1 - \theta = \frac{N_u}{n}\left(1 + \xi\left(\frac{y-u}{\beta}\right)\right)^{-\frac{1}{\xi}}.$$

$$\hat{q}_\theta = u + \frac{\beta}{\xi}\left[\left((1-\theta)\frac{n}{N_u}\right)^{-\xi} - 1\right] \tag{14}$$

Thus the θth quantile value at risk of $F(y)$ can be estimated by

$$VaR_\theta = u - \frac{\beta}{\xi}\left[1 - \left[\frac{n}{N_u}(1-\theta)\right]^{-\xi}\right] \tag{15}$$

Equation (15) is an analytical tool which can be used for assessing riskiness of trading activities therein, Value at risk (VaR). Consequently, the conditional expectation of loss given that the loss is beyond VaR level is the expected shortfall.

5. Results and discussion

We focus on the distribution of daily gasoline price changes in Kenya from July 1, 2006 to June 20, 2009 (data from Kenya National Oil Corporation). Since the price changes are so high we chose logarithmic changes instead of simple net returns. A problem of using simple net returns is that prices are bounded from below and that this makes the return distribution skewed for large positive and

negative returns. We focus only on the negative tail of the distribution even though the same concept applies to positive tails.
Table 1, reports some statistics on the price changes series. The high volatility is confirmed as evidenced by the very high excess kurtosis. The very small P-value of the Pormanteau test (Ljung) together with visual inspection of figure 1 indicates a high degree of volatility clustering effect.

Table 1 Descriptive statistics of daily gasoil price changes in Kenya.

Statistic	Value
Mean	-7.958×10^{-5}
Standard deviation	0.209
Skewness	0.289
Kurtosis	23.379
Ljung P-Value	2.20×10^{-16}

To get better tail estimates, we first filter the data to capture some of the most important dependencies and thereafter apply the ordinary extreme value technique. The advantage is that independent and identical assumption behind the EVT based tail quantile estimator is less likely to be violated.
To pre filter the time series we combine the Autoregressive moving average (ARMA) with the simplest generalized autoregressive conditional heteroscedastic (GARCH (1, 1)) model. Thus we have the following model

$$r_t = a_0 + a_1 r_{t-1} - b_1 e_{t-1} + e_t.$$

$$\sigma_t^2 = \varphi_0 + \varphi_1 e_{t-1}^2 + \varphi_2 \sigma_{t-1}^2 \qquad (16)$$

where σ_t^2 is the conditional variance of e_t and $e_t = \sigma_t \varepsilon_t$ with $\varepsilon_t \sim N(0,1)$. Let the unconditional EVT quantiles of the residual distribution be $\hat{q}_\theta$, the conditional tail quantiles of our original rate of change distribution will be given by

$$q_{t,\theta} = a_0 + a_1 r_{t-1} - b_1 e_{t-1} + \sigma_t \hat{q}_\theta \qquad (17)$$

In implementing POT method, we rely on a reasonable choice of the threshold that is approximately 5.5%. We produce a one step a head forecast of the conditional mean volatility using equation (16) and use these values to scale up our residual quantiles with equation (17). The summary statistics are shown in table 2. From table 3, with 5% probability, the daily rate of change of gasoil price could be as low as -0.624% and given that the rate of change is less than -0.624 the average rate of change value is -1.404%. Likewise, with 1% probability the daily rate of change of gasoil price could be as low as -1.3818%, and given this rate of change, the average rate of change value is -2.1870.

6. Conclusion

In this paper we have shown that quantile regression posses a more flexible appealing properties enough to capture the underlying complex dependence structure of a time series data. We have also illustrated how extreme value theory can be used to model tail related risk measures depending on data availability, frequency, desired time horizon and the level of complexity one is willing to introduce in the model. It can also be seen from the empirical study that this method can be used to determine the threshold by quantification.

7. Acknowledgements

We would like to thank The Kenya National Council for Science and Technology for their grant support towards this research. We would also like to thank The Kenya National Oil Corporation for enabling us access the data used for this study. Finally we wish to thank the reviewers and the editors for their comments and suggestion.

Table 2 ARMA-GARCH parameters, statistics of the standard residuals as well as POT parameters.

AR-GARCH parameters	
a_0	***
a_1	-0.0472 (0.0576)***
b_1	-0.6168 (0.0263)
$\varphi_0 \times 10^{-3}$	1.0576 (0.2733)
φ_1	0.1469 (0.0120)
φ_2	0.8758 (0.0090)
Standardized residuals descriptive statistics	
Mean	0.03256
Standard deviation	1.01179
Skewness	0.00122
Kurtosis	10.60099
Ljung P-value	0.97
POT parameters	
ξ	0.04635 (0.0393)
β	0.61109 (0.0493)
u	0.055

**** Not Significant parameter estimates. Figures in parenthesis are standard errors*

Table 3 Risk measures computed via Peak over threshold measure at 5%, 1% and 0.1% probability.

Probability	quantile	Shortfall
0.95	0.624	1.404
0.99	1.3818	2.1870
0.999	3.2574	4.7766

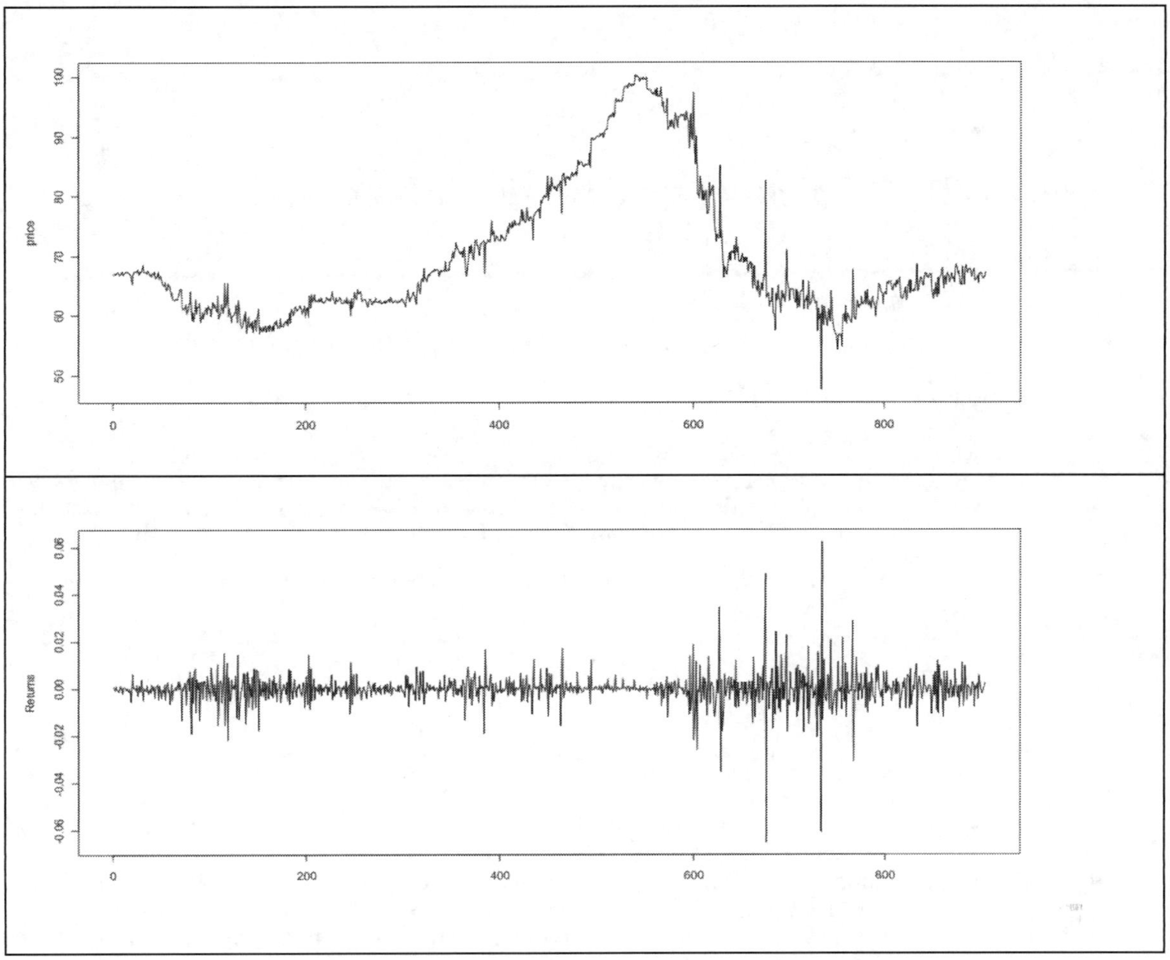

Figure1 Gasoil price changes (%) superimposed on the time series trend of gasoil daily prices.

8. References

[1] Byström H N E 2005 Extreme value theory and extremely large electricity price changes *Int. Rev. of Eco. Fin.* **14** (1) 41-55

[2] Sebitosi AB 2010 Is the South African electricity tariff model conducive to an energy efficient economy *En. for S. Dev. J.* **14** 315-319

[3] Bunn D W and Karakatsani N 2003 *Forecasting Electricity Prices* (London: London Business School)

[4] Danielsson J and deVries C G 1997 Tail index and quantile estimation with very high frequency data *J. Emp. Fin.* **4** 241-257

[5] Mbugua L, Mwita P and Mwalili S 2011 Application of nonparametric methods in studying energy consumption *Rwanda J.* **23** 7-20

[6] Koenker R W and Basset G W 1978 Regression quantiles *Econometrica* **46** 33-50

[7] Embrechts P, Klüppelberg C and Mikosch T 1997 *Modelling Extremal Events for Insurance and Finance* (Berlin: Springer)

[8] Pickands J 1975 Statistical inference using extreme order statistics *Ann. of stat.* **3** 119-131

[9] McNeil AJ and Frey R 2000 Estimation of tail-related risk measures for heteroscedastic financial time series an extreme approach *J. of Emp. Fin.* **7** 271-300

Measures of self-blocking system with infinite space

Jyh-Bin Ke and Yu-Li Tsai

National Chung-Hsing University, Applied Mathematics, Taichung, Taiwan

jbke@amath.nchu.edu.tw

Abstract. In this paper, we study an infinite series queueing system with self-blocking phenomenon. Poisson arrivals and exponential service times are assumed. We develop the structured generator matrix to compute steady-state probabilities of the self-blocking system with infinite space by matrix-geometric method. The stability condition of the system is obtained in closed-form. We also present some performance measures including mean number in the system, and blocking probability, etc. The characteristics of the system with different service order are discussed as well.

1. Introduction

Traditionally, most of the published studies of series queues deal with the type of problems in which customers must be served by a series of service stations. This kind of systems is known as "tandem queues". In this paper, we investigate a special self-blocking queueing system which consists of only two service stations, as shown in Figure 1. In this system, a customer is forced to enter station-2 directly if both service stations are available. A completed service is defined as when a customer is fully served at either station-1 or station-2. It is apparent that we can use this self-blocking queueing system to model the operations in a gas station or other similar queueing systems. This kind of configuration in the gas station is popular in the metropolitan area due to the land use restriction.

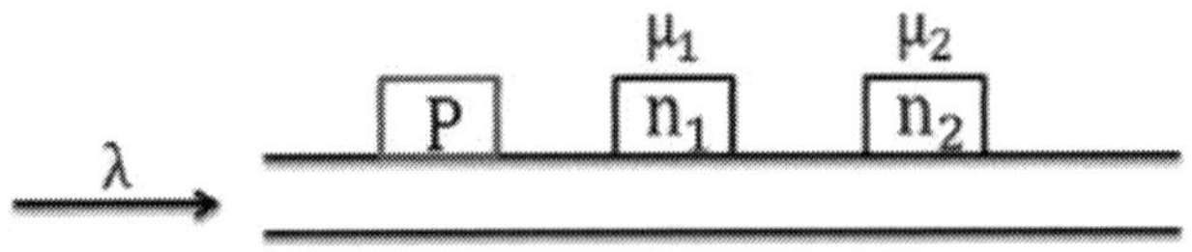

Figure 1. Scheme of self-block queueing system.

In the field of tandem queues, Hunt [1] first studied tandem queues with blocking phenomenon in sequential arrays of waiting lines with Poisson arrivals and exponential service times. Avi-Itzhak [2] provided an analysis about a two-stage tandem queue with blocking, arbitrary input and regular service times. The waiting time distributions and other characteristics of such queueing system can be obtained by some properties. Avi-Itzhak and Yadin [3] investigated a queueing system consisting of a sequence of two service stations with no intermediate queue. They derived the moment generating functions of the steady-state queueing times as well as the generating functions of the steady-state number of customers in the various parts of the system under assumptions of Poisson process of arrivals and arbitrarily distributed service times. Neuts [4] presented a series system with a finite intermediate waiting room. He demonstrated that this system can be studied in terms of an imbedded semi-Markov process.

Performance and approximation analysis of tandem queues can be found in the following works. Langaris and Conolly [5] conducted an analysis on the waiting time of a two-stage queueing system with blocking. The survey of closed queueing networks with blocking was provided by Onvural [6]. Gomez-Corral [7] focused on the embedded process at departures of a tandem queueing system with

Published under licence by IOP Publishing Ltd

blocking and Markovian arrival process. Gomez-Corral and Martos [8] studied the influence of several flows of signals on the performance evaluation of a two-stage tandem system with blocking.

In this paper, we investigate the performance measures of a system with two service units operating independently in series simultaneously. The phenomenon of self-blocking occurs at the scenario when the station-1 has completed the service, but the station-2 is still serving for the other customer. We study the impacts of the performance measures of the system by switching different service orders. Therefore, we can suggest a better disposition strategy for the self-blocking queueing system according to our research. In addition, we derive the stability condition of the system with infinite capacity.

2. Problem formulation and notations

The assumptions of the model are described as follows. Suppose that a queueing system which consists of two independent service units in series operating simultaneously, as shown in Figure 1. Suppose that each customer enters to the system in accordance with Poisson process with mean arrival rate λ. Each customer is forced to move to the station-2 if both service stations are available. The service is complete when a customer is fully served at either station-1 or station-2. This means that the customer can depart the system when the service is performed completely at either station. The main characteristic of the service process in this queueing system is self-blocking, which occurs at the scenario when the station-1 has completed the service, but the station-2 is still serving for the other customer. A queue with infinite capacity is allowed in front of station-1. We also assume that no waiting space exists at station-2. It is assumed that each service station can serve only one customer at a time and the service rate is independent of the number of customers. The service is performed on the first come first serve (FCFS) discipline. In addition, the time to serve a customer of station-1 and station-2 is exponentially distributed with mean service time $1/\mu_1$ and $1/\mu_2$, respectively.

λ Mean arrival rate of Poisson arrivals

μ_1 Mean service rate of the station-1

μ_2 Mean service rate of the station-2

P_{n_1,n_2,n_3}, The steady-state probability P_{n_1,n_2,n_3} of n_1 customers in the station-2 and n_2 customers in the station-1 and n_3 customers in the queue.

3. Steady-state equations and structured generator matrix

The quasi-birth-death process of the self-blocking queueing system with infinite capacity is shown in Figure 2. The infinite infinitesimal steady-state equations of the self-blocking queueing system are given by:

$$\lambda P_{0,0,0} = \mu_1 P_{0,1,0} + \mu_2 P_{1,0,0} + \mu_2 P_{1,b,0}, \quad (1)$$

$$(\lambda + \mu_2) P_{1,0,0} = \lambda P_{0,0,0} + \mu_1 P_{0,1,1} + \mu_2 P_{1,b,1}, \quad (2)$$

$$(\lambda + \mu_1 + \mu_2) P_{1,1,0} = \lambda P_{1,0,0} + \mu_1 P_{0,1,2} + \mu_2 P_{1,b,2}, \quad (3)$$

$$(\lambda + \mu_2) P_{1,b,0} = \mu_1 P_{1,1,0}, \quad (4)$$

$$(\lambda + \mu_1) P_{0,1,0} = \mu_2 P_{1,1,0}, \quad (5)$$

$$(\lambda + \mu_1 + \mu_2) P_{1,1,i+1} = \lambda P_{1,1,i} + \mu_1 P_{0,1,i+3} + \mu_2 P_{1,b,i+3}, \quad i \geq 0, \quad (6)$$

$$(\lambda+\mu_2)P_{1,b,i+1}=\lambda P_{1,b,i}+\mu_1 P_{1,1,i+1}, \qquad i\geq 0, \tag{7}$$

$$(\lambda+\mu_1)P_{0,1,i+1}=\lambda P_{0,1,i}+\mu_2 P_{1,1,i+1}, \qquad i\geq 0. \tag{8}$$

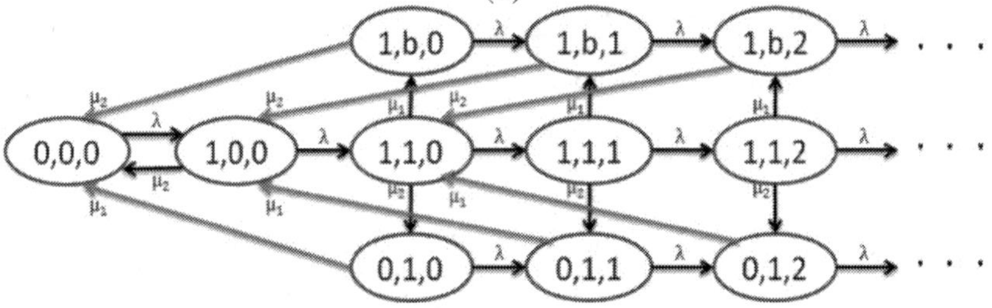

Figure 2. The quasi-birth-death process of the self-block system with infinite capacity.

We construct the structured generator matrix according to the quasi-birth-death process of the system. The structured generator matrix can be written in the following block form:

$$Q=\begin{bmatrix} B_{0,0} & B_{0,1} & 0 & 0 & 0 & 0 & L \\ B_{1,0} & A_1 & A_0 & 0 & 0 & 0 & L \\ B_{2,0} & A_2 & A_1 & A_0 & 0 & 0 & L \\ 0 & A_3 & A_2 & A_1 & A_0 & 0 & L \\ 0 & 0 & A_3 & A_2 & A_1 & A_0 & L \\ 0 & 0 & 0 & A_3 & A_2 & A_1 & L \\ M & M & M & M & M & M & O \end{bmatrix}_g \tag{9}$$

where

$$B_{0,0}=\begin{bmatrix} -\lambda & \lambda \\ \mu_2 & -(\lambda+\mu_2) \end{bmatrix}, \quad B_{1,0}=\begin{bmatrix} 0 & 0 \\ \mu_2 & 0 \\ \mu_1 & 0 \end{bmatrix}, \quad B_{2,0}=\begin{bmatrix} 0 & 0 \\ 0 & \mu_2 \\ 0 & \mu_1 \end{bmatrix}, \quad B_{0,1}=\begin{bmatrix} 0 & 0 & 0 \\ \lambda & 0 & 0 \end{bmatrix}, \quad A_0=\begin{bmatrix} \lambda & 0 & 0 \\ 0 & \lambda & 0 \\ 0 & 0 & \lambda \end{bmatrix},$$

$$A_1=\begin{bmatrix} -(\lambda+\mu_1+\mu_2) & \mu_1 & \mu_2 \\ 0 & -(\lambda+\mu_2) & 0 \\ 0 & 0 & -(\lambda+\mu_1) \end{bmatrix}, \quad A_2=\begin{bmatrix} 0 & 0 & 0 \\ 0 & 0 & 0 \\ 0 & 0 & 0 \end{bmatrix}, \text{ and } A_3=\begin{bmatrix} 0 & 0 & 0 \\ \mu_2 & 0 & 0 \\ \mu_1 & 0 & 0 \end{bmatrix}_g$$

3.1. Matrix-geometric method and stability conditions

Let us define $\mathbf{P}=[\mathbf{P}_0,\mathbf{P}_1,\mathbf{P}_2,...]$ as the steady-state probability vector corresponding to the structured generator matrix Q. Therefore, we can yield $\mathbf{P}$ by solving $\mathbf{P}Q=\mathbf{0}$, while obeying the normalization condition, $\mathbf{P1}=1$. The global balance equations of the quasi-birth-death process are considered:

$$\mathbf{P}_0 B_{0,0}+\mathbf{P}_1 B_{1,0}+\mathbf{P}_2 B_{2,0}=\mathbf{0}, \tag{10}$$

$$\mathbf{P}_0 B_{0,1}+\mathbf{P}_1 A_1+\mathbf{P}_3 A_3=\mathbf{0}, \tag{11}$$

$$\mathbf{P}_i A_0+\mathbf{P}_{i+1} A_1+\mathbf{P}_{i+3} A_3=\mathbf{0}, \qquad i\geq 1. \tag{12}$$

A constant rate matrix R can be introduced that leads to the matrix-geometric equation:

$$\mathbf{P}_i = \mathbf{P}_{i-1}R = \mathbf{P}_1 R^{i-1}, \qquad i > 1. \tag{13}$$

Eq.(12) can be simplified to

$$A_0 + RA_1 + R^3 A_3 = 0. \tag{14}$$

The normalization condition can be written as:

$$\sum_{i=0}^{\infty} \mathbf{P}_i \mathbf{1} = 1. \tag{15}$$

Using some derived results referred in Neuts [4], we can obtain an important normalization condition equation constrained to Eqs. (14) and (15):

$$\mathbf{P}_0 \mathbf{1} + \mathbf{P}_1 (I - R)^{-1} \mathbf{1} = 1. \tag{16}$$

We evaluate the rate matrix, R, iteratively by successive substitution. To obtain $\mathbf{P}_i$, Eqs. (10), (11) and (16) are taken into account

$$(\mathbf{P}_0, \mathbf{P}_1) \begin{pmatrix} B_{0,0} & B^*_{0,1} & \mathbf{1} \\ B_{1,0} + RB_{2,0} & (A_1 + R^2 A_3)^* & (I - R)^{-1} \mathbf{1} \end{pmatrix} = (\mathbf{0}, 1). \tag{17}$$

where $(.)^*$ indicates that the last column of the included matrix is removed to avoid linear dependency. The removed column is replaced by the normalization condition. Equation (17) is solved for evaluating $\mathbf{P}_0$ and $\mathbf{P}_1$.

In order to confirm whether the quasi-birth-death process is tractable by the matrix-geometric method, the stability of the queueing system must be verified by checking the stability condition of Neuts [4]:

$$\mathbf{P}_A A_0 \mathbf{1} < \mathbf{P}_A A_2 \mathbf{1} + 2(\mathbf{P}_A A_3 \mathbf{1}), \tag{18}$$

where $\mathbf{P}_A$ is the steady-state probability vector corresponding to the generator matrix A.
Thus, the stability conditions can be concluded as the following inequalities:

$$\rho < \frac{2}{1 + \frac{\mu_1}{\mu_2} + \frac{\mu_2}{\mu_1}} \overset{\text{for } \mu_1 = \mu_2}{=} \frac{2}{3} \tag{19}$$

where $\rho \equiv \frac{\lambda}{\mu_1 + \mu_2}$ g

3.2. Performance measures

Performance measures of the self-blocking queueing system with infinite capacity are defined as:

(1) Mean number of customers in the system

$$L = (P_{1,0,0} + P_{0,1,0} + P_{1,b,0}) + \sum_{n=2}^{\infty} (P_{1,b,n-1} + P_{1,1,n-2} + P_{0,1,n-1})\, n_g \tag{20}$$

(2) Mean waiting time in the system (Little's Law)

$$W = \frac{L}{\lambda}. \tag{21}$$

(3) Blocking probability of the customer in the station-1

$$P_b = \sum_{n=0}^{\infty} P_{1,b,n} \overset{\mu_1=\mu_2}{=} \frac{1}{3}\mathbf{P}_1(1-R)^{-1}\cdot\mathbf{1} \tag{22}$$

4. Numerical results

We perform numerical experiments in this section. We first present how stability conditions change as the ratio of the service rate of the station-1 to the service rate of the station-2 varies from 0.1 to 10, as shown in Figure 3. We observe that the order of service stations doesn't affect the stability condition which can be verified by the symmetric figure to the ratio $\mu_1/\mu_2 = 1$. It is noted that when the service rate of the station-1 is equal to the service rate of the station-2, the stability condition for the self-blocking queueing system must be less than $2/3$. Next, we inspect the trends of blocking probability and mean number of customers in the system as the mean arrival rate increases. Therefore, we fix the system parameters at $\mu_1 = 1$, $\mu_2 = 1$ and vary the values of ρ from 0.01 to 0.66, as shown in Figure 4. We observe that the mean number of customers in the system increases slowly for low values of ρ, then it increases very rapidly as ρ approaches to 0.66. The increasing trends of blocking probability as a function of ρ are also shown in Figure 4.

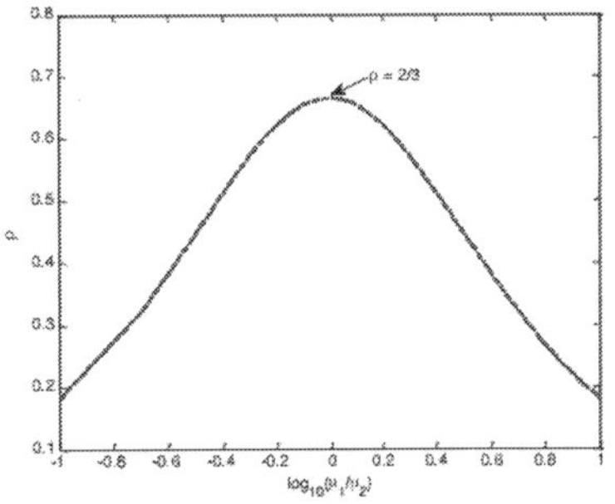

Figure 3. Stability condition curve of this queueing system.

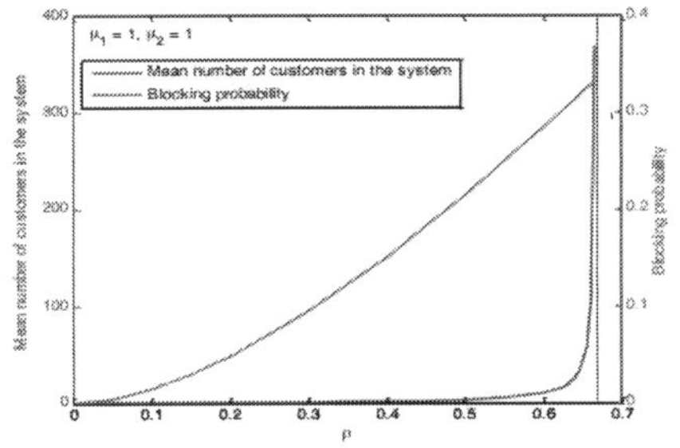

Figure 4. Mean number of customers in the system and blocking probability of this system

Next, we study the impact of different service orders that cause the distinct performances of the self-blocking queueing system. Comparisons with different service orders on mean waiting time in the system and mean service time of the system and blocking probability are given in the following special case: setting different service orders for $\mu_1 = 2,\ \mu_2 = 1$, $\mu_1 = 1,\ \mu_2 = 2$ and vary the values of ρ from 0.01 to 0.56.

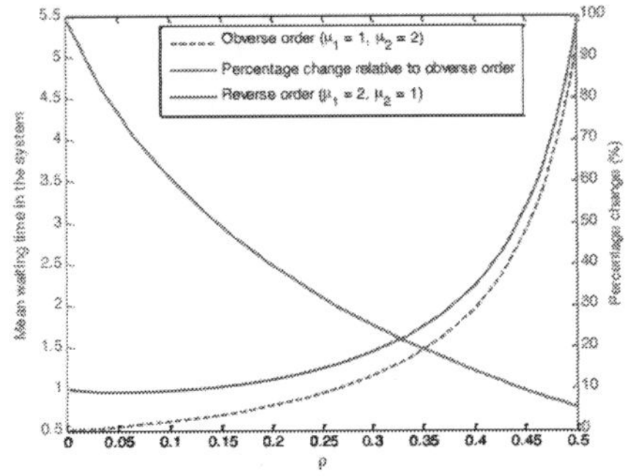

Figure 5. Mean waiting time in the system with different service orders.

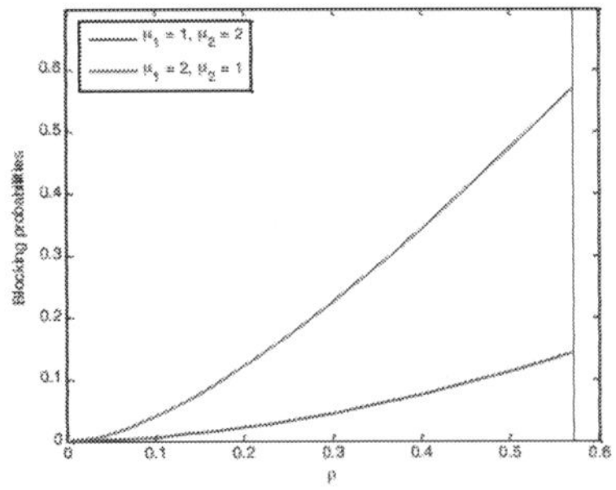

Figure 6. Blocking probability with different service orders.

Intuitively, if we set the higher service rate for the station-2, the performance on blocking probability of the self-blocking queueing system is going to be less than that of higher service rate of the station-1. We define service order called *obverse order*, if we dispose higher service rate for the station-2 (i.e. $\mu_2 > \mu_1$). Otherwise, if the service rate of the station-2 is lower than that of the station-1, it is called *reverse order* (i.e. $\mu_2 < \mu_1$).

We observed that the performances of the system placed in the obverse order on mean waiting time in the system, mean service time of the system and blocking probability are less than that of the system placed in the reverse order, as shown in Figure 5 ~ Figure 6, respectively. Percentage change relative to the obverse order on mean waiting time of the system is presented in the red line of Figure 5. From Figure 6 of the case we have studied above, it is noted that if we want to optimize the performances of the self-blocking queueing system, it is better to dispose the obverse order for the self-blocking system. We can increase the efficiency for the self-blocking system which means maintain lower performances on the mean number of customers in the system, mean waiting time in the system, mean service time of the system and blocking probability by disposing the service order with obverse way.

5. Conclusions

In this paper, we study a series type queueing system with self-blocking phenomenon. We have applied matrix-geometric method to evaluate steady-state probabilities of the system with infinite capacity. We have also derived stability conditions of the system for both service stations with same service rates or with different service rates.

Numerical results of performance measures of the self-blocking queueing system indicate that it is better to adjust the service station-2 with higher service rate than that of the service station-1 in order to effectively reduce the mean number of customers in the system, mean waiting time in the system, and blocking probability of the self-blocking queueing system.

References

[1] G. C. Hunt 1965 Sequential arrays of waiting lines. *Operations Research* **4** 674–83.
[2] B. Avi-Itzhak 1965 A sequence of service stations with arbitrary input and regular service times. *Management Science* **11** 565–71.
[3] B. Avi-Itzhak, M. Yadin 1965 A sequence of two servers with no intermediate queue. *Management Science* **11** 553–64.
[4] M. F. Neuts 1968 Two queues in series with a finite, intermediate waiting room. *Journal of Applied Probability* **5** 123–42.
[5] C. Langaris, B. Conolly 1984 On the waiting time of a two stage queueing system with blocking. *Journal of Applied Probability* **21**628-38.
[6] R. O. Onvural 1990 Survey of closed queueing networks with blocking. *ACM Computing Surveys* **22** 83-121.
[7] A. Gomez-Corral 2002 A tandem queue with blocking and Markovian arrival process. *Queueing Systems* **41** 343-70.
[8] A. Gomez-Corral, M.E. Martos 2006 Performance of two-stage tandem queues with blocking: The impact of several flows of signals. *Performance Evaluation* **63** 910-38.

Infinite capacity multi-server queue with second optional service channel

Jau-Chuan Ke[1], Chia-Huang Wu [2], and Wen Lea Pearn[2]

[1]Department of Applied Statistics National Taichung University of Science and Technology ,Taichung, Taiwan, ROC

jauchuan@nutc.edu.tw

[2]Department of Industrial Engineering and Management National Chiao Tung University, Taiwan, ROC

duckboy614583@gmail.com, wlpearn@mail.nctu.edu.tw

Abstract. This paper deals with an infinite-capacity multi-server queueing system with a second optional service (SOS) channel. The inter-arrival times of arriving customers, the service times of the first essential service (FES) and the SOS channel are all exponentially distributed. A customer may leave the system after the FES channel with probability (1-θ), or at the completion of the FES may immediately require a SOS with probability θ $(0 \leq \theta \leq 1)$. The formulae for computing the rate matrix and stationary probabilities are derived by means of a matrix analytical approach. A cost model is developed to determine the optimal values of the number of servers and the two service rates, simultaneously, at the minimal total expected cost per unit time. Quasi-Newton method are employed to deal with the optimization problem. Under optimal operating conditions, numerical results are provided in which several system performance measures are calculated based on assumed numerical values of the system parameters.

1. Introduction

There are R channels (servers) that provide the first essential service (FES) as well as the second optional service (SOS) to arriving customers. The service times of FES and the SOS are exponentially distributed with mean $1/\mu_1$ and$1/\mu_2$, respectively. As soon as the FES of a customer is completed, he may leave the system with probability (1-θ) or, opt for the SOS provided by the same server with probability θ $(0 \leq \theta \leq 1)$. Each channel can serve only one customer at a time and it also provides only one of FES or SOS at a time. It is assumed that customers arrive according to a Poisson process with parameter λ and FCFS discipline is adopted. Customers who upon entry into the channel facility, find that all channels are busy need to wait in the queue until a channel becomes available. The various stochastic (arrival or service) processes involved in the system are independent of each other.

In many real service systems, one encounters numerous examples of the queueing situations where all arrivals require the main service and only some may request the subsidiary service provided by the channel. Analytic steady-state solutions of an M/M/R queue with a second optional channel

IC-MSQUARE 2012: International Conference on Mathematical Modelling in Physical Sciences IOP Publishing
Journal of Physics: Conference Series **410** (2013) 012115 doi:10.1088/1742-6596/410/1/012115

have not been found. This motivates us to investigate an infinite capacity M/M/R queueing system with a SOS channel.

2. Markov chain model

For an infinite capacity M/M/R queueing system with second optional service (SOS) channel. The states of the system are described by the pair $(i, j), i = 0,1,2,\ldots$ and $j = 0,1,2,\ldots,R$, where i denotes the number of customers in the FES channel (including customers waiting in the queue), and j denotes the number of customers in the SOS channel. If $(i + j) \leq R$, i.e., there are available servers, the customers upon to the server will get service immediately. If $(i + j) > R$, i.e., all servers are busy, the new arriving customer must wait in the queue until a server becomes available. We define the following notations in steady-state:

> $P_{i,j} \equiv$ probability that there are i customers in the FES channel and there are j customers in the SOS channel, where $i = 0,1,2,\ldots$ and $j = 0,1,2,\ldots,R$.

The infinitesimal generator $\mathbf{Q}$ of the quasi birth-and-death (QBD) process describing the M/M/R queueing system with SOS channel is of the block-tridiagonal form (see Neuts (1981)):

$$
\mathbf{Q} = \begin{array}{c} \\ l(0) \\ l(1) \\ l(2) \\ \vdots \\ l(R-1) \\ l(R) \\ l(R+1) \\ \vdots \end{array}
\begin{array}{c}
\begin{array}{cccccccccc} l(0) & l(1) & l(2) & \cdots & \cdots & \cdots & l(R-2) & l(R-1) & l(R) & l(R+1)\ \cdots \end{array} \\
\left[\begin{array}{ccccccccc}
\mathbf{A}_0 & \mathbf{B} & & & & & & & \\
\mathbf{C}_1 & \mathbf{A}_1 & \mathbf{B} & & & & & & \\
 & \mathbf{C}_2 & \mathbf{A}_2 & \mathbf{B} & & & & & \\
 & & \ddots & \ddots & \ddots & & & & \\
 & & & & \mathbf{C}_{R-1} & \mathbf{A}_{R-1} & \mathbf{B} & & \\
 & & & & & \mathbf{C}_R & \mathbf{A}_R & \mathbf{B} & \\
 & & & & & & \mathbf{C}_R & \mathbf{A}_R & \mathbf{B} \\
 & & & & & & \ddots & \ddots & \ddots
\end{array}\right]
\end{array}. \tag{1}
$$

Each entry of the matrix $\mathbf{Q}$ is a square matrix of order $R+1$ listed as follows:

$$\mathbf{B} = \lambda \mathbf{I}, \tag{2}$$

$$
\mathbf{A}_i = \begin{bmatrix}
a_{i,0} & & & & \\
\mu_2 & a_{i,1} & & & \\
 & 2\mu_2 & a_{i,2} & & \\
 & & \ddots & \ddots & \\
 & & & R\mu_2 & a_{i,R}
\end{bmatrix}, \ i = 0,\ldots,R \tag{3}
$$

$$
\mathbf{C}_i = \begin{bmatrix}
c_{i,0} & d_{i,0} & & & & \\
 & c_{i,1} & d_{i,1} & & & \\
 & & c_{i,2} & d_{i,2} & & \\
 & & & \ddots & \ddots & \\
 & & & & c_{i,R-1} & d_{i,R-1} \\
 & & & & & 0
\end{bmatrix}, \ i = 1,\ldots,R \tag{4}
$$

where $\mathbf{I}$ is the identity matrix of order $R+1$, and

$$a_{i,j}=\begin{cases}-(\lambda+i\mu_1+j\mu_2), & 1\le i+j\le R,\\ -[\lambda+(R-j)\mu_1+j\mu_2], & i+j>R.\end{cases} \tag{5}$$

$$c_{i,j}=\begin{cases}i(1-\theta)\mu_1, & 1\le i+j\quad R,\ \le\\ (R-j)(1-\theta)\mu_1, & i+j>R.\end{cases} \tag{6}$$

$$d_{i,j}=\begin{cases}i\theta\mu_1, & 1\le i+j\le R,\\ (R-j)\theta\mu_1, & i+j>R.\end{cases} \tag{7}$$

Let $\mathbf{F}=\mathbf{C}_R+\mathbf{A}_R+\mathbf{B}$, and $\mathbf{X}=[x_0,x_1,\ldots,x_R]$ are the irreducible generator and the invariant vector corresponding to $\mathbf{F}$. Then $\mathbf{F}$ satisfies the linear equations

$$\mathbf{XF}=\mathbf{0} \text{ and } \mathbf{Xe}=1, \tag{8}$$

where $\mathbf{0}$ and $\mathbf{e}$ are column vector with dimension $R+1$ and all elements equal to zero and one, respectively. By Theorem 3.1.1 of Neuts (1981), the standard drift condition is

$$\mathbf{XBe}<\mathbf{XC_R e}, \tag{9}$$

which is the necessary and sufficient condition for stability of the QBD-$\mathbf{Q}$ process. Solving equation (8) to obtain $\mathbf{X}$ as

$$x_i=\binom{R}{i}\left(\frac{\theta\mu\theta\mu}{\mu_2}\right)^i\left(1+\frac{1}{\mu_2}\right)^{-R},\quad i=0,1,\ldots,R. \tag{10}$$

Substituting $\mathbf{B}$ and $\mathbf{C}_R$ into equation (9) yields

$$\mu_1(R-L_2)>\lambda, \tag{11}$$

where

$$\begin{aligned}L_2&=x_1+2x_2+\ldots+Rx_R\\ &=\sum_{i=1}^{R}ix_i=\sum_{i=1}^{R}i\binom{R}{i}\left(\frac{\theta\mu\theta}{\mu_2}\right)^i x_0=\frac{R\ \mu\theta}{\mu_2}x_0=\frac{R\ \mu\theta\mu}{\mu_2}\left(1+\frac{1}{\mu_2}\right)^{-R}\end{aligned} \tag{12}$$

denotes the expected number of customers in the SOS channel. It is note that if $\theta=0$ or $\mu_2\to\infty$(i.e., L_2=0), equation (11) could be reduced to the stability condition for the ordinary M/M/R queueing system without SOS channel.

Algorithm: Linear Progression Algorithm

INPUT $\mathbf{B}$, $\mathbf{A}_R$, $\mathbf{C}_R$, $\mathbf{e}=[1,\ldots,1]^{\mathrm{T}}$, $\mathbf{I}$ is the identity matrix, and the tolerance δ.

OUTPUT approximate solution $\mathbf{T}$

Step 1 $\mathbf{G}=(-\mathbf{A}_R)^{-1}\mathbf{C}_R$

Step 2 while $\|\mathbf{e}-\mathbf{Ge}\|\ge\delta$ do Steps 3-4

Step 3 set $\mathbf{H}=\mathbf{A}_R+\mathbf{BG}$

Step 4 and $\mathbf{G}=(-\mathbf{H})^{-1}\mathbf{C}_R$

Step 5 set $\mathbf{T} = \mathbf{B}(-\mathbf{H})^{-1}$

Step 6 OUTPUT

3. Probability computation

Algorithm : Recursive Solver

INPUT $\mathbf{R}$, $\mathbf{B}$, $\mathbf{A}_0$, $\mathbf{A}_1,\ldots,\mathbf{A}_R$, $\mathbf{C}_1,\mathbf{C}_2,\ldots,\mathbf{C}_R$, $\mathbf{T}$, and $\mathbf{H}$.

OUTPUT approximate solution $\mathbf{P}_0,\mathbf{P}_1,\mathbf{P}_2,\ldots$

Step 1 set $\phi_1 = \mathbf{C}_1(-\mathbf{A}_0)^{-1}$

Step 2 for $i = 2$ to R do Step 3

Step 3 set $\phi_i = \mathbf{C}_i[-(\phi_{i-1}\mathbf{B}\quad \mathbf{A}_{i-1})]^{-1}$

Step 4 end

Step 5 set $\phi_{R+1} = \mathbf{C}_R[-(\phi_R\mathbf{B}+\ \mathbf{A}_R)]^{-1}$

Step 6 for $k = 0$ to $R-1$ do Step 7

Step 7 set $\mathbf{\Phi}_k = \prod_{i=R}^{k+1} \phi_i$

Step 8 end

Step 9 Solving $\mathbf{P}_R\mathbf{T}[\phi_{R+1}\mathbf{B}+\mathbf{H}] = 0$, $\mathbf{P}_R[\sum_{k=1}^{R}\prod_{i=R}^{k}\phi_i + \mathbf{I} + \mathbf{T}(\mathbf{I}-\mathbf{T})^{-1}]\mathbf{e}\quad 1$

Step 10 for $i = 0$ to $R-1$ do Step 11

Step 11 set $\mathbf{P}_i = \mathbf{P}_R\mathbf{\Phi}_i$

Step 12 end

Step 13 for $i = R+1$ to infinity do Step 14

Step 14 set $\mathbf{P}_{i+1} = \mathbf{P}_i\mathbf{T}$

Step 15 end

Step 16 OUTPUT

References

[1] M. Neuts 1981, Matrix-Geometric Solutions in Stochastic Models, John Hopkins University Press, Baltimore.

Controlling arrival for the machine repair problem with switching failure

Fu-Min Chang, Jau-Chuan Ke and Cheng-Hwai Liou

Chaoyang University of Technology, Department of Finance, Taichung, Taiwan

fmchang@cyut.edu.tw, jauchuan@nutc.edu.tw, jhliou@nutc.edu.tw

Abstract. This study investigates the warm-standby machine repair problem which involves a controlling arrival policy and switching failure probability. It involves operating machines with S warm standbys and one single server. The failure times and repair times are assumed to follow an exponential distribution. For such system, some system performance measures are derived and a steady-state expected cost function per unit time is developed. By using Quasi-Newton method followed direct search method, we can find the joint optimal parameter values at maximum profit such that the availability constraint is satisfied.

1. Introduction

Controllable queueing models aims to find the optimal operating policy, which are rules for turning the server on and off that result in the lowest long-run cost. The pioneering work in the issue of controlling the arrivals, also called *F*-policy, was first investigated by Gupta [1]. The *F*-policy aims to control the arrival process when service control is not possible through *N*-policy. Ke and Wang [2] analyzed machine repair problems with constant balking probability, negative exponential distributed reneging, and unreliable servers. The system steady-state availability, MTTF, and some system performance measures were presented. Wang, Ke, and Ke[3] investigated the profit analysis of the M/M/R machine repair problem with balking, reneging, and standby switching failures. A comprehensive and exhaustive discussion of machine repair problems was given by Haque and Armstrong[4].

In this paper, we study the issue for the F-policy M/M/1 /K machine repair system consisting of M operating machines with S warm standbys. In this system, each of the operating machine fails independent of the state of the others and has an exponential time-to-failure distribution with parameter λ. Whenever an operating machine breaks downs, it is instantly replaced by a standby machine. Each of the available standby machine fails independent of the state of all the others and has an exponential time-to-failure distribution with parameter α $(0 \le \alpha \le \lambda)$. Whenever a machine fails, it is immediately sent to the repair facility where repair work is provided in the order of breakdowns, with a time–to-repair which is exponential distribution with parameter μ. We consider that there is always the possibility of failures during the switching process from standby state to operating state. We assume that there is a significant probability q of a switching failure. The server requires an exponential startup time with parameter γ to start allowing failed machines in the queue. Arriving failed machines form a single waiting line based on the order of their arrivals; that is, they are queued according to the first-come, first-served (FCFS) policy. We assume that when a standby machine moves into an operating state, its characteristics will be that of an operating machine. If a machine repair system reaches its finite capacity K ($K = M + S < \infty$), no further failed machines are allowed to enter the system until enough machines who are already in the system have been repaired so that the number of failed machines decreases to a predetermined threshold F ($0 \le F \le K-1$).

2. Steady-state equations and matrix-geometric solutions

The states of the presnted system are described by the pair (i,n), where i represents server's state and n represents the number of failed machines in the system, respectively. Let i=0 denote the failed machine is not allowed to enter into the system, i=1 denote the failed machine is allowed to enter into the system. Thus $(i,n) \in \{(i,n): i=0;\ n=0,\ 1,2,L\ ,K\} \cup \{(i,n): i=1;\ n=0,1,\ 2,L-,K\ \ 1\}$.The failure rate λ_n for this system is given by

$$\lambda_n = \begin{cases} M\lambda(1-q)+(S-n)\alpha & n=0,1,..S-1 \\ (K-n)\lambda & n=S,...K-1 \\ 0 & otherwise \end{cases}$$

In steady-state, let $P_0(n)$ denotes the probability that there are n failed machines in the system when the arrivals are not allowed to enter into the system, where $n = 0,\ 1,\ 2,L\ ,K$. We also let $P_1(n)$ denotes the probability that there are n failed machines in the system when the arrivals are allowed to enter into the system, where $n = 0,\ 1,\ 2,L\ ,K-1$.

Referring to the state-transition rate diagram for the F-policy M/M/1/K machine repair system with standby switching failures for F≤S case, the equations of the machine repair model are constructed as follows:

$$\gamma P_0(0) = \mu P_0(1), \qquad (1)$$

$$(\mu+\gamma)P_0(n) = \mu P_0(n+1), \qquad 1 \le n \le F \quad (2)$$

$$\mu P_0(n) = \mu P_0(n+1), \qquad F+1 \le n \le K-1 \quad (3)$$

$$\mu P_0(K) = \lambda P_1(K-1), \qquad (4)$$

$$(M\lambda + S\alpha)P_1(0) = \gamma P_0(0) + \mu P_1(1), \qquad (5)$$

$$[M\lambda + (S-1)\alpha + \mu]P_1(1) = \gamma P_0(1) + [M\lambda(1-q) + S\alpha]P_1(0) + \mu P_1(2), \qquad (6)$$

$$[M\lambda + (S-n)\alpha + \mu]P_1(n) = \gamma P_0(n) + [M\lambda(1-q) + (S-n+1)\alpha]P_1(n-1) + \mu P_1(n+1) + \sum_{i=0}^{n-2} \Phi_{n-1-i} P_1(i) \quad 2 \le n \le F \quad (7)$$

$$[M\lambda + (S-n)\alpha + \mu]P_1(n) = [M\lambda(1-q) + (S-n+1)\alpha]P_1(n-1) + \mu P_1(n+1) + \sum_{i=0}^{n-2} \Phi_{n-1-i} P_1(i) \quad F+1 \le n \le S \quad (8)$$

$$[(M-1)\lambda + \mu]P_1(S+1) = M\lambda P_1(S) + \mu P_1(S+2) + \sum_{i=0}^{S-1} \Theta_i P_1(i) \qquad (9)$$

$$[(K-n)\lambda + \mu]P_1(n) = (K-n+1)\lambda P_1(n-1) + \mu P_1(n+1) \qquad S+2 \le n \le K-2 \quad (10)$$

$$(\lambda + \mu)P_1(K-1) = 2\lambda P_1(K-2) \qquad (11)$$

Similar to F≤ S case, we can obtain the following steady-state equations for F>S case.

$$\gamma P_0(0) = \mu P_0(1), \qquad (12)$$

$$(\mu+\gamma)P_0(n) = \mu P_0(n+1), \qquad 1 \le n \le F \quad (13)$$

$$\mu P_0(n) = \mu P_0(n+1), \qquad F+1 \le n \le K-1 \quad (14)$$

$$\mu P_0(K) = \lambda P_1(K-1), \qquad (15)$$

$$(M\lambda + S\alpha)P_1(0) = \gamma P_0(0) + \mu P_1(1), \qquad (16)$$

$$[M\lambda + (S-1)\alpha + \mu]P_1(1) = \gamma P_0(1) + [M\lambda(1-q) + S\alpha]P_1(0) + \mu P_1(2), \qquad (17)$$

$$[M\lambda+(S-n)\alpha+\mu]P_1(n)=\gamma P_0(n)+[M\lambda(1-q)+(S-n+1)\alpha]P_1(n-1)+\mu P_1(n+1)+\sum_{i=0}^{n-2}\Phi_{n-1-i}P_1(i) \quad 2\le n\le S \tag{18}$$

$$[(M-1)\lambda+\mu]P_1(S+1)=\gamma P_0(S+1)+M\lambda P_1(S)+\mu P_1(S+2)+\sum_{i=0}^{S-1}\Theta_i P_1(i) \tag{19}$$

$$[(K-n)\lambda+\mu]P_1(n)=\gamma P_0(n)+(K-n+1)\lambda P_1(n-1)+\mu P_1(n+1) \quad S+2\le n\le F \tag{20}$$

$$[(K-n)\lambda+\mu]P_1(n)=(K-n+1)\lambda P_1(n-1)+\mu P_1(n+1) \quad F+1\le n\le K-2 \tag{21}$$

$$(\lambda+\mu)P_1(K-1)=2\lambda P_1(K-2) \tag{22}$$

Since the closed-form probability solutions of equations (1)-(22) are too complicated to develop explicit expressions by using a recursive method. Hence we use matrix-geometric methods to analysis this problem. We find that the equations (1)-(22) of the present model can be expressed in the matrix form. Then the steady-state probabilities, $P_0(n), P_1(n)$,can be computed by using the following normalizing conditions:

$$\sum_{n=0}^{K}P_0(n)+\sum_{n=0}^{K-1}P_1(n)=1. \tag{23}$$

3. System performance measures and cost analysis

Using the steady-state probabilities derived in the previous section, we compute some system performance measures. The assumptions and notations are defined as follows: L denotes the expected number of failed machines in the system; P_B denotes the probability that the server is in the busy state; P_I denotes the probability that the server is in the idle state; P_{bl} denotes the probability that the system is blocked. *E(O)* denotes the expected number of operating machines in the system; *E(S)* denotes the expected number of standby machines in the system; *M.A.* denotes machine availability(the fraction of the total time that the machines are working). *S.R.* denotes the average switching failure rate;*W* denotes the expected waiting time in the system. The expressions for L, P_B, P_I, P_{bl} , $E(O)$, $E(S)$,*MA*, *S.R.* and *W* are given by:

$$L=\sum_{n=0}^{K}nP_0(n)+\sum_{n=0}^{K-1}nP_1(n),\ P_B=\sum_{n=1}^{K}P_0(n)+\sum_{n=1}^{K-1}P_1(n),\ P_I=1-P_B,\ P_{bl}=\sum_{n=0}^{K}P_0(n),$$

$$E(O)=M-[\sum_{n=1}^{M}nP_0(S+n)+\sum_{n=1}^{M-1}nP_1(S+n)]\ \ E(S)=\sum_{n=0}^{S}(S-n)P_0(n)+\sum_{n=0}^{S}(S-n)P_1(n),$$

$M.A.=1-\dfrac{L}{K}$, $S.R.=\sum_{n=1}^{S}M\lambda qP_1(n-1)$, Let *W* denote the expected waiting time in the system. By using the Little's formula we obtain $W=L/\lambda_e$, where the effective arrival rate in the system is $\lambda_e=\sum_{n=0}^{K-1}\lambda_n P_1(n)$.

We construct a total expected cost function per unit time for the F-policy M/M/1/K repair system with standby switching failures, in which *F*, *S* and μ are decision variables. The discrete variables *F* and *S* are required to be natural numbers, and the continuous variable μ is positive numbers. Our main objective is to search the optimal values of (F,S,μ), say (F^*,S^*,μ^*) so as to minimize the total expected cost.

We define the following cost elements: C_h denotes the holding cost per unit time when one failed machine joins the system; C_l denotes the cost per unit time of a failed machine after all standbys are exhausted; C_2 denotes the cost per unit time when one machine is acting as a warm standby state; C_3 denotes the cost per unit time when the server is idle; C_4 denotes the cost per unit time when the server is busy; C_5 denotes fixed cost for each lost failed machine when the system is blocked; C_6 denotes the cost per unit time of providing the service rate μ ; C_w denotes the waiting cost per unit time when one machine is waiting for services. Using the definitions of these elements listed above, the total expected cost function per unit time is determined by:

$$TC(F,S,\mu)=C_hL+C_1[M-E(O)]+C_2E(S)+C_3P_I+C_4P_B+C_5\lambda_eP_{bl}+C_6\mu+C_WW$$

It is extremely difficult to develop the useful analytical results for the optimum value (F^*,S^*,μ^*), because that the values F and S are discrete quantities, μ is continuous quantity, and the cost function is highly complex. To determine the optimal value (F^*,S^*,μ^*), we apply the direct search method to obtain optimal value. Firstly, we obtain optimal value (F,S,μ^*) when F and S are initially fixed. Next, based on this solution, we then use direct search method to obtain the optimal value (F^*,S^*,μ^*), which minimize the cost function $TC(F,S,\mu)$. For fixed values F and S, the direct search algorithm is applying to search μ until the minimum value of $TC(F,S,\mu)$ is achieved, say (F,S,μ^*) and constraints are satisfied. In practice, the server rate could be adusted to further minimize the total cost. It is assumed that the value of service rate has the upper bound of μ_U The investment budget is restricted by a given budget C. After obtaining the joint optimal value μ^* of the continuous variable μ, we use direct search method again to obtain the optimal value (F^*,S^*) such that the expected cost function $TC(F,S,\mu^*)$ attains a minimum value, say $TC(F^*,S^*,\mu^*)$.

4. Conclusions

In this paper, we studied an F-policy M/M/1/K machine repair system with standby switching failures, and obtain the steady-state analytic solutions. The stationary probability vectors were obtained using matrix-geometric analysis and the technique of matrix recursive. We have provided an efficient method to determine the optimum repair policy, including the optimal threshold, the number of standbys and the optimal server rate simultaneously, in order to minimize the expected cost function when system availability is maintained at given level, and calculated various system performance measures under optimal operating conditions. This research presents an extension of the Markovian model theory and the analysis of the model will provide a useful performance evaluation tool.

References

[1] Gupta, S.M. (1995). Interrelationship between controlling arrival and service in queueing systems. Computers and Operations Research, 22, 1005–1014.

[2] Ke, J. C., & Wang, K. H. (1999). Cost analysis of the M/M/R machine repair problem with balking, reneging, and server breakdowns. Journal of the Operational Research Society, 50(3), 275–282.

[3] Wang, K. H., Ke, J. B., & Ke, J. C. (2007). Profit analysis of the M/M/R machine repair problem with balking, reneging, and standby switching failures. Computers and Operations Research, 34(3), 835 - 847.

[4] Haque, L., & Armstrong, M. J. (2007). A survey of the machine interference problem. European Journal of Operational Research, 179, 469–482.

Symmetric Path Fitted Variational Integrators

D.S. Vlachos

University of Peloponnese, Department of Computer Science and Technology, Laboratory of Mathematical Modeling and Complex Systems, Terma Karaiskaki str., 22100 Tripolis, Greece

E-mail: dvlachos@uop.gr

O.T. Kosmas

University of Erlangen-Nuremberg, Department of Mechanical Engineering, Konrad-Zuse-Strae 3-5, Erlangen 91052, Germany

E-mail: odykosm@uop.gr

Abstract. Recently, the benefits of variational integrators have been combined with efficient high order techniques. On the other hand, a special set of high order methods are the symmetric ones, those who preserve time reversal symmetry and show improved behavior in long term integration. In the present work, we will introduce a systematic way to construct symmetric variational integrators. The idea is to apply the variational principle not in a set of intermediate points but to a set of parameters that characterize a symmetric orbit between starting and ending points. The estimated symmetric orbit may be a polynomial or a general function that sometimes is indicated by the nature of the problem to solve. The results show excellent behavior in long term integration and acceleration of the method when special functions are used.

1. Introduction

It is well known that the dynamics of seemingly unrelated conservative systems in mechanics, physics, biology, and chemistry fit the Hamiltonian formalism ([1]). Included among these are particle, rigid body, ideal fluid, solid, and plasma dynamics. An important property of the Hamiltonian flow or solution to a Hamiltonian system is that it preserves the Hamiltonian and the symplectic form (see, for example, [2]). A key consequence of symplecticity is that the Hamiltonian flow is phase-space volume preserving (Liouville's theorem). Since analytic expressions for the Hamiltonian flow are rarely available, approximations based on discretization of time are used. A numerical integration method which approximates a Hamiltonian flow is called symplectic if it discretely preserves a symplectic 2-form to within numerical round off ([3, 4]) and standard otherwise. By ignoring the Hamiltonian structure, a standard method often introduces spurious dynamics.

Symplectic integrators can be derived by a variety of ways including Hamilton-Jacobi theory, symplectic splitting, and variational integration techniques. Early investigators, guided by Hamilton-Jacobi theory, constructed symplectic integrators from generating functions which approximately solve the Hamilton–Jacobi equation. The symplectic splitting technique is based on the property that symplectic integrators form a group, and thus, the composition of symplectic-preserving maps is also symplectic. The idea is to split the Hamiltonian into terms whose flow can be explicitly solved and then compose these individual flows in such a fashion that the composite flow is consistent and convergent with the Hamiltonian flow being simulated. On the other hand, variational integration techniques determine integrators from a discrete Lagrangian and associated discrete variational principle. The discrete Lagrangian can be designed to inherit the symmetry associated with the action of a Lie group, and hence by a discrete Noether's theorem, these methods can also preserve momentum invariants (for a discussion of the above statements, see [1]).

Variational integration theory derives integrators for mechanical systems from discrete variational principles ([5, 6, 7]). Variational principles have been successfully applied to partial differential equations and to stochastic systems as well. In the general theory, discrete analogs of the Lagrangian, Noether's theorem, the Euler–Lagrange equations, and the Legendre transform can be easily obtained. Moreover, variational integrators can readily incorporate holonomic constraints (via Lagrange multipliers) and non-conservative effects (via their virtual work). The algorithms derived from this discrete principle have been successfully tested in infinite and finite-dimensional conservative, dissipative, smooth and non-smooth mechanical systems.

In general, the accuracy is not the terminus of the application of variational integrators, but rather their ability to discretely preserve essential structure of the continuous system and in computing statistical properties of larger groups of orbits, such as in computing Poincaré sections or the temperature of a system (see, for example. On the other hand, high accuracy can be obtained using special designed methods as it is explained in [5, 6, 7]. In the case of time reversible systems, it has been found that both accuracy and long term behavior is dramatically improved when using methods that are symmetric, which means that they preserve the time reversal symmetry.

In the present work, we will introduce a systematic way to construct symmetric variational integrators. The idea is to apply the variational principle not in a set of intermediate points but to a set of parameters that characterize a symmetric orbit between starting and ending points. The estimated symmetric orbit may be a polynomial or a general function that sometimes is indicated by the nature of the problem to solve. The results show excellent behavior in long term integration and acceleration of the method when special functions are used.

2. Symmetric Path Fitted Methods

2.1. Symmetric Methods

Conservative mechanical systems are reversible in the sense that if we invert the initial direction of the velocity and keep the initial position, the solution trajectory remains the same (only the direction of motion is changed). In general, let ρ be an invertible transformation in the phase space of a system described by the differential equation $\dot{y} = f(y)$. Then, the differential equation and the vector field $f(y)$ are called ρ-reversible if

$$\rho f(y) = -f(\rho y) \tag{1}$$

On the other hand, a map $\Phi(y)$ is called ρ-reversible if

$$\rho \circ \Phi = \Phi^{-1} \circ \rho \tag{2}$$

In the case of a numerical method Φ_h, the condition 2 takes the form

$$\rho \circ \Phi_h = \Phi_{-h} \circ \rho \tag{3}$$

Finally, a method is symmetric if (setting $\rho = -id$, where id is the identity map)

$$\Phi_h = \Phi_{-h}^{-1} \tag{4}$$

or when interchanging y_0 with y_1 and h with $-h$ the resulting method is the same. Symmetric methods have several benefits-in long term integration. One of them is that their order is even (since both the method and the adjoint one must have exactly the same truncation error).

2.2. Path Fitted Variational Integrators

In the case of path fitted variational integrators, we have a parametric estimation of the orbit, and the variational principles applied in order to optimize the values of the free parameters. Thus, considering the Lagrangian of the system

$$L = L(q^i, \dot{q}^i, t), \qquad i = 1, .., n \tag{5}$$

where $2 \cdot n$ is the dimension of the configuration space, we assume that the path between q_k^i at time t_k and q_{k+1}^i at time t_{k+1} can be parametrized as

$$q^i(t) = g_k^i\left(t, q_k^i, x^{i,j}\right)\ ,\ j = 1, 2, \ldots, K \tag{6}$$

where K is the number of free parameters. Moreover, we can assume that the parameter $x^{i,1}$ is the final point of the orbit. Thus, we have the following conditions

$$g_k^i\left(t_k, q_k^i, x_k^{i,j}\right) = q_k^i \tag{7}$$

$$g_k^i\left(t_{k+1}, q_k^i, x_k^{i,j}\right) = x_k^{i,1} \tag{8}$$

As it has been described above, by an appropriate choice of quadrature scheme we can break up the action integral into pieces, which we denote by $S_{d,k}$. Replacing now in the Lagrangian equ. 6, we have

$$S_{d,k} = S_{d,k}\left(q_k^i, x_k^{i,j}\right) \tag{9}$$

The condition in equ. 8 can be obtained using Lagrange multipliers λ_k^i and thus we can redefine the discrete Lagrangian as

$$S_d = \sum_{k=0}^{N-1} S_{d,k}(q_k^i, x_k^{i,j}) - \sum_{k=0}^{N-2}\sum_{i=1}^{n} \lambda_k^i (x_k^{i,1} - q_{k+1}^i) \tag{10}$$

Then by demanding $\delta S_d = 0$ and fixing points at $t = t_0$ and $t = t_N$ we can derive the equations that the unknown variables must follow

$$\begin{aligned} \frac{\partial S_{d,k}}{\partial q_k^i} + \lambda_{k-1}^i &= 0 \\ \frac{\partial S_{d,k}}{\partial x_k^{i,1}} - \lambda_k^i &= 0 \\ \frac{\partial S_{d,k}}{\partial x_k^{i,J}} &= 0, \quad j = 2, 3, \ldots, K \end{aligned} \tag{11}$$

It is easy to verify that the Lagrange multipliers λ_k^i are the momenta p_k^i. Then, the system of K equations

$$\begin{aligned} \frac{\partial S_{d,k}}{\partial q_k^i} + p_k^i &= 0 \\ \frac{\partial S_{d,k}}{\partial x_k^{i,j}} &= 0, \quad j = 2, 3, \ldots, K \end{aligned} \tag{12}$$

is solved to give the K unknowns $x_k^{i,j}$, $j = 1, 2, \ldots, K$. Finally, for the computation of the p_{k+1}^i and q_{k+1}^i we have

$$\begin{aligned} p_{k+1}^i &= \frac{\partial S_{d,k}}{\partial x_k^{i,1}} \\ q_{k+1}^i &= x_k^{i,1} \end{aligned} \tag{13}$$

2.3. Polynomial fitting

Let $h = t_{k+1} - t_k$. Then we can define the polynomial function

$$g_k^i\left(t, q_k^i, x_k^{i,j}\right) = q_k^i\left(1 - \frac{t}{h}\right) + x_k^{i,1} \cdot \frac{t}{h} + P\left(t, x_k^{i,2}, x_k^{i,3}, \ldots, x_k^{i,K}\right) \tag{14}$$

where $P\left(t, x_k^{i,2}, x_k^{i,3}, \ldots, x_k^{i,K}\right)$ is a polynomial in t which depends on $x_k^{i,2}, x_k^{i,3}, \ldots, x_k^{i,K}$ and is symmetric in the interchange of t with $h - t$. Moreover, $P(0) = P(h) = 0$.

It is easy to verify that interchanging q_k^i with q_{k+1}^i and t with $h - t$ (we assume here that t represents $t - t_k$ so it varies fro 0 to h), the path remains the same and this holds for the resulting method iff $P(t) = P(h - t)$. Setting $t^{(1)} = t$ and $t^{(2)} = h - t$ we can write the polynomial P in 2-variables as $P\left(t^{(1)}, t^{(2)}\right)$. Thus, P must be a symmetric polynomial. In general, let $x_1, \ldots, x_n$ be variables, and denote for $k \geq 1$ by $p_k(x_1, \ldots, x_n)$ the k-th power sum:

$$p_k(x_1, \ldots, x_n) = \sum\nolimits_{i=1}^{n} x_i^k = x_1^k + \cdots + x_n^k, \tag{15}$$

and for $k \geq 0$ denote by $e_k(x_1, \ldots, x_n)$ the elementary symmetric polynomial that is the sum of all distinct products of k distinct variables, so in particular

$$\begin{aligned}
e_0(x_1, \ldots, x_n) &= 1, \\
e_1(x_1, \ldots, x_n) &= x_1 + x_2 + \cdots + x_n, \\
e_2(x_1, \ldots, x_n) &= \sum\nolimits_{i<j} x_i x_j, \\
e_n(x_1, \ldots, x_n) &= x_1 x_2 \cdots x_n, \\
e_k(x_1, \ldots, x_n) &= 0, \quad \text{for } k > n.
\end{aligned}$$

Then, according to the Newton's identities

$$k \cdot e_k(x_1, \ldots, x_n) = \sum_{i=1}^{k} (-1)^{i-1} e_{k-i}(x_1, \ldots, x_n) p_i(x_1, \ldots, x_n), \tag{16}$$

and thus we can write

$$P\left(t, x_k^{i,2}, x_k^{i,3}, \ldots, x_k^{i,K}\right) = \sum_{j=2}^{K+1} x^{i,j}\left(t^j + (h - t)^j\right) \tag{17}$$

with

$$x^{i,K+1} = -\sum_{j=2}^{K} x^{i,j} \tag{18}$$

in order to have $P(0) = 0$.

2.4. Trigonometric fitting

In the case of oscillatory problems, trigonometric or phase fitting can improve dramatically the behavior of the integrators (se, for example, [5, 6]). The analysis is the same as in the polynomial fitting, except that now we are dealing with the phase function. In detail, consider the path

$$q_k^i(t) = A\cos\left(f_k^i(t)\right) + B\sin\left(f_k^i(t)\right) \tag{19}$$

where $f(t)$ is a polynomial similar to that in equ. 14. Thus,

$$f_k^i(t) = f_0^i\left(1 - \frac{t}{h}\right) + f_1^i\frac{t}{h} + \sum_{j=2}^{K+1} f_j^i\left(t^j + (h-t)^j\right) \tag{20}$$

Both constants A, B are calculated be the equations

$$q_k^i(t_k) = q_{k+1}^i$$
$$q_k^i(t_{k+1}) = x_k^{i,1}$$

3. The 2-body problem

We now turn to the study of two objects interacting through a central force. The most famous example of this type, is the Kepler problem (also called the two-body problem) that describes the motion of two bodies which attract each other. In the solar system the gravitational interaction between two bodies leads to the elliptic orbits of planets and the hyperbolic orbits of comets.

If we choose one of the bodies as the center of our coordinate system, the motion will stay in a plane. Denoting the position of the second body by $\mathbf{q} = (q_1, q_2)^T$, the Lagrangian of the system takes the form (assuming masses and gravitational constant equal to 1)

$$L(\mathbf{q}, \dot{\mathbf{q}}, t) = \frac{1}{2}\dot{\mathbf{q}}^T\dot{\mathbf{q}} + \frac{1}{|\mathbf{q}|} \tag{21}$$

The initial conditions are taken

$$\mathbf{q} = (1-\epsilon, 0)^T\ ,\ \dot{\mathbf{q}} = \left(0, \sqrt{\frac{1+\epsilon}{1-\epsilon}}\right)^T \tag{22}$$

where ϵ is the eccentricity of the orbit. In the first experiment, we take the eccentricity $\epsilon = 0.5$ and constant step size $h = 0.05$ and plot the relative error in the energy during one period for several number of path parameters K.

In our experiment, we take the eccentricity $\epsilon = 0.95$ and use an adaptive time step control in order to keep the relative error in energy smaller than 10^{-7}. Table 1 shows the number of integration steps needed for one period. The order of approximation is again clearly increases with increasing K as the mean step size needed for the same error in energy increases from $1.8 \cdot 10^{-3}$ for $K = 4$ to 0.1 for $K = 8$.

Table 1. Number of integration steps

S	No of Steps
2	$> 10^4$
3	$> 10^4$
4	3526
5	460
6	421
7	181
8	142

4. Conclusions

The use of symmetric paths as estimators for trajectories in the construction of variational principles, produces integrators with excellent behavior in energy for long term integration. Moreover, a generalization of the method proposed, can easily applied in order to construct high order methods. The results show that in both stiff problems (like the 2-body one with high eccentricity) and many body one, the symmetric variational integrator can produces accurate and stable solutions. Moreover, the method can be easily adapted in order to use it for the construction of trigonometric fitted or in general function fitting symmetric variational integrators for special problems (like oscillatory ones).

References

[1] N. Rabee, Ph.D. thesis, California Institute of Technology Pasadena, California (2007).
[2] J. Marsden and T. Ratiu, Springer Texts in Applied Mathematics, 17 (1999).
[3] R. Ruth, IEEE Transactions on Nuclear Science, 30, 2669 (1983).
[4] K. Feng, J. Comp. Math, 4, 279 (1986).
[5] O.T. Kosmas and D.S. Vlachos, Journal of Computational and Applied Mathematics, 236 (10), 2632 (2012)
[6] O. Kosmas and D.S. Vlachos,, Computer Physics Communications, 181 (3), 562 (2010).
[7] D.S. Vlachos and O. Kosmas, arXiv:0904.0112 (2009).

Evaluation of muscle force predictions using optimization theory

Yunus Ziya Arslan[1], Motoshi Kaya[2], Walter Herzog[3]

[1]Department of Mechanical Engineering, Faculty of Engineering, Istanbul University, Istanbul, Turkey

[2]Department of Physics, Graduate School of Science, University of Tokyo, Tokyo, Japan

[3]Human Performance Laboratory, The University of Calgary, Calgary, AB, Canada

yzarslan@istanbul.edu.tr

Abstract. Prediction of muscle forces using optimization based models of muscle coordination is an active research area in biomechanics. Theoretical calculation of individual muscle forces depends on solving the redundancy problem. In a musculoskeletal model, redundancy arises since the number of muscles in the model exceeds the number of degrees-of-freedom present. One of the widely used methods to solve this problem is to formulate a physiologically sound cost function and optimize this function subject to mechanical equality and inequality constraint equations. In this study, force predictions obtained from different optimization-based models were compared with those obtained from experimentally measured individual muscle forces recorded during a variety of movement conditions. Advantages and limitations of the tested models were discussed.

1. Introduction

Analyzing the force distribution problem among muscles has attracted many investigators over years [1, 2]. Understanding the muscle coordination can give insight into how the motor control and body mechanics interact for force generation during various types of movement tasks. Force distribution problem arises in a musculoskeletal system since the number of muscles crossing a joint is more than the number of rotational degree of freedom.

One common approach to the force distribution problem is to formulate a static optimization model [3, 4]. The approach proposed by Crowninshield and Brand [5] is by far the most used optimization model in the literature. In this model, the cost function takes the form of a weighted sum of muscle forces raised to some non-linear power. The weighting is done using constant values for the physiological cross-sectional area.

In recent years, some cost functions, which take into account the contractile conditions of muscles that change during movement such as muscle velocity and length, have been proposed [6, 7]. Since both the instantaneous contractile conditions and physiological properties of muscle are incorporated into these optimization strategies, realistic force predictions can be obtained using these cost functions.

In this study, muscle force predicted from optimization strategies proposed by Crowninshield and Brand [5], Dul et al. [8] and Arslan [7] were compared with those obtained from measured muscle forces. Merits and shortcomings of the optimization strategies were discussed.

2. Musculoskeletal model of the cat hindlimb

Contractile conditions of muscles have a profound effect on force sharing among synergistic muscles which becomes particularly apparent when some muscles in a synergistic group are driven to their contractile limits [9]. Optimization strategies compared in this study were tested in the musculoskeletal model of the cat hindlimb shown in Figure 1. The synergistic soleus (SO) and medial gastrocnemius (MG) were modeled as straight line muscles.

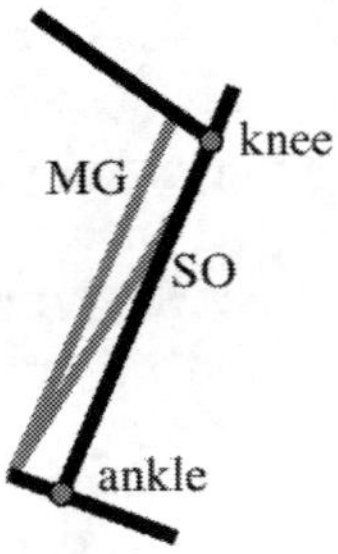

Figure 1. Schematic representation of the cat hindlimb musculoskeletal system.

3. Optimization models

3.1. Objective Function I

The objective function proposed by Crowninshield and Brand [5] is given below.

$$\text{Minimize} \left\{ \sum_{i}^{n} \left(\frac{f_i}{PCSA_i}\right)^3 \right\} \tag{1}$$

subject to

$$h - \sum_{i=1}^{n} \boldsymbol{d}\, \boldsymbol{f}^T = 0, \quad f_i \geq 0, \quad i = 1,2. \tag{2}$$

where f_i, $PCSA_i$, h, $\boldsymbol{d}$ denote the unknown muscle forces of SO and MG, physiological cross sectional area, ankle joint moment and moment arm vector at the ankle, respectively.

3.2. Objective Function II

The optimization strategy proposed by Dul et al. [8] is

$$\text{Maximize} \left\{ \text{Minimum} \left\{ a_i \frac{f_i}{F_{i\max}} 100 \right\}^{p_i} \right\} \tag{3}$$

subject to equation (2).

In the cost function, $F_{i\max}$ is the maximum isometric muscle force, a_i and p_i are the muscle parameters depending on the percentage of slow-twitch fibers.

3.3. Objective Function III

The function proposed by Arslan [7] is given below.

$$\text{Minimize}\left\{\sum_{i}^{n}\frac{(f_i v_i)^2+\omega(f_i l_i)^2}{F_i(v_i)F_i(l_i)S_i}\right\} \quad (4)$$

subject to equation (2).

In the equation, $F_i(v_i)$, $F_i(l_i)$ and S_i are maximal muscle forces at instantaneous muscle contraction velocity v_i, maximal muscle force at instantaneous muscle length l_i and the percentage of slow-twitch muscle fibers, respectively. ω was assumed constant and equal to 1 (in unit of $1/s^2$).

This optimization strategy was designed to punish forces in muscles with non-favorable contractile conditions and enhance forces in muscles with favorable contractile conditions. Minimizing the proposed objective function enables the minimization of the instantaneous muscle power ($f_i v_i$) and the restriction of the lengthening of muscle ($f_i l_i$). By incorporating $F_i(v_i)$ and $F_i(l_i)$ as weighting factors into the cost function, the force-velocity and force-length properties of the synergistic muscles become crucial determinants in the determination of the unknown muscle forces [6]. $F_i(v_i)$ and $F_i(l_i)$ partly depend on the PCSA, hence PCSA was included in the cost function in an indirect way. The percentage of slow-twitch muscle fiber S_i, which reflects muscle endurance, was incorporated into the objective function as a weighting factor [8].

4. Experimental protocol

Five cats were trained to perform level and 45° upslope walking [10]. "E" shaped tendon force transducers were used to measure the forces produced by SO and MG muscles during a variety of locomotion tasks. Five reflective markers were placed over the hip, knee, ankle, metatarsophalangeal (MP) joint, and toe of the hindlimb to obtain knee, ankle, and MP joint angles. A motion analysis system was used to record the positions of the markers. Ground reaction forces were recorded using force platforms. Muscle-tendon lengths were calculated using the joint kinematics and the tendon travel technique. Changes in muscle lengths were calculated as the first time derivative of muscle-tendon length. Since SO and MG are not active during swing phase, only the stance phase of the step cycle was analyzed. All procedures were approved by the Life Sciences Animal Ethics Committee of the University of Calgary.

5. Results

In Figure 2, comparison of the representative force sharing loops obtained from the different cost functions with the experimentally measured forces from cat SO and MG muscles are given for upslope (Figure 2a) and level walking (Figure 2b). It can be clearly seen from the figures that typical force patterns obtained from Cost Function III are more similar to the actual muscle forces than those obtained by Cost Function I and II.

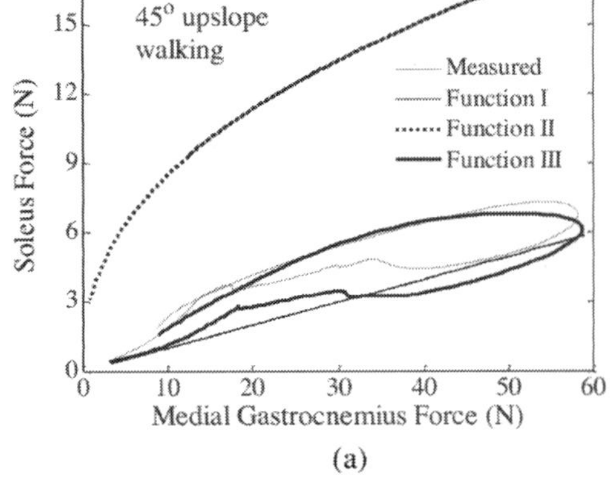

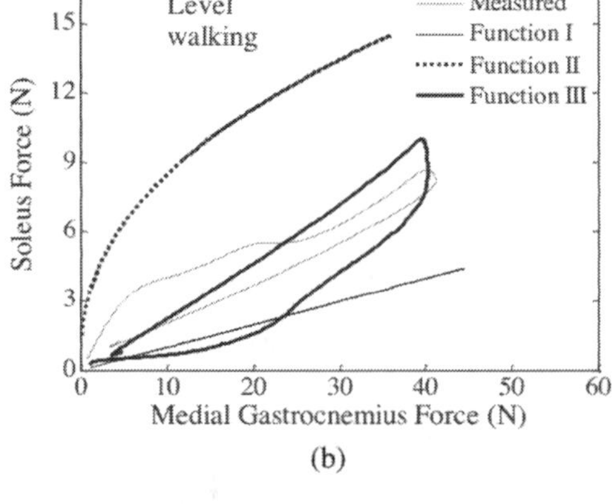

Figure 2. Experimentally and theoretically obtained force sharing loops among cat calf muscles.

6. Discussion

Contractile conditions of muscles have a profound effect on force sharing among synergistic muscles which becomes particularly apparent when some muscles in a synergistic group are driven to their contractile limits [9, 11]. Since it is computationally efficient and does not require contractile information about the target muscles, the cost functions proposed by Crowninshield and Brand [5] and Dul et al. [8] has been extensively used in many studies [1, 2, 3, 4]. However, during movement, contractile conditions of muscle change continuously and thus these cost function fail in the prediction of basic force-sharing properties, such as loop-type force sharing behavior between synergistic muscles in a one degree of freedom model and simultaneous increase in force in one agonist and decrease in another [1]. Furthermore these cost functions often focus on specific tasks rather than a generalized prediction of muscle forces across tasks of different intensity and metabolic cost [12].

The cost function proposed by Arslan [7] is sensitive to changes in contractile conditions of muscles. This optimization strategy made better predictions of cat SO and MG forces for a variety of locomotor tasks compared to Cost Function I and II. It was concluded from the obtained results that realistic individual muscle force predictions are possible when accounting for the instantaneous contractile conditions.

References

[1] Herzog W 1996 Force-sharing among synergistic muscles: Theoretical considerations and experimental approaches *Exerc. Sport. Sci. Rev.* **24** 173-202

[2] Erdemir A, McLean S, Herzog W and van den Bogert AJ 2007 Model-based estimation of muscle forces exerted during movements *Clin. Biomech.* **22(2)** 131-154

[3] Crowninshield RD and Brand RA 1981a The prediction of forces in joint structures: Distribution of intersegmental resultants *Exerc. Sport. Sci. Rev.* **9** 159-181

[4] Tsirakos D, Baltzopoulos V and Bartlett R 1997 Inverse optimization: Functional and physiological considerations related to the force-sharing problem *Crit. Rev. Biomed. Eng.* **25(4-5)** 371-407

[5] Crowninshield RD and Brand RA 1981b A physiologically based criterion of muscle force prediction in locomotion *J. Biomech.* **14(1)** 793-801

[6] Schappacher-Tilp G, Binding P, Braverman E and Herzog W 2009 Velocity-dependent cost function for the prediction of force sharing among synergistic muscles in a one degree of freedom model *J. Biomech.* **42(5)** 657-660

[7] Arslan YZ 2009 Mechanics and force optimization analysis of the musculoskeletal system, PhD Thesis, Istanbul University

[8] Dul J, Johnson GE, Shiavi R, Townsend MA and 1984 Muscular synergism-II. A minimum-fatigue criterion for load sharing between synergistic muscles *J. Biomech.* **17(9)** 675-684

[9] Hodgson JA 1983 The relationship between soleus and gastrocnemius muscle activity in conscious cats - a model for motor unit recruitment?, *J. Physiol.* **337** 553-562

[10] Kaya M 2003 Coordination of cat hindlimb muscles during voluntary movements, PhD Thesis, University of Calgary

[11] Kaya M, Leonard TR and Herzog W 2008 Premature deactivation of soleus during the propulsive phase of cat jumping *J. R. Soc. Interface* **5(21)** 415-426

[12] Nelson FE and Roberts TJ 2008 Task-dependent force sharing between muscle synergists during locomotion in turkeys *J. Exp. Biol.* **211** 1211-1220

Multi-step numerical methods derived using discrete Lagrangian integrators

D.S. Vlachos[1] and O.T. Kosmas[2]

[1] Department of Computer Science and Technology, University of Peloponnese, Greece
[2] Chair of Applied Dynamics, University of Erlangen-Nuremberg, Germany

E-mail: `dvlachos@uop.gr`

Abstract. On the basis of the variational integrators theory, we initially examine the possibility of deriving multi-step numerical methods. Then, we propose an integration technique that approximates the action integral within one time interval by using appropriate expressions for the relevant configurations and velocities. These approximations depend on a specific number of known configurations defined at previous time nodes. Multi-step numerical methods can finally be deduced, by defining, as usually, the Lagrange function as a weighted sum over the discrete Lagrangians corresponding to each of the curve segments and using the discrete Euler-Lagrange equations.

1. Introduction and motivation

Numerical integration of ordinary differential equations with oscillatory solutions has been the subject of intensive research in the past few decades. Special cases of such ordinary differential equations are often met in real problems, such as the N-body problem. For highly oscillatory problems, standard nonspecialized methods may require a huge number of steps to track the oscillations. One of the possible ways to obtain a more efficient integration process, is to construct numerical methods with an increased algebraic order, even though the implementation of high algebraic order meets several difficulties [1].

On the other hand, variational integrators are relatively new tools to model dynamical systems. In particular, they constitute an alternative to continuous ordinary differential equations to simulate mechanical systems. They can be considered as a class of integrators [1, 2] that preserve (or nearly preserve, depending on the particular integrator) fundamental physical observables like energy and momentum [3, 4, 5]. They have traditionally been used to study conservative systems (e.g., celestial mechanics), but modern variational integrators also include forcing and dissipation. In fact, the results with external forcing often have better accuracy than those of classical methods.

In the present contribution, a new way to derive multi-step numerical methods is investigated. Based on the variational integration schemes, the proposed technique approximates the action integral within one time interval by using expressions for configurations and velocities that depend on a specific number of known configurations at previous time nodes. By defining the Lagrange function as a weighted sum over the discrete Lagrangians corresponding to the each of the curve segments and using the discrete Euler-Lagrange equations, multi-step numerical methods can be defined. The application of these methods may be tested in several numerical problems.

2. Discrete mechanics and variational integrators

When discretizing a mechanical system, whose continuous Lagrangian is defined as $L : TQ \to R$, on the time interval $[0, T]$, a finite number of time nodes $0 = t_0 < ... < t_N = T$ derives the sequence of configurations $q_0, ..., q_N$, where $q_k \approx q(t_k)$ for $N \in \mathbb{N}$. The discrete Lagrange function $L_d : Q \times Q \to \mathbb{R}$ can be defined by the approximation of the action integral along the curve segment between q_k and q_{k+1}, i.e. $L_d(q_k, q_{k+1}) \approx \int_{t_k}^{t_{k+1}} L(q, \dot{q})dt$. The discrete action sum corresponding to the above discrete Lagrangians then reads

$$S_d(q_0, ..., q_N) = \sum_{k=0}^{N-1} L_d(q_k, q_{k+1}) \approx \int_0^T L(q, \dot{q})dt \tag{1}$$

Applying Hamilton's actions principle, $\delta S_d(q_0, ..., q_N) = 0$, by requiring the endpoints to be fixed, i.e. $\delta q_0 = \delta q_N = 0$, discrete Euler-Lagrange equations are obtained as

$$D_2 L_d(q_{k-1}, q_k) + D_1 L_d(q_k, q_{k+1}) = 0, \qquad k = 1, ..., N-1 \tag{2}$$

where $D_1 L_d$ and $D_2 L_d$ denote the partial derivatives of L_d with respect to the first and second arguments, see e.g. [4].

3. Deriving multi-step methods from discrete Lagrangian integrators

Following the steps of [4, 5, 6] to derive multi-step numerical schemes using variational integrators, we use intermediate configuration $q^j \in [q_k, q_{k+1}]$ at time $t^j \in [t_k, t_{k+1}]$ for $j = 0, ..., S$ ($S \in \mathbb{N}$) by expressing $t^j = t_k + C^j h$ for $C^j \in [0, 1]$ such that $C^0 = 0, C^S = 1$ for $h = t_{k+1} - t_k$. Under these assumptions the general expression for q^j can then be written as

$$q^j = f_{k-\sigma}(t^j) q_{k-\sigma} + ... + f_{k+1+\sigma}(t^j) q_{k+1+\sigma}, \qquad \sigma \in \mathbb{N} \tag{3}$$

where $2\sigma + 2$ represents the number of configuration points needed for the above approximation and

$$\begin{aligned} f_{k-\sigma}(t^j) = 0, &\quad \text{for} \quad j = 0, j = S \quad \text{and} \quad \sigma \neq 0 \\ f_k(t_k) = f_{k+1}(t_{k+1}) = 1, &\quad f_k(t_{k+1}) = f_{k+1}(t_k) = 1 \end{aligned} \tag{4}$$

The velocity obtained for the above expression of the intermediate point can be written as

$$\dot{q}^j = \dot{f}_{k-\sigma}(t^j) q_{k-\sigma} + ... + \dot{f}_{k+1+\sigma}(t^j) q_{k+1+\sigma}, \qquad \sigma \in \mathbb{N} \tag{5}$$

($\dot{f}$ denotes, as usually, time derivative). It is worth notting that the phase-fitted discrete Lagrangian integrators of Ref. [6] are a special case of the above expressions for $\sigma = 0$ and

$$f_k(t^j) = \frac{\sin\left(u - \frac{t^j - t_k}{h} u\right)}{\sin u}, \qquad f_{k+1}(t^j) = \frac{\sin\left(\frac{t^j - t_k}{h} u\right)}{\sin u} \tag{6}$$

($u \in \mathbb{R} - \{k\pi, k \in \mathbb{Z}\}$). It is now clear that the latter functions satisfy the assumptions of Eqs. 4. For these functions the expressions for intermediate points are

$$\begin{aligned} q^j &= \frac{\sin\left(u - \frac{t^j - t_k}{h} u\right)}{\sin u} q_k + \frac{\sin\left(\frac{t^j - t_k}{h} u\right)}{\sin u} q_{k+1} \\ \dot{q}^j &= \frac{1}{h} \left[- \frac{u \cos\left(u - \frac{t^j - t_k}{h} u\right)}{\sin u} q_k - \frac{u \cos\left(\frac{t^j - t_k}{h} u\right)}{\sin u} q_{k+1} \right] \end{aligned} \tag{7}$$

4. Numerical schemes derived using Catmull-Rom splines

Following the above assumptions, we employ the cubic Hermitian spline interpolation on a single unit interval to obtain points $q^j \in [q_k, q_{k+1}]$ from a linear combination of the form [7]

$$q^j = a_1 q_k + a_2 p_k + a_3 q_{k+1} + a_4 p_{k+1} \tag{8}$$

where

$$\begin{aligned} a_1 &= 2(c^j)^3 - 3(c^j)^2 + 1, \qquad a_2 = (c^j)^3 - 2(c^j)^2 + c^j, \\ a_3 &= -2(c^j)^3 + 3(c^j)^2, \qquad a_4 = (c^j)^3 - (c^j)^2. \end{aligned} \tag{9}$$

If we chose the tangents to be given by $p_k = (q_{k+1} - q_{k-1})(2h)^{-1}$, a Catmull-Rom spline interpolation of the interval can be obtained as [7]

$$q^j = f_{k-1} q_{k-1} + f_k q_k + f_{k+1} q_{k+1} + f_{k+2} q_{k+2} \tag{10}$$

where

$$\begin{aligned} f_{k-1} &= -\frac{a_2}{2h}, \qquad f_k = a_1 - \frac{a_4}{2h}, \\ f_{k+1} &= \frac{a_2}{2h} + a_3, \qquad f_{k+2} = \frac{a_4}{2h}. \end{aligned} \tag{11}$$

Using the latter expression for q^j the discrete Lagrangian is of the same form, i.e. it depends only on the endpoints $[q_k, q_{k+1}]$. From Eq. (10) the expression for $\dot{q}^j$ can be functional cast in the form

$$\dot{q}^j = \dot{f}_{k-1} q_{k-1} + \dot{f}_k q_k + \dot{f}_{k+1} q_{k+1} + \dot{f}_{k+2} q_{k+2} \tag{12}$$

where

$$\begin{aligned} \dot{f}_{k-1} &= \frac{-3(c^j)^2 + 4c^j - 1}{2h^2}, \qquad \dot{f}_k = \frac{12(c^j)^2 h - 12c^j h - 3(c^j)^2 - 2(c^j)}{2h^2}, \\ \dot{f}_{k+1} &= \frac{3(c^j)^2 - 4c^j - 12(c^j)^2 h + 12c^j h + 1}{2h^2}, \qquad \dot{f}_{k+2} = \frac{3(c^j)^2 - 2c^j}{2h^2}. \end{aligned} \tag{13}$$

By combining Eqs. (10) and (12), with Eqs. (9) and (9) the discrete Lagrangian reads

$$L_d(q_{k-1}, q_k, q_{k+1}, q_{k+2}) \approx h \sum_{j=1}^{S} W_j L(q(t_k + c^j h), \dot{q}(t_k + c^j h)). \tag{14}$$

Using the Hamilton's principle we can now compute the variations of the action sum S providing that the boundary points q_0,q_1,q_2 and q_{N-2},q_{N-1},q_N are kept fixed. Furthermore, by demanding the variations of the action to be zero for any choice of δq_k with $\delta q_0 = \delta q_1 = \delta q_2 = \delta q_N = \delta q_{N+1} = \delta q_{N+2} = 0$, we obtain

$$\begin{aligned} D_1 L_d(q_k, q_{k+1}, q_{k+2}, q_{k+3}, h) &+ D_2 L_d(q_{k-1}, q_k, q_{k+1}, q_{k+2}, h) + \\ D_3 L_d(q_{k-2}, q_{k-1}, q_k, q_{k+1}, h) &+ D_4 L_d(q_{k-3}, q_{k-2}, q_{k-1}, q_k, h) = 0 \end{aligned} \tag{15}$$

The latter equations are the discrete Euller-Lagrange equations stemming from the discrete Lagrangian Eq. (14) and must hold for each value of k.

We can now use the expressions (10) and (12) to derive the discrete Lagrangian of Eq. (14) for some specific physical problems. For the resulting Lagrangian, the discrete Euller–Lagrange equations of (15) will be used.

5. Simple harmonic oscillator

We first consider the simple harmonic oscillator described by the Lagrangian

$$L = \frac{1}{2}\dot{q}^2 - \frac{1}{2}\omega q^2 \tag{16}$$

The interpolation technique of Eqs. (10) defines the discrete Lagrangian

$$L_d(q_{k-1}, q_k, q_{k+1}, q_{k+2}, h) = \frac{h}{2}\left[\sum_{j=0}^{S} W_j\left(\dot{q}^j\right)^2 - \omega^2 \sum_{j=0}^{S} W_j\left(q^j\right)^2\right] \tag{17}$$

Subsequently the discrete Euler-Lagrange equations (15) give the variational integrator (see Ref. [4])

$$P_1 q_{k+3} + P_2 q_{k+2} + P_3 q_{k+1} + P_4 q_k + P_3 q_{k-1} + P_2 q_{k-2} + P_1 q_{k-3} = 0 \tag{18}$$

where

$$\begin{aligned} P_1 &= \sum_{j=0}^{S} W_j\left(\dot{f}_{k-1}\dot{f}_{k+2} - \omega^2 f_{k-1} f_{k+1}\right) \\ P_2 &= \sum_{j=0}^{S} W_j\left[\dot{f}_{k-1}\dot{f}_{k+1} + \dot{f}_k\dot{f}_{k+1} - \omega^2(f_{k-1}f_{k+1} + f_k f_{k+1})\right] \\ P_3 &= \sum_{j=0}^{S} W_j\left[\dot{f}_{k-1}\dot{f}_k + \dot{f}_k\dot{f}_{k+1} + \dot{f}_{k+1}\dot{f}_{k+2} - \omega^2(f_{k-1}f_k + f_k f_{k+1} + f_{k+1}f_{k+2})\right] \\ P_4 &= \sum_{j=0}^{S} W_j\left[\dot{f}^2_{k-1} + \dot{f}^2_k + \dot{f}^2_{k+1} + \dot{f}^2_{k+2} - \omega^2(f^2_{k-1} + f^2_k + f^2_{k+1} + f^2_{k+2})\right] \end{aligned} \tag{19}$$

For the above integrator since the first six points must be known, any numerical scheme can be used as starting method.

6. Summary and conclussions

In the present we have investigated the possibility of deriving multi-step numerical schemes by combining variational integrators theory and spline interpolation techniques. A special case of the proposed technique, that uses Catmull-Rom spline interpolationhas been derived for the numerical solution of physical problems with oscillating solution.

References

[1] E. Hairer, Ch. Lubich, G. Wanner, Geometric Numerical Integration. Structure-Preserving Algorithms. Springer, Berlin, 2002.

[2] B. Leimkuhler, S. Reich, Simulating Hamiltonian Dynamics. Cambridge Univ. Press, 2004.

[3] J. Wendlandt, J. Marsden, Mechanical integrators derived from a discrete variational principle. Physica D 106, 223–246, 1997.

[4] J. E. Marsen and M. West, Discrete mechanics and variational integrators, Acta Numer., vol. 10, 357–514, 2001.

[5] C. Kane, J. Marsden, M. Ortiz, Symplectic-energy-momentum preserving variational integrators. Journal of Mathematical Physics 40, 3353–3371, 1999.

[6] O. T. Kosmas, D. S. Vlachos, Phase-fitted discrete Lagrangian integrators. Computer Physics Communications 181, 562–568, 2010.

[7] E. Catmull, R. Rom, A class of local interpolating splines, Computer Aided Geometric Design, Academic Press, New York,1974.

Fractional exclusion statistics in systems with localized states

G. A. Nemnes[1,2] and D. V. Anghel[1]

[1] "Horia Hulubei" National Institute for Physics and Nuclear Engineering (IFIN-HH), 077126 Magurele-Ilfov, Romania

[2]University of Bucharest, Faculty of Physics, "Materials and Devices for Electronics and Optoelectronics" Research Center, P.O. Box MG-11, 077125 Magurele-Ilfov, Romania

Abstract. We develop a model based on the fractional exclusion statistics to describe systems with localized states. The local distribution of the energy levels is captured in the formalism by including the positions in the definition of the species. The particle distributions on the energy axis, as well as in the real space are determined for test-case systems with a peak/dip profile in the local density of states.

1. Introduction

The fundamentals of the fractional exclusion statistics (FES) were introduced by Haldane in his seminal work [1], while the thermodynamical properties of the ideal FES gas were calculated by Wu [2] and Isakov [3]. The formalism was amended recently to include the change of the FES parameters at the change of the particle species [4, 5, 6, 7].

The fractional exclusion statistics is a generalization of the Pauli exclusion principle and in its simplest form represents an interpolation between the Fermi and Bose statistics.

A number of physical systems have been shown to exhibit FES. Amongst them are the quasiparticle excitations at lowest Landau level in the fractional quantum Hall effect, spinon excitations in a spin-$\frac{1}{2}$ quantum antiferomagnet [1, 2], Bose and Fermi systems described in the thermodynamic Bethe ansatz [3, 8, 9, 10], excitations [1] or motifs of spins [11, 12] in spin chains, elementary volumes obtained by coarse-graining in the phase-space of a system [3, 13, 14, 4, 15, 16], etc.

In a more recent study [17], the transition rates in an ideal FES gas were established. The proposed stochastic method generalizes a previous approach for ideal Fermi and Bose gases [18]. The transition rates include the step and the acceptance probabilities. The former component of the transition rate, namely the step probability, is defined in accordance with the generalized exclusion principle. The method can be applied to study non-equilibrium phenomena in interacting Fermi and Bose systems, including disordered, glassy systems, considered from the perspective of ideal FES systems.

Here we extend the FES formalism in order to describe systems with localized states. Concrete applications reside in the area of semiconductor physics (e.g. Coulomb glasses), as well as condensed matter physics (e.g. bosons trapped in optical lattices). In our model the species are organized in order to include the classical degrees of freedom, which are the positions. In section

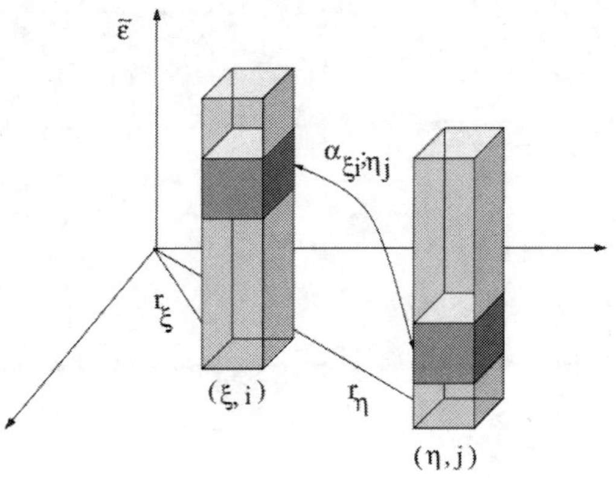

Figure 1. Partitioning of the real space and the energy axis. The statistical parameters $\alpha_{\xi i;\eta j}$ relate the species (ξi) and (ηj).

2 we present the main features of the formalism and the method of calculation, which includes the specification of the statistical parameters and a discussion of the numerical method employed. Section 3 contains the results obtained in the case of the non-uniform systems considered, which are systems with a peak/dip profile in the local density of states.

2. Formalism and Method

The system consists of a countable set of sites, randomly located at positions $\mathbf{r}_I$ in an s-dimensional lattice of total volume Ω. Each site may accommodate maximum one particle (fermion). In the absence of particle-particle interaction the energy of the particle on the site is ϵ_I.

We define the species as the elementary volumes in the $(s+1)$-dimensional position-energy volume, $\Omega \times \epsilon$. The species centered at $\mathbf{r} \times \epsilon$ will be denoted by $\delta\Omega_{\mathbf{r}} \times \delta\epsilon$. If the density of sites is $\sigma(\mathbf{r}, \epsilon)$, then in the species $\delta\Omega_{\mathbf{r}} \times \delta\epsilon$ we have $G_{\delta\Omega_{\mathbf{r}},\delta\epsilon} = \sigma(\mathbf{r}, \epsilon)\delta\Omega_{\mathbf{r}}\delta\epsilon$ states and $N_{\delta\Omega_{\mathbf{r}},\delta\epsilon}$ particles. For thermodynamic calculations we shall assume that both, $G_{\delta\Omega_{\mathbf{r}},\delta\epsilon}$ and $N_{\delta\Omega_{\mathbf{r}},\delta\epsilon}$ are much bigger than 1 and the range of ϵ will be from 0 to ∞ in any of the elementary volumes, $\delta\Omega_{\mathbf{r}}$.

The interaction energy between the particles is only position dependent and will be denoted by $V(|\mathbf{r}_\xi - \mathbf{r}_\eta|)$.

The quasiparticle energies will be defined following Ref. [15, 20], as

$$\tilde{\epsilon}_{\mathbf{r}_I} = \epsilon_{\mathbf{r}_i} + \sum_{\tilde{\epsilon}_{\mathbf{r}_J} < \tilde{\epsilon}_{\mathbf{r}_I}} V(|\mathbf{r}_I - \mathbf{r}_J|) n_{\mathbf{r}_J}. \tag{1}$$

where $n_{\mathbf{r}_I}$ is the occupation of the site located at $\mathbf{r}_I$.

In the quasi-continuous case, assuming that $\tilde{\epsilon}_{\mathbf{r}}(\epsilon) \geq \tilde{\epsilon}_{\mathbf{r}'}(\epsilon')$ if $\epsilon \geq \epsilon'$, Eq. (1) becomes

$$\tilde{\epsilon}_{\mathbf{r}}(\epsilon) = \epsilon_{\mathbf{r}} + \int_\Omega \mathrm{d}^s\mathbf{r} \int_0^\epsilon \sigma(\mathbf{r}, \epsilon') V(|\mathbf{r} - \mathbf{r}'|) n_{\mathbf{r}',\epsilon'}, \tag{2}$$

where $n_{\mathbf{r},\epsilon}$ is the (average) particle population, defined in any elementary volume as $n_{\mathbf{r},\epsilon} \equiv N_{\delta\Omega_{\mathbf{r}},\delta\epsilon}/G_{\delta\Omega_{\mathbf{r}},\delta\epsilon}$.

The definition (2) leads to the density of sites in the $\Omega \times \tilde{\epsilon}$ space, $\tilde{\sigma}(\mathbf{r}, \tilde{\epsilon}) = \sigma(\mathbf{r}, \epsilon)(\mathrm{d}\tilde{\epsilon}_{\mathbf{r}}(\epsilon)/\mathrm{d}\epsilon)^{-1}$, and to the FES parameters [15, 20]:

$$\alpha_{\xi i;\eta j} = \delta\Omega_\xi \left[\delta_{ij}\sigma(\mathbf{r}_\xi, \epsilon_i) + \theta(i-j)\delta\epsilon_i \frac{\partial\sigma(\mathbf{r}_\xi, \epsilon_i)}{\partial\epsilon_i}\right] \times V(|\mathbf{r}_\xi - \mathbf{r}_\eta|), \tag{3}$$

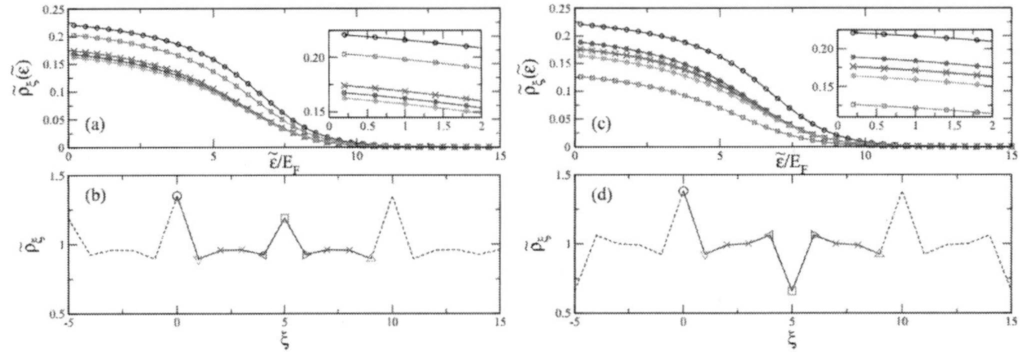

Figure 2. Particle densities for a peak-peak system (a-b) and for a peak-dip system (c-d). The symbols indicate the index of the species. The insets represent enlarged sections of the main plots.

where the first doublet, (ξi), specifies the species in which the number of states changes, whereas the second doublet, (ηj), specifies the species in which the number of particle changes; $\theta(k)$ is the step function, $\theta(k > 0) = 1$ and $\theta(k \leq 0) = 0$. Note that the parameters $\alpha_{\xi i;\eta j}$ obey the general rules of Ref. [5] and the ansatz of Ref. [16].

The populations, $n_{\xi i}$, are obtained by solving the non-linear system of equations for fermions [16],

$$0 = \beta(\mu - \tilde{\epsilon}_i) + \ln \frac{1 - n_{\xi i}}{n_{\xi i}} + \sum_{\eta j} \alpha_{\eta j;\xi i} \ln[1 - n_{\eta j}]. \tag{4}$$

Note that the populations $n_{\xi i}$ are in general different functions of energy for each site ξ.

The particle density as a function of quasi-particle energy can be expressed as $\tilde{\rho}(\mathbf{r}_\xi, \tilde{\epsilon}_i) \equiv \tilde{\rho}_{\xi i} = n_{\xi i}\tilde{\sigma}_{\xi i}$ and the spatially integrated density of particles in our discrete approach is $\tilde{\rho}_\xi = \sum_i \tilde{\rho}_{\xi i}$.

The solution of the nonlinear system is found iteratively using gradient descent starting from an initial guess solution. Our implementation used GSL for this purpose.

3. Applications

We consider a one dimensional interacting system of fermions with periodic boundary conditions. The local density of states is assumed to be independent on energy, $\sigma(\mathbf{r}, \epsilon) \equiv \sigma(\mathbf{r})$, and the particle-particle interaction is the screened Coulomb potential,

$$V(r; \gamma, \lambda) = \kappa \frac{\exp(-r/\lambda)}{r^\gamma},, \tag{5}$$

with $\gamma = 1$ and $\lambda = 3R_\xi$ being a finite screening length – R_ξ is the size of the species ξ. In order to remove the singularity at the origin, we take $V(r) = V(R_0) \equiv V_0$ for $r < R_0 = R_\xi$. Furthermore we set $V(R_\xi) = 1$.

For numerical calculations we partition the physical space into $N_R = 10$ segments ("volumes") and the energy axis into $N_E = 50$ segments. This leads to a number of $N_R \times N_E = 1000$ species. The system contains $N = N_R E_F \sigma$ particles, where the $E_F = 1$ is the Fermi energy. In this way we set up the energy scale. The following results are obtained at $k_B T = 1$.

We analyze two systems with non-uniform density of states. The first system is assumed to have two peaks in the otherwise constant density of states, i.e. $\sigma_{\xi_1=0} = 2.0$, $\sigma_{\xi_2=5} = 1.5$, while $\sigma_\xi = 1.0$ for the rest of the species. The results for the particle distributions on the energy axis,

$\tilde{\rho}_{\xi i}$, and in the positions space, $\tilde{\rho}_{\xi}$ are plotted in Fig. 2(a) and (b), respectively. One can see that the particles tend to accumulate in the two species with higher density of states. The first order neighboring species of both ξ_1 and ξ_2 exhibit a drop in the particle density, due to a larger repulsion induced by the particle accumulations at the two sites. Because of this, although the two groups of four species between the peaks have the same density of states ($\sigma_{\xi} = 1.0$), the particle densities are different.

Next we consider a system with a peak and a dip in the local density of states, i.e. $\sigma_{\xi_1=0} = 2.0$, $\sigma_{\xi_2=5} = 0.5$, while $\sigma_{\xi} = 1.0$ for the rest of the species. The results are plotted in Figs. 2(c) and 2(d). The major difference appears at the species with the lowest density of states. As expected, there is a smaller number of particles in species ξ_2 than in the rest of the species and because of this the density of particles rises from the species $\xi_1 + 1$ to $\xi_2 - 1$, as the particles are repelled from the accumulation at ξ_1.

4. Conclusions

We presented a general method based on the fractional exclusion statistics to extract the equilibrium properties of interacting systems with localized states. The formalism was illustrated on two test-cases of systems with non-uniform density of states – systems with peak/dip profiles in the local density of states.

The method may be extended to a variety of interacting Fermi and Bose systems and may represent an useful tool in calculating the thermodynamical properties of nonuniform interacting particle systems.

Acknowledgments

The work was supported by the Romanian National Authority for Scientific Research projects PN-II-ID-PCE-2011-3-0960 and PN09370102/2009. The travel support from the Romania-JINR Dubna collaboration project Titeica-Markov and project N4063 are gratefully acknowledged.

References

[1] Haldane F D M, 1991 *Phys. Rev. Lett.* **67** 937
[2] Wu Y S, 1994 *Phys. Rev. Lett.* **73** 922
[3] Isakov S B, 1994 *Phys. Rev. Lett.* **73** 2150
[4] Anghel D V, 2007 *J. Phys. A: Math. Theor.* **40** F1013, arXiv:0710.0724.
[5] Anghel D V, 2009 *Europhys. Lett.* **87** 60009, arXiv:0906.4836
[6] Anghel D V, 2010 *Phys. Rev. Lett.* **104** 198901
[7] Wu Y S, 2010 *Phys. Rev. Lett.* **104** 198902
[8] Bernard D and Wu Y S, 1995 *New Developments on Integrable Systems and Long-Ranged Interaction Models*, edited by M. L. Ge and Y. S. Wu (World Scientic, Singapore), p. 10, cond-mat/9404025.
[9] Potter G G, Müller G, and Karbach M, 2007 *Phys. Rev.* E **75** 61120
[10] Potter G G, Müller G, and Karbach M, 2007 *Phys. Rev.* E **76** 61112
[11] Liu D, Lu P, Müller G, and Karbach M, 2011 *Phys. Rev.* E **84** 021136
[12] Liu D, Vanasse J, Müller G, and Karbach M, 2012 *Phys. Rev.* E **85** 011144
[13] Sutherland B, 1997 *Phys. Rev.* B **56** 4422
[14] Iguchi K and Sutherland B, 2000 *Phys. Rev. Lett.* **85** 2781
[15] Anghel D V, 2008 *Phys. Lett.* A **372** 5745, arXiv:0710.0728.
[16] Anghel D V, 2010 *Europhys. Lett.* **90** 10006, arXiv:0909.0030.
[17] Nemnes G A and Anghel D V, 2010 *J. Stat. Mech.* P09011
[18] Guastella I, Bellomonte L and Sperandeo-Mineo R M, 2009 *J. Stat. Mech.* P02021
[19] Anghel D V, 2012, arXiv:1204.4064.
[20] Anghel D V, 2012 *Phys. Lett.* A **376** 892
[21] Tsallis C, 2002 *Chaos, Solitons and Fractals* **13** 371
[22] Anghel D V, 2002 *J. Phys. A: Math. Gen.* **35** 7255

Fractional exclusion statistics: the concept and some applications

Dragoş-Victor Anghel

Horia Hulubei National Institute of Physics and Nuclear Engineering, Măgurele, Jud. Ilfov, Romania

E-mail: dragos@theory.nipne.ro

Abstract. I present briefly the concept of fractional exclusion statistics (FES) and I analyze two models of interacting particle systems. After I show that the models are equivalent by transforming one into the other by redefining some terms in the Hamiltonians, I calculate the heat capacity of the system and compare it with the heat capacity of an ideal Bose or Fermi gas. I show that under certain conditions the heat capacities may not be equal. I show that the interacting particle system may always be described as an ideal gas obeying FES and I show the method to construct such a gas.

1. Introduction

Considerable insight may be gained into the microscopic and macroscopic physical properties of a system of interacting particles if this can be described in a certain approximation as a gas of ideal quasiparticles. The difficulty in finding such a description is due to the fact that in general the interaction between the particles cannot be completely removed in the quasiparticle formulation or the physical properties of the quasiparticle gas do not follow closely enough the physical properties of the original system. A method by which a system of interacting particles may be described as an ideal gas of particles with identical physical properties is provided by the fractional exclusion statistics (FES).

The concept of FES was introduced by Haldane [1] and the thermodynamic properties of FES systems were calculated mainly by Wu [2] and Isakov [3]. A FES system consists of a countable number of species of particles, indexed here by capital letters, $I, J = 0, 1, \ldots$. Each species consists of a finite number of single-particle states, G_I, and particles, N_I – G_I will also be called the *dimension* of the species I. The total number of particles in the system is $N = \sum_I N_I$.

The *key* characteristics of FES is that the dimensions of the species depend on the particle populations, namely if N_I changes by δN_I, for any I, then G_J changes by $\delta G_J = -\alpha_{JI}\delta N_I$ for any J. The coefficients α_{IJ} are called the FES parameters. If I denote by $G_I^{(0)}$ the dimensions of the species in the absence of any particles in the system ($N = 0$), then for a given particle distribution, say $\{N_I\}$, the species dimensions are

$$G_I = G_I^{(0)} - \sum_J \alpha_{IJ} N_J. \quad (1)$$

If the particles are bosons, then the total number of micro-configurations in the system is [2, 4]

$$W^{(-)}(\{N_I\}) = \prod_I \frac{(G_I^{(0)} + N_I - 1 - \sum_J \alpha_{IJ} N_J)!}{N_I!(G_I^{(0)} - 1 - \sum_J \alpha_{IJ} N_J)!} = \prod_I \frac{(G_I + N_I - 1)!}{N_I!(G_I - 1)!}, \tag{2a}$$

whereas if they are fermions, it is

$$W^{(+)}(\{N_I\}) = \prod_I \frac{(G_I^{(0)} - \sum_J \alpha_{IJ} N_J)!}{N_I!(G_I^{(0)} - N_I - \sum_J \alpha_{IJ} N_J)!} = \prod_I \frac{G_I!}{N_I!(G_I - N_I)!}. \tag{2b}$$

In general, by the superscripts $(-)$ and $(+)$ I shall denote Bose and Fermi systems, respectively. If the numbers $\{G_I\}$, $\{G_I^{(0)}\}$ and $\{N_I\}$ are integers whereas α_{IJ}'s are real, then the numbers of the type $G_I^{(0)} - \sum_J \alpha_{IJ} N_J$ that appear in the Eqs. (2) are defined up to an arbitrary constant of the order 1. This choice has no physical relevance in the thermodynamic limit, when all the number involved are much bigger than 1. Nevertheless, in mesoscopic systems such a constant cannot be neglected in rigorous calculations. Then the physical quantities that remain an integers are G_I and N_I [5, 6, 7].

Denoting $\beta = 1/(k_B T)$, where T is the temperature of the system, introducing the notation $n_I \equiv N_I/G_I$, and associating an energy ϵ_I and a chemical potential, μ_I $(I = 0, 1, \ldots)$ to each of the species in the system, one can write the grandcanonical partition function,

$$\mathcal{Z}^{(\pm)} = W^{(\pm)} \prod_I \exp\left[\beta(\epsilon_I - \mu_I) n_I G_I\right], \tag{3}$$

which, if it is maximized with respect to the n_I's leads to the equations for the equilibrium populations [4],

$$\beta(\mu_I - \epsilon_I) + \ln \frac{[1 \mp n_I^{(\pm)}]}{n_I^{(\pm)}} = \mp \sum_J \alpha_{JI} \ln[1 \mp n_J^{(\pm)}]. \tag{4}$$

In Eq. (4) the upper signs stand for fermions and the lower ones for bosons. For ideal Bose and Fermi gases $\alpha_{IJ} = 0$ for any I and J and Eq. (4) renders the well-known Fermi (for upper signs) and Bose (for lower signs) populations, as expected.

2. FES in general interacting particle systems

The FES is the general method by which interacting particle systems can be described as ideal gases of quasiparticles. Let me exemplify it here by two generic models.

2.1. The models

In the *first model* the total energy of the system is equal to the sum of the energies of the quasiparticles, namely

$$E(\{n_{\mathbf{i}}\}) \equiv \sum_{\mathbf{i}} \tilde{\epsilon}_{\mathbf{i}} n_{\mathbf{i}}, \tag{5}$$

where $\mathbf{i}$ are sets of quantum numbers, each set defining a quasiparticle state; $\tilde{\epsilon}_{\mathbf{i}}$ and $n_{\mathbf{i}}$ are the quasiparticle energy and the population of the state $\mathbf{i}$, respectively. The entropy of the system has the typical form,

$$S^{(\pm)}(T,\mu) = \mp k_B \sum_{\mathbf{i}} \left\{ \left[1 \mp n_{\mathbf{i}}^{(\pm)}(T,\mu)\right] \log\left[1 \mp n_{\mathbf{i}}^{(\pm)}(T,\mu)\right] \pm n_{\mathbf{i}}^{(\pm)}(T,\mu) \log\left[n_{\mathbf{i}}^{(\pm)}(T,\mu)\right] \right\}, \tag{6}$$

where again the upper signs are for fermions, the lower ones are for bosons; by $n_{\mathbf{i}}^{(\pm)}(T,\mu)$ I denote the equilibrium population of the state $\mathbf{i}$, at temperature T and chemical potential

μ. If $U^{(\pm)}(T,\mu) \equiv \sum_{\rm i} \tilde{\epsilon}_{\rm i} n_{\rm i}(T,\mu)$ and $N^{(\pm)}(T,\mu) \equiv \sum_{\rm i} n_{\rm i}(T,\mu)$ are the (equilibrium) internal energy and total particle number of the system, then the equilibrium populations, $n_{\rm i}^{(\pm)}(T,\mu)$, are determined by minimizing the grandcanonical potential,

$$\Omega^{(\pm)}(T,\mu) = U^{(\pm)}(T,\mu) - TS^{(\pm)}(T,\mu) - \mu N^{(\pm)}(T,\mu), \tag{7}$$

with respect to $n_{\rm i}^{(\pm)}$'s. This procedure leads to the populations

$$n_{\rm i}^{(\pm)}(T,\mu) = \left[e^{\beta(\tilde{\epsilon}_{\rm i}^{(1)}-\mu)} \pm 1\right]^{-1}, \tag{8}$$

where

$$\tilde{\epsilon}_{\rm i}^{(1)} = \tilde{\epsilon}_{\rm i} + \sum_{\rm j} \frac{\partial \epsilon_{\rm j}}{\partial n_{\rm i}} n_{\rm j} \tag{9}$$

are the *new* quasiparticle energies.

We observe that the populations (8) are not the ideal Bose or Fermi gas populations, with the energies $\tilde{\epsilon}_{\rm i}$, but they depend on the more complicated new quasiparticle energies, $\tilde{\epsilon}_{\rm i}^{(1)}$. Moreover,

$$\tilde{E}^{(1)}(\{n_{\rm i}\}) \equiv \sum_{\rm i} \tilde{\epsilon}_{\rm i}^{(1)} n_{\rm i} = \sum_{\rm i} \tilde{\epsilon}_{\rm i} n_{\rm i} + \sum_{\rm i}\sum_{\rm j} \frac{\partial \tilde{\epsilon}_{\rm j}}{\partial n_{\rm i}} n_{\rm i} n_{\rm j} \equiv E(\{n_{\rm i}\}) + \frac{1}{2}\sum_{\rm i}\sum_{\rm j} \tilde{V}_{\rm ij} n_{\rm i} n_{\rm j} \tag{10}$$

where I have identified

$$\tilde{V}_{\rm ij} \equiv 2\partial \epsilon_{\rm j}/\partial n_{\rm i}. \tag{11}$$

In the *second model* we start with a total energy of the system of the Fermi liquid theory form [8],

$$E(\{n_{\rm i}\}) = \sum_{\rm i} \epsilon_{\rm i} n_{\rm i} + \frac{1}{2}\sum_{\rm ij} V_{\rm ij} n_{\rm i} n_{\rm j}, \tag{12}$$

where $\epsilon_{\rm i}$ is the free particle energy and $V_{\rm ij}$ characterizes the particle-particle interaction energy.

The procedure for finding the physical properties of this system is similar to the procedure employed for the previous model. We write the entropy like in Eq. (6), the grandcanonical potential like in Eq. (7) and determine the equilibrium particle distribution by minimizing $\Omega^{(\pm)}$ with respect to the populations. The equilibrium population has again the expression (8), but with a quasiparticle energy,

$$\tilde{\epsilon}_{\rm i}^{(2)} = \frac{\partial E}{\partial n_{\rm i}} = \epsilon_{\rm i} + \sum_{\rm j} V_{\rm ij} n_{\rm j}, \tag{13}$$

The sum of the quasiparticle energies is

$$\tilde{E}^{(2)}(\{n_{\rm i}\}) \equiv \sum_{\rm i} \tilde{\epsilon}_{\rm i}^{(2)} n_{\rm i} = \sum_{\rm i} \epsilon_{\rm i} n_{\rm i} + \sum_{\rm i}\sum_{\rm j} V_{\rm ij} n_{\rm i} n_{\rm j} \equiv E(\{n_{\rm i}\}) + \frac{1}{2}\sum_{\rm i}\sum_{\rm j} V_{\rm ij} n_{\rm i} n_{\rm j}. \tag{14}$$

We observe that the *model 1* and *model 2* are very similar to each other. They can be viewed as completely equivalent if we redefine the quasiparticle energies of model 2 as

$$\tilde{\epsilon}_{\rm i} \equiv \epsilon_{\rm i} + \frac{1}{2}\sum_{\rm j} V_{\rm ij} n_{\rm j}. \tag{15}$$

With this redefinition, the total energy of the system in the *model 2* has the same expression (5) as in *model 1*, whereas the identification (11) makes the expressions for the total quasiparticle energies, $\tilde{E}^{(1)}$ and $\tilde{E}^{(2)}$, similar. Now we can discuss only *model 1* and the results may be immediately extended to *model 2*.

2.2. Transforming the interacting particle systems into ideal particle systems

Let us first check if the thermodynamic properties of these interacting particle systems are those of ideal Bose or Fermi gases. For this let me calculate the heat capacity, $C_V^{(\pm)} \equiv (\partial U^{(\pm)}/\partial T)_{N=\mathrm{const.}} = T \equiv (\partial S^{(\pm)}/\partial T)_{N=\mathrm{const.}}$, by using the formula

$$\frac{C_V^{(\pm)}}{T} \equiv \left(\frac{\partial S^{(\pm)}}{\partial T}\right)_N = \left(\frac{\partial S^{(\pm)}}{\partial T}\right)_\mu - \left(\frac{\partial S^{(\pm)}}{\partial \mu}\right)_T \left(\frac{\partial N^{(\pm)}}{\partial T}\right)_\mu \left(\frac{\partial N^{(\pm)}}{\partial \mu}\right)_T^{-1}, \tag{16}$$

with

$$\frac{\partial S^{(\pm)}}{\partial T} = k_\mathrm{B} \sum_\mathrm{i} \ln\left(\frac{1-n_\mathrm{i}^{(\pm)}}{n_\mathrm{i}^{(\pm)}}\right) \frac{\partial n_\mathrm{i}^{(\pm)}}{\partial T}, \quad \frac{\partial S^{(\pm)}}{\partial \mu} = k_\mathrm{B} \sum_\mathrm{i} \ln\left(\frac{1-n_\mathrm{i}^{(\pm)}}{n_\mathrm{i}^{(\pm)}}\right) \frac{\partial n_\mathrm{i}^{(\pm)}}{\partial \mu}, \tag{17a}$$

$$\frac{\partial N^{(\pm)}}{\partial T} = \sum_\mathrm{i} \frac{\partial n_\mathrm{i}^{(\pm)}}{\partial T}, \quad \frac{\partial N^{(\pm)}}{\partial \mu} = \sum_\mathrm{i} \frac{\partial n_\mathrm{i}^{(\pm)}}{\partial \mu}. \tag{17b}$$

Since $n_\mathrm{i}^{(\pm)}(T,\mu)$ is given by Eq. (8), taking into account that $\tilde{\epsilon}_\mathrm{i}^{(2)}$ depends on both, T and μ, due to its dependence on $\{n_\mathrm{i}\}$, then the temperature variation of $n_\mathrm{i}^{(\pm)}$ should be calculated from the system

$$\frac{\partial n_\mathrm{i}^{(\pm)}(T,\mu)}{\partial(k_B T)} = \frac{\beta^2(\tilde{\epsilon}_i^L-\mu)e^{\beta(\tilde{\epsilon}_i^L-\mu)}}{\left[e^{\beta(\tilde{\epsilon}_i^L-\mu)}+1\right]^2}\left[1-\frac{1}{\beta(\tilde{\epsilon}_i^L-\mu)}\sum_j V_\mathrm{ij}\frac{\partial n_\mathrm{j}^{(\pm)}(T,\mu)}{\partial(k_B T)}\right], \tag{18a}$$

$$\frac{\partial n_\mathrm{i}^{(\pm)}(T,\mu)}{\partial \mu} = \frac{\beta e^{\beta(\tilde{\epsilon}_i^L-\mu)}}{\left[e^{\beta(\tilde{\epsilon}_i^L-\mu)}+1\right]^2}\left[1-\sum_j V_\mathrm{ij}\frac{\partial n_\mathrm{j}^{(\pm)}(T,\mu)}{\partial \mu}\right]. \tag{18b}$$

From Eqs. (16), (17) and (18) we conclude that the heat capacity of the system has the temperature dependence of an ideal Bose or Fermi gas if the expressions in the square brackets in Eqs. (18) are equal (or close enough) to 1.

If the expressions in the parentheses in Eqs. (18) differ significantly from 1, then the system is not *thermodynamically equivalent* (i.e. does not have the same heat capacity and entropy) [9] with an ideal Bose or Fermi system.

To show how we can transform the gas of *model 1* or *model 2* into an ideal FES gas, let us use the continuous limit and define the density of states (DOS), $\tilde{\sigma}(\tilde{\epsilon})$. We follow the procedure of Ref. [10] and coarse-grain the $\tilde{\epsilon}$ axis into elementary, fixed intervals, say $\delta\tilde{\epsilon}_I \equiv \tilde{\epsilon}_{I+1} - \tilde{\epsilon}_I$. Each interval represents a species of particles, like the ones described in Section 1. Each species contains $G_I = \tilde{\sigma}(\tilde{\epsilon})\delta\epsilon_I$ states and $N_I \equiv n(\tilde{\epsilon}_I)\tilde{\sigma}(\tilde{\epsilon})\delta\epsilon_I$ particles, where $n(\tilde{\epsilon}_I) \equiv N_I/G_I$. Since the energies $\tilde{\epsilon}_\mathrm{i}$ depend on the populations, this implies that $\tilde{\sigma}(\tilde{\epsilon})$ depends also on the populations, so by including a number, δN_I of particles into the species I, the DOS along the whole $\tilde{\epsilon}$ axis may change. But since the intervals $\delta\tilde{\epsilon}_I$ are fixed and if we fix also the number of particles in each interval (except for the variation δN_I), then the variation of DOS changes the number of states in the species and this gives rise to FES. The FES parameters are [10]

$$\alpha_{\tilde{\epsilon}_J\tilde{\epsilon}_I} = -\delta\tilde{\epsilon}_J \frac{d}{d\tilde{\epsilon}}\left[\frac{\delta\tilde{\epsilon}_J}{\delta\tilde{\rho}(\tilde{\epsilon}_I)}\tilde{\sigma}(\tilde{\epsilon}_J)\right], \tag{19}$$

where $\tilde{\epsilon}_J$ is the species where it is observed the change in the number of states, after the inclusion of the particles in the species I.

3. Conclusions

After a brief introduction of the fractional exclusion statistics (FES) formalism, I analyzed the thermodynamic properties of two types of interacting particle systems. I showed that the systems are equivalent from the thermodynamics perspective and they can be transformed one into the other by a redefinition of some terms in the Hamiltonian. I also showed that such interacting particle systems may not always be *thermodynamically equivalent* [9] with a Bose or a Fermi ideal gas, but they can always be described as ideal gases of FES quasiparticles. I also showed the procedure for finding such an ideal FES gas description.

Acknowledgments

Discussions with dr. Gerhard Müller and dr. Alexandru Nemneş are gratefully acknowledged. The work was supported by the Romanian National Authority for Scientific Research CNCS-UEFISCDI projects PN-II-ID-PCE-2011-3-0960 and PN09370102/2009. The travel support from the Romania-JINR Dubna collaboration project Titeica-Markov and project N4063 are gratefully acknowledged.

References

[1] Haldane F D M 1991 *Phys. Rev. Lett.* **67** 937
[2] Wu Y S 1994 *Phys. Rev. Lett.* **73** 922
[3] Isakov S B 1994 *Phys. Rev. Lett.* **73** 2150
[4] Anghel D V 2010 *EPL* **90** 10006 arXiv:0909.0030
[5] Liu D, Vanasse J, Müller G and Karbach M 2012 *Phys. Rev. E* **85** 011144
[6] Liu D, Lu P, Müller G and Karbach M 2011 *Phys. Rev. E* **84** 021136
[7] Lu P, Liu D, Müller G and Karbach M 2012 *Cond. Matt. Phys.* **15** 13001
[8] Pines D and Noziéres P 1966 *The Theory of Quantum Liquids* (W. A. Benjamin, Inc., New York, Amsterdam)
[9] Lee M H 1997 *Phys. Rev. E* **55** 1518
[10] Anghel D V 2012 Fractional exclusion statistics – the method to describe interacting particle systems as ideal gases arXiv:1207.6534, to appear in Physica Scripta T

IC-MSQUARE 2012: International Conference on Mathematical Modelling in Physical Sciences IOP Publishing
Journal of Physics: Conference Series **410** (2013) 012122 doi:10.1088/1742-6596/410/1/012122

Convolution/deconvolution of generalized Gaussian kernels with applications to proton/photon physics and electron capture of charged particles

W. Ulmer
Klinikum München-Pasing, Department of Radiation Therapy and MPI of Physics, Göttingen, Germany

Email: waldemar.ulmer@gmx.net

Abstract. Scatter processes of photons lead to blurring of images produced by CT (computed tomography) or CBCT (cone beam computed tomography). Multiple scatter is described by, at least, one Gaussian kernel. In various tasks, this approximation is crude; we need two/three Gaussian kernels to account for long-range tails (Landau tails), which appear in Molière scatter of protons, energy straggling and electron capture of charged particles passing through matter and Compton scatter. The ideal image (source function) is subjected to Gaussian convolution to yield a blurred image. The inverse problem is to obtain the source image from a detected image. Deconvolution methods of linear combinations of two/three Gaussian kernels with different parameters s_0, s_1, s_2 can be derived via an inhomogeneous Fredholm integral equation of second kind (IFIE2) and Liouville - Neumann series (LNS) to provide the source function ρ. Scatter functions s_0, s_1, s_2 are best determined by Monte-Carlo. An advantage of LNS is given, if the scatter functions s_0, s_1, s_2 depend on coordinates. The convergence criterion can always be satisfied with regard to the above mentioned cases. A generalization is given by an analysis of the Dirac equation and Fermi-Dirac statistics leading to Landau tails applied to Bethe-Bloch equation (BBE) of charged particles and electron capture.

1. Introduction

Multiple scatter can be treated with, at least, one single Gaussian kernel [1], which we abbreviate by K(s, u − x), but it may refer to more space dimensions. The ideal image (source function ρ without any blurring) is subjected to a Gaussian convolution in order to yield an image function ϕ (blurred image):

$$\phi = \int K(s,u-x)\cdot\rho(u)du \text{ with } K = \frac{1}{s\cdot\sqrt{\pi}}\cdot\exp(-(u-x)^2/s^2) \quad (1)$$

The magnitude of the parameter s represents a measure of the blurring strength that as $s \to 0$ the kernel K tends to the δ-distribution and ϕ becomes identical with ρ. In many situations the restriction to one Gaussian kernel represents a crude approximation, and we need a linear combination of Gaussian kernels with K_g as a resulting convolution kernel to account for long-range tails, which appear in the Molière multiple scatter theory of protons with accounting for Landau tails [2, 3] or in Compton scatter [1]:

$$\left.\begin{aligned} K_g &= c_0\cdot K(s_0,u-x)+c_1\cdot K(s_1,u-x)+c_2\cdot K(s_2,u-x) \\ \phi &= \int K_g(s_0,s_1,s_2,c_0,c_1,c_2,u-x)\cdot\rho(u)du. \end{aligned}\right\} \quad (2)$$

In every case, the parameters in Eq. (2) have to satisfy $c_0+c_1+c_2 = 1$, $c_0 > c_1$, $c_0 > c_2$ and $s_0 < s_1$, $s_0 < s_2$. The restriction to two Gaussian kernels results by setting $c_2 = 0$. If $c_2 \neq 0$, $c_1 < 0$ can also be valid, but $K_g \geq 0$

must exist. The previous study [1] of the inverse task of K_g requires $s_1 \wedge s_2 > s_0 \cdot \sqrt{2}$. This restriction may lead to critical cases (proton dosimetry, image processing with CBCT). The LNS method can circumvent this restriction, since it only needs that $s_1 \wedge s_2 > s_0$ is satisfied. The inverse task of Eq. (1) is to determine the ideal source image from a determined image. If the scatter parameters are known, we are able to calculate the idealistic source ϱ by $K^{-1}(s, u - x)$. A possible representation [1] of K^{-1} is given by:

$$\left.\begin{aligned} &K^{-1}(s,u-x) = \sum_{n=0}^{N} c_n(s) \cdot H_{2n}\left(\tfrac{u-x}{s}\right) \cdot K(s,u-x) \text{ and } \rho(x) = \int K^{-1}(s,u-x) \cdot \phi(u)\,du. \\ &N \to \infty;\ c_n = (-1)^n \cdot s^{2n} / (2^n \cdot n!);\ (n = 0, 1, \ldots, \infty) \end{aligned}\right\} \quad (3)$$

Due to many applications of $K^{-1}(s, u-x)$ it represents is a proven tool circumventing ill-posed aspects [1, 4 – 7]. H_{2n} refer to Hermite polynomials of even order. The coefficients c_n of the two-point Hermite polynomials of K^{-1} are determined by a Lie series expansion (in practice: $N < \infty$). K and K^{-1} can be derived as Green's functions [5, 7]. The intention of this study is to derive $K_g^{-1}(s_0, c_0, s_1, c_1, s_2, c_2, u - x)$ of a linear combination of two/three Gaussian convolution kernels according to Eq. (2) and to give some examples via an IFIE2 and related LNS.

2. Methods

The basic formulas of all calculations are the following two operator functions, defined as Lie series:

$$O^{\pm 1} = \exp\left(\mp \tfrac{1}{4} \cdot s^2 \cdot \tfrac{d^2}{dx^2}\right) = 1 + \sum_{n=1}^{\infty} \tfrac{s^{2n}}{n! \cdot 4^n} \cdot (\mp 1)^n \cdot d^{2n} / dx^{2n}. \qquad (4)$$

O^{+1} and O^{-1} obey the following relation $O^{-1} \cdot O^{+1} = O^{+1} \cdot O^{-1} = 1$ (unit operator). In three dimensions, we have to substitute d^2/dx^2 by the 3D Laplace operator Δ. We point out that the operator O^{-1}results from an operator calculus of the Boltzmann canonical ensemble $exp(-H/k_B \cdot T)$ connected to a free particle Schrödinger equation $H = -(\hbar^2/2m) \cdot \Delta$. If ϱ(x) represents a source and ϕ(x) an image function, the following relationships have to be valid:

$$\phi(x) = O^{-1} \cdot \rho(x) \textit{ and } \rho(x) = O \cdot \phi(x). \qquad (5)$$

$$\phi(x) = \int K(s,u-x) \cdot \rho(u)du \textit{ and } \rho(x) = \int K^{-1}(s,u-x) \cdot \phi(u)du. \ (6)$$

The equivalence between Eq. (5) and Eq. (6) has been previously shown [1]. In Eq. (5) the class of permitted functions is C^∞, whereas in Eq. (6) we can consider the spaces either L_1 or L_2.

The differential operator formulation of Eq. (2) is:

$$\phi(x) = O_g^{-1} \cdot \rho(x) = [c_0 \cdot O_0^{-1}(s_0) + c_1 \cdot O_1^{-1}(s_1) + c_2 \cdot O_2^{-1}(s_2)] \cdot \rho(x). \qquad (7)$$

In every case, the condition $O_g^{-1} \cdot O_g^{+1} = 1$ has to be satisfied. According to a previous study [7] we obtain:

$$\left.\begin{aligned} &O_g = [c_0 \cdot O_0^{-1} + c_1 \cdot O_1^{-1} + c_2 \cdot O_2^{-1}]^{-1} . \phi \\ &[c_0 \cdot O_0^{-1} + c_1 \cdot O_1^{-1} + c_2 \cdot O_2^{-1}] \cdot \rho \Rightarrow \tfrac{1}{c_0} \cdot O_0 \cdot [1 + \tfrac{c_1}{c_0} \cdot O_0 \cdot O_1^{-1} + \tfrac{c_2}{c_0} \cdot O_0 \cdot O_2^{-1}]^{-1} \cdot \phi = \rho. \\ &\tfrac{1}{c_0} \cdot O_0 \cdot \phi = [1 + \tfrac{c_1}{c_0} \cdot O_0 \cdot O_1^{-1}] \cdot \rho \end{aligned}\right\} \quad (8)$$

We use the following abbreviation: $f(x,y,z) = \frac{1}{c_0} \cdot \int K_0^{-1}(s_0, \vec{u} - \vec{x}) \cdot \phi(\vec{u}) d^3u$. (9)

Function f incorporates the inhomogeneous part of the Fredholm integral equation of second kind (IFIE2).

In order to derive an alternative method to solve the inverse problem of a linear combination of Gaussian convolutions, we consider Eq. (8) with regard to two kernels (the generalization to $c_2 \neq 0$ is stated elsewhere [8]). By that, we obtain the desired formula, which can be transformed to an IFIE2:

$$\left.\begin{array}{l} f(\vec{x}) = \rho(\vec{x}) + \frac{c_1}{c_0} \cdot \int \rho(\vec{u}) \cdot K_f(\sigma, \vec{u} - \vec{x}) d^3u \\ \sigma^2 = s_1^2 - s_0^2 \end{array}\right\}. \tag{10}$$

The LNS procedure consists of iterated procedures, and the resolving kernel K_{res} is given by:

$$\left.\begin{array}{l} K_{fn)}(\vec{u}-\vec{x}) = \int\int .... \int K_f(\sigma,\vec{u}-\vec{u}_1) \cdot K_f(\sigma,\vec{u}_1-\vec{u}_2) \cdot .... \cdot K_f(\sigma,\vec{u}_{n-1}-\vec{x}) d^3u_1 d^3u_2 ....d^3u_{n-1} \\ K_{res}(\vec{u}-\vec{x},\lambda) = \sum_{n=0}^{L} \lambda^n \cdot K_{fn+1)}(\vec{u}-\vec{x}) \ (L \to \infty) \end{array}\right\} \tag{11}$$

The solution of the integral equation becomes:

$$\rho(\vec{x}) = \int K_{res}(\vec{u}-\vec{x},\lambda) \cdot f(\vec{u}) d^3u \ . \tag{12}$$

The application of BBE for the determination of the electronic stopping power is established for the passage of protons through homogeneous media. A particular importance of BBE appears in Monte-Carlo calculations to simulate charged projectile particles along the track. This equation reads:

$$\left.\begin{array}{l} -dE(z)/dz = (K/\mathrm{v}^2) \cdot [\ln(2m\mathrm{v}^2/E_I) - \ln(1-\beta^2) + a_{shell} + a_{Barkas} + a_0 \mathrm{v}^2 + a_{Bloch}] \\ K = (Z\rho/A_N) \cdot 8\pi q^2 e_0^4 / 2m \end{array}\right\}. \tag{13}$$

Parameters and details of Eq. (13) are given in [4, 8]. The integration of BBE provides the range formula:

$$R_{CSDA} = \sum_{n=1}^{N} a_n E_0^n \quad (N \Rightarrow \infty) \tag{14}$$

In practical problems N = 4 is sufficient. The inversion can be obtained via Gompertz functions (N = 5):

$$E_0 = R_{csda} \sum_{i=1}^{N} c_k \exp(-\lambda_k R_{csda}) \ (N \to \infty) \text{ and } E(z) = (R_{csda} - z) \sum_{i=1}^{N} c_k \exp(-\lambda_k (R_{csda} - z)) \tag{15}$$

The stopping power is determined by dE(z)/dz and yields:

$$\left.\begin{array}{l} S(z) = dE(z)/dz = -E(z)/(R_{CSDA} - z) + \sum_{k=1}^{N} \lambda_k E_k(z) \ (N \to \infty) \\ E_k(z) = c_k (R_{CSDA} - z) \cdot \exp[-\lambda_k (R_{CSDA} - z)] \end{array}\right\}. \tag{16}$$

Due to electron capture formula (16) is only applicable to protons, whereas for heave carbons we have to account for $q^2(E)$, which can be described by generalized convolutions, based on Fermi-Dirac statistics. With regard to our task Dirac equation and Fermi-Dirac statistics to describe the particle motion are an adequate starting-point (E_F: energy of the Fermi edge, d_s: density of states):

$$\left.\begin{array}{l} H_D = c\vec{\alpha}\,\vec{p} + \beta mc^2; \ \hat{H} = H_D - E_F \\ f_F(\hat{H}) = \frac{1}{1+\exp[(H_{D-E_F})/E_{ex}]} \cdot d_s(H_D) \end{array}\right\} \tag{17}$$

The application of Eq. (17) to BBE finally provides the stopping power S_E and $q^2(E)$:

$$\left.\begin{array}{l} S_E = N_f \cdot \exp(-(E_n(k) - E_{Average,n})^2 / 2\sigma_E(n)^2) \cdot \sum_{l=0}^{\infty} b_l(n, mc^2) \cdot (E_n(k)/2E_{ex})^l \\ q^2(E) = q_0^2[erf(E/s_E) - A \cdot (1 - E/E_0) \cdot (1 - (E/E_0) \cdot \exp(-E^2/s_E^2)] \\ s_E^2 = q_0^3 \cdot \pi \cdot m^2c^4(1 + 1/M^2c^4);\; A = q_0^2 \cdot \pi^2 \cdot mc^2 / Mc^2 \\ Rcsda = \alpha(E_0 \cdot N / q_{eff}^2) + \beta(E_0 \cdot N / q_{eff}^2)^2 + \gamma(E_0 \cdot N / q_{eff}^2)^3 + \delta(E_0 \cdot N / q_{eff}^2)^4 \\ \mathrm{E_0: initial\ energy/per\ nucleon;\ N: nucleon\ number} \end{array}\right\} \quad (18)$$

Please note that the parameters [4] must be modified: $\alpha = 0.0069465598$; $\beta = 0.0008132157$; $\gamma = -0.00000121069$; $\delta = 0.000000001051$. If $N = 1$ and $q_{eff} = 0.995$ the above formula is valid for protons.

3. Applications

Parameters necessary for calculations can be found in [4, 8]. With regard to LNS in image processing we consider an image recorded by CBCT (Figures 1 – 3), 140 kV X-rays.

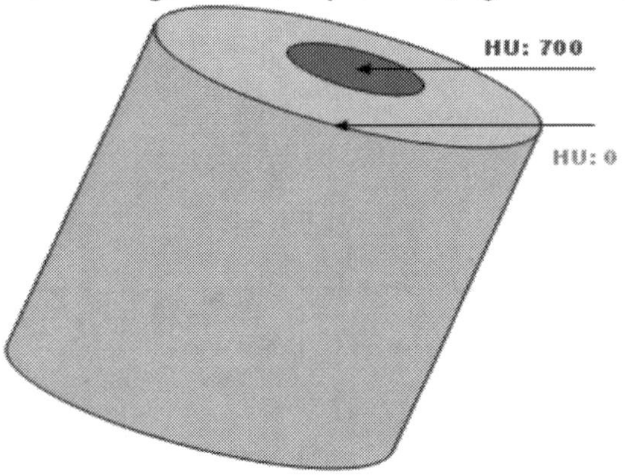

Figure 1. Phantom (left side) for a CBCT check.

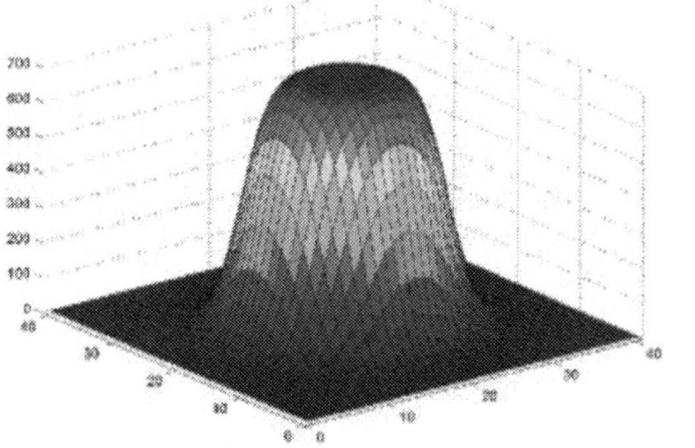

Figure 2. Measured 2D dose distribution (in Hounsfield values).

The result of deconvolution via LNS is shown in Figure 3; details are given in [8].

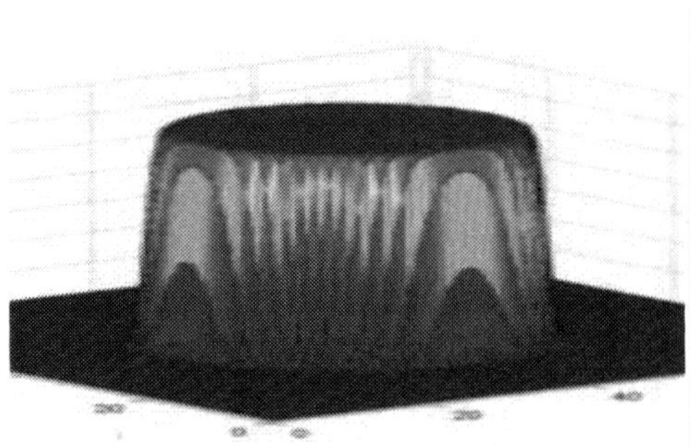

Figure 3. Deconvolution of Figure2.

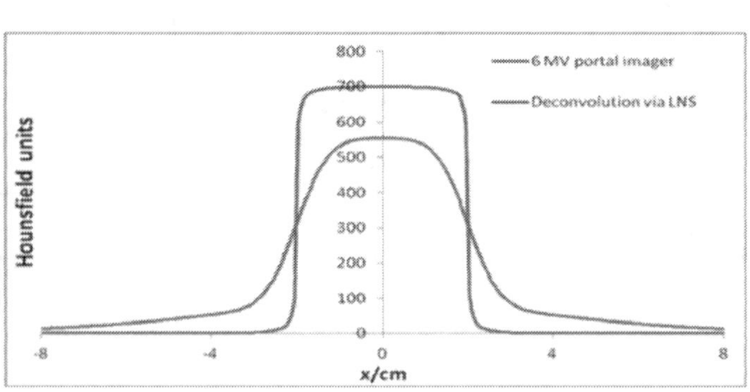

Figure 4. Identical task in the MV domain (6 MeV, portal imager, IMRT).

The LNS procedure will now be applied to proton physics (IMPT) and the generalized convolution (Fermi-Dirac statistics) to electron capture. The field-size for IMPT amounts to 4.8 x 4.8 mm², which we have used due to available data of the HCL. It turned out by many studies that in proton therapy scanning methods have a preferred importance, since broad beams a rather difficult to handle due to the varying range of targets and range shifts resulting from heterogeneity of patient tissue [6]. The importance of the deconvolution of very narrow proton beams is demonstrated by Figure 5, which is closely related to

Figure 6. With regard to scanning beams in proton radiotherapy and IMPT the deconvolution can provide reliable information on the necessity of superposition of neighboring proton beamlets to avoid underdose in DOIs or fluence modulation in IMPT. We also present results of calculations for protons, He and C ions; the initial energy amounts to 400 MeV/nucleon. This appears to be a reasonable restriction with regard to therapeutic conditions. Thus Figure 8 shows that at the end of the projectile track all charged ions nearly behave in the same manner, whereas Figure 7 is restricted to C ions, which start with different initial energies E_0.

4. Discussion

The LNS procedure is powerful with regard to the inverse problem of linear combinations of Gaussian convolution kernels. The application of the Fermi-Dirac statistics (instead of Boltzmann) can be handled with linear combinations of shifted Gaussian kernels [4]. Thus desired back calculations can also be carried out with the LNS-procedure, e.g. the calculation of $q^2(E)$ of measured Bragg curves of heavy carbons. By considering the energy dependence of the projectile particle charge $q^2(E)$, the LET of the stopping power S(E) of charged particles can be accounted for in dose calculations.

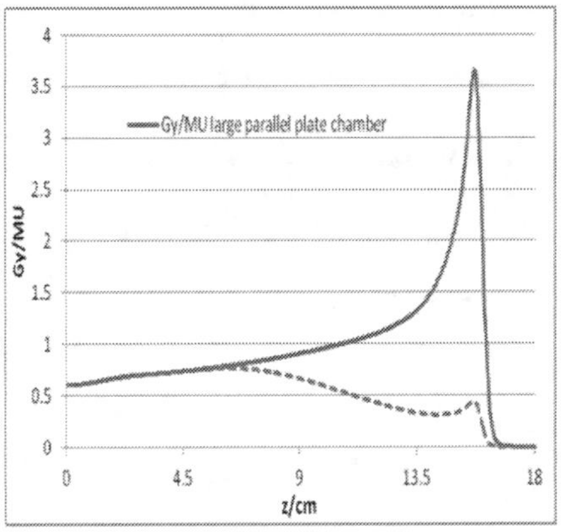

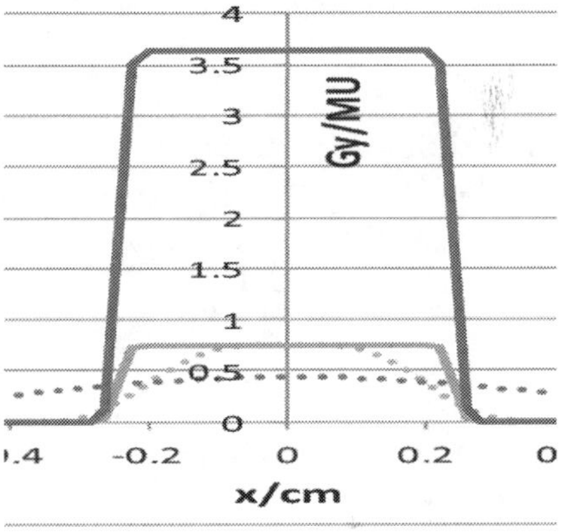

Figure 5. Stopping-power of 158 MeV protons (HCL). **Figure 6.** Transverse profiles at z =6 cm and Bragg peak.

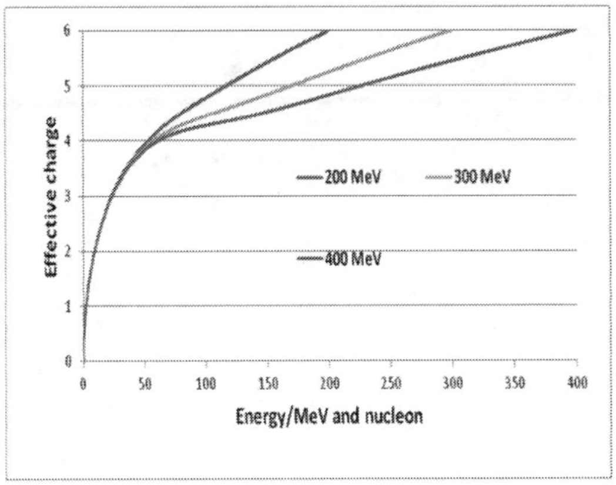

Figure 7. Charge q(E) of C ions; E_0 = 200, 300 and 400 MeV/nucleon.

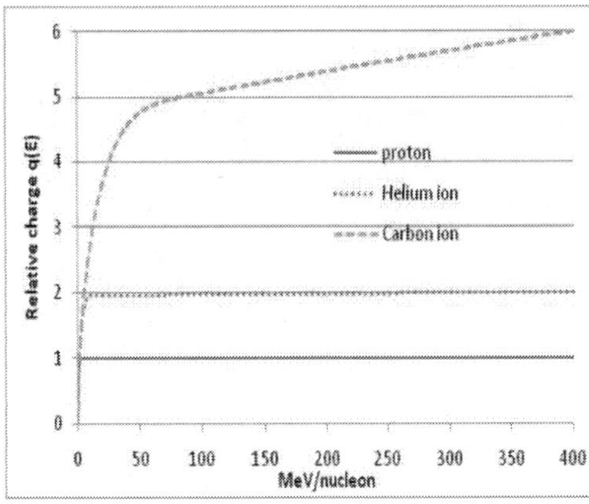

Figure 8. q(E) for 3 nuclei, E_0 = 400 MeV/nucleon

References

[1] Ulmer W 2010 *Inverse Problems* **26** 085002.
[2] Molière G 1955 *Z. Naturforschung* ***10a*** 177.
[3] Bethe H A 1953 *Phys. Rev.* ***89*** 1256.

[4] Ulmer W and Matsinos E 2011 *European Physics Journal (ST)* **190** 1.
[5] Ulmer W and Kaissl W 2003 *Phys. Med. Biol.* **48** 707.
[6] Bäumer C and Farr J 2011 *Med. Phys.* ***38*** 2904.
[7] Ulmer W 2012 Deconvolution of a linear combination of Gaussian kernels by Liouville-Neumann series applied to an integral equation of second kind with applications to radiation physics/image processing. *An Introductory Guide to the Emerging Areas of Digital Image Processing (Edited by: A. Mishra)*. iConcept Press. ISBN: 978-14775548-0-7.
[8] Ulmer W 2012 http://arxiv.org/abs/1208.6223

Exotic Lepton Flavour Violating Processes in the Presence of Nuclei

D.K. Papoulias[1] and T.S. Kosmas[2]
Division of Theoretical Physics, University of Ioannina, GR 45110 Ioannina, Greece
E-mail: [1]`dimpap@cc.uoi.gr` [2]`hkosmas@uoi.gr`

Abstract.
The discovery of neutrino oscillations indicates the existence of massive neutrinos in contrast to the massless neutrinos predicted by the Standard Model. One of the simplest extensions of the SM obtained by adding a heavy right-handed neutrino singlet, N_R, per neutrino generation is the Seesaw mechanism. Within the context of this mechanism, flavour changing neutral current neutrino-nucleus reactions of the type ${}_AN_Z\,(\nu_\alpha, \nu_\beta)\,{}_AN_Z^*$ and ${}_AN_Z\,(\tilde{\nu}_\alpha, \tilde{\nu}_\beta)\,{}_AN_Z^*$, with $\alpha \neq \beta$, are predicted to occur. In this contribution, motivated by the extensive studies (theoretical and experimental) of the LFV in $\mu^- \to e^-$ conversion in nuclei, we investigate FCNC in neutrino-nucleus reactions. From a nuclear theory point of view, the Donnelly-Walecka model for cross sections calculations is employed. To this purpose, the single-particle transition matrix elements are evaluated from a Mathematica code developed in this work. Neutrino-nucleus reactions have important impact in Astrophysics and hence a detailed study of such exotic processes is of significant importance.

1. Introduction

It is well known that neutrinos interact very rarely with matter through the weak interaction of the Standard Model (SM) [1]. Such reactions proceed either by charged-current processes mediated by $W^\pm$ boson exchange or through neutral-current processes mediated by a neutral Z boson [1, 2] (see Fig. 1). At nuclear-level, these processes are represented by the following reactions

$$\nu_\alpha(\tilde{\nu_\alpha}) + (A, Z) \to \alpha^\mp + (A, Z \pm 1), \tag{1}$$

$$\nu_\alpha(\tilde{\nu_\alpha}) + (A, Z) \to \nu_\alpha(\tilde{\nu_\alpha}) + (A, Z)^*, \tag{2}$$

where $\alpha = e, \mu, \tau$, is the flavour index and the asterisk (*) indicates an excited final nuclear state. So far, the ν-nucleus reactions (1) and (2) have been extensively studied both theoretically and experimentally, however further research is still needed [3, 4]. Many questions related to the absolute neutrino mass, the total number of neutrino flavours, the Dirac or Majorana nature of neutrinos and others remain still open [4].

Even though neutrinos interact very weakly with matter, they provide a very useful tool for Astrophysical searches since they can escape even from the interior of distant stars [5, 6]. So far, many efforts have been done for a better understanding of the Sun's nature, the Supernova explosion and other Astrophysical phenomena through ν-nucleus interaction probes [1].

Recently, non-standard ν-nucleus reactions, that may play significant role to Supernova physics, have been considered [4, 5, 6]. Among these, the flavour changing neutral current

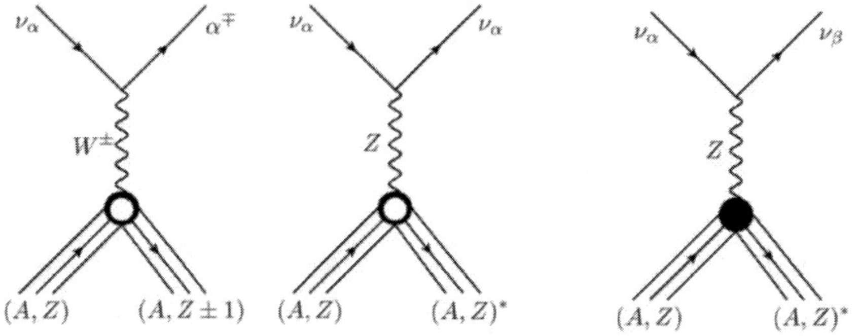

Figure 1. A set of Feynman diagrams for SM charged-current (left), SM neutral-current (middle) and FCNC (right), processes in the presence of nuclei. The non-standard physics enters in the complicated vertex •.

(FCNC) reactions

$$\nu_\alpha + (A, Z) \to \nu_\beta + (A, Z)^*, \tag{3}$$

with different flavour indices, i.e $\alpha \neq \beta$, posses prominent position. Such processes however, have not been studied in detail from a nuclear theory point of view. In addition, the lepton flavour violating (LFV) muon to electron conversion in nuclei is presently intensively investigated [7, 8, 9, 10, 11, 12]. Our aim is to deeper explore the aforementioned exotic reactions, predicted to occur within the context of particle models beyond the SM like the Seesaw mechanism [10], by performing realistic nuclear structure calculations.

2. FCNC neutrino-nucleus and other LFV processes

FCNC ν-nucleus processes constitute an attractive scenario to probe new physics both from a theoretical and an experimental nuclear physics point of view. It also provides a new challenge to increase the sensitivity of the relevant branching ratios. Our purpose is to systematically study and understand the unknown new physics entering the complex vertex of Fig. 1.

The quark-level interaction Hamiltonian will be constructed within the context of various models beyond the SM. In the present work, inspired from the methods used in Refs. [7, 10] we adopt the Seesaw mechanism as reference model. The parameters that give rise to LFV entering the quark-level interaction Lagrangian can be constrained by the upper limits on the branching ratio derived from $\mu^- \to e^-$ conversion experiments, like the mu2e and Project X at Fermilab, USA [11] and the COMET at J-PARC, Japan [12] at a sensitivity up to $R_{\mu^- e^-} = 10^{-18}$.

In previous comprehensive $\mu^- \to e^-$ conversion studies, it has been claimed that the reformulation of the quark-level effective Lagrangian yields the effective nucleon-level Lagrangian written in terms of the nucleon isospin operators as [7, 8, 9, 10]

$$\begin{aligned} \mathcal{L}^N_{eff} = G_\alpha \Big[\sum_{A,B} j^A_\mu \left(\alpha^{(0)}_{A,B} J^{B\mu}_{(0)} + \alpha^{(3)}_{A,B} J^{B\mu}_{(3)} \right) + \sum_{C,D} j^C \left(\alpha^{(0)}_{C,D} J^D_{(0)} + \alpha^{(3)}_{C,D} J^D_{(3)} \right) \\ + j_{\mu\nu} \left(\alpha^{(0)}_T J^{\mu\nu}_{(0)} + \alpha^{(3)}_T J^{\mu\nu}_{(3)} \right) \Big], \qquad \alpha = \text{ph}, \text{nph}. \end{aligned} \tag{4}$$

Here, the index α refers to the photonic (α = ph) and non-photonic (α = nph) contribution [10]. The components of the isoscalar, $J_{(0)}$, and isovector, $J_{(3)}$, nucleon currents are defined as

$$J^{V\mu}_{(k)} = \bar{N}\gamma^\mu \tau_k N, \quad J^{A\mu}_{(k)} = \bar{N}\gamma^\mu\gamma_5\tau_k N, \quad J^S_{(k)} = \bar{N}\tau_k N, \quad J^P_{(k)} = \bar{N}\gamma_5\tau_k N, \quad J^{\mu\nu}_{(k)} = \bar{N}\sigma^{\mu\nu}\tau_k N,$$

with $k = 0, 3$. Each component is separately treated, however not all of them should be taken into account. In fact, the pseudoscalar and tensor nucleon components are negligible. Therefore, at nuclear-level the effective interaction Hamiltonian H_{eff}, takes the well-known current-current form [7, 10].

The nuclear calculations for partial transition rates are based on the matrix elements (ME) of the interaction Hamiltonian between the initial and any final nuclear states, i.e.

$$\langle f|H_{eff}|i\rangle = \frac{G_F}{\sqrt{2}} l^\mu \int d^3x\, e^{-i\mathbf{q}\cdot\mathbf{x}} \langle f_{nucl}|\hat{\mathcal{J}}_\mu(\mathbf{x})|i_{nucl}\rangle, \tag{5}$$

(G_F is the weak interaction coupling constant). The magnitude of the three momentum transfer $|\vec{q}|$, is obtained from the kinematics of the reaction [1, 13] and the ME of the leptonic current has been written as

$$\langle l_f|j_\mu|l_i\rangle = l_\mu\, e^{-i\mathbf{q}\cdot\mathbf{x}}. \tag{6}$$

Hence, Eq. (5) implies that one needs to compute the ME of the hadronic current density.

FCNC in the ν-nucleus reactions (3) may be predicted by the same particle physics theories beyond the SM that predict the exotic $\mu^- \to e^-$ conversion in nuclei. The same holds also for their description within the framework of a nuclear structure method (Shell-model, RPA, QRPA, Fermi-Gas models, etc.) [7].

3. Brief Description of the Basic Nuclear Operators

In the present work, we employ the Donnelly-Walecka method, that describes in a unified way any semi-leptonic process in the presence of nuclei [1, 2]. It has been proved, that the multipole expansion of the hadronic current density ME, leads to eight irreducible tensor operators written in terms of the weak nucleon form factors F_x, $x = 1, 2, A, P$, as

$$M^{Coul}_{JM}(q\mathbf{r}) = F_1 M^J_M(q\mathbf{r}), \quad iM^{Coul\,5}_{JM}(q\mathbf{r}) = \frac{q}{M}\left[F_A \Omega^J_M(q\mathbf{r}) + \frac{1}{2}\left(F_A + \omega F_P\right) \Sigma''^J_M(q\mathbf{r})\right], \tag{7}$$

$$L_{JM}(q\mathbf{r}) = \frac{\omega}{q} M^{coul}_{JM}(q\mathbf{r}), \quad -iL^5_{JM}(q\mathbf{r}) = \left[F_A - \frac{q^2}{2M} F_P\right] \Sigma''^J_M(q\mathbf{r}), \tag{8}$$

$$T^{el}_{JM}(q\mathbf{r}) = \frac{q}{M}\left[F_1 \Delta'^J_M(q\mathbf{r}) + \frac{1}{2}\left(F_1 + 2MF_2\right) \Sigma^J_M(q\mathbf{r})\right], \quad -iT^{el\,5}_{JM}(q\mathbf{r}) = F_A \Sigma'^J_M(q\mathbf{r}), \tag{9}$$

$$iT^{mag}_{JM}(q\mathbf{r}) = \frac{q}{M}\left[F_1 \Delta^J_M(q\mathbf{r}) - \frac{1}{2}\left(F_1 + 2MF_2\right) \Sigma'^J_M(q\mathbf{r})\right], \quad T^{mag\,5}_{JM}(q\mathbf{r}) = F_A \Sigma^J_M(q\mathbf{r}). \tag{10}$$

On the right-hand side of Eqs. (7)-(10), we notice that seven (due to CVC theory) linearly independent new operators, namely $M^{coul}_{JM}, \Sigma^J_M, \Sigma'^J_M, \Sigma''^J_M, \Delta^J_M, \Delta'^J_M$ and Ω^J_M, that do not carry form-factor dependence, may be defined (see Ref. [2]). For the reduced ME of the latter operators (they enter the description of the cross section for any semi-leptonic process in the presence of nuclei), a compact formalism has been developed, which reads [2]

$$\langle j_1||T^J_i||j_2\rangle = e^{-y} y^{\beta/2} \sum_{\mu=0}^{n_{max}} \mathcal{P}^{i,\,J}_\mu y^\mu, \qquad i = 1, \cdots, 7. \tag{11}$$

T^J_i, $i = 1, \cdots, 7$, represent any of the above operators, $n_{max} = (N_1 + N_2 - \beta)/2$ and $N_i = 2n_i + \ell_i$ indicates the harmonic oscillator quanta of the $i-$th level. The well-defined quantum numbers in our notation are $|(n_1\ell_1)j_1\rangle \equiv |j_1\rangle$. For an analytical description of the aforementioned formalism the reader is referred to [2].

4. Results and Discussion

As a first step of our calculational procedure, we have constructed an efficient modern Mathematica code to compute the reduced ME for all basic operators T_i^J and for any configuration $(j_1, j_2)J$, using the analytical expressions of Eq. (11). As expected, the coefficients obtained in Ref [2, 13], coincide with those of the present work. As an example, the coefficients $\mathcal{P}_\mu^{6,\,J}$ for the reduced ME of the operator $T_6^J \equiv \Delta_M'^J$ are tabulated in Table 1. We mention however, that in Ref. [13], only configurations for which $N_1 - N_2 \leq 3$ are included, while in our code there is not any similar restriction. We conclude that all geometrical coefficients $\mathcal{P}_\mu^J$ are simple rational numbers for the diagonal elements (not shown here), or square roots of rational numbers for the non-diagonal elements. The double differential cross sections of processes (3) from which the nuclear calculations start, includes reduced ME of the form $|\langle j_1||T_i^J||j_2\rangle|^2$, hence their evaluation is crucial. We currently perform realistic nuclear structure calculations for coherent cross sections that depend on the nuclear form factors for protons and neutrons. The appropriate code, provides improved results, by a refined version of the fractional occupation probabilities method of Ref. [14]. These calculations are in progress and will appear soon.

Table 1. The coefficients $\mathcal{P}_\mu^{6,\,J}$ for the the $\langle j_1||\Delta_M'^J||j_2\rangle$.

$(n_1\ell_1)_{j_1} - (n_2\ell_2)_{j_2}$	J	$\mu = 0$	$\mu = 1$	$\mu = 2$	$\mu = 3$	$\mu = 4$
$0p_{1/2} - 0s_{1/2}$	1	$\sqrt{\frac{1}{6}}$				
$0p_{3/2} - 0s_{1/2}$	1	$-\sqrt{\frac{1}{3}}$	0			
$0d_{5/2} - 0s_{1/2}$	2	$-\sqrt{\frac{3}{5}}$	0			
$0f_{5/2} - 0p_{1/2}$	2	$-\sqrt{\frac{7}{5}}$	$\sqrt{\frac{20}{63}}$	0		
$0d_{7/2} - 0p_{1/2}$	4	$\sqrt{\frac{8}{63}}$	0			
$1p_{3/2} - 1s_{1/2}$	1	$-\sqrt{\frac{5}{9}}$	$\sqrt{\frac{4}{45}}$	$-\sqrt{\frac{4}{45}}$	0	
$0d_{3/2} - 0f_{7/2}$	3	$-\sqrt{\frac{16}{175}}$	$\sqrt{\frac{4}{45}}$	$-\sqrt{\frac{16}{1575}}$	0	
$1f_{5/2} - 2p_{3/2}$	2	0	$\sqrt{\frac{8}{7}}$	$-\sqrt{\frac{288}{343}}$	$\sqrt{\frac{2888}{27783}}$	$-\sqrt{\frac{32}{27783}}$

References

[1] V. Tsakstara and T.S. Kosmas 2011, *Phys. Rev.* **C 83** 054612; 2011 **C 84**, 064620 and references therein
[2] V. Ch. Chasioti and T. S. Kosmas 2009, *Nucl. Phys. A* **829**, 234
[3] Y. Kuno and Y. Okada 2001, *Rev. Mod. Phys.* **73** 151
[4] S.P. Das, F.F. Deppisch, O. Kittel, and J.W.F. Valle 2012, *Phys. Rev.* **D 86** 055006 and references therein
[5] P. S. Amanik, G. M. Fuller and B. Grinstein 2005, *Astropart. Phys.* **24**, 160
[6] P. S. Amanik and G. M. Fuller 2007, *Phys. Rev.* **D 75**, 083008
[7] T.S. Kosmas 2001, *Nucl. Phys.* **A 683** 443
[8] A. Faessler, T.S. Kosmas, S. Kovalenko and J.D. Vergados 2000 *Nucl. Phys.* **B 587** 25
[9] T.S. Kosmas, S. Kovalenko and I. Schmidt 2001 *Phys. Lett.* **B 511** 203; 2001 **B 519** 78
[10] F. Deppisch, T.S. Kosmas and J.W.F. Valle 2006, *Nucl. Phys.* **B 752**, 80
[11] R. Bernstein and G. Kribs, talk at 2012 Project X Physics Study (PXPS12), Fermilab
[12] Y.G. Cui et al. (COMET Collaboration), KEK Report 2009-10
[13] T. W. Donnelly and W.C. Haxton (1979), *At. Data Nucl. Data Tables* **23**, 103
[14] T.S. Kosmas and J.D. Vergados 1992, *Nucl. Phys.* **A 536** 72

IC-MSQUARE 2012: International Conference on Mathematical Modelling in Physical Sciences IOP Publishing
Journal of Physics: Conference Series **410** (2013) 012124 doi:10.1088/1742-6596/410/1/012124

Electron-capture and its role to explosive neutrino-nucleosynthesis

P.G. Giannaka and T.S. Kosmas

Theoretical Physics Division, University of Ioannina, GR-45110 Ioannina, Greece

E-mail: pgiannak@cc.uoi.gr

Abstract. In explosive nucleosynthesis, among other reactions, prominent position posses the electron capture by nucleons and nuclei. This process plays important role in the core-collapse of massive stars by modifying the quantity Y_e. From a nuclear theory point of view, this process may be studied by using the same nuclear methods (e.g. the quasi-particle random phase approximation, QRPA, employed in the present work) with the one-body charge changing nuclear processes (β-decay modes, charged-current electron-neutrino absorption by nuclei, etc). In this work we concentrate on a detailed study of the orbital e-capture by medium-heavy nuclei (Fe region) which play fundamental role in the evolution of massive stars and in the so called explosive neutrino nucleosynthesis. We study extensively the cross sections and rates of e-capture process by "iron group peaked nuclei", isotopes that are important for investigating the neutrino nucleosynthesis and other related phenomena.

1. Introduction

The recent studies on presupernova scenarios indicate that e^--capture on nucleons and nuclei is an important process, mostly in the early stage of the collapse of massive stars. Refinements of the methods used for calculating electron capture on heavy nuclei during the collapse phase shows that the capture on free protons dominates [1, 2], in contrast to the findings of previous works. Captures on nuclei produce neutrinos with rather small energies, making the inclusion of inelastic neutrino-nucleus reactions in supernova simulations more important. Even though capture on free protons is favored compared to that on nuclei, due to the significantly lower Q-value (difference between the nuclear mass of the initial and final state), the number of free protons divided by the total number of nucleons (Y_p) is quite low ($Y_p \sim 10^{-6}$ in the $15M_{\odot}$ presupernova model) [2, 3].

Obviously, electron capture reduces the ratio Y_e, favoring nuclear composition with more neutron-rich and heavier nuclei, including those with $N > 40$, which dominate the matter composition for densities larger than about $10^{10} g\, cm^{-3}$. Also recently it was shown that some models for electron capture on nuclei are not realistic, because the Pauli blocking of the GT transitions overestimates the correlations and temperature effects [2].

In this work, we study in detail the orbital electron-capture on medium-heavy nuclei, process represented by reaction

$$(A, Z) + e^- \rightarrow (A, Z-1)^* + \nu_e \qquad (1)$$

Even thought the investigation of this process started decades ago, more accurate transition rates are needed due to its significant importance in core-collapse supernova dynamics and other astrophysical phenomena. To this purpose, in the present work we have chosen to study this process in a set of isotopes which are important from an astrophysical point of view that belong to the "iron group peaked nuclei" [4]. We focus on extensive cross sections calculations following the Donnelly-Walecka multipole decomposition method [7]. For the computation of the many-body nuclear wave functions we utilize the quasi-particle random phase approximation (or QRPA for short) as described below [6, 7, 8].

2. The Donnelly-Walecka multipole decomposition method

In this paper, the method of studying orbital e^--capture process on nuclei, compared to previous similar works, is improved in the following: 1) In the level where use of compact analytical expressions for the required reduced matrix elements of all basic multipole operators is made. 2) In the level of constructing the nuclear Hamiltonian where the Bonn C-D potential (instead of the Bonn C used up to now) for the realistic two-body nuclear forces is employed [6, 7, 8].

The nuclear calculations start by writing down the weak interaction Hamiltonian $\hat{H}_w$ which in a current-current form describes all the semi-leptonic processes in the presence of nuclei. By denoting the leptonic current as j_μ^{lept} and the hadronic current as $\hat{\mathcal{J}}_\mu$, $\hat{H}_w$ is written as:

$$\hat{\mathcal{H}}_w = \frac{G \cos\theta_c}{\sqrt{2}} j_\mu^{lept}(\mathbf{x})\hat{\mathcal{J}}_\mu(\mathbf{x}) \tag{2}$$

where $G = 1.1664 \times 10^{-5}\, GeV^{-2}$ and θ_c is the Cabibbo angle ($\cos\theta_c = 0.974$) [7]. The hadronic current, which is of interest from a nuclear theory point of view, consists of the polar-vector and the axial-vector components which can be written by using isovector, axial-vector and pseudoscalar form factors (they depend on the four-momentum transfer q_μ^2) and read:

$$\hat{\mathcal{J}}_\mu = g_V \hat{J}_\mu^V - g_A \hat{J}_\mu^A \tag{3}$$

In Eq. (3), g_V (g_A) represent the coupling of the polar (axial) vector interaction (at $q_\mu^2 = 0$, static parameters).

One of the main goals of this effort is to construct an advantageous code for calculating electron-capture cross sections in various currently interesting nuclear isotopes. This code is based on the present formalism and the QRPA that uses realistic Bonn C-D residual interaction [7].

3. Evaluation of the nuclear wave functions within QRPA.

It is well known that the nucleus consists of protons and neutrons which, in a rather rough approximation can be considered as independent particles, attracted by the nuclear center through a central strong force. This attraction can be described by a mean field, as for example the Woods-Saxon potential. In our case, use of a Woods-Saxon potential with Coulomb corrections and a spin orbit part is made.

The two-nucleon correlations, also known as the residual two-body interaction, are necessary to be included in a reliable nuclear Hamiltonian. In our case they are assumed to be described by the Bonn C-D potential. The initially evaluated bare two-body nucleon-nucleon matrix-elements of the latter interaction refer to a specific isotope of mass number A. In the specific isotope (A, Z) studied, a renormalization of the two-body interaction is required. Towards this purpose we use four multiplicative parameters that renormalize the bare residual interaction. The first two $g_{pair}^{p,n}$, known as pairing parameters for protons (p) and neutrons (n) renormalize the monopole interaction, the third g_{pp} tunes the particle-particle channel and the fourth g_{pm} renormalizes the particle-hole interaction.

Table 1. Parameters for the interaction of proton pairs, g^p_{pair}, and neutron pairs, g^n_{pair}. They are fixed in such way that the corresponding experimental gaps, Δ_p^{exp} and Δ_n^{exp}, to be reproduced.

	Abundance(%)	*b (h.o.)*	g^n_{pair}	g^p_{pair}	S_n (MeV)	S_p (MeV)	Δ_n^{exp} (MeV)	Δ_n^{theor} (MeV)	Δ_p^{exp} (MeV)	Δ_p^{theor} (MeV)
^{96}Ru	5,540	2,158	0,987	0,835	10,694	7,344	1,0824	1,0821	1,4955	1,4954
^{102}Pd	1,020	2,178	0,978	0,958	10,568	7,806	1,3094	1,3085	1,4947	1,4938

^{96}Ru			^{102}Pd		
State	g_{ph}	g_{pp}	State	g_{ph}	g_{pp}
0^+	0.403	0.781	0^+	0.377	0.907
2^+	0.579	1.189	2^+	0.671	1.350
4^+	0.905	0.546	4^+	1.040	0.322
6^+	1.085	1.195	6^+	1.108	0.274

Table 2. Strength parameters for the particle-particle (g_{pp}) and particle-hole (g_{ph}) interaction for various multipolarities.

The ground state of the nucleus in question, is obtained in the context of the Bardeen Cooper Schrieffer (BCS) theory. The solution of the relevant BCS equations gives the probability amplitudes V and U for each single particle level to be occupied or unoccupied, respectively, and the quasi-particle energies.

Subsequently, we construct the excited states of the studied isotope by solving the QRPA equations, which in matrix form are written as

$$\begin{pmatrix} \mathcal{A} & \mathcal{B} \\ -\mathcal{B} & -\mathcal{A} \end{pmatrix} \begin{pmatrix} X^m \\ Y^m \end{pmatrix} = \Omega^m_{J^\pi} \begin{pmatrix} X^m \\ Y^m \end{pmatrix} \tag{4}$$

Their solution is an eigenvalue problem, which gives us the X and Y amplitudes for forward and backward scattering, respectively, as well as the QRPA excitation energies [8]. In the Donelly-Walecka method the solution of the QRPA equations is obtained separately for each multipole set of states $|J^\pi\rangle$.

4. Results and discussion

In Table 1 the values of the BCS parameters are listed through which we construct the wave function for the initial (ground) state. The parameters determined for the final states, of the isotope in question are shown in Table 2.

After the determination of the particle-particle and particle-hole parameters, we proceed by checking the reproducibility of the energy spectrum of the studied isotopes. In Figure 1, as a concrete example we show the energy spectrum of the ^{96}Ru isotope. For comparison, the experimental spectrum is also shown. As can be seen, the agreement is good in the low-lying spectrum.

In the final step we receive the results for cross-sections of the process which, as is known, are proportional to the square of the relevant matrix elements [7, 8]. The required nuclear matrix elements are calculated in the context of the QRPA [7].

The model space chosen in our QRPA calculations includes harmonic oscillator levels as follows: for the ^{96}Ru isotope (14 orbits) $0d_{5/2}$, $1s_{1/2}$, $0d_{3/2}$, $0f_{7/2}$, $1p_{1/2}$,$1p_{3/2}$, $0f_{5/2}$, $0g_{9/2}$, $1d_{5/2}$, $0g_{7/2}$, $1d_{3/2}$, $2s_{1/2}$, $0h_{9/2}$, $0h_{11/2}$, and for the ^{102}Pd (11 orbits) $0f_{7/2}$, $1p_{1/2}$,$1p_{3/2}$, $0f_{5/2}$, $0g_{9/2}$, $1d_{5/2}$, $0g_{7/2}$, $1d_{3/2}$, $2s_{1/2}$, $0h_{9/2}$, $0h_{11/2}$ [7].

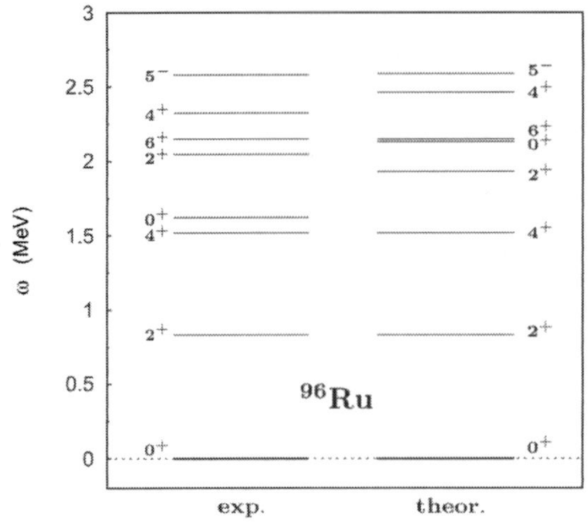

Figure 1. Comparison of the QRPA (theoretical) excitation spectrum (resulting from the solution of the QRPA eigenvalue problem) with the low-lying (up to about 3 MeV) experimental one. The agreement is good.

The reliability of the present calculations, could come out of the comparison of our electron capture cross sections with those evaluated in other similar processes, e.g the corresponding charged-current neutrino nucleus reactions, i.e. the particle conjugate process of the electron capture.

5. Summary and Conclusions

As discussed in the Introduction, during the presupernova and collapse phase, electron captures on nuclei (and in the late stage also on free protons) plays an important role, as the nuclear β-decay during silicon burning does. Electron captures become increasingly possible as the density in the star's center is increased. It is accompanied by an increase of the chemical potential (Fermi energy) of the degenerate electron gas and it reduces the electron-to-baryon ratio Y_e of the matter composition.

In this work, we use an advantageous numerical approach, constructed by our group recently, to calculate all basic multipole transition matrix elements needed for obtained e-capture cross sections. The required nuclear wave functions are obtained within the context of the QRPA using realistic two-body forces. Results for the cross sections are expected to be received soon.

5.1. Acknowledgments

This research has been co-nanced by the European Union (European Social Fund ESF) and Greek national funds through the Operational Program Education and Lifelong Learning" of the National Strategic Reference Framework (NSRF) - Research Funding Program: Heracleitus II. Investing in knowledge society through the European Social Fund.

References

[1] Juodagalvis A , K Langanke et al 2005 *Nucl. Phys.* A **747** 87
[2] Langanke K, Martinez-Pinedo G, et al 2003 *Phys. Rev. Lett.* **90** 241102.
[3] Langanke K, Martinez-Pinedo G, et al 2003 *Rev. Mod. Phys.* **75**.
[4] Frohlich C, Martinez-Pinedo G,et al 2006 *Phys. Rev. Lett.* **96** 142502.
[5] Kosmas T S and Oset E 1996 *Phys. Rev.* C **53** 1409.
[6] Donnelly T W and Peccei R D 1979 *Phys. Rep.* **50** 1.
[7] Chasioti V C and Kosmas T S 2009 *Nucl. Phys.* A **829** 234.
[8] Tsakstara V and Kosmas T S 2011 *Phys. Rev.* C **83** 054612.

Artificial γ-ray imaging of a galactic, relativistic astrophysical jet

T Smponias[1], **T S Kosmas**[1]

[1] Theoretical Physics Section, University of Ionnina, 45110, Greece

E-mail: hkosmas@uoi.gr

Abstract. The jet in a microquasar stellar system is simulated with a relativistic hydrocode (PLUTO) and then it is imaged, in the γ-rays wave band, with a line-of-sight method, including both emission and self-absorption. A synthetic image is produced which allows to better estimate the system's physical properties. The calculation procedure in our method has been simplified by exploiting the ability to de-couple the hydrodynamical from the radiative quantities in the computations.

1. Introduction

Microquasars (MQ) are a class of X-ray binaries comprising a close binary stellar system, where a collapsed object (neutron star or black hole) orbits together with a main sequence companion, in close proximity. An accretion stream carries matter from the main sequence star towards the compact object, leading to the formation of an accretion disk around the collapsed star [12]. The main reason for the formation of the disk is conservation of angular momentum, which prevents matter, originally part of a rotating star, to fall directly to the compact object, thus losing angular momentum instantaneously. From the close vicinity of the black hole, or neutron star for that purpose, twin relativistic jets are ejected, fed by accretion disk matter [12] and pointed perpendicular to the plane of the accretion disk.

The jets in MQ are sometimes highly relativistic e.g. GRS 1915+105 [4], or mildy relativistic, like SS433 ([1],[10],[3]). In some cases, apparent superluminal motion is also exhibited (e.g. GRS 1915+105). The jet ejection can be episodic, or a more continuous flow. Strong correlations tend to appear among emissions in different wavelengths, during various phases of system behaviour [9].

In the present work we concentrate on emission from the MQ jet of SS433, an object that has been repeatedly observed over the past three and a half decades. The basic features of the system are very briefly summarized below. The system consists of a main sequence supergiant and a collapsed stellar remnant, orbiting at close distance with a period of around 13 days. Relativistic (0.26c) jets are precessing around an axis inclined at 79 degrees in relation to the direction to the Earth, the precession angle being roughly 20 degrees and the precession period about 160 days [3]. The jets have been verified, from optical spectroscopy, to mainly consist of proton flow ([1], [3]), yet no definite such data exist as of today for other MQ systems (see however [6]).

The ejected plasma can be approximately treated as a fluid flow, assuming a mild magnetic field permeats the jet matter [5]. The field provides a coupling of the jet particles at smaller

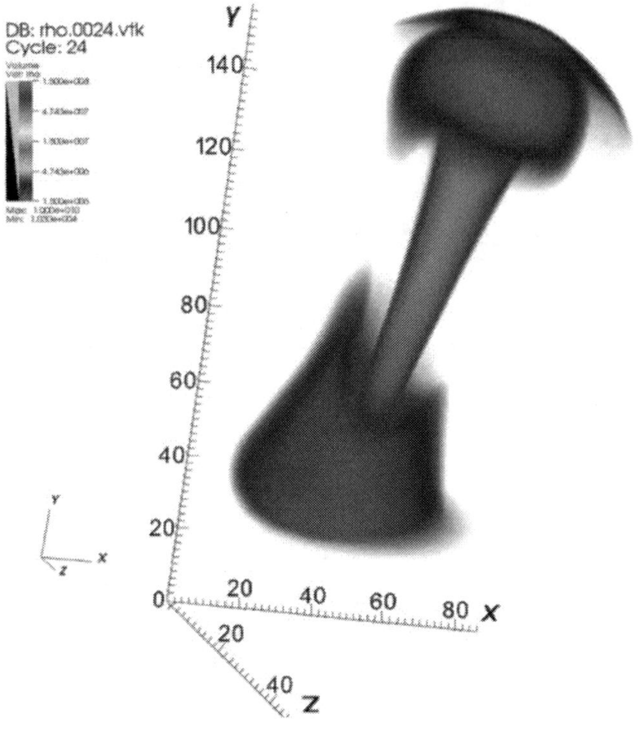

Figure 1. A 3-D density plot of the jet moving through the stellar wind, with a resolution of 300*500*300 (XYZ respectively) cartesian homogeneous grid. The accretion disk wind in the model has been crossed by the jet and matter from it can be seen left over around the jet base. The spatial scale (10^{10} cm per unit length) is the same for all axes.

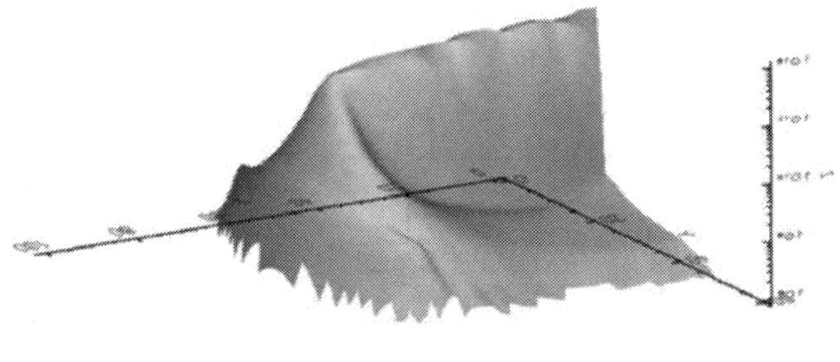

Figure 2. A plot of the γ-ray intensity (z-axis, arbitrary units) of the artificial image (formed on x-y plane) of the jet system, at the same time instant as in Fig. 1. A small amount of self-absorption has been included, in a simplified manner (proportional to the local emission coefficient at each point). Fast and dense matter contributes the most to emission, while absorption is maximum where the column density is largest.

scales, therefore making possible a macroscopically continuous flow, where the jet plasma now behaves as an ideal gas. This approximation opens the possibility of modelling the jet with numerical hydrodynamics codes, allowing the investigation of physical conditions in the jet. Trial and error leads to a set of initial/boundary conditions (IBC's) that, when inserted into the models, lead to synthetic images that are then compared to actual observed properties. If a match is achieved, then this means that the initial input (IBC) was correct.

This paper attempts to model a microquasar jet using a 3-dimensional (3D) relativistic hydrodoce and then, a line-of-sight (LOS) radiative transfer code [13] that produces artificial γ-ray synthetic images of the hydrocode model system. Meanwhile, the emission and absorption coefficients for creating the synthetic images are provided by a separate routine, written in Mathematica [14], in order to deal with the complexity of the relevant mathematical expressions.

2. Methods

2.1. Calculating the emission and absorption coefficients

In order to calculate the -self-absorbed- emission from the simulated jet and its surrounding winds, one needs to obtain the γ-ray emission and absorption coefficients at every location. This can be achieved through the solution of the equation of radiative transfer, along multiple parallel lines of sight intersecting the volume of the jet-wind system. The full details of this calculation shall be found in [14], based on the formalism of [16], [7], [15] and [13]. More specifically, the one dimensional (no scattering sideways) equation of radiative transfer is solved for each LOS. This equation, in the case of time independent radiative transfer, and also in the absence of sideways scattering reduces to:

$$[\hat{\Omega} \cdot \nabla + k_\nu] I_\nu = j_\nu. \tag{1}$$

where I_ν is the local intensity, at frequency ν, Ω is the solid angle, j_ν denotes the local emission coefficient and k_ν the local absorption coefficient. Along a line of sight that subtends a solid angle $\Delta\Omega$, the above expression becomes

$$\Delta\Omega \cdot \frac{dI_\nu}{dl} = j_\nu - k_\nu I_\nu \tag{2}$$

The two indices are, in general, dependent on the properties of the jet matter. Consequently, in order to calculate the emission from a given system, such as the microquasar jet, one also needs the description of the geometrical and dynamical conditions of the system in question. The emission and absorption coefficients for γ-rays are obtained from the proton-proton interaction for emission and from a number of processes (γ-γ interaction with ambient soft photons $\gamma\gamma \rightarrow e^+e^-$, photopion production $\gamma N \rightarrow \pi_i\gamma$, photopair production $\gamma N \rightarrow Ne^+e^-$) for absorption [15]. A simplified absorption process, with the relevant coefficient set proportional to the emission coefficient, shall be considered for this paper.

2.2. Numerical hydrodynamic simulations using the PLUTO code

The jet was represented as a 3-dimensional relativistic flow of ideal gas within the framework of the PLUTO code [11], originating from the vicinity of the compact object and then crossing the accretion disk wind and the stellar wind of the companion. The domain of the simulation is Cartesian XYZ with a resolution of 300*500*300 for this specific run. The algorithm used (piecewise-linear method: [8], [2]) includes linear interpolation in cells and a dimensionally split method of characteristic tracing to advance the solution in time at each cell interface. The simulation is let to evolve until the jet, injected at the one side of the computational box, begins to exit from the other side. Although the binary companion lies outside the domain, its stellar wind is still present, with a $\frac{1}{r^2}$ inverse dependence on the distance from the (not-included-yet assumed-to-exist at a specific but remote location) companion star. The accretion disk wind is represented in a simplified manner, using a cylindrical shape wind construct, with a density decreasing as proportional to $\frac{1}{y^2}$, where y is parallel to the jet precession axis.

2.3. Line of sight integration

In order to calculate the emission and absorption of γ-rays from the jet and from the surrounding winds, line of sight integration is performed. The 3-dimensional volume of a snapshot from the hydrocode, is transferred, in the form of 3-D data arrays, to the imaging program (LOS code) in IDL. Then, the data volume is crossed by many parallel lines of sight (LOS's), all set parallel to a preset direction, i.e. oriented towards the Earth. Each LOS constitutes a 1-D vector array, whose elements are selected from the 3-D array, as the LOS path is drawn through the volume. The emission and absorption coefficients, ϵ_ν and κ_ν respectively, are also calculated along the way. Finally, when a LOS has been completed, its elements forming a 1-D array, the equation of radiative transfer (no sideways scattering is allowed) is solved along the given LOS, using the obtained emission and absorption coefficients. The result, for each LOS, is the intensity from that LOS and it is assigned as the radiative intensity of the pixel where the LOS meets the plane of observation, i.e. the plane where the synthetic image is formed. This way, each pixel of the artificial image is assigned the intensity from a LOS.

The emission and absorption coefficients, ϵ_ν and κ_ν are also calculated along the way. Finally, when a LOS has been completed, its elements forming a 1-D array, the equation of radiative transfer (no sideways scattering allowed) is solved along the given LOS, using the obtained emission and absorption coefficients.

3. Results

A computer simulation study of the SS433 jet was performed, using the PLUTO hydrocode and the in-house radiative transfer codes ([13], [14]). The size of the computational grid for PLUTO was set to 300*500*300 equal sized cells, while the imaging was performed at a resolution 5 times less (60*100*60). The jet speed is set to u=0.26c and the jet enters the grid at the center of the xz plane. The accretion disk wind and stellar companion wind are represented in the model as ambient matter, that is traversed by the jet on its way outwards, along the positive Y-direction. In Figure 1 we can see a snapshot of the jet density, where the jet has crossed the accretion disk wind matter, part of which is now piling up around the jet base.

Then, the model system was artificially imaged, in γ-rays, using the LOS code (Figure 2). The emission coefficient was set to be proportional to the hydrodynamical density ρ, times the square of the local velocity u^2, at each point, while the absorption coefficient was simply set proportional to the emission coefficient, as a first attempt to include the absorption. As can be seen, the fast moving matter contributes more to the emission. The absorption, on the other hand, is generally stronger where the column density is higher (i.e. where the LOS has to cross more matter along its way through the system).

4. Conclusions

The simplified emission and absorption model used here is based on the fact that in the formalism of Ref. [15] the hydrodynamical quantities can be de-coupled from the radiative ones, during the calculation of γ-ray emission. Therefore, the same radiative calculation can be applied all over the model system and then the result should be multiplied by the density, velocity, etc. This approximation, however, is not always (e.g. for neutrino emission) valid.

References

[1] Begelman, M C, Hatchett, S P, McKee, C F, Sarazin, C L, & Arons, J 1980, Ap. J., 238, 722
[2] Collela, P, SIAM Journal for Sci. Stat. Computing, Vol.6 No.1 (1985), pp. 104-117
[3] Fabrika, S 2004, Astrophysics and Space Physics Reviews, 12, 1
[4] Fender, R P, Garrington, S T, McKay, D J, et al. 1999, MNRAS, 304, 865
[5] Ferrari, A 1998, Ann. Rev. Astron. Astrophys, 36, 539
[6] Heinz, S 2008, International Journal of Modern Physics D, 17, 1947
[7] Kelner, S R, Aharonian, F A, & Bugayov, V V 2006, Phys. Rev. D, 74, 034018
[8] van Leer, B, 1979, J. Comput. Phys. 32, 101-136
[9] Malzac, J, & Belmont, R 2009, MNRAS, 392, 570
[10] Margon, B 1984, Ann. Rev. Astron. Astrophys, 22, 507
[11] Mignone, A, Bodo, G, Massaglia, S, et al. 2007, Ap. J. Supp., 170, 228
[12] Mirabel, I F, & Rodríguez, L F 1999, Ann. Rev. Astron. Astrophys, 37, 409
[13] Smponias, T, & Kosmas, T S 2011, MNRAS, 412, 1320
[14] Smponias, T & Kosmas, T, 2012, MNRAS, to be submitted
[15] Reynoso, M M, Romero, G E, & Christiansen, H R 2008, MNRAS, 387, 1745
[16] Romero, G E, Torres, D F, Kaufman Bernadó, M M, & Mirabel, I F 2003, Astron. & Astrophys., 410, L1

IC-MSQUARE 2012: International Conference on Mathematical Modelling in Physical Sciences IOP Publishing
Journal of Physics: Conference Series **410** (2013) 012126 doi:10.1088/1742-6596/410/1/012126

Application of simulation techniques in supernova physics

V. Tsakstara[1], T.S. Kosmas[1]

[1]Theoretical Physics Section, University of Ioannina, GR 45110 Ioannina, Greece

E-mail: vtsaksta@cc.uoi.gr

Abstract. In supernova (SN) physics, the responses of nuclear neutrino-detectors to SN neutrino-spectra, are studied by convoluting original theoretical cross sections with well known SN (anti)neutrino-energy distributions (such distributions are the two-parameter Fermi-Dirac and Power-Law distributions). Also, the interpretation of SN neutrino signals, created at various nuclear ν-detectors, is explored by applying simulation techniques in low-energy $\widetilde{\nu}_e$ anti-neutrino spectra of boosted β-radioactive ^6He ions (beta-beam neutrinos). In the present paper, we employ simulation techniques to analyze original SN anti-neutrino signals by using synthetic beta-beam spectra (defined as linear combinations of boosted beta-beam spectra of ^6He). The quality of the fits, obtained by using the MERLIN optimization package, is in general good. From a nuclear theory point of view, the resulted nuclear responses reflect the effectiveness of some detector materials as SN neutrino detectors (COBRA, CUORE, ICARUS experiments).

1. Introduction

The convolution (or folding) needed in the present work, from a mathematical point of view, is a well known basic operation applied in many areas of research (physics, mathematics, engineering, informatics, etc.) to extract information from images [1]. This operation takes two functions, f and g, and produces a third function C that, in a sense, represents the amount of overlap between f and a translated version of g. Thus, for two continuous one-dimensional (simplest case) integrable functions f and g, the convolution is defined by the integral [1]

$$C(t) \equiv (f \star g)(t) = \int_{\alpha}^{\beta} f(t')g(t-t')dt' . \quad (1)$$

One can notice that, the function g in the integrand is shifted. Equation (1) shows that, the convolution constitutes a particular kind of integral transform which blends one function with another and expresses the overlap of the function g as it is shifted over the function f. The integration range in Eq. (1), depends on the domain on which f and g are defined and it could be a finite range, but often we take $\alpha = -\infty$ and $\beta = +\infty$ by extending either periodically or with zeros the integration range in both directions. It is worth mentioning the specific type of convolution which is defined between spectral distributions very oftenly used in physics, as we do in the present paper where we utilize Fermi-Dirac (FD) and Power-law (PL) distributions (many times 3-dimensional convolutions are needed) [1, 2].

In this paper, through the energy-spectrum, $\eta(\varepsilon_\nu)$, of a SN neutrino source, represented by Fermi-Dirac (FD), Power-Law (PL), Maxwell-Boltzman (MB), etc. distributions, theoretical ν-nucleus cross sections (of neutral- and charged-current reactions) computed in the context of a nuclear model, are connected via the convolution procedure with physical observables measured in ν-detection experiments. Then, the signals created at the nuclear detectors can be analysed by utilizing the convolution (folding) method described below.

2. Convolution (folding) procedure for ν-detection studies

For the purposes of the present work, convoluted double differential cross sections as well as convoluted single differential and flux-averaged total (theoretical) cross sections are needed.

In the case of the original double differential ν-nucleus cross section, $d^2\sigma(\theta,\omega,\varepsilon_\nu)/d\Theta d\omega$, where $d\Theta = sin\theta d\theta$ [3], the convolution with a spectral ν-energy distribution, $\eta(\varepsilon_\nu)$, is defined by the expression [2]

$$\left[\frac{d^2\sigma(\theta,\omega)}{d\Theta d\omega}\right]_{fold} = \int_\omega^\infty \frac{d^2\sigma(\theta,\omega,\varepsilon_\nu)}{d\Theta d\omega}\eta(\varepsilon_\nu)d\varepsilon_\nu\,, \tag{2}$$

where the energy ω in the lower limit of the integral denotes that incoming neutrinos with energy ε_ν cause nuclear transitions for which the excitation energy ω satisfies the condition $\omega \leq \varepsilon_\nu$ [3].

The SN neutrinos studied here, have rather broad energy distribution, $\eta(\varepsilon_\nu) \equiv dN_\nu(\varepsilon_\nu)/d\varepsilon_\nu$, where N_ν denotes the number of neutrinos reaching the detector, characteristic of the considered neutrino or anti-neutrino type. The distributions $\eta(\varepsilon_\nu)$ are usually normalized as

$$\int_0^\infty \eta(\varepsilon_\nu)d\varepsilon_\nu = 1\,. \tag{3}$$

A realistic energy distribution for SN neutrinos results by inserting the chemical potential μ, in the black body shape distribution. Then, the well-known two-parameter Fermi-Dirac (FD) distribution is obtained which is written as [2]

$$\eta_{FD}(T,n_{dg};\varepsilon_\nu) = \left[\int_0^\infty \frac{x^2}{e^{x-n_{dg}}+1}dx\right]^{-1} \frac{(\varepsilon_\nu^2/T^3)}{1+e^{(\varepsilon_\nu/T-n_{dg})}}\,, \tag{4}$$

where T (in MeV) and $n_{dg} = \mu/T$, are the neutrino temperature and the known as degeneracy parameter, respectively.

Alternatively, the SN-ν energy spectrum can also be described with the analytically simpler two-parameter Power-Law (PL) energy distribution of the form [2]

$$\eta_{PL}(\langle\varepsilon_\nu\rangle,\alpha;\varepsilon_\nu) = \frac{(\alpha+1)^{\alpha+1}}{\Gamma(\alpha+1)}\frac{\varepsilon_\nu^\alpha}{\langle\varepsilon_\nu\rangle^{\alpha+1}}e^{-(\alpha+1)(\varepsilon_\nu/\langle\varepsilon_\nu\rangle)}\,, \tag{5}$$

where $\langle\varepsilon_\nu\rangle$ is the average neutrino energy. The parameter α adjusts the width of the distribution (pinching parameter). In Fig. 1(a) we show some FD and PL neutrino spectral distributions.

Recently, some accelerated (in storage rings) β-radioactive nuclei have been proposed as sources of laboratory ν-beams (beta-beam neutrinos) [4, 5]. Such facilities may produce pure beam neutrinos in which the possible neutrino flavors are either the ν_e (for β^+-decaying ions) or the $\tilde{\nu}_e$ (for β^--decaying ions). The laboratory ν-beams are used to search for standard and non-standard neutrino physics at low and intermediate energies (ν-nucleus interactions, neutrino properties, neutrino oscillations, etc.) and measure ν-nucleus scattering cross sections [4].

In the laboratory-frame, the energy distribution of β-beam neutrinos, assuming that the decaying nuclei are moving with velocity $u = c\sqrt{\gamma^2-1}/\gamma$, reads [5]

$$\eta_\gamma(\varepsilon_\nu) = \frac{ln2}{m_e^5(ft)}\frac{F(\pm Z,\varepsilon_e)E_e p_e \varepsilon_\nu^2}{2\gamma^3(1+u)(1-u^2)} \tag{6}$$

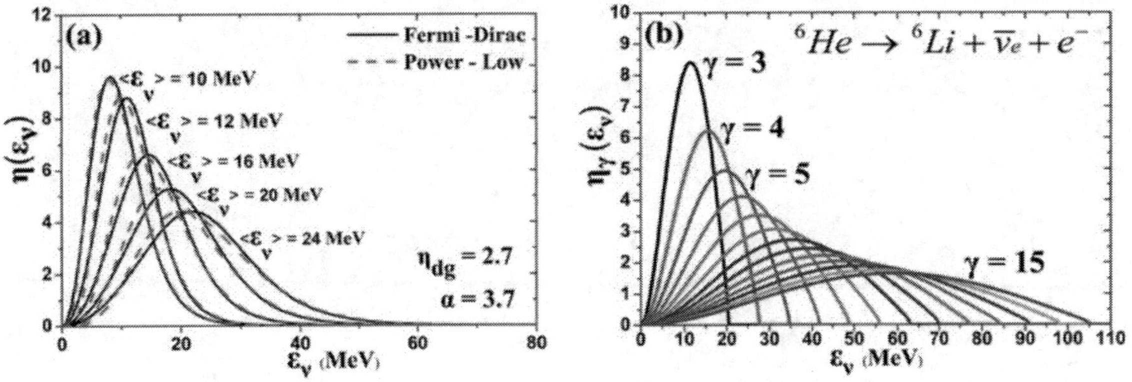

Figure 1. (a) Equivalent Fermi-Dirac (FD) and Power-law (PL) ν-energy distributions for various values of their parameters. (b) Normalized β-beam anti-neutrino spectra originating from boosted radioactive ^{6}He nuclei. The boost factors are integers between γ=3 and γ=15.

(γ denotes the known Lorentz factor and c the speed of light). In the latter expression, m_e, E_e and p_e are the mass, energy and momentum, respectively, of the outgoing e^- (or e^+) of the reaction [see Fig. 1(b)]. (ft) is the known ft-value and $F(\pm Z, \varepsilon_e)$ denotes the Coulomb correction function (Fermi function) which accounts for the electromagnetic interaction between the emitted e^- (or e^+) and the charge distribution of the daughter nucleus (final state interaction) [2].

By exploiting low-energy β-beam ν-spectra of the form of Eq. (6), corresponding to different boost velocities (γ-factors), one could construct normalized synthetic neutrino energy distributions $\eta_{bb}(\varepsilon_\nu)$ given by linear combinations as

$$\eta_{bb}(\varepsilon_\nu) = \sum_{j=1}^{N} \alpha_j \eta_{\gamma_j}(\varepsilon_\nu) \,, \tag{7}$$

where N is the number of different γ boosted spectra included in the synthetic spectrum [2, 4].

The convolution of the theoretical cross sections [with ν-spectra of the form of Eqs. (4)-(7)], provides estimates of nuclear-detector responses to energy-distributions of the observed neutrinos. The features of the arriving at the detector SN-ν flux are encoded in the nuclear response of the detector which theoretically could be reproduced by convoluted cross sections calculations [1, 3]. As an example, using Eqs. (6) and (7), we apply the convolution method to simulate SN-ν signals recorded at a nuclear detector with detector medium the ^{66}Zn isotope [5].

3. Application of simulation techniques in SN-ν detection

The distributions of synthetic spectra $\eta_{bb}(\varepsilon_\nu)$, may be used to fit original SN-ν energy spectra, $\eta_{SN}(\varepsilon_\nu)$, reaching terrestrial ν-detectors. This may be achieved by adjusting the weight parameters α_j of Eq. (7) through a minimization procedure which determines the participation of each of the γ_j components in the linear combination $\eta_{bb}(\varepsilon_\nu)$ [4]. The best fit of $\eta_{bb}(\varepsilon_\nu)$ to an original SN-ν spectrum $\eta_{SN}(\varepsilon_\nu)$, representing the signal recorded at the ν-detector, results by minimizing the integral [4, 5]

$$\mathcal{Q} = \int_0^\infty |\sum_{j=1}^{N} \alpha_j \eta_{\gamma_j(\varepsilon_\nu)} - \eta_{SN}(\varepsilon_\nu)| d\varepsilon_\nu \,. \tag{8}$$

For the simulations performed in the present paper, we employ various SN-ν spectral distributions. The minimization of the integral in Eq. (8), was achieved by utilizing the optimization algorithm MERLIN [2, 6].

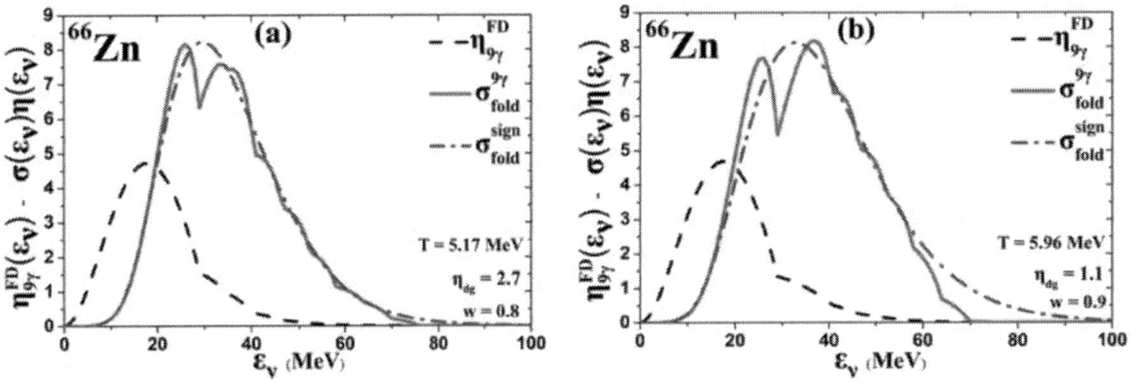

Figure 2. Folded total cross-section (in 10^{-40} cm^2) $\sigma^{9\gamma}_{fold}$ (full line) for ^{66}Zn. They resulted by using a synthetic spectrum with 9 Lorentz factors $\gamma = 5, 6, \ldots, 13$. The assumed SN-$\nu$ signal, σ^{sign}_{fold} (dash-dotted line), is well reproduced [w = 0.8 in (a) and w = 0.9 in (b), see text]. The illustrated normalized original SN-neutrino spectrum, $\eta_{FD}(\varepsilon_\nu)$, is in 10^{-2} MeV^{-1} [2].

In Fig. 2 (full line), the folded total cross-section of a signal, $\sigma^{sign}_{fold}(\varepsilon_\nu)$, recorded at the ^{66}Zn detector [5], is compared with the corresponding (fitted) folded cross section, $\sigma^{bb}_{fold}(\varepsilon_\nu)$. The latter is obtained by using the synthetic spectrum $\eta_{bb}(\varepsilon_\nu)$ with nine components (N=9), i.e. beta-beam anti-neutrino spectra with Lorentz factors γ= 5, 6, ..., 13 (see dashed line of Fig. 2). The assumed original SN-ν spectra that create the signal on the detector, are described by two-parameter FD distributions. Their parametrization is described in Fig. 2, where we give also the width parameter w defined in Eq. (19) of Ref. [2], namely w = 0.8 Fig. 2(a), and w = 0.9 Fig. 2(b). As can be seen, in all cases the peak of the synthetic spectrum, $\eta_{bb}(\varepsilon_\nu)$, is located at lower energies as compared to the folded total cross-section $\sigma^{bb}_{fold}(\varepsilon_\nu)$. Concerning the quality of the fitting in Fig. 2, even thought the MERLIN package is a very effective algorithm [6], it is not very good due to the fact that the curves for $\gamma = 3, \ldots, 7$ have noticeably different shapes than those for $\gamma > 7$. We mention that, in Ref. [4] authors utilized a weighted factor of the order of 2 for values $\gamma > 7$ which improves significantly the quality of the fit.

4. Conclusions

We concentrated on the convolution method used in nuclear and astro-nuclear physics studies and, in particular, in the investigation of nuclear responses of various neutrino detectors to the energy-spectra of SN-ν. By using the minimization procedure utilized recently to fit the synthetic beta-beam spectra on original SN-ν spectra we employed various boosted beta-beam spectra of β-radioactive ^{6}He ions to simulate specific SN-ν scenarios.

References

[1] Tsakstara V, Kosmas T S, Sinatkas J, Chasioti V C, Divari P C 2007 *Proc. Int. Conf. on Computational Methods in Science and Engineering (Corfu)* vol 2 (AIP) p 1383 and references therein.
[2] Tsakstara V and Kosmas T S 2011 *Phys. Rev.* C **84** 064620 and references therein.
[3] Tsakstara V and Kosmas T S 2011 *Phys. Rev.* C **83** 054612.
[4] Jachowicz N, McLaughlin G C, Volpe C 2008 *Phys. Rev.* C **77**, 055501.
[5] Tsakstara, V, Kosmas T S, Wambach J 2011 *J. Prog. Part. Nucl. Phys.* **66** 424.
[6] Lagaris I E, Likas A, Fotiadis D 1997 *Comput. Phys. Commun.* **104** 1; Kosmas T S, Lagaris I E 2002 *J. Phys.* G **28** 2907.

Exact wave functions of bound μ^- for calculating ordinary muon capture rates

I.S. Kardaras [1], V.N. Stavrou [1], I.G. Tsoulos [2] and T.S. Kosmas [1]

[1] Theoretical Physics Section, University of Ioannina, GR 45110 Ioannina, Greece
[2] Department of Computer Science, University of Ioannina, GR 45110 Ioannina, Greece

E-mail: iokard@uoi.gr, vstavrou@cc.uoi.gr, hkosmas@cc.uoi.gr

Abstract. The goal of the present contribution is twofold: (i) To compute exact wave functions for a muon bound in the extended Coulomb potential of a muonic atom by solving the Dirac equation within the context of genetic algorithms and neural network techniques using experimental finite-size charge-densities for the attracting nucleus. (ii) To calculate partial and total rates of the ordinary muon capture in various muonic atoms. In contrast to the majority of previous realistic calculations for μ^--capture rates, in our present work we utilize the above mentioned exact wave functions for a muon orbiting at the 1s and 2p atomic orbits. The required many-body nuclear wave functions are obtained by diagonalizing the eigenvalue problem of the quasi-particle random phase approximation (QRPA).

1. Introduction

As is well known, when negative muons, produced by the decay of slow pions (e.g. in a meson factory at Fermilab, USA, or at JPARC, Japan), slow down in matter, it is possible for them to be captured in atomic orbits. Afterwards, fast electromagnetic cascades bring the muon of the muonic atom down to the innermost (1s or 2p) orbits [1, 2]. The bound in such an orbit muon may disappear, either by decay, known as muon decay in orbit and represented by the reaction

$$\mu_b^- \to e^- + \tilde{\nu}_e + \nu_\mu \, , \qquad (1)$$

or by capture by the nucleus the main channel of which is the ordinary muon capture represented by [3, 4]

$$\mu_b^- + (A, Z) \to (A, Z-1)^* + \nu_\mu \, . \qquad (2)$$

We mention that, some of the most sensitive recent experiments (at Fermilab and JPARC) search extensively for additional channels that involve muon number non-conservation predicted to occur in modern gauge theories [5]. The most important exotic possibilities of this kind are: (i) The neutrinoless muon to electron conversion

$$\mu_b^- + (A, Z) \to (A, Z) + e^- \, , \qquad (3)$$

which violates the muon number and is allowed by lepton conservation (this is searched by the mu2e experiment at Fermilab and the COMMET experiment at JPARC), and (ii) the muon to positron conversion

$$\mu_b^- + (A, Z) \to (A, Z-2) + e^+ \qquad (4)$$

Published under licence by IOP Publishing Ltd

which violates both the muon number and the lepton number as well.

In the present paper we focus our attention on process (2) which has been the subject of extensive experimental and theoretical investigations started on the early 50's. Recently, however, the interest for this process has been revived due to its important role in astrophysics and especially in supernova dynamics [3]. For terrestrial experiments studying the muon capture by a nuclear target (A,Z), a competitive channel is the radiative muon capture [1, 2]

$$\mu_b^- + (A,Z) \to (A, Z-1)^* + \nu_\mu + \gamma \tag{5}$$

It should be stressed that the ordinary muon capture Eq. (2) is one of the most useful channels for testing the nuclear structure models, as the low-energy electron scattering on nuclei.

2. Theoretical Framework - Brief description of the Formalism

2.1. The interaction Hamiltonian of the muon capture process

In the context of the current-current interaction hypothesis, the effective interaction Hamiltonian for the ordinary muon capture takes the form [3, 4]

$$\hat{H}_{eff} = \int d^3x \hat{\mathcal{H}}_{eff}(\mathbf{x}) = \frac{G}{\sqrt{2}} \int d^3\mathbf{x} \hat{\mathcal{J}}_\lambda^-(\mathbf{x})(j^+)^\lambda(\mathbf{x}) \tag{6}$$

where j_λ^+ denotes the lepton charge-changing current

$$j_\lambda^+ = \bar{\psi}_{\nu_\mu}\gamma_\lambda(1-\gamma_5)\tau_+\psi_\mu \tag{7}$$

(ψ_μ, ψ_{ν_μ} are the lepton spinors), and J_λ^- represents the hadronic charge-changing current written as

$$J_\lambda^- = \bar{u}_n(P')[F_V\gamma_\lambda + F_M\sigma_{\lambda\nu}q_\nu + F_A\gamma_\lambda\gamma_5 + F_P\gamma_5 q_M]\tau_- u_p(P) \tag{8}$$

In the latter equation, $u_{p,n}(P)$ represent the proton or neutron spinor. The quantities F_V, F_M, F_A, F_P are the well known nucleon form factors for the polar-vector, magnetic, axial-vector and pseudoscalar components of the hadronic current, respectively. All these form factors are functions of the four-momentum transfer q_μ^2 given by the kinematics of the reaction. In our convention [6, 7] we write

$$q_\mu^2 \equiv q_\mu q^\mu = q_0^2 - \mathbf{q}^2 \,. \tag{9}$$

The magnitude of the three momentum transfer $q = |\mathbf{q}|$, which enters the nuclear matrix elements, can be approximated by [3]

$$q \equiv \nu \simeq m_\mu - \varepsilon_b + E_i - E_f. \tag{10}$$

In the latter equation, m_μ is the muon rest mass, ϵ_b is the muon binding energy at the muonic atom, and E_i, E_f represent the energies of the initial (ground) and final nuclear states, respectively. In Eq. (10) we neglect the nuclear recoil and assume that the initial momentum of the bound muon is negligible.

2.2. Partial and total muon capture rates

In our nuclear structure calculations, the even-even nuclear systems are supposed to be spherically symmetric with an initial (ground) state $|J_i^{\pi_i}\rangle = |0^+\rangle$. After the μ^- capture, the studied nucleus is left in a final excited state $|J_f^{\pi_f}\rangle$ having angular momentum J_f and parity π_f. Using natural units ($\hbar = c = 1$), the partial capture rate for a transition from the ground state to an excited state is written as [3, 4]

$$\sigma_{i\to f} = \frac{\Omega(\varepsilon_f)^2}{2\pi} \sum_{s_f,s_i} \frac{1}{(2J_i+1)} \sum_{M_f,M_i} |\langle f|\hat{H}_{eff}|i\rangle|^2 \tag{11}$$

where ε_f is the energy of the outgoing ν_μ and Ω stands for the quantization volume. After replacing $\hat{H}_{eff}$ in Eq. (11) with its expression from Eq. (6) and elaborating on the result, the partial muon capture rate $\sigma_{i\to f}$ reads [3, 4]

$$\sigma_{if} = \frac{G^2(\varepsilon_f)^2}{2\pi}\frac{1}{2J_i+1}\left\{\sum_{J=0}^{\infty}|\langle J_f\|\Phi^\mu(\hat{\mathcal{M}}_J - \hat{\mathcal{L}}_J)\|J_i\rangle|^2 + \sum_{J=1}^{\infty}|\langle J_f\|\Phi^\mu(\hat{\mathcal{T}}_J^{el} - \hat{\mathcal{T}}_J^{mag})\|J_i\rangle|^2\right\} \quad (12)$$

The operators $\hat{\mathcal{M}}_J$ (Coulomb or charge), $\hat{\mathcal{L}}_J$ (longitudinal), $\hat{\mathcal{T}}_J^{el}$ (transverse electric) and $\hat{\mathcal{T}}_J^{mag}$ (transverse magnetic) contain polar-vector and axial-vector components (see e.g. Refs. [6, 7]). The nuclear wave functions for the initial $|J_i\rangle$ and the final $|J_f\rangle$ states, in the present work are constructed within the context of the quasi-particle random phase approximation (QRPA) [6, 7]. In Eq. (12) Φ^μ stands for the muon wave function. Usually, in the allowed muon capture process, an average value for the 1s state, i.e. $\langle\Phi^\mu_{1s}\rangle^2$, is taken out of the integrals of Eq. (12) [1, 3]. This is given in terms of the effective nuclear charge Z_{eff} which sees the bound muon [5]. In the present work, however, exact wave functions for the bound μ^- orbiting at the innermost atomic orbits of a muonic atom are obtained by solving the Schroedinger and Dirac equations as is described below [5].

3. Computation of the exact wave function for a bound μ^-

3.1. Solving the Schroedinger equation

For non-relativistic bound muons we can use only the large component of the muon spinor (the small one is assumed to be zero) in the calculations of the ordinary muon capture rates. In such cases, one solves the Schroedinger equation which for the reduced radial muon wave function $u(r) = r\Phi^\mu(r)$ is written as [5]

$$-\frac{\hbar^2}{2\mu}\frac{d^2}{dr^2}u(r) + V(r)u(r) = Eu(r)\,. \quad (13)$$

where V(r) represents the extended Coulomb potential originating from the finite-size nuclear charge-density (E is the energy eigenvalue) and μ denotes the reduced mass of the muon-nucleus system. The latter equation is solved by assuming that $u(r)$ can be written as a superposition of appropriately parametrized sigmoid functions and applying the same techniques as in solving the Dirac equation (see below) [5, 8].

3.2. Solving the Dirac equation

In the present contribution we consider both radial components of the bound muon wave function, the large, f(r), and the small, g(r). They are given from the solution of the Dirac equations [3, 9]

$$\frac{d}{dr}f(r) + \frac{1}{r}f(r) = \frac{1}{\hbar}[\mu c^2 - E + V(r)]g(r) \quad (14)$$

$$\frac{d}{dr}g(r) - \frac{1}{r}g(r) = \frac{1}{\hbar}[\mu c^2 + E - V(r)]f(r)\,, \quad (15)$$

The solution of these equations is obtained by utilizing neural network techniques and genetic algorithms in order to optimize the parametric expressions discussed in Sect. 3.1. In Fig. 1, the large and small components of the Dirac spinor for a muon orbiting in the Φ_{1s} (left) and Φ_{2p} (right) orbits of a ^{28}Si muonic atom, are shown. For comparison, the wave function obtained by solving the Schroedinger equation (13) is also shown in the case of the Φ_{1s} wave function.

4. Construction of the many body nuclear wave functions

The second basic ingredient in our present study is the construction of the initial and final nuclear states entering the capture rates of Eqs. (11) and (12). We deduce them within the QRPA, the main features of which are outlined in Refs. [6, 7].

The nuclear ground state of even-even isotopes is reliably calculated from the solution of the BCS equations obtained with the single-particle energies of the effective field of a Coulomb-corrected Woods-Saxon potential plus the pairing interaction (monopole) part of a one-meson exchange Bonn CD two-body potential. Similarly, the final (excited) nuclear states of an even-even isotope are computed by solving the QRPA equations as described e.g. in Refs. [6, 7]. Results for partial and total muon capture rates in a set of nuclei (^{28}Si, ^{64}Zn, ^{98}Mo, ^{124}Sn, ^{156}Gd, ^{208}Pb, etc.), obtained as discussed above, will be published elsewhere [9].

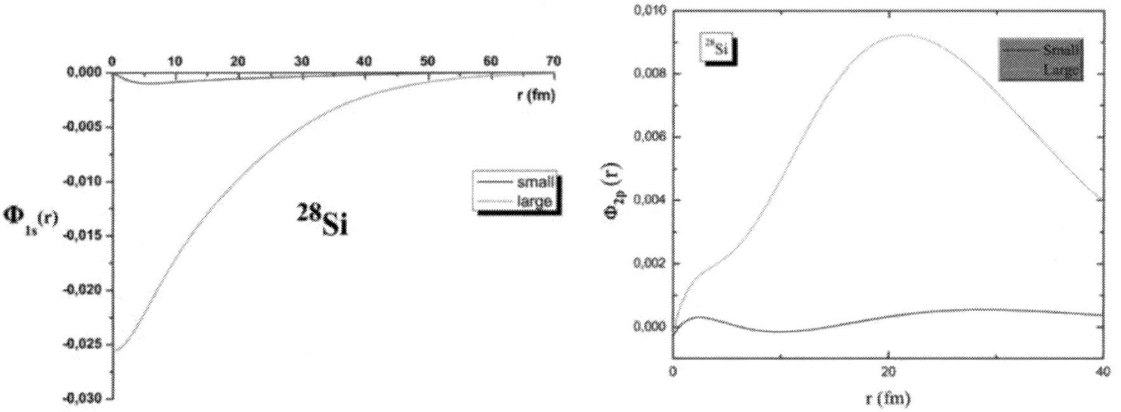

Figure 1. Wave function Φ^{μ}_{1s} (left) and Φ^{μ}_{2p} (right) for a bound muon in the ^{28}Si nucleus calculated as described in the text.

5. Summary

The ordinary μ^- capture by atomic nuclei is studied in currently interesting nuclei (^{28}Si, ^{64}Zn, ^{98}Mo, ^{124}Sn, ^{156}Gd, ^{208}Pb, etc.) [9]. The nuclear wave functions are computed within the context of the QRPA by employing the Bonn CD residual interaction that is a realistic description of the two-body nuclear forces. For the reduced matrix elements of the relevant tensor multipole operators, coming out of the Donnelly-Walecka projection method, we employ the computational approach developed by our group recently [8]. The required muon wave functions are obtained by solving the Dirac equation using neural network techniques and genetic algorithms. As a special case in our present work, we illustrate the Φ_{1s} and Φ_{2p} wave functions for a muon bound in the ^{28}Si nucleus. Results for the capture rates are expected to be published soon.

References

[1] Measday D F 2001 Phys. Rep. **354** 243.
[2] Suzuki T, Measday D F, and Roalsvig J P 1987 Phys. Rev. **C 35** 2212.
[3] Zinner N T, Laganke K, Vogel P Phys. Rev. **C 74** 024326, and References therein.
[4] Marketin T , Paar N, Niksic T, and Vretenar 2009 Phys. Rev. **C 79** 0546323.
[5] Kosmas T S and Lagaris I E 2002 J. Phys. **G 28** 2907.
[6] Chasioti V C and Kosmas T S 2009 Nucl. Phys. **A 829** 234-252.
[7] Tsakstara V and T S Kosmas 2011 Phys. Rev. **C 83** 054612.
[8] Tsoulos I G 2008 Appl. Math. and Comput. **203** 598.
[9] Kardaras I S 2010 MSc. Thesis, Univ. of Ioannina; Kardaras I S and Kosmas T S, in preparation.

Spectral triplets, emergent geometry and entropy in Moyal plane

B. Chakraborty[a] **and F. G. Scholtz** [b]

[a]S. N. Bose National Centre for Basic Sciences,JD Block, Sector III, Salt Lake, Kolkata-700098, India

[b]National Institute for Theoretical Physics, Private Bag X1, Matieland, Stellenbosch 7602, South Africa

Abstract. It was shown by Doplicher et.al. that the measurement of spacetime intervals of the order of Planck length scale is operationally impossible, as the process of measurement invariably gives rise to a black hole formation. This can be avoided by postulating non-vanishing commutation relations between the coordinates, which are now promoted to the level of operators. Formulation of quantum mechanics in these kinds of spaces through the introduction of Hilbert spaces of Hilbert-Schmidt operators is then shown to allow the construction of spectral triplets a la Connes naturally. The computation of spectral distance between pure and mixed states is then shown to exhibit a deep connection between entropy and geometry.

1. Introduction

Ever since Connes developed the mathematical basis of Noncommutative geometry (NCG) [1], it was mostly applied to the so-called "almost commutative" (AC) spaces to provide a unified framework of the Standard model of particle physics [1]. This structure is believed to be adequate to describe physics up to the GUT scale. But still at a higher energy scale, like in the vicinity of Planck scale, one has to consider more serious non-commutative spaces and adopt the prescription of Doplicher et. al [2] where the spacetime coordinates are promoted to the level of operators satisfying certain non-vanishing commutation relations like $[\hat{x}_\mu, \hat{x}_\nu] = i\theta_{\mu\nu}$ as one of the possibilities, so that the gravitational instabilities arising out of the measurement processes to localise an event can be avoided. This naturally raises the question whether the formulation of noncommutative quantum mechanics as set out in [3, 4], which entails a representation of the quantum system on the Hilbert space of Hilbert-Schmidt operators, does not naturally encode geometric information in the context of Connes NCG where the notion of a spectral triplet is introduced ? Our investigation in this direction indeed shows that the answer is in the affirmative, besides giving a much simpler algorithm to compute the spectral distance in an operatorial and a basis independent approach, so that we can avoid using any star product and any ambiguities that may result there-from[5]. Furthermore, in contrast to certain other recent computations [6, 7], carried out using the Moyal star product, we do not restrict our analysis to the classical configuration space $\mathcal{H}_c$ only; we extend this to the above mentioned Hilbert space of Hilbert-Schmidt operators, called here the quantum Hilbert space $\mathcal{H}_q$, which has a natural tensorial structure $\mathcal{H}_q = \mathcal{H}_q \otimes \mathcal{H}_q$. This offers rather tantalizing possibilities in terms of modifications of the implied geometry and the possible statistical underpinnings thereof.

2. Quantum mechanics on the non-commutative plane

In two dimensions the non-commutative plane is defined through the commutation relations

$$[\hat{x_i}, \hat{x_j}] = i\theta_{ij}. \tag{1}$$

One constructs standard creation and annihilation operators $b^\dagger$ and $b = \hat{x}_1 + i\hat{x}_2/\sqrt{2\theta}$ satisfying $[\hat{b}, \hat{b}^\dagger] = 1$, and view the non-commutative plane i.e. the classical configuration space $\mathcal{H}_c$(counter part of $\mathcal{R}^2$ in commutative quantum mechanics) as a boson Fock space spanned by the eigenstate $|n\rangle$ of the radial operator $b^\dagger b$:

$$\mathcal{H}_c = span\{|n\rangle = \frac{1}{\sqrt{n!}}(b^\dagger)^n|0\rangle\}. \tag{2}$$

Next we introduce the quantum Hilbert space in which the physical states of the system and the non-commutative Heisenberg algebra are to be represented. This is taken to be the set of all bounded trace-class operators (the Hilbert-Schmidt operators) over $\mathcal{H}_c$ and we refer to it as the quantum Hilbert space,$\mathcal{H}_q$,

$$\mathcal{H}_q = \{\psi : \mathrm{tr}_c(\psi^\dagger\psi) < \infty\}, \tag{3}$$

where the subscript c refers to tracing over $\mathcal{H}_c$. Elements of $\mathcal{H}_q$ are denoted by a round bracket $\psi \equiv |\psi)$ and the inner product is defined as $(\phi|\psi) = \mathrm{tr}_c(\phi^\dagger\psi)$. We reserve $\dagger$ to denote hermitian conjugation on the classical Hilbert space, while $\ddagger$ denotes hermitian conjugation on the quantum Hilbert space. Note that $\mathcal{H}_q$ has a natural tensor product structure in that it can be viewed as the tensor product of $\mathcal{H}_c$ and its dual, i.e., $\mathcal{H}_q = \mathcal{H}_c \otimes \mathcal{H}_c^\star$. We can therefore also write the elements of $\mathcal{H}_q$ in the form $|\psi, \phi) = |\psi\rangle\langle\phi|$. We shall use this notation quite often below and refer to ψ as the left hand and ϕ as the right hand sector.

A unitary representation of the non-commutative Heisenberg algebra

$$[\hat{x}_i, \hat{x}_j] = i\theta_{ij}, \quad [\hat{x}_i, \hat{p}_j] = i\delta_{ij}, \quad [\hat{p}_i, \hat{p}_j] = 0 \tag{4}$$

is obtained by the following action:

$$\hat{X}_i\psi = \hat{x}_i\psi, \quad \hat{P}_i\psi = \frac{1}{\theta}\epsilon_{ij}[\hat{x}_j, \psi]. \tag{5}$$

Here we use capital letters to distinguish operators acting on the quantum Hilbert space from those acting on the classical Hilbert space. It is also useful to introduce the following quantum operators $B, B^\ddagger, \hat{P}, \hat{P}^\ddagger$, where

$$B = \frac{1}{\sqrt{2\theta}}\left(\hat{X} + i\hat{Y}\right), \quad \hat{P} = \hat{P}_x + i\hat{P}_y \tag{6}$$

These operators act as follow

$$\begin{aligned} B\psi(\hat{x}, \hat{y}) &= b\psi(\hat{x}, \hat{y}), \quad B^\ddagger\psi(\hat{x}, \hat{y}) = b^\dagger\psi(\hat{x}, \hat{y}), \\ P\psi(\hat{x}, \hat{y}) &= -i\sqrt{\frac{2}{\theta}}[b, \psi(\hat{x}, \hat{y})], \quad P^\ddagger\psi(\hat{x}, \hat{y}) = i\sqrt{\frac{2}{\theta}}[b^\dagger, \psi(\hat{x}, \hat{y})]. \end{aligned} \tag{7}$$

The interpretation of this quantum system now proceeds as for a standard one. The only modification required is that position measurement must now be interpreted in the context of a weak measurement (Positive Operator Valued Measure) rather than a strong (Projective Valued Measurement). The essence of the construction is based on the minimal uncertainty

Journal of Physics: Conference Series **410** (2013) 012128 doi:10.1088/1742-6596/410/1/012128

normalised coherent states on $\mathcal{H}_c$ given by $|z\rangle = e^{-z\bar{z}/2}e^{zb^\dagger}|0\rangle$ where $z = \frac{1}{\sqrt{2\theta}}(x+iy)$ is a dimensionless complex number. These states provide an overcomplete basis on the non-commutative configuration space. Corresponding to these states one constructs a state in quantum Hilbert space as follows

$$|z,z) = |z\rangle\langle z|, \quad B|z,z) = z|z,z), \tag{8}$$

which leads to the natural interpretation of (x,y) as the dimensionful position coordinates. These states provide an overcomplete set on the quantum Hilbert space enabling us to obtain the operators in terms of the orthonormal basis $|n\rangle$

$$\pi_z = \frac{1}{2\pi\theta}|z,z)e^{\overleftarrow{\partial_{\bar{z}}}\overrightarrow{\partial_z}}(z,z| = \frac{1}{2\pi\theta}\sum_n |z,n)(z,n| \tag{9}$$

forming a set of complete, positive, but non-orthogonal and unormalized projection operators [8], i.e.,

$$\int dxdy\; \pi_z = 1_q \quad (\psi|\pi_z|\psi) \geq 0, \forall\psi, \quad \pi_z\pi_w \neq \delta(z-w), \quad \pi_z^2 \propto \pi_z. \tag{10}$$

They therefore provide a Positive Operator Valued Measure (POVM) that can be used to give a consistent probability interpretation by assigning the probability of finding the outcome of a position measurement to be (x,y), given that the system is described by the density matrix ρ, to be $P(x,y) = \mathrm{tr_q}(\pi_z\rho)$, which is completely insensitive to the right sector.

3. Spectral triplets on classical configuration space

We introduce the spectral triplet $(\mathcal{A},\mathcal{H},D)$ where $\mathcal{A} = \mathcal{H}_q$, $\mathcal{H} = \mathcal{H}_c \otimes C^2$, the action of elements of $a \in \mathcal{A}$ on $\mathcal{H}$ is defined through the representation $\pi(a)(|\psi\rangle,|\phi\rangle) = (a|\psi\rangle, a|\phi\rangle)$ and the Dirac operator is defined as $D = -i\sigma^i P_i$ with $\sigma^1 = \begin{pmatrix} 0 & 1 \\ 1 & 0 \end{pmatrix}$,$\sigma^2 = \begin{pmatrix} 0 & -i \\ i & 0 \end{pmatrix}$ being the first two Pauli matrices. Then in the commutator $[D,\pi(a)]$ the Dirac operator can effectively be identified as,

$$D = -i\sqrt{\frac{2}{\theta}}\begin{pmatrix} 0 & b^\dagger \\ b & 0 \end{pmatrix}. \tag{11}$$

States ω are positive linear functionals of norm one over $\mathcal{A}$. The Connes' spectral distance between two states is then defined by

$$d(\omega,\omega') = \sup_{a\in B}|\omega(a)-\omega'(a)|, B = \{a \in \mathcal{A} : ||[D,\pi(a)]||_{\mathrm{op}} \leq 1\}, ||A||_{\mathrm{op}} = \sup_{\phi\in\mathcal{H}}\frac{||A\phi||}{||\phi||}. \tag{12}$$

Exploiting the close link between states and density matrices (taken to be hermitian, semi-positive and normalised) ω and ρ_ω respectively, any state over $\mathcal{A}$ can be uniquely written as [9] $\omega(a) = \mathrm{tr_c}(\rho_\omega a)$. Note that $\rho_\omega \in \mathcal{A} = \mathcal{H}_q$. If ω is a pure state, so is ρ_ω, i.e. it is of the form $\rho_\omega = |\psi\rangle\langle\psi|$ for some $|\psi\rangle \in \mathcal{H}_c$.

Using this we can write the Connes' distance function as

$$d(\rho,\rho') = \sup_{a\in B}|\mathrm{tr_c}((d\rho)a)| \tag{13}$$

where $d\rho = \rho - \rho'$ and we use the notation $d\rho$ to indicate that we are interested in infinitesimal changes. Using the Schwarz inequality $|(d\rho,a)| \leq ||d\rho||_{\mathrm{tr}}||a||_{\mathrm{tr}}$,this allows us to write Connes' distance function, in a more tractable form as,

$$d(\rho,\rho') = \frac{\mathrm{tr_c}(d\rho)^2}{||[D,\pi(d\rho)]||_{\mathrm{op}}}. \tag{14}$$

Let us apply this to the example in [6] where the distance between harmonic oscillator states is calculated. In terms of the harmonic oscillator basis $|n\rangle$,introduced in (2), we can write $a = \sum_{m,n=0} a_{m,n}|m\rangle\langle n|$. The states are taken as $\omega_m(a) = a_{m,m}$ with corresponding pure state density matrix $\rho_m = |m\rangle\langle m|$. We then take $d\rho = \rho_{m+1} - \rho_m$. The calculation is now straightforward:

$$[D, d\rho] = -i\sqrt{\frac{2}{\theta}} \begin{pmatrix} 0 & \left[b^\dagger, d\rho\right] \\ [b, d\rho] & 0 \end{pmatrix}, \tag{15}$$

A simple calculation yields the diagonal forms of

$$\begin{aligned} ||\, [b, d\rho]\, ||_{\text{op}}^2 &= ||4(m+1)|m+1\rangle\langle m+1| + m|m\rangle\langle m| + (m+2)|m+2\rangle\langle m+2|||_{\text{op}} = 4(m+1), \\ ||\left[b^\dagger, d\rho\right]||_{\text{op}}^2 &= ||(m+2)|m+1\rangle\langle m+1| + 4(m+1)|m\rangle\langle m| + m|m-1\rangle\langle m-1|||_{\text{op}} = 4(m+1) \end{aligned}$$

From this we obtain $||[D, d\rho]||_{\text{op}} = 2\sqrt{\frac{2(m+1)}{\theta}}$ and

$$d(m+1, m) = \sqrt{\frac{\theta}{2(m+1)}}, \tag{16}$$

which agrees precisely with the result of [6].

Next we calculate the distance between two coherent harmonic oscillator states. Here the pure state density matrix are $\rho = |z\rangle\langle z|$. Then

$$\begin{aligned} d\rho &= |z+dz\rangle\langle z+dz| - |z\rangle\langle z| \\ &= \left(b^\dagger - \bar{z}\right)|z\rangle\langle z|dz + |z\rangle\langle z|\,(b-z)\,d\bar{z} + O(dz^2, d\bar{z}^2). \end{aligned} \tag{17}$$

This can be written more simply as

$$d\rho = d\bar{z}|\tilde{0}\rangle\langle\tilde{1}| + dz|\tilde{1}\rangle\langle\tilde{0}| \tag{18}$$

where we have introduced new bosons $\tilde{b} = b - z$, $\tilde{b}^\dagger = b^\dagger - \bar{z}$ and the vacuum $|\tilde{0}\rangle = |z\rangle$ for which $\tilde{b}|\tilde{0}\rangle = 0$, $|\tilde{1}\rangle = \tilde{b}^\dagger|\tilde{0}\rangle$. From this

$$[b, d\rho] = dz|\tilde{0}\rangle\langle\tilde{0}| - \sqrt{2}d\bar{z}|\tilde{0}\rangle\langle\tilde{2}| - dz|\tilde{1}\rangle\langle\tilde{1}| \tag{19}$$

and proceeding as before, we get

$$||\, [b, d\rho]^\dagger\, ||_{\text{op}}^2 = ||\left[b^\dagger, d\rho\right]^\dagger||_{\text{op}}^2 = 3d\bar{z}dz \tag{20}$$

so that $||[D, d\rho]||_{\text{op}} = \sqrt{\frac{6d\bar{z}dz}{\theta}}$ and

$$d(z+dz, z) = \sqrt{\frac{2\theta}{3}}\sqrt{d\bar{z}dz}, \tag{21}$$

yielding the Euclidean distance. One can quickly check that replacing the operator norm in (16) and (21) by the trace norm one recovers, up to numerical factors, the same results. In what follows we shall often replace the operator norm with the trace norm and write

$$d(\rho, \rho') = \frac{\text{tr}_c(d\rho)^2}{||[D, \pi(d\rho)]||_{\text{tr}}}. \tag{22}$$

4. Spectral triplets on quantum Hilbert space

Unlike in the previous section, a generic state $|\psi,\phi) = |\psi\rangle\langle\phi|\epsilon\mathcal{H}_q$ or its corresponding pure state density matrix is given by $\rho_q = |\psi,\phi)(\psi,\phi|$. The family of such states with fixed $|\psi\rangle$ have many-to-one relation with the points $|\psi\rangle$ (associated with the density matrix $|\psi\rangle\langle\psi|$) in $\mathcal{H}_c$.

On the other hand, as discussed in section 2 and further below, in non-commutative quantum mechanics a particle maximally localized at a point (x, y), without any prior knowledge of the right hand sector, is described by the mixed state density matrix $\rho = \sum_n |z,n)(z,n|$ $(z = x+iy)$. When the right hand sector is also specified the particle is described by a pure state density matrix $\rho = |z,n)(z,n|$, but note that there are infinitely many choices for the right hand sector given a point (x, y) in classical configuration space. In this section, we would like to compute the various distances between pure / mixed states. To this end we introduce a further spectral triplet $(\mathcal{A}, \mathcal{H}, D)$ on quantum Hilbert space where $\mathcal{A} = B(\mathcal{H}_q)$, i.e., the bounded operators on $\mathcal{H}_q$, $\mathcal{H} = \mathcal{H}_q \otimes C^2$, the action of elements of $a \in \mathcal{A}$ on $\mathcal{H}$ is defined through the representation $\pi(a)(|\psi), |\phi)) = (a|\psi), a|\phi))$ and the (effective) Dirac operator is defined as

$$D = -i\sqrt{\frac{2}{\theta}} \begin{pmatrix} 0 & B^\ddagger \\ B & 0 \end{pmatrix}. \tag{23}$$

Note that the Dirac operator defined here acts only on the left sector as it does not involve any momentum operators.

Using the same Dirac operator we can compute the spectral distance between the states $\rho_q(m,\phi) = |m,\phi)(m,\phi|$ and $\rho_q(m+1,\phi') = |m+1,\phi')(m+1,\phi'|$ where ϕ' is different and orthogonal to ϕ, i.e., $\langle\phi|\phi'\rangle = \delta_{\phi,\phi'}$. Introducing $d\rho_q = |m+1,\phi')(m+1,\phi'| - |m,\phi)(m,\phi|$, we proceed as before to see that for $\phi = \phi'$ we recover the earlier result (16), as if ϕ is merely a spectator. But for $\phi \neq \phi'$ we get $d(\rho(m+1,\phi'), \rho(m,\phi)) = \sqrt{\frac{2\theta}{m+2}}$. This shows that $d(\rho(m+1,\phi), \rho(m,\phi)) < d(\rho(m+1,\phi'), \rho(m,\phi))$ and demonstrates that, despite the fact that the Dirac operator only acts on the left hand sector, the distance between two states depends on the right hand sector. One may contemplate an even more general situation by taking the right hand sector to be a statistical mixture that changes from point to point, i.e., we can consider the following density matrices

$$\rho_q(m) = \sum_n p_n(m)|m,n)(m,n|, \qquad \rho_q(z,\bar{z}) = \sum_n p_n(z,\bar{z})|z,n)(z,n| \tag{24}$$

subject to the respective conditions

$$\sum_n p_n(m) = 1, \forall m \quad , \quad \sum_n p_n(z,\bar{z}) = 1, \forall z \tag{25}$$

where $|m\rangle$ and $|z\rangle$ are as above, while $|n\rangle$ is an arbitrary orthonormal basis in classical configuration space and p_n a set of position dependent probabilities. By comparing the averages and fluctuations of $B_L^\ddagger B_L$ and $\hat{X}_i$ in the respective states $\rho_q(m)$ and $\rho_q(z,\bar{z})$, we see that $\rho_q(m)$'s are localised at a fixed radial distance $\sqrt{\theta(2m+1)}$ and $\rho_q(z,\bar{z})$ are maximally localised at (x,y).

Let us first compute the Connes' distance function for the density matrices (24), using (22). Setting $d\rho_q = \rho_q(m+1) - \rho_q(m)$ and replacing operator norm with trace norm (14)(which affects the result at most upto a constant), one obtains

$$d(m+1,m) = \frac{\sqrt{\theta}}{2}\frac{\sum_n \left(p_n^2(m+1) + p_n^2(m)\right)}{\sqrt{\sum_n \left((2m+3)p_n^2(m+1) + (2m+1)p_n^2(m) + 2(m+1)p_n(m+1)p_n(m)\right)}}. \tag{26}$$

At this stage, it is not clear a priori what criterions should be used to choose the probability distributions. However, we can investigate about the following two possibilities, which present themselves in an obvious manner. And for that we proceed one by one.Let us first ask which choice of probabilities actually minimizes the distance function. Starting with (26), the distance between two points m_i and m_f is given by

$$d(m_f, m_i) = \sum_{m=m_i}^{m_f-1} d(m+1, m). \tag{27}$$

Carrying out the minimization by the method of Lagrange's multipliers, we find that all the probabilities must be equal say $(1/\Omega)$ if Ω is the cut-off for the sum over n and independent from m and n. In this case the distance function reduces to

$$d(m+1, m) = \frac{\sqrt{\theta}}{\sqrt{6\Omega}} \frac{1}{\sqrt{m+1}}, \tag{28}$$

which only differs from (16) by a probability dependent global scale factor. Note that the operators π_z introduced in (9) are density matrices with this property.

Next we explore the geometry emerging from the second perspective. Here, we would like to fix the probabilities from the condition of local thermal equilibrium i.e. local average energy is fixed and the local entropy associated with the density matrices (24)is optimized. To do this we introduce a local entropy associated with the density matrices (24) in the usual manner

$$S(m) = -\sum_n p_n(m) \log p_n(m), \qquad S(z, \bar{z}) = -\sum_n p_n(z, \bar{z}) \log p_n(z, \bar{z}). \tag{29}$$

A simple calculation shows that maximizing this local entropy leads, as above, to equally distributed probabilities. This, of course, corresponds to the $T \to \infty$ limit where only entropy plays a role and all states are assigned equal probabilities.

Associating an energy scale ϵ_n with the right hand states labeled by $|n\rangle$ and require the local average energy to be fixed on $E(m)$, maximization of the local entropy yields a local Boltzman distribution

$$p_n(m) = \frac{e^{-\beta(m)\epsilon_n}}{Z(\beta(m))}, \qquad Z(\beta(m)) = \sum_n e^{-\beta(m)\epsilon_n}, \tag{30}$$

where $\beta(m)$ is the local inverse temperature. From this the distance function (26) can be computed. In particular, if we assume the local average energy and hence the temperature to be independent of m, we obtain

$$d(m+1, m) = \frac{\sqrt{\theta}}{\sqrt{6}} \frac{\sqrt{Z(2\beta)}}{Z(\beta)} \frac{1}{\sqrt{m+1}}, \tag{31}$$

which again only differs from (16) by a temperature dependent global scale factor.

This implied modified geometry for mixed states becomes even more explicit in the case of the continuous coherent state basis, which we consider next.

The Connes' distance in the coherent state simplifies considerably with a slight redefinition of the Dirac operator. We take the Dirac operator, written in terms of $\tilde{b}$ and $|\tilde{0}\rangle = |z\rangle$ to be

$$D = -i\sqrt{\frac{2}{\theta}} \begin{pmatrix} 0 & \left[B^\ddagger, 2\pi\theta\pi_z\right] \\ \left[B, 2\pi\theta\pi_z\right] & 0 \end{pmatrix} = -i\sqrt{\frac{2}{\theta}} \begin{pmatrix} 0 & \sum_n |\tilde{1}, n)(\tilde{0}, n| \\ \sum_n |\tilde{0}, n)(\tilde{1}, n| & 0 \end{pmatrix}. \tag{32}$$

Note that the Dirac operator is still insensitive to the right hand sector.

The computation now follows the same route as before and one obtains

$$d(z+dz,z) = \frac{\sqrt{\theta}}{2}\sqrt{2\sum_n\left(p_n(z)^2 + \frac{\partial p_n(z)}{\partial\bar z}\frac{\partial p_n(z)}{\partial z}\right)d\bar z dz + \sum_n\left(\left(\frac{\partial p_n(z)}{\partial\bar z}\right)^2 d\bar z^2 + \left(\frac{\partial p_n(z)}{\partial z}\right)^2 dz^2\right)}. \tag{33}$$

This clearly exhibits a modified geometry. Assuming that p_n is just a function of the dimensionful radial coordinate $r = \sqrt{\frac{\bar z z}{2\theta}}$ one obtains

$$d(z+dz,z) = \frac{1}{2\sqrt{2}}\sqrt{\sum_n\left(2p_n(r)^2 + p_n'(r)^2\right)}\sqrt{dr^2 + r^2 d\phi^2}, \tag{34}$$

which is related to the Euclidean distance by a probability dependent conformal transformation. We can of course again compute the probabilities from the condition of local equilibrium. In the $T \to \infty$ limit, the probabilities are obtained by maximizing (29), which yields equal, r-independent probabilities and subsequently, up to a global scale, an Euclidean geometry. If we constrain the local average energy to $E(r)$, the temperature becomes position dependent and the probabilities again follow a Boltzman distribution

$$p_n(r) = \frac{e^{-\beta(r)\epsilon_n}}{Z(\beta(r))}, \tag{35}$$

yielding a conformal scale factor. If the local average energy is position independent, so is the temperature, and we obtain an Euclidean geometry with a global, temperature dependent, scale factor

$$d(z+dz,z) = \frac{\sqrt{Z(2\beta)}}{2Z(\beta)}\sqrt{dr^2 + r^2 d\phi^2}. \tag{36}$$

As the prefactor is bounded by $0 < \frac{\sqrt{Z(2\beta)}}{2Z(\beta)} \leq \frac{1}{2}$, reaching its minimum at $T \to \infty$ and its maximum at $T \to 0$ this suggest, quite remarkably, that distances expand as the global temperature is lowered.

5. Conclusions

The formulation of non-commutative quantum mechanics on the Hilbert space of Hilbert-Schmidt operators naturally encodes the notion of spectral triplets in the sense of Connes and thus the notion of geometry. The geometry can be made explicit on the level of the classical configuration space and the quantum Hilbert space. In the former case, as in commutative quantum mechanics, there is a one-to-one correspondence between pure state density matrices and points in the classical configuration space. In the latter case this no longer holds due to the tensor product structure of the quantum Hilbert space and the distance function depends on information encoded in the right hand sector, even though the Dirac operator only acts on the left hand sector. It then becomes natural to define a local entropy. The condition of local equilibrium then fully determines the distance function and, subsequently, the geometry. When the temperature is independent of position, Euclidean geometry results with a scale determined by the global temperature. Remarkably the scale increases as the temperature is lowered, leading to expanding distances at lower temperature.

References

[1] A. Connes, Non-commutative Geometry, Academic Press, 1994.
[2] S.Doplicher, K.Fredenhagen and J.E.Roberts, Comm.Math.Phys. 172,187 (1995).
[3] F.G. Scholtz, B.Chakraborty, J Govaerts and S.Vadya, J.Phys. A40,14581(2007).

[4] F.G.Scholtz, L.Gouba, A.Hafver, C.M.Rohwer J.Phys.A 42,175303 (2009).
[5] P. Basu, B.Chakraborty and F.G. Scholtz, J. Physics. A.44,285204(2011).
[6] E. Cagnache, F. D'Andrea, P Martinetti and J-C Wallet, Jnl. Geom. and Phys. 61, 1881 (2011).
[7] P. Martinetti and L Tomassini, arxiv:1110.6164 [math-ph].
[8] C. M. Rohwer, "Additional degrees of freedom associated with position measurements in non-commutative quantum mechanics", arXiv:1206.1242 [hep-th].
[9] O. Bratelli and D.W. Robinson, Operator Algebras and Quantum Statistical Mechanics, Second edition, Vol 1., Springer, 1987.

IC-MSQUARE 2012: International Conference on Mathematical Modelling in Physical Sciences IOP Publishing
Journal of Physics: Conference Series **410** (2013) 012129 doi:10.1088/1742-6596/410/1/012129

The entropy of dense non-commutative fermion gases

J N Kriel[1] **and F G Scholtz**[1,2]

[1] Institute of Theoretical Physics, University of Stellenbosch, Stellenbosch 7600, South Africa
[2] National Institute for Theoretical Physics (NITheP), 7600 Stellenbosch, South Africa

Abstract. We investigate the properties of two- and three-dimensional non-commutative fermion gases with fixed total z-component of angular momentum, J_z, and at high density for the simplest form of non-commutativity involving constant spatial commutators. Analytic expressions for the entropy and pressure are found. The entropy exhibits non-extensive behaviour while the pressure reveals the presence of incompressibility in two, but not in three dimensions. Remarkably, for two dimensional systems close to the incompressible density, the entropy is proportional to the square root of the system size, i.e., for such systems the number of microscopic degrees of freedom is determined by the circumference, rather than the area (size) of the system. From a commutative perspective these results can also be interpreted in terms of an effective excluded volume. Possible generalizations to three dimensions, which restore rotational symmetry, are briefly discussed.

1. Introduction

The realization that gravitational instabilities make the absolute localization of events impossible, which in turn may be captured through a non-commutative space-time structure [1], has led to considerable interest in the formulation of non-commutative quantum field theories [2], quantum mechanics [3] and the possible physical consequences of non-commutativity.

One aspect of non-commutative systems that has received very little attention in the literature is the thermodynamics. The first attempt to understand the thermodynamics of such systems was made in [4]. However, the computation could still only be carried out numerically and thus the thermodynamic limit could not be probed reliably. The results of these calculations were quite striking and showed strong deviations from commutative behaviour at high densities and temperatures. The most important feature is the appearance of incompressibility at high densities for fixed angular momentum. This reflects the effective excluded volume implied by the non-commutative length scale. Under these conditions several other thermodynamic features also deviate strongly from the commutative case and, in particular, the entropy is no longer extensive.

This sets the scene for the present paper, which aims at understanding the thermodynamics of non-commutative fermion gases in two and three dimensions analytically, particularly at high densities and low temperature.

2. The two-dimensional non-commutative fermion gas

We begin by summarizing the key results of [4, 5]. Consider a system of N spinless particles of mass m_0 which are confined to a disc with area $A = \pi R^2$. The particle coordinates satisfy $[x, y] = i\theta$ $(\theta > 0)$ and we set $R^2 = \theta(2M + 1)$ with M a positive integer. The latter serves as a dimensionless measure of the system size. The single particle energies are then

$E_{r,m} = \hbar^2 x_{r,m}/(\theta m_0)$ with $x_{r,m}$ the zeros of the Laguerre polynomial $L^m_{M+1}(x)$ $(m \geq 0)$ or $L^{|m|}_{M+m+1}(x)$ $(-M \leq m < 0)$. Here m is an angular momentum label while r labels the $M+1$ $(m \geq 0)$ or $M+m+1$ $(-M \leq m < 0)$ states within a specific angular momentum sector. Note that m is bounded from below by $-M$ and so there are only a finite number of negative angular momentum states.

The q-potential of the grand canonical ensemble is

$$q(A, \beta, \mu, \omega) = \sum_{m=-M}^{\infty} \sum_{r} \log[1 + e^{-\beta(E_{r,m} - \mu - \hbar\omega m)}]. \tag{1}$$

We set $E_0 = \hbar^2/(\theta m_0)$ and introduce the dimensionless parameters $\tilde{\beta} = E_0\beta$, $\tilde{\mu} = \mu/E_0$ and $\tilde{\omega} = \hbar\omega/E_0$ in terms of which the q-potential becomes

$$q(M, \tilde{\beta}, \tilde{\mu}, \tilde{\omega}) = \sum_{m=-M}^{\infty} \sum_{r} \log[1 + e^{-\tilde{\beta}(x_{r,m} - \tilde{\mu} - \tilde{\omega} m)}]. \tag{2}$$

The central thermodynamic quantities can be calculated from q using

$$N = \frac{1}{\tilde{\beta}} \frac{\partial q}{\partial \tilde{\mu}}, \qquad \frac{L}{\hbar} = \frac{1}{\tilde{\beta}} \frac{\partial q}{\partial \tilde{\omega}}, \qquad \frac{S}{k} = q - \tilde{\beta} \frac{\partial q}{\partial \tilde{\beta}} \tag{3}$$

and we also define dimensionless measures of the density and pressure by $\tilde{\rho} = N/(2M+1)$ and $\tilde{P} = q/(\tilde{\beta}(2M+1))$.

In the thermodynamic limit (2) can be replaced by

$$q(M, \tilde{\beta}, \tilde{\mu}, \tilde{\omega}) = \int_{-M}^{\infty} dm \int_{x_-(m)}^{x_+(m)} dx\ D(x, m) \log[1 + e^{-\tilde{\beta}(x - \tilde{\mu} - \tilde{\omega} m)}] \tag{4}$$

where $D(x, m)$ is the asymptotic density of the zeros of the Laguerre polynomial $L^m_{M+1}(x)$ $(m \geq 0)$ or $L^{|m|}_{M+m+1}(x)$ $(-M \leq m < 0)$ in the limit where M and m tend to infinity in a fixed ratio. It is known that $D(x, m)$ is given by [6]

$$D(x, m) = \frac{\sqrt{4Mx - (m-x)^2}}{2\pi x} \tag{5}$$

for x in the interval on which the argument of the radical is non-negative.

The low density limit of (4) can be explored quite easily and one finds that it yields precisely the commutative density of states and thermodynamics. More interesting is the high density limit, which we investigate next.

To see the effects of non-commutativity it is necessary to consider densities which scale with the system size. This will render the chemical potential extensive and ensure that $c = \tilde{\mu}/M$ remains finite as M tends to infinity. A more appropriate measure of the density is therefore $\nu \equiv \tilde{\rho}/M \approx N/(2M^2)$, which will be kept fixed when taking $M \to \infty$. We also define the scaled angular momentum by $\ell \equiv L/(M^3\hbar)$.

The integrals defining the q-potential can be approximated at low temperatures using the Sommerfeld expansion [7]. The final result, up to linear order in temperature, reads

$$q(M, b, c, w) = \frac{M^2}{6b} \left[\frac{c\,(b^2c^2 + \pi^2)\,\Theta(-c)}{w^2} - \frac{(w-1)^2\,(3b^2c^2 + \pi^2) - 3b^2c(w-1)w^2 + b^2w^4}{(w-1)^3} \right] \tag{6}$$

where $\Theta(\cdot)$ is the step function. This result is valid for $w < 1$ and $c \geq w^2/(w-1)$.

Transforming from (b, c, w) back to $(\tilde{\beta}, \tilde{\mu}, \tilde{\omega})$ yields the desired expression for $q(M, \tilde{\beta}, \tilde{\mu}, \tilde{\omega})$ in (2) in the high density, low temperature limit.

Our next task is to combine (3) and (6) and solve for $\tilde{\mu}$ and $\tilde{\omega}$ as functions of the angular momentum and number of particles. We find that for solutions to exist, the angular momentum is bounded from below by $-M^3/6$; a reflection of the finite number of negative angular momentum states. A further, more striking feature, is that for each $\ell > -1/6$ there exists a maximum density $\tilde{\rho}_c = M\nu_c$ with

$$\nu_c = (3 + 2\sqrt{18\ell + 3})/12. \tag{7}$$

No such restriction is present in the commutative case and this is a clear indication that non-commutativity results in an excluded area effect at high densities. We investigate the thermodynamic properties close to this critical density in the next section.

In general the exact expressions for $\tilde{\mu}$ and $\tilde{\omega}$ are quite cumbersome and so we will focus only on the physically interesting region close to the critical density. Here we find

$$\frac{\tilde{\mu}}{M} = 1 - \frac{7 + 24\ell - 24\nu(6\nu - 1)}{24\left[(8\nu + 2)(\nu - \nu_0)(\nu_c - \nu)\right]^{1/2}} \qquad \tilde{\omega} = 1 - \frac{1}{4}\left[\frac{8\nu + 2}{(\nu - \nu_0)(\nu_c - \nu)}\right]^{1/2} \tag{8}$$

where $\nu_0 \equiv \left(3 - 2\sqrt{18\ell + 3}\right)/12$. Note that as ν approaches ν_c these quantities diverge: $\tilde{\mu} \to \infty$ and $\tilde{\omega} \to -\infty$.

Substituting the expressions for $\tilde{\mu}$ and $\tilde{\omega}$ in (8) into the relations (3) yields expressions for the entropy and pressure of the non-commutative gas close to the critical density:

$$\frac{S}{Mk} = \left[\frac{\nu_c - \nu_0}{1 + 4\nu_c}\right]^{1/2} \frac{2^{3/2}\pi^2}{3\tilde{\beta}} \sqrt{\nu_c - \nu} \quad \text{and} \quad \frac{\tilde{P}}{M^2} = \left[\frac{1 + 4\nu_c}{\nu_c - \nu_0}\right]^{1/2} \frac{(1 + 48\nu_c^2)}{96\sqrt{2}} \frac{1}{\sqrt{\nu_c - \nu}}. \tag{9}$$

In terms of the dimensionless system size M and particle density $\tilde{\rho} \approx N/(2M)$ this implies, for example, that at $\ell = 0$

$$S \sim \sqrt{M}\sqrt{M\nu_c - \tilde{\rho}} \quad \text{and} \quad \tilde{P} \sim M^{5/2}/\sqrt{M\nu_c - \tilde{\rho}}. \tag{10}$$

We see that as ν approaches ν_c the entropy vanishes as the square root of $\nu_c - \nu$, indicating that the "tight packing" limit has been reached and only a single microstate is accessible to the system. This coincides with a divergence in the pressure which signals that the system is incompressible at the critical density. Due to their entropic nature these phenomenon are expected to persist at higher temperatures. In particular, the value of ν_c and the $\sqrt{\nu_c - \nu}$ dependence of the entropy and pressure should remain unaffected provided that the system is sufficiently close to its maximum density.

The next question we address is the following: How will the entropy of systems close to the incompressible density ν_c scale with system size? The physical motivation for this question is that we would like to know how the entropy of a very dense gas, close to its maximum density, behaves as a function of the size or number of particles when more particles are added and the size of the object increased, but in such a way that the density is always kept very close to the maximal density. For this purpose consider a system containing N particles with a minimum size M_c given by $M_c^2 = N/(2\nu_c)$. The density can then be expressed as $\nu = N/(2M^2) = \nu_c M_c^2/M^2$. The limit that interest us is when the system size exceeds the minimum size by a small fixed area $2\pi\theta\Delta M$. Then $M = M_c + \Delta M$ with ΔM a non-negative integer such that $\Delta M << M_c$ and $\nu \approx \nu_c(1 - \frac{2\Delta M}{M_c})$, i.e., the density is close to the incompressible density. Crucial here is that the quantization of the system size implies that $\Delta M \sim 1$. With ΔM held fixed we now consider the dependence of the entropy on the system size $M \approx M_c$. Substituting in (9) the entropy close to the incompressible point reads

$$S(M_c) \sim M\sqrt{\nu_c}\sqrt{1-\frac{M_c^2}{M^2}} \approx \sqrt{2\nu_c \Delta M}\sqrt{M_c}. \tag{11}$$

As $R^2 \approx 2M\theta$ it follows that the entropy is proportional to the circumference, rather than the area, of the system. When $M >> M_c$, i.e. far from the incompressible point, normal extensive behaviour is recovered.

3. The three-dimensional non-commutative fermion gas

We assume the following commutation relations in three dimensions

$$[x_i, x_j] = i\theta_{i,j} \tag{12}$$

with θ_{ij} a real anti-symmetric matrix. Through an appropriate choice of coordinates it is always possible to restrict the non-commutativity to two of the spacial coordinates with the third coordinate being commutative. We consider a cylindrical geometry with radius R and length L with its central axis orientated along the z-direction. The x- and y-directions are taken to be non-commutative and at a fixed z the problem therefore reduces to the two-dimensional case considered earlier. In the z-direction we have the regular one dimensional free particle problem with vanishing boundary conditions. In the notation of (2) the q-potential reads

$$q_{3D}(M, \tilde{\beta}, \tilde{\mu}, \tilde{\omega}, \tilde{\gamma}) = \sum_{n=1}^{\infty}\sum_{m,r} \log[1 + e^{-\tilde{\beta}(x_{r,m}+\tilde{\gamma}n^2-\tilde{\mu}-\tilde{\omega}m)}] \tag{13}$$

$$= \sum_{n=1}^{\infty} q_{2D}(M, \tilde{\beta}, \tilde{\mu} - \tilde{\gamma}n^2, \tilde{\omega}) \tag{14}$$

where $\tilde{\gamma} = \gamma/E_0$ and $\gamma = \hbar^2\pi^2/(2m_0L^2)$. In the thermodynamic limit R and L will tend to infinity in a fixed ratio. This can be made explicit by writing $\tilde{\gamma} = G/M$ where $G = (\pi/2)^2(R/L)^2$ is a dimensionless constant defining the shape of the cylinder. As is evident from (14) the q_{3D}-potential is just a sum of two-dimensional potentials with shifted chemical potentials. The sum terminates automatically once $\tilde{\mu} - \tilde{\gamma}n^2$ drops below $M\tilde{\omega}^2/(\tilde{\omega} - 1)$, as discussed after equation (6). Under the same assumptions that led to q_{2D} in (6) we can now express q_{3D} as

$$q_{3D}(M, \tilde{\beta}, \tilde{\mu}, \tilde{\omega}, \tilde{\gamma}) = \int_0^{n_+} \mathrm{d}n \; q_{2D}(M, \tilde{\beta}, \tilde{\mu} - \tilde{\gamma}n^2, \tilde{\omega}) \tag{15}$$

where $n_+ > 0$ is such that $\tilde{\mu} - \tilde{\gamma}n_+^2 = M\tilde{\omega}^2/(\tilde{\omega} - 1)$. This integral is straightforward to perform since q_{2D} is a piecewise polynomial in its third argument.

In three dimensions it is convenient to use $\nu = N/M^3$ and $\ell = L/(\hbar M^4)$ as measures of the density and angular momentum. Combining (15) and (3) leads to a set of equations for $\tilde{\mu}$ and $\tilde{\omega}$ which can be solved numerically for given ν and ℓ. Unlike the two-dimensional case there is no maximum density at which the entropy vanishes and pressure diverges, i.e. no incompressible point. This reflects the fact that the excluded *area* effect is restricted to the $x - y$ plane and does not place any restriction on how the commutative modes in the z-direction are occupied. However, the effects of non-commutativity are still present and responsible for modifying the scaling behaviour of the entropy as we progress from low to high densities. We find that S grows like $\nu^{1/3}$ at low densities and decreases like $1/\nu$ at high densities. In these two limits we are able to solve for $\tilde{\mu}$ and $\tilde{\omega}$ when $\ell = 0$ to find

$$\frac{S}{kM^2} = \left[\frac{\pi^6}{18G\tilde{\beta}^3}\right]^{1/3} \nu^{1/3} \quad \text{and} \quad \frac{S}{kM^2} = \frac{(4.2397\ldots)}{G\tilde{\beta}}\frac{1}{\nu} \tag{16}$$

for low and high densities respectively. In terms of the dimensionless volume $\tilde{V} = V/\theta^{3/2}$ and density $\tilde{\rho} = N/\tilde{V}$ these expressions become

$$\frac{S}{k} = \frac{1}{\tilde{\beta}}\left(\frac{\pi}{6}\right)^{2/3} \tilde{V}\tilde{\rho}^{1/3} \quad \text{and} \quad \frac{S}{k} = \frac{\eta G^{2/3}}{\tilde{\beta}}\frac{\tilde{V}^{7/3}}{\tilde{\rho}}. \tag{17}$$

with $\eta \approx 6.4757 \times 10^{-4}$. The low density result agrees exactly with that of a commutative fermion gas [7]. It is independent of both the shape of the cylinder (i.e. G) as well as the non-commutative parameter θ. In contrast, the high density result depends on both these parameters and also exhibits an unusual non-extensive volume dependence. Furthermore, this entropy is inversely proportional to the particle density as is the case for a one-dimensional ideal gas. This observation that some aspects of the high density thermodynamics mimic that of a one-dimensional system is also observed in the relation between pressure and energy density. For a non-interacting gas in d dimensions it holds quite generally that $P = (2/d)(E/V)$. We indeed find that $P = (2/3)(E/V)$ and $P = 2(E/V)$ in the low and high densities limits respectively.

These results hint that this type of non-commutativity and constraint on J_z lead to a high density limit that behaves as a commutative one dimensional system. This counters our expectations for the real world and we can therefore only assume that this form of non-commutativity, the symmetry breaking it implies and the constraint on J_z are inappropriate. A natural alternative is to change the non-commutative relations from the simple form (12) to those of a fuzzy sphere $[x_i, x_j] = i\theta\epsilon_{ijk}x_k$. In this way R^3 can be realized as an "onion structure" with quantized radius and the two dimensional program of [4] can in principle be executed to compute the spectrum of a three dimensional well.

4. Conclusions

The thermodynamic behaviour of two- and three-dimensional non-commutative gases was investigated analytically in the high density region. Strong deviations from conventional commutative behaviour was found. In particular an incompressible point was found in two dimensions and the entropy exhibits in both cases non-extensive behaviour. More profoundly, for two dimensional systems with density close to the incompressible density the entropy exhibits a very particular dependence on the system size in that it is proportional to the square root of the system size, i.e. the circumference rather than the area. Incompressibility is absent in three dimensions and the scaling of entropy at high density is unconventional, showing rather a crossover to one dimensional commutative thermodynamics. This is probably due to the particular form of non-commutativity used here in three dimensions and the symmetry breaking it implies. In particular the constraint on J_z, rather than J, is problematic and cannot be implemented correctly in the current framework.

5. Acknowledgements

This work was supported under a grant of the National Research Foundation of South Africa.

References

[1] Doplicher S, Fredenhagen K and Roberts J E 1995 *Commun. Math. Phys.* **172** 187
[2] Douglas M R and Nekrasov N A 2001 *Rev. Mod. Phys.* **73** 97

Journal of Physics: Conference Series **410** (2013) 012129 doi:10.1088/1742-6596/410/1/012129

[3] Scholtz F G, Gouba L, Hafver A and Rohwer C M 2009 *Jnl. Phys. A* **42** 175303
[4] Scholtz F G and Govaerts J 2008 *Jnl. Phys. A* **41** 505003
[5] Scholtz F G, Chakraborty B, Govaerts J and Vaidya S 2007 *Jnl. Phys. A* **40** 14581
[6] Dette H and Studden W J 1995 *Constructive Approximation* **11** 227
[7] Pathria R K 1972 *Statistical Mechanics* (Oxford: Pergamon)

IC-MSQUARE 2012: International Conference on Mathematical Modelling in Physical Sciences IOP Publishing
Journal of Physics: Conference Series **410** (2013) 012130 doi:10.1088/1742-6596/410/1/012130

Gravothermal instability with a cosmological constant in the canonical ensemble[1]

Minos Axenides[1], George Georgiou[1] and Zacharias Roupas[1,2]

[1]Institute of Nuclear and Particle Physics, N.C.S.R. Demokritos, GR-15310 Athens, Greece
[2]Physics Department, National Technical University of Athens, GR-15780, Athens, Greece

E-mail: roupas@inp.demokritos.gr

Abstract. We present here how the gravothermal or Antonov's instability, which was originally formulated in the microcanonical ensemble, is modified in the presence of a cosmological constant and in the canonical ensemble. In contrast to the microcanonical ensemble, there is a minimum, and not maximum, radius for which metastable states exist. In addition this critical radius is decreasing, and not increasing, with increasing cosmological constant. The minimum temperature for which metastable states exist is decreasing with increasing cosmological constant, while above some positive value of the cosmological constant, there appears a second critical temperature. For lower temperatures than the second critical temperature value, metastable states reappear, indicating a typical reentrant phase transition. The two critical temperatures merge when the cosmological density equals one half the mean density of the system.

1. Introduction

Antonov in 1962 [1] discovered an instability that later became known as Antonov's instability or gravothermal catastrophe [2]. It signified the beginning of the study of thermodynamics of self-gravitating systems [3, 4, 5, 6] and the nowadays rapidly developing field of statistical mechanics with long-range interactions [7, 8].

In the original formulation of Antonov's instability [1, 2], the system consists of particles (stars) bound by a spherical shell with insulating and perfectly reflecting walls. The particles interact only with Newtonian gravity (no cosmological constant or any relativistic effects are present) and the number of particles and the energy of the system are constant, i.e. the system is studied in the microcanonical ensemble. It was found that the entropy of the system has no global maximum and that for $ER/GM^2 < -0.335$ no equilibrium states exist. For $ER/GM^2 > -0.335$ there exist local entropy maxima, i.e. metastable states, only for density contrast below the critical value $(\rho_0/\rho_R)_{cr} = 709$, where ρ_0, ρ_R are the central and edge densities, respectively.

In gravity the ensembles are not equivalent. In the microcanonical ensemble, stable equilibria with negative specific heat do exist. This negative specific heat region is replaced by a phase transition in the canonical ensemble [3]. The Antonov system in the canonical ensemble (with no cosmological constant), is studied in detail by Chavanis [4]. He found that the free energy does not have a global minimum and equilibria exist only for temperatures greater than the

[1] Text of a talk given at the workshop *Gravity, Quantum and Black Holes* in Budapest, in September 2012. Details and extensions of the results, reported briefly here, will appear elsewhere.

critical dimensionless temperature $(RT/GM)_{cr} = 0.40$ which corresponds to dimensionless inverse temperature $(GM\beta/R)_{cr} = 2.52$. These equilibria are stable only for density contrast less than the critical value $(\rho_0/\rho_R)_{cr} = 32.1$ (see Ref. [9], as well).

We studied the gravothermal instability with a cosmological constant Λ in the microcanonical ensemble in Ref. [10]. It was found that for a fixed energy, the maximum radius (which equals $R_{cr} = (-0.335/E)GM^2$ with no Λ) is increasing with increasing Λ, while for $\Lambda > 0$ there appears a second critical radius which is decreasing and the two radii merge at some marginal Λ. This was identified as a reentrant behaviour (see Ref. [11, 12, 13, 14] for details on reentrant phase transitions in statistical systems with long-range interactions). In the present work, the system is placed in a heat bath and the walls are non-insulating, i.e. we study the canonical ensemble. In contrast to the microcanonical one, we find that the critical radius R_{cr}, for a fixed temperature and mass, is decreasing with increasing cosmological constant. In addition the instability occurs for $R < R_{cr}$ in the canonical ensemble, in contrast to the microcanonical one. This is evident in case $\Lambda = 0$, since then there exist no equilibria if $GM\beta/R > 2.52 \Rightarrow R < GM\beta/2.52$ for a fixed temperature and mass. A reentrant phase transition appears in the canonical ensemble, when one examines the system with respect to the temperature for a fixed radius. There appear two critical temperatures for $\Lambda > 0$. For the intermediate temperature values, no thermodynamic equilibria are possible.

2. Analysis and results

In the Newtonian limit, the Poisson equation is modified [15, 16] in the presence of a cosmological constant Λ as:

$$\nabla^2\phi = 4\pi G\rho - 8\pi G\rho_\Lambda \tag{1}$$

with $\rho_\Lambda = \frac{\Lambda c^2}{8\pi G}$. Thus, the gravitational potential can be decomposed to two parts

$$\phi = \phi_N + \phi_\Lambda \tag{2}$$

where ϕ_N is the Newtonian and ϕ_Λ the cosmological potential, that are given by

$$\phi_N = -G\int \frac{\rho(r')}{|\vec{r}-\vec{r}'|}d^3\vec{r}' \ , \ \phi_\Lambda = -\frac{4\pi G}{3}\rho_\Lambda r^2 \tag{3}$$

For convenience we call the Newtonian limit of de Sitter ($\rho_\Lambda > 0$) and Anti-de Sitter ($\rho_\Lambda < 0$) spaces [17] just dS and AdS, respectively.

We consider a self-gravitating gas of N particles with unity mass inside a spherical non-insulating shell inside a heat bath and restrict only to spherical symmetric configurations. We work in the mean field approximation, for which the N-body distribution function is replaced by the one body distribution function $f(\vec{r},\vec{v})$. The Boltzmann entropy is defined as $S = -k\int f \ln f d^3\vec{r}d^3\vec{p}$. The Helmholtz free energy is equal to $F = E - TS$. Equivalently one can work with the Massieu function [4, 9] $J = -F/T$ that gives

$$J = S - \frac{1}{T}E \tag{4}$$

The maximization of J with constant T and the maximization of S with constant E (and constant M in both cases), with respect to perturbations $\delta\rho$, are the same [4] to first order in $\delta\rho$, i.e. give the same equilibria, described by the Maxwell-Boltzmann distribution function

$$f = \left(\frac{\beta}{2\pi}\right)^{\frac{3}{2}} \rho(r)e^{-\frac{1}{2}\beta v^2} \tag{5}$$

where

$$\rho(r) = \rho_0 e^{-\beta(\phi-\phi(0))} \tag{6}$$

This is proved in [4] without a cosmological constant, and holds in the presence of Λ as well, with the difference being that ϕ does not satisfy Poisson equation now, but equation (1). The two ensembles have the same equilibria. However, the second variations of J and S are different. Therefore, what is different in the two ensembles is the turning point at which an instability sets in. This causes a great qualitative difference for the two ensembles.

Introducing the dimensionless variables $y = \beta(\phi - \phi(0))$, $x = r\sqrt{4\pi G\rho_0\beta}$ and $\lambda = 2\rho_\Lambda/\rho_0$, and using equation (6), equation (1) becomes

$$\frac{1}{x^2}\frac{d}{dx}\left(x^2\frac{d}{dx}y\right) = e^{-y} - \lambda \tag{7}$$

called the Emden-Λ equation. Let us call $z = R\sqrt{4\pi G\rho_0\beta}$ the value of x at R. In order to generate the series of equilibria needed to study the stability of the system, the Emden-Λ equation has to be solved with initial conditions $y(0) = y'(0) = 0$, keeping M constant and for various values of the parameters ρ_Λ, β, ρ_0. This is a rather complicated problem, since, unlike $\Lambda = 0$ case, while solving for various z, mass is not automatically preserved, because of the mass scale $M_\Lambda = \rho_\Lambda\frac{4}{3}\pi R^3$ that Λ introduces. A suitable λ value has to be chosen at each z. We define the dimensionless mass

$$m \equiv \frac{M}{2M_\Lambda} = \frac{3}{8\pi}\frac{M}{\rho_\Lambda R^3} = \frac{\bar{\rho}}{2\rho_\Lambda} \tag{8}$$

where $\bar{\rho}$ is the mean density of matter. Calling $z = R\sqrt{4\pi G\rho_0\beta}$ the value of x at R, equation (8) can also be written as $m = 3B/\lambda z^2$ where $B = GM\beta/R$ is the dimensionless inverse temperature. It can be calculated by integrating the Emden-Λ equation, to get:

$$B(z) = zy'(z) + \frac{1}{3}\lambda z^2 \tag{9}$$

We developed an algorithm to solve equation (7) for various values of λ, z keeping m fixed. From equation (8) it is clear that solving for various fixed m can be interpreted as solving for various ρ_Λ and/or R for a fixed M.

We find that in dS case there exist multiple series of equilibria for a given cosmological constant, while in AdS case there is just one series likewise $\Lambda = 0$ case. Studying the series $\beta(\frac{\rho_0}{\rho_R})$ for both AdS and dS, we determined the critical values of radii and temperature beyond which there are no equilibria. These critical values are plotted in Figures 1 and 2, with respect to ρ_Λ, where the unshaded region is the region of instability in both figures.

In Figure 1 the radius denoted R_A is the minimum radius for which thermodynamic equilibria exist in the canonical ensemble. R_H is the radius of the homogeneous solution which exists only in dS and is defined by equation $\rho = 2\rho_\Lambda = const$, i.e. $R_H = (3M/(8\pi\rho_\Lambda))^{\frac{1}{3}}$. In the shaded region of Figure 1 there can always be found metastable states. For AdS case ($\rho_\Lambda < 0$), these have always monotonically decreasing density ($\rho_0 > \rho_R$) and suffer a transition to unstable equilibria at a critical density contrast $(\rho_0/\rho_R)_{cr}$ whose value depends on ρ_Λ. For dS case and for $R < R_H$, the metastable states suffer a transition to unstable equilibria likewise AdS, but the density is not necessarily monotonic, neither does $\rho_0 > \rho_R$ always hold. For each ρ_Λ there exists a tower of solutions with qualitative (and quantitative of course) different density functions. In the region $R > R_H$, all metastable states have $\rho_0 < \rho_R$. The ones, whose density is a monotonic function of R do not suffer any transition to unstable equilibria, while all others series have a turning point.

In Figure 2 we see that in AdS the critical temperature is decreasing for increasing cosmological constant, while in dS there appears a reentrant phase transition. For small positive

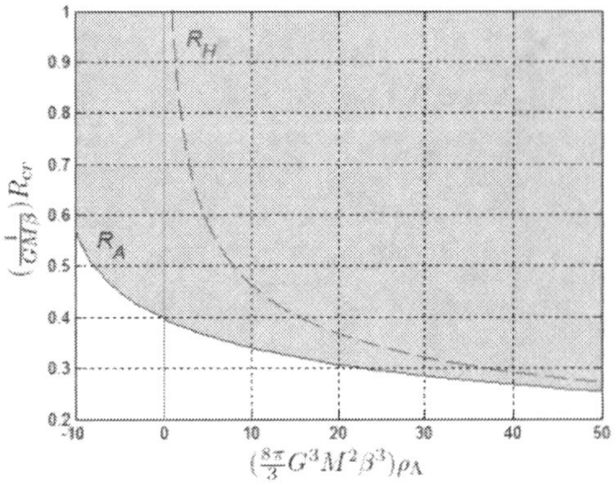

Figure 1. The critical radius w.r.t. ρ_Λ for fixed M, β.

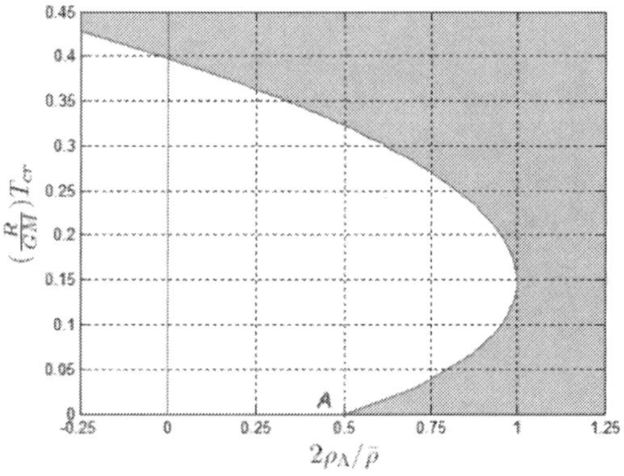

Figure 2. The critical temperature w.r.t. ρ_Λ for fixed M, R.

values of ρ_Λ, compared to the mean density $\bar{\rho}$, the system cannot stabilize for low temperatures. It needs higher temperature values to obtain a significant pressure gradient to balance gravity. As ρ_Λ is increasing, the cosmological repelling force increases, enforcing the total outward pushing force and hence, enabling the system to stabilize at lower temperatures. Point A, denotes the marginal value $\rho_\Lambda^{min} = \bar{\rho}/4$ for which a dynamically static equilibrium is allowed. This limiting configuration corresponds to a singular solution at which all mass is concentrated at the edge. It can easily be calculated: $GM/2R^2 = \frac{8\pi G}{3}\rho_\Lambda^{\min} R \Rightarrow \rho_\Lambda^{min} = 3M/16\pi R^3 = \bar{\rho}/4$. For $\rho_\Lambda > \rho_\Lambda^{min}$, solutions with $T = 0$ (static dynamical equilibria) do exist and as ρ_Λ increases the mass is allowed to be closer to the center. All thermodynamic equilibria below the lower critical branch (low temperatures) have increasing density ($\rho_0 < \rho_R$). For fixed R, M and a fixed ρ_Λ with $\bar{\rho}/4 < \rho_\Lambda < \bar{\rho}/2$, as we increase the temperature beginning from zero, we pass from a region of metastable states with $\rho_0 < \rho_R$ to a region of no equilibria, at a first critical temperature T_1. We have to increase the temperature even more to reach a new region of metastable states at a second critical temperature T_2. For $T > T_2$ the metastable states have decreasing density $\rho_0 > \rho_R$ if $R < R_H$. This is a reentrant phase transition similar to the ones found in other statistical systems with long-range interactions [11].

References

[1] V.A. Antonov, *Vest. Leningrad Univ.* **7**, 135, (1962)
[2] D. Lynden-Bell and R. Wood, *MNRAS* **138**, 495, (1968)
[3] T. Padmanabhan, *Phys. Rep.* **188**, 285, (1990)
[4] P.H. Chavanis, *A&A* **381**, 340, (2002)
[5] J. Katz, *Found. Phys.* **33**, 223, (2003)
[6] P.H. Chavanis, *Int. J. Mod. Phys. B*, **20**, 3113 (2006)
[7] D. Lynden-Bell, *Physica A* **263**, 293, (1999)
[8] A. Campa, T. Dauxois and S. Ruffo, *Phys. Rep.* **480**, 57 (2009)
[9] J. Katz, *MNRAS*, **183**, 765, (1978)
[10] M. Axenides, G. Georgiou and Z. Roupas, *Preprint* arXiv:1206.2839[astro-ph.CO]
[11] F. Staniscia, P.H. Chavanis, De Ninno and D. Fanelli, *Phys. Rev. E* **80**, 021138, (2009)
[12] C.K. Thomas and H.G. Katzgraber, *Phys. Rev. E* **84**, 040101(R), (2011)
[13] T. Dauxois, P. de Buyl, L. Lori and S. Ruffo, *J. Stat. Mech.* P06015, (2010)
[14] F. Staniscia, P.H. Chavanis and G. De Ninno, *Phys. Rev. E* **83**, 051111 (2011)
[15] S.R. Green and R.M. Wald, *Phys. Rev. D* **85**, 063512, (2012)
[16] M. Axenides, E.G. Floratos and L. Perivolaropoulos, *Mod. Phys. Lett.* **A15**, 1541, (2000)
[17] G.W. Gibbons and C.E. Patricot, *Class. Quant. Grav.* **20**, 5225, (2003)

Analytical approaches to 2D CDT coupled to matter

Max R Atkin [a] **and Stefan Zohren** [b c]

[a] Fakultät für Physik, Universität Bielefeld, Postfach 100131, D-33501 Bielefeld, Germany
[b] Department of Physics, Pontifica Universidade Católica do Rio de Janeiro, Rua Marquês de São Vicente 225, 22451-900 Gávea, Rio de Janeiro, Brazil
[c] Rudolf Peierls Centre for Theoretical Physics, 1 Keble Road, Oxford OX1 3NP, UK.

E-mail: [a] `matkin@physik.uni-bielefeld.de`, [b] `zohren@fis.puc-rio.br`

Abstract. We review some recent results by Ambjørn et al. (1202.4435) and the authors (1202.4322,1203.5034) in which multicritical points of the CDT matrix model were found and in a particular example identified with a hard dimer model. This identification requires solving the combinatorial problem of counting configurations of dimers on CDTs.

1. Introduction

Casual Dynamical Triangulation (CDT) together with its predecessor Dynamical Triangulation (DT) are approaches to giving meaning to the formal expressions appearing in a path integral quantisation of gravity. The idea is to regulate the path integral by approximating the geometries appearing in the integration by discrete triangulations, thereby replacing the path integral over geometries with a summation over triangulations (s.o.t) each weighted by a discrete form of the original action. Once the s.o.t is computed the regularisation is removed by taking a scaling limit around a critical point described by a continuum theory.

Much work was done in the early 90s on DT in $D = 2, 3, 4$ dimensions. For $D = 2$ the s.o.t could be formulated in terms of a hermitian matrix model [1]. These models possess critical points whose continuum theory is gravity interacting with matter. For $D > 2$ the work was numerical in nature and much less promising; no second order critical point was found at which a continuum theory could be extracted [2]. This is due to the preference the theory has for the geometry to undergo spatial topology change.

To remedy this, CDT only sums over triangulations containing a preferred time slicing w.r.t which spatial topology change is forbidden [3]. For CDT with $D > 2$, there are indications of a critical surface on which a continuum theory of extended de-Sitter spacetimes exists (see [4] for a review). There are also interesting results suggesting that in the UV the number of dimensions of the theory reduces to two [5] prompting suggestions that the continuum theory should be identified with a theory known as Horava-Lifshitz gravity [6]. For $D = 2$ the theory without matter has been solved using a variety of methods. In particular, a generalised CDT allowing some topology change has a matrix model formulation [7], opening the possibility of introducing matter and allowing a string worldsheet interpretation; a causal string field theory (CSFT) [8].

In principle adding matter to CDT is easy; we simply add extra degrees of freedom to the triangulation. This at least allows numerical investigations to be conducted [9]. Here however, by "adding matter" we mean treating such a matter sector analytically. Of course since no analytical

Published under licence by IOP Publishing Ltd

treatments of pure CDT for $D > 2$ exists our discussion is confined to two dimensions. There are a number of reasons to be interested in adding matter analytically. Recent numerical work [10] has indicated that when the central charge of the matter exceeds one, the two-dimensional theory undergoes a transition to a phase with a three-dimensional geometry corresponding to a de-Sitter space similar to that seen in the $D > 2$ simulations. Having an analytic understanding of this result would give insight into the transition appearing in the higher dimensional theories [4]. Secondly, due to the lack of matter the CSFT corresponds to strings in zero dimensions. It would be interesting to study CSFT in more complicated backgrounds, hopefully leading to an understanding of how CSFTs differ from usual SFTs [11]. Indeed there has been recent progress in embedding Horava-Lifshitz gravity into string theory [12]. One might suspect that there exists more connections between CDT and string theory and having analytic understanding of matter will be an excellent tool for investigating this.

2. Multicritical CDT from a matrix model

The matrix model approach to (C)DT is based on the fact that triangulations are in bijection [1] with Feynman diagrams of matrix models of the form,

$$Z = \int [d\phi] e^{-N \operatorname{Tr} V(\phi)} = \int_{-\infty}^{\infty} \prod_{i=1}^{N} d\lambda_i \prod_{j \neq i} (\lambda_j - \lambda_i)^2 e^{-N \sum_{i=1}^{N} V(\lambda_i)}, \tag{1}$$

where ϕ is an $N \times N$ hermitian matrix, $[d\phi]$ is the standard measure on such matrices and $V(\phi) = \sum_{k=0}^{\infty} t_k \phi^k / k$ is known as the potential. We therefore identify Z with the s.o.t with an action determined by the t_k. The advantage of the matrix model approach is that well developed techniques for computing Z exist [1]. Indeed, in the second equality we have written the integral in terms of the eigenvalues λ_i of ϕ. Such an integral can be viewed as a gas of charged particles in a potential and may be evaluated by orthogonal polynomial methods. Once Z is computed, we take a scaling limit about a second order phase transition to recover a continuum theory. Such a transition occurs when the eigenvalues of ϕ begin to spill over a confining barrier. By tuning the parameters appearing in V one can obtain critical points describing the continuum limit of DT coupled to $(2, 2p+1)$ conformal minimal matter [1].

In the last few year it was realised [7] that pure CDT may also be obtained as a scaling limit of the matrix model (1) if one uses the potential $V_{\mathrm{CDT}}(\phi) = \frac{1}{\beta}\left(-g\phi + \phi^2/2 - g\phi^3/3\right)$. In [7] it is shown that β acts as a coupling for spatial topology change. To forbid topology change we require $\beta = 0$, which is clearly singular in the above action, partially explaining why it took so long to find a matrix model for CDT. The insight of [7] was to make an scaling ansatz for the variables of,

$$g = g^* - a^2\Lambda \qquad \text{and} \qquad \beta = a^3 g_s, \tag{2}$$

where a is lattice length which goes to zero in the scaling limit and hence $\beta \to 0$. This leads to a *continuum theory* in which there is a coupling for spatial topology change. By moving the term $(\lambda_j - \lambda_i)^2$ in (1) in to the potential we see that β controls the strength of repulsion between eigenvalues. By scaling β to zero we allow the eigenvalues to collect at the bottom of the potential. A similar transition, in which eigenvalues *appear* in a new minima of the potential, was studied in [13] in which the scaling limit about the point where the eigenvalues first appear is again described by a matrix model. Here, we find a similar phenomenon if we scale ϕ in (1) as $\phi = \phi^* + a\Phi$. This results in the scaled partition function [7],

$$\tilde{Z}_{\mathrm{CDT}} = \int [d\Phi] e^{-\frac{N}{g_s}\left(2\Lambda\Phi - \frac{1}{6}\Phi^3\right)}. \tag{3}$$

We stress here that one can compute continuum CDT amplitudes from (3) or by first computing the discrete amplitude using (1) and then taking the limit with (2). To consider matter one follows DT and looks for a non-trivial scaling limit for higher order V in which $\beta \to 0$. Following [7], in [14] and later [15] it was realised that by choosing the potential and scaling ansatz to be of the form,

$$V_{\mathrm{mCDT}}(\phi) = -g\phi + \frac{1}{2}\phi^2 - g\sum_{k=1}^{m+1} \frac{t_k}{k}\phi^k \tag{4}$$

$$\phi = \phi^* + a\Phi, \qquad g = g^* - a^m\Lambda \qquad \text{and} \qquad \beta = a^{m+1}g_s, \tag{5}$$

we obtain the scaled partition function,

$$\tilde{Z}_{\mathrm{mCDT}} = \int [d\Phi] e^{-\frac{N}{g_s}\mathrm{Tr}\left[(-\partial_g\partial_\phi V^*)\Lambda\Phi + \frac{1}{(m+1)!}(\partial_\phi^{m+1}V^*)\Phi^{m+1}\right]}. \tag{6}$$

The critical point $(\{t_k^*\}, g^*, \phi^*)$ is defined such that $V^{(n)}(\{t_k^*\}, g^*, \phi^*) = 0$ for $n \leq m$. There are a number of interesting observations about (6). Firstly, the continuum theory is again a matrix model; this should be compared with [13] and it would be interesting to understand the relation between these two results. Note also that the scaling limit does not change N and hence N is a free parameter. It was argued for the pure case $m = 2$ [7] and for $m > 2$ [15], that N can be identified with a coupling that distinguishes the splitting and joining of universes *in the continuum theory*. If we do not wish to distinguish these processes we can set $N = 1$ thereby giving a simple expression for the entire summation of the topological expansion. Finally, we identify (6) as a degenerate form of the generalised Kontsevich model. Such a model has appeared in non-critical string theory before as the partition function of the continuum theory of a stack of FZZT branes [18]. The continuum limit of mth order multi-critical CDT is therefore equivalent, at least mathematically, to this. Such an observation was also made in [16] for the case of pure CDT. It would of course be highly interesting to understand if there exists a physical explanation for this coincidence.

3. Identification of $m = 3$ with hard dimers

Identifying the continuum theory for $m > 2$ is difficult as we have no continuum formulation of the gravity sector. This is in contrast with DT in which the continuum gravity sector is Liouville theory and for multicritical points it is known the continuum theory is a tensor product of Liouville theory with a minimal model. Progress on this problem was made in [14, 17] where $m = 3$ was identified with a hard dimer model as is also the case for DT with a quartic interaction. However in DT it is clear how hard dimers, which are objects occupying two adjacent triangles such that no two dimers occupy the same triangle, can be mapped to squares; see Figure 1(a). For CDT such a mapping is no longer valid.

For $D = 2$ the partition function is the generating function for CDTs of a given size containing a given number of dimers. The idea of [17] was to explicitly solve a hard dimer model on CDT by use of a bijection between CDT and labelled tree graphs; this is shown in Figure 1(b). Unlabelled trees are easy to count since any sub-graph is also a tree, allowing a recursive equation for the generating function of the number of trees of a given size to be found and solved. Counting the labelled trees in correspondence with hard dimers is a difficult task; the bijection removes links and hence local interactions in the CDT can become non-local in the tree, thereby preventing the trees from being defined recursively.

In [17] enough non-local interactions were removed to allow an analytical treatment but not so many as to prevent new critical behaviour. This can be seen in Figure 1 (a); we only allow a restricted class of dimers to appear. Note that the type-3 dimers are non-local, however they can be accounted for by noting there exists a bijection between trees containing such dimers and distinct trees in which they are replace by type-2 dimers (see [17] for details). Dimers of type

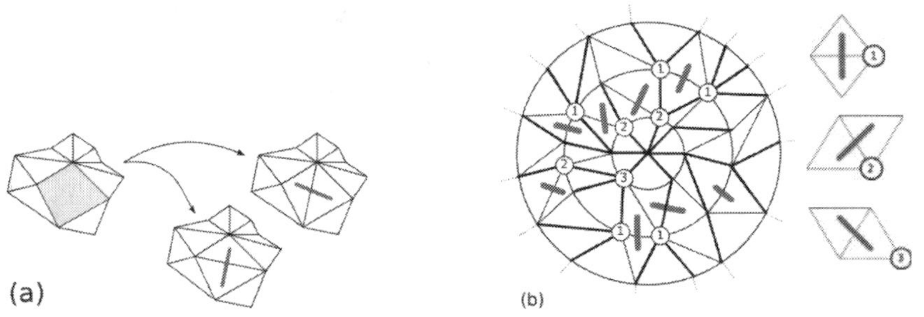

Figure 1. **(a)** The mapping between squares and hard dimers. This mapping does not respect the causal structure of CDT. **(b)** A bijection between hard dimers in CDT and labelled trees. The tree is shown in thickened lines and the labels are added using the rules on the right.

one and two only affect dimers which are adjacent to them in the tree. Hence trees containing only these can be counted recursively. The critical exponents of the model were computed in [17] and found to agree with the values predicted from the matrix model.

4. Summary

We have given an overview of the recent progress in analytically analysing matter coupled to CDT together with a number of examples of why such work is important. We have seen that it is a useful tool for connecting CDT with string theory and will hopefully help in clarifying the manner in which the de-Sitter phase of higher dimensional CDT arises. A natural continuation of this work is to generalise the results here to any (p, q) minimal model with the first target being the Ising model. The unitary models are particularly interesting as one may compare to the numerical work of [9].

References

[1] J. Ambjørn, B. Durhuus, T. Jonsson, *Quantum geometry. A statistical field theory approach.* Cambridge University Press, UK, 1997; P. H. Ginsparg, G. W. Moore, *TASI 92*, pp. 277–470. 1993. hep-th/9304011; P. Di Francesco, P. H. Ginsparg, J. Zinn-Justin, *Phys. Rept.* **254** (1995) 1–133, hep-th/9306153.
[2] P. Bialas, Z. Burda, A. Krzywicki and B. Petersson, Nucl. Phys. B **472** (1996) 293 [hep-lat/9601024].
[3] J. Ambjørn and R. Loll, Nucl. Phys. B **536** (1998) 407 [hep-th/9805108].
[4] J. Ambjørn, A. Görlich, J. Jurkiewicz and R. Loll, 1203.3591 [hep-th].
[5] J. Ambjorn, J. Jurkiewicz and R. Loll, Phys. Rev. Lett. **95** (2005) 171301 [hep-th/0505113].
[6] P. Horava, Phys. Rev. D **79** (2009) 084008 [arXiv:0901.3775 [hep-th]].
[7] J. Ambjørn, R. Loll, Y. Watabiki, W. Westra, and S. Zohren, *Phys. Lett.* **B665** (2008) 252–256, 0804.0252 [hep-th]; *Phys.Lett.* **B670** (2008) 224–230, 0810.2408 [hep-th]; *Acta Phys. Polon.* **B39** (2008) 3355, 0810.2503 [hep-th].
[8] J. Ambjørn, R. Loll, W. Westra and S. Zohren, JHEP **0712**, 017 (2007) [arXiv:0709.2784 [gr-qc]]; J. Ambjorn, R. Loll, Y. Watabiki, W. Westra and S. Zohren, JHEP **0805** (2008) 032 [arXiv:0802.0719 [hep-th]].
[9] J. Ambjørn, K. N. Anagnostopoulos and R. Loll, Phys. Rev. D **60** (1999) 104035 [hep-th/9904012]. J. Ambjørn, K. N. Anagnostopoulos, R. Loll and I. Pushkina, Nucl. Phys. B **807** (2009) 251 [arXiv:0806.3506 [hep-lat]].
[10] J. Ambjørn, A. T. Görlich, J. Jurkiewicz and H. G. Zhang, Nucl. Phys. B **863** (2012) 421 [arXiv:1201.1590 [gr-qc]].
[11] N. Ishibashi and H. Kawai, Phys. Lett. B **314** (1993) 190 [hep-th/9307045].
[12] T. Griffin, P. Horava and C. M. Melby-Thompson, JHEP **1205** (2012) 010 [arXiv:1112.5660 [hep-th]].
[13] Eynard B 2005 J. Stat. Mech. P07005 [math-ph/0605064]. Tom Claeys. [arXiv:0711.2609].
[14] J. Ambjørn, L. Glaser, A. Görlich and Y. Sato, Phys. Lett. **B712**, 109 (2012) 1202.4435.
[15] M. R. Atkin and S. Zohren, 1203.5034 [hep-th].
[16] J. Ambjørn, R. Loll, Y. Watabiki, W. Westra, and S. Zohren, *Acta Phys.Polon.* **B40** (2009) 3479–3507, 0911.4208 [hep-th].
[17] M. R. Atkin and S. Zohren, Phys. Lett. **B712**, 445 (2012) 1202.4322 [hep-th].
[18] A. Hashimoto, M. Huang, A. Klemm and D. Shih, JHEP **0505** (2005) 007, hep-th/0501141.

Towards a loop space description of non-linear sigma model

Partha Mukhopadhyay
The Institute of Mathematical Sciences, C.I.T. Campus, Taramani, Chennai 600113, India
E-mail: parthamu@imsc.res.in

Abstract. Non-linear sigma model with target space M can be described as a single particle quantum mechanics in the corresponding free loop space LM. We first discuss a formal description of this loop space quantum mechanics (LSQM) using the general coordinates in LM. Then we consider a semi-classical limit where the string wavefunction is localized on the submanifold of vanishing loops. The semi-classical expansion is related to the tubular expansion of LSQM around this submanifold. We develop the mathematical framework required to compute the effective dynamics on the submanifold in the Born-Oppenheimer sense at leading order in α' expansion. In particular, we show that the linearized tachyon effective equation is reproduced correctly with divergent terms all proportional to the Ricci scalar of M.

1. Introduction and summary

This talk will be based on [1, 2, 3]. We will consider a closed bosonic string moving in a curved target space M[1]. The relevant classical Lagrangian in unit gauge is given by,

$$L = \frac{1}{2}\int_0^{2\pi} \frac{d\sigma}{2\pi}\, G_{\alpha\beta}(Z(\sigma))[\dot{Z}^\alpha(\sigma)Z^\beta(\sigma) - \partial Z^\alpha(\sigma)\partial Z^\beta(\sigma)] , \tag{1.1}$$

where $G_{\alpha\beta}$ is the metric on M. A $\dot{}$ and a ∂ denote time and space (σ) derivatives respectively. The configuration space of our system is given by the collection of all small loops which can be entirely contained in a single convex neighborhood [8] in M.

The worldsheet theory can be viewed as the worldline description of a single-particle mechanical system where the particle moves in the free loop space LM corresponding to M,

$$LM = C^\infty(S^1, M) . \tag{1.2}$$

In §2 we first discuss the formal structure of the corresponding quantum mechanics [1, 2]. Then in §3 we discuss a semi-classical limit [3] where the string wavefunction is localized on the submanifold of vanishing loops (which is isomorphic to M) in LM. The corresponding semi-classical expansion is related to the tubular expansion [9, 10] of the theory around this submanifold. We also discuss the linearized tachyon effective equation at leading order and the underlying mathematical framework for our computation. Finally, we conclude in §4 with future directions.

[1] See [4, 5] for the study of non-linear sigma model in Lagrangian framework using background field method. We wish to study a loop space description which is a natural set-up for Hamiltonian framework [6, 7].

2. Loop space quantum mechanics

A string embedding $Z^\alpha(\sigma)$ corresponds to a point in LM. The general coordinates of this point is given by[2],

$$z^a = \int_0^{2\pi} \frac{d\sigma}{2\pi}\, Z^\alpha(\sigma) e^{-i\mathrm{a}\sigma} \ . \tag{2.3}$$

The Lagrangian in (1.1) takes the following form in the above general coordinates,

$$L \ = \ \frac{1}{2} g_{ab}(z) \dot{z}^a \dot{z}^b - V \ , \quad V = \frac{1}{2} g_{ab}(z) v^a(z) v^b(z) \ , \tag{2.4}$$

where the metric g and a vector field v in LM are given by,

$$g_{ab}(z) \ = \ \int_0^{2\pi} \frac{d\sigma}{2\pi}\, G_{\alpha\beta}(Z(\sigma)) e^{i(\mathrm{a}+\mathrm{b})\sigma} \ , \quad v^a(z) = \int_0^{2\pi} \frac{d\sigma}{2\pi}\, \partial Z^\alpha(\sigma) e^{-i\mathrm{a}\sigma} \ . \tag{2.5}$$

Such relations, which relate geometric quantities in LM to those in M taken as local fields on S^1, follow certain systematics which is discussed in detail in [1, 3][3]. The above Lagrangian is the standard non-linear system of a particle in an arbitrary curved space which can be quantized following DeWitt in [12]. In particular, generally covariant position space representation of operators can be constructed in scalar states. For example, the Hamiltonian is given by: $\langle\chi|H|\psi\rangle = \int dw\ \chi^*(z)[-\frac{\hbar^2}{2}\mathcal{D}^2 + V]\psi(z)$, where $dw = dz\sqrt{g}$ is the invariant measure, $\hbar = \alpha'$ and $\mathcal{D}^2$ is the Laplacian in loop space. In our case, such a quantization is formal because of the infinite dimensionality of LM (UV divergences are present in the form of infinite dimensional traces). Nonetheless, an interesting result is obtained by applying the above procedure to the Virasoro generators and then computing their algebra [1]. The final result is,

$$\langle\chi| \left\{ \begin{array}{l} [L_\mathrm{a}, L_\mathrm{b}] = (\mathrm{a}-\mathrm{b})\hbar L_{\mathrm{a}+\mathrm{b}} \\ [\tilde{L}_\mathrm{a}, \tilde{L}_\mathrm{b}] = (\mathrm{a}-\mathrm{b})\hbar \tilde{L}_{\mathrm{a}+\mathrm{b}} \\ [L_\mathrm{a}, \tilde{L}_\mathrm{b}] = \hbar^2 \mathcal{A}_{\mathrm{ab}} \end{array} \right\} |\psi\rangle \ , \quad \forall \mathrm{a}, \mathrm{b} \in \mathbb{Z} \ , \tag{2.6}$$

where $\mathcal{A}_{\mathrm{ab}}$ is an *operator anomaly* term which is linear in the Ricci tensor of M. A generalization of DeWitt's analysis in [12] to higher rank tensor states has been obtained in [2]. As consistency would require, the analysis leads to the same result for the above algebra, with scalar states replaced by tensorial ones.

3. A semi-classical limit

It is of interest to understand how to make sense of the infinite dimensional structure of LSQM as discussed in the previous section. As a first step towards this direction we study a semi-classical limit ($\hbar = \alpha' \to 0$) where the wavefunction localizes on $M \hookrightarrow LM$. The general idea,

[2] We adopt the following convention for an infinite-dimensional coordinate index. It is given by a lower case Latin alphabet, which is associated to a pair containing the corresponding Greek alphabet (a target space index) and an integer, denoted by the same Latin alphabet in text format. For example, $a \to (\alpha, \mathrm{a})$, $b \to (\beta, \mathrm{b})$. We will also adopt a similar association between such a pair and the corresponding upper case Latin alphabet when the integer is non-zero, i.e. $A \to (\alpha, \mathrm{a})$, $B \to (\beta, \mathrm{b})$ etc. only when $\mathrm{a}, \mathrm{b} \neq 0$.

[3] Some geometric facts related to v are as follows: it is a Killing vector field. This induces an isometry of LM, which corresponds to the reparametrization invariance and therefore, is present irrespective of the property of M. Notice that v vanishes on the submanifold $M \hookrightarrow LM$ of vanishing loops. This situation is similar to the consideration of Kobayashi's theorem in [11] (in finite dimensions), which claims that the space of fixed points of an isometry is a totally geodesic submanifold of even co-dimension. We will find in next section that $M \hookrightarrow LM$ is indeed totally geodesic. Although, this has infinite number of transverse directions, from the discussion below eq.(2.3), it is clear that for every transverse index $A \to (\alpha, \mathrm{a})$, there is a pair $\bar{A} \to (\alpha, -\mathrm{a})$.

which goes along the line of what is called *constrained quantum systems* in the literature [13], is as follows. Given that V vanishes, more importantly, minimizes on the submanifold[4], one finds a suitable $\hbar$-dependent rescaling of the theory, in particular, of the transverse coordinates such that in the semi-classical limit the potential deepens heavily causing the localization. Moreover the transverse coordinates become *fast* with respect to the longitudinal ones in the Born-Oppenheimer sense. The goal is to find the effective theory of the longitudinal fluctuations on the submanifold order by order in α'-expansion. The technical/conceptual challenges are as follows: (1) Finding the precise definition of the rescaled theory. Once this is understood, each term in the semi-classical expansion of, say, the Hamiltonian can be written down in terms of the tubular expansion coefficients of the loop space metric. (2) To understand tubular expansion of the loop space metric. (3) Divergences are still present in general. How to interpret them?

3.1. Semi-classical expansion

In [3] the first question was answered by demanding the correct flat space limit. The rescaled Hamiltonian (up to next to leading order in $\hbar$) tunrs out to be,

$$\begin{aligned}\mathcal{H} &= \mathcal{H}^{\perp}_{flat} + \hbar\Delta\mathcal{H} + O(\hbar^{3/2}) \ , \\ \Delta\mathcal{H} &= -\frac{1}{2}(\nabla^{\alpha} + i\bar{\omega}^{\alpha AB}\ \Lambda_{AB})(\nabla_{\alpha} + i\bar{\omega}_{\alpha}{}^{CD}\ \Lambda_{CD}) - \frac{1}{4}\bar{r}_{\parallel} - \frac{1}{12}\bar{r}_{\perp} \\ &\quad + \frac{1}{6}\bar{r}^{ABCD}\ \Lambda_{AB}\ \Lambda_{CD} + \frac{1}{6}\sum_{A,B,C,D} \epsilon_{ABCD}\,\bar{r}_{ACDB}\mathrm{y}^{A}\mathrm{y}^{C}\mathrm{y}^{D}\mathrm{y}^{B} \ , \end{aligned} \tag{3.7}$$

where $\mathcal{H}^{\perp}_{flat}$ is precisely the oscillator part of the flat space Hamiltonian with the correct normal ordering constant, ∇ is the covariant derivative in M, ω and r_{abcd} are spin connection and Riemann tensor of LM respectively with a bar indicating that the quantities are being computed on the submanifold. ϵ_{ABCD} is a number involving $|\mathrm{a}|, \cdots, |\mathrm{d}|$ and y^A is the rescaled transverse coordinate[5]: $\mathrm{y}^A = \sqrt{\frac{|\mathrm{a}|}{\hbar}}y^A$, y^A being the Fermi normal coordinate (FNC) [9, 10]. Finally,

$$\bar{r}_{\parallel} = \bar{r}^{\alpha B}{}_{\alpha B} \ , \quad \bar{r}_{\perp} = \bar{r}^{AB}{}_{AB} \ , \quad \Lambda_{AB} = -\frac{i}{2}(\eta_{AC}y^{C}\frac{\partial}{\partial y^{B}} - \eta_{BC}y^{C}\frac{\partial}{\partial y^{A}}) \ , \tag{3.8}$$

where,

$$\eta_{AB} = \eta_{\alpha\beta}\delta_{\mathrm{a+b},0} \ . \tag{3.9}$$

Notice that Λ_{AB} is the angular momentum operator in the transverse space and $\bar{\omega}_{\alpha AB}$ is analogous to a non-abelian Berry connection [14].

The quadratic tachyon effective action is obtained by computing the expectation value of the above Hamiltonian in the state[6]: $\psi(z) = T(x)\chi(y)$, where $T(x)$ is the tachyon field in M and $\chi(y)$ is the wavefunction for the oscillator ground state in flat space. This gives the following linearized effective equation at leading order: $[-\nabla^2 + m_T^2 + V_{eff}]T = 0$, where m_T is the correct tachyon mass and,

$$V_{eff} = -\frac{1}{2}\bar{r}_{\parallel} - \frac{1}{6}\bar{r}_{\perp} + \sum_{B,D}(\beta_{BD}\bar{r}^{BD}{}_{BD} + \gamma_{BD}\bar{\omega}^{\alpha BD}\bar{\omega}_{\alpha BD}) \ , \tag{3.10}$$

β_{BD}, γ_{BD} being certain numbers involving $|\mathrm{b}|$, $|\mathrm{d}|$.

[4] This is not true when M is Lorentzian. However, this is the standard problem of negative norm states and should be solved with the help of ghosts in the usual manner.

[5] Notice that non-zero transverse coordinates correspond to non-zero loops and therefore should have upper case Latin indices according to the rules mentioned in footnote 2.

[6] The coordinate system in LM, which was implicit in eqs(3.7), is given by $z^a = (x^\alpha, y^A)$, x being the general coordinates on the submanifold.

3.2. Tubular expansion in loop space and divergence in tachyon effective equation

The final results have so far been written in terms of certain tensors of LM evaluated on the submanifold. How do we write them entirely in terms of intrinsic geometric data of M? The above results are obtained from the FNC expansion of g up to quadratic order which, for a general embedding, is given by [3][7],

$$\begin{aligned} g_{\alpha\beta} &= G_{\alpha\beta} + \bar{s}_{\alpha\beta C}y^C + (\bar{\omega}_\alpha{}^\gamma{}_C\bar{\omega}_{\beta\gamma D} + \bar{\omega}_\alpha{}^B{}_C\bar{\omega}_{\beta BD} + \bar{r}_{\alpha CD\beta})y^C y^D + O(y^3) , \\ g_{\alpha B} &= \bar{\omega}_{\alpha BC}y^C + \frac{2}{3}\bar{r}_{\alpha CDB}y^C y^D + O(y^3) , \\ g_{AB} &= \eta_{AB} + \frac{1}{3}\bar{r}_{ACDB}y^C y^D + O(y^3) , \end{aligned} \tag{3.11}$$

where $\bar{s}$ is the second fundamental form [15]. The expansion coefficients (barred objects), which are tensors of the ambient space evaluated on the submanifold, carry information about the extrinsic properties of the embedding. In a generic situation they can be chosen somewhat independently of the intrinsic properties of the submanifold. However, the only independent geometric data that are used to construct LM are those of M. Therefore understanding the tubular expansion in loop space means knowing all the expansion coefficients[8] in terms of the intrinsic geometric data of M. This has been determined (up to an arbitrary real constant q) in [3] for the metric expansion in (3.11) up to quadratic level. The results are given by eq.(3.9) and,

$$\begin{aligned} &\bar{\omega}_{\alpha\beta D} = 0(\Rightarrow \bar{s}_{\alpha\beta D} = 0) , \quad \bar{\omega}_{\alpha BD} = E_{(\beta)\gamma}(x)\partial_\alpha E_{(\delta)}{}^\gamma(x)\delta_{\mathrm{b+d,0}} , \\ &\bar{r}_{\alpha DE\beta} = 2qR_{\alpha(\delta\eta)\beta}(x)\delta_{\mathrm{d+e,0}} , \quad \bar{r}_{\alpha BDE} = 0 , \quad \bar{r}_{ABDE} = R_{(\alpha\beta\delta\eta)}(x)\delta_{\mathrm{a+b+d+e,0}} , \end{aligned} \tag{3.12}$$

where $R_{\alpha\beta\gamma\delta}$ and $E^{(\alpha)}{}_\beta$ are the Riemann tensor and the vielbein of M respectively. The first equation implies [15] that $M \hookrightarrow LM$ is totally geodesic as expected (see footnote 3). Using eqs(3.9, 3.12) in (3.10) one finds,

$$V_{eff} \propto R(x) , \tag{3.13}$$

where R is the Ricci scalar of M and the proportionality constant is divergent. This shows that *the tachyon effective equation at leading order is correctly reproduced up to divergent terms all proportional to the equation of motion for the background metric.*

3.3. The mathematical argument

The general procedure of arriving at eqs.(3.12) uses a basic mathematical structure that is relevant for multi-particle dynamics in curved space. The method of finding the centre of mass (CM) would be as follows. Given a multi-particle configuration, erect a tangent space T_xM such that x lies in the neighborhood of that configuration. Then find the preimage of the configuration in T_xM under the exponential map $\exp_x$. If the sum of the position vectors in T_xM vanishes, then x is the CM. Therefore all possible configurations can be described on the tangent bundle TM in the following way. The CM always lies on the zero section $TM_0(\cong M)$ whereas any given configuration lies entirely on the corresponding fibre with one constraint that the average position on the fibre is at the origin. The exponential map being a diffeomorphism, this configuration in TM is actually diffeomorphic to the original configuration whose correct description is given in $M \times M$ where the CM lies on the diagonal submanifold $\Delta(\cong M)$. If $\Phi : TM \to M \times M$ is the relevant bundle map, then we may write,

$$\Phi : (x, \hat{y}) \to (\exp_x(\hat{y}), \exp_x(-\hat{y})) . \tag{3.14}$$

[7] Lower case symbols are used to denote tensors of the ambient space at a generic point $z = (x, y)$.

[8] Closed form expressions for the all-order results for vielbein have been computed in [10] for a generic embedding.

This induces a Riemannian structure from $M \times M$, which has a simple direct product structure, to TM such that $TM_0 \hookrightarrow TM$ admits a tubular neighborhood which is diffeomorphic to that of $\Delta \hookrightarrow M \times M$. The relevant tubular expansion coefficients are all related to intrinsic geometric data of M which can be calculated by exploiting the coordinate transformation in (3.14).

In the context of a string this overall picture remains the same with the multi-particle configuration on a fibre being replaced by a loop $\hat{Y}^\alpha(\sigma)$ with the constraint: $\int_0^{2\pi} \hat{Y}^\alpha(\sigma) = 0$. The submanifold $M \hookrightarrow LM$ is the same as TM_0 and the transverse coordinates (FNC) y^A are given by the Fourier modes of $\hat{Y}^\alpha(\sigma)$ following the definition similar to (2.3). Therefore the Riemannian structure in LM is directly related to that of TM as discussed above.

4. Future directions

Many avenues need to be explored with the new approach described here and its supersymmetric counterpart. The study of the semi-classical limit discussed here needs to be extended to the massless modes. Among other things it is important to understand the pattern of leading order divergences. The implication of the Born-Oppenheimer approach to the study of low energy effective theory in general should be explored. The leading order analysis of the DeWitt-Virasoro algebra in (2.6) and its possible relation to the tensor representation worked out in [2] should also be investigated.

What is the best way to handle the divergences at sub-leading order is an important question. This may require one to understand the tubular expansion in loop space in the sense described below eq.(3.11) more completely.

It may be interesting to explore if the string path integral can be understood as a tubular expansion around particle worldline in M. Perhaps the question of interest is if the interpretation of this type of tubular geometry can in any way be helpful to understand α'-corrections in string theory.

[1] P. Mukhopadhyay, arXiv:0912.3987 [hep-th]; P. Mukhopadhyay, Pramana **76**, 407 (2011) [arXiv:1010.0930 [hep-th]].
[2] P. Mukhopadhyay, arXiv:1004.2396 [hep-th].
[3] P. Mukhopadhyay, arXiv:1202.2735 [hep-th].
[4] See for example reviews:
C. G. Callan and L. Thorlacius ; A. A. Tseytlin, Int. J. Mod. Phys. A **4**, 1257 (1989).
[5] J. Honerkamp, Nucl. Phys. B **36**, 130 (1972); D. Friedan, Phys. Rev. Lett. **45**, 1057 (1980); D. H. Friedan, Annals Phys. **163**, 318 (1985). L. Alvarez-Gaume, D. Z. Freedman and S. Mukhi, Annals Phys. **134**, 85 (1981).
[6] E. Witten, J. Diff. Geom. **17**, 661 (1982); E. Witten, PUPT-1050.
[7] E. Frenkel, A. Losev and N. Nekrasov, Nucl. Phys. Proc. Suppl. **171**, 215 (2007) [hep-th/0702137]; E. Frenkel, A. Losev and N. Nekrasov, arXiv:0803.3302 [hep-th].
[8] R. M. Wald, "General Relativity," *Chicago, Usa: Univ. Pr. (1984) 491p*
[9] A. Gray, "Tubes," Second Edition (Progress in Mathematics). Addison-Wesley Publishing Company, New York, 1990.
[10] P. Mukhopadhyay, arXiv:1203.1151 [gr-qc].
[11] S. Kobayashi, "Fixed points of isometries," Nagoya Math. J. **13** (1958), 63-68.
[12] B. S. DeWitt, Phys. Rev. **85**, 653 (1952).
[13] R. A. Marcus, J. Chem. Phys. **45**, 44934499 (1966); H. Jensen, H. Koppe, Ann. Phys. **63**, 586591 (1971); R. C. T. da Costa, Phys. Rev. A **25**, 28932900 (1982); P. Maraner, J. Phys. A **63**, 2939-2951 (1995); K. A. Mitchell, Phys. Rev. A **63**, 042112 (2001); G. F. Dell'Antonio and L. Tenuta, J. Phys. A **37**, 5605-5624 (2004); J. Wachsmuth, S. Teufel, arXiv:0907.0351v3 [math-ph]; J. Wachsmuth, S. Teufel, Phy. Rev. A **82**, 022112 (2010).
[14] M. V. Berry, Proc. R. Soc. London, Ser. A **392**, 45 (1984).
B. Simon, Phys. Rev. Lett.**51**, 2167 (1983).
[15] S. Kobayashi and K. Nomizu. "Foundations of Differential Geometry II," Wiley, New York, 1st edition, 1969.

Third quantization: modeling the universe as a 'particle' in a quantum field theory of the minisuperspace

S J Robles Pérez

Instituto de Física Fundamental, Consejo Superior de Investigaciones Científicas, Serrano 121, 28006 Madrid (SPAIN) and Estación Ecológica de Biocosmología, Pedro de Alvarado 14, 06411 Medellín (SPAIN).

E-mail: `salvarp@imaff.cfmac.csic.es`

Abstract. The third quantization formalism of quantum cosmology adds simplicity and conceptual insight into the quantum description of the multiverse. Within such a formalism, the existence of squeezed and entangled states raises the question of whether the complementary principle of quantum mechanics has to be extended to the quantum description of the whole space-time manifold. If so, the *particle* description entails the consideration of a multiverse scenario and the *wave* description induces us to consider as well correlations and interactions among the universes of the multiverse.

1. Introduction

The so-called third quantization formalism was initially developed [1, 2, 3] to quantum mechanically represent the fluctuations of the space-time which, at the Planck length, provide it with a *foam* structure [4, 5], in which virtual black holes, wormholes and baby universes cohabit. The main aim was to find an explanation of a vanishing cosmological constant [2, 6], which constituted the standard cosmological model at that time, and to study the effects that the space-time foam would produce in the coherence properties of matter fields [7, 8, 9] as well as in the effective value of fundamental constants [10].

On the other hand, a quantum description of the multiverse within the third quantization formalism has received some criticisms [11]. The most remarkable of them are that the scale factor cannot generally be taken as a time variable provided that it is not a monotonic function of time, and that the third quantization does not add anything to what it is already described within the usual quantum cosmological approach, whether this is given by the Wheeler-DeWitt equation or by the path integral formalism. However, as we shall try to briefly expose in this letter, it does add.

The scale factor cannot generally be taken as a time variable of the minisuperspace in a similar way as a well-defined time variable cannot be given in a general curved space-time. However, for universes with high degree of symmetry, the scale factor can be seen as a formal time-like variable of the minisupermetric that defines the geometrical structure of the corresponding minisuperspace. Furthermore, for large parent universes, the third quantization adds: i) simplicity of the model with respect to other approaches; ii) well-known standard procedures that can be applied to the quantum description of the universe; and, iii) conceptual insight of the quantum description of both the universe and the multiverse. For instance, we shall see that the existence in the multiverse of quantum states having no classical analogue, like entangled and squeezed states, would no longer be associated to non-locality features because the concepts of locality or non-locality are meaningless in the quantum multiverse.

Nowadays, the multiverse has reached a wider acceptance [12], and the third quantization formalism supplies us with a quantum description of the multiverse that parallels that of a quantum field theory in a curved space-time. It also provides us with novel approaches for customary problems of quantum cosmology, like the boundary conditions, the problem of the cosmological constant, and the arrow of time, among others, and a framework in which new phenomena that were not contemplated so far may come out.

2. Third quantization formalism

The third quantization formalism consists of considering the wave function of the universe as a field to the quantized that propagates upon the minisuperspace. For instance, for a closed homogeneous and isotropic space-time minimally coupled to n scalar fields, $\vec{\varphi} \equiv (\varphi_1, \ldots, \varphi_n)$, which represent the matter content of the universe, the Wheeler-DeWitt equation can be written as [13]

$$\left(-\frac{\hbar^2}{\sqrt{-G}} \partial_A (\sqrt{-G} G^{AB} \partial_B) + \mathcal{V}(q^A) \right) \phi(q^A) = 0, \tag{1}$$

where, $\{q^A\} \equiv \{a, \vec{\varphi}\}$, are the configuration variables of the minisuperspace, the potential $\mathcal{V}(q^A)$ is given by

$$\mathcal{V}(q^A) = a^3 \Lambda - a + a^3 (V(\varphi_1) + \ldots + V(\varphi_n)),$$

being $V(\varphi_i)$ the potential of the scalar field φ_i, and G^{AB} is the inverse of the minisupermetric

$$G_{AB} = \mathrm{diag}(-a, a^3, \ldots, a^3), \tag{2}$$

with determinant, $G = -a^{3n+1}$. In Eq. (1), $\phi(q^A) \equiv \phi(a, \vec{\varphi})$ is the wave function of the universe being considered. The Lorentzian signature of the minisupermetric (2) allows us to consider the scale factor as a formal time-like variable of the minisuperspace. This has not to be confused with a physical time variable in terms of 'clocks and rods' measured by any observer. The relation between the scale factor and the Friedmann time, t, has to be found *a-posteriori.*

We shall consider two particular cases: a massless scalar field in a de-Sitter space-time and a slow-varying field in a closed FRW space-time. For the massless scalar field, φ, the Wheeler-DeWitt equation (1) explicitly reads

$$\ddot{\phi} + \frac{\dot{\mathcal{M}}}{\mathcal{M}} \dot{\phi} - \frac{1}{a^2} \phi'' + \omega^2(a) \phi = 0, \tag{3}$$

where, $\phi \equiv \phi(a, \varphi)$ is the wave function of the universe, with $\dot{\phi} \equiv \frac{\partial \phi}{\partial a}$ and $\phi' \equiv \frac{\partial \phi}{\partial \varphi}$, $\dot{\mathcal{M}} \equiv \frac{\partial \mathcal{M}}{\partial a}$ with $\mathcal{M} = a$, and $\omega(a) = \frac{a}{\hbar}\sqrt{a^2\Lambda - 1}$. Following the analogy between the third quantization formalism and a quantum field theory in a curved space-time, the wave function ϕ can be promoted to an operator that can be decomposed in normal modes as

$$\hat{\phi} = \frac{1}{\sqrt{2\pi}} \int dk \left(e^{ik\varphi} A_k(a) \hat{c}_k + e^{-ik\varphi} A_k^*(a) \hat{c}_k^\dagger \right), \tag{4}$$

where $\hat{c}_k$ and $\hat{c}_k^\dagger$ are the usual annihilation and creation operators of the harmonic oscillator with mass and frequency $\mathcal{M}(a)$ and $\omega(a)$, respectively, evaluated on the boundary hypersurface Σ_0, for which $a = a_0$ and $\varphi = \varphi_0$. They represent annihilation and creation of universes, respectively, within the framework of the third quantization formalism. The probability amplitudes, $A_k(a)$ and $A_k^*(a)$, satisfy the equation of a damped harmonic oscillator with a mode-dependent frequency given by

$$\omega_k = \frac{1}{\hbar}\sqrt{a^4\Lambda - a^2 + \frac{\hbar^2 k^2}{a^2}}, \tag{5}$$

where the last term in the radicand of Eq. (5) is a quantum correction that does not appear in the classical picture. For $k = 0$, there is a Lorentzian region for values, $a > a_t \equiv \frac{1}{\sqrt{\Lambda}}$, and an Euclidean region for values, $a_t > a > 0$. The transition hypersurface $\Sigma_t \equiv \Sigma(a_t)$ corresponds then to the appearance of time [13]. For values $k_m > k > 0$, where [14] $k_m = \frac{4}{27\hbar^2\Lambda^2}$, before reaching the collapse the Euclidean instanton finds a new Lorentzian region (see Ref. [14]). Then, following a reasoning that parallels that proposed in Refs. [15, 16, 17], two instantons can then be joined forming a double instanton that would give rise to an entangled pair of universes with a composite quantum state given by [14]

$$\phi_{I,II} = \int dk \left(e^{ik(\varphi_I + \varphi_{II})} A_{I,k}(a) A_{II,k}(a)\, \hat{b}^\dagger_{I,k} \hat{b}^\dagger_{II,k} + e^{-ik(\varphi_I+\varphi_{II})} A^*_{I,k}(a) A^*_{II,k}(a)\, \hat{b}_{I,k}\, \hat{b}_{II,k} \right), \tag{6}$$

where $\varphi_{I,II}$ are the scalar fields of each single universe, labelled by I and II, respectively.

3. The 'particle' definition of the universe

Let us now consider the case of a slow-varying field in a closed FRW space-time, for which the Wheeler-DeWitt equation (1) reads

$$\ddot{\phi} + \frac{\dot{\mathcal{M}}}{\mathcal{M}}\dot{\phi} + \omega_{sv}^2(a,\varphi)\phi = 0, \tag{7}$$

where, $\omega_{sv}(a,\varphi) = \frac{a}{\hbar}\sqrt{a^2 V(\varphi) - 1}$. Eq. (7) is formally the equation of a harmonic oscillator defined on minisuperspace. Different representations can be chosen to quantum mechanically describe the universes. For instance, we could use the constant operators given in Eq. (4). However, in terms of such representation, the number operator, $\hat{N}_0 \equiv \hat{c}_k^\dagger \hat{c}_k$, is not an invariant operator and its eigenvalues depend on the value of the scale factor.

We would expect that, for a given model of the multiverse, the number of universes would not depend on the value of the scale factor of a particular single universe provided that we are not considering interaction terms in the Hamiltonian. Thus, it seems appropriate imposing the boundary condition that the number of universes of the multiverse is represented by an invariant operator. In the case of a slow-varying field, the Hamiltonian for which the Heisenberg equations of motion give rise to Eq. (7), is given by

$$\hat{H} = \frac{1}{2\mathcal{M}}\hat{P}_\phi^2 + \frac{\mathcal{M}\omega_{sv}^2}{2}\hat{\phi}^2, \tag{8}$$

where $\hat{P}_\phi$ is the third quantized momentum conjugated to the wave function operator $\hat{\phi}$. Then, an invariant representation can be found by following the method developed by Lewis [18] and others [19, 20, 21]. It is given by the operators,

$$\hat{b}(a) \equiv \sqrt{\frac{1}{2\hbar}}\left(\frac{1}{R}\hat{\phi} + i(R\hat{P}_\phi - \dot{R}\hat{\phi})\right), \tag{9}$$

$$\hat{b}^\dagger(a) \equiv \sqrt{\frac{1}{2\hbar}}\left(\frac{1}{R}\hat{\phi} - i(R\hat{P}_\phi - \dot{R}\hat{\phi})\right), \tag{10}$$

Journal of Physics: Conference Series **410** (2013) 012133 doi:10.1088/1742-6596/410/1/012133

where, $R = \sqrt{\phi_1^2 + \phi_2^2}$, being ϕ_1 and ϕ_2 two independent solutions of the Wheeler-de Witt equation (7). Then, $\hat{N} \equiv \hat{b}^\dagger \hat{b}$, is an invariant operator fulfilling the boundary condition of the multiverse. However, in terms of the invariant representation, the Hamiltonian (8) turns out to formally be the Hamiltonian of a parametric amplifier that, in quantum optics, is associated with the generation of entangled pairs of photons [22]. Similarly, we can interpret that the Hamiltonian (8) is associated, in the representation given by Eqs. (9-10), with the creation and annihilation of entangled pairs of universes in a multiverse scenario [14].

However, for an observer inhabiting a large parent universe, for which $a \gg 1$, the appropriate representation of the universe is given by the asymptotic representation which is described in terms of the usual creation and annihilation operators of the harmonic oscillator with mass $\mathcal{M}(a)$ and frequency $\omega(a, \varphi)$. It turns out to be that [14] the vacuum state of the multiverse becomes, in the asymptotic representation, a two-mode squeezed state that represents an entangled pair of universes. The quantum state of each single universe of the entangled pair turns out to be then given by a thermal state, which is indistinguishable from a classical mixture and whose thermodynamical properties can be computed. In particular, the entropy and energy of entanglement may provide us with a time variable and with a vacuum energy, respectively, for each single universe [14, 23].

4. Wave-particle duality in the multiverse

The existence of squeezed and entangled states in the multiverse raises the question of whether Bell's inequalities can be violated by such states in the multiverse, too. If so, however, such violation could not be related to non-locality features because there is no space-time among the universes of the quantum multiverse. It would be rather related to the interdependence of the quantum states that represent different universes.

The violation of classical inequalities by entangled and squeezed states in quantum optics is fundamentally related to the concept of complementary of the quantum theory [24]. The extension of the complementary principle to the quantum multiverse would imply that the quantum state of the whole space-time manifold has to be complementary described in terms of *particles and waves*. In terms of *particles* means that, in an appropriate representation, the universes have to be considered as individual entities, giving rise to a multiverse scenario. In terms of *waves* implies that interference effects among universes should be contemplated as well. Such a vision of the multiverse opens the door for novel approaches to customary problems in quantum cosmology. For instance, the thermodynamics of entanglement provides us with a new tool for studying the thermodynamical properties of a single universe and the related features of the vacuum energy and the arrow of time.

Furthermore, the consideration of the multiverse as a collective system also supplies us with unexpected phenomena. For instance, in Ref. [25] it is studied the quantum state of a multiverse made up of de-Sitter universes all of them with the same value, Λ, of their cosmological constants. Even considering no interaction among the universes, the energy spectrum of the collective system splits into two energy levels, which correspond respectively to two normal modes, one of which corresponds to a state of a very small value of the energy. Then, the universes could initially be created in the excited level, which would provide us with a large value of the vacuum energy in the initial state of the universe, and decay afterwards into the ground state that entails a very small value of their effective cosmological constants. The effect is even more evident when a 'nearest' interaction is considered [25]. The spectrum splits then into a large number of energy levels, being the ground state a state with a very small value of the vacuum energy. The reasoning made in Ref. [25] is not conclusive but it plainly shows that the logic of the multiverse is rather different than the logic of a set of universes, being these taken individually, showing that the multiverse is much more that the mere sum of its parts.

Acknowledgments

To the memory of Prof. Pedro González Díaz, who always encouraged us to adopt a fearless and an open-minded attitude as well in science as in everyday life.

References

[1] McGuigan M. Third quantization and the Wheeler-DeWitt equation. Phys Rev D. 1988;38:3031.
[2] Rubakov VA. On third quantization and the cosmological constant. Phys Lett B. 1988;214:503–507.
[3] Strominger A. Baby Universes. In: Coleman S, Hartle JB, Piran T, Weinberg S, editors. Quantum Cosmology and Baby Universes. vol. 7. World Scientific, London, UK; 1990. .
[4] Wheeler JA. On the nature of quantum geometrodynamics. Ann Phys. 1957;2:604–614.
[5] Hawking SW. Spacetime foam. Nucl Phys B. 1978;144:349–362.
[6] Coleman S. Why there is nothing rather than something? A theory of the cosmological constant. Nucl Phys B. 1988;310:643–668.
[7] Coleman S. Black holes as red herrings: Topological fluctuations and the loss of quantum coherence. Nucl Phys B. 1988;307:867.
[8] González-Díaz PF. What is really being lost in wormholes? Phys Rev D. 1990;42(12):3983–3996.
[9] González-Díaz PF. Regaining quantum incoherence for matter fields. Phys Rev D. 1992;45:499.
[10] Giddings SB, Strominger A. Loss of incoherence and determination of coupling constants in quantum gravity. Nucl Phys B. 1988;207:854–866.
[11] Vilenkin A. Approaches to quantum cosmology. Phys Rev D. 1994;50:2581–2594.
[12] Carr B, editor. Universe or Multiverse. Cambridge University Press, Cambridge, UK; 2007.
[13] Kiefer C. Quantum gravity. Oxford University Press, Oxford, UK; 2007.
[14] Robles-Pérez S, González-Díaz PF. Quantum entanglement in the multiverse. [arXiv:11114128]. 2011;.
[15] Barvinsky AO, Kamenshchik AY. Cosmological Landscape From Nothing: Some Like It Hot. JCAP. 2006;0609:014.
[16] Barvinsky AO, Kamenshchik AY. Cosmological landscape and Euclidean quantum gravity. J Phys A. 2007;40:7043–7048.
[17] Barvinsky AO. Why there is something rather than nothing (out of everything)? Phys Rev Lett. 2007;99:071301.
[18] Lewis HR, Riesenfeld WB. An Exact Quantum Theory of the Time-Dependent Harmonic Oscillator and of a Charged Particle in a Time-Dependent Electromagnetic Field. J Math Phys. 1969;10:1458–1473.
[19] Pedrosa IA. Comment on "Coherent states for the time-dependent harmonic oscillator". Phys Rev D. 1987;36:1279.
[20] Dantas CMA, Pedrosa IA, Baseia B. Harmonic oscillator with time-dependent mass and frequency and a perturbative potential. Phys Rev A. 1992;45:1320.
[21] Vergel DG, Villaseñor JS. The time-dependent quantum harmonic oscillator revisited: Applications to quantum field theory. Ann Phys. 2009;324:1360.
[22] Scully MO, Zubairy MS. Quantum optics. Cambridge University Press, Cambridge, UK; 1997.
[23] Robles-Pérez S. Entanglement arrow of time in the multiverse. [arXiv:12035774]. 2012;.
[24] Reid MD, Walls DF. Violations of classical inequalities in quantum optics. Phys Rev A. 1986;34:1260–1276.
[25] Alonso-Serrano A, Bastos C, Bertolami O, Robles-Pérez S. Interacting universes and the cosmological constant. (in preparation). 2012;.

Journal of Physics: Conference Series **410** (2013) 012134 doi:10.1088/1742-6596/410/1/012134

Two-dimensional quantum dilaton gravity and the quantum cosmological constant

Simone Zonetti[1] **and Jan Govaerts**[1,2,3]

[1] Centre for Cosmology, Particle Physics and Phenomenology (CP3), Institut de Recherche en Mathématique et Physique (IRMP), Université catholique de Louvain, Chemin du Cyclotron 2, B-1348 Louvain-la-Neuve, Belgium

[2] International Chair in Mathematical Physics and Applications (ICMPA-UNESCO Chair), University of Abomey-Calavi, 072 B. P. 50, Cotonou, Republic of Benin

[3] Fellow of the Stellenbosch Institute for Advanced Study (STIAS), 7600 Stellenbosch, South Africa

E-mail: simone.zonetti@uclouvain.be, jan.govaerts@uclouvain.be

Abstract. We address the cosmological constant problem in the context of two-dimensional dilaton-Maxwell gravity, coupled to an arbitrary number of scalar matter fields. We are able to quantize the model non-perturbatively; we determine that the realization of the classical symmetries at the quantum level provides a mechanism that fixes the value of the cosmological constant.

1. Introduction

The cosmological constant problem is one of the long-standing issues of modern physics. While we can measure the value of the cosmological constant with great accuracy, we are not able to calculate it in a coherent theoretical framework and the theoretical predictions in quantum field theory are radically different from observations. This disagreement is a hint of the difficult conciliation of Quantum Mechanics and General Relativity in a theory of quantum gravity. Current approaches to the cosmological constant problem, in particular, do not account for the quantum nature of the gravitational interaction and rely on perturbative calculations. In this paper we discuss how quantum gravity can determine the value of the cosmological constant, labelled Λ, and we address the issue in the simplified scenario of two-dimensional dilaton-Maxwell gravity. In Section 2 we discuss a general idea on how quantum gravity, in any space-time dimensions, can provide a mechanism that fixes the cosmological constant. In Section 3 we detail how this ideas apply in the specific case of two-dimensional dilaton-Maxwell gravity, while in Section 4 we briefly discuss our results.

2. The cosmological constant in a quantum theory of gravity

Two interesting examples of a cosmological constant determined by the symmetries in quantum gravity are given by one-dimensional gravity [1] and a two-dimensional toy model [2], both in the presence of scalar matter . We can generalize the principles contained in these works to the case of quantum gravity in any number of space-time dimensions d.

In a completely general way the diffeomorphism invariance, in the Hamiltonian formulation,

provides classical constraint equations, some of which will also include the cosmological constant, $\mathcal{H}^\mu(\Lambda, \dots) = 0$, with $\mu = 1, \dots, d$. Additional gauge symmetries provide additional constraints. When turned into quantum constraint operators, these classical equations become conditions that determine the physical states of the model, $\hat{\mathcal{H}}^\mu(\Lambda)|\psi_{phys}\rangle = 0$, in a way dependent on the value of the cosmological constant and spatial coordinates. This same condition can be seen as a way to determine the value of the cosmological constant required for a given quantum state to be physical. Once a basis is chosen in Hilbert space, one can solve the set of equations $\langle\psi_{phys}|\hat{\mathcal{H}}^\mu(\Lambda)|\psi_{phys}\rangle = 0$ in the parameter space spanned by the the cosmological constant itself, the (complex) coordinates which cover the specific (sub)space of states considered and possibly the undetermined parameters of the model. In order to do so the dependence on spatial coordinates has to be integrated out, for instance looking at the Fourier modes of each equation. For each μ this reduces the number of equations from $d-1$ non countable infinities (an equation per spatial point) to $d-1$ countable ones (one equation per each Fourier mode). Moreover the commutation relations among the different modes might drastically reduce the number of independent equations. A notable example is the Weyl symmetry in string theory, which requires only three modes of the Virasoro generators to annihilate physical states. This is true in any number of dimensions, as long as diffeomorphism invariance holds. In particular, due to the specific form of the cosmological constant term in the gravitational action, Λ will appear linearly in the constraints. There will then be at most one value of the cosmological constant that allows a given quantum state to be physical.
To turn the formal constraints equations into something able to provide an actual result for Λ it is of course necessary to have fully quantized the theory, so that the explicit form and algebra of $\hat{\mathcal{H}}^\mu$ are known. Therefore we have to turn our attention to those models of gravity that we are able to quantize, also in the presence of additional fields.

3. Two-dimensional dilaton gravity and the quantized cosmological constant

Two-dimensional generalized dilaton theories represent a very interesting class of models, widely studied in the last two decades. They proved to be very useful in the understanding of classical and quantum gravity, allowing to face conceptual issues also relevant to higher dimensions. Most of the results are well summarized in [3, 4]. In order to generalize the results of [2], we consider the most general action principle of dilaton gravity in the presence of a Maxwell field:

$$S_{DM} = \int_{\mathcal{M}} dx^2 \sqrt{-g} \left(XR - U(X) X_{,\mu} X^{,\mu} - 2V(X) - \frac{1}{4} G(X) F_{\mu\nu} F^{\mu\nu} \right) , \tag{1}$$

where X is the dilaton, U, V and G are arbitrary functions of X, and $F_{\mu\nu}$ is the usual field strength for the vector gauge field A_μ. We will consider all fields to vanish at infinity so no boundary terms appear. In [5] it was realized that, with specific restrictions on the U and G potentials, there exists a dual formulation of (1) in terms of two decoupled Liouville fields on flat space-time. In particular the dual action principle takes the form:

$$S_{dual} = \int d^2x \sqrt{-g_\flat} \left[\frac{1}{2} \left(Z_{,\mu} Z^{,\mu} - Y_{,\mu} Y^{,\mu} \right) - 2\Lambda e^{Z/\xi} - e^{-Y/\xi} \frac{\vartheta}{2\Lambda} F_{\mu\nu} F^{\mu\nu} + \xi \left(Z - Y \right) R_\flat \right] , \tag{2}$$

where Z and Y are two scalar fields, Λ is the cosmological constant, ϑ is an arbitrary constant and ξ plays the role of a coupling constant. The space-time metric gives a line element $dx^2 = -\lambda_0 \lambda_1 dt^2 + (\lambda_0 - \lambda_1) dt\, ds + ds^2$, which is in fact pure gauge. The subscript $\flat$ indicates gravitational quantities that are calculated with respect to this metric. The inclusion of scalar matter is trivial and it is simply done by adding an arbitrary number D of terms in the form:

$$S_\phi = -\frac{1}{2} \int d^2x \sqrt{-g_\flat} \phi_{,\mu} \phi^{,\mu} \tag{3}$$

To deal with the gauge invariances it is useful to look at the BRST formulation of the model [6]; after the introduction of ghost degree of freedom and the BRST charge, the BRST extended constraints are readily calculated and, as expected from the properties of the duality, it is easy to check that their algebra is first class. In particular the (smeared) Poisson brackets read:

$$\begin{aligned}\{L^{\pm}(f), L^{\pm}(g)\} &= \pm L^{\pm}(g_{,s}f - f_{,s}g) \ , \\ \{L^{\pm}(f), L^{\emptyset}(g)\} &= 0 \ , \\ \{L^{\pm}(f), L^{\mp}(g)\} &= -\tfrac{1}{4\Lambda_G}\left(e^{Y/\xi}\Pi_1 L^{\emptyset}\right)(fg) \ ,\end{aligned} \tag{4}$$

where $\pm$ label the two Virasoro generators and $\emptyset$ indicates the constraint related to the $U(1)$ gauge invariance. As expected these brackets are vanishing on the constraints surface. In the following we will choose to work in the conformal gauge for the gravitational sector, $\lambda_0 = \lambda_1 = 1$, and the Coulomb gauge in the $U(1)$ sector, $A_0 = A_{1,s} = 0$.

In order to quantize this model in the canonical formalism we can compactify the space dimension to a circle, in a way equivalent to the standard box quantization in quantum field theory. We can set the compactification scale to 1 with a suitable choice of units. Since Newton's constant is dimensionless in two dimensions, we are always entitled to this choice. In the following we are also considering $c = \hbar = 1$. Therefore we consider $s \in [0, 2\pi)$, so that all classical fields will be required to be periodic along the compact dimension, i.e. for any field F we have $F(t,0) = F(t,2\pi)$. Quantization follows the usual prescriptions of Dirac's approach to constrained systems and field operators are expressed in terms of creation/annihilation operators. Normal ordering is chosen when needed for composite operators. The decoupling between Z, Y and scalar matter fields plays a fundamental role in allowing us to quantize the fields, and therefore the classical constraints, by working in each sector separately. In particular the two BRST extended Virasoro constraints are a sum of terms from each field $L^{\pm} = L^{\pm,Z} + L^{\pm,Y} + L^{\pm,g} + \sum_{n=1}^{D} L^{\pm,\phi}$, where g labels the contribution from the ghosts. We will require the classical symmetries to be unbroken at the quantum level: the quantum operators corresponding to the L's will have to exhibit a closed algebra. In particular we will make sure that no central term appears in the commutators. The $L^{\emptyset}$ constraint has identically vanishing brackets with the two Virasoro generators and itself, hence the corresponding operator is quantum mechanically trivial. The Virasoro algebra, on the other hand will exhibit a central extension. The ghosts and scalar matter degrees of freedom reproduce the usual results known from string theory. Looking at their Fourier modes the contributions to the total central extension are $c_g = -\frac{13}{6}r^3 - \frac{1}{6}r$ and $c_\phi = \frac{1}{12}r^3 + \frac{1}{6}r$. The two Liouville fields on the other hand require the presence of a quantum correction to the coupling constant ξ which appears in front of the term containing $R_\flat$ in (2) [7] and in particular one has $\xi_Z = \xi_Y = \xi - \frac{1}{8\pi\xi}$. Interestingly, as Z and Y appear with opposite signs in (2), the central extensions of their Virasoro contributions cancel each other. The total quantum Virasoro algebra will therefore contain a central extension $c = \frac{1}{12}(D-26)\,r^3 + \frac{D+1}{6}r$.

By fixing $D = 26$ we can eliminate the cubic contribution, while by shifting the zero modes as $\hat{L}_0^{\pm} \to \hat{L}_0^{\pm} - 27/12$ we can reabsorb the linear term, giving a Virasoro algebra free of central extensions. The quantum constraint equations (dropping the ghost contributions by virtue of BRST invariance) are therefore given by:

$$\begin{aligned}\langle \hat{L}^+ + \hat{L}^- \rangle &= -\tfrac{1}{2}\langle P_Z^2 + Z_{,s}^2\rangle + \tfrac{1}{2}\langle P_Y^2 + Y_{,s}^2\rangle + \textstyle\sum \tfrac{1}{2}\langle P_i^2 + \phi_{i,s}^2\rangle + 2\xi_Z\langle Z_{,ss}\rangle - 2\xi_Y\langle Y_{,ss}\rangle + \\ &\quad + \Lambda\left[2\langle e^{Z/\xi}\rangle + \tfrac{1}{4\Lambda_G}\langle e^{Y/\xi}P_{A_1}^2\rangle\right] - 27/6 \\ \langle \hat{L}^+ - \hat{L}^- \rangle &= \langle P_Z Z_{,s}\rangle - 2\xi_Z\langle P_{Z,s}\rangle + \langle P_Y Y_{,s}\rangle - 2\xi_Y\langle P_{Y,s}\rangle + \textstyle\sum\langle P_i\phi_{i,s}\rangle \\ \langle L^{\emptyset}\rangle &= \langle P_{A_1,s}\rangle \ ,\end{aligned} \tag{5}$$

where the P's are conjugate momenta. These equations can be solved for the cosmological constant and its explicit value calculated with a suitable choice for the representation of the states and the operators themselves.

4. Results and outlook

A first interesting result is that the Z field contributes to Λ with a sign opposite with respect to the contributions of the Y and scalar matter fields. This allows excitations of the Z field to provide cancellations which can in principle generate arbitrarily small values for Λ, in contrast with the usual QFT picture in which Λ is naturally of the order of the Planck scale.
Furthermore, as Λ appears linearly in the quantum constraints equations, once a quantum state is required to be physical, only one specific value of the cosmological constant is allowed, as it is expected from the considerations made in Section 2. On the other hand for a given value of Λ different quantum states can be physical.
In addition a series of quantum corrections appear: besides the shift of the zero modes of the Virasoro generators, which determines a non vanishing and purely quantum cosmological constant in the vacuum, the coupling constants ξ_Z and ξ_Y require additional contributions. These corrections will influence the value of Λ, suggesting that quantum gravity might play a very important role in building up the value of the cosmological constant.
With a suitable choice of representation for quantum states and operators it is also possible to study the spectrum of the cosmological constant for the lower excitations of this model. To this purpose the natural choice for the states are Fock excitations, while we can employ a coherent states representation for the operators [8]. With this combination it is possible to turn (5) into gaussian integrals over complex variables and to solve them explicitly for Λ in terms of the parameters that define the quantum states (see [2, 9] for a detailed description of the procedure applied to a slightly different model). Let us stress that the resulting value of the cosmological constant is determined in a completely non-perturbative way and to our knowledge this is the first case, although two-dimensional, in which the quantum nature of the gravitational interaction is taken into account.

Acknowledgments

SZ benefits from a PhD research grant of the Institut Interuniversitaire des Sciences Nucléaires (IISN, Belgium). This work is supported by the Belgian Federal Office for Scientific, Technical and Cultural Affairs through the Interuniversity Attraction Pole P6/11.

References

[1] Govaerts J 2004 *Proc. 3rd International Workshop on Contemporary Problems in Mathematical Physics, 1-7 November 2003, Cotonou (Benin), eds. J. Govaerts, M.N. Hounkonnou and A.Z. Msezane* (World Scientific (Singapore)) pp 244–272 (*Preprint* hep-th/0408022)
[2] Govaerts J and Zonetti S 2011 *Class. Quantum Grav.* **28** 185001 (*Preprint* 1102.4957v1)
[3] Grumiller D, Kummer W and Vassilevich D V 2002 *Phys. Rept.* **369** 327–430 (*Preprint* hep-th/0204253)
[4] Grumiller D and Meyer R 2006 *Turk.J.Phys.* **30** 349–378 (*Preprint* hep-th/0604049)
[5] Zonetti S and Govaerts J 2012 *J. Phys. A: Math. Theor. 45 (2012) 042001* (*Preprint* 1111.1612)
[6] Govaerts J 1991 *Hamiltonian Quantisation and Constrained Dynamics* (*Leuven Notes in Mathematical and Theoretical Physics* vol 4) (Leuven University Press)
[7] Curtright T L and Thorn C B 1982 *Phys. Rev. Lett.* **48** 1309–1313
[8] Govaerts J 2008 Klauder's phase space coherent state path integral and the quantisation of action-angle variables unpublished notes
[9] Zonetti S 2012 *Two-dimensional quantum dilaton gravity and the quantized cosmological constant* Ph.D. thesis Academie Louvain - Université Catholique de Louvain

Hoop Conjecture, Minimal Length and Black Hole Formation in the Asymptotically Safe Scenario of Quantum Gravity

Sayandeb Basu[1], **David Mattingly**[2]

[1]Physics Department, University of the Pacific, 3601 Pacific Avenue, Stockton CA 95211, USA
[2]Physics Department, University of New Hampshire, Durham, New Hampshire, 03824, USA

E-mail: [1]`sbasu@pacific.edu`

E-mail: [2]`david.mattingly@unh.edu`

Abstract. Black hole dominance, assumed to be a fairly ubiquitous feature of any theory of quantum gravity, amounts to that any observer trying to perform a localized experiment on ever smaller length scales will ultimately be thwarted by the formation of a trapped surface within the spatial domain of the experiment. The argument based on Thorne's hoop conjecture, conjointly leads to a fundamental length scale in physics. Black hole dominance also suggests that ordinary field theory cannot be used to describe quantum gravity in the extreme UV, contrary to implications of asymptotic safety. We re-examine black hole dominance in an asymptotically safe scenario, in the presence of higher curvature terms an with running couplings, by modifying a proof of Thorne's hoop conjecture. We find that the proof falls apart, and along with it, so does the argument for a mandatory formation of a trapped surface inside the domain of the experiment. However, neither is there a contrary proof that local trapped surfaces do not form. Instead in this approach whether an observer can perform local measurements in arbitrary small regions of spacetime depends on the specific values of the couplings near the UV fixed point. In this sense there is no all pervasive *local* version of the minimal length argument. However, we argue that one trapped surface must still form outside an experiment, when the domain of this experiment is localized to scales much smaller than the Planck length. This enshrouding horizon then prevents any information from reaching observers at infinity, thus retaining a vestige of "asymptotic darkness" for them.

1. Introduction

Black hole dominance (or asymptotic darkness), of the spectrum of particle states near the Planck scale is often invoked to argue against the non-renormalizability of gravity. A renormalizable field theory at a certain scale is a perturbation of a CFT by relevant operators with finite couplings. But In d-dimensions, black hole thermodynamics would yield an energy scaling $\rho_{\rm BH} = e^{E^{\frac{d-2}{d-1}}}$, in black hole sates, while the density of states of a CFT scales as $\rho_{\rm CFT} = e^{E^{\frac{d-1}{d}}}$ [1]. Strictly speaking proving the conjecture requires a quantum theory of gravity. However, the argument rests on Thorne's hoop conjecture (HC) [2] , and quantum mechanics: If an amount of energy or matter Es compressed into a region of characteristic size L, a trapped surface of size L forms if $L < R_S = 2GE$. If L scales according to $L \propto 1/E$, then a trapped surface forms for any experiment at energies $E > E_{Planck}$. The HC is unproven. However it finds strong support from numerical work on particle scattering [3] which vindicates the Aichelburg-Sexl result [4], and other work on Planck scale scattering [5, 6].

Asymptotic darkness also implies that the Planck length is the minimum observable length [7] for *any* observer. Consider simple General Relativity and a static, spherically symmetric experiment. The solution at radii $r > L$ is Schwarzschild, which has trapped surfaces at every radius from L out to $2GL^{-1}$. Therefore there is no observer at radius L to infinity that could receive any signal from the experiment once it is scaled down past L_{Planck}, which becomes a fundamental measurement cutoff for any observer. Thus, using HC, one can argue for black hole dominance, and the appearance of a minimal length. We query, first, if this continues to be true in asymptotic safety scenarios of quantum gravity, where gravity *is* supposed to be a field theory based upon the metric all the way to the extreme UV. Skeptics object that the hoop conjecture is an artefact of IR gravity, and hence black holes may not form at all in the UV. But notice that if one ramps the energy scale of an experiment, so that $E \gg E_{\text{Planck}}$, a large black hole will eventually form; large black holes have low horizon curvature scales, and therefore IR gravity should be a perfectly valid description. Hence, momentarily postponing the debate of what happens in the far UV, within asymptotic safety, compatibilty with GR, must be restored in the low energy limit, for a similar experiment. The UV/IR connection implied by the formation of large black holes, has long been discussed as a distinctive feature of high scale quantum gravity [8, 9], and recently led to ideas of UV "classicalization" of non-renormalizable theories [10].

We now revisit and parallel the usual minimum length arguments by incorporating one feature, that of running couplings implied by asymptotic safety, into a proof due to Bizon et al [11] for a minimum length under certain conditions. We frequently refer to the authors' research article [12] for certain details we have omitted here.

2. Hoop Theorem for Spherical Symmetry

Black holes can form in myriad ways, so black hole dominance is a rather vague statement. The hoop conjecture, implies that a black hole will form, regardless of the details, if we concentrate enough matter into a small volume of space. So to be concrete, we choose a specific bahevior for gravity, spacetime toplogy, and matter distribution as follows **1.** We assume gravity is asymptotically safe, **2.** the spacetime topology allows a foliation by spacelike surfaces, Σ_t, labeled by a time function t, and outward normal n^a. **3.** A spherical ball of matter Ω, of *proper* radius L and density ρ on a maximal slice Σ_0 with boundary $\partial\Omega$ and spacelike unit normal r^a $(r \cdot n = 0)$ **4.** The system is momentarily stationary. The reason for these choces are the following. First the event horizon is a global object, and its sharp location in quantum gravity is *prima facie* suspect. This motivates reformulating black hole thermodynamics, and the hoop conjecure in terms of local trapping/apparent horizons [11, 13, 14, 15, 16]. Spherical symmetry and stationarity are used for simplicity, but note that the momentarily stationary assumption intuitively matches the idea of scaling a static system down in some frame.

The proof uses the Hamiltonian constraint of General Relativity, which for our purposes is written as

$$^3R + K^2 - K_{ab}K^{ab} = \kappa_0{}^{-1}\left(H_\kappa + \frac{1}{2}\rho\right), \quad \rho = T_m^{ab} n_a n_b;\ \kappa_0{}^{-1} = 16\pi G. \tag{1}$$

Here H_κ denotes the contribution from the higher curvature terms, which appear in a general truncation of the 4d effective action for gravity *á la* asymptotic safety [20, 21, 22]

$$I = \int d^4x\sqrt{-g}\big(-\Lambda(\mu) + \kappa_0(\mu)R + \kappa_1(\mu)R^2 + \kappa_2(\mu)R_{ab}R^{ab} + L_m \dots\big). \tag{2}$$

The coupling constants, κ_I ($I = 1, 2, 3, \ldots$) of the higher derivative terms in eq. 2 are implicitly absorbed in H_κ above. One could question the use of the classical Hamiltonian constraint in quantum gravity, but we underscore that we are asking how asymptotic safety modifies the *standard* minimum length argument, which is based on just such a semiclassical picture, and as a first attempt, we apply the predictions of Asymptotic Safety to an otherwise mathematically precise argument.

To proceed we choose a standard gauge adapted to spherical symmetry for the spatial metric on Σ_0, namely $q_{ab} = \varphi^4 \delta_{ab}$. This in turn implies that the 3-Ricci scalar, ${}^3R = -8\varphi^{-5}\nabla^2\varphi$. Further for a slice where everything is momentarily stationary, the extrinsic curvature $K_{ab} = 0$; These restrictions on the canonical data (q_{ab}, K^{ab}) simplifies the constraint equation 1 to

$$-8\varphi^{-5}\nabla^2\varphi = {\kappa_0}^{-1}\left(H_\kappa + \frac{1}{2}\rho\right) \tag{3}$$

This implies that $\nabla^2\varphi \leq 0$ (or $\frac{d\varphi}{dr} \leq 0$) if the higher curvature satisfy the positive energy condition as well as the matter contribution [17, 18, 19]. Now, any 2-sphere inside Ω is *not* a trapped surface if the expansion $\theta = r^a_{;a} > 0$. with $r^a{}_{;a} = (r^2\varphi^6)^{-1}\frac{d(r^2\varphi^4)}{dr}$, this means

$$\theta_r = \partial_r\left(r\varphi^2\right) > 0\;, \tag{4}$$

Integrating eq. 3 over Ω and simplifying for spherical symmetry obtains

$$-32\pi\int_0^{r_0} dr\varphi\partial_r(r^2\partial_r\varphi) = \frac{1}{2\kappa_0}E + \kappa_0^{-1}\int_\Omega dv H_\kappa. \tag{5}$$

where $E = \int\sqrt{q}d^3xT_m^{ab}n_an_b$ is the proper energy of matter inside Ω. A partial integration of the term on the left hand side then yields

$$16\pi L - 16\pi r\partial_r(r\varphi^2)\Big|_{d\Omega} + 16\pi\int dr\left(r\varphi\partial_r\varphi + \varphi^{-1}r(\partial_r\varphi)\partial_r(r\varphi^2)\right) = \kappa_0^{-1}\left(\frac{E}{2} + \int_\Omega dv H_\kappa\right). \tag{6}$$

where $L = \int_0^{r_0}\varphi^2 dr$ is the proper radius of Ω. If there are no trapped surfaces in Ω and $\partial\Omega$ is not a trapped surface, then using $\frac{d\varphi}{dr} \leq 0$ and $\theta_r > 0$, as above, all terms except the first on the left side of eq. 6 are negative. Therefore we have

$$16\pi L > \kappa_0^{-1}\left(\frac{E}{2} + \int_\Omega dv H_\kappa\right). \tag{7}$$

as the condition for*no* trapped surfaces in Ω.

For the case of pure General Relativity, i.e with $H_\kappa = 0$ and no running couplings, eq. 7 implies that as we increase E and therefore decrease L as $L \sim \frac{1}{E}$, a trapped surface must eventually form insideΩ, as the left hand side decreases and the right side increases, thereby violating the inequality at some point. This happens when $E = 32\pi\kappa_0 L \approx 32\pi\kappa_0 E^{-1}$, i.e at $E \approx \sqrt{\kappa_0} = E_{\rm Pl}$.

2.1. Implications of Asymptotic Safety

The reader is referred to Sec. IIIA of Ref.[12] for details on the behavior of running coupling constants. We will here summarize the essential point. In the far UV the behavior of the running Newton constant is dominated by the Non-Gaussian fixed point g_* and given by [23]

$$\frac{1}{16\pi\kappa_0(E)} = G(E) \simeq \frac{g_*}{E^2} \tag{8}$$

Momentarily confining ourselves to the Einstein Hilbert+matter truncation, we find from eq. 7 that there are no trapped surfaces in Ω if

$$L > \frac{g_*}{2E^2}E \tag{9}$$

Using the scaling $L \sim \frac{1}{E}$, factors of E drop out from both sides, and gives

$$g_* < 2 \tag{10}$$

for no trapped surfaces to be found in Ω. This shows that upto $O(1)$ numbers arising from order of magnitude estimates/scaling arguments etc, there is some value of coefficients for which the standard minimum length arguments fail. Interestingly, g_* in many truncations have been found to be close to 2. We quote here the values obtained by Benedetti et al [24], who consider truncations with minimally coupled scalar field and find

$$g_* = 0.860 \quad \text{(For Einstein-Hilbert truncation)} \tag{11}$$

$$g_* = 2.279 \quad \text{(For } R^2 + C^2 \text{ truncation)} \tag{12}$$

Consider now the impact of higher curvature terms with running couplings on our argument. Returning to eq. (7) we find that no trapped surfaces form inside Ω if

$$L > \frac{g_*}{E^2}\left(\frac{E}{2} + \int dv H_\kappa\right) \tag{13}$$

With L replaced by E^{-1} eq. (13) becomes

$$E^{-1} > \frac{g_*}{E^2}\left(\frac{E}{2} + \int H_\kappa dv\right) \tag{14}$$

We now observe that the integral involving the higher curvature terms is dimensionally an energy and so the integral must be a combination of E multiplied by the dimensionless coefficients $\bar{\kappa}_I$. We therefore have,

$$E^{-1} > g_* \left((2E)^{-1} + E^{-1}\bar{I}(\bar{\kappa}_I)\right) \tag{15}$$

where $\bar{I}(\bar{\kappa}_I)$ is a dimensionless number whose details depend on both the unknown κ_I and the precise solution for the metric in the presence of thehigher curvature terms with our given stress tensor. It suffices for us that it is a linear function of $\bar{\kappa}_I$. In asymptotic safety the $\bar{\kappa}_I$ coefficients remain finite as $E \to \infty$, and hence I is also finite. Therefore as $E \to \infty$ there is no proof that a trapped surface is formed, as each side scales the same way with E, and so the argument for a minimum length no longer necessarily holds (at least in this formulation). Whether a trapped surface can still form is sensitive to truncation details, coefficients and the actual value of g_*.

3. The Infrared Limit and General Relativity

While the question of whether or not a trapped surface forms inside Ω for Planck sized experiments becomes muddled with asymptotic safety, in the limit of energy $E \gg E_{Pl}$ there still must be a trapped surface at large radius, as this corresponds to a large black hole with low horizon curvature. To reach this spacetime picture, we can analyze the "horizon" structure which emerges by considering a background distribution of matter withh total energy E_B in Ω, and ask when a probe experiment at a length scale L_E (or equivalent energy scale E_E) encounters a trapped surface. Since we are away from the fixed point dominated regime we now use a specific form of $G(E)$ [25, 26],

$$G(E) = \frac{g_*}{E^2 + g_* M_{Pl}^2} \tag{16}$$

return to equation (7), but modify the scenario. We define

$$F(E_E) = L_E - G(E_E)\left(\frac{E_E + E_B}{2} + \int dv H_\kappa\right) \tag{17}$$

.

If $F(E) > 0$, then there is no trapped surface in the region. Assume that g_* is such that there are no trapped suface in Ω as $E_E \to \infty$.Given this, as we take E_E towards the IR eventually $F(E_E)$ may cross zerosignifying that at some radius L_E, there is a trapped surface. As we move towards even lower energies, F may cross zero again. Hence there can be in generel multiple horizons. We can locate these horizons by setting $F(E) = 0$. A straightforward algebra [12], then gives, in the case of $E_B \gg E_{\mathrm{pl}}$, two roots,

$$L_{E+} = \frac{E_B J_B}{M_{Pl}^2} \tag{18}$$

$$L_{E-} = \frac{1 - g_* J_E}{g_* E_B J_B}. \tag{19}$$

where $J_E = 1/2 + I_E(\bar{\kappa}_I)$ and $J_B = 1/2 + I_B(\bar{\kappa}_I)$, encapsulate the effect of the higher curvature terms. L_{E+} is $R \approx GE_B$, i.e. it corresponds to the Schwarzschild radius for a large black hole. To get an intuitive picture, set the higher curvature terms to zero. In that case $J_E = J_B = \frac{1}{2}$, the inner horizon location becomes $L_{E-} = (2 - g_*)/(g_* E_B)$. In the previous section, for just one experiment, our limit was roughly $g_* < 2$ to avoid the minimum length argument. Here we have the additional background energy E_B in addition to E_E, so we must further decrease g_* to avoid a trapped surface forming in the domain of an experiment. In other words, if we chose $E_E = E_B$ for example, then there is twice the energy density as we had in the previous section in a region of size $L_E = L_B$ so g_* would need to be less than one to prevent formation of a trapped surface. If we consider $g_* < 1$, such that this bound is satisfied, then we see that the inner horizon is reached when $L_E \approx 2/(g_* E_B)$. Thus the following picture emerges: We have a background energy density $E_B >> M_{Pl}$ in a region of radius $L_B = 1/E_B$. We probe the system at length scales smaller than L_B and we see no trapped surface. As we approach the length scale $(2 - g_*)L_B/g_*$ we see a trapped surface form. This behavior persists until we reach the large radius GE_B, at which point we return to semiclassical physics outside an ordinary Schwarzschild horizon. Thus in this picture, local observers performning local measurements see no trapped surfaces, but they cannot transmit any measurements to asymptotic observers, for whom, there is still a minimum length.

In sum therefore, our findings indicate that within asymptotic safety, running of couplings conspire such that the necessity of forming a trapped surface at the scale of a high energy experiment vanishes. Whether trapped surfaces form is a matter of delicate details about truncations and values of fixed points. Hence, it is of general interest to explore such issues with realistic matter content. Significantly compatibilty with General Relativity in the IR can be restored. The difference between local measurements and asymptotic information which emerges in this discussion warrants caution about blanket statements regarding the status of minimum length within Asymptotic Safety.

References

[1] Shomer A 2007 *Preprint* 0709.3555 [hep-th]
[2] Thorne K S 1972 *Magic Without Magic* ed J Klauder (W. H. Freeman, San Francisco, 1972).
[3] Choptuik M, Pretorius F 2010 *Phys. Rev. Lett* **104** 111101 (*Preprint* 0908.1780 [gr-qc])
[4] Aichelburg P, Sexl R U 1971 *Gen. Rel. Grav* **2** 303
[5] Giddings S B, Srednicki M 2008 *Phys. Rev.* D **77** 085025
[6] Banks T, Fischler W 1999 *Preprint* hep-th/9906038

[7] Calmet X, Graesser M and Hsu S D H 2005 *Int. J. Mod. Phys.* D **14** 2195
[8] Aharony O, Banks T 1999 *Journal of High Energy Physics*JHEP, 9903(1999)016
[9] Dvali G 2011 *Phys. Rev* D **84** (2011)
[10] Dvali G 2011 *Journal of High Energy Physics* JHEP1108(2011)108
[11] Bizon P, Malec E and O'Murchadha N 1988 *Phys. Rev. Lett.* **61** 1147
[12] Basu S, Mattingly D 2010 *Phys. Rev.* D **82** 124017
[13] Ashtekar A, Krishnan B 2004 *Living Rev.Rel.* **7** 10 *eprint* gr-qc/0407042
[14] Ashtekar A, Beetle C and Fairhurst S 1999 *Class.Quant.Grav.* **16**:L1-L7 *eprint* gr-qc/9812065
[15] Flanagan E 1992 *Phys. Rev.* **D** 15 1429
[16] Barrabés C et al 2004 *Phys. Rev.* **D** 69 19501(R)
Yoshino H, Nambu Y 2003 *Phys. Rev.* **D** 67 024009
Chiba T et al 1994 *Class. Quant. Grav.* **11** 431
[17] Strominger A *Phys. Rev.* D **30**, 2257 (1984)
[18] Deser S, Tekin B *Phys. Rev.* D **67**, 084009 (2003)
[19] Deser S, Tekin B *Phys. Rev.* D **75** 084032(2007)
[20] Weinberg S 1979 *Genneral Relativity: An Einstein Centenary Survey* ed S W Hawking and W. Israel (Cambridge University Press, Cambridge 1979)
[21] Weinberg S 2009 *eprint*0908.1964 [hep-th]
[22] Neidermaier M, Reuter M 2006, *Living. Rev. Relativity* **9** 5
[23] Narain G, R. Percacci 2009 *Acta Phys. Polon* B **40** 3439 *eprint* 0910. 5390 [hep-th].
[24] Benedetti D, Machado P and Sauressig F 2010 *Nucl. Phys.* B **824** 168
[25] Bonanno A, Reuter M 2000 *Phys.Rev.*D **62** 043008
[26] Hewett J, Rizzo *Journal of High Energy Physics*JHEP, 0712(2007)009

Journal of Physics: Conference Series **410** (2013) 012136 doi:10.1088/1742-6596/410/1/012136

A Metric for Testing the Nature of Black Holes

Tim Johannsen[1]

Department of Physics and Astronomy, University of Waterloo, 200 University Avenue West, Waterloo, ON, N2L 3G1, Canada
Perimeter Institute for Theoretical Physics, 31 Caroline Street North, Waterloo, ON, N2L 2Y5, Canada
Physics Department, University of Arizona, 1118 East 4th Street, Tucson, AZ 85721, USA

Abstract.
In general relativity, astrophysical black holes are uniquely described by the Kerr metric. Observational tests of the Kerr nature of these compact objects and, hence, of general relativity, require a metric that encompasses a broader class of black holes as possible alternatives to the usual Kerr black holes. Several such Kerr-like metrics have been constructed to date, which depend on a set of free parameters and which reduce smoothly to the Kerr metric if all deviations vanish. Many of these metrics, however, are valid only for small values of the spin or small perturbations of the Kerr metric or contain regions of space where they are unphysical hampering their ability to properly model the accretions flows of black holes. In this paper, I describe a Kerr-like black hole metric that is regular everywhere outside of the event horizon for black holes with arbitrary spins even for large deviations from the Kerr metric. This metric, therefore, provides an ideal framework for tests of the nature of black holes with observations of the emission from their accretion flows, and I give several examples of such tests across the electromagnetic spectrum with current and near-future instruments.

1. Introduction

By virtue of the no-hair theorem, black holes in general relativity are fully and uniquely characterized by only three parameters: their masses, spins, and electrical charges (Israel 1967, 1968; Carter 1971, 1973; Hawking 1972; Robinson 1975; Mazur 1982). These objects are described by the Kerr-Newman metric (Newman et al. 1965), which, in the case of astrophysical black holes, reduces to the Kerr metric (Kerr 1963), since any net charge quickly neutralizes.

The no-hair theorem, then, provides a means to test general relativity by verifying that black holes are indeed Kerr black holes. Several such tests have been suggested to date, which are based on observations of either gravitational waves (see Hughes 2010 for a review), electromagnetic emission from accretion flows (Johannsen & Psaltis 2010a), pulsar-black hole binaries (Wex & Kopeikin 1999), the ephemeredes of stars around the supermassive black hole Sgr A* at the galactic center (Will 2008), or of the quasar OJ287 (Valtonen et al. 2011).

While tests in the regime of weak gravitational fields can rely on a parameterized post-Newtonian formulation (e.g., Will 1993), strong-field tests require a careful modeling of the black hole spacetime itself in terms of a parametric deviation from the Kerr metric. These Kerr-like spacetimes depend on one or more free parameters in addition to the mass and spin, and signatures of possible violations of the no-hair theorem can be studied as a function of the

[1] CITA National Fellow

deviation parameters. Several parametric deviations have been constructed so far (e.g., Manko & Novikov 1992; Collins & Hughes 2004; Glampedakis & Babak 2006; Yunes & Pretorius 2009; Vigeland & Hughes 2010; Vigeland, Yunes, & Stein 2011; Johannsen & Psaltis 2011b; Yagi, Yunes & Tanaka 2012).

In this paper, I review the Kerr-like black hole spacetime constructed by Johannsen & Psaltis (2011b), which is regular everywhere outside of the event horizon even for high values of the spin and large deviations from the Kerr metric. This property makes this metric an ideal framework for strong-field tests the no-hair theorem with observations of the accretion flows of black holes.

2. A Suitable Metric for Tests of the No-Hair Theorem with Observations of Black Hole Accretion Flows

The Kerr metric is a stationary, axisymmetric, asymptotically flat, vacuum solution of the Einstein field equations and, due to the no-hair theorem, the only spacetime in general relativity that has all of these characteristics. Kerr-like metrics retain as many of these properties in order to closely mimic the observational appearance of Kerr black holes. All of them include the Kerr metric as the limiting case if the deviation parameters are set to zero.

By construction, parametric deviations from the Kerr metric are usually stationary, axisymmetric, and asymptotically flat. If such a metric is also a vacuum solution in general relativity, it either harbors a naked singularity or is plagued with pathological regions in the exterior domain where causality is violated (see Johannsen et al. 2012a). Some Kerr-like metrics are valid only for small or intermediate values of the black hole spin (e.g., Glampedakis & Babak 2006; Yunes & Pretorius 2009), while others are based on an expansion in the deviation parameters in order to remain a vacuum solution of the Einstein equations under certain conditions (e.g., Vigeland & Hughes 2010). Depending on the desired application, additional metric properties can be important, such as its Petrov type (Vigeland et al. 2011) or violations of parity (Yunes & Pretorius 2009; Yagi et al. 2012).

The metric by Johannsen & Psaltis (2011b) in Boyer-Lindquist-like coordinates is given by the line element

$$\begin{aligned} ds^2 &= -[1+h(r,\theta)]\left(1-\frac{2Mr}{\Sigma}\right)dt^2 - \frac{4aMr\sin^2\theta}{\Sigma}[1+h(r,\theta)]dtd\phi + \frac{\Sigma[1+h(r,\theta)]}{\Delta + a^2\sin^2\theta h(r,\theta)}dr^2 \\ &\quad +\Sigma d\theta^2 + \left[\sin^2\theta\left(r^2+a^2+\frac{2a^2Mr\sin^2\theta}{\Sigma}\right) + h(r,\theta)\frac{a^2(\Sigma+2Mr)\sin^4\theta}{\Sigma}\right]d\phi^2, \end{aligned} \tag{1}$$

where

$$\Sigma \equiv r^2 + a^2\cos^2\theta, \quad \Delta \equiv r^2 - 2Mr + a^2, \quad h(r,\theta) \equiv \sum_{k=0}^{\infty}\left(\epsilon_{2k} + \epsilon_{2k+1}\frac{Mr}{\Sigma}\right)\left(\frac{M^2}{\Sigma}\right)^k. \tag{2}$$

This metric is a stationary, axisymmetric, and asymptotically flat parametric deviation that describes an actual black hole for all values within the range of spins $|a| \leq M$ without relying on an expansion in the deviation parameters ϵ_k and that is regular, i.e., it is free of any pathologies outside of the event horizon. Possible nonzero lowest-order deviations ϵ_k, $k = 0, 1, 2$, of the Kerr metric are ruled out by current solar-system tests of gravity and by the requirement that the above metric reduces to Newtonian gravity far from the black hole. For simplicity, I will allow the leading-order coefficient ϵ_3 to be the only nonvanishing deviation parameter.

With this choice, the shape of the event horizon depends on the values of the spin and the parameter ϵ_3. Figure 1 shows the location of the horizon in the xz-plane of a black hole with a value of the spin $a = 0.75M$, where $x \equiv \sqrt{r^2+a^2}\sin\theta$ and $z \equiv r\cos\theta$. While the event horizon of a Kerr black hole is spherical, the horizon of a black hole in this metric is prolate for positive

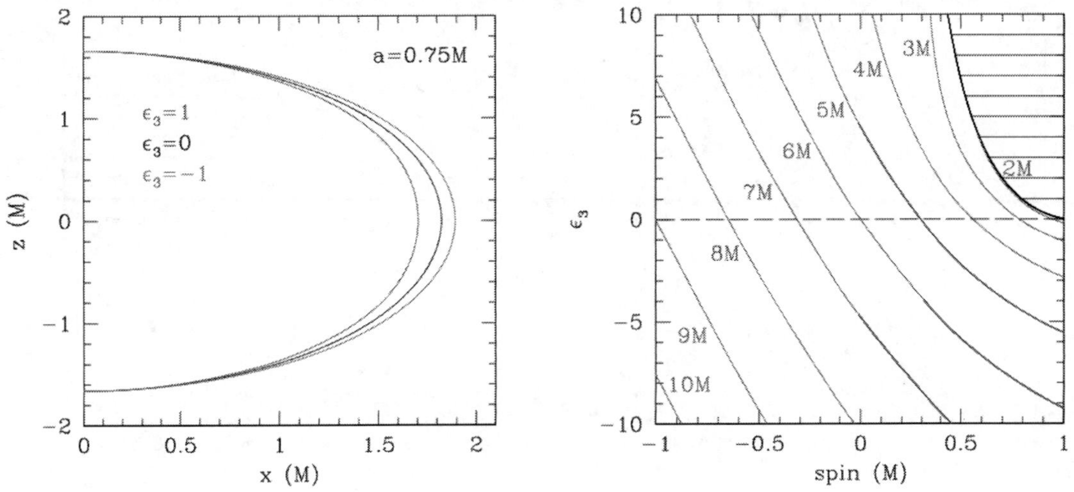

Figure 1. Left: Event horizon of a black hole with spin $a = 0.75M$ for several values of the parameter ϵ_3 that measures the deviation from the Kerr metric. For values of the parameter $\epsilon_3 > 0$ the shape of the event horizon is prolate, while for values of the parameter $\epsilon_3 < 0$ the shape of the event horizon is oblate. Right: Contours of constant radius of the ISCO for values of the spin $-1 \leq a/M \leq 1$ and of the deviation parameter $-10 \leq \epsilon_3 \leq 10$. The ISCO radius decreases for increasing values of the spin and the parameter ϵ_3. The shaded region marks the part of the parameter space where the event horizon is not entirely closed (only shown for the case $a > 0$; Johannsen & Psaltis 2011b).

values of the parameter ϵ_3 and oblate for negative values of the parameter ϵ_3. For high values of the spin and values of the parameter ϵ_3 exceeding a maximum bound as shown in Figure 1, the central object turns into a naked singularity.

The properties of this spacetime make it ideally suited for tests of the no-hair theorem that involve observations of accretion flows. The innermost stable circular orbit (ISCO) is often taken to be the inner edge of geometrically thin accretion disks and can be arbitrarily close to the event horizon in the case of rapidly spinning black holes. Since relativistic effects are most prominent in the immediate vicinity of black holes, it is critical that this region is properly modeled by a parametric deviation. Since the metric by Johannsen & Psaltis (2011b) is regular everywhere outside of the horizon, the electromagnetic emission of accretion flows can be modeled self-consistently. Figure 1 also shows the location of the ISCO as a function of the spin and the parameter ϵ_3. For increasing values of the spin and the deviation parameter, the radius of the ISCO approaches the event horizon.

3. Astrophysical Tests

Accretion flows of black holes allow us to observe potential violations of the no-hair theorem across the electromagnetic spectrum (Johannsen & Psaltis 2010a). Observables include (a) the direct imaging of the event horizon with Sgr A* (Johannsen & Psaltis 2010b; Johannsen 2012; Johannsen et al. 2012b) with the *Event Horizon Telescope*, a planned global very-long baseline interferometer (Doeleman et al. 2009a, b; Fish et al. 2009), (b) relativistically broadened iron lines (Johannsen & Psaltis 2012), (c) quasi-periodic variability (Johannsen & Psaltis 2011a;

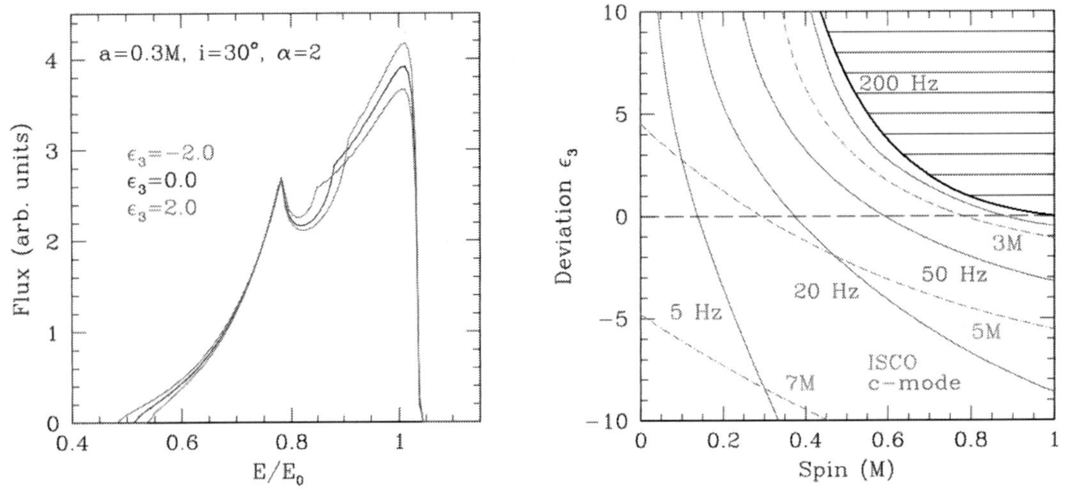

Figure 2. Left: Iron line profiles for a black hole with spin $a = 0.3M$, inclination $\theta_0 = 30°$ and emissivity index $\alpha = 2$ for several values of the parameter ϵ_3. Violations of the no-hair theorem manifest predominantly in the blue-shifted peak and at low energies. Right: C-mode frequency contours for a $10M_\odot$ black hole as a function of the spin and the parameter ϵ_3. These frequencies increase with increasing values of the spin and decreasing values of the parameter ϵ_3. Contours of constant ISCO radius (dashed lines) are shown for comparison (Johannsen & Psaltis 2012).

Bambi 2012a), (d) accretion disk spectra (Bambi & Barausse 2011; Krawczynski 2012), (e) X-ray polarization (Krawczynski 2012), and (f) jets (Bambi 2012b).

In Figure 2 I plot profiles of iron lines with the characteristic double-horn shape, which is produced by relativistic effects, for a black hole with a value of the spin $a = 0.3M$ and several values of the parameter ϵ_3. The accretion disk is assumed to be geometrically thin and optically thick with a radial extent ranging from the ISCO to $r = 15M$. The emission is isotropic with an emissivity $\propto r^{-\alpha}$, $\alpha = 2$, and the disk is viewed at an inclination angle of $i = 30°$. Violations of the no-hair theorem manifest as significant alterations of the line shape.

Figure 2 also shows frequency contours of so-called corrugation modes (c-modes; Silbergleit et al 2001) as a viable model of the quasi-periodic variability observed in both stellar-mass and supermassive black holes (e.g., Remillard & McClintock 2006). The effects of potential violations of the no-hair theorem on iron line profiles and quasi-periodic variability may be observed with upcoming or future X-ray missions such as *Astro-H*, *ATHENA* and *LOFT*.

Tests of the no-hair theorem and of general relativity that are based on observations of the accretion flows of black holes require a carefully designed parametric deviation from the Kerr spacetime. The metric by Johannsen & Psaltis (2011b) provides the necessary framework and is ideally suited for the description of rapidly-spinning black holes, which are mostly found in nature (Brenneman & Reynolds 2009; McClintock et al. 2011). Observational tests of the no-hair theorem can be realized in the near future.

References

[1] Bambi C and Barausse E 2011 *ApJ* **731** 121

[2] Bambi C 2012a, arXiv:1205.6348
[3] ———. 2012b *Phys. Rev. D* **85** 043002
[4] Brenneman L W and Reynolds C S 2009 *ApJ* **702** 1367
[5] Carter B 1971 *Phys. Rev. Lett* **26** 331
[6] ———. 1973 in *Black Holes* eds. DeWitt & DeWitt (New York: Gordon and Breach)
[7] Collins N A and Hughes S A 2004 *Phys. Rev. D* **69** 124022
[8] Doeleman S S, Fish V L, Broderick A E and Loeb A 2009a *ApJ* **695** 59
[9] Doeleman S S et al. 2009b *Astro2010: The Astronomy and Astrophysics Decadal Survey*, Science White Papers, **68**
[10] Fish V L, Broderick A E, Doeleman S S and Loeb A 2009 *ApJ* **692** L14
[11] Glampedakis K and Babak S 2006 *Class. Quantum Grav.* **23** 4167
[12] Hawking S W 1972 *Commun. Math. Phys.* **25** 152
[13] Hughes S A 2010, arXiv:1002.2591
[14] Israel W 1967 *Phys. Rev.* **164** 1776
[15] ———. 1968 *Commun. Math. Phys.* **8** 245
[16] Johannsen T and Psaltis D 2010a *ApJ* **716** 187
[17] ———. 2010b *ApJ* **718** 446
[18] ———. 2011a *ApJ* **726** 11
[19] ———. 2011b *Phys. Rev. D* **83** 124015
[20] ———. 2012, arXiv:1202.6069
[21] Johannsen T 2012 *Adv. Astron.* **2012** 1
[22] Johannsen T et al. 2012a, in preparation
[23] Johannsen T, Psaltis D, Gillessen S, Marrone D P, Özel F, Doeleman S S and Fish V L 2012b, arXiv:1201.0758
[24] Kerr R P 1963 *Phys. Rev. Lett.* **11** 237
[25] Krawczynski H 2012, *ApJ* **754**, 133
[26] Manko V S and Novikov I D 1992 *Class. Quantum Grav.* **9** 2477
[27] Mazur P O *J. Phys. A* **15** 3173
[28] McClintock J E, Narayan R, Davis S W, Gou L, Kulkarni A, Orosz J A, Penna R F, Remillard R A and Steiner J F *Class. Quantum Grav.* **28** 4009
[29] Newman E T, Couch K, Chinnapared K, Exton A, Prakash A and Torrence R 1965 *J. Math. Phys.* **6** 918
[30] Remillard R A and McClintock J E 2006 *ARA&A* **44** 49
[31] Robinson D C 1972 *Phys. Rev. Lett.* **34** 905
[32] Silbergleit A S, Wagoner R V and Ortega-Rodríguez M 2001 *ApJ* **548** 335
[33] Valtonen M J, Mikkola S, Lehto H J, Gopakumar A, Hudec R and Polednikova J 2011 *ApJ* **742** 22
[34] Vigeland S J and Hughes S A 2010 *Phys. Rev. D* **81** 024030
[35] Vigeland S J, Yunes, N and Stein L C 2011 *Phys. Rev. D* **83** 104027
[36] Wex N and Kopeikin S M 1999 *ApJ* **514** 388
[37] Will C M 1993 *Theory and Experiment in Gravitational Physics* (Cambridge: Cambridge University Press)
[38] ———. 2008 *ApJ* **674** L25
[39] Yagi K, Yunes N and Tanaka T 2012 arXiv:1206.6130
[40] Yunes N and Pretorius F 2009 *Phys. Rev. D* **79** 084043

Near horizon geometry, Brick wall model and the Entropy of a scalar field in the Reissner-Nordstrom black hole backgrounds

Kaushik Ghosh[1]
Department of Physics, St. Xavier's College, 30, Mother Teresa Sarani, Kolkata-700016, India.

Abstract.
In this article we will find the entropy of a scalar field in the Reissner-Nordstrom black hole backgrounds using the brick wall model of t' Hooft. We will use the semi-classical WKB approximation. We will consider the modes which are globally stationary so that the WKB quantization rule used in the brick wall model remains valid. In the Schwarzschild black hole this consideration had led to a new expression of the entropy different from the conventional expression which is inversely divergent in the brick wall cut-off parameter and in terms of a proper distance cut-off parameter, is proportional to the area of the event horizon. The new expression of the scalar field entropy obtained in this article is logarithmically divergent in the brick wall cut-off parameter and is not proportional to the area of the black hole event horizon. For the extremal Reissner-Nordstrom black hole background the entropy of the scalar field is again divergent in the brick wall cut-off parameter and vanishes if the temperature of the Hawking radiation and the black hole is taken to be zero. We will next consider the entropy for a thin shell of matter field surrounding the black hole horizon. When expressed in terms of a covariant cut-off parameter, the entropy of a thin shell of matter field surrounding the horizon in the non-extreme Reissner-Nordstrom black hole background is given by an expression proportional to the area of the black hole horizon. We will briefly explain the significance of this result.

1. Introduction

The thermodynamical aspects of the black holes was first established by Bekenstein [1]. He obtained an expression for the entropy of the black holes. This expression was proportional to the area of the horizons of the black holes. Hawking obtained the exact expression for the entropy by considering the behaviour of matter fields in the black hole backgrounds [2]. The entropy of a black hole, considered as a thermodynamical system, was found out to be $\frac{A}{4}$. Here A is the horizon surface area. The thermodynamical properties of a minimally coupled matter fields in black hole backgrounds was discussed by 't Hooft [3]. The black hole is assumed to be in thermal equilibrium with surrounding matter. For a static black hole the rate of particles radiated by the black hole should be equal to the rate of absorption of matter by the black hole. 't Hooft [3] assumed that the matter field wave function is vanishing near the horizon as well as at infinity, i.e, for a large value of the radial coordinate. Since then a lot of works had been done on the entropy of matter fields in the black hole backgrounds using the brick-wall model [4,5,6,7,8,9,10,11,12 13, 14].

In a previous article [15] we had discussed a few aspects of the entropy of a scalar field in the Schwarzschild black hole background using the brick-wall model of 't Hooft [3] with a proper consideration of the centrifugal potential. In the brick wall model of 't Hooft the semi-classical

[1] E-address: ghosh_kaushik06@yahoo.co.in

WKB quantization rule is applied to the radial part of the matter field solution for the purpose of the counting of the states. We considered the fact that the modes for which the WKB quantization rule, used in the brick wall model, remains to be valid should be stationary throughout the range of the radial variable. We found that, with this consideration, the entropy of a scalar field is different than that obtained in [3]. The expression of entropy obtained by 't Hooft is inversely divergent in the brick wall cut-off parameter and in terms of a proper distance cut-off parameter, is proportional to the area of the event horizon. The new expression of the scalar field entropy obtained in [15] is logarithmically divergent in the brick wall cut-off parameter and is not proportional to the area of the black hole event horizon. We also note that the term logarithmically divergent in the brick wall cut-off parameter is independent of the mass of the black hole although the leading order infrared divergent term is dependent on the mass of the black hole.

In the present article we will find the entropy of a scalar field in the Reissner-Nordstrom and the extremal Reissner-Nordstrom black hole background. We will use the brick wall model of t' Hooft together with the counting of the states followed in [15]. The expression for the entropy of a massless scalar field, obtained in this article, in the non-extreme Reissner-Nordstrom black hole background is similar to that of the Schwarzschild black hole and is logarithmically divergent in the brick wall cut-off parameter. For the extremal Reissner-Nordstrom black hole background the entropy of the scalar field is again divergent in the brick wall cut-off parameter. In this case the divergent part contains a term inversely dependent on the brick wall cut-off parameter and a term logarithmically dependent on the same. The matter field entropy vanishes if the temperature of the Hawking radiation and the black hole is taken to be zero.

In section:IV we will consider the entropy of a thin shell of massless scalar field surrounding the black hole event horizon [3]. The thickness of the thin shell is taken to be the same as the brick wall cut-off parameter. In the flat space-time the entropy of a thin shell of scalar field, obtained by using the WKB approximation, is proportional to the volume of the thin shell in the limit the thickness is small compared to the radius of the inner surface. The internal energy and the entropy vanish when the thickness vanishes. In the black hole backgrounds we will find that the entropy of a thin shell of massless scalar field surrounding the horizon following the procedure of section:III. We will find that, in terms of a proper distance cut-off parameter, the scalar field entropy for a thin shell surrounding the black hole horizon is proportional to the area of the black hole event horizon. We will find that the free energy and the entropy are inversely dependent on the coordinate cut-off parameter and is more strongly divergent in comparison to the earlier case of a scalar field occupying the complete manifold. This is expected because for a thin shell of matter field surrounding the horizon, the upper limit of the angular quantum number is restricted by eqn (16) and when the thickness of the matter shell is small compared to the radius of the horizon, a greater number of modes contribute to the free energy and the entropy. This is associated with the vanishing of the centrifugal potential at the horizon. Similar situation will remain valid with the Schwarzschild black hole [24]. It is remarkable that the entropy is given by a universal expression *proportional to the area of the horizon*. The expressions of the matter field entropy obtained in this section are important in the context of explaining the black hole entropy in terms of the near horizon states [20] and also for the entanglement entropy approach to explain the black hole entropy [21].

2. Brick-wall model and the Entropy of a scalar field in the Reissner-Nordstrom black hole background

The radial part of the wave equation of a massive scalar field in the Schwarzschild black hole background is given by the following expression:

$$(1-\frac{2M}{r})^{-1}E^2\psi(r)+\frac{1}{r^2}\partial_r[r(r-2M)\partial_r\psi(r)]-[\frac{l(l+1)}{r^2}+m^2]\psi(r)=0 \qquad (1)$$

Here l is the angular momentum quantum number and E is the energy.

We follow the procedure of t' Hooft to calculate the entropy of a scalar field using the brick wall model and the WKB quantization rule [3]. We consider massless fields but the discussions can also be extended for massive fields. The boundary conditions on the scalar field are the following:

$$\psi(2M+h), \psi(L) = 0 \tag{2}$$

where $h << 2M$ and $L >> 2M$. Here h is the brick-wall cut-off parameter. We consider the solutions of the wave-equation obtained by the WKB approximation. The solutions are assumed to be of the form $\rho(r)\exp[iS(r)]$. The solutions are stationary w.r.t the radial variable throughout the spatial manifold and the amplitude is assumed to be a slowly varying function of the radial coordinate r. These solutions are also important for the analysis of the Hawking radiation.

We now consider the free-energy calculation using a proper density of states. Let us first consider the WKB quantization condition together with the brick wall boundary condition. We have the following condition:

$$\pi n = \int_{r_1}^{L} dr k(r,l,E) \tag{3}$$

Here $r_1 = 2M + h$ and $k(r,l,E)$ is the radial wave number. We assume that for a given set of values for E, l, r_1 and L there exist one solution of the above integral equation. The upper value of l is restricted by the condition that $k(r)$ should remain real throughout the complete region of integration. We have,

$$k(r) = \frac{1}{V(r)}\sqrt{E^2 - \frac{V(r)}{r^2}l(l+1)} \tag{4}$$

Here we are considering a massless scalar field. At $r = 2M+h$ with $h \to 0$, l can be varied freely. However for a state for which $k(r,E,l)$ is real throughout the complete region from $r = 2M+h$ to $r = L$ the upper limit of l is restricted by the maximum value of $\frac{V(r)}{r^2}$ which occurs at $r = 3M$ for the Schwarzschild black hole. The upper limit of l is given by the following expression,

$$l(l+1) = 27E^2M^2 \tag{5}$$

The expression for the free energy of a massless scalar field is given by the following expression [17]:

$$\pi\beta F = -\beta\int_0^\infty \frac{dE}{[\exp(\beta E) - 1]}\int_{2M+h}^{L}\frac{dr}{V(r)}\int dl(2l+1)\sqrt{E^2 - \frac{V(r)}{r^2}[l(l+1)]} \tag{6}$$

Here $V(r) = 1 - \frac{2M}{r}$. The range of E integration is from zero to infinity although one should be considering the back reaction to the metric. The range of r integration is from $2M+h$ to L where $L >> 2M$. The l integration in eqn.(6) for the free energy introduces a factor $\frac{r^2}{V(r)}$ in the radial integral. We can take the upper limit of l integration to be $l(l+1) = s(27E^2M^2)$, where s can be taken to be arbitrarily close to one but not one so that $k(r)$ is real for all values of r. The product $s\frac{V(r)}{r^2}(27E^2M^2)$ is less than E^2 for all values of r. We first do the l -integration. The free energy is given by the following expression:

$$\pi\beta F = -\frac{2}{3}\beta\int\frac{dE}{[\exp(\beta E) - 1]}\int\frac{r^2dr}{V^2(r)}[E^3 - \{E^2 - s(27E^2M^2)\frac{V(r)}{r^2}\}^{\frac{3}{2}}] \tag{7}$$

We can do a binomial expansion in s to find out the leading divergent term in the brick-wall cut-off parameter. The maximum value of the term $27M^2\frac{V(r)}{r^2}$ is one. We thereafter do the radial integration and the energy integration respectively. It can easily be seen the term associated with the linear order term in s is the most divergent term in the brick-wall cut-off parameter h. The entropy is given by $S = \beta^2\frac{dF}{d\beta}$. The entropy associated with the linear order term in $'s'$ is given by,

$$S = \frac{27sI}{128\pi^4}[\frac{L}{M} + 2\ln(\frac{L}{h})] \tag{8}$$

Where $I = \int_0^\infty \frac{x^3 dx}{(\exp x - 1)} = \frac{\pi^4}{15}$. The divergent part of the entropy associated with the brick-wall cut-off parameter is logarithmically divergent in the cut-off parameter. The divergent behaviour is expected as the constant-time foliations intersect at the horizon and can be related with infinite red shift. We have assumed $h << 2M$ and $L >> 2M$. Here β is given by the usual expression $\beta = 8\pi M$ and h is the brick-wall cut-off parameter. We find that with this expression of β the divergent part of the entropy in terms of the brick-wall cut-off parameter is independent of the horizon surface area. We also note that the term logarithmically divergent in the brick wall cut-off parameter is independent of the mass of the black hole although the leading order infrared divergent term is dependent on the mass of the black hole. The higher order terms in s are not divergent in h and vanish in the limit $L \to \infty$. We will later discuss the situation with the massive fields.

In passing we note that in the original reference, [3], the upper limit of the l integration in equation (6) was taken so that k is zero. This makes the upper limit dependent on r and many solutions do not have a part stationary throughout the complete range of r and we can not apply the WKB quantization rule for those modes. The scalar field entropy is given by the following expression:

$$S = \frac{8\pi^3}{45h} 2M (\frac{2M}{\beta})^3 \tag{9}$$

Unlike eqn(8) this expression is inversely dependent on h. In terms of a coordinate invariant cut-off, given by eqn (16), this expression is proportional to the area of the horizon and is independent of the mass or charge of a particular black hole.

We now consider the entropy of a massless scalar field in the non-extreme Reissner-Nordstrom black hole background. The procedure remains the same and only the metric function $V(r) = (1 - \frac{2M}{r})$ is replaced by the new metric function $V(r) = (1 - \frac{2M}{r} + \frac{Q^2}{r^2})$, where Q is the charge of the black hole and we have $M > |Q|$. The black hole horizon is at $r = r_+ = M + \sqrt{(M^2 - Q^2)}$. In this case finding out the maximum value of $\frac{V(r)}{r^2}$ is little cumbersome and this occurs at,

$$\tilde{r} = \frac{1}{2}[3M + \sqrt{(9M^2 - 8Q^2)}] \tag{10}$$

which is outside the event horizon. Proceeding in the same way as above we obtain the following expression for the most significant part of the scalar field entropy:

$$S = \frac{4\pi^3 s}{15\beta^3}[L + r_+ ln(\frac{L}{r_+}) + \frac{{r_+}^2}{r_+ - r_-} ln(\frac{r_+}{h})][\frac{\tilde{r}^2}{V(\tilde{r})}] \tag{11}$$

Where $r_- = M - \sqrt{(M^2 - Q^2)}$. We thus find that the divergent part of the scalar field entropy in the non-extreme Reissner-Nordstrom black hole background is similar to that of the Schwarzschild black hole and in particular the near horizon contribution is again logarithmically divergent in the brick wall cut-off parameter. The expression reduces to that in the Schwarzschild black hole in the limit $Q = 0$. In the next section we will find that the expression of the entropy of a thin shell of matter field surrounding the horizon is given by an expression proportional to the horizon surface area with the proportionality constant being the same for both the Scwarzschild black hole and the non-extreme Reissner-Nordstrom black hole.

We now consider the extremal Reissner-Nordstrom black hole in detail. In this case we have,

$$V(r) = (1 - \frac{M}{r})^2 \tag{12}$$

The maximum value of $\frac{V(r)}{r^2}$ occurs at $r = 2M$ and is given by $\frac{1}{16M^2}$. The divergent part of the free energy, both ultraviolet and infrared, is given by,

$$F = -\frac{16\pi^3}{15}\frac{M^2}{\beta^4}[(\frac{M^2}{h}) + M ln(\frac{L}{h}) + L] \tag{13}$$

Here $'s'$ is taken to be one. Hence the divergent part of the massless scalar field entropy is given by the following expression:

$$S = \frac{64\pi^3}{15}\frac{M^2}{\beta^3}[(\frac{M^2}{h}) + Mln(\frac{L}{h}) + L] \tag{14}$$

The temperature of the Hawking radiation and hence that of the extremal Reissner-Nordstrom black hole is zero and hence both the free-energy and the entropy of the scalar field vanish. However it is well-known from the imaginary-time Euclidean sector analysis [18,19] that the temperature can be taken to be arbitrary as there is no conical singularity to avoid in the corresponding Euclidean manifold. In this case the scalar field entropy is divergent. The topology of the extremal Reissner-Nordstrom black hole is different from that of the non-extreme black holes and the corresponding expressions for the scalar field entropies are also different.

3. Entropy of a thin shell of scalar field surrounding the black hole event horizon

We now consider a thin shell of matter field surrounding the outer horizon in the non-extreme Reissner-Nordstrom black hole background. The boundary conditions on the scalar field are the following:

$$\psi(r_+ + h), \psi(r_+ + 2h) = 0 \tag{15}$$

We again assume $h << r_+$. We can proceed similar to the preceding section. Within the region of interest the maximum of $\frac{V(r)}{r^2}$ occurs at $r = r_+ + 2h$ and to apply the WKB quantization rule, the angular quantum number, l, should be less than the value obtained from the following equation:

$$N(N+1) = \frac{E^2(r_+)^4}{2h(r_+ - r_-)} \tag{16}$$

As before we consider the upper limit of the l -integration to be given by sN with $s < 1$. With this choice, $k(r)$ is real throughout the region of interest. In this case the free energy calculation is a little cumbersome. Unlike the previous cases, we will have near horizon contributions from the higher order terms in $'s'$. These terms are inversely dependent on the brick wall parameter and we have,

$$F = -[2sln(2) - \tfrac{2}{3}\sum(-1)^p \frac{(\frac{3}{2}!)s^p}{(\frac{3}{2}-p)!p!}\frac{(2^{p-1}-1)2^{2-p}}{(p-1)}] \\ [\frac{\pi^3}{60\beta^4 h}\frac{(r_+)^6}{(r_+ - r_-)^2}] \tag{17}$$

Where $p \geq 2$, is an integer, and $h << r_+$. In this case the integral for the free energy, for $s < 1$, picks divergent contribution only from the near horizon region and can be associated with the brick wall cut-off h. With the h-dependent part explicitly factored out, the remaining terms within the third parentheses is finite and can be shown to be greater than $ln[4\exp(3/2)]$ by a small value.

The brick wall cut-off parameter is a radial parameter and we can replace it by the covariant cut-off parameter ϵ given by the following expression:

$$\epsilon = \int_{r_+}^{(r_+ + h)} \frac{dr}{\sqrt{V(r)}} \tag{18}$$

For $h << r_+$ we have,

$$\epsilon^2 = (\frac{4{r_+}^2}{r_+ - r_-})h \tag{19}$$

In terms of this covariant cut-off parameter we have the following expression for the scalar field entropy,

$$S = \eta \frac{4\pi^3}{15\beta^3} [\frac{{r_+}^8}{(r_+ - r_-)^3}] \frac{1}{\epsilon^2} \tag{20}$$

Where $'\eta'$ is the numerical factor within the third parentheses of the equation (17). $'\beta'$ is the inverse of the Hawking temperature and is given by, $\beta = \frac{4\pi {r_+}^2}{r_+ - r_-}$. With this value of $'\beta'$, the scalar field entropy becomes proportional to the area of the event horizon and is given by the following expression:

$$S = \frac{\eta}{960\pi} \frac{A}{\epsilon^2} \tag{21}$$

Where $'A'$ is the area of the black hole event horizon. Thus we have a universal expression for the matter field entropy which is proportional to the area of the event horizon. If we now equate this expression to the Bekenstein-Hawking entropy [3], we find that the value of the covariant cut-off parameter becomes independent of the mass of the black hole and is given by the following expression:

$$\epsilon = \sqrt{\frac{\eta}{240\pi}} \tag{22}$$

This expression is different from the corresponding expression obtained by 't Hooft,

$$\epsilon = \sqrt{\frac{1}{90\pi}} \tag{23}$$

Similar situation will also remain valid for the Schwarzschild black hole background [24]. Thus the value of $'\epsilon'$ given by eqn(22) is a universal constant, independent of the black hole parameters. It is same for both the non-extreme Reissner-Nordstrom black hole and the Schwarzschild black hole. The value is expressed in terms of the Planck length. The situation for the extreme Reissner-Nordstrom black hole is not much illuminating and the scalar field entropy vanishes when Hawking temperature is taken to be zero.

4. conclusion

To conclude, we have considered the entropy of a scalar field in the the Reissner-Nordstrom black hole backgrounds. We have used the brick wall model of 't Hooft. The scalar field is vanishing near the horizon as well as far away from the horizon. The density of states is calculated using the WKB approximation. We found that the entropy associated with the solutions which are stationary throughout the region of interest so that the WKB quantization rule is valid is different from that obtained by 't Hooft. The expression for the entropy is divergent in the brick wall cut-off parameter and this ill-behavedness is associated with the intersection of the constant-time foliations at the horizon which is a fixed point set of the time-like Killing vector. The scalar field entropy is not proportional to the area of the black hole event horizon. It will be interesting to compare the discussions in this article with the calculations of the matter field entropy made in the imaginary-time Euclidean manifold [19]. The Euclidean sector topology of the extremal Reissner-Nordstrom black hole is different from that of the non-extreme Reissner-Nordstrom black hole and leads to a vanishing entropy for the black hole [17]. The entropy of the matter field also vanishes if we choose the temperature to be zero. We also considered the entropy of a thin shell of scalar field surrounding the black hole horizon. We have obtained an expression for the matter field entropy which is proportional to the horizon surface area. This is valid for both the Schwarzschild and the non-extreme Reissner-Nordstrom black holes and the proportionality constants are the same. The expressions of the thin shell matter field entropy obtained in the article are important in the context of explaining the black hole entropy in terms of the near horizon states [20] and also for the entanglement entropy approach to explain the black hole entropy [21,22]. The fact that the solutions of the matter field may not contain a global oscillatory term for all values of $'l'$ is also important to discuss the scattering and absorption of matter fields in black hole backgrounds. These discussions may also be important to consider backscattering and to find out the luminosity of Hawking radiation [19]. The logarithmically divergent expression of the matter field entropy may be significant for the optical metric analysis of the black hole thermodynamics [13].

References

[1] Bekenstein J 1973 *Phys. Rev. D* **7** 2333
[2] Hawking S W 1975 *Commun. Math. Phys.* **43** 199
[3] 't Hooft G 1985 *Nucl. Phys. B* **256** 727
[4] Susskind L and Uglum J 1994 *Phys. Rev. D* **50** 2700
[5] Ghosh A and Mitra P 1995 *Phys. Lett. B* **357** 295
[6] Ichinose I and Satoh Y 1995 *Nucl. Phys. B* **447** 34
[7] Solodukhin S N 1995 *Phys. Rev. D* **51** 68
[8] Cai R G and Zhang Y Z 1996 *Mod. Phys. Lett. A* **11**
[9] Cognola G 1998 *Phys. Rev. D* **57** 6292
[10] Ghosh K 2003 *Phys. Rev. D* **60** 124027
[11] Ghosh K 2012 *A Few Comments on Classical Electrodynamics: SupplementIX* **V153** http://www.arxiv.org/physics/0605061
[12] Bombelli L, Kaul R K, Lee J and Sorkin R 1986 *Phys. Rev. D* **34** 373
[13] de Alwis S P and Ohta N 1995 *Phys. Rev. D* **52** 3529
[14] Setare M R 2006 *Int. J. Mod. Phys. A* **21** 6183-6190
[15] Ghosh K 2009 *Nucl. Phys. B* **814** 212-216
[16] Gibbons G W and Hawking S W 1979 *Commun. Math. Phys.* **66** 291
[17] Tolman R C 1979 *The principles of Statistical Mechanics* (Dover Publications, Inc. New York)
[18] Hawking S W, Horowitz G T and Ross S F 1995 *Phys. Rev. D* **51** 4302
[19] Hawking S W 1977 *Commun. Math. Phys.* **56** 133
[20] Birmingham D, Gupta K S and Sen S 2001 *Phys.Lett. B* **505** 191
[21] Srednicki M 1993 *Phys. Rev. Let.* **71** 666
[22] Birrel N D and Davies P C W 1982 *Quantum Fields in Curved Space* (Cambridge University Press)
[23] Ghosh K 1999 *Phys. Rev. D* **60** 104003
[24] Ghosh K 2012 *Communicated*

IC-MSQUARE 2012: International Conference on Mathematical Modelling in Physical Sciences IOP Publishing
Journal of Physics: Conference Series **410** (2013) 012138 doi:10.1088/1742-6596/410/1/012138

Thermodynamic curvature: pure fluids to black holes

George Ruppeiner

Division of Natural Sciences, New College of Florida, Sarasota, Florida 34243-2109

E-mail: ruppeiner@ncf.edu

Abstract. Thermodynamics unavoidably contains fluctuation theory, expressible in terms of a unique thermodynamic information metric. This metric produces an invariant thermodynamic Riemannian curvature scalar R which, in fluid and spin systems, measures interatomic interactions. Specifically, $|R|$ measures the size of organized fluctuating microscopic structures, and the sign of R indicates whether the interactions are effectively attractive or repulsive. R has also been calculated for black hole thermodynamics for which there is no consensus about any underlying microscopic structures. It is hoped that the physical interpretation of R in fluid and spin systems might offer insight into black hole microstructures. I give a brief review of results for R in black holes, including stability, the sign of R, $R = 0$, diverging $|R|$, and various claims of "inconsistencies" in thermodynamic metric geometry.

1. Introduction

Thermodynamics [1] has a reputation as being a science complete in its basic laws. Less well known is thermodynamic fluctuation theory which follows from thermodynamics [2]. This fluctuation theory may be represented by a unique thermodynamic information Riemannian metric [3], resulting in an invariant Riemannian thermodynamic curvature scalar R. Physically, R reveals information about interatomic interactions, and I illustrate here with recent calculations of R in pure fluids with thermodynamic data from the NIST fluid database [4].

As was shown by Bekenstein [5], Hawking [6], and others, thermodynamics may be extended to black holes. Logically, thermodynamic fluctuation theory, including the physical interpretation of R developed in fluid and spin systems, should extend likewise. This paper includes a tabulation of calculations of R from general relativistic black hole solutions, for which no underlying microscopic models are known. I also discuss various claims of "inconsistencies" concerning divergences of black hole heat capacities and cases with $R = 0$.

2. Pure fluids

The basic structure in thermodynamic fluctuation theory in pure fluids is an infinite environment in some reference state "0", characterized by two independent intensive thermodynamic variables, and containing an open subsystem with constant volume V. This subsystem exchanges energy and particles with its environment, and its thermodynamic state fluctuates about an equilibrium characterized by maximum entropy. The probability of a fluctuation away from equilibrium is given by Einstein's famous formula [2]: probability $\propto \exp\left[-V(\Delta\ell)^2/2\right]$, where $(\Delta\ell)^2$ is the invariant, positive definite, thermodynamic information metric. Expressed in the pair of independent thermodynamic variables x^1 and x^2: $(\Delta\ell)^2 = g_{\mu\nu}\Delta x^\mu \Delta x^\nu$, where $\Delta x^\alpha \equiv (x^\alpha - x^\alpha_0)$ denotes the difference between the thermodynamic variables x^α of the subsystem and their equilibrium values x_0^α. Denote by S, X^1, and X^2 the entropy, internal energy, and particle number, respectively, of the subsystem. X^1 and X^2 are each conserved quantities, and S is additive between the subsystem and its environment. If $x^\alpha = X^\alpha$, then:

$$g_{\alpha\beta} = -\frac{1}{k_B V}\frac{\partial^2 S}{\partial X^\alpha \partial X^\beta}, \tag{1}$$

where k_B is Boltzmann's constant [3]. $g_{\alpha\beta}$ transforms as a second-rank tensor.

The thermodynamic metric Eq. (1) leads to an invariant Riemannian thermodynamic curvature scalar R. $R = 0$ for the ideal gas [7], and $|R|$ diverges as the correlation volume ξ^3 near the critical point [3, 7, 8]. R is thus a measure of effective interatomic interactions. Model calculations reveal that the sign of R corresponds to whether interactions are effectively attractive ($R < 0$) or repulsive ($R > 0$); see the tabulation in [9]. The connection of $|R|$ to fluctuating structure size has been established directly by means of a covariant thermodynamic fluctuation theory [3].

Most recently [4], R has been calculated in pure fluids using data based on experiment. Pure fluids have interatomic interaction potentials of the Lennard-Jones type, strongly repulsive at short range, attractive at long range, and with a minimum at a distance where the atoms in the condensed liquid and solid phases reside. A fluid typically has density smaller than that of a condensed state, and we thus expect an average attractive interatomic interaction, and negative R. Negative R is indeed the norm in pure fluids. Figure 1 shows R for hydrogen along the coexistence curve, with: 1) R in both phases diverging to $-\infty$ at the critical point, 2) $|R|$ in both phases decreasing as the triple point is approached, with liquid phase values approaching the order of a molecular volume, and 3) a region of small positive R in the condensed liquid phase near the triple point as the liquid organizes into a solid-like structure.

One may now pose a puzzle: which atomic arrangements produce the thermodynamic curvatures seen in Fig. 1? Figure 2 suggests a solution encompassing most fluid situations: Fig. 2a shows widely spaced atoms where attractive interactions form loose fluctuating clusters of volume $|R| \propto \xi^3$. Such structures are typical of the critical point regime. Fig. 2b shows a tight cluster of atoms in the vapor phase prevented from collapse by repulsive interactions. Such clusters are present in water, but not in hydrogen [4]. Fig. 2c shows a haphazard arrangement of atoms in a condensed liquid state. Fig. 2d shows an organized solid-like state. The liquid phase in Fig. 1 shows a change in sign of R corresponding to a transition $c \to d$.

Results for R in pure fluids are thus associated with simple microscopic patterns. Black hole physics does not have as well developed a theoretical microscopic structure. However, the calculation of R, and insights from fluid or spin models, might lend some guidance in this difficult domain.

3. Black holes

Black hole solutions to general relativity have well-defined thermodynamic properties. The role of the conserved thermodynamic variables is played by mass, angular momentum, and charge (M, J, Q). Logically, black hole thermodynamics leads to thermodynamic fluctuation theory and an information metric Eq. (1) (set $V = 1$, and use (M, J, Q) for the X^α's [10]). This information metric produces the thermodynamic curvature R [11, 12]. A tentative attempt to physically interpret R for black holes has been made in terms of the spatial size of correlated fluctuations on the event horizon [10]. Perhaps the puzzle solving tried above in pure fluids can eventually lead to insight about possible black hole microstructures.

Debated in black hole thermodynamics has been the possibility raised by Davies [13] that the curve of diverging heat capacity $C_{J,Q} \equiv T(\partial S/\partial T)_{J,Q}$ in the Kerr-Newman black hole corresponds to a phase transition. Since diverging heat capacities are a feature of second-order phase transitions in fluid and spin systems, Davies' association would appear logical.

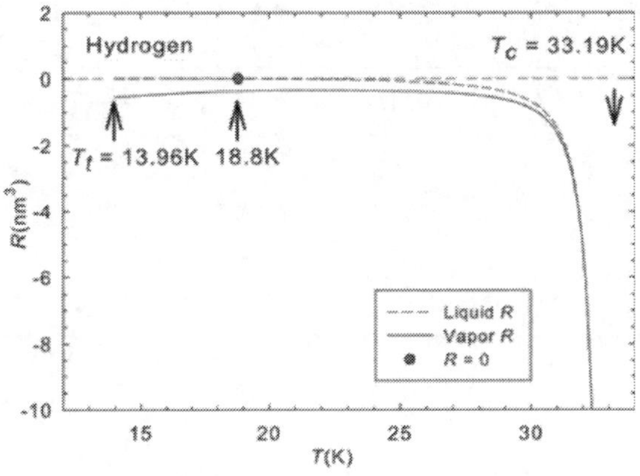

Figure 1. R for hydrogen along the coexistence curve in the liquid and vapor phases from the triple point to the critical point, with temperatures $T = T_t$ and $T = T_c$.

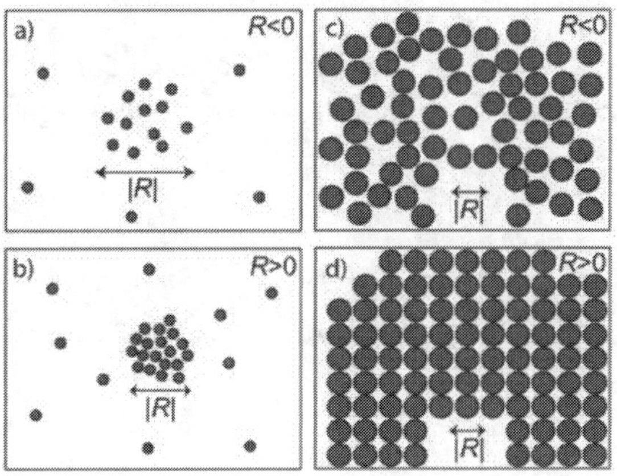

Figure 2. a) Atoms pulled together by attractive interactions; b) atom cluster held up by repulsive interactions; c) disorganized liquid phase; and d) organized solid phase.

However, in ordinary thermodynamics the connection between thermodynamic anomalies and phase transitions gets support from microscopic models. Without such models there can be no assurance of any phase transition. For black holes, if some heat capacity diverges, how could we be sure that we have not just made an inappropriate choice of thermodynamic variables, which reveals infinities with no really fundamental significance? What would we make of curves where one heat capacity diverges, but the other heat capacities stay regular? What if each heat capacity diverges along its own curve, as happens in the Kerr-Newman black hole [10]? Which curve corresponds to a true phase transition? These questions identify difficulties associating diverging heat capacities with phase transitions. It seems simpler to look at diverging $|R|$ to indicate phase transitions, since R has a unique status in identifying microscopic order, which in fluid and spin systems goes to the heart of the issue of phase transitions.

Black hole solutions with R identically zero have also given rise to debate; see [17] for a review. Examples of such solutions are the BTZ and the Reissner-Nordström [12] black holes. If R measures in some sense the range of interactions, then one might expect $|R|$ to be large for black hole thermodynamics, reflecting the gravitational forces present in these objects. But such reasoning need not obtain. In a classical black hole, the gravitating particles have collapsed to a central singularity, shrinking the interactions between them to zero volume. The statistics underlying the thermodynamics might reside on the event horizon, where unknown constituents might interact with each other by forces perhaps not gravitational. In this scenario, gravity might merely be a nonstatistical force holding the assembly together, and a result $R = 0$, where the unknown constituents move independently of each other, might not be so unreasonable.

Table 1 summarizes a number of black hole calculations of R. Tabulated are the spatial (+ time) dimensions, whether or not there are regimes of thermodynamic stability (positive definite metric matrix Eq. (1)), the sign of R (or an indication "0" if R is identically zero), whether or not there are places where the sign of R changes through zero, which might indicate a Hawking-Page phase transition [16], and whether or not there are divergences $|R| \to \infty$ with $T \neq 0$. Most solutions here with R not identically zero diverge to $\pm\infty$ at the extremal curve $T \to 0$. I omit string theory black hole solutions for simplicity. Table 1 shows a progression from the elegant BTZ black hole, stable for all states, and with identically zero R, to the older solutions Kerr, Reissner-Nordström, and Kerr-Newman, none of which rise to the level of stability. (Omitted are cases of the Kerr-Newman black hole in which one of the parameters is held fixed, while the

Table 1. Properties of R for black holes solutions from general relativity.

Name of solution	Dimension	Variables	Stable	Sign of R	$R = 0$	$\frac{\lvert R\rvert\to\infty}{T\neq 0}$
BTZ [12]	$2+1$	(M, J)	yes	0	-	no
Kerr [12]	$3+1$	(M, J)	no	$+$	no	no
Reissner-Nordström[12]	$3+1$	(M, Q)	no	0	-	no
Kerr-Newman [10, 14]	$3+1$	(M, J, Q)	no	$+$	no	no
K-AdS [15]	$3+1$	(M, J)	yes	$-$	no	yes
RN-AdS [12]	$3+1$	(M, Q)	yes	$\pm$	yes	yes
KN-AdS [16]	$3+1$	(M, J, Q)	yes	$\pm$	yes	yes
Black Hole [18]	$4+1$	(M, J)	no	$+$	no	no
Small Black Ring [18]	$4+1$	(M, J)	no	$\pm$	yes	yes
Large Black Ring [18]	$4+1$	(M, J)	yes	$-$	no	yes

others fluctuate; such do show instances of stability [10]). A richer structure, including stability, R zero crossing, and R divergences, clearly from either increasing the number of dimensions, adding an asymptotic AdS scenario, and/or more complex topology. Too intricate to tabulate here are results for the Myers-Perry black holes [19].

Acknowledgments

I thank George Skestos for travel support, and Bhupendra Tiwari for productive discussions.

References

[1] Callen H 1985 *Thermodynamics and an Introduction to Thermostatistics* (New York: Wiley)
[2] Landau L and Lifshitz E 1977 *Statistical Physics* (New York: Pergamon)
[3] Ruppeiner G 1995 Riemannian geometry in thermodynamic fluctuation theory *Rev. Mod. Phys.* **67** 605-59
[4] Ruppeiner G 2012 Thermodynamic curvature from the critical point to the triple point *Phys. Rev.* E **86** 021130
[5] Bekenstein J 1974 Generalized second law of thermodynamics in black-hole physics *Phys. Rev.* D **9** 3292-3300
[6] Hawking S 1976 Black holes and thermodynamics *Phys. Rev.* D **13** 191-7
[7] Ruppeiner G 1979 Thermodynamics: a Riemannian geometric model *Phys. Rev.* A **20** 1608-13
[8] Johnston D, Janke W, and Kenna R 2003 Information geometry, one, two, three (and four), *Acta Phys. Pol.* B **34** 4923–37
[9] Ruppeiner G 2010 Thermodynamic curvature measures interactions Am. J. Phys. **78** 1170-80
[10] Ruppeiner G 2008 Thermodynamic curvature and phase transitions in Kerr-Newman black holes *Phys. Rev.* D **78** 024016
[11] Ferrara S, Gibbons G, and Kallosh R 1997 Black holes and critical points in moduli space *Nuc. Phys.* B **500**, 75
[12] Åman J, Bengtsson I, and Pidokrajt N 2003 Geometry of black hole thermodynamics *Gen. Relativ. Gravit.* **35**, 1733-43
[13] Davies P 1977 The thermodynamic theory of black holes *Proc. R. Soc.* A **353** 499-521
[14] Mirza B and Zamaninasab M 2007 Ruppeiner geometry of RN black holes: flat or curved? *J. High Energy Phys.* **6** 059
[15] Sahay A, Sarkar T, and Sengupta G 2010 On the Thermodynamic Geometry and Critical Phenomena of AdS Black Holes *Preprint* 1004.1625
[16] Sahay A, Sarkar T, and Sengupta G 2010 Thermodynamic geometry and phase transitions in Kerr-Newman-AdS black holes *JHEP* **1004** 118
[17] Medved A 2008 A commentary on Ruppeiner metrics for black holes *Mod. Phys. Lett.* A **23** 2149-61
[18] Arcioni G and Lozano-Tellechea E 2005 Stability and critical phenomena of black holes and black rings *Phys. Rev.* D **72** 104021
[19] Åman and Pidokrajt N 2010 On explicit thermodynamic functions and extremal limits of Myers-Perry black holes *Preprint* 1004.5550

IC-MSQUARE 2012: International Conference on Mathematical Modelling in Physical Sciences IOP Publishing
Journal of Physics: Conference Series **410** (2013) 012139 doi:10.1088/1742-6596/410/1/012139

Universal Property of Quantum Gravity implied by Bekenstein-Hawking Entropy and Boltzmann formula

Hiromi Saida

Department of Physics, Daido University, Nagoya 457-8530, Japan

E-mail: saida@daido-it.ac.jp

Abstract. We search for a universal property of quantum gravity. By "universal", we mean the independence from any existing model of quantum gravity (such as the super string theory, loop quantum gravity, causal dynamical triangulation, and so on). To do so, we try to put the basis of our discussion on theories established by some experiments. Thus, we focus our attention on thermodynamical and statistical-mechanical basis of the black hole thermodynamics: Let us assume that the Bekenstein-Hawking entropy is given by the Boltzmann formula applied to the underlying theory of quantum gravity. Under this assumption, the conditions justifying Boltzmann formula together with uniqueness of Bekenstein-Hawking entropy imply a reasonable universal property of quantum gravity. The universal property indicates a repulsive gravity at Planck length scale, otherwise stationary black holes can not be regarded as thermal equilibrium states of gravity. Further, in semi-classical level, we discuss a possible correction of Einstein equation which generates repulsive gravity at Planck length scale.

1. Introduction

We investigate the basis of black hole thermodynamics in order to search for universal properties of quantum gravity, where "universal" means the independence from any existing model of quantum gravity. The basic concept of black hole thermodynamics is as follows: *Because a stationary black hole emits the thermal radiation (Hawking radiation), the black hole is in a thermal equilibrium state of gravity. And, usually, the micro-scopic origin of black hole is expected to be the quantum states of gravity.* This concept depends on the thermal radiation theory which indicates that the source of "Planck spectrum" is a thermal equilibrium object.

By accepting the black hole thermodynamics, it is reasonable to require that the quantum gravity satisfies the presupposition of thermal radiation theory, otherwise we can not regard black holes as thermal equilibrium states by observing the Hawking radiation. Here, the presupposition of thermal radiation theory is that the ordinary thermodynamics and quantum statistical mechanics work well on any thermal equilibrium object (at least in laboratory system). Therefore, it is natural to require that *the basic idea of statistical mechanics is applicable to thermal equilibrium states of not only laboratory systems but also quantum gravity (black holes).* Here, the basic idea of statistical mechanics is, for example, the Boltzmann formula, $S_{BH} = \ln \Omega_{BH}$, where S_{BH} is the Bekenstein-Hawking entropy and Ω_{BH} is the number of states of black hole counted by the underlying quantum gravity.

From the above, we adopt the following two suppositions for our purpose:

Supposition 1: A stationary black hole is a thermal equilibrium state composed of micro-states of underlying quantum gravity.

Supposition 2: Basic formalism of statistical mechanics (e.g. Boltzmann formula, grand partition function and so on) is applicable to underlying quantum gravity so as to describe the thermal equilibrium states (black holes).

The issues for each supposition are as follows:

Issue on Supposition 1 (Uniqueness of BH entropy): Note that, if the Boltzmann formula, $S_{BH} = \ln \Omega_{BH}$, holds for black holes, then it defines the Bekenstein-Hawking entropy S_{BH} *uniquely* in micro-scopic point of view. This means that the Boltzmann formula can never work well unless the uniqueness of entropy is proven in macro-scopic point of view of thermodynamics (independent of any micro-scopic theory) [1, 2, 3]. In ordinary thermodynamics, the uniqueness of entropy is proven by some basic axioms of thermodynamics [1, 2]. However, a basic axiom of ordinary thermodynamics do not hold in black hole thermodynamics [4, 5, 6]. Then, it is not necessarily manifest whether the entropy S_{BH} is unique or not in black hole thermodynamics. Hence, we have to prove the uniqueness of S_{BH} before the discussion of Boltzmann formula (see Sec.2).

Issue on Supposition 2 (Justification of Boltzmann formula): Once the uniqueness of S_{BH} is proven in macro-scopic point of view, we can proceed to discussions on Boltzmann formula. Note that the validity of Boltzmann formula in ordinary quantum statistical mechanics is justified by some intrinsic properties of Hilbert space of ordinary quantum mechanics. Those intrinsic properties appear in Ruelle-Tasaki theorem and Dobrushin theorem [3, 7, 8] (see Sec.3). *Thus, if S_{BH} is given by the Boltzmann formula, we can expect that those theorems of ordinary quantum mechanics are related with some property of underlying quantum gravity. Some universal property of quantum gravity may be extracted from those theorems* (see Sec.4).

2. Uniqueness Theorem of Bekenstein-Hawking Entropy

In the rigorous axiomatic formulation of ordinary thermodynamics [1, 2], the basic axioms consist of not only the "four laws of thermodynamics" but also some more requirements. One of the axioms of ordinary thermodynamics modified in black hole thermodynamics is the classification requirement of state variables; all thermodynamic state variables are classified into two categories, extensive variables (e.g. energy, entropy and so on) and intensive variables (e.g. temperature, pressure and so on). However, as explained in detail in ref.[4, 6], the state variables in black hole thermodynamics can not be classified into the two categories, but classified into *three* categories. Hence, various theorems (including the uniqueness of entropy) in ordinary thermodynamics, whose proof uses the classification of state variables, should be re-proven in black hole thermodynamics. Ref.[4] gives an axiomatic proof to the uniqueness of S_{BH}:

Uniqueness Theorem of Entropy: Let K be an extensive variable which increases along irreversible adiabatic processes. Then there exist two constants $a\,(>0)$ and b such that $K = aS_{BH} + b$, where S_{BH} is the Bekenstein-Hawking entropy. (See ref.[4] for the proof.)

Once this theorem is proven, we can proceed to discussions on Boltzmann formula. In this sense, we can regard the conclusion in Sec.4 as a result implied by the uniqueness of S_{BH}.

3. Conditions Justifying Boltzmann Formula

Consider the ordinary quantum mechanics in this section. For simplicity, consider the system of N identical particles in a region of volume V with no external field, and:

- Interaction potential : $\Phi(\vec{x}_1, \cdots, \vec{x}_N) = \sum_{j=1}^{\infty} \phi^{(j)}(\vec{x}_{i_1}, \cdots, \vec{x}_{i_j})$, where $i_j \in (1, \cdots, N)$.
- Energy eigen value : $E_k(V, N)$, where $k = 0, 1, 2, \cdots$ and $E_k \le E_{k+1}$ ("=" for degeneracy).
- Number of states : $\Omega(V, N; U) :=$ "Number of eigen states satisfying $E_k \le U$" $= \max_{E_k \le U} k$.

For the first, let us show the Ruelle-Tasaki theorem which clarifies the sufficient conditions for the validity of Boltzmann formula [4, 7, 3]:

Ruelle-Tasaki Theorem: Suppose the following two conditions of $\Phi(\vec{x}_1, \cdots, \vec{x}_N)$:

Condition A: Arbitrary j-particle interaction, $\phi^{(j)}$, becomes negative for sufficiently large distribution of j particles. That is, there exists a constant $r_A\,(>0)$, such that $\phi^{(j)}(\vec{x}_{i_1}, \cdots, \vec{x}_{i_j}) \le 0$ for $r_A \le \min_{k,l=1,\cdots,j} |\vec{x}_{i_k} - \vec{x}_{i_l}|$.

Condition B: Interaction potential Φ is bounded below. That is, there exists a constant $\phi_B\,(>0)$, such that $\Phi(\vec{x}_1, \cdots, \vec{x}_N) \ge -N\,\phi_B$.

Then, the limit, $\sigma(\varepsilon, \rho) := \lim_{t.l.} \dfrac{\ln \Omega(V, N; U)}{V}$, exists uniquely, where $\lim_{t.l.}$ is thermodynamic limit defined by $V \to \infty$ with fixing $\rho := N/V$ and $\varepsilon := U/V$ at constant values.

By this theorem, the thermodynamics limit of $\ln \Omega(V, N; U)$ is defined well. Furthermore, it can be also proven that $\sigma(\varepsilon, \rho)$ is concave about its arguments (ε, ρ) and monotone increasing about ε, and remains constant along *reversible* adiabatic processes [3]. These behaviors are some characteristic properties of entropy already known in thermodynamics. Then, in statistical mechanics, it is assumed that $\ln \Omega(V, N; U)$ is equal to the entropy (the Boltzmann formula).

Obviously, the above conditions A and B are the sufficient conditions for the validity of Boltzmann formula. Furthermore, a system, which holds Boltzmann formula but violates conditions A or B, has not been found experimentally. *Hence, at least in laboratory systems, it is reasonable to require that any physical system satisfies the conditions A and B.*

Next, let us show the Dobrushin theorem which clarifies the necessary condition for the existence of thermal equilibrium states [4, 7, 8]:

Dobrushin theorem: Consider the case satisfying following two presuppositions:

Presupposition C: The $j(\neq 2)$-particle interactions disappear, and the total interaction potential is a sum of two-particle interactions, $\Phi(\vec{x}_1, \cdots, \vec{x}_N) = \sum_{1\le i<j\le N} \phi^{(2)}(\vec{x}_i, \vec{x}_j)$.

Presupposition D: $\phi^{(2)}(\vec{x}, \vec{y}) \propto |\vec{x} - \vec{y}|^{-\alpha}$ $(\alpha > 0)$ for sufficiently large $|\vec{x} - \vec{y}|$.

Under these presuppositions, if the ground partition function can be defined uniquely (i.e. if thermal equilibrium states exist), then the following inequality holds,

$$I_V^{(2)} := \frac{1}{V^2} \int\!\!\int_V \mathrm{d}^3x_1\, \mathrm{d}^3x_2\, \phi^{(2)}(\vec{x}_1, \vec{x}_2) \ge 0\,. \tag{1}$$

Obviously, $I_V^{(2)} \ge 0$ is the necessary condition for the existence of thermal equilibrium states. Note that the presuppositions C and D are natural when we consider the gravity.

4. Conclusion and Discussion

Our supposition 2 in Sec.1 implies that the ordinary quantum mechanics and quantum gravity shares the same properties which justify the Boltzmann formula and existence of thermal equilibrium states. The sufficient conditions A, B and the necessary condition $I_V^{(2)} \ge 0$ are expected to be such shared properties [1]. The typical form of the potential is shown in Fig.1. Note that, these three conditions (A, B and $I_V^{(2)} \ge 0$) are for the interaction potential.

On the other hand, it is not clear whether the full quantum gravity is expressed by using the interaction potential Φ or Hamiltonian (or Lagrangian). However, it seems to be certain that a semi-classical expression of quantum gravity is described by an effective field equation of spacetime metric (with some auxiliary field if needed). Hence, our suggestion is as follows:

[1] So far, as a fact, no counter-example to conditions A and B are found experimentally. It may be reasonable to extend this fact to quantum gravitating systems, so that quantum gravity also satisfies those conditions.

Implication from the above: If the effective field equation of quantum gravity at semi-classical level is expressed by a correction term $f_{\mu\nu}$ added to Einstein equation such as

$$R_{\mu\nu} - \frac{1}{2} R\, g_{\mu\nu} + f_{\mu\nu} = \frac{8\pi G}{c^4} T_{\mu\nu} \,, \tag{2}$$

(where $T_{\mu\nu}$ is the energy-momentum tensor, $g_{\mu\nu}$ the metric, $R_{\mu\nu}$ the Ricci tensor, and $R := R_\mu{}^\mu$), then $f_{\mu\nu}$ should generate the repulsive gravity at short (e.g. Planck) length scale, and a necessary property of $f_{\mu\nu}$ is given in Eq.(3) below.

Let us note that the Raychaudhuri equation holds for arbitrary timelike/null congruence in arbitrary spacetime. For simplicity, consider the local hypersurface orthogonal (i.e. vanishing vorticity) geodesic congruence. Then, the Raychaudhuri equation becomes, $\mathrm{d}\theta/\mathrm{d}\lambda = -R_{\mu\nu}u^\nu u^\nu - \sigma^2 - \theta/p$, where λ is the affine parameter, u^μ the tangent of geodesic, $\theta := u^\mu{}_{;\mu}$ is the expansion (the rate of change of sectional area of congruence, d[area]/dλ), σ^2 the square of shear tensor, and $p = 3$ or 2 for respectively timelike or null geodesics. Note that, in the general relativity ($f_{\mu\nu} = 0$), the so-called *energy condition* (e.g. the null energy condition, $T_{\mu\nu}u^\mu u^\nu \geq 0$ for arbitrary null vector u^μ) gives $-R_{\mu\nu}u^\mu u^\nu \leq 0$ by the Einstein equation and results in $\mathrm{d}\theta/\mathrm{d}\lambda \leq 0$ by the Raychaudhuri equation, which indicates the attractive nature of gravity. On one hand, if there arises a case of $-R_{\mu\nu}u^\nu u^\nu \geq 0$, then $\mathrm{d}\theta/\mathrm{d}\lambda \geq 0$ may be possible by the Raychaudhuri equation, which indicates the repulsive gravity.

Then, when we require the repulsive gravity in Planck length scale, the correction term $f_{\mu\nu}$ should give the inequality $-R_{\mu\nu}u^\nu u^\nu \geq 0$ in Planck length scale. Here, by Eq.(2), we find, $-R_{\mu\nu}u^\nu u^\nu = -(8\pi G/c^4)\left[T_{\mu\nu}u^\mu u^\nu - (1/2)T\,u^2\right] + \left[f_{\mu\nu}u^\mu u^\nu - (1/2)f\,u^2\right]$, where $T = T_\mu{}^\mu$ and $f = f_\mu{}^\mu$. Hence, the inequality $-R_{\mu\nu}u^\nu u^\nu \geq 0$ turns to be,

Necessary property in Planck scale: $$f_{\mu\nu}u^\mu u^\nu - \frac{1}{2} f\,u^2 \;\geq\; \frac{8\pi G}{c^4}\left[T_{\mu\nu}u^\mu u^\nu - \frac{1}{2} T\,u^2\right] \tag{3}$$

If a suitable form of $f_{\mu\nu}$ is possible, the repulsive gravity may be realized with preserving the traditional energy conditions (e.g. dominant, weak and strong energy conditions).

In this manuscript, any existing model of quantum gravity is not used. (See ref.[4] for detail.) Therefore, the above implication can be regarded as a universal property of quantum gravity.

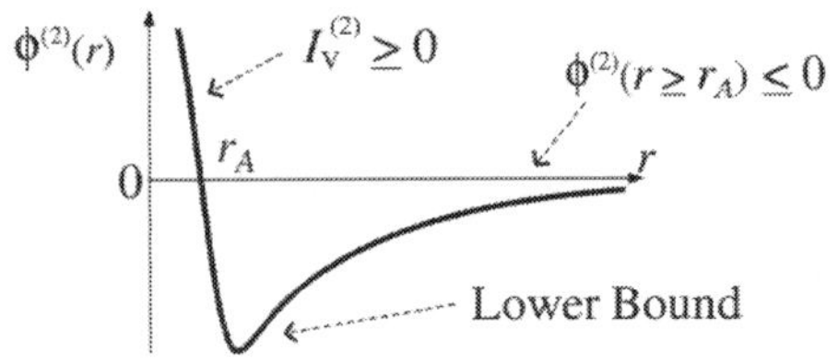

Figure 1. Typical form of the effective potential of quantum gravity at semi-classical level. Thermal equilibrium states of gravity (black holes) may not exist unless a repulsive region appears, where r_A may be Planck scale.

References

[1] E.H.Lieb and J.Yngvason, *The Physics and Mathematics of the Second Law of Thermodynamics*, Phys.Rep.310 (1999) 1–96.

[2] H.Tasaki, *Thermodynamics (Netus-Rikigaku)*, Baifu-kan Publ., 2000. [Published in Japanese]

[3] H.Tasaki, *Statistical Mechanics (Tokei-Rikigaku)*, Baifu-kan Publ., 2008. [Published in Japanese]

[4] H.Saida, *Universal Property of Quantum Gravity implied by Uniqueness Theorem of Bekenstein-Hawking Entropy*, Entropy 13 (2011) 1611–1647. (Invited as a feature (and refereed) paper for the special issue *black hole thermodynamics* edited by J.Bekenstein in the journal *Entropy*.)

[5] J.W.,Jr.York, *Black-hole thermodynamics and the Euclidean Einstein action*, Phys.Rev.D33 (1986) 2092–2099

[6] H.Saida, *To what extent is the entropy-area universal?*, Prog.Theor.Phys.122 (2010) 1515–1552.

[7] D.Ruelle, *Statistical Mechanics –Rigorous Results–*, Imperial College Press and World Scientific Publ., 1999.

[8] R.L.Dobrushin, *Investigation of the Conditions of the Asymptotic Existence of the Configuration Integral of the Gibbs Distribution*, Teorija Verojatn. i ee Prim.9 (1964) 626–643.

IC-MSQUARE 2012: International Conference on Mathematical Modelling in Physical Sciences IOP Publishing
Journal of Physics: Conference Series **410** (2013) 012140 doi:10.1088/1742-6596/410/1/012140

Horizon thermodynamics in pregeometry

Lorenzo Sindoni
Max Planck Institute for Gravitational Physics (Albert Einstein Institute), Am Mühlenberg 1, 1, 14476 Potsdam, Germany
E-mail: sindoni@aei.mpg.de

Abstract. The discovery of the various thermodynamical aspects of classical gravity has led to a picture in which Einstein equations might be seen as an equation of state, relating the dynamics of null hypersurfaces to thermodynamic relations for a system at or close to equilibrium. The explanation of such a fact is expected to come from a complete theory of quantum gravity. In absence of it, I will use a pregeometric model of emergent gravity to discuss the role of Wheeler's boundary of a boundary principle in this specific problem. The goal is to show the relevance of the equilibration of the microscopic degrees of freedom in establishing the thermodynamical status of the semiclassical gravity limit, and how the equilibration of the system might lead to additional physical effects. The role of phase transitions (conjectured in certain combinatorial models proposed to define a quantum theory of gravity, namely tensor models and group field theories) will be also considered in this light, showing how criticality and deviations from it relate to these ideas.

One of the most striking facts is that Einstein's equations can be derived starting from a thermodynamic point of view, associating thermodynamical nature to local acceleration horizons and to their dynamics as a result of the stress energy tensor of matter fields [1].

The objective of this contribution is to discuss horizon thermodynamics in a class of models designed to give the metric tensor as a composite field of some other microscopic, pregeometric degrees of freedom. We will further stress the necessity of an equilibration condition in order for the thermodynamical interpretation to make sense. While many ideas are certainly not new, the objective is to make them as explicit as possible in a concrete case.

The models that we are considering have been dubbed "Pregeometry" [2]. We briefly sketch the idea underlying the models, referring to the literature for a detailed discussion [3]. Reduced to the bone, they are the quantization of the Polyakov action of a bosonic or fermionic membrane of assigned dimension (for us it will be 4) embedded in a flat (pseudo-)Riemannian manifold of equal or higher dimension. The general idea is that the metric tensor as we know it is just a sort of meson field, *i.e.* a composite operator. We will work within a field theory model, and we will denote as $\hat{\psi}_A$ $(A = 1, ...N)$ the field operators associated to the microscopic degrees of freedom, where A is a label denoting collectively possible further fine grained details of the field (internal symmetries)

Using the Lagrange multipliers $\gamma^{\mu\nu}$ and the auxiliary fields $\phi_{\mu\nu}$, the partition function for the system can be written as

$$Z = \int \mathcal{D}\phi\mathcal{D}\gamma\mathcal{D}\psi \exp\left(-\int d^4x \left\{ (\det \phi_{\mu\nu})^{1/2} + \sqrt{\gamma}\gamma^{\mu\nu}(\phi_{\mu\nu} - \eta^{AB}\partial_\mu\psi_A\partial_\nu\psi_B) \right\} \right), \quad (1)$$

and, following standard techniques, one can associate to it a generating functional for the metric as a composite operator, such that

$$g_{\mu\nu} = \frac{1}{Z}\frac{\delta Z[K]}{\delta K^{\mu\nu}}\bigg|_{K=0}. \tag{2}$$

Given the quadratic dependence of the bare action of the fields ψ, we can easily integrate them away to get an explicit effective action for the composite metric tensor.

These models are not complete since they are nonrenormalizable. Indeed, standard QFT calculations show that the effective action for the composite metric will indeed be given in terms of the divergent contributions (which are cutoff regulated) of the heat kernel expansion of the one loop effective action. We can also evaluate transition amplitudes between assigned states of the fields ψ_A at two spacelike boundaries of the spacetime manifold.

By construction, in such a model we obtain effective quantum equations of motion of the form

$$\mathscr{G}^{\kappa}_{\mu\nu}(g_{\alpha\beta}) = 8\pi G_0 T^{\kappa}_{\mu\nu}, \tag{3}$$

with a conserved stress energy tensor for the matter fields, coupled only to the metric degrees of freedom, G_0 a coupling constant, and $\mathscr{G}$ a covariantly conserved rank two tensor obtained from the effective action of the composite metric tensor. The action will depend on the boundary conditions imposed on the functional integral (or, better, the boundary states), collectively labeled by κ.

The thermodynamical behavior of such models can be easily inferred by means of a generalization of Cartan's moment of rotation and the formalization of the field equation in terms of the boundary of a boundary principle advocated by Wheeler (see [5] for an extensive discussion). We construct the vector valued 3-forms:

$$\mathscr{G}^{(k)\mu}{}_{\nu}\mathbf{e}_{\mu} * dx^{\nu}, \qquad T^{\mu}_{\nu}\mathbf{e}_{\mu} * dx^{\nu}, \tag{4}$$

where $*$ is the Hodge star operator, and we integrate the equations on a portion H of a cylindrical horizon between an initial and a final slice, once we have projeted the vector valued forms onto the generator of the null surface,

$$\int_H \mathscr{G}^{(\kappa)\mu}{}_{\nu}\xi_{\mu} * dx^{\nu} = 8\pi G_{(\kappa)} \int_H T^{\mu}_{\nu}\xi_{\mu} * dx^{\nu}. \tag{5}$$

The RHS is just proportional to the heat flow that enters the thermodynamic reasoning of [1]. The analysis of the LHS should lead to the identification of the entropy function as it is coming from the microscopic theory, by means of the Hodge decomposition theorem of a generic differential form into an exact, coexact and harmonic components

$$\mathscr{G}_{(k)}(\xi) = \mathscr{G}^{(k)\mu}{}_{\nu}\xi_{\mu} * dx^{\nu} = \mathrm{d}_{\mathrm{ext}}\alpha + \delta\beta + \gamma. \tag{6}$$

When ξ is a Killing vector field, the coexact part of this form is zero and the heat flow is associated to the difference of a certain state function between the initial and final slices:

$$\int_{\partial H} \alpha_k(\xi) + \int_H (\delta\beta_{(k)}(\xi) + \gamma_{(k)}(\xi)) \propto \delta Q. \tag{7}$$

This relation, which is understood, after [1], as a Clausius-like relation connecting heat fluxes with what can be interpreted as changes of gravitational entropy of horizons (in the case of black holes, for instance), suggests to identify the equilibrium entropy terms as

$$S_k(\sigma_i;\xi) \propto \int_{\sigma_i} \alpha_{(k)}(\xi), \tag{8}$$

the proportionality factor to be determined once the temperature is identified. Notice that this treatment has the same root of the analysis of Wald [6]. The advantage is that now we are able to highlight the source of the nonequilibrium terms, and the general structure, in order to be able to treat more general models.

These models easily support a (non-equilibrium [7]) thermodynamical interpretation and can predict the various terms appearing in a Clausius-like relation, including the internal entropy production terms, which are related to the coexact part of $\mathscr{G}_{(\kappa)}(\xi)$. This interpretation is ultimately possible because the effective equations for the metric can be put in a form that naturally conforms to Wheeler's boundary of a boundary principle.

This thermodynamical reading of the effective dynamics of the composite metric tensor is rather formal: the effective action for the metric fully depends on the assigned boundary data. This will become manifest from the fact that states with the same boundary metrics will evolve differently, since the latter do not exhaust the information encoded in the label κ. Furthermore, no semiclassical limit has ever been invoked in the discussion: the thermodynamical interpretation seems to hold even in the deep quantum regime. An equilibration condition, possibly in the form of an H-theorem, is then required.

Equilibrium is deeply related to a notion of time, a delicate subject to be discussed in theories with reparametrization invariance. We will assume the existence of a clock, such that it is possible to define the condition of equilibrium as a condition of stationarity of macroscopic variables under Hamiltonian evolution. We will then assume to work with a Schrödinger equation for the full density matrix of the deparametrized system,

$$i\hbar\frac{\partial}{\partial t}\hat{\rho} = [\hat{\rho}, \hat{H}], \tag{9}$$

where t is the time associated to the clock and $\hat{H}$ is the resulting Hamiltonian.

Decoherence will be the key to equilibration, leading to an equilibrium density matrix uniquely controlled by macroscopic geometrical data. The ideas that we are going to discuss here go along the lines of the attempts made to establish thermalization in quantum statistical mechanics [8].

The full Hilbert space will be decomposed in the tensor product of three Hilbert spaces,

$$\mathscr{H}_{\text{tot}} = \mathscr{H}_{\text{micro}} \otimes \mathscr{H}_{\text{grav}} \otimes \mathscr{H}_{\text{matter}}, \tag{10}$$

where $\mathscr{H}_{\text{micro}}$ is associated to the invisible microscopic configurations, $\mathscr{H}_{\text{grav}}$ is associated to the degrees of freedom giving the metric (for which we will use greek indices α, β, γ), and, finally, $\mathscr{H}_{\text{matter}}$ is associated to the matter degrees of freedom used to probe geometry.

Given the full density matrix for a bipartite system, we split its density matrix as follows:

$$\rho^{ai}_{bj} = \rho^{a}_{b}(\mathcal{A})\rho^{i}_{j}(\mathcal{B}) + \rho^{ai}_{bj}(\text{ent}), \tag{11}$$

where $\rho(\mathcal{A})$ and $\rho(\mathcal{B})$ are the reduced density matrices and $\rho(\text{ent})$ encodes the entanglement between the two subsystems. Using successive splittings as in (11) for our tripartite system, we can compute the effective equations for the gravitational and matter degrees of freedom, once the microscopic ones are traced away,

$$i\hbar\frac{\partial}{\partial t}\rho^{\alpha}_{\beta}(\text{grav}) = [H_{\text{eff}}(\text{grav}), \rho(\text{grav})]^{\alpha}_{\beta} + i\hbar\Xi^{\alpha}_{\beta}(\text{grav}), \tag{12}$$

where Ξ is associated to the effect of the entanglement, while the one for matter fields is similar in structure. As expected, decoherence is appearing in both gravity and matter sectors. However, the strength of the decoherence effects might be different in the two cases, since it ultimately depends on the entanglement with the microscopic degrees of freedom, to which matter degrees of freedom are decoupled, as we have assumed[1]. In order to equilibration to take place we need that $\Xi(\text{grav})$ has a large effect on the matrix $\rho(\text{grav})$.

In the interaction picture, we might expect that the density matrix $\hat{\rho}_{\text{int}}(\text{grav})$ approaches an asymptotic equilibrium configuration, $\hat{\rho}_0(\text{grav})$:

$$\hat{\rho}_{\text{int}}(\text{grav}) = \hat{\rho}_{\text{int}}(\text{grav})(t=0) + \int^t dt' \hat{\Xi}_{\text{int}}(\text{grav}) \to \hat{\rho}_0(\text{grav})(t). \tag{13}$$

Whether this happens or not strongly depends on the underlying dynamics, which influences also the kind of information about the initial state that is lost and the one that is preserved. If we want that the equilibration preserves the initial assignement of the spatial metric tensor g_{ij} (and, analogously, for the extrinsic curvature), we need

$$\Delta g_{ij} = \Delta t_{eq} \frac{\partial g_{ij}}{\partial t} = \frac{\partial}{\partial t} \mathrm{Tr}\left(\hat{\rho}(\text{grav})\hat{g}_{ij}\right) \approx 0, \tag{14}$$

so that the metric data do not change much during equilibration. It is only when this condition is met that the metric can be treated realistically as a thermodynamic variable, given that it really specifies uniquely the equilibrium density matrix at fixed macroscopic conditions. It is not unreasonable to expect that the equilibration time has to be of the order of the Planck time, for gravity. This is far from obvious and exceeds the limits of the present discussion [9].

Let us stress that equilibration is not the only path to the understanding of gravitational thermodynamics. Group field theories [10] promise to generate combinatorially the path integral for the gravitational field in terms of the Feynman diagrams for certain quantum/statistical systems, possessing an interpretation in terms of discrete geometries. To recover GR the continuum and semiclassical limit, phase transitions for the GFT partition functions are required (despite the lack of conclusive results, this is supported by various evidences).

A partial analysis [11] suggests that part of the geometrical data (*e.g.* a global scale factor), their dynamics and the associated density matrix for the microscopic and macroscopic degrees of freedom should depend only on the way in which the phase transition is approached, *i.e.* on its universality class. While this is certainly an attractive alternative to equilibration, it is not clear if it is the correct way, even though it might lead to a solution of old naturalness problems [12].

Acknowledgments
I would like to thank Goffredo Chirco and Arnau Riera for extensive discussions.

References

[1] Jacobson T 1995 *Phys. Rev. Lett.* **75** 1260
[2] Akama K, 1978 *Prog. Theor. Phys.* **60** 1900
[3] Sindoni L 2012 *SIGMA* **8** 027
[4] Sexty D and Wetterich C 2012 *Preprint* arXiv:1208.2168
[5] Kheyfets A 1986 *Found. Phys.* **16** 483
[6] Wald R M 1993 *Phys. Rev. D* **48** 3427
[7] Guedens R, Jacobson T and Sarkar S 2012 *Phys. Rev.* D **85** 064017
[8] Reimann P 2010 *New J. Phys.* **12** 055027
[9] Sindoni L 2012 On horizon thermodynamics for a composite metric, to appear
[10] Oriti D 2011 The microscopic dynamics of quantum space as a group field theory *Preprint* arXiv:1110.5606
[11] Sindoni L 2011 Gravity as an emergent phenomenon: a GFT perspective *Preprint* arXiv:1105.5687
[12] Finazzi S, Liberati S and Sindoni L 2012 *Phys. Rev. Lett.* **108** 071101

[1] Obviously, decoupling does not imply absence of entanglement.

IC-MSQUARE 2012: International Conference on Mathematical Modelling in Physical Sciences IOP Publishing
Journal of Physics: Conference Series **410** (2013) 012141 doi:10.1088/1742-6596/410/1/012141

Color Superconductors, Fermions in Reissner-Nordström-anti-de Sitter Spacetimes and Zero Central Charge Supersymmetry[1]

V K Oikonomou

Max Planck Institute for Mathematics in the Sciences
Inselstrasse 22, 04103 Leipzig, Germany

E-mail: `Vasilis.Oikonomou@mis.mpg.de`

Abstract. Two fermionic systems that have an underlying supersymmetric structure are studied, a color superconductor model and Dirac fermion in a Reissner-Nordström-anti-de Sitter gravitational background. In the chiral limit of both systems, an $N = 2$ with zero central charge $d = 1$ quantum algebra, underlies both systems.

During the last years, the relation between gravity and condensed matter physical systems has been considerably scrutinized with many valuable findings [1]. The experimental verification [2] of many topological states of condensed matter, made the motivation for further studying this relation quite stronger. In view of these correspondenses, and owing to the fact that these topological states of matter appear when ultra high quark densities are deployed, a question arises. Can color superconductivity be materialized in ultra high quark densities and is a color superconductor (CS hereafter) a topological insulator [3]. These conditions amount to condensed matter physics of QCD, something that can be true in neutron stars. The answer is in the affirmative as shown in [3]. In relation to this QCD related cooper pair condensation, color superconductivity might be relevant to a condensation that fermionic fields experience in evolved AdS gravitational backgrounds. There exists a similar pattern of physical phenomena in a holographic superconducting phase transition [1] studied in [4], but for an Reissner-Nordström-anti-de Sitter black hole (RN-AdS BH hereafter) spacetime, and the results where consistent with the holographic superconducting phase transition. In [4] the order parameter is a charged Dirac fermion directly coupled to an abelian gauge field. In this fashion, this model of a charged Dirac fermion in an RN-AdS BH background, could be a simple model of CS. The approach of the authors of [4] is done by computing the quasinormal modes of the charged Dirac fermion field in the aforementioned curved background. The quasinormal modes [5] describe long lasting period of damped gravitational waves oscillations and due to the few physical parameters of the black hole namely, the mass, angular momentum and charge of the black holes, these modes make the spacetime parameters identification easier. In addition these modes are used to AdS/CFT correspondence [6] theories. Due to the fact that the charged Dirac fermion in the background of an RN-AdS BH could model simply color superconductivity, we exploit in this paper that in

[1] Prepared for the International Conference on Mathematical Modeling in Physical Sciences September 2012 Budapest, Hungary

both the aforementioned systems there exists a hidden $N = 2$, $d = 1$ supersymmetric algebra [7, 8] (SUSY QM hereafter). Both systems are studied in their chiral limits $m = 0$. The very own existence of the SUSY QM algebra in the gravitational fermionic system is guaranteed by the existence of quasinormal modes, while for the other system supersymmetry is related to localized fermionic solutions around the vortex boundary of the CS. For details and similar structures in alternative backgrounds see [13] and [10, 11, 9] respectively. The SUSY QM algebras are very important since the supercharges could be related to harmonic superspace [12], or to $d = 1$ supersymmetric sigma models. It is intriguingly interesting that the color superconductors and the fermionic system in RN-AdS BH are linked via the same supersymmetric underlying pattern. We are interested in zero mode localized fermionic solutions around vortices, with a non-trivial pairing gap (for details see [13, 3]). These solutions correspond to the zero modes of the Bogoliubov-de Genne equation:

$$\begin{pmatrix} -i\mathbf{a}\partial + \beta m - \mu & e^{i\theta}|\Delta(r)| \\ e^{-i\theta}|\Delta(r)| & i\mathbf{a}\partial - \beta m + \mu \end{pmatrix} \Phi(x) = 0 \tag{1}$$

The pairing gap is $\Delta(x) = e^{i\theta}|\Delta(r)|$ with $\lim_{r\to\infty}|\Delta(r)| > 0$.. The solution to the above equation (??) is:

$$\Phi(r, \theta, z) = e^{ip_z z}\phi_{p_z}(r, \theta) \tag{2}$$

In the case $m = 0$ and $p_z = 0$ the exponentially localized solutions around the vortex are equal to [3]:

$$\varphi_R = \frac{e^{i\frac{\pi}{4}}}{\sqrt{\lambda}} \begin{pmatrix} J_0(\mu r) \\ ie^{i\theta}J_1(\mu r) \\ 0 \\ 0 \\ e^{-i\theta}J_1(\mu r) \\ -iJ_0(\mu r) \\ 0 \\ 0 \end{pmatrix} e^{-\int_0^r|\Delta(r')|dr'}, \quad \varphi_L = \frac{e^{-i\frac{\pi}{4}}}{\sqrt{\lambda}} \begin{pmatrix} 0 \\ 0 \\ J_0(\mu r) \\ -ie^{i\theta}J_1(\mu r) \\ 0 \\ 0 \\ e^{-i\theta}J_1(\mu r) \\ iJ_0(\mu r) \end{pmatrix} e^{-\int_0^r|\Delta(r')|dr'} \tag{3}$$

The above are classified according to their γ_5 parity. This fermionic system, which is formed by the zero modes of the Bogoliubov-de Genne equation, can constitute an $N = 2$ supersymmetric quantum mechanics algebra. Indeed, the Bogoliubov-de Genne equation can be written as:

$$D\Phi(x) = 0 \tag{4}$$

with D and $D^\dagger$ equal to:

$$D = \begin{pmatrix} -i\mathbf{a}\partial + \beta m - \mu & e^{i\theta}|\Delta(r)| \\ e^{-i\theta}|\Delta(r)| & i\mathbf{a}\partial - \beta m + \mu \end{pmatrix}, \quad D^\dagger = \begin{pmatrix} i\mathbf{a}\partial - \mu & e^{i\theta}|\Delta(r)| \\ e^{-i\theta}|\Delta(r)| & -i\mathbf{a}\partial + \mu \end{pmatrix}, \tag{5}$$

The above matrix can be used to built a supersymmetric algebra. The supercharges of the SUSY QM algebra, Q and $Q^\dagger$ can be defined in terms of D and $D^\dagger$ as,

$$Q = \begin{pmatrix} 0 & D \\ 0 & 0 \end{pmatrix}, \quad Q^\dagger = \begin{pmatrix} 0 & 0 \\ D^\dagger & 0 \end{pmatrix} \tag{6}$$

Additionally, the quantum Hamiltonian of the SUSY QM system is,

$$H = \begin{pmatrix} DD^\dagger & 0 \\ 0 & D^\dagger D \end{pmatrix} \tag{7}$$

It is easy to see that the supercharges (6) and the Hamiltonian and (7), satisfy the following relations:

$$\{Q,Q^{\dagger}\} = H \ ,Q^2 = 0,\ Q^{\dagger^2} = 0,\ \{Q,W\} = 0,\ W^2 = I,\ [W,H] = 0 \tag{8}$$

Supersymmetry is unbroken. Indeed the operators D and $D^{\dagger}$ defined in relation (5) are Fredholm for the localized solutions around the vortex. The vector space ker$D$ is given by the solutions of the equation $D\Phi = 0$, namely relation (4), while the localized left and right handed solutions of the equation $D^{\dagger}\Phi = 0$ are given by:

$$\varphi_R' = \frac{e^{i\frac{\pi}{4}}}{\sqrt{\lambda}} \begin{pmatrix} e^{i\theta}J_1(\mu r) \\ iJ_0(\mu r) \\ 0 \\ 0 \\ J_0(\mu r) \\ -ie^{-i\theta}J_1(\mu r) \\ 0 \\ 0 \end{pmatrix} e^{-\int_0^r |\Delta(r')|dr'},\ \varphi_L' = \frac{e^{-i\frac{\pi}{4}}}{\sqrt{\lambda}} \begin{pmatrix} 0 \\ 0 \\ e^{i\theta}J_1(\mu r) \\ -iJ_0(\mu r) \\ 0 \\ 0 \\ J_0(\mu r) \\ ie^{-i\theta}J_1(\mu r) \end{pmatrix} e^{-\int_0^r |\Delta(r')|dr'} \tag{9}$$

Likewise, as in the D operator case, the operator $D^{\dagger}$ is Fredholm with the two solutions constituting the space ker$D^{\dagger}$. Since the two operators D and $D^{\dagger}$ are Fredholm, the Fredholm index of the operator D is given by:

$$\text{ind}D = \dim \text{ker}D - \dim \text{ker}D^{\dagger} = \dim \text{ker}D^{\dagger}D - \dim \text{ker}DD^{\dagger} \tag{10}$$

Hence, the Witten index of the corresponding SUSY QM algebra is given by:

$$\Delta = -\text{ind}D \tag{11}$$

Based on the fact that ker$D$ = ker$D^{\dagger}$, as we found previously, the Witten index of the SUSY QM algebra is zero and since $n_- = n_+ \neq 0$, supersymmetry is unbroken. Following the same line of reasoning as above and using the RN-AdS BH spacetime formalism, it is proved [13] that we can construct two supersymmetric algebras N_1 and N_2. Indeed, N_1 is constructed in terms of the supercharges that are in turn constructed in terms of

$$D_{RN} = \begin{pmatrix} \partial_r - \frac{i}{f}(\omega + qA_t) & \frac{i}{r\sqrt{f}}k_x \\ -\frac{i}{r\sqrt{f}}k_x & \partial_r + \frac{i}{f}(\omega + qA_t) \end{pmatrix},\ D_{RN}^{\dagger} = \begin{pmatrix} \partial_r + \frac{i}{f}(\omega^* + qA_t) & \frac{i}{r\sqrt{f}}k_x \\ -\frac{i}{r\sqrt{f}}k_x & \partial_r - \frac{i}{f}(\omega^* + qA_t) \end{pmatrix} \tag{12}$$

The supercharges of the $N = 2$ algebra Q_{RN} and $Q_{RN}^{\dagger}$ are equal to:

$$Q_{RN} = \begin{pmatrix} 0 & D_{RN} \\ 0 & 0 \end{pmatrix},\ Q_{RN}^{\dagger} = \begin{pmatrix} 0 & 0 \\ D_{RN}^{\dagger} & 0 \end{pmatrix} \tag{13}$$

in the case of N_1, while N_2 is constructed from,

$$Q_{RN'} = \begin{pmatrix} 0 & D_{RN'} \\ 0 & 0 \end{pmatrix},\ Q_{RN'}^{\dagger} = \begin{pmatrix} 0 & 0 \\ D_{RN'}^{\dagger} & 0 \end{pmatrix} \tag{14}$$

with

$$D_{RN'} = \begin{pmatrix} \partial_r - \frac{i}{f}(-\omega - qA_t) & -\frac{i}{r\sqrt{f}}k_x \\ \frac{i}{r\sqrt{f}}k_x & \partial_r + \frac{i}{f}(-\omega - qA_t) \end{pmatrix} \tag{15}$$

These operators are not Fredholm and a regularized version of the Witten index is needed. The heat-kernel regularized index [8], both for the operator D_{RN}, denoted $\mathrm{ind}_t D_{RN}$ and for the Witten index, Δ_t, which are defined as:

$$\mathrm{ind}_t D_{RN} = \mathrm{Tr}(-We^{-tD_{RN}^{\dagger}D_{RN}}) = \mathrm{tr}_{-}(-We^{-tD_{RN}^{\dagger}D_{RN}}) - \mathrm{tr}_{+}(-We^{-tD_{RN}D_{RN}^{\dagger}}) \tag{16}$$
$$\Delta_t = \lim_{t\to\infty} \mathrm{ind}_t D_{RN}$$

Since $\mathrm{ker}D_{RN} \neq 0$, the zero modes equation for the operator $D_{RN}^{\dagger}$ can yield two results, either that $\mathrm{ker}D_{RN}^{\dagger} \neq 0$ or that $\mathrm{ker}D_{RN}^{\dagger} = 0$. If the second is true, then the Witten index is different than zero, $\Delta_t \neq 0$, hence supersymmetry is unbroken. In the first case, $\mathrm{ker}D_{RN}^{\dagger} \neq 0$, it can either be that $\mathrm{ker}D_{RN}^{\dagger} = \mathrm{ker}D_{RN}$ or that $\mathrm{ker}D_{RN}^{\dagger} \neq \mathrm{ker}D_{RN}$. In both cases supersymmetry is unbroken (for details see [13]).

The fermionic system in the Reissner-Nordström-anti-de Sitter black hole background has two $N = 2$ supersymmetric quantum mechanics algebras. The two superalgebras are invariant under the transformations:

$$Q'_{RN} = e^{-ia}Q_{RN}, \qquad Q'^{\dagger}_{RN} = e^{ia}Q^{\dagger}_{RN} \tag{17}$$
$$Q'_{RN'} = e^{-ia'}Q_{RN'}, \qquad Q'^{\dagger}_{RN'} = e^{ia'}Q^{\dagger}_{RN'}$$

Hence, each quantum system is invariant under a global-$U(1)$ symmetry. Correspondingly, the total system is invariant under an $U(1) \times U(1)$ symmetry. Each of the aforementioned $U(1)$ symmetries is a Hilbert states symmetry namely of the spaces, $\mathcal{H_{RN}}^{+}$, $\mathcal{H_{RN}}^{-}$ and $\mathcal{H_{RN'}}^{+}$, $\mathcal{H_{RN'}}^{-}$. Let ψ_{RN}^{+} and ψ_{RN}^{-} denote the Hilbert states corresponding to the spaces $\mathcal{H}_{RN}^{+}$ and $\mathcal{H}_{RN}^{-}$. Then, the $U(1)$ transformation of the states is equal to,

$$\psi'^{+}_{RN} = e^{-i\beta_{+}}\psi^{+}_{RN}, \qquad \psi'^{-}_{RN} = e^{-i\beta_{-}}\psi^{-}_{RN} \tag{18}$$

with $a = \beta_{+} - \beta_{-}$. In the same vain, for the spaces $\mathcal{H_{RN'}}^{+}$, $\mathcal{H_{RN'}}^{-}$ we have,

$$\psi'^{+}_{RN'} = e^{-i\beta'_{+}}\psi^{+}_{RN'}, \qquad \psi'^{-}_{RN'} = e^{-i\beta'_{-}}\psi^{-}_{RN'} \tag{19}$$

with $\psi^{+}_{RN'}$ and $\psi^{-}_{RN'}$ the Hilbert states of the spaces $\mathcal{H_{RN'}}^{+}$, $\mathcal{H_{RN'}}^{-}$.

The two fermionic systems we presented are believed to be interrelated, with the fermionic system in curved background being a promising candidate for describing the color superconductor. Thus, the supersymmetries we found, indicate that under certain very general assumptions, these two systems have an underlying supersymmetric structure, of $N = 2$, $d = 1$ type. However, the supersymmetries of the two spaces have different Hilbert space structure, since in the color superconductor case the operators are Fredholm and in the other case they are not.

References

[1] Gubser Steven S, *Phys.Rev.* D **78**, 065034 (2008)
[2] Koenig M et al., *Science* **318**, 766 (2007)
[3] Nishida Y, *Phys.Rev.*D **81**, 074004 (2010)
[4] Rong-Gen Cai, Zhang-Yu Nie, Bin Wang, Hai-Qing Zhang, *Preprint* arXiv:1005.1233
[5] Kokkotas K D, Schmidt, B G, *Living Rev. Rel.* **2** , 2 (1999)
[6] Kanti, P, *Int. J. Mod. Phys.* A **19**, 4899 (2004)
[7] Thaller Bernd, *The Dirac Equation* Springer 1992
[8] Junker G. , *Supersymmetric Methods in Quantum and Statistical Physics*, Springer, 1996
[9] Oikonomou V K, *Preprint* arXiv: 1204.2395
[10] Oikonomou V K, *Gen. Rel. Grav.*, **44**, 1285 (2012)
[11] Oikonomou V K, *Mod. Phys. Lett.* A **25**,2611 (2010)
[12] Galperin A S, Ivanov E A, Ogievetsky V I, Sokatchev E S, *Harmonic Superspace*, Cambridge University Press 2001
[13] Oikonomou V K, *Preprint* arXiv:1208.0740

A note on the Newman-Unti group and the BMS charge algebra in terms of Newman-Penrose coefficients

Glenn Barnich and Pierre-Henry Lambert

Physique Théorique et Mathématique, Université Libre de Bruxelles and International Solvay Institutes, Campus Plaine C.P. 231, B-1050 Bruxelles, Belgium

E-mail: gbarnich@ulb.ac.be, Pierre-Henry.Lambert@ulb.ac.be

Abstract. The symmetry algebra of asymptotically flat spacetimes at null infinity in four dimensions in the sense of Newman and Unti is revisited. As in the Bondi-Metzner-Sachs gauge, it is shown to be isomorphic to the direct sum of the abelian algebra of infinitesimal conformal rescalings with $\mathfrak{bms}_4$. The latter algebra is the semi-direct sum of infinitesimal supertranslations with the conformal Killing vectors of the Riemann sphere. Infinitesimal local conformal transformations can then consistently be included. We work out the local conformal properties of the relevant Newman-Penrose coefficients, construct the surface charges and derive their algebra.

1. Introduction

The definitions of asymptotically flat four dimensional space-times at null infinity by Bondi-Van der Burg-Metzner-Sachs [1, 2] (BMS) and Newman-Unti (NU) [3] in 1962 merely differ by the choice of the radial coordinate. Such a change of gauge should not affect the asymptotic symmetry algebra if, as we contend, this concept is to have a major physical significance.

The problem of comparing the symmetry algebra in both cases is that, besides the difference in gauge, the very definitions of these algebras are not the same. Indeed, NU allow the leading part of the metric induced on Scri to undergo a conformal rescaling. When this generalization is considered in the BMS setting, it turns out that the symmetry algebra is the direct sum of the BMS algebra $\mathfrak{bms}_4$ [4] with the abelian algebra of infinitesimal conformal rescalings [5], [6]. There are two novel and independent aspects in this computation.

- The first concerns the fact that the BMS algebra in 4 dimension involves the conformal Killing vectors of the unit, or equivalently, the Riemann sphere and can consistently accommodate infinitesimal local conformal transformations. The symmetry algebra $\mathfrak{bms}_4$ then involves two commuting copies of the non centrally extended Virasoro algebra, called superrotations in [7], and simultaneously the supertranslations generators are expanded in Laurent series. The standard, globally well-defined symmetry algebra $\mathfrak{bms}_4^{\text{glob}}$ consists in restricting to the globally well defined conformal Killing vectors of the sphere which correspond to infinitesimal Lorentz transformation, while the supertranslation generators are expanded into spherical harmonics.
  This local versus global versions of the symmetry algebra are of course not related to the BMS gauge choice, but will also occur in alternative characterizations of the asymptotic symmetry algebra where the conformal Killing vectors of the sphere play a role. Examples of this are the geometrical approach of Geroch [8] based on Penrose's definition of null infinity [9] and also, as we will explicitly discuss in this paper, the asymptotic symmetries in the NU framework.
- The second aspect is related to the modified Lie bracket that should be used when the vector fields parametrising infinitesimal diffeomorphisms depend explicitly on the metric. Indeed, when using the modified Lie bracket, the space-time vectors realize the asymptotic symmetry algebra everywhere in the bulk and furthermore, even on Scri, this bracket is needed to disentangle the algebra when conformal rescalings of the induced metric on Scri are allowed. Similarly, in the context of the AdS/CFT correspondence, this bracket allows one to realize the asymptotic symmetry algebra in the bulk and to disentangle the symmetry algebra at infinity when considering transformations that leave the Fefferman-Graham ansatz invariant only up to conformal rescaling of the boundary metric [10]. From a mathematical point of view, the modified Lie bracket is the natural bracket of the Lie algebroid that is associated to any theory with gauge invariance [11].

What we will do in this paper is to re-derive from scratch the asymptotic symmetry algebra in the NU framework by focusing on metric aspects and on the two novel features discussed above. As expected, the symmetry algebra is again the direct sum of $\mathfrak{bms}_4$ with the abelian algebra of infinitesimal conformal rescalings of the metric on Scri and thus coincides, as it should, with the generalized symmetry algebra in the BMS approach. A related analysis of asymptotic symmetries in the NU context from the point of view of Scri and emphasizing global issues instead can be found in [12], [13].

Even though the results presented here are not really surprising in view of those in the BMS framework and the close relation between the NU and BMS approaches, the exercise of working out the details is justified because the NU framework is embedded in the context of the widely used Newman-Penrose formalism [14] so that explicit formulae in this context are directly relevant in many applications, see e.g. the review article [15].

As a first application, we study the transformation properties of the Newman-Penrose coefficients parametrizing solution space in the NU approach. Our main focus is on the inhomogeneous terms in the transformation laws that contain the information on the central extensions of the theory. We then discuss the associated surface charges by following the analysis in the BMS gauge [16] and briefly compare with standard expressions that can be found in the literature. The algebra of these charges is derived and shown to involve field dependent central charges in the case of $\mathfrak{bms}_4$ which vanish for $\mathfrak{bms}_4^{\text{glob}}$.

2. NU metric ansatz for asymptotically flat spacetimes

The metric ansatz of NU is based on a family of null hypersurfaces labelled by the first coordinate, $x^0 \equiv u = \text{const}$. The second coordinate $x^1 \equiv r$ is chosen as an affine parameter for the null geodesic generators l^μ of these hypersurfaces, so that $l^\mu = -\delta^\mu_r$. Up to a change of signature from $(+,-,-,-)$ to $(-,+,+,+)$, a renumbering of the indices and the tetrad transformation that makes P_{NU} real, the line element considered in section 4 of NU [3] can be written as

$$ds^2 = W du^2 - 2 dr du + g_{AB}(dx^A - V^A du)(dx^B - V^B du) , \qquad (1)$$

with associated inverse metric

$$g^{\mu\nu} = \begin{pmatrix} 0 & -1 & 0 \\ -1 & -W & -V^B \\ 0 & -V^A & g^{AB} \end{pmatrix} , \qquad (2)$$

where

$$g_{AB} dx^A dx^B = r^2 \bar{\gamma}_{AB} dx^A dx^B + r C_{AB} dx^A dx^B + o(r) , \qquad (3)$$

with $\bar{\gamma}_{AB}$ conformally flat. Below, we will use standard stereographic coordinates $\zeta = \cot \frac{\theta}{2} e^{i\phi}, \bar{\zeta}$, $\bar{\gamma}_{AB} dx^A dx^B = e^{2\tilde{\varphi}} d\zeta d\bar{\zeta}$, $\tilde{\varphi} = \tilde{\varphi}(u, x)$.

In addition, the choice of origin for the affine parameter of the null geodesics is fixed through the requirement that the term proportional to r^{-2} in the expansion of the spin coefficient $-\rho = D_\rho l_\nu m^\rho \bar{m}^\nu$ is absent.

When expressed in terms of the metric, one finds

$$\rho = -\frac{1}{4} g^{AB} g_{AB,r} = -\frac{1}{4} \partial_r \ln |g| = -r^{-1} + \frac{1}{4} C^A_A r^{-2} + o(r^{-2}) , \qquad (4)$$

where $g = \det g_{\rho\nu}$ and the index has been raised with the inverse of $\bar{\gamma}_{AB}$. The requirement is thus equivalent to the condition

$$C^A_A = 0\,. \tag{5}$$

In the following we denote by $\bar{D}_A$ the covariant derivative with respect to $\bar{\gamma}_{AB}$ and by $\bar{\Delta}$ the associated Laplacian and by $\bar{R}$ the scalar curvature. In complex coordinates $\zeta, \bar{\zeta}$, $C_{\zeta\bar{\zeta}} = 0$ and we define for later convenience $C_{\zeta\zeta} = e^{2\widetilde{\varphi}}c$, $C_{\bar{\zeta}\bar{\zeta}} = e^{2\widetilde{\varphi}}\bar{c}$. Finally,

$$V^A = O(r^{-2}), \qquad W = -2r\partial_u\widetilde{\varphi} + \bar{\Delta}\widetilde{\varphi} + O(r^{-1})\,, \tag{6}$$

where $\bar{\Delta}\widetilde{\varphi} = 4e^{-2\widetilde{\varphi}}\partial\bar{\partial}\widetilde{\varphi}$ with $\partial = \partial_\zeta, \bar{\partial} = \partial_{\bar{\zeta}}$.

The more restrictive fall-off conditions in [3] are relevant for integrating the field equations but play no role in the discussion of the asymptotic symmetry algebra.

3. Asymptotic symmetries in the NU approach

The infinitesimal NU transformations can be defined as those infinitesimal transformations that leave the form (2) and the fall-off conditions (3)-(6) invariant, up to a rescaling of the conformal factor $\delta\widetilde{\varphi}(u, x^A) = \widetilde{\omega}(u, x^A)$. In other words, they satisfy

$$\mathcal{L}_\xi g^{uu} = 0, \quad \mathcal{L}_\xi g^{uA} = 0, \quad \mathcal{L}_\xi g^{ur} = 0, \tag{7}$$

$$\partial_r\Big[\frac{1}{\sqrt{|g|}}\partial_\rho(\sqrt{|g|}\xi^\rho)\Big] = o(r^{-2})\,, \tag{8}$$

$$\begin{aligned} \mathcal{L}_\xi g^{rA} = O(r^{-2}), \quad \mathcal{L}_\xi g^{AB} = -2\widetilde{\omega} g^{AB} + O(r^{-3}), \\ \mathcal{L}_\xi g^{rr} = 2r\partial_u\widetilde{\omega} + 2\widetilde{\omega}\bar{\Delta}\widetilde{\varphi} - \bar{\Delta}\widetilde{\omega} + O(r^{-1})\,. \end{aligned} \tag{9}$$

Equations (7) are equivalent to

$$\partial_r\xi^\nu = g^{\nu\rho}\partial_\rho\xi^u \iff \begin{cases} \partial_r\xi^u = 0\,, \\ \partial_r\xi^A = \partial_B\xi^u g^{BA}\,, \\ \partial_r\xi^r = -\partial_u\xi^u - \partial_A\xi^u V^A\,, \end{cases} \tag{10}$$

and are explicitly solved by

$$\begin{cases} \xi^u = f, \\ \xi^A = Y^A + I^A, \quad I^A = -\partial_B f \int_r^\infty dr' g^{AB}, \\ \xi^r = -r\partial_u f + Z + J, \quad J = \partial_A f \int_r^\infty dr' V^A, \end{cases} \tag{11}$$

with $\partial_r f = 0 = \partial_r Y^A = \partial_r Z$. Equation (8) then implies

$$Z = \frac{1}{2}\bar{\Delta} f\,. \tag{12}$$

The first equation of (9) requires $\partial_u Y^A = 0$, the second that Y^A is a conformal Killing vector of $\bar{\gamma}_{AB}$, which amounts to

$$Y^\zeta \equiv Y = Y(\zeta), \quad Y^{\bar{\zeta}} \equiv \bar{Y} = \bar{Y}(\bar{\zeta}), \tag{13}$$

in the coordinates $\zeta, \bar{\zeta}$, and also that

$$\partial_u f = f\partial_u \widetilde{\varphi} + \frac{1}{2}\widetilde{\psi}, \tag{14}$$

with $\psi = \bar{D}_A Y^A$, or more explicitly in $\zeta, \bar{\zeta}$ coordinates, $\psi = \partial Y + \bar{\partial}\bar{Y} + 2Y\partial\widetilde{\varphi} + 2\bar{Y}\bar{\partial}\widetilde{\varphi}$, and $\widetilde{\psi} = \psi - 2\widetilde{\omega}$. Finally, the last equation of (9) implies

$$2(\partial_u Z + Z\partial_u \widetilde{\varphi}) = Y^A \partial_A \bar{\Delta}\widetilde{\varphi} + \psi\bar{\Delta}\widetilde{\varphi} + 2\partial_A f \bar{\gamma}^{AB}\partial_B \partial_u \widetilde{\varphi} + f\bar{\Delta}\partial_u \widetilde{\varphi} - \bar{\Delta}\widetilde{\omega}, \tag{15}$$

which is identically satisfied when taking the previous relations into account.

One approach is to consider that (14) fixes $\widetilde{\omega}$ in terms of f and Y, $\widetilde{\omega} = \frac{1}{2}\psi + f\partial_u\widetilde{\varphi} - \partial_u f$. Consider Scri, the space $\mathscr{I}$ with coordinates $u, \zeta, \bar{\zeta}$ and metric

$$ds^2_{\mathscr{I}} = 0du^2 + e^{2\widetilde{\varphi}} d\zeta d\bar{\zeta}. \tag{16}$$

The NU algebra is then defined as the commutator algebra of the vector fields

$$\xi = f\frac{\partial}{\partial u} + Y^A \frac{\partial}{\partial x^A}, \tag{17}$$

with $f = f(u, x^A)$ arbitrary and $Y^A(x)$ conformal Killing vectors of a conformally flat metric in 2 dimensions, or equivalently, the algebra of conformal vector fields of the degenerate metric (16).

This is not the symmetry algebra of asymptotically flat spacetimes in the sense of NU however. Indeed, $\widetilde{\varphi}$ is arbitrary, it can for instance be considered as the finite ambiguity related to Penrose's conformal approach [9, 17, 18] to null infinity. One can then interpret $\widetilde{\varphi}$ as part of the background structure, or in other words, of the gauge fixing [8], and compute the asymptotic symmetries for a fixed choice of $\widetilde{\varphi}$, i.e., $\widetilde{\omega} = 0$ in the formulae above, or ask the more general question of how the asymptotic symmetries depend on changes in $\widetilde{\varphi}$ by an arbitrary infinitesimal amount $\widetilde{\omega}$. In both cases, one has to consider (14) as a differential equation for f. As we now show, the symmetry algebra will then be isomorphic to the trivially extended $\mathfrak{bms}_4$ algebra by the abelian algebra of infinitesimal conformal rescalings, as it should, and as a consequence, the Poincaré algebra is embedded therein in a natural way. Furthermore, there is a natural realization of the asymptotic symmetry algebra on an asymptotically flat 4 dimensional bulk spacetime. Note also that, for $\widetilde{\omega} = 0$, equation (14) has been interpreted from the point of view of Penrose's conformal approach to null infinity in [12] following [19] and related to the preservation of null angles, which is the standard way [9, 17, 20, 21] to recover the BMS algebra from geometrical data on Scri.

The general solution for (14) reads

$$f = e^{\widetilde{\varphi}}\big[\widetilde{T} + \frac{1}{2}\int_0^u du' e^{-\widetilde{\varphi}}\widetilde{\psi}\big], \quad \widetilde{T} = \widetilde{T}(\zeta, \bar{\zeta}), \tag{18}$$

and the general solution to equations (7)-(9) defining the asymptotic symmetries is given by ξ^ρ as in (11) where Z,Y^A,f satisfy (12), (13), (18) with $\widetilde{\omega}$ arbitrary. Asymptotic Killing vectors thus depend on $Y^A, \widetilde{T}, \widetilde{\omega}$ and the metric, $\xi = \xi[Y, \widetilde{T}, \widetilde{\omega}; g]$.

For such metric dependent vector fields, consider on the one hand the suitably modified Lie bracket taking the metric dependence of the spacetime vectors into account,

$$[\xi_1, \xi_2]_M = [\xi_1, \xi_2] - \delta^g_{\xi_1}\xi_2 + \delta^g_{\xi_2}\xi_1, \tag{19}$$

where $\delta^g_{\xi_1}\xi_2$ denotes the variation in ξ_2 under the variation of the metric induced by ξ_1, $\delta^g_{\xi_1} g_{\mu\nu} = \mathcal{L}_{\xi_1} g_{\mu\nu}$.

Consider on the other hand the extended $\mathfrak{bms}_4$ algebra, i.e., the semi-direct sum of the algebra of conformal Killing vectors of the Riemann sphere with the abelian ideal of infinitesimal supertranslations, trivially extended by infinitesimal conformal rescalings of the conformally flat degenerate metric on Scri. More explicitly, the commutation relations are given by $[(Y_1, \widetilde{T}_1, \widetilde{\omega}_1), (Y_2, \widetilde{T}_2, \widetilde{\omega}_2)] = (\widehat{Y}, \widehat{\widetilde{T}}, \widehat{\widetilde{\omega}})$ where

$$\begin{cases} \widehat{Y}^A = Y_1^B\partial_B Y_2^A - Y_2^B\partial_B Y_1^A, \\ \widehat{\widetilde{T}} = Y_1^A\partial_A\widetilde{T}_2 - Y_2^A\partial_A\widetilde{T}_1 + \frac{1}{2}(\widetilde{T}_1\partial_A Y_2^A - \widetilde{T}_2\partial_A Y_1^A), \\ \widehat{\widetilde{\omega}} = 0\,. \end{cases} \tag{20}$$

It thus follows that

Theorem 3.1. *The spacetime vectors $\xi[Y, \widetilde{T}, \widetilde{\omega}; g]$ realize the extended $\mathfrak{bms}_4$ algebra in the modified Lie bracket,*

$$\Big[\xi[Y_1, \widetilde{T}_1, \widetilde{\omega}_1; g], \xi[Y_2, \widetilde{T}_2, \widetilde{\omega}_2; g]\Big]_M = \xi[\widehat{Y}, \widehat{\widetilde{T}}, \widehat{\widetilde{\omega}}; g]\,, \tag{21}$$

in the bulk of an asymptotically flat spacetime in the sense of Newman and Unti.

Note in particular that for two different choices of the conformal factor $\widetilde{\varphi}$ which is held fixed, $\widetilde{\omega} = 0$, the asymptotic symmetry algebras are isomorphic to $\mathfrak{bms}_4$, which is thus a gauge invariant statement.

Proof. The proof follows closely the one in [6] for the BMS gauge. In order to be self-contained we recall the different steps here. In a first stage, one shows that on $\mathscr{I}$, the vectors fields $\bar{\xi}[Y,\widetilde{T},\widetilde{\omega};\widetilde{\gamma}]$ given in (17) with f as in (18) realize the extended $\mathfrak{bms}_4$ algebra in terms of the modified Lie bracket. Indeed, this is obvious for the A components which do not depend on the metric so that the modified bracket reduces to the standard Lie bracket for these components. For the u component, taking into account that

$$\delta^g_{\bar{\xi}_1} f_2 = \widetilde{\omega}_1 f_2 + \frac{1}{2}e^{\widetilde{\varphi}}\int_0^u du' e^{-\widetilde{\varphi}}[-\widetilde{\omega}_1(\psi_2 - 2\widetilde{\omega}_2) + 2Y_2^A\partial_A\widetilde{\omega}_1],$$

we have $[\bar{\xi}_1,\bar{\xi}_2]^u_M|_{u=0} = e^{\widetilde{\varphi}}|_{u=0}\widehat{T}$. Direct computation then shows that $\partial_u([\bar{\xi}_1,\bar{\xi}_2]^u_M) = \widehat{f}\partial_u\widetilde{\varphi} + \frac{1}{2}\bar{D}_A\widehat{Y}^A$ with $\widehat{f}$ given by (18) with $\widetilde{T},Y,\widetilde{\omega}$ replaced by their hatted counterparts, implying the result for the u component.

For the spacetime vectors, direct computation gives $[\xi_1,\xi_2]^u_M = [\bar{\xi}_1,\bar{\xi}_2]^u_M = \widehat{f}$. Using the defining property (10), one then finds that $\partial_r([\xi_1,\xi_2]^\rho_M) = g^{\rho\nu}\partial_\nu\widehat{f}$. For the A components the result then follows from the one on $\mathscr{I}$, $\lim_{r\to\infty}[\xi_1,\xi_2]^A_M = \widehat{Y}^A$. This is due to the fact that I^A goes to zero at infinity, that the non-vanishing term at infinity does not involve the metric and that the correction term in the bracket does not change the asymptotic behaviour. Finally, for the r component, we still need to check that the r independent component of $[\xi_1,\xi_2]^r_M$ is given by $\frac{1}{2}\bar{\Delta}\widehat{f}$, which follows by direct computation. □

For completeness, let us also stress here that, if one focuses on local properties and expands the conformal Killing vectors $Y^A\partial_A$ and the infinitesimal supertranslations T in Laurent series,

$$l_n = -\zeta^{n+1}\frac{\partial}{\partial\zeta},\quad \bar{l}_n = -\bar{\zeta}^{n+1}\frac{\partial}{\partial\bar{\zeta}},\quad n\in\mathbb{Z}, \tag{22}$$

$$\widetilde{T}_{m,n} = \zeta^m\bar{\zeta}^n,\quad m,n\in\mathbb{Z}, \tag{23}$$

the commutation relations for the complexified $\mathfrak{bms}_4$ algebra read

$$\begin{aligned} &[l_m,l_n] = (m-n)l_{m+n},\quad [\bar{l}_m,\bar{l}_n] = (m-n)\bar{l}_{m+n},\quad [l_m,\bar{l}_n] = 0,\\ &[l_l,T_{m,n}] = (\frac{l+1}{2} - m)T_{m+l,n},\quad [\bar{l}_l,T_{m,n}] = (\frac{l+1}{2} - n)T_{m,n+l}. \end{aligned} \tag{24}$$

The $\mathfrak{bms}_4$ algebra contains as subalgebra the Poincaré algebra, which we identify with the algebra of exact Killing vectors of the Minkowski metric equipped with the standard Lie bracket. It is spanned by the generators

$$l_{-1},\, l_0,\, l_1,\quad \bar{l}_{-1},\, \bar{l}_0,\, \bar{l}_1,\quad \widetilde{T}_{0,0},\, \widetilde{T}_{1,0},\, \widetilde{T}_{0,1},\, \widetilde{T}_{1,1}\,. \tag{25}$$

Non trivial central extensions of the algebra (24) have been studied in [7]: the computation of $H^2(\mathfrak{bms}_4)$ reveals that there are only the standard ones for the Virasoro algebra extending the first two commutation relations.

4. Explicit relation between the NU and the BMS gauges

The definition of asymptotically flat space-times in the BMS approach [1], [2], [4] as reviewed in [5], [6], amounts to replacing $g_{uu} = 1/g^{uu} = -1$ by

$$g_{uu} = 1/g^{uu} = -e^{2\beta}, \qquad \beta = O(r^{-2}) \tag{26}$$

in (1) and (2) while imposing the additional requirement that

$$\det g_{AB} = r^4 \det \bar{\gamma}_{AB} \,. \tag{27}$$

Both definitions then differ just by a choice of radial coordinate. Indeed, replacing the radial coordinate by a function of the 4 coordinates preserves the zeros in (1) and (2) (see e.g. the discussion in [22]). Furthermore, to first non trivial order in r, the determinant condition leads to the same restriction (5) as the choice of the origin of the affine parameter. It follows that the relation between the two radial coordinates does not involve constant terms and is of the form

$$r' = r + O(r^{-1})\,. \tag{28}$$

More explicitly, starting from the NU approach, BMS coordinates are obtained by defining the new radial coordinates as [23]

$$r_{\rm BMS} = \left(\frac{\det g_{AB}}{\det \bar{\gamma}_{AB}}\right)^{\frac{1}{4}}. \tag{29}$$

Conversely, starting from the BMS approach with radial coordinate r, NU coordinates are obtained by changing the radial coordinate to

$$r_{\rm N} = r - \int_r^\infty dr'(e^{2\beta} - 1)\,. \tag{30}$$

These changes of coordinates only affect lower order terms in the asymptotic expansion of the metric that play no role in the definition of asymptotic symmetries and explains a posteriori why the asymptotic symmetry algebras in both approaches are isomorphic.

We will now work out the explicit relation between the free data characterizing asymptotic solution space in both approaches. The inverse metric in the BMS gauge (as discussed in [6]) is given by

$$g^{\mu\nu}_{BMS} = \begin{pmatrix} 0 & -e^{-2\beta} & 0 \\ -e^{-2\beta} & -e^{-2\beta}\frac{V}{r} & -e^{-2\beta}U^B \\ 0 & -e^{-2\beta}U^A & g^{AB} \end{pmatrix}. \tag{31}$$

$$g_{AB} = r^2\bar{\gamma}_{AB} + rC_{AB} + \frac{1}{4}\bar{\gamma}_{AB}C^C_D C^D_C + O(r^{-1}), \tag{32}$$

For simplicity, we assume here that there is no trace-free part D_{AB} at order 0 and that the conformal factor is time-independent, $\partial_u \widetilde{\varphi} = 0$, in which case the news tensor is simply $N_{AB} = \partial_u C_{AB}$ and $f = T + \frac{1}{2}u\widetilde{\psi}$ with $T = e^{\widetilde{\varphi}}\widetilde{T}$. Writing

$$C_{\zeta\zeta} = e^{2\widetilde{\varphi}}c, \quad C_{\bar{\zeta}\bar{\zeta}} = e^{2\widetilde{\varphi}}\bar{c}, \quad C_{\zeta\bar{\zeta}} = 0, \tag{33}$$

we have

$$\begin{aligned}
\beta &= -\frac{1}{4}r^{-2}c\bar{c} + O(r^{-4}),\\
U^\zeta &= -\frac{2}{r^2}e^{-4\widetilde{\varphi}}\partial(e^{2\widetilde{\varphi}}\bar{c}) - \frac{2}{3r^3}\Big[N^\zeta - 4e^{-4\widetilde{\varphi}}\bar{c}\bar{\partial}(e^{2\widetilde{\varphi}}c)\Big] + O(r^{-4}),\\
\frac{V}{r} &= 4e^{-2\widetilde{\varphi}}\partial\bar{\partial}\widetilde{\varphi} + r^{-1}2M + O(r^{-2}),
\end{aligned} \tag{34}$$

which implies in particular that

$$r_{\mathrm{N}} = r + \frac{c\bar{c}}{2r} + O(r^{-3})\,. \tag{35}$$

The angular momentum and mass aspects $N^\zeta = N^\zeta(u,\zeta,\bar{\zeta}), M = M(u,\zeta,\bar{\zeta})$ satisfy the evolution equations

$$\partial_u M = -\frac{1}{8}N^A_B N^B_A + \frac{1}{8}\bar{\Delta}\bar{R} + \frac{1}{4}\bar{D}_A\bar{D}_C N^{CA}, \tag{36}$$

$$\begin{aligned}
\partial_u N_A = \partial_A M + \frac{1}{4}C^B_A\partial_B\bar{R} + \frac{1}{16}\partial_A\big[N^B_C C^C_B\big] - \frac{1}{4}\bar{D}_A C^C_B N^B_C\\
- \frac{1}{4}\bar{D}_B\big[C^B_C N^C_A - N^B_C C^C_A\big] - \frac{1}{4}\bar{D}_B\big[\bar{D}^B\bar{D}_C C^C_A - \bar{D}_A\bar{D}_C C^{BC}\big].
\end{aligned} \tag{37}$$

Consider now the "eth" operators [24] defined here for a field η^s of spin weight s according to the conventions of [25] through

$$\eth\eta^s = P^{1-s}\bar{\partial}(P^s\eta^s), \qquad \bar{\eth}\eta^s = P^{1+s}\partial(P^{-s}\eta^s)\,, \qquad P = \sqrt{2}e^{-\widetilde{\varphi}}\,, \tag{38}$$

where $\eth, \bar{\eth}$ raise respectively lower the spin weight by one unit and satisfy

$$[\bar{\eth}, \eth]\eta^s = \frac{s}{2}\bar{R}\,\eta^s\,. \tag{39}$$

The spin weights of the various quantities are summarized in table 1. Note that the P used here differs from the one used in [3], which we will denote by P_N below. It also no longer denotes the particular function $\frac{1}{2}(1+\zeta\bar{\zeta})$, contrary to the notation used in [6, 16]. In the current conventions, the particular value of P adapted to the unit sphere is $\frac{1}{\sqrt{2}}(1+\zeta\bar{\zeta})$.

In order to compare with the notation used in [3], we use $\zeta = x^3 + ix^4$. With $x'^\alpha = u, r_{\mathrm{N}}, x^3, x^4$ and $x^\mu = u, r, \zeta, \bar{\zeta}$, computing $g^{\alpha\beta}_{\mathrm{N}}(x') = -\Big(\frac{\partial x'^\alpha}{\partial x^\mu}g^{\mu\nu}_{\mathrm{BMS}}\frac{\partial x'^\beta}{\partial x^\nu}\Big)(x(x'))$, where the overall minus sign takes the change of signature into account, then gives the following dictionary by comparing with [3]:

$$P_N = \frac{1}{\sqrt{2}}e^{-\widetilde{\varphi}} = \frac{1}{2}P\,, \quad \nabla = 2\bar{\partial}\,, \quad \mu^0 = -P^2\partial\bar{\partial}\ln P = \frac{1}{2}\bar{\Delta}\widetilde{\varphi} = -\frac{1}{4}\bar{R}\,,$$
$$\Psi^0_2 + \bar{\Psi}^0_2 = -2M - \partial_u(c\bar{c})\,, \quad \sigma^0 = \bar{c}\,, \qquad \omega^0 = \bar{\eth}\sigma^0\,, \tag{40}$$
$$\Psi^0_1 = -PN_{\bar{\zeta}} - \sigma^0\eth\bar{\sigma}^0 - \frac{3}{4}\eth(\sigma^0\bar{\sigma}^0)\,.$$

For convenience, let us also use

$$\Psi^0_3 = -\eth\dot{\bar{\sigma}}^0 - \frac{1}{4}\bar{\eth}\bar{R}, \qquad \Psi^0_4 = -\ddot{\bar{\sigma}}^0\,. \tag{41}$$

In these terms,

$$\dot{\Psi}^0_3 = \eth\Psi^0_4, \quad \dot{\Psi}^0_2 = \eth\Psi^0_3 + \sigma^0\Psi^0_4, \qquad \dot{\Psi}^0_1 = \eth\Psi^0_2 + 2\sigma^0\Psi^0_3\,. \tag{42}$$

Indeed, the first equation holds by definition and the assumed time-independence of P. The evolution equation (36) is equivalent to the real part of the second equation. Taking into account the on-shell relation of the NU framework,

$$\Psi^0_2 - \bar{\Psi}^0_2 = \bar{\eth}^2\sigma^0 - \eth^2\bar{\sigma}^0 + \bar{\sigma}^0\dot{\sigma}^0 - \sigma^0\dot{\bar{\sigma}}^0\,, \tag{43}$$

we find

$$M = -\Psi^0_2 - \sigma^0\dot{\bar{\sigma}}^0 + \frac{1}{2}\bar{\eth}^2\sigma^0 - \frac{1}{2}\eth^2\bar{\sigma}^0\,, \tag{44}$$

in terms of which (36) is fully equivalent to the second equation of (42) and (37) is equivalent to the last equation of (42), in agreement with [3].

5. Transformation laws of the NU coefficients characterizing asymptotic solutions

Let $\mathcal{Y} = P^{-1}\bar{Y}$ and $\bar{\mathcal{Y}} = P^{-1}Y$. The conformal Killing equations and the conformal factor then become

$$\eth\bar{\mathcal{Y}} = 0 = \bar{\eth}\mathcal{Y}, \qquad \psi = (\eth\mathcal{Y} + \bar{\eth}\bar{\mathcal{Y}})\,. \tag{45}$$

It follows for instance that

$$\bar{\eth}\eth\mathcal{Y} = -\frac{\bar{R}}{2}\mathcal{Y}, \quad \eth^2\psi = \eth^3\mathcal{Y} - \frac{1}{2}\bar{\mathcal{Y}}\eth\bar{R}, \quad \bar{\eth}\eth\psi = -\frac{1}{2}[\eth(\bar{R}\mathcal{Y}) + \bar{\eth}(\bar{R}\bar{\mathcal{Y}})]\,. \tag{46}$$

Using the notation $S = (Y, \widetilde{T}, \widetilde{\omega})$, we have $-\delta_S\bar{\gamma}_{AB} = 2\widetilde{\omega}\bar{\gamma}_{AB}$ for the background metric and

$$[-\delta_S, \bar{\eth}]\eta^s = -\widetilde{\omega}\bar{\eth}\eta^s + s\bar{\eth}\widetilde{\omega}\eta^s, \quad [-\delta_S, \eth]\eta^s = -\widetilde{\omega}\eth\eta^s - s\eth\widetilde{\omega}\eta^s\,. \tag{47}$$

To work out the transformation properties of the NU coefficients characterizing asymptotic solution space, one needs to evaluate the subleading terms in $\mathcal{L}_\xi g_N^{\alpha\beta}$ on-shell. This can also be done by translating the results from the BMS gauge, which yields

$$\begin{aligned}
-\delta_S\sigma^0 &= [f\partial_u + \mathcal{Y}\eth + \bar{\mathcal{Y}}\bar{\eth} + \frac{3}{2}\eth\mathcal{Y} - \frac{1}{2}\bar{\eth}\bar{\mathcal{Y}} - \widetilde{\omega}]\sigma^0 - \eth^2 f\,,\\
-\delta_S\dot{\sigma}^0 &= [f\partial_u + \mathcal{Y}\eth + \bar{\mathcal{Y}}\bar{\eth} + 2\eth\mathcal{Y} - 2\widetilde{\omega}]\dot{\sigma}^0 - \frac{1}{2}\eth^2\widetilde{\psi}\,,\\
-\delta_S\Psi_4^0 &= [f\partial_u + \mathcal{Y}\eth + \bar{\mathcal{Y}}\bar{\eth} + \frac{1}{2}\eth\mathcal{Y} + \frac{5}{2}\bar{\eth}\bar{\mathcal{Y}} - 3\widetilde{\omega}]\Psi_4^0\,,\\
-\delta_S\Psi_3^0 &= [f\partial_u + \mathcal{Y}\eth + \bar{\mathcal{Y}}\bar{\eth} + \eth\mathcal{Y} + 2\bar{\eth}\bar{\mathcal{Y}} - 3\widetilde{\omega}]\Psi_3^0 + \eth f\Psi_4^0\,,\\
-\delta_S\Psi_2^0 &= [f\partial_u + \mathcal{Y}\eth + \bar{\mathcal{Y}}\bar{\eth} + \frac{3}{2}\eth\mathcal{Y} + \frac{3}{2}\bar{\eth}\bar{\mathcal{Y}} - 3\widetilde{\omega}]\Psi_2^0 + 2\eth f\Psi_3^0,\\
-\delta_S\Psi_1^0 &= [f\partial_u + \mathcal{Y}\eth + \bar{\mathcal{Y}}\bar{\eth} + 2\eth\mathcal{Y} + \bar{\eth}\bar{\mathcal{Y}} - 3\widetilde{\omega}]\Psi_1^0 + 3\eth f\Psi_2^0\,.
\end{aligned} \tag{48}$$

Following for instance the terminology in [26] section 3, but now for general infinitesimal transformations $\zeta' = \zeta + \epsilon Y(\zeta)$, $\bar{\zeta}' = \bar{\zeta} + \epsilon\bar{Y}(\bar{\zeta})$ instead of those associated to linear fractional transformations on the sphere and also considering $\bar{\zeta}$ as the holomorphic coordinate instead of ζ, a field η has spin weight s and conformal weight w if it transforms as

$$-\delta_{Y,\bar{Y}}\eta = \big[Y^A\partial_A + \frac{s}{2}(\bar{\partial}\bar{Y} - \partial Y) - \frac{w}{2}\psi\big]\eta\,. \tag{49}$$

A tensor density of rank $s \geqslant 0$ and weight n transforms as

$$-\delta_{Y,\bar{Y}}A_{\bar{\zeta}\ldots\bar{\zeta}} = \big[Y^A\partial_A + s\bar{\partial}\bar{Y} + n(\partial Y + \bar{\partial}\bar{Y})\big]A_{\bar{\zeta}\ldots\bar{\zeta}}\,. \tag{50}$$

while for rank $s \leqslant 0$ and weight n, we have

$$-\delta_{Y,\bar{Y}}A_{\zeta\ldots\zeta} = \big[Y^A\partial_A - s\partial Y + n(\partial Y + \bar{\partial}\bar{Y})\big]A_{\zeta\ldots\zeta}\,. \tag{51}$$

It then follows that a tensor density of weights (s, n) defines a field of weights $(s, -(2n+|s|))$ and conversely, a field of weights (s, w) defines a tensor density of weights $(s, -\frac{1}{2}(w+|s|))$. For $s \geqslant 0$, this is done through $\eta = A_{\bar{\zeta}\ldots\bar{\zeta}}P^{2n+s}$ and $A_{\bar{\zeta}\ldots\bar{\zeta}} = P^w\eta$. For $s \leqslant 0$, we have $\eta = A_{\zeta\ldots\zeta}P^{2n-s}$ and $A_{\zeta\ldots\zeta} = P^w\eta$. Note that complex conjugation gives rise to opposite spin weight and rank but leaves the conformal and density weights unchanged. Alternatively, (49) can be written as

$$-\delta_{\mathcal{Y},\bar{\mathcal{Y}}}\eta = \big[\mathcal{Y}\eth + \bar{\mathcal{Y}}\bar{\eth} + \frac{s-w}{2}\,\eth\mathcal{Y} - \frac{s+w}{2}\,\bar{\eth}\bar{\mathcal{Y}}\big]\eta\,. \tag{52}$$

When focusing on $T = 0 = \widetilde{\omega}$ at the surface $u = 0$ and on the homogeneous part of the transformations, this gives the weights summarized in tables 1, 2. These tables are extended to the Lie algebra elements, which are passive in all our computations, by writing $[Y, \widetilde{T}] = -\delta_{Y,\bar{Y}}\widetilde{T}$ and $[Y, Y']^A = -\delta_{Y,\bar{Y}}Y'^A$.

Table 1. Spin and conformal weights

	σ^0	$\dot{\sigma}^0$	Ψ_4^0	Ψ_3^0	Ψ_2^0	Ψ_1^0	$\mathcal{Y}$	T
s	2	2	-2	-1	0	1	-1	0
w	-1	-2	-3	-3	-3	-3	1	1

Table 2. Rank and density weights

	$P^{-1}\sigma^0$	$P^{-2}\dot{\sigma}^0$	$P^{-3}\Psi_4^0$	$P^{-3}\Psi_3^0$	$P^{-3}\Psi_2^0$	$P^{-3}\Psi_1^0$	$\tilde{Y}$	$\widetilde{T}$
s	2	2	-2	-1	0	1	-1	0
n	$-\frac{1}{2}$	0	$\frac{1}{2}$	1	$\frac{3}{2}$	1	-1	$-\frac{1}{2}$

6. Surface charge algebra

In this section, $\widetilde{\omega} = 0$ so that $f = T + \frac{1}{2}u\psi$ and we use the notation $s = (\mathcal{Y}, \bar{\mathcal{Y}}, T)$ for elements of the symmetry algebra, which is given in these terms by $[s_1, s_2] = \widehat{s}$ where

$$\widehat{\mathcal{Y}} = \mathcal{Y}_1 \eth \mathcal{Y}_2 - (1 \leftrightarrow 2), \qquad \widehat{\bar{\mathcal{Y}}} = \bar{\mathcal{Y}}_1 \bar{\eth} \bar{\mathcal{Y}}_2 - (1 \leftrightarrow 2),$$
$$\widehat{T} = (\mathcal{Y}_1 \eth + \bar{\mathcal{Y}}_1 \bar{\eth}) T_2 - \frac{1}{2}\psi_1 T_2 - (1 \leftrightarrow 2)\,. \tag{53}$$

The translation of the charges, the non-integrable piece due to the news and the central charges computed in [16] gives here

$$Q_s[\mathcal{X}] = -\frac{1}{8\pi G}\int d^2\Omega^\varphi \Big[\big(f(\Psi_2^0 + \sigma^0 \dot{\bar{\sigma}}^0) + \mathcal{Y}(\Psi_1^0 + \sigma^0 \eth \bar{\sigma}^0 + \frac{1}{2}\eth(\sigma^0 \bar{\sigma}^0))\big) + \text{c.c.}\Big],$$
$$\Theta_s[\delta\mathcal{X}, \mathcal{X}] = \frac{1}{8\pi G}\int d^2\Omega^\varphi\, f\big[\dot{\bar{\sigma}}^0 \delta\sigma^0 + \text{c.c.}\big]\,, \tag{54}$$
$$K_{s_1,s_2}[\mathcal{X}] = \frac{1}{8\pi G}\int d^2\Omega^\varphi \Big[\big(\frac{1}{4} f_1 \eth f_2 \bar{\eth} R + \frac{1}{2}\bar{\sigma}^0 f_1 \eth^2 \psi_2 - (1 \leftrightarrow 2)\big) + \text{c.c.}\Big]\,.$$

Note that one could also write the charges $Q_s[\mathcal{X}]$ by allowing for the additional terms $(\frac{1}{2}\eth^2\bar{\sigma}^0 - \frac{1}{2}\bar{\eth}^2\sigma^0)$ in the first parenthesis since these terms cancel with the corresponding terms in the complex conjugate expression. Note also that not Ψ_2^0 but only $\Psi_2^0 + \bar{\Psi}_2^0$ is free data on-shell because of the relation (43).

We recognize all the ingredients of the surface charges described in [27], which in turn have been related there to previous expressions in the literature and, in particular, to the twistorial approach of Penrose [28]. More precisely, up to conventions, $Q_{0,0,T}$ agrees with Geroch's linear super-momentum [8] $Q_{gn} + \overline{Q}_{gn}$, as given in equation (A1.12) of [27]. The angular (super-)momentum that we get is

$$Q_{\mathcal{Y},0,0} = -\frac{1}{8\pi G}\int d^2\Omega^\varphi\, \mathcal{Y}\Big[\Psi_1^0 + \sigma^0\eth\bar\sigma^0 + \frac{1}{2}\eth(\sigma^0\bar\sigma^0) - \frac{u}{2}\eth\big(\Psi_2^0 + \bar\Psi_2^0 + \partial_u(\sigma^0\bar\sigma^0)\big)\Big]. \quad (55)$$

It differs from Q_{η_c} given in equation (4) of [27] by the explicitly u-dependent term of the second line. It thus has a similar structure to Penrose's angular momentum as described in equations (11), (12), and (17a) of [27] in the sense that it also differs by a specific amount of linear supermomentum, but the amount is different and explicitly u-dependent,

$$Q_{\mathcal{Y},0,0} = Q^{u=0}_{\mathcal{Y},0,0} + \frac{1}{2}uQ_{0,0,\eth\mathcal{Y}}\,. \quad (56)$$

The main result derived in [16] states that

- if one is allowed to integrate by parts,

$$\int d^2\Omega^\varphi\, \eth\eta^{-1} = 0 = \int d^2\Omega^\varphi\, \bar\eth\eta^1, \quad (57)$$

  where $d^2\Omega^\varphi = \frac{2d\zeta\wedge d\bar\zeta}{iP^2}$,
- if one defines the "Dirac bracket" through

$$\{Q_{s_1}, Q_{s_2}\}^*[\mathcal{X}] = -\delta_{s_2}Q_{s_1}[\mathcal{X}] + \Theta_{s_2}[-\delta_{s_1}\mathcal{X}, \mathcal{X}], \quad (58)$$

then the charges define a representation of the $\mathfrak{bms}_4$ algebra, up to a field dependent central extension,

$$\{Q_{s_1}, Q_{s_2}\}^* = Q_{[s_1,s_2]} + K_{s_1,s_2}, \quad (59)$$

where K_{s_1,s_2} satisfies the generalized cocycle condition

$$K_{[s_1,s_2],s_3} - \delta_{s_3}K_{s_1,s_2} + \text{cyclic}(1,2,3) = 0\,. \quad (60)$$

The representation theorem contained in equations (59) and (60) can be verified directly in the present context by starting from (54), (43) and using the properties (39), (57) of $\eth$, the evolution equations (42), the conformal Killing equations (45), the $\mathfrak{bms}_4$ algebra (53) and the transformation laws (48).

Several remarks are in order:

- Integrations by parts are justified for regular functions on the sphere and thus for $\mathfrak{bms}_4^{\text{glob}}$ and regular solutions. In the case of Laurent series more care is needed, see e.g. [29]. We will address this question elsewhere.
- For the globally well-defined $\mathfrak{bms}_4^{\text{glob}}$ algebra on the sphere, the central charge K_{s_1,s_2} vanishes.

- The non-conservation of the charges follows by taking $s_2 = (0,0,1)$ and $s_1 = s$. Indeed, since $\frac{d}{du}Q_s = \frac{\partial}{\partial u}Q_s - \delta_{(0,0,1)}Q_s$, the equality of the right hand sides of (58) and (59) gives

$$\frac{d}{du}Q_s = -\frac{1}{8\pi G}\int d^2\Omega^{\varphi}\left[\dot{\bar{\sigma}}^0(-\delta_s\sigma^0) + \frac{1}{4}\eth f\bar{\eth}\bar{R} + \frac{1}{2}\bar{\sigma}^0\eth^2\psi + \text{c.c.}\right]. \qquad (61)$$

For $s = (0,0,1)$, this gives the standard Bondi-Sachs mass loss formula,

$$\frac{d}{du}Q_{0,0,1} = -\frac{1}{8\pi G}\int d^2\Omega^{\varphi}\left[\dot{\bar{\sigma}}^0\dot{\sigma}^0 + \text{c.c.}\right]. \qquad (62)$$

It also follows that the standard $\mathfrak{bms}_4^{\text{glob}}$ charges are all conserved on the sphere in the absence of news.

To the best of our knowledge, except for the previous analysis in the BMS gauge, the above representation result does not exist elsewhere in the literature. A more detailed discussion of its implications, a detailed comparison with results in the literature as well as a self-contained derivation of the $\mathfrak{bms}_4$ transformation laws in the context of the Newman-Penrose formalism will be given elsewhere.

Acknowledgments

The authors thank Cédric Troessaert for useful discussions. G.B. is Research Director of the Fund for Scientific Research-FNRS (Belgium). This work is supported in part by the Belgian Federal Science Policy Office through the Interuniversity Attraction Pole P6/11, by IISN-Belgium, by "Communauté française de Belgique - Actions de Recherche Concertées" and by Fondecyt Projects No. 1085322 and No. 1090753.

References

[1] Bondi H, van der Burg M G and Metzner A W 1962 *Proc. Roy. Soc. Lond. A* **269** 21
[2] Sachs R K 1962 *Proc. Roy. Soc. Lond. A* **270** 103
[3] Newman E T and Unti T W J 1962 *Journal of Mathematical Physics* **3** 891–901 URL `http://link.aip.org/link/?JMP/3/891/1`
[4] Sachs R K 1962 *Phys. Rev.* **128** 2851–2864
[5] Barnich G and Troessaert C 2010 *Phys. Rev. Lett.* **105** 111103 (*Preprint* 0909.2617)
[6] Barnich G and Troessaert C 2010 *JHEP* **05** 062 (*Preprint* 1001.1541)
[7] Barnich G and Troessaert C 2010 *Proceedings of Science CNCFG* **010** (*Preprint* 1102.4632)
[8] Geroch R 1977 *Symposium on the asymptotic structure of space-time* ed Esposito P and Witten L (Plenum, New York) pp 1–105
[9] Penrose R 1963 *Phys. Rev. Lett.* **10** 66–68
[10] Imbimbo C, Schwimmer A, Theisen S and Yankielowicz S 2000 *Class. Quant. Grav.* **17** 1129–1138 (*Preprint* hep-th/9910267)
[11] Barnich G 2010 *AIP Conf. Proc.* **1307** 7–18 (*Preprint* 1010.0899)

[12] Foster J 1978 *Journal of Physics A: Mathematical and General* **11** 93 URL http://stacks.iop.org/0305-4470/11/i=1/a=012

[13] Foster J 1987 *International Journal of Theoretical Physics* **26**(11) 1107–1124 ISSN 0020-7748 10.1007/BF00669365 URL http://dx.doi.org/10.1007/BF00669365

[14] Newman E and Penrose R 1962 *Journal of Mathematical Physics* **3** 566–578 URL http://link.aip.org/link/?JMP/3/566/1

[15] Newman E P and Tod K P 1980 *General Relativity and Gravitation. 100 Years after the Birth of Albert Einstein. Volume 2* ed Plenum Press pp 1–36

[16] Barnich G and Troessaert C 2011 *JHEP* **1112** 105 (*Preprint* 1106.0213)

[17] Penrose R 1964 *Relativity, groups and topology: Les Houches 1963* ed C DeWitt B D (Gordon and Breach) pp 563–584

[18] Penrose R 1965 *Proc. Roy. Soc. Lond.* **A284** 159

[19] Tamburino L A and Winicour J H 1966 *Phys. Rev.* **150** 1039

[20] Penrose R 1974 *Group theory in non-linear problems* ed Barut A O (Reidel Publishing Company, Dodrecht, Holland) pp 1–58

[21] Schmidt B, Walker M and Sommers P 1975 *General Relativity and Gravitation* **6**(5) 489–497 ISSN 0001-7701 10.1007/BF00762453 URL http://dx.doi.org/10.1007/BF00762453

[22] Fletcher S J and Lun A W C 2003 *Classical and Quantum Gravity* **20** 4153–4167 URL http://stacks.iop.org/0264-9381/20/4153

[23] Kroon J A V 1999 *General Relativity and Gravitation* **31** 1219–1224 URL http://dx.doi.org/10.1023/A:1026712421739

[24] Newman E T and Penrose R 1966 *J. Math. Phys.* **7** 863–870 URL http://link.aip.org/link/?JMP/7/863/1

[25] Penrose R and Rindler W 1986 *Spinors and Space-Time, Volume 2: Spinor and Twistor Methods in Space-Time Geometry* (Cambridge University Press)

[26] Held A, Newman E T and Posadas R 1970 *Journal of Mathematical Physics* **11** 3145–3154 URL http://link.aip.org/link/?JMP/11/3145/1

[27] Dray T and Streubel M 1984 *Classical and Quantum Gravity* **1** 15–26 URL http://stacks.iop.org/0264-9381/1/15

[28] Penrose R 1982 *Proceedings of the Royal Society of London. A. Mathematical and Physical Sciences* **381** 53–63 (*Preprint* http://rspa.royalsocietypublishing.org/content/381/1780/53.full.pdf+html) URL http://rspa.royalsocietypublishing.org/content/381/1780/53.abstract

[29] Saidi E H and Zakkari M 1990 Harmonic distributions, Diff(S2), and Virasoro algebra Tech. Rep. IC/90/257 ICTP

IC-MSQUARE 2012: International Conference on Mathematical Modelling in Physical Sciences IOP Publishing
Journal of Physics: Conference Series **410** (2013) 012143 doi:10.1088/1742-6596/410/1/012143

Relative locality and gravity's weight on worldline fuzziness

Giovanni Amelino-Camelia, Valerio Astuti and Giacomo Rosati

Dipartimento di Fisica, Università "La Sapienza" and Sez. Roma1 INFN, P.le A. Moro 2, 00185 Roma, Italy

E-mail: amelino@roma1.infn.it

Abstract. We use the example of the much-studied κ-Minkowski noncommutative spacetime for illustrating a novel approach toward the analysis of the possible implications of spacetime noncommutativity. Our starting point is the proposal that spacetime noncommutativity is most naturally introduced within the manifestly-covariant formulation of quantum mechanics. This allows us to obtain a crisp characterization of the relativity of spacetime locality present in κ-Minkowski theories. And we also develop a novel description of how κ-Minkowski noncommutativity affects the fuzziness of worldlines.

1. Spacetime noncommutativity and covariant formulation of quantum mechanics

κ-Minkowski noncommutative spacetime[1, 2, 3, 4, 5],

$$[x_j, x_0] = i\ell x_j \ , \qquad [x_j, x_k] = 0 \ , \tag{1}$$

has attracted very intense interest over the last twenty years, but the issues we are interested in remain largely unexplored. We would like to characterize a class of physical effects linked with the description of spacetime fuzziness codified by κ-Minkowski noncommutativity, and it is at this point not even clear which of our current theories could make room for this noncommutativity. Evidently classical mechanics cannot be our choice, since its formalization provides no room for noncommutativity of coordinates. And even giving a formulation of κ-Minkowski spacetime noncommutativity in a quantum-mechanics setup is not straightforward. This is due to the fact that in κ-Minkowski the time coordinate should be an operator that does not commute with the spatial-coordinate operators, but in the standard setup of quantum mechanics we are not in the situation of time being described by an operator that commutes with the spatial-coordinate operators: in the standard setup of quantum mechanics time is not an observable at all, it just plays the role of evolution parameter.

We believe that it was indeed because of this mismatch between the nature of time in quantum mechanics and the properties of the κ-Minkowski time coordinate that progress did not materialize for formulating some of the observable spacetime consequences of κ-Minkowski that we are here interested in. We here summarize and offer an overall perspective on our results reported in Refs. [6, 7, 8] whose main starting point is recent progress on a manifestly-covariant formulation of quantum mechanics, which matured significantly over the last decade [9, 10, 11, 12]. In this powerful reformulation of quantum mechanics both the spatial coordinates and the time coordinate play the same type of role. And there is no "evolution", since dynamics is codified in a constraint, just in the same sense familiar for the covariant formulation of classical mechanics (see, *e.g.*, chapter 4 of Ref. [13]). Spatial and time coordinates are well-defined operators on a "kinematical Hilbert space", which is just an ordinary Hilbert space of normalizable wave functions [11]. And on the kinematical Hilbert space one has standard momenta π_0,π_1 conjugate to the spacetime coordinates q_0, q_1 [11]

$$[\pi_0, q_0] = i\ , \quad [\pi_1, q_1] = -i\ , \quad [\pi_1, q_0] = [\pi_0, q_1] = 0\ , \tag{2}$$

specializing to the case of a 2D spacetime.

For free massless particles the physical Hilbert space will be characterized by states satisfying the constraint

$$\mathcal{H}\psi \equiv (\pi_0^2 - \pi_1^2)\psi = 0$$

where ψ is then a state on the physical Hilbert space.

We focus on 2D κ-Minkowski, so that its noncommutativity is fully encoded in

$$[x_1, x_0] = i\ell x_1\ , \tag{3}$$

and we observe [6, 7, 8] that this 2D-κ-Minkowski defining commutator (3) is satisfied by posing a relationship between κ-Minkowski coordinates and the phase-space observables of the covariant formulation of quantum mechanics (the ones of Eq. (2), here viewed merely as formal auxiliary operators [6]):

$$x_1 = e^{\ell\pi_0} q_1\ , \quad x_0 = q_0\ . \tag{4}$$

Besides allowing a meaningful description of the noncommutativity of the κ-Minkowski time coordinate, this setup leads straightforwardly [6, 8] to a description of translation transformations in κ-Minkowski:

$$\mathcal{T}_{a^\mu} \triangleright f(x_0, x_1) \longleftrightarrow -ia^\mu \left[\pi_\mu, f(q_0, q_1 e^{\ell\pi_0})\right]\ , \tag{5}$$

One can then study [6, 8], within this setup, the properties of boost transformations in κ-Minkowski, thereby obtaining two other strongly-characterizing results. The first result [6, 8] concerns the form of the operator that should enforce the Hamiltonian constraint for free particles

$$\mathcal{H}_\ell = \left(\frac{2}{\ell}\right)^2 \sinh^2\left(\frac{\ell\pi_0}{2}\right) - e^{-\ell\pi_0}\pi_1^2\ . \tag{6}$$

and the second result [6, 8] concerns the measure for integration over momenta, needed for evaluating scalar products when working in the "momentum representation":

$$d\pi_0 d\pi_1 \longrightarrow d\pi_0 d\pi_1 e^{-\ell\pi_0} \tag{7}$$

2. Fuzzy points, translation transformations and relative locality

Evidently within our description a point will be identified with a state in the kinematical Hilbert space that gives rather well determined values to x_0 and x_1. A class of states which is well suited for exploring the properties of κ-Minkowski fuzziness is the one of gaussian states on our kinematical Hilbert space. We denote them by $\Psi_{\overline{q}_0,\overline{q}_1}(\pi_\mu; \overline{\pi}_\mu, \sigma_\mu)$ and they are specified by functions of the variables π_μ parametrized by $\overline{\pi}_\mu$, σ_μ, and $\overline{q}_\mu$:

$$\Psi_{\overline{q}_0,\overline{q}_1}(\pi_\mu; \overline{\pi}_\mu, \sigma_\mu) = Ne^{-\frac{(\pi_0-\overline{\pi}_0)^2}{4\sigma_0^2} - \frac{(\pi_1-\overline{\pi}_1)^2}{4\sigma_1^2}} e^{i\pi_0\overline{q}_0 - i\pi_1\overline{q}_1}, \tag{8}$$

where N is a normalization constant. Of course, our main focus of attention will be on establishing how the κ-Minkowski scale ℓ affects the results, since this is going to be our indicator of the difference between classical Minwkoswki spacetime and κ-Minkowski.

Following the analysis we reported in Ref. [6] one finds that the κ-Minkowski scale ℓ turns out to play a particularly significant role in the properties of the coordinate x_1,for which we find

$$\langle x_1 \rangle = \langle q_1 \rangle \left\langle e^{\ell \pi_0} \right\rangle = \overline{q}_1 e^{\ell \overline{\pi}_0} e^{-\frac{\ell^2 \sigma_0^2}{2}} , \quad \delta x_1 = e^{\ell \overline{\pi}_0} \left[\frac{1}{4\sigma_1^2} + \overline{q}_1^2 \left(1 - e^{-\ell^2 \sigma_0^2} \right) \right]^{1/2} , \tag{9}$$

whereas for the κ-Minkowski time coordinate x_0 one has

$$\langle x_0 \rangle = \overline{q}_0 - i\frac{\ell}{2} , \qquad \delta x_0 = \frac{1}{2\sigma_0} \tag{10}$$

(comments on the imaginary contribution $-i\ell/2$ to $\langle x_0 \rangle$ are given in Ref. [6]).

Our next task is to act with a translation transformation on our gaussian state and see what changes as a result. For the expectation values here of interest this is equivalently done by acting with the translation transformation on the operators x_1 and x_0. Following again our work in Ref. [6] one finds that

$$\langle \mathcal{T}_{a^\mu} \triangleright x_0 \rangle = \overline{q}_0 - a_0 - i\frac{\ell}{2} , \quad \delta \left(\mathcal{T}_{a^\mu} \triangleright x_0 \right) = \frac{1}{2\sigma_0} , \tag{11}$$

and

$$\langle \mathcal{T}_{a^\mu} \triangleright x_1 \rangle = (\overline{q}_1 - a_1)\, e^{\ell \overline{\pi}_0} e^{-\frac{\ell^2 \sigma_0^2}{2}} , \quad \delta(\mathcal{T}_{a^\mu} \triangleright x_1) = e^{\ell \overline{\pi}_0} \left[\frac{1}{4\sigma_1^2} + (\overline{q}_1 - a_1)^2 \left(1 - e^{-\ell^2 \sigma_0^2} \right) \right]^{1/2} . \tag{12}$$

The interpretation here of course is such that the x_0, x_1 are operators characterizing the distance of a given (fuzzy) point from the frame origin of some observer Alice, and then $\mathcal{T}_{a^\mu} \triangleright x_0$, $\mathcal{T}_{a^\mu} \triangleright x_1$ are the operators that characterize the distance of that point from the frame origin of an observer Bob, purely translated with respect to Alice. And accordingly one can deduce the relation between the mean values and uncertainties in positions among two distant observers in relative rest by comparing (10) to (11) and comparing (9) to (12). This we did in detail in Ref. [6]. We here summarize the main message of that analysis, focusing on the case of two fuzzy points in κ-Minkowski as described by two distant observers: one of the points is near observer Alice, while the other one is near observer Bob, purely translated with respect to Alice. By comparing (10) to (11) and comparing (9) to (12) one then finds two main features [6]:
(i) the same point appears to be more fuzzy to a distant observer than to a nearby observer,
(ii) the point at Alice is not described as being at Alice in the coordinatization of spacetime of observer Bob, and *vice versa* the point at Bob is not described as being at Bob in the coordinatization of spacetime of observer Alice.
The second feature, (ii), is essentially already known from previous studies of relative locality in the classical limit [14, 15]: one can have consistently relativistic theories where pairs of points found to be coincident by a nearby observer (or, as in the case here considered, a point found to coincide with the origin of that observer) are instead described as noncoincident if one uses the inferences about those points by a distant observer. Feature (i) was established for the first time in our Ref. [6], and is a feature of relative locality for the fuzziness of points in a quantum spacetime. The emerging picture is fully relativistic (though in the sense of deformed relativistic symmetries [16, 17, 18, 19]): all observers in κ-Minkowski perceive their origin as the point of lowest fuzziness and attribute to distant points fuzziness proportional to the distance.

3. Fuzziness of κ-Minkowski worldlines

In the previous section we worked on the kinematical Hilbert space. For our next (and here last) task we must progress to the level of the physical Hilbert space. We want to now use κ-Minkowski noncommutativity as a way to model the influence of quantum-gravity effects on the fuzziness of worldlines. This should be viewed in the context of attempts to describe the dynamics of matter particles as effectively occurring in an "environment" of short-distance quantum-gravitational degrees of freedom: for propagating particles with wavelength much larger than the Planck length, when it may be appropriate to integrate out these quantum-gravitational degrees of freedom, the main residual effect of short-distance gravity could indeed be an additional contribution to the fuzziness of worldlines, which one could model with spacetime noncommutativity.

For working on the physical Hilbert space we follow (as we did in Ref. [8]) the prescription adopted in Ref. [11]: we obtain the needed feature of invariance of physical observables under the action of the constraint $\mathcal{H}_\ell$ by introducing a new scalar product that projects all the orbit of the gauge transformation generated by $\mathcal{H}_\ell$ on the same state. For massless particles, on which we here focus, this amounts to inserting the Hamiltonian constraint as follows

$$\langle\psi|\phi\rangle_{\mathcal{H}_\ell} = \langle\psi|\delta\left(\mathcal{H}_\ell\right)\Theta(\pi_0)|\phi\rangle \tag{13}$$

where $\langle\rangle$ is the scalar product on the kinematical Hilbert space and $\Theta(\pi_0)$ specifies a restriction [11] to positive-energy solutions of the constraint.

The next hurdle we must face concerns the identification of an observable suitable for the characterization of the fuzziness of the worldline. The apparently obvious choices, x_1 and x_0, are actually not suitable for this task, since they are not self-adjoint operators on our physical Hilbert space (in particular they do not commute with $\mathcal{H}_\ell$). We propose to remedy this by focusing on the following "intercept operator" $\mathcal{A}$:

$$\mathcal{A} = e^{\ell\pi_0}\left(q_1 - \mathcal{V}q_0 - \frac{1}{2}[q_0, \mathcal{V}]\right) \tag{14}$$

where $\mathcal{V}$ is short-hand for $\mathcal{V} \equiv (\partial\mathcal{H}_\ell/\partial\pi^0)^{-1}\partial\mathcal{H}_\ell/\partial\pi^1$.

One may notice that $\mathcal{A}$ is describable as an ℓ-deformed Newton-Wigner operator [20]. And it is well known that within special-relativistic quantum mechanics there is no better estimator of localization than the Newton-Wigner operator (it can only be questioned for localization comparable to the Compton wavelength of the particle [20], but this merely conceptual limit of ideal localization is not relevant for our purposes here [8]).

Let us focus, for conceptual clarity, on the analysis of the properties of $\mathcal{A}$ for the case of $\Psi_{0,0}$, *i.e.* for $\overline{q}_0 = 0, \overline{q}_1 = 0$. One then easily finds that

$$\langle\Psi_{0,0}|\mathcal{A}|\Psi_{0,0}\rangle_{\mathcal{H}_\ell} = 0$$

so this is a case where the particle intercepts the observer Alice in her origin.

The fuzziness of this intercept, which reflects the fuzziness of the worldline [7, 8] described by $\Psi_{0,0}$, is characterized by [8]

$$\delta\mathcal{A}^2_{[\ell]} = \left(\langle\Psi_{0,0}|\mathcal{A}^2|\Psi_{0,0}\rangle_{\mathcal{H}_\ell}\right)_{[\ell]} \approx \ell\langle\pi_0\rangle\sigma^{-2}/2 \tag{15}$$

where for simplicity we focused on the leading ℓ-dependent contributions (index $[\ell]$) and we assumed that σ_1 is small enough, in comparison to σ_0, $\bar{\pi}_1$, to allow a saddle point approximation in the π_1 integration; then σ (without indices) is the effective gaussian width after the saddle point approximation in π_1: $\sigma^{-2} \equiv \sigma_1^{-2} + <\mathcal{V}>^2\,\sigma_0^{-2}$.

In the interpretation we proposed in Ref. [8] we describe Eq. (15) as the fuzziness of the worldline "at Alice" (at the point of crossing the origin of Alice's reference frame). It is then interesting to establish whether observers reached by the particle at cosmological distances from Alice will observe bigger fuzziness. We characterize such observers as those who are connected to Alice by a pure translation, so that for them the state of the particle is Ψ_{a_0,a_1}, and are, like Alice, such that $<\mathcal{A}>=0$, *i.e.* $\langle\Psi_{a_0,a_1}|\mathcal{A}|\Psi_{a_0,a_1}\rangle_{\mathcal{H}_\ell}=0$. Unsurprisingly this requires $a^1=\langle\mathcal{V}\rangle a^0$. And for these observers one finds

$$\delta\mathcal{A}^2_{[\ell]}=\left(\langle\Psi_{a^0,\langle\mathcal{V}\rangle a^0}|\mathcal{A}^2|\Psi_{a^0,\langle\mathcal{V}\rangle a^0}\rangle_{\mathcal{H}_\ell}\right)_{[\ell]}\approx\left(\frac{\ell\langle\pi_0\rangle}{2\sigma^2}+\ell^2a_0^2\sigma^2\right) \tag{16}$$

From this we see that the approach we developed in Refs. [6, 7, 8] leads to the first example of a quantum-spacetime picture (and of an interpretation of relativistic quantum mechanics in such a spacetime) providing the main ingredient of scenarios [21, 22, 23, 24, 25, 26] motivating the idea that quantum-gravity effects could affect worldline fuzziness in ways that inevitably lead to an increase of this fuzziness as the particle propagates. One can in fact interpret our observer Alice, the observer on the worldline for whom the fuzziness of the intercept takes the minimum value, as the observer at the source (where the particle is produced), and then the intercept of the particle worldline with the origin of the reference frames of observers distant from Alice (where the particle could be detected) has bigger uncertainty.

As we stressed in Refs. [7, 8] this formalization of gravity's contribution to worldline fuzziness could be relevant for an ongoing phenomenological effort aimed at finding experimental evidence of this sort of spacetime-fuzziness effects [21, 22, 23, 24, 25, 26].

References

[1] Majid S and Ruegg H 1994 *Phys. Lett.* B **334** 348
[2] Lukierski J, Ruegg H and Zakrzewski W J 1995 *Ann. Phys.* **243** 90
[3] Amelino-Camelia G and Majid S 2000 *Int. J. Mod. Phys.* A **15** 4301 (*Preprint* arXiv:hep-th/9907110)
[4] D'andrea F 2006 *J. Math. Phys.* **47** 062105 (*Preprint* arXiv:hep-th/0503012)
[5] Meljanac S and Kresic-Juric S 2011 *Int. J. Mod. Phys.* A **26** 3385 (*Preprint* arXiv:1004.4647 [math-ph])
[6] Amelino-Camelia G, Astuti V and Rosati G 2012 *Preprint* arXiv:1206.3805 [hep-th]
[7] Amelino-Camelia G, Astuti V and Rosati G 2012 *Preprint* arXiv:1207.2509 [gr-qc]
[8] Amelino-Camelia G, Astuti V and Rosati G 2012 Predictive description of Planck-scale-induced spacetime fuzziness *Preprint* Sapienza University of Rome
[9] Halliwell J J 2001 *Phys. Rev.* D **64** 04408 (*Preprint* arXiv:gr-qc/0008046)
[10] Gambini R and Porto R A 2001 *Phys. Rev.* D **63** 105014
[11] Reisenberger M and Rovelli C 2002 *Phys. Rev.* D **65** 125016 (*Preprint* arXiv:gr-qc/0111016)
[12] Freidel L, Girelli F and Livine E R 2007 *Phys. Rev.* D **75** 105016
[13] Henneaux M and Teitelboim C 1992 *Quantization of Gauge Systems* (Princeton: Princeton University Press)
[14] Amelino-Camelia G, Matassa M, Mercati F and Rosati G 2011 *Phys. Rev. Lett.* **106** 071301 (*Preprint* arXiv:1006.2126 [gr-qc])
[15] Amelino-Camelia G, Freidel L, Kowalski-Glikman J and Smolin L 2011 *Phys. Rev.* D **84** 084010 (*Preprint* arXiv:1101.0931 [hep-th])
[16] Amelino-Camelia G 2002 *Int. J. Mod. Phys.* D **11** 35 (*Preprint* arXiv:gr-qc/0012051)
[17] Kowalski-Glikman J 2001 *Phys. Lett.* A **286** 391 (*Preprint* arXiv:hep-th/0102098)
[18] Magueijo J and Smolin L 2003 *Phys. Rev.* D **67** 044017 (*Preprint* arXiv:gr-qc/0207085)
[19] Amelino-Camelia G 2010 *Symmetry* **2** 230 (*Preprint* arXiv:1003.3942 [gr-qc])
[20] Newton T D and Wigner E P 1949 *Rev. Mod. Phys.* **21** 400
[21] Amelino-Camelia G 1999 *Nature* **398** 216
[22] Ng Y J, van Dam H and Christiansen W A 2003 *Astrophys. J.* **591** L87 (*Preprint* arXiv:astro-ph/0302372)
[23] Lieu R and Hillman L W 2003 *Astrophys. J.* **585** L77
[24] Ragazzoni R, Turatto M and Gaessler W 2003 *Astrophys. J.* **587** L1
[25] Christiansen W A, Ng Y J and van Dam H 2006 *Phys. Rev. Lett.* **96** 051301
[26] Tamburini F, Cuofano C, Della Valle M and Gilmozzi R 2011 *Astron. Astrophys.* **533** A71

IC-MSQUARE 2012: International Conference on Mathematical Modelling in Physical Sciences IOP Publishing
Journal of Physics: Conference Series **410** (2013) 012144 doi:10.1088/1742-6596/410/1/012144

Recent results on a non-minimal coupling between curvature and matter

Jorge Páramos

Instituto de Plasmas e Fusão Nuclear, Instituto Superior Técnico, Universidade Técnica de Lisboa, Avenida Rovisco Pais 1, 1049-001 Lisboa, Portugal

E-mail: `paramos@ist.edu`

Abstract. This work presents a review of recent findings from the consideration of a non-minimal coupling between matter and geometry, namely the possibility of mimicking dark matter in clusters and the description of gravitational collapse — thus adding to the wide range of phenomena already covered by the theory.

1. Introduction

Dark gravity is the collective name under which a growing number of researchers are tackling several outstanding issues in contemporary Cosmology, namely dark matter and dark energy: instead of assuming that Einstein's General Relativity (GR) is valid at the relevant energy scales and focusing on the search for the missing matter or energy contribution, this alternative posits instead that the dark sector of our Universe merely reflect deviations of gravity from GR, so that no real dark matter or dark energy species are in principle required.

Several phenomenological attempts to generalize GR have surfaced, positing putative low-energy modifications of Einstein's theory that should be derived from a yet unknown high-energy, fundamental theory of gravitation: in particular, $f(R)$ theories substitute the linear scalar curvature R term in the Einstein-Hilbert action with a function $f(R)$ [1]. This may be generalized by considering a non-minimal coupling with matter [2],

$$S = \int \left[\kappa f_1(R) + f_2(R)\mathcal{L}\right] \sqrt{-g} d^4x \quad , \quad \kappa = \frac{c^4}{16\pi G} \quad . \tag{1}$$

This work presents a view of recent results obtained with this framework [3, 4], developed in collaboration with O. Bertolami, C. Bastos and P. Frazão. Other results [5, 6, 7, 8, 9, 10, 11] were discussed in Ref. [12].

2. The model

Variation with respect to the action Eq. (1) yields the modified Einstein field equations,

$$2\left(\kappa F_1 + F_2\mathcal{L}\right) G_{\mu\nu} = 2\left(\nabla_\mu\nabla_\nu - g_{\mu\nu}\Box\right)\left(\kappa F_1 + F_2\mathcal{L}\right) - \left[\kappa(F_1 R - f_1) + F_2\mathcal{L}R\right] g_{\mu\nu} + f_2 T_{\mu\nu} \quad , \tag{2}$$

where $F_i(R) \equiv f_i'(R)$. One recovers GR by setting $f_1(R) = R$ and $f_2(R) = 1$. Its trace is

$$\kappa(F_1 R - 2f_1) + F_2\mathcal{L}R = \frac{1}{2} f_2 T - 3\Box\left(\kappa F_1 + F_2\mathcal{L}\right) \quad , \tag{3}$$

where T is the trace of the energy-momentum tensor. The Bianchi identities may be used to derive the non-(covariant) conservation of the energy-momentum tensor,

$$\nabla_\mu T^{\mu\nu} = \frac{F_2}{f_2}\left(g^{\mu\nu}\mathcal{L} - T^{\mu\nu}\right)\nabla_\mu R \ . \tag{4}$$

This may be rewritten as an extra force imparted on test particles, so that its trajectory will deviate from geodesical motion. Thus, the Equivalence Principle may be broken if the *r.h.s.* of the last equation varies significantly for different matter distributions [2, 13].

3. Cluster dark matter mimicking

One first reviews a mechanism that mimics the dark matter at cluster scales [3]. In order to isolate the effect of the non-minimal coupling, one sets $f_1(R) = 1$ and a power-law $f_2(R) = 1 + (R/R_n)^n$. Inserting this into Eq. (3), together with the Lagrangian density $\mathcal{L} = -\rho$ and the energy-momentum tensor $T_{\mu\nu} = \rho U_\mu U_\nu$ for non-relativistic dust yields

$$R = (1-2n)\left(\frac{R}{R_n}\right)^n \frac{\rho}{2\kappa} - 3n\Box\left[\left(\frac{R}{R_n}\right)^n \frac{\rho}{\kappa R}\right] \ . \tag{5}$$

If the terms stemming from the non-minimal coupling dominate, an exact solution is found when the last term vanishes; interpreting the additional curvature obtained as a mimicked "dark matter", one obtains a direct translation between visible and mimicked dark matter,

$$\rho_{dm} \equiv 2\kappa R = \rho_n\left[(1-2n)\frac{\rho}{\rho_n}\right]^{1/(1-n)} \quad , \quad \rho_n \equiv 2\kappa R_n \ . \tag{6}$$

A more evolved study of Eq. (5) shows that its most general solution exhibits small oscillations around Eq. (6): this makes the gradient term actually dominate Eq. (5), and yields a perturbative $f_2(R) \ll 1$, which guarantees that the weak, strong, null and dominant energy conditions are satisfied and prevents the appearance of Dolgov-Kawasaki instabilities [14].

In order to fit the dark matter density profiles of several clusters (A133, A262, A383, A478, A907, A1413, A1795, A1991, A2029, A2390, RX J1159+5531, MKW 4, USGC S152 and A586, which are almost virialized and spherical clusters [15]), one numerically solves Eq. (5) with ρ given by available curves for the visible matter density, and fit the resulting ρ_{dm} to the reported dark matter profiles. A best fit is obtained for a value $n = 0.2$, as depicted in Fig. 1 for the Abell A907 cluster. The other clusters also yield an adequate fit [3], showing that the outlined mechanism, first proposed in the context of galactic dark matter [8], is also useful at this scale.

This fitting procedure yields an upper bound for $r_n \equiv R_n^{-1/2}$, based upon the two following considerations: firstly, it is found that the dark matter density profiles reported in Ref. [15] are impossible to mimic if the non-minimal coupling is non-perturbative, so that $f_2(R) \sim 1$ — which numerically translates into $r_{0.2} \ll 1$ Mpc. A second, more stringent bound is derived from the physical constraint that the extra force arising from the non-conservation of the energy-momentum tensor Eq. (4) is much smaller than the Newtonian gravitational force: this leads to the stronger condition $r_{0.2} \ll 51$ pc. With the latter constraint and the positive exponent $n = 0.2$, the non-minimal coupling is much smaller than unity, $f_2(R) \sim 1 + 10^{-8}$.

4. Gravitational Collapse

One now discusses the features of the gravitational collapse of a spherical body of homogeneous dust linearly coupled to gravity [4], thus generalizing the familiar Oppenheimer-Snyder scenario (OS). One considers $f_2(R) = 1 + \epsilon R/\kappa$ and a trivial $f_1(R) = R$, in order to highlight the effect of the former. Compatibility between a non-minimally coupled preheating mechanism and Starobinsky inflation [11] dictates that the coupling strength is of the order $10^9 < \epsilon < 10^{13}$.

The Lagrangean density of matter appears explicitly in the modified field Eq. (2): this has prompted the question as to what is the right form for this quantity, as discussed in Ref. [7] — where it was argued that $\mathcal{L} = -\rho$, not $\mathcal{L} = p$, is the correct one. Nevertheless, one asserts the physical impact of the two forms by writing $\mathcal{L} = -\alpha\rho$: the choice $\mathcal{L} = -\rho$ corresponds to $\alpha = 1$, while $\mathcal{L} = p$ is obtained by setting $\alpha = 0$ (as the pressure vanishes in the dust distribution). The parameter α is thus a binary variable, considerably simplifying several calculations.

Introducing the above expression, together with the energy-momentum tensor for dust and the adopted forms for $f_1(R)$ and $f_2(R)$ into the modified field Eq. (2), one eventually obtains

$$\frac{1}{6}\kappa a^2\rho = \left(\kappa^2 - \epsilon\rho\right) k + \left[\kappa^2 + (3\alpha - 1)\epsilon\rho\right] (\dot{a})^2 + \epsilon(\alpha - 1)a\ddot{a}\rho \ , \tag{7}$$

where $a(t)$ is scale factor and k the spatial curvature of the FRW describing the homogeneous collapse: the latter is given by $k = k_0/[1 - (1 + \alpha)\epsilon_0/2]$, with k_0 the value found in GR and $\epsilon_0 \equiv \epsilon\rho_0/\kappa^2$. Since the initial density is much smaller than the typical density of neutron stars, $\rho_0 \ll \rho_N \sim 10^{18}\ \mathrm{kg/m^3}$, this parameter is vanishingly small, $\epsilon_0 \ll 10^{-62}$.

Using Eq. (3) and recalling that α is a binary variable, one integrates Eq. (4) to find

$$\rho(t) = \rho_0 \left(\frac{\kappa + \epsilon R}{\kappa + \epsilon R_0}\right)^{\alpha - 1} \left(\frac{a_0}{a}\right)^3 = \frac{\rho_0 a^{-3}}{1 + \epsilon(1 - \alpha)\frac{\rho_0}{2\kappa^2}\left(a^{-3} - 1\right)} \ . \tag{8}$$

Combining Eq. (7) and (8), one obtains the differential equation

$$\dot{a} = -\sqrt{k(1 - a)\frac{a^2 + (a + 1)\alpha\epsilon_0}{a^3 + 2\alpha\epsilon_0}} \quad , \quad a(t_0) = 1 \ . \tag{9}$$

Analysis of Eq. (9) and Eq. (8) shows that, in the $\mathcal{L} = -\rho \rightarrow \alpha = 1$ scenario, the gravitational collapse deviates (very weakly, as $\epsilon_0 \ll 1$) from GR due to the more evolved dynamics; this is depicted in Fig. 2 for larger values of ϵ_0, for clarity. The usual dependence of the density on the scale factor $\rho \sim a^{-3}$ is maintained, and a point-like singularity with infinite density reached.

The $\mathcal{L} = p \rightarrow \alpha = 0$ scenario is much more interesting: although the dynamics of the scale factor are qualitatively the same as in OS collapse, the modified dependence for the density yields a geometric point-like singularity with a final finite density $\rho \rightarrow \rho_f = 2\kappa^2/\epsilon$. Since ϵ is very large, this falls well below the Planckian domain, $\rho \rightarrow \rho_f \ll M_P^4$ — although still many orders of magnitude above the typical density of neutron stars.

To match the interior FRW metric with the outer Schwarzschild metric, one inspects the behaviour of the extrinsic curvature K_{ab} across the surface of the body, which reads

$$K_{ab}^+ = \left(1 - \frac{\epsilon\alpha}{\kappa^2}\rho\right) K_{ab}^- + \frac{\epsilon\alpha}{\kappa^2}\rho K^- h_{ab} \ . \tag{10}$$

In the $\mathcal{L} = p \rightarrow \alpha = 0$ scenario, the extrinsic curvature is continuous, and this matching shows that the mass M of the spherical body (as measured from the gravitational potential away from it) according to $M = M_0/(1 - \epsilon_0/2)$, where M_0 is the usual gravitational mass obtained in GR. As this depends on the initial density ρ_0, different event horizons arise from the collapse of stars with the same mass, but distinct radius — thus breaking the no-hair theorem.

The alternative description $\mathcal{L} = -\rho \rightarrow \alpha = 1$ is problematic, as the matching of the inner and outer spacetimes is unfeasible unless unnatural extra terms are added to the boundary action. This could be related to the non-vanishing effective pressure that arises due to the non-minimal coupling — *i.e.* a non-vanishing $p_{eff} \sim G_{rr} \neq 0$— reminiscent of the similar matching problem found in the gravitational collapse of a homogeneous sphere with pressure in GR.

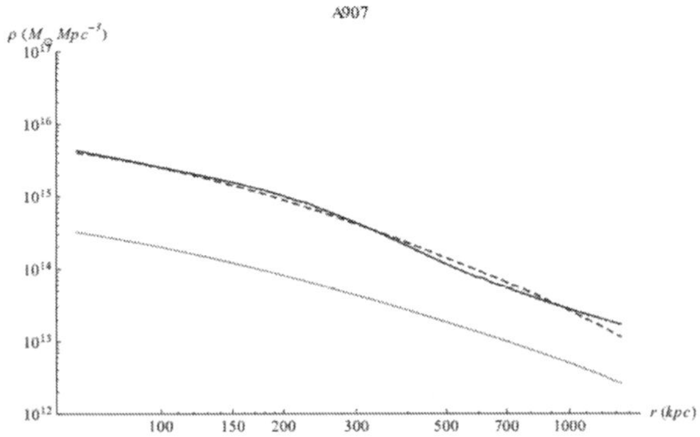

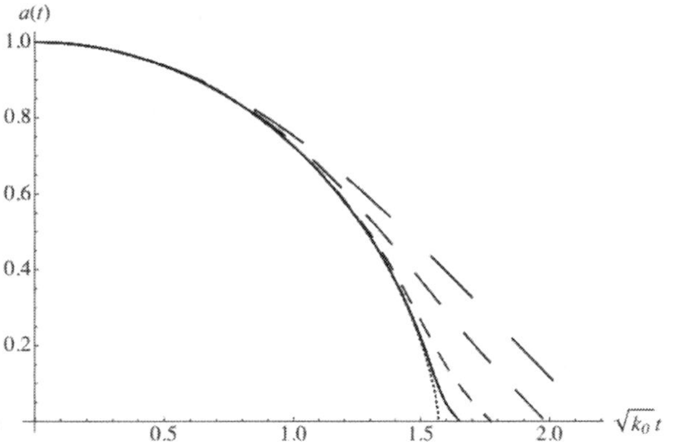

Figure 1. Dark (full) and visible (gray) matter density, with obtained mimicked dark matter density superimposed (dashed).

Figure 2. Evolution of the scale factor for different values of $\epsilon_0 = [10^{-3},\ 10^{-2},\ ,10^{-1},\ 1]$ (full line, small, medium and large dash); dotted indicates $\epsilon_0 = 0$.

5. Conclusions and Outlook

A non-minimal coupling between geometry and matter covers a broad spectrum of applications in an elegant, natural way,as these and other results have shown. The different forms adopted may be regarded as an approximation to a more evolved forms for $f_2(R)$, each valid in a particular regime: early *vs.* late time, central *vs.* long range, *etc.* — hinting at a Laurent series expansion for the latter. Yet untested terms of the above series may be evaluated by studying other phenomena and environments, where distinct curvatures and densities are at play.

Acknowledgments

This work was developed in the context of the First International Conference on Mathematical Modeling in Physical Sciences, 3-7 September 2012, Budapest, Hungary. I wish to thank the organization, specially Elias Vagenas, for the invitation and hospitality.

Bibliography

[1] A. D. Felice and S. Tsujikawa, *Living Rev. Rel.* **13**, 3 (2010).
[2] O. Bertolami, C. Boehmer, T. Harko and F. Lobo, *Phys. Rev.* **D 75**, 104016 (2007).
[3] O. Bertolami, P. Frazão and J. Páramos, *Phys. Rev.* **D 86**, 044034 (2012).
[4] J. Páramos and C. Bastos, arXiv:1207.6258 [gr-qc].
[5] O. Bertolami and J. Páramos, *Phys. Rev.* **D 77**, 084018 (2008).
[6] O. Bertolami and J. Páramos, *Class. Quantum Gravity* **25**, 245017 (2008).
[7] O. Bertolami, F. S. N. Lobo and J. Páramos, *Phys. Rev.* **D 78**, 064036 (2008).
[8] O. Bertolami and J. Páramos, *JCAP* **1003**, 009 (2010).
[9] O. Bertolami, P. Frazão, and J. Páramos, *Phys. Rev.* **D 81**, 104046 (2010).
[10] O. Bertolami and J. Páramos, *Phys. Rev.* **D 84**, 064022 (2011).
[11] O. Bertolami, P. Frazão and J. Páramos, *Phys. Rev.* **D 83**, 044010 (2011).
[12] J. Páramos, arXiv:1111.2740 [gr-qc].
[13] T. P. Sotiriou and V. Faraoni, *Class. Quantum Gravity* **25**, 5002 (2008).
[14] O. Bertolami and M. C. Sequeira, *Phys. Rev.* **D79**, 104010 (2009).
[15] A. Vikhlinin *et al.*, *Ap. J.* **640**, 691 (2006).

Non-perturbative analysis of the gravitational energy in Hořava Theory

Alvaro Restuccia[1], Jorge Bellorín[2], Adrián Sotomayor[3]

[1] Departamento de Física, Universidad de Antofagasta, Antofagasta, Chile
[1] Departamento de Física, Universidad Simón Bolívar, Caracas, Venezuela
[3] Departamento de Matemáticas, Universidad de Antofagasta, Antofagasta, Chile

E-mail: arestu@usb.ve,jorgebellorin@usb.ve,asotomayor@uantof.cl

Abstract. We perform a non-perturbative analysis of the constraints of the Hořava Gravitational theory. In distinction to Einstein gravity the theory has constraints of the first class together with second class ones. We analyze the consequences of having to impose second classes constraints at any time in the quantum formulation of the theory. The second class constraints are formulated as strongly elliptic partial differential equations allowing a global analysis on the existence and uniqueness of the solution. We discuss the possibility of formulating the theory in terms of a master action with first class constraints only. In this case the Hořava theory would correspond to a gauged fixed version of the master theory. Finally we obtain , using the non-perturbative solution of the constraints, the explicit expression of the gravitational energy. It is, under some assumptions, always positive and the solution of Hořava field equations at minimal energy is the Minkowski metric.

1. Introduction

Recently, Hořava [1] proposed a gravitational theory with Lifshitz-like anisotropic scaling at short distances:

$$t \to b^z t, \quad x \to bx.$$

The formulation breaks the relativistic symmetry at short distances with the idea of regaining it at large distances. The benefit would be to obtain a power counting renormalizable theory of gravity. The anisotropic scaling between time and space allows to include in the action high enough spatial derivatives which contribute to the interactions and to the propagators improving the UV properties of the theory. In order to obtain a renormalizable theory z should be equal to the number of spatial dimension and all terms compatible with the gauge symmetry should be included in the action.

In distinction to General Relativity, Hořava Gravity is restricted not only by first class contraints but also by second class constraints which are potentially dangerous because they may introduce non-localities in the quantum formulation of the theory. Directly related to this analysis and an important aspect of any gravitational theory is to determine the gravitational energy and to established its positivity. The positive mass theorem plays a fundamental role in General Relativity. It has firstly proved in [2, 3, 4] that for asymptotically flat space-times the total energy-momentum in General Relativity is well defined, it is greater or equal to zero and it vanishes for flat space-time. We will discuss the existence and uniqueness of the solution to

Published under licence by IOP Publishing Ltd

the constraints and will prove the positivity of the gravitational energy for the Hořava theory in the large distance regime. Moreover the expression for the gravitational energy we will obtain remains valid even when one includes all the interacting terms corresponding to the $z = 3$ Hořava theory.

2. The Hořava action

Hořava theory is formulated on a foliated manifold $\mathcal{M} = \Sigma \times \mathbb{R}$ where Σ is a three dimensional Riemann manifold which we will assume to be complete, connected and asymptotically flat.

The theory is expressed in terms of an ADM formulation where the metric is given by

$$ds^2 = \left(-N^2 + N_i N^i\right) dt^2 + 2N_i dx^i dt + g_{ij} dx^i dx^j. \tag{1}$$

The extrinsic curvature of Σ, denoted K_{ij}, satisfies

$$K_{ij} = \frac{1}{2N}\left(\dot{g}_{ij} - \nabla_i N_j\right) \tag{2}$$

where the lapse N is assumed to be different from zero.

The Hamiltonian of Hořava theory is generically given by

$$\int dt d^3x \sqrt{g} N \left(G^{ijkl} K_{ij} K_{kl} + V(g, N)\right) \tag{3}$$

where

$$G^{ijkl} = \frac{1}{2}\left(g^{ik}g^{jl} + g^{il}g^{jk} - \lambda g^{ij}g^{kl}\right), \tag{4}$$

λ is a dimensionless parameter which may be included because each term is separately a tensor under spatial diffeomorphism the gauge symmetry of the theory. More precisely the above action is invariant under diffeomorphism on Σ and reparametrization on t:

$$\widetilde{t} = f(t), \quad \widetilde{x} = g(x). \tag{5}$$

The symmetry (5) may be enlarged by considering anisotropical conformal gauge transformations, which are relevant in the formulation of conformal gravity [Hořava, 2012][5]. We will only be concerned with transformations (5).

In the potential $V(g, N)$ one should include all possible local, up to $z = 3$ terms, compatible with the gauge symmetries of the theory.

Typical $z = 3$ terms are $\nabla_k R_{ij} \nabla^k R^{ij}$, $R\Delta R$, $R^{ij}\Delta R_{ij}$ which contribute to the interactions but also modify the propagators improving the UV behaviour of the theory. Other $z = 3$ terms such as R^3, $RR_{ij}R^{ij}$ are pure interacting terms.

3. The hamiltonian in the large distance regime

The most general Hamiltonian describing the large distance regime in Hořava theory is given by [6, 7]

$$H = \int d^3x \left(N\mathcal{H} + N_i \mathcal{H}^i + \sigma\phi\right) + E_{ADM} - 2\alpha\phi_N \tag{6}$$

$$\mathcal{H} \equiv G_{ijkl} \frac{\pi^{ij}\pi^{kl}}{\sqrt{g}} + \sqrt{g}\left(-R + \alpha\left(2\nabla_i a^i + a_i a^i\right)\right) \tag{7}$$

$$\mathcal{H}^i \equiv -2\nabla_j \pi^{ji} + \phi \partial^i N \tag{8}$$

$$\phi_N \equiv \oint d\Sigma_i \partial_i N, \tag{9}$$

where we have assumed that $\lambda \neq \frac{1}{3}$. For $\lambda = \frac{1}{3}, G^{ijkl}$ is not invertible. The theory with $\alpha = 0$ for any value of λ is equivalent to General Relativity [8, 9]. The terms which depends on N and their derivatives in the potential were introduced in [10].

As in General Relativity the ADM energy in included as the boundary term. This is necessary in order to obtain the equation of motion under variations δg_{ij} asymptotically of order $O(r^{-1})$ [11]. Similarly, the flux of N at spatial infinity, ϕ_N, cancels a non zero contribution coming from $\delta(2\alpha N \nabla_i a^i)$ for variations of N with asymptotic decay $\delta N = O(r^{-1})$.

$\mathcal{H}^i = 0$ is a first class constraint related to the generator of spatial diffeomorphisms. $\mathcal{H} = 0$ is a second class constraint. In order to preserve it we obtain a PDE for the Lagrange multiplier σ,

$$\partial_i \left(N \sqrt{g} g^{ij} \partial_j \left(\sigma / N \right) \right) = \partial_i \left(-\gamma N^2 \nabla^i \pi + N^2 G^i_{klm} \pi^{lm} \right). \tag{10}$$

The Dirac procedure ends at this stage.

4. The non-perturbative analysis of the constraints

In order to analize the constraints $\mathcal{H} = 0$ and the equation for the Lagrange multiplier we consider the conformal transformation

$$\begin{array}{c} g_{ij} = e^{\varphi} \widehat{g}_{ij} \\ \pi^{ij} e^{-\varphi} \widehat{\pi}^{ij} \end{array} \tag{11}$$

where the Riemannian metric $\widehat{g}_{ij}$ satisfies $\det \widehat{g}_{ij} = 1$.

We also consider the decomposition of $\pi^{ij} = \widehat{\pi}^{ij}_T + \frac{1}{3} \widehat{g}^{kj} \widehat{\pi}$ where $\widehat{g}^{ij}$ is the inverse of $\widehat{g}_{ij}$ and $\widehat{\pi}^{ij}_T$ is traceless: $\widehat{g}_{ij} \widehat{\pi}^{ij} = 0$.

The transformation $(g_{ij}, \pi^j) \rightarrow \left(\widehat{g}_{ij}, \widehat{\pi}^{ij}_T, \varphi, \widehat{\pi} \right)$ is canonical:

$$\int_\Sigma d^3x \pi^{ij} \dot{g}_{ij} = \int_\Sigma d^3x \left(\widehat{\pi}^{ij}_T \dot{\widehat{g}}_{ij} + \widehat{\pi} \dot{\varphi} \right). \tag{12}$$

In the new variables the constraint $\mathcal{H} = 0$ becomes

$$\partial_i \left(e^{\beta\varphi} \widehat{g}^{ij} \partial_j e^{\frac{\xi}{2\alpha}} \right) + G e^{\frac{\xi}{2\alpha}} = 0 \tag{13}$$

where

$$G = -\frac{e^{\beta\varphi}}{4\alpha} \left[e^{-2\varphi} \left(\widehat{\pi}^{ij}_T \widehat{\pi}_{Tij} + (3 - 9\lambda)^{-1} \right) \widehat{\pi}^2 - \widehat{R} - \beta \partial_i \varphi \partial^i \varphi \right] \tag{14}$$

and $\xi \equiv \alpha \ln |N| + \varphi$.

We notice that G depends on the canonical pairs $\left(\widehat{g}_{ij}, \widehat{\pi}^{ij}_T \right)$ and $(\varphi, \widehat{\pi})$. It does not depend on N. Equation (13) is a linear partial differential equation on $e^{\frac{\xi}{2\alpha}}$. It is strongly elliptic. It is convenient to consider $e^{\frac{\xi}{2\alpha}} = 1 + u$ and to define a suitable space of functions for u.

We can prove the following proposition [12]:

Proposition 1 *Given an asymptotically flat Riemannian manifold with C^∞ metric g_{ij} and momenta π^{ij}, with the asymptotic behaviour*

$$\begin{array}{c} g_{ij} = \delta_{ij} + O(r^{-1}) \\ \pi^{ij} = O(r^{-2}), \end{array}$$

then the asymptotic solution for (13) exists, it is C^∞ and the asymptotic behaviour is

$$u = O(r^{-1}).$$

We are then motivated to introduce the space $\widehat{C}^1$:

$$\widehat{C}_1(\Sigma) = \left\{u \in C^1(\Sigma) : u = O(r^{-1}) \text{ when } r \to \infty\right\}.$$

For u and v in $\widehat{C}_1(\Sigma)$ we define the bilinear functional

$$(u, v) = \int_\Sigma d^3x \left(e^{\beta\varphi}\widehat{g}^{ij}\partial_i u \partial_j v + Guv\right).$$

If $G \geq 0$ on Σ then (u, v) defines an internal product in $\widehat{C}_1(\Sigma)$. We denote $\widehat{\mathcal{H}}^1$ the Hilbert space obtained by the completion of $\widehat{C}_1(\Sigma)$ with respect to the norm induced by the above internal product.

The assumption $G \geq 0$ will be essential to prove existence and uniqueness for the solution of the constraint. If G is not positive the operator

$$\mathcal{O} \equiv -\partial_i \left(e^{\beta\varphi}\widehat{g}^{ij}\partial_j \cdot\right) + G\cdot$$

may have a nontrivial kernel K. In that case the solution u to the constraint

$$\mathcal{O}u = -G \tag{15}$$

exists if and only if G is orthogonal to K. Even if this condition is satisfied the solution would not be unique.

We can prove the following propositions [12].

Proposition 2 *Given a complete, connected, asymptotically flat Riemannian manifold Σ and momenta π^{ij} satisfying $G \geq 0$, there always exist in $\widehat{\mathcal{H}}^1$ a unique weak solution to the constraint (15).*

Proposition 3 *Under the previous assumptions and considering a C^∞ metric g_{ij} and momenta π^{ij}, the weak solution is C^∞.*

Proposition 4 *Under the previous assumptions the solution satisfies $1 + u \geq 0$. Hence we may identify*

$$e^{\frac{\xi}{2\alpha}} \equiv 1 + u \geq 0.$$

Proposition 5 *Under the previous assumptions the solution for the Lagrange multiplier σ exists and is unique. The asymptotic behaviour is*

$$\sigma = O(r^{-2})$$

when $r \to \infty$.

These propositions prove under above assumptions, the existence and uniqueness of the solution to the constraint $\mathcal{H} = 0$ and of the equation for the Lagrange multiplier.

We can finally obtain the following results concerning the gravitational energy of the Hořava gravity.

The gravitational energy is given by the flux of the ξ field:

$$E = -2 \oint d\Sigma_i g^{ij} \partial_j \xi.$$

Under the previous assumptions and for $\alpha \leq 0$ the gravitational energy E is positive.

The energy E has a minimum for $\xi = 0$. At the minimum and using the field equations for Hořava gravitational theory we obtain

$$\begin{gathered} g_{ij} = \delta_{ij}, \pi^{ij} = 0 \\ N = 1 \end{gathered} \tag{16}$$

in the gauge $N^i = 0$.

It is important to distinguish this space-time metric from one obtained by performing an anisotropic conformal transformation on a Lifchitz space-time. If fact, the field equations for the action we are considering are invariant under an isotropic conformal transformation (exactly the same one that leave invariant Einstein equations for General Relativity). The metrics are equal but the symmetries of the underlying spaces are different.

5. Conclusions

We presented a non-perturvatibe analysis of the existence and uniqueness of the second class contraints of the Hořava Gravity in the large distance regime where $z = 3$ interacting terms and their contributions to the propagator are not relevant. The action we considered includes all the $z = 1$ possible contributions (without the cosmological term). We provide a sufficient condition ensuring the existence and uniqueness of the solution to the constraints. After the elimination of : a) the lapse N, from the solution for the ξ field, and its conjugate momentum from the second class contraints, b) the gauge field N^i and the longitudinal part of π^{ij} from the first class constraint, we are then left with the physical degrees of freedom in terms of the conjugate pairs $\left(\pi_T^{ij}, \hat{g}_{ij}\right)$ and $(\hat{\pi}, \varphi)$. There is one additional degree of freedom with respect to General Relativity. The interesting point of our presentation is that the gravitational energy of the theory is directly expressed as the flux of the ξ field whose solution is obtained directly from the second class constraint $\mathcal{H} = 0$. We gave a sufficient condition ensuring the positivity of the gravitational energy for Hořava theory.

Although the second class contraints can be qualitatively solved and the solution renders a positive gravitational energy still the goal of a renormalizable field theory has not been reached. The reason is that the complicated (when $z = 3$ contributions are included) second class constraints have to be imposed at any time and the possibility to explicitly solve them without introducing non-localities is far for been obtained. An alternative approach to handle this problem arises from the structure of the metric of Poisson bracket of the second class constraint. Their contribution to the measure of the path integral is through the square root of its determinant. In general, even with the inclusion of $z = 3$ terms in the hamiltonian, the square root reduces only to the bracket $\{\mathcal{H}, \phi\}$. This contribution is similar to the one of a theory with first class constraint only. In that case the bracket is evaluated between the first class constraint and its gauge fixing condition. This suggest the existence of a master action with first class constraints only from which the Hořava theory would arise as a gauge fixed of the master theory. In that case a BRST quantization would then solve the problem.

Acknowledgments

A. R. and A. S. are partially supported by Project Fondecyt 1121103, Chile.

References

[1] P. Hořava, Phys. Rev. D 79, 084008 (2009).
[2] S. Deser and C. Teitelboim, Phys. Rev. Lett. 39, 249 (1977).
[3] R. Schoen and S-T Yau, Commun. Math. Phys. 79, 47 (1981); 79, 231 (1981).
[4] E. Witten, Commun. Math. Phys. 80, 381 (1981).
[5] T. Griffin, P. Hořava, C. M. Melby-Thompson, JHEP 1205, 010 (2012).
[6] W. Donnelly and T. Jacobson, Phys. Rev. D 84, 104019 (2011).
[7] Bellorin and A. Restuccia, Phys. Rev. D 84, 104037 (2011).
[8] J. Bellorin and A. Restuccia, Int. J. Mod. Phys. D 21, 1250029-1 (2012).
[9] J. Bellorin and A. Restuccia, Phys. Rev. D 83, 044003 (2011).
[10] D. Blas, O. Pujolas, and S. Sibiryakov, Phys. Rev. Lett. 104, 181302 (2010).
[11] T. Regge and C. Teitelboim, Ann. Phys. (N.Y.) 88, 286 (1974).
[12] J. Bellorin, A. Restuccia and A. Sotomayor, Phys. Rev. D 85.124060 (2012).

Neutrino Oscillations at the Intensity Frontier: The NOvA Experiment

Athanasios Hatzikoutelis
Physics Department
University of Tennessee Knoxville
USA

E-mail: a.hatzikoutelis@utk.edu

Abstract. The "NuMI Off-Axis electron-neutrino Appearance" (NOvA) is a second generation, long- baseline, neutrino oscillation, experiment. It is made of two detectors, a large Far detector (14 ktons) and a similar Near detector (222 tons), both made of mostly active scintillator and separated by 810 km. Along with the 700 kW NuMI-beam upgrade (a prelude to the Intensity Frontier), it will be the leading neutrino experiment at Fermilab. In the wake of the recent measurement of the θ_{13} mixing angle, NOvA is positioned to see evidence of the neutrino mass hierarchy, possibly to resolve the θ_{23} octant ambiguity, and begin the study of the CP violation at the lepton sector. The experiment is under construction. The design and potential of this experiment is presented here along with the current status.

1. Introduction

The last two years have been an exceptional time for neutrino physics. Several experiments have measured for the first time a non-vanishing value for the mixing angle θ13. These are accelerator driven electron-neutrino appearance measurement like the Japan-based T2K [1] and the USA-based MINOS [2], or nuclear reactor driven anti-neutrino disappearance ones, like the China-based Daya Bay [3], with strong evidence of reactor electron antineutrino disappearance consistent with neutrino oscillations and the France-based Double Chooz [4] with the most recent measurements to date. They progressively defined the value of θ_{13} to within a non-zero value-band around 9° with various confidence levels each (based on their systematics and the assumptions in the models). This large value makes it more possible to explore CP-violation and the mass hierarchy of the neutrinos. Now, we can address some of the compelling questions of neutrino physics and the lepton sector. What is the neutrino mass hierarchy? Do neutrino oscillations violate CP symmetry? Is the θ_{23} mixing angle 45° (maximal)? Do these findings indicate new symmetries or new selection rules, forces, sectors?

The Physics community has come together in several instances over the past few years trying to decide on the path for the next generation and further in the future experiments. Work on the Physics Frontiers [5] of Intensity, Energy and the Cosmic frontier, will provide answers, or better yet, will over- constrain the above parameters in order to determine any new physics beyond the Standard Model.

2. The NOvA Experiment

The second generation of the long baseline, accelerator based, neutrino oscillation experiments, is made up of three experiments; the MINOS from Fermilab at 790km baseline, the T2K from JPARC at 293 km, and the NOvA Experiment, also from Fermilab at 810 km. The NOvA experiment aims to lead in the measurement of appearance of electron neutrinos and anti-neutrinos from the NuMI (Neutrinos at the Main Injector) neutrino beam from Fermilab (Fermi National Accelerator Lab in Batavia, Illinois, USA). The NOvA neutrinos are produced by a 120 GeV proton accelerator complex at 400 kW with a scheduled upgrade to 700 kW. After the protons fall on a graphite target, a secondary beam of pions and kaons is generated. In order to focus this beam, we use magnetic horns. By passing current through these horns, the charged-particles of the secondary-beam can focus forward. They can also control the consistency neutrino or anti-neutrino) of the tertiary beam by the direction of the current. That is the method used to create the anti-neutrino beam. Further downstream, the experiment is made of two functionally identical detectors positioned off-axis by 14 mrad. This angle yields for the resulting neutrino beam, a much tighter energy distribution in the spectrum, compared to an on-axis orientation (fig.1). The reduced flux is peaked about 2GeV within 20% in energy. For this peak energy, the first oscillation maximum falls at the distance of the NOvA-Far detector.

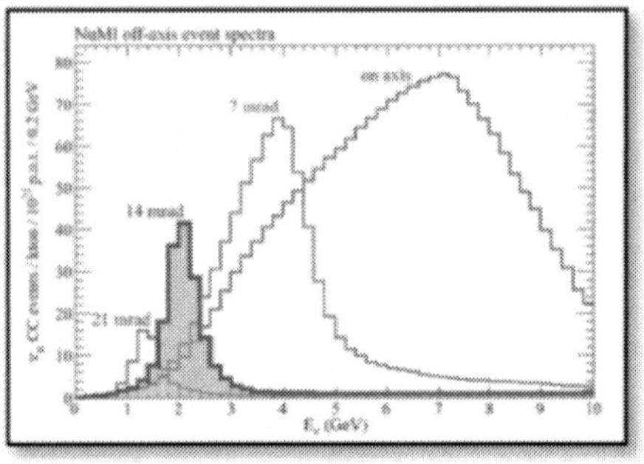

Figure 1. Energy spectrum of the off axis (14 mrad) technique yielding a narrow band around 2 GeV for the NOvA neutrino beam.

2.1. NOvA Detector Design

The basic design of a NOvA detector cell is long extruded PVC (plastic) tubes/cells, 4x6 cm in cross-section, with a length of 15 m for the Far and 4 m for the Near detector. Each is filled with a mineral oil liquid scintillator, instrumented with a single wavelength shifting fiber, and read out by an Avalanche Photo-Diode (APD). This provides a 64% active-material and low-Z calorimeter with a radiation length of X0~38 cm and tracking capabilities. This technology is optimal for electromagnetic shower reconstruction as well as muon tracking.

The near detector design calls for about 18,000 such cells, totaling 222 tons, positioned at a distance of 990 m from the NuMI beam target. The Far detector is much larger, numbering about 370,000 cells for 14 ktons, and at a distance of 810 km located in Ash River, Minnesota.

The detectors are instrumented by regions. They are synchronized and readout at high speed, continuously and with no dead time. The threshold is set to that of a half of a minimum ionizing particle energy deposition at the far end of the longest cell. This is equivalent to an energy threshold of about 6-8 MeV. All data is buffered in a large computing farm and triggered on asynchronously by the NOvA data acquisition and the Accelerator Trigger Systems. In figure 2, there are event displays of some indicative, simulated event topologies of the Near detector response to the prominent signatures of the NOvA experiment. A 48% reconstruction efficiency of 2 GeV electron-neutrinos from charge-current quasi- elastic scattering and a 1% rate of fake neutral-current interactions, with the proton recoil visible to few MeV, make the NOvA detection system an exceptionally good detector for its size.

2.2. The NOvA measurement

The NOvA experiment is designed to measure independently four probabilities of oscillation between the muon and electron neutrinos as they travel from the Near detector site to the Far detector 810 km away. We will combine the measured probabilities to evaluate the θ13 and θ23. Through these measurements, we can determine the mass hierarchy and the octant of the θ23, providing we have enough statistics (3-year data run that can restrict the 1σ and 2σ curves as in figure 3). Depending on the combination of the hierarchy and CP-violation, we can resolve them up to 38% of the phase space. This means we may be able to see the first evidence of CP violation in the lepton sector.

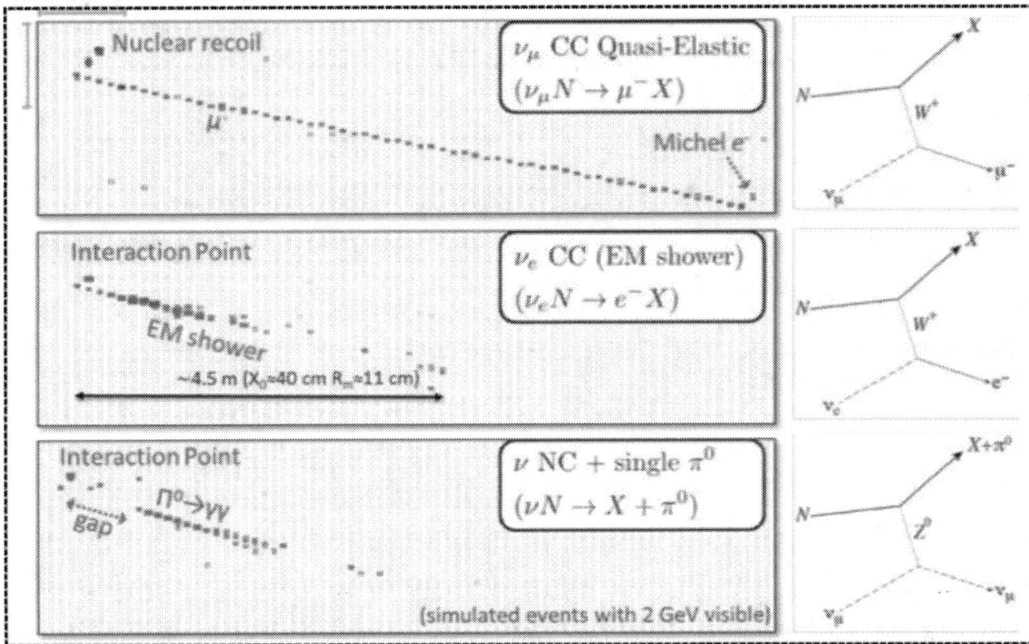

Figure 2. Event topologies of simulated events in the Far detector for various signature interactions of 2 GeV neutrinos. Muon tracks are well differentiated from showers for reconstruction.

Accepting the value for the $\theta_{13} \sim 9^\circ$, we can parameterize the bi-probabilities of appearance measured at the NOvA experiment in the plot of figure 3 between assumptions of θ_{23} and all the values of the CP-violating parameter. This particular plot assumes that the θ_{23} is maximal. A NOvA measurement is represented by a star in the plots. The curves around it represent 1σ and 2σ resolution after a 3 year run with neutrinos and anti-neutrinos each. The measurement (star) will fall on one of the eclipses representing the CP violation case and the mass hierarchy. Depending on that point, the NOvA may be able to resolve the hierarchy with up to 2σ resolution and indicate the extent of CP violation (see fig. 3). This is the most optimistic of the scenarios. If the measurement (star) falls in some point in an eclipse for which the resolution circles overlap with larger parts of these eclipses, then a bound can be set. If the θ_{23} is sufficiently non-maximal, its octant may be resolved easier than the mass hierarchy. There are several other scenarios that span the phase space but their presentation is not for the short present document.

2.3. NDOS prototype

The Near Detector On the Surface (NDOS) is the prototype of a NOvA Near detector. The NDOS was built for component production, installation and integration experience. It has been completed since May 2011. It is positioned at a crossing of the beams from both the Main Injector (110 mrad off axis) and the Booster accelerators (on-axis but 23o rotated). It is also situated 100 m above the beams (i.e. on the surface). Until May 2012, it was doing physics runs when the NuMI beam was shut down for the upgrade. It is planned to pick up again after NuMI beam resumes in April 2013. In the meantime, it is used for engineering runs. These serve to evaluate hardware and software technology tests. The data acquisition and the detector control systems (hardware and software) have been using the NDOS as a life size test- stand for their developments. APD technologies are currently being evaluated before being widely deployed on the Far detector in the winter of 2012. The track and vertex reconstruction software have been using the NuMI and Booster beams to create extensive libraries of good events. Cosmic muons are also used but for background rejection algorithms. All these signals are also used to calibrate the detector. By measuring the attenuation length in the cells and the spectrum of Michel electrons, we get exceptional resolution and low readout thresholds.

3. Current Status of NOvA

The Far detector laboratory has been completed and the NOvA project got beneficial occupancy since April 2011. It is located at Ash River, Minnesota. It is on the surface with 3 m earth equivalent overburden above the detector hall. The detector construction has started as of late summer of 2012. Construction of the full 14 ktons detector is expected to be completed in 2014. The Near detector cavern is being prepared and the detector completion is expected by 2014 as well. The NuMI beam upgrades are on schedule to begin delivering the new beam power by the accelerator restart in Spring of 2013.

4. Summary

The next decade will be the time that some of the most important questions of neutrino physics can be answered. The NOvA experiment will be in the position to answer some of them. It has the opportunity to resolve the mass hierarchy and the octant of θ23, if it runs at least for 3 years on neutrino and anti-neutrino beams. Most importantly, it could show some first evidence of CP violation in the lepton sector. The experiment is well prototyped by now and well understood. It is also on track in construction for the Far and Near detectors and the beam upgrades will provide it with the best chance to lead the way in understanding the lepton sector and beyond.

References

[1]. K. Abe et al., "T2K" PRL. 107, 041801 (2011), and PR D 85, 031103(R) (2012).
[2]. P. Adamson et al., "MINOS" PRL. 107, 181802 (2011), hep-ex/1108.0015v1.
[3]. F. P. An et al, "Daya Bay" PRL 108, 171803(2012), or arXiv:1203.1669 [hep-ex].
[4]. Y. Abe et al, "Double Chooz " PRD 86, 052008 (2012) or arXiv:1207.6632v1[hep-ex] 27 July 2012,and DC first indications in PRL 108, 131801(2012) or arXiv: 1112.6353[hep-ex] 13 March 2012.
[5]. Neutrino Experiments at the Intensity frontier: http://if-neutrino.fnal.gov/neutrinos.pdf

Neutral-current neutrino-nucleus reactions and their impact to supernova physics

V Tsakstara[1], T S Kosmas[1], A A Dzhioev[2,3], A I Vdovin[2]

[1] Theoretical Physics Section, University of Ionnina, 45110, Creece
[2] Bogoliubov Laboratory of Theoretical Physics, JINR, Dubna, 141980, Russia

E-mail: vtsaksta@cc.uoi.gr

Abstract. We study neutral-current neutrino-nucleus reactions in nuclei that are relevant for supernova (SN) simulations and for terrestrial experiments aiming at neutrino astrophysics as well as ν-nucleus scattering cross sections measurements. Such studies allow us to improve estimates of nuclear responses to low energy neutrinos in light of the operation of nuclear ν-detectors with very-low threshold and very high sensitivity. The adopted ν-energy range is extended to rather high energies (up to 100 MeV) so as to consider allowed and forbidden multipole contributions to cross sections. Both contributions are calculated within the quasi-particle random phase approximation by using realistic two-body forces (Bonn CD potential) for the residual interaction of the nuclear Hamiltonian. As a special application the ^{56}Fe isotope is chosen due to its significant role in SN physics and ν-detection.

1. Introduction

Neutrinos are neutral particles (without electric charge and magnetic moment) having very small masses (according to the recent experimental evidences). Since they are not deflected from electromagnetic fields and interact very weakly gravitationally, they can travel cosmological large distances without interaction, so as they may escape the very dense surrounding deepest in the interior of massive stars [1]. This property makes them almost ideal messenger-particles of distant stars, a feature that allows us to look into the innermost part of the stars. However, the very low ν-interaction cross-section is simultaneously the biggest disadvantage for their detection that can be probed through measuring Cherenkov light from secondary particles or through the de-excitation products of ν-induced excitations of nuclear ν-detectors in extremely sensitive (mostly underground) experiments. Such probes, however, need significant event rates and, therefore, huge volume detectors have to be constructed [2, 3, 4]. On the other hand, from supernova (SN) physics it is known that, the death of massive (with mass $M \geq 8 - 10M_{\odot}$) stars is marked through very energetic explosions known as core-collapse supernovae. In a typical SN, an energy of about 10^{53} ergs is released (within 10 seconds after the explosion) in the form of neutrinos which (except their oscillation on propagation) travel to Earth undistorted.

In the present research plan, we explore the role of ν-nucleus reactions in understanding open questions relevant to SN physics. Our strategy is as follows: First, we consider neutral-current (NC) ν-nucleus scattering assuming that the nuclear target is at zero temperature, for some isotopes contents of promising ν-detection materials that are quite advantageous (Fe,

[3] Present address: Physics Department, Université libre de Bruxelles, Campus Plaine, CP231, Blvd du Triomphe

Ar, Mo, Te, etc.): (i) for measuring rare events (double-beta decay, etc. [5, 6]) and (ii) for potential use in studying neutrino physics issues [2, 3, 4]. Second, we treat charged-current (CC) interactions of neutrinos with the above nuclei. The relevant cross sections are by about an order of magnitude larger than the corresponding for NC ones. Third, we consider nuclear responses to SN-neutrinos focusing on the possibility of including in these estimations contributions coming from particle decays of ν-induced excited states lying above the separation energies for protons and neutrons [4]. In studying detector responses to SN neutrinos, particle decays cause extra signals which are important to be included in the study.

As a concrete nuclear system in this paper, we examine the nucleus ^{56}Fe, which plays a significant role in SN physics involving neutrinos. We note that, for the iron group isotopes in SN environment the important thermal effects on NC ν-nucleus scattering have been studied [7, 8].

2. Neutrino-nucleus cross sections at zero temperature

By applying the Donnelly-Walecka multipole decomposition method [2], the double differential cross section of NC ν-nucleus scattering reads

$$\frac{d^2\sigma_{i\to f}}{d\Omega d\omega}(\phi,\theta,\omega,\varepsilon_i) = \delta(E_f - E_i - \omega)\frac{2G^2\varepsilon_f^2 \cos^2(\theta/2)}{\pi(2J_i+1)}\left[\mathcal{C}_V + \mathcal{C}_A - \mathcal{C}_{VA}\right], \qquad (1)$$

The δ-function in the right hand side of the latter equation denotes the energy conservation, so as $\omega = \varepsilon_\nu - \varepsilon_f \approx E_f - E_i$, where ω is the excitation energy of the nucleus, E_i and E_f represent the energy of the initial (ground) and final states of the studied nucleus, respectively. ε_ν (ε_f) is the incoming (outgoing) neutrino energy.

The term $\mathcal{C}_V$ ($\mathcal{C}_A$) in Eq. (1) is a summation over the contributions coming from the polar-vector (axial-vector) operators written in terms of the matrix elements of the eight multipole operators $\widehat{M}_J^{(5)}$, $\widehat{L}_J^{(5)}$, $\widehat{T}_J^{el(5)}$ and $\widehat{T}_J^{mag(5)}$, where the superscript '5' refers to the four axial-vector components of the hadronic current. The interference term $\mathcal{C}_{VA}$ in Eq. (1) contains the product of transverse polar-vector and transverse axial-vector matrix elements [2] For normal parity transitions, $\mathcal{C}_{VA}$ contains contributions of $\widehat{T}_J^{el}$ and $\widehat{T}_J^{mag5}$ operators while for abnormal parity ones $\mathcal{C}_{VA}$ contains matrix elements of $\widehat{T}_J^{mag}$ and $\widehat{T}_J^{el5}$. Total cross sections, σ, are subsequently calculated by integrating (numerically) Eq. (1) as is described in Refs. [2, 3, 4].

2.1. Determination of the QRPA model parameters

The initial (ground) state ($|0^+\rangle$) of the even-even ^{56}Fe isotope has been computed by solving the BCS equations. In Table 1, we tabulate the values of the pairing parameters ($g_{pair}^{p,n}$) and the energy gaps ($\Delta_{p,n}^{th}$) for protons (p) and neutrons (n) determined at the BCS level for this isotope. The first parameters renormalize the pairing force and adjust the gaps $\Delta_{p,n}^{th}$ to the empirical ones, $\Delta_{p,n}^{exp}$, obtained through the application of the well known three point formulae [4].

Table 1. Parameters $g_{pair}^{p,n}$, for protons (p) and neutrons (n), respectively, determining the BCS pairing interactions, and values of the energy gaps, $\Delta_{p,n}^{th}$. The latter are adjusted to reproduce the empirical energy gaps, $\Delta_{p,n}^{exp}$ for ^{56}Fe (b is the h.o. size parameter).

Isotope	b (fm)	g_{pair}^{n}	g_{pair}^{p}	S_p	S_n	Δ_p^{exp}	Δ_p^{th}	Δ_n^{exp}	Δ_n^{th}
^{56}Fe	1.996	0.901	0.849	10.184	11.197	1.568	1.568	1.366	1.363

The final (excited) states $|J_f^{\pi_f}\rangle$ of the isotope in question have been calculated by solving the QRPA equations [2]. The model space used consists of the major shells with $\mathcal{N} = 2, 3, 4\ \hbar\omega$ (inert core ^{16}O) which include 14 single particle orbits. The excited states $|J_f^{\pi_f}\rangle$ have been checked by comparing the low-lying QRPA spectrum with the corresponding experimental excitations by adjusting accordingly the parameters that renormalize the particle-particle (g_{pp}) and particle-hole (g_{ph}) channels of the residual two-body matrix elements entering the QRPA matrices [2].

3. Results and Discussion

In the present contribution, as a fist step towards the above purposes, we apply the QRPA method to calculate NC ν-^{56}Fe scattering cross sections focusing on the study of the behavior of the $\sigma(\omega, \varepsilon_\nu = const., T = 0)$ in various leading sets of multipole states (up to $J^\pi = 6^\pm$). The assumed incoming ν-energy range is $0 \leq \varepsilon_\nu \leq 100$ MeV.

The results of such original state-by-state QRPA calculations for the most important multipolarities are illustrated in Fig. 1(a,b,c), for the positive parity transitions 0^+, 1^+ and 2^+, and in Fig. 1(d), for the negative parity transitions 1^-. In this figure, the cross sections refer to $\varepsilon_\nu = 30$ MeV, but such plots could be also made for other values of ε_ν. As it is, in general, expected, for inelastic scattering of SN neutrinos, the leading multipoles, which yield the most significant cross sections, are the $J^\pi = 1^+$ and $J^\pi = 1^-$. We have chosen to show $\sigma(\omega, \varepsilon_\nu = 30\ \text{MeV})$, because $\varepsilon_\nu = 30$ MeV is rather close to the mean energies of SN ν_x neutrinos and $\tilde{\nu}_x$ anti-neutrinos. Recent SN neutrino simulations suggest mean energies $\langle\varepsilon_{\nu_x}\rangle = \langle\varepsilon_{\tilde{\nu}_x}\rangle =$ 22 - 25 MeV [3].

As discussed in Ref. [2], the 1^- multipole contributions contain spurious admixtures which in the present work are approximately removed. The maximum peak appears in the case of the 1^+ multipole, in which the axial-vector pieces L_1^5 and T_1^{el5} dominate over the vector piece T_1^{mag}. Since the momentum transfer q is rather small for SN-neutrino scattering, the axial-vector

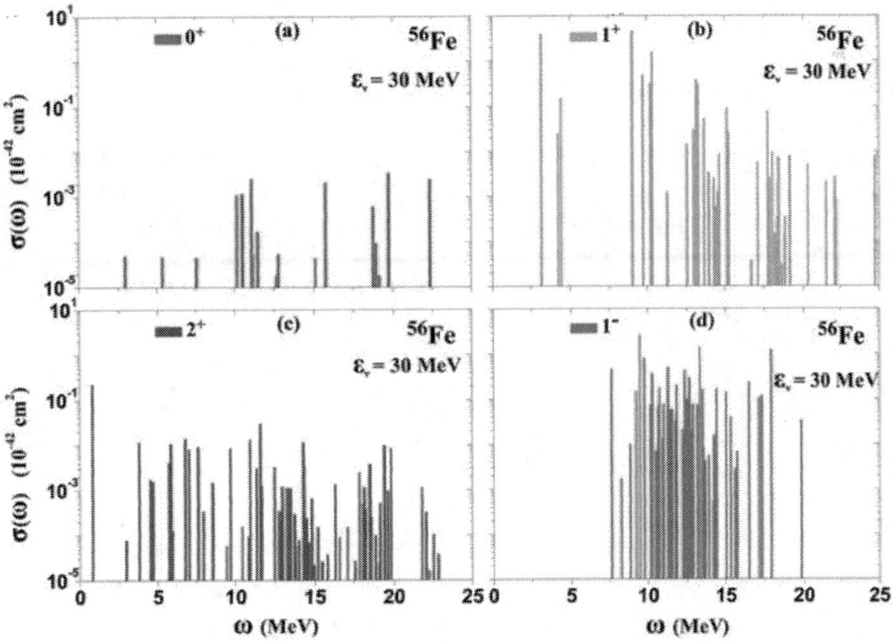

Figure 1. NC ν-^{56}Fe scattering cross sections $\sigma(\omega)$, as a function of the nuclear excitation energy ω for: 0^+ (a), 1^+ (b), 2^+ (c) and 1^- (d) transitions ($\varepsilon_\nu = 30$ MeV).

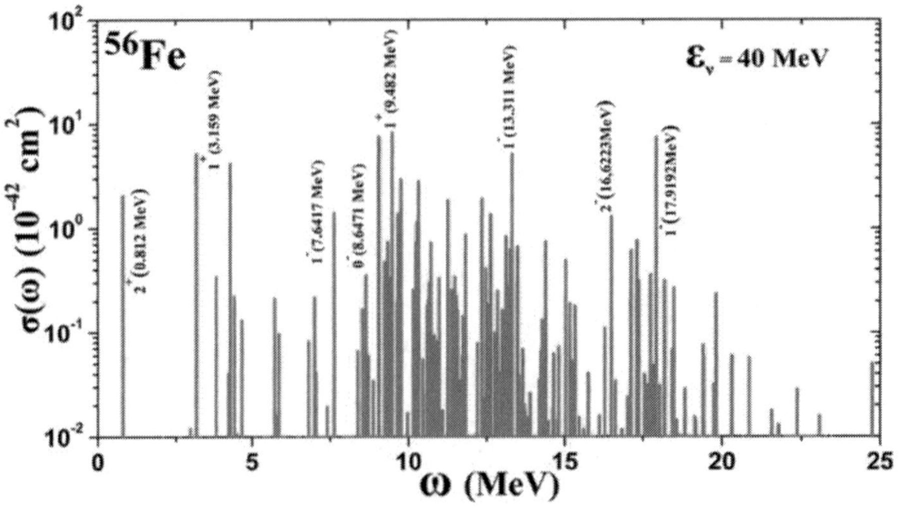

Figure 2. Cross section $\sigma(\omega)$ as a function of the excitation energy ω for the nucleus ^{56}Fe. The incoming ν-energy was $\varepsilon_\nu = 40$ MeV. The most pronounced peaks belong to $J = 1$ transitions.

matrix elements dominate over the vector ones (even by an order of magnitude or more).

In Fig. 2, the variation of $\sigma(\omega, \varepsilon_\nu = ct.)$ for the reaction $^{56}Fe(\nu, \nu')^{56}Fe^*$, is demonstrated. As can be seen, $\sigma(\omega)$ presents some characteristic clearly pronounced peaks at various excitation energies ω and specifically for transitions $J^\pi = 1^+, 1^-$, but also for $J^\pi = 0^+, 2^+$. The maximum peak corresponds to a $J^\pi = 1^+$ (ground state) transition at $\omega = 9.482$ MeV and a second one corresponds to a 1^- transition at $\omega = 13.311$ MeV in agreement with the findings of Refs. [2, 4].

4. Summary and Conclusions

Neutral-current ν-nucleus reactions cross sections in nuclei that are important for SN simulations and for terrestrial experiments aiming at astrophysical ν-detection, are required. A great number of open issues in supernova physics may be unraveled by future galactic (or extragalactic) SN explosions which nowadays may well be analyzed with very high statistics by Earth experiments.

In this work, we focused on cross section calculations of the reaction $^{56}Fe(\nu, \nu')^{56}Fe^*$ (at zero temperature). At finite temperature in stellar environment, thermal population of the excited states may enhance the weak interaction rates and cross sections at low ν-energies. Thermal effects for NC ν-scattering on Fe group isotopes are important to understand SN dynamics as discussed in Refs. [8, 9, 10]

References

[1] Langanke K and Martínez-Pinedo 2003 *Rev. Mod. Phys.* **75** 819.
[2] Tsakstara V and Kosmas T S 2011 *Phys. Rev.* C **83** 054612 and references therein.
[3] Tsakstara V and Kosmas T S 2011 *Phys. Rev.* C **84** 064620.
[4] Tsakstara V and Kosmas T S 2012 *Phys. Rev.* C **86** 044618.
[5] H. Ejiri, Phys. Rep. **338**, 265 (2000).
[6] K. Zuber, Phys. Lett. **B 519**, 1 (2001); ibid **B 571**, 148 (2003).
[7] Dzhioev A A, Vdovin A I, Ponomarev V Y, Wambach J, Langanke K, and Martínez-Pinedo G 2010 *Phys. Rev.* C **81** 015804.
[8] Dzhioev A A, Vdovin A I, Ponomarev V Y, Wambach J 2011 *Phys. At. Nucl.* **74** 1162
[9] Dzhioev A A, Vdovin A I, Tsakstara V, Kosmas T S *Journal of Physics: Conference Series* , submitted
[10] Dzhioev A A, Tsakstara V, Kosmas T S, Vdovin A I, Ponomarev V Yu, and Wambach J Phys. Rev. C to be submitted.

Nilpotence in Physics: the case of Tsallis entropy

A J Creaco[1], **N Kalogeropoulos**[2,3]

[1] Science Department, BMCC - The City University of New York, 199 Chambers St., New York, NY 10007, USA

[2] Pre-Medical Department, Weill Cornell Medical College in Qatar, Education City, PO Box 24144, Doha, Qatar

E-mail: [1] acreaco@bmcc.cuny.edu, [2,3] nik2011@qatar-med.cornell.edu

Abstract. In an attempt to understand the Tsallis entropy composition property, we construct an embedding of the reals into the set of 3×3 upper triangular matrices with real entries. We explore consequences of this embedding and of the geometry of the ambient 3×3 Heisenberg group. This approach establishes the polynomial growth of the volume of phase space of systems described by the Tsallis entropy and provides a general framework for understanding Abe's formula in terms of the Pansu derivative between Riemmanian spaces.

1. Introduction

The Harvda/Charvat [1], Daróczy [2], Tsallis [3] entropy (henceforth Tsallis entropy) is a single parameter family of entropic functionals introduced in the Statistical Mechanics community in 1988 [3]. It has attracted considerable interest since that time, for several reasons. One is that it claims to describe phenomena having long range spatial and temporal correlations, something that the Boltzmann/Gibbs/Shannon (BGS) entropy is, a priori, unsuitable to do [4] (and references therein). Examples are long-range interactions such as the gravitational interaction applied, in particular, in astrophysical or general relativistic contexts and electrodynamics as applied in charged plasmas, for instance. Another reason is that the Tsallis entropy appears to be suitable for describing some out of equilibrium phenomena. As it befits a concept claiming a foundational role in any part of Physics, as well as a wide-ranging set of applications, the Tsallis entropy has, inevitably, also attracted some criticism (see e.g. [5], [6]). Even though its scope of applicability, if any, is currently unclear, its contribution in providing a viable alternative to the BGS entropy which gives rise to a reasonable, and well-developed, thermodynamic formalism is undeniable.

2. Additivity and extensivity

Consider a set having a sample space Ω. As examples of physical interest, Ω can be a countable set with a measure, such as the cardinality of subsets, and a metric, or it can be a Riemannian manifold or, more generally, a metric measure (mm) space. We will follow the notation of the Riemannian case in the sequel, but the results can be straightforwardly extended to the discrete case or further generalized to mm spaces. Let the statistical behavior of the aspects of the system under study be given by the probability function $p : \Omega \to [0, 1]$. Assume that p has

[3] Corresponding Author

no atoms and let its Radon-Nikodym derivative with respect to $dvol_\Omega$ be indicated by ρ. The Tsallis entropy of p is defined [3],[4] as

$$S_q[\rho] = k_B \frac{1}{q-1}\left\{1 - \int_\Omega [\rho(x)]^q \, dvol_\Omega\right\} \tag{1}$$

Here k_B is Boltzmann's constant that will be set equal to one, for simplicity, in what follows. The parameter $q \in \mathbb{R}$, where $\mathbb{R}$ indicates the set of the real numbers, is called nonextensive or entropic parameter. It is straightforward to check that

$$\lim_{q\to 1} S_q = S_{BGS} \tag{2}$$

Consider two occurrences $\Omega_1, \Omega_2 \subset \Omega$ that are independent, in the conventional sense of the word, namely $\rho_{\Omega_1 * \Omega_2} = \rho_{\Omega_1} \cdot \rho_{\Omega_2}$ where $\Omega_1 * \Omega_2$ stands for the combined system resulting from the interactions of Ω_1 and Ω_2. The Tsallis entropy is then

$$S_q[\rho_{\Omega_1 * \Omega_2}] = S_q[\rho_{\Omega_1}] + S_q[\rho_{\Omega_2}] + (1-q) S_q[\rho_{\Omega_1}] S_q[\rho_{\Omega_2}] \tag{3}$$

According to (3), the Tsallis entropy is not additive, in the conventional sense of the word. However this is not an impediment to formulating a valid thermodynamic formalism [4] (and references therein). Additivity is quite distinct from extensivity, although in the usual BGS entropy case this distinction is somewhat blurred. The important property that any thermodynamic potential should have is extensivity, appropriately defined. This amounts to requiring that the thermodynamic potentials are proportional to the number of effective degrees of freedom of the system. Clearly, additive systems are extensive, but the converse is not necessarily true. Property (3) above is, arguably, the most distinct property of the Tsallis entropy when compared to its BGS counterpart. In order to restore the manifest additivity in (3), a generalized addition $\oplus_q$ was defined [7],[8] by

$$x \oplus_q y = x + y + (1-q)xy, \quad \forall\, x, y \in \mathbb{R} \tag{4}$$

This is a usual process in Physics, where new objects are defined to naturally reflect properties of some operations: consider the role of tensors, for instance, in General Relativity, the role of covariant derivatives in Gauge Theories and Gravity, appropriately defined superfields in supersymmetric and supergravitational models etc. Using (4) the generalized independence/additivity of the Tsallis entropy can be rewritten in the more "covariant" form

$$S_q[\rho_{\Omega_1 * \Omega_2}] = S_q[\rho_{\Omega_1}] \oplus_q S_q[\rho_{\Omega_2}] \tag{5}$$

3. BGS vs. Tsallis entropies: hyperbolic metrics

In an attempt to compare the concept of independence as encoded through the BGS and the Tsallis entropies, we set up in [9] a coordinate system whose one axis was parametrized by $\mathbb{R}$ and whose other axis was parametrized by its Tsallis entropy induced deformation [10]. By using a semi-direct product construction, we found that the resulting plane has a metric of constant sectional curvature which is given by

$$k = -[\log(2-q)]^2 \tag{6}$$

If the Tsallis entropy composition property is a feature of phase space of a system, as opposed to being emergent (e.g ultrametrics in the case of spin glasses) then it induces a hyperbolic structure on phase space. Hence someone can claim that the Tsallis entropy is the "hyperbolic analogue" of the "Euclidean" BGS entropy [9]. Then the Cartan-Hadamard theorem guarantees that the Tsallis entropy is universal, if one also accepts the rest of the Shannon/Khitchine/Abe/Santos/Suyari axioms. The conclusion that we reached can be extended to the case of interacting systems via convex potentials, even when the systems are described tby different values of the entropic parameters, in which case, the resulting system is described by an induced $CAT(k)$ structure rather than by one induced by the hyperbolic plane. As a result of this formalism, we were able to establish in [11] that if a system's thermodynamic behavior is described by the Tsallis entropy and if the composition property relfects a property of its phase space, then the largest Lyapunov exponent of the underlying dynamical system should be zero. The converse is not necessarily true though. Hence all such systems described by the Tsallis entropy exhibit "weak chaos". Moreover under the above assumptions we were able to show in [12] why one should use the escort distribution

$$\rho_{esc} = \frac{1}{N}\,\rho^q, \qquad N = \int_\Omega \rho^q \; dvol_\Omega \tag{7}$$

rather than ρ itself in maximizing and in deriving the moments of the quantities of interest stemming from the Tsallis entropy.

4. BGS vs Tsallis entropies: sub-Riemannian structures

The goal of the construction of this section is the same as of the previous one: to compare the usual and the generalized additions in an attempt to elucidate the meaning and implications of the Tsallis entropy composition. Here, we construct a matrix the non-trivial elements of which give rise to the ordinary and generalized additions, upon matrix multiplication. Then we explore geometric consequences of this construction. To begin with, we perform a formal re-scaling of the Tsallis entropy

$$\tilde{S}_q[\rho] \equiv (1-q) S_q[\rho] \tag{8}$$

Then $\tilde{S}_q$ obeys a modified additivity which induces a generalized addition, mimicking the steps leading to (4), given by

$$x \tilde{\oplus}_q y = x + y + xy \tag{9}$$

Next, we define the map $\mathcal{S}: \mathbb{R} \rightarrow \mathfrak{U}^{3\times 3}$ defined by

$$\mathcal{S}(x) \;=\; \begin{pmatrix} 1 & x & x \\ 0 & 1 & x \\ 0 & 0 & 1 \end{pmatrix} \tag{10}$$

Here $\mathfrak{U}^{3\times 3}$ stands for the set of upper triangular matrices with real entries having unit elements in their diagonal. Upon ordinary matrix multiplication, we get

$$\mathcal{S}(x)\;\mathcal{S}(y) \;=\; \begin{pmatrix} 1 & x+y & x\tilde{\oplus}_q y \\ 0 & 1 & x+y \\ 0 & 0 & 1 \end{pmatrix} \tag{11}$$

This is a very desirable form since the elements in the secondary diagonal are additive, whereas the "corner" term follows the generalized additivity (10). So, the embedding (11) allows someone to compare the ordinary and the generalized additions within the same structure.

A drawback of (11) is that it is not a homomorphism, nor is the image of $\mathbb{R}$ under $\mathcal{S}$ a subgroup of $\mathfrak{U}^{3\times 3}$. The latter is bothersome, on physical grounds, as we wish to have a structure that allows the composition of elements, such as a group or, more generally, a category. If we wish to continue working with groups, we are lead to consider not just with the image of $\mathbb{R}$ under $\mathcal{S}$, but the whole ambient space $\mathfrak{U}^{3\times3}$. This group is easily checked to be 2-step nilpotent and its Lie algebra is easily seen to be generated by the matrices

$$X = \begin{pmatrix} 0 & 1 & 0 \\ 0 & 0 & 0 \\ 0 & 0 & 0 \end{pmatrix}, \quad Y = \begin{pmatrix} 0 & 0 & 0 \\ 0 & 0 & 1 \\ 0 & 0 & 0 \end{pmatrix}, \quad Z = \begin{pmatrix} 0 & 0 & 1 \\ 0 & 0 & 0 \\ 0 & 0 & 0 \end{pmatrix} \tag{12}$$

which obey $[X, Y] = Z$ with all other commutators being zero. This is exactly the 3-dimensional Heisenberg algebra. Hence someone can claim that the Tsallis entropy can be viewed as the "quantum analogue" of the BGS entropy, understanding however that this is just a formal analogy and not claiming any deeper physical significance to it, at least at this stage. The effective "Planck's constant" in this description, is a function of q.

From a geometric viewpoint, the embedding (11) reduces the study of the Tsallis entropy composition (10) into the framework of the geometry of the Heisenberg groups, or more generally brings it in the far wider context of sub-Riemannian (Carnot-Carathéodory) geometry [13], [14]. This framework allows us to draw some far-reaching conclusions. One of them is that if a system is described by the Tsallis entropy and if the composition property is not emerging, but is follows directly from a corresponding property of the phase space, then the phase space volume of the system grows in a polynomial (power-law) fashion. This stems from a continuum analogue of a theorem due to Milnor/Wolf/Titis/Guivar'ch/Bass/Gromov stating that a finitely generated group has polynomial growth if and only if it is virtually nilpotent. The desired result can be reached by taking the "continuum limit", or by applying Kanai's discretization results for manifods with lower Ricci curvature bounds. The conclusion is an inverse, in a sense, of a result of Tsallis et al. [4] in the discrete case, and is applicable for both discrete and continuous systems. Another implication of this construction is that we can see Abe's formula involving the Jackson derivative of a probability distribution as a special case of the Pansu derivative between sub-Riemannian spaces. Details and more concrete explanations of the statements of this paragraph will be given in a forthcoming work.

Acknowledgments

We wish to thank the Organizing Committee of IC-MSQUARE 2012 for inviting us to present our work in the mini-workshop "Gravity, Quantum, and Black Holes" and for establishing a pleasant and stimulating venue for the effective presentation and fruitful exchange of ideas.

References

[1] 1 Harvda J and Charvat F 1967 *Kybernetika* **3** 30
[2] 2 Daróczy Z 1970 *Inf. Comp./Inf. Contr.* **16** 36
[3] 3 Tsallis C 1988 *J. Stat. Phys.* **52** 479
[4] 4 Tsallis C 2009 *Introduction to Nonextensive Statistical Mechanics: Approaching a Complex World* (New York, NY: Springer)
[5] 5 Gross D H E 2002 *Physica A* **305** 99
[6] 6 Grassberger P 2008 Comment on "Central limit behavior in deterministic dynamical systems" *Preprint* arXiv:0809.1406
[7] 7 Nivanen L, Le Mehauté A and Wang Q A 2003 *Rep. Math. Phys.* **52** 437
[8] 8 Borges E P 2004 *Physica A* **340** 95
[9] 9 Kalogeropoulos N 2012 *Physica A* **391** 3435
[10] 10 Kalogeropoulos N 2012 *Physica A* **391** 1120
[11] 11 Kalogeropoulos N Vanishing largest Lyapunov exponent and Tsallis entropy *Preprint* arXiv:1203.2707
[12] 12 Kalogeropoulos N Escort distributions and Tsallis entropy *Preprint* arXiv:1206.5127
[13] 13 Bellaiche A and Risler J-J 1996 *Sub-Riemannian Geometry* (Basel: Birkhäuser)
[14] 14 Montgomery R 2006 *A Tour of Subriemannian Geometries, Their Geodesics and Applications* (Providence, RI : Amer. Math. Soc.)

Journal of Physics: Conference Series **410** (2013) 012149 doi:10.1088/1742-6596/410/1/012149

A topological extension of GR: Black holes induce dark energy

M Spaans[1]
Kapteyn Astronomical Institute, University of Groningen, 9700AV Groningen, The Netherlands

E-mail: spaans@astro.rug.nl

Abstract. A topological extension of general relativity is presented. The superposition principle of quantum mechanics, as formulated by the Feynman path integral, is taken as a starting point. It is argued that the trajectories that enter this path integral are distinct and thus that space-time topology is multiply connected. Specifically, space-time at the Planck scale consists of a lattice of three-tori that facilitates many distinct paths for particles to travel along. To add gravity, mini black holes are attached to this lattice. These mini black holes represent Wheeler's quantum foam and result from the fact that GR is not conformally invariant. The number of such mini black holes in any time-slice through four-space is found to be equal to the number of macroscopic (so long-lived) black holes in the entire universe. This connection, by which macroscopic black holes induce mini black holes, is a topological expression of Mach's principle. The proposed topological extension of GR can be tested because, if correct, the dark energy density of the universe should be proportional the total number of macroscopic black holes in the universe at any time. This prediction, although strange, agrees with current astrophysical observations.

1. Quantum mechanics, gravity and topology

The success of quantum mechanics and Einstein gravity is overwhelming. Hence, one should take great care in preserving the beautiful features of both, while pursuing the mystery of dark energy. A topological extension of GR, which allows for quantum aspects, is discussed here, following [1].

The Feynman path integral provides a very elegant formulation of quantum mechanics. In it, the motion of matter, like a particle world line, derives from a superposition of wave amplitudes along many different paths. To physically construct a world line it thus seems reasonable to assume that nature knows about all the different paths that particles can follow. To distinguish such paths, one can use topology. After all, when one takes paths as the central objects in quantum mechanics, then one must have a means to identify them, whatever the erratic trajectories that particles can follow. Indeed, many of the paths that enter the Feynman path integral are continuous, but not necessarily differentiable. Topology allows for the continuous deformation of space-time. Therefore, one can construct distinct paths through space-time by assigning a multiply connected character to space-time. Specifically, [1] proposes a lattice of three-tori embedded in 4 dimensions to create many

[1] To whom any correspondence should be addressed.

topologically distinct paths between any two space-time points, and thus facilitate the Feynman path integral, as follows. A three-torus is a solid cube with its faces identified through the 4th dimension. A lattice of three-tori can be constructed by connecting each face of a three-torus to a face of a neighboring three-torus through a 3D junction. Wave amplitudes can then freely travel along the lattice junctions and pick up phase shifts along the 3 loops comprising any three-torus, $T^3=S^1\times S^1\times S^1$. The local isotropy and flatness of the three-torus allows one to define relativistic quantum mechanics in small space-time patches, in the usual manner, while the global multiply connected nature of the lattice of three-tori implements the superposition principle using space-time itself.

The Einstein equation has many symmetries, but it is not invariant under local conformal (scale changing) transformations. As a result, one finds large metric fluctuations when one goes to small spatial scales. In fact, at the Planck scale of $l_P\sim10^{-33}$ cm, these fluctuations in space-time become huge (even singular) and space-time should take the form of a quantum foam with Planck mass ($m_P\sim10^{-5}$ g) black holes (BHs), popping out off and into the vacuum [2]. Space-time thus enjoys the presence of mini BHs at the Planck scale, i.e., is also multiply connected when one considers Einstein gravity at the Planck scale. A fundamental problem with the quantum foam is that it fluctuates violently on a Planck time of $t_P\sim10^{-43.5}$ sec and is therefore difficult to conceive as a stable structure.

Nevertheless, as a towards a theory of quantum gravity, and using the freedom to choose the (global) topology of space-time within GR, one can think of a Planck scale space-time that comprises a lattice of three-tori with an unstable foam of mini black holes attached to it. However, this model is quite heuristic and without dynamical content. This can be improved as follows.

One first notes that any topological space-time manifold can be constructed by adding together three-dimensional building blocks called prime manifolds. These primes (e.g., handles, three-tori, three-spheres, etc.) lend their name from the property that they themselves cannot be written as the connected sum of smaller three-dimensional units. Furthermore, interactions among those primes can never change their intrinsic properties, like the number of closed loops they contain. Finally, following the superposition principle of quantum mechanics, any property of Planck scale topology must be carried by the linear sum of these primes. One can then conclude that a plausible linear basis set consists of the non-chiral prime manifolds $S^1\times S^2$, S^3 and T^3, where 1) the non-irreducible handle manifold appears as part of the large scale solution to the Einstein equation, i.e. the Schwarzschild BH solution; 2) the topologically trivial three-sphere is a unit element required for the large scale vacuum limit; 3) the three-torus is confined to the Planck scale and implements the superposition principle.

To implement topology driven dynamics on this basis set of primes, one introduces a loop algebra with operators that create (T*) and destroy (T) the S^1 loops contained in a prime P. The central topological object here is the loop because quantum mechanical superposition deals with paths and any pair of paths, with the same starting and end point, is homeomorphic to a loop. Similarly, the presence of a handle $S^1\times S^2$ requires a loop and allows a pair of paths in the form of a loop that has one branch going though the handle and the other around it. One can define topological dynamics through

$$T^*P = P\times S^1 \quad \text{and} \quad TP = nQ, \qquad (1)$$

with n the number of closed loops in P while Q denotes P with a loop contracted to a point. Details on the loop algebraic equations of motion are provided in [1], but it is sufficient to realize that the action of the scalar operator N=(TT*+T*T) amounts to

$$NT^3 = 7T^3, \quad NS^1\times S^2 = 3S^1\times S^2 \quad \text{and} \quad NS^3 = S^3. \qquad (2)$$

Hence, the operator N defines the multiplicity of the primes and as such allows one to build a global topological manifold with heptaplets, triplets and singlets of the primes T^3, $S^1 \times S^2$ and S^3, respectively, under the repeated action of N. That is, a single BH (three-torus) *induces* a triplet (heptaplet) of BHs (three-tori). For the Planck scale, three-tori one can thus construct a lattice of three-tori, $L(T^3)$, but BHs carry mass and are subject to Hawking evaporation. Therefore, N induces triplets of BHs with a success that depends on the longevity of existing BHs, and this allows for topology change.

The quantum foam stability problem that plaques GR at the Planck scale is then solved automatically since macroscopic (so long lived) BHs induce extra pairs of mini BHs every Planck time under the action of the construction operator N. Macroscopic BHs are topologically indistinguishable from mini BHs, but are stable against Hawking evaporation, unlike their mini BH counterparts. Hence, an ensemble of macroscopic BHs provides a naturally stable quantum foam. Furthermore, topological induction of mini BHs extends Mach's principle: Einstein's interpretation dictates that the global distribution of matter determines the (changes in) local geometry, and vice versa. Wheeler's local quantum foam, with its non-trivial topology in the form of mini BHs, now attached to the lattice of three-tori, is a quantum expression of GR. Therefore, following Einstein's ideas for geometry, matter that is globally in the form of macroscopic BHs should determine the (changes in) Planck scale topology, and vice versa [1].

The spectrum of the operator N is discrete, as one would expect because topology, with its invariance under continuous deformations of space-time, cares only about countable quantities (topological invariants). Hence, the number of macroscopic BHs (*not* their space density) in the entire universe, a global quantity, determines the local occurrence of mini BHs. To be more quantitative about this, one must compute with what probability δ Wheeler's quantum foam enjoys a mini BH in every Planckian volume of a three-torus, given a specific number of macroscopic BHs. The topological extension of Mach's principle then states that the total number of macroscopic BHs in the universe determines the properties of the quantum foam on $L(T^3)$. Hence, δ is *linearly* proportional to the number of macroscopic BHs in the entire three-volume of the universe, for any given time slice of thickness t_P.

Because topology allows one to identify all space-time points through the underlying lattice of three-tori, violations in causality do not occur as nature acquires information on the total number of macroscopic BHs in the entire universe, so in all of three-space, for every time slice. The induced mini BHs constitute a dark energy density because their creation as three-dimensional objects requires an increase in four-volume to embed them. In all, the dark energy density, $\Lambda(z)$, is linearly proportional to the total number of macroscopic BHs in the entire three-volume of the universe, $N_{BH}(z)$, at a given redshift z. In this, redshift is a distinctive measure of time, in the usual 3+1 decomposition for the evolution of the three-geometry and three-topology of space-time. I.e., redshift labels different global time slices in the geometrical and topological dynamics of three-space. Algebraically,

$$\Lambda(z) = \delta(z)\, m_P / l_P^3, \qquad (3)$$

with

$$\delta(z) = N_{BH}(z)\; l_P^3 / l_I^3, \qquad (4)$$

and l_I is the size of the universe when the first BH forms that exists for longer than a Hubble time. I.e., l_I is *frozen in* when the quantum foam stabilizes and one can speak of a universe with a specific global topology in the sense of Mach. Each mini BH pair, which for an external observer seems to appear spontaneously in some region of the quantum foam, contributes about m_P gram of mass per Planckian volume. Equation (4) shows that, for N_{BH} independent of redshift, any local observer concludes that the dark energy density Λ in g cm^{-3} remains constant as the universe expands. In all, one has the local space-time quantum foam that Wheeler envisioned, but globally constrained by the total number of macroscopic BHs in the entire universe.

2. Testing the Connection between Dark Energy and Macroscopic Black Holes

As discussed above, the dark energy density of our universe is linearly proportional to the total number of macroscopic BHs in three-space at any redshift. Very practically, this means that if the number of BHs in the entire universe at redshift 1 is twice higher than at redshift 2, then Λ in g cm^{-3} doubles from z=2 to z=1. In [3] a detailed calculation on the time dependence of the dark energy density is presented. Here, a simple estimate is provided. The bulk of the stars, and thus stellar BHs, in the universe appears to be present as early as z=1, when a $(1+z)^4$ decline ensues in the cosmic star formation rate [4]. Hence, since BH formation follows high-mass (>8 $M_\odot$) star formation, one expects to find an effectively constant Λ for z<1, when the bulk of the BHs in the universe is in place. Conversely, given the rapidity with which massive stars/BHs are produced around z=1-3, this epoch should exhibit a strong decrease in the number of BHs, and thus Λ, from z=1 to z=3.

These qualitative expectations can be quantified as follows. For a local observer now, it is natural to consider $\Lambda(z)$ normalized to its present value $\Lambda(0)$, where the latter corresponds to the current number of BHs in the observable universe. The ratio $\Lambda(z)/\Lambda(0)$ can now be expressed analytically, in terms of dN_{BH}/dt for observers that measure the rate of macroscopic BH formation in their comoving volume, as

$$\Lambda(z)/\Lambda(0) = \int_y^z dt(z)\, dN_{BH}/dt \,/ \int_y^0 dt(z)\, dN_{BH}/dt \approx (1+z)^{-0.36} \text{ for } z<1, \quad (5)$$

for a flat WMAP7 cosmology, BH formation starting at some large redshift y (~30-1000) but at a very modest pace, and the usual cosmic time interval dt(z). Although a redshift coordinate is used, there is no pertinent dependence on coordinate system because of the normalization in equation (5). Any observer, whatever its coordinate system, finds the dimensionless ratio of two numbers to be covariant. It is assumed that the observer can derive the temporal change in the total number of BHs, as three-space evolves, because he lives in a representative comoving volume that enjoys the same evolution in BH formation as the whole universe.

One finds numerically, using the data in [4], that $\Lambda(z)/\Lambda(0)$ roughly scales as $(1+z)^{-0.36}$ for z<1. I.e., $\Lambda(z)$ changes by less than ~30% more recently than a redshift of unity. Furthermore, a rapid decline in the dark energy density by a factor of 5 occurs from z=1-3, as one moves to before the peak in the star formation history of the universe. The value of the equation of state parameter $w=p/\varrho$ (for pressure p and density ϱ) is estimated to be $w= -1.061^{+0.069}_{-0.068}$ in [5], for z<1. These data are consistent with equation (5), because a constant w= -1.1 corresponds to a change in $\Lambda(z)/\Lambda(0)$ of about 30%, for z<1. Conversely, because the bulk of the stellar mass BHs appears not to have been formed before z=1, one expects to have w< -1 for z>1.

References
[1] Spaans M 1997 *Nuc. Phys. B* **492** 526
[2] Wheeler J A 1957 *Ann. of Phys.* **2** 604
[3] Aykutalp A and Spaans M 2012 *Preprint arXiv:1201.6372*
[4] Hopkins A M and Beacom J F 2006 *ApJ* **651** 142
[5] Sullivan M *et al.* 2011 *ApJ* **737** 102

Acknowledgements
The author is grateful to Elly van 't Hof and Aycin Aykutalp for discussions on space-time topology and black hole formation.

Fermion Particle Production as a Dynamical Casimir Effect inside a Three Dimensional Sphere

Mohammad Reza Setare and Asrin Seyedzahedi

Faculty of Science, Department of Physics, University of Kurdistan, Pasdaran Ave. Sanandaj, Iran

E-mail: rezakord@ipm.ir

Abstract. In this paper we analyze the problem of fermion creation as a dynamical Casimir effect inside a three dimensional sphere. We present an appropriate wave function which satisfies the Dirac equation in this geometry with MIT bag model boundary condition. We consider the radius of the sphere to have dynamics and introduce the time evolution of the quantized field by expanding it over the instantaneous basis. We explain how we can obtain the average number of particles created. In this regard we find the Bogoliubov coefficients. We consider an oscillation and determine the coupling conditions between different modes that can be satisfied depending on the cavity's spectrum. Assuming the parametric resonance case we obtain an expression for the mean number of created fermions in each mode of an oscillation.

1. Introduction

The Casimir effect is one of the most interesting manifestations of the nontrivial properties of the vacuum state in a quantum field theory (see, e.g., [1, 2, 3, 4] and references therein) and can be viewed as a polarization of the vacuum due to the boundary conditions. A new phenomenon, namely the quantum creation of particles known as the dynamical Casimir effect occurs when the geometry of the system varies in time. In two dimensional spacetime and for conformally invariant fields the problem with dynamical boundaries can be mapped to the corresponding static problem and hence allows a complete study (see [4, 5] and references therein). In higher dimensions the problem is much more complicated and only partial results are available. The vacuum stress induced by uniform acceleration of a perfectly reflecting plane has been considered in [6]. The corresponding problem for a sphere expanding in the four-dimensional spacetime with constant acceleration has been investigated by Frolov and Serebriany [7, 8] in the perfectly reflecting case and by Frolov and Singh [9] for semi-transparent boundaries. For more general cases of motion by vibrating cavities the problem of particle and energy creation has been considered on the base of various perturbation methods [10, 11, 12, 13, 14, 20, 15, 16]. It has been shown that a gradual accumulation of small changes in the quantum state of the field could result in a significant observable effect. A new application of the dynamical Casimir effect has recently appeared in connection with the suggestion by Schwinger [17] that the photon production associated with changes in the quantum electrodynamic vacuum state arising from a collapsing dielectric bubble could be relevant for sonoluminescence.
Particle creation from the quantum scalar vacuum by expanding or contracting spherical shell with Dirichlet boundary conditions have been considered in [18]. In another paper the case has

been considered when the sphere radius performs oscillation with a small amplitude and the expression are derived for the number of created particles to the first order of the perturbation theory [19].

In order to obtain the number of produced fermions, the organization of this paper is as follows, in sec. II we present a normalized eigenfuntion for a massive spinor field inside a spherical shell of finite radius. Considering that the fermionic field obeys the MIT bag condition on the spherical shell, explicit behavior for boundary dependent term is exhibited. We consider the radius of the sphere to have dynamics. Then we introduce the time evolution of the quantized field by expanding it over the instantaneous basis in sec. III. Section IV is devoted to the derivation of Bogoliubov coefficients. Following the steps given in [20] we arrive at an infinite set of coupled differential equations for the coefficients of the expansion. We explain how we can obtain the average number of particles created after the end of the motions. The mean number of created particles depends on the Bogoliubov coefficient. We consider an oscillation with small amplitudes of oscillations and determine the coupling conditions between different modes that can be satisfied depending on the cavity's spectrum. Assuming the parametric resonance case we obtain an expression for the mean number of created fermions in each mode of an oscillation and in sec. VI. we present our conclusion.

2. Eigenfunctions for a Spinor Field

The Dirac Hamiltonian inside the sphere is

$$\hat{H}_D = \hat{\alpha} \cdot \hat{p} + \hat{\beta} m, \tag{1}$$

with

$$\hat{\alpha} = \begin{pmatrix} 0 & \hat{\sigma} \\ \hat{\sigma} & 0 \end{pmatrix}. \tag{2}$$

Since the angular momentum operator and the parity operator commute with the Hamiltonian, states have definite energy, angular momentum and parity. Let us write the four-components spinor field ψ in terms of two-components ones as

$$\psi(\vec{x}, t) = \begin{pmatrix} \zeta_{jlm} \\ \chi_{jl'm} \end{pmatrix} \exp(-i\omega t), \tag{3}$$

and it can be seen easily that

$$\begin{aligned} (\hat{\sigma}.\hat{p})\chi_{jl'm} &= \omega\ \zeta_{jlm}, \\ (\hat{\sigma}.\hat{p})\zeta_{jlm} &= \omega\ \chi_{jl'm}, \end{aligned} \tag{4}$$

where ω is the frequency of the modes. The angular parts of the spinors are the standard spinor spherical harmonics Ω_{jlm} whose explicit form is given in [21]

$$\begin{aligned} \zeta_{jlm} &= f(r)\Omega_{jlm} \\ \chi_{jl'm} &= g(r)\Omega_{jl'm} \end{aligned} \tag{5}$$

where j specifies the value of the total angular momentum, m its projection, $l = j \pm \frac{1}{2}$, and $l' = 2j - l$. By taking into account these relations, we obtain the following set of differential equations for the radial functions

$$\begin{aligned} f'(r) + \frac{1+\kappa}{r} f(r) - (\omega + m) g(r) &= 0 \\ g'(r) + \frac{1-\kappa}{r} g(r) + (\omega - m) f(r) &= 0 \end{aligned} \tag{6}$$

where we use the notation

$$\kappa = \{ \begin{array}{c} -(l+1),\ j = l + 1/2 \\ l,\ j = l - 1/2 \end{array} . \tag{7}$$

These lead to second order differential equations for the separate functions

$$\begin{aligned} f''(r) + \frac{2}{r} f'(r) + \left[p^2 - \frac{\kappa(\kappa+1)}{r^2} \right] f(r) = 0 \\ g''(r) + \frac{2}{r} g'(r) + \left[p^2 - \frac{\kappa(\kappa-1)}{r^2} \right] g(r) = 0 \end{aligned} \tag{8}$$

with the solutions

$$f(r) = N\ j_l(pr),\ g(r) = \mp i\ N \sqrt{\frac{\omega - m}{\omega + m}}\ j_{l'}(pr) \tag{9}$$

where $p = \sqrt{\omega^2 - m^2}$. In these solutions $j_l(pr)$ represents the spherical Bessel function of order l. As a result for a given j we have two types of eigenfunctions with different parities corresponding to $j = l \pm 1/2$. Due to the orthogonality conditions the normalization constant reads $N = p\sqrt{(\omega + m)/(\omega\pi)}$. In order to guaranty complete confinement of the fermionic field inside the sphere it is needed to impose a boundary condition. The MIT bag model boundary condition is a proper boundary condition. Imposing the prevalent form of this boundary condition on the eigenfunction we have

$$(1 + i\hat{r} \cdot \vec{\gamma})\psi(\mathbf{x})\Big|_{r=a} = 0, \tag{10}$$

where a is the sphere radius and the $\hat{r}$ is the outward-pointing normal to the sphere. In terms of spinors ζ_{jlm} and $\chi_{jl'm}$ this condition becomes

$$\zeta_{jlm} + i(\vec{\sigma} \cdot \hat{r})\ \chi_{jl'm} = 0,\ r = a \tag{11}$$

and leads to the following quantization condition for the momentum

$$j_l(pa) = \pm \sqrt{\frac{\omega - m}{\omega + m}}\ j_{l\pm1}(pa),\ \ \pm\ \texttt{for}\ j = l \pm 1/2. \tag{12}$$

Let us denote the roots to Eq. (12) by $\Lambda_{n\kappa} = pa$ where n is the number of the root and κ is the quantum number introduced by Eq. (7).

3. Expansion of Dirac field over the instantaneous basis

The Fourier expansion of the Dirac field for an arbitrary moment of time, in terms of creation and annihilation operators, can be written as

$$\psi(\mathbf{x}, t) = \sum_n a_n^{in} u_n(\mathbf{x}, t) + b_n^{\dagger in} v_n(\mathbf{x}, t), \tag{13}$$

where a_n^{in} and $b_n^{\dagger in}$ are the annihilation and creation operators correspond to the particles in the 'in' region. By index 'in' we mean times in which the system is static and therefore the spherical shell has a constant radius a. The mode functions $u_n(x, t)$ form a complete orthonormal set of solutions of the wave equation with the MIT bag model boundary condition. For the static cavity each field mode is determined by the Eq. (12) and the mode functions $u_n(x,t)$ have the form of the static solution mentioned in Eq. (3, 5, 9).

Now consider that the radius of sphere varies as a function of time $a(t)$, which has constant initial value a and after a finite time modulation stops and the radius takes its initial value. When the modulation begins the boundary condition varies with time. To satisfy this time-dependent boundary condition we expand the mode functions with respect to an instantaneous basis [22] for particles

$$u_n(\mathbf{x},t) = \sum_k (i\gamma^\nu \partial_\nu + m) Q_k^{(n)}(t)\varphi_k(\mathbf{x},t) \tag{14}$$

and a similar instantaneous basis for anti particles where

$$\varphi_k(\mathbf{x},t) = N \begin{pmatrix} j_l(kr)\Omega_{jlm} \\ \mp i\sqrt{\frac{\omega-m}{\omega+m}}\, j_{l'}(kr)\Omega_{jl'm} \end{pmatrix} \tag{15}$$

and the initial conditions are

$$\begin{aligned} Q_k^{(n)}(0) &= \frac{1}{\sqrt{2\omega_n}}\,\delta_{n,k} \\ \dot{Q}_k^{(n)}(0) &= -i\sqrt{\frac{\omega_n}{2}}\,\delta_{n,k}. \end{aligned} \tag{16}$$

When the modulation accurse we can write the Fourier expansion of the Dirac field one more time, but this time with index 'out' to indicate this new region of time with new annihilation and creation operators

$$\psi(\mathbf{x},t) = \sum_n a_n^{out} u_n(\mathbf{x},t) + b_n^{\dagger out} v_n(\mathbf{x},t). \tag{17}$$

Since the expansion in Eq.(17) for the field modes must be a solution of the wave, we insert it in Dirac equation equation. Considering that the φ_k's form a complete and orthogonal set of solutions of the wave equation and they depend on t only through $a(t)$, we obtain a set of coupled equations for $Q_k^{(n)}(t)$

$$\ddot{Q}_k^{(n)}(t)\varphi_k(\mathbf{x},t) + \omega_k{}^2(t)Q_k^{(n)}(t)\varphi_k(\mathbf{x},t) = -2\dot{Q}_k^{(n)}(t)\dot{\varphi}_k(\mathbf{x},t) - Q_k^{(n)}(t)\ddot{\varphi}_k(\mathbf{x},t) \tag{18}$$

where $\omega_k^2 = m^2 + |\vec{k}|^2$.

4. Derivation of Bogoliubov coefficients

We now want to obtain the number of fermions produced after the modulation. The mean number of particles produced in the mode k is the average value of the number operator $a_k^{\dagger out} a_k^{out}$ with respect to the initial vacuum state. Therefor we are interested in times after the modulation. In this region, the radius takes its initial value and the right hand side in Eq.(18) vanishes and it reduces to the following simple form

$$(\partial_t{}^2 + \omega_p{}^2)Q_p^{(n)}(t) = 0, \tag{19}$$

the solution reads

$$Q_p^{(n)}(t) = A_p^{(n)} e^{i\omega_p t} + B_p^{(n)} e^{-i\omega_p t}, \tag{20}$$

with $A_p^{(n)}$ and $B_p^{(n)}$ constant coefficients to be determined. The creation and annihilation operators for particles and anti-particles in the 'in' and 'out' regions obey the usual anticommutation relations. Using the Bogoliubov canonical transformation one can expand the 'in' operators in terms of the 'out' operators

$$\begin{aligned} a_n^{in} &= \alpha_n^* a_n^{out} - \beta_n b_n^{\dagger out}, \\ b_n^{\dagger in} &= \beta_{-n}^* a_{-n}^{out} + \alpha_{-n} b_{-n}^{\dagger out}. \end{aligned} \tag{21}$$

Substituting Eq. (20) in Eq. (14) and then in Eq. (17), and considering Eq. (21) we expand φ_k for out region in terms of the creation and annihilation operators in the in region. After some calculations and by means of Eq. (4) eventually we get

$$\alpha_n^* = \frac{B_p^{(n)}}{N} \begin{pmatrix} 2m - \omega_p & \omega_p \\ \omega_p & 2m + \omega_p \end{pmatrix}. \tag{22}$$

Using the relation between the Bogoliubov coefficients for fermionic fields

$$|\beta_p|^2 = 1 - |\alpha_p|^2, \tag{23}$$

therefore the mean number of particles produced in the mode $\vec{p}$ through an arbitrary modulation of the single fermion mode is the average value of the number operator with respect to the initial vacuum state (defined through $a_p^{in}|0_{in}\rangle = 0$)

$$\langle N_p \rangle = |\beta_p|^2 = 1 - 2\pi |B_p^{(n)}|^2. \tag{24}$$

Up to this point the equations are valid for an arbitrary variation of radius of the sphere. The only assumption was that $a(0) = a$. Our purpose is to investigate the number of fermion created inside the sphere, therefore we look for harmonic oscillation of the radius which could enhance that number by means of resonance effects for some specific external frequencies. So we consider the following oscillations

$$a(t) = a(1 + \varepsilon \sin(\Omega t)), \tag{25}$$

where the amplitude of oscillation is very small and we can content ourselves with its first order and Eq.(18) for a massless case takes the form [23]

$$\begin{aligned} \ddot{Q}_p^{(n)}(t) + {\omega_p}^2 Q_p^{(n)}(t) &= 2\varepsilon {\omega_p}^2 \sin(\Omega t) Q_p^{(n)}(t) \\ &+ \frac{2{\omega_p}^2}{\pi} \varepsilon \Omega \cos(\Omega t) \sum_k g_{pk} \dot{Q}_k^{(n)}(t) \\ &- \frac{{\omega_p}^2}{\pi} \varepsilon \Omega^2 \sin(\Omega t) \sum_k g_{pk} Q_k^{(n)}(t), \end{aligned} \tag{26}$$

where

$$g_{pk} = -a(t) \int_0^{a(t)} d^3x \; \varphi_p^* (\mathbf{x}, t) \frac{\partial \varphi_k(\mathbf{x}, t)}{\partial a}. \tag{27}$$

Since $\varepsilon \ll 1$ it is natural to assume that the solution of Eq. (26) is of the form

$$Q_p^{(n)}(t) = A_p^{(n)}(t) \; e^{i\omega_p t} + B_p^{(n)}(t) \; e^{-i\omega_p t}, \tag{28}$$

functions $A_p^{(n)}(t)$ and $B_p^{(n)}(t)$ varying slowly with time. We insert Eq. (28) into the Eq. (26) to obtain differential equations for $A_p^{(n)}(t)$ and $B_p^{(n)}(t)$. After neglecting their second derivatives and multiplying equation by $e^{\pm i\omega t}$ we average over the fast oscillations, we have

$$\frac{dA_p^{(n)}}{d\tau} = -\frac{\omega_p}{2}B_p^{(n)}\delta(2\omega_p - \Omega) + \sum_k(-\omega_k + \frac{\Omega}{2})\delta(-\omega_p - \omega_k + \Omega)\frac{\Omega\omega_p}{2\pi}g_{pk}B_k^{(n)}$$
$$+\sum_k\left[(\omega_k + \frac{\Omega}{2})\delta(\omega_p - \omega_k - \Omega) + (\omega_k - \frac{\Omega}{2})\delta(\omega_p - \omega_k + \Omega)\right] \times \frac{\Omega\omega_p}{2\pi}g_{pk}A_k^{(n)}, \tag{29}$$

$$\frac{dB_p^{(n)}}{d\tau} = -\frac{\omega_p}{2}A_p^{(n)}\delta(2\omega_p - \Omega) + \sum_k(-\omega_k + \frac{\Omega}{2})\delta(-\omega_p - \omega_k + \Omega)\frac{\Omega\omega_p}{2\pi}g_{pk}A_k^{(n)}$$
$$+\sum_k\left[(\omega_k + \frac{\Omega}{2})\delta(\omega_p - \omega_k - \Omega) + (\omega_k - \frac{\Omega}{2})\delta(\omega_p - \omega_k + \Omega)\right] \times \frac{\Omega\omega_p}{2\pi}g_{pk}B_k^{(n)} \tag{30}$$

where $\tau = \varepsilon t$ is a time scale. Now we shall solve equations (29) and (30). Depending on the radius of the static sphere and the frequency spectrum of the modes, there are different kinds of solutions. If we consider that the frequency of the boundary is twice the frequency of some unperturbed mode namely $\Omega = 2\omega_p$, in this condition if $\omega_k - \omega_p = \Omega$ the resonant mode p will be coupled to some other mode k. Let us now suppose that the coupling conditions $|\omega_p \pm \omega_k| = \Omega$ are not fulfilled. In this case and for a massless field the equations (29) and (30) reduces to

$$\frac{dA_p^{(n)}}{d\tau} = -\frac{\omega_p}{2}B_p^{(n)}, \tag{31}$$
$$\frac{dB_p^{(n)}}{d\tau} = -\frac{\omega_p}{2}A_p^{(n)}. \tag{32}$$

Satisfying the initial condition mentioned in Eq. (16), the solutions of these coupled equations are

$$A_p^{(n)}(\tau) = -\frac{1}{\sqrt{2\omega_p}}\sinh(\frac{\omega_p}{2}\tau), \tag{33}$$
$$B_p^{(n)}(\tau) = \frac{1}{\sqrt{2\omega_p}}\cosh(\frac{\omega_p}{2}\tau). \tag{34}$$

By using Eq. (24) the average number of produced fermions in the mode p is

$$\langle N_p\rangle = 1 - 2\pi\cosh^2(\frac{\omega_p}{2}\tau), \tag{35}$$

where for our massless case $\omega_p = \Lambda_{n\kappa}/a$ and $\Lambda_{1,-1} = 2.04$, $\Lambda_{2,-1} = 5.40$, etc. For each κ there are two roots corresponding to $l' = l \pm 1$.

5. Conclusion

We have investigated the particle creation for a Dirac field in a three dimensional sphere with the MIT bag model boundary condition. We have considered the radius of the sphere to modulate during a finite time interval and when modulation stops the radius takes its initial value. In this regard we have used the Bogoliubov coefficients to obtain the number of fermions created during the motion. We have taken into account the usual parametric resonance case ($\Omega = 2\omega_p$) and find the number of fermion created in this system.

5.1. Acknowledgments
We acknowledge helpful discussions with Prof. Robert Mann on Eq. (14). Also we thank Prof. Diego Dalvit for comment on Eq. (26). Also we thank the referee for good observations and kind guidelines.

References
[1] Mostepanenko V M and Trunov N N 1997 *The Casimir Effect and Its Applications* (Clarendon, Oxford).
[2] Plunien G, Muller B, and Greiner W 1986, Phys. Rep. **134**, 87.
[3] Lamoreaux S K 1999 Am. J. Phys. **67**, 850.
[4] Bordag M, Mohidden U and Mostepanenko V M 2001 Phys. Rep. **353**, 1.
[5] Birrell N D and Davies P C W 1982 *Quantum Fields in Curved Space* (Cambridge University Press).
[6] Candelas P and Deutsch D 1977 Proc. Roy. Soc. A**354**, 79.
[7] Frolov V P and Serebriany 1979 J. Phys. A: Math. Gen. **12**, 2415.
[8] Frolov V P and Serebriany 1980 J. Phys. A: Math. Gen. **13**, 3205.
[9] Frolov V and Singh D 1999 Class. Quantum Grav.**16**, 3693.
[10] Calucci G 1992 J. Phys. A: Math. Gen.**25**, 3873.
[11] Jauregui R, Villarreal C and Hacyan S 1995 Mod. Phys. Lett A**10**,7.
[12] Sassaroli E, Srivastava Y N and Widom A 1994 Phys. Rev. A**50**, 1027.
[13] Dodonov V V and Klimov A B 1996 Phys. Rev. A**53**, 2664.
[14] Lambrecht A, Jaekel M T and Reynaud S 1996 Phys. Rev. Lett. **77**, 615.
[15] Schutzhold R, Plunien G and Soff G 1998 Phys. Rev. A**57**, 2311.
[16] Dodonov V V 1998 J. Phys. A: Math. Gen. **31**, 9835.
[17] Schwinger J 1993 *Proc. Nat. Acad. Sci.* 90, 985, 2105, 4505, 7285; 91, 6473 (1994).
[18] Setare M R and Saharian A A 2001 Mod.Phys.Lett. A**16**,927.
[19] Setare M R and Saharian A A 2001 Mod.Phys.Lett. A**16**,1269.
[20] Law C K 1995 Phys. Rew. A. **51**, 2537.
[21] Berestetskii V B, Lifshits E M and Pitaevskii L P 1982 *Quantum Electrodynamics* (Pergamon Press, Oxford).
[22] Fuentes I, Mann R B, Martin-Martinez E and Moradi S 2010 Phys. Rev. D **82**, 045030.
[23] Crocce M, Dalvit Diego A R and Mazzitelli Francisco D 2001 Phys. Rev. A **64**, 013808.

IC-MSQUARE 2012: International Conference on Mathematical Modelling in Physical Sciences IOP Publishing
Journal of Physics: Conference Series **410** (2013) 012151 doi:10.1088/1742-6596/410/1/012151

Unitary evolution and uniqueness of the Fock quantization in flat cosmologies

G A Mena Marugán[1], D Martín-de Blas[1] and L Castelló Gomar[2]

[1] Instituto de Estructura de la Materia, IEM-CSIC, Serrano 121, 28006 Madrid, Spain.
[2] Facultad de C.C. Físicas, Universidad Complutense de Madrid, 28040 Madrid, Spain.

E-mail: mena@iem.cfmac.csic.es, daniel.martin@iem.cfmac.csic.es

Abstract. We study the Fock quantization of scalar fields with a time dependent mass in cosmological scenarios with flat compact spatial sections. This framework describes physically interesting situations like, e.g., cosmological perturbations in flat Friedmann-Robertson-Walker spacetimes, generally including a suitable scaling of them by a background function. We prove that the requirements of vacuum invariance under the spatial isometries and of a unitary quantum dynamics select (a) a unique canonical pair of field variables among all those related by time dependent canonical transformations which scale the field configurations, and (b) a unique Fock representation for the canonical commutation relations of this pair of variables. The proof is generalizable to any compact spatial topology in three or less dimensions, though we focus on the case of the three-torus owing to the especially relevant implications.

1. Introduction

The Fock quantization of linear fields in curved spacetimes (either physical backgrounds or auxiliary spacetimes in effective descriptions) is subject to ambiguities which affect the physical predictions. The canonical commutation relations (CCR's) of the fieldlike system admit an infinite number of nonequivalent Fock representations [1]. In simple cases, like e.g. in Minkowski spacetime, one can remove this ambiguity and select a preferred representation by imposing that the vacuum be invariant under the spacetime symmetries. But for less symmetric backgrounds (or when the curved spacetime is only an auxiliary structure) one cannot appeal to this criterion, and other vacuum properties must be required in order to pick out a unique Fock representation.

In this work we will consider the case of scalar fields in cosmological spacetimes which have spatial sections with the topology of a three-torus. We focus our study on the case of three-torus topology owing to its physical relevance, for instance in the study of cosmological perturbations around flat Friedmann-Robertson-Walker (FRW) spacetime with compact sections [2]. The analysis is nonetheless generalizable to other compact spatial sections [3] (see e.g. [2, 4] for three-spheres). We will show that one can select a unique class of unitarily equivalent Fock quantizations by means of a requirement which, furthermore, is itself stable under unitary transformations. This requirement includes the invariance under the spatial isometries. However, this invariance does not suffice to fix the Fock representation, essentially because, in cosmological scenarios, no timelike symmetry is available. The requirement is then completed by demanding instead that the dynamics be unitary in the quantum theory.

On the other hand, in cosmological settings like those that we want to study, it is natural to consider time dependent linear canonical transformations in which the field is scaled [4, 5]. In these transformations, the dynamics is modified (owing to the time dependent changes) and, in general, a new fieldlike system is reached. This introduces a new kind of ambiguity when facing the Fock quantization. Nonetheless, we will show that our requirement of spatial symmetry invariance and unitary evolution fixes this freedom completely as well.

Before proving these results, let us emphasize the importance of determining a unique Fock quantum description for fields in cosmological spacetimes, since this uniqueness provides robustness to the quantum predictions and makes them significant.

2. Fock quantization with unitary evolution

Let us begin by considering the quantization of a real scalar field φ with a time dependent mass $s(t)$ propagating in a globally hyperbolic static spacetime, $\mathbb{I}\times\Sigma$, where $\mathbb{I}$ is a closed time interval and the spatial sections Σ have the topology of a three-torus. A system of this type describes, for instance, the gauge-invariant energy density perturbations around an FRW spacetime after a suitable field scaling (in conformal time and in isotropic and adiabatic conditions) [4], as well as the ultraviolet behavior of the scalar perturbations after gauge fixing and rescaling [6]. The phase space of the system is given by $(\varphi, P_\varphi) = (\varphi_{|t_0}, \sqrt{h}\dot{\varphi}_{|t_0})$. The dot stands for the time derivative, t_0 is an initial time, and h is the determinant of the spatial metric, h_{ab}. The Hamilton field equations are $\dot{\varphi} = h^{-1/2}P_\varphi$ and $\dot{P}_\varphi = h^{1/2}\left[\Delta\varphi - s(t)\varphi\right]$, where Δ is the standard Laplace-Beltami (LB) operator for the three-torus metric. These field equations are invariant under the spatial isometries resulting from the composition of rigid rotations in each of the three spatial directions $i = 1,2,3$, $T_{\vec{\alpha}} = T_{\alpha_1}\circ T_{\alpha_2}\circ T_{\alpha_3}$, where $T_{\alpha_i}: \theta_i \to \theta_i + \alpha_i$ and $\alpha_i \in S^1$.

We make a Fourier decomposition of the scalar field using a basis of real eigenfunctions of the LB operator:

$$\varphi(t,\vec{\theta}) = \frac{1}{\pi^{3/2}}\sum_{j=0}^{3}\sum_{\vec{n}_j}\left[q_{\vec{n}_j}(t)\cos(\vec{n}_j\cdot\vec{\theta}) + x_{\vec{n}_j}(t)\sin(\vec{n}_j\cdot\vec{\theta})\right]. \tag{1}$$

Here, $\vec{n}_0 = (n_1,n_2,n_3)$ with $n_i \in \mathbb{N}$, and $\vec{n}_i$ is obtained from $\vec{n}_0$ by flipping the sing of n_i. A similar expansion of the field momentum leads to modes determined by the coefficients $p_{\vec{n}_j} = \dot{q}_{\vec{n}_j}$ and $y_{\vec{n}_j} = \dot{x}_{\vec{n}_j}$. Then, the equations of motion for the field modes coincide in each eigenspace of the LB operator, with eigenvalue $-\omega_n^2 = -|\vec{n}_j|^2$ (the subindex $n \in \mathbb{N}$ designates here the ordering in the sequence of different eigenvalues):

$$\ddot{q}_{\vec{n}_j} + \left[\omega_n^2 + s(t)\right]q_{\vec{n}_j} = 0, \qquad \ddot{x}_{\vec{n}_j} + \left[\omega_n^2 + s(t)\right]x_{\vec{n}_j} = 0. \tag{2}$$

The possible Fock representations of the CCR's are totally characterized by the choice of a complex structure (CS) J [1], namely, of a real, canonical linear map on phase space whose square is minus the identity and which is compatible with the symplectic structure, Ω (containing the information about the canonical Poisson brackets), in the sense that the bilinear map $\Omega(J\cdot,\cdot)$ is positive definite.

As starting point, we choose the CS which would be naturally associated with a free massless field. This CS, J_0, is determined by the spatial metric, and hence its invariance under the three-torus isometries is warranted. The corresponding annihilationlike variables are

$$a_{\vec{n}_j} = \frac{1}{\sqrt{2\omega_n}}(\omega_n q_{\vec{n}_j} + ip_{\vec{n}_j}), \qquad \tilde{a}_{\vec{n}_j} = \frac{1}{\sqrt{2\omega_n}}(\omega_n x_{\vec{n}_j} + iy_{\vec{n}_j}). \tag{3}$$

For simplicity, we ignore the zero modes, which can be quantized separately (possibly by nonstandard methods). The creationlike variables $\{(a^*_{\vec{n}_j}, \tilde{a}^*_{\vec{n}_j})\}$ are the complex conjugate of (3). The action of J_0 is diagonal in these variables: $J_0(a_{\vec{n}_j}) = ia_{\vec{n}_j}$, $J_0(a^*_{\vec{n}_j}) = -ia^*_{\vec{n}_j}$, and likewise for $\tilde{a}_{\vec{n}_j}$ and $\tilde{a}^*_{\vec{n}_j}$. Note that these variables are precisely those promoted to annihilation and creation operators in the Fock representation defined by J_0.

The time evolution of the annihilationlike variables is given by the Bogoliubov transformation

$$a_{\vec{n}_j}(t) = \alpha_n(t,t_0)a_{\vec{n}_j}(t_0) + \beta_n(t,t_0)a^*_{\vec{n}_j}(t_0), \tag{4}$$

where the alpha and beta functions coincide in each eigenspace of the LB operator, and $|\alpha_n(t,t_0)|^2 - |\beta_n(t,t_0)|^2 = 1$. This transformation is unitary in the Fock representation defined by J_0 if and only if the set of beta functions, $\{\beta_{\vec{n}_j}(t,t_0)\}$, is square summable. Taking into account the eigenspace degeneracy, g_n, the considered condition amounts to the square summability of $\{g_n|\beta_n(t,t_0)|^2\}$. On the other hand, the asymptotic behavior when $n \to \infty$ (provided that the timelike function $s(t)$ has an integrable first derivative in each compact time subinterval [2]) is

$$\alpha_n(t,t_0) = e^{-i\omega_n(t-t_0)} + \mathcal{O}\left(\omega_n^{-1}\right), \qquad \beta_n(t,t_0) = \mathcal{O}\left(\omega_n^{-2}\right). \tag{5}$$

The variation of g_n with n in the three-torus case is quite involved, because of accidental degeneration. Nonetheless, one can prove that the sequence $\{g_n/\omega_n^4\}$ is summable if so is $\{D_n/n^4\}$, where D_n is the number of eigenstates with eigenvalue in the interval $(n-1, n]$. It is easy to realize [7] that $D_n = \mathcal{O}(n^2)$ for large n and, therefore, the evolution is indeed unitarily implementable with respect to the CS J_0.

3. Uniqueness results

Let us briefly sketch the proof that the requirement of invariance under the spatial symmetries and of a unitary quantum dynamics indeed selects a unique family of (unitarily equivalent) Fock representations, as well as a preferred canonical pair of field variables.

3.1. Fock representation of the CCR's

First, we characterize the complex structures, J, that are invariant under the group of transformations $T_{\vec{\alpha}}$. These transformations only mix modes with the same labels $\vec{n}_j$ (and the mixing is independent for each value of the label). As a consequence, any invariant CS must be block diagonal, with 4×4 blocks, in the considered basis on phase space. Furthermore, imposing the invariance and compatibility of the symplectic structure with the CS, we conclude that every invariant CS J can be obtained from J_0 by means of a transformation of the type $J = KJ_0K^{-1}$, where K is block diagonal, with 4×4 blocks of the form (see [7] for details):

$$K_{\vec{n}_j} = \begin{pmatrix} \mathcal{K}_{\vec{n}_j} & \mathbf{0} \\ \mathbf{0} & \mathcal{K}_{\vec{n}_j} \end{pmatrix}; \qquad \mathcal{K}_{\vec{n}_j} = \begin{pmatrix} \kappa_{\vec{n}_j} & \lambda_{\vec{n}_j} \\ \lambda^*_{\vec{n}_j} & \kappa^*_{\vec{n}_j} \end{pmatrix}. \tag{6}$$

Here, $\kappa_{\vec{n}_j}$ and $\lambda_{\vec{n}_j}$ are complex numbers satisfying $|\kappa_{\vec{n}_j}|^2 - |\lambda_{\vec{n}_j}|^2 = 1$, so that K is a canonical transformation. The evolution $\mathcal{U}(t,t_0)$ is unitarily implementable in the representation given by J if and only if $K^{-1}\mathcal{U}(t,t_0)K$ is unitarily implementable in the representation given by J_0, for all instants of time. As before, unitarity amounts to the square summability of the corresponding, new beta functions, which are

$$\beta^J_{\vec{n}_j}(t,t_0) = (\kappa^*_{\vec{n}_j})^2\beta_n(t,t_0) - \lambda^2_{\vec{n}_j}\beta^*_n(t,t_0) + 2i\kappa^*_{\vec{n}_j}\lambda_{\vec{n}_j}\Im[\alpha_n(t,t_0)]. \tag{7}$$

The symbol $\Im[\cdot]$ denotes the imaginary part. It is easy to realize that, if $\{\beta^J_{\vec{n}_j}(t,t_0)\}$ is square summable, then the same must happen with $\{\lambda_{\vec{n}_j}\Im[\alpha_n(t,t_0)]\}$ [2]. Owing to the asymptotic (ultraviolet) behavior of the functions $\alpha_n(t,t_0)$ (and making an average in time, together with a convenient use of Lusin's theorem [2]), one arrives then to the conclusion that $\{|\lambda_{\vec{n}_j}|^2\}$ must be square summable. But this immediately guarantees that the representations given by J and J_0 are unitarily equivalent (since the transformation K is then implemented unitarily).

3.2. Field description

As the final step in our analysis, we consider now the most general form of a time dependent, linear canonical transformation which scales the field configuration:

$$\phi = f(t)\varphi, \qquad P_\phi = f(t)^{-1}P_\varphi + g(t)\sqrt{h}\varphi. \tag{8}$$

The real functions $f(t)$ and $g(t)$ are assumed to be twice differentiable, and $f(t)$ nonvanishing. Besides, without loss of generality, we can set initially $f(t_0) = 1$ and $g(t_0) = 0$ [4].

One can show that a unitary evolution with respect to a CS which is invariant under the spatial symmetries is only possible when $f(t) = 1$ and $g(t) = 0$ $\forall t \in \mathbb{I}$ [7]. In order to prove this result, we consider the field dynamics of the new canonical pair and compute the alpha and beta functions for the CS J_0:

$$\bar{\alpha}_n = f_+(t)\alpha_n + f_-(t)\beta_n^* + i\frac{g(t)}{2\omega_n}[\alpha_n + \beta_n^*], \quad \bar{\beta}_n = f_+(t)\beta_n + f_-(t)\alpha_n^* + i\frac{g(t)}{2\omega_n}[\beta_n + \alpha_n^*], \tag{9}$$

where $2f_\pm(t) = f(t) \pm 1/f(t)$. Then, the beta functions $\bar{\beta}^J_{\tilde{n}_j}$ for any other invariant CS, J, can be obtained as in expression (7). Once more, if the evolution is to be unitary, these beta functions have to be square summable at all times. Hence, in particular, the beta functions must tend to zero for large ω_n. Recalling the asymptotic behavior (5), one can then show that this vanishing limit is possible only if $f_-(t)$ is the zero function, a fact which implies that $f(t) = 1$ [7]. Moreover, by studying the behavior for large ω_n when $f(t)$ is the unity, one can prove that the square summability of the beta functions, needed for the unitarity of the dynamics, requires that $g(t)$ vanish identically [7]. In this way, the field description of the system turns out to be completely determined.

4. Conclusions

We have considered the Fock quantization of scalar fields in (possibly effective) cosmological spacetimes whose spatial sections have a three-torus topology. We have shown that the requirement of invariance under the spatial symmetries and of a unitary dynamics picks out a preferred canonical pair of variables for the field description –among all those related by time dependent canonical transformations which include a scaling of the field configuration. Besides, for such a canonical pair, the introduced requirement turns our to select a unique family of unitarily equivalent Fock representations of the CCR's. As a result, the ambiguity in the Fock quantization coming from the scaling of the field or from a choice of vacuum is removed. This allows one to reach robust physical predictions in the quantum theory for linear scalar fields with quadratic potentials, propagating in some types of nonstationary scenarios, which include the relevant case of compact flat FRW cosmologies.

Acknowledgments

This work was supported by the research grant MINECO FIS2011-30145-C03-02. D. M-dB has been supported by CSIC and the European Social Fund under the grant JAEPre_09_01796.

References

[1] Wald R M, *Quantum Field Theory in Curved Spacetime and Black Hole Thermodynamics* (Chicago University Press, Chicago, 1994).
[2] Cortez J, Mena Marugán G A, Olmedo J, and Velhinho J M, Phys. Rev. D **81**, 044037 (2010).
[3] Cortez J, Mena Marugán G A, Olmedo J, and Velhinho J M, Classical Quantum Gravity **28**, 172001 (2011).
[4] Cortez J, Mena Marugán G A, and Velhinho J M, Phys. Rev. D **83**, 025002 (2011).
[5] Mukhanov V, *Physical Foundations of Cosmology* (Cambridge University Press, Cambridge, England, 2005).
[6] Fernández-Méndez M, Mena Marugán G A, Olmedo J, and Velhinho J M, Phys. Rev. D **85**, 103525 (2012).
[7] Castelló Gomar L, Cortez J, Martín-de Blas D, and Mena Marugán G A, *in preparation.*

Low energy physics and properties of extra space

Sergey G. Rubin
National Research Nuclear University "MEPhI" , Moscow, Russia
E-mail: sergeirubin@list.ru

Abstract. The mechanism of low energy physics formation in the framework of multidimensional gravity is discussed. It is shown that a wide set of parameters of a primary theory could lead to the observable Universe. Quantum fluctuations of extra space metric and its consequent classical evolution play an important role in this process.

1. Introduction

"A theory of everything" (TOE) is usually considered as a final goal. It means an existence of some universal lagrangian containing fields, symmetries and parameters (coupling constants) that should describe all phenomena in the Universe. But even if the goal was achieved main questions listed above remain unanswered:

- What is the origin of the fields? Why just these fields are realized in the Nature?
- Why just these symmetries are realized but not others?
- Why the parameters have specific values?
- Why our space has specific number of dimensions (4, 10, 11, 26)?
- The situation is complicated by the known problem: the physical parameters acquire their observable values which are extremely fine tuned.

One may conclude that a TOE can not be a real final theory due to its internal reasons.

The alternative approach - multiverse - is also known. In this case it is supposed that many theories (universes) could be realized in Nature and we accidentally live in some space domain (Universe) described by one of the theory. This way is partly declared by the string theory. Main objection to this approach can be expressed shortly in one phrase: "If everything is possible, it is not a science". Let me suggest some example as a counterargument. Indeed, it would be strange if somebody would explain the observed Earth mass just saying that a lot of planets with different masses do exist. Does it mean that any scientific researches in this direction are meaningless? Not at all. Instead we develop the theory of planet formation and some other scientific directions like the geology, the geophysics.

2. Formation of low energy physics

Our aim is to propose the approach that connects observable physical parameters and initial parameters of nonlinear gravity acting in $D-$ dimensional. It will be shown that a universe similar to ours can be formed for a wide range of initial parameters. In case of success, answers to the questions itemized above would be quite evident. The main tools are multidimensional gravity and quantum theory. The latter is needed to justify a quantum creation of manifolds with various metrics. For our purposes we need only the same fact of a nonzero probability of any metric formation in a small space region.

The essence of the approach is as follows. Suppose that a sufficient number of extra dimensions does exist. A metric of a whole space M_0 is chosen in the form

$$M_0 = M_4 \times M. \tag{1}$$

Here M_4 is our 4-dim space and M is an extra space. It is supposed that every 4-dim lagrangian, e.g. the Standard Model lagrangian is deduced from a primary one that acts in the whole space M_0. Let this lagrangian contains initial parameters a_{in}. The observed values of the parameters a_{obs} of the Standard Model are their functions

$$a_{obs} = f(a_{in}, g_M). \tag{2}$$

Here g_M is a metric tensor of the extra space. A form of the function f is a separate problem which was discussed partly in [1]. The internal metric g_M is usually fixed by low energy physics what leads to strict connection between two sets a_{in} and a_{obs}. Thus, in spite of many interesting applications have been performed, see e.g. [3, 4], the problem of origin of the same parameters a_{in} remains. Let us make a next step and suppose that the geometry has the form

$$M_0 = M_4 \times M \times K, \tag{3}$$

where K is an another extra space with a metric g_K. Formula (2) is transformed into the two-step expression

$$a_{in} = f_1(b_{in}, g_K), \quad a_{obs} = f(a_{in}, g_M), \tag{4}$$

where b_{in} is a new set of initial parameters. For the first sight nothing has been changed. In fact we have obtain additional freedom - an internal metric g_K may be varied freely. The set a_{in} can be obtained from a variety of sets b_{in} by fitting the metric g_K. We will refer to the set a_{in} as a secondary set and to the set b_{in} as a primary one. Our nearest goal is to elaborate some production mechanism of various metrics g_K. It seems a trivial task because quantum fluctuations claimed produce any form of metric. The problem becomes much less simple if one takes into account a classical motion of an extra space metric g_K just after its nucleation. As a result, a set of final metrics $g_K(t \to \infty)$ appears to be quite limited [1]. This strongly reduces a number of possibilities for the metric variation in (4).

As was shown in [1] the metrics could evolve classically to some stationary configuration $g_{stat}(y) = const$ that does not depend on an initial metric. This way gives rise a large but limited number of an initial sets a_{in} of parameters that lead to the universe similar to ours. The problem of fine tuning would be hardly solved.

Another wide class of theories could be obtained if classical equations for the metric g_K have finite-dimensional attractors. In this case a behavior of the metric at large times depends on an arbitrary initial metric and form a continuous set. Thus we come to infinite set of effective theories.

So far we have implicitly assumed that the structure of the space is $T \times M_{D_0-1}$, i.e. we do not distinguish an extra space and the observable 3-dim space of our Universe. In fact, the huge difference in their sizes has to be explained. As was shown in [2], the reason of their difference is laid in initial conditions. If an initial metric satisfies the conditions

$$M_0 = M_1 \times K_1; \quad R_M \ll R_K, \tag{5}$$

where R_M, R_K are the Ricci scalar of the main space and the extra space correspondingly, their following evolution is different. Relaxation time in the extra space K is much smaller so that an evolution of metric of the main space M proceeds at (almost) stationary extra metric. Initial conditions dictate a shape of the extra space while the latter influences a dynamics of the main space.

3. Explicit form of secondary parameters

To illustrate the above, consider a toy model with the specific pure gravitational action in a D_0-dim space

$$S = \int d^{D_0} z\sqrt{G} F(R_{M_0}, \{b\}); \quad F(R, \{b\}) = \sum_{n=1}^{N} b_n R^n. \tag{6}$$

Here R_{M_0} is the Ricci scalar and $\{b\} = b_1, b_2, ...b_N$.

We follow only those geometries that represent a direct product

$$U = M_4 \times M \times K \tag{7}$$

and satisfy the inequality written in (5). A classical motion of the metric g_K as a function in the extra space M is of interest and we will omit any mentioning about our 3-dim space.

Metric

$$ds^2 = dt^2 - g_{ab}(t, y)dy^a dy^b - e^{2\beta_1(t)}\gamma_{1,ij}(z)dz_1^i dz_1^j - e^{2\beta_2(t)}\gamma_{2,ij}(z)dz_2^i dz_2^j. \tag{8}$$

where $g_{ab}(t, y)$ is the metric in M, $\gamma_{1,2;ij}(z)$ are positively defined internal metrics of the extra space $K = K_1 \times K_2$ and $e^{2\beta_{1,2}(t)}$ are scaling factors (see [5, 3]) of the spaces $K_{1,2}$. Also, $D = dimM, d = dimK$.

Evolution of the extra space metric is governed by classical equations for the functions $\beta_{1,2}(t)$. As was shown in [2], the effective action for the field β has the form

$$S = \frac{1}{2}V[d_1] \int d^D y\sqrt{|G^{(D)}|} \{\text{sign} F' \cdot [R_4 + K] - 2V(\phi)\} \tag{9}$$

$$K_E = \frac{1}{d}\left(\partial\sigma + \frac{F''}{F'}\partial\phi\right)^2 + \left(\frac{F''}{F'}\right)^2 (\partial\phi)^2 + \sum_i d_i(\partial\beta_i)^2, \tag{10}$$

$$-2V_E(\phi_i) = \mathrm{e}^{-\sum_i \beta_i d_i}|F'|^{-d_0/d}F(\phi), \tag{11}$$

$$\phi_{1(2)}(t) := (d_{1(2)} - 1)\mathrm{e}^{-2\beta_{1(2)}(t)}, \quad \phi := d_1\phi_1 + d_2\phi_2. \tag{12}$$

As was shown in [6] the lines $\phi_{1,2}^* = 0$ are attractors of the classical equations for the fields $\phi_{1,2}$. We are interested in solutions of the form $\phi_1 \to 0; 0 < \phi_2 < \infty$.

Under these assumptions the Ricci scalar

$$R = R_M + \phi_1(t) + \phi_2(t) \tag{13}$$

is easily obtained. After some algebra with formulas (13) and (6) the reduced action can be obtained

$$S = \int d^{D_M} y\sqrt{g_M} F(R_M, \{a\}), \tag{14}$$

where the intermediate set of parameters a is connected to a set of primary parameters b

$$a_n(t) = \int d^{D_K} z\sqrt{g_K}\frac{F^{(n)}(\phi_1(t) + \phi_2(t), \{b\})}{n!}. \tag{15}$$

g_M, g_K are determinants of the metric tensors of spaces M and K. This formula may be considered as the connection between the primary set b and the metric g_K of extra space K. Variety of the sets b can be obtained by continuous variation the metric g_K at fixed intermediate parameters a. This illustrates the hypothesis:

In the framework of multidimensional gravity, the observable set of effective parameter values could be obtained by a continuous set of initial parameter values.

Formula (15) indicates that all physical parameters vary with time. Their modern values represent the asymptotes at time tends to infinity.

In this paper we discuss the idea of "inverse landscape". By this is meant that quantum fluctuations of extra space metric and their subsequent classical evolution could lead to observable values of physical parameters in wide range of initial parameter values of a primary lagrangian. In this case the fine tuning problem seems solvable.

[1] Rubin S G and Zinger A S 2012 *GERG*, 44, 9 2283
[2] Bronnikov K A and Rubin S G 2006 *Phys. Rev.* D **73** 124019
[3] Bronnikov K A, Konoplich R V and Rubin S G 2007 *Class. Quantum Grav.* **24**, 1261; gr-qc/0610003
[4] Bolokhov S V, Bronnikov K A and Rubin S G 2011 *Phys.Rev.* D **84**:044015; arXiv:1011.2828
[5] Carroll S M et al. 2002 *Phys.Rev.* D **66**:024036; hep-th/0110149
[6] Bronnikov K A, Rubin S G, Svadkovsky I V 2010 *Phys.Rev.* D **81**:084010

How fast is the wave function collapse?

A. Yu. Ignatiev

Theoretical Physics Research Institute, Melbourne 3163, Australia

E-mail: a.ignatiev@ritp.org

Abstract. Using complex quantum Hamilton-Jacobi formulation, a new kind of non-linear equations is proposed that have almost classical structure and extend the Schrödinger equation to describe the collapse of the wave function as a finite-time process. Experimental bounds on the collapse time are of order 0.1 ms to 0.1 ps and the areas where sensitive probes of the possible collapse dynamics can be done include Bose-Einstein condensates, ultracold neutrons or ultrafast optics.

1. Introduction

Recently there has been considerable interest in the fundamental principles of quantum mechanics, including its theoretical, experimental, and applied aspects [1, 2, 3], the physical meaning and reality of the wave function being a current focus of attention [4]. Among the least understood features of quantum mechanics is the collapse of the wave function (also known as the reduction of the wave packet)—"the most interesting point of the entire theory" [5].

In particular, it is not known whether the collapse is instantaneous or takes a finite time [6].

This issue was first raised by von Neumann, and to this day it remains a difficult and unsolved task, notwithstanding the progress in constructing the theories of dynamical collapse [7, 8]. Besides foundational interest, this question is also of utmost importance for quantum computing and quantum communications [9] because the finite collapse time could set the ultimate limit to the speed of quantum computation and information processing. Also, it would shed new light on the basic issues such as the Many-Worlds interpretation of quantum mechanics.

Recent years have also seen rapid growth of attention to a promising novel approach to quantum mechanics: complex quantum Hamilton-Jacobi (CQHJ) formulation [10]; one of its main purposes is to bring classical and quantum mechanics as close to each other as possible. An important and interesting question is whether or not the new formulation can give us new insights into fundamental long-standing issues such as the quantum measurement problem.

This paper proposes a new family of 'collapsible' Schrödinger equations, i.e., non-linear equations describing the collapse of the wave function as a dynamical phenomenon with a certain time-scale occurring during the final stages of the measurement process. We then mention briefly the experimental consequences of the suggested approach [11].

2. Classical mechanics in the light of quantum

Usually, we ask ourselves if quantum mechanics can be recast into a more classical form. But we can also turn this question upside down and ask if classical mechanics can be given a quantum 'look and feel'.

For example, we can ask if anything resembling the wave function f can be introduced into classical mechanics.

The first obstacle that we meet is this: If we have two one-particle systems and make out of them one two-particle systems, then wave functions must be *multiplied*. But in classical mechanics, the lagrangians, hamiltonians, or actions must be *added*, not multiplied.

There is an easy way out of this difficulty, though [12]. We just need to take an additive quantity, like the action S, and exponentiate it thus making the resulting quantity multiplicative:

$$f = \exp(iS/\hbar). \tag{1}$$

(An arbitrary constant had to be introduced for dimensional reasons, and the natural choice for it is the Planck constant $\hbar$.)

Next, what is the classical 'momentum operator'? Recalling that according to the Hamilton-Jacobi theory $\mathbf{p} = \nabla S$, we obtain:

$$\mathbf{p}(\mathbf{r}, t) = \frac{\hbar}{i}\frac{\nabla f}{f}. \tag{2}$$

3. Derivation of the CQHJ equation

We start with the ordinary Schrödinger equation for a particle of mass m in the potential $V(\mathbf{r})$:

$$i\hbar\psi_t = -\frac{\hbar^2}{2m}\Delta\psi + V\psi. \tag{3}$$

(Throughout the paper, the subscript t denotes the partial derivative with respect to time, $f_t \equiv \partial f/\partial t$.)

Motivated by the previous discussion, let us introduce a new variable $\mathbf{p}(\mathbf{r}, t)$:

$$\mathbf{p} = \frac{\hbar}{i}\frac{\nabla\psi}{\psi}. \tag{4}$$

Our next goal is to obtain a closed dynamical equation for $\mathbf{p}$. Differentiating Eq. (4) with respect to time, we find:

$$\mathbf{p}_t = \frac{\hbar}{i}\frac{\nabla\psi_t}{\psi} - \mathbf{p}\frac{\psi_t}{\psi}. \tag{5}$$

In deriving the second term, we have used the definition (4) rewritten in the form

$$\nabla\psi = \frac{i}{\hbar}\mathbf{p}\psi. \tag{6}$$

It is convenient to start ψ-elimination from the second term of Eq. (5). For this purpose, we first find $\Delta\psi$, again using (6):

$$\Delta\psi = \nabla\left(\frac{i}{\hbar}\mathbf{p}\psi\right) = \frac{i}{\hbar}\left(\psi\nabla\cdot\mathbf{p} + \mathbf{p}\nabla\psi\right). \tag{7}$$

Plugging this into the Schrödinger equation (3) and multiplying both sides by $i\mathbf{p}/(\hbar\psi)$, we obtain:

$$-\mathbf{p}\frac{\psi_t}{\psi} = \frac{1}{2m}\mathbf{p}\nabla\cdot\mathbf{p} + \frac{i}{2m\hbar}\mathbf{p}p^2 + \frac{i}{\hbar}\mathbf{p}V. \tag{8}$$

Next, we need to eliminate ψ from the first term of Eq. (5), so we substitute the right-hand side of Eq. (7) into the Schrödinger equation (3) instead of $\Delta\psi$, then take the gradient of both parts and divide both of them by ψ, which yields:

$$-i\hbar\frac{\nabla\psi_t}{\psi} = \frac{i\hbar}{2m}\nabla(\nabla\cdot\mathbf{p}) - \frac{1}{2m}\mathbf{p}\nabla\cdot\mathbf{p} - \frac{1}{2m}\nabla p^2 - \frac{i}{2\hbar m}p^2\mathbf{p} - \nabla V - \frac{i}{\hbar}\mathbf{p}V. \tag{9}$$

We now add the this formula, part-by-part, with Eq. (8) and then use Eq. (5) to finally obtain:

$$\mathbf{p}_t = -\nabla V - \frac{1}{2m}\nabla p^2 + \frac{i\hbar}{2m}\nabla(\nabla \cdot \mathbf{p}). \tag{10}$$

So it turns out that the complete elimination of the wave function is indeed possible, and Eq. (10) gives a new dynamical equation in a closed form.

This equation plays the role of the Schrödinger equation in the CQHJ approach [10].

A slight rearrangement gives this equation an elegant form:

$$\mathbf{p}_t = -\nabla(V + \frac{p^2}{2m} + \frac{\hat{\mathbf{p}}\mathbf{p}}{2m}), \tag{11}$$

where $\hat{\mathbf{p}} = -i\hbar\nabla$.

This suggests introducing the 'quantum Hamiltonian'

$$H = V + \frac{p^2}{2m} + \frac{\hat{\mathbf{p}}\mathbf{p}}{2m}, \tag{12}$$

so we can rewrite our Eq. (10) in a 'canonical form'

$$\mathbf{p}_t = -\nabla H. \tag{13}$$

4. Non-linear equations for collapse

The satisfactory description of the measurement process is likely to require a non-linear extensions of the Schrödinger dynamics. If we are looking for such equations, they should combine two features: (a) when the quantum system is isolated, they should go over to the standard Schrödinger equation; (b) they should include the non-linear interaction between the system and apparatus.

From the conventional perspective, it is hard to see which form such equations should take. By contrast, the CQHJ approach can lead us to a solution in the following way:

First, we replace the condition (a) by an equivalent: when the quantum system is isolated, the new equations should go over to the CQHJ equation, which is conveniently rewritten in yet another form

$$\mathbf{p}_t = \mathbf{F} - \nabla(\frac{p^2}{2m} + \frac{\hat{\mathbf{p}}\mathbf{p}}{2m}), \tag{14}$$

where $\mathbf{F} = -\nabla V$. Despite being non-linear in terms of $\mathbf{p}$, this equation is, of course, linear in terms of ψ.

Second, we need to satisfy condition (b). A natural way to do that is to add an extra 'collapsing' force $\mathbf{F}_c$ which effectively describes the 'measuring' interaction. The key point here is that the extra force must be *non-potential*, because otherwise the resulting equation would have the usual Schrödinger form (3) with V replaced by $V + V_c$ and therefore would still be linear in ψ.

Thus, finally, one of our non-linear 'collapsible Schrödinger equations' takes the form

$$\mathbf{p}_t = \mathbf{F}_c - \nabla H. \tag{15}$$

Due to its homogeneity, this equation automatically conserves probability. More precisely, the homogeneity here means the invariance of our equation under the transformation $\psi \to N(t)\psi$ where $N(t)$ is an arbitrary time-dependent scale factor.

5. Experimental consequences

It can be shown [11] that the collapsible Schrödinger equations can lead to remarkable consequences. Instead of the traditionally assumed instantaneous collapse, we now have the possibility that it is a continuous physical process taking a finite, non-zero time to proceed. This feature offers an exciting opportunity of experimental tests. Due to the limited space, we refer the intersted reader to Ref. [11] for more details, including the experimental bounds on the collapse time (of order 0.1 ms to 0.1 ps) and its convenient dimensionless measure which helps to identify the areas where sensitive probes of the possible collapse dynamics can be done. Examples are experiments with Bose-Einstein condensates, ultracold neutrons or ultrafast optics.

Our approach is not to be confused with the well-known dynamical collapse theories of Ghirardi-Rimini-Weber and Pearle (CSL) where localization is occurring spontaneously and continuously, even in the absence of a measuring device, due to a stochastic noise the physical source of which is yet to be clarified [7]. Unlike CSL, our collapsible Schödinger equations can be stochastic, noiseless.

Also, it is quite distinct from the earlier models of the Bohm-Bub type [8] in that it may *not* require any hidden variables [11].

The author is thankful to L. Ignatieva, V. A. Kuzmin, and M. E. Shaposhnikov for reading the manuscript and helpful discussions. Valuable comments of M. G. Albrow, R. Dörner, and F. Thaheld are gratefully acknowledged.

References

[1] see, e.g., S. L. Adler, *Quantum Theory as an Emergent Phenomenon* (Cambridge University Press, Cambridge, 2004); G. C. Ghirardi, A. Rimini, T. Weber, Phys. Rev. D **34**, 470 (1986); P. Pearle, Phys. Rev. A **39**, 2277 (1989); S. Weinberg, arXiv:1109.6462 [quant-ph]; J. S. Bell, Speakable and unspeakable in quantum mechanics (Cambridge University Press, Cambridge, 1989); W. H. Zurek, Physics Today **44**, 36 (1991); E. Joos and H. D. Zeh, Z. Phys. B **59**, 223 (1985); A. E. Allahverdyan, R. Balian, and T. M. Nieuwenhuizen, arXiv:1107.2138 [quant-ph]; F. H. Thaheld, BioSystems **92**, 114 (2008); A. Yu. Ignatiev, Rad. Phys. Chem. **75**, 2090 (2006); for a comprehensive bibliography see A. Cabello arXiv: quant-ph/0012089.

[2] A. J. Leggett, Science **307**, 871 (2005); J. Phys.: Cond. Mat. **14**, R415 (2002).

[3] G. 't Hooft, Class. Quant. Gravity, **16**, 3263 (1999); AIP Conf. Proc. **957**, 154 (2007).

[4] R. Colbeck and R. Renner, Phys. Rev. Lett. 108, 150402 (2012); M. F. Pusey, J. Bartlett, and T. Rudolph, Nature Physics, doi:10.1038/nphys2309.

[5] E. Schrödinger, Naturwissenschaften, 23, 807 (1935), English translation in: J. A. Wheeler and W. H. Zurek (eds.) *Quantum Theory and Measurement* (Princeton University Press, Princeton, N. J., 1983).

[6] N. Bohr, Nature **121**, 580 (1928); J. von Neumann, *Mathematische Grundlagen der Quantenmechanik* (Julius Springer-Verlag, Berlin, 1932) [English transl.: Princeton University Press, Princeton, N.J., 1955]; V. B. Braginsky and F. Ya. Khalili, *Quantum Measurement* (Cambridge University Press, Cambridge, 1992).

[7] For a recent review see, e.g., A. Bassi *et al.*, arXiv:1204.4325 [quant-ph].

[8] D. Bohm and J. Bub, Rev. Mod. Phys. **38**, 453 (1966); J. H. Tutsch, Rev. Mod. Phys. **40**, 232 (1968); P. M. Pearle, Phys. Rev. D **13**, 857 (1976); arXiv:quant-ph/0611211, 0611212 and references therein; A. N. Grigorenko, J. Phys. A: Math. Gen. **28**, 1459 (1995); T. Durt, Helv. Phys. Acta **72**, 356 (1999).

[9] M. A. Nielsen and I. L. Chuang, *Quantum computation and quantum information* (Cambridge University Press, Cambridge, 2010); R. Jozsa, arXiv:quant-ph/0508124.

[10] R. A. Leacock and M. J. Padgett, Phys. Rev. Lett. **50**, 3 (1983); Phys. Rev. D **28**, 2491 (1983); M. V. John, Found. Phys. Lett., **15**, 329, (2002); Ann. Phys. **324**, 220 (2009); arXiv:1104.3197; C.-D. Yang, Ann. Phys. **319**, 399 (2005); *ibid.* **319**, 444 (2005); Y. Goldfarb, I. Degani, and D. J. Tannor, J. Chem. Phys. **125**, 231103 (2006); Á. S. Sanz, F. Borondo, and S. Miret-Artés, J. Phys.: Cond. Mat. **14**, 6109 (2002); C. C. Chou *et al.*, Phys. Rev. Lett. **102**, 250401 (2009); Ann. Phys. **325**, 2193 (2010); B. Poirier, Phys. Rev. A **77**, 022114 (2008); C. C. Chou and R. E. Wyatt, Phys. Lett. A, **374**, 2608 (2010).

[11] A. Yu. Ignatiev, arXiv:1204.3373 [quant-ph].

[12] E. Schrödinger, Ann. Physik, **79**, 361 (1926).

Aspects of dynamical dimensional reduction in multigraph ensembles of CDT

Georgios Giasemidis[1], **John F. Wheater**[1] **and Stefan Zohren**[2,1]

[1] Rudolf Peierls Centre for Theoretical Physics, 1 Keble Road, Oxford OX1 3NP, UK
[2] PUC-Rio, R. Marquês de São Vicente 225, Rio de Janeiro, Brazil

E-mail: giasemidis@physics.ox.ac.uk, j.wheater1@physics.ox.ac.uk, zohren@fis.puc-rio.br

Abstract. We study the continuum limit of a "radially reduced" approximation of Causal Dynamical Triangulations (CDT), so-called multigraph ensembles, and explain why they serve as realistic toy models to study the dimensional reduction observed in numerical simulations of four-dimensional CDT. We present properties of this approximation in two, three and four dimensions comparing them with the numerical simulations and pointing out some common features with 2+1 dimensional Hořava-Lifshitz gravity.

1. Introduction

The phenomenon of the spectral dimensional reduction has attracted special interest within the quantum gravity community during the last seven years. It was first observed in computer simulations of the CDT (Causal Dynamical Triangulation) approach to quantum gravity [1] where a scale dependent spectral dimension varying from 4 in the IR to 2 in the UV limit was observed [2]. The running of the spectral dimension serves as a dynamical mechanism that might regulate the theory at short scales so that it might possesses a non-Gaussian fixed point as suggested by Weinberg [3]. Thereafter several authors confirmed a similar result from different approaches to quantum gravity, e.g. asymptotically safe Quantum Einstein Gravity [4, 5], Hořava-Lifshitz gravity [6, 7], spin foams [8, 9, 10, 11], multi-fractional space-times [12, 13]. The asymptotic safety scenario and Hořava-Lifshitz gravity share further characteristics with CDT and their relation is an active research field [14, 15, 16]. Besides the contributions from different approaches, within the CDT approach, analytical computation explaining the numerical results seemed out of reach due to the complexity of the problem. To remedy this, the authors recently introduced a radial approximation of CDT, a so-called multigraph ensemble, and showed that it serves as realistic toy model to study dynamical reduction of the spectral dimension in CDT [17, 18]. In the following we review the results of [17, 18] and comment on the relation between the radial approximation of CDT and three-dimensional Hořava-Lifshitz gravity.

2. The multigraph approximation: Motivation and validity

The causal constraint in $(d+1)$-dimensional CDT requires the topology to be $I \times \Sigma^d$ where d is the dimension of spatial hyper-surfaces, which are connected by $(d+1)$-simplices. The foliation structure separates the edges in two types; space-like (blue) and time-like (red) (see Fig.1 for an example in $2+1$ dimensions). Consider the mapping which acts on rooted causal triangulations by collapsing spatial hyper-surfaces into single points (blue) and retaining all the time-like edges (red) (Fig. 1). The resulting graph is a rooted multigraph M with $L_n \geq 1$, $n \geq 0$, number of

Published under licence by IOP Publishing Ltd

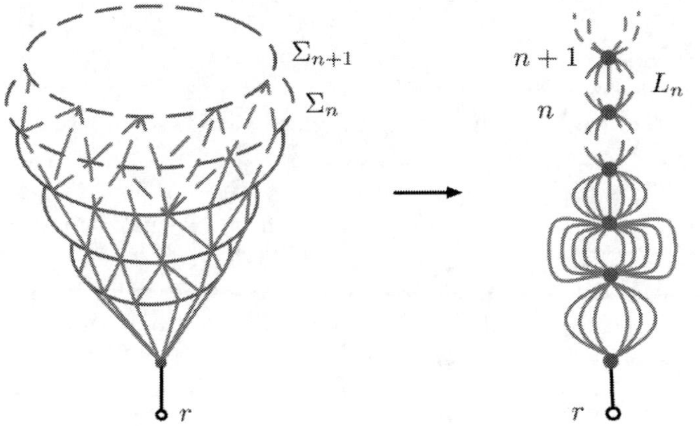

Figure 1. Illustration of how to obtain a multigraph from a causal triangulation.

edges connecting vertices n and $n+1$. Note that the multigraph ensemble inherits its measure from the measure of the causal triangulation ensemble.

We study the spectral dimension d_s on graphs by considering random walks on them and determining the asymptotic behaviour of the first diverging k-th derivative of generating function of the return probabilities to the root, $p_G(t)$ (refer to [17, 18] for details),

$$Q_G^{(k)}(x) \equiv \left(\sum_{t=0}^{\infty} p_G(t)(1-x)^{t/2}\right)^{(k)} \sim x^{-(k+1)+d_s/2}, \qquad x \to 0, \qquad k \geq 0, \tag{1}$$

for $2k \leq d_s < 2(k+1)$. Denoting by M_n the multigraph obtained from M by removing all the vertices and edges from the root to vertex $n-1$ and relabelling the remaining multigraph and setting $\eta_n \equiv Q_{M_n}/L_n$, the generating function satisfies the recursion relation [19]

$$\eta_{M_n}(x) = \eta_{M_{n+1}}(x) + \frac{1}{L_n} - xL_n\eta_{M_n}(x)\eta_{M_{n+1}}(x). \tag{2}$$

which is the starting point to determine the spectral dimension.

The reason why multigraphs serve as realistic models that share the same spectral dimension with CDT is twofold; firstly it is proven [19] that recurrent multigraphs bound above the spectral dimension of two-dimensional CDT and further this bound is believed to be tight. Secondly, in a rigorous proof [19] it was shown that the number of space-like links in a slice of a two-dimensional CDT remains finite for finite time. This indicates that a random walker on a CDT spends only a finite time on (finite) spatial hyper-surfaces which implies that the spectral dimension is not affected by random walks on spatial surfaces. This argument suggests that the multigraph approximation does not effect the spectral dimension of CDT.

Having motivated the use of multigraphs, we now discuss the properties that determine the spectral dimension on these graphs. Rigorous results for multigraph ensembles with recurrent measures related to two-dimensional CDT [19, 17] and transient measures [17] indicate that the spectral dimension is determined by the volume and resistance growth (considering the graph as an electrical network where each edge has resistance one). The knowledge of these two quantities is sufficient to determine the spectral dimension of multigraph ensembles. In the next section we apply this intuition to "radially reduced" higher-dimensional CDT ensembles.

3. Higher-dimensional models and results

Due to the absence of an analytical model for higher-dimensional CDT we propose a measure μ of the corresponding multigraph ensembles by imposing three assumptions [17, 18], which are closely related to volume and resistance growth. The first assumption concerns the connectivity,

$$\langle L_N \rangle_{\mu(\nu_D)} \simeq \nu_D N^{3-\epsilon} + N. \tag{3}$$

Here ν_D is the inverse bare Newton's constant and $D = d+1$ the topological dimension. In the following we choose $\epsilon = 1$ for $D = 3$ while for $D = 4$, ϵ is arbitrarily small and is introduced for purely mathematical reasons [17]. The assumption (3) implies for the volume growth $\underline{c}N \langle L_N \rangle_{\mu(\nu_D)} \leq \langle B(N) \rangle_{\mu(\nu_D)} \leq \bar{c}N \langle L_N \rangle_{\mu(\nu_D)}$ and is in agreement with computer simulations of four-dimensional CDT.

The second assumption bounds from above the resistance from the vertex N to infinity and controls the downward fluctuations in connectivity while the third ansatz controls the upward fluctuations in connectivity. In particular, these two assumptions read

$$R(N) \leq \frac{N}{\langle L_N \rangle_{\mu(\nu_D)}} \psi_+(\sqrt{\nu_D} N^{\frac{2-\epsilon}{2}}), \tag{4}$$

$$L_N \leq \langle L_N \rangle_{\mu(\nu_D)} \psi(\sqrt{\nu_D} N^{\frac{2-\epsilon}{2}}) \tag{5}$$

for $N > N_0 > 0$ and almost all graphs of the ensemble. Here $\psi(x)$, $\psi_+(x)$ are diverging and slowly varying functions at $x = 0$ and $x = \infty$. Note that (3)-(5) are higher-dimensional analogues of the analytical two-dimensional model. The description of the higher-dimensional model $(2 \leq d_s \leq 4)$ requires transient multigraphs with finite resistance $R(N)$. Therefore the spectral dimension is extracted from $Q'(x)$, which diverges for $x \to 0$.

The scale dependent spectral dimension depends on a length scale which is set by ν_D. Special attention has to be paid to define the correct scaling limit of the discrete random graph model in which both the walk length as well as the characteristic length scale diverge in a "double scaling limit" [17, 20],

$$|\tilde{Q}'(\xi, G_{(D)})| \equiv \lim_{a \to 0} (a/G_{(D)}) \langle |Q'_M(x = a\xi)| \rangle_{\mu(\nu_D = a^{1-\epsilon/2}/G_{(D)})}. \tag{6}$$

In particular, one obtained for $D = 3$ and $D = 4$ respectively,

$$|\tilde{Q}'(\xi, G_{(3)})| \sim \begin{cases} \xi^{-1}, & \xi >> G_{(3)}^{-2}, \\ \xi^{-1/2}, & \xi << G_{(3)}^{-2}, \end{cases} \qquad |\tilde{Q}'(\xi, G_{(4)})| \sim \begin{cases} \xi^{-1}, & \xi >> G_{(4)}^{-1}, \\ \xi^{-\epsilon/2}, & \xi << G_{(4)}^{-1}, \end{cases} \tag{7}$$

where $G_{(3)}$ and $G_{(4)}$ are the renormalised Newton's constants in three and four dimensions respectively. This result implies that for $D = 3, 4$ one has $d_s^\infty = D$ at large distances while $d_s^0 = 2$ at short scales, which agrees with the numerical results in three [21, 22] and four-dimensional CDT [2, 22]. Although $G_{(3)}^2$ and $G_{(4)}$ set a scale on the *duration* of the walk, it is their square root which corresponds to the *length extent* on the graph and is identified with the Planck length in three and four dimensions respectively.

We can apply a Tauberian theorem to $\langle |Q'(x)| \rangle_{\mu(\nu_D)}$ to find the average return probability density. Scaling the latter we determine the probability density $P(\sigma)$ for continuous diffusion time σ. In [18] we pursued it for $D = 4$ and showed that the functional form of $P(\sigma)$ is consistent with the one conjectured from computer simulations in [2]. Here we give the analogous result for $D = 3$. In this case one has

$$P(\sigma) \equiv \lim_{a \to 0} a^{-1} \langle p(t = \lfloor \sigma/a \rfloor) \rangle_{\mu(\nu_3)} \simeq \frac{1}{\sigma^{3/2} \left(1 + \sqrt{2} G_{(3)}/\sqrt{\sigma}\right)} \tag{8}$$

which implies that the scale dependent spectral dimension is given by

$$D_s(\sigma) \equiv -2 \frac{d \log P(\sigma)}{d \log \sigma} = 3 - \frac{1}{1 + \sqrt{\sigma/(2G_{(3)}^2)}} = \begin{cases} 2, & \sigma \to 0, \\ 3, & \sigma \to \infty. \end{cases} \tag{9}$$

Journal of Physics: Conference Series **410** (2013) 012154 doi:10.1088/1742-6596/410/1/012154

We note that, while having the correct limits, this result is slightly different than the rational or exponential fit suggested in [21, 22] to explain the numerical data. It would be interested so to test how well the data fits in the functional form (9) of the spectral dimension.

4. Discussion

We reviewed the results of [17, 18] regarding a radial reduction of CDT, the multigraph approximation, which can be used to study the phenomenon of scale dependent spectral dimension analytically. Besides discussing previous results on the four-dimensional case, we also applied this approximation to three-dimensional CDT.

The three-dimensional model provides an interesting playground to study the relations between different approaches to Quantum Gravity [16, 23]. One possibility to relate the results from the multigraph approximation with other approaches is to follow the techniques of [23] to find the dispersion relation $\omega(k)$ which corresponds to (9). In [24] a dispersion relation was introduced which originates from the foliation-defining scalar of 2+1 dimensional Hořava-Lifshitz gravity and which is given by $\omega(k)^2 = Ak^2\frac{1+Bk^2+Ck^4}{1+Dk^2}$. It is readily seen that it has the same asymptotic behaviour as the dispersion relation $\omega(k)$ derived from (9) when $k \to 0$ and $k \to \infty$. Trying to link those two results might shed further light on the relationship between Hořava-Lifshitz gravity and CDT(-like) models.

Acknowledgments

This work was supported by the A.G. Leventis Foundation, the A.S. Onassis Foundation grant F-ZG/097/2010-2011, EPSRC grant EP/I01263X/1 and STFC grant ST/G000492/1.

References

[1] Ambjørn J and Loll R 1998 *Nucl.Phys.* **B536** 407–434 (*Preprint* hep-th/9805108)
[2] Ambjørn J, Jurkiewicz J and Loll R 2005 *Phys.Rev.Lett.* **95** 171301 (*Preprint* hep-th/0505113)
[3] S. Weinberg, in General Relativity, S. Hawking and W. Israel, eds., p. 790-831, 1980.
[4] Litim D F 2004 *Phys.Rev.Lett.* **92** 201301 (*Preprint* hep-th/0312114)
[5] Lauscher O and Reuter M 2005 *JHEP* **0510** 050 (*Preprint* hep-th/0508202)
[6] Hořava P 2009 *Phys.Rev.* **D79** 084008 (*Preprint* arXiv:0901.3775 [hep-th])
[7] Hořava P 2009 *Phys.Rev.Lett.* **102** 161301 (*Preprint* arXiv:0902.3657 [hep-th])
[8] Modesto L 2009 *Class.Quant.Grav.* **26** 242002 (*Preprint* arXiv:0812.2214 [gr-qc])
[9] Caravelli F and Modesto L 2009 (*Preprint* arXiv:0905.2170 [gr-qc])
[10] Magliaro E, Perini C and Modesto L 2009 (*Preprint* arXiv:0911.0437 [gr-qc])
[11] Modesto L and Nicolini P 2010 *Phys.Rev.* **D81** 104040 (*Preprint* arXiv:0912.0220 [hep-th])
[12] Calcagni G 2011 *Phys.Lett.* **B697** 251–253 (*Preprint* arXiv:1012.1244 [hep-th])
[13] Calcagni G 2012 (*Preprint* arXiv:1205.5046 [hep-th])
[14] Reuter M and Saueressig F 2011 *JHEP* **1112** 012 (*Preprint* arXiv:1110.5224 [hep-th])
[15] Ambjørn J, Gorlich A, Jordan S, Jurkiewicz J and Loll R 2010 *Phys.Lett.* **B690** 413–419 (*Preprint* arXiv:1002.3298 [hep-th])
[16] Anderson C *et al.* 2012 *Phys.Rev.* **D85** 044027 (*Preprint* arXiv:1111.6634 [hep-th])
[17] Giasemidis G, Wheater J F and Zohren S 2012 *J. Phys. A: Math. Gen.* **45** 355001 (*Preprint* arXiv:1202.6322 [hep-th])
[18] Giasemidis G, Wheater J F and Zohren S 2012 *Phys.Rev.* **D** in press (*Preprint* arXiv:1202.2710 [hep-th])
[19] Durhuus B, Jonsson T and Wheater J F 2006 *J. Stat. Phys.* **139** 859–881 (*Preprint* arXiv:0908.3643 [math-ph])
[20] Atkin M R, Giasemidis G and Wheater J F 2011 *J.Phys.A* **A44** 265001 (*Preprint* arXiv:1101.4174 [hep-th])
[21] Benedetti D and Henson J 2009 *Phys.Rev.* **D80** 124036 (*Preprint* arXiv:0911.0401 [hep-th])
[22] Kommu R 2012 *Class.Quant.Grav.* **29** 105003 (*Preprint* arXiv:1110.6875 [gr-qc])
[23] Sotiriou T P, Visser M and Weinfurtner S 2011 *Phys.Rev.* **D84** 104018 (*Preprint* arXiv:1105.6098 [hep-th])
[24] Sotiriou T P, Visser M and Weinfurtner S 2011 *Phys.Rev.Lett.* **107** 131303 (*Preprint* arXiv:1105.5646 [gr-qc])

Theory of Excited States of Finite Systems in Coulomb External Potential

Á. Nagy

Department of Theoretical Physics, University of Debrecen, Debrecen, Hungary

E-mail: anagy@madget.atomki.hu

Abstract.

Recently a theory of excited states of Coulomb systems (P. W. Ayers, M. Levy and Á. Nagy, Phys. Rev. A **85**, 042518 (2012)) has been been put forward. The talk will present and develop this new theory. It will be shown that the Coulomb density determines the Hamiltonian and the degree of excitation. The definition of a single, universal functional which is enough to describe Coulomb systems is presented. The availability of the theory is discussed.

1. Introduction

Nowadays excitation energies are frequently calculated with time-dependent density functional theory(See, e. g. [1, 2, 3]), though there are time-independent theories, too. The first rigorious theory was the subspace theory of Theophilou [4] which was enlarged to an ensemble theory by Gross et al [5]. These approaches have, however, the disadvantage that their application to highly excited states is rather complicated and they cannot be used in case of core excitations. There also exist theories for a single excited state. The Levy-Nagy theory [6], for example, is a bifunctional theory, that is, the excited-state energy is a functional not only of the excited-state density, but also of the external potential (or the ground-state density). The Nagy theory of a single excited state [7, 8], on the other hand, is based on Kato's theorem and valid for Coulomb external potential. Recently, the latter has been generalized [9]. The talk will present and develop this new theory.

2. Cusp condition for excited states

The ground-state electron density is sufficient in principle to determine all molecular properties. This can be simply understood following Bright Wilson's [10] argument: A well-known theorem of quantum mechanics, Kato's theorem [11] leads to [12]

$$Z_\beta = -\frac{1}{2n(r)} \frac{\partial \bar{n}(r)}{\partial r}\bigg|_{r=R_\beta} , \qquad (1)$$

where the partial derivatives are taken at the nuclei β and $\bar{n}(r)$ is the angular average of the density. So the cusps of the density tell us where the nuclei are (R_β) and what the atomic numbers Z_β are. On the other hand, the integral of the density gives us the number of electrons:

$$N = \int n(\mathbf{r}) d\mathbf{r} . \qquad (2)$$

Kato's theorem is valid not only for the ground state but also for the excited states. So, if the density n_k of the k-th electron states is known the Hamiltonian $\hat{H}$ is also in principle known and its eigenvalue problem

$$\hat{H}\Psi_i = (\hat{T} + \hat{V} + \hat{V}_{ee})\Psi_i = E_i\Psi_i \tag{3}$$

can be solved, where

$$\hat{T} = \sum_{j=1}^{N}(-\frac{1}{2}\nabla_j^2) , \qquad \hat{V}_{ee} = \sum_{i=1}^{N-1}\sum_{j=i+1}^{N}\frac{1}{|\mathbf{r}_i - \mathbf{r}_j|} , \qquad \hat{V} = \sum_{i=1}^{N}\sum_{\beta=1}^{M} -Z_J/|\mathbf{r}_i - \mathbf{R}_\beta| , \tag{4}$$

are the kinetic energy, the electron-electron and the electron-nucleon operators, respectively.

There are certain special cases, however, where Eq. (1) does not determine the atomic number. The simplest example is the $2p$ orbital of the hydrogen atom. In this case the spherical average of the derivative of the density is zero and the value of the density

$$n_{2p}(r) = cr^2e^{-Zr} \tag{5}$$

is also zero at the nucleus. It means that in this case Kato's theorem does not give the atomic number. Similar cases occur in those highly excited atoms, ions or molecules, in which there are no s-electrons. There exist, however, a more general cusp expression for the density, that can be applied even in these special cases. The corresponding relations for the wave functions were derived by Pack and Brown [13]. (Further works concerning the cusp of the density are Ref. [14, 15, 16].) The cusp relations for highly excited states for the density were also derived [17, 18]:

$$\left.\frac{\partial\bar{\eta}_l(r)}{\partial r}\right|_{r=R_\beta} = -\frac{2Z}{l+1}\eta_l(\mathbf{R}_\beta) , \tag{6}$$

where

$$\eta_l(\mathbf{r}) = \frac{n(\mathbf{r})}{r^{2l}} \tag{7}$$

and l is the smallest integer for which η_l is not zero at the nucleus. In the example mentioned above Eq. (7) leads to

$$\eta_{2p}(r) = \frac{n_{2p}}{r^2} = ce^{-Zr} \tag{8}$$

and the new cusp relation has the form:

$$-2Z\eta_{2p}(0) = 2\eta_{2p}'(0) . \tag{9}$$

So we can again readily obtain the atomic number from the electron density. (Further useful cusp relations can be found in [19].) The argument above proves

Theorem 1: The density of a Coulomb system determines the external potential .

3. Theory of excited states in Coulomb external potential

Theorem 2: The excited state density n_k determines not only the Hamiltonian but the eigenvalue E_k, too.

To prove this theorem, suppose that we have two wave function Ψ_k and Ψ_k' for which the density is the same: $n_k = n'_k$. It is well known, that the asymptotic behaviour of the density is governed by the ionization energy I [20, 21]. Consequently,

$$I_k = E_k - E_1^{(N-1)}, \tag{10}$$

where $E_1^{(N-1)}$ is the ground-state energy of the system after an electron is removed. Similarly, we can write that $I'_k = E'_k - E_1^{(N-1)}$. As the density is the same $n_k = n'_k$, so is the ionization energy $I_k = I'_k$. Therefore $E_k = E'_k$. So, for a non-degenerate state the density uniquely determines both the energy and the wave function.

Eq. (10) is valid only if

$$\varepsilon = E_1^{(N-1)} - E_k > 0, \tag{11}$$

For highly excited states ε might be negative. A more general expression valid for any excited state can be found in [21]: With the nomenclature of Wigner [22] the eigenfunctions of the Hamiltonian $\hat{H}^{Coul}$ can be classified by the irreducible representations $D^{(l)}, l = 0, 1, ..., [N/2]$. If the eigenfunction belongs to the irreducible representation $D^{(l)}$ of the symmetric group S_N, it can be ionized into $N-1$-electron states which transform according to the irreducible representations $D^{(l)}$ or $D^{(l-1)}$ of the symmetric group S_{N-1}. Denoting by $E_{1,l}^{(N-1)}$ the lowest eigenvalue of the Hamiltonian H^{N-1} and defining

$$\tilde{E}_1^{(N-1)} = min\left(E_{1,l}^{(N-1)}, E_{1,l-1}^{(N-1)}\right), \tag{12}$$

the decay of the wave function is determined by

$$\varepsilon_{gen} = \tilde{E}_1^{(N-1)} - E_k. \tag{13}$$

The present theory is valid provided this quantity is positive, that is, for a bound state.

Define now the functional F.

$$F[n] = E - \int n(\mathbf{r})v(\mathbf{r})d\mathbf{r} \tag{14}$$

is valid for any stationary state of any Coulomb system. As there is no known way to decide (without constructing the external potential and solving the Schrödinger equation), whether a given density is Coulombic it would be better to define a functional F for all electron densities.

First, F is defined as a functional of a trial density n and a Coulomb density n^{Coul} corresponding to the the kth state of some Coulomb Hamiltonian

$$F[n, n^{Coul}] = \min_{\substack{\Psi \to n \\ \{\langle \Psi | \Psi_j^{Coul}[n^{Coul}] \rangle = 0\}_{j=1}^{k-1}}} \langle \Psi | \hat{T} + \hat{V}_{ee} | \Psi \rangle . \tag{15}$$

The minimization is done with the constraints that each Ψ yield $n(\mathbf{r})$ and is simultaneously ortogonal to the first $k-1$ states of the Coulomb system specified by $n^{Coul}(\mathbf{r})$. Then a universal functional $F[n]$ can be constructed as follows. Assume that there exists a unique Coulomb density that is closest to the (non-Coulomb) density n. (The best measure for "closest" is not detailed here.) If there are several Coulomb densities from the same 'distance' from n, the one yielding to the smallest F (Eq. (15)) is selected:

$$F_\epsilon^{Coul}[n] = \min_{n^{Coul}} F[n, n^{Coul}]; \quad ||n^{Coul} - n|| \le \epsilon. \tag{16}$$

ϵ is supposed to be large enough to ensure the existence of at least one stationary state Coulomb density in the distance smaller than ϵ. With ϵ_{min} denoting the smallest possible value of ϵ,

$$F^{Coul}[n] = F^{Coul}_{\epsilon_{min}}[n]. \tag{17}$$

This procedure should be done in the "vicinity" of all Coulomb densities $n_k(\mathbf{r})$. In this way the functional F^{Coul} is defined for any density. Supposing that this functional is functionally differentiable we are led to the the Euler equation

$$v^{Coul}([n], \mathbf{r}) = -\frac{\delta F^{Coul}[n]}{\delta n} \tag{18}$$

up to a constant. The theory presented here is for non-degerate states. The generalization for degerate states will be published elsewhere. An extension to the Kohn-Sham scheme is under progress.

Acknowledgments
The work is also supported by the TAMOP 4.2.1/B-09/1/KONV-2010-0007 and the TAMOP 4.2.2/B-10/1-2010-0024 projects. The project is co-financed by the European Union and the European Social Fund. Grant OTKA No. K 100590 is also gratefully acknowledged.

References

[1] H. Appel, E.K.U. Gross and K. Burke, Phys. Rev. Lett. **90**, 043005 (2010).
[2] M. E. Casida, J. Mol. Struct., Theochem **914**, 3 (2009).
[3] M. Petersilka, U. J. Gossmann and E.K.U. Gross, Phys. Rev. Lett . **76**, 1212 (1996).
[4] A.K. Theophilou, J. Phys. C **12** 5419 (1978).
[5] L.N. Oliveira, E.K.U. Gross and W. Kohn, Phys. Rev. A **37**, 2805, 2809, 2821 (1988).
[6] M. Levy and Á. Nagy, Phys. Rev. Lett. **83**, 4361 (1999). Á. Nagy and M. Levy, Phys. Rev. A **63**, 2502 (2001).
[7] Á. Nagy, Int. J. Quantum. Chem. **69**, 247 (1998).
[8] Á. Nagy, Int. J. Quantum. Chem. **70**, 681 (1998).
[9] P. W. Ayers, M. Levy and Á. Nagy, Phys. Rev. A **85**, 042518 (2012).
[10] N. C. Handy, in *Quantum Mechanical Simulation Methods for Studying Biological Systems* Eds. D. Bicout and M. Field (Springer–Verlag, Heidelberg,1996) p.1.
[11] T. Kato, Commun. Pure Appl. Math. **10**, 151 (1957).
[12] E. Steiner, J. Chem. Phys. **39**, 2365(1963); N. H. March, *Self-consistent fields in atoms* (Pergamon, Oxford, 1975).
[13] R. T. Pack and W. B. Brown, J. Chem. Phys. **45**, 556 (1966).
[14] F. J. Gálvez, J. Porras, J. C. Angulo and J. S. Dehesa, J. Phys. B. **21**, L271 (1988); J. C. Angulo, J. S. Dehesa and F. J. Gálvez, Phys. Rev. A **42**, 641 (1990); erratum Phys. Rev. A **43**, 4069 (1991); J. C. Angulo and J. S. Dehesa, Phys. Rev. A **44**, 1516 (1991); F. J. Gálvez and J. Porras, Phys. Rev. A **44**, 144 (1991); J. Porras and F. J. Gálvez, Phys. Rev. A **46**, 105 (1992).
[15] R. O. Esquivel, J. Chen, M. J. Stott, R. P. Sagar and V. H. Smith, Jr., Phys. Rev. A **47**, 936 (1993); R. O. Esquivel, R. P. Sagar, V. H. Smith, Jr., J. Chen and M. J. Stott, Phys. Rev. A **47**, 4735 (1993);
[16] J. S. Dehesa, T. Koga and E. Romera, Phys. Rev. A **49**, 4255 (1994); T. Koga, Theor. Chim. Acta **95**, 113 (1997); T. Koga and H. Matsuyama, Theor. Chim. Acta **98**, 129 (1997); J. C. Angulo, T. Koga, E. Romera and J. S. Dehesa, THEOCHEM **501-502**, 177 (2000).
[17] Á Nagy and K. D. Sen, J. Phys. B **33**, 1745 (2000).
[18] P. W. Ayers, Proc. Natl. Acad. Sci. **97**,1951 (2000).
[19] Á Nagy and K. D. Sen, Chem. Phys. Lett. **332**, 154 (2000); J. Chem. Phys. **115**, 6300 (2001).
[20] C.-O. Almbladh and U. von Barth, Phys. Rev. B **31**, 3231 (1985); N. C. Handy, M. T. Marron and H. J. Silverstone, Phys. Rev. **180**, 45 (1969); M. M. Morrell, R. G. Parr and M. Levy, J. Chem. Phys. **62**, 549 (1975); H. J. Silverstone, Phys. Rev. A **23**, 1030 (1981).
[21] R. Ahlrichs, J. Chem. Phys. **64**, 2706 (1976); M. Hoffmann-Ostenhof and T. Hoffmann-Ostenhof, Phys. Rev. A **16**,1782 (1977).
[22] E. P. Wigner, Group theory and its application to quantum mechanics of atomic spectra (Academic Press, New York, 1959).

Journal of Physics: Conference Series **410** (2013) 012156 doi:10.1088/1742-6596/410/1/012156

Random Walk Model for the Growth of Monolayer in Dip Pen Nanolithography

H. Kim, S. Ha, and J. Jang[1]
Department of Nanomaterials Engineering, Pusan National University, Miryang 627-706, Republic of Korea

E-mail: jkjang@pusan.ac.kr

Abstract. By using a simple random-walk model, we simulate the growth of a self-assembled monolayer (SAM) pattern generated in dip pen nanolithography (DPN). In this model, the SAM pattern grows mainly via the serial pushing of molecules deposited from the tip. We examine various SAM patterns, such as lines, crosses, and letters by changing the tip scan speed.

1. Introduction

Dip-pen nanolithography (DPN) emerged as a general tool to fabricate nanopatterns on various substrates, including metallic, insulating, and semiconducting substrates [1-3]. In DPN, an atomic force microscope (AFM) tip functions as a source for continuously creating nanodroplets of molecules that subsequently spread to form a self-assembled monolayer (SAM) on a substrate. The molecular transport from the tip to the substrate is mediated by the water meniscus which naturally forms between the tip and substrate under ambient conditions [4-6].

Little is known about the molecular mechanism and time scale of the SAM growth in DPN. In this respect, theory and modelling have proven to be useful [4, 7-10]. Given molecules are designed to chemisorb to the substrate, Jang et al.[8] proposed a diffusion model of DPN by assuming that molecules are irreversibly trapped once they reach the chemisorbing sites on the substrate. Molecules can diffuse on top of other molecules. This model fails to reproduce the dendritic SAM patterns observed in DPN using dodecylamine [11] or polyethylene glycol [12] on mica. Molecular dynamics (MD) simulation [7] showed that a SAM pattern mainly grows via the serial pushing mechanism (Figure 1a) in which a molecule in the upper layer (open circle in the second layer in the figure) pushes a molecule on the substrate out of its original position, and the molecule pushed out in turn pushes out another molecule nearby. Such a consecutive push propagates and reaches the periphery, and then the pushing stops. This serial pushing was taken to have a finite directional coherence length, N_d, defined as the number of consecutive pushing events which has the same direction (see Figure 1b). By simply varying N_d, this model can reproduce various SAM patterns, ranging from an isotropic to a dendritic pattern. The serial pushing model also captured the essential features of an MD simulation [10]. Previously, the serial pushing model was applied for the case where an AFM tip is fixed in position. However, the SAM patterns in DPN are generally made by moving the tip on a substrate. Then, the outcome of DPN depends on the moving speed of tip. Herein, we simulate various

[1] To whom any correspondence should be addressed.

line-based SAM patterns and study how DPN is affected by the tip scan speed and the directional coherence length.

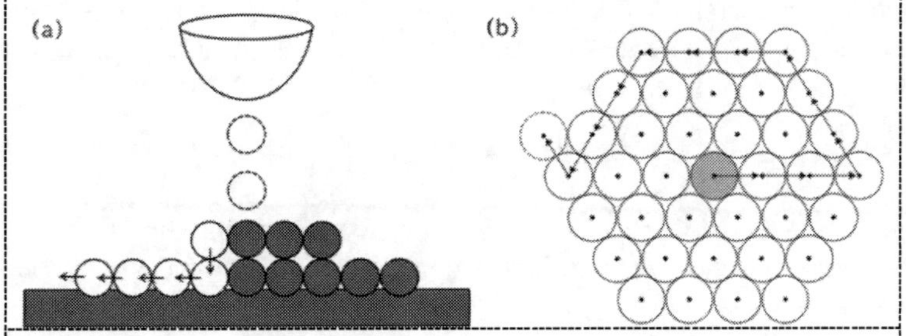

Figure 1. (a) Serial pushing model. A molecule (an open circle in the second layer) pushes the molecule below it (one of the open circles in the first layer), and the molecule pushed out in turn pushes its neighbor out of place. This pushing propagates (drawn as left arrows) toward the periphery of the monolayer until there is no molecule left to push out. (b) Schematic diagram for serial pushing events with the directional coherence length N_d of 3. The serial pushing changes its direction after every 3 consecutive pushing events.

2. Details of Model and Simulation

Random walk simulations are performed by using a three dimensional lattice and a discrete time step. The three dimensional lattice was constructed by vertically stacking two dimensional trigonal lattices. The lattice spacing of the trigonal lattice is identified as the distance between the nearest neighbor molecules in the SAM of octadecanethiol on a gold surface, $l = 0.5385$ nm [7, 10]. Every molecule belonging to of the layers above the substrate executes a three dimensional random walk. If a radom walk of such a molecule happens to be at the position of a molecule below it, this random walk is taken to push the molecule below out of place. Then the molecule pushed out in turn pushes its neighbor out of place. This push-induced movement laterally propagates toward the periphery until there are no molecules left to push out. This sequence of push-induced movements occurs without time delay. Unlike the molecules in the upper layers, a molecule on the bare surface is allowed to move only if it is pushed by one of its nearest neighbors, above or beside it.

If a molecule initially pushes out its neighboring molecule, the subsequent $(N_d - 1)$ movements take the same direction as the initial pushing direction. The direction of the $(N_d + 1)$th molecular displacement becomes random, and then, the subsequent $(N_d - 1)$ movements are taken to occur in the same direction as the $(N_d + 1)$th displacement. This series of push-induced movements continues until it stops at the periphery (where there are no molecules left to push out).

The time step of simulation Δt is set by following Jang et al [8]. Theory of diffusion shows Δt and l satisfy $4D\Delta t = l^2$, where D is the diffusion constant of molecule. Here, estimating the value of D is rather tricky as D refers to the molecular diffusion on top of other molecules (not the surface diffusion on a bare surface). Previously, D was estimated for eicosanethiol in the context of the microcontact printing, as $D = 7\times10^{-8} cm^2 s^{-1}$ [13]. Using this value of D, we get $\Delta t = 10.4$ ns.

In the present work, the tip scan speed v is reported in terms of a dimensionless value, v^*, defined as $v^* = v/(l/\Delta t)$. For a slow tip scan speed, $v^* < 1$, the tip moves by l for every $1/v^*$ time steps. If the displaced tip is not exactly at one of the lattice points, the tip position is taken to be at the lattice point nearest to the tip position. We consider three tip scan speeds, v^*=0.25, 0.5, and 1.

Another parameter needed for the present simulation is the dripping rate of molecules from the tip, n. Here, the dripping rate is reported in terms of a reduced rate, $n^* = n\Delta t$, which represents the number of molecules dropped per simulation time step. Throughout the present work, n^* is fixed to be 5.

3. Results and Discussion

We first examine how the SAM patterns are affected by varying the directional coherence length of serial pushing. In Figure 2, we show cross patterns generated by using a tip scan speed v^* of 0.25. The horizontal line is drawn first and the vertical line next. As N_d varies as 0 (a), 10 (b), 20 (c), and 30 (d), the periphery of the cross SAM pattern becomes more rugged. Compared to the case where the tip is fixed in position, the effects of changing N_d are small. A dendritic pattern with self-replicating branches was found for high values of N_d in the case of a fixed tip. However, we did not find any dendritic pattern for a moving tip.

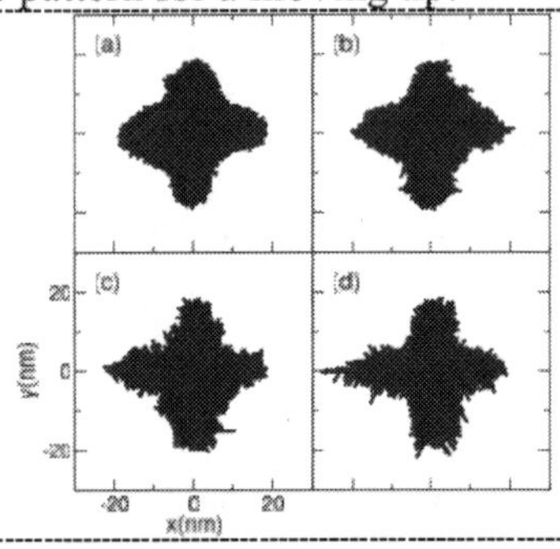

Figure 2. Cross SAM patterns generated by scanning a tip. The N_d of serial pushing is varied as 0 (a), 10 (b), 20 (c), and 30 (d). The tip scan speed v^* and the molecular dripping rate n^* are fixed to be 0.25 and 5, respectively.

We move on to study the effects of tip scan speed on the SAM patterns. By varying v^* as (a) 0.25, (b) 0.5, and 1 (c), we inspect the SAM line pattern (Figure 3). N_d is fixed to be 30 for all the cases. As the tip scans faster, the width of SAM line becomes narrow. Note also the SAM line pattern gets rougher in its periphery as the tip scans slower, even though the same N_d value was used for all the cases in Figure 3.

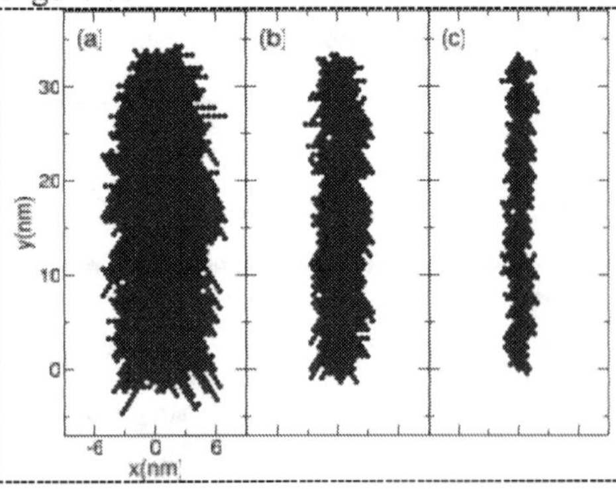

Figure 3. The SAM lines patterned by using the tip scan speeds v^*s of 0.25 (a), 0.5 (b), and 1 (c). The directional coherence length (N_d) and the molecular dripping rate (n^*) are 30 and 5, respectively.

4. Conclusion

The growth of a SAM pattern in dip-pen nanolithography was simulated by using a simple random walk model. In this model, a SAM pattern primarily grows via the serial pushing of molecules initiated by molecules dropping from an AFM tip (source). This serial pushing is taken to have a finite length for its direction called directional coherence length. To model various SAM patterns made of lines, we have implemented the movement of tip in the present simulation. For these line-based SAM patterns, the tip scan speed plays a major role in determining the outcome of DPN. As the tip scan speed increases, the SAM patterns become narrower and less rugged in their peripheries. The effect of directional coherence length is relatively small compared to the case of a tip fixed in position.

References

[1] Piner R D, Zhu J, Xu F, Hong S and Mirkin C A 1999 "Dip-Pen" Nanolithography *Science* **283** 661-3
[2] Mirkin C A 2007 The Power of the Pen: Development of Massively Parallel Dip-Pen Nanolithography *ACS Nano* **1** 79-83
[3] Salaita K, Wang Y and Mirkin C A 2007 Applications of dip-pen nanolithography *Nat Nano* **2** 145-55
[4] Kim H, Saha L C, Saha J K and Jang J 2010 Molecular simulation of the water meniscus in dip-pen nanolithography *Scanning* **32** 2-8
[5] Rozhok S, Piner R and Mirkin C A 2002 Dip-Pen Nanolithography:□ What Controls Ink Transport? *The Journal of Physical Chemistry B* **107** 751-7
[6] Giam L R, Wang Y and Mirkin C A 2009 Nanoscale Molecular Transport: The Case of Dip-Pen Nanolithography† *The Journal of Physical Chemistry A* **113** 3779-82
[7] Heo D M, Yang M, Kim H, Saha L C and Jang J 2009 Tip Dependence of the Self-Assembly in Dip-Pen Nanolithography *J. Phys. Chem. C* **113** 13813-8
[8] Jang J, Hong S, Schatz G C and Ratner M A 2001 Self-assembly of ink molecules in dip-pen nanolithography: A diffusion model *J. Chem. Phys.* **115** 2721-9
[9] Kim H and Jang J 2009 Serial Pushing Model for the Self-Assembly in Dip-Pen Nanolithography *J. Phys. Chem. A* **113** 4313-9
[10] Kim H, Schatz G C and Jang J 2010 Simplistic Model for the Dendritic Growth of a Monolayer in Dip Pen Nanolithography *J. Phys. Chem. C* **114** 1922-7
[11] Manandhar P, Jang J, Schatz G C, Ratner M A and Hong S 2003 Anomalous Surface Diffusion in Nanoscale Direct Deposition Processes *Physical Review Letters* **90** 115505
[12] Rivas-Cardona J A and Banerjee D 2007 Microfluidic device for delivery of multiple inks for dip pen nanolithography *Journal of Micro/Nanolithography, MEMS, and MOEMS* **6** 033004-
[13] Delamarche E, Schmid H, Bietsch A, Larsen N B, Rothuizen H, Michel B and Biebuyck H 1998 Transport Mechanisms of Alkanethiols during Microcontact Printing on Gold *The Journal of Physical Chemistry B* **102** 3324-34

Acknowledgments
This study was supported by National Research Foundation Grants funded by the Korean Government (MEST) (No. 2011-0027445 and No. 2011-0027696). J.J. wishes to thank the Korea Institute of Science and Technology Information for the use of the PLSI supercomputing resources.

Dimer adsorption on square nanotubes with first- and second-neighbor interactions

A J Phares[1] and D W Grumbine Jr[2]

[1] Department of Physics, Villanova University, Mendel Science Center, Villanova, PA 19085-1699, USA

[2] Department of Physics, St. Vincent College, 300 Fraser Purchase Rd., Latrobe, PA 15650-4580, USA

E-mail: alain.phares@villanova.edu

Abstract. The thermodynamic properties of dimer adsorption on square M sites wide terraces, with first- and second-neighbor adsorbate-adsorbate interaction energies, were recently numerically obtained using an extension of the transfer matrix technique developed by the authors in 1993. The rank of the transfer matrix increases rapidly with size M, and required the development of an efficient algorithm for its construction at any temperature of the system, and for all possible values of the adsorbate-adsorbate and adsorbate-substrate interaction energies. This article presents the non-trivial generalization of this algorithm for the study of dimer adsorption problems on a square lattice nanotube surface, with M sites on its circular normal section.

1. Introduction: Problem setup

Molecules in a medium (a gas or a solution) are adsorbed on a square-lattice nanotube surface (the substrate) as dimers bound to two first-neighbor sites, at thermodynamic equilibrium, with no overlapping. Pairwise interactions between dimers include first- and second-neighbors. The nanotube is viewed as vertical, assembled by wrapping on a vertical cylinder a square-lattice terrace, with M sites in its horizontal width, such that the 1st site on the right is a first-neighbor with the Mth site on the left. The parameters are: the absolute temperature T; the chemical potential energy μ of the molecules in the medium; the adsorbate-substrate interaction energy V_0, and the adsorbate-adsorbate first- and second-neighbor energies, V and W . There are four relevant activities, with k_B indicating Boltzmann's constant:

$$x = \exp[(\mu + V_0)/k_BT] \ , \ y = \exp[V_0/k_BT] \ , \ z = \exp[W/k_BT] \ . \tag{1}$$

Statistical thermodynamics provides all of the adsorption properties from the knowledge of the partition function Z associated with the system, which depends in this case on the three activities identified above.

Our approach is to construct a transfer matrix that depends on the activities, and whose eigenvalues provide the partition function of the system, from which all of its thermodynamic properties can then be computed. The set, $\{\theta_0, \theta, \beta\}$, of the statistical averages of the coverage of the surface, θ_0, of the numbers per site of first-neighbors, θ, and second-neighbors, β, describe the occupational state of the system. Phases correspond to dimers locked into a given distribution,

Published under licence by IOP Publishing Ltd

$\{\theta_0, \theta, \beta\}$, in a perfect order (zero entropy) or partial order (non-zero entropy), over a relatively wide range of values of μ. Phases gradually disappear with increasing temperature. All possible phases of a given adsorption system occur at relatively low temperatures. The temperature-energy phase space associated with the adsorption system is 4-dimensional with variables $(\mu+V_0)$, V, W, and T.

The transfer matrix method provides all of the phases that exist in the temperature-energy phase diagram which may be used as a guide in analyzing experimental data. Phases observed experimentally correspond to a specific temperature-energy region of the phase diagram, leading to information on adsorbate-adsorbate and adsorbate-substrate interaction energies, and on the range of values of T and μ for which the phases may be observed. The phase diagram also provides the conditions under which transitions between phases may occur.

2. Transfer matrix

The energy of the system relies on the count of the number of dimers on the lattice, the number of dimer ends that are first- and second-neighbors. The counting proceeds along each strip of M horizontal sites, and these are added for all of the strips in the vertical length of the nanotube. A dimer is counted as a member of a strip when both of its ends are in that strip, or when its other end is in the strip above (if below, it belongs to the lower strip). The numbers of first- and second-neighbors belonging to a given strip are those between dimers both members of that strip, or one belonging to that strip and the other to the strip below.

There is a one-to-one correspondence between a matrix element of the transfer matrix and the states of two consecutive strips, referred to as "lower" and "upper" strips [1, 2, 3]. The row-index of an element is associated with the state of the lower strip, and the column-index with the state of the upper strip. A vertical dimer belonging to the lower strip cannot have its upper end in the upper strip if that site is occupied by one end of a dimer since overlapping is excluded. A matrix element is zero when overlapping between the states of the lower and upper strips occurs; otherwise it is of the form $x^a y^b z^c$, where a, b, and c are the numbers of dimers, and the numbers of first- and second-neighbors, respectively, following the counting procedure described above.

A site may be in one of five states indicated by a number α. This number is (0) if the site is vacant. If the site is occupied by one end of a dimer, its state is indicated as (1) when its other end is in the site below, (2) when its other end is above, (3) when its other end is to its left, or (4) when to its right. The problem consists of locating the zero elements and, for every non-zero element, of determining exponents a, b, and c.

The state of the ith site in a strip of M sites is denoted α_i, randomly picking one of the sites to be the first. Let $D(M)$ be the number of states of the strip, with no dimer with its ends in the 1st and M^{th} sites; thus excluding $(\alpha_1, \alpha_M) = (4, 3)$, which is the case for a terrace. In each of the cases $\alpha_i = (0), (1)$, or (2), the number of states is $D(M-1)$; and, when $\alpha_i = (3)$, it is $D(M-2)$ since $(\alpha_2, \alpha_{M-1}) = (4,3)$ is excluded. Therefore, $D(M) = 3D(M-1) + D(M-2)$. The number of states with $(\alpha_1, \alpha_M) = (4, 3)$ is $D(M-2)$. Thus the total number, $R(M)$, of states is given by the recursive relations:

$$R(M) = 3D(M-1) + 2D(M-2) \; ; \; D(M) = 3D(M-1) + D(M-2) \; . \tag{2}$$

For $M = 10$, $R(10) = 154451$ while for a terrace $D(10) = 141481$.

3. Recursive construction of the transfer matrix

The row-indices of the transfer matrix $\mathbf{NT}(M)$ are associated with an state of the lower strip given by the list $\{\alpha_i\}$; similarly, the column-indices of $\mathbf{NT}(M)$ are associated with an state of the upper strip given by the list $\{\beta_i\}$. Thus a (row, column)-matrix element is indicated as the $(\{\alpha_i\}, \{\beta_i\})$-matrix element.

Block-matrices are extracted from the transfer matrix $\mathbf{NT}(M)$ by selecting a number of $(\{\alpha_i\}, \{\beta_i\})$-matrix elements. The selection is based on considering the states in the lower and upper strips of M sites from $i = 1$ to $(M - k)$ in a given state, allowing all of the remaining k sites to be in all of the possible states. In this manner, one identifies, for every k, eight families of block-matrices indicated as $\mathbf{T}$, $\mathbf{P}$, $\mathbf{K}$, $\mathbf{L}$, $\mathbf{T}'$, $\mathbf{P}'$, $\mathbf{K}'$, and $\mathbf{L}'$. The unprimed block-matrices exclude the first sites in the lower and upper strips to be both in the state (4); and the primed block-matrices correspond to the case where both of these sites are in state (4). There are 6 members in each of the unprimed families and 2 members in each of the primed families, for a total of 32 block-matrices for every k, indicated as $\mathbf{T}(n + 2m; k)$, $\mathbf{P}(n + 2m; k)$, $\mathbf{K}(n + 2m; k)$, $\mathbf{L}(n + 2m; k)$, $\mathbf{T}'(n; k)$, $\mathbf{P}'(n; k)$, $\mathbf{K}'(n; k)$, and $\mathbf{L}'(n; k)$, where n may take on the values 1 or 2, and m the values 1, 2, or 3. The unprimed matrices with $m = 0$ (a total of 8) are the only ones required to construct the transfer matrix for terraces.

Using the standard Kronecker-delta notation, matrices $\mathbf{T}(n+2m; k)$, $\mathbf{P}(n+2m; k)$, $\mathbf{T}'(n; k)$, and $\mathbf{P}'(n; k)$ are $D(k - \delta_{2m}) \times D(k - \delta_{1m})$ matrices; matrices $\mathbf{K}(n + 2m; k)$ and $\mathbf{K}'(n; k)$ are $D(k-\delta_{2m}) \times D(k-1-\delta_{1m})$; and matrices $\mathbf{L}(n+2m; k)$, and $\mathbf{L}'(n; k)$ are $D(k-1-\delta_{2m}) \times D(k-\delta_{1m})$ matrices. One can then show that the unprimed (primed) matrices for a given k are obtained from the the unprimed (primed) matrices for lower values of k, thus setting up a recursive construction. The recursion relations for the unprimed matrices are found as:

$$\mathbf{T}(n{+}2m; k) = \begin{pmatrix} \mathbf{T}(1+2m; k-1) & 0 & \begin{matrix} xz^{n-1}\times \\ \mathbf{P}(1+2m; k-1) \end{matrix} & \begin{matrix} xz^{n-1}\times \\ \mathbf{K}(1+2m; k-1) \end{matrix} \\ \mathbf{T}(2+2n; k-1) & 0 & \begin{matrix} xz^{n-1}\times \\ \mathbf{P}(2+2m; k-1) \end{matrix} & \begin{matrix} xyz^{n-1}\times \\ \mathbf{K}(2+2m; k-1) \end{matrix} \\ 0 & \begin{matrix} z^{n-1}\times \\ \mathbf{P}(2+2m; k-1) \end{matrix} & 0 & 0 \\ \mathbf{L}(1+2m; k-1) & 0 & \begin{matrix} xyz^{n-1}\times \\ \mathbf{L}(1+2m; k-1) \end{matrix} & \begin{matrix} xy^2z^{n+1}\times \\ \mathbf{k}(2+2m; k-2) \end{matrix} \end{pmatrix} \tag{3}$$

$$\mathbf{P}(n{+}2m; k) = \begin{pmatrix} \mathbf{T}(1+2m; k-1) & 0 & \begin{matrix} xyz^{n-1}\times \\ \mathbf{P}(1+2m; k-1) \end{matrix} & \begin{matrix} xyz^{n-1}\times \\ \mathbf{K}(1+2m; k-1) \end{matrix} \\ z\mathbf{T}(2+2m; k-1) & 0 & \begin{matrix} xy^2z^{n}\times \\ \mathbf{P}(2+2m; k-1) \end{matrix} & \begin{matrix} xy^2z^{n}\times \\ \mathbf{K}(2+2m; k-1) \end{matrix} \\ 0 & \begin{matrix} yz^{n}\times \\ \mathbf{P}(2+2m; k-1) \end{matrix} & 0 & 0 \\ z\mathbf{L}(1+2m; k-1) & 0 & \begin{matrix} xy^2z^{n}\times \\ \mathbf{L}(1+2m; k-1) \end{matrix} & \begin{matrix} xy^3z^{n+2}\times \\ \mathbf{P}(2+2m; k-2) \end{matrix} \end{pmatrix} \tag{4}$$

$$\mathbf{K}(n + 2m; k) = \begin{pmatrix} z^{n-1}\mathbf{P}(1+2m; k-1) \\ yz^{n}\mathbf{P}(2+2m; k-1) \\ 0 \\ yz^{n}\mathbf{L}(2+2m; k-1) \end{pmatrix} \tag{5}$$

$$\mathbf{L}(n + 2m; k) = \begin{pmatrix} z^{n-1}\mathbf{T}(2+2m; k-1) & 0 & xy^nz^n\mathbf{P}(2+2m; k-1) & xy^nz^n\mathbf{K}(2+2m; k-1) \end{pmatrix} \tag{6}$$

There are conditions that must be met at the M-end of the lower and upper strips, when $k = 0$ or 1. These initial conditions for the $\mathbf{T}$- and $\mathbf{P}$-families are:

$$\mathbf{T}(n; 0) = \mathbf{P}(n; 0) = 1 \; ; \qquad \mathbf{T}(n+2; 1) = \begin{pmatrix} z^{n-1} \\ yz^{n-1} \\ 0 \end{pmatrix} \; ; \qquad \mathbf{P}(n+2) = \begin{pmatrix} yz^{n-1} \\ y^2z^{n-1} \\ 0 \end{pmatrix} \; ;$$
$$\mathbf{T}(n+4; 1) = \begin{pmatrix} 1 & 0 & xyz^{n-1} \end{pmatrix} \; ; \quad \mathbf{P}(n+4; 1) = \begin{pmatrix} z & 0 & xy^2z^n \end{pmatrix} . \tag{7}$$

There are no $\mathbf{K}(n+2m;0)$ or $\mathbf{L}(n+2m;0)$ matrices, and $\mathbf{K}(n+4;1) = yz^n$; $\mathbf{L}(n+2;1) = y^n z^n$. More involved conditions must be met on the other end of the two strips to account for first- and second-neighbors across the boundary between the 1st and Mth sites.

There are two members in each of the primed families of matrices, for which $m = 0$ and $n = 1$ or 2. The recursion relations among the primed matrices are the same as those among their unprimed counterparts (where m is set to zero). Here the 1st and Mth sites are restricted to be in the states $(\alpha_1, \alpha_M) = (\beta_1, \beta_M) = (4, 3)$. The only conditions to be satisfied are at $k = 0$ or 1: there are no $\mathbf{K}'(n;0)$ or $\mathbf{L}'(n;0)$, and $\mathbf{T}'(n;0) = z^{n-1}$, $\mathbf{P}'(n;0) = yz^n$, and

$$\mathbf{K}'(n;1) = \begin{pmatrix} yz^n \\ y^2 z^{n-1} \\ 0 \end{pmatrix} , \quad \mathbf{L}'(n;1) = \begin{pmatrix} z^n & 0 & xy^{n+1}z^{n+2} \end{pmatrix} . \tag{8}$$

The transfer matrix $\mathbf{NT}(M)$ is divided into 5×5 block-matrices. The row-indices (or column-indices) of a block-matrix correspond to all of the possible states of the lower strip (or upper strip) restricting α_1 (or β_1) to be in this order: (0), (1), (2), (3), or (4), leading to

$$\mathbf{NT}(M) = \begin{pmatrix} \mathbf{T}(1;M-1) & 0 & x\mathbf{P}(1;M-1) & x\mathbf{K}(1;M-1) & x\mathbf{P}(3;M-2) \\ \mathbf{T}(2;M-1) & 0 & xy\mathbf{P}(2;M-1) & xy\mathbf{K}(2;M-1) & xyz\mathbf{P}(4;M-2) \\ 0 & \mathbf{P}(2;M-1) & 0 & 0 & 0 \\ \mathbf{L}(1;M-1) & 0 & xy\mathbf{L}(2;M-1) & xy^2z^2\mathbf{P}(2;M-1) & xyz\mathbf{L}(4;M-1) \\ \mathbf{T}(6;M-1) & 0 & xyz\mathbf{P}(6;M-1) & xyz\mathbf{K}(6;M-1) & xy^2z^2\mathbf{Q}(M-1) \end{pmatrix} \tag{9}$$

The $\mathbf{Q}$-block in the lower right corner of $\mathbf{NT}(M)$ is a square matrix of rank $D(M-2)$ and is obtained recursively in terms of the families of primed matrices as follows:

$$\mathbf{Q}(M-2) = \begin{pmatrix} \mathbf{T}'(1;M-3) & 0 & xyz\times \mathbf{P}'(1;M-3) & xyz\times \mathbf{K}'(1;M-3) \\ z\mathbf{T}'(2;M-3) & 0 & xy^2z^2\times \mathbf{P}'(2;M-3) & xy^2z^2\times \mathbf{K}'(2;M-3) \\ 0 & yz^2\times \mathbf{P}'(2;M-3) & 0 & 0 \\ z\mathbf{L}'(1;M-1) & 0 & xy^2z^2\times \mathbf{L}'(2;M-2) & xy^3z^4\times \mathbf{P}'(2;M-4) \end{pmatrix} \tag{10}$$

The algorithm that results from this mathematical construction makes it possible to study dimer adsorption on nanotube surfaces with increasing diameter.

4. Discussion

We have written C language code to perform this algorithm which is similar to the code we wrote for dimers adsorbed on square terraces [1, 2]. The results for nanotubes will be valuable for comparison with experimental results. We plan to extend the algorithm for co-adsorption of two species in a manner similar co-adsorption of several monomer species [3].

References

[1] Phares A J, Wunderlich F J, Curley J D, and Grumbine Jr D W 1993 *Journal of Physics A* **26** 6847
[2] Phares A J, Pasinetti P M, Grumbine Jr D W, and Wunderlich F J 2011 *Physica B: Condensed Matter* **406** 1096
[3] Phares A J 2011 *Langmuir* **27** 8105

Dynamics of irradiation: from molecules to nano-objects and from material science to biology

P. Wopperer[1,2], P. M. Dinh[1], P. G. Reinhard[2], E. Suraud[1]

1) Laboratoire de Physique Théorique, Université P. Sabatier, 118 Rte de Narbonne, F-31062 Toulouse cedex, France
2) Institut für Theoretische Physik II, Universität Erlangen-Nürnberg, Staudstr. 7, D-91058 Erlangen, Germany

E-mail: `suraud@irsamc.ups-tlse.fr`

Abstract. We give a brief review of theoretical and computational tools developed for the description and analysis of irradiation dynamics of cluster and molecules. We illustrate the capabilities of the method on the demanding example of C_{60} irradiated by various laser fields.

1. Introduction

Irradiation of matter constitutes a key tool in physics, chemistry, and biology, for analyzing structural and dynamical properties of atoms, molecules, clusters and bulk material. Lasers offer here an especially flexible and powerful instrument [1] with enormous technological progress reached in the manipulation of laser light. Collisions with charged projectiles are also used as sources of short electromagnetic pulses but require access to dedicated facilities. Generally speaking radiation damage [2] is of high current interest for example in connection with biological tissues ("human controlled" as in a medical context or "natural" when referring to earth or space radiations). In all cases, the immediate electronic response of the irradiated system plays a key role as the doorway to all subsequent dynamical scenarios. The basic optical response corresponding to electronic oscillations, defines the coupling between irradiation and matter from gentle to strong perturbations [3]. Equally important in energetic cases is electron transport, particularly electron emission. As important examples one can cite the many studies on irradiation of clusters by short and intense laser pulses [3], especially through energy (Photo Electron Spectra, PES [4]) and, more recently, angle-resolved [5] distributions of emitted electrons (Photo Angular Distributions, PAD). Secondary electrons in DNA damage [6] also provide a remarkable example. The long term societal importance of the latter studies is obvious, especially in relation to oncology. The microscopic understanding of underlying mechanisms is still in its infancy and requires dedicated efforts, particularly from a modeling point of view in order to better understand dedicated ongoing experiments.

The analysis of electronic emission from a finite system is thus a key issue in a wide range of physical, chemical and biological processes. Electrons are usually the first constituents to respond to an electromagnetic pulse. Strong excitations lead to immediate ionization of the system, often with dramatic long-time effects (dissociation, Coulomb explosion, ...). It implies electronic transport and possible indirect effects on neighboring species. Dissociative Electron Attachment (DEA) provides a typical example in biological systems where primarily emitted

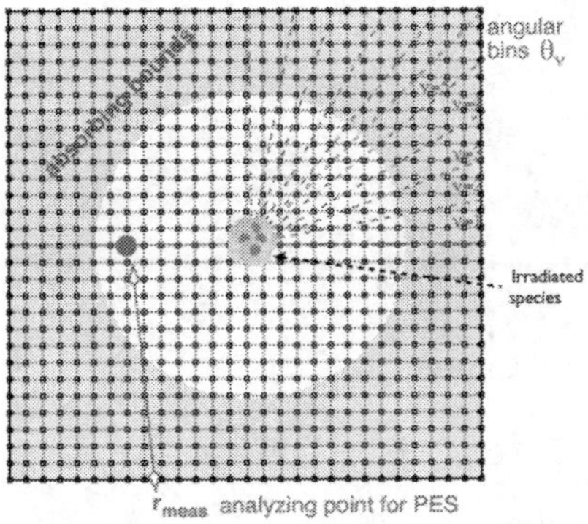

Figure 1. Schematic view of the computation of observables from electron emission, on the basis of a Cartesian coordinate space grid. The absorbing layer is indicated by yellow shading. The PAD is collected in angular bins as indicated. The PES are deduced from collecting phase information at measuring points $\mathbf{r}_{meas}$, placed short before the absorbing bounds.

electrons are attached to target biological molecules which finally leads to the break up of the latter. Emitted electrons may also ultimately provide a valuable insight into reaction pathways, for example in terms of PES and PAD. Electrons thus provide the first response at short time scales, which is more or less quickly coupled to other degrees of freedom. They are finally useful probes along the whole dynamical process, especially when emitted and properly recorded. We shall illustrate the capabilities of our approach on such typical electronic signals.

The theoretical description of irradiation scenarios with electronic emission has made remarkable progress in the past [3] and, at the same time raises challenging questions for further developments. Electron response was analyzed since long [3, 7, 8] especially in the linear domain, leading to a large corpus of theoretical investigations of the optical response of clusters and molecules [3, 8, 9]. But only very few of these go as far as to consider in a routine way non-linear excitations leading to electronic emission [7]. We have developed over the last 15 years elaborate dynamical tools to simulate such non-linear situations in simple metal clusters [7], in metal clusters in contact with an environment [10] and in organic systems [11].

2. Formal and computational framework

The formal framework is Time Dependent Density Functional Theory (TDDFT, [12]) in real time, for the description of electronic dynamics, coupled to classical Molecular Dynamics (MD) for ionic degrees of freedom. The TDDFT is used at the level of the Time-Dependent Local-Density Approximation (TDLDA), augmented with an averaged self-interaction correction (SIC) and absorbing boundary conditions. Coupling to ions is mediated by local pseudo-potentials following standard parametrizations thereof [7]. Laser excitation is described by an external classical laser field with a finite temporal envelope. We can also consider collisions with bypassing highly charged ions. The latters are then described as classical particles (when highly charged) or dressed with quantum electrons when one considers neutral or little charged projectiles [11]. We use standard coordinate-space techniques as outlined in detail in [7, 8] to represent electronic wave functions. Ions are treated as classical particles following standard MD propagation.

In order to illustrate the capabilities of this method we consider examples of electronic signals. The analysis of electron emission requires absorbing boundary conditions, as indicated in Figure 1. Absorption is performed after each time step. After having performed a standard time step from t to $t+\delta t$ we apply a mask function $\mathcal{M}$ (active in an absorbing margin $R_{\rm in} < |\mathbf{r}| < R_{\rm out}$) to the preliminary single electron wavefunctions ϕ_α, which removes gradually any flow outwards, see [14] for details. The PES delivers the kinetic energies of emitted electrons and can be evaluated

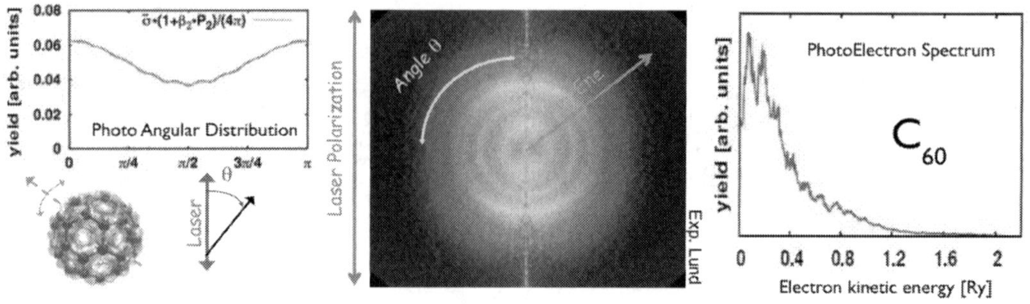

Figure 2. Example of application for the analysis of PES and PAD in C_{60}. An example of experimental PESPAD spectrum is shown in the middle panel, data taken from [17]. A typical PES is shown in upper right panel and a PAD is shown in upper left panel with a small scheme detailing the orientation problem (left lower panel).

for each state separately from the phase oscillations of the single-electron wavefunction at an analyzing point $\mathbf{r}_{\text{meas}}$ near the absorbing bounds, see Figure 1 and [15]. The protocol $\phi_\alpha(\mathbf{r}_{\text{meas}}, t)$ is then Fourier transformed to $\tilde{\phi}_\alpha(\mathbf{r}_{\text{meas}}, \omega)$ and the spectrum $\mathcal{P}(\omega) \propto |\tilde{\phi}_\alpha(\mathbf{r}_{\text{meas}}, \omega)|^2$ is translated to a kinetic-energy spectrum of emitted electrons by identifying $\hbar\omega = \varepsilon_{\text{kin}}$, thanks to the absorbing bounds which leave only outgoing waves in its vicinity. The angular distributions (PAD) are evaluated in angular bins (Figure 1) by collecting all probability removed by the absorption step and accumulating it in the bin to which $\mathbf{r}$ belongs. Again that can be done for each wavefunction separately, which finally delivers an energy and angular resolved analysis of electronic emission. Note that the calculation of PAD requires extra caution due to the compulsory angular averaging (orientation of cluster/molecule with respect to laser polarization), but this can be handled rather efficiently [16].

3. Example of application

We illustrate our approach on the case of C_{60}, which consists of 60 C atoms arranged in a highly symmetric configuration (fullerene, [18]) as illustrated in lower left panel of Figure 2. The Carbon $1s$ electrons are considered as core electrons and accounted for by the pseudopotential. The four $1p$ electronsshell of each C atom provide the 240 dynamically active electrons of C_{60}, with a high ionization potential (IP) of 7.58 eV . The spectral span of the occupied valence states is also rather large ranging over about 20 eV from -27 eV up to the IP [17]. To resolve the s.p. spectrum by a one-photon process, one takes a photon with frequency just above the lowest bound state at $\omega_{\text{las}} = 33$ eV. The associated PAD is shown in the upper left panel of of Figure 2. The PES is illustrated in the upper right panel, this time for a photon frequency of 1.55 eV, deeply inside the multi-photon regime of emission. Comparison with experimental data, are extremely encouraging, leading to qualitative and even quantitative general agreement.

Computations of PES and PAD on such a large system as C_{60} represent an enormous task requiring a sizable computational effort. These computations are feasible as the underlying developed code package is highly parallelized, each processor accounting for a small number of single electron wave functions. Most of the electronic time propagation can be indeed performed processor per processor, the major communication consisting in exchanging the electron density which is computed as a sum over the wave function norms, thus implying full exchange between processors. Still, there remains some space for improvement in terms of computational performance. First attempts in that direction concern the use of more efficient FFT packages such as FFTW. The use of GPU at some places also provides interesting possibilities of speedup. A concerted program of optimization of our code package is underway implying several universities in Europe and China. After less than one year the first results (FFT, GPU) have already allowed significant gain in computational cost, still without exploiting all possible applications of GPU.

4. Conclusions

We have briefly reviewed the theoretical description of irradiation dynamics focusing on PES and PAD. Numerics is based on a Cartesian coordinate-space grids with absorbing boundary. PAD are simply computed from gathering the lost particles in angular segments while PES are obtained from recording the time evolution of each single electron wavefunction at a few selected measuring points near the absorbing bounds. We have been able to perform calculations on the large C_{60} cluster. The results for C_{60} are particularly interesting as several measurements on this large cluster are available and even more are to be expected in near future. Accordingly, more calculations for C_{60} are underway. We can cite as example the impact of plasmon frequency on PAD and PES characteristics, following either laser irradiation or collisions by a highly charged ion. Improving computational performances is also an important aspect which gathers joined efforts from several groups. The next envisioned step is to couple the fully microscopic (quantum) description (as discussed here) to large scale MD descriptions of an environment thus providing an original multi-scale description of irradiation dynamics and of its impact on environment.

This work was supported by the Deutsche Forschungsgemeinschaft, project RE 322/10-1, Institut Universitaire de France, and Agence Nationale de la Recherche.

References

[1] U. Keller, 2003, Nature, **424**, 831 (2003)
[2] K. Tokesi et al, Proceedings RADAM08, AIP Proceedings, **1080** (2008)
[3] Th. Fennel et al, Rev. Mod. **82**, 1793 (2010)
[4] E. E. B. Campbell et al, Phys. Rev. Lett. **84**, 2128 (2000)
[5] C. Bartels et al, Science **323**, 1323 (2009)
[6] B. Boudaiffa, P. Cloutier, D. Hunting, M. A. Huels, and L. Sanche, Science **287**, 1658 (2000)
[7] F. Calvayrac, P.-G. Reinhard, E. Suraud, and C. A. Ullrich, Phys. Rep. **337**, 493 (2000)
[8] P.-G. Reinhard and E. Suraud, Introduction to Cluster Dynamics, Wiley (New York) 2003
[9] http://www.tddft.org/programs/octopus/
[10] P. M. Dinh, P.-G. Reinhard, and E. Suraud, Phys. Rep. **485**, 43 (2010)
[11] Z. P. Wang et al, Intern. J. Quant. Chem. **111**, 480 (2010)
[12] M. A. Marques et al, Lect. notes in Physics, **837** (2012)
[13] S. Kümmel, M. Brack, P.G. Reinhard, Euro. Phys. J. D **9**, 149 (1999)
[14] P.–G. Reinhard et al, Phys. Rev. E, **73**, 036709 (2006)
[15] A. Pohl, P.–G. Reinhard, and E. Suraud, Phys. Rev. Lett. , **84**, 5090 (2000).
[16] P. Wopperer et al Phys. Lett. A, **375**, 39 (2010)
[17] M. Kjellberg et al, Phys. Rev. A **81**, 023202 (2010)
[18] W. I. F. David et al, Nature **353**, 147 (1991)

Visualization of quantum electronic and lattice fluctuations by using the resonating Hartree-Fock method

Norikazu Tomita[1], Shuji Watanabe[1], Keisuke Shida[1], H. Gomi[2], and A. Takahashi[2]

[1]Yamagata University, 1-4-12 Kojirakawa, Yamagata, 990-8560, Japan
[2]Nagoya Institute of Technology, Gokiso, Showa, Nagoya, 488-8555, Japan

E-mail: tomita@sci.kj.yamagata-u.ac.jp

Abstract. Large quantum fluctuations in the strongly correlated electron systems as well as electron-phonon coupled systems are visualized by the resonating Hartree-Fock (HF) method). We show that ,in the two-dimensional Hubbard model, doped holes form topological defects called polarons. It is also shown that the quantum fluctuations drastically change the lattice structures in the Su-Schrieffer-Heeger model.

1. Introduction

Visualization of large quantum fluctuations has been a challenging issue in not only condensed matter physics but also nuclear physics. Recent progress in computer science has made it possible to describe the many-body effects efficiently. Density matrix renormalization group (DMRG)[1], quantum[2] and variational[3] Monte Carlo simulations are well established numerical methods. Furthermore, we have even the exact solution, such as Lieb-Wu solution for the one-dimensional Hubbard model. In spite of such progress or even the exact solutions, we cannot usually see physics behind the many-body effects intuitively.

In this paper, we introduce an alternative numerical approach, called a resonating Hartree-Fock (HF) method, in which a many-fermion wave function is constructed by superposition of non-orthogonal Slater determinants (S-dets). The orbitals in all the S-dets, as well as the superposition coefficients, are simultaneously optimized. As we mention details in the following sections, the resonating HF method describes the many-body effects efficiently. In addition, this method can make it possible to reveal physics behind the complicated many-body effects by analyzing the structures of the S-dets generating the resonating HF wave function. We apply the method to the two-dimensional (2D) Hubbard model on a square lattice. More recently, we have developed to describe the quantum lattice fluctuations by superposition of coherent states of phonons. By using this approach, we investigate the quantum lattice fluctuations in Su-Schrieffer-Heeger (SSH) model. The paper is organized as follows, In §2, we introduce the resonating HF method and coherent state representation of phonons. Results and discussions are given in §3. Brief summary is given in §4.

2. Method and model

Generally speaking, a many-fermion wave function is exactly constructed by superposition of S-dets. In spite of recent progress in computer ability, however,it is impossible to obtain the exact wave functions for more than 30 interacting fermions. When we approximate the exact wave function by a single S-det, it is called a HF approximation. To include the large correlations, the resonating HF method has been developed, where the wave function is constructed by superposition of non-orthogonal S-dets.[4, 5] Non-orthogonality is important because one S-det includes the complete excitation effects from other S-dets automatically. As a result, large correlation effects can be included quite efficiently. A wave function is explicitly given by

$$|\Psi>=\sum_{n=1}^{N_S} C_n \sum_G P^G|\psi_n>, \tag{1}$$

where N_S denotes the number of S-dets generating the resonating HF wave function. The molecular orbitals of all the constituting S-dets $|\psi_n>$, as well as the superposition coefficients C_n, are variationally determined. To incorporate the correlation effects from symmetry broken states, the DODS (different orbitals for different spins)-type S-dets are employed in the resonating HF method. Therefore, we adopt symmetry projections for each constituting S-det to recover the original symmetry of the system, which are symbolically denoted by P^G in eq. (1).

We should note that one of the most important features of the resonating HF method is that we can visualize quantum fluctuations by analyzing the structures of S-dets. In the next section, the method is applied to the 2D Hubbard model, whose Hamiltonian is given by

$$H=-\sum_{<i,j>\sigma}^{N} t(i,j)(a_{i,\sigma}^{\dagger}a_{j,\sigma}+a_{j,\sigma}^{\dagger}a_{i,\sigma})+U\sum_{i}^{N} n_{i,\uparrow}n_{i,\downarrow}, \tag{2}$$

where $t(i,j)=t$ for the electron hopping to the nearest-neighbor sites, while $t(i,j)=t'$ for the hopping to the next nearest-neighbor sites. U and N represent an on-site Coulomb repulsion and a system size, respectively.

Recently, we have extended the method to include the quantum lattice fluctuation effects. A coherent state representation of phonons is used for this purpose, which is given by

$$|z>=e^{-\frac{1}{2}\mathbf{z}^2}e^{\mathbf{z}\cdot\mathbf{b}^{\dagger}}|0>, \tag{3}$$

where $\mathbf{b}^{\dagger}=(b_1^{\dagger},b_2^{\dagger},...)$ represents the set of creation operators of phonons. $\mathbf{z}=(z_1,z_2,....)$ are determined by the lattice structure, such as

$$z_k=\sqrt{\frac{m\omega_k}{2N\hbar}}\sum_{n=0}^{N-1}<z|q_n|z>\exp\{-\frac{2\pi ink}{N}\}, \tag{4}$$

where q_n denotes the lattice displacement from the equilibrium position. Thus, the extended form of the resonating HF wave function is written by

$$|\Psi>=\sum_{n=1}^{N_S} C_n \sum_G P^G|\psi_n>|z_n>. \tag{5}$$

Ideally, we should optimize these z_n's of the coherent states, and in fact, we can carry out the optimization when we employ a single coherent state. However, at the present stage, it is complicated to optimize plural coherent states. Therefore, in the following calculations, we put the coherent states by hand. We choose the coherent states from the optimized result for a single coherent state. On the other hand, the S-dets are optimized by the resonating HF procedures. In the next section, we apply the method to the Su-Schrieffer-Heeger (SSH) model for the 1D electron-phonon coupled system. SSH Hamiltonian is given as

$$H = -t\sum_{l\sigma}^{N}\{1+\alpha(q_{l+1}-q_l)\}(a_{l,\sigma}^{\dagger}a_{l+1,\sigma}+a_{l+1,\sigma}^{\dagger}a_{l,\sigma})+\sum_{l}^{N}\{\frac{1}{2m}p_l^2+\frac{k}{2}(q_{l+1}-q_l)^2\}. \quad (6)$$

In this model, the lattice displacement affects the electron transfer of the first term. We will investigate the quantum lattice fluctuation effects on this model beyond the adiabatic and classical approximation.

3. Results and Discussion

One of the most interesting problems in the copper-oxides is what carriers are. To answer the question, we calculate the resonating HF wave function for the 2D doped Hubbard model with the parameters of $U/t = 8$ and $t'/t = -0.32$. Figure 1 shows the structures of a typical Slater determinants generating the resonating HF wave function for $N = 16 \times 16$ and $N_e = 208$. We can see many dips, where the anti-ferromagnetic ordering is suppressed and the charge density wave ordering is raised. These are kinds of topological defects and called polarons. As also seen in Fig.1, polarons have net charges and spins. Other S-dets, which are not shown here to save space, also have many polarons with different relative configurations. Our results indicate that the doped holes become polarons in the 2D Hubbard model.[6] Another interesting question is whether an attractive interaction works between polarons. Figure 2 shows the HF and projected HF energies of the states having two polarons. We change distance between polarons, which is shown as a horizontal axis. For the projected HF state, symmetry projections are adopted to the HF S-dets and no further orbital optimizations are carried out. From Fig.2, we can see that the attractive interaction works between polarons, when we partially incorporate the electron correlation effects by adopting the symmetry projections. Thus, the electron correlation effects cause the attractive interaction between polarons. Our results strongly suggest that doped holes become polarons and they can be carriers of superconductors.

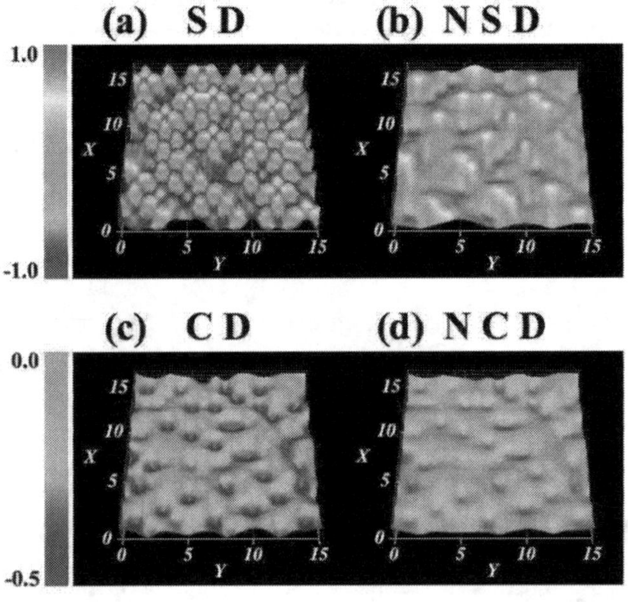

Figure 1. Structures of a typical S-det for $N = 256$ and $N_e = 208$.

Next, we investigate the lattice fluctuation effects in the SSH model. We show in Fig.3 x-dependence of the lowest state energies. The system size is $N = 98$. Black circles represent the energy of the state with single coherent state of phonons having the uniform dimerization. On the other hand, a red circle represents the energy of the lowest state with 5 coherent states of phonons. We do not have enough space to show the lattice structures, but soliton-like defects, which convert the dimerization pattern from long-short-long-short to short-long-short-long or vise versa, do not stabilize the ground state. This indicates that the dimerization is very strong in the half-filled ground state. In the case of doped system with $N_e = 90$ electrons (or 8 holes), a single coherent state of phonons indicates the equidistant lattice, as shown by black circles in Fig.3(b). On the other hand, when we increase the number of coherent states, the dimerized lattice becomes still the lowest state. But, its dimerization amplitude is very small, as shown by a red circle in Fig.3(b).

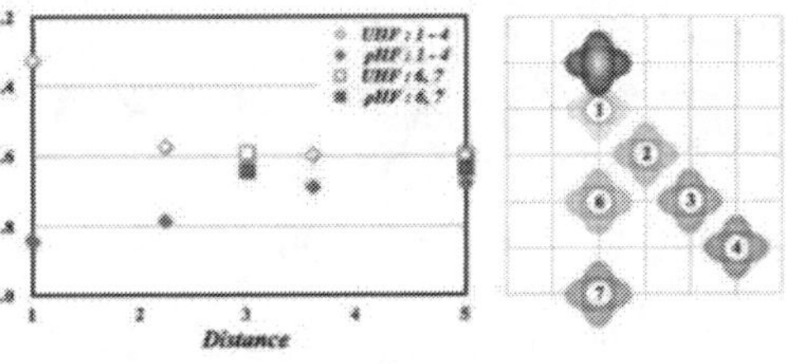

Figure 2. Energy of UHF and pHF states with a single pair of polarons.

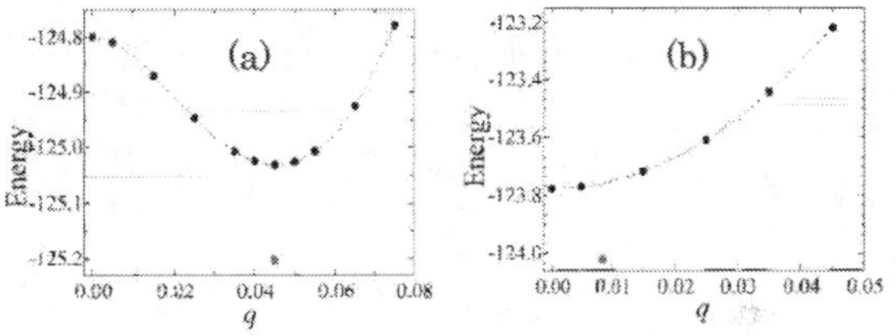

Figure 3. Dimerization amplitude(q)-energy relation for the half-filled (a) and 8% doped (b) systems.

When the number of electrons is reduced to $N_e = 86$, the ground state has the equidistant lattice, even if we increase the number of coherent states. In the classical approximation, the soliton lattice is always the lowest state in any doping regimes. Therefore, the present results show that quantum fluctuations destabilizes the soliton lattice state.

4. Summary

We have visualized the quantum fluctuations in the 2D doped Hubbard model and SSH model by using the resonating HF method. Quantum lattice fluctuation effects have been described by superposition of coherent states of phonons. We have shown that doped holes form polarons in the 2D Hubbard model. It has been suggested that an attractive interaction works between two polarons when the electron correlation is partially incorporated within the framework of the projected HF approximation. These results have strongly suggested that polarons could be carriers of high-temperature superconductors.

Then, it has been shown that the dimerization is very strong in the SSH model at a half-filling. On the other hand, quantum fluctuations due to lattice solitons stabilize the ground state energies in the lightly doped systems, though the dimerization decreases with the increase of doping. Finally, the ground state has an equi-distant lattice beyond 12% doping.

This work was supported by Grant-in-Aid for Scientific Research No.23540359, MEXT, Japan.

References

[1] S. R. White, Phys. Rev. Lett. **69**, 2863(1992).
[2] *Quantum Monte Carlo Method in Condensed Matter Physics*, edited by M. Suzuki(World Scientific, Singapore, 1993).
[3] H. Yokoyama and H. Shiba, J. Phys. Soc. Jpn. **56**, 3582(1987).
[4] H. Fukutome, Prog. Theor. Phys. **80**, 417(1988).
[5] N. Tomita, Phys. Rev. **B69**, 045110(2004).
[6] N. Tomita and S. Watanabe, Phys. Rev. Lett. **103**, 116401(2009).

Oscillations at the Si/electrolyte contact: Discretization of phase oscillators

J Grzanna and H J Lewerenz
Helmholtz Zentrum Berlin for Materials and Energy,
Hahn-Meitner Platz 1, D-14109 Berlin, Germany

E-mail: grzanna@helmholtz-berlin.de

Abstract. The origin of sustained current oscillations at the Si/electrolyte contact is not fully understood. Oscillatory functions are regarded which describe the oscillating oxide thickness at the silicon electrode. We consider an initially vanishing two-dimensional time dependent function which oscillates between a minimum and a maximum oxide thickness at each location of the electrode. The function is continuous except at single points of the electrode at which the oxide thickness drops deeply due to the formation of nanopores in the oxide. The oscillatory function is represented by a set of infinite (infinitesimal) oscillators. The mathematical model is based on the fact that it is sufficient to register the oscillators only one time per i-th cycle at their minimum or when the phase of the oscillator equals $i\ 2\pi$. In phase-space representation, the passing of the phase trough the $i\ 2\pi$ planes leads to oscillator density functions $p_i(t)$ which define the (differential) number of oscillators passing their minimum at the i-th time at the time t. Two consecutive oscillator density functions are connected by an integral equation representing a Markov process. Together with a local model for the oxide microstructure, a fit of the model parameter to the measured current is possible. The result is that the existence of two types of oxides (with different nanopore densities) can explain sustained current oscillations and, further, it is possible to calculate the mean nanopore distance in both types of oxide.

1. Introduction

Oscillatory processes are observed in almost any field of science. The oscillatory behaviour of silicon electrodes has been reported about two decades ago [1]. In photo-electrochemistry and electrochemistry, oscillatory behaviour has been extensively studied on Si photo-electrodes [2]. Several models were developed based on self-oscillating domains [3], oxide-induced interfacial stress [4] and the so-called current bursts [5]. The observation of nano-dimensioned pores, fluctuating with the phase of the oscillating (photo)current of Si electrodes immersed in dilute ammonium fluoride solutions motivated the further development of the original stress model, also to spatial resolution, by applying cellular automates [6-8]. In section 2, the basics of the oscillatory process are presented. The discretization of the used phase oscillators is illustrated in section 3 and finally the fit to measured data is discussed in section 4.

2. The oscillatory process

The relation between the oxide evolution at the electrode and the considered two-dimensional time dependent oscillating function $d(t,x,y)$ is shown in figure 1 in which d denotes the oxide thickness, t

the time, x,y the spatial coordinates at the electrode, and i the cycle (Si layer) number. The basic process at the surface of the Si electrode is characterised by a cyclic oxidation (figure 1b) of a small silicon layer (with a thickness of ca. 10 nm) and the following removal of the oxide by etching (figure 1c-d). The oxide thickness evolves not uniformly at the electrode due to cracks and nanopores at the oxide (figure 1b-d) which result from stress due to the lattice mismatch between silicon and its oxide. Hence the oxidation can proceed at different layers (cycles) at the same time. The introduced function d reflects the dynamics of the oxide thickness and oscillates between a minimum and maximum value (figure 1, bottom). At the left hand side of a single minimum value of d, the oscillation process is in a different (neighbouring) cycle relative to the right hand side (figure 1d).

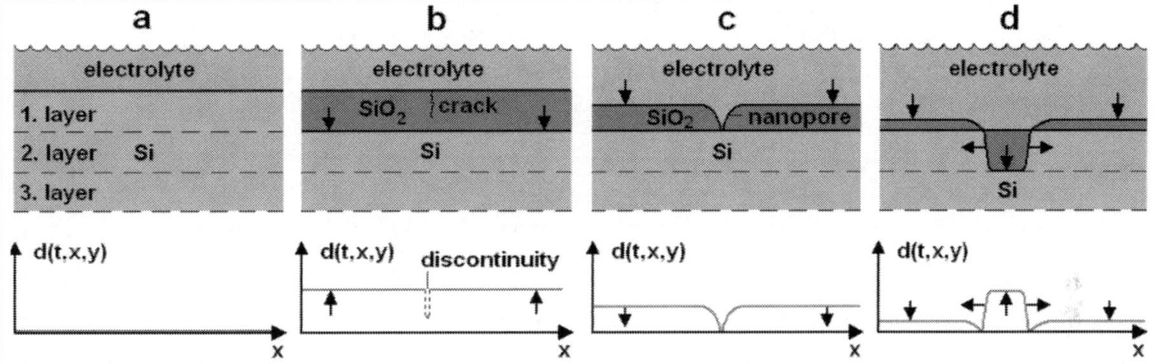

Figure 1. Relation between the oxide thickness evolution at the electrode at the top and the oscillating function $d(t,x,y)$ at the bottom.

The mathematical model is based on the fact that it is sufficient to register the time s at which d equals the minimum value at the location (x,y). At this time s (which depends on (x,y)), the oxide starts to grow (figure 1a, d) and during the (constant) oxide growth time an anodic current $E_i(t-s)$, whose duration is short with regard to the oxide etching time, flows [4, 7]. Then, the total current can be calculated by superposition of all elementary current peaks $E_i(t-s)$ [4].

3. Discretization of phase oscillators

The oscillating function d is described by a set of infinite oscillators. Each oscillator is dedicated one-to-one to an infinitesimal point $(x,y)\in\Omega$ of the electrode surface domain $\Omega\subseteq R^2$. With

$$d(t,x,y)=\frac{\cos(\varphi(t,x,y)+\pi)+1}{2}(d^{\max}-d^{\min})+d^{\min} \quad \text{for } i2\pi\le\varphi(t,x,y)\le(i+1)2\pi, i=0,1,2,... \quad (1)$$

the oscillatory process is described by a two-dimensional time depending phase function $\varphi(t,x,y)$. Due to equation (1), φ propagates monotone increasing (with variable shape) in the phase-space $\Phi=\{(x,y,\varphi)\,|\,(x,y)\in\Omega, \varphi\in R_+\}$. If an oscillator passes its minimum at the i–th time, the phase equals $i\,2\pi$. Due to the above mentioned basics, it is sufficient to register the time s at which the phase passes the so-called snap-shot screens $S_i=\{(x,y,i\,2\pi)\,|\,(x,y)\in\Omega\}$ that are planes in Φ with a phase difference of 2π. The latter corresponds to a discretization of the phase (oscillators) because only the multiple of 2π is considered and not the continuum. The passing of the phase through each snap-shot screen S_i is described by a density function $p_i(t)$ of oscillators which defines the differential number of oscillators passing their minimum simultaneously at the time t at the i-th time. For the calculation of $p_i(t)$, firstly the intersection $\Gamma_i(t) = \{(x,y) \mid (x,y) \in \Omega \text{ and } \varphi(t,x,y) = i\,2\pi\}$

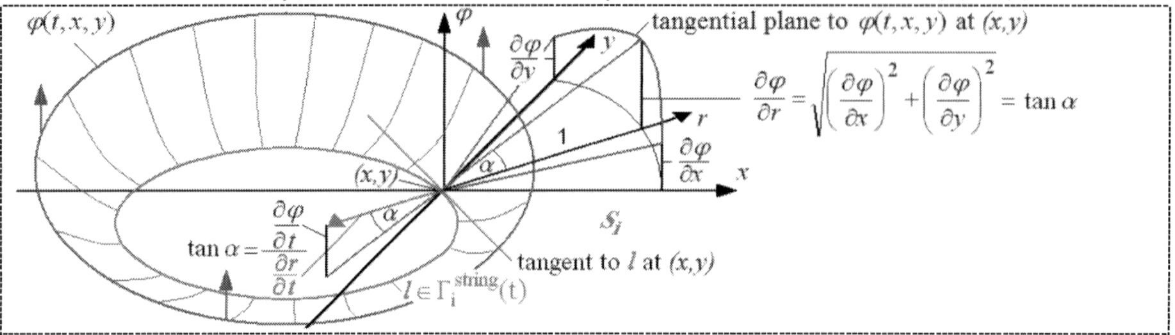

Figure 2. The passing of the phase $\varphi(t,x,y)$ through S_i (here the *(x,y)* plane) is shown.

between the phase φ and S_i is determined. $\Gamma_i(t) = \Gamma_i^{string}(t) + \Gamma_i^{area}(t)$ subdivides into strings (contour lines of φ) and to areas at which the phase equals $i\,2\pi$. Normally such areas does not exist (all $\Gamma_i^{area}(t)$ are empty) but we consider here also the possibility that some few $\Gamma_i^{area}(t_j)$ are not empty for $j = 1(1)m$. Assuming the phase φ passes completely the $i\,2\pi$-screen S_i during the time interval $[t_i^{\min}, t_i^{\max}]$, we get the factorization $\Omega = \bigcup_{t \in [t_i^{\min}, t_i^{\max}]} \Gamma_i(t)$. Secondly, the incremental velocity of a string is calculated. The string of intersection between φ and S_i moves at the $i\,2\pi$-screen with the incremental velocity of

$$\frac{\partial r}{\partial t}(t,x,y) = \frac{\partial \varphi}{\partial t}\left(\frac{\partial \varphi}{\partial r}\right)^{-1} = \frac{\partial \varphi}{\partial t}\left(\left(\frac{\partial \varphi}{\partial x}\right)^2 + \left(\frac{\partial \varphi}{\partial y}\right)^2\right)^{-1/2} \quad \text{for } (x,y) \in \Gamma_i^{string}(t) \tag{2}$$

to the (here) opposite direction of the normal r at the point (x,y) of the string $l \in \Gamma_i^{string}(t)$ (figure 2). Thirdly, we integrate the velocity of the strings and add (only in the case of $m > 0$) weighted Dirac functions and get

$$p_i(t) = \frac{1}{N_i} \int_{\Gamma_i^{string}(t)} \frac{\partial r}{\partial t}(t,x,y)\, dx\, dy + \sum_{j=1}^{m} F_i(t)\, \delta(t - t_j) \quad \text{for } t \in [t_i^{\min}, t_i^{\max}] \tag{3}$$

at which N_i is a factor to fulfill $\int_{t_i^{\min}}^{t_i^{\max}} p_i(t)\, dt = \int_\Omega d\Omega$ (each oscillator correlates to one $(x,y) \in \Omega$) and $F_i(t)$ denotes the area of $\Gamma_i^{area}(t)$. $p_{i-1}(t)$ and $p_i(t)$ are connected by the two-dimensional (in t normalized) function $q_{i,s}(t-s)$ which denotes the probability density for the period $t-s$ of oscillators starting their $(i-1)$-th cycle at the time s. We get the Markov process [4] (figure 3)

$$p_i(t) = \int_{t_{i-1}^{\min}}^{t_{i-1}^{\max}} p_{i-1}(s)\, q_{i,s}(t-s)\, ds \quad \text{for } t \in [t_i^{\min}, t_i^{\max}] \tag{4}$$

$q_{i,s}(t-s)$ is calculated as $p_i(t)$ with the exception that Ω is replaced by $\Gamma_{i-1}(s)$ in the definition of $\Gamma_i(t)$ and the factor N_i makes sure the normalization of $q_{i,s}(t-s)$ in the t direction [6].

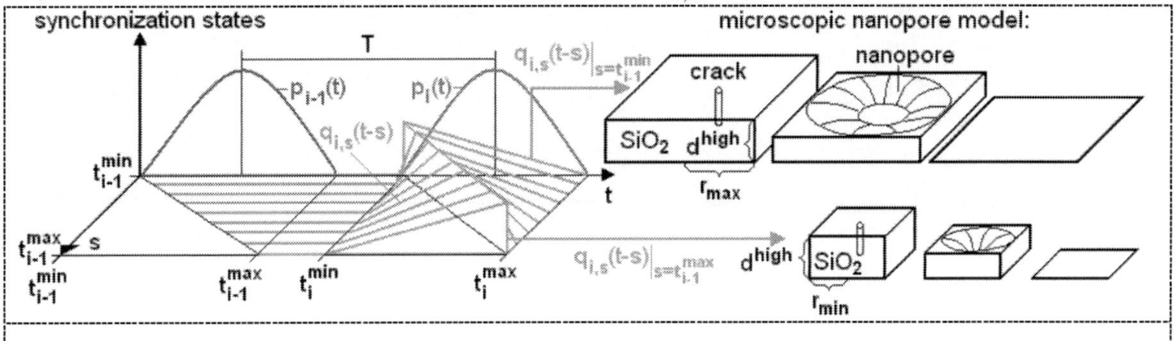

Figure 3. A sketch related to equation (4) (left) and the microscopic nanopore model (right).

4. Modelling sustained current oscillations

The simulation procedure entails:
i) find a kernel $q_{i,s}$ which fulfills equation (4) with $p_i(t) = p_{i-1}(t-T)$ (T – oscillation period) and
ii) fit the total current $I(t) = \sum_{i=0}^{\infty} \int_{t_i^{\min}}^{t_i^{\max}} p_i(s)\, E_i(t-s)\, ds$ of the model to the measured current.

Figure 3 shows the case of sustained oscillations for what $t_{i-1}^{\max} - t_{i-1}^{\min} = t_i^{\max} - t_i^{\min}$ and $q_{i,s}(t-s) = 0$ for all t with $t < t_i^{\min}$ and $t_i^{\max} < t$ are necessary conditions. Here, the shape of q is deduced from the normalized Greens function of the Helmholtz differential equation [8]. From the shape of q and the microscopic nanopore model it follows that two types of oxide (I and II) exist. We consider the s dependent oxide micro structure evolution (spatial temporal mean) during a cycle (figure 3, right). The maximum oxide life time $t_i^{\max} - t_{i-1}^{\min}$ is determined by the maximum oxide thickness $d^{high}(< r_{\max})$ in the case of the earlier grown oxide I with some few cracks and nanopores per area unit (see figure 3, top right). The shorter life time $t_i^{\max} - t_{i-1}^{\max}$ of the later grown oxide II can be realized only by more cracks and nanopores per unit area. In this case, the oxide life time is determined by a smaller half pore distance $r_{\min}(< d^{high})$ (see figure 3, bottom right) due to three dimensional etching in nanopores. Hence, the microscopic nanopore model allows (by simulating the etching) to determine the relation between the oxide lifetime t_i^{max}-s and the mean nanopore distance $2r(s)$ for $t_{i-1}^{\min} \le s \le t_{i-1}^{\max}$.

5. Conclusions

The mathematical concept to describe the oscillatory process at the Si/electrolyte contact is based on the discretization of phase oscillators which are registered at the multiple of 2π. An integral equation describes the sequence of the resulting phase oscillator densities $p_i(t)$ at $i2\pi$ as a Markov process. From the kernel $q_{i,s}(t-s)$ of the integral equation, it follows that two types of oxide with different nanopore density can explain sustained oscillations. In addition, the mean nonopore distance of the different oxide types can be calculated using the microscopic nanopore model.

References

[1] Gerischer H, Lübke M 1988 *Ber. Bunsenges. Phys. Chem.* **92** 573
[2] Aggour M, Giersig M and Lewerenz H J 1995 *J. Electroanal. Chem.* **383** 67
[3] Ozanam F, Blanchard N and Chazalviel J N 1993 Electrochim. Acta 38 1627
[4] Grzanna J, Jungblut H and Lewerenz H J 2000 J. Electroanal. Chem. 486 181+190
[5] Foca E, Carstensen J and Föll H 2007 J. Electroanal. Chem. 603 175

[6] Grzanna J, Jungblut H and Lewerenz H J 2007 Phys. Stat. Sol. (a) 204 1245
[7] Grzanna J, Notz T and Lewerenz H J 2009 Phys. Stat. Sol. (c) 6 1639
[8] Grzanna J, Notz T, Stempel T and Lewerenz H J 2011 ECS Transactions 33 127

Maximum entropy reconstruction of the configurational density of states from microcanonical simulations

Sergio Davis

Grupo de Nanomateriales, Departamento de Física, Facultad de Ciencias, Universidad de Chile, Casilla 653, Santiago, Chile

E-mail: sdavis@gnm.cl

Abstract. In this work we develop a method for inferring the underlying configurational density of states of a molecular system by combining information from several microcanonical molecular dynamics or Monte Carlo simulations at different energies. This method is based on Jaynes' Maximum Entropy formalism (MaxEnt) for Bayesian statistical inference under known expectation values. We present results of its application to measure thermodynamic entropy and free energy differences in embedded-atom models of metals.

1. Introduction

The Maximum Entropy (MaxEnt) principle, proposed by Jaynes in 1957 [1] and widely used in Bayesian probability, allows the construction of the most unbiased probabilistic models that reproduce certain information. Recently, Caticha [2] has proposed a Maximum Relative Entropy principle to update a probabilistic model under known information from a previous state of knowledge. In this work we apply this idea to the microcanonical distribution of configurational degrees of freedom, in order to update a previous guess of the configurational density of states (CDOS) with information from several microcanonical simulations at different energies.

2. Microcanonical distribution

We will consider a classical system of $6N$ degrees of freedom (3N momenta, denoted collectively by $\mathbf{p}$, 3N coordinates denoted by $\mathbf{r}$), with Hamiltonian

$$\mathcal{H} = \frac{\mathbf{p}^2}{2m} + \Phi(\mathbf{r}). \tag{1}$$

The probability of the system having phase space coordinates $(\mathbf{r}, \mathbf{p})$ at fixed total energy E is given by [3, 4]

$$P(\mathbf{r}, \mathbf{p}; E) = \frac{1}{\Omega(E)}\delta(E - \mathcal{H}(\mathbf{r}, \mathbf{p})), \tag{2}$$

where

$$\Omega(E) = \frac{1}{h^{3N}} \int d\mathbf{r}d\mathbf{p}\delta(E - \mathcal{H}(\mathbf{r},\mathbf{p})) \quad (3)$$

is the density of states having energy between E and $E + dE$, and h is Planck's constant. From this density of states, the microcanonical entropy is obtained by using Boltzmann's formula, $S(E) = k_B \ln \Omega(E)$.

Given that the dependence of the Hamiltonian on $\mathbf{p}$ is fully known, those degrees of freedom can be removed from the problem by integrating them explicitly [5, 6, 7, 8]. The probability distribution function for the configurational degrees of freedom takes the form

$$P(\mathbf{r}; E) = \frac{1}{\eta(E)}\Theta(E - \Phi(\mathbf{r}))\sqrt{E - \Phi(\mathbf{r})}^{3N-2}, \quad (4)$$

where

$$\eta(E) = \int d\mathbf{r}\Theta(E - \Phi(\mathbf{r}))\sqrt{E - \Phi(\mathbf{r})}^{3N-2} \quad (5)$$

is proportional to $\Omega(E)$ and $\Theta(x)$ is Heaviside's step function.

Using Eq. 4 we can define the microcanonical expectation value of any configurational quantity $A(\mathbf{r})$ as

$$\langle A \rangle_E = \frac{1}{\eta(E)} \int d\mathbf{r}\Theta(E - \Phi(\mathbf{r}))\sqrt{E - \Phi(\mathbf{r})}^{3N-2} A(\mathbf{r}). \quad (6)$$

3. Determination of the Density of States

From Eq. 6, it can be deduced that

$$P(\phi; E) = \langle \delta(\Phi(\mathbf{r}) - \phi) \rangle = \frac{1}{\eta(E)}\Theta(E - \phi)\sqrt{E - \phi}^{3N-2}\mathcal{D}(\phi) \quad (7)$$

where

$$\mathcal{D}(\phi) = \int d\mathbf{r}\delta(\Phi(\mathbf{r}) - \phi) \quad (8)$$

is the configurational density of states (CDOS). If we have access to $\mathcal{D}(\phi)$ we can compute the expectation of any function of ϕ at an arbitrary energy E as

$$\langle A(\Phi) \rangle_E = \frac{1}{\eta(E)} \int_{\phi_0}^{E} d\phi\mathcal{D}(\phi)\sqrt{E - \phi}^{3N-2} A(\phi). \quad (9)$$

Our aim is to reconstruct $\mathcal{D}(\phi)$ from several known configurational averages at different energies. Suppose we have a prior estimate of the density of states, $\mathcal{D}_0(\phi)$ and incorporate knowledge of the following expectation value

$$\langle f(\Phi) \rangle_E = F(E), \quad (10)$$

then the most unbiased estimate of $\mathcal{D}(\phi)$ is obtained by maximization of the relative entropy

$$S[\mathcal{D}, \mathcal{D}_0] = -\int d\phi P(\phi; E) \ln \Big(\frac{P(\phi; E)}{P_0(\phi; E)}\Big) \quad (11)$$

under the constraint given in Eq. 10, where P and P_0 are the probability density functions using $\mathcal{D}$ and $\mathcal{D}_0$, respectively. This leads to the following updating rule

$$\mathcal{D}(\phi) = \mathcal{D}_0(\phi)\exp(\lambda f(\phi)), \quad (12)$$

where λ is a Lagrange multiplier to be determined by imposing the validity of the constraint itself, condition which can be written as

$$\frac{\partial}{\partial \lambda} \ln Z(\lambda) = F(E), \tag{13}$$

where

$$Z(\lambda) = \int_{\phi_0}^{E} d\phi \sqrt{E-\phi}^{3N-2} \mathcal{D}_0(\phi) \exp(\lambda f(\phi)). \tag{14}$$

Now, consider the case $f(\phi) = \Theta(b-\phi)\Theta(\phi - a)$, where a and b form an arbitrary interval so that $f(\phi) = 1$ if $a < \phi < b$, and $f(\phi) = 0$ otherwise. Then, $F(E)$ is just the probability $P(a < \Phi < b|E)$, that is, the probability of $\Phi(\mathbf{x})$ falling into the interval $[a, b]$. The normalization constant $Z(\lambda)$ is given by

$$Z(\lambda) = G(\phi_0, a) + G(b, E) + \exp(\lambda)G(a, b) \tag{15}$$

where

$$G(x, y) = \int_{x}^{y} d\phi \sqrt{E-\phi}^{3N-2} \mathcal{D}_0(\phi). \tag{16}$$

Eq. 13 thus connects λ with $P(a < \Phi < b|E)$ in the following form,

$$P(a < \Phi < b|E)\Big(G(\phi_0, a) + G(b, E) + \exp(\lambda)G(a, b)\Big) = \exp(\lambda)G(a, b), \tag{17}$$

therefore

$$\exp(\lambda) = \frac{P(a < \Phi < b|E)}{1 - P(a < \Phi < b|E)} \frac{(G(\phi_0, a) + G(b, E))}{G(a, b)}. \tag{18}$$

Consider now a discretization of ϕ in n intervals, $\phi = i\Delta\phi$, where $\Delta\phi = (E - \phi_0)/n$, and i an integer index between 0 and $n-1$ (thus the condition $\phi < E$ is imposed implicitly, as $k < n$). To further simplify, we measure ϕ so that $\phi_0 = 0$ and $\Delta\phi = 1$. We can then approximate, for sufficiently large n,

$$G(a, b) = \sum_{i=a}^{b} \sqrt{E-i}^{3N-2} d_i^{(0)} \tag{19}$$

where $d_i^{(0)} = \mathcal{D}_0(i)$. Taking $a = k$ and $b = k + 1$, we have

$$d_k = \frac{N_k(E)}{N - N_k(E)} \frac{\sum_{i \neq k} \sqrt{E-i}^{3N-2} d_i^{(0)}}{\sqrt{E-k}^{3N-2}}. \tag{20}$$

where $N_k(E)$ is a histogram of potential energies at total energy E.

This is a relatively simple rule for incorporating information from a microcanonical simulation at total energy E into a prior guess of a discretized CDOS. The only input needed from the simulation is the histogram $N_k(E)$. This updating rule is guaranteed to converge: when it does ($d_k = d_k^{(0)}$), we have

$$P_k(E) = \frac{N_k(E)}{N} = \frac{\sqrt{E-k}^{3N-2} d_k}{\sum_{i=0}^{n-1} \sqrt{E-i}^{3N-2} d_i}, \tag{21}$$

which is just a discretized version of Eq. 7, and the denominator is an estimate of $\eta(E)$, the full (kinetic plus configurational) density of states.

In practice, we may start with a constant CDOS as initial guess, $d_k^{(0)} = 1$, and iteratively combine the information from several microcanonical simulations at different energies, each one only updating the CDOS for the potential energies actually observed.

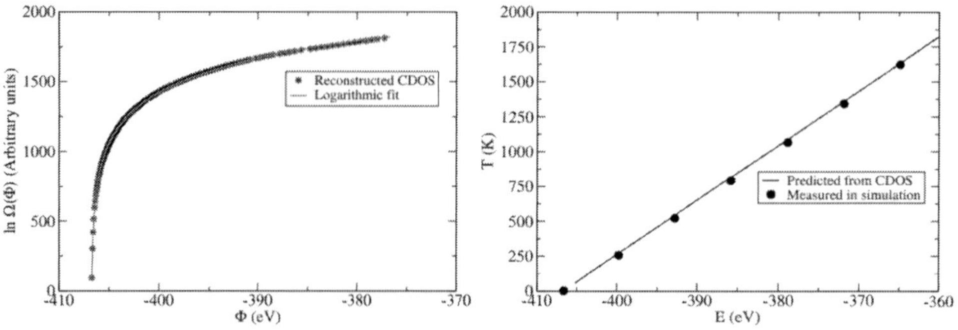

Figure 1. Left panel, reconstructed configurational density of states (CDOS) for Au, 108 atoms. Right panel, predicted caloric curve from CDOS and actual simulation values.

4. Results

Figure 1 (left panel) shows the CDOS computed using this procedure for a Au system composed of 108 atoms, simulated at 24 different energies (corresponding to applying initial kinetic energies between 6K and 6000K to the ideal face-centered cubic structure) via molecular dynamics with the Sutton-Chen potential [9]. This CDOS agrees remarkably with a simple model

$$\mathcal{D}(\phi) = \Big(a(\phi - \phi_0) + 1\Big)^b, \tag{22}$$

with fitted parameters a =43.764 eV^{-1}, b =134.297, and the energy of the ideal crystalline structure ϕ_0 =-406.715 eV. The right panel of figure 1 shows the predicted values of temperature as a function of total energy, together with the measured values from several simulations. These predicted temperatures are derived from the CDOS by computing, according to Eq. 9, the expectation value

$$T = \Big\langle \frac{2(E - \Phi)}{3N} \Big\rangle_E. \tag{23}$$

5. Final Remarks

We have developed a method for obtaining the configurational density of states from microcanonical simulations, based on maximization of the relative configurational entropy under known expectation values. The formula (Eq. 20) resembles the histogram reweighting methods for the canonical ensemble [10].

Acknowledgments

SD acknowledges FONDECYT grant 3110017.

References

[1] Jaynes E T 1957 *Physical Review* **106** 620–630
[2] Caticha A and Giffin A 2006 (*Preprint* `arXiv:physics/0608185v1`)
[3] Khinchin A I 1949 *Mathematical Foundations of Statistical Mechanics* (Dover Publications Inc.)
[4] Huang K 1987 *Statistical Mechanics* (Wiley)
[5] Severin E S, Freasier B C, Hamer N D, Jolly D L and Nordholm S 1978 *Chem. Phys. Lett.* **57** 117
[6] Pearson E M, Halicioglu T and Tiller W A 1985 *Phys. Rev. A* **32** 3030–3039
[7] Ray J R 1991 *Phys. Rev. A* **44** 4061–4064
[8] Davis S 2011 *Phys. Rev. E* **84** 050101(R)
[9] Sutton A P and Chen J 1990 *Phil. Mag. Lett.* **61** 139
[10] Ferrenberg A M and Swendsen R H 1989 *Phys. Rev. Lett.* **63** 1195–1198

Solution of linearized Ginzburg-Landau problem for mesoscopic superconductors by conformal mapping

Paulo J. Pereira[1], **Victor V. Moshchalkov**[1], **Liviu F. Chibotaru**[2]

[1] INPAC-Institute for Nanoscale Physics and Chemistry, Katholieke Universiteit Leuven, Celestijnenlaan 200 D, B-3001 Leuven, Belgium

[2] Division of Quantum and Physical Chemistry and INPAC, Institute for Nanoscale Physics and Chemistry, Katholieke Universiteit Leuven - Celestijnenlaan, 200F, B-3001 Leuven, Belgium

Abstract. We present a new method for the solution of linearized Ginzburg-Landau problem for mesoscopic superconducting nanostructures of arbitrary shapes in applied magnetic field. The method is based on the conformal mapping of the analytical solution for the disk and uses a specially designed superconducting gauge for the vector potential corresponding to the magnetic field. As a demonstration of the methods accuracy, we calculate the distribution of the order parameter in superconducting regular polygons and compare the obtained solutions with the available numerical results. We further consider an example of irregular polygon and show the evolution of the vortex patterns in function of the geometry of samples boundary. The obtained results will be compared with available experimental data on mesoscopic and nanoscopic superconductors.

The linearized Ginzburg-Landau (LGL) equation describes the distribution of the superconducting order parameter (Ψ) during the nucleation of the superconductivity, i.e. close to the normal-superconducting phase boundaries of the $T_c(H_{c2})$ diagram [1]. It can also be used to solve the non-linear Ginzburg-Landau problem[2, 3].

The LGL problem is defined by the equations

$$\frac{1}{2m}(-i\hbar\vec{\nabla} - 2e\vec{A})^2\Psi = \alpha\Psi \tag{1}$$

$$(-i\hbar\vec{\nabla} - 2e\vec{A})|_{n.b.}\Psi = 0, \tag{2}$$

where $\vec{A}$ is the vector potential corresponding to the applied magnetic field ($\vec{H} = \nabla \times \vec{A}$), Ψ is the superconducting order parameter, α is the condensation energy density (in the linearized Ginzburg-Landau is an eigenvalue to be determined) and $n.b.$ means normal to the boundary. The most common method to solve Ginzburg-Landau (GL) problem in superconductivity (the linearized GL equation, non-linear GL equation, the GL equation coupled with the magnetic field equation, the time dependent GL, the two component GL) is the finite difference method . However comparison with analytical and semi-analytical methods applied to the LGL equation has shown that the finite difference method needs a great number of points (with grids of at least 201 by 201 points [4]) to become sufficiently accurate to obtain vortex features like anti-vortex in the square [5]. Still, analytical and semi-analytical methods are only applicable, to the best of our knowledge, to a few geometries like the circle [6, 7], square [5], rectangle [8], equilateral triangle [9] and annular [10] geometries.

Most of the newly accessible bounded geometries by semi-analytical methods were analyzed by a superconducting gauge method [11]. Recently, a numeric method [12], not so efficient as the mentioned gauge method, was developed to solve the LGL problem for bounded geometries

Published under licence by IOP Publishing Ltd

using the finite element method and the superconducting gauge approach. The solutions for this problem were found for different geometries, including some highly symmetric, like the square, triangle, pentagon and five point stars.

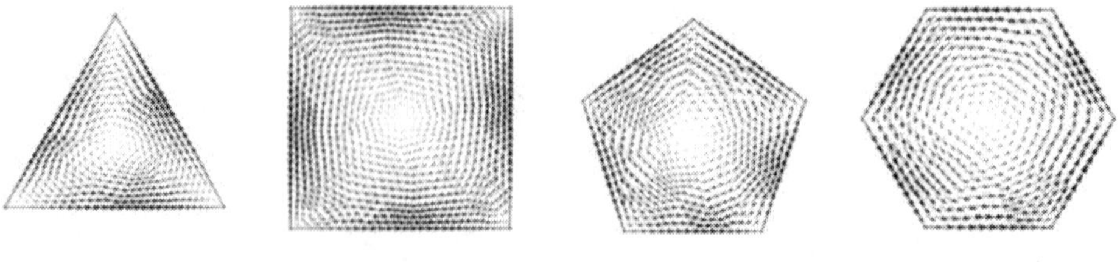

Figure 1. Plots of the vector potential with the superconducting gauge fixed by the method described in the text.

We propose to solve the LGL problem in a generic polygon with the eigenfunctions of the Laplacian operator (with Neumann boundary condition), defined on the circle, as basis functions. We also propose to fix the gauge of the vector potential, that reduces the boundary condition of the Ginzburg-Landau problem to the Neumann boundary condition, by solving Laplace equation with a general Neumann boundary condition (that can be found by the Dini integral formula [13]).

First, we calculated the eigenfunctions of the Laplacian operator in the circle (with Neumann boundary condition). These well known solutions are

$$\begin{aligned} \Psi_{(l,n)}(r,\theta) &= J_l(u'_{(l,n)}r)e^{il\theta} \\ J_l'(u'_{(l,n)}) &= 0, \end{aligned} \tag{3}$$

where $u'_{(l,n)}$ is the n zero of the l Bessel function derivative and $J_l(u'_{(l,n)}r)$ is the l Bessel function rescaled to match the Neumann boundary conditions. The solutions for the circle (auxiliary space) were mapped onto the general polygon (original space) by the SC mapping. We have represented the functions with a mesh in the original space. Afterwards, this set of functions was orthogonalized via the Löwding orthogonalization process. Then, the new functions were used as a basis for the LGL problem. To apply the functions as a basis for the LGL equation, the superconducting gauge needs, first, to be fixed using the gauge condition[11], $\vec{A}'|_{n.b.} = 0$, which means $\vec{\nabla}|_{n.b.}S = -\vec{A}|_{n.b.}$, where $\vec{A}' = \vec{A} + \vec{\nabla}S$.

If we want to satisfy gauge condition while keeping the Coulomb gauge, the function S must be a solution of the problem:

$$\Delta S = 0 \tag{4}$$

$$\vec{\nabla}|_{n.b.}S = -\vec{A}|_{n.b.}, \tag{5}$$

where $\vec{A}$ is a general vector potencial. In general, this problem is only solvable when

$$\int_{boundary} \vec{A}|_{n.b.} = 0, \tag{6}$$

where the integration is made over the boundary. This condition is fulfilled if the original vector potential is in the Coulomb gauge, $\vec{\nabla}.\vec{A} = 0$. To solve the new problem, we have mapped the equation back to the auxiliary space (circle). The solution to this problem is given by the Dini integral formula [13]

$$\begin{aligned} S(u(r,\theta),v(r,\theta)) = \frac{1}{4\pi}\int_{\phi=0}^{2\pi} -|w'(z)|\,\vec{A}(u,v)|_{n.b.} \\ \log\left(r^2 - 2r\cos(\theta-\phi)+1\right)d\phi, \end{aligned} \tag{7}$$

for the unitary circle, where ϕ is a integration angle. This method makes it possible to fix the superconducting gauge for any magnetic field distribution, such that the vector potential is tangential to the boundary.

Fig. 1 and Fig. 2 show the vector potential and the distribution of the order parameter for the triangle, square, pentagon and hexagon geometry using this method and Fig. 3 below presents the same quantities but for a deformed hexagon.

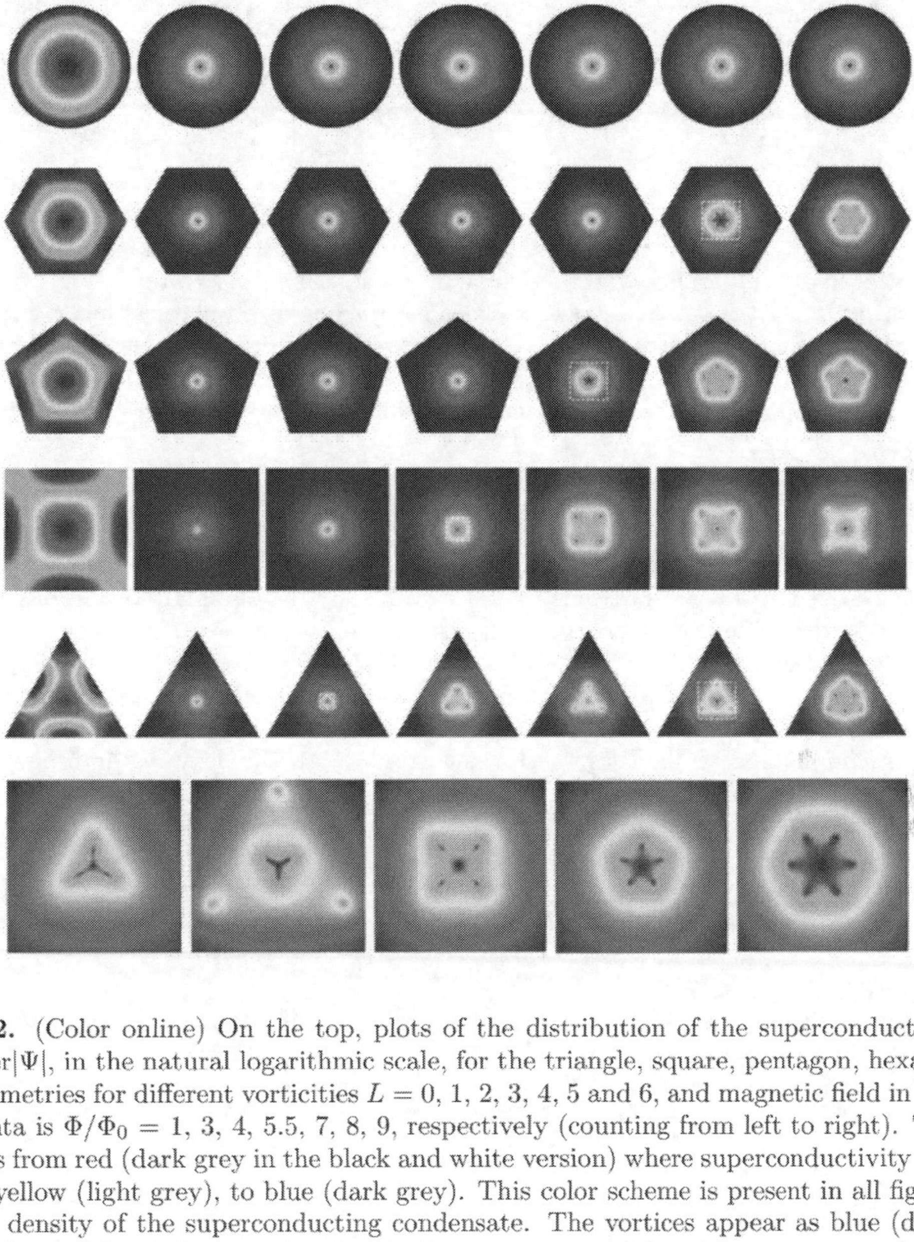

Figure 2. (Color online) On the top, plots of the distribution of the superconducting order parameter$|\Psi|$, in the natural logarithmic scale, for the triangle, square, pentagon, hexagon, and circle geometries for different vorticities $L = 0, 1, 2, 3, 4, 5$ and 6, and magnetic field in magnetic flux quanta is $\Phi/\Phi_0 = 1, 3, 4, 5.5, 7, 8, 9$, respectively (counting from left to right). The color scale goes from red (dark grey in the black and white version) where superconductivity is strong, through yellow (light grey), to blue (dark grey). This color scheme is present in all figures that show the density of the superconducting condensate. The vortices appear as blue (dark grey) spots in the middle of the sample. On the bottom, zooms (of anti-vortex patterns) from the areas enclosed by white dashed boxes, corresponding from left to right, to the pairs (geometry, number of vortices, vorticity number) = (triangle, 3, 2) ; (triangle, 6, 5) ; (square, 4, 3) ; (pentagon, 5, 4) ; (hexagon, 6, 5)

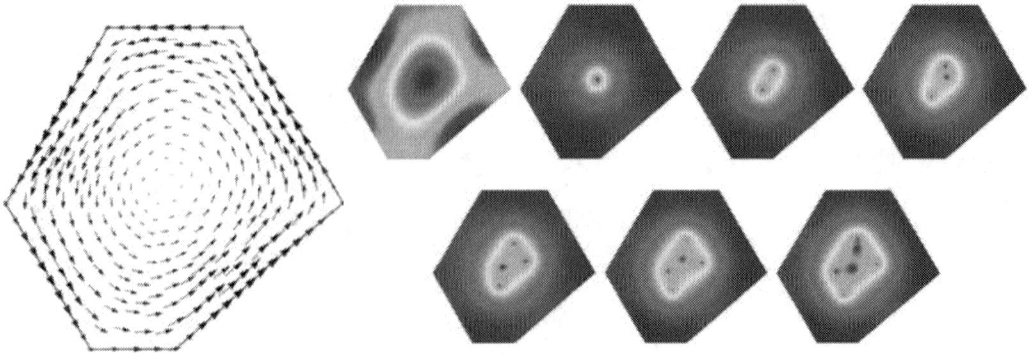

(a) (b)

Figure 3. Fig. 3a : Vector potential with the superconducting gauge fixed (the vector potential lines are parallel to the boundary), as described in the text. Fig. 3b (Color online) : Distribution of the superconducting order parameter, in the natural logarithmic scale, in the deformed hexagon for vorticities $L = 0, 1, 2, 3, 4, 5$ and 6 and magnetic field in magnetic flux quanta $\Phi/\Phi_0 = 1, 3, 4, 5.5, 7, 8, 9$, respectively (counting from left to right, from top to bottom). The vortices appear as blue (or in black and white version as dark gray) spots in the middle of the sample.

1. Conclusions

The proposed method extends naturally the previously developed superconducting gauge method. The use of basis functions that already satisfy the boundary conditions is a key element present in the original method and the extended one. These basis functions are different for each of these methods, but only few number of basis functions are needed to get a very good approximation of the solution in both methods. These methods were compared for the case of the triangle and the square geometries. The solutions of the extended method match the ones of the original method for these geometries. With this new method, new geometries are now accessible to semi-analytical treatment. The application of the latter in the present work allowed to observe certain common features shared by the solutions of LGL equation calculated for regular polygons. The solutions of LGL equation for different polygons presented here have demonstrated that the proposed method is efficient (uses only a few basis functions) and fairly accurate. Moreover it can be equally applied to irregular polygons with comparable efficiency.

Acknowledgments

The authors gratefully acknowledge financial support of Methusalem program at the KU Leuven and the FWO grants of Flemish Science Foundation.

[1] D. Saint-James, E. J. Thomas, G. Sarma, Type II superconductivity, (Pergamon Press, Oxford, 1969)
[2] V. V. Moshchalkov, X. G. Qiu, and V. Bruyndoncx, Phys. Rev. B **55**, 11793, (1997)
[3] L. F. Chibotaru, G. Teniers, A. Ceulemans, and V. V. Moshchalkov, Phys. Rev. B **70**, 094505 (2004)
[4] J. Bonča and V. V. Kabanov, Phys. Rev. B 65, 012509 (2001); private communication
[5] L. F. Chibotaru, A. Ceulemans, V. Bruyndoncx, and V. V. Moshchalkov, Nature (London) **408**, 833 (2000)
[6] D. Saint-James, Phys. Lett. **15**, 13 (1965)
[7] R. B. Dingle, Proc. R. Soc. London Ser. A **211**, 500 (1952)
[8] G. Teniers, L.F. Chibotaru, A. Ceulemans and V.V. Moshchalkov, Europhys. Lett., **63**,296 (2003)
[9] L. F. Chibotaru, A. Ceulemans, V. Bruyndoncx, and V. V. Moshchalkov, Phys. Rev. Lett. **86**, 1323 (2001)
[10] V. Bruyndoncx, L. Van Look, M. Verschuere, V. V. Moshchalkov, Phys. Rev. B **60**, 10468 (1999)
[11] L. F. Chibotaru, A. Ceulemans, M. Morlele, G. Teniers, C. Carballeira and V. V. Moshchalkov, J. Math. Phys. **46**, 095108 (2005)
[12] S. Kim, M. Gunzburger, J. Peterson and C. Hu, Commun. Comput. Phys., **6**, 673 (2008)
[13] L. V. Kantorovich and V. I. Krylov, Approximate Methods of Higher Analysis, (Interscience Publishers Inc., New York, 1964)

Material quality assessment of silk nanofibers based on swarm intelligence

Bruno Brandoli Machado, Wesley Nunes Gonçalves, Odemir Martinez Bruno

Physics Institute of São Carlos, University of São Paulo, Av. Trabalhador São Carlense, 400 - 13560-970 - São Carlos, SP, Brasil

E-mail: brandoli@icmc.usp.br,wnunes@ursa.ifsc.usp.br,bruno@ifsc.usp.br

Abstract. In this paper, we propose a novel approach for texture analysis based on artificial crawler model. Our method assumes that each agent can interact with the environment and each other. The evolution process converges to an equilibrium state according to the set of rules. For each textured image, the feature vector is composed by signatures of the live agents curve at each time. Experimental results revealed that combining the minimum and maximum signatures into one increase the classification rate. In addition, we pioneer the use of autonomous agents for characterizing silk fibroin scaffolds. The results strongly suggest that our approach can be successfully employed for texture analysis.

1. Introduction

The silk fibroin is a protein extracted from cocoons of *Bombyx mori* silkworm. It has been widely used in biomedical applications due to its high capacity to suture tissues, to regenerate bones and its biocompatibility with several types of human cells used in prosthesis [1]. In the last years, researchers have proposed to improve the nanoscale features of silk fibroin by adding glycerol. Though the presence of glycerol can provide better material properties, it also can alter the silk fibroin molecules interactions, damaging the result in its surface. Therefore, texture analysis methods emerge as a powerful tool for determining the suitable concentration of glycerol.

Many methods for texture description have been proposed in the literature. These methods are based on *statistical analysis* of the spatial distribution (e.g., co-occurrence matrices [2] and local binary pattern [3]), *stochastic models* (e.g., Markov random fields [4]), *spectral analysis* (e.g., Fourier descriptors [5], Gabor filters [6, 7]), *complexity analysis* (e.g., fractal dimension [8]), *agent-based model* (e.g., deterministic tourist walk [9, 10]). Although there are effective texture methods, they do not capture the richness of patterns of the silk fibroin scaffolds.

In this paper, we present a methodology for classifying surface properties of silk fibroin by means of texture analysis. The texture description approach proposed here is based on the artificial crawler model [11, 12]. We propose a new rule of movement that not only moves artificial crawler agents toward higher intensity, as well as to lower ones. We confirm that this strategy increases the discriminatory power and outperforms the state-of-the-art method.

This paper is organized as follows. Section 2 details the original artificial crawler model. Section 3 presents our approach to characterize textured images. Section 4 discusses the results of the experiments. Finally, conclusions are given in Section 5.

2. The Original Artificial Crawler Model

The first artificial crawler (ACrawler) model was developed in [11, 12]. Let us consider an image that assigns to each pixel $p = (x_p, y_p)$ an intensity $I(p) \in [0, 255]$. Each pixel holds a neighborhood set $\eta(p) = \{q \mid dist(p, q) \leq \sqrt{2}\}$, where $dist$ is the Euclidean distance.

At each time t, an agent i is characterized by two attributes: (1) the level of energy e_i^t and (2) the spatial position in the image ρ_i^t. First, n agents are born with identical energy ϵ. Such energy can either wax or wane their lifespan according to energy consumption and influence of the environment. On images, the environment is treated as a 3D surface with different altitudes that correspond to grey values in z-axis of the images. Higher intensities supply nutrients to the agents, while lower altitudes correspond to the land. The algorithm consists of a set of rules:

(i) **Born**: Each agent i is born with the same energy; $\forall_i, e_i^0 = \epsilon$.

(ii) **Survival threshold**: An agent i dies if its energy is below the threshold; $\forall_{t,i}$, if $e_i^t \leq e_{min}$ then i dies.

(iii) **Movement**:

$$\forall_i : e_i^t > e_{min}, \rho_i^{t+1} = f(\rho_i^t)$$

$$f(\rho) = \begin{cases} \rho_i^t, & \text{if (a) is satisfied} \\ \rho_{max}^t, & \text{if (b) is satisfied} \\ \rho_m^t, & \text{if (c) is satisfied} \end{cases}$$

(a) Agents settle down if the grey level of its 8-neighbors are lower than itself.
(b) Agents move if there exist one of its 8-neighbors (ρ_{max}^t) with higher intensity.
(c) If there exist more than one neighbor with higher intensity, an agent moves to the pixel that already was occupied (ρ_m^t).

(iv) **Energy consumption**: Each time t consumes one unity of energy; $\forall_i : e_i^t > e_{min}, e_i^{t+1} = e_i^t - e_{unity}$.

(v) **Law of the jungle**: An agent with higher energy eats up another with lower one; $\forall_{i,j} : \rho_i^{t+1} = \rho_j^{t+1}, e_{max}\{e_i^t, e_j^t\} = \max\{e_i^t, e_j^t\}$.

(vi) **Gain of energy**: It up to dates the energy absorption from the environment, where λ is a rate of absorption over the pixel $I(\rho_i^t)$; $\forall_i, e_i^{t+1} = e_i^t + \lambda I(\rho_i^t)$.

(vii) **Limit of energy**: It bounds the maximum energy e_{max}; $\forall_{t,j} : e_j^t \geq e_{max}, e_j^{t+1} = e_{max}$.

To quantify the multi-agent system, a curve of live agents at each time is obtained:

$$\varphi = [\psi(0), \psi(1), \ldots, \psi(t_{max})] \tag{1}$$

where $\psi(t)$ is the number of live agents at time t and t_{max} is maximum iteration.

3. A Novel Approach with Artificial Crawler to Texture Analysis

The artificial crawler model described above consists of moving agents to a neighbor pixel with the highest intensity. Despite the promising results, this idea does not extract all the richness of textural pattern. Our approach differs from the original ACrawler model in terms of movement: each agent is not only able to move to the higher altitudes as well as to lower ones. It allows the model to extract the details present in peaks and valleys of the images.

First, the agents move to higher intensities as the original artificial crawler method. Thus, the artificial crawlers are performed using this rule and the curve φ_{max} is obtained. Throughout the paper, this rule of movement will be referred as max. We can observe that the original artificial crawler method only models the peaks of a textured image. To obtain a robust and effective texture representation, we propose to move artificial crawlers toward lower intensities – this rule of movement will be referred throughout the paper as to *min*. In our approach, artificial crawlers are randomly placed in the image with initial energy ϵ. Then, the conditional movement step is modified as follows:

(*a*) Agents settle down if the grey level of its 8-neighbors are higher than itself.

(*b*) Agents move to a specific pixel if there exist one of its 8-neighbors (ρ^t_{min}) with lower intensity.

(*c*) If there exist more than one neighbor with lower intensity, an agent moves to the pixel that already was occupied (ρ^t_m).

The multi-agent systems using the rule of movement *min* is characterized as the original method by using the number of live agents at each time. Considering that now we have two rules of movement, the final feature vector of our approach is composed by the concatenation of φ_{max} and φ_{min} according to:

$$\varphi = [\varphi_{max}, \varphi_{min}] \tag{2}$$

4. Experimental Results

In this section, we demonstrate the effectiveness of our approach. We first outline details of the experimental setup, and then, experiments carried out on two datasets are discussed: Brodatz and silk fibroin.

For classification, we adopted the Linear Discriminant Analysis (LDA) using ten-fold cross validation [13]. The number of agents placed on the pixels of the image was initial set to 1000 with a coverage rate of 10%, varying from 1000 to 35000. In our experiments, all agent was born with an initial energy ϵ of 10 units and the loss for each iteration consumes 1 unit of energy. The absorption rate was set to 0.01 in terms of the current pixel. For the survival threshold and the upper bound of energy were set to 1 and 12 units, respectively.

Experiment 1: First, we perform an analysis of our method on the Brodatz dataset (40 classes with 10 samples) [14]. The results of the proposed method are compared with existing methods in Table 1. It is observed that the our method outperforms the state-of-the-art. The highest classification rate of 98.25%(±1.69) was obtained by our method, which is followed by a classification rate of 95.25%(±3.43) obtained by the Gabor filter, one of the most traditional texture analysis method.

Experiment 2: In this experiment, we present a comparative study of our approach to assess the quality of the silk fibroin scaffolds. The potential of the silk fibroin is enhanced by including glycerol solutions (usually 0% to 10% with step of 2.5%) during scaffold formation [1]. The silk fibroin dataset contains 5 classes with 10 samples. Figure 1 shows three samples for each concentration. In the silk fibroin dataset, our method achieved highest classification rates when compared with traditional texture analysis methods. The experimental results, presented in Table 1, shows that our method achieved a classification rate of 96%(±8.43). These experimental results indicate that our method is consistent and can be applied in real-world applications.

5. Conclusion

In this paper we presented a novel approach based on artificial crawler for texture classification. We have demonstrated how the feature vector can be improved by combining *min* and *max* curves, instead of using only the strategy for the maximum of intensity of the pixels. Although traditional methods of texture analysis have provided satisfactory results, the approach proposed here has proved to be superior for characterizing textures from a popular benchmark and from the silk fibroin scaffolds analysis.

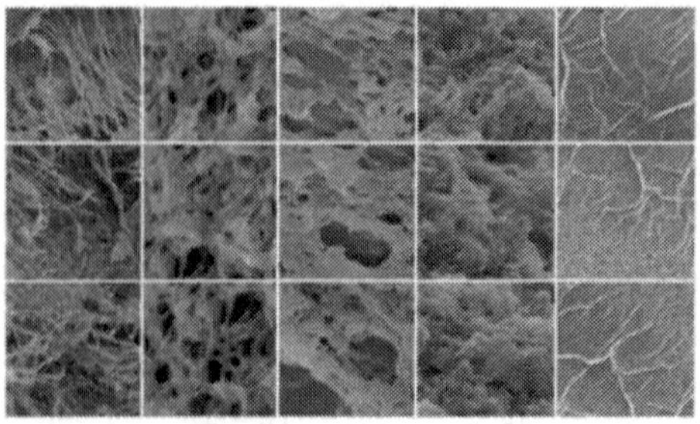

Figure 1. Samples for each glycerol concentration. The first column corresponds to 0% of concentration, the second 2.5%, and so on up to 10%.

Table 1. Experimental results for texture methods in the Brodatz and Silk Fibroin datasets.

Method	Brodatz	Silk Fibroin
Fourier descriptors [5]	86.50 (±6.58)	78.00 (±22.01)
Co-occurrence matrices [2]	91.25 (±2.65)	94.00 (±9.66)
Original artificial crawler [11]	93.00 (±5.50)	84.00 (±15.78)
Gabor filter [7]	95.25 (±3.43)	62.00 (±19.44)
Proposed method	**98.25** (±1.69)	**96.00** (±8.43)

Acknowledgements
BBM and WNG were supported by FAPESP under grants 2011/02918-0 and 2010/08614-0, respectively. OMB was supported by CNPq grants 306628/2007-4 and 484474/2007-3.

References
[1] Shenzhou L, Xiaoqin W, Qiang L, Xiaohui Z, Jonathan A K, Neha U, Omenetto F and Kaplan D L 2010 *Biomacromolecules* **11** 143–150 ISSN 1064-5462
[2] Haralick R M, Shanmugam K and Dinstein I 1973 *IEEE Transactions on Systems, Man and Cybernetics* **3** 610–621
[3] Kashyap R L and Khotanzad A 1986 *IEEE Trans. Pattern Anal. Mach. Intell.* **8**(4) 472–481 ISSN 0162-8828
[4] Cross G R and Jain A K 1983 *IEEE Trans. Pattern Anal. Mach. Intell.* **5** 25–39
[5] Azencott R, Wang J P and Younes L 1997 *IEEE Trans. Pattern Anal. Mach. Intell.* **19**(2) 148–153 ISSN 0162-8828
[6] Machado B B, Gonçalves W N and Bruno O M 2011 Enhancing the texture attribute with partial differential equations *ACIVS* pp 337–348
[7] Gabor D 1946 *Journal of Institute of Electronic Engineering* **93** 429–457
[8] Bruno O M, de Oliveira Plotze R, Falvo M and de Castro M 2008 *Information Sciences* **178**(12) 2722–2733 ISSN 0020-0255
[9] Backes A R, Gonçalves W N, Martinez A S and Bruno O M 2010 *Pattern Recogn.* **43**(3) 685–694 ISSN 0031-3203
[10] Gonçalves W N, Backes A R, Martinez A S and Bruno O M 2012 *Expert Syst. Appl.* **39** 11818–11829
[11] Zhang D and Chen Y Q 2004 Classifying image texture with artificial crawlers *International Conference on Intelligent Agent Technology* IAT '04 (Los Alamitos, CA, USA: IEEE Computer Society) pp 446–449 ISBN 0-7695-2101-0
[12] Zhang D and Chen Y Q 2005 *IJPRAI* **19** 355–374
[13] Fukunaga K 1990 *Introduction to statistical pattern recognition* 2nd ed (San Diego, CA, USA: Academic Press) ISBN 0-12-269851-7
[14] Brodatz P 1966 *Textures: A Photographic Album for Artists and Designers* (New York: Dover Publications)

On some nonclassical fullerenes with several heptagonal rings

Francisco J. Sánchez-Bernabe [1]

[1] Department of Mathematics, Universidad Autónoma Metropolitana-Iztapalapa, San Rafael Atlixco 186, Colonia Vicentina, México, D.F., 09340, Mexico

E-mail: `fjsb@xanum.uam.mx`

Abstract. Several works on nonclassical fullerenes with heptagons have mainly considered the case with just one heptagon. In this context the isolated pentagon rule is not satisfied. The study of nonclassical fullerenes is important because some of them are more stable than the corresponding classical isomers with the same number of pentagonal bonds. We present several nonclassical fullerenes with pentagons, hexagons and two, three, or more heptagons.

1. Introduction

Classical fullerenes are those carbon cage molecules with exactly 12 pentagons and $n/2 - 10$ hexagons, Patrick Fowler and David Manolopoulos [1]. All classical fullerenes satisfy the so called isolated pentagon rule IPR, Harold Kroto [2]. On the other hand, nonclassical fullerenes are those carbon cage molecules embedded with one or more squares or heptagons. In this last case, sometimes pentagon-pentagon adjacencies appear, and the most stable structure allows for the least pentagon-pentagon adjacencies, which is known as the pentagon adjacency penalty rule PAPR, Eleanor Campbell *et al* [3]. Actually, the IPR or PAPR has been an efficient criterion for explaining the stability of fullerenes.

2. Background

An amount of work has been done to study nonclassical fullerenes. For instance, Andrés Ayuela *et al* [4] show theoretical evidence for the existence of a nonclassical fullerene C_{62} with one heptagonal, 13 pentagonal and 19 hexagonal rings. Jie An *et al* [5] study the isomers of fullerene C_{26} composed of square, pentagonal, hexagonal, and heptagonal faces. Yuan-Zhi Tan *et al* [6] consider the fullerene C_{68} which contains one heptagonal ring. Furthermore, Li-Hua Gan *et al* [7] study fullerenes C_{46}, C_{48}, C_{50}, and C_{52}, some of them composed of one heptagonal ring.

3. Calculations

We have obtained the graphs of our results by running the V0.3 version of (Carbon Generator) CaGe software, Gunnar Brinkmann [8], [9]. Schlegel diagrams [10] are also provided for each considered fullerene. Our first example of nonclassical fullerene contains 68 carbons with two heptagonal rings, 14 pentagons, and 20 hexagons. Therefore, this polyhedron has a total number of 36 faces, 102 edges, and of course 68 vertexes. It is shown the 2-dimensional representation (Schlegel diagram) of this fullerene in Figure 1, and its 3-dimensional graph in Figure 2.

Journal of Physics: Conference Series **410** (2013) 012164 doi:10.1088/1742-6596/410/1/012164

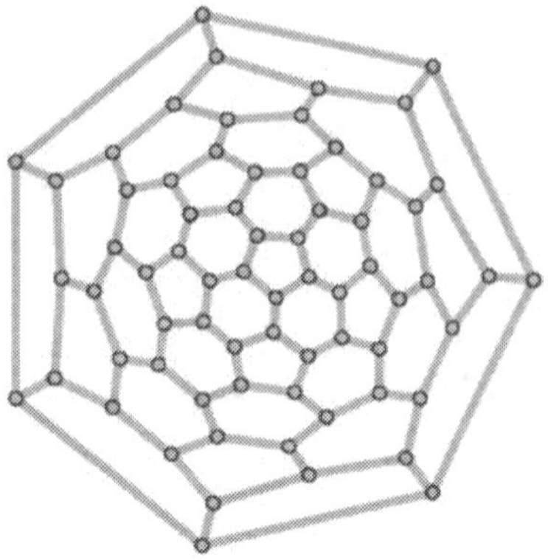

Figure 1. Schlegel diagram for the 68 carbons fullerene.

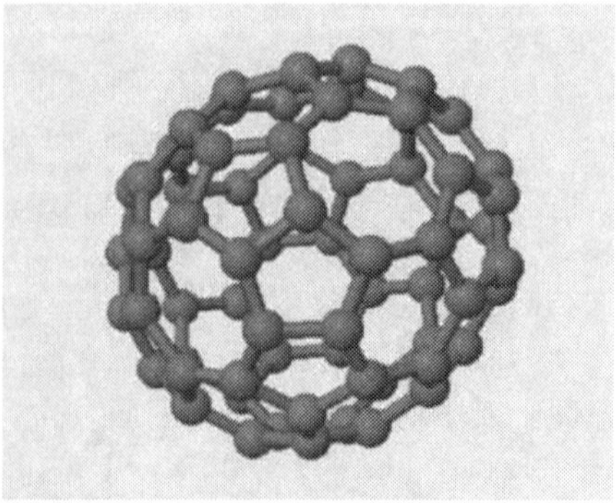

Figure 2. Fullerene with 68 carbons, 2 heptagons, 20 hexagons, and 14 pentagons.

The second example of nonclassical fullerene has 80 carbons, three heptagonal rings, 15 pentagons, and 24 hexagons. This polyhedron has a total number of 42 faces, 120 edges, and of course 80 vertexes. The Schlegel diagram of this fullerene is shown in Figure 3, and its 3-dimensional graph in Figure 4.

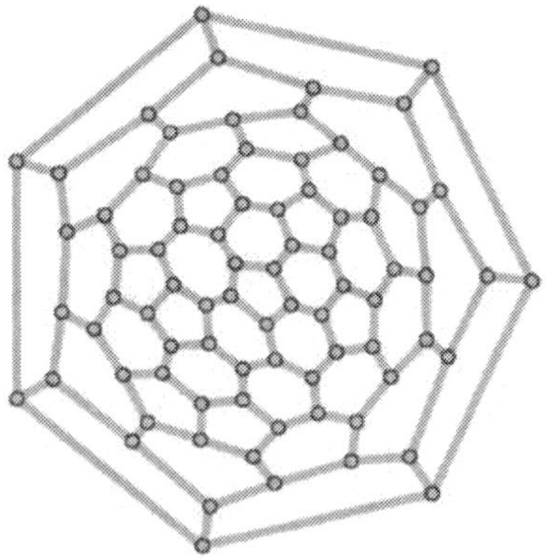

Figure 3. Schlegel diagram for the 80 carbons fullerene.

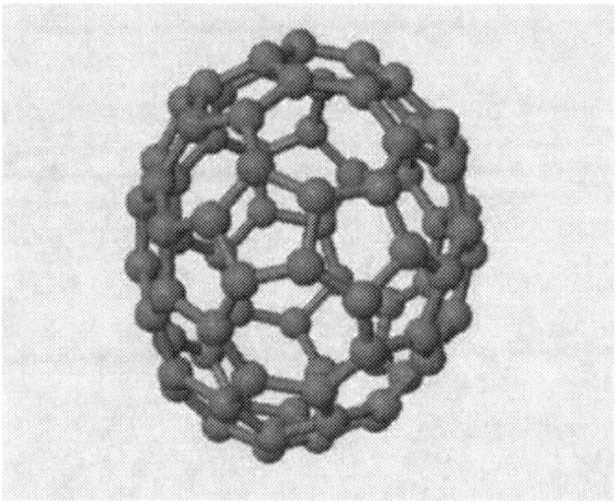

Figure 4. Fullerene with 80 carbons, 3 heptagons, 24 hexagons, and 15 pentagons.

Next, the third example of nonclassical fullerene contains 82 carbons with four heptagonal rings, 16 pentagons, and 23 hexagons. Therefore, this polyhedron has a total number of 43 faces, 123 edges, and of course 82 vertexes. It is shown the 2-dimensional representation of this fullerene in Figure 5, and its 3-dimensional graph in Figure 6.

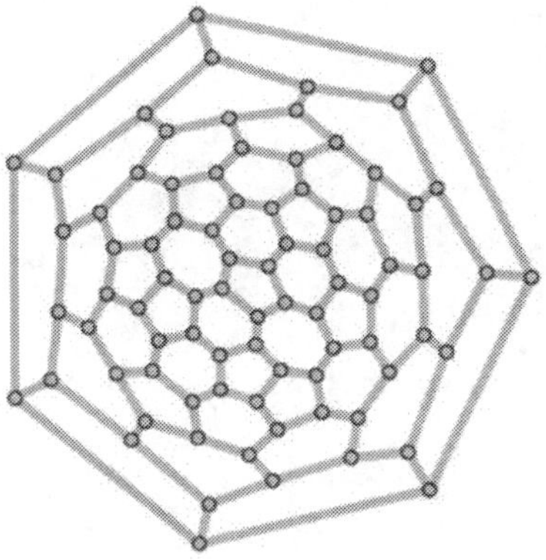

Figure 5. Schlegel diagram for the 82 carbons fullerene.

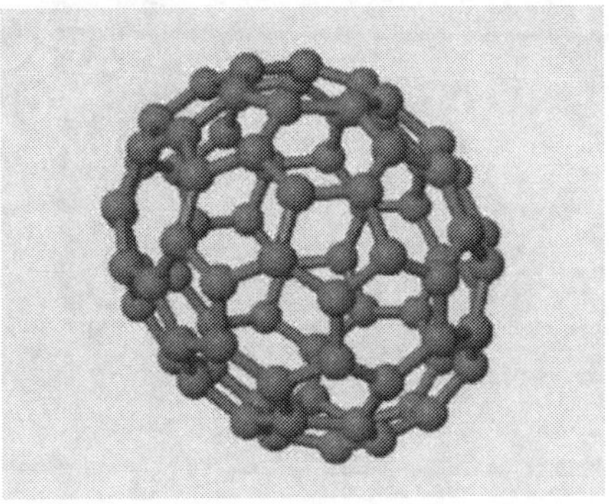

Figure 6. Fullerene with 82 carbons, 4 heptagons, 23 hexagons, and 16 pentagons.

Finally, the fourth example of nonclassical fullerene contains 76 carbons with six heptagonal rings, 18 pentagons, and 16 hexagons. Therefore, this polyhedron has a total number of 40 faces, 114 edges, and of course 76 vertexes. The 2-dimensional representation of this fullerene is shown in Figure 7, and its 3-dimensional graph in Figure 8.

Obtained results are summarized in Table 1.

Table 1. Properties of considered fullerenes.

carbons	edges	heptagons	hexagons	pentagons
68	102	2	20	14
80	120	3	24	15
82	123	4	23	16
76	114	6	16	18

4. Conclusions

We have considered four nonclassical fullerenes with two, three, four, and six heptagonal rings. In the first case, the 68 carbons fullerene contains 2 heptagonal rings. One of them is surrounded by 4 pentagons next to each other, and one pentagon lies between 2 hexagons. The other heptagonal ring is surrounded by a couple of pentagons, which lies between two hexagons; the border is

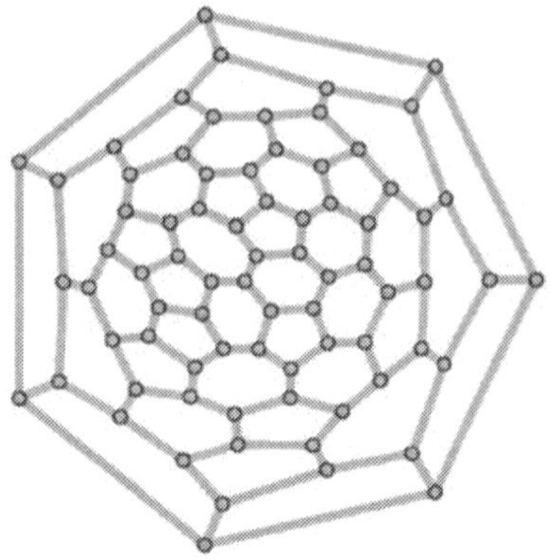

Figure 7. Schlegel diagram for the 76 carbons fullerene.

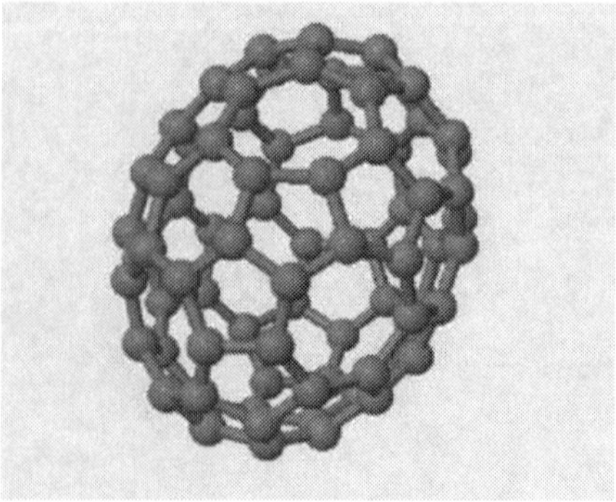

Figure 8. Fullerene with 76 carbons, 6 heptagons, 16 hexagons, and 18 pentagons.

completed by one hexagon in the middle of 2 pentagons. With respect to the 80 carbons case, the structure of the pentagons and hexagons surrounding each one of the 3 heptagonal rings follows one of the patrons of the 68 carbons fullerene. A similar situation occurs with the 82 carbons case. But, the 76 carbon fullerene, presents two new cases: a couple of pentagons, then a hexagon, followed by another couple of pentagons, and the boundary of this heptagonal ring is completed by 2 hexagons next to each other. The other type boundary surrounding a heptagonal ring is: a couple of hexagons, which lies between two pentagons; the border is completed by one pentagon in the middle of 2 hexagons.

Acknowledgments
The author thanks Berenice Sánchez and Óscar del Valle for installation of CaGe software.

References
[1] Fowler P W and Manolopoulos D E 1995 *An Atlas of Fullerenes* (Oxford: Clarendon Press)
[2] Kroto H W 1986 The stability of the fullerenes C_n with $n = 24, 28, 32, 36, 50, 60$ and 70 *Nature* **329** 529-531
[3] Campbell E E B, Fowler P W, Mitchell D and Zerbetto F 1996 *Chem. Phys. Lett.* **250** 544
[4] Ayuela A, Fowler P W, Mitchell, Schmidt R, Seifert G and Zerbetto F 1996 *J. Phys. Chem.* **1000** 15634-36
[5] An J, Gan L-H, Zhao J-Q and Li R 2010 *J. Chem. Phys.* **132** 154304
[6] Tan Y-Z, Chen R-T, Liao Z-J, Li J, Zhu F, Lu X, Xie S-Y, Li J, Huang R-B and Zheng L-S 2011 Carbon arc production of heptagon-containing fullerene[68] *Nature Comm. 2* NCOMMS1431(2011)420
[7] Gan L-H, Zhao J-Q and Hui Q 2010 *J. Comp. Chem.* **31** 1715
[8] Brinkmann G, Delgado O, Friedrichs S, Lisken A, Peeters A and Cleemput N V 2010 A virtual environment for studying some special classes of plane graphs - an update *Match Comm. Math. Comp. Chem.* **63** 533
[9] Brinkmann G, Dress A W M 1997 A constructive enumeration of fullerenes *J. Algorithms* **23** 345
[10] Coxeter H S M 1973 *Regular Polytopes* (New York: Dover)

Casimir forces for inhomogeneous planar media.

C. Xiong[1], T.W. Kelsey[1*], S.A. Linton[1] and U. Leonhardt[2]
[1]School of Computer Science, University of St Andrews, KY16 9SX, UK
[2]School of Physics and Astronomy, University of St Andrews, KY16 9SS, UK
chun.xiong@yahoo.cn, {twk,sl4}@st-andrews.ac.uk, ulf@st-andrews.ac.uk

Abstract. Casimir forces arise from vacuum fluctuations. They are fully understood only for simple models, and are important in nano- and microtechnologies. We report our experience of computer algebra calculations towards the Casimir force for models involving inhomogeneous dielectrics. We describe a methodology that greatly increases confidence in any results obtained, and use this methodology to demonstrate that the analytic derivation of scalar Green's functions is at the boundary of current computer algebra technology. We further demonstrate that Lifshitz theory of electromagnetic vacuum energy can not be directly applied to calculate the Casimir stress for models of this type, and produce results that have led to alternative regularisations. Using a combination of our new computational framework and the new theory based on our results, we provide specific calculations of Casimir forces for planar dielectrics having permittivity that declines exponentially. We discuss the relative strengths and weaknesses of computer algebra systems when applied to this type of problem, and describe a combined numerical and symbolic computational framework for calculating Casimir forces for arbitrary planar models.

1. Introduction

Casimir forces result from zero-point vacuum fluctuations confined between two dielectric materials [1]. Although these forces were predicted theoretically in the 1940s, empirical evidence confirming the theory has only been obtained in recent years [2, 3]. Lifshitz theory [4] is a theoretical approach to the calculation of Casimir forces, in which the Green's tensor for the electric field is used to derive electromagnetic stress and energy density.

The standard planar model is to have two plates, L and R, of uncharged dielectric materials separated in the x direction by a few micrometers. The materials have permittivities $\varepsilon_L(x, i\xi)$ and $\varepsilon_R(x, i\xi)$, depending on displacement and frequency ξ, which completely describe the media since we assume that there is no magnetic response (we enforce $\mu_L(x, i\xi) = 1 = \mu_R(x, i\xi)$ for the magnetic permeabilities involved). The gap between the plates, C, is either a quantum vacuum or third dielectric with $\varepsilon_C(x, i\xi)$ equal to a constant; we consider such a model to be *homogeneous*. For this model the Casimir force can be both calculated analytically and measured empirically [5].

In this paper we consider *inhomogeneous* models where the permittivity of the central region, $\varepsilon_C(x, i\xi)$, varies with x. In particular, we explore the applicability of Lifshitz theory to inhomogeneous models. The Lifshitz regularisation process described in Section 2 was derived with homogeneous media in mind, we therefore explore the possibility that this could be a confounding factor in attempts to calculate Casimir forces in the inhomogeneous case.

2. The calculation of Casimir stress in planar media

Stresses on objects in electromagnetic fields are given by Maxwell's stress tensor, in which $\hat{\mathbf{E}}$ and $\hat{\mathbf{H}}$ are respectively the electric and magnetic fields, $\hat{\mathbf{B}}$ is the magnetic induction and $\hat{\mathbf{D}}$ is the electric displacement.

$$\hat{\sigma} = \hat{\mathbf{E}} \otimes \hat{\mathbf{D}} + \hat{\mathbf{B}} \otimes \hat{\mathbf{H}} - \frac{1}{2}(\hat{\mathbf{E}} \cdot \hat{\mathbf{D}} + \hat{\mathbf{B}} \cdot \hat{\mathbf{H}})\mathbb{I}_3 \tag{1}$$

For stationary electromagnetic fields, the divergence of the Maxwell's stress tensor gives the force density $\hat{\mathbf{f}}$,

$$\hat{\mathbf{f}} = \nabla \cdot \hat{\sigma}. \tag{2}$$

The expectation values (also known as correlation functions) for the tensor products in Equation (1) are related to the retarded Green's function as follows:

$$\langle \hat{\mathbf{E}}(\mathbf{r}, t) \otimes \hat{\mathbf{D}}(\mathbf{r}', t) \rangle = -\frac{\hbar}{\pi c^2} \int_0^\infty d\xi \varepsilon(\mathbf{r}, i\xi)\xi^2 \mathbf{G}(\mathbf{r}, \mathbf{r}', i\xi), \tag{3}$$

$$\langle \hat{\mathbf{B}}(\mathbf{r}, t) \otimes \hat{\mathbf{H}}(\mathbf{r}', t) \rangle = \frac{\hbar}{\pi} \int_0^\infty d\xi \frac{1}{\mu(\mathbf{r}, i\xi)} \nabla \times \mathbf{G}(\mathbf{r}, \mathbf{r}', i\xi) \times \overleftarrow{\nabla}' \tag{4}$$

The notation $\times\overleftarrow{\nabla}'$ denotes a curl on $\mathbf{r}'$ in $\mathbf{G}(\mathbf{r}, \mathbf{r}', i\xi)$ from the right. $\mathbf{G}(\mathbf{r}, \mathbf{r}', i\xi)$ is the retarded Green's tensor for the vector potential in a Coulomb gauge, and is defined as the solution of the following inhomogeneous electromagnetic wave equation

$$\nabla \times \frac{1}{\mu} \nabla \times \mathbf{G}(\mathbf{r}, \mathbf{r}', i\xi) + \varepsilon \frac{\xi^2}{c^2} \mathbf{G}(\mathbf{r}, \mathbf{r}', i\xi) = \delta(\mathbf{r} - \mathbf{r}')\mathbb{I}_3. \tag{5}$$

The Green's function should always obey the reciprocity relation:

$$\mathbf{G}(\mathbf{r}, \mathbf{r}', i\xi) = \mathbf{G}(\mathbf{r}', \mathbf{r}, -i\xi). \tag{6}$$

We are considering planar dielectrics, for which the permittivity $\varepsilon(\mathbf{r}, i\xi) = \varepsilon(x, i\xi)$ and magnetic permeability $\mu(\mathbf{r}, i\xi) = \mu(x, i\xi)$, i.e. depend only on the x-coordinate.
The Fourier-transformed Green's function $\mathbf{G}(x, x', u, v, i\xi)$ is given by the Fourier-transformed wave equation:

$$\nabla \times \frac{1}{\mu(x, i\xi)} \nabla \times \mathbf{G}(x, x', u, v, i\xi) + \varepsilon(x, i\xi)\frac{\xi^2}{c^2} \mathbf{G}(x, x', u, v, i\xi) = \delta(x - x'). \tag{7}$$

The Casimir force depends only on the $xx-$component of Maxwell's stress tensor because the force density is also independent of y and z. In the limit $\mathbf{r} \to \mathbf{r}'$, the result for σ_{xx} is

$$\sigma_{xx} = -\frac{\hbar c}{4\pi^2} \int_0^\infty du \int_0^\infty d\kappa u \left(\frac{1}{\mu}(w^2 - \partial_x \partial_{x'})\tilde{g}_{Es} + \frac{1}{\varepsilon}(w^2 - \partial_x \partial_{x'})\tilde{g}_{Ms} \right) |_{v=0, \mathbf{x}'=\mathbf{x}}, \tag{8}$$

with

$$w = \sqrt{u^2 + v^2 + \varepsilon\mu\kappa^2}, \quad \kappa = \frac{\xi}{c},$$
$$\tilde{g}_{Es} = \tilde{g}_E - \mu\tilde{g}_0, \quad \text{and} \quad \tilde{g}_{Ms} = \tilde{g}_M - \varepsilon\tilde{g}_0.$$

In Lifshitz theory, $\tilde{g}_{Es}$ and $\tilde{g}_{Ms}$ are the regularized electric and magnetic Green's functions (where regularisation involves subtraction of the the relevant divergent part). $\tilde{g}_0$ is the infinite contribution from the retarded Green's function in a space with homogeneous medium:

$$\tilde{g}_0 = -\frac{1}{2w} e^{(-w|x-x'|)}. \tag{9}$$

Equation (7) reduces to

$$\frac{d^2}{dx^2}\tilde{g}(x) - (u^2 + v^2 + \varepsilon\kappa^2)\tilde{g}(x) = \delta(x - x') \tag{10}$$

where $\tilde{g}$ denotes either the electric or magnetic Green's function, and in which none of the left-hand parameters depends on x. The general solution involves trigonometric functions and the Heaviside function; specific solutions are easily obtained from the boundary conditions.

3. Casimir forces for inhomogeneous planar media

To obtain the Casimir force for a planar dielectric model, the sequence of calculations is now :

(i) Calculate the scalar Green's functions – Equation (7) – for the permittivity of the specific media under consideration (recalling the modelling assumption $\mu(x, i\xi) = 1$ described in Section 1). These calculations can be performed either numerically of analytically, but should incorporate two checks for internal consistency:
   (a) The limits of the Green's functions from the left and from the right should be equal;
   (b) The reciprocity relation – Equation 6 – should hold. Clearly, if either check fails then any further calculations will be in gross error. If both succeed, then there is no guarantee of correctness, although confidence in correctness is greatly increased.

(ii) Perform the corrected regularisation – Equation 7 – in order to remove (most of) the infinite parts from the above scalar Green's functions. The theory behind these new results is beyond the scope of this paper, but it should be noted that, in general, infinite components remain (at the boundaries in the examples we have presented here). It may be that these discontinuities are artefacts of the new regularisation theory that do nor represent physical truth, or it could be that Casimir stresses do tend to infinity at certain well-defined points. More research is needed into this intriguing possibility. In any event, the corrected regularisation is straightforward to compute as the parameters in the subtractions do not depend on x.

(iii) Solve the double integral – Equation (8) – to obtain the stress tensor σ_{xx}. This is the hard part: we have obtained numerical solutions only for models involving exponential decay. Our approach is to make two simplifications:
   (a) Instead of integrating from zero to ∞, we integrate from zero to a "large" finite value which, for practical purposes, is determined by how much time we are prepared to wait for a solution;
   (b) We search for approximations of values for the double integral at fixed points in the x region, followed by curve-fitting to obtain a smooth approximation of the stress, thereby reducing the problem to a single integration performed many times.

   Even after these simplifications the resulting PDEs can not, in general, be solved using the numeric libraries of either Mathematica or Maple. Moreover, to our knowledge, no library of numeric PDE solvers (such as NAG) contains routines that solve our specific type of PDE. The main difficulties arise from the nature of the problem itself: there are, for example, three regular and one irregular singular points in the PDE of $gMs(x, x')$.

(iv) The divergence of σ_{xx} is the theoretically predicted Casimir force per unit volume – Equation (2). This is trivial to compute given a smooth function that approximates the Casimir stress.

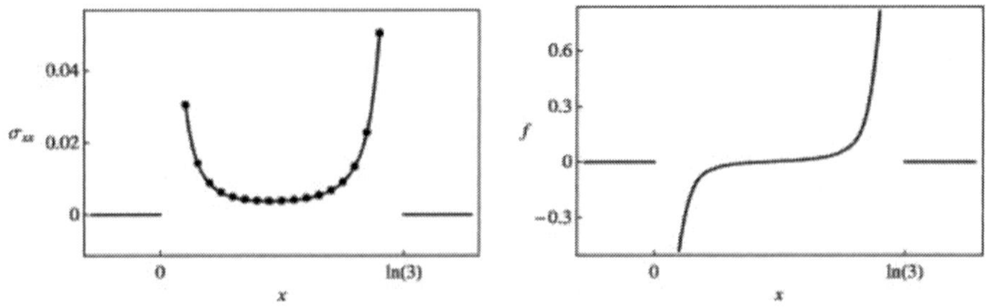

Figure 1. The Casimir stress (8) and the Casimir force per unit volume for the permittivity model $\varepsilon(x) = 3e^{-x}$ (with no magnetic response) from $x = 0$ to $x = 0.1$. In this case $\epsilon_R = log_e(3)$.

4. Conclusions

We have reported the first (to our knowledge) calculations of Casimir forces for inhomogeneous permittivity models for planar dielectrics. Intermediate calculations suggested problems with the standard regularisation process: using these insights, new theoretical regularisation results have been derived [6], and we have incorporated and tested this new theory. We have described the computational framework used to obtain Casimir forces, and highlighted the work still needed before efficient computation for arbitrary models becomes a reality.

Our findings suggest that the calculation of scalar Green's functions for arbitrary inhomogeneous media is at the boundary of the current capabilities of Maple and Mathematica. Using Maple we can get satisfactory results for exactly one model, and believe we could increase the number of such models if the modified Heun function implementation within Maple were to be improved. Mathematica is less useful for these calculations, as no Heun function implementation is present in the current system. However, we have successfully replicated Maple results using Mathematica, indicating that the lengthy and complex Mathematica expressions produced as intermediate output are completely correct.

Future avenues of research include (i) the testing of any proposed alternative regularisation using our methodology of comparing results from two systems for both the the right and left limits, (ii) the development of a combined numeric-symbolic framework that agrees with the symbolically derived results presented here for the exponential model, and which can be used to calculate Casimir forces for those inhomogeneous model that lie beyond the current analytic capabilities of Maple and Mathematica.

Casimir forces remain difficult, both theoretically and in terms of practical computation. By predicting Casimir forces for more complicated models than simple constant-permittivity dielectrics, we can investigate the poorly-understood effects and interactions of forces at nano-scales. Moreover, our results form hypotheses for applied physics investigations: can the forces that we predict be measured in practical experiments? This multi-disciplinary approach involves investigations that are at, or beyond, the the current limits of computational mathematics, theoretical physics and applied physics.

[1] Casimir H B G 1948 *Proc. K. Ned. Akad. Wet.* **51** 793–795
[2] Lamoreaux S K 1997 *Physical Review Letters* **78** 5–8
[3] Hertlein C, Helden L, Gambassi A, Dietrich S and Bechinger C 2008 *Nature* **451** 172–175 ISSN 0028-0836
[4] Dzyaloshinskii I E, Lifshitz E M and Pitaevskii L P 1961 *Advances in Physics* **10** 165–209
[5] Bordag M, Klimchitskaya G L, Mohideen U and Mostepanenko V M 2009 *Advances in the Casimir Effect* (Oxford)
[6] Philbin T, Xiong C and Leonhardt U 2010 *Annals of Physics* **325** 579 – 595 ISSN 0003-4916

Fluorescence Computed Tomography with Polychromatic Source Data

Eduardo Miqueles & Alvaro R.De Pierro⋆

Department of Applied Mathematics, University of Campinas, São Paulo, Brazil

⋆ Supported by FAPESP grant No 09/15844-4, CNPq grants No 476825/2006-0 and 304820/2006-7

E-mail: `miqueles@ime.unicamp.br`, `alvaro@ime.unicamp.br`

Abstract. Fluorescence computed tomography is a synchrotron imaging technique aiming at reconstructing the fluorescence emission within a sample object. For a polychromatic source hitting the object, the amount of fluorescence detected is defined by a linear equation. For the monochromatic case, the operator is a Generalized Attenuated Radon Transform (GART). The main goal is to reconstruct the density function, given the sinogram data and the weight function. An eficient iterative algorithm for the inversion of the GART was presented recently by the authors. This inversion can only be performed if the weight function is previously known, which means that $\mu = \mu(\cdot, \varepsilon)$ and λ are also known. For monochromatic XFCT (acronym for x-rays fluorescence computed tomography), the determination of λ is a dificult task, and we have considered the approximation $\lambda \approx \mu$, which is valid for low energies ranging from 3Kev to 10Kev. So, for solving our problem, the first step is to find μ given the polychromatic sinogram. There are different approaches for this in the literature. Recently, an elegant and efficient method for solving this problem was introduced, using a fixed point algorithm. Opposite to this, where $\mu(\cdot, \varepsilon)$ needs to be computed for all $\varepsilon \in E$, we claim that the integral of $\mu(\cdot, \varepsilon)$ for all ε has a physical meaning and provides a good aproximation for the solution. Also we present fast algorithm for computations

1. Introduction

$$p(\eta) = \int_E \int_{\Omega(\eta)} S(\varepsilon) f(\mathbf{x}) W_{\mu_\varepsilon, \lambda}(\mathbf{x}, \eta) d\mathbf{x} d\varepsilon$$

Equation 1: XFCT polychromatic sinogram p, given the emission f and linear attenuation coefficients μ and λ. Weight function W can be found in [3].

Polychromatic x-ray fluorescence tomography is a new imaging technology, which is able to be implemented on a synchrotron device. In this paper we give a brief mathematical formalism to this problem and propose an approximated algorithm for the solution. When the source is monochromatic, several algorithms have been proposed, and a description can be found in [2] and [3]. Using the same physical approach, but now taking into account many energy levels for the incoming x-ray, the yield of fluorescence detected in the polychromatic case is given by the sinogram p defined in Eq.1. In that equation, $\Omega(\eta)$ determines the x-ray path (parameterized by η) and $\mathbf{x} \in \Omega$. The positive function S describes the spectrum of energy, varying on the interval $E = [\varepsilon_{\min}, \varepsilon_{\max}]$, for the sample under investigation. W is a weight function containing λ and μ_ε (see [2] for exact definitions) where $\lambda = \lambda(\mathbf{x})$ represents the fluorescence attenuation coefficient for rays emanating from within the sample at pixel $\mathbf{x}$, $\mu_\varepsilon = \mu_\varepsilon(\mathbf{x})$ is the transmission attenuation of the sample, for the incident ray, which is dependent on the energy ε. For the monochromatic case, i.e., $S(\varepsilon) = \delta(\varepsilon - \bar{\varepsilon})$, the operator is a Generalized Attenuated Radon Transform (GART) [2]. The main goal is to reconstruct the density function $f = f(\mathbf{x})$ from Eq.1, given the sinogram data $p(\eta)$ and the weight function W. It is straightforward to note that Eq.1 can be rewritten as Eq.2 just after the interchange of the integrals.

$$p(\eta) = \int_{\Omega(\eta)} f(\mathbf{x})\omega(\mathbf{x},\eta)d\mathbf{x} \qquad \text{(a)}$$

$$\omega(\mathbf{x},\eta) = \int_E W_{\mu_\varepsilon,\lambda}(\mathbf{x},\eta)d\varepsilon \qquad \text{(b)}$$

Equation 2: (a) Generalized Attenuated Radon Transform (GART) (b) Weight function for polychromatic XFCT.

An eficient iterative algorithm for the inversion of Eq.2.(a) was presented in [3]. This inversion can only be performed if the weight function is previously known, which means that $\mu_\varepsilon = \mu(\cdot,\varepsilon)$ and λ are also known. For monochromatic XFCT, the determination of λ is a dificult task, and we have considered the approximation $\lambda \approx \mu_\varepsilon$ with good results. Function μ_ε obeys the polychromatic transmission model at Eq.3, with $\mathscr{R}$ standing for the classical Radon Transform. So, for solving Eq.1, the first step is to find μ_ε given the polychromatic sinogram $r(\eta)$. There are different approaches for this in the literature.

$$r(\eta) = \int_E S(\varepsilon)e^{-\mathscr{R}\mu_\varepsilon(\eta)}d\varepsilon$$

Equation 3: Model for function $\mu_\varepsilon = \mu(\cdot,\varepsilon)$ with $S(\varepsilon)$ the energy spectrum.

In [4], an elegant and efficient method for solving Eq.3 was introduced, using a fixed point algorithm. Since we have an underdetermined problem, the method in [4] is based on assuming that the transmission map μ is linear combination of known attenuation maps, with known coefficients depending on ε. **Remark:** In fact, we have to solve Eq.3 for each function μ_ε, given the sinogram $q(\eta) = -\log r(\eta)$; this is more realistic according to Lambert-Beer law.

2. Decomposition at Energy Levels

Each chemical element from the periodic table has a linear mass attenuation coefficient that depends explicitly on the frequency of the incident eletromagnetic wavefront exciting him. Plotting this dependency at different energy levels ε (i.e. different frequencies), we obtain a plot similar to the one presented in Fig.1 for Calcium. In this figure, we clearly see that for energies in the neighborhood of $\varepsilon = 4$ there is a jump for this attenuation indicating a K-level energy for the chemical element. We denote this function as $U_Z(\varepsilon)$. In a general setting, where an object described by attenuation μ is composed with more than just one element, this association has to be done at each *pixel* $\mathbf{x}$. Also, for a fixed energy level $\varepsilon \in E$, the geometry of the object determined by the map $\mu_\varepsilon = \mu_\varepsilon(\mathbf{x})$ is usually refered as *a phantom*. We assume that we have a discrete energy set $E = \{\varepsilon_0, \ldots, \varepsilon_N\}$.

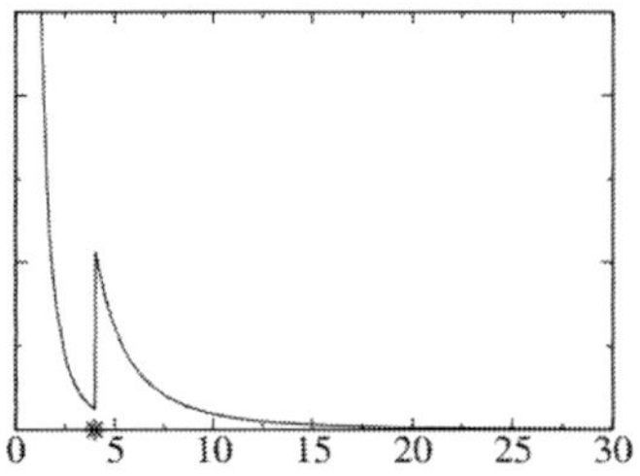

Figure 1: $U_Z = U_Z(\varepsilon)$ for `Calcium` in the range [1Kev,30Kev].

Using the same approach of [4] we describe the phantom μ_ε as a linear combination of the constituent elements $\{\mathsf{Z}_0, \ldots, \mathsf{Z}_J\}$ with Z_j describing the atomic number. Therefore $\mu_\varepsilon(\mathbf{x}) = \sum_{j=1}^{J} U_{\mathsf{Z}_j}(\varepsilon)v_j(\mathbf{x})$ where each function U_Z is previously known for each element[1]. We use $J = 7$ and Fig.2 as a reference to set v_j. Region R_4 is the main body of the object[2] with air within R_8. Hence, $v_j(\mathbf{x}) = \mathbf{1}_{\mathsf{R}_j}(\mathbf{x})$ and there is Zinc, Nickel, Manganese, Iron, Potassium, Chlorine and Copper within regions $\{\mathsf{R}_5, \mathsf{R}_6, \mathsf{R}_7, \mathsf{R}_3, \mathsf{R}_1, \mathsf{R}_2\}$, respectively. We present in Fig.3, three phantoms $\mu_\varepsilon(\cdot)$, each one at a different energy level ε. For simplicity, the images depicted in Fig.3 are displayed without a scaling bar, althoug Fig.3.(c) has lower intensity compared to Fig.3.(a). Indeed, U_Z is a decreasing function of ε.

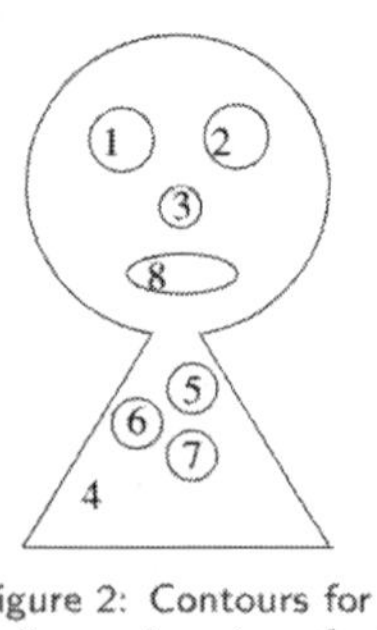

Figure 2: Contours for indicator functions $\{v_j\}$ at region R_j.

[1] `http://www.nist.gov/index.html`
[2] We use $\mathbf{1}_A(\mathbf{x})$ as the indicator function over the set A.

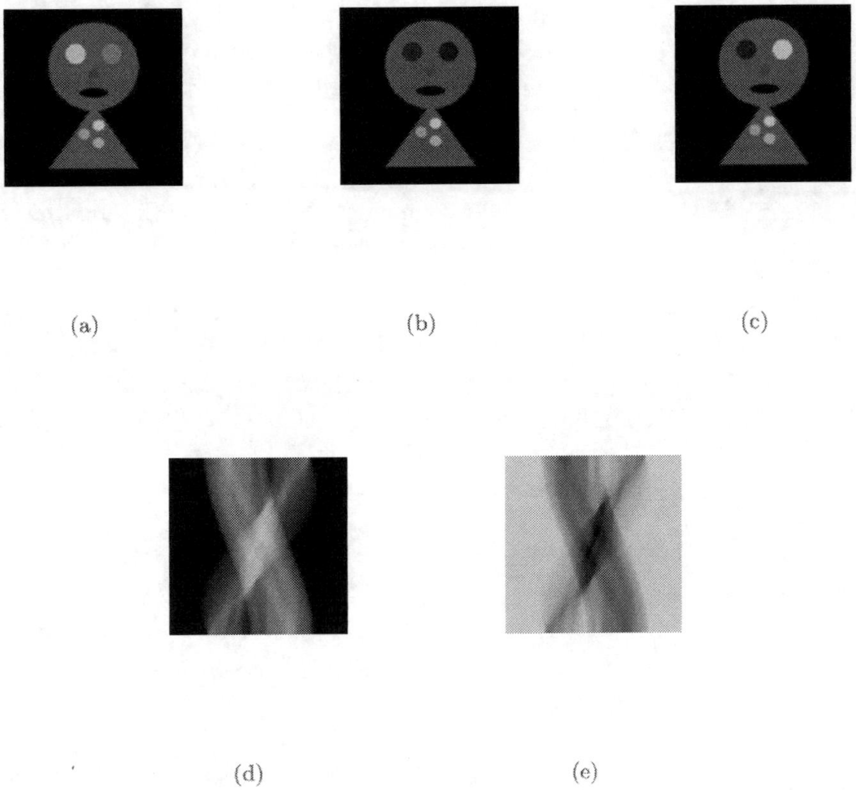

(a) (b) (c)

(d) (e)

Figure 3: (a)-(c): Functions $\mu(\varepsilon_{60}, \mathbf{x})$, $\mu(\varepsilon_{180}, \mathbf{x})$ and $\mu(\varepsilon_{380}, \mathbf{x})$ respectively; where $\varepsilon_{60} = 4$Kev, $\varepsilon_{180} = 10$Kev and $\varepsilon_{380} = 20$Kev (d) Sinogram $q(\eta) = -\log r(\eta)$ (e) Sinogram $r(\eta)$ (See Eq.3)

3. Reconstruction Algorithm

We begin by noticing that the classical filtered back-projection algorithm, for the inverse of the Radon transform $\mathscr{R}^{-1}$, when applied to the sinogram $q(\eta) = -\log r(\eta)$, gives rise to a reconstructed map[3] $\tilde{\mu}(\mathbf{x})$, very similar to the original phantom μ_ε, for a given ε. This is shown in Fig.4, where we clearly see that the constrast within each region is averaged by the back-projection method. Therefore, it is reasonable to imagine the map μ_ε as a distortion of $\tilde{\mu}$, with an error that can be reduced iteratively, as in [3]. In a discrete setting, we want to reconstruct functions $\{\mu_{\varepsilon_1}, \dots, \mu_{\varepsilon_N}\}$ obeying Eq.4 for a given $r(\eta)$, as in Fig.3.(e). Our aim is to find μ_{ε_j} from r, for each $j = 1, \dots, N$.

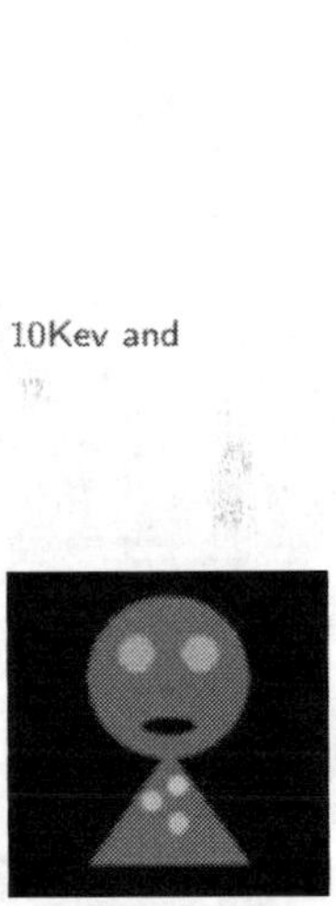

Figure 4: Map function $\mathscr{R}^{-1}[-\log r]$ with r given by Eq.3

$$\mathcal{P}[\mu_{\varepsilon_1}, \dots, \mu_{\varepsilon_N}](\eta) \equiv \sum_{j=1}^{N} S(\varepsilon_j) e^{-\mathscr{R}\mu_{\varepsilon_j}(\eta)} \Delta\varepsilon_j = r(\eta)$$

Equation 4: Discrete model for r from Eq.3. Given r, we want to recover function μ_{ε_j}, $\forall\, j$

Ignoring ε_i for $i \neq j$, we rewrite Eq.4 simply as $\mathcal{P}[\mu_{\varepsilon_j}] = r$. This is an underdetermined problem because the operator $\mathcal{P}$ depends on the other energies ε_i for $i \neq j$. We propose an approximation for $\mathcal{P}$ in order to detach our equation into N easier subproblems.

[3] A general phantom lies within a L^2-space of compactly supported functions, denoted as $\mathbf{U}$.

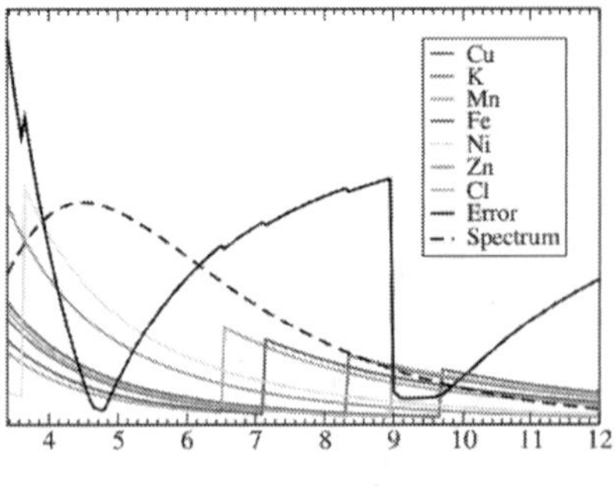

Figure 5: Error analysis. See text for details.

Returning to the continuum, and fixing an $\varepsilon \in E$, recall that $\mathcal{P}(\mu_\varepsilon)$ is determined by the integral in Eq.3. From the Mean Value Theorem, there exists $\tau \in E$ such that $\mathcal{P}(\mu_\varepsilon) = \exp[-\mathscr{R}\mu_\tau(\eta)]$. The value for τ is theoretical, although we discovered - after several numerical simulations - that τ lies in the neighborhood of $\varepsilon^* = \operatorname{argmax}\{S(\varepsilon), \varepsilon \in E\}$; this is shown in Fig.5. Indeed, the solid line represents the error curve $\| \exp[-\mathscr{R}\mu_\varepsilon] - r\| / \|r\|$ for $3 \leq \varepsilon \leq 12$. We can clearly observe two local minima, the first one being the most important and very close to the maximum point of S (displayed in dashed line), i.e., near $\varepsilon = 4.5$Kev). We will assume throughout this paper the approximation $\tau = \varepsilon^*$. The nonincreasing functions U_Z are also depicted in Fig.5. Using the linear combination from $\mu_\varepsilon(\mathbf{x})$ from previous Section, we see that $\frac{\partial \mu_\varepsilon}{\partial \varepsilon} < 0$ since $U_Z'(\varepsilon) < 0$. This is also true in general, i.e., the higher the energy, the lower the attenuation. From this fact, and also from numerical experiments, we observed that $\frac{\partial \mu_\varepsilon(\mathbf{x})}{\partial \varepsilon} \approx -C\mu_\varepsilon(\mathbf{x})$ for some constant C. Let us assume $C = 1$. It follows from the mean value theorem that for $\ell \in (0,1)$

$$\mu_\varepsilon(\mathbf{x}) - \mu_\tau(\mathbf{x}) = \frac{\partial \mu_\rho(\mathbf{x})}{\partial \varepsilon}(\varepsilon - \tau), \quad \rho = (1-\ell)\varepsilon + \ell\tau \quad \Rightarrow \quad \mu_\varepsilon(\mathbf{x}) - \mu_\tau(\mathbf{x}) \approx -\mu_\varepsilon(\mathbf{x})(\varepsilon - \tau) \tag{4}$$

Using Eq.4 and the linearity of $\mathscr{R}$, we approximate $\mathcal{P}$ using the following sequence: $e^{-\mathscr{R}\mu_\tau} = e^{-\mathscr{R}_\varepsilon} e^{\mathscr{R}(\mu_\varepsilon - \mu_\tau)}$ implying $e^{-\mathscr{R}\mu_\tau} \approx e^{-\mathscr{R}\mu_\varepsilon} e^{-(\varepsilon-\tau)\mathscr{R}\mu_\varepsilon}$ and therefore $e^{\mathscr{R}\mu_\tau} \approx e^{[1+(\varepsilon-\tau)]\mathscr{R}\mu_\varepsilon}$. Hence,

$$\mathcal{P}[\mu_\varepsilon] \approx \mathbb{P}[\mu_\varepsilon] = e^{-g(\varepsilon)\mathscr{R}\mu_\varepsilon}$$

with $g(\varepsilon) = 1 + (\varepsilon - \tau)$. Our new system of equation is $\mathbb{P}[\mu_\varepsilon] \approx r$, or equivalently, $\mathcal{H}[\mu_\varepsilon] \equiv \mathbb{P}[\mu_\varepsilon] - r \approx 0$. We use Newton's method to solve $\mathcal{H}[\mu_\varepsilon] \approx 0$ in order to find an approximate solution μ_ε. It is straightforward that the Fréchet derivative of $\mathcal{H}$ at $u \in \mathbf{U}$, denoted by $\mathcal{H}'[u]$, is the linear transformation $\mathcal{H}'[u] = -g(\varepsilon)\mathbb{P}[u]\mathscr{R}$. This yields $\mathcal{H}'[u] \cdot h = -g(\varepsilon)\mathbb{P}[u]\mathscr{R}h$ for any $h \in \mathbf{U}$. The reconstruction method, for each energy level $\varepsilon \in E$, is depicted in Algorithm 1, within the above box.

(1) $\mu_\varepsilon^{(k+1)}(\mathbf{x}) = \mu_\varepsilon^{(k)}(\mathbf{x}) + z^{(k)}(\mathbf{x}), \quad \varepsilon \in E$

(2) $\mathcal{H}'\left[\mu_\varepsilon^{(k)}\right] \cdot z^k = -\mathcal{H}\left[\mu_\varepsilon^{(k)}\right]$

(3) $z^{(k)} = \frac{1}{g(\varepsilon)}\mathscr{R}^{-1}\left[\mathcal{H}[\mu_\varepsilon^{(k)}]/\mathbb{P}[\mu_\varepsilon^{(k)}]\right]$

$(\star)$ $\mu_\varepsilon^{(k+1)} = \mu_\varepsilon^{(k)} + \frac{1}{g(\varepsilon)}\mathscr{R}^{-1}\left[1 - e^{g(\varepsilon)\mathscr{R}\mu_\varepsilon^{(k)}} r\right]$

Algorithm 1: Sequence of steps to develop the reconstruction algorithm $(\star)$, for a fixed ε. See text for details.

We emphasize that the final equation $(\star)$ at Algorithm 1, is equivalent to a fixed point method for the Fredholm equation $u(\mathbf{x}) = \bar{u}(\mathbf{x}) + \mathscr{D}u(\mathbf{x})$ associated to $\mathbb{P}[u] = r$, where $\bar{u} = \mathbb{J}^{-1}[r]$, $\mathscr{D}u = \mathbb{J}^{-1}(r - \mathbb{P}[u])$ and $\mathbb{J} = \mathcal{H}'(u)$. The convergence is guaranteed whenever $\mathbb{J}^{-1}$ is a good approximation for the inverse of $\mathbb{P}$.

4. Numerical Results

For our simulations, we have considered a 256×256 phantom μ_ε, as in Fig.2 and a polychromatic sinogram $q = -\log(r)$ obtained with 300 views and 300 rays at each view. The results are shown below. The stopping criterion used for Algorithm 1 was $\|\mu_\varepsilon^{(k+1)} - \mu_\varepsilon^{(k)}\| \leq \sigma$ with $\sigma = 10^{-3}$.

Simulations have been performed using `raft` [1], a general purpose reconstruction package. The approximated results are much better than the straightforward inversion by Filtered Back Projection from the data. Moreover, the method is more general

Acknowledgements: We thank *Carlos Alberto Pérez*, from the National Laboratory of Synchrotron Light, Brazil, for useful discussions and providing spectrum data and mass attenuation coefficients. The first author was supported by FAPESP Grant No 09/15844-4, and the second by CNPq Grants No 476825/2006-0 and 304820/2006-7

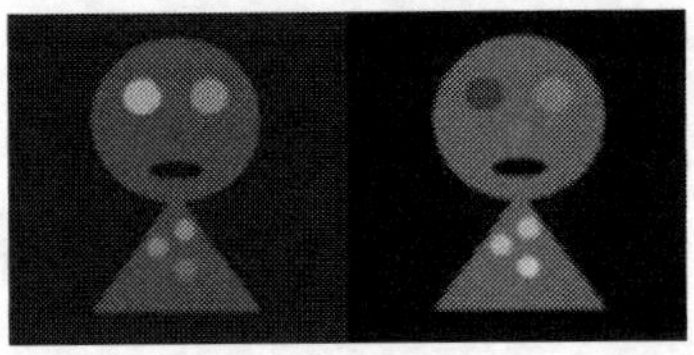

(a)

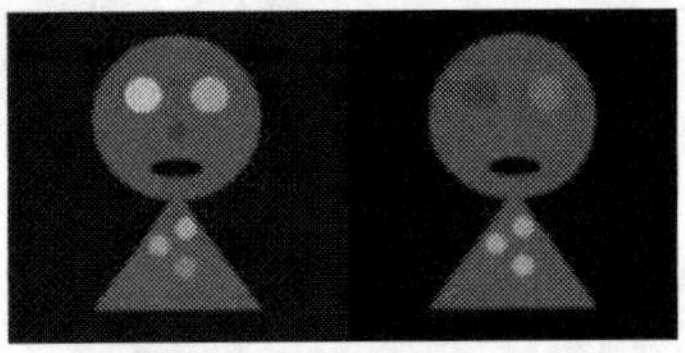

(b)

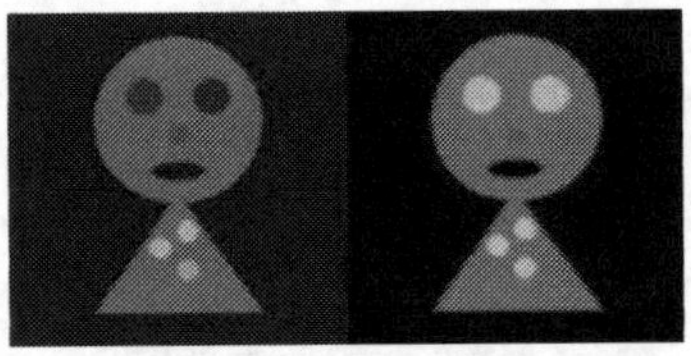

(c)

Figure 6: Comparison between original phantom μ_ε (left) and reconstructed $\mu_\varepsilon^{(k)}$ (right). (a) $\varepsilon = \varepsilon_{100} = 5.95$Kev and $k = 30$ iterations. (b) $\varepsilon = \varepsilon_{80} = 4.95$Kev and $k = 15$ iterations. (c) $\varepsilon = \varepsilon_{200} = 10.95$Kev and $k = 3$

than the one suggested in [4], that needs additional hypotheses on the behaviour of the attenuation coefficients. This makes it applicable to lower K-binding energy values.

References

[1] E.Miqueles, A.R.De Pierro, *C-library raft: Reconstruction algorithms for tomography. Applications to X-ray fluorescence tomography*, Computer Physics Commun. 182(12): 2661-2673, 2011.

[2] E.Miqueles, A.R.De Pierro, *Exact analytic reconstruction in x-ray fluorescence CT and approximated versions* , Physics in Medicine and Biology, Vol.55, No.4, 2010.
[3] E.Miqueles, A.R.De Pierro, *Iterative reconstruction in X-ray fluorescence tomography based on Radon inversion*, IEEE Trans.Medical Imaging, Vol.30, No.2, February 2011.
[4] C.Yan, R.Whalen, G.Beaupré, S.Yen, S.Napel, *Reconstruction Algorithm for Polychromatic CT Imaging: Application to Beam Hardening Correction*, Trans.Med.Imaging, Vol. 9, No.1, January 2000.

Journal of Physics: Conference Series **410** (2013) 012167 doi:10.1088/1742-6596/410/1/012167

Fast Continuous Iterative Methods in X-Rays Fluorescence Computed Tomography

Eduardo Miqueles & Alvaro R.De Pierro⋆

Department of Applied Mathematics, University of Campinas, São Paulo, Brazil

⋆ FAPESP grant No 09/15844-4, CNPq grants No 476825/2006-0 and 304820/2006-7

E-mail: `miqueles@ime.unicamp.br`, `alvaro@ime.unicamp.br`

Abstract. Fluorescence computed tomography is a synchrotron imaging technique that reconstructs the fluorescence emission within a sample object. For a monochromatic source hitting the object, the amount of fluorescence detected is given as a linear equation. Iterative methods based on the inversion of the Radon transform were introduced. These methods were compared with the Expectation Maximization algorithm, implemented in a continuous setting. This implementation provided better quality results, but with higher computational cost. Recently, a faster OS-EM algorithm was applied in XFCT, in a discrete setting. In this manuscript we further improve on previous results by considering a relaxed version of OS-EM, in a continuous setting (faster implementation per iteration).

1. Introduction

Fluorescence computed tomography is a synchrotron imaging technique that reconstructs the fluorescence emission within a sample object. For a monochromatic source hitting the object, the amount of fluorescence detected is given by p as the following equation

$$p(\eta) = \int_{\Omega(\eta)} f(\mathbf{x}) W_{\mu,\lambda}(\mathbf{x},\eta) d\mathbf{x} \tag{1}$$

where $\Omega(\eta)$ determines the path of the incoming wave (parameterized by η) and $\mathbf{x} \in \Omega$. W is a weight function containing λ and μ_ε (see [2] for exact definitions) where $\lambda = \lambda(\mathbf{x})$ represents the fluorescence attenuation coefficient for rays emanating from within the sample at the pixel $\mathbf{x}$, $\mu_\varepsilon = \mu_\varepsilon(\mathbf{x})$ is the transmission attenuation of the sample, for the incident ray, which is dependent on the energy ε. The operator $\mathscr{R}_W f(\eta)$ defined by (1) is a Generalized Attenuated Radon Transform (GART) [2]. The main goal is to reconstruct the density function $f = f(\mathbf{x})$ from (1), given the sinogram data $p(\eta)$ and the weight function W.

Variable $\eta = (t,\theta)$ stands for a polar coordinate system, where $|t| \leq 1$ and $\theta \in [0, 2\pi]$, as depicted in Figure 1. Also, $\xi = (\cos\theta, \sin\theta)$ is a direction vector.

2. RAMLA

In [3], iterative methods based on the inversion of the Radon transform were introduced. These methods were compared with the Expectation Maximization algorithm, implemented in a continuous setting. This implementation provided better quality results, but with higher

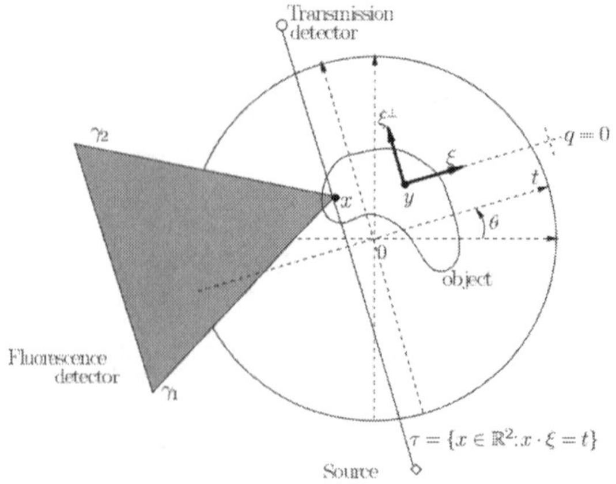

Figure 1: Geometry for XFCT experiments

computational cost. In [5] a faster OS-EM algorithm was applied in XFCT, in a discrete setting. In this manuscript we further improve on previous results [3, 5] by considering a relaxed version of OS-EM (see [1]), in a continuous setting (faster implementation per iteration), that is given by RAMLA (*Row-action maximum likelihood algorithm*). Such an algorithm is defined below by **A**:

$$f^{(k,i+1)}(\mathbf{x}) = \mathbf{A}[f^{(k,i)}](\mathbf{x}) \tag{2}$$

$$\equiv f^{(k,i)}(\mathbf{x}) + \lambda_k f^{(k,i)}(\mathbf{x}) \mathscr{B}_W^{(i)} p^{(k,i)}(\mathbf{x}) \tag{3}$$

$$= f^{(k,i)}(\mathbf{x}) \left[1 + \lambda_k\right] \mathscr{B}_W^{(i)} p^{(k,i)}(\mathbf{x}) \tag{4}$$

with λ_k a positive relaxation parameter tending to zero sublinearly and

$$p^{(k,i)}(t,\theta) = \frac{p(t,\theta)}{\mathscr{R}_W f^{(k,i)}(t,\theta)} - 1. \tag{5}$$

The operator $\mathscr{B}_W^{(i)}$ is a partial Backprojection operator, integrating over I_i, defined by

$$\mathscr{B}_W^{(i)} p(\mathbf{x}) = \int_{\theta \in I_i} W(\mathbf{x},\theta) p(\mathbf{x} \cdot \xi, \theta) d\theta \tag{6}$$

Here, $[0, 2\pi] = \bigcup_{i=0}^{N-1} I_i$, defines a partition of the interval $[0, 2\pi]$. In practice, $[0, 2\pi]$ is represented by a grid $\{\theta_0, \theta_1, \theta_2, \ldots, \theta_{N-1}\}$. A schematic representation for the algorithm **A** follows from Table 1.

#	Iter. $k=0$	Iter. $k=1$	Iter. $k=2$	
Begin Cycle	$f^{(0,0)} = f^{(0)}$	$f^{(1,0)} = f^{(1)}$	$f^{(2,0)} = f^{(2)}$	...
Cycle	$f^{(0,1)} = \mathbf{A}[f^{(0,0)}]$ $f^{(0,2)} = \mathbf{A}[f^{(0,1)}]$ $\vdots$ $f^{(0,N-1)} = \mathbf{A}[f^{(0,N-2)}]$	$f^{(1,1)} = \mathbf{A}[f^{(1,0)}]$ $f^{(1,2)} = \mathbf{A}[f^{(1,1)}]$ $\vdots$ $f^{(1,N-1)} = \mathbf{A}[f^{(1,N-2)}]$	$f^{(2,1)} = \mathbf{A}[f^{(2,0)}]$ $f^{(2,2)} = \mathbf{A}[f^{(2,1)}]$ $\vdots$ $f^{(2,N-1)} = \mathbf{A}[f^{(2,N-2)}]$	...
End Cycle	$f^{(1)} = f^{(0,N-1)}$	$f^{(2)} = f^{(1,N-1)}$	$f^{(3)} = f^{(2,N-1)}$	...

Table 1: RAMLA Algorithm **A**, from Eq.(2)

We use the following partition for the interval $[0, 2\pi]$

$$I_i = [i + T] = \{j \in \mathbb{Z} : j \equiv (i + t) \bmod N, \ t \in T\} \tag{7}$$

Figure 2: Phantom $f(\mathbf{x})$ used for reconstruction at left and sinogram $p(\eta)$ at right.

with $T \subset \mathbb{Z}$. In this article, we use:

(a) $T = \emptyset$. Here, the partial backprojection (6) reduces to $\mathscr{B}_W^{(i)} p(\mathbf{x}) = W(\mathbf{x}, \theta_i) p(\mathbf{x} \cdot \xi_i, \theta_i)$ yielding (2) as the Twomey [1] formula:

$$f^{(k,i+1)}(\mathbf{x}) = f^{(k,i)}(\mathbf{x}) + \lambda_k f^{(k,i)}(\mathbf{x}) W(\mathbf{x}, \theta_i) \left[\frac{p(\mathbf{x} \cdot \xi_i, \theta_i)}{\mathscr{R}_W f^{(k,i)}(t, \theta)} - 1 \right] \tag{8}$$

(b) $T = \{0, 3, 6\}$. In this case $I_i = \{i, [i+3], [i+6]\}$.

(c) $T = \{0, 3, 6, 9, 12, 15\}$. In this case $I_i = \{i, [i+3], [i+6], [i+9], [i+12], [i+15]\}$.

Using (b) and (c) at (6), we obtain the following iteration formula,

$$f^{(k,i+1)}(\mathbf{x}) = f^{(k,i)}(\mathbf{x}) + \frac{\lambda_k}{\mathsf{card}(I_i)} f^{(k,i)}(\mathbf{x}) \sum_{\ell \in I_i} W(\mathbf{x}, \theta_\ell) \left[\frac{p(\mathbf{x} \cdot \xi_\ell, \theta_\ell)}{\mathscr{R}_W f^{(k,i-1)}(\mathbf{x} \cdot \xi_\ell, \theta_\ell)} - 1 \right]. \tag{9}$$

We denote $I_i^{(a)}$, $I_i^{(b)}$ and $I_i^{(c)}$ as the index set from equation (7) with subset T described as above, at (a), (b) and (c). Also, card denote the cardinality of a set.

3. Numerical Results

The relaxation parameter used for algorithm (2) was $\lambda_k = \frac{1}{k}$. The results were generated using constant attenuation functions μ and λ for weight function $W_{\mu,\lambda}$ at (1) (see [3] for a precise definition of W). The phantom used for reconstruction is due to Golosio [6], depicted in Figure 2. The numerical simulations were carried out without noise, as shown in the sinogram from Figure 2. We use a starting point for algorithm (2) as $f^{(0)}(\mathbf{x}) = \alpha$, $\forall\, \mathbf{x}$, where $\alpha^{-1} = \int_N p(\eta) d\eta$ and $N = [0, 2\pi] \times [-1, 1]$. Also, the discretized version of $f(\mathbf{x})$ has dimension 64×64 and $\eta = (t, \theta) \in N$ varies within a meshgrid with dimensions 100×180; i.e., 180 views and 100 rays at each view.

The results are displayed in the following sequence:

(I) Partial backprojection (6), using $I_0^{(a)}$ and $I_0^{(c)}$, are presented in Figure 3. **(II)** Reconstruction using the expectation maximization algorithm, see [3], for this particular data is presented in Figure 4, depicting iterates

Figure 3: Partial backprojection at $i = 0$ using two diferent angle sets: $I_0^{(a)}$ and $I_0^{(c)}$. See text for details.

$f^{(3)}$, $f^{(5)}$ and $f^{(7)}$. **(III)** Reconstruction with (2) and $I_i^{(a)}$ for the partial backprojection (6) is presented in Figure 5, depicting iterates $f^{(3)}$, $f^{(5)}$ and $f^{(7)}$. **(IV)** Reconstruction with (2) and $I_i^{(b)}$ for the partial backprojection (6) is presented in Figure 6, depicting iterates $f^{(3)}$, $f^{(5)}$ and $f^{(7)}$. **(V)** Reconstruction with (2) and $I_i^{(c)}$ for the partial backprojection (6) is presented in Figure 7, depicting iterates $f^{(3)}$, $f^{(5)}$ and $f^{(7)}$.

† Simulations have been performed using `raft`, a general purpose reconstruction package [4]

4. References

[1] J.A. Browne and A.R De Pierro, *A row-action alternative to the EM algorithm for maximizing likelihoods in emission tomography*, IEEE Trans. Medical Imaging, V15, 4, 687-699, October 1996.

[2] E.Miqueles and A.R.De Pierro, *Exact analytic reconstruction in x-ray fluorescence CT and approximated versions* , Physics in Medicine and Biology, Vol.55, No.4, 2010.

[3] E.Miqueles and A.R.De Pierro, *Iterative reconstruction in X-ray fluorescence tomography based on Radon inversion*, Trans.Medical Imaging, Vol.30, No.2, February 2011.

[4] E.Miqueles, A.R.De Pierro, *C-library raft: Reconstruction algorithms for tomography. Applications to X-ray fluorescence tomography*, Computer Physics Commun. 182(12): 2661-2673, 2011.

[5] Q. Yang, B. Deng, W. Lv, F. Shen, R. Chen, Y. Wang, G. Du, F. Yan, T. Xiaoa and H. Xua, *Fast and accurate X-ray fluorescence computed tomography imaging with the ordered-subsets expectation maximization algorithm*, J. Syncrothron Rad.,19, 210-215, 2012.

[6] B.Golosio, A.Simionovici, A.Somogyi1, L.Lemelle, M.Chukalina, A.Brunetti, *Internal elemental microanalysis combining x-ray fluorescence, Compton and transmission tomography*, J. Appl. Phys. 94, 145, 2003.

Figure 4: From left to right: EM iterates 3, 5, 7 and function f.

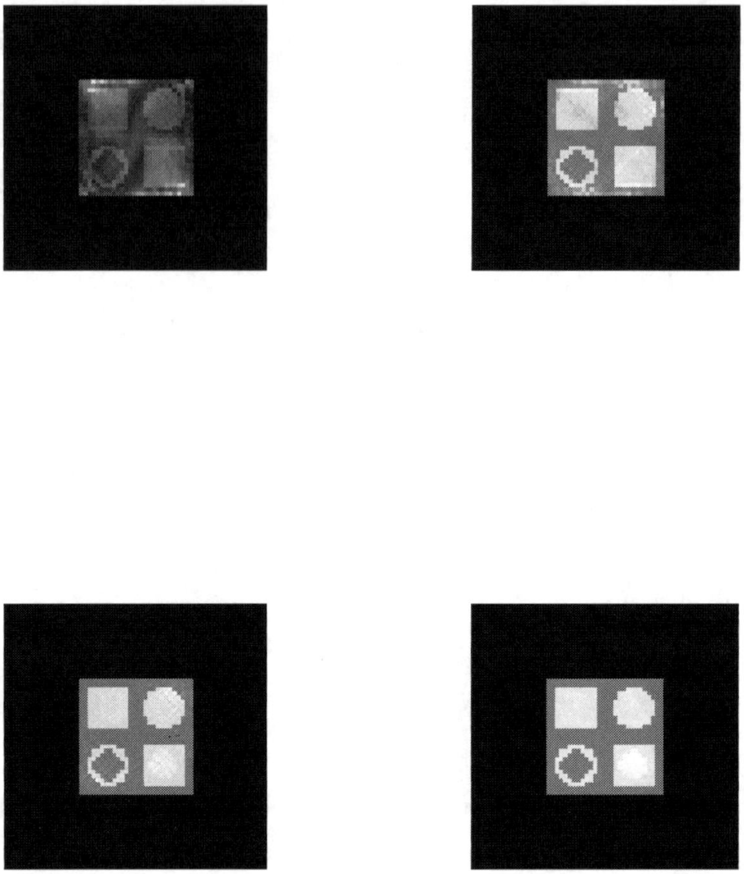

Figure 5: From left to right: RAMLA iterates 3, 5, 7 and function f. Here, $I_i = I_i^{(a)}$

Figure 6: From left to right: RAMLA iterates 3, 5, 7 and function f. Here, $I_i = I_i^{(b)}$

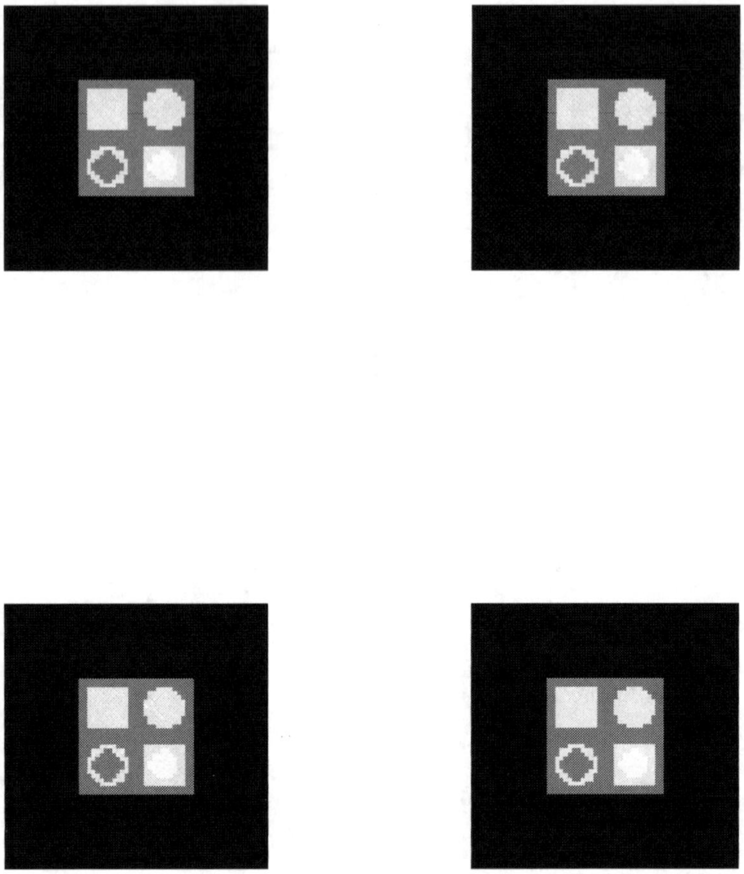

Figure 7: From left to right: RAMLA iterates 3, 5, 7 and function f. Here, $I_i = I_i^{(c)}$

A theoretical model of cytosolic calcium elevation following wounding in urothelial cell monolayers

Peter A. Appleby[1], Saqib Shabir[2], Jennifer Southgate[2], Dawn Walker[1]

[1] Department of Computer Science, Kroto Research Institute, University of Sheffield, Sheffield, UK
[2] Jack Birch Unit for Molecular Carcinogenesis, Department of Biology, University of York, York, UK

E-mail: p.appleby@sheffield.ac.uk

Abstract. Scratch wounding of a urothelial cell monolayer triggers a number of events including the release of soluble, diffusible signalling factors and mechanical stimulation of cells at the wound edge. These events cause a sustained elevation in cytosolic calcium concentration in the cells surrounding the wound and a transient rise in those further away. The precise form of this calcium transient is believed to play a central role in determining the subsequent response of individual cells and ultimately leads to a co-ordinated, population-level response that rapidly closes the wound. Here we present a framework for modelling the initial phases of this process. We combine a PDE model of diffusion in the extracellular medium and an ODE model of calcium signalling that has been tailored to represent urothelial cells. The ODE model is capable of generating a wide range of calcium transients, including spikes, bursts, oscillations and sustained elevations in the cytosolic calcium concentration. In multi-cell simulations of scratch wounding in a perfusion flow we find that the spatial position of the cells relative to the wound site leads to distinct classes of calcium response, with cells proximal to the wound exhibiting a sustained elevation and cells distal to the wound exhibiting a more transient elevation. We compare these results to existing experimental data and generate a number of novel predictions that could be used to test the model experimentally.

1. Introduction

Calcium signalling is known to play a central role in a wide variety of cellular processes. A key mechanism is the release of calcium from the endoplasmic reticulum. Typically, a small elevation in cytosolic calcium concentration ($[Ca^{++}]_i$) triggers further release of calcium via ryanodine- and IP_3-receptor activated calcium channels, leading to a rapid and much larger elevation in $[Ca^{++}]_i$. This is followed by a rapid falling phase, so the overall pattern is that of a sharp calcium spike. A number of modelling studies have examined the mechanisms underlying calcium transients (for review, see [1]). These models are capable of reproducing a variety of experimentally-observed calcium transients [2, 3, 4, 5, 6]. In our own work we have integrated and extended these models to produce a theoretical model of calcium signalling in epithelial cells. Our hypothesis is that the calcium transient, defined as the precise temporal profile of the variation in $[Ca^{++}]_i$, directs subsequent cell behaviour [7]. Our model is capable of reproducing calcium transients observed in normal human urothelial cells grown in monolayer cell culture [8].

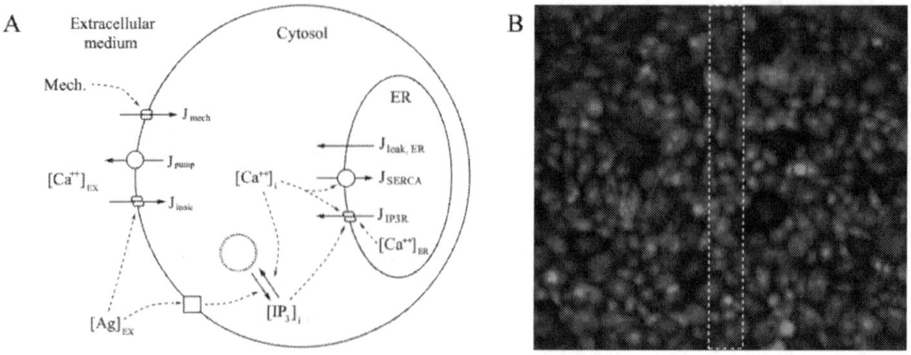

Figure 1. **(A)** Diagram of the extended calcium signalling model. Leak currents are represented by straight arrows, pumped currents by arrows overlayed with circles, and channel currents by arrows overlayed with two ellipses. **(B)** An example of a cultured normal human urothelial cell monolayer. Cells have been loaded with fura red-AM and 5μM fluo4-AM (which produces the green fluorescence) and imaged prior to wounding using a spinning disk confocal microscope. Scratch wounding is performed by removing a strip of cells using a sharpened glass pipette (white dashed box).

Here we extend our model to describe calcium transients in a simulated urothelial cell monolayer following scratch wounding.

2. Model

We build upon the one-pool model of ER-dependent calcium signalling described in [8]. This model relates the cytosolic, extracellular and ER calcium concentrations via a system of ODEs and contains five key components: leak currents, SERCA pumping, plasma membrane Ca^{++} ATPase (PMCA) pumping, and the ionic and P2Y signalling pathways. Writing the cytosolic calcium concentration as $[\mathrm{Ca}^{++}]_{\mathrm{i}}$, the ER calcium concentration as $[\mathrm{Ca}^{++}]_{\mathrm{ER}}$, and the extracellular calcium concentration as $[\mathrm{Ca}^{++}]_{\mathrm{EX}}$, and denoting the various calcium currents as J_x, the governing equations of the model are

$$\frac{d\,[\mathrm{Ca}^{++}]_{\mathrm{EX}}}{dt} = 0, \tag{1}$$

$$\frac{d\,[\mathrm{Ca}^{++}]_{\mathrm{ER}}}{dt} = -J_{\mathrm{leak,ER}} + J_{\mathrm{SERCA}} - J_{\mathrm{IP3R}}, \tag{2}$$

$$\frac{d\,[\mathrm{Ca}^{++}]_{\mathrm{i}}}{dt} = J_{\mathrm{leak,ER}} - J_{\mathrm{pump}} - J_{\mathrm{SERCA}} + J_{\mathrm{ionic}} + J_{\mathrm{IP3R}}, \tag{3}$$

where all concentrations are defined relative to the total cell volume. The full mathematical details of this model and its implementation are given in [8] where we show that, given an appropriate choice of parameters, it can reproduce calcium transients observed in cultured urothelial cell monolayers that have been stimulated by application of the purinergic agonist ATP. Here we consider a virtual wounding experiment where cells are removed from a cell layer by scratch wounding. During scratch wounding soluble diffusible signalling factors such as ATP are released from the wound site and cells near the wound undergo mechanical deformation [9]. The mechanical stimulus is believed to activate ionic channels in the cell membrane and we add a new term, $J_{\mathrm{mech}}(t)$, to Eq. 3 to represent this

$$J_{\mathrm{mech}}(t) = f_{\max}\,[\mathrm{H}(x_1, x) - \mathrm{H}(x_2, x)]\; g_{\max}\,[\mathrm{H}(t_1, t) - \mathrm{H}(t_2, t)]\,[\mathrm{Ca}^{++}]_{\mathrm{EX}}, \tag{4}$$

where x is spatial location of the cell, $x_{1,2}$ determine the spatial extent of the mechanical deformation, $t_{1,2}$ determine the duration of activation of the ionic channels and $f_{\max}$ and $g_{\max}$

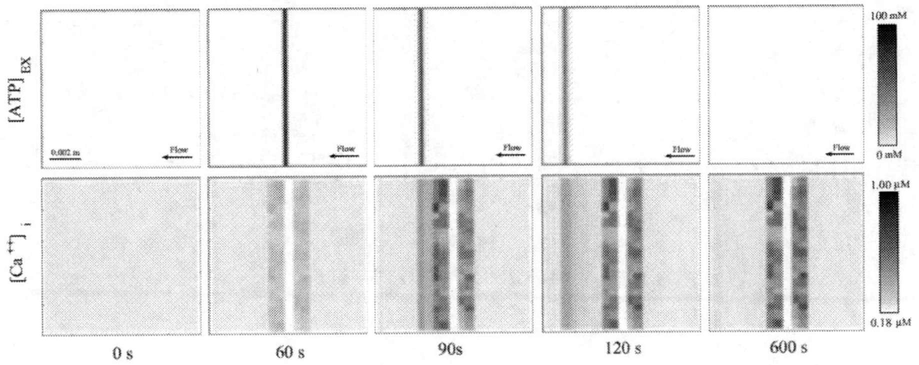

Figure 2. Multi-cell simulations of wounding in a perfusion chamber. The top row shows the extracellular ATP concentration, the bottom row shows cytosolic calcium concentrations where each cell is represented by a square 10 pixels wide and assigned a grey scale value that corresponds to its cytosolic calcium concentration. Wounding is simulated at $t = 60$ s by the removal of the central column of cells, triggering release of ATP and mechanical deformation of cells proximal to the wound. As the ATP is carried over the monolayer individual cells are excited and undergo a transient elevation in $[Ca^{++}]_i$. Cells distal to the wound site rapidly return to basal cytosolic calcium concentration. Cells proximal to the wound show a sustained calcium elevation due to the influence of the ionic current $J_{mech}(t)$.

are scaling factors. A sketch of this model is shown in Fig. 1A. We also include a model of diffusion where the concentration of an extracellular signalling factor, ϕ, evolves according to

$$\frac{\partial \phi}{\partial t} = D \bigtriangledown^2 \phi, \tag{5}$$

where D is the diffusion coefficient and $\bigtriangledown^2$ the Laplace operator. We incorporate a uniform flow of the extracellular medium at a fixed speed s = 5x10^{-5} m/s. This corresponds to an experimental set up where the extracellular medium flows over the cell layer at a constant rate due to a perfusion flow. Solutions of the diffusion equation are calculated using standard techniques from the literature [10].

3. Results

We simulate a uniform layer of cells of total area 0.01m^2, which represents a confluent layer of cultured urothelial cells similar to that illustrated in Fig. 1B. The cytosolic calcium concentration of each cell is governed by the ODE model described above using the default parameter set given in [8]. The parameters J_{ionic}, J_{mech}, and J_{IP3R} contain dependencies on channel numbers and we expect them to vary from cell to cell. We therefore scale each of these parameters by a Gaussian probability distribution with a mean of 1 and standard deviation of 0.2. We set $x_1 = 0.001$, $t_1 = 1$ min and $t_2 = 10$ min. We assume the mechanical stimulation caused by wounding is symmetrical around the wound site, so that $x_1 = -x_2$.

Figure 2 shows the evolution of the extracellular ATP concentration and cytosolic calcium concentrations of the cell population during a virtual wounding experiment. Initially all of the cells have a resting cytosolic calcium concentration of 0.2 mM and no ATP is present in the extracellular medium. Following scratch wounding there is a brief elevation in ATP at the wound site, which is rapidly carried over the cell layer by the perfusion flow. Cells distal to the wound site undergo a transient elevation in cytosolic calcium concentration. Cells proximal to the wound exhibit a sustained increase in cytosolic calcium concentration due to the influence of the ionic current J_{mech} that persists after the ATP has dissipated. Figure 3 shows the time course of cytosolic calcium concentration for cells grouped into proximal (within 1 mm of the wound edge) and distal (greater than 1 mm from the wound edge) populations. Proximal cells exhibit a characteristic rise in calcium concentration followed by a sustained plateau. The magnitude

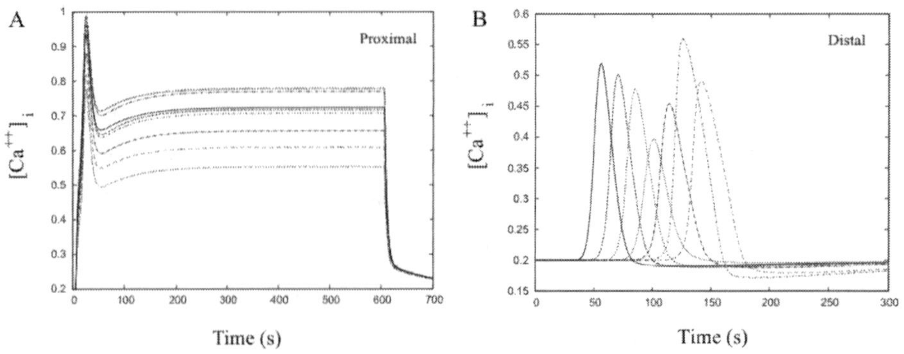

Figure 3. Representative time courses of $[Ca^{++}]_i$ grouped according to proximal and distal populations in Fig. 2. **(A)** Proximal cells exhibit an initial peak followed by a sustained elevation in cytosolic calcium concentration. **(B)** Distal cells exhibit a much more transient response, with a single peak followed by a return to baseline. Cells are activated sequentially as the perfusion flow carries the ATP over the cell layer. Note the different timescales of these responses. The distal cells return to baseline within 60 s, while the proximal cells remain activated for approximately 600 s.

of the plateau varies from cell to cell but the overall shape is preserved. Distal cells exhibit a characteristic single-spike response that lasts around ten seconds, with an onset time that is dependent on the spatial distance of the cell from the wound site.

4. Discussion

In this paper we have examined a theoretical model of cytosolic calcium transients in urothelial cells following scratch wounding. We observe two distinct classes of response within the cell population. Cells proximal to the wound site exhibit a sustained elevation in calcium, while cells distal to the wound site exhibit a more transient, single-spike behaviour lasting a few seconds. Both of these results compare favourably to existing experimental data [7]. A number of experimental tests of the model can be conceived. Direct mechanical stimulation of the cells should activate the ionic channels that mediate J_{mech}, leading to a sustained elevation in $[Ca^{++}]_i$ without the need for wounding. Alternatively, scratch wounding a monolayer in the presence of a P2Y antagonist should strongly attenuate the distal response but preserve the proximal response. Performing these experiments would be a direct test of the assumptions behind the model. Our model provides the framework for a detailed exploration of how calcium transients in individual cells direct subsequent behavioural responses.

Acknowledgments

PAA and SS are supported by Wellcome Trust grant 090577/Z/09/Z. JS is supported by York Against Cancer.

[1] Schuster S, Marhl M and Hfer T 2002 *Eur J Biochem* **269** 1333–1355
[2] Young G W D and Keizer J 1992 *Proc Natl Acad Sci U S A* **89** 9895–9899
[3] Atri A, Amundson J, Clapham D and Sneyd J 1993 *Biophys J* **65** 1727–1739 URL `http://dx.doi.org/10.1016/S0006-3495(93)81191-3`
[4] Goldbeter A, Dupont G and Berridge M J 1990 *Proc Natl Acad Sci U S A* **87** 1461–1465
[5] Borghans J M, Dupont G and Goldbeter A 1997 *Biophys Chem* **66** 25–41
[6] Lemon G, Gibson W G and Bennett M R 2003 *J Theor Biol* **223** 93–111
[7] Shabir S and Southgate J 2008 *Cell Calcium* **44** 453–464 URL `http://dx.doi.org/10.1016/j.ceca.2008.02.008`
[8] Appleby P, Shabir S, Southgate J and Walker D 2012 *BMC Sys. Bio.* **(in review)**
[9] Churchill G C, Atkinson M M and Louis C F 1996 *J Cell Sci* **109 (Pt 2)** 355–365
[10] Morton K W and Mayers D 2005 *Numerical Solution Of Partial Differential Equations: An Introduction* (Cambridge University Press)

Three-dimensional modeling and highly refined mesh generation of the aorta artery and its tunics

J A Cazotto[1], L A Neves[2], J M Machado[2], J C Momente[2], Y Shiyou[4], M F Godoy[3], G F D Zafalon[2], A R Pinto[2], C R Valêncio[2]

[1]IGCE, Department of Statistics, Applied Mathematics and Computation (DEMAC), São Paulo State University (UNESP), Rio Claro, SP, Brazil.
[2]Department of Computer Science and Statistics, São Paulo State University, São José do Rio Preto, SP, Brazil.
[3]Interdisc. Center for the Study of Chaos and Complexity (NUTTECC), Faculty of Medicine of São José do Rio Preto (FAMERP), São José do Rio Preto, SP, Brazil.
[4]College of Electrical Engineering, Zhejiang UniversityHangzhou, People Republic of China.

E-mail: jessica.cazotto@gmail.com

Abstract. This paper describes strategies and techniques to perform modeling and automatic mesh generation of the aorta artery and its tunics (adventitia, media and intima walls), using open source codes. The models were constructed in the Blender package and Python scripts were used to export the data necessary for the mesh generation in TetGen. The strategies proposed are able to provide meshes of complicated and irregular volumes, with a large number of mesh elements involved (12,000,000 tetrahedrons approximately). These meshes can be used to perform computational simulations by Finite Element Method (FEM).

1. Introduction

Computational fluid dynamics (CFD) methods based on three-dimensional vessel representations have recently shown to provide prognostically relevant hemodynamic data [1]. However, the automatic mesh generation of a given geometric domain is one of the most complex software problems which must be solved, mainly considering a high degree of refinement and obeying the solid boundaries conveniently. The assessment of clinically relevant parameters may depend on the histological features represented, which is not a trivial task. The models presented are restricted to the cases when histological features or free use conditions are necessary [2]. Models considering histological results may provide more clinically relevant information, and when built in open source codes they allow modifications without restriction. In this context, there are only a few examples of open source codes available, specially those dedicated to the solid modeling and automatic constrained tetrahedralization, which may be used to investigate three-dimensional vessel models based on FEM simulations [2].

In this context, we describe strategies to generate tetrahedral meshes of the aorta considering adventitia, media and intima walls. The strategies here proposed were based on open source code integration for vessel modeling and automatic mesh generation with a high degree of refinement.

2. Methodology

The aorta was constructed with macro and microscopic features. The macroscopic features are those which define the aorta path and ramifications. The model considers a structure with four segments: ascending aorta, aortic arch, thoracic aorta and abdominal aorta. The considered ramifications were the common carotid arteries, subclavian arteries, brachiocephalic artery, renal arteries and common iliacs. The dimensions were defined by a cardiologist. The microscopic features were dimensions of the adventitia and media tunic, with approximately 0.8 mm of thickness each, and the intima tunic with 130 μm of thickness. The proposed model was built with the Blender package (maintained by Blender Foundation) [3], an integrated system of tools and available under a dual license (BL / GNU General Public License). This package includes support to Python script language, which is based on a high level programming language model. We represented the aorta with the Blender package and used Python script to export the domain, which was used in the TetGen [4]. TetGen generates tridimensional meshes from the Delaunay algorithm [4], which is one of the most popular and robust algorithms for domain discretization.

The model was built with strategies that ensured the mesh generation with TetGen. The first strategy was the using of quadrilateral faces to discretize domains without bifurcation. The second was to consider triangular faces to represent regions with bifurcation. The triangular faces provide appropriate representations of regions with acute angles, like those occurring in bifurcation regions. For the aorta, the quadrilateral faces were used to define the ascending aorta, aortic arch, brachiocephalic artery, common carotid arteries, subclavian arteries, thoracic aorta, abdominal aorta and common iliac arteries. The triangular faces were used to represent the abdominal aorta bifurcation to the common iliac arteries and to the renal arteries, the aortic arch bifurcations to the left common carotid artery, to the left subclavian artery and to the brachiocephalic artery and the brachiocephalic bifurcations to the right common carotid artery and to the right subclavian artery. The histological particularities of the structure were represented with properties that enable highly refined meshes. These properties were face distortion minimization and density control of vertices. The first indicates that quadrilateral faces can be modified to fill regions with acute angles, since the internal angles respect values between 45 and 135 degrees as shown in figure 1. These limits avoided errors in the mesh generator stage. For a quadrilateral face not included in this interval, the region was discretized with a triangular element. A triangular element (triangular face) can be equilateral, isosceles or scalene which might be seen in figure 2a. The second property used was the density control of vertices, which defines the number of nodes in a region. The increasing number of vertices allows a better representation of curved regions, smoothing the direction transition, respecting the features defined on the first property and the orthogonality of the faces (figure 2b).

The aorta was stored into Blender files (*.blend*), which are a kind of container designated as *datablock*. An *object datablock* contains information about geometric operations applied into a given solid model. A *mesh datablock* stores information about vertices, edges, faces and holes, which resulted from the operations assigned by an *object datablock*. The integration of the models constructed by Blender, with the mesh generator Tetgen, was accomplished by Python scripts [5]. The *script* operates into objects generated by the Blender package through four iterative structures. These structures read the data contained in *mesh* and *object datablocks,* and translate the information about vertices, faces, holes and attributes accordingly. The processed information is an ASCII file (*.poly*), properly written in the formats required by TetGen. Since the iteration loops in [5] export only the active solid, some adaptation is necessary to provide the meshing of mixed solids. The performed adaptation executes the process for each member in the group of existing solids and generates a single ASCII file *(.poly*), composed by 4 parts: an indexed list of point coordinates; a list of solid faces; a list of volume holes; and, a list of attributes or boundaries (constraints).

The aorta was constructed and meshed in a computer with processor Intel® CoreTM i5, model 2410M, with 2,3 GHz, 4 GB of RAM and with a dedicated video card (AMD Radeon HD 6470M). The operating system used was the Windows 7 with 64 bits, running the Blender version 2.49b and the mesh generator TetGen version 1.4.

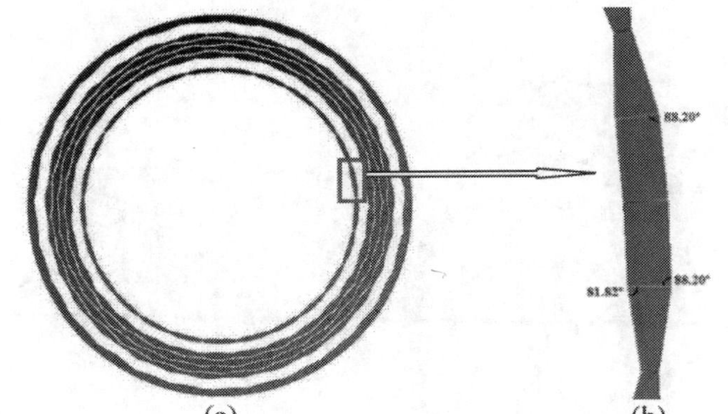

(a) (b)

Figure 1. Illustration of the quadrilateral faces used for the adventitia tunic representation (a), and the corresponding angles with values between 45 and 135 degrees (b).

(a) (b)

Figure 2. Illustration of the triangular faces (equilateral and isosceles triangles) used for the bifurcations representation (a), and the increase in the concentration of vertices to the aortic arch (b).

3. Results

The aorta was built using the concept of integration mentioned in section 2 and it is shown in Figure 3. Figure 3a shows the aorta that was constructed with the information from the ASCII file (.poly). The generated mesh of the adventitia tunic is showed in Figure 3b, which was obtained with 3,741,764 tetrahedral elements in approximately 16 minutes. The mesh of the media tunic was built in 4 parts. The part which requested most time was generated in approximately 49 minutes and has 7,930,965 tetrahedral elements. Finally, the intima tunic was generated in 79 minutes and has 11,778,860 elements. For instance, a transversal cut applied in the plane x,y, which is presented in the Figure 4, allows more detailed information about the adventitia and media walls histology meshed. Notice these blood vases comprise distinct layers of tissues with different material properties, and therefore a good domain mesh must represent these layers conveniently.

4. Conclusions

We have introduced strategies for the integration of data structures and vessel representations, based on the open source packages Blender and Tetgen. We have constructed the aorta artery that is a complex example, considering the abdominal aorta, the aortic arch and the brachiocephalic artery, as well as the bifurcations of these vessels. All these structures were generated successfully with the corresponding tunics. The adventitia tunic has 3,741,764 elements. Due to the topology and thickness of the media and the intima tunics, they have almost 8,000,000 and 12,000,000 elements, respectively. The approach here presented may already give a contribution for FEM simulations of several phenomena in blood vases, using highly refined tetrahedral meshes.

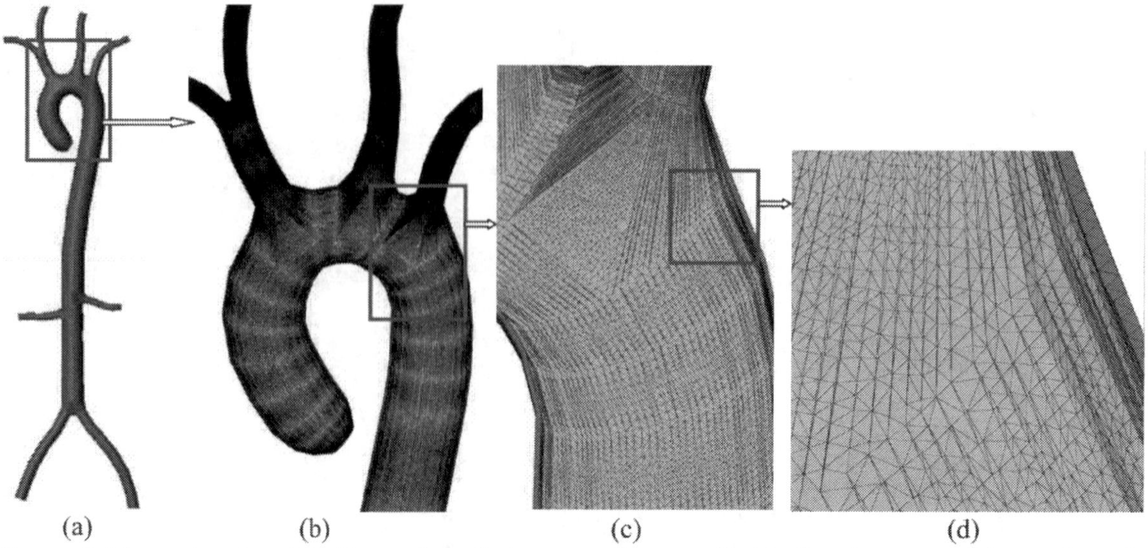

Figure 3. The aorta artery (a), illustration of a transversal cut applied in the plane z, x of the aortic arch (b), and an enlargement to a more detailed view ((c) and (d)).

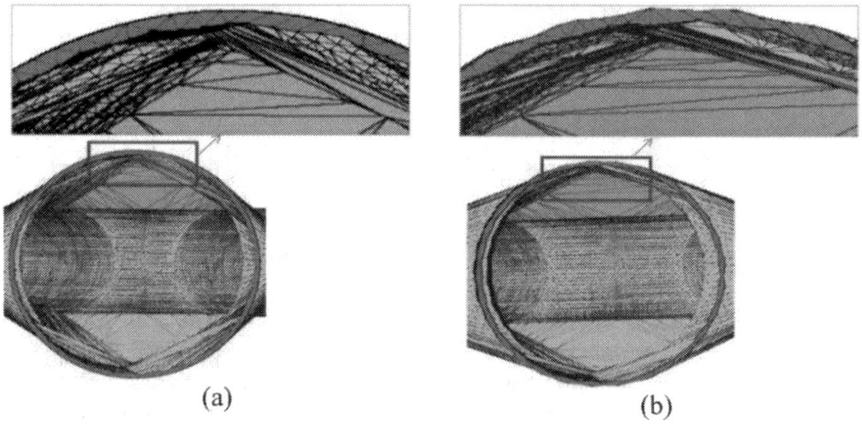

Figure 4. A transversal cut applied in the plane x,y of the adventitia tunic (a) and a part of the media tunic (b).

Acknowledgments
This study was supported by UNESP/PROPe (Pró-Reitoria de Pesquisa) and part by FAPESP.

References

[1] Hayase H, Tokunaga K, Nakayama T, Sugiu K, Nishida A, Arimitsu S, Hishikawa T, Ono S, Ohta M and Date I 2011 Computational fluid dynamics of carotid arteries after carotid endarterectomy or carotid artery stenting based on postoperative patient-specific computed tomography angiography and ultrasound flow data *Neurosurgery* **68** 1096–101

[2] Milašinović D, Ivanovic M, Tengg-Kobligk H, Böckler D and Filipović N 2008 Software tools for generating CFD simulation models of blood flow from CT images, and for postprocessing *Journal of the Serbian Society for Computational Mechanics* **2** 51–58

[3] Blender Foundation. [Online]. Available in: http://www.blender.org/.Accessed on: May 2012

[4] TetGen, User's Manual. Available in: http://tetgen.berlios.de/files/tetgen-manual.pdf Accessed on: May 2012

[5] Pedroso D 2007 TetGen export. [Online]. Available in: http://cvs.savannah.gnu.org/viewvc/*checkout*/mechsys/mechsys/src/py_scripts/tetgen_export.py

Journal of Physics: Conference Series **410** (2013) 012170 doi:10.1088/1742-6596/410/1/012170

Mathematical Modeling for Tomography in Domains with Reflecting Obstacles

Kamen Lozev

3161 S. Sepulveda Boulevard, Apt. 307, Los Angeles, CA 90034

E-mail: kamen@ucla.edu

Abstract. This work develops new numerical methods for the solution of the tomography problem in domains with reflecting obstacles. We compare the solution's performance for Lambertian reflection, for classical tomography with ubroken rays and for specular reflection. Our numerical method using Lambertian reflection improves the solution's accuracy by an order of magnitude compared to classical tomography with ubroken rays and for tomography in the presence of a specularly reflecting obstacle the numerical method improves the solution's accuracy approximately by a factor of three times. We present efficient new algorithms for the solution's software implementation and analyze the solution's performance and effectiveness.

1. Introduction

Let $f(x)$ be a continuous function in Ω, where $\Omega = \Omega_0\backslash\overline{\Omega_1}$ and Ω_0 is a compact convex set in $\mathbb{R}^2$ with a smooth boundary and Ω_1 a convex obstacle with a smooth boundary such that $\overline{\Omega_1} \subset \Omega_0 \subset \mathbb{R}^2$. Consider $\partial\Omega_0$ as the observation boundary of Ω_0 and the points of this boundary as transmitters and receivers of ray signals or ray solutions of the wave equation

$$u_{tt} - c^2(x)\Delta u = 0 \tag{1}$$

where $c(x) > 0$ is the variable speed of sound in Ω and $u|_{\partial\Omega_1} = 0$.

The data for the tomography problem are all integrals $\int_\gamma f(l)dl = C_\gamma$ where γ are rays in Ω that have both of their end points in $\partial\Omega_0$. One end point is a transmitter and the other end point of each of the rays a receiver. The classical tomography problem is to find $f(x)$ in Ω knowing the values of all integrals C_γ where γ are all straight line segments or unbroken rays in Ω. In the case of a domain Ω_0 without an obstacle $\Omega_1 \subset \Omega_0$, this problem is widely studied theoretically and numerically [1, 2, 3, 4]. When there are obstacles present, this problem is much less studied. Key theoretical work is done in [5, 6, 7, 8] for domains with one obstacle and with both broken rays, i.e. rays reflecting at the obstacle, and unbroken rays. A key result from [5] is that the tomography problem in the presence of one reflecting obstacle is well-posed. In other words, if an obstacle $\Omega_1 \subset \Omega_0$ is present then we have the well-posed problem of recovering $f(x)$ in $\Omega_0\backslash\Omega_1$ from the set C_γ where γ are all broken and unbroken rays in the domain starting and ending at the observation boundary. This problem is called the Broken Ray Tomography Problem[5].

In a basic tomography setup transmitters and receivers of wave signals are placed at the domain's boundary $\partial\Omega_0$. Ray signals are generated by the transmitters and received by the receivers. Travel times for signal propagation from transmitters to receivers are measured and these travel times $T(A,B)$ are the values of line integrals of a function $f(x) = \frac{1}{c(x)}$ where $c(x) > 0$ is the speed of sound at point $x \in \Omega_0 \backslash \Omega_1$. This measurement procedure gives the C_γ data for solving the Tomography Problem by relating signal travel times to the values of line integrals of $f(x)$. Given sufficient data, $f(x)$ and from here the velocity $c(x)$ are computed with tomographic reconstruction algorithms [1, 2, 3, 4]. The tomographic algorithms presented in the next section require and compute the ray path γ of each ray and, in order to compute these ray paths, we consider reflection models at $\partial\Omega_1$ and models of the speed of sound in Ω.

The theory of broken ray tomography from [5, 6, 7, 8] is based on a reflection model at $\partial\Omega_1$ that is mirror-like i.e. the angle of incidence is equal to the angle of reflection. In this paper, we consider a Lambertian reflection model in which incident rays are reflected at the obstacle in all possible directions and present results that show that the broken ray tomography reconstruction error is smaller when we consider Lambertian reflection.

We consider a mathematical model of the speed of sound

$$c(x) = c_o + \epsilon(x) \tag{2}$$

where $x \in \Omega$ that models the speed of sound as a continuous function close to a known constant c_o. It is shown in [9] that for sufficiently small $\epsilon(x)$ waves propagate along the known geodesics of $c_o(x)$ when

$$c(x) = c_o(x) + \epsilon(x) \tag{3}$$

where $c(x)$, $c_o(x)$ and $\epsilon(x)$ are continuous functions in Ω. The acoustic geodesics for constant speed of sound $c_o(x) = c_o$ are straight lines when there is no obstacle. Therefore, for the model 2, we consider two cases. In the first case, $\gamma = \tilde{\gamma_1}$ is an unbroken ray composed of a straight line segment $\tilde{\gamma_1}$. In the second case of a broken ray, $\gamma = \tilde{\gamma_1} \bigcup \tilde{\gamma_2}$ is the union of two straight line segments that intersect at a reflection point at the obstacle.

For unbroken rays the travel time or time of flight is

$$T(A,B) = \int_\gamma \frac{ds}{c_o + \epsilon(x(s))} = \int_{\tilde{\gamma_1}} \frac{ds}{c_o + \epsilon(x(s))}$$

and this model leads to the classical tomography problem with $f(x) = \frac{1}{c_o+\epsilon(x)}$.

For broken rays and a known obstacle, we know that the acoustic wave $u(x)$ propagates along the known straight line segments $\tilde{\gamma_1}$ and $\tilde{\gamma_2}$. γ_1 and γ_2 are known because in addition to the time of flight, our data measurement procedure gives the end points of the ray γ and its initial velocity, which in turn imply the reflection point of γ at the known obstacle Ω_1. Then

$$T(A,B) = \int_\gamma \frac{ds}{c_o + \epsilon(x(s))} = \int_{\tilde{\gamma_1}} \frac{ds}{c_o + \epsilon(x(s))} + \int_{\tilde{\gamma_2}} \frac{ds}{c_o + \epsilon(x(s))}$$

and, when data of this type is added to the set of measurements for the time of flight for unbroken rays, this gives the set C_γ for the Broken Ray Tomography Problem with $f(x) = \frac{1}{c_o+\epsilon(x)}$. In other words, the data set C_γ for the Broken Ray Tomography Problem contains the travel times of all broken and unbroken rays in the domain that start and end at the observation boundary.

2. Numerical Solution of the Broken Ray Tomography Problem for Lambertian Reflection

The first algorithm for the Broken Ray Tomography Problem is presented in [10] for a known obstacle Ω_1, specular reflection and the model 2 for the speed of sound. This work extends the first numerical solution of the Broken Ray Tomography Problem and develops an algorithm for finding the velocity structure of Ω for Lambertian reflectance at $\partial\Omega_1$. The broken ray tomography problem for an obstacle with Lambertian reflectance is solved with the following algorithm that constructs the finite set of broken and unbroken rays and computes the associated ray travel times by numerical integration or from ray travel time data.

Require: Domain Ω_0
Require: Obstacle $\Omega_1 \subset \Omega_0$
Require: Finite set of receiver points $R = (x_r, y_r)$ on $\partial\Omega_0$
Require: Finite set of transmitter points $T = (x_t, y_t)$ on $\partial\Omega_0$
Require: Finite set of obstacle boundary points $H = (x_h, y_h)$ on $\partial\Omega_1$
Require: Number of broken rays n_b.
Require: Number of unbroken rays n_u.
  {Algorithm for reconstructing $f(x, y)$ in Ω in the presence of obstacle Ω_0}
  initialize an empty list L that will contain all broken and unbroken rays
  for $p = 1 \to n_b$ **do**
    generate a random broken ray r from input data R, T, H, Ω_0 and Ω_1 {a broken ray is a unique triple $((x_r, y_r), (x_t, y_t), (x_h, y_h))$ that does not intersect the obstacle except at the reflection point (x_h, y_h)}
    add r to L
  end for
  for $p = 1 \to n_u$ **do**
    generate a random unbroken ray r from input data R, T, H, Ω_0 and Ω_1 {an unbroken ray is a unique pair $((x_r, y_r), (x_t, y_t))$ that does not intersect the obstacle}
    add r to L
  end for
  randomize L as a preprocessing step before starting the Kaczmarz method
  for all rays r in L **do**
    Compute travel time p_r for ray r. In numerical simulations, this is obtained via nimerical integration in Ω_0 along r. Store p_r in list P.
  end for
  Reconstruct $f(x, y)$ by the Kaczmarz method [11] for the linear system associated with rays r in L and corresponding travel times p_r in P.

As input to the above algorithm give n_b and n_u to be equal or approximately equal to the maximum number of broken and unbroken rays for the finite input sets R, T and H, or alternatively, during ray generation, generate new rays until all rays with end points in the input sets are generated. In addition, again in order to approximate the requirements of the theory of broken ray tomography for inclusion of all broken and unbroken rays, provide as input as many transmitter and receiver points on the observation boundary as possible. Ideally, the transmitter and receiver points are uniformly distributed on the boundary and sufficiently close to each other so that each set approximates the set of all points on the observation boundary. The above algorithm uses Lambertian reflection by considering broken rays as random triples $((x_r, y_r), (x_t, y_t), (x_h, y_h))$ and uses all rays in the domain.

3. Experimental Results

In order to show the effectiveness of the numerical solution of the broken ray tomography problem with Lambertian reflection, a Java implementation of the above algorithm compares the reconstructed values of $f(x, y)$ in Ω_0 with the known values of $f(x, y)$ for the same test function $f(x, y) = K\sqrt{(x - x_0)^2 + (y - y_0)^2}$, where (x_0, y_0) is the center of the computation grid, and test environment as in [10]. The following table compares the reconstruction error for classical tomography without reflection and broken ray tomography with Lambertian reflection.

Experiment	ART Error	ART Iterations	BRTL Error	BRTL Iterations
1	1.269592e-004	54293	1.272610e-005	32635
2	1.300814e-004	54204	5.272083e-006	40557
3	1.478464e-004	45923	7.530267e-006	37251
4	1.642026e-004	42480	1.110909e-005	33104
5	1.927561e-004	23037	1.163345e-005	33424
6	1.985439e-004	33705	1.896151e-005	28522
7	8.991362e-005	88190	1.401511e-005	30061
8	1.641201e-004	40717	1.955676e-005	29687
9	1.089771e-004	64932	7.102756e-006	40126
10	1.934379e-004	32899	2.149935e-005	28723
Average	1.516838e-004	48038	1.294065e-005	33409

Table 1. Error and number of iterations for broken ray tomography with Lambertian reflection(BRTL) at the boundary of the reflecting obstacle for a fixed number of 126050 rays. The average error for BRTL is 1.294065e-005 and the average number of iterations for finding a solution is 33409, and are shown in the right two columns of the table. The results for classical tomography with the ART method with the same number of rays are shown in the left two columns of the table. The average error for ART is 1.516838e-004 and the average number of iterations is 48038.

The results show that the reconstruction error of the new numerical solution of the broken ray tomography problem using Lambertian reflection is approximately three times smaller compared to an average reconstruction error using specular reflection of 3.874848e-005 by the algorithm from [10] for the same test function, domain and obstacle and an order of magnitude smaller than the reconstruction error for classical tomography in the presence of a reflecting obstacle. The new method also appears to be faster compared to classical tomography with ART. The order of the rays in both the ART and BRTL tests in this work is randomized before applying the Kaczmarz method therefore the speed of convergence difference is due to the use of reflection. We will report further results on the speed of convergence.

4. Acknowledgements
I would like to thank Professor Gregory Eskin for his continuous guidance.

References
[1] Natterer F 1986 *The Mathematics of Computerized Tomography* (Stuttgart: B G Teubner)
[2] Kak A C and Slaney M 1988 *Principles of Computerized Tomographic Imaging* (New York: IEEE Press)
[3] Natterer F and Wübbeling F 2001 *Mathematical methods in image reconstruction* (Philadelphia: SIAM)
[4] Faridani A 2003 URL `http://people.oregonstate.edu/~faridana/preprints/AFintroct.pdf`
[5] Eskin G 2004 *Inverse Problems* **20** 1497–1516
[6] Eskin G 2008 *J. Math. Phys.* **49** 022105–022105–18
[7] Eskin G 2006 *Phase Space Analysis of Partial Differential Equations* ed Bove, Colombini and Del Santo D
[8] Eskin G 2005 *J. Phys.: Conf. Ser. 12.* **23**
[9] Romanov V G 1974 *Integral Geometry and Inverse Problems for Hyperbolic Equations* (Berlin Heidelberg New York: Springer-Verlag)
[10] Lozev K 2012 *International Journal of Tomography and Statistics* **20** 1–23
[11] Kaczmarz S 1937 *Bulletin de l'Academie Polonaise des Sciences et Letters* **A35** 355–357

Algorithms for Shape and Trajectory Reconstruction of Obstacles in Domains with Variable Speed of Sound

Kamen Lozev
3161 S. Sepulveda Boulevard, Apt. 307, Los Angeles, CA 90034
E-mail: kamen@ucla.edu

Abstract. We present new parallel algorithms for shape and trajectory reconstruction of moving obstacles using reflected rays. In contrast to tomography where the focus of the reconstruction method is to recover the velocity structure of the domain, the shape and trajectory reconstruction procedure directly finds the shape and trajectory of the obstacle. The rays are curves determined by the variable speed of sound and initial conditions and we develop ultrasonic ray models based on a system of differential equations. The method can achieve high-resolution and computational efficiency.

1. Introduction

Consider an ultrasonic wave, or signal, described by the wave equation

$$u_{tt} - c^2(x, y, z)\Delta u = 0 \tag{1}$$

where $c(x, y, z) > 0$ is the variable speed of sound in $\Omega(t) = \Omega_0 \backslash \overline{\Omega_1}(t)$ and

$$u|_{\partial\Omega_1(t)} = 0$$

for $t > 0$ and $\overline{\Omega_1(t)} \subset \Omega_0 \subset \mathbb{R}^3$, where $\Omega_1(t)$ is a convex moving obstacle with a smooth boundary in a bounded domain Ω_0.

We consider an environment without caustics and look for solutions of the form

$$u(x, y, z, t) = \sum_{j=0}^{\infty} A_j(x, y, z) \frac{e^{i\omega(W(x,y,z)-t)}}{\omega^j} \tag{2}$$

where the eikonal function $W(x, y, z) = const$ defines a surface of constant phase. These solutions of the ray equation are called rays or ray solutions. Suppose that for all t we are given all integrals $\int_\gamma f(l)dl = C_\gamma$ where γ are broken rays in $\Omega(t)$ such that for each point $P \in \partial\Omega_1(t)$ there is at least one broken ray γ reflecting at P. A broken ray is a ray reflecting at the obstacle and starting and ending at the observation boundary $\partial\Omega_0$.

Let $f(x) = \frac{1}{c(x,y,z)} > 0$ in Ω. Then as we know C_γ correspond to signal travel times in a medium with speed of sound $c(x, y, z)$. The shape and trajectory reconstruction problem is to find $\partial\Omega_1(t)$ given the sets $C_\gamma(t)$ where $\gamma \in \Omega(t)$.

Consider $\partial\Omega_0$ as the observation boundary and signal transmitters and receivers with known locations along the boundary. Transmitters send ray signals with known initial zenith and azimuth angles at known transmission times. Receivers receive reflected ray signals and record the time when the signal was received. The combined data leads to a set of data points

$$B_k = (x_l, y_l, z_l, x_r, y_r, z_r, \phi_k, \theta_k, t_k, \xi_k)$$

where where ϕ_k and θ_k are the initial incident and azimuth angles of the ray from the transmitter, x_l, y_l, z_l are the coordinates of the transmitter endpoint of the ray, and x_r, y_r, z_r are the coordinates of the receiver endpoint of the ray, t_k is the time of flight for the signal and ξ_k is a frequency of the signal. We present algorithms for the shape and trajectory reconstruction problem for finite travel times and variable speed of sound derived from the equations for the Shooting Method for two-point seismic ray tracing[JG].

$$\frac{dx}{dt} = c(x,y,z)\sin\phi\cos\theta \tag{3}$$

$$\frac{dy}{dt} = c(x,y,z)\sin\phi\sin\theta \tag{4}$$

$$\frac{dz}{dt} = c(x,y,z)\cos\phi \tag{5}$$

$$\frac{\partial\phi}{dt} = -\cos\phi(\frac{\partial c}{\partial x}\cos\theta + \frac{\partial c}{\partial y}\sin\theta) + \frac{\partial c}{\partial z}\sin\phi \tag{6}$$

$$\frac{\partial\theta}{dt} = \frac{1}{\sin\phi}(\frac{\partial c}{\partial x}\sin\theta - \frac{\partial c}{\partial y}\cos\theta) \tag{7}$$

This system of equations derived from the eikonal equation has wide applications in seismology and is used in algorithms for seismic ray tracing [SK]. The system of equations 3 is for a Cartesian coordinate system where $\vec{R}(t) = (x(t), y(t), z(t))$ is the ray position vector, $\phi(t)$ is the incident angle of the ray direction vector with the z axis and $\theta(t)$ is the azimuth angle that the projection of the ray direction vector makes with the positive x axis.

We consider that the speed of sound $c(x, y, z)$ is known inside Ω. We will reconstruct the position of the reflection point P given the positions of a transmitter L and receiver S, the incident and azimuth angles of the transmitted ray with respect to a Cartesian coordinate system centered at the transmitter, and the time of flight of the signal. We know the initial position $(x(0), y(0), z(0))$ and therefore we know the velocity of the ray at $t = 0$. We know the angle and initial speed of wave propagation for the transmitted signal at time $t = 0$. Then, knowing the initial conditions, we can find numerically the signal path from the transmitter L through the reflection point P for a given travel time τ_k. Let the travel time from the transmitter L to the reflection point P be τ_k and the total travel time from the transmitter L to the receiver S through the reflection point P be t_k where $t_k > \tau_k$.

From symmetry, we can imagine that the signal received at the receiver S is transmitted from the receiver S and that it arrives at the reflection point P in time $t = t_k - \tau_k$ because its travel time is $t_k - \tau_k$. The ray path from S to P is again described by the above system of equations for seismic ray tracing, however we do not know the initial angles for the signal from the receiver S.

In order to reconstruct the intersection point P_k of rays starting from the transmitter and receiver for a given data point

$$B_k = (x_l, y_l, z_l, x_r, y_r, z_r, \phi_k, \theta_k, t_k, \xi_k)$$

containing the measured values for the signal, we step through a discrete set of values for τ_k. Starting to trace a ray path from L, at each time step, each new point on the path is a candidate reflection point P. At this stage, this is basic initial value ray tracing. Then we apply the two point seismic ray tracing shooting method to see if we can reach from receiver S candidate point P for the remainder of our time budget. If we can then we have found the reflection point P for this data point. Otherwise, we continue with the next time step of the initial value ray tracing and check the next candidate point P. We repeat this until we find a reflection point P or exhaust our time budget t_k. For example, let

$$\tau_{ks} = \frac{t_k n_s}{N_r}$$

where N_r is an integer that specifies the time-step resolution and n_s is an integer that specifies the number of the time step and such that $0 \leq n_s < N_r$. Each time τ_{ks} corresponds to a unique point P_{ks} on the ray path starting from the transmitter L such that P_{ks} can be reached from L in time τ_{ks}. Next, for each time step corresponding to time τ_{ks} to reach candidate P_{ks} from L, we step through the range of initial angles for the signal starting from the receiver S and check whether the curve from S will intersect P_{ks} in time $t_k - \tau_{ks}$. Alternatively, we can step through a range of reflection angles at the point P_{ks} that is found for the given τ_{ks} and check whether the reflected signal will intersect S in time $t_k - \tau_{ks}$.

2. Reconstruction algorithms

The input to the following algorithm is the speed of sound $c(x)$ for the domain Ω and a set of data points or ray coordinates corresponding to broken rays. The output is a unique set of points in $\mathbb{R}^3$. The points from the output are the reflection points reconstructed from the input data.

Require: Set of broken ray data points $B_k = (x_l, y_l, z_l, x_r, y_r, z_r, \phi_k, \theta_k, t_k, \xi_k)$
Require: Speed of sound $c(x)$ for domain Ω
  {Algorithm for Shape and Trajectory Reconstruction of Moving Obstacles}
  {Estimated time complexity is $O(T^2 A)$ where T is the number of discretization points for the time of flight, and A is the number of discretization points for the angle space}
  for all data points B_k **do**
    $h_k = \frac{t_k}{N_r}$
    $L = (X_0, Y_0, Z_0) = (x_l, y_l, z_l)$ set this initial position to be position of transmitter
    $S = (aX_0, aY_0, aZ_0) = (x_r, y_r, z_r)$ set this initial position to be position of receiver
    $\Phi_0 = \phi_k$
    $\Theta_0 = \theta_k$
    $T_0 = 0$
    $aT_0 = 0$
    for $s = 0 \to N_r - 1$ **do**
      {Compute the next point on the ray from the transmitter by fourth order Runge-Kutta step and the ray tracing system 3}
      $X_{s+1} = RK4_X(h_k, T_s, X_s, Y_s, Z_s, \Phi_s, \Theta_s)$
      $Y_{s+1} = RK4_Y(h_k, T_s, X_s, Y_s, Z_s, \Phi_s, \theta_s)$
      $Z_{s+1} = RK4_Z(h_k, T_s, X_s, Y_s, Z_s, \Phi_s, \Theta_s)$
      $\Phi_{s+1} = RK4_\Phi(h_k, T_s, X_s, Y_s, Z_s, \Phi_s, \Theta_s)$
      $\Theta_{s+1} = RK4_\Theta(h_k, T_s, X_s, Y_s, Z_s, \Phi_s, \Theta_s)$
      $T_{s+1} = T_s + h_k$
      $P_{s+1} = (X_{s+1}, Y_{s+1}, Z_{s+1})$ point on solution of ray tracing equations with initial values for transmitter that is at time T_{s+1} away from the transmitter L
      if $!(P_{s+1} \in \Omega_0)$ **then**
        There must be a measurment error. Continue with next data point B_k
      end if

for all initial angles $a\Phi_0, a\Theta_0$ in discretized angle space of the receiver **do**
for $p = 0 \to N_r - 1$ **do**
{Compute the next point on the ray from the receiver by fourth order Runge-Kutta step and the ray tracing system 3}
$aX_{p+1} = RK4_X(h_k, aT_p, aX_p, aY_p, aZ_p, a\Phi_p, a\Theta_p)$
$aY_{p+1} = RK4_Y(h_k, aT_p, aX_p, aY_p, aZ_p, a\Phi_p, a\Theta_p)$
$aZ_{p+1} = RK4_Z(h_k, aT_p, aX_p, aY_p, aZ_p, a\Phi_p, a\Theta_p)$
$a\Phi_{p+1} = RK4_\Phi(h_k, aT_p, aX_p, aY_p, aZ_p, a\Phi_p, a\Theta_p)$
$a\Theta_{p+1} = RK4_\Theta(h_k, aT_p, aX_p, aY_p, aZ_p, a\Phi_p, a\Theta_p)$
$aT_{p+1} = aT_p + h_k$
$P_{\alpha_{p+1}} = (aX_{p+1}, aY_{p+1}, aZ_{p+1})$ point on solution of ray tracing equations with initial angles $a\Phi_0$ and $a\Theta_0$ and initial position S, that is time aT_{p+1} away from S
if $!(P_{\alpha_{p+1}} \in \Omega_0)$ **then**
Exit this for loop and continue with next pair of initial angles $a\Phi_0, a\Theta_0$ from outer for loop
end if
if $distance(P_{s+1}, P_{\alpha_{p+1}}) < \epsilon_1$ and $|T_{s+1} + aT_{p+1} - t_k| < \epsilon_2$ **then**
$P_k = P_{s+1}$ {Solution for current data point B_k found. Continue with next data point B_{k+1}}
end if
if $T_{s+1} + aT_{p+1} > t_k + \epsilon_2$ **then**
{We are over the travel time budget t_k. Continue looking for a solution with the next set of initial angles $a\Phi_0, a\Theta_0$.}
end if
end for
end for
end for
end for

RK4 stands for a fourth order Runge-Kutta method although other time-dependent numerical methods can be used as well. The above algorithm leads to a class of new algorithms when data structures are used to store the rays for faster processing. For example, when rays from the receiver in its set of directions are stored and looked up in a data structure such as an array, the performance of the algorithm can be improved to $O(TA^2)$ where T is the number of discretization points for the time of flight, and A is the number of discretization points for the angle space. These algorithms are very suitable for parallelization and their parallelized implementation is key for efficient real-time processing, performance and computational efficiency. For example, the computation for tracing the rays from the receiver can be parallelized in multiple threads. When the above algorithm is run on a set of points $\{B_k\}$ from one sampling time interval T_k, the algorithm reconstructs the shape of the obstacle during this sampling interval. In order to reconstruct the trajectory of the obstacle, the algorithm is run on the data points for each of the sampling intervals. The method can achieve high resolution because it can process a very large number of distinct points on the obstacle's boundary.

3. Acknowledgements

I would like to thank Professor Gregory Eskin and Professor James Ralston.

4. References

[JG] Julian, B.R. and D. Gubbins *Three-dimensional seismic ray tracing J. Geophys., 43, 95-113, 1977*
[SK] M.S. Sambridge and B.L.N.Kennet *Boundary value ray tracing in a heterogeneous medium: a simple and versatile algorithm* Geophys. J. Int., 101, 157-168, 1990

AUTHOR INDEX

AUTHOR INDEX

AUTHOR INDEX

AUTHOR INDEX